Paul H. Glaser

The Alpine Flora
of the Rocky
Mountains

The Alpine Flora of the Rocky Mountains

VOLUME I

The Middle Rockies

Richard W. Scott

UNIVERSITY OF UTAH PRESS
Salt Lake City

© 1995 by the University of Utah Press

All rights reserved

Printed on acid-free paper

LIBRARY OF CONGRESS CATALOGING-IN-PUBLICATION DATA

Scott, Richard W. (Richard Walter), 1941–
    The alpine flora of the Rocky Mountains / Richard W. Scott.
       p.   cm.
    Includes bibliographical references and index.
    Contents: v. 1. The Middle Rockies
    ISBN 0-87480-482-5 (alk. paper)
    1. Mountain plants—Rocky Mountains—Identification.
2. Mountain plants—Rocky Mountains—Pictorial works.   I. Title.
QK139.S36   1996
581.978—dc20   96-6190
                                           CIP

Contents

Preface vii
Acknowledgments ix

Introduction 1

The Alpine Zone 3
    Timberlines 3
    The Alpine Landscape and Geomorphic Processes 3
    The Alpine Environment 5
    Adaptations of Alpine Plants 7

The Middle Rocky Mountains 11
    Geography 11
    Drainage Basins 12
    Formation of the Ranges 13
    Glaciation: The Pleistocene and the Holocene 13
    The Mountain Ranges 16

The Flora 23
    Characteristics 23
    Taxonomic Treatment 27

The Middle Rocky Mountain Alpine Flora 31
    Key to the Families 31
    Adiantaceae 35
    Apiaceae 38
    Aspleniaceae 60
    Asteraceae (Compositae) 67
    Betulaceae 201
    Boraginaceae 203
    Brassicaceae 217
    Callitrichaceae 265
    Campanulaceae 266
    Caprifoliaceae 269
    Caryophyllaceae 272
    Celastraceae 302
    Chenopodiaceae 304
    Crassulaceae 308
    Cupressaceae 314
    Cyperaceae 317
    Ericaceae 381
    Fabaceae (Leguminosae) 395
    Fumariaceae 429
    Gentianaceae 431
    Geraniaceae 445

Grossulariaceae 448
Hydrocharitaceae 453
Hydrophyllaceae 455
Hypericaceae (Clusiaceae) 458
Isoetaceae 459
Juncaceae 460
Lamiaceae (Labiatae) 473
Liliaceae 475
Linaceae 485
Lycopodiaceae 487
Onagraceae 489
Ophioglossaceae 500
Orchidaceae 501
Papaveraceae 504
Pinaceae 506
Plantaginaceae 514
Poaceae 515
Polemoniaceae 589
Polygonaceae 604
Portulacaceae 624
Primulaceae 631
Ranunculaceae 640
Rosaceae 672
Rubiaceae 709
Salicaceae 712
Saxifragaceae 728
Scrophulariaceae 753
Selaginellaceae 793
Valerianaceae 796
Violaceae 800

Appendix 1: Glossary of Descriptive Alpine Terminology 805
Appendix 2: Glossary of Botanical Terms 809
Appendix 3: Authors of Accepted Species Names 830
Appendix 4: Chromosome Numbers of Alpine Plants in the Middle Rocky Mountain Flora 845
Bibliography 864
Index of Common Names 871
Index of Latin Names 879

Preface

THIS MANUAL HAD ITS beginning in my early interest in the mountains: their distribution, characteristics, environments, and biota. The alpine zone in particular has always intrigued me, especially the alpine plants, which grow and thrive in a seemingly marginal and harsh environment.

The first detailed regional treatment of alpine plants within the Middle Rocky Mountains was my M.S. thesis at the University of Wyoming, "The Alpine Flora of Northwestern Wyoming," which was distributed as a special publication of the Rocky Mountain Herbarium. I later expanded the flora to include the entire state of Wyoming, and then added regions of southern Montana and northeastern Utah to complete the entire Middle Rocky Mountain region. Included in this latest version are the Medicine Bow Mountains of southeastern Wyoming, even though, according to some authors, they are technically not part of the Middle Rocky Mountains. Thus, the flora serves a dual purpose: first, it deals with the alpine plants within the political boundaries of the state of Wyoming; second, it treats the more natural physiographic unit, the Middle Rocky Mountains, which encompasses the states of Montana, Wyoming, and Utah, and a very small bit of Idaho.

The taxa included in this manual are based on specimens collected throughout the Middle Rocky Mountain region by myself, students, colleagues, and others. Collections represented by specimens in regional herbaria with Middle Rocky Mountain holdings are also included. Where reports of other investigators indicate the presence of a species in the alpine zone of a particular mountain range, and I have no record of a specimen, I use the phrase "reported by" to show the basis for its inclusion and to distinguish it from those that are documented by herbarium specimens. The inventory is, of course, by no means complete. I am continuing to add to the documentation and would appreciate receiving new information as it becomes available.

At first, my plans were to complete a field manual with diagnostic keys that could be used to identify all the alpine vascular plant species found in the Middle Rocky Mountains from timberline to their upper elevational limits. Later, distribution maps for each species were developed as field documentation became available. Finally, I added a general, but condensed, overview of the characteristics of the alpine zone, along with line drawings to illustrated each of the species treated in the manual. The University of Utah Press has now agreed to publish a multivolume series to cover all the ranges of the Rocky Mountains and the associated Colorado Plateau. I find this very exciting, since the alpine zone is composed of diverse and extreme environments that are important for understanding vascular plant adaptations and their biological, biogeographical, and evolutionary significance. The alpine flora is predicted by many authors as one of the very first to possibly reflect structural changes as a result of global warming. This lends some urgency to the documentation of species and their distribution patterns in the vast expanse of the Rocky Mountains.

This treatment of the Middle Rockies, which began in northwestern Wyoming, thus becomes volume 1 of a three-volume set covering the alpine zone of the entire Rocky Mountain region. I hope that all users will consider the treatment useful, and that the experienced botanist will not find it too brief, nor the amateur too technical.

Acknowledgments

A LARGE NUMBER OF PEOPLE helped make my job of collecting data, synthesizing information, and writing easier. My deepest appreciation is extended to all. Some who deserve special mention are Bob Lichvar, Rob Kirkpatrick, Stuart Markow, Charmaine Refsdal, and especially Walt Fertig, who shared their master's thesis records from the Gros Ventre, southern Absaroka, western Teton, northern Uinta, and western Wind River ranges.

Students and staff at the Rocky Mountain Herbarium, University of Wyoming, extended numerous courtesies during my use of the facility. Professor Ron Hartman, Curator, and Ernie Nelson, Herbarium Manager, were particularly helpful, allowing me to use species distribution information from their expanding GIS (geographic information system). Many curators of other herbaria throughout the region also provided access to their collections along with unlimited cooperation and much help.

Assistance was provided in the field by Robin and Mabel Jones, Dennis Van den Bos, Lee Wickel, Gary Weldon, and Mike Bynum. Mike Bynum tested a number of the keys in the field, and Yurika Hasegawa tested them in the Central Wyoming College Herbarium. Hollis Marriott provided specimens and distribution information for selected species from the Wyoming Natural Diversity Database.

Several people were involved in the typing and mechanics of the manuscript. Debbie Shaver started it, and was followed by Tina Leonhardt, Cheryl Portwood, Lisa Ferguson, Katherine Matlack, and Judi Parham, finishing with the much-appreciated efforts of Patricia Hatle-McCoy and Lynn Kinter, who typed and retyped what we always thought was *the final* revision. Karen Aurand organized the scientific and common name indexes.

Lisa Erven, reference librarian at Central Wyoming College, pounded the electronic pathways of the Internet in search of old literature, new literature, and a multitude of sources for interlibrary loans.

The illustrations of plant species are a mixture of previously published and original line drawings. Thanks is due to the University of Washington Press and Stanford University Press for permission to use material from their publications. Walt Fertig supplied a number of original drawings made especially for this volume, and two Kaye Thorne drawings were furnished through the combined efforts of Duane Atwood and Al Winward.

In my transition to modern GIS techniques of mapping, the distribution maps were constructed by the older technique of applying stick-on dots. Thanks to Michelle Van Vleet, Josh Fraley, and Christine Dixon, who assisted with the process of positioning 60,519 of them.

The editors at the University of Utah Press deserve special mention for their skillful approach to making the transition from manuscript to finished book a smooth process and productive experience. Their helpful attitudes were appreciated.

Last and most importantly, I wish to acknowledge and thank my wife Bev for her patience, field assistance, and support. Her companionship and stimulating discussion made even the longest and toughest of field trips a fun experience.

Introduction

MOUNTAIN PLANTS HAVE been called orophytes by some botanists (Love et al., 1971). It is a term descriptive of both their habitat and their environmental requirements, the Greek root *oro* meaning "mountain," and *phyte*, "a plant." As a group, orophytes are indicators of environments characterized by long snowy winters, short growing seasons, abundant summer precipitation, low temperatures, high light intensity, and frequent strong winds. The group can be further broken into subgroups based on the specific environments that occur at different elevations. This manual treats only one subgroup of the total assemblage of orophytes, the alpine plants of the Middle Rocky Mountains. These are the hyperorophytes, and they occur from timberline, which forms the lower limit of the alpine zone, to high elevations and mountain summits. The environments found at these high elevations are among the most severe on earth.

The alpine flora includes a number of plants long recognized by botanists, gardeners, and others for their hardiness, adaptability, and often showy, insect-pollinated flowers. These species have a rich tradition as rock garden plants (Williams, 1986), and they have been generating enthusiastic interest for many years, beginning in the Alps (Rambert, 1869; Schröter & Schröter, 1904) and later in the United States (Duft & Moseley, 1989; Strickler, 1990) and elsewhere.

Alpine plants are frequently described as dwarfed and mat forming. Many are, but a close examination of the entire assemblage of species reveals a wide spectrum of differences, not only of life-forms but also of flower types, adaptive strategies, and regional biogeographic groups. Some groups consist of wide-ranging arctic and alpine species, and some are arctic disjuncts with a sporadic distribution pattern in the Rocky Mountain alpine zone. Some are widely distributed alpine species found in all mountain ranges, and some are narrow alpine endemics restricted to relatively small geographic areas or specific habitats distributed over a wider area. Some groups include plants of lower elevations that reach their upper limits in the alpine zone. In spite of the dramatic environmental boundary indicated by regional timberline at the lower limit of the alpine zone, a number of these latter species are not restricted by it and easily cross it. As a result, the alpine flora is an often surprising mixture of diverse life-forms and regional biogeographic groups.

The Alpine Zone

Timberlines

THE ALPINE ZONE of the Middle Rocky Mountains occupies the area between timberline and the summits of the highest peaks. At some sites there is a distinct boundary, with conifers on the lower elevational side and low shrub or herbaceous vegetation on the upper side. At other sites the boundary is very indistinct due to the gradual shortening and thinning of arborescent growth forms. It is not uncommon for small, isolated clumps (or islands) of conifers to form. In places where the ecotone is very broad and gradual, these islands become more elongate downwind and more deformed as the altitude increases. Such clumps are referred to as krummholz, krupelkiefer, or elfinwood; their forms and movement are described by Marr (1977) and Wardle (1968). At the highest elevations krummholz is represented by very low clumps restricted to the lee sides of boulders and small topographic depressions that collect winter snow.

The variations in timberline characteristics have led to a complex terminology that describes site-specific characteristics of the ecotone. Thus, *forest limit* refers to the upper limit of closed-canopy forest. *Tree limit* is the upper limit of arborescent growth forms, which are commonly said to be either 2 m or 3 m in height. *Tree species limit* is the uppermost krummholz limit. Ilmari Hustich (1966) described five separate limits, the lowest of which he termed *economic forest-line,* closed-canopy forest that can be commercially cut. Above this is the *physiognomic forest line,* the limit of open-canopy forest. The *tree-line* is the absolute upper altitudinal limit of trees, regardless of species. *The tree species line* is the upper limit represented by any-sized individuals of a particular species. Finally, the *historic tree line* indicates earlier advances of forest and tree-line species under different climatic regimes.

Timberline is defined here as the upper limit of more or less continuous arborescent growth forms 3 m or more in height. This definition most closely agrees with the concept of tree limit, or Hustich's physiognomic forest line. The elevation of timberline varies with both microsite and macrosite conditions. At the macrosite level there is a broad correlation between timberline elevation and latitude. Generally, the higher the latitude, the lower the timberline elevation in the Northern Hemisphere. Thus, timberline increases from sea level in polar regions (arctic timberline) to 4115 m (13,500 ft) at 25° north latitude in Mexico (Swan, 1968). In the Middle Rocky Mountains, mean timberline (Figure 1) varies over nearly 6 degrees of latitude (39° 45' to 45° 37.5' N) from about 2682 m (8800 ft) in the Beartooth Range of southwestern Montana to about 3293 m (10,800 ft) in the Uinta and central Wasatch Ranges of Utah.

The Alpine Landscape and Geomorphic Processes

D. L. Blackstone (1971) noted that "landscape is the ultimate product of a long series of events extending back in geologic time almost to the inception of the earth as a planet." Landscape is also a major factor influencing the distribution of biota, and the alpine landscape of the Middle Rocky Mountains is no exception. It owes its overall character both to the older geologic events referred to by Blackstone and to the relatively recent Pleistocene glaciations that sculpted and deposited the surfaces in the region, both above and below timberline. Glacial features are common here. Features that are especially sig-

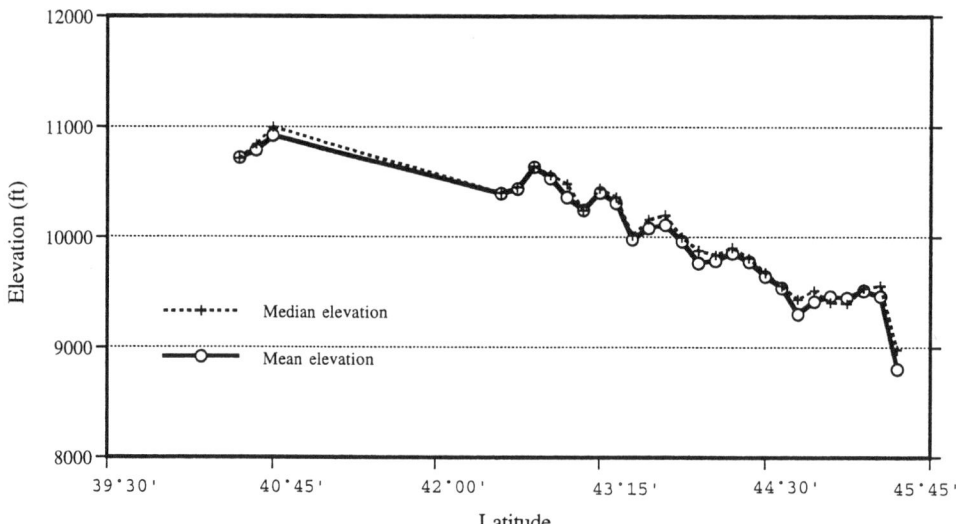

Figure 1. Timberline elevations along 110 degrees W Longitude. Data points represent mean and median elevations determined from random samples along latitudinal transects intersecting the line of longitude.

nificant in terms of alpine plant distribution are lateral and end moraines, which are deposits around the edges of glaciers. The relative ages of such deposits are often evident from the spectrum of vascular plant species colonizing and occupying their substrates. Glacial cirques—steep-walled, armchairlike basins—still remain long after the retreat and disappearance of Pleistocene ice. Some are presently occupied by more recent Neoglacial ice remnants. Most have a northern or northeasterly orientation and provide a specialized (usually cool to cold and shady) habitat for plant species. For example, the only record of *Lycopodium selago* L. in the alpine (or any other) zone of the Middle Rocky Mountains is from a north-facing cirque wall in the Teton Range. Other glacial features that directly or indirectly provide habitat for alpine vascular plants are outwash, kettle lakes associated with moraines, and tarns associated with cirques.

Following Pleistocene glaciation, factors such as cold temperatures in both summer and winter, low evaporation rate, and snowiness (Russell, 1933), interacted to control major alpine geomorphic processes and their resultant effect on landscapes. These processes are nivation, solifluction, and frost action.

Nivation is the denudation that occurs around snowfields and snowbanks when sediments are removed from the substrate by meltwater, resulting in a depression or shallow basin. Continued nivation results in an ever-deepening basin, which allows more and more snow to accumulate during the winter. The persistence of this accumulated snow has a profound effect on the distribution of plant species, and late in the growing season nivation basins are readily detected by changes in vegetation patterns. Nivation basins can be the first step in the formation of mountain glaciers and the resulting steep-walled glacial cirques that are particularly common on lee slopes of all Middle Rocky Mountain ranges (Flint, 1957; Anderson 1975).

Solifluction involves the slow downslope creep of water-saturated soils and rocks over permafrost, creating stairlike lobes of varying sizes that extend across slopes. Changes in slope and exposure brought about by solifluction create microenvironments often detectable by local changes in the species composition of vegetation mats. A variation of the solifluction lobe that is nearly level and sorted by creep to fine materials upslope and coarse materials on the downslope rim is the altiplanation terrace, particularly well developed and evident in the Absaroka, Beartooth, Uinta, and Wind River Ranges. Similar processes on level or near-level sites have created features such as frost boils, polygons, and circles. On shallow slopes, elongation due to gravity creep results in stone stripes, nets, and garlands. Many of these features are evidenced by corresponding vegetation patterns.

Frost action, or frost wedging, is the weathering process by which rocks are split and wedged apart by continuing freeze-thaw cycles. On steep and vertical slopes this process

Table 1. Average summer air temperature conditions at the alpine vegetation-atmosphere boundary, 10 cm or less above the ground surface (degrees C, June–August)

Mountain Range	Means	Maxima	Minima	Source
Presidential				
1917 m (6288 ft)	9–12 (D)	15–20 (D)	4–8 (D)	Bliss, 1966
Sierra Nevada				
3540 m (11,611 ft)	—	14–18 (D)	5–8 (D)	Chabot & Billings, 1972
Beartooth				
3242 m (10,635 ft)	—	16–27 (W)	-4–0 (W)	P.L. Johnson &
3140 m (10,300 ft)*	—	13–18 (W)	-3–1 (W)	Billings, 1962
Medicine Bow				
3354 m (11,000 ft)	8–13 (M)	16–24 (M)	-3–7 (M)	Bliss, 1956
Wind River				
3383 m (11,100 ft)	5–9 (D)	10–16 (D)	0–3 (D)	Scott, 1992, unpub.

NOTE: D = daily, W = weekly, and M = monthly.
*Shelter temperatures at 4 ft above the ground surface.

creates aprons of coarse debris (talus) or fine debris (scree) at the foot of the slopes. Snow accumulation at the base of such slopes causes the frost-wedged debris to collect in a morainelike arc on the outer perimeters of the snow, forming a protalus rampart. Frost wedging on shallow slopes and nearly level slopes results in a jumble of rocks known as felsenmeer, a German term that translates literally as "sea of rocks." An area of stabilized felsenmeer colonized by vegetation mats is an alpine fellfield.

The Alpine Environment

The alpine zone is frequently described as a harsh and severe environment. Microclimates are of extreme importance in controlling the distribution patterns of organisms here. Because of the high elevation of the alpine zone in mid-latitude mountain ranges, relatively small changes in slope and exposure result in microclimatic modifications of large magnitude and corresponding changes in the distribution patterns of organisms, particularly plants.

Overall, the alpine environment is characterized by low winter and variable summer temperatures, high wind velocities, and intense light. Total visible light radiation is more intense, as are short and ultraviolet wavelengths (Caldwell et al., 1980). The range in daily mean radiation values is 12–17 MJ/m² on Mount Washington in the Presidential Range of the northeastern United States, 21–25 MJ/m² in the Rocky Mountains, and 25–28 MJ/m² in the Sierra Nevada of California. Maximum daily values are 33.5–35.6 MJ/m² in the Presidential Range (Bliss, 1966). Clear summer days in the Wind River Range of Wyoming typically receive as much as 31 MJ/m² total daily radiation above timberline in the middle of July. The high elevation and clear air give the alpine zone the maximum photoperiod for photosynthesis. As in the lowlands, surface and near-surface temperatures vary with exposure, so that south-facing slopes are generally warmest and north-facing slopes are generally coolest. Temperatures are also much warmer near the ground, resulting in large lapse rates (Geiger, 1965). In the plant growth zone at about 10 cm (4 in) or less above the ground surface, moderate daily air temperatures are the rule (Table 1).

The reflectance from predominantly light-colored granite, granodiorite, quartzite, or dolomite surfaces common in the alpine zone and the corresponding strong reradiation of heat make freezing temperatures common, even at timberline, more so early and late in the growing season. For example, during a 122-day sampling period in summer

Figure 2. Growing season precipitation at different elevations in the Wind River Mountains, Wyoming. Spruce Island (S.I.) is at timberline (3231 m - 10,600 ft); Blueberry Glade (B.G.) is subalpine at 2916 m (9,600 ft). See Figure 3 for locations and elevations.

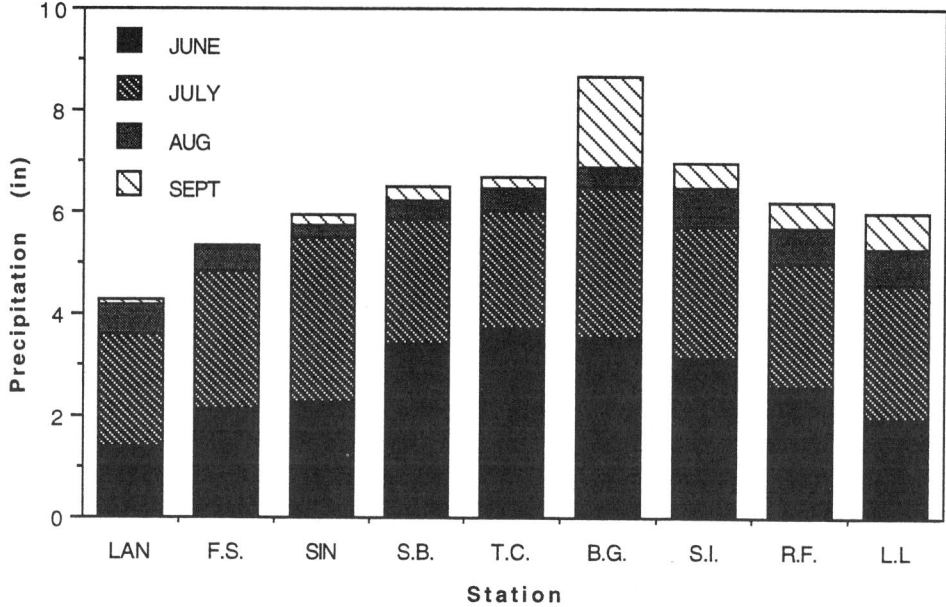

Figure 3. Location of stations on a microclimatological profile in the Wind River Mountains, Wyoming. Values on the x-axis are lat./long. coordinates projected onto the longitudinal scale.

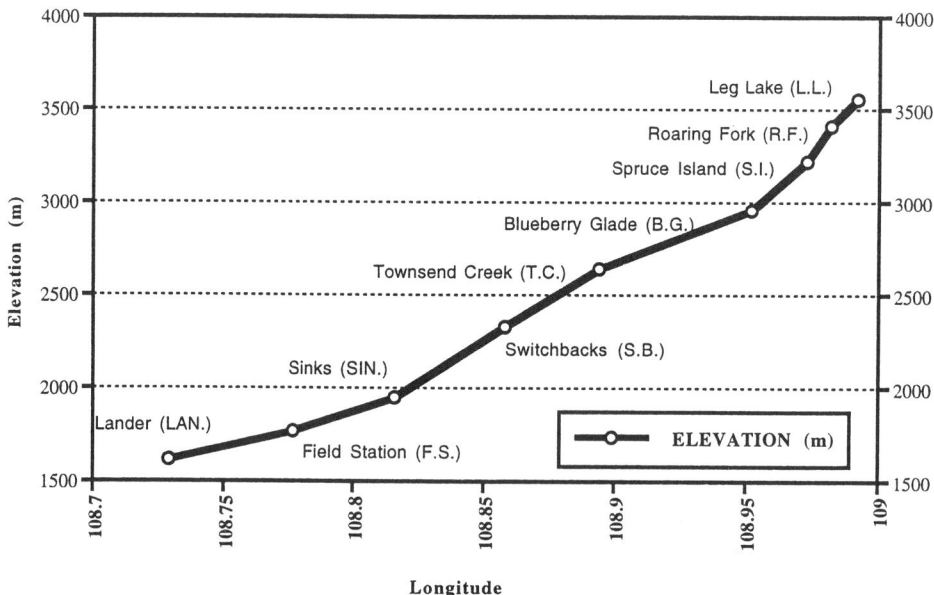

1992 (June–October), microclimate stations in the Wind River Range recorded at or below freezing temperatures 26 nights at 3231 m (10,600 ft), 50 nights at 3383 m (11,100 ft), and 53 nights at 3536 m (11,600 ft). There were 9 days at or above 20°C at 3231 m (10,600 ft), 2 at 3383 m (11,100 ft), and none at 3536 m (11,600 ft).

Wind plays two major roles in the alpine zone, one biological and one physical. Wind acts on plants' physiological systems by affecting their heat and water retention, and ultimately their assimilation of energy and resultant growth. Physically, wind affects the distribution and depth of snow cover, resulting in a summer mosaic of growing seasons and conditions reflected in community patterns, composition, and appearance (Harshberger, 1929; Billings & Bliss, 1959; Bliss, 1963; Scott, 1974; Helm, 1982). The few wind studies that have been conducted over an entire year have shown winter winds to be generally two to five times greater in velocity than summer winds (Bliss, 1985). At timberline, wind speed and direction are primary influences on the symmetry of trees (Woolridge et al., 1992). The elevation of timberline is influenced both directly and indirectly by wind: directly by the erosion of cuticular surfaces (Hadley & Smith, 1986, 1987); and indirectly through the desiccation of unripened shoot tissues (Wardle, 1971). At elevations above timberline, summer winds also affect gas exchange and other physiological processes (Caldwell, 1970).

The prevailing storm tracks in the Middle Rocky Mountains are from the southsouthwest, and the westerly, southerly, and southwesterly flanks of these ranges thus receive more precipitation than the opposite sides as a result of the rain shadow effect. In addition, precipitation apparently decreases with elevation in the alpine zone (Price, 1981). Some workers think this phenomenon might be no more than sampling error resulting from the higher wind velocities at higher elevations. Snow fence wind deflectors installed around precipitation gauges on Niwot Ridge, Colorado, increased the recorded precipitation above previous annual records (Barry, 1973). However, our studies in the Wind River Range with similar (although admittedly small) wind deflectors show a distinct summer precipitation maximum in the vicinity of timberline with corresponding decreases above and below (Figures 2 and 3).

Adaptations of Alpine Plants

The physical, or abiotic, factors operating in the alpine zone have selected for distinctive characteristics and forms that allow plants to survive and reproduce in this unique environment. Some of these characteristics are morphological features such as fleshy leaves, evergreen leaves, thick cuticles, epidermal hairs, and dark scales (Bliss, 1962). The subterranean organs of at least some species of alpine plants also exhibit adaptive features. Daubenmire (1941), for example, measured root systems in the Medicine Bow Range and found proliferation of branch roots near the ground line, but no consistency of storage function or morphology. Webber and May (1977) showed that ratios between aboveground biomass and belowground biomass are associated with growth form. Shrub communities had a ratio of 1:3, taprooted fellfield communities had a ratio of 1:14, and graminoid meadows had the maximum ratio, 1:25.

Many alpine plants overwinter with both leaf and flower buds in an advanced stage of development, and growth may thus occur as soon as conditions become favorable. Some alpine plants can begin growth early in the season because they have the ability to concentrate carbohydrate reserves in the zone near the soil surface. Mooney and Billings (1960) described this relationship between carbohydrates and growth for the three alpine species *Polygonum bistortoides* Pursh, *Saxifraga rhomboidea* Greene, and *Geum rossii* (R. Br.) Ser. (as *G. turbinatum*) from the Medicine Bow Range of Wyoming. Carbohydrate reserves were high in the fall after summer photosynthesis; development occurred under winter snows, resulting in some depletion of reserves; in the spring, accelerated growth resulted in a sharp drop in carbohydrates and ended with flowering; and summer photosynthesis replaced depleted reserves from flowering until dormancy occurred in the fall. Some alpine species initiate growth even under the snow. Two Middle Rocky Mountain species that undergo significant vegetative development below snowbanks (subnivean development) and flower sometimes within hours of snowmelt are *Erythronium grandiflorum* Pursh and *Caltha leptosepala* DC. Cold resistance, freezing tolerance, and early development in plants such as *C. leptosepala* appear to be linked to the presence of nonstructural carbohydrate reserves (Bliss, 1985). However, the underlying metabolic processes and the controls on their synthesis and release remain poorly known. Hamerlynck and Smith (1994) found no evidence of photosynthetic carbon assimilation in *E. grandiflorum,* despite substantial subnivean growth and development, until postemergence from snowbanks. Their studies indicate that plants depend on stored energy reserves for preemergent developmental events while they are protected from large temperature fluctuations during the early growing season.

If snow persists long enough into the growing season, vascular plants may be eliminated altogether and replaced by cryptogams. Two species of vascular plants in the

Middle Rocky Mountains that are persistent in late snow beds are *Androsace septentrionalis* L. and *Claytonia megarhiza* (Gray) Parry ex S. Wats. Their adaptive strategies are quite different. *Androsace septentrionalis* is an annual that can grow, flower, and set seed in a short period. *Claytonia megarhiza* is a perennial with a large, fleshy taproot for storing the energy needed for rapid growth following snowmelt. The perennial habit is an adaptive feature and is much more common in alpine plants than the annual habit. Alpine perennials are either woody shrubs or nonwoody graminoids and herbs.

By definition, trees do not occur in the alpine zone; at least, arborescent forms of tree species do not commonly occur above timberline. Alpine woody plants, then, are represented by erect or prostrate shrubs and krummholz conifer formations that become more stunted and deformed as elevation increases.

Most alpine shrubs are deciduous and are confined to moist areas such as stream banks, lakeshores, and sites protected by winter snow accumulation. Such sites are generally snow-free by mid-June. Members of the genus *Salix* are easily the most common of all alpine shrubs; 13 species are represented in numerous plant communities of the Middle Rocky Mountain flora. Erect shrubs such as *Salix glauca* L. and, to a lesser extent, *S. planifolia* Pursh form distinct belts above timberline. At higher elevations, well above timberline, shrubby plants such as *S. arctica* Pall. and *S. reticulata* L. are distinctly prostrate and confined to the few centimeters just above warm soil or rock surfaces. A few alpine shrubs are evergreen. As with deciduous shrubs, the erect forms, such as *Phyllodoce empetriformis* (Sw.) D. Don and *Kalmia microphylla* (Hook.) A. Heller, are confined to moist stream banks and lakeshores near timberline. As elevation increases, the prostrate growth form of species such as *Dryas octopetala* L. is selected over the erect form.

Nonwoody plants are conveniently described as either graminoid or herbaceous. Graminoid plants include grasses (*Poa* spp., *Festuca* spp., *Elymus* spp., *Deschampsia* spp., and others), sedges (*Carex* spp.), and rushes (*Juncus* spp.). All are common and widespread in a wide variety of microenvironments in the alpine zone, particularly on mesic slopes and moist to wet meadows. Most have extensive underground structures such as rhizomes or fibrous roots that allow for overwinter energy storage. Growth thus occurs at an accelerated rate as the soil warms following snowmelt and the accumulated carbohydrate reserves are used.

Herbaceous plants employ a similar adaptive strategy in the rosette and cushion growth forms, which permit survival in two of the most severe and least-exploited alpine microenvironments: exposed ridgetops lacking winter snow cover and late snow beds with a very short growing season. The rosette form is characterized by a few, often broad, leaves lying near the ground and originating from the crown of a thick taproot or dense cluster of fibrous roots. *Claytonia megarhiza* (Gray) Parry ex S. Wats. is such a plant. It occupies exposed sites up to the upper limits of vascular plants in the Wind River Range, which has the highest elevations of the Middle Rocky Mountains. The stored energy reserves allow rapid development and flowering following melting of late snow beds, while the horizontal leaves at ground level are in the zone of lowest wind velocity and maximum temperature. Cushion growth forms of *Silene acaulis* (L.) Jacq., *Saxifraga oppositifolia* L., *Minuartia* spp., *Phlox pulvinata* (Wherry) Cronq., and *Trifolium nanum* Torr. provide a similar adaptive advantage.

Annuals are few in alpine floras, with reports ranging from 10 (Major and Taylor, 1977) to 4 (Jackson and Bliss, 1982) or fewer depending on range and location. Considering all ranges and all species in the Middle Rocky Mountain alpine flora, 13 species are annuals. These are *Androsace septentrionalis* L., *Chenopodium capitatum* (L.) Asch., *Collinsia parviflora* Lindl., *Collomia linearis* Nutt., *Gentianella tenella* (Rottb.) Boerner, *Gentianopsis detonsa* (Rottb.) Ma, *Gymnosteris parvula* A. Heller, *Koenigia islandica* L., *Polygonum douglasii* Greene, *P. kelloggii* Greene, *P. minimum* S. Wats., *P. sawatchense* Small, and *Spergularia rubra* (L) J. & K. Presl. Eight annual species are also weakly biennial: *Minuartia dawsonensis* (Britt.) House, *Draba crassifolia* Grah., *D.*

praealta Greene, *Gentiana prostrata* Haenke ex Jacq., *Gentianella amarella* (L.) Boerner, *Poa annua* L., *Rorippa curvipes* Greene, and *Spraguea umbellata* Torr. Both groups grow primarily on open rocky ridgetops, wet gravelly stream banks, old frost boils and polygons, and scree slopes. Rarely are they found mixed in perennial vegetation mats or associations. Since they complete their life cycles in a compressed period, one would assume that they would be ideally adapted to the short alpine growing season. Lending support to this idea, Went (1953) showed that numerous alpine annuals occur in the Sierra Nevada of California. He attributed this to the desertlike conditions there: short growing season, high solar radiation, restricted moisture supply, and daily temperature extremes. However, Bliss (1985), among others, explained that most annuals are adapted to environments with higher temperatures and drier growing seasons than those commonly found in the alpine zone, and are consequently rare to uncommon in most alpine ranges.

Transitional between the annual and perennial habits are biennials. Plants that are biennial or weakly perennial in the Middle Rocky Mountain flora are *Machaeranthera canescens* (Pursh) Gray, *Descurainia incana* (Bernh. ex Fisch. & Mey.) Dorn, *D. torulosa* Roll., *Frasera speciosa* Dougl. ex Griseb., and *Ipomopsis aggregata* (Pursh) V. Grant. All appear to be restricted to the timberline ecotone or the lowest part of the alpine zone regardless of the mountain range.

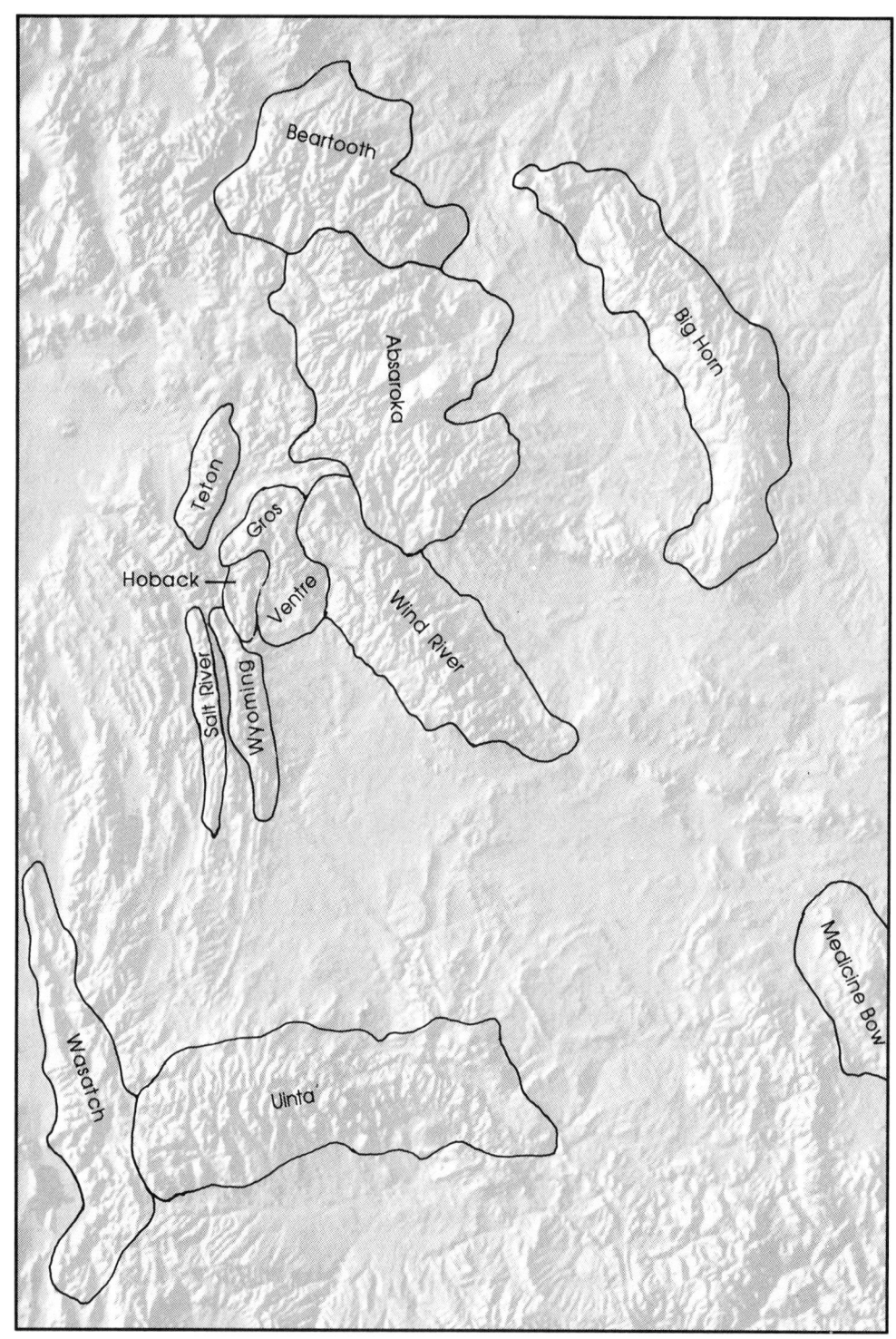

Figure 4. Ranges of the Middle Rocky Mountains.

The Middle Rocky Mountains

Geography

CORDILLERA ("rope" or "string") is a Spanish term used to describe mountain range systems that consist of more or less parallel chains. Thus, the system of mountain ranges that forms the western backbone of North America makes up the North American Cordillera. Actually, there are several discrete or nearly discrete chains of mountain ranges. Some are coastal, and others, located at varying distances inland, are continental. All can be grouped into two major chains that form the total system: the coastal Sierra Nevada–Cascade–Coastal Canadian complex and the continental Rocky Mountain complex, which is more or less continuous from Alaska to Mexico. Within the Rocky Mountain complex south of Alaska, it is convenient to recognize the Northern Rockies, the Middle Rockies, the Southern Rockies, and the associated Colorado Plateau.

This manual deals with the flora of the alpine zone of the Middle Rocky Mountains, one of the major physiographic provinces of North America. The province extends from southwestern Montana through Wyoming into northeastern Utah and includes that area from north to south between 45° 37.5' and 39° 45' north latitude, and east to west between 106° 30' and 112° west longitude. This forms a rough rectangle 428 km by 669 km (266 mi by 416 mi). If the Medicine Bow Mountains of southeastern Wyoming (traditionally placed in the Southern Rockies) are added, the eastern limit is moved slightly to 106° 00' west longitude, increasing the width of the rectangle to 486 km (302 mi), which gives a total map area of 325,134 km² (125,632 mi²). Within the traditionally recognized Middle Rockies are 11 major mountain ranges with alpine zones plus numerous smaller and lower ranges. In southeastern Wyoming there is a small, isolated alpine crest known as the Snowy Range, a unit of the much larger Medicine Bow Range of Wyoming and northern Colorado. Its inclusion brings the number of alpine ranges treated here to 12 (Figure 4). A treatment that includes the Medicine Bows is not entirely without precedent. At least one work (Richmond, 1986) shows the southeastern boundary of the Middle Rocky Mountains as falling between the southern Medicine Bow Range and the northern tip of the Front Range of Colorado. The Middle Rockies, then, according to this concept and including the Medicine Bow Range, occupy 53,147 km² (20,520 mi²) of the total map area of 325,134 km² (125,632 mi²).

One very distinct characteristic of the Middle Rocky Mountain ranges is their diversity in terms of formation, geology, and orientation. A second is their discontinuity; most form isolated environmental islands, at least with respect to the upper elevational zones, and certainly in the alpine zones.

The alpine ranges of the Middle Rocky Mountains, arranged roughly from north to south, are Beartooth, Absaroka, Big Horn, Teton, Gros Ventre, Hoback, Wind River, Wyoming, Salt River, Medicine Bow, Uinta, and Wasatch. Some are separated from one another by the cold desert of intermontane basins and form islands at low elevations (1372 m, or 4500 ft). Such islands include not only the alpine zone but also middle elevations and the foothill zones. Other ranges are separated at the 2438-m (8000-ft) level, and only the upper montane, subalpine, and alpine zones are included. If closed contour lines are used to outline mountain ranges and to separate each range from all the others,

Figure 5. Areas of Middle Rocky Mountain ranges at the 2438 m (8,000 ft) level.

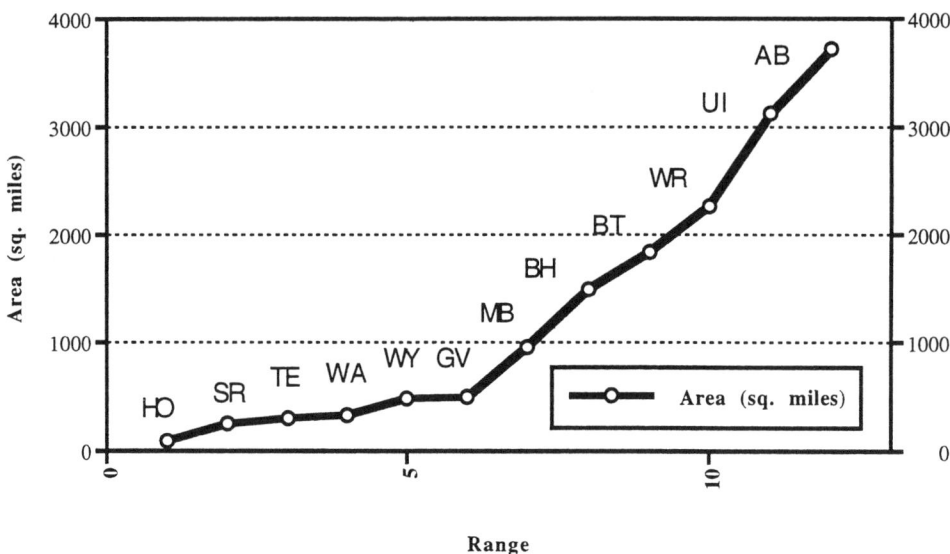

then the 2438-m (8000-ft) level is the lowest elevation that provides consistent separation. The area of each range, based on separation at this altitude, is shown in Figure 5.

Drainage Basins

Between and surrounding the margins of the Middle Rocky Mountain ranges are large intermontane basins and major river valleys. All the major rivers of the western United States have their headwaters in the Middle Rocky Mountains. Some of these flow through the intermontane basins and then around or beside the ranges, but most flow through the mountain ranges as superimposed, or superposed, streams and rivers. Their canyons and valleys have been formed by downcutting from earlier Tertiary erosional surfaces that once buried, or nearly buried, the ranges. Widespread general uplift (epierogenesis) during the late Tertiary led to a quickening of streams and increased erosion. As the sediments were eroded away and the buried ranges were exposed, the streams cut down to their present-day beds, allowing them to flow across the mountain crests rather than around or beside them.

At the northwest corner of the Middle Rocky Mountains, the Yellowstone River drains the Yellowstone Park area and flows north to the Missouri River, forming not only the boundary between the Beartooths and the Gallatin Range but the dividing line between the Middle and Northern Rockies.

The low elevations of the very large Big Horn Basin north of the nonalpine Owl Creek and Bridger Mountains separate the Absaroka and Beartooth Ranges from the Big Horn uplift to the east. The Big Horn uplift is cut across the northern portion by the superposed Big Horn River delineating the Pryor Mountains, of wild horse fame, to the north in Montana, and the Big Horns in Wyoming to the south. The east side of the Big Horns is bounded by the Powder River Basin, with a prominent gap to the south along the edge of the Great Plains. This gap, north of the Laramie and Medicine Bow Ranges, is the eastern edge of the Wyoming Basin, which is composed of the smaller Shirley Basin, Laramie Basin, Great Divide Basin, and Washakie Basin. The Medicine Bows can be divided into northern and southern segments (Houston et al., 1968). The southern segment, in this manual, is the southeastern corner of the Middle Rockies, joining with the Southern Rockies at the northern edge of the Front Range of Colorado.

In the interior of the Middle Rockies, the Wind River Basin separates the Absarokas and the Wind Rivers, in which are the headwaters of the Wind River–Big Horn River system. To the west, the Green River Basin separates the Wind River Range from the overthrust Hoback, Wyoming, and Salt River Ranges. This large basin is bisected by the Green River, which at its headwaters separates the Wind River Range from the Gros Ventre Range to the northwest. South and southwest of the Green River Basin are the Uinta Range (the southern border of the Middle Rockies) and the Wasatch Range (the western border) on the edge of the Salt Lake Valley. The Green River is superposed on the eastern edge of the Uinta Range, where it forms the spectacular Canyon of Ladore.

The Wasatch Front, on the western edge of the Middle Rockies, is equally spectacular as it rises steeply above the Salt Lake Valley. Farther north along the western edge is the Teton range, which is bordered on the west by Pierre's Hole; on the east by Jackson Hole, which separates it from the Gros Ventre Range; on the south by the superposed Snake River; and on the north by the southern limits of the Yellowstone-Absaroka volcanic field. These major drainages, intermontane basins, and valleys are indicated in Figure 6.

Formation of the Ranges

The concepts of plate tectonics provide a framework to describe and explain the dynamic state of the earth's surface. Large and small crustal plates move, collide, drift, or break apart, and in these processes become locally deformed. Some plates remain stable for long periods and then begin to move, causing deformations that change their surface characteristics and morphology. In this process mountain ranges are formed, particularly at and near plate margins.

The Middle Rocky Mountains occupy only a small portion of the very large North American Plate. Early geologic time (Figure 7), from approximately 570 million years ago (Ma) until about 70 Ma in this part of the North American Plate, was generally a time of deposition. Sedimentary layers in shallow lake basins and inland seas built up into thick deposits over earlier Precambrian rocks. In the late Cretaceous and early Tertiary periods, a time of intense deformation and construction within the North American Plate, known as the Laramide Orogeny, resulted in the folding and uplift of these deposits. In some areas magmas found their way to the surface, ultimately becoming extensive areas of volcanic rock. Although most of the intense tectonic activity was over by the late Eocene and early Oligocene, some continues today, and there is evidence that ranges such as the Tetons are still being uplifted, or at least were until recently.

The highest summits of the Middle Rocky Mountains today are considered to be former projections above the old Rocky Mountain Peneplain (Atwood & Atwood, 1938), which was formed during the mid-Tertiary period. The landscape at that time was one of monadnock peaks along divides and crestlines surrounded by sedimentary deposits. The sediments of this low-relief landscape have long since been removed by erosion, leaving high remnant surfaces in many of the mountain ranges. Wind River Peneplain, Gilbert Peak Surface, Cross Lake Surface, and Medicine Bow Peneplain are only a few of the local names that have been applied to remnants of this regional surface. Thus, the story since the end of the mid-Tertiary has been one of basin filling and erosional sculpting of the ranges by wind, water, and most recently by Pleistocene ice.

Glaciation: The Pleistocene and the Holocene

The Pleistocene and Holocene epochs make up the Quaternary period, which began 1.65 Ma (Richmond, 1986). Pleistocene ice events had profound effects on the mountain landscapes and on the kinds and numbers of habitats available to organisms. Plants, being nonmotile, were particularly vulnerable to the effects of climatic changes associated with episodes of continental glaciation.

The Pleistocene in the Northern Hemisphere was characterized by at least four major ice advances and recessions (see Richmond and Fullerton, 1986, for a detailed discussion of these). The last two (measured in thousands of years, or ka) were the Illinoian (302–132 ka) and the Wisconsin (122–10 ka), which can be divided into the Eowisconsin,

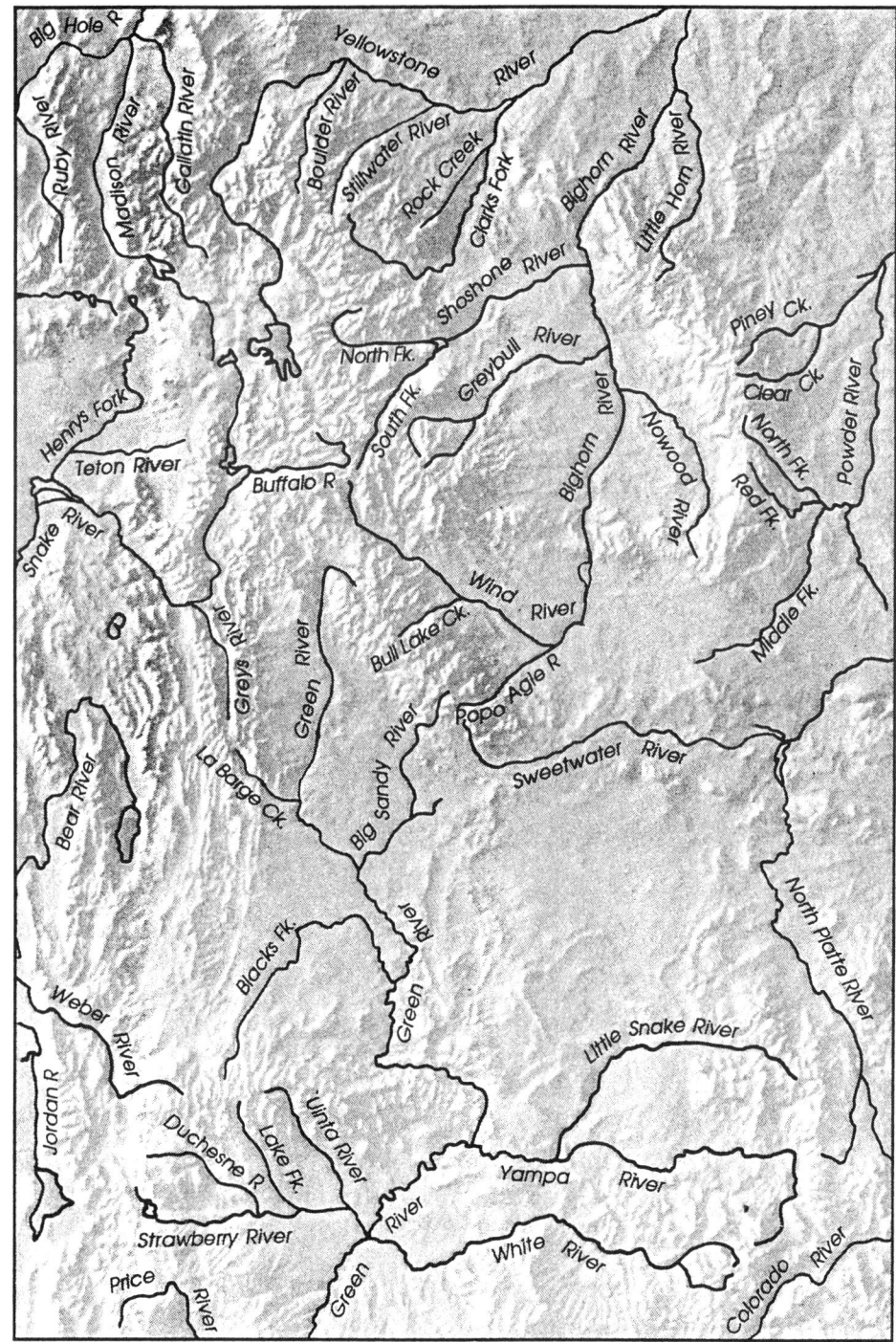

Figure 6. Major drainages and tributaries of the Middle Rocky Mountains.

the Early Wisconsin, the Middle Wisconsin, and the Late Wisconsin. Lobes of the continental ice sheet flowed south during the Illinoian and Wisconsin to about 47° north latitude in Montana. In the Middle Rocky Mountains, south of the ice sheet margin, there were several large ice centers and ice caps. Large valley glaciers advanced from these and flowed down to the plains (Figure 8). The most recent major advances from the mountains are the Bull Lake (Illinoian) and the Pinedale (Wisconsin). In addition, relatively small moraines at upper elevations in the Wind River Range, estimated to be about 12,000 years old, represent a very late Wisconsin advance that has been named the

Figure 7. The geologic time scale (after Knight, 1990).

EON	ERA	PERIOD	EPOCH	TIME*
PHANEROZOIC	Cenozoic	Quaternary	Holocene Pleistocene	-- 0.01 -- 1.6
		Tertiary	Pliocene Miocene Oligocene Eocene Paleocene	-- 5.3 -- 23.7 -- 36.6 -- 57.8 -- 66.4
	Mesozoic	Cretaceous Jurassic Triassic		-- 144 -- 208 -- 245
	Paleozoic	Permian Pennsylvanian Mississippian Devonian Silurian Ordovician Cambrian		-- 286 -- 320 -- 360 -- 408 -- 438 -- 505 -- 570
PRECAMBRIAN	Proterozoic			-- 2,500
	Archean			-- 4,600

* In millions of years before present; not to scale.

Donald Creek Advance (Richmond, 1984). The Bull Lake and Pinedale Advances apparently occurred in all the Middle Rocky Mountain ranges. There is evidence of Donald Creek moraines in other ranges, particularly the Uintas. The first two advances resulted in glacial sculpting of cirques, peaks, and ridges at higher elevations and deposition of large moraines at lower elevations. Glacial scour created numerous lakes at all elevations, although the best developed are in foothills canyons and adjacent valleys and basins.

The periglacial climate of the intermontane basins between the ranges must have been prevailingly cold during the late Pleistocene. The occurrence of fossil frost wedges and nonsorted polygons (Mears, 1981) indicates the presence of permafrost and with it a cold, arid, windy, steppe-tundra environment 10–13°C (18–23°F) colder than today. The basins, or portions of them, could have served as refugia for alpine plants excluded from the mountain ranges by ice caps and ice centers. This would have been a system very similar to the Alaskan and Yukon refugia described by Hultén (1937), from which arctic-alpine plants migrated back into surrounding mountain ranges when the Pleistocene ice receded.

The end of the Pleistocene and beginning of the Holocene (or Recent) epoch occurred at roughly 12,000–10,000 B.P. (before the present), although the exact time varies from one location in North America to another. For purposes of uniformity, Hopkins (1975) assigned the Pleistocene-Holocene boundary an arbitrary age of 10,000 B.P. Since that time, minor glacial advances have occurred, reaching maxima around 5000 B.P. during the "Neoglacial" (Porter and Denton, 1967). A recent maximum during the period 1550–1850 A.D. (300–100 B.P.) has been named the Little Ice Age or Lesser Ice Age (Matthes, 1942). These Post-Wisconsin Neoglacial advances, or stades, created relatively fresh glaciated surfaces and moraines, mostly at higher elevations, which, as habitat, have considerable influence on alpine plant distribution. Judging from the degree of weathering, lichen colonies, carbon-14 dates, and the amount of vegetation cover, the Neoglacial moraines appear to range in age from 8000 to 100 B.P. Four stades that are reasonably

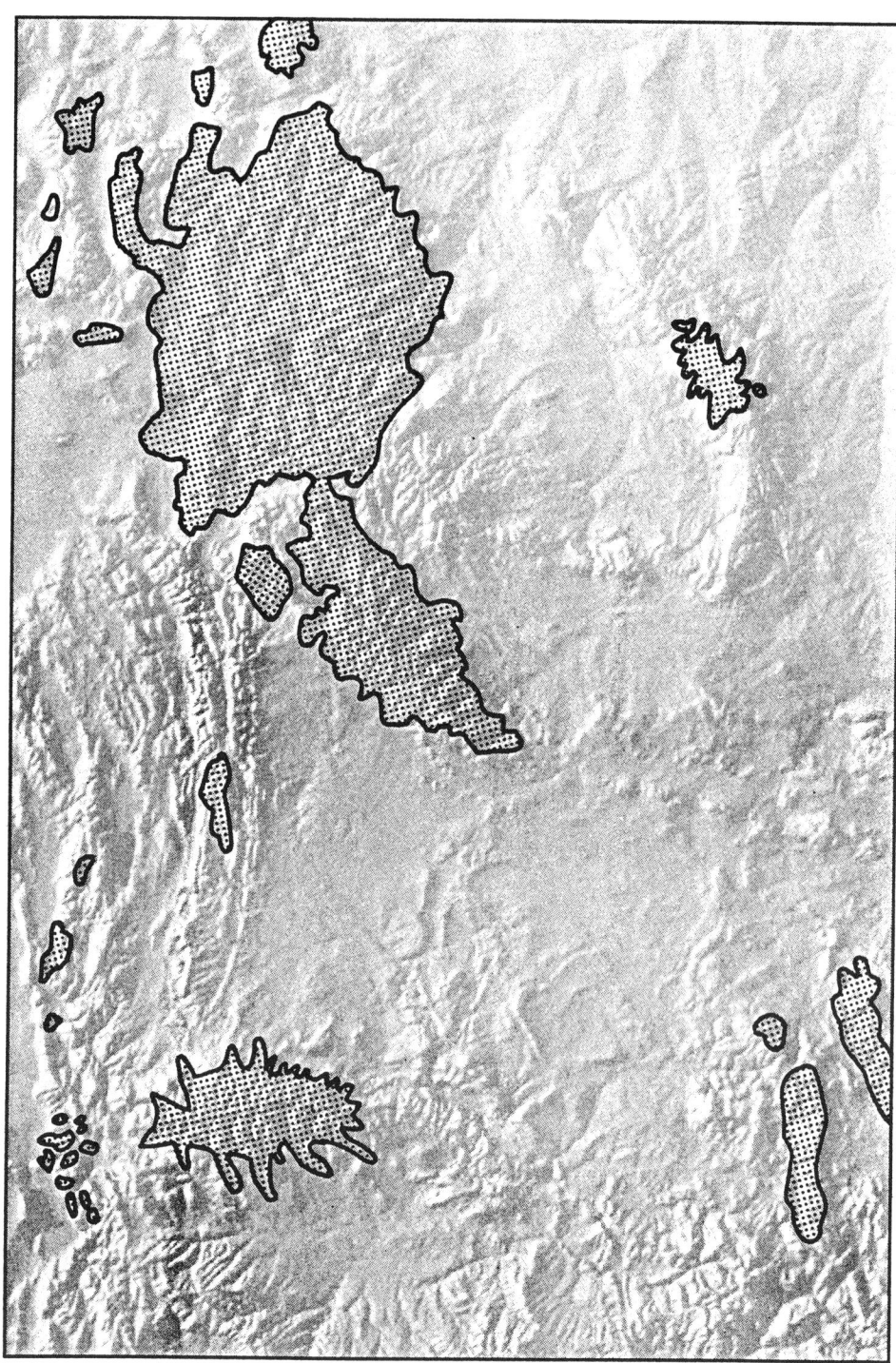

Figure 8. Pleistocene ice distribution in the Middle Rocky Mountains. Areas covered include the approximate boundaries of both Bull Lake and Pinedale advances (modified from Montagne, 1972).

well represented throughout the Middle Rocky Mountains are indicated in Table 2. Moraines of the Gannett Peak Stade form fresh, poorly vegetated surfaces close to alpine cirque walls and near the snouts of existing glaciers.

The Mountain Ranges

Based on general structure and method of formation there are four types of ranges in the Middle Rocky Mountains (Figure 9). First, along the western edge, are the highly folded

Table 2. Stades of the Middle Rocky Mountains

Stade	Age (B.P.)	Time*
Temple Peak	8000–6200	Early Neoglacial
Indian Basin	5000–3000	Middle Neoglacial
Audubon	2000–1000	Late Neoglacial
Gannett Peak	300–100	Little Ice Age

*The term *Neoglacial* is used in the broadest sense; Heuberger (1974) pointed out that glacial advances have been nearly continuous since the Wisconsin, and Neoglacial may not describe a single time period of glacial maxima, as originally intended.

and thrust-faulted sedimentary ranges that make up the Overthrust Belt; these include the Wyoming, Salt River, and Hoback Ranges. Second, to the south, is the fault-blocked Wasatch Range, and to the north its smaller twin, the Teton Range. North and east is the third type, the large area of mostly horizontal, stratified volcanic deposits that make up the Absaroka Range. The fourth type lies east and south of the Absaroka volcanics and includes the folded and faulted anticlines of the Beartooth, Big Horn, Gros Ventre, Wind River, Uinta, and Medicine Bow Ranges. All of the latter ranges are bounded on one or more sides by downfolded structures (synclines) and/or fault blocks that are manifested as intermontane or marginal basins. The locations and characteristics of the alpine ranges of the Middle Rocky Mountains, in roughly north to south order, are listed below.

Beartooth Range

The Beartooth Range, the highest range in Montana, occupies areas of both southern Montana and northern Wyoming. It is bounded on the west by the valley of the Yellowstone River. On the southern border it is separated from the Absarokas by the Lamar River and Clark's Fork, both tributaries of the Yellowstone River. The range is an anticlinal thrust wedge that was thrust to the northeast during the Laramide Orogeny. Foose (1960, *in* Eardley, 1962) calculated the thrusting at as much as 3048 m (10,000 ft). Subsequent erosion has exposed a large expanse of Precambrian rock, which makes up a significant portion of the range. The range's extensive alpine topography is the result of Pleistocene climate and glaciation, which created U-shaped valleys, glacial cirques, and fellfields. Small cirque glaciers still exist, and numerous lakes occupy basins excavated by ice.

The Beartooth Range is composed of three northwest-trending belts, each made up of one or more rock units (Mueller et al., 1988). The southwestern Beartooths on the Yellowstone Park boundary, or the South Snowy Block, are Precambrian crystalline rocks overlain in places by Paleozoic sediments and Tertiary volcanics. Along the northern edge of the South Snowy Block is a broad zone of volcanic deposits originating from a vent system beginning in the vicinity of Cooke City, Montana, and extending northwesterly through the entire range. North of this volcanic band, which has been called the Cooke City Sag (Foose et al., 1961), is the largest expanse of Precambrian granitic and metamorphic rocks in the range. It includes the North Snowy Block on the northwest and the main Beartooth massif (Reid et al., 1975) extending southeasterly to Clark's Fork of the Yellowstone in Wyoming. Sedimentary rocks occur along the margins of this portion of the Beartooth uplift. In a few places, such as Clay Butte and Beartooth Butte, sedimentary rocks belonging to the Devonian Beartooth Butte formation may be found above timberline. The highest peak, Granite Peak, 3901 m (12,799 ft), is also the highest in Montana.

Absaroka Range

The Absaroka Range occupies a significant area of northwestern Wyoming, including a major portion of Yellowstone Park. In this treatment the northern Absarokas include the area north of the North Fork of the Shoshone River in Wyoming that extends to the

Figure 9. Structure of Middle Rocky Mountain ranges and associated basins: 1a) anticlines; 1b) asymmetric anticline with thrust fault; 2) fault block; 3) volcanic plateau; 4) overthrust.

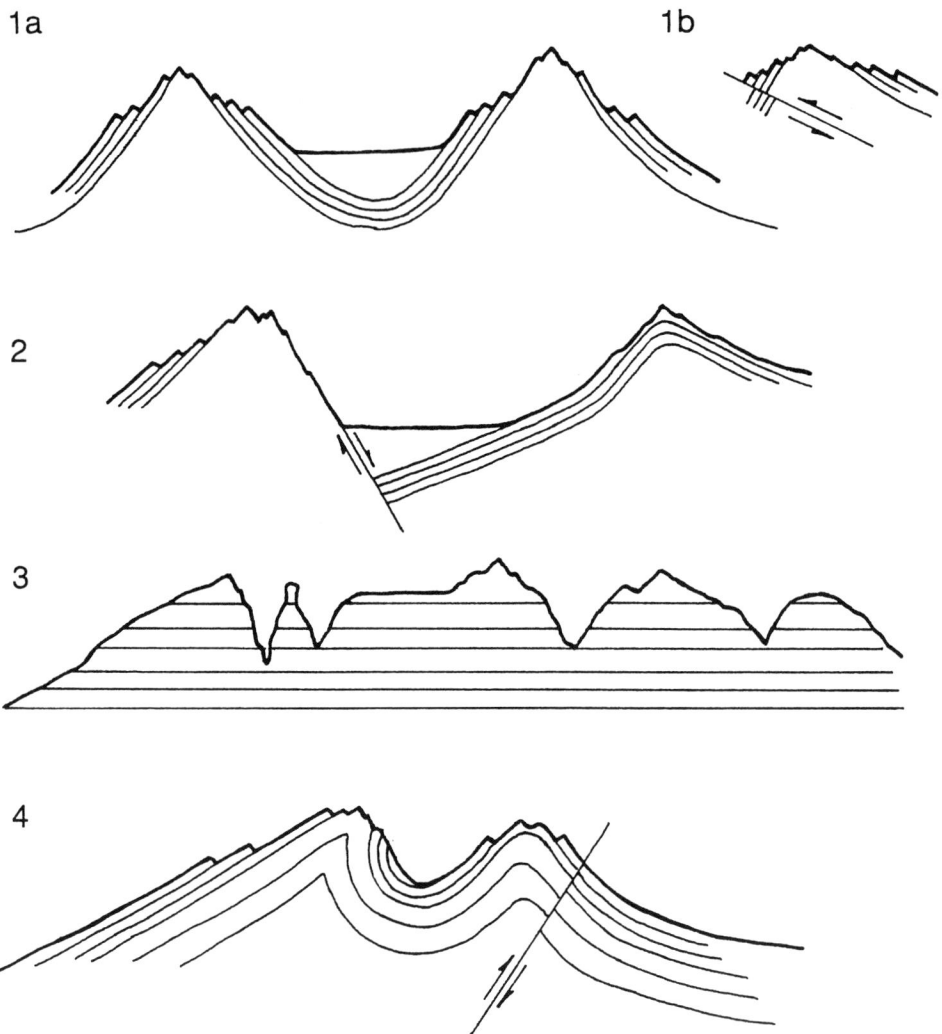

southern Montana border east and south of the Yellowstone River. South of the North Fork of the Shoshone River is the southern Absaroka Range, the "Shoshone Range" of earlier authors (Fenneman, 1931), which is separated from the Gros Ventre and Wind River Ranges to the south and southeast by the timbered, low mountains of the Mount Leidy Highlands. The Absarokas are a volcanic plateau and differ from all other Middle Rocky Mountain ranges in being a depositional feature rather than an uplift. The rocks are mostly Eocene age (49–44 Ma), although a few younger rocks are known to have been deposited as recently as 480,000 ± 60,000 B.P. at the southern limits of the range (J. D. Love & Love, 1983). Very evident, especially in the southern Absarokas, are horizontal strata made up of ash layers, conglomerates, and breccias. All such strata have varying amounts of ash and are very soft; as a consequence, the range is highly dissected, with uplands of rolling plateau remnants separated by very narrow and deep river valleys. The uplands in these areas are reminiscent of the arctic tundra or the Scottish moors. Some northern portions of the Absarokas developed on basalts, which lend a different morphology of high, sharp peaks, such as Sunlight and Stinkingwater Peaks, south of the Montana border. There are no large, flat-topped plateaus in the northeastern portions of the range as there are in the south. Near the extreme southern rim of the Absaroka volcanics is the almost completely buried Washakie Range of Cretaceous age (J. D. Love, 1939), an extension of the nonalpine Owl Creek Mountain and Bridger Mountain uplift which connects the southern Big Horns with the Absarokas. The summits of Black

Mountain and Crow Mountain north of U.S. Highway 287 between Riverton and Dubois are part of this range. The Washakie Range is significant botanically because of the isolated sedimentary rock lenses that are being uncovered as the overlying Absaroka volcanics are eroded away. Few lakes occur in the Absarokas due to the soft, porous volcanic rocks. The highest peak, Franc's Peak, is 4009 m (13,153 ft).

Big Horn Range

The Big Horn Range is a curved, nearly J-shaped structure trending northwest to southeast. It is a very complex, asymmetric anticline that is thrust (and therefore steepened) toward the east, with gentler slopes at some locations on the west, interspersed with extremely steep faces. Foothills are lacking for the most part. The range can be broken roughly into three sections based on morphology; the north and south ends are steeper on the west side due to asymmetrical uplift, and the central section is steeper on the east, again due to asymmetry of uplift accompanied by some overthrusting. Much of the crest of the range is made up of Paleozoic sedimentary rocks, although there is a core of Precambrian igneous and metamorphic rocks that is particularly well exposed in the central Cloud Peak region. This region was glacierized during the Pleistocene and exhibits well-developed high alpine topography. The north and south portions with Paleozoic rocks (mainly limestones and dolomites) above timberline are characterized by gentle slopes and rolling plateaus and hills. To the north the Big Horns grade into the nonalpine Pryor Mountains of Montana. To the south, at the tip of the J, they grade into the Bridger Mountains, which, along with the Owl Creek Mountains, form a connection with the southern Absaroka Range. The highest peak is Cloud Peak, at 4013 m (13,167 ft).

Gros Ventre Range

Between Jackson Hole to the west and the Wind River Range to the east is the Gros Ventre Range. It trends northwest to southeast, forming a small oval with a connection to the Hoback Range on the southwest side. To the north, the nonalpine Mount Leidy Highlands separate the Gros Ventre from the southern Absarokas. This range is an asymmetrical anticline of mostly Paleozoic sedimentary rocks with a few small, isolated Precambrian exposures of igneous and metamorphic rocks. Thrust faulting and folding have occurred along the southwest margin, causing the block to be much steeper on that side than along the northeastern margin, where it dips under the buried Washakie Range at the margin of the southern Absarokas. The range is one of lower summits, as demonstrated by its highest mountain, Doubletop Peak, at 3561 m (11,682 ft).

Teton Range

Probably the most spectacular of the Middle Rocky Mountain ranges, the Tetons are a fault block (normal fault) range bordered on the east by the valley of Jackson Hole. The valley is the downthrown block, with the range itself being the upthrown block. This upthrown fault block was lifted more to the east, where it forms the crest of the range, the block itself being almost undeformed. The fault is nearly vertical, resulting in a steep eastern face and gentler slopes to the west. The younger, overlying Paleozoic sediments have been almost completely eroded away from the crest, leaving most of the higher summits composed of Precambrian rocks, mostly metamorphic schists and gneisses. At least one sedimentary cap remains; there is a thin layer of Cambrian Flathead Sandstone on the summit of Mount Moran. The range was glaciated during the Pleistocene, resulting in steep, jagged topography now occupied by a few small cirque glaciers. Sharp mountain peaks are separated by steep-sided, U-shaped valleys. Moraines left from the ice advances that created these valleys are prominent features in Jackson Hole near the base of the range. Most of the moraines are easily detected by vegetation differences. Big sage and various grass species grow on the well-drained glacial outwash of the valley, while lodgepole pine occupies the more poorly sorted (and moisture-retaining) moraines. To the south, the lower and less spectacular segment of the Teton uplift in Wyoming and adjacent Idaho is designated the Snake River Range by some authors (Bonney & Bonney, 1977). This area, approximately 35 by 27 km (22 by 17 mi) at the longest and widest points, connects the Tetons with the Salt River and Wyoming Ranges of the Overthrust Belt. The area is treated here as the southern portion of the Teton Range with a margin formed by the canyon of the superposed Snake River. The highest peak in this section is Mt. Baird Peak, at 3060 m (10,040 ft). The highest peak in the entire range is the Grand Teton, at 4197 m (13,770 ft).

Wind River Range

Although not the largest range, the Wind River Range has one of the most extensive alpine areas in the coterminous United States. It is the largest discrete mountain mass in Wyoming (Blackstone, 1971) and is an asymmetric, anticlinal thrust wedge that has been thrust to the southwest, its steepest side. As a result, Precambrian rocks are exposed down to the adjoining valley level. The shallower northeast flank is characterized by sedimentary hogback ridges and steep rock wedges known as flatirons. The Continental Divide trends northwest-southeast and forms the crest along most of the length of the range. It has an extensive Precambrian core of igneous and metamorphic rocks with younger Paleozoic and Mesozoic sedimentary rocks exposed on the northeast flanks. Some of these sedimentary rocks, particularly limestones and dolomites, occur above timberline in the northern part of the range. The extensive Wind River Fault on the southwest side of the range results in very steep flanks lacking sedimentary formations. Above this fault is a broad, benchlike structure at about 2865 m (9400 ft), the Cross Lake Surface, which contains numerous glacial lakes and is possibly a peneplain remnant (Atwood & Atwood, 1938). The northern portion of the range is glacierized, with several large valley glaciers and many smaller cirque and niche glaciers. The alpine topography is extensive and spectacular, with numerous features such as cirques, horn peaks, aretes, and moraines. Lakes, especially glacial lakes, are abundant everywhere in the range. Pleistocene glaciers built extensive moraine systems near the mouths of canyons at the base of the range, and a number of these have dammed streams to form deep, narrow foothills lakes. The highest summit, Gannett Peak, at 4208 m (13,804 ft), is the highest mountain in the Middle Rocky Mountains.

Overthrust Ranges

The Wyoming overthrust ranges treated here are three small, parallel ranges that are more or less linear in shape, have limited and somewhat disjunct alpine areas, and have summits lower than all the other alpine ranges of the Middle Rocky Mountains. They have been both folded and often overturned, and forcibly thrust eastward along fault planes. They are characterized by steep slopes with Paleozoic sedimentary rocks above timberline and Mesozoic rocks in the valleys. The alpine zone consists of extensive scree and talus slopes, sharp ridges, and generally unstable terrain. Precambrian rocks are lacking. Lakes and ponds are few and totally lacking above timberline.

Hoback Range. The Hoback Range, a small, oval range at the northern tip of the Wyoming Range, is defined by the Hoback River to the southeast, east, and north; the Snake River to the north; and the Little Greys River to the west. The highest summit is Hoback Peak, at 3311 m (10,862 ft).

Wyoming Range. The Wyoming Range borders the Green River Basin on the west and runs north-south along the east side of the Greys River. It is a steep, thrust-faulted range with the highest elevations and most extensive alpine areas of the three overthrust ranges. The highest mountain, Wyoming Peak, is 3463 m (11,363 ft).

Salt River Range. The Salt River Range borders the Greys River on the west and runs north-south, paralleling the Wyoming Range to the east. Like the Wyoming Range, it is steep and thrust-faulted. Mount Fitzpatrick, at 3325 m (10,907 ft), is the highest mountain.

North and west of the Salt River Range are three small, rugged, and spectacular ranges. The Snake River Range of Wyoming and Idaho connects the Salt Rivers with the

Tetons to the north. Northwestward, the Snake River Range becomes the Big Hole Mountains of Idaho. Across the Snake River valley directly to the west is the Caribou Range, also in Idaho. All are characterized by fairly extensive, discontinuous subalpine balds that resemble alpine tundra and contain a number of alpine species. The summit elevations are all less than 3048 m (10,000 ft), with the exception of Mt. Baird of the Snake River Range, at 3056 m (10,025 ft). Because of their low summits, strong subalpine character, and lack of continuity with adjacent alpine ranges, the two Idaho ranges are not considered to be alpine. The Snake River Range is treated here as a southern extension of Teton Range, occupying a 35-km (22-mi) segment between Teton Pass and the canyon of the superposed Snake River.

Medicine Bow Range

The Medicine Bow Range occupies the western edge of the Laramie Basin in southeastern Wyoming. It is an anticlinal range, asymmetrically thrust to the east over the western margin of the Laramie Basin (Eardley, 1962) during the Laramide Orogeny. The rocks of the range are Precambrian igneous and metamorphic, primarily gneisses and quartzites. Paleozoic and Mesozoic rocks are exposed on the lower flanks. Alpine terrain is concentrated in a relatively small area known as the Snowy Range, which has been identified as a monadnock rising above the surrounding peneplainal surface of the northern Medicine Bows (Atwood & Atwood, 1938). This peneplain occurs as a rather broad, flattish surface at an elevation of about 2743 m (9000 ft). A second cluster of alpine mountains in the southern portion of the range is called the Rawah Peaks (Blackstone, 1971). The Medicine Bow Mountains, specifically the Snowy Range, were extensively modified by Pleistocene glaciation (Houston et al., 1968). Alpine features such as cirques, moraines, and protalus ramparts are evident near the summit crests. The highest mountain, Medicine Bow Peak, is 3659 m (12,006 ft).

Wasatch Range

The Wasatch Range trends north-south in northeastern Utah to form the southwestern edge of the Middle Rocky Mountains. In the central portion of the range, the Wasatch Front forms the eastern edge of the Salt Lake Valley. Like the Tetons, this is a fault block range with the upthrown block rising steeply and asymmetrically from the downthrown valley block. In the north the Wasatch Range blends into the nonalpine Bear River Range, which connects it with the Wyoming ranges of the Overthrust Belt. In the south it terminates just south of Mount Nebo, east of the town of Nephi. The Wasatch Range is a complex block of folded and faulted Paleozoic and Mesozoic sedimentary rocks with some Precambrian rocks locally exposed through erosion. Limestones are abundant, even above timberline. The range is dissected by deep canyons and valleys that trend east-west. Glacial features, including small but well-developed cirques, are evident in the central portion of the range. Moraines occur in many of the lower or foothills canyons. Generally the west flank is much steeper than the more gradually sloping east side of the fault block. Steep, unstable scree slopes are widespread and common. The major peaks, and thus most of the alpine areas, are located along the western edge. The highest mountain is 3636-m (11,928-ft) Mount Nebo at the extreme southern tip of the range. Mount Timpanogos, in the center of the range, at 3581 m (11,750 ft), is cited in some older publications as the highest in the range.

Uinta Range

The Uinta Range occupies an east-west band in northeastern Utah south of the Wyoming border. It is one of the few ranges, and the largest one in North America, with an east-west orientation. It is a broad, flat-topped anticline, slightly asymmetric to the north, the steepest flank. Erosion has removed the overlying younger sediments to expose the Precambrian core of igneous and metamorphic rocks, mainly quartzites. Paleozoic and Mesozoic limestones, sandstones, and shales are still present on the lower flanks of the range. Pleistocene glaciation was intense here, and nearly all the central surfaces were covered by ice. Cirques in various stages of development, from nivation basins to compound systems, may be observed, along with U-shaped valleys, moraines, and horn peaks. Glacial lakes are abundant. Because of the flatness of the anticline, many of the ridges left between glacial cirques have nearly the same elevation and appearance as the crest of the range. This gives rise to a characteristic topography of flat-topped ridges separated by cirque basins. The Uinta Range forms the southeastern margin of the Middle Rocky Mountains, separating them from the Colorado Plateau to the south. Like the Wind River Range, it has an extensive and large alpine area along the crest. The highest summit, Kings Peak, is 4123 m (13,528 ft).

The Flora

Characteristics

ALL THE FLOWERING PLANTS and vascular cryptogams known to occur above timberline in the Middle Rocky Mountains are described in this manual, as are the timberline conifers. The flora consists of 700 taxa including 55 subspecies and 314 varieties and belonging to 51 families, 204 genera, and 609 species (Table 3). Thirteen of the species belong to the Division Pteridophyta, the vascular cryptogams or fern allies; and 596 species are in the Division Spermatophyta. The spermatophytes are represented by the Subdivision Gymnospermae, the gymnosperms, with 8 species in the Class Coniferae, the conifers. In the Subdivision Angiospermae, the flowering plants, there are 133 species in the Class Monocotyledoneae, the monocots; and 455 species in the Class Dicotyledoneae, the dicots. The 3 largest families are Asteraceae (108 species), Poaceae (55 species) and Cyperaceae (54 species). Thirteen families are represented by a single genus and species.

Table 3. Middle Rocky Mountain alpine taxa

Taxon	Genera	Species	Subspecies	Varieties	Total Taxa
DIVISION PTERIDOPHYTA					
Adiantaceae	2	2	0	0	2
Aspleniaceae	5	6	0	3	6
Isoetaceae	1	1	0	0	1
Lycopodiaceae	1	1	0	0	1
Ophioglossaceae	1	1	1	0	1
Selaginellaceae	1	2	0	1	2
Total: 6	11	13	1	4	13
DIVISION SPERMATOPHYTA					
Subdivision Gymnospermae, Class Coniferae					
Cupressaceae	1	2	0	2	2
Pinaceae	4	6	0	1	6
Total: 2	5	8	0	3	8
Subdivision Angiospermae, Class Monocotyledonae					
Callitrichaceae	1	1	0	0	1
Cyperaceae	4	54	5	22	62
Hydrocharitaceae	1	1	0	0	1
Juncaceae	2	11	1	4	11
Liliaceae	6	9	0	2	9
Orchidaceae	2	2	0	3	3
Poaceae	18	55	9	25	66
Total: 7	34	133	15	56	153
Subdivision Angiospermae, Class Dicotyledonae					
Apiaceae	8	17	3	1	17
Asteraceae	29	108	8	56	126
Betulaceae	1	1	0	0	1
Boraginaceae	5	11	0	3	12

Taxon	Genera	Species	Subspecies	Varieties	Total Taxa
Brassicaceae	10	37	1	31	46
Campanulaceae	1	2	0	1	2
Caprifoliaceae	2	2	1	3	3
Caryophyllaceae	9	24	1	12	28
Celastraceae	1	1	0	0	1
Chenopodiaceae	2	3	0	2	3
Crassulaceae	1	5	0	2	5
Ericaceae	6	12	2	3	12
Fabaceae	5	26	5	7	30
Fumariaceae	1	1	0	0	1
Gentianaceae	5	10	1	1	10
Geraniaceae	1	2	0	0	2
Grossulariaceae	1	4	1	1	4
Hydrophyllaceae	1	2	0	2	2
Hypericaceae	1	1	1	0	1
Lamiaceae	1	1	0	1	1
Linaceae	1	1	0	1	1
Onagraceae	3	8	0	6	9
Papaveraceae	1	1	0	0	1
Plantaginaceae	1	1	0	0	1
Polemoniaceae	6	11	2	6	14
Polygonaceae	5	17	2	12	23
Portulacaceae	3	5	0	4	5
Primulaceae	4	7	0	6	9
Ranunculaceae	7	27	3	14	33
Rosaceae	11	29	2	20	35
Rubiaceae	2	2	0	0	2
Salicaceae	2	14	2	8	16
Saxifragaceae	6	21	3	15	24
Scrophulariaceae	9	34	0	18	39
Valerianaceae	1	3	1	1	3
Violaceae	1	4	0	4	4
Total: 36	154	455	39	251	526
TOTAL OF ALL DIVISIONS: 51	204	609	55	314	700

Table 4. The number of taxa in the circumpolar arctic flora and the Middle Rocky Mountain alpine flora. Arctic flora from Polunin (1959).

	Taxonomic Group	Proportion[a]
Circumpolar flora		
Families	66	1.0
Genera	230	0.287
Species	892	0.074
Middle Rocky Mountain flora		
Families	51	1.0
Genera	204	0.250
Species	609	0.084

[a]The proportion was calculated by dividing the number of families by the number of taxa in each taxonomic rank.

Figure 10. Sizes of alpine floras in the Middle Rocky Mountains.

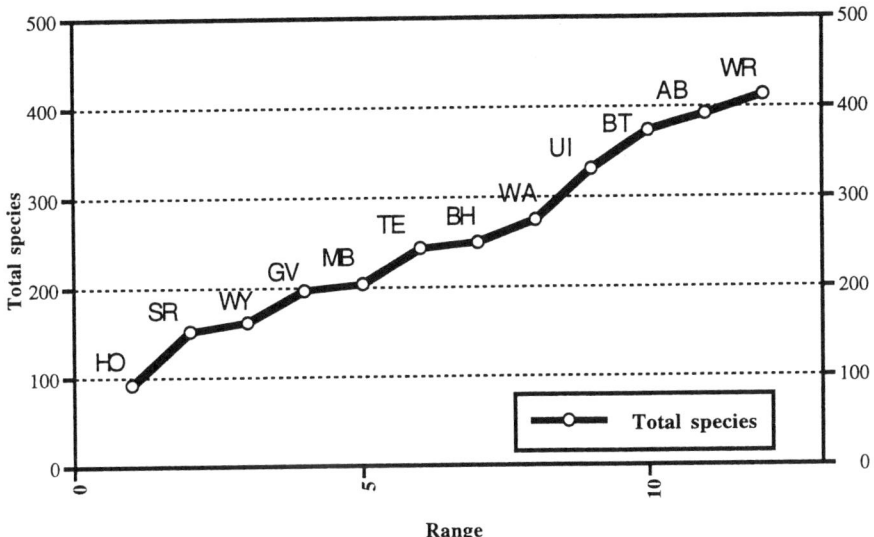

The largest genera, each with more than 10 species, are *Carex* (47), *Erigeron* (21), *Potentilla* (16), *Senecio* (15), *Draba* (15), *Salix* (13), *Poa* (12), *Saxifraga* (12), and *Ranunculus* (11). Of the 204 genera present, 96 are represented by a single species.

Both arctic and alpine vegetation units are often very similar in appearance and share similar plant growth forms. Both are referred to as tundra. Both regions have at least some species in common at the local microenvironmental level.

A comparison of the Middle Rocky Mountain alpine flora with the circumpolar arctic flora also demonstrates some similarity in the percentages of genera and species present (Table 4). Both regions are either above or north of climatic treeline, but the arctic flora, which occupies a much larger area, has 5 more families, 26 more genera, and 283 more species. Such figures have to be interpreted with caution, however, since the species concepts of different authors have considerable influence on the numbers of taxa said to be in common (or not in common). A similar numerical relationship appears between individual ranges in the Middle Rocky Mountains.

The number of species present, or the species richness, varies from one mountain range to the next. It seems to be best associated with the areal extent of the range; the larger the range, the greater the species richness. MacArthur and Wilson (1967) attributed this to the greater habitat space available in larger areas. The proximity of a range to other high mountain ranges also contributes to species richness, as does mountain mass (Price, 1981; Arno & Hammerly, 1984). One would thus expect small alpine floras in small ranges and proportionately larger floras in larger ranges, especially those in close proximity. This holds reasonably true in the case of Middle Rocky Mountain alpine ranges (Figure 10). The largest ranges with correspondingly large alpine areas are the Absaroka, Beartooth and Wind River Ranges, which also have the greatest number of species: 392, 374, and 413, respectively. The smallest range in areal extent, the Hoback, has only 92 species, fewer than any other range.

A second characteristic of Rocky Mountain alpine floras in general and the Middle Rockies in particular is the presence of disjunct species. Among the species that occur once or only a few times in the Middle Rocky Mountains are *Antennaria monocephala* DC., *Erigeron lanatus* Hook., *Saussurea weberi* Hult., *Draba fladnizensis* Wulf., *D. glabella* Pursh, *D. porsildii* Mulligan, *Parrya nudicaulis* (L.) Regel, and *Potentilla nana* Willd. ex Schlecht. Some are arctic species that have their southernmost distributional limits in the Middle Rockies, and some overlap into Middle Rocky Mountain ranges from adjacent Southern or Northern Rocky Mountain regions.

Among the species endemic to one or more mountain ranges of the Middle Rocky

Mountains, or with a restricted distribution pattern that includes a portion of the area or seems to be centered on the Middle Rockies, are *Angelica grayi* (Coult. & Rose) Coult. & Rose, *Cymopteris evertii* Hartman & Kirkpatrick, *Erigeron garrettii* A. Nels., *Erigeron goodrichii* Welsh, *Arabis williamsii* Roll., *Descurainia torulosa* Roll., *Lesquerella garrettii* Pays., *L. paysonii* Roll., *L. utahensis* Rydb., *Thlaspi parviflorum* A. Nels., and *Astragalus molybdenus* Barneby. Species usually associated with the plains or lower mountain zones that reach their maximum elevational limit in the alpine zone, usually near or just above timberline, include *Chrysothamnus viscidiflorus* (Hook.) Nutt., *Eriophyllum lanatum* (Pursh) Forbes, *Monolepis nuttalliana* (Schultes) Greene, *Nothocalais nigrescens* (Henders.) A. Heller, and *Oenothera cespitosa* Fraser. In the Wasatch Range, the shrubs *Cercocarpus ledifolius* Nutt. ex T. & G. and *Holodiscus dumosus* (Nutt.) A. Heller are members of the latter group.

The distribution patterns of cosmopolitan weeds such as *Cerastium arvense* L., *Taraxacum officinale* G.H. Weber ex Wiggers, and *Tragopogon dubius* Scop. vary from widespread to restricted in the Middle Rocky Mountain alpine zone. Finally, a few lowland oddities such as *Populus balsamifera* L. are included in the flora based on collections from elevations above timberline. *Populus tremuloides* Michx. has been observed forming timberline stands on south-facing rocky slopes in the northern Wind River Range.

The alpine flora thus consists of many elements and groups of species that have affinities with other geographical areas and elevational zones. In general, and regardless of the affinity, the number of vascular plant species and the number of groups represented in the flora decrease with elevation, and no vascular plants occur on the very highest summits. The 11 species that have been documented at the highest known elevations for vascular plants in the Middle Rocky Mountains, just above 4084 m (13,400 ft), are *Draba ventosa* Gray and *Smelowskia calycina* (Steph. ex Willd.) Mey. (Brassicaceae), *Minuartia obtusiloba* (Rydb.) House and *Sagina saginoides* (L.) Karst. (Caryophyllaceae), *Carex rupestris* Allioni (Cyperaceae), *Trifolium nanum* Torr. (Fabaceae), *Poa lettermannii* Vasey (Poaceae), *Claytonia megarhiza* (Gray) Parry ex S. Wats. (Portulacaceae), and *Saxifraga cespitosa* L., *S. chrysantha* Gray, and *S. flagellaris* Willd. ex Sternb. (Saxifragaceae). To my knowledge no vascular plant species occurs in the narrow band above 4110 m (13,481 ft) in the three highest ranges—the Tetons, Uintas, and Wind Rivers, which have maximum elevations of 4197 m (13,771 ft), 4123 m (13,528 ft), and 4208 m (13,804 ft), respectively.

Species not included in this flora that might be expected based on their alpine occurrence in nearby mountain ranges, or near timberline within our area, are (ordered by family, genus, and species) Aspleniaceae: *Woodsia oregana* D.C. Eat.; Asteraceae: *Arnica alpina* (L.) Olin & Ladau, *Aster campestris* Nutt., *A. engelmannii* (D.C. Eat.) Gray, *Erigeron coulteri* Porter, *E. divergens* T. & G., *Haplopappus lanuginosus* Gray, *Senecio crocatus* Rydb., *S. cymbalaria* Pursh; Brassicaceae: *Descurainia sophia* (L.) Webb ex Prantl, *Draba streptocarpa* Gray, *Streptanthus cordatus* Nutt.; Campanulaceae: *Campanula parryi* Gray, *C. scabrella* Engelm.; Caryophyllaceae: *Stellaria americana* (Porter ex Robins.) Standley; Cyperaceae: *Kobresia simpliciuscula* (Wahl.) Mack.; Ericaceae: *Cassiope mertensiana* (Bong.) D. Don; Fabaceae: *Lupinus sericeus* Pursh; Hydrophyllaceae: *Phacelia lyallii* (Gray) Rydb.; Liliaceae: *Stenanthium occidentale* Gray, *Xerophyllum tenax* (Pursh) Nutt.; Linaceae: *Linum kingii* S. Wats.; Polygonaceae: *Erigonum caespitosum* Nutt., *E. chrysops* Rybd.; Pinaceae: *Larix lyallii* Parl.; Primulaceae: *Primula angustifolia* Torr.; Ranunculaceae: *Delphinium glaucum* S. Wats.; Rosaceae: *Dryas drummondii* Richards. ex Hook., *Luetkea pectinata* (Pursh) Kuntze, *Spiraea splendens* Baumann ex Koch; Santalaceae: *Comandra umbellata* (L.) Nutt.; Saxifragaceae: *Saxifraga lyallii* Engl.; Scrophulariaceae: *Castilleja linariifolia* Benth., *Penstemon alpinus* Torr., *P. ellipticus* Coult. & Fish., *P. leonardii* Rydb.; and Valerianaceae: *Valeriana sitchensis* Bong. Of these, *Stellaria americana* (Porter ex Robins.) Standley (Caryophyllaceae), *Kobresia simpliciuscula* (Cyperaceae), *Cassiope mertensiana* (Bong.) D. Don (Ericaceae), *Xerophyllum tenax*

(Pursh) Nutt. (Liliaceae), and *Saxifraga lyallii* Engl. (Saxifragaceae) have recently been collected from timberline and the lower alpine zone of the northern margin of the Beartooth Range (Evert, *pers. comm.*).

Taxonomic Treatment

Organization

The taxonomic treatment begins with a dichotomous key to all the families. The families are then presented in alphabetical order for ease of location, and each family is also organized alphabetically by genus and species. Dichotomous keys are provided for each taxonomic rank, ultimately leading to the species treatment. Infraspecific taxa are treated within the species.

Chromosome numbers

Chromosome numbers are listed in Appendix 4. These have been obtained from floras, manuals, monographs, and specialized treatments, e.g. Ornduff, 1967–68; Mosquin & Hayley, 1966; Bolkovskikh, 1969; A. Löve et al., 1971; Moore, 1973; Goldblatt, 1981, 1984, 1985, 1988; Goldblatt & Johnson, 1990, 1991. They provide a valuable tool, among many such tools in modern taxonomy, for the establishment of species limits.

Nomenclature

Each species treatment begins with the accepted name, followed by the author, place of publication, date of publication, and common name. My abbreviations of authors' names generally follow the recommendations of Brummitt and Powell (1992). I have tried to adhere to the International Code of Botanical Nomenclature (Greuter et al., 1988) in the determination of accepted species names. Each accepted species name is followed by synonyms at the specific level as used in monographs, manuals, and floras of regional application, or which have a relation to the alpine flora of the Middle Rocky Mountains. I have attempted to include all the species synonyms that apply to the alpine plants of the Middle Rocky Mountains, but infraspecific synonymy is not included. I generally followed Kartesz and Kartesz (1980) and Kartesz (1994) as a guide to synonyms, but I obtained numerous additions from regional floras, monographs of taxa, and the Harvard Gray Herbarium Index of New World Plants, now easily accessible through the Internet. The synonyms, alphabetized by genus and species, follow the accepted species name. In cases where an accepted specific epithet has been proposed in a different rank or taxon and was later transferred, that original name is listed at the beginning of the alphabetical list. I generally followed Beetle (1970) for common names, although I also consulted numerous regional floras and manuals. In a very few cases I simply anglicized the Latin binomial to obtain a common name. Since common names are in such widespread use, especially by the lay public and wildflower enthusiasts, one is provided for each species.

Author's names

Appendix 3 contains a brief bibliography of each author of an accepted name. The intent is to provide a short characterization of the person, his or her life span, a representative position he or she held during his or her lifetime, and a representative work, if applicable. Following Brummitt and Powell (1992), the use of "fl." (floruit) after an author's name is in lieu of a birth date and represents any single date when the author is known to have published. I compiled the list through searches of Internet library databases such as CARL (Colorado Alliance of Research Libraries), VICTOR (University of Maryland system), and MELVYL (University of California System); from bibliographic works (e.g., Jackson, 1964; Stafleu & Cowan, 1976); and from various floras and manuals. In the interest of brevity, multiple works or revisions by the authors are represented by a single citation. For example, J. G. C. Lehmann published *Monographia generis Potentillarum* in 1820, issued a supplement in 1835, and revised it with illustrations in 1856 as *Revisio Potentillarum,* but I include only the first publication. Lengthy Latin titles of early works have been abbreviated to the first descriptive words.

Species Description

Each species description is based on my observations of plants in the field, herbarium specimens and references from regional manuals and floras. Each description is organized as follows: habit of the plant; underground parts; stems; leaves; inflorescences; flower characteristics in the order of sepals, petals, androecium, gynoecium; fruits; and seeds. If parts or structures were not available for description, they were omitted. While I made a concerted attempt to be as quantitative as possible in the descriptions, I include few measurements. Alpine plants are dependent on and responsive to microclimates, and in many cases the sizes of parts and structures are a function of the habitat; that is, of the microclimate. Measurements produce a range that may be misleading, or at least not very helpful, especially if the taxon also has a wide distribution in lowland habitats, as many do. I thus made the decision to omit most measurements from the species descriptions. Although I frequently examined a large number of herbarium specimens of each species, the descriptions are based on the few that seemed to be "typical" specimens, especially those typical of the alpine zone. In the case of some rare plants, my description is based on a single specimen and referenced with the originally published description or descriptions in regional floras or journals.

Distribution

The species description is followed by a short habitat description. The terms used in alpine ecology and geomorphology to describe environments and habitats are listed in Appendix 1. Next, the Middle Rocky Mountain distribution is given by mountain range. If the species occurs in approximately half or fewer of the ranges, the ranges are listed by name; if in more than half, the general distribution is given instead. (The ranges can be ascertained directly from the distribution maps.) The Middle Rocky Mountain distribution is followed by the North American and/or worldwide distribution. The distribution pattern begins in the northwestern corner of North America and continues to the south and east. Thus, a species that occurs in Montana, Wyoming, Idaho, Oregon, California, Washington, and Utah would be described as "Washington south and east to California, Utah, Wyoming, and Montana."

Comments

Comments on infraspecific taxa and how to distinguish them follow the distribution information.

Distribution Maps

The maps show the distribution pattern of each species in the alpine flora. No attempt is made to distinguish patterns of infraspecific taxa. The dots are based on herbarium specimens and, in a very few cases, field observations when no specimens were available. Published distribution information based on herbarium specimens has also been incorporated. The dots represent both alpine and nonalpine specimens; no attempt has been made to distinguish the two. Locations for the dots were determined by use of the U.S. Geological Survey Geographic Names Information System (GNIS). The maps thus show the overall range of each species in the Middle Rocky Mountains. The strictly alpine distribution patterns of obligate high alpine species, such as *Papaver kluanense* D. Löve, are reflected on the maps. Wide-ranging species of lowlands, uplands, and the alpine zone, such as *Taraxacum officinale* Wiggers, have alpine distribution patterns that can be determined only from the mountain range description in the accompanying species treatment. It is likely that a species that occurs in a mountain range will someday be observed in the alpine zone, but I have not made that assumption in this treatment. Thus, if I have not seen a herbarium specimen to document the alpine occurrence of a species, it is not considered part of the alpine flora of the range, even if it occurs at lower elevations in that range. Some areas on the maps are blank (most notably Colorado). These fall outside the boundaries of the Middle Rocky Mountains and will be treated in future volumes.

Concepts of Taxa

In this flora I treat family, genus, species, subspecies, and variety. I have used a conservative approach to the concept of taxa both above the species level and below it. Families and genera are treated in the traditional sense, as groups of similar entities based on anatomical and morphological characteristics. The species is recognized as an interbreeding unit that is reproductively isolated from other such units, with (ideally) morphologic distinctiveness. Thus, a "good" species should be separated from other species by a gap in the variation of observable traits and a corresponding barrier to interbreeding. Unfortunately, this is not always the case when barriers to interbreeding are weak. Gaps between species are sometimes very narrow and hard to discern. A conservative treatment that gives broad limits to species sometimes solves or minimizes this problem with respect to use of keys and identification, but still recognizes the variation—only at the infraspecific, rather than the specific, rank.

Taxonomists include both "lumpers," who recognize broad limits to species, and "splitters," who would describe many species within these broad limits. In the treatment of most groups I categorize myself as a "lumper," and I have taken a correspondingly conservative approach to species definition. Where many workers might use the species to describe variation within a group, I rely more on the subspecies and the variety for this purpose. Of the two, subspecies is the higher rank and is used to describe more strongly differentiated geographical differences; variety is used to describe minor geographical or habitat differences within species. In this flora, I use subspecies in 21 families to describe 55 infraspecific taxa, and variety in 36 families to describe 314 infraspecific taxa. At the same time, I have tried to follow the lead of specialists in certain groups without making new nomenclatural changes. For example, in the alpine zone the *Poa sandbergii* complex has been represented by *P. ampla, P. canbyi, P. gracillima, P. incurva, P. juncifolia, P. sandbergii,* and *P. secunda,* but I followed Kellogg (1985) in recognizing a single variable species, for which the name *Poa secunda* Presl has priority, with no clearly defined infraspecific taxa. On the other hand, I used the name *Carex macloviana* D'Urv. to describe the complex that includes *C. haydeniana, C. microptera,* and *C. pachystachya* in the alpine zone. These taxa are treated as species by some, but here as varieties of *C. macloviana* using names already available.

The Middle Rocky Mountain Alpine Flora

Key to the Families

1. Plants reproducing by spores, without flowers or seeds; plants often fernlike or mosslike
   2. Leaves grasslike (linear-subulate), clustered on a short stem; plants aquatic or amphibious — Isoetaceae
   2. Leaves not grasslike; stems elongate; plants terrestrial
      3. Plants fernlike; leaves well developed, 2 cm or more long
         4. Sporangia in a terminal panicle above the sterile leaf — Ophioglossaceae
         4. Sporangia on the margins or dorsal surface of ordinary or sometimes specialized leaves
            5. Sori wholly or partially covered by the reflexed (inrolled) leaf margins; petioles with singular vascular bundles — Adiantaceae
            5. Sori not covered by the reflexed leaf margins; petioles with 2 or more vascular bundles, at least at the base — Aspleniaceae
      3. Plants mosslike; leaves small, often scalelike, 1 cm or less long
         6. Sporophytes homosporous, the spores minute; leaves 5 mm long, erect to spreading — Lycopodiaceae
         6. Sporophytes heterosporous, the megasporangia usually containing 4 large megaspores, the microsporangia with numerous small microspores; leaves 2–4 mm long, appressed — Selaginellaceae
1. Plants reproducing by seeds, with or without flowers; habit various
   7. Seeds naked on a carpellary scale, not enclosed in an ovary; leaves linear, subulate, or scalelike, mostly evergreen; stigmas none; stems woody (Gymnospermae)
      8. Cones fleshy, berrylike; leaves subulate or scalelike, opposite or whorled — Cupressaceae
      8. Cones woody; leaves needlelike, alternate, or in fascicles of 2–5 — Pinaceae
   7. Seeds enclosed within an ovary; leaves various, mostly deciduous; stigmas present; stems woody or herbaceous (Angiospermae)
      9. Leaves mostly parallel-veined; flowers 3- or 6-merous; vascular bundles of stem not arranged in a ring; embryo with a single cotyledon (Monocotyledonae)
         10. Plants submerged aquatics — Hydrocharitaceae
         10. Plants terrestrial, often of wet habitats, but not submerged
            11. Plants grasslike; perianth of chaffy bristles or scales, or lacking, or of cottony hairs; underground parts consisting of fibrous roots and rhizomes
               12. Perianth of 6 conspicuous, chaffy segments; fruit a 3-valved capsule — Juncaceae

12. Perianth inconspicuous or lacking, or of cottony hairs; fruit an akene or grain
   13. Culms not jointed, usually solid and triangular; leaves 3-ranked; fruit an akene — Cyperaceae
   13. Culms jointed, usually hollow and terete; leaves 2-ranked; fruit a grain — Poaceae
11. Plants seldom grasslike; perianth of petaloid segments; underground parts consisting of bulbs, corms, or fleshy roots
   14. Flowers regular, hypogynous — Liliaceae
   14. Flowers irregular, epigynous — Orchidaceae
9. Leaves mostly net-veined; flowers 4- or 5-merous; vascular bundles of stem arranged in a ring; embryo with two cotyledons (Dicotyledoneae)
   15. Flowers in catkins; plants woody
      16. Fruit a capsule; seeds comose; plants dioecious — Salicaceae
      16. Fruit a winged or wingless nutlet; seeds not comose; plants monoecious — Betulaceae
   15. Flowers not in catkins; plants mostly herbaceous
      17. Corolla none, the perianth consisting of a single series, or lacking
         18. Stamens numerous, more than 10 — Ranunculaceae
         18. Stamens 10 or fewer
            19. Leaves opposite
               20. Plants aquatic, perianth lacking — Callitrichaceae
               20. Plants terrestrial, perianth of a single series — Caryophyllaceae
            19. Leaves alternate or basal
               21. Ovary 1-carpellate; fruit an akene or utricle
                  22. Sheathing stipules present — Polygonaceae
                  22. Sheathing stipules absent — Chenopodiaceae
               21. Ovary 2-carpellate; fruit a capsule — Scrophulariaceae
      17. Corolla present, the perianth of 2 series differing in texture, size, or color
         23. Corolla of separate petals
            24. Flowers epigynous
               25. Inflorescence a simple or compound umbel — Apiaceae
               25. Inflorescence a raceme, or the flowers axillary
                  26. Flowers 4-merous; ovary 2- or 4-celled; herbs — Onagraceae
                  26. Flowers 5-merous; ovary 1-celled; shrubs — Grossulariaceae
            24. Flowers hypogynous or perigynous
               27. Corolla irregular
                  28. Stamens numerous — Ranunculaceae
                  28. Stamens 10 or fewer
                     29. Ovary simple; none of the petals spurred; fruit a legume — Fabaceae
                     29. Ovary compound; at least 1 petal spurred; fruit a capsule
                        30. Petals 4; stamens 6 — Fumariaceae
                        30. Petals 5; stamens 5 — Violaceae
               27. Corolla regular

31. Leaves punctate with blackish purple glands in ours — Hypericaceae
31. Leaves lacking punctate glands
    32. Flowers perigynous, the stamens inserted on the hypanthium
        33. Leaves opposite; flowers 4-merous — Celastraceae
        33. Leaves alternate; flowers 5-merous
            34. Stipules present; stamens mostly more than 10 — Rosaceae
            34. Stipules lacking; stamens mostly 10 or fewer — Saxifragaceae
    32. Flowers hypogynous, the stamens inserted on the receptacle
        35. Stamens numerous, more than twice as many as the petals (or sepals if petals lacking)
            36. Plants with milky or colored sap; petals 4 — Papaveraceae
            36. Plants lacking milky or colored sap; petals more than 5, or lacking
                37. Pistils 10 or more; fruit an aggregate of follicles or akenes — Ranunculaceae
                37. Pistil 1; fruit a capsule or berry
                    38. Leaves simple; fruit a capsule — Portulacaceae
                    38. Leaves ternately compound; fruit a berry — Ranunculaceae
        35. Stamens twice as many, or fewer than twice as many as the petals
            39. Leaves thick, fleshy
                40. Sepals 2; stipules usually present; fruit a capsule — Portulacaceae
                40. Sepals 5; stipules usually lacking; fruit a group of follicles — Crassulaceae
            39. Leaves thin, not fleshy
                41. Petals 4; stamens 6, 4 long and 2 short; fruit a silique or silicle — Brassicaceae
                41. Petals 5 or sometimes lacking; stamens 5 or 10; fruit a capsule
                    42. Leaves palmately lobed or divided — Geraniaceae
                    42. Leaves entire or toothed
                        43. Leaves opposite — Caryophyllaceae
                        43. Leaves alternate, at least above — Linaceae
23. Corolla of united petals
    44. Flowers in an involucrate head; corollas tubular or ligulate or both — Asteraceae
    44. Flowers not as above
        45. Stamens more numerous than the lobes of the corolla
            46. Flowers irregular, the ovary 1-celled — Fabaceae
            46. Flowers regular or nearly so, the ovary 3- to several-celled — Ericaceae

45. Stamens not more numerous than the lobes
 of the corolla
  47. Flowers epigynous
    48. Leaves alternate   Campanulaceae
    48. Leaves opposite or whorled
      49. Plants shrubby   Caprifoliaceae
      49. Plants herbaceous
        50. Leaves exstipulate; calyx reduced
            to an epigynous ring, becoming
            plumose in fruit; fruit an akene   Valerianaceae
        50. Leaves stipulate; calyx more or
            less conspicuous, not becoming
            plumose in fruit; fruit a schizocarp
            in ours, covered with hooked
            bristles   Rubiaceae
  47. Flowers hypogynous
    51. Stamens the same number as the corolla
        lobes and opposite them   Primulaceae
    51. Stamens alternate with the corolla lobes
        or fewer
      52. Flowers irregular
        53. Plants with more or less square
            stems and opposite, aromatic
            leaves; fruit a 1-seeded nutlet   Lamiaceae
        53. Plants lacking square stems; leaves
            alternate, opposite, or basal, not
            aromatic; fruit a capsule   Scrophulariaceae
      52. Flowers regular
        54. Ovary deeply 4-lobed   Boraginaceae
        54. Ovary not deeply 4-lobed
          55. Corolla scarious, inconspicuous;
              plants scapose with flowers in
              dense, bracteate spikes   Plantaginaceae
          55. Corolla petaloid, conspicuous;
              plants lacking the above
              combination of characters
            56. Style 3-cleft; ovary 3-celled Polemoniaceae
            56. Style entire or 2-cleft; ovary
                1-celled
              57. Leaves alternate; stamens
                  exerted   Hydrophyllaceae
              57. Leaves opposite; stamens
                  included   Gentianaceae

Adiantaceae
Maidenhair Family

Sporophytes from short to elongate, branched or unbranched, rhizomes with brownish scales. Leaves (fronds) evergreen or deciduous, with 1 to several times pinnately compound blades, these elevated on slender, very often dark and wiry petioles (stipes), with a single vascular bundle. Fertile blades sometimes dissimilar to the sterile blades in size and shape. Sori marginal or submarginal, elongate, covered by the indusium-like recurved leaf margin. Sporangia stalked, with a well-developed vertical annulus, often exserted. Spores all alike, tetrahedral. Gametophytes very small, green, cordate to reniform, glabrous.

KEY TO THE GENERA OF ADIANTACEAE

1. Sterile and fertile leaves dissimilar, the fertile longer with narrower pinnules   *Cryptogramma*
1. Sterile and fertile leaves all alike or nearly so, the pinnules similar   *Pellaea*

Cryptogramma R. Br.
Rockbrake

Small, evergreen or deciduous ferns of damp, rocky situations. Fronds dimorphic, the sterile blades spreading, glabrous, shorter and broader than the longer, more erect, brownish, fertile, blades. Petioles slender, usually wiry, with a single vascular bundle, glabrous to hairy, and usually scaly at the base. Sori marginal or submarginal, on the thickened ends of the veins, without true indusia, but covered by the revolute margins of the leaf segments. Spores tetrahedral.

Cryptogramma acrostichoides R. Br.
in Richards., *Bot. App. Frankl.
Journ.* 754, 767, 1823.

American Rockbrake

Acrostichum crispum (L.) Vill.
Allosorus acrostichoides (R. Br.) Spreng.
Allosorus crispus (L.) Bernh. ex Spreng.
Blechnum crispum (L.) Hartm.
Cryptogramma crispa (L.) R. Br. ex Hook.
Gymnogramma acrostichoides (R. Br.) Presl
Osmunda crispa L.
Phorolobus acrostichoides (R. Br.) Fee
Phorolobus crispus (L.) Desv.
Pteris crispa (L.) All.
Struthiopteris crispa (L.) Wallr.

Densely tufted ferns from short, scaly rhizomes covered by persistent stipe bases. Sterile fronds evergreen: stipes light brown to greenish; blades ovate-lanceolate, 2–3-pinnate, the ultimate segments rounded, finely serrate. Fertile fronds exceeding the sterile ones, 2–3-pinnate, the segments linear to linear-oblong, with revolute to recurved margins nearly meeting beneath. Sori covering the ventral surface of the fertile pinnules.

Moist or dry rock ledges, cliffs, talus, scree, and occasionally felsenmeer in most alpine areas of our region. Not yet known from above timberline in the Absaroka, Gros Ventre, Hoback, and Wyoming ranges. Circumboreal; Alaska to Ontario, south to California, New Mexico, Colorado, Nebraska, and Michigan.

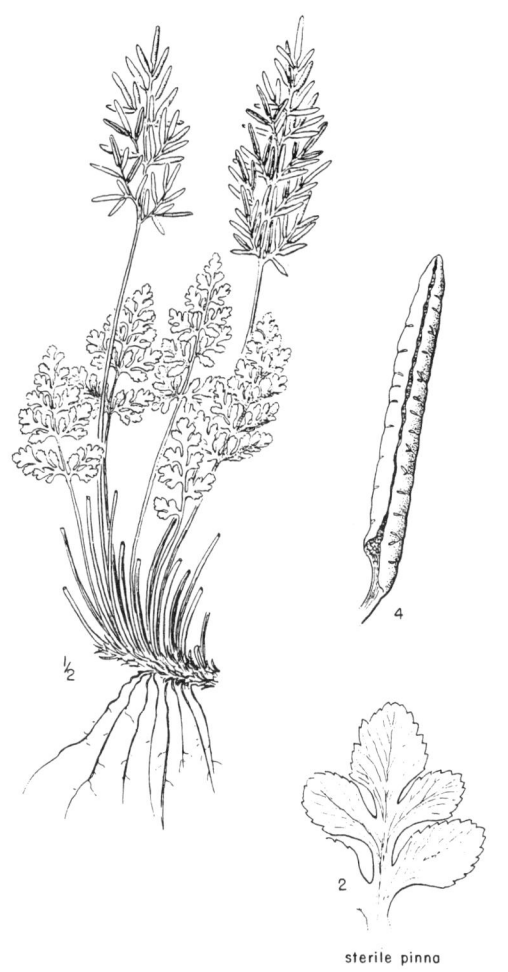

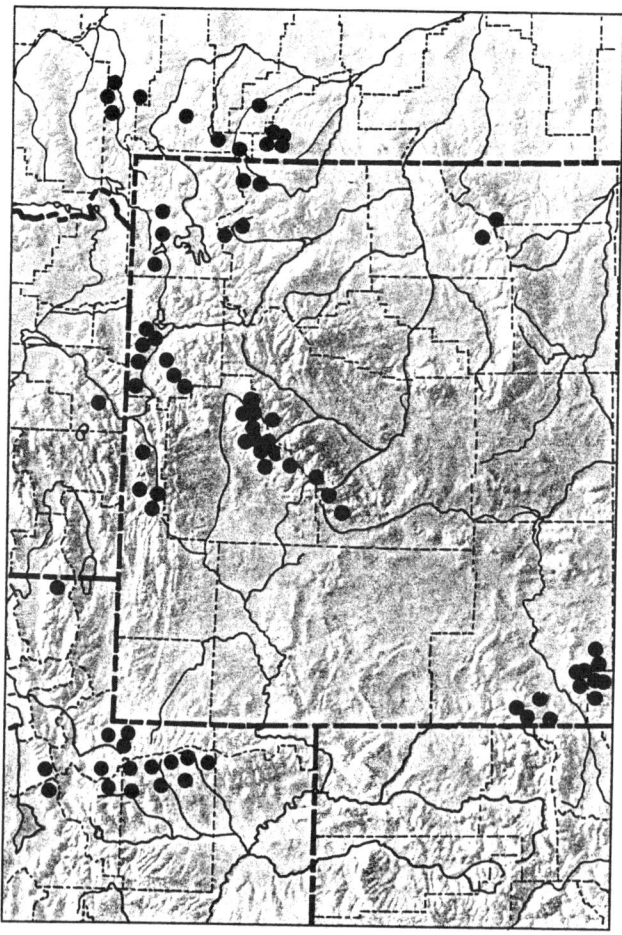

Pellaea Link
Cliffbrake

Small, mostly evergreen, xeromorphic ferns from short, thick, densely brown-scaly rhizomes. Fronds 1–2-pinnate, glabrous to somewhat hairy; stipes wiry, dark reddish, brownish, or blackish, with a singular vascular bundle; pinnae oblong to nearly orbicular, or bifid, the margins revolute. Sori elongate, marginal, confluent or nearly so produced on the thickened ends of the veins; indusia lacking, the sori partially covered by the revolute leaf margins. Sporangia long-stalked and often conspicuously exerted. Spores tetrahedral.

Pellaea breweri D.C. Eat.
Proc. Am. Acad. 6:555, 1865.

Brewer Cliffbrake

Allosoris breweri (D.C. Eat.) Kuntze

Densely tufted, lithophilous ferns from short, branched rhizomes covered by brownish scales and numerous old stipe bases. Fronds pinnate, oblong to oblong-lanceolate; stipes reddish brown, transversely grooved at the base, the rachis green to yellowish brown; pinnae glabrous, opposite or alternate, the lower unequally bifid and asymmetrical, resembling a butterfly wing, the upper entire and lanceolate. Sori elongate, confluent or nearly so, partially covered by the revolute leaf margins.

Rocky ledges, cliffs, and talus, usually on limestone. In the alpine zone known only from the Big Horn, Gros Ventre, Salt River, and Wyoming Ranges. Washington to Wyoming, southward to California, Utah, and Colorado.

Apiaceae
(Umbelliferae)
Parsnip Family

Annual or perennial herbs with alternate or basal, mostly compound leaves, the petioles often dilated and sheathing. Flowers umbellate in ours, epigynous, 5-merous, perfect, regular or rarely irregular at the margins of umbellets. Sepals 5, very small, sometimes obsolete. Petals 5, distinct, the tips inflexed. Stamens 5, alternate with the petals. Pistil 1, the ovary 2-celled, with 1 ovule per cell, the styles 2. Fruit of 2 mericarps, each usually 5-ribbed, the walls with or without oil tubes in the intervals between the ribs.

KEY TO THE GENERA OF APIACEAE

1. Leaves all simple, entire, linear, lanceolate or oblanceolate — *Bupleurum*
1. Leaves compound
    2. Leaves 1–3 times compound, with well-defined, broad leaflets
        3. Leaflets 3, 10–30 cm wide — *Heracleum*
        3. Leaflets more than 3, much less than 10 cm wide
            4. Fruits clavate, hispid, in open umbels with few rays — *Osmorhiza*
            4. Fruits oblong to orbicular, not hispid, in compact umbels with many rays
                5. Fruits dorsally flattened; petioles of leaves dilated and strongly sheathing — *Angelica*
                5. Fruits laterally flattened; petioles not dilated and strongly sheathing — *Ligusticum*
    2. Leaves pinnately dissected, usually lacking well-defined, broad leaflets
        6. Plants stout, mostly more than 4 dm tall — *Ligusticum*
        6. Plants slender, mostly less than 3 dm tall
            7. Flowers white; fruits flattened dorsally — *Cymopterus*
            7. Flowers yellow; fruits flattened laterally or dorsally
                8. Plants less than 1 dm tall, with creeping rootstocks — *Oreoxis*
                8. Plants usually more than 2 dm tall, with slender or tuberously thickened taproots
                      9. Fruits winged dorsally; involucel of prominent lanceolate bracts; base of stems covered by persistent leaf bases — *Cymopterus*
                      9. Fruits winged laterally; involucel of oblanceolate to elliptic bracts; base of stems lacking persistent leaf bases — *Lomatium*

Angelica L.
Angelica

Tall, stout, perennial herbs, with 1 to several stems from a taproot. Leaves ternate-pinnate to pinnately compound, with serrate, dentate, or cleft leaflets; petioles dilated and sheathing, the uppermost sometimes lacking blades. Flowers white, purplish, or brownish, in a loose compound umbel; involucre present or lacking; involucel of narrow bractlets in ours. Calyx teeth minute or lacking. Stylopodium conic. Fruit oblong to orbicular, strongly flattened dorsally, the ribs winged both dorsally and laterally in ours; oil tubes few to numerous; carpophore bifid to the base.

KEY TO THE SPECIES OF ANGELICA

1. Flowers purplish or brownish; ovaries and fruit glabrous — *A. grayi*
1. Flowers white to light pink; ovaries and fruit scabrous — *A. roseana*

Angelica grayi (Coult. & Rose) Coult. & Rose
Contr. U.S. Natl. Herb. 7:154, 1900.

Gray's Angelica

Selinum grayi Coult. & Rose

Stout, perennial herb to 6 dm tall. Leaves pinnate to ternate-pinnate; leaflets serrate, the petioles dilated, at least on cauline leaves. Flowers purplish brown, in compound umbels. Fruits glabrous, flattened dorsally; dorsal ribs narrowly winged; lateral ribs broadly winged.

Seen only from the Medicine Bow Mountains. Apparently endemic to southern Wyoming and western Colorado.

Angelica roseana Henders.
Contr. U.S. Natl. Herb. 5:201, 1899.

Rose Angelica

Rompelia roseana (Henders.) Koso.-Pol.

Stout, single- or multistemmed, perennial herb to 6 dm tall. Leaves ternate-pinnate; leaflets serrate or dentate, often incised, the petioles swollen at the base. Flowers white to pink in compound umbels. Fruits scabrous, dorsally flattened; ribs winged, the lateral ones more strongly and prominently than the dorsal.

Rocky fellfields of the Big Horn, Teton, Uinta, and Wasatch Ranges; reported from the Kirwin area of the Absaroka Range (Evert, 1983). Montana and Idaho south to Colorado and Utah.

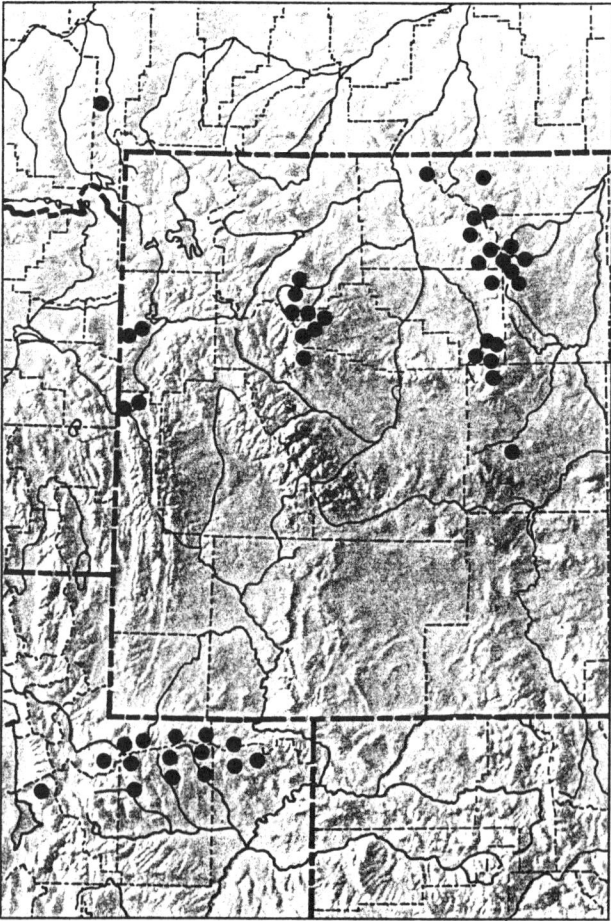

Bupleurum L.
Thorowax

Ours are low, perennial herbs with glabrous or often glaucous, simple, mostly basal, entire leaves. Flowers small, yellow or sometimes purplish, in a loose, compound umbel; the involucre of conspicuous foliaceous bracts or none, the involucel of ovate bractlets. Calyx teeth obsolete or lacking. Stylopodium prominent, depressed-conic, the styles short, somewhat reflexed. Fruit oblong to orbicular, flattened laterally, the ribs filiform; oil tubes present or lacking; carpophore bifid to the base.

Bupleurum americanum Coult. & Rose
Rev. N. Am. Umbell. 115, 1888.

American Thorowax

Bupleurum angulosum Cham. & Schlecht.
Bupleurum angulosum H. & A.
Bupleurum purpureum Blank.
Bupleurum ranunculoides Hook.

Perennial, multistemmed herb from a taproot and branched caudex. Leaves entire, linear, lanceolate, or oblanceolate, prominently nerved. Flowers yellow, in a compound umbel, the umbellets compact and rounded. Fruits oblong, flattened laterally.

Common in meadows and fellfields of the northern and western ranges of the Middle Rocky Mountains. Alaska south to Montana, Idaho, and Wyoming.

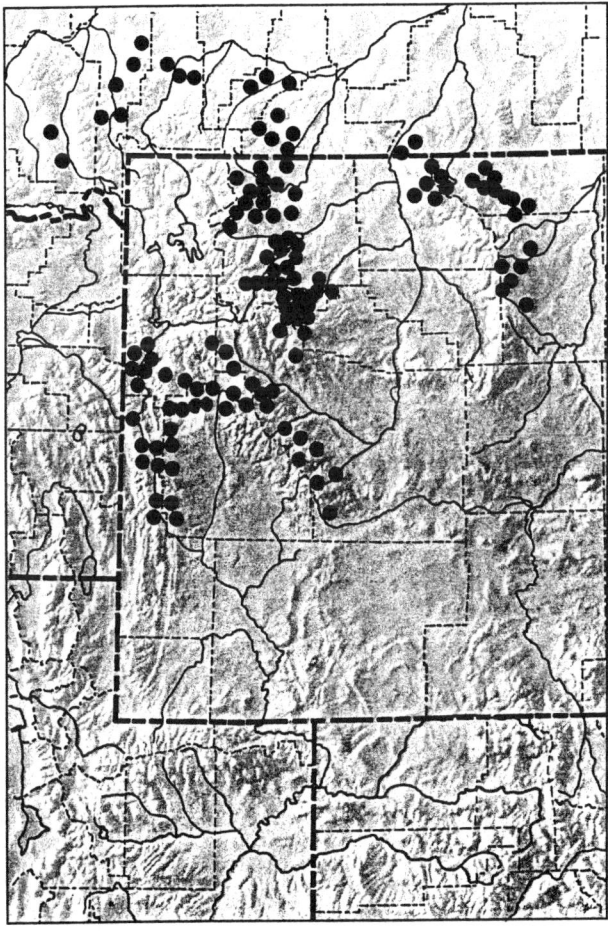

Cymopterus Raf.
Springparsley

Low, acaulescent or caulescent, pubescent or glabrous, perennial herbs from a stout taproot; pseudoscapes often present. Leaves bipinnate to pinnately or ternately decompound, the ultimate divisions narrow; the bases sometimes persistent on the crown of the taproot. Flowers yellow, white, or purple, in loose or compact, compound umbels, the involucre of scarious or green bracts or lacking, the involucel of ovate to linear bractlets, on one side of the umbellet. Calyx teeth prominent to obsolete, sometimes unequal. Stylopodium none. Fruit oblong to ovoid, flattened dorsally, the lateral and dorsal ribs winged; oil tubes 1 to several in the intervals, several on the commissure; carpophore bifid to the base.

KEY TO THE SPECIES OF *CYMOPTERUS*

1. Plants with a conspicuous pseudoscape separating the root crown and origin of leaves — *C. longipes*
1. Plants lacking a pseudoscape, the leaves cauline or arising directly from the root crown
    2. Flowers white; calyx lobes more or less scarious, blunt, less than 0.9 mm long
        3. Ovaries and immature fruit glabrous; mature fruit glabrous, the dorsal and intermediate wings thin, 0.5–3.5 mm broad; carpophore present — *C. nivalis*
        3. Ovaries and immature fruit puberulent; mature fruit muricate, the dorsal and intermediate wings stout, 0.2–0.5 mm broad; carpophore lacking — *C. evertii*
    2. Flowers yellow; calyx lobes green, acute, more than 1 mm long
        4. Plants acaulescent; leaves oblong; involucel of conspicuous green bractlets usually exceeding the umbellets — *C. longilobus*
        4. Plants caulescent; leaves ovate; involucel of bractlets not exceeding the umbellet — *C. terebinthinus*

Cymopterus evertii Hartman & Kirkpatrick
Brittonia 38:420, 1986.

Evert's Springparsley

Tufted, perennial herb from a stout taproot and few- to many-branched caudex, the branches covered by persistent leaf bases. Leaves pinnate-pinnatifid, bipinnate, bipinnate-pinnatifid, to tripinnate. Flowers in a subcompact, compound umbel, the rays unequal; involucre lacking; involucel of linear, lanceolate, or elliptic, entire to erose, acute bractlets, about equaling the flowers. Sepals triangular to lanceolate, subequal; petals white to cream, obovate. Fruits terete to dorsally flattened, elliptic to suborbicular, muricate, the ribs short, stout, and rounded.

Rocky, open slopes and ridges of Carter Mountain in the Absaroka Range, with nonalpine locations known from lower elevations in the Big Horn Basin of southeastern Park County, Wyoming (Hartman & Kirkpatrick, 1986), and from Uintah County, Utah. *Cymopterus evertii* is closely related to *C. nivalis* but differs by characters mentioned in the species key.

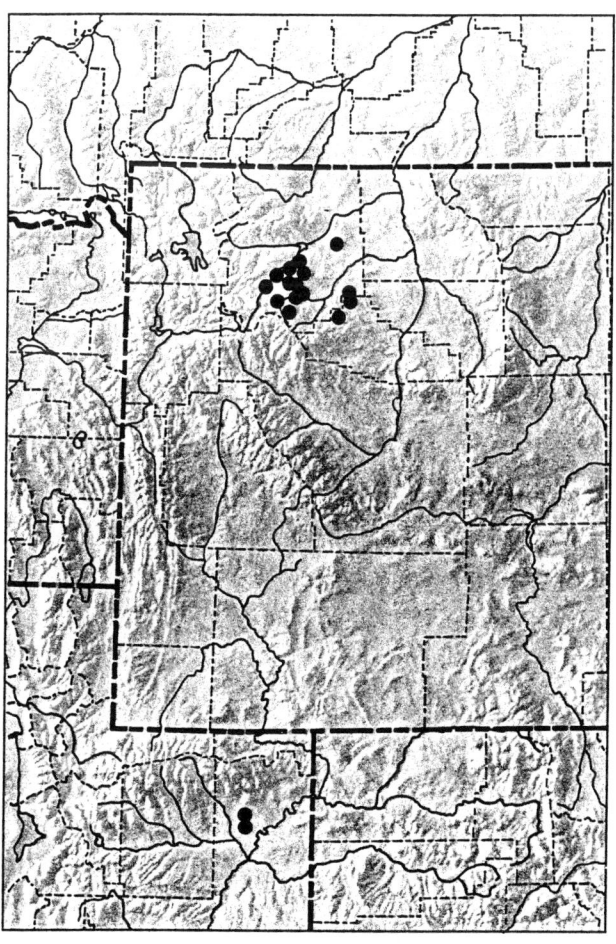

Cymopterus longilobus (Rydb.) W.A. Weber
Phytologia 33:105, 1976.

Henderson Springparsley

Pseudopteryxia longiloba Rydb.
Cymopterus hendersonii (Coult. & Rose) Cronq.
Pseudocymopterus hendersonii Coult. & Rose
Pseudopteryxia hendersonii (Coult. & Rose) Rydb.
Pteryxia hendersonii (Coult. & Rose) Math. & Const.

Acaulescent, perennial herb from a stout taproot and branched caudex, densely covered by persistent leaf bases. Leaves numerous, bipinnate, or occasionally tripinnate. Flowers in a compound umbel, the rays unequal; involucre lacking; involucel of conspicuous green bractlets, usually exceeding the umbellet. Fruits ovoid-oblong, the ribs narrowly winged.

Common on fellfields, Neoglacial moraines, and sheltered rocky ledges in most alpine ranges of the Middle Rocky Mountains. Not yet known from above timberline in the Big Horn and Medicine Bow Ranges. Idaho and Montana south to Nevada and New Mexico.

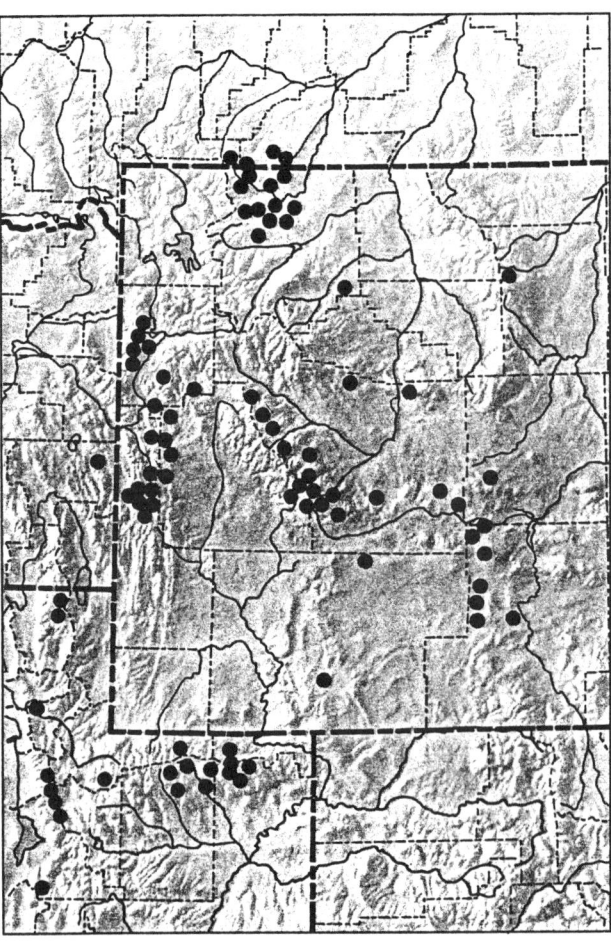

Cymopterus longipes S. Wats.
Bot. King Exped. 124, 1871.

Longstalk Springparsley

Aulospermum angustum Osterh.
Aulospermum longipes (S. Wats.) Coult. & Rose
Cogswellia lapidosa (M.E. Jones) Rydb.
Cymopterus lapidosus (M.E. Jones) M.E. Jones
Lomatium lapidosum (M.E. Jones) Garrett
Peucedanum lapidosum M.E. Jones

Perennial herb with a conspicuous pseudoscape. Leaves pinnately dissected, the leaflets pinnately lobed. Flowers yellow or white; involucre lacking; involucel of linear bractlets. Fruits oblong; ribs winged, the dorsal much narrower than the lateral.

Dry slopes of the Gros Ventre, Hoback, Salt River, Teton, Wind River, and Wyoming Ranges. Wyoming and Idaho south to Utah and Colorado.

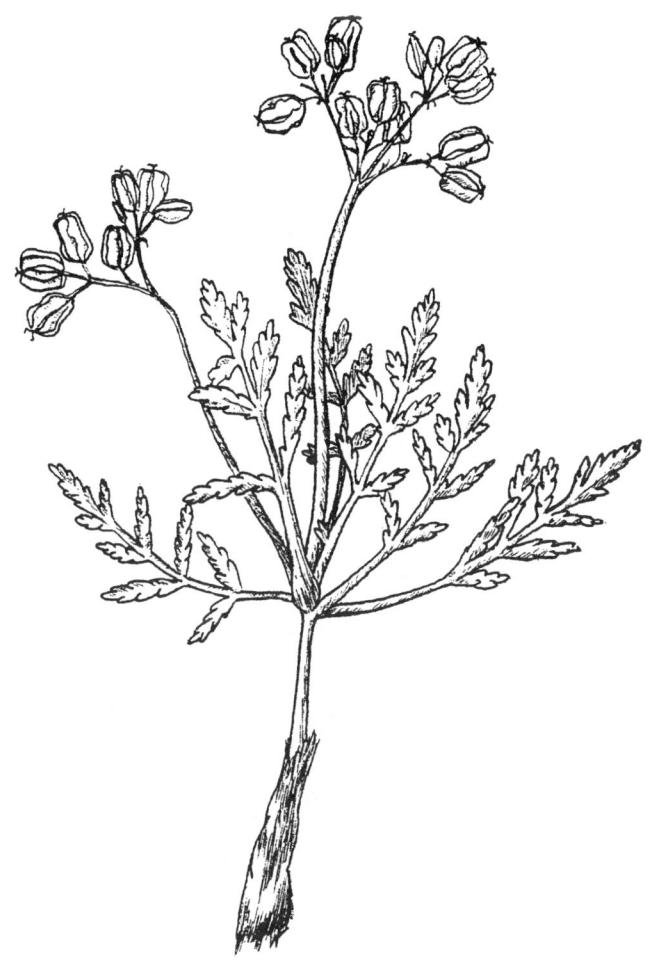

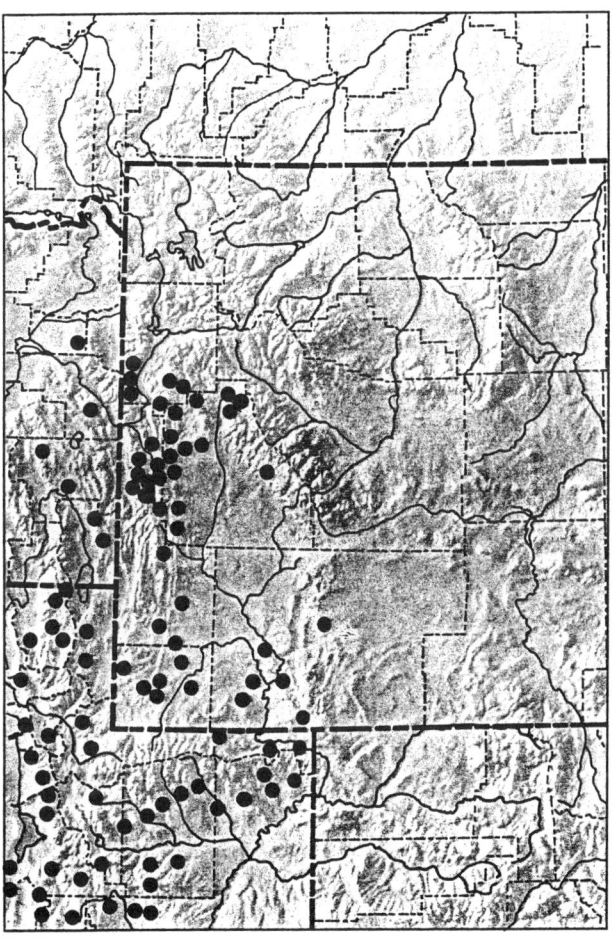

Cymopterus nivalis S. Wats.
Bot. King Exped. 123, 1871.

Snowbank Springparsley

Cymopterus bipinnatus S. Wats.
Cynomarathrum macbridei A. Nels.
Pseudocymopterus bipinnatus (S. Wats.) Coult. & Rose
Pseudocymopterus nivalis (S. Wats.) Rydb.
Pseudoreoxis bipinnatus (S. Wats.) Rydb.
Pseudoreoxis nivalis (S. Wats.) Mathias

Low, acaulescent, caespitose perennial from a stout taproot; caudex branched and covered by persistent leaf bases. Leaves basal, finely dissected into numerous glaucous segments. Flowers white in a dense, compact umbel; involucre lacking. Fruits ovoid-oblong, narrowly winged, flattened dorsally.

Seen only from rocky, south-facing limestone slopes in the northern Wind River Range. Montana to Oregon and south to Nevada, Idaho, and Wyoming

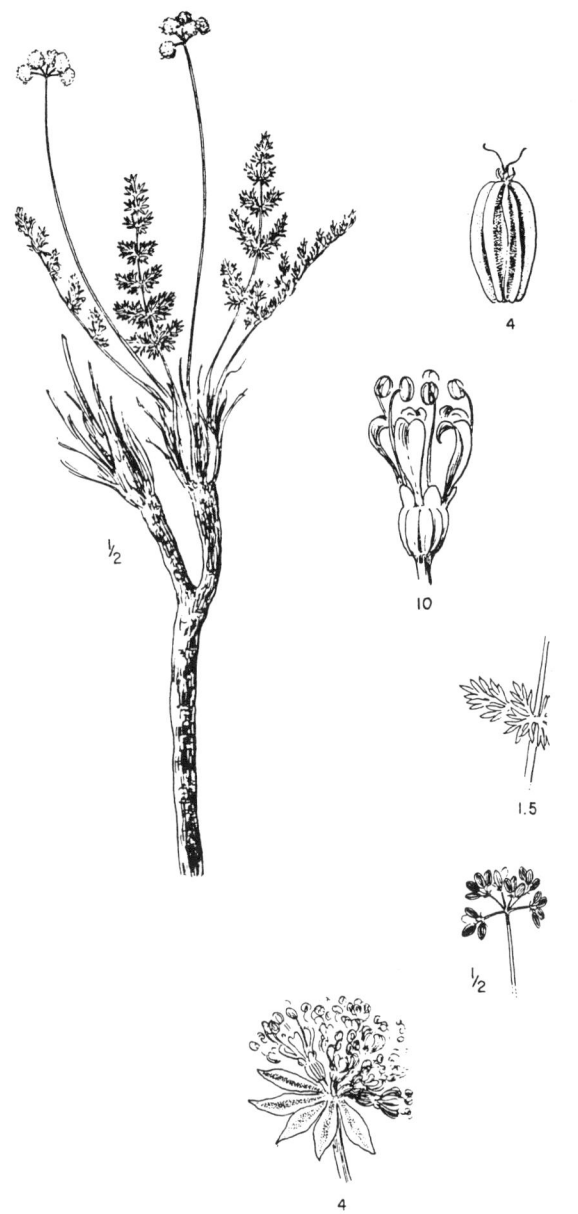
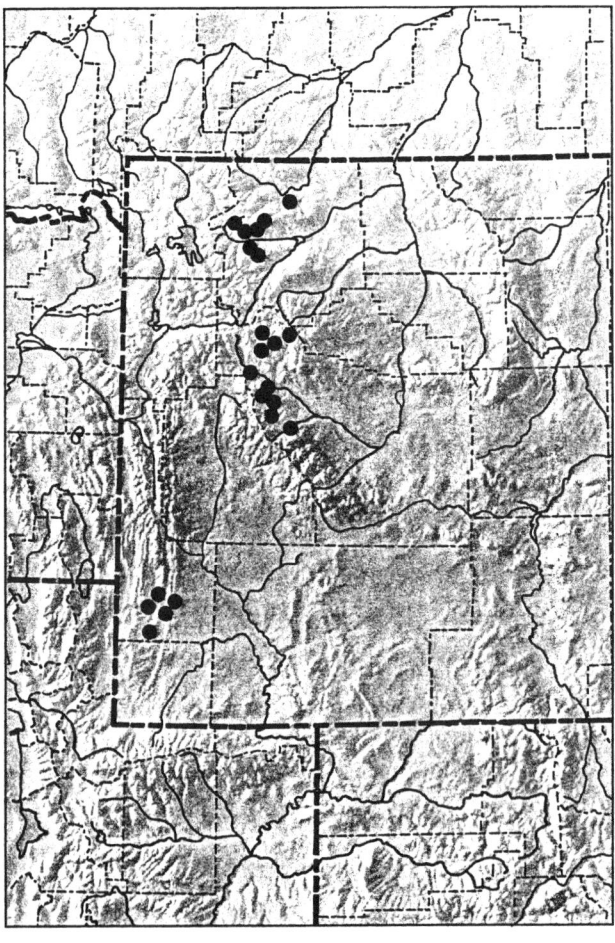

Cymopterus terebinthinus (Hook.) T. & G.
Fl. N. Am. 1:624, 1840.

Turpentine Springparsley

Selinum terebinthinum Hook.
Cymopterus calcareus M.E. Jones
Pteryxia terebinthacea (Hook.) Nutt.
Pteryxia terebinthina (Hook.) Coult. & Rose

Perennial herb from a taproot and branched caudex, often with a few persistent leaf bases. Leaves basal and cauline on the lower stem, ternate-pinnate, finely dissected. Flowers yellow, in a compound umbel; rays unequal; involucre lacking; involucel of bractlets not exceeding the umbellet. Fruits oblong to narrowly ovoid; ribs winged, both dorsal and lateral wings about the same length.

In the alpine zone known from the Beartooth, Big Horn, Hoback, and Wind River Ranges. Wyoming to Washington and south to Nevada and Colorado.

Middle Rocky Mountain alpine plants with small, dense inflorescences are var. *albiflorus* (T. & G.) M.E. Jones.

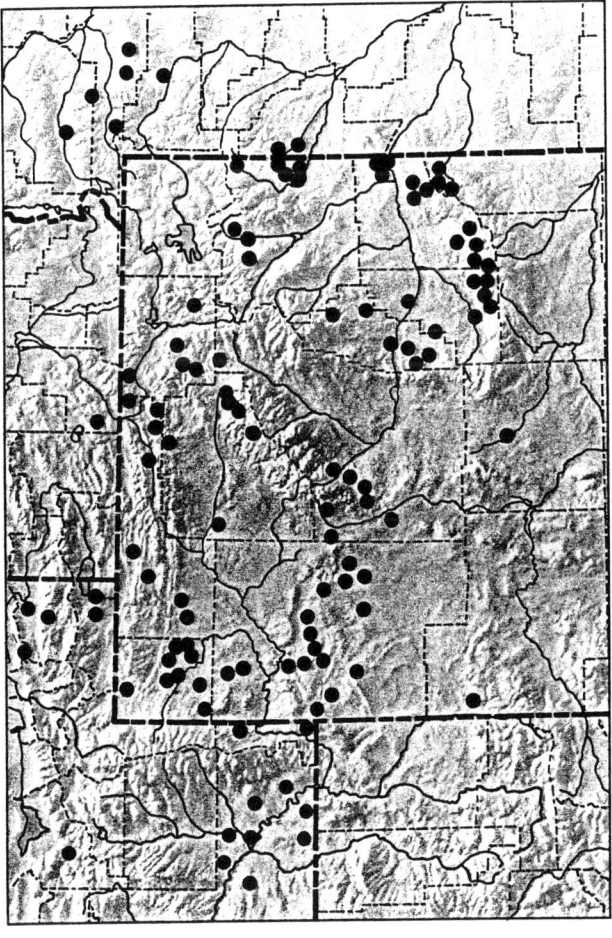

Heracleum L.
Cowparsnip

Stout, pubescent, biennial or perennial herbs from a taproot or fascicled, fibrous roots. Leaves large, ternately compound, the leaflets 10–30 cm broad, ovate to orbicular, cordate at the base, the margins serrate and lobed, the upper cauline leaves with sheathing petioles conspicuously dilated at the base. Flowers white, in a loose, terminal or axillary, compound umbel, the involucre of 5–10 deciduous bracts, the involucel of numerous slender bractlets. Calyx teeth minute or obsolete. Outer flowers of the umbellets irregular, with enlarged, often bifid corolla lobes. Stylopodium conic, the styles short or recurved. Fruit orbicular to elliptic, strongly flattened dorsally, often pubescent, the ribs filiform; oil tubes large, extending only partway from the stylopodium to the base of the mericarp, solitary in the intervals, 2–4 on the commissure; carpophore bifid to the base.

Heracleum sphondylium L.
Sp. Pl., 249, 1753.

Hogweed Cowparsnip

Heracleum douglasii DC.
Heracleum lanatum Michx.
Heracleum maximum Bartr.
Heracleum montanum Schleich.
Pastinaca lanata (Michx.) Koso.-Pol.
Selinum sphondylium (L.) E.H. Krause
Sphondylium lanatum (Michx.) Greene
Sphondylium vulgare S.F. Gray

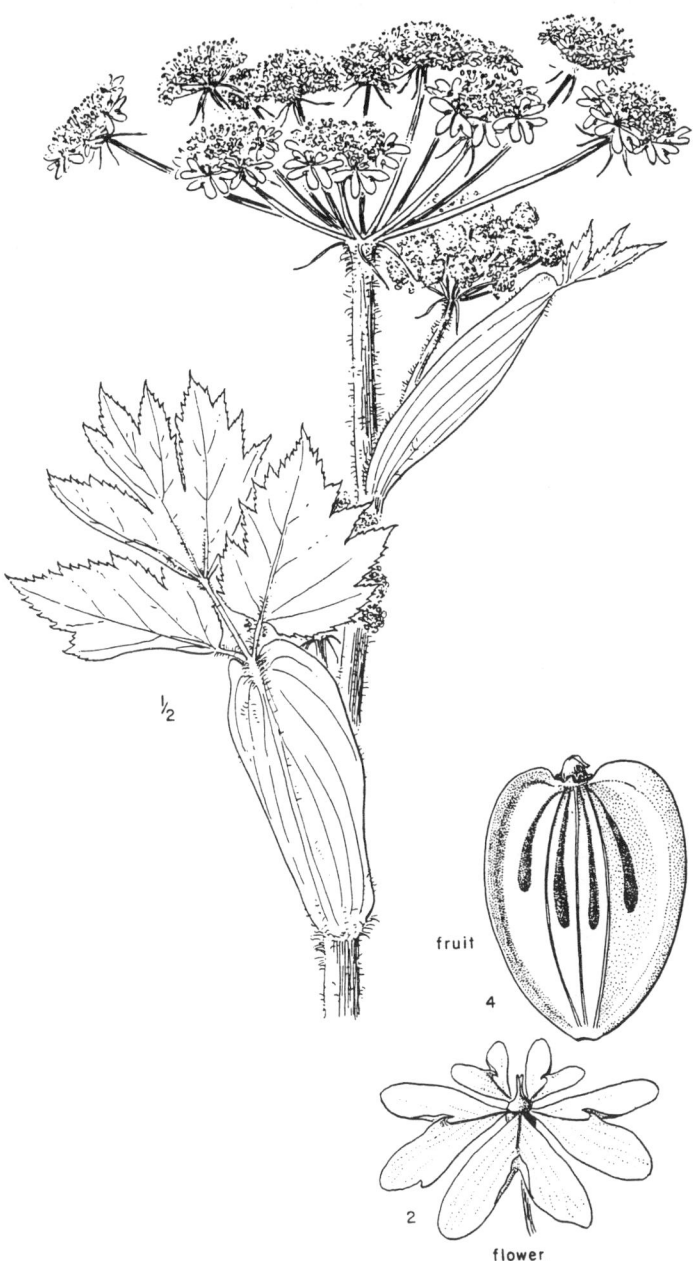

Tall, stout, perennial herb to 1.5 m, from a taproot or fascicled roots. Leaves ternate; leaflets up to 30 cm wide, coarsely doubly-serrate, palmately lobed, the lower surfaces tomentose to villous; petioles with swollen bases. Flowers white, in compound umbels up to 20 cm wide; rays unequal. Involucre of deciduous lanceolate bracts; involucel similar to involucre. Fruits obovate to obcordate, glabrous to sparsely pubescent.

Known from above timberline only in rocky places and on moist stream banks in the northern Wind River Range, although common in the subalpine and montane zones throughout the Middle Rocky Mountains. Transcontinental in the northern United States and Canada, and south in the West to California and Arizona; Siberia.

Brummitt (1971) provided a convincing argument for combining the North American plants that have been called *H. lanatum* Michx. with the European *H. sphondylium* L. to form a single, polymorphic circumboreal species. Following this treatment, North American plants with ternate leaves are ssp. *montanum* (Schleich. ex Gaud.) Briq. ex Schinz & R. Keller.

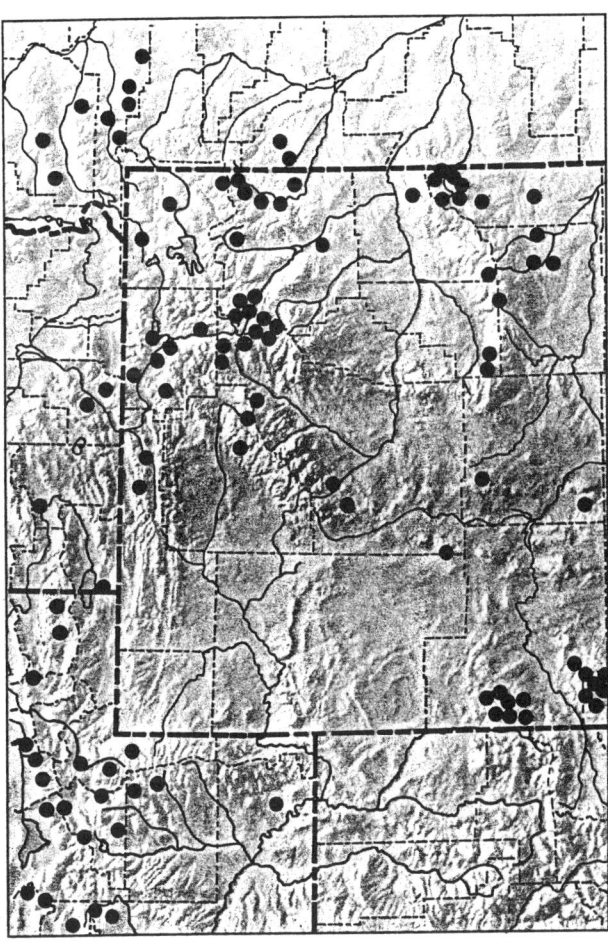

Ligusticum L.
Lovage

Tall, stout, perennial herbs from taproots. Leaves basal and cauline, ternate to ternate-pinnate; leaflets pinnately dissected. Flowers white to pinkish, in flat-topped, compound umbels; involucre and involucel mostly lacking, sometimes of reduced, inconspicuous bracts and bractlets. Calyx teeth evident or minute and obscure. Stylopodium low-conical. Fruit oblong, flattened laterally to subterete in cross section, glabrous, the ribs narrowly winged; carpophore bifid to the base.

KEY TO THE SPECIES OF *LIGUSTICUM*

1. Leaves pinnatifid, the ultimate segments lanceolate, ovate, or oblong, usually more than 3 mm wide ... *L. porteri*
1. Leaves dissected, the ultimate segments linear to very narrowly lanceolate, less than 3 mm wide
    2. Plants mostly less than 5 dm tall, appearing acaulescent; umbels 1–2; cauline leaf reduced or lacking ... *L. tenuifolium*
    2. Plants mostly more than 5 dm tall, distinctly caulescent; umbels 2–5; cauline leaves 1 or more, well developed ... *L. filicinum*

Ligusticum filicinum S. Wats.
Proc. Am. Acad. 11:140, 1876.

Fernleaf Lovage

Ligusticum apiifolium Benth. & Hook.
Ligusticum scopulorum Parry

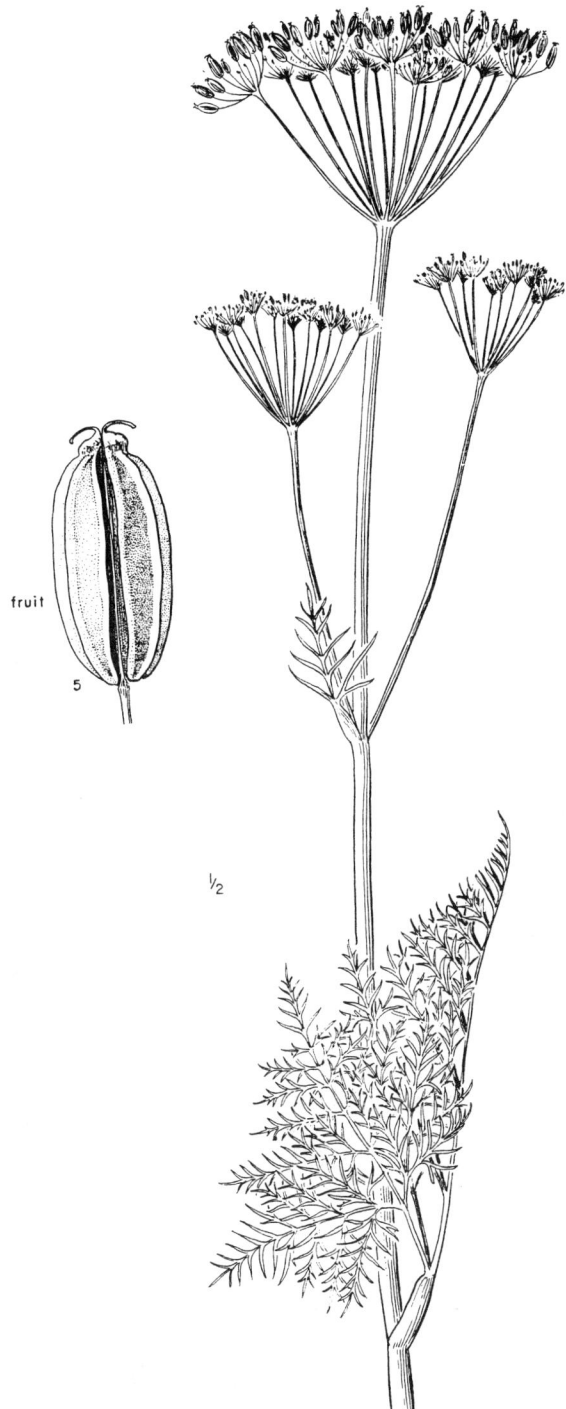

Glabrous to scaberulous perennial herb from a somewhat fleshy taproot. Leaves 3–4-pinnately dissected, the ultimate segments linear to narrowly lanceolate; petioles dilated at the bases. Umbels compound, 1 to several, terminal or in the upper leaf axils; flowers white to greenish; rays approximately equal; involucre and involucel lacking. Fruit oblong, the ribs narrowly winged.

Subalpine slopes and meadows; occurring at timberline in the southern Absarokas near Togwotee Pass, and the Teton and Wind River Ranges. Montana, Idaho, Utah, Wyoming, and Colorado.

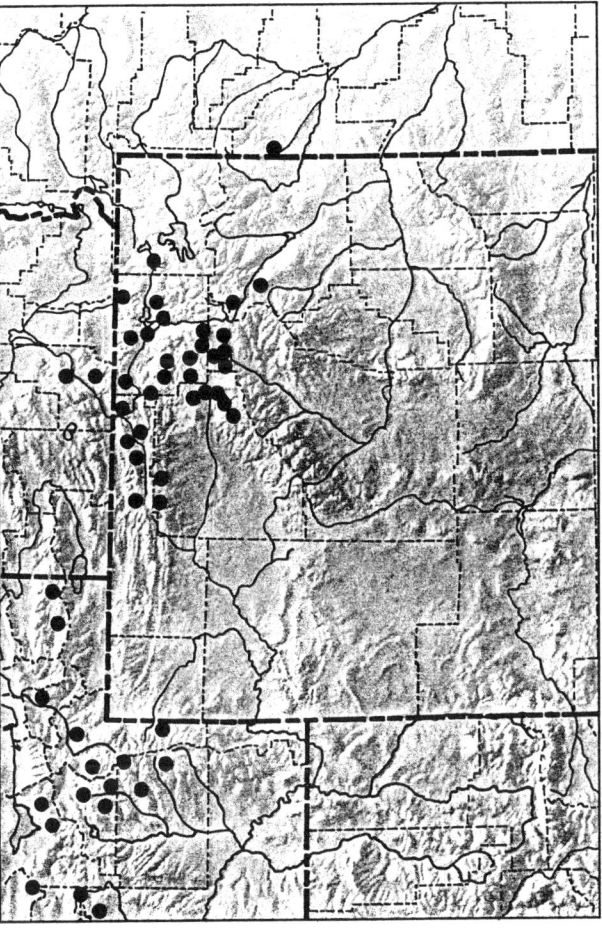

Ligusticum porteri Coult. & Rose
Rev. N. Am. Umbell. 86, 1888.

Porter Lovage

Ligusticum affine A. Nels.
Ligusticum goldmanii Coult. & Rose
Ligusticum madrense Rose
Ligusticum nelsonii Coult. & Rose
Ligusticum simulans Coult. & Rose

Tall, stout, glabrous to puberulent, caulescent, perennial herb, to 1 m tall. Leaves 1–3 times ternately-pinnately compound, the leaflets ovate, deeply incised. Flowers white in flat-topped, compound umbels; rays approximately equal; involucre and involucel wanting. Fruit oblong, winged, 5–8 mm long.

In the alpine zone, known only from tundra of the Medicine Bow Mountains. Wyoming to Nevada and south to Colorado and New Mexico.

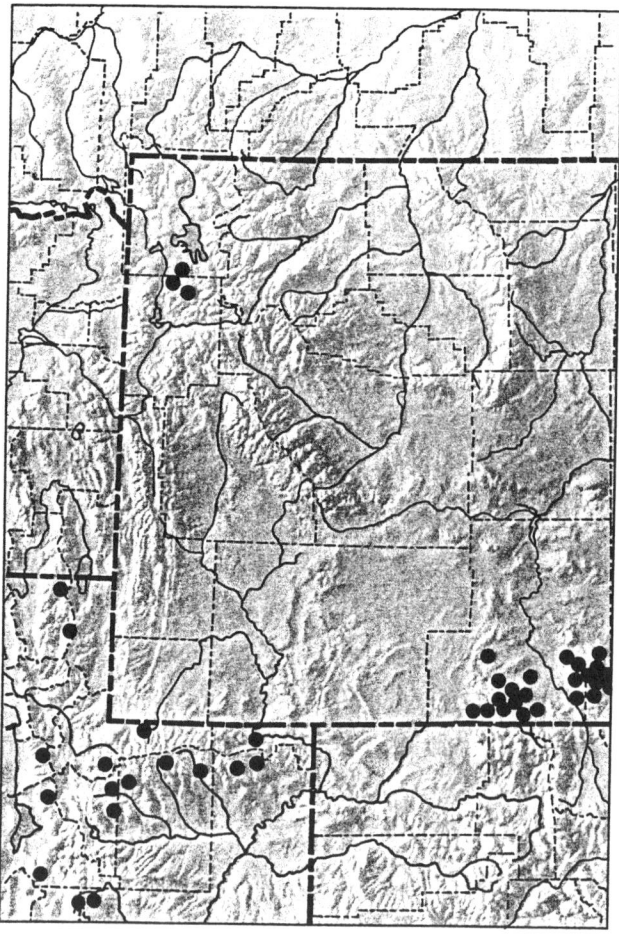

Ligusticum tenuifolium S. Wats.
Proc. Am. Acad. 14:293, 1879.

Slenderleaf Lovage

Ligusticum oreganum Coult. & Rose

Scapose or subscapose perennial from a taproot and unbranched caudex covered by old leaf bases; scapes or stems mostly glabrous, often scaberulous in the inflorescence. Basal leaves pinnately dissected, the ultimate segments narrowly lanceolate to linear; cauline leaf single, reduced, ternately dissected, short-petiolate to sessile. Flowers white or tinged with lavender, in 1 or 2, occasionally 3, compound, terminal umbels; rays approximately equal; involucre lacking or of narrow, linear bractlets. Fruit oblong, the ribs narrowly winged.

Meadows and moist stream banks of the Uinta Range. Oregon to Montana, Wyoming, Utah, and Colorado.

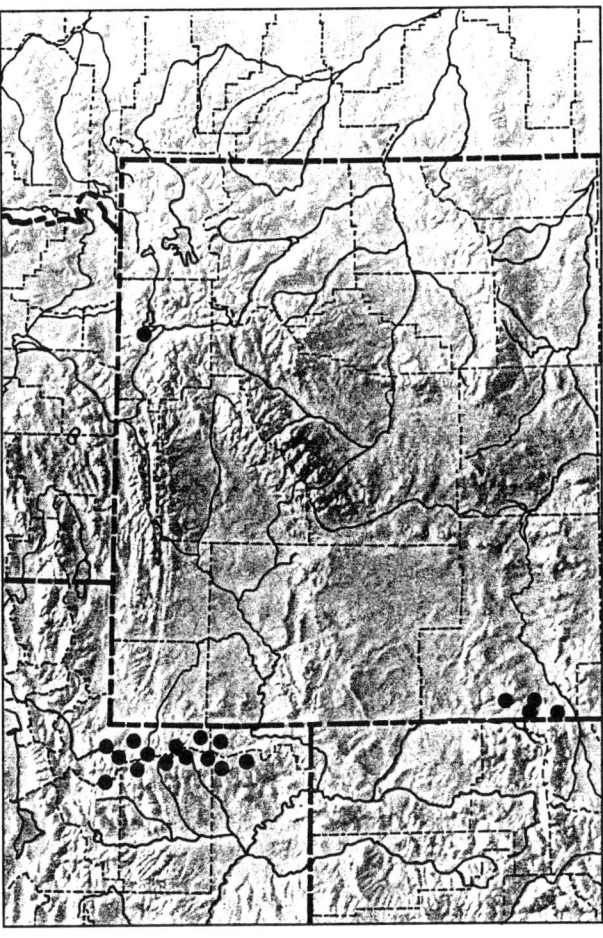

Lomatium Raf.
Biscuitroot

Low, glabrous to pubescent, acaulescent to caulescent, perennial herbs from fleshy or tuberous taproots. Leaves mostly basal, variously compound or decompound. Flowers white or yellow, in loose, compound umbels, these usually terminal; involucre lacking or inconspicuous; involucel of filiform to obovate, conspicuous bractlets, rarely lacking. Calyx teeth obsolete or minute. Fruit linear to orbicular, dorsally flattened, the dorsal ribs evident to obsolete, the lateral ribs thin or corky-winged; oil tubes solitary to numerous in the intervals, 2 or more on the commissure; carpophore bifid to the base; stylopodium none.

KEY TO THE SPECIES OF *LOMATIUM*

1. Leaves with few ultimate divisions, the divisions commonly more than 10 mm long
   2. Plants puberulent; leaves mostly ternate — *L. triternatum*
   2. Plants glabrous; leaves mostly pinnate — *L. graveolens*
1. Leaves with numerous ultimate divisions, the divisions mostly less than 6 mm long — *L. cous*

Lomatium cous (S. Wats.) Coult. & Rose
Contr. U.S. Natl. Herb. 7:214, 1900.

Cous Biscuitroot

Peucedanum cous S. Wats.
Cogswellia circumdata M.E. Jones
Cogswellia cous (S. Wats.) M.E. Jones
Cogswellia montana (Coult. & Rose) M.E. Jones
Lomatium circumdatum (S. Wats.) Coult. & Rose
Lomatium montanum Coult. & Rose
Lomatium purpureum A. Nels.
Peucedanum circumdatum S. Wats.
Peucedanum montanum (Coult. & Rose) Blank.

Scapose or occasionally short caulescent, glabrous to pubescent, perennial herb from a tuberously thickened taproot. Leaves ternate-pinnate, the ultimate divisions oblanceolate to obovate. Flowers yellow, in compound umbels; rays unequal; involucre wanting; involucel of broadly oblanceolate to obovate bractlets. Fruits oblong to oval, laterally winged, the dorsal ribs winged or not.

Dry, rocky ridges and slopes of the Absaroka, Beartooth, Big Horn, Gros Ventre, Hoback, and Wind River Ranges. Oregon to Montana and south to Idaho and Wyoming.

Scaberulous, yellow-flowered plants with involucels lacking or of inconspicuous linear-lanceolate, attenuate bractlets have been called *L. attenuatum* Evert and might be expected above timberline on dry volcanic or limestone lithosols in the North Fork of the Shoshone River drainage of the central Absaroka Range.

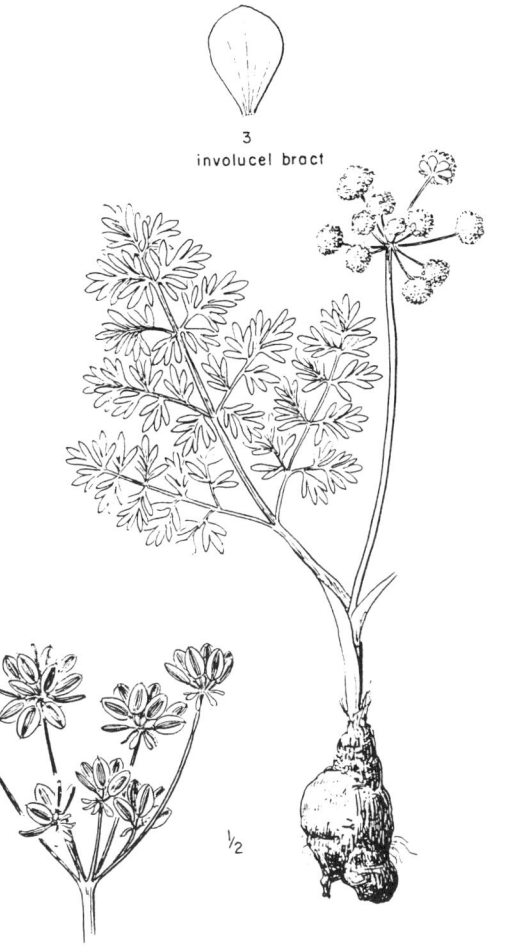

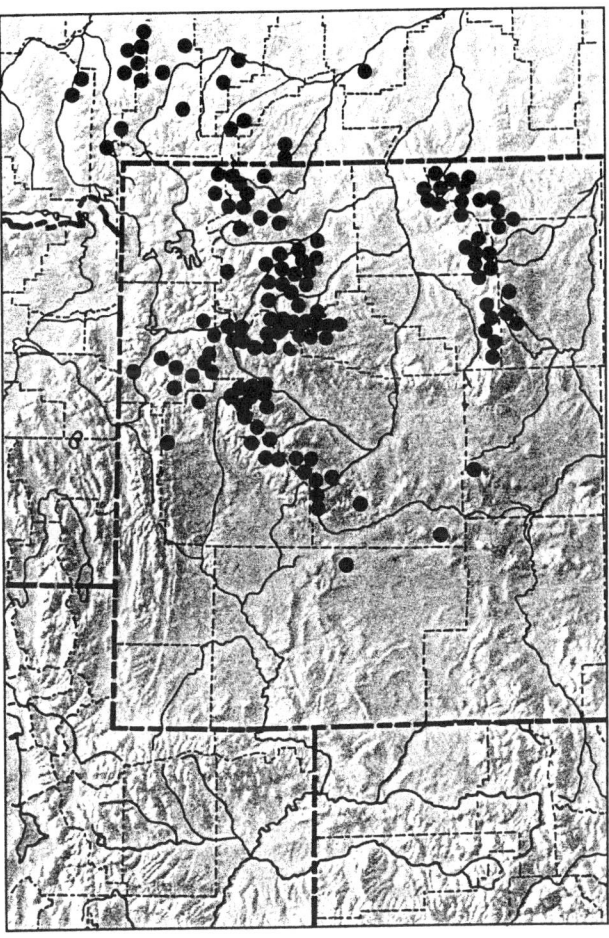

Lomatium graveolens (S. Wats.) Dorn & Hartman
Madroño 35:70–71, 1988.

King Biscuitroot

Peucedanum graveolens S. Wats.
Cynomarathrum alpinum (S. Wats.) Coult. & Rose
Lomatium alpinum (S. Wats.) Macbr.
Lomatium kingii (S. Wats.) Cronq.
Peucedanum kingii S. Wats.

Strongly aromatic, glabrous, scapose perennial herb from branched, woody caudex, often covered with persistent leaf bases. Leaves 1-2-pinnate, sometimes ternate-pinnate, the ultimate division remote, linear; petioles with dilated sheaths. Flowers in compound umbels; bractlets of the involucel linear. Calyx teeth somewhat scarious; petals yellow, often fading to white. Fruits narrowly oblong, laterally winged, the dorsal ribs filiform.

Scree and talus above timberline in the Wyoming Range; to be expected from above timberline in the Wasatch Range on similar substrates, especially limestone. Wyoming, Utah, and Nevada.

Alpine records from the Wyoming Range are var. *graveolens*.

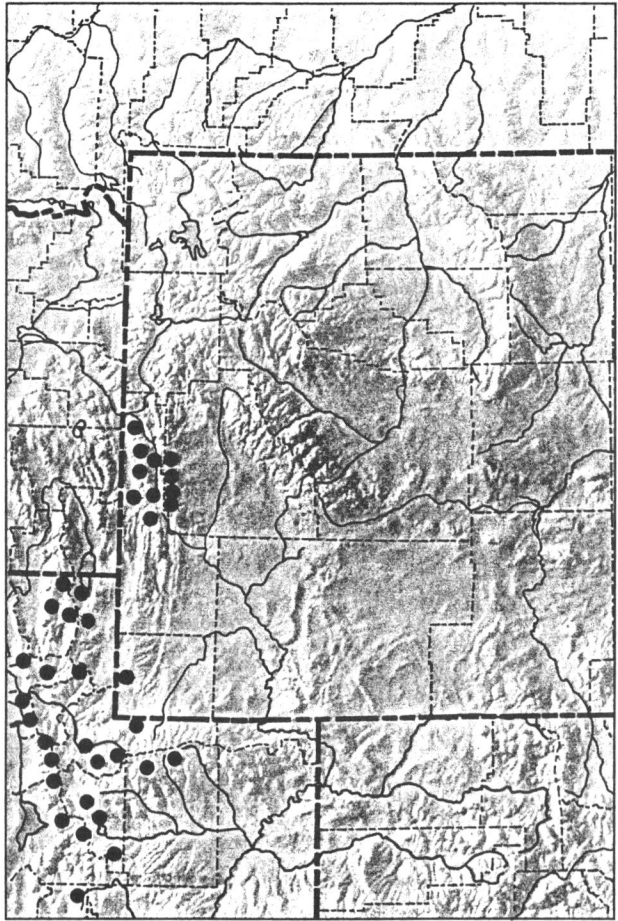

Lomatium triternatum (Pursh) Coult. & Rose
Contr. U.S. Natl. Herb. 7:227, 1900.

Nineleaf Lomatium

Seseli triternatum Pursh
Cogswellia alata (Coult. & Rose) Coult. & Rose
Cogswellia anomala (M.E. Jones ex Coult. & Rose) M.E. Jones
Cogswellia brevifolia (Coult. & Rose) M.E. Jones
Cogswellia leptophylla (Hook.) Rydb.
Cogswellia platycarpa (Torr.) M.E. Jones
Cogswellia robustior (Coult. & Rose ex Holz) Coult. & Rose
Cogswellia simplex (Nutt. ex S. Wats.) M.E. Jones
Cogswellia triternata (Pursh) M.E. Jones
Eulophus triternatus (Pursh) Nutt.
Lomatium alatum (Coult. & Rose) Coult. & Rose
Lomatium anomalum M.E. Jones ex Coult. & Rose
Lomatium brevifolium (Coult. & Rose) Coult. & Rose
Lomatium platycarpum (Torr.) Coult. & Rose
Lomatium robustius (Coult. & Rose ex Holz) Coult. & Rose
Lomatium simplex (Nutt. ex S. Wats.) Macbr.
Peucedanum nuttallii Walp.
Peucedanum simplex Nutt. ex S. Wats.
Peucedanum triternatum (Pursh) Nutt. ex T. & G.

Scapose or subscapose, hirtellous-puberulent, perennial herb from an elongate taproot with a simple or few-branched crown; stems or scapes simple, erect, usually solitary. Leaves basal or nearly so, ternately compound or ternate-pinnately compound, the ultimate divisions linear, 1–10 cm long; bases dilated, sheathing. Flowers yellow, sometimes fading to white, in compound umbels, the rays spreading or ascending, unequal in length; involucre wanting; involucel of linear or filiform bractlets. Fruits oblong to broadly elliptic or suborbicular, glabrous, the wings narrower than the body; dorsal ribs filiform.

The single alpine collection is from fine scree slopes in the southern Absaroka Range. British Columbia and Alberta south to California, Utah, and Colorado.

The alpine collection with narrow, oblong fruits is ssp. *triternatum*.

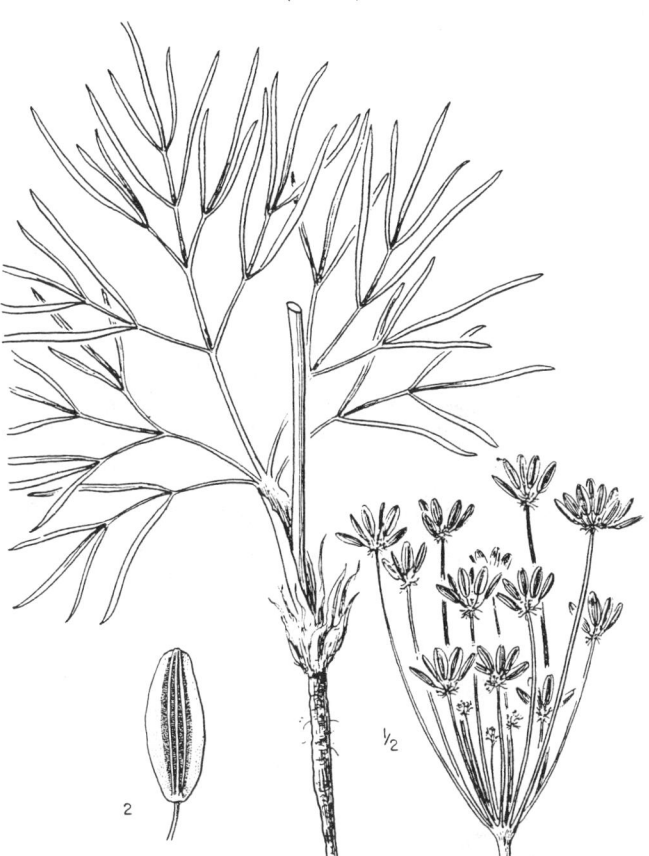

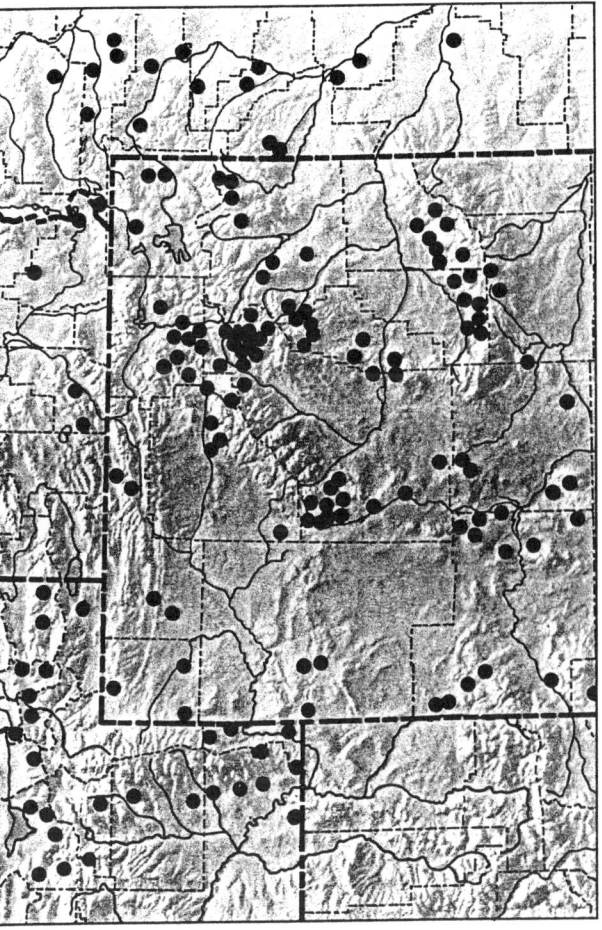

Oreoxis Raf.
Oreoxis

Low, scapose perennials from slender roots. Leaves once- or twice-pinnately compound. Flowers white to yellow in compound umbels; involucre usually lacking; bractlets of involucel mostly on one side (dimidiate), usually exceeding the umbellet. Calyx teeth evident. Stylopodium lacking. Fruit oblong to ovoid, flattened laterally, winged; oil tubes 1 to several in the intervals, 2 to several on the commissure; carpophore lacking.

Oreoxis alpina (Gray) Coult. & Rose
Contr. U.S. Natl. Herb. 7:144, 1900.

Alpine Oreoxis

Cymopterus alpinus Gray

Low, acaulescent, perennial herb less than 1 dm tall. Leaves once- or twice-pinnate, the divisions linear. Flowers yellow, in a compound umbel; involucre lacking, sometimes represented by a single bract; involucel of conspicuous linear bractlets. Fruit oblong, pubescent, winged.

Alpine tundra in the Medicine Bow and Uinta Ranges. Southern Wyoming south to Colorado, Utah, New Mexico, and Arizona. This small genus is endemic to the southern Rocky Mountains and occurs in our region only in the Medicine Bow and Uinta Ranges.

Alpine plants that are glabrous or nearly so are distinguished from the montane ssp. *puberulenta* W.A. Weber as ssp. *alpina*.

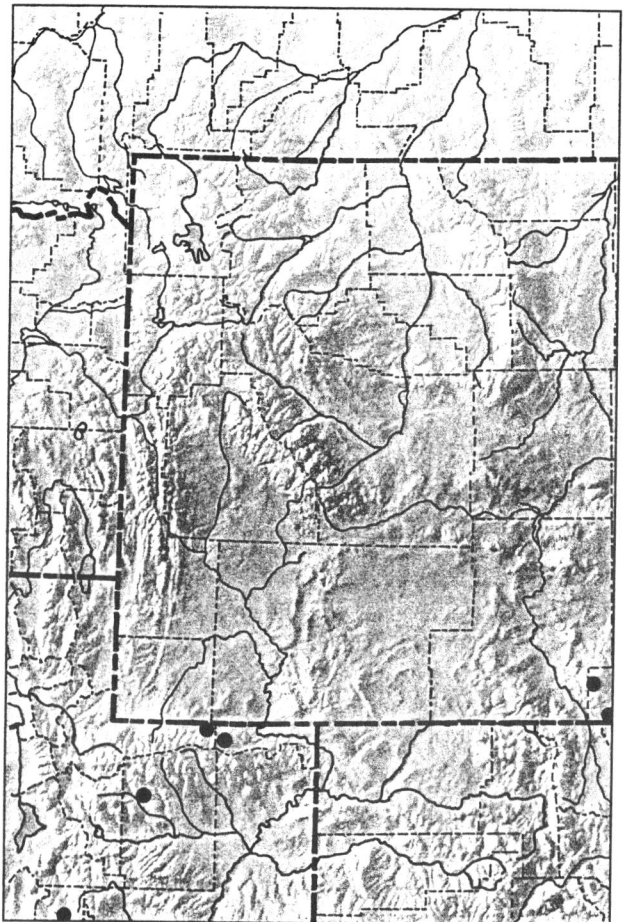

Osmorhiza Raf.
Sweetcicely

Caulescent, perennial herbs from stout taproots; stems simple or branched from near the base. Leaves basal and cauline, ternate or pinnate, the leaflets toothed to ternately parted. Flowers white to greenish in ours, in open terminal or axillary compound umbels; involucre lacking, or of 1 to a few leaflike bracts; involucel lacking, or of several narrow, reflexed bractlets. Calyx teeth obsolete or minute. Stylopodium conic to depressed. Fruit clavate to linear-oblong, terete or subterete in cross section, glabrous or hispid, the ribs narrow; carpophore bifid above.

Osmorhiza depauperata Phil.
Anal. Univ. Chile 85:726, 1894.

Small Sweetcicely

Washingtonia obtusa Coult. & Rose
Osmorhiza obtusa (Coult. & Rose) Fern.

Caulescent, perennial herb from a slender, unbranched taproot; stems simple, solitary. Leaves basal and cauline, glabrate, 1–2-ternate, the leaflets thin, ovate to deltoid, toothed to ternately parted; basal leaves long-petiolate, the bases dilated; cauline leaves short-petiolate, becoming sessile above. Umbels few to several, open and long-pedunculate at maturity. Involucre and involucel lacking, or rarely of 1 or 2 small bracts or bractlets. Sepals lacking; petals greenish white; stylopodium somewhat conic to depressed; styles very short, inconspicuous. Fruits clavate, hispid, the apex acute to obtuse, lacking a subapical constriction.

Timberline and krummholz margins in the Wind River Range. Alaska south in western North America to California, New Mexico, Colorado, and South Dakota.

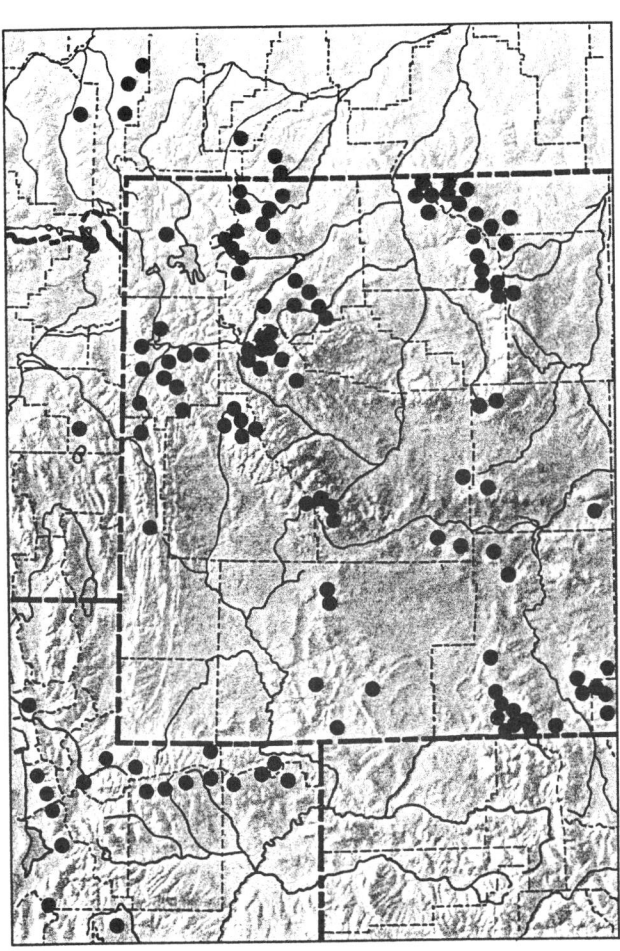

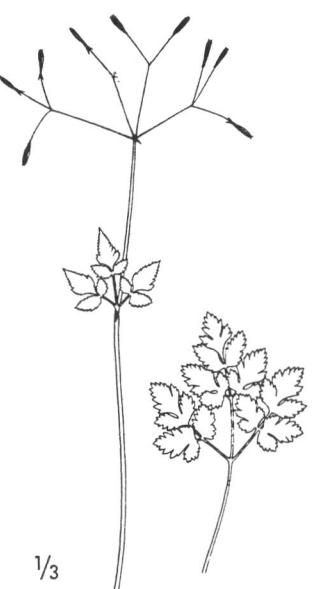

⅓

Aspleniaceae
Spleenwort Family

Sporophytes from scaly, creeping, horizontal or erect rhizomes. Leaves (fronds) circinately coiled when young, commonly with simple, pinnatifid, or decompound blades, these elevated on slender to stout petioles (stipes) with 2 to several vascular bundles, at least below, the blades similar to only slightly dimorphic. Sori on the lower leaf surface, orbicular to oblong, naked or covered by a membranous indusium. Sporangia stalked, with a well-developed vertical annulus. Spores all alike, bilateral. Gametophytes green, cordate, glabrous to glandular or hairy.

KEY TO THE GENERA OF ASPLENIACEAE

1. Sori elongate
    2. Leaves 1-pinnately compound — *Asplenium*
    2. Leaves 2–3-pinnately compound — *Athyrium*
1. Sori round or nearly so
    3. Indusium lacking — *Athyrium*
    3. Indusium present, sometimes inconspicuous
        4. Pinnae spinulose-toothed, or at least mucronate; indusium superior — *Polystichum*
        4. Pinnae lacking marginal spines; indusium partly or wholly inferior
            5. Old leaf bases persistent, dark, wiry; indusium inferior, peltate, cleft into radiating segments — *Woodsia*
            5. Old leaf bases lacking; indusium hoodlike, attached under the margin of the sorus — *Cystopteris*

Asplenium L.
Spleenwort

Small, tufted, evergreen ferns from short, creeping, scaly rhizomes. Fronds simple, or once-pinnate with irregularly toothed to subentire leaflets. Petioles wiry, glabrous to sparsely scaly, greenish to blackish, with 2 vascular bundles, at least below. Sori oblong to linear, borne on the veins; indusia hyaline, glabrous, entire, attached laterally on the outer edge and opening toward the middle of the leaflet, sometimes inconspicuous when sporangia mature. Spores bilateral.

Asplenium trichomanes-ramosum L.
Sp. Pl. 1082, 1753.

Green Spleenwort

Asplenium viride Huds.

Small, slender, spreading fern from a short rhizome covered with blackish scales. Roots brown, fibrous. Fronds pinnately compound, glabrous or sparsely brownish-glandular; stipe dark brown, the rachis green; leaflets opposite or alternate, rhombic to ovate, at least the lower petiolate, the margins coarsely toothed. Sori straight, oblong to elliptic, the indusium pale, undulate-margined.

Rocky places, cliffs and ledges near timberline, especially on limestone of the Big Horn, Teton, Uinta, and Wasatch ranges. Circumboreal; Alaska to Newfoundland, southward to New York, Wisconsin, Colorado, Utah, and Nevada.

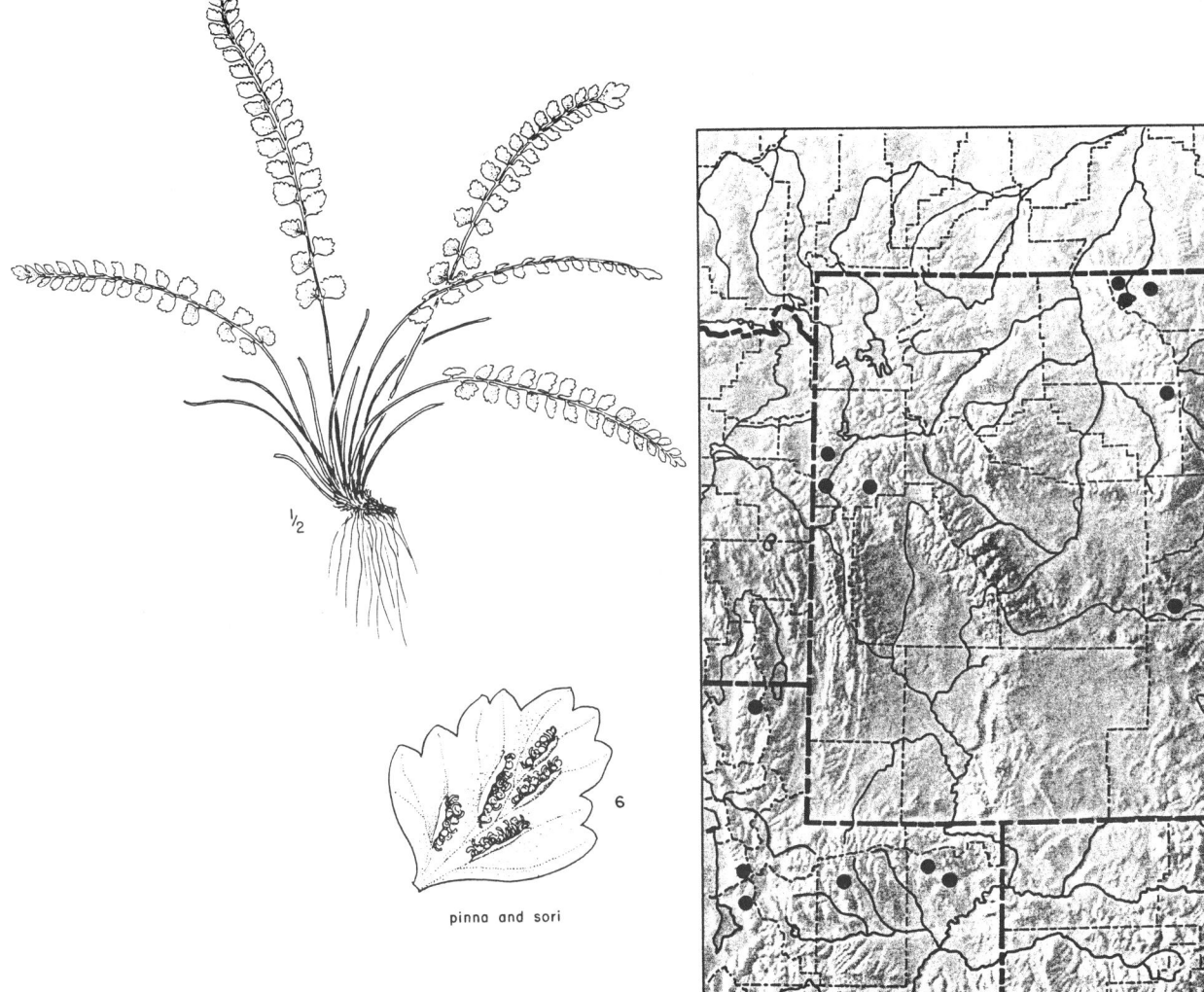

Athyrium Roth.
Lady Fern

Medium-sized, tufted ferns from scaly, short-creeping rhizomes, the scales entire, fibrous. Fronds erect to spreading, long-stipitate, the blades commonly 2–4-pinnate, glabrous to sparsely hairy or scaly; petioles stout, with 2 vascular bundles, at least below. Sori elongate and lunate, or orbicular; indusia lacking, or hyaline and attached laterally along the vein on the outer edge and opening toward the middle of the pinnule, the margin entire to erose or fimbriate. Spores bilateral.

KEY TO THE SPECIES
ATHYRIUM

1.	Indusium lacking; sori orbicular	*A. alpestre*
1.	Indusium present; sori lunate	*A. filix-femina*

Athyrium alpestre (Hoppe) Clairville
Man. Herbor. Suisse, 301. 1811.

Alpine Lady Fern

Aspidium alpestre Hoppe
Athyrium americanum (Butters) Maxon
Athyrium distentifolium Tausch ex Opiz
Polypodium alpestre (Hoppe) Spreng.
Pseudathyrium alpestre (Hoppe) Newm.

Tufted ferns from short rhizomes scaly with old leaf bases. Fronds 2–4-pinnate, glabrous, broadly lanceolate, with straw-colored stipes; pinnae obliquely inserted, narrowly deltoid, becoming crowded above; pinnules pinnatifid, the margins shallowly serrate. Sori round, numerous on the pinnules; indusia lacking.

Wet to moist places, often on talus near timberline and in the lower alpine zone of the Beartooth, Medicine Bow, and Teton ranges. Alaska, southward to California, Nevada, and Colorado.

North American plants are var. *americanum* Butters.

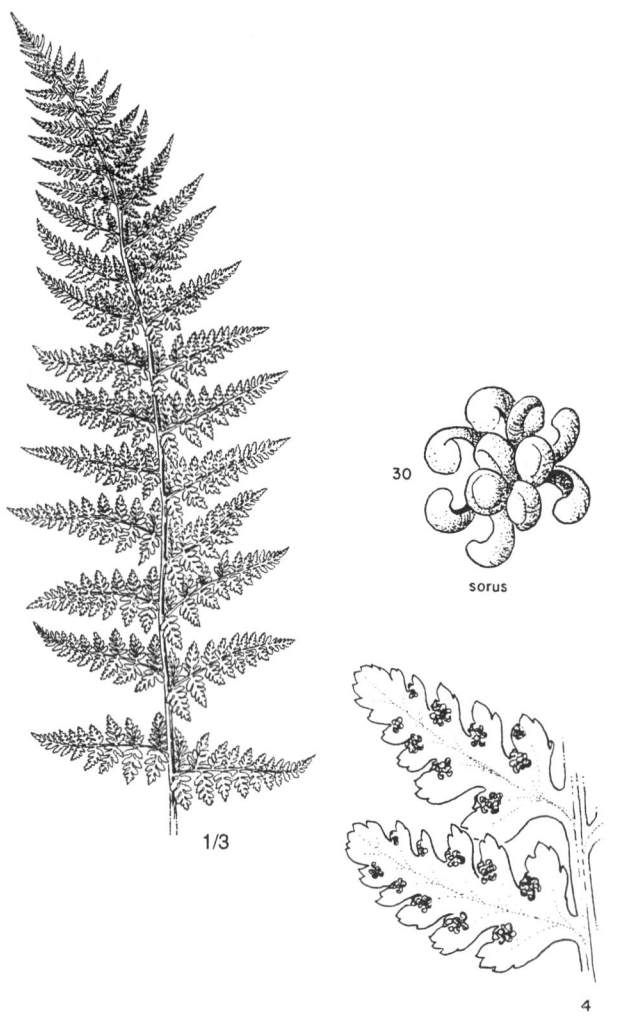

62 Aspleniaceae

Athyrium filix-femina (L.) Roth ex Mertens
Rom. Arch. Bot. 2(1):106. 1799

Common Lady Fern

Polypodium filix-femina L.
Aspidium angustum Willd.
Aspidium filix-femina (L.) Swartz
Asplenium cyclosorum (Rupr.) Fern.
Asplenium filix-femina (L.) Bernh.
Athyrium angustum (Willd.) K. Presl
Athyrium asplenioides (Michx.) A.A. Eat.
Cyathea filix-femina (L.) Bertol.
Cystopteris filix-femina (L.) Coss. & Germ
Nephrodium asplenioides Michx.

Tufted ferns from a short, stout, scaly rhizome covered by persistent leaf bases. Fronds 2–3-pinnate, glabrous or sparsely puberulent on the midveins; blades elliptic with straw-colored to brown stipes; pinnae alternate, lanceolate, reduced upwards; pinnules lanceolate to elliptic, the margins deeply incised to serrate, the bases mostly decurrent on the rachis. Sori lunate or horseshoe-shaped; indusia with the free margin ciliate, erose, or subentire.

Moist places in the Middle Rocky Mountains, mostly at lower elevations, but occurring above timberline in the Tetons. Circumboreal; most of North America.

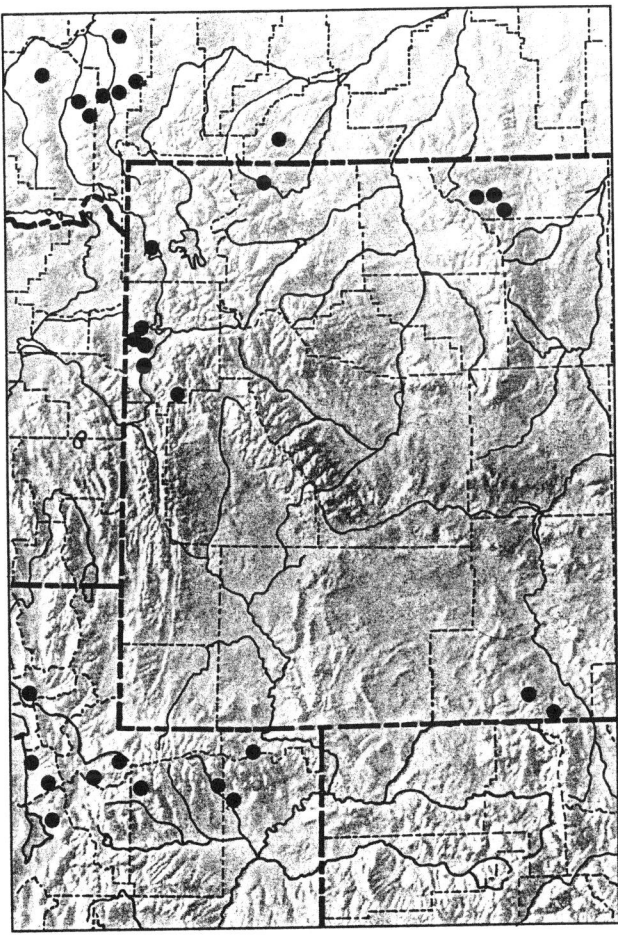
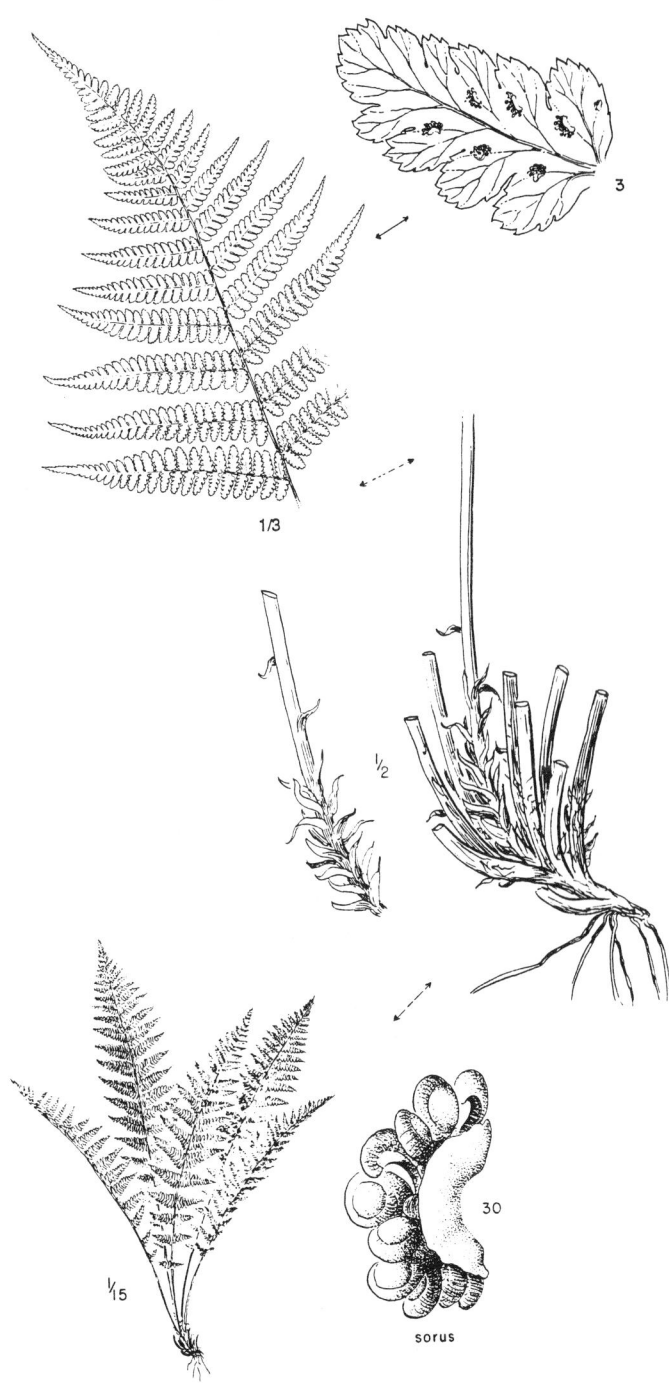

Cystopteris Bernh.
Bladder Fern

Small, tufted, delicate ferns from slender, scaly rhizomes, the scales lanceolate to ovate. Fronds bipinnate to decompound, the fertile and sterile blades similar in size and shape. Petioles slender, scaly at base, with 2 vascular bundles. Sori orbicular, attached to veins midway between the margin and midrib of the blade. Indusium hyaline, hood-like, laterally attached, deciduous or concealed by the sporangia at maturity. Spores bilateral, and spiny in ours.

Cystopteris fragilis (L.) Bernh.
Schrad. Neu. Journ. Bot. 1(2):26, 1806.

Brittle Bladder Fern

Polypodium fragile L.
Aspidium fragile (L.) Swartz
Athyrium fragile (L.) Spreng.
Cyathea fragilis (L.) J.E. Smith
Cystopteris dickieana Sim
Cyclopteris fragilis (L.) S.F. Gray
Cystea fragilis (L.) J.E. Smith
Filix fragilis (L.) Underwood

Small, loosely tufted ferns from short, creeping rhizomes crowded with old petiole bases. Fronds deciduous, 2-pinnate, or nearly so; stipes brown, at least below, often with a few scales at the base; blades lanceolate, elliptic, or narrowly ovate; glabrous or sometimes glandular; pinnules deltoid to lanceolate, pinnatifid or toothed, decurrent and offset on the stipe. Sori small, round, borne on the veins of the pinnules; indusia hood-shaped, the margins entire or erose, often deciduous.

Damp, or at least protected, ledges, cliffs, and talus of nearly all our alpine ranges. Not yet known from above timberline in the Hoback Range. Circumboreal; Alaska to Newfoundland, southward to New York, Missouri, Texas, and California.

Mountain plants are the typical var. *fragilis*.

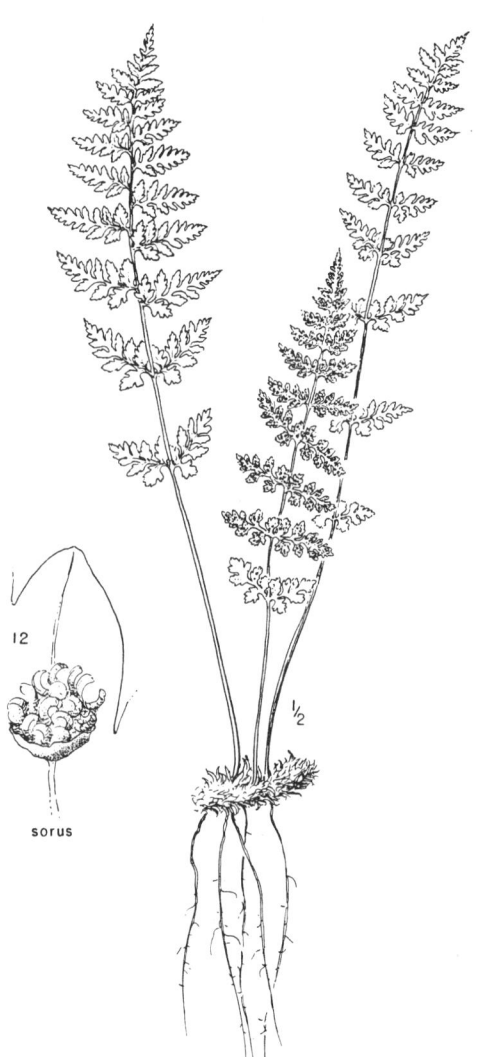

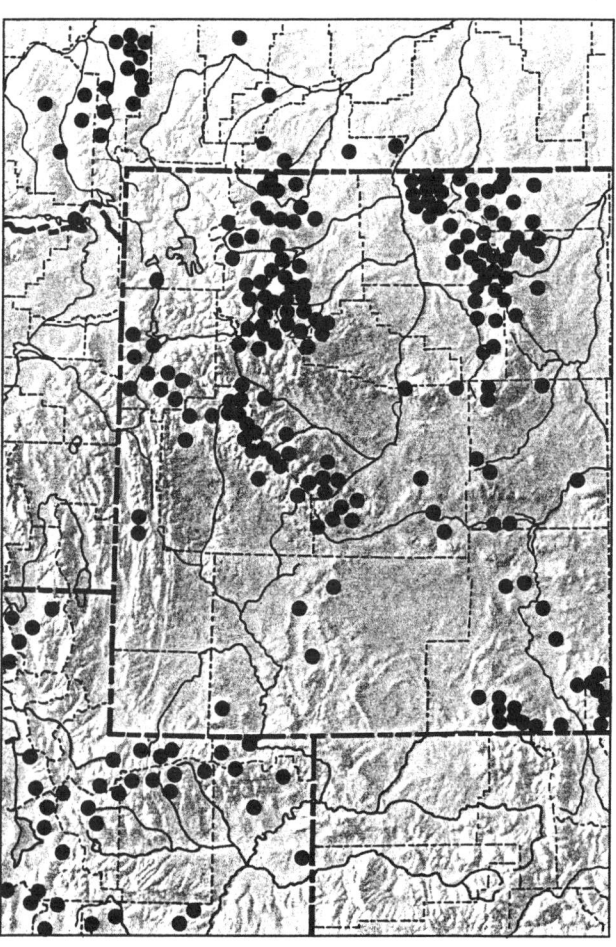

Polystichum Roth
Shield Fern, Holly Fern

Tufted evergreen ferns from short, stout, creeping to somewhat erect, brown-scaly rhizomes. Fronds scaly, especially on the rachis, the scales often dimorphic; blades pinnate, the pinnae sessile or nearly so, asymmetrical at the base, often with sharply toothed margins; petioles stout, shorter than the blades, with 4 or 5 vascular bundles. Sori orbicular, borne on the veins in one or more rows on each side of the midrib. Indusium peltate, the margin entire, ciliate, or erose. Spores bilateral.

Polystichum lonchitis (L.) Roth
Tent. Fl. Germ. 3(1):71, 1800.

Holly Fern

Polypodium lonchitis L.
Aetopteron lonchitis (L.) House
Aspidium lonchitis (L.) Swartz
Dryopteris lonchitis (L.) Kuntze
Hyopeltis lonchitis (L.) Todaro

Loosely tufted ferns from short, stout, erect to ascending rhizomes; scales reddish brown to chestnut. Leaves several, evergreen, pinnate, linear to oblanceolate, short-stipitate, the rachis scaly. Pinnae alternate, the lower deltoid and equilateral, the middle and upper asymmetrically toothed at base, lanceolate, falcate; margins unevenly spinulose-toothed. Sori round, in 1 or 2 rows on the middle and upper pinnae, medial; indusia round, the margins erose-dentate.

Ledges, cliffs, and coarse talus near timberline and in the lower alpine zone of the Beartooth, Teton, and Wasatch ranges. Circumboreal; south in western North America to California, Utah, and New Mexico.

Polystichum scopulinum (D.C. Eaton) Maxon, the Rock Holly Fern, might be encountered at timberline in similar habitats. It differs from *P. lonchitis* by having pinnae conspicuously cleft near the base, with 1–3 lobes, and teeth mucronate, but not spinulose as in *P. lonchitis*.

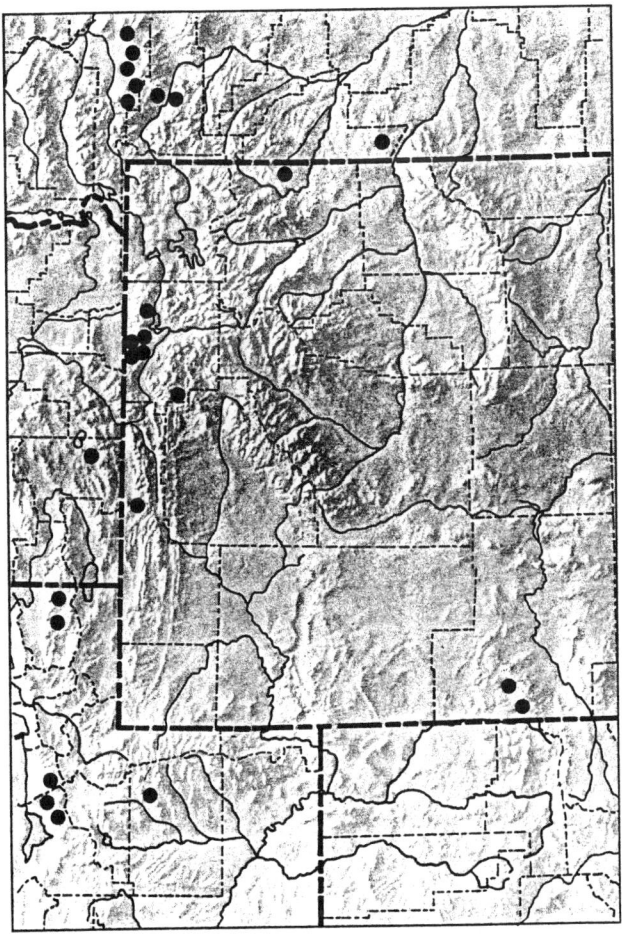

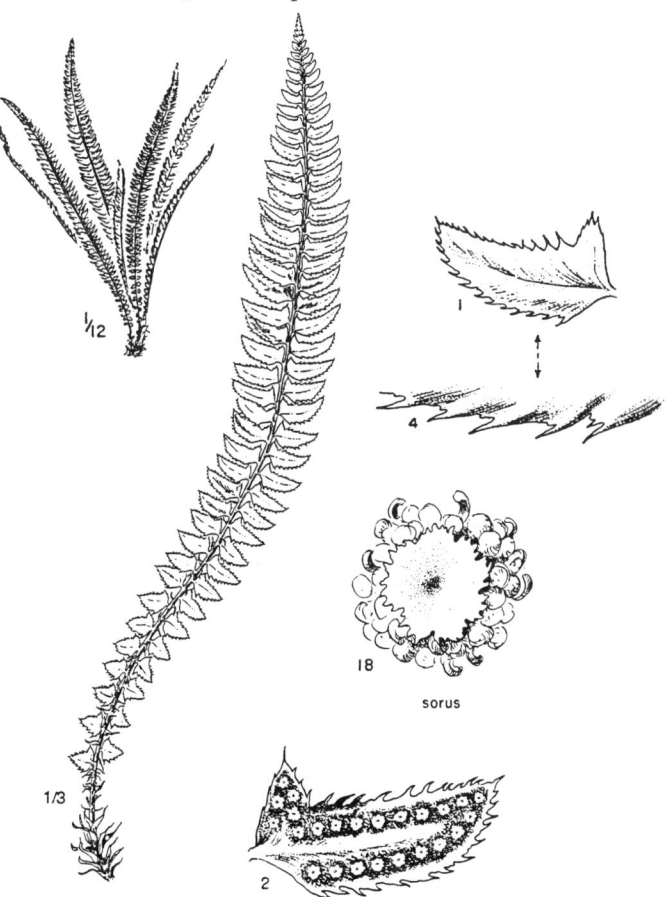

Woodsia R. Br.
Cliff Fern

Small to medium-sized, tufted ferns from short-creeping rhizomes covered by persistent petiole bases and brownish scales. Roots fibrous, brown, clustered on the rhizome. Leaves 1–2-pinnate; pinnae triangular, the pinnules crenate or dentate; petioles slender, wiry, shorter than the blades, persistent, with 2 vascular bundles at least at the base. Sori orbicular, borne on the veins; indusium inferior, peltate, at maturity splitting into spreading, mostly unequal, radial scalelike or filiform segments, often concealed by the sporangia. Spores bilateral.

Woodsia scopulina D.C. Eat.
Can. Nat. II, 2:91, 1865.

Rocky Mountain Woodsia

Physematium scopulinum (D.C. Eat.) Trev.

Small to medium-sized ferns, often forming tufts, from short, slender rhizomes; scales few, entire, brown with a dark stripe. Leaves several to numerous, mostly bipinnate, whitish-pilose and glandular below and on the rachis, glabrous or pilose and glandular above; hairs flattened, septate; pinnae alternate to nearly opposite, mostly lanceolate, the pinnules oblong to elliptic, obtuse, crenate-serrate or lobed; petioles reddish brown, at least below, whitish-pilose and glandular to glabrate. Sori round, medial; indusia cleft into narrow, often hairlike, stellate segments.

Rocky places near timberline in the lower alpine zone of the Absaroka, Big Horn, Medicine Bow, Teton, Uinta, Wasatch, and Wind River Ranges, and might be expected on cirque walls, talus, and moraines in other ranges of our region. Alaska to Quebec, southward to North Carolina, Tennessee, Oklahoma, New Mexico, Arizona, and California.

Plants of the lower alpine zone are the typical var. *scopulina*.

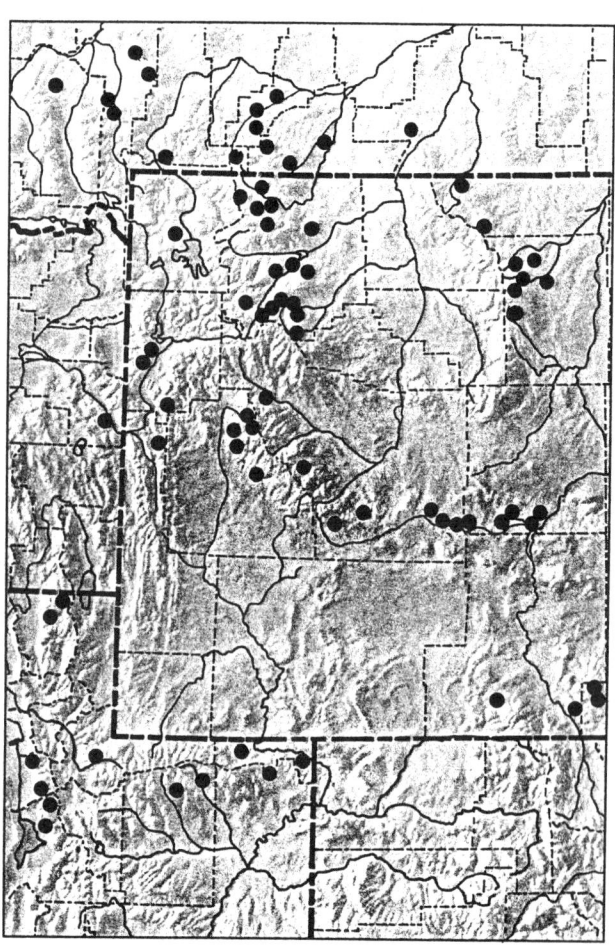

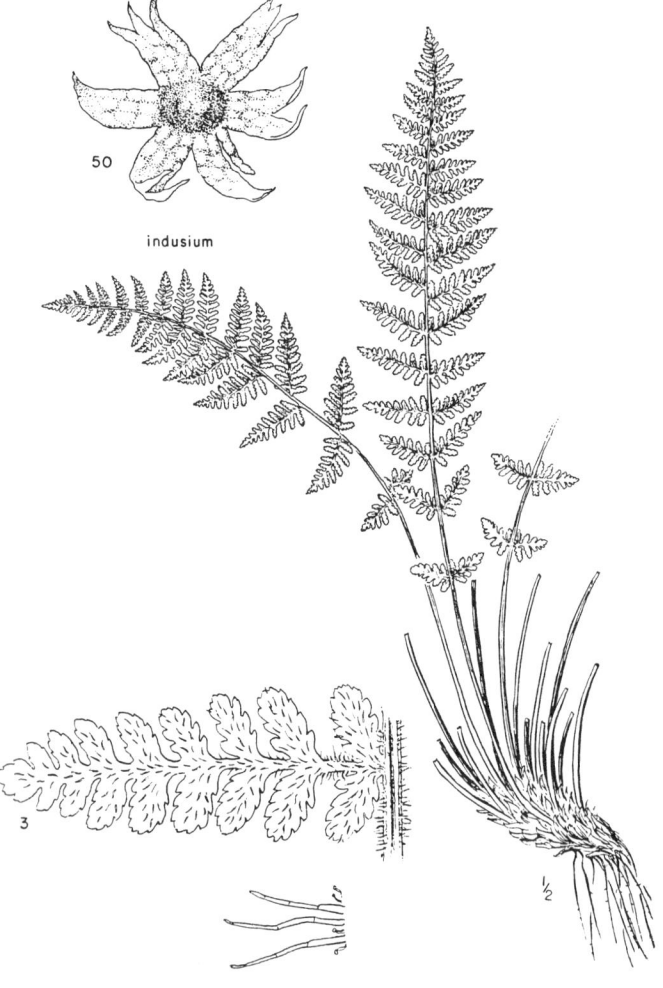

Asteraceae
(Compositae)
Aster Family

Herbs, shrubs, or semishrubs with simple and entire to variously divided, exstipulate leaves. Inflorescence a heterogamous or homogamous head subtended by an involucre of 1 or more rows of variously textured bracts which surround a dry or fleshy, chaffy or naked receptacle. Flowers epigynous, perfect, unisexual or sterile, regular or irregular, commonly 5-merous, all tubular, the central tubular and the marginal irregular and ligulate, or all ligulate. Stamens 5, alternate with the lobes of the corolla in regular flowers, inserted on the corolla tube, the filaments or, usually, the anthers united into a tube. Calyx (pappus) lacking or developing late and crowning the summit of the mature akene, of hairs, bristles, scales, or awns. Pistil 1, the ovary 1-celled, of 2 united carpels, the style 2-cleft at the apex, sometimes becoming elongate and beaklike in fruit. Fruit an akene.

KEY TO THE GENERA OF ASTERACEAE

1. Corollas all ligulate, perfect, irregular; juice milky
    2. Heads several to many; plants leafy-stemmed
        3. Pappus white; plants less than 10 cm tall in ours — *Crepis*
        3. Pappus brown; plants more than 20 cm tall in ours — *Hieracium*
    2. Heads solitary on naked peduncles; plants scapose
        4. Pappus of plumose bristles; involucral bracts in 1 series, more than 2 cm long — *Tragopogon*
        4. Pappus of capillary bristles or scales; involucral bracts in more than 1 series, less than 2 cm long
            5. Akenes not beaked; pappus of slender scales — *Nothocalais*
            5. Akenes beaked; pappus of capillary bristles
                6. Leaves mostly deeply pinnatifid or laciniate; akenes spinulose-muricate at the apex, 4–5-ribbed — *Taraxacum*
                6. Leaves mostly entire or shallowly toothed; akenes not spinulose-muricate at the apex, 10-ribbed — *Agoseris*
1. Corollas both tubular and ligulate or all tubular; juice not milky
    7. Heads discoid, the corollas all tubular
        8. Pappus of capillary to plumose bristles
            9. Leaf margins and involucral bracts spine-tipped — *Cirsium*
            9. Leaf margins and involucral bracts lacking spines
                10. Plants shrubby; flowers yellow
                    11. Involucral bracts acuminate, aligned to form vertical ranks; disk flowers 5–7 — *Chrysothamnus*
                    11. Involucral bracts acute, not forming vertical ranks; disk flowers 10–25 — *Haplopappus*
                10. Plants herbaceous; flowers white, pink, or purple
                    12. Pappus of plumose bristles; corollas purple — *Saussurea*
                    12. Pappus of capillary bristles; corollas white or pink
                        13. Plants often mat formers; stems with a basal tuft of leaves, the cauline leaves reduced upward — *Antennaria*
                        13. Plants not forming mats; stems lacking a basal tuft of leaves, the cauline leaves scarcely reduced upward — *Anaphalis*
        8. Pappus of scales or lacking
            14. Heads solitary on naked peduncles in ours — *Chaenactis*
            14. Heads several to numerous
                15. Leaves mostly grayish green, pubescent, but lacking dolabriform hairs; heads in elongate spikes, racemes, or panicles — *Artemisia*

15. Leaves green, glabrate, with sparse, dolabriform hairs; heads in flat-topped corymbs .... *Sphaeromeria*
7. Heads radiate, the marginal corollas ligulate
    16. Pappus of hyaline or chaffy scales
        17. Plants shrubby; ray flowers 5 mm or less long .... *Gutierrezia*
        17. Plants herbaceous; ray flowers 6 mm or more long
            18. Involucral bracts fewer than 10, each subtending, or nearly subtending, an individual ray flower .... *Eriophyllum*
            18. Involucral bracts more than 10, not subtending individual ray flowers
                19. Plants aromatic and glandular-pubescent; akenes elongate, at least 4 times longer than wide .... *Hulsea*
                19. Plants not aromatic, glabrous to arachnoid or woolly; akenes short, 2–3 times longer than wide .... *Hymenoxys*
    16. Pappus partly or wholly of capillary bristles, bristlelike scales, or lacking
        20. Leaves opposite, not all basal .... *Arnica*
        20. Leaves alternate or all basal
            21. Flowers yellow
                22. Involucral bracts essentially uniseriate, equal, narrow, usually with only a few shorter ones at the base .... *Senecio*
                22. Involucral bracts in 2 or more distinct series, equal or imbricate, sometimes broad and somewhat leafy
                    23. Involucral bracts arranged in vertical ranks; ray flowers mostly 3 or fewer .... *Petradoria*
                    23. Involucral bracts not in vertical ranks; ray flowers more than 5
                        24. Pappus of 2 distinct series, the outer series of much shorter scales or bristles than the inner .... *Heterotheca*
                        24. Pappus of 1 series, the bristles equal or unequal
                            25. Heads solitary to few, the involucre 6–12 mm high; pappus bristles unequal; plants usually with taproots .... *Haplopappus*
                            25. Heads several to many, in racemes or panicles, the involucre 4–6 (–11) mm high; pappus bristles equal; plants with fibrous roots arising from a rhizome or short, branched caudex .... *Solidago*
            21. Flowers white, lavender, blue, or purple
                26. Pappus of disk flowers wholly or partly of capillary bristles
                    27. Leaves and involucral bracts spinulose-tipped .... *Machaeranthera*
                    27. Leaves and involucral bracts lacking spinulose tips
                        28. Involucral bracts subequal or more or less imbricate, often green in part, but neither definitely leafy nor with chartaceous base and herbaceous green tip; style branches lanceolate or broader, acute to obtuse, 0.5 mm long or less, or obsolete .... *Erigeron*

28. Involucral bracts either subequal and the outer leafy or, more commonly, evidently imbricate, with chartaceous base and evident green tip, sometimes chartaceous throughout; style branches lanceolate or narrower, acute or acuminate, ordinarily longer than 0.5 mm *Aster*
26. Pappus of the disk flowers lacking or of bristlelike scales
    29. Leaves entire; rays numerous, more than 14 *Townsendia*
    29. Leaves pinnatifid or pinnately dissected; rays few, commonly 3–5 *Achillea*

Achillea L.
Yarrow

Erect, aromatic, perennial herbs with alternate, finely pinnately dissected leaves. Heads small, usually radiate, in a dense corymbose inflorescence, the involucral bracts imbricate in several series, with hyaline or scarious margins, the receptacle chaffy. Ray flowers few, 5–12, white in ours, pistillate, and fertile. Disk flowers several to many, perfect, and fertile. Akenes oblong, flattened, glabrous, the pappus lacking.

Achillea millefolium L.
Sp. Pl., 899, 1753.

Common Yarrow

Achillea alpicola (Rydb.) Rydb.
Achillea angustissima Rydb.
Achillea arenicola A. Heller
Achillea asplenifolia sensu auctt. non Vent.
Achillea borealis Bong.
Achillea californica Pollard
Achillea eradiata Piper
Achillea fusca Rydb.
Achillea gigantea Pollard
Achillea gracilis Raf.
Achillea lanulosa Nutt.
Achillea laxiflora Pollard & Cockll.
Achillea megacephala Raup
Achillea nigrescens (E. Mey.) Rydb.
Achillea occidentalis (DC.) Raf. ex Rydb.
Achillea pacifica Rydb.
Achillea puberula Rydb.
Achillea rosea Desf.
Achillea subalpina Greene
Achillea tomentosa Pursh

Aromatic, rhizomatous, fibrous-rooted perennial; stems erect to ascending, villous, simple to branched above. Leaves alternate, sometimes with a basal tuft, linear to oblong, 2–3-pinnately dissected, the ultimate segments linear, acute to subulate, only the lower petiolate, the cauline leaves reduced upward. Heads numerous, mostly in flat-topped, corymbose cymes. Involucral bracts villous to glabrate, oblong, scarious, with a greenish keel and mostly brownish to blackish margins. Ray flowers mostly 5, white or sometimes pink; disk flowers 10–30, white to cream, pappus lacking. Receptacle chaffy. Akenes oblong, flattened, glabrous.

Fellfields, meadows, slopes, stream banks, especially in the lower alpine zone near timberline in all Middle Rocky Mountain ranges. Circumboreal; widely distributed in North America.

Alpine plants of short stature and having involucral bracts with dark brownish to blackish margins are referable to ssp. *lanulosa* (Nutt.) Piper, var. *alpicola* (Rydb.) Garrett. They apparently intergrade with and at times are indistinguishable from the introduced ssp. *millefolium*.

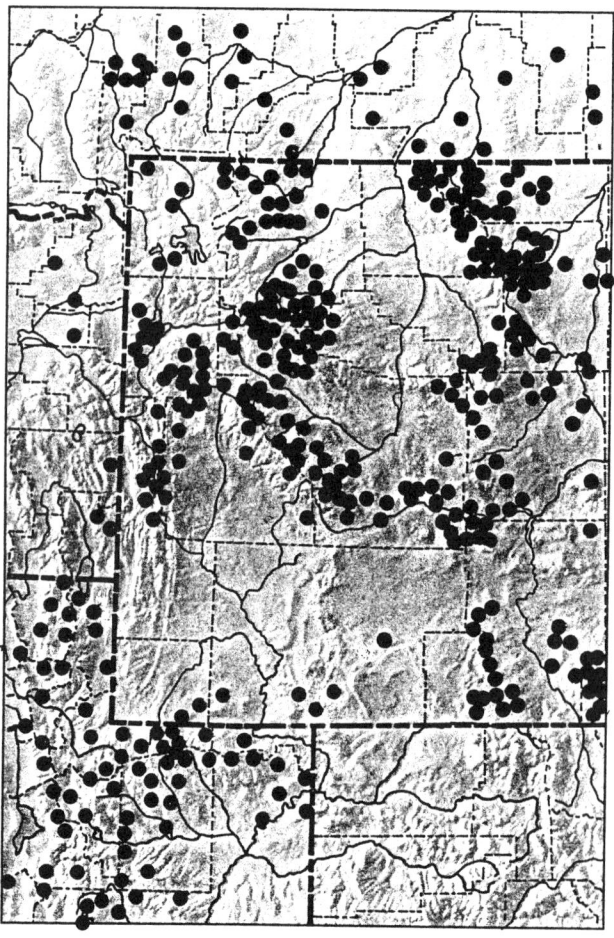

½

Agoseris Raf.
False Dandelion

Scapose, perennial herbs with milky juice, from taproots, and with sessile, entire to toothed leaves. Heads solitary and terminal on long peduncles, the bracts of the involucre in several series, the outer ones shorter and broader, the receptacles flat, naked. Flowers all ligulate, yellow or orange, sometimes drying pink or purple. Akenes oblong to linear, 10-ribbed, tapering into a smooth beak bearing a pappus of numerous white, capillary bristles.

KEY TO THE SPECIES OF
AGOSERIS

1. Corollas orange; akenes with a beak more than half as long as the body *A. aurantiaca*
1. Corollas yellow; akenes with a beak less than half as long as the body *A. glauca*

Agoseris aurantiaca (Hook.) Greene
Pittonia 2:177, 1891.

Orange Agoseris

Troximon aurantiacum Hook.
Agoseris carnea Rydb.
Agoseris confinis Greene
Agoseris gracilens (Gray) Kuntze
Agoseris gracilenta (Gray) Greene
Agoseris graminifolia Greene
Agoseris greenei (Gray) Rydb.
Agoseris howellii Greene
Agoseris longirostris Greene
Agoseris nana Rydb.
Agoseris purpurea (Gray) Greene
Agoseris rostrata Rydb.
Agoseris subalpina G.N. Jones
Macrorhynchus purpureus Gray
Macrorhynchus troximoides T. & G.
Troximon gracilens Gray
Troximon montanum (Osterh.) A. Nels.
Troximon purpureum (Gray) A. Nels.

Scapose, glabrous to somewhat villous, perennial herb from a simple to branched caudex and taproot; scapes villous-tomentose, at least below the inflorescence. Leaves oblong to oblanceolate, entire or with a few teeth or lobes, glabrate to villous, rounded to acuminate. Heads single. Involucral bracts imbricate, narrowly lanceolate, acuminate, the outer ones often obtuse. Flowers orange, drying purple or pink. Akenes striate, tapering abruptly to a slender, obscurely striate beak more than half as long as to longer than the body.

Meadows, stream banks, and fellfields of the Absaroka, Beartooth, Big Horn, Teton, Uinta, and Wasatch Ranges. Alaska and Yukon Territory south to California, Utah, and New Mexico.

Two weakly differentiated varieties are recognized as follows:

1. Involucral bracts imbricate, broad, obtuse, mottled with irregular purple blotches . . . var. *purpurea* (Gray) Cronq.
1. Involucral bracts not imbricate, or only slightly imbricate, narrow, acute, dotted with fine purple spots
. . . var. *aurantiaca*

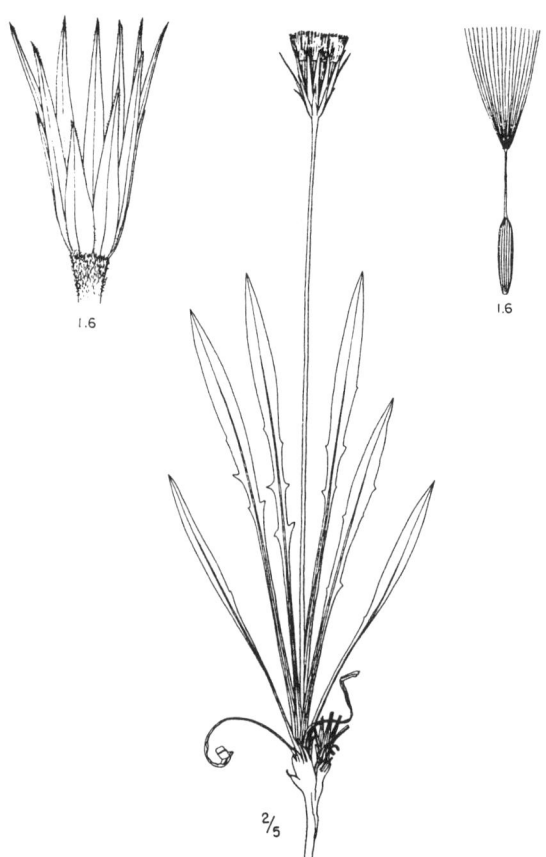

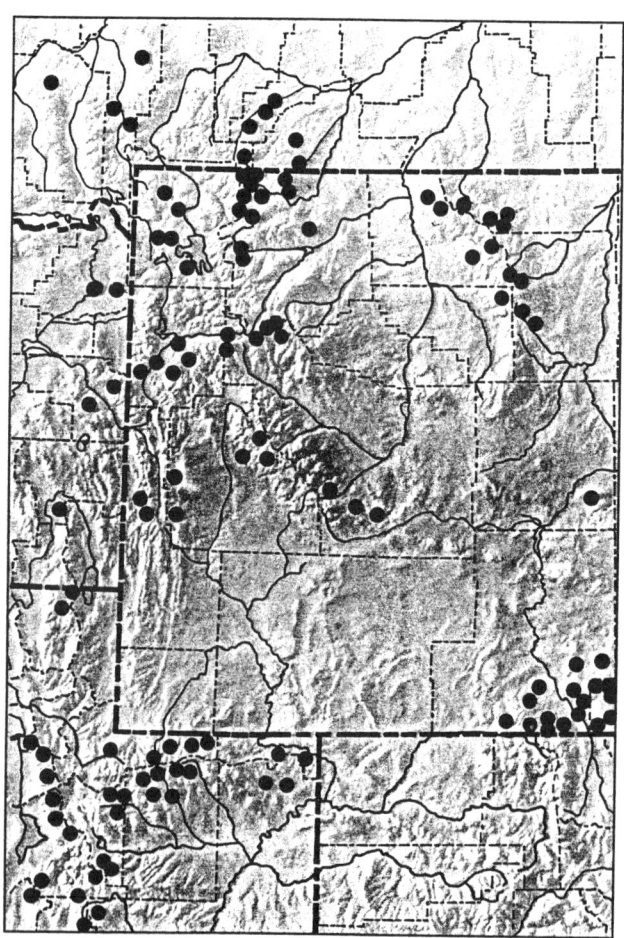

Agoseris glauca (Pursh) Raf.
Atlantic J. 6:39, 1833.

Pale Agoseris

Troximon glaucum Pursh
Agoseris agrestis Osterh.
Agoseris altissima Rydb.
Agoseris arachnoidea Rydb.
Agoseris arizonica Greene
Agoseris aspera Rydb.
Agoseris attenuata Rydb.
Agoseris caudata Greene
Agoseris dasycarpa Greene
Agoseris deas leonis Greene
Agoseris eisenhoweri B. Boivin
Agoseris isomeris Greene
Agoseris lacera Greene
Agoseris laciniata (Nutt.) Greene
Agoseris lanulosa Greene
Agoseris lapathifolia Greene
Agoseris leontodon Rydb.
Agoseris longula Greene
Agoseris maculata Rydb.
Agoseris microdonta Greene
Agoseris monticola Greene
Agoseris parviflora (Nutt.) D. Dietr.
Agoseris procera Greene
Agoseris pubescens Rydb.
Agoseris pumila (Nutt.) Rydb.
Agoseris rosea (Nutt.) D. Dietr.
Agoseris roseata Rydb.
Agoseris scorzoneraefolia (Schrad.) Greene
Agoseris taraxacifolia (Nutt.) D. Dietr.
Agoseris taraxacoides Greene
Agoseris turbinata Rydb.
Agoseris vestita Greene
Agoseris villosa Rydb.
Ammogeton scorzoneraefolius Schrad.

Macrorhynchus glaucus (Pursh) D.C. Eat.
Macrorhynchus laciniatus (Nutt.) T. & G.
Stylopappus laciniatus Nutt.
Troximon arachnoideum (Rydb.) A. Nels.
Troximon arizonicum Greene
Troximon parviflorum Nutt.
Troximon pubescens (Rydb.) A. Nels.
Troximon pumilum Nutt.
Troximon roseum Nutt.
Troximon taraxacifolium Nutt.
Troximon villosum (Rydb.) A. Nels.

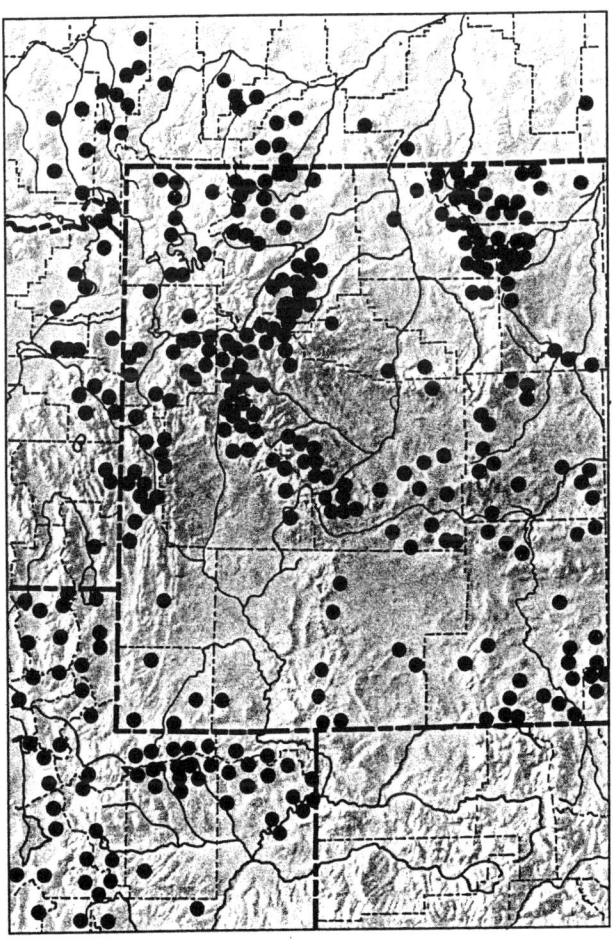

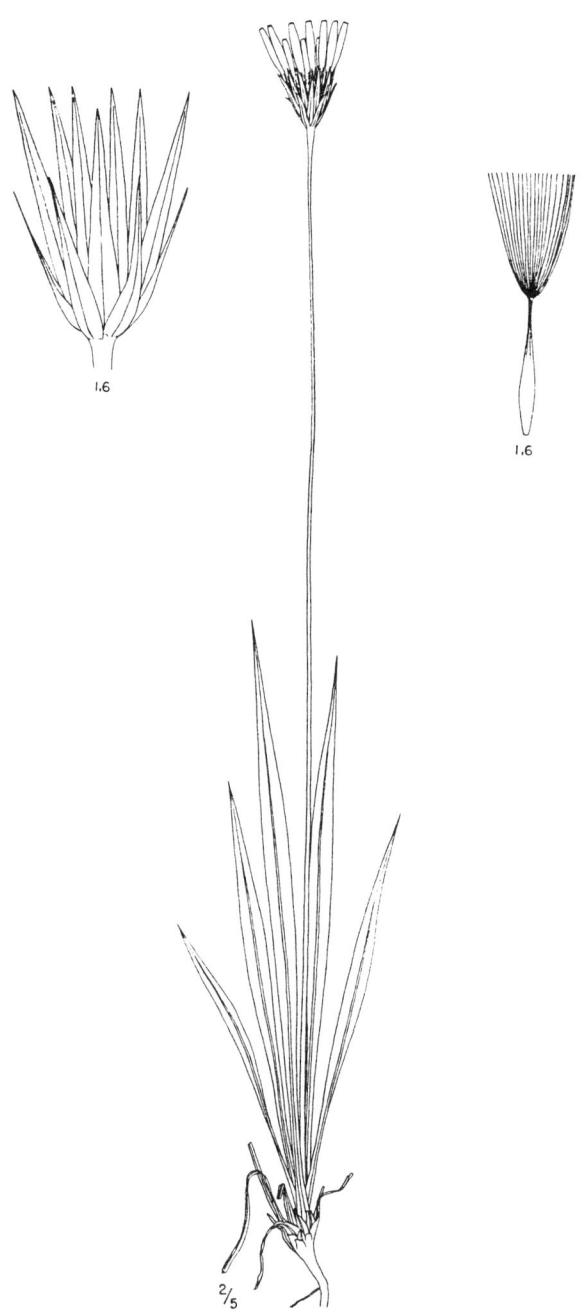

Perennial from a taproot and a simple to branched caudex; scapes glabrous or sometimes woolly. Leaves linear-lanceolate to oblanceolate, entire to toothed or laciniate-pinnatifid, glabrous or pubescent. Heads solitary. Involucral bracts lanceolate, acute to acuminate, glabrous to somewhat villous, sometimes purple spotted. Flowers yellow, sometimes drying pink or purplish. Akenes tapering gradually to a stout, striate beak less than half as long as the body.

Meadows, streambanks, willow thickets, and fellfields of all alpine ranges. Alaska and Yukon Territory south in western North America to California, Arizona, and New Mexico.

The three varieties which occur in the alpine zone of the Middle Rocky Mountains may be distinguished as follows:

1. Plants mostly less than 15 cm tall: involucral bracts purple or with a broad, purple midrib
   . . . var. *cronquistii* S.L. Welsh
1. Plants mostly more than 20 cm tall: involucral bracts green or purple-spotted
   2. Plants glabrous, sometimes cillate on the lower part of the leaves . . . var. *glauca*
   2. Plants pubescent, at least on the involucre
      . . . var. *dasycephala* (T. & G.) Jeps.

Anaphalis DC.
Pearly Everlasting

Tomentose, perennial, dioecious or polygamodioecious herbs from rhizomes in ours. Stems erect, with alternate, entire leaves. Heads corymbose, several to many, the pistillate ones often with a few staminate or perfect flowers in the center. Involucral bracts scarious, imbricate in several series. Receptacle naked, mostly convex. Staminate flowers tubular with undivided styles, the anthers caudate. Pistillate flowers narrowly tubular, the styles bifid. Perfect flowers sterile with undivided styles. Akenes glabrous, the pappus of capillary bristles.

Anaphalis margaritacea (L.) B. & H.
Gen. Pl. 2:303, 1873.

Pearly Everlasting

Gnaphalium margaritaceum L.
Anaphalis angustifolia Rydb.
Anaphalis lanata (A. Nels.) Rydb.
Anaphalis occidentalis (Greene) A. Heller
Anaphalis sierrae A. Heller
Anaphalis subalpina (Gray) Rydb.
Antennaria margaritacea (L.) R. Br.
Nacrea lanata A. Nels.

Erect, rhizomatous, mostly solitary-stemmed, tomentose perennial, the pubescence often becoming rusty with age. Leaves alternate, linear to lanceolate, sessile, tomentose below, glabrous or at least less pubescent above; margins often revolute, the lower soon deciduous. Heads numerous in a short, corymbose inflorescence. Involucral bracts imbricate in several series, dry, scarious, pearly white, sometimes with a dark basal spot. Receptacle naked. Akenes brown to olive, papillate, the pappus of capillary bristles.

Forest margins and rocky meadows at middle elevations in the mountains. Reported from the lower alpine zone of the Teton Range by Spence and Shaw (1981). Eastern Asia; North America, south to North Carolina, Kansas, New Mexico, Arizona, and California.

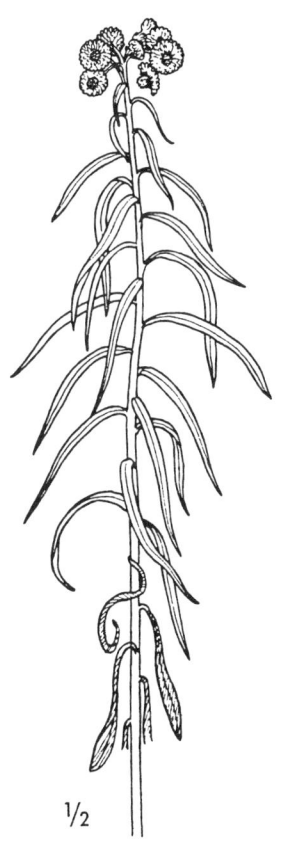

½

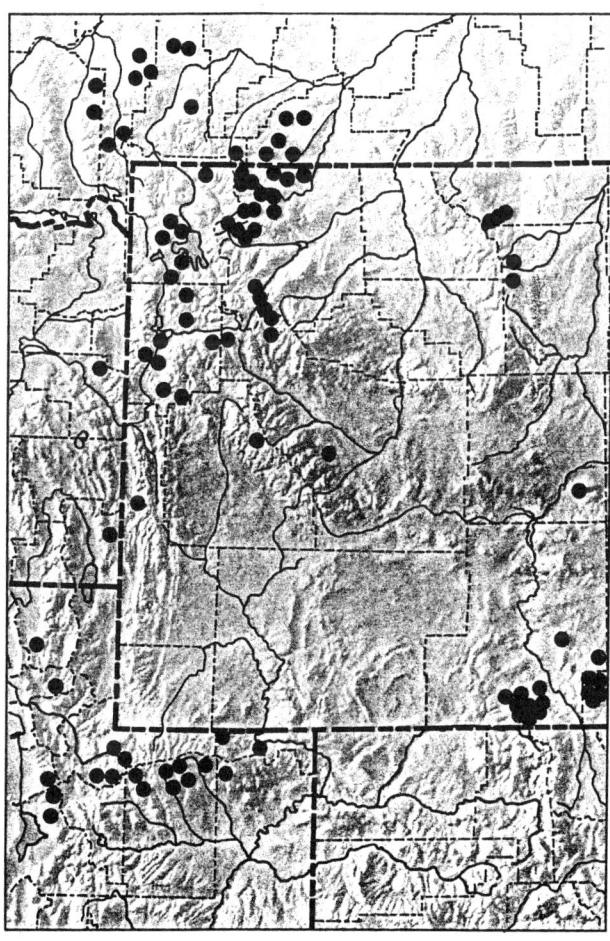

Antennaria Gaertn.
Pussytoes

Low, perennial, dioecious herbs, often stoloniferous, with alternate or mostly basal, entire leaves. Heads many-flowered, in capitate or corymbose clusters, or sometimes solitary, the involucral bracts in several series, imbricate, the tips scarious, white or colored, the receptacle flat or convex, naked. Staminate flowers tubular, with caudate anthers, an undivided style, and a rudimentary ovary and pappus. Pistillate flowers tubular-filiform, the style 2-cleft, the pappus of numerous capillary bristles united at the base. Akenes terete, usually glabrous.

KEY TO THE SPECIES OF *ANTENNARIA*

1. Heads solitary; tips of upper cauline leaves scarious, resembling the involucral bracts — *A. monocephala*
1. Heads 2 or more; tips of upper cauline leaves usually not resembling the involucral bracts
    2. Plants without stolons, mostly not mat forming
        3. Involucre blackish, the inner bracts sometimes narrowly white-tipped — *A. lanata*
        3. Involucre white or pinkish, sometimes with a dark spot at the base — *A. anaphaloides*
    2. Plants with numerous leafy stolons, often forming large, dense mats
        4. Terminal portion of involucral bracts white or pink
            5. Involucral bracts with a dark brown or black spot at the base of the scarious terminal portion; basal leaves narrowly oblanceolate — *A. corymbosa*
            5. Involucral bracts not darkened at the base of the terminal scarious portion; basal leaves oblanceolate to spatulate or broader — *A. microphylla*
        4. Terminal portion of the involucral bracts blackish, brownish, or greenish
            6. Involucral bracts with a dark blackish or greenish tip, usually acute — *A. alpina*
            6. Involucral bracts with a pale brown tip, usually obtuse
                7. Leaves glandular — *A. aromatica*
                7. Leaves lacking glands — *A. umbrinella*

Antennaria alpina (L.) Gaertn.
Fruct. 2:410, 1791.

Alpine Pussytoes

Gnaphalium alpinum L.
Antennaria alaskana Malte
Antennaria angustifolia Ekman
Antennaria arenicola Malte
Antennaria atriceps Fern.
Antennaria austromontana E. Nels.
Antennaria bayardii Fern.
Antennaria bocheria Pors.
Antennaria borealis Greene
Antennaria brevistyla Fern.
Antennaria brunnescens Fern.
Antennaria cana (Fern. & Wieg.) Fern.
Antennaria candida Greene
Antennaria canescens (Lange) Malte
Antennaria chlorantha Greene
Antennaria compacta Malte
Antennaria confusa Fern.
Antennaria crymophila Pors.
Antennaria densa Greene
Antennaria densifolia Pors.
Antennaria ekmaniana Pors.
Antennaria foggii Fern.
Antennaria friesiana (Trautv.) Ekman
Antennaria macounii Greene
Antennaria media Greene
Antennaria megacephala Fern.
Antennaria modesta Greene
Antennaria neoalaskana Pors.

Antennaria pedunculata Pors.
Antennaria pulchella Greene
Antennaria pulvinata Greene
Antennaria reflexa E. Nels.
Antennaria rousseaui Pors.
Antennaria scabra Greene
Antennaria sornborgeri Pors.
Antennaria stolonifera Pors.
Antennaria subcanescens Ostenf. ex Malte
Antennaria tomentella E. Nels.
Antennaria vexillifera Fern.

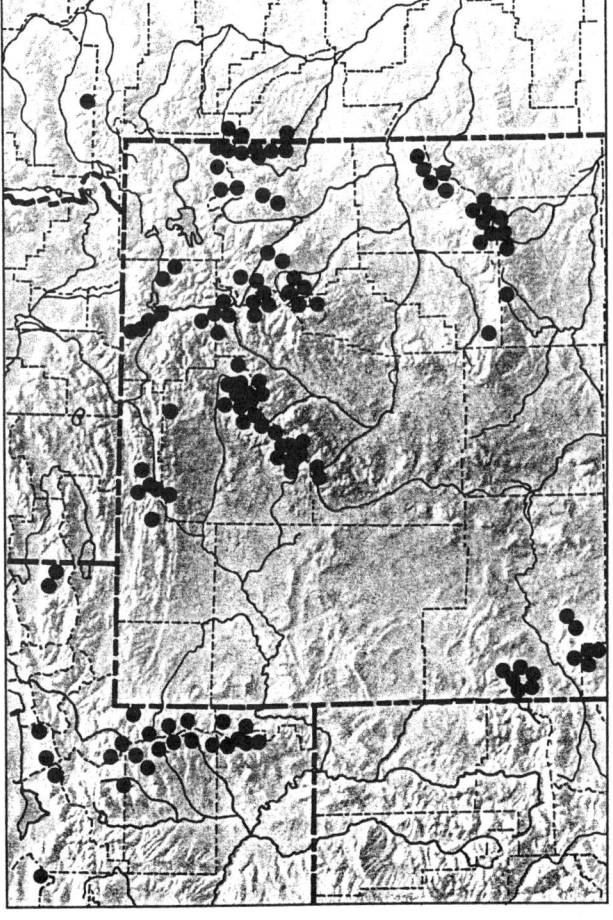

Mat-forming perennial with leafy stolons, the stolons decumbent-spreading; stems erect, tomentose. Basal leaves rosulate, oblanceolate to spatulate, mostly obtuse, tomentose on both sides or glabrate above; cauline leaves several, reduced upward, linear to linear-oblong. Heads 3–7, in a dense, subcapitate cyme; pistillate involucres lanate below, the tips scarious, dark brownish to greenish black, usually slender, acute; staminate involucres often with whitish tips. Akenes glabrous.

Fellfields in most alpine ranges of the Middle Rocky Mountains. Not yet known from above timberline in the Hoback and Salt River Ranges. Circumpolar; south in western North America to California, Arizona, and New Mexico.

Antennaria alpina represents a large, widespread species complex with considerable variation in morphology and reproduction. Chromosome numbers also vary and are represented by at least diploid through octoploid sets (Bayer, 1990). As treated here, Middle Rocky Mountain plants are var. *media* (Greene) Jeps.

Antennaria anaphaloides Rydb.
Mem. N.Y. Bot. Gard. 1:409, 1900.

Pearly Pussytoes

Tufted, perennial herb from a simple to branched caudex; stems 1 to several, erect or ascending, tomentose. Basal leaves petiolate, the blades lanceolate, entire, acute, with 3 usually evident veins, tomentose on both sides; cauline leaves linear, reduced upward and becoming sessile, tomentose as the basal ones. Heads numerous in a usually open, corymbose cyme. Involucres tomentose, at least at the base, the bracts brown with obtuse, often erose, white, scarious tips. Akenes glabrous.

Meadows near timberline in the Beartooth, Big Horn, and Wind River Ranges. British Columbia, Alberta, and Saskatchewan south to Oregon, Nevada, Utah, and Colorado.

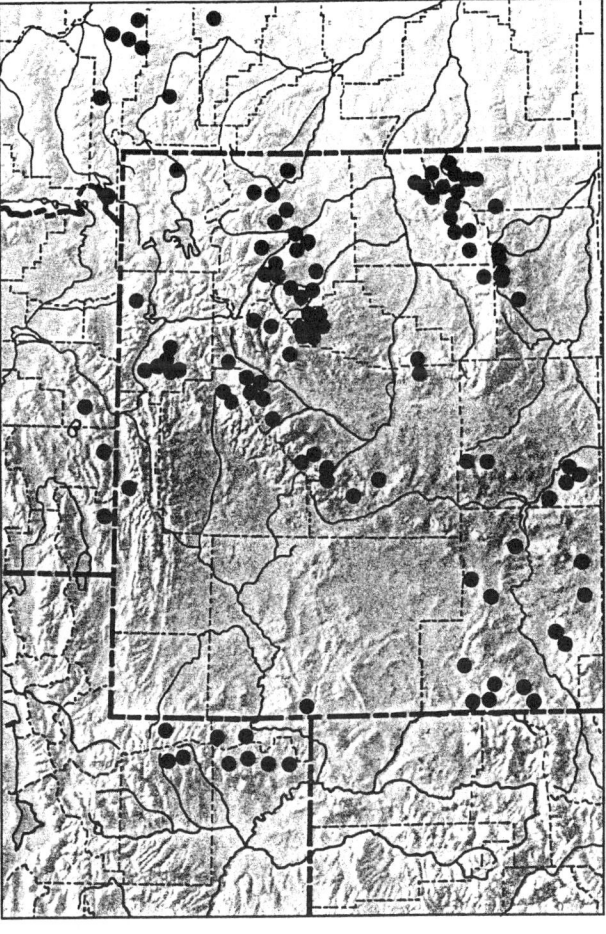

Antennaria aromatica Evert
Madroño 31:109, 1984.

Aromatic Pussytoes

Mat-forming, stoloniferous, aromatic, perennial herb from a branched, woody caudex covered with old leaf bases; stems erect, glandular-tomentose, exceeding the basal leaves. Basal leaves oblanceolate to spatulate, tomentose on both surfaces, glandular, often somewhat mucronate; cauline leaves linear-lanceolate to oblanceolate, glandular, tomentose as the basal ones, the upper sometimes with a scarious, brownish, acute tip. Heads 2–5 in a compact corymbose or subcapitate cyme. Involucres tomentose at base, the bracts usually glandular, the tips mostly acute, erose, scarious, and greenish or brownish, the base of the scarious portion often darker than the tip. Akenes tuberculate.

Scree, talus, and rocky crevices, usually on limestone, near timberline in the Absaroka, Beartooth, Big Horn, Gros Ventre, and Wind River Ranges. Alberta south to Montana and Wyoming. Reported from Colorado (E.L. Hartman & Rottman, 1988; Siems & Neely, 1990).

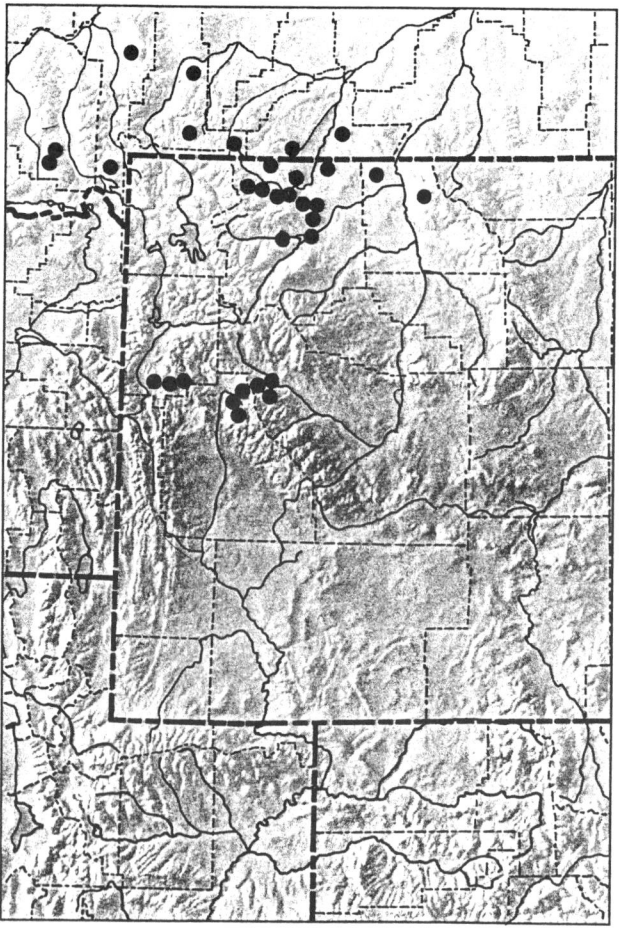

Antennaria corymbosa E. Nels.
Bot. Gaz. 27:212, 1899.

Flattop Pussytoes

Antennaria hygrophila Greene
Antennaria nardina Greene

Loose, mat-forming, tomentose, stoloniferous, perennial herb, the stolons slender, leafy; stems 1 to several, erect to ascending. Basal leaves petiolate, the blades spatulate to narrowly oblanceolate, acute, mostly 1-nerved, tomentose on both sides; cauline leaves linear. Heads several, in a compact cyme. Involucres tomentose at the base, the bracts obtuse or the inner ones sometimes acute, with a conspicuous dark brown to black spot at the base of each white to cream terminal portion. Akenes puberulent.

Fellfields and talus of the Absaroka, Beartooth, Uinta, Wasatch, and Wind River Ranges. Alberta and Saskatchewan south to Oregon, California, Utah, and Colorado.

2/3

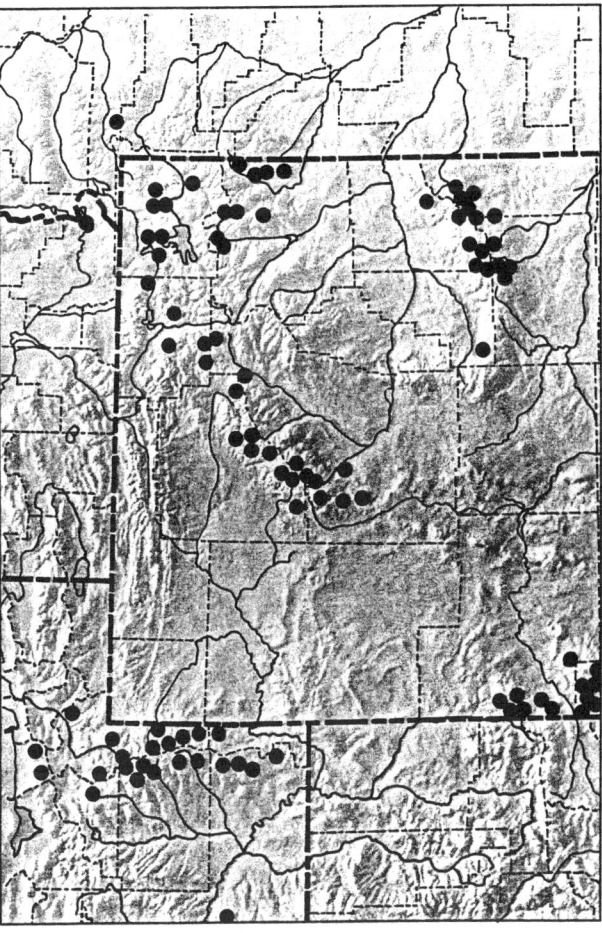

Antennaria lanata (Hook.) Greene
Pittonia 3:288, 1898.

Woolly Pussytoes

Tufted, tomentose, perennial herb from a simple to branched caudex. Leaves mostly sessile, the basal leaves erect, oblanceolate, usually with 3 prominent veins; cauline leaves reduced upward, narrowly oblanceolate to linear. Heads several in compact cymes. Involucres tomentose near the base, the bracts brown or greenish black below, whitish and scarious near the mostly acute tip.

Moist meadows and lakeshores of the Absaroka, Beartooth, and Wind River Ranges. British Columbia and Alberta south to Oregon and Wyoming.

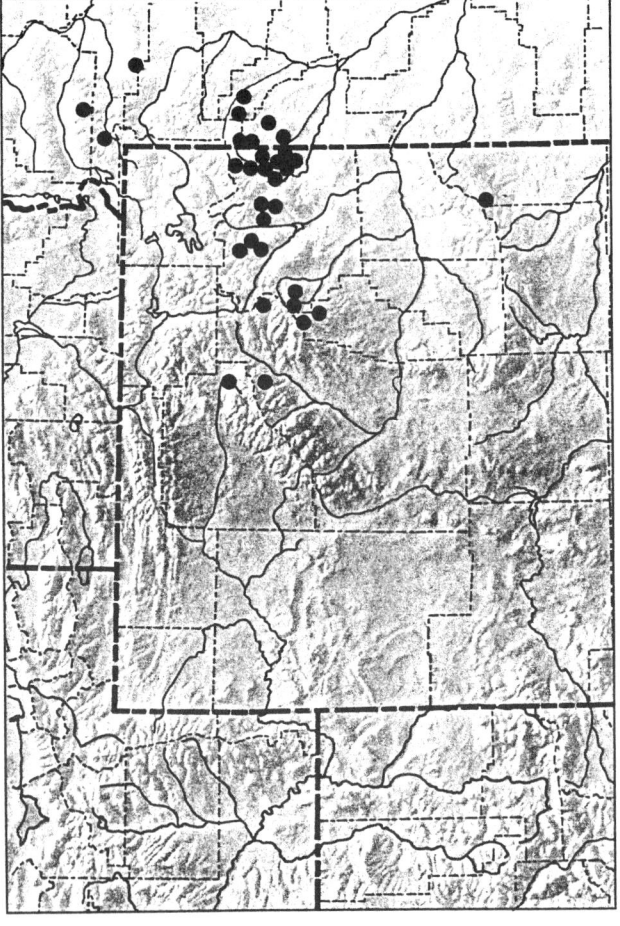

Antennaria microphylla Rydb.
Bull. Torr. Club 24:303, 1897.

Rosy Pussytoes

Antennaria acuminata Greene
Antennaria alborosea A.E. & M.P. Pors.
Antennaria angustifolia Rydb.
Antennaria arida E. Nels.
Antennaria bracteosa Rydb.
Antennaria breitungii Pors.
Antennaria concinna E. Nels.
Antennaria elegans Pors.
Antennaria erigeroides Greene
Antennaria foliacea Greene
Antennaria formosa Greene
Antennaria groenlandica M.P. Pors.
Antennaria hansii Kern.
Antennaria hendersonii Piper
Antennaria imbricata E. Nels.
Antennaria incarnata Pors.
Antennaria laingii Pors.
Antennaria leontopodioides Cody
Antennaria leuchippii M.P. Pors.
Antennaria nitida Greene
Antennaria oxyphylla Greene
Antennaria parvifolia Greene
Antennaria rosea Greene
Antennaria scariosa E. Nels.
Antennaria solstitialis Greene
Antennaria sorida Greene
Antennaria speciosa E. Nels.
Antennaria subviscosa Fern.

Mat-forming, stoloniferous, perennial herb with creeping, leafy stolons; stems erect, tomentose. Basal leaves usually petiolate, the blades oblanceolate to obovate, acute to obtuse, tomentose on both surfaces, occasionally glabrate with age; cauline leaves linear to narrowly oblanceolate, reduced upward. Heads several to many in compact to subcapitate cymes. Involucres tomentose at base, the bracts oblong, obtuse to acute, brownish to greenish below, the terminal scarious portion deep pink to white. Akenes glabrous.

Fellfields, meadows, and stream banks in nearly all alpine ranges of the Middle Rocky Mountains. Not yet known from above timberline in the Medicine Bow Range. Alaska south in western North America to California, New Mexico, Nebraska, and the Dakotas.

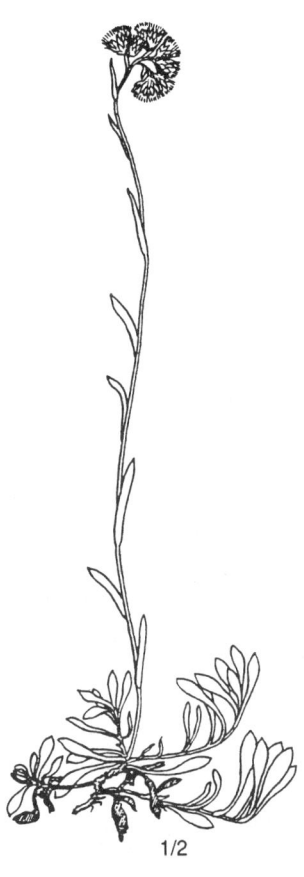

1/2

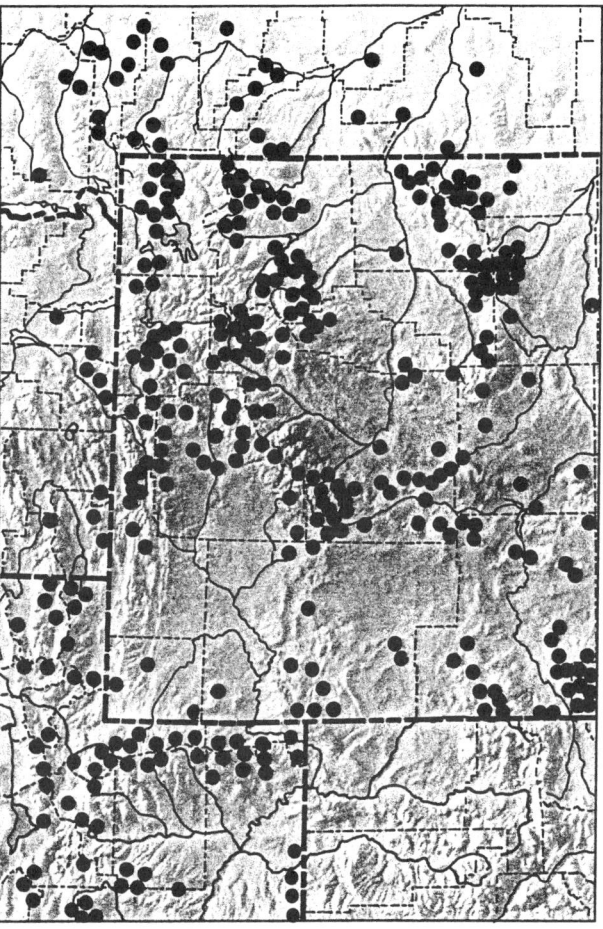

Antennaria monocephala DC.
Prodromus 6:269, 1837.

Singlehead Pussytoes

Antennaria angustata Greene
Antennaria burwellensis Malte
Antennaria congesta Malte
Antennaria exilis Greene
Antennaria fernaldiana Polunin
Antennaria glabrata (J. Vahl) Greene
Antennaria hudsonica Malte
Antennaria nitens Greene
Antennaria philonipha Pors.
Antennaria pygmaea Fern.
Antennaria shumaginensis Pors.
Antennaria solitaria Rydb.
Antennaria tansleyi Polunin
Antennaria tweedsmurii Polunin

Mat-forming, tomentose, perennial herb with short stolons; stems erect to ascending. Basal leaves oblanceolate to spatulate, tomentose beneath, glabrous to glabrate above, 1-nerved, the tips short-mucronate; cauline leaves linear, at least the upper ones with brown, scarious tips. Heads solitary, erect. Involucre tomentose at base, the bracts greenish below, often darker in the middle, brown and scarious at the tip. Akenes glabrous.

Fellfields and patterned ground of the Absaroka, Beartooth, Big Horn, Teton, and Wind River Ranges. Siberia; Alaska south to Wyoming.

Rocky Mountain alpine plants with short stolons and leaves glabrate to thinly tomentose above are ssp. *monocephala*, var. *monocephala*.

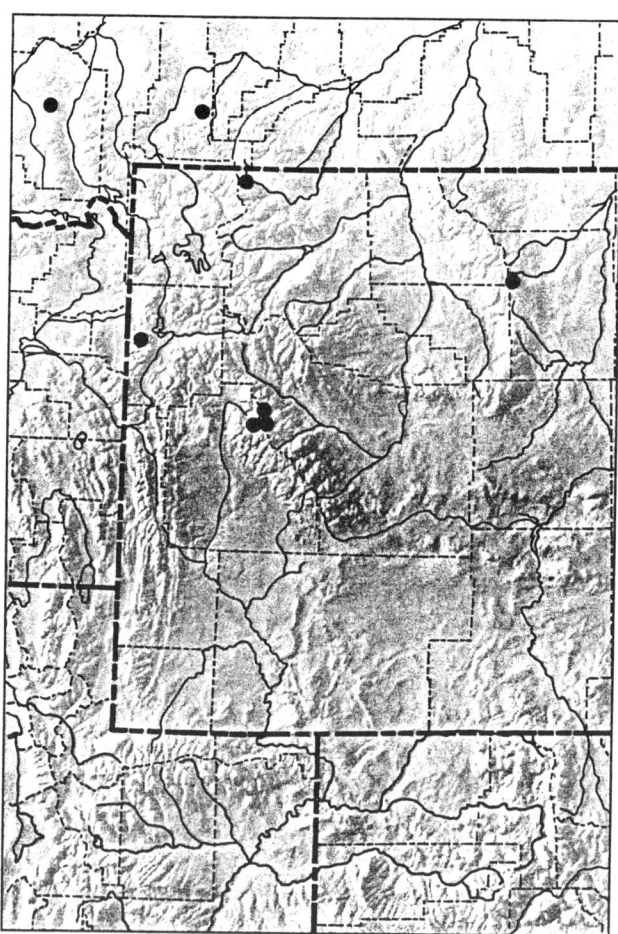

Antennaria umbrinella Rydb.
Bull. Torr. Club 24:302, 1897.

Brown Pussytoes

Antennaria aizoides Greene
Antennaria albescens (E. Nels.) Rydb.
Antennaria albicans Fern.
Antennaria columnaris Fern.
Antennaria confinis Greene
Antennaria ellyae Pors.
Antennaria flavescens Rydb.
Antennaria fusca E. Nels.
Antennaria gormanii St. John
Antennaria intermedia (Rosenv.) M.P. Pors.
Antennaria isolepis Greene
Antennaria lanulosa Greene
Antennaria maculata Greene
Antennaria mucronata Greene
Antennaria pallida E. Nels.
Antennaria peasei Fern.
Antennaria sansonii Greene
Antennaria sedoides Greene
Antennaria straminea Fern.
Antennaria viscidula Rydb.
Antennaria wiegandii Fern.

Mat-forming, tomentose, stoloniferous perennial, the stolons leafy. Basal leaves oblanceolate to spatulate, densely tomentose on both sides, the tips acute to somewhat mucronate; cauline leaves linear to linear-oblong. Heads 3–8 in a dense, terminal, subcapitate cyme. Involucres lanate-tomentose below, the bracts dark brown to blackish green at the base, the tips scarious, brownish to whitish, mostly obtuse. Akenes glabrous.

Fellfields and meadows, especially near timberline, in nearly all alpine ranges of the Middle Rocky Mountains. Not yet known from above timberline in the Medicine Bow Range. Alaska and Canada; south in western North America to California, Arizona, and Colorado.

Antennaria umbrinella apparently intergrades with *A. alpina* at elevations above timberline and with *A. microphylla* below timberline.

3/4

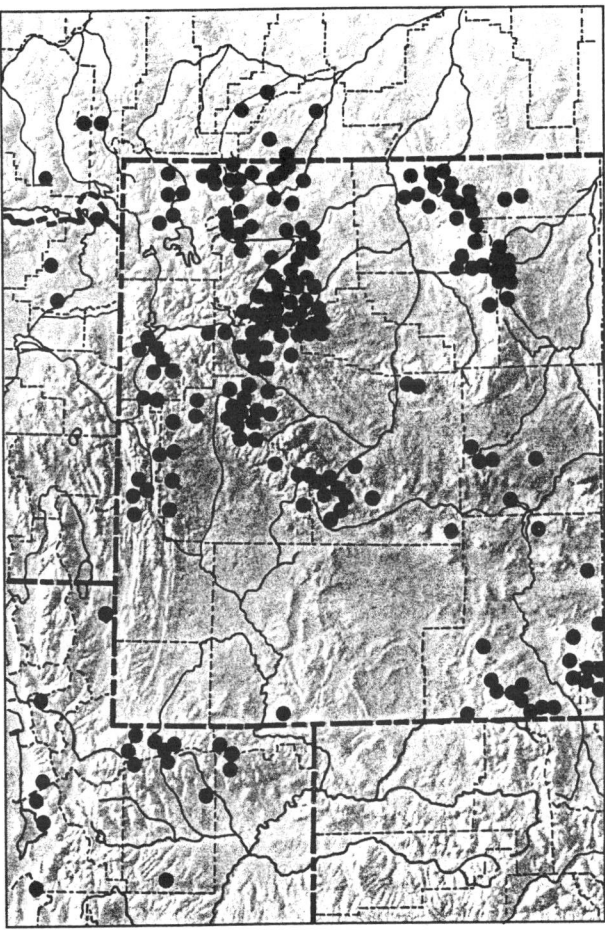

Arnica L.
Arnica

Perennial herbs from fibrous roots or rhizomes, with simple, opposite, entire or toothed leaves. Heads radiate in ours, large and showy, solitary to several, the involucral bracts equal, linear, in 1 or 2 series, the receptacle flat or convex, naked. Ray flowers yellow or yellow-orange, pistillate, rarely lacking. Disk flowers yellow, perfect, and fertile. Akenes linear, cylindric, 5–10-nerved, the pappus of white or tan, barbellate or subplumose, capillary bristles.

KEY TO THE SPECIES OF *ARNICA*

1. Cauline leaves in 5–10 pairs — *A. longifolia*
1. Cauline leaves in 2–4 pairs
    2. Pappus plumose to subplumose, tan; basal leaves neither conspicuous nor tufted — *A. mollis*
    2. Pappus barbellate, white; basal leaves usually conspicuous
        3. Leaves relatively broad, ovate to cordate, toothed
            4. Akenes glabrous, at least at the base; involucre glabrous, occasionally with a few long hairs; basal leaves narrowly ovate — *A. latifolia*
            4. Akenes pubescent or glandular; involucre pubescent with long, spreading hairs; basal leaves ovate to cordate — *A. cordifolia*
        3. Leaves relatively narrow, lanceolate, oblanceolate, spatulate, or elliptic, entire
            5. Ray flowers 10–20; old leaf bases with axillary tufts of tan to brown woolly hairs; lower cauline leaves petiolate — *A. fulgens*
            5. Ray flowers less than 10; old leaf bases lacking tan or brown woolly hairs in axillary tufts; lower cauline leaves sessile — *A. rydbergii*

Arnica cordifolia Hook.
Fl. Bor.-Am. 1:331, 1834.

Heartleaf Arnica

Arnica abortiva Greene
Arnica andersonii Piper
Arnica austinae Rydb.
Arnica chionophila Greene
Arnica evermannii Greene
Arnica hardinae St. John
Arnica humilis Rydb.
Arnica macrophylla Nutt.
Arnica ovalis Rydb.
Arnica parvifolia Greene
Arnica pumila Rydb.
Arnica subcordata Greene
Arnica whitneyi Fern.

Caulescent, glandular-puberulent to villous perennial from creeping, slender, nonscaly rhizomes often with clusters of leaves on short shoots; stems solitary or clustered. Cauline leaves in 2–4 pairs, the lower leaves long-petiolate, cordate, coarsely serrate, dentate or entire, puberulent to pilose; upper leaves sessile, ovate to lanceolate, the margins toothed to entire. Heads 1–3, turbinate to campanulate; involucres glandular, pilose at the base, the bracts lanceolate to oblanceolate, acute to acuminate. Ray flowers yellow; disk flowers pilose on the tube. Akenes hirsute, often glandular; pappus barbellate, white or whitish.

Borders of timberline krummholz formations in the Absaroka, Uinta, Wasatch, and Wind River Ranges. Alaska and Yukon Territory south to California, Arizona, New Mexico, South Dakota, and Michigan.

Dwarfed plants of the subalpine and lower alpine zone mostly less than 20 cm tall, with leaves only slightly cordate at the base are var. *pumila* (Rydb.) Maguire. Also see comments under *A. mollis*, below.

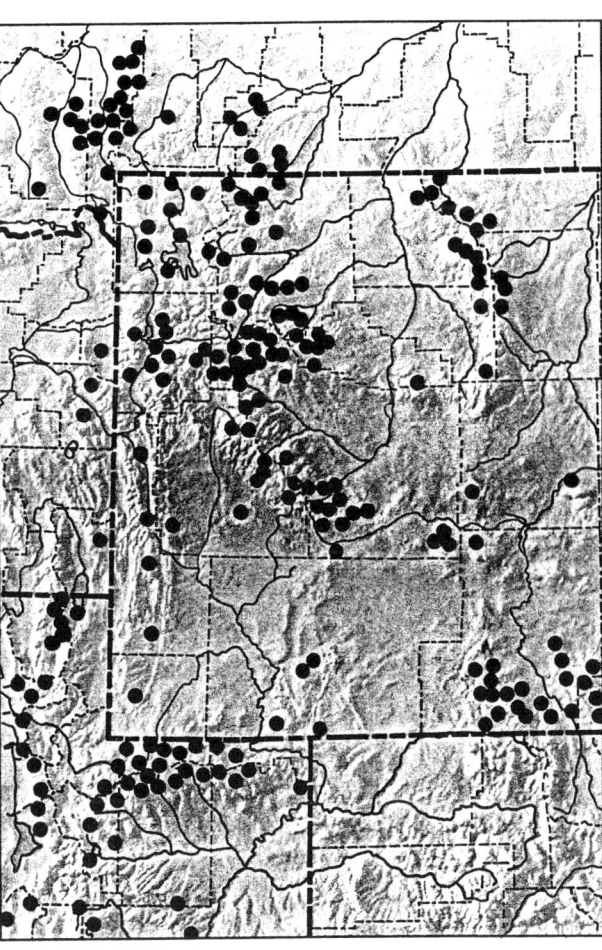

Arnica fulgens Pursh
Fl. Am. Sept., 527, 1814.

Orange Arnica

Arnica monocephala Rydb.
Arnica pedunculata Rydb.

Somewhat tufted perennial from a short, thick, scaly rhizome with numerous fibrous roots; stems glandular, often puberulent, with 4–6 pairs of leaves. Basal leaves and persistent petioles with axillary tufts of brown woolly hair; blades oblanceolate to elliptic, prominently 3–5-nerved, the margins entire or nearly so. Cauline leaves reduced and sessile upward, lanceolate or linear. Heads 1–3, mostly solitary, hemispheric, the peduncle apex pilose and glandular below. Involucre glandular, villous, the bracts lanceolate to narrowly elliptic, somewhat rounded at the tip. Ray flowers orange-yellow; disk flowers with some glands and usually a few spreading white hairs. Akenes hirsute; pappus barbellate and white, cream, or light brown.

Timberline meadows of the Absaroka, Beartooth, Wind River, and Wyoming Ranges. British Columbia, Alberta, and Saskatchewan south to California, Utah, Colorado, and the Dakotas.

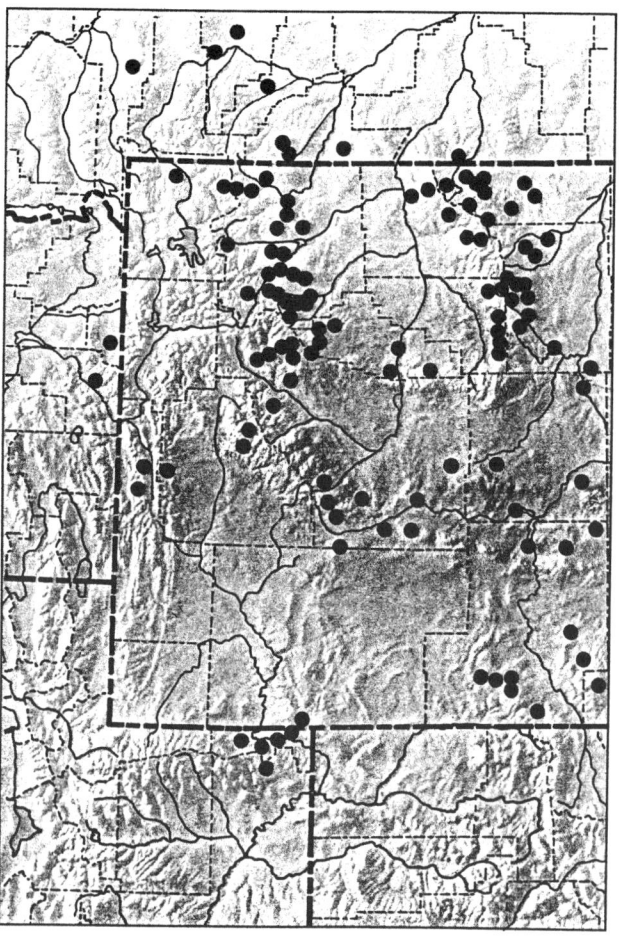

Arnica latifolia Bong.
Mém. Acad. Imp. Sci. St. Pétersb., Sér. 6, Sci. Math. 2:147, 1832.

Broadleaf Arnica

Arnica aphanactis Piper
Arnica aprica Greene
Arnica betonicaefolia Greene
Arnica columbiana A. Nels.
Arnica eriopoda Gand.
Arnica flodmanii Rydb.
Arnica gracilis Rydb.
Arnica grandifolia Greene
Arnica granulifera Rydb.
Arnica jonesii Rydb.
Arnica laevigata Greene
Arnica latucina Greene
Arnica leptocaulis Rydb.
Arnica menziesii Hook.
Arnica multiflora Greene
Arnica ovalifolia Greene
Arnica platyphylla A. Nels.
Arnica puberula Rydb.
Arnica teucrifolia Greene
Arnica ventorum Greene

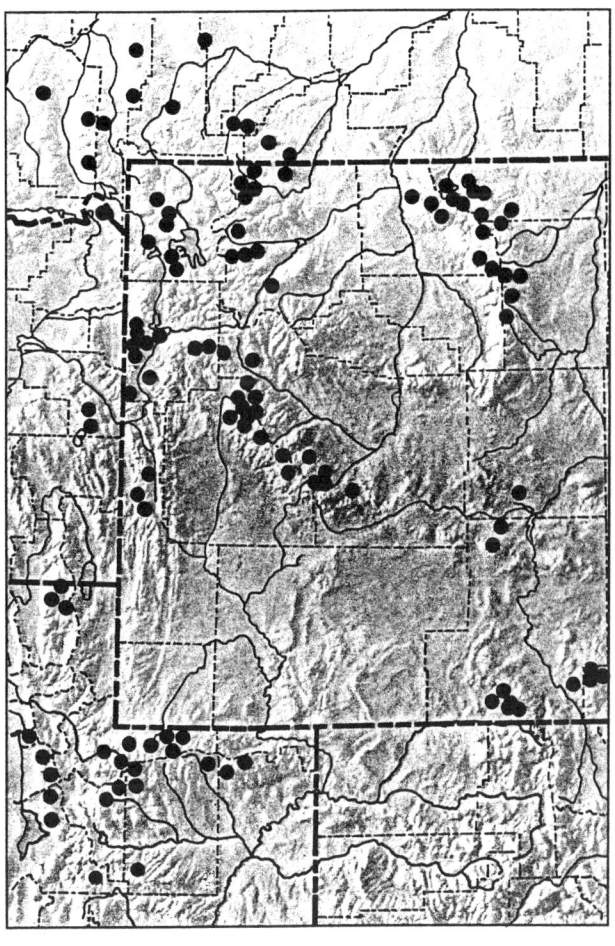

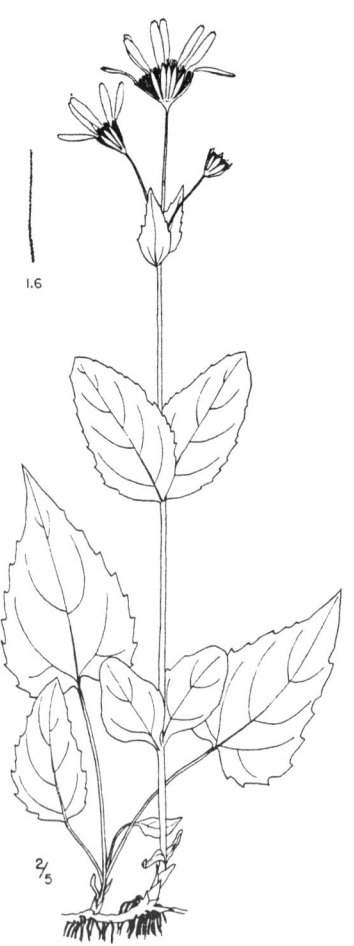

Glandular-puberulent to pilose perennial from short to somewhat creeping, scaly rhizomes with fibrous roots; stems simple or branched, erect to ascending, with 2–4 pairs of leaves. Basal leaves petiolate, the blades ovate to cordate, withered by flowering time, or produced on short shoots separate from the stems. Cauline leaves petiolate, becoming sessile and reduced upward; blades ovate to lanceolate, the margins serrate-dentate to entire. Heads 1–3, sometimes 5 or more, turbinate to subhemispheric, the peduncle apex villous and often glandular below. Involucre glandular, usually pilose at the base, the bracts lanceolate to oblanceolate, acute. Ray flowers yellow; disk flowers pubescent on the tube. Akenes glabrous to puberulent or glandular above; pappus white, barbellate.

Talus, stream banks, scree slopes, and fellfields in most alpine ranges of the Middle Rocky Mountains. Not yet known from above timberline in the Hoback, Medicine Bow, Salt River, Wasatch, and Wyoming Ranges. Alaska and Yukon Territory south to California, Utah, and Colorado.

Two varieties occur in the alpine zone: the low, tufted var. *gracilis* (Rydb.) Cronq. (of higher elevations) with 3–9 heads, involucres 7–13 mm high, and leaves less than 2.5 cm wide; and var. *latifolia* (near timberline), which is usually taller, erect, with 1–3 heads, involucres 10–18 mm high, and leaves 1.5–8 cm wide. Also see comments under *A. mollis,* below.

Arnica longifolia D.C. Eat. in S. Wats. *Bot. King Exped.* 5:186, 1871.

Longleaf Arnica

Arnica arcana A. Nels.
Arnica caudata Rydb.
Arnica myriadenia Piper
Arnica polycephala A. Nels.

Tufted, glandular to puberulent perennial from a branched caudex and short rhizomes; stems erect to decumbent-ascending, with 5–10 pairs of leaves. Basal leaves lacking. Cauline leaves lanceolate to elliptic, acute to acuminate, the lower petiolate, becoming reduced and sessile upward, with 3–5 primary veins, the margins entire or nearly so. Heads 1 to many, turbinate to campanulate, the peduncle apex sparsely villous below. Involucre pilose and glandular, the bracts lanceolate to oblong, acute to acuminate, ciliate. Ray flowers yellow; disk flowers short-pubescent on the tube. Akenes glandular to glabrate; pappus barbellate, light brown to straw-colored.

Widespread in wet talus, scree, and rocky streamsides in most alpine ranges of the region. Not yet known from above timberline in the Hoback, Medicine Bow, Uinta, and Wasatch Ranges. British Columbia and Alberta south to California, Nevada, Utah, and Colorado.

Rocky Mountain plants with a short-barbellate pappus (rather than subplumose) have been called ssp. *genuina* Maguire.

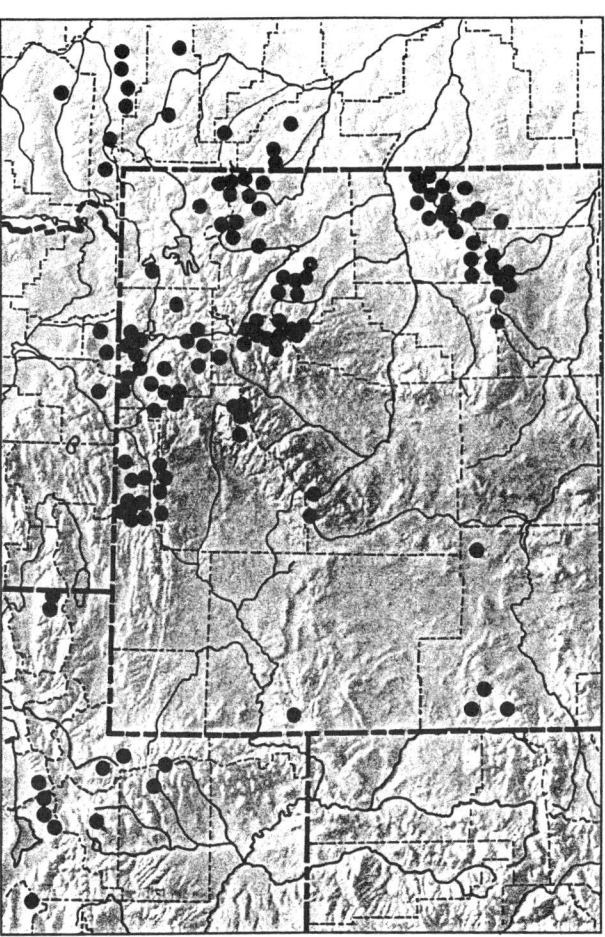

Arnica mollis Hook.
Fl. Bor.-Am. 1:331, 1834.

Hairy Arnica

Arnica amplifolia Rydb.
Arnica arachnoidea Rydb.
Arnica coloradensis Rydb.
Arnica crocea Greene
Arnica crocinea Greene
Arnica merriami Greene
Arnica ovata Greene
Arnica rivularis Greene
Arnica scaberrima Greene
Arnica sylvatica Greene
Arnica subplumosa Greene

Pubescent, often glandular perennial from a short, branched rhizome with numerous fibrous roots; stems erect, with 3–4 pairs of leaves. Lower leaves sessile or petiolate, the upper reduced and sessile; blades narrowly ovate, elliptic, or obovate below, becoming elliptic to lanceolate above, the margins denticulate to entire. Heads solitary or few, campanulate to hemispheric, the peduncle apex sparsely to moderately pilose below. Involucre villous-pilose, glandular, the bracts lanceolate to elliptic, acute to somewhat acuminate. Ray flowers yellow. Akenes pubescent to glandular; pappus subplumose to plumose, light brown.

Meadows, moist fellfields, and stream banks in nearly all alpine ranges of the Middle Rocky Mountains. Not yet known from above timberline in the Gros Ventre and Hoback Ranges. Alaska, Yukon Territory, MacKenzie (Northwest Territory), south to California, Nevada, Utah, and Colorado.

Hybrids between *A. mollis*, *A. cordifolia*, *A. latifolia*, and *A. rydbergii* are sometimes called *A. diversifolia*, the Sticky Arnica. Such specimens are distinguished from typical *A. mollis* by their longer petioles on the lower cauline leaves, which have broadly ovate blades and truncate to subcordate bases. Typical *A. mollis* has lower cauline leaves sessile to short-petiolate, narrowly ovate, elliptic, or obovate, with cuneate bases.

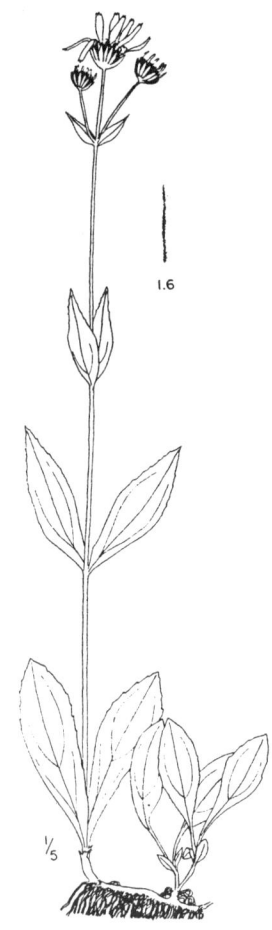

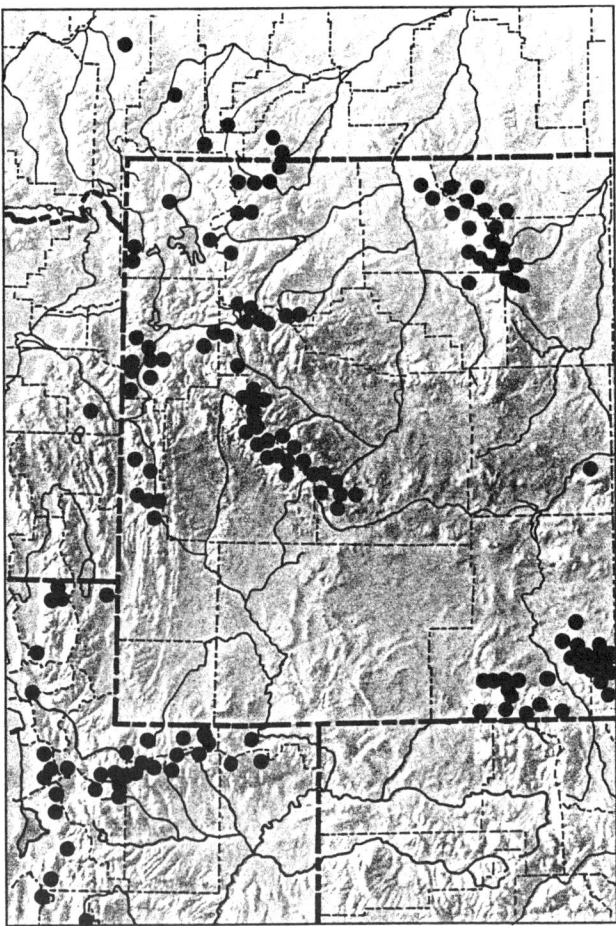

Arnica rydbergii Greene
Pittonia 4:36, 1899.

Rydberg Arnica

Arnica aurantiaca Greene
Arnica caespitosa A. Nels.
Arnica cascadensis Greene
Arnica lasiosperma Greene
Arnica sulcata Rydb.
Arnica tenuis Greene

Somewhat tufted, pilose-glandular perennial from scaly rhizomes; stems erect or ascending, with 3–4 pairs of leaves. Basal leaves petiolate, the blades oblanceolate to spatulate, often produced on separate shoots; cauline leaves becoming sessile and reduced upward, the blades ovate, obovate, or lanceolate, 3–5-nerved; margins entire to denticulate. Heads 1–3, turbinate-campanulate, the peduncle apex villous, glandular below. Involucre pilose and glandular to glabrate, the bracts lanceolate, acute, ciliate. Ray flowers yellow. Akenes villous with short hairs; pappus barbellate, white.

Timberline meadows and fellfields in most alpine ranges of the Middle Rocky Mountains. Not yet known from above timberline in the Hoback, Teton, Uinta, and Wasatch Ranges. British Columbia and Alberta south to California, Utah, and Colorado.

See comments about hybrids under *A. mollis,* above.

Arnica angustifolia M. Vahl, Alpine Arnica, will key here. The first record from the Middle Rocky Mountains has recently been collected from limestones on Arrow Mountain in the northern Wind River Range. It is distinguished from *A. rydbergii* by dense, woolly-villous pubescence and white-villous, often purple-tinged involucral bracts 10–15 mm long.

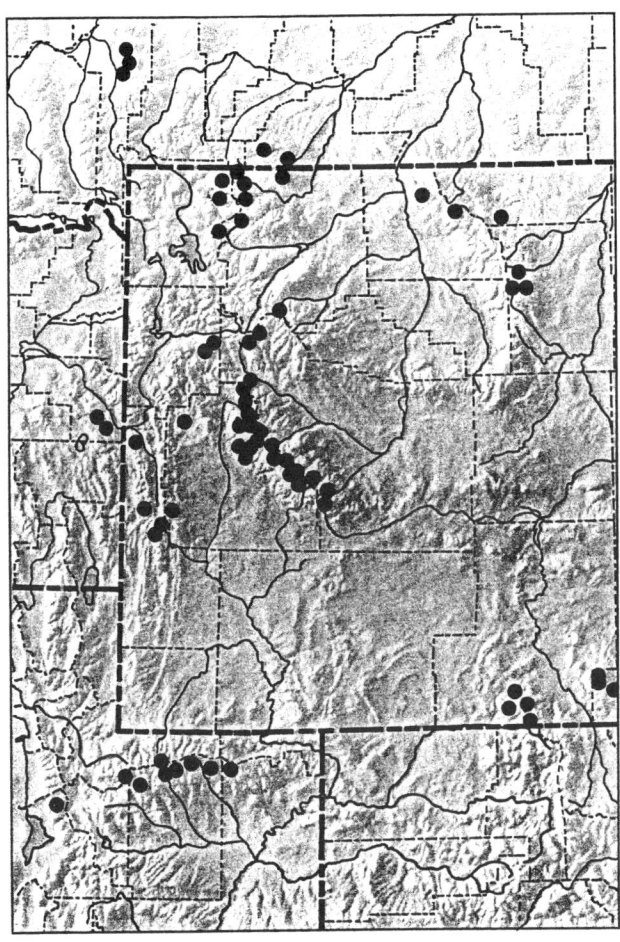

Artemisia L.
Sagebrush, Wormwood

Annual, biennial, or perennial herbs or shrubs, usually strongly aromatic, with alternate, entire to variously dissected leaves. Heads small, commonly paniculate but sometimes in spikes or racemes, the involucral bracts dry, at least the margins scarious, imbricate in 2–4 series, the receptacle flat or somewhat hemispheric, naked or pubescent. Flowers all tubular, all perfect, or the outer flowers pistillate and the inner ones fertile or sterile. Akenes usually glabrous, the pappus lacking or sometimes a short crown.

KEY TO THE SPECIES OF *ARTEMISIA*

1. Plants shrubs or subshrubs, woody, at least at the base
    2. Receptacle pubescent with long hairs between the flowers; leaves 2–3 times ternately divided; plants subshrubs forming mats — *A. frigida*
    2. Receptacle glabrous; leaves 3-toothed to irregularly lobed at the apex; plants shrubs, not mat forming — *A. tridentata*
1. Plants herbaceous perennials
    3. Disk flowers at center of inflorescence sterile; basal leaves 2–3-pinnatifid, rarely ternate — *A. campestris*
    3. Disk flowers at center of inflorescence fertile; basal leaves entire to pinnatifid
        4. Receptacle with numerous hairs between the flowers — *A. scopulorum*
        4. Receptacle glabrous
            5. Leaves mostly cauline, white-tomentose to glabrous; plants mostly rhizomatous
                6. Leaves entire to coarsely pinnatifid; involucre pubescent — *A. ludoviciana*
                6. Leaves finely 1–3-pinnately dissected; involucre glabrous to glabrate — *A. michauxiana*
            5. Leaves mostly basal, or at least the basal ones best developed, sparsely villous to glabrous; plants from a branched caudex — *A. norvegica*

Artemisia campestris L.
Sp. Pl., 846, 1753.

Common Sagewort

Artemisia borealis Pallas
Artemisia bourgeauana Rydb.
Artemisia camporum Rydb.
Artemisia canadensis Michx.
Artemisia caudata Michx.
Artemisia forwoodii S. Wats.
Artemisia maccallae Rydb.
Artemisia pacifica Nutt.
Artemisia richardsoniana Bess.
Artemisia ripicola Rydb.
Artemisia scouleriana (Bess.) Rydb.
Artemisia spithamea Pursh
Oligosporus borealis (Pallas) Poljakov
Oligosporus campestris (L.) Cass.
Oligosporus pacifica (Nutt.) Poljakov

Glabrous to sparsely pubescent, biennial to perennial herb from a taproot and simple, broad caudex; stems 1 to several. Basal leaves long-petiolate, 2–3-pinnatifid, the ultimate divisions often ternate with linear to oblong segments, glabrous or silky-canescent; cauline leaves pinnatifid and short-petiolate, becoming reduced, ternate and sessile upward, pubescent as the basal ones. Heads numerous in a narrow, spikelike panicle, somewhat pendulous with age. Involucre glabrous to villous, the bracts with hyaline margins. Marginal flowers pistillate, fertile; center flowers sterile; receptacle glabrous. Akenes glabrous.

Meadows and fellfields of the lower alpine zone in the Absaroka and Beartooth Ranges. Circumboreal; south in western North America to Oregon, Utah, New Mexico, Arizona, and Texas.

North America plants that occur near timberline in the Middle Rocky Mountains are ssp. *borealis* (Pallas) Hall & Clem., var. *scouleriana* (Bess.) Cronq.

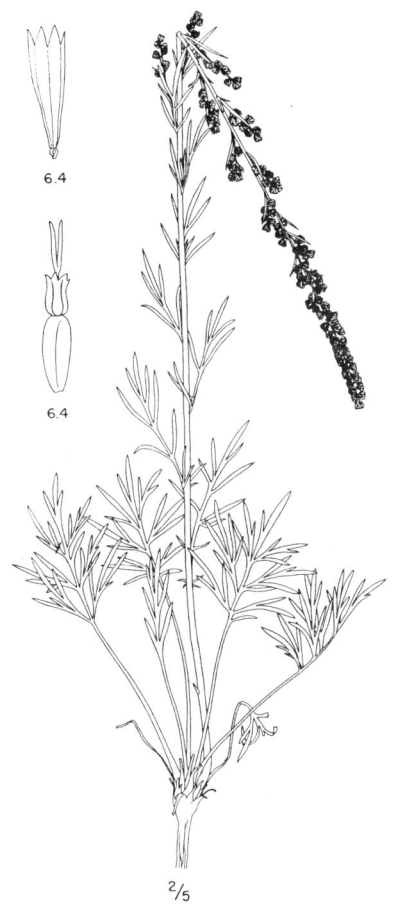

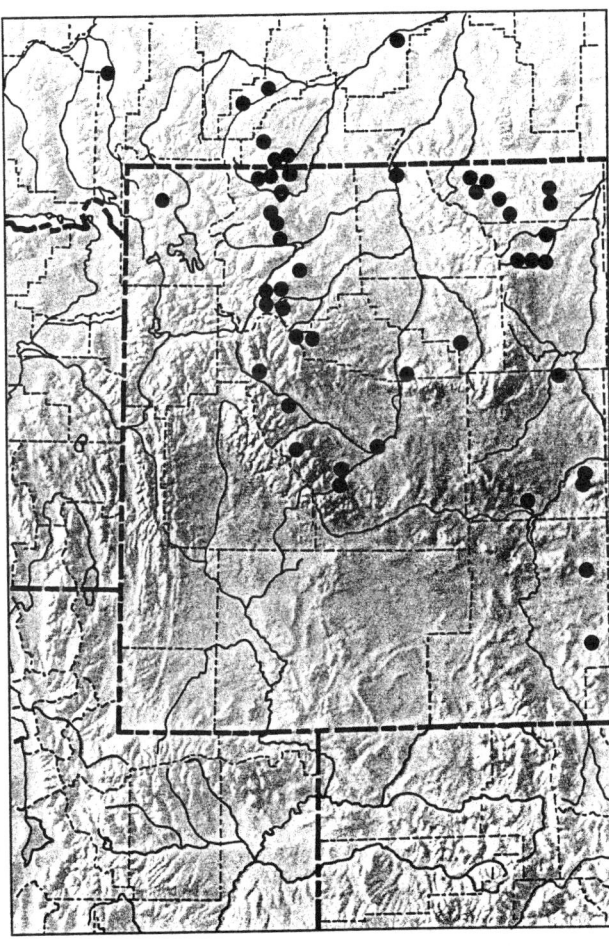

Artemisia frigida Willd.
Sp. Pl. 3:1838, 1804.

Fringed Sagewort

Absinthium frigidum (Willd.) Bess.

Spreading, mat-forming, whitish-tomentose perennial from a woody caudex; stems simple to branched, erect. Basal leaves crowded, petiolate, 2–3-pinnatifid into linear segments, tomentose; cauline leaves becoming reduced and sessile upward. Heads numerous in a narrow, racemelike panicle, sometimes nodding with age. Involucre pilose-tomentose, the bracts with scarious, brownish margins. Marginal flowers pistillate, fertile; center flowers perfect, fertile; receptacle villous, with long hairs between the flowers. Akenes glabrous.

This common lowland species occurs in rocky habitats of the lower alpine zone of the Absaroka, Beartooth, Uinta, Wasatch, Wind River, and Wyoming Ranges. Circumboreal; south in western North America to Washington, Idaho, Utah, Arizona, and New Mexico.

3/4

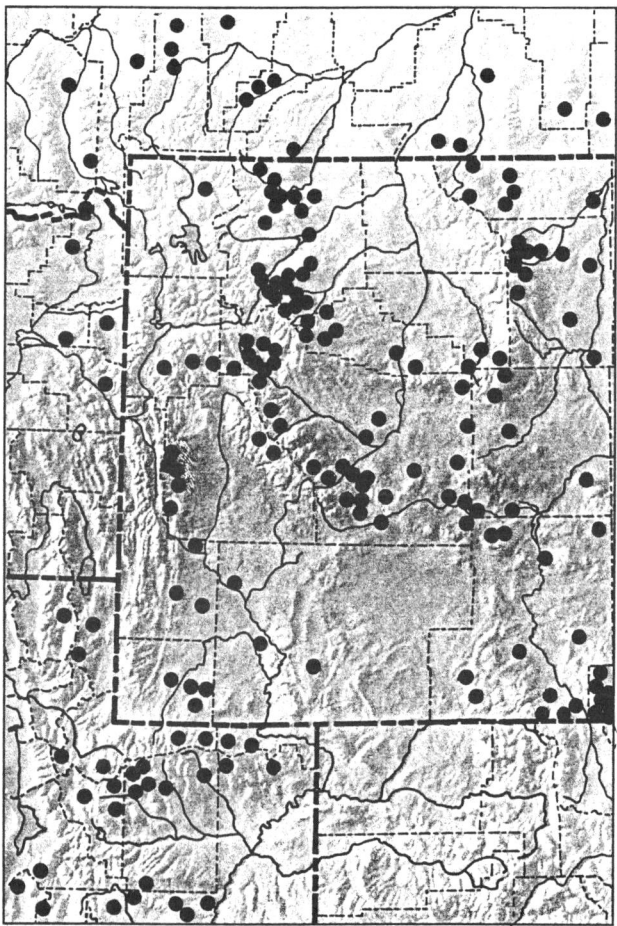

Artemisia ludoviciana Nutt.
Gen. Pl. 2:143, 1818.

Prairie Sagewort

Artemisia albula Wooton
Artemisia argophylla Rydb.
Artemisia atomifera Piper
Artemisia brittonii Rydb.
Artemisia candicans Rydb.
Artemisia cuneata Rydb.
Artemisia diversifolia Rydb.
Artemisia falcata Rydb.
Artemisia floccosa Rydb.
Artemisia flodmanii Rydb.
Artemisia gnaphalodes Nutt.
Artemisia gracilenta A. Nels.
Artemisia herriottii Rydb.
Artemisia incompta Nutt.
Artemisia latiloba Rydb.
Artemisia lindheimeriana Scheele
Artemisia mexicana Willd. ex Spreng.
Artemisia microcephala Wooton
Artemisia pabularis (A. Nels.) Rydb.
Artemisia paucicephala A. Nels.
Artemisia platyphylla Rydb.
Artemisia potens A. Nels.
Artemisia pudica Rydb.
Artemisia purshiana Bess.
Artemisia rhizomata A. Nels.
Artemisia serrata Nutt.
Artemisia silvicola Osterh.
Artemisia underwoodii Rydb.
Oligosporus mexicanus (Willd. ex Spreng.) Less.

Erect, tomentose to glabrate, perennial herb from a long-creeping rhizome; stems simple or branched above. Leaves tomentose to glabrate, very variable in shape, simple to pinnatifid, the margins entire to irregularly toothed or lobed. Heads numerous in a narrow to open panicle. Involucre tomentose to glabrous, the bracts with broad, scarious margins. Marginal flowers pistillate, fertile; center flowers perfect, fertile; receptacle glabrous. Akenes glabrous.

Scree, talus, and exposed rocky ridges of most alpine ranges of the Middle Rocky Mountains. Not yet known from above timberline in the Beartooth, Gros Ventre, and Medicine Bow Ranges. British Columbia to Ontario, south in the West to California, Mexico, and Arkansas.

Alpine plants with pinnately parted to pinnatifid leaves that are gray-pubescent on both surfaces and flowers in a narrow, spicate panicle are var. *latiloba* Nutt.

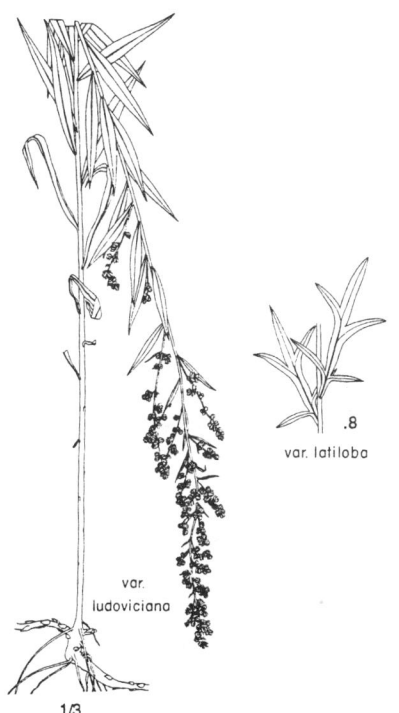

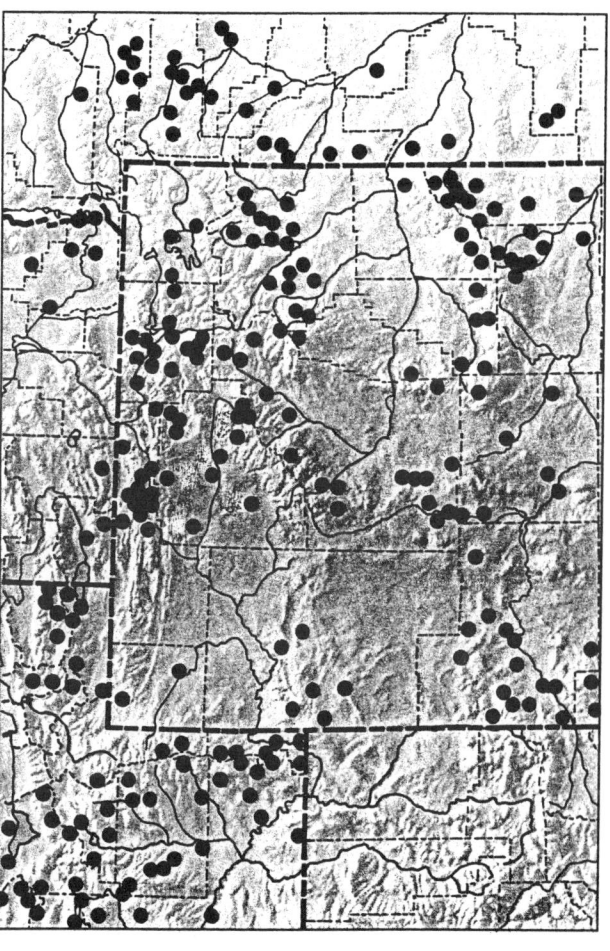

Artemisia michauxiana Bess. in Hook.
Fl. Bor.-Am. 1:324, 1834.

Michaux Sagewort

Artemisia discolor Dougl. ex Bess.
Artemisia graveolens Rydb.
Artemisia subglabra A. Nels.
Artemisia tenuis Rydb.

Erect, perennial herb from a woody caudex, rhizome, or occasionally a taproot; stems simple, glabrous to tomentose. Leaves petiolate, mostly green-glabrate above, tomentose beneath, 2-pinnatifid, the ultimate segments very short, narrow, and toothed; upper leaves reduced, sessile, 1-pinnatifid to entire. Heads several to numerous, erect to reflexed in an elongate, narrow, spikelike inflorescence. Involucre glabrous or nearly so, the bracts with brownish, scarious margins. Marginal flowers pistillate, fertile; center flowers perfect, fertile; receptacle glabrous. Akenes glabrous.

Talus, scree, ledges, and other rocky places in the Absaroka, Beartooth, Uinta, Wasatch, and Wind River Ranges. Yukon Territory south to California, Idaho, Utah, and Wyoming.

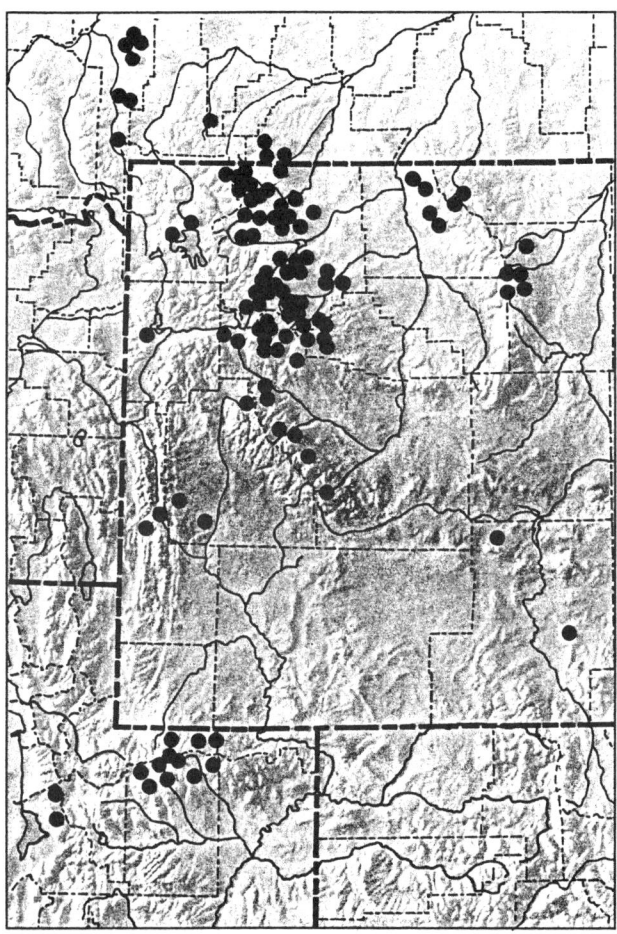

Artemisia norvegica Fries
Nov. Fl. Suec., 56, 1817.

Spruce Sagewort

Artemisia arctica Less.
Artemisia chamissoniana Bess. in Hook.
Artemisia longipedunculata Rudolphi ex Bess.
Artemisia saxicola Rydb.

Erect, glabrous to villous perennial from a branched caudex, rhizome, or occasionally a taproot; stems simple, single to clustered, often reddish purple near the base. Basal leaves petiolate, 2-pinnately dissected, the ultimate divisions linear, lanceolate, acute; cauline leaves reduced upward, becoming sessile. Heads several, large, nodding, in a narrow raceme or racemelike panicle. Involucres villous, or glabrate with age, the bracts with dark brownish to blackish, scarious margins. Marginal flowers pistillate, fertile; center flowers perfect, fertile; receptacle glabrous. Akenes glabrous.

Talus and fellfields of the Absaroka, Gros Ventre, Teton, Uinta, and Wind River Ranges. Circumboreal; south in western North America to California, Idaho, Utah, and Colorado.

North American plants are ssp. *saxatalis* (Bess.) Hall & Clements; they have relatively smaller heads and longer, narrower, pinnatifid leaves than the typical European ssp. *norvegica*.

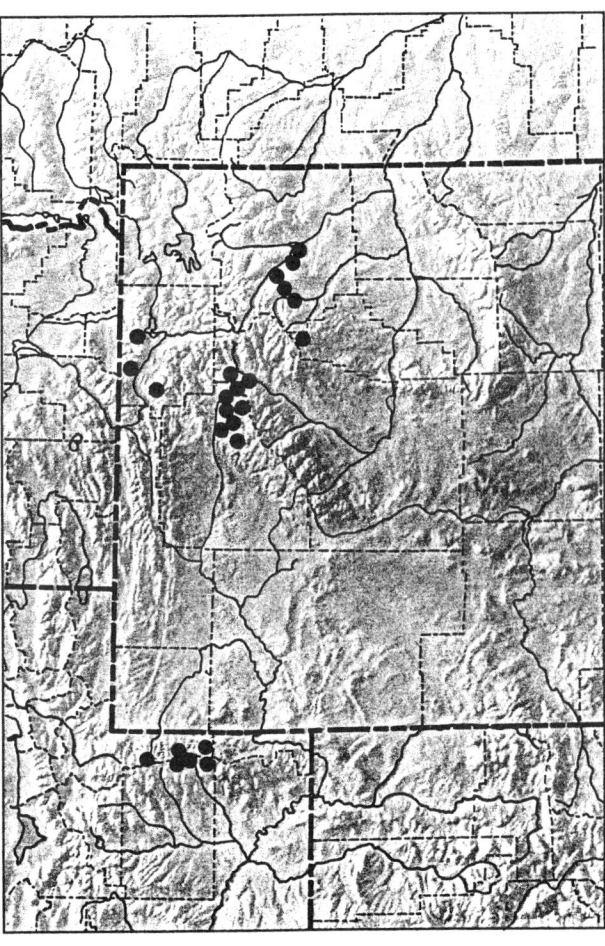

Artemisia scopulorum Gray
Proc. Acad. Phila. 1863:66, 1863.

Alpine Sagewort

Tufted, perennial herb from a simple or, mostly, branched caudex often covered with old leaf bases, and somewhat fleshy roots; stems erect, simple, 1 to several, sparsely tomentose to somewhat villous, often reddish. Basal leaves petiolate, bipinnatifid, the segments linear to oblong-lanceolate, canescent; cauline leaves few, reduced upward, 1–2-pinnatifid, canescent. Heads several to numerous in a spike or raceme, the lower heads often nodding. Involucres villous, the bracts with prominent blackish or dark brown, scarious margins. Marginal flowers pistillate, fertile; center flowers perfect, fertile, villous; receptacle long-villous with white hairs. Akenes glabrous.

Widespread in fellfields and on talus, scree, and rocky stream banks of most alpine ranges in the Middle Rocky Mountains. Not yet known from above timberline in the Hoback, Salt River, and Wyoming Ranges. Montana, Wyoming, Utah, Colorado, and New Mexico.

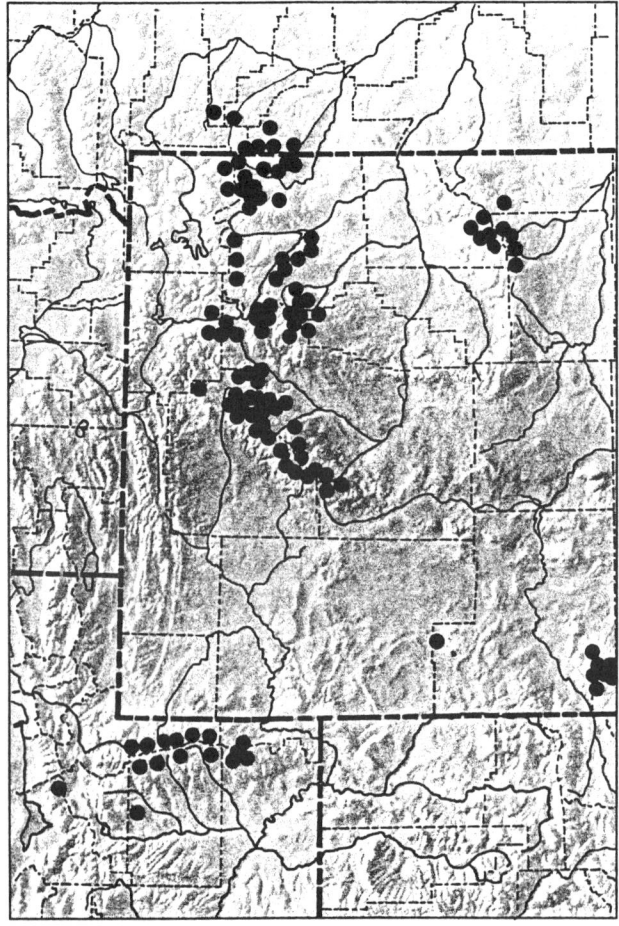

Artemisia tridentata Nutt.
Trans. Am. Phil. Soc., II, 7:398, 1841.

Big Sagebrush

Artemisia angusta Rydb.
Artemisia angustifolia (Gray) Rydb.
Artemisia parishii Gray
Artemisia rothrockii Gray
Artemisia spiciformis Osterh.
Artemisia vaseyana Rydb.
Seriphidium tridentatum (Nutt.) W.A. Weber

Erect, much-branched, aromatic shrub; bark shredded-fibrous; young twigs and leaves silvery-canescent, the leaves persistent, cuneate, 3–5-toothed or cleft at the apex; upper leaves becoming linear, entire. Heads numerous, nearly sessile in panicles or (occasionally) racemes. Involucre canescent to tomentose, sometimes glabrate, the bracts green with scarious margins. Flowers all perfect, fertile; receptacle glabrous. Akenes glabrous.

Timberline slopes of the Absaroka, Big Horn, Gros Ventre, Salt River, and Wasatch Ranges. British Columbia and Alberta south to California, New Mexico, and Nebraska.

Plants of mountain slopes that are less than 1 m tall, with some leaves usually more than 1.5 cm long, and inflorescences overtopping the foliage are var. *vaseyana* (Rydb.) B. Boivin.

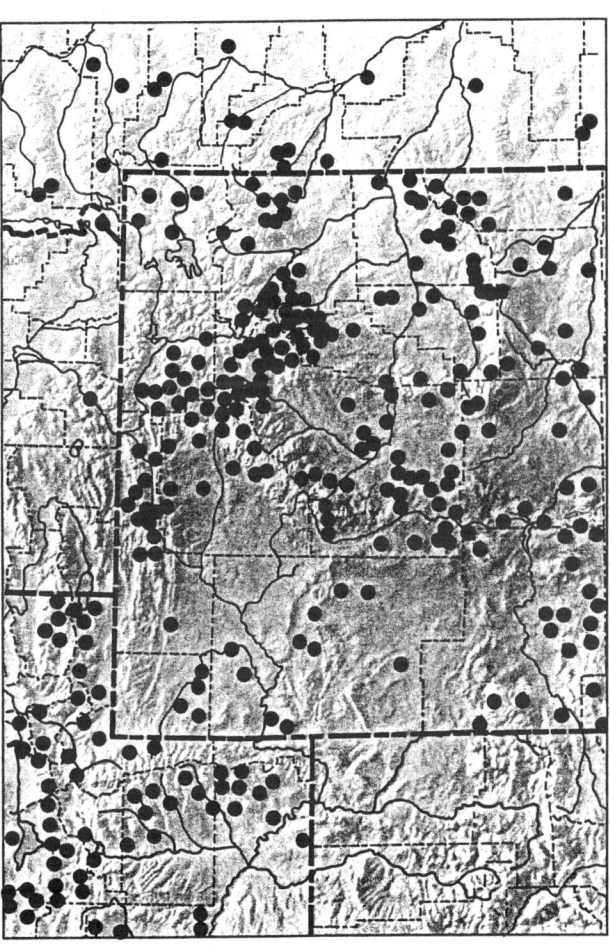

Aster L.
Aster

Perennial or rarely annual herbs with alternate, entire or toothed leaves. Heads radiate or rarely discoid, solitary or numerous and in racemes, panicles, or corymbs; the involucral bracts usually herbaceous at the tip and chartaceous below, or sometimes chartaceous throughout, often imbricate, the receptacle flat, naked. Ray flowers white, pink, blue, or purple, pistillate. Disk flowers yellow or reddish purple, numerous, perfect. Akenes somewhat flattened, mostly several-nerved, the pappus of numerous equal or subequal, capillary bristles.

KEY TO THE SPECIES OF ASTER

1. Plants less than 13 cm tall, from taproots; leaves mostly basal, linear or linear-elliptic to narrowly oblanceolate — *A. alpigenus*
1. Plants usually more than 17 cm tall, from fibrous roots; leaves mostly cauline, oblanceolate to narrowly obovate
    2. Leaves with shallow teeth; involucral bracts purple-tipped, often purple-margined — *A. sibiricus*
    2. Leaves entire; involucral bracts with or without purple margins and tips
        3. Involucral bracts with broad, scarious margins, only the midrib and tip green — *A. glaucodes*
        3. Involucral bracts green, the scarious margins inconspicuous or lacking
            4. Involucral bracts glandular — *A. integrifolius*
            4. Involucral bracts lacking glands
                5. Inflorescence subtended by 1 or more foliaceous bracts below those of the receptacle; stems mostly decumbent, 17–23 cm tall; middle cauline leaves mostly more than 1 cm wide — *A. foliaceus*
                5. Inflorescence lacking foliaceous bracts below those of the receptacle; stems mostly erect, 20–30 cm tall; middle cauline leaves mostly less than 1 cm wide — *A. spathulatus*

Aster alpigenus (T. & G.) Gray
Proc. Am. Acad. 8:389, 1872.

Alpine Aster

Aplopappus alpigenus T. & G.
Aster andersonii (Gray) Gray
Aster elatus (Greene) Cronq.
Aster haydenii Porter
Aster pulchellus D.C. Eat.
Erigeron andersonii Gray
Oreastrum alpigenum (T. & G.) Greene
Oreastrum andersonii (Gray) Greene
Oreastrum elatus Greene
Oreastrum haydenii (Porter) Rydb.
Oreostemma alpigenum (T. & G.) Greene
Oreostemma andersonii (Gray) Greene
Oreostemma haydenii (Porter) Greene

Subscapose perennial from a simple or branched caudex and usually a taproot; stems 1 to several, simple, glabrate to puberulent. Basal leaves oblanceolate to linear-elliptic, entire, erect to somewhat spreading; cauline leaves reduced upward, linear to linear-lanceolate. Heads solitary on the stems; involucre hemispheric, imbricate, the bracts usually purplish in part and chartaceous at the base. Ray flowers deep purple to lavender, rarely white; disk flowers yellow. Akenes glabrous in ours.

Fellfields and rocky meadows of nearly all of the northern alpine ranges. Not yet known from the Hoback Range. Washington south to California, Idaho, and Wyoming.

Wyoming and southern Montana plants 13 cm tall or less, with akenes glabrous, at least below, and with linear to linear-elliptic leaves are var. *haydenii* (Porter) Cronq.

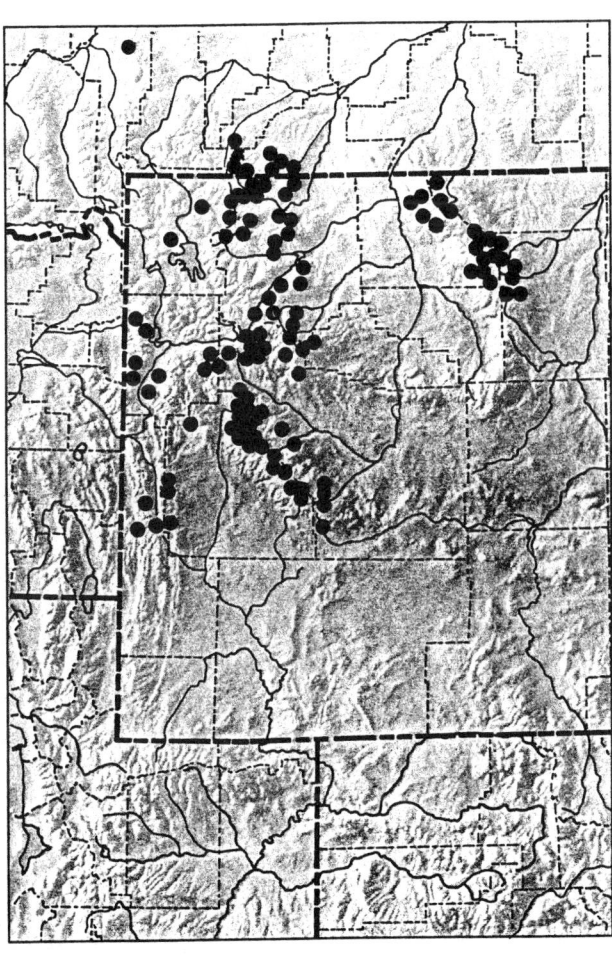

Aster foliaceus Lindl. ex DC.
Prodromus 5:228, 1836.

Leafybract Aster

Aster amplissimus Greene
Aster apricus (Gray) Rydb.
Aster burkei (Gray) T.J. Howell
Aster canbyi (Gray) Vasey ex Rydb.
Aster ciliomarginatus Rydb.
Aster cusickii Gray
Aster diabolicus Piper
Aster eriocaulis Rydb.
Aster frondeus (Gray) Greene
Aster glastifolius Greene
Aster hendersonii Fern.
Aster incertus A. Nels.
Aster kootenayi A. Nels. & Macbr.
Aster phyllodes Rydb.
Aster subspicatus Nees
Aster tweedyi Rydb.
Aster vaccinus Piper

Caulescent perennial from a branched, creeping rhizome or caudex; stems 1 to several, erect to somewhat decumbent at the base, usually reddish purple, glabrous to villous, the hairs often in lines below the leaf bases. Lower leaves mostly petiolate, the blades oblanceolate to obovate, entire or nearly so, the margins ciliate; upper leaves sessile, reduced in size, lanceolate to ovate, entire. Heads 1 to several, showy, in a corymbose inflorescence; involucre hemispheric, the outer bracts green, foliaceous; the inner bracts smaller than the outer, imbricate, whitish or yellowish at the base, acute to obtuse. Ray flowers purple, bluish, rose, or lavender; disk flowers yellow. Akenes pubescent.

Meadows and willow thickets of the lower alpine zone in most of the Middle Rocky Mountain ranges. Not yet known from above timberline in the Salt River, Teton, and Wyoming Ranges. Alaska south to California, New Mexico, and Colorado.

Two varieties occur at timberline and higher in the mountains. They may be distinguished as follows:
1. Stems decumbent or ascending, mostly less than 23 cm tall; involucral bracts purplish on the margins and tips . . . var. *apricus* Gray
1. Stems erect, mostly more than 23 cm tall; involucral bracts green on the margins and tips . . . var. *parryi* (Eaton) Gray

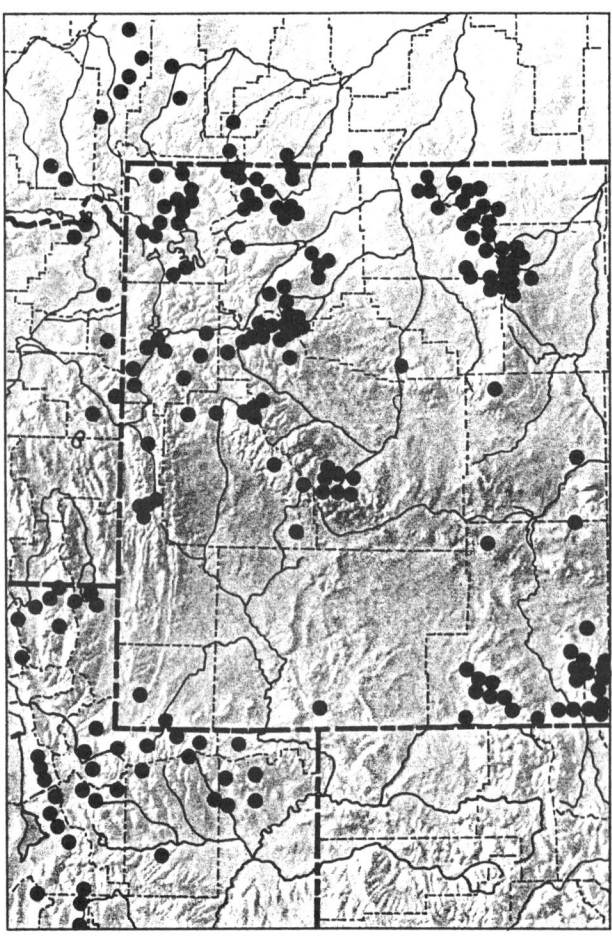

A. f. var. apricus

Aster glaucodes Blake
Proc. Biol. Soc. Wash. 35:174, 1922.

Blueleaf Aster

Aster glaucus (Nutt.) T. & G.
Aster wasatchensis (M.E. Jones) Blake
Eucephalus formosus Greene
Eucephalus glaucus Nutt.

Perennial from rhizomes; stems simple to branched, glabrous to puberulent or glandular in the inflorescence. Leaves sessile, lanceolate or oblong, entire, glabrous and glaucous, reticulate-veined, the lower ones early deciduous. Heads few to many in a short, corymbose inflorescence; involucre campanulate to hemispheric, the bracts chartaceous with a greenish midvein and greenish to purplish, usually acute tip. Ray flowers white, lavender, or pinkish; disk flowers yellowish. Akenes pubescent to glabrate.

Margins of meadows at timberline in the Absaroka, Uinta, Wasatch, and Wind River Ranges. Wyoming south to Utah, Arizona, and Colorado.

Mountain plants lacking glandular pubescence in the inflorescence are var. *glaucodes*.

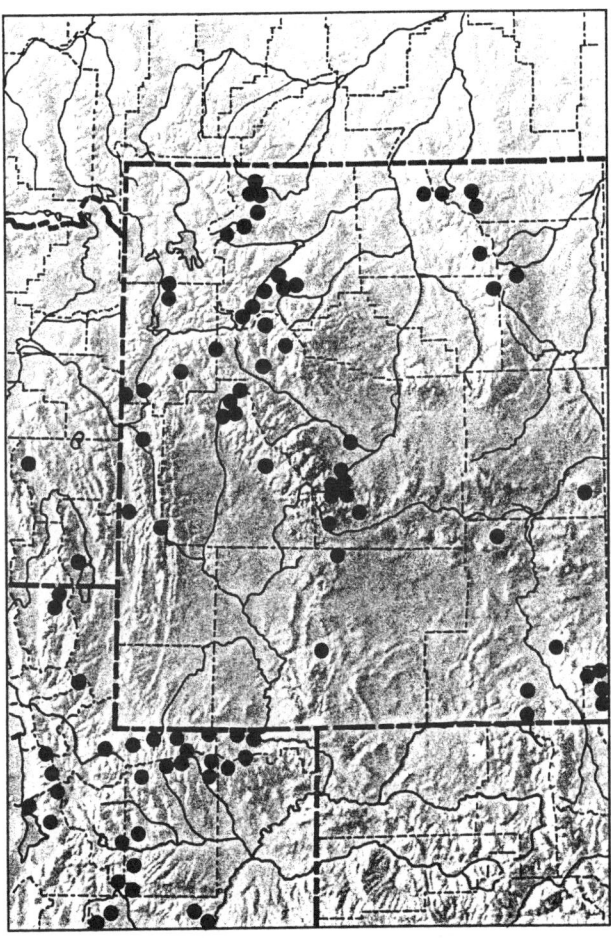

Aster integrifolius Nutt.
Trans. Am. Phil. Soc., II, 7:291, 1840.

Thickstem Aster

Aster amplexifolius Rydb.

Erect, perennial herb from a rhizome or short caudex and fibrous roots; stems mostly solitary, simple, glandular-villous. Lower leaves petiolate, the blades elliptic to oblanceolate, glandular-villous, ciliate, the margins entire; cauline leaves reduced upward, sessile and clasping, lanceolate to somewhat oblong, the margins entire. Heads showy, few to several, in corymbose inflorescences; involucre glandular, the bracts subequal, hardly imbricate, the outer bracts somewhat leafy, the inner ones narrowly lanceolate, purplish. Ray flowers purple, violet, or pinkish; disk flowers yellow.

Timberline margins of the Salt River and Wind River Ranges. Washington south and east to Idaho, Utah, Wyoming, and Montana.

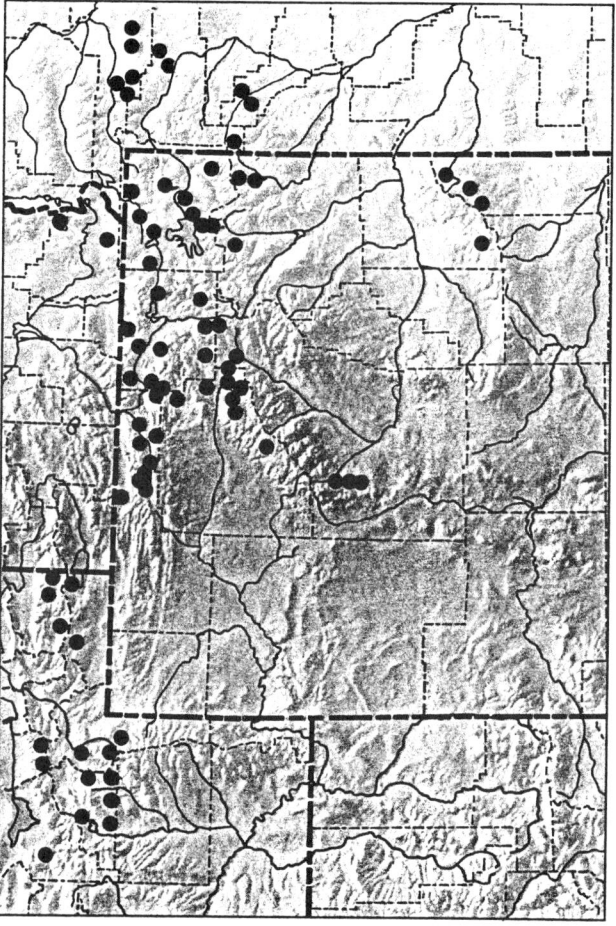

Aster sibiricus L.
Sp. Pl., 872, 1753.

Siberian Aster

Aster bakerensis St. John
Aster espenbergensis Nees
Aster meritus A. Nels.
Aster montanus Rich.
Aster pygmaeus Lindl.
Aster richardsonii Spreng.
Aster subintegerrimus (Trautv.) Ostenf.

Erect to ascending perennial from simple to branched, elongate rhizomes; stems 1 to several, short-villous, simple or branched, usually reddish purple. Leaves sessile to short-petiolate, the blades lanceolate, elliptic, or oblanceolate, entire to irregularly serrate, glabrous above, glabrate to villous below, the tips acute to acuminate; basal leaves early deciduous and smaller than the middle ones. Heads solitary to few and cymose; involucre short-villous, the bracts loosely imbricate, chartaceous below, green-tipped, the margins often purplish, ciliate. Ray flowers purple; disk flowers yellowish. Akenes pubescent.

Timberline margins of the Absaroka and Beartooth Ranges. A nonalpine collection has been reported from the Uinta Range in Summit County, Utah (Albee et al., 1988). Eurasia; Alaska and Yukon Territory south to Oregon, Idaho, Utah, and Wyoming.

Middle Rocky Mountain plants as described above with few heads and involucral bracts that are loosely imbricate and lack squamose tips are var. *meritus* (A. Nels.) Raup.

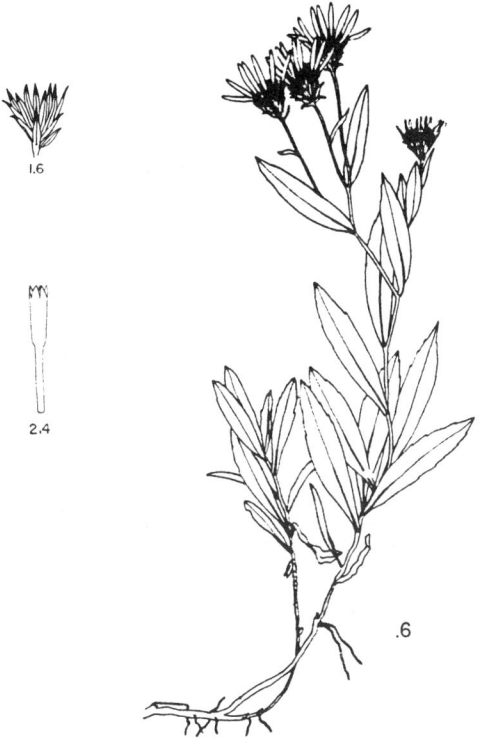

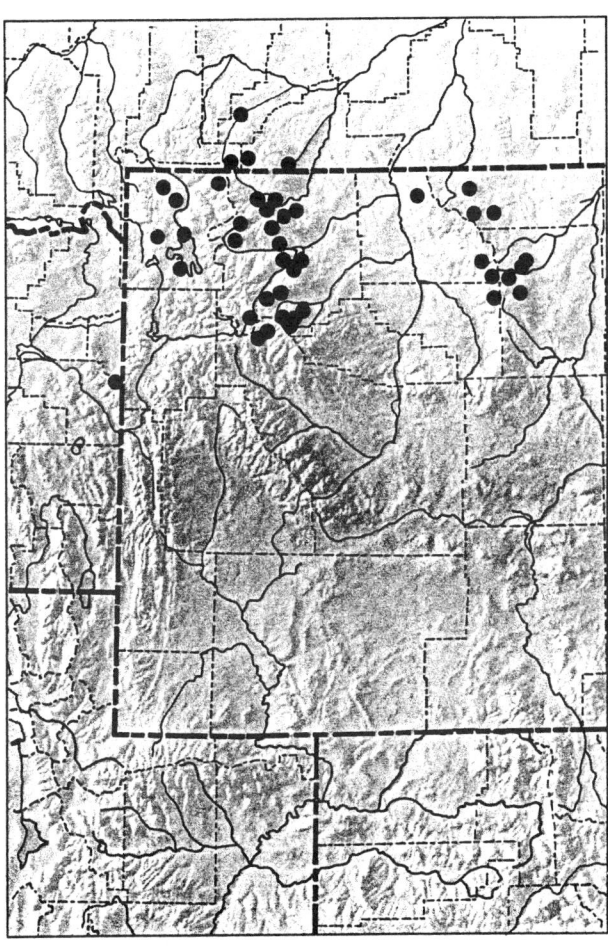

Aster spathulatus Lindl. in Hook.
Fl. Bor.-Am. 2:8, 1834.

Western Aster

Aster andinus Nutt.
Aster delectabilis Hall
Aster delectus Piper
Aster fremontii (T. & G.) Gray
Aster misellus Piper
Aster occidentalis (Nutt.) T. & G.
Aster paludicola Piper
Aster subspathulatus Rydb.
Aster vallicola Greene
Aster williamsii Rydb.
Aster yosemitanus (Gray) Greene
Tripolium occidentale Nutt.

Erect, perennial herb from a short rhizome or caudex and fibrous roots; stems several, loosely villous at least above, often reddish. Lower leaves petiolate, the blades oblanceolate, mostly glabrous, the margins entire, ciliate; upper leaves reduced, sessile, and lanceolate, linear, or narrowly elliptic, entire. Heads showy, 1 to several in a cymose panicle; involucre hemispheric, the bracts mostly acute, occasionally obtuse, linear to linear-oblong, the outer ones mostly green with chartaceous bases, the inner ones mostly chartaceous with green or purplish tips. Ray flowers purple, violet, or bluish; disk flowers yellowish. Akenes pubescent.

Willow thickets and timberline meadows of the Absaroka, Beartooth, Big Horn, Gros Ventre, and Wind River Ranges. British Columbia south to California, Utah, and Colorado.

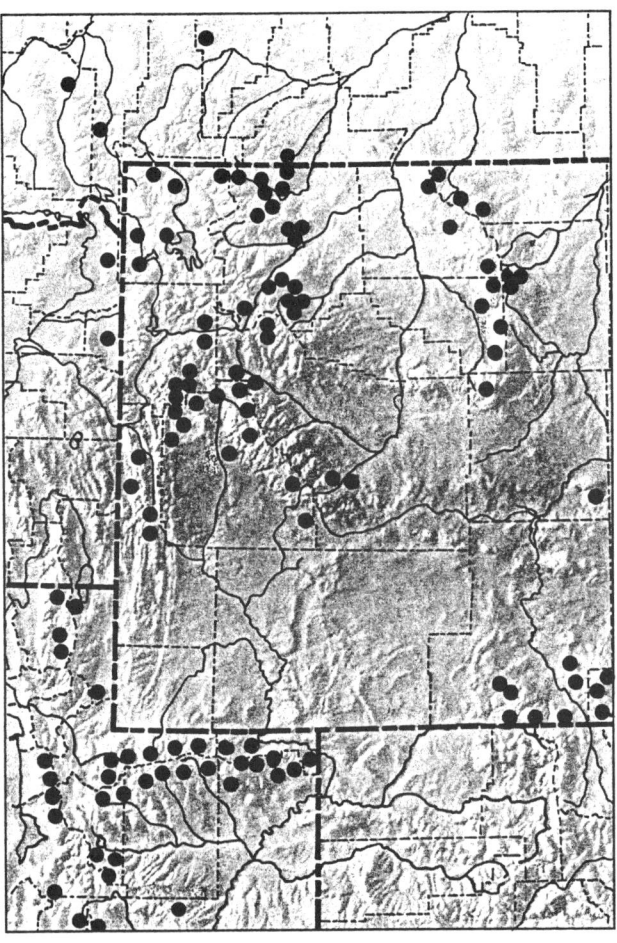

Chaenactis DC.
Dusty Maiden

Annual, biennial, or perennial herbs from taproots, with alternate, entire to pinnately or irregularly dissected leaves. Heads solitary and terminal in ours, the involucral bracts narrow, equal or imbricate, the receptacle flat, naked. Flowers all tubular, pink in ours. Anthers sagittate. Style branches externally hairy nearly to the base. Akenes clavate or terete and somewhat compressed, the pappus of few to numerous hyaline scales.

KEY TO THE SPECIES OF *CHAENACTIS*

1. Plants 10 cm or less tall; internodes and stems very short, the leaves appearing basal or nearly so; heads solitary, sometimes 2 — *C. alpina*
1. Plants more than 10 cm tall; internodes visible, the leaves definitely cauline; heads few to several in a corymb — *C. douglasii*

Chaenactis alpina (Gray) Jones
Proc. Calif. Acad. Sci., II, 5:699, 1895.

Alpine Dusty Maiden

Chaenactis leucopsis Greene
Chaenactis miniscula Greene
Chaenactis rubella Greene

Dwarfed, subscapose perennial from a stout taproot or short-branched caudex; stems very short, terminated by the much longer peduncles, which exceed the leaves. Leaves sparsely to densely tomentose, petiolate, the blades 2-pinnatifid. Heads solitary, sometimes 2, the peduncles glandular or tomentose; involucre glandular to tomentose, the bracts narrowly oblanceolate, often purplish. Disk corollas white or lavender, glandular to tomentose; pappus scales oblanceolate. Akenes pubescent.

Talus and scree of all western alpine ranges of the Middle Rocky Mountains. Not yet known from above timberline in the Big Horn and Medicine Bow Ranges. Oregon south and east to California, Idaho, Utah, Colorado, Wyoming, and Montana.

Alpine varieties may be distinguished as follows:
1. Involucres and peduncles glandular . . . var. *alpina*
1. Involucres and peduncles tomentose
 . . . var. *leucopsis* (Greene) Stockwell

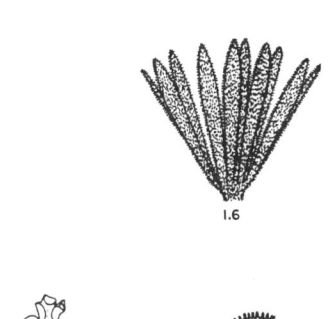

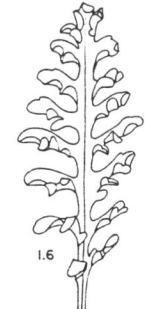

Chaenactis 109

Chaenactis douglasii (Hook.) Hook. & Arn.
Bot. Beechy Voy., 354, 1838.

Douglas Dusty Maiden

Hymenopappus douglasii Hook.
Chaenactis achilleaefolia Hook. & Arn.
Chaenactis angustifolia Greene
Chaenactis brachiata Greene
Chaenactis cheilanthoides Greene
Chaenactis cineria Stockwell
Chaenactis humilis Rydb.
Chaenactis imbricata Greene
Chaenactis panamintensis Stockwell
Chaenactis rubricaulis Rydb.
Chaenactis suksdorfii Stockwell
Macrocarpus achilleaefolius (Hook. & Arn.) Nutt.
Macrocarpus douglasii (Hook.) Nutt.

Erect biennial or perennial from a taproot or branched caudex; stems tomentose, sometimes glandular above, mostly solitary, simple, occasionally branched above. Leaves mostly tomentose, 1–3-pinnatifid, reduced upward; basal rosette present early, becoming withered with age. Heads few to several, corymbose, the peduncles glandular; involucre glandular to tomentose, the bracts narrow, oblong to linear, obtuse. Disk corollas white, cream, or pink, glandular-puberulent at least below; pappus scales linear to narrowly oblanceolate. Akenes pubescent.

Rocky slopes and coarse, sandy soils, mostly at lower elevations; becoming alpine in the Absaroka and Wasatch Ranges. British Columbia south to California, Arizona, Colorado, and North Dakota.

Middle Rocky Mountain plants of higher elevations that are mostly perennial, multistemmed, and often with a branched caudex are var. *montana* M.E. Jones.

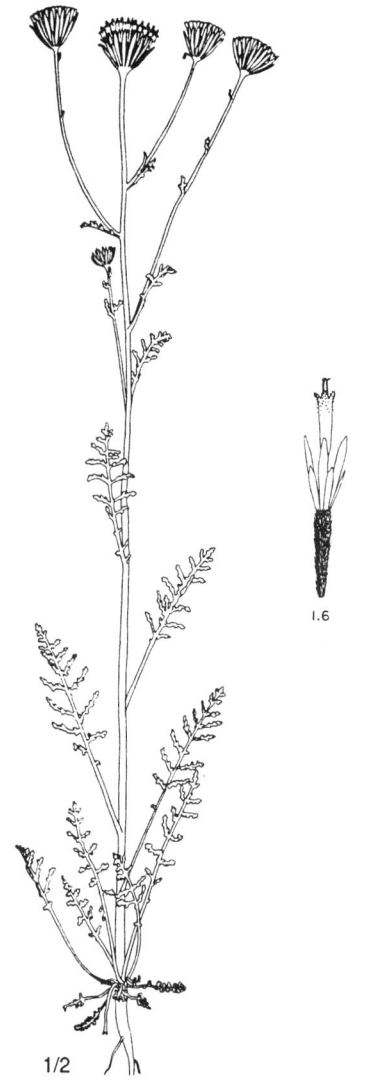

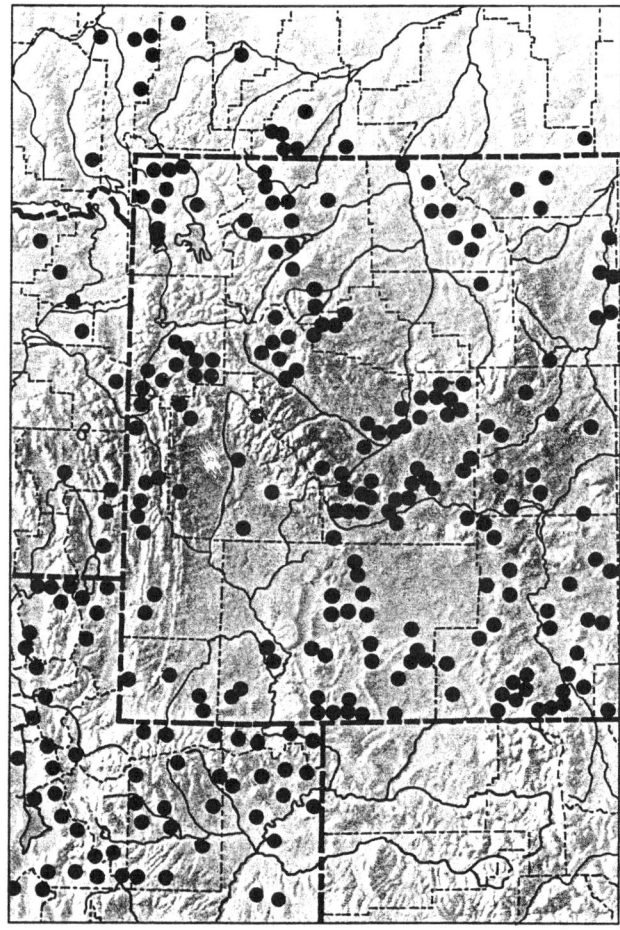

Chrysothamnus Nutt.
Rabbitbrush

Shrubs with glabrous, puberulent, or densely tomentose stems, usually with taproots. Leaves alternate, linear to elliptic, sessile or nearly so, entire. Heads discoid, numerous, 5–20-flowered, mostly in panicles or corymbs. Involucral bracts imbricate, often forming vertical ranks, chartaceous or coriaceous, the tips sometimes green-herbaceous; receptacle naked. Disk flowers yellow, perfect; style branches flattened with elongate, pubescent appendages. Akenes slender, terete to somewhat angled, pubescent to nearly glabrous; pappus of numerous white or whitish capillary bristles.

Chrysothamnus viscidiflorus (Hook.) Nutt.
Trans. Am. Phil. Soc., II, 7:324, 1840.

Douglas Rabbitbrush

Crinitaria viscidiflora Hook.
Aster viscidiflorus (Hook.) Kuntze
Bigelowia douglasii Gray
Bigelowia glauca K. Schum.
Bigelowia lanceolata (Nutt.) Gray
Bigelowia viscidiflora (Hook.) DC.
Chrysothamnus axillaris Keck
Chrysothamnus douglasii (Gray) Clements & Clements
Chrysothamnus elegans Greene
Chrysothamnus glaucus A. Nels.
Chrysothamnus humilus Greene
Chrysothamnus lanceolatus Nutt.
Chrysothamnus latifolius (D.C. Eat.) Rydb.
Chrysothamnus leucocladus Greene
Chrysothamnus linifolius Greene
Chrysothamnus marianus Rydb.
Chrysothamnus puberulus (D.C. Eat.) Greene
Chrysothamnus pumilus Nutt.
Chrysothamnus serrulatus (Torr.) Rydb.
Chrysothamnus stenolepis Rydb.
Chrysothamnus stenophyllus (Gray) Greene
Chrysothamnus tortifolius (Gray) Greene
Linosyris lanceolata (Nutt.) T. & G.
Linosyris pumila (Nutt.) Gray
Linosyris serrulata Torr. in Stansb.
Linosyris viscidiflora (Hook.) T. & G.

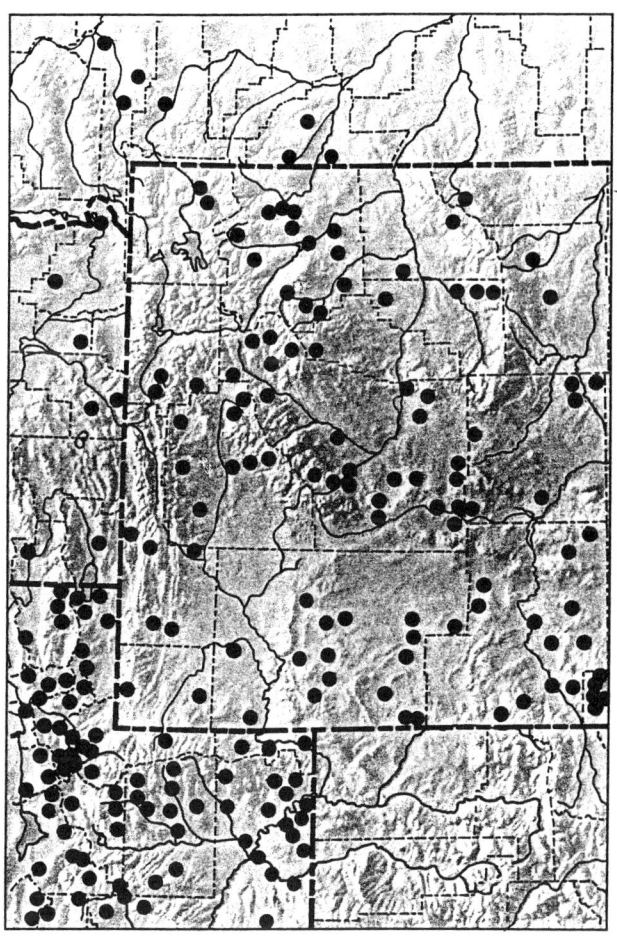

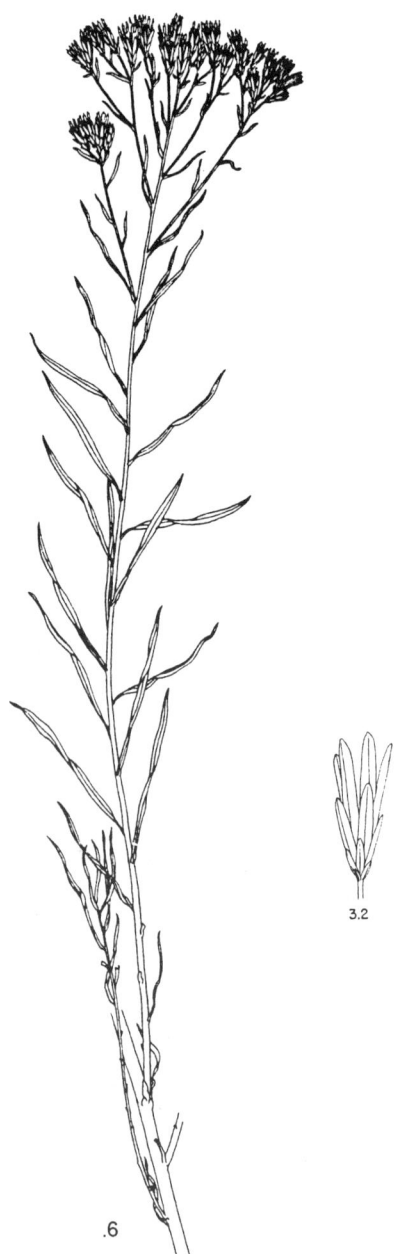

Low shrub, less than 5 dm tall in the alpine zone, with erect to spreading branches; twigs glabrous to puberulent, but lacking a dense, whitish tomentum. Leaves linear, oblong, or narrowly lanceolate, and glabrous to puberulent, usually viscid, 1–5-nerved. Heads numerous, in terminal cymes. Involucre glandular or puberulent, the bracts obtuse to acute, with narrow, hyaline margins. Disk corollas usually 5, yellow. Akenes villous; pappus of whitish capillary bristles.

Scree slopes and other rocky places in the Wasatch and northern Wind River Ranges. British Columbia south and east to California, Arizona, New Mexico, and South Dakota.

Alpine specimens with puberulent twigs and the leaves 3–5-nerved and flattened, rather than twisted, are var. *lanceolatus* (Nutt.) Greene.

Cirsium Mill.
Thistle

Coarse, biennial or perennial herbs with alternate, spinulose, pinnatifid leaves. Heads large, solitary or few to several, in racemes or panicles, the involucral bracts subequal, in several series, at least some of them spine-tipped, the receptacle flat, bristly. Flowers purple or whitish, all tubular, the corollas deeply 5-cleft. Anthers caudate at the base, the filaments papillose-hairy. Style with a thickened, hairy ring. Akenes obovate to oblong, glabrous, usually somewhat flattened, the pappus of numerous plumose bristles that are united at the base and fall together.

KEY TO THE SPECIES OF *CIRSIUM*

1. Leaves glabrous beneath, or with a few multicellular hairs — *C. eatonii*
1. Leaves tomentose beneath
   2. Involucral bracts pectinate-ciliate, the hairs apparently cross-connecting between the bracts — *C. subniveum*
      3. Heads mostly in dense clusters, the peduncles short and hidden by the upper bracts; leaves toothed to shallowly lobed — *C. hookerianum*
      3. Heads mostly in loose clusters, the peduncles elongate and evident; leaves parted more than halfway to the midrib — *C. subniveum*
   2. Involucral bracts glabrous to pubescent, but not pectinate-ciliate, hairs, if present, not cross-connecting — *C. pulcherrimum*

Cirsium eatonii (Gray) Robins.
Rhodora 13:240, 1911.

Eaton Thistle

Cnicus eatonii Gray
Carduus leiocephalus A. Heller
Carduus olivescens Rydb.
Carduus polyphyllus Rydb.
Carduus tweedyi Rydb.
Cirsium olivescens (Rydb.) Petr.
Cirsium polyphyllum (Rydb.) Petr.
Cirsium tweedyi (Rydb.) Petr.

Perennial, often tufted, glabrous to arachnoid herb from a taproot and simple or branched caudex, often crowned with old leaf bases. Basal leaves oblong to elliptic, pinnatifid, spiny-margined, green and glabrous to glabrate on both sides; cauline leaves reduced upward, with margins like the basal ones. Heads few and scattered in upper leaf axils or sometimes in a terminal, spiciform cluster; involucres arachnoid, villous, or tomentose, the bracts lanceolate, attenuate; outer bracts with spinose margins, the dorsal ridges not developed. Disk corollas pink, purplish, or ochroleucous.

Ledges, talus, and scree of most of the western alpine ranges. Not yet known from above timberline in the Salt River and Wyoming Ranges. Montana, Idaho, Wyoming, and Utah.

The 2 alpine varieties may be distinguished by the following characters (*C. eatonii* var. *murdockii* S.L. Welsh, is endemic to the Uinta Range).

1. Corollas pink or purplish; involucral bracts white-tomentose, mostly lacking multicellular hairs
     . . . var. *eatonii*
1. Corollas ochroleucous; involucral bracts gray or brown-villous with multicellular hairs
     . . . var. *murdockii* S.L. Welsh

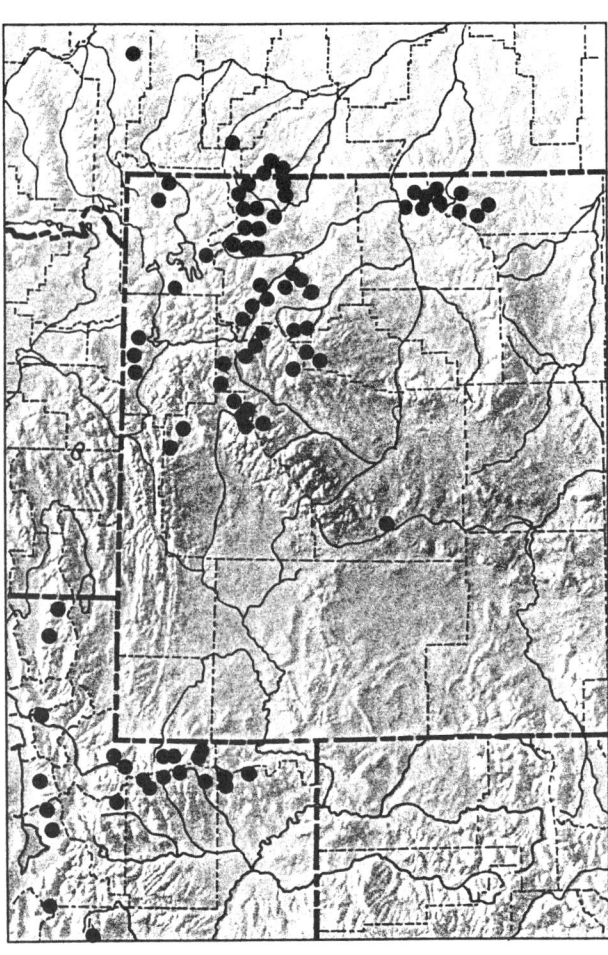

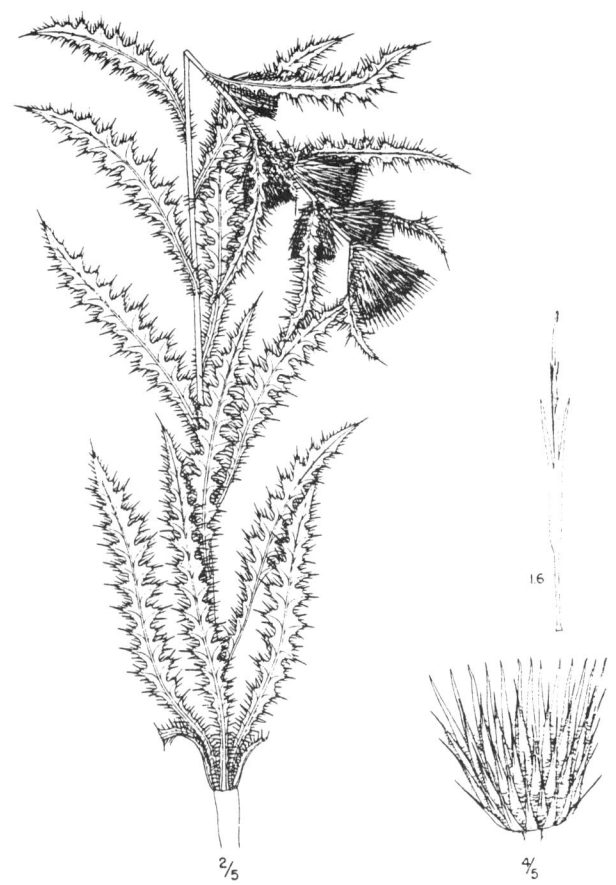

114   Asteraceae

Cirsium hookerianum Nutt.
Trans. Am. Phil. Soc., II, 7:418, 1841.

White Thistle

Carduus butleri Rydb.
Carduus hookeriana (Nutt.) A. Heller
Cirsium butleri (Rydb.) Petr.
Cnicus hookeriana (Nutt.) Gray

Erect, glabrate to arachnoid-tomentose, biennial to weakly perennial herb; stems simple, or rarely branched, succulent below and tapering to a slender tip. Leaves oblanceolate to narrowly oblong, becoming linear upward, decurrent at the base, greenish-glabrate to villous above, white-tomentose below, the margins toothed or shallowly lobed with short spines. Heads few, in a dense terminal spiciform cluster, or sometimes scattered in the upper leaf axils; involucre arachnoid-tomentose, the bracts lanceolate, spine-tipped. Disk corollas white to ochroleucous.

Timberline meadows, rocky slopes, and krummholz margins. Reported from the Beartooth Range by Lessica (1993). British Columbia and Alberta south to Washington, Idaho, and northern Wyoming.

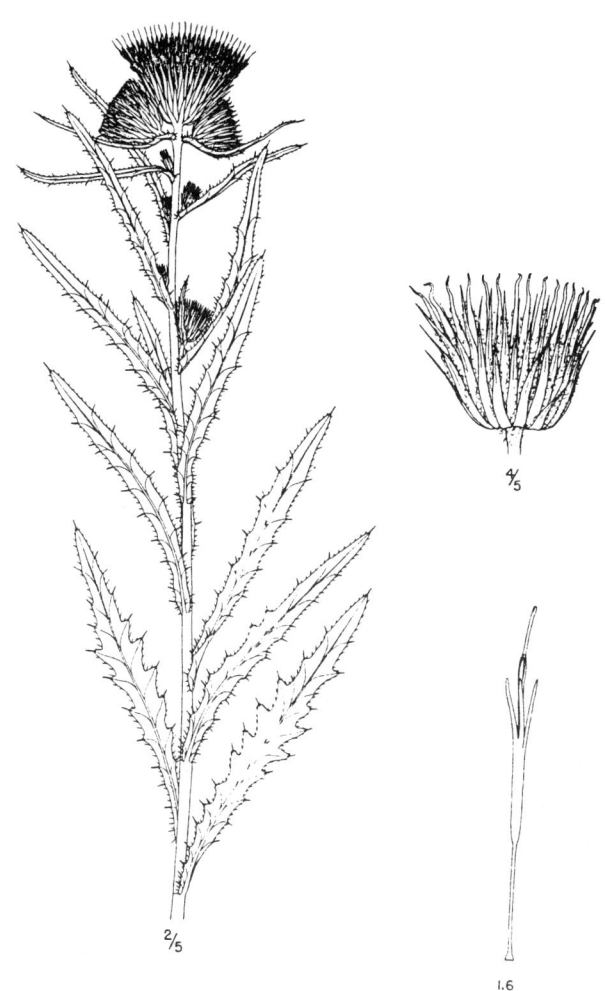

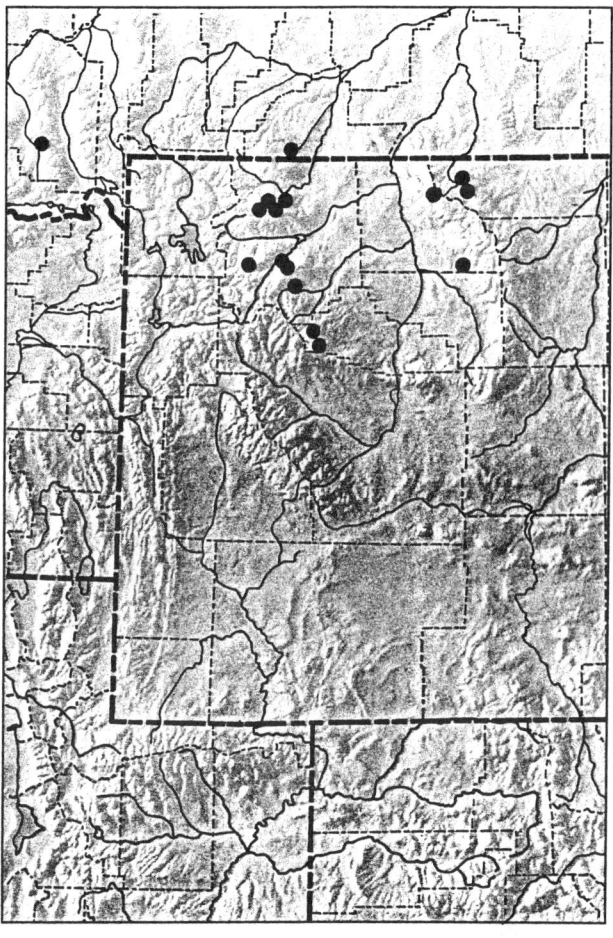

Cirsium pulcherrimum (Rydb.) K. Schum.
Just's Bot. Jaresb. 29:566, 1903.

Showy Thistle

Carduus pulcherrimus Rydb.

Erect, floccose perennial herb from a taproot and simple caudex; stems simple or branched. Leaves oblong to elliptic, sinuate-pinnatifid, spiny-margined, green and glabrous to floccose above, white-tomentose below, reduced upward; lower leaves with clasping bases; upper leaf bases decurrent on the stem. Heads few, terminating the branches; involucres glandular, the bracts lanceolate, spine-tipped, the margins pilose. Disk corollas pink, purple, or whitish.

Rocky timberline meadows in the Big Horn Mountains. Montana, Wyoming, and northeastern Utah.

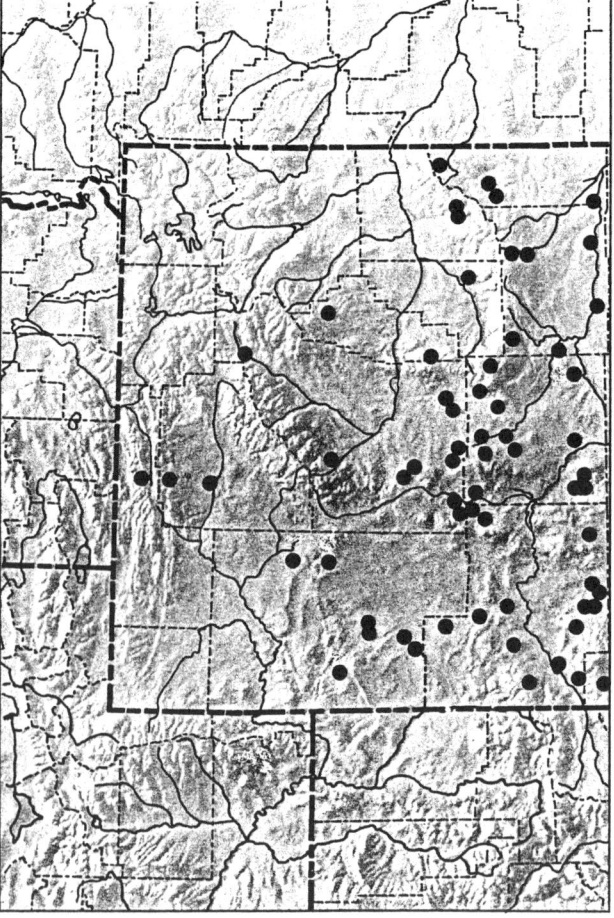

Cirsium subniveum Rydb.
Fl. Rocky Mtns., 1006, 1917.

Snow Thistle

Carduus nevadensis Greene
Cirsium davisii Cronq.
Cirsium humboldtense Rydb.
Cirsium nevadensis (Greene) Petr.
Cirsium wallowense M. Peck

Tomentose to villous perennial from a taproot; stems simple or branched. Leaves elliptic to oblong, spiny-margined, pinnatifid, tomentose beneath, glabrate and greener above, the bases decurrent. Heads few, terminating the branches; involucres tomentose, the bracts ovate-lanceolate, with tomentose margins that appear to connect the bracts, spine-tipped, the dorsal ridges developed and glandular. Disk corollas pink, purplish, or whitish.

Rocky places in the Absaroka, Salt River, Wind River, and Wyoming Ranges near timberline. Oregon south to Idaho, Utah, and western Wyoming.

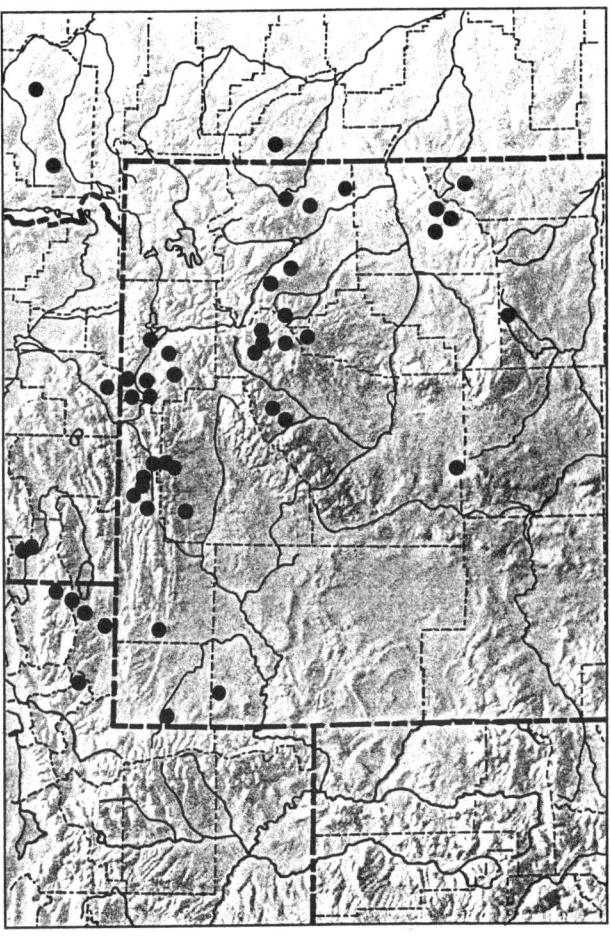

Crepis L.
Hawksbeard

Annual or perennial, commonly leafy-stemmed herbs with milky juice, from woody roots, with basal and cauline, alternate, toothed to pinnatifid leaves. Ours have few to many heads on mostly branched stems, the principal bracts of the involucre in 1 or 2 series, the outer bracts shorter, few to many, the receptacle flat, naked or ciliate. Flowers all ligulate, yellow, the involucre usually purple or purplish in ours. Akenes oblong, terete, mostly 10-ribbed, beakless or sometimes with a short beak, the pappus of numerous white, capillary bristles.

KEY TO THE SPECIES OF *CREPIS*

1. Plants less than 10 cm tall; heads shorter than or equaling the leaves; flowers less than 12 per head — *C. nana*
1. Plants more than 15 cm tall; heads much exceeding the leaves; flowers more than 20 per head — *C. runcinata*

Crepis nana Richards.
Bot. App. Frankl. J., 746, 1823.

Dwarf Hawksbeard

Askellia nana (Richards.) W.A. Weber
Barkhausia nana (Richards.) DC.
Hieraciodes nanum (Richards.) Kuntze
Prenanthes pygmaea Ledeb.
Youngia nana (Richards.) Rydb.

Low, glabrous, caespitose, perennial herb, mostly less than 10 cm tall, from a slender taproot, often with a short, stout, erect caudex; stems several to numerous, often much branched, erect to ascending. Basal leaves petiolate, the blades ovate, spatulate or obovate, glabrous, entire to toothed or occasionally lyrate-pinnatifid; cauline leaves, if present, similar to the basal ones, reduced upward. Heads few to many, axillary on short peduncles, few-flowered; involucres cylindrical, the outer bracts lanceolate, unequal and much shorter than the 10 oblong, glabrous, often purplish inner ones. Ray flowers yellow, often drying pinkish. Akenes light yellowish brown, ribbed, constricted into a short, broad beak; pappus white.

Rocky stream banks, scree, and talus of the Absaroka, Beartooth, Uinta, Wasatch, and Wind River Ranges. Circumpolar; Alaska south in western North America to California, Utah, and Colorado.

Middle Rocky Mountain alpine plants are ssp. *nana* and ssp. *ramosa* Babcock. They are distinguished as follows:

1. Plants less than 10 cm tall, the leaves exceeding the inflorescences . . . ssp. *nana*
1. Plants more than 10 cm tall, the leaves much shorter than the inflorescences . . . ssp. *ramosa*

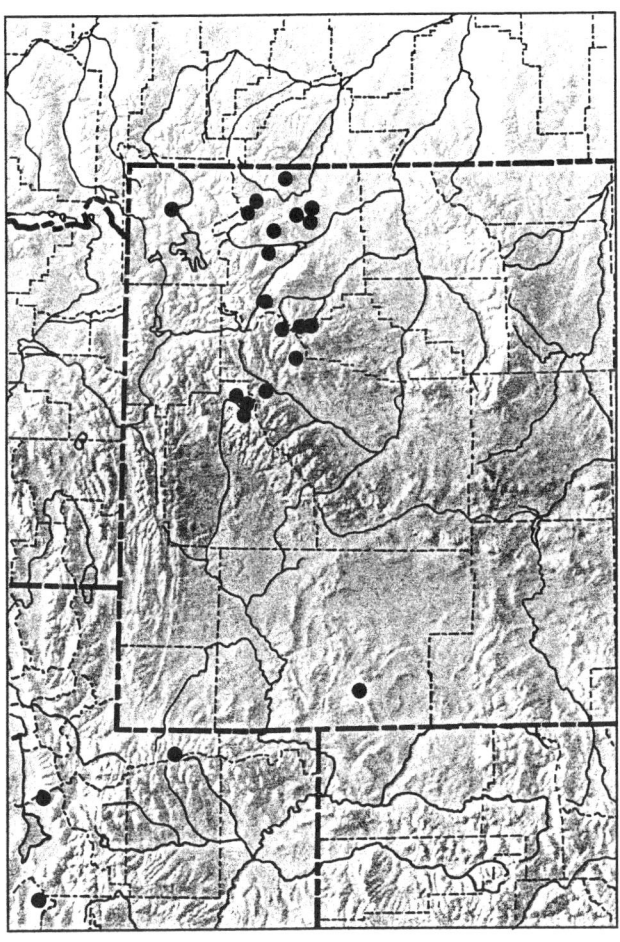

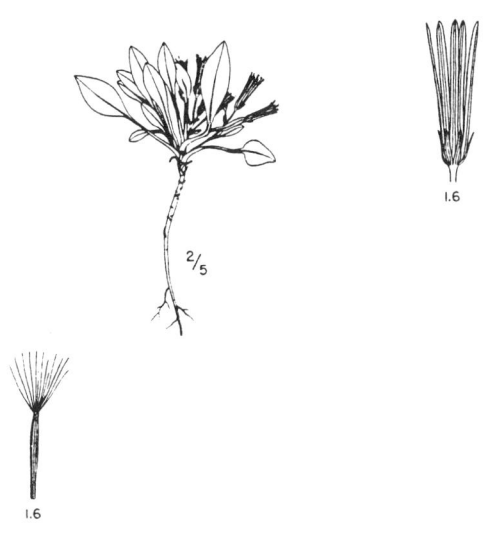

Crepis runcinata (E. James) T. & G.
Fl. N. Am. 2:487, 1843.

Meadow Hawksbeard

Hieracium runcinatum E. James
Crepidium glaucum Nutt.
Crepidium runcinatum (E. James) Nutt.
Crepis alpicola (Rydb.) A. Nels.
Crepis andersonii Gray
Crepis barberi Greenm.
Crepis denticulata Rydb.
Crepis glauca (Nutt.) T. & G.
Crepis glaucella Rydb.
Crepis neomexicana Woot. & Standl.
Crepis obtusissima Greene
Crepis pallens Greene
Crepis perplexans Rydb.
Crepis petiolata Rydb.
Crepis platyphylla Greene
Crepis riparia A. Nels.
Crepis tomentulosa Rydb.
Hieraciodes runcinatum (E. James) Kuntze
Psilochaenia runcinata (E. James) Löve & Löve

Subacaulescent, rosulate, glabrous to hispid, perennial herb mostly more than 15 cm tall, from a taproot or branched caudex; stems 1–3. Basal leaves petiolate, the blades oblanceolate to obovate, entire to runcinate-toothed; cauline leaves lacking or much reduced. Heads 3–30 in a corymbose inflorescence; involucres campanulate, glabrous to hispid, sometimes with glandular hairs, the outer bracts unequal, much shorter than the narrowly lanceolate inner ones. Ray flowers yellow, often fading to white. Akenes brownish, ribbed, contracted into a short, broad beak; pappus of white capillary bristles.

Timberline meadows and krummholz margins of the Absaroka and Beartooth Ranges. Washington east to Minnesota and south to California, Utah, New Mexico, and Nebraska.

The 2 mountain varieties are distinguished as follows:
1. Involucral bracts with glandular hairs
   . . . var. *runcinata*
1. Involucral bracts lacking glandular hairs
   . . . var. *glauca* (Nutt.) B. Boivin

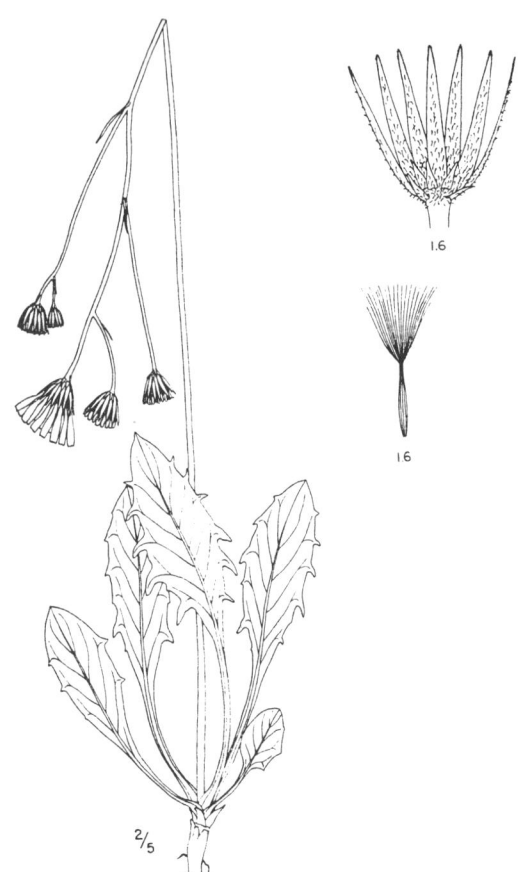

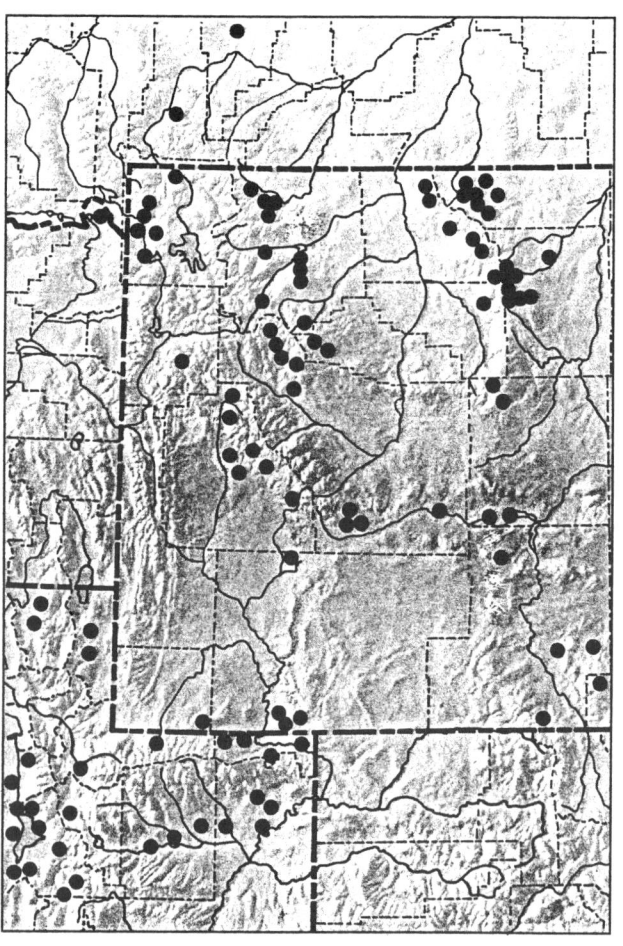

Erigeron L.
Fleabane, Daisy

Annual, biennial, or perennial herbs with alternate or basal, entire or pinnatifid leaves. Heads radiate or rarely discoid, few or solitary, the involucral bracts narrow, varying from herbaceous and subequal to scarcely herbaceous and evidently imbricate, the receptacle usually flat, naked. Ray flowers white, pink, blue, or purple, pistillate. Disk flowers yellow, numerous, perfect. Akenes flattened, 2- to many-nerved, the pappus of capillary, often brittle bristles, with or without a short outer series of bristles or scales.

KEY TO THE SPECIES OF *ERIGERON*

1. Leaves pinnatifid, parted, or lobed
    2. Caudex of several long, slender rhizomelike branches; leaves flabellate, 3-lobed, the lobes broad, ovate — *E. flabellifolius*
    2. Caudex stout, unbranched or with a few stout branches; leaves trifid to pinnately divided, the lobes narrow, linear
        3. Leaves pinnately lobed — *E. pinnatisectus*
        3. Leaves trifid to ternately dissected — *E. compositus*
1. Leaves entire
    4. Rays narrow, 0.5 mm or less wide, erect, nearly the same length as the pappus — *E. acris*
    4. Rays mostly more than 0.6 mm wide, spreading, exceeding the pappus
        5. Plants (10–) 20–40 cm tall; stems bearing several well-developed leaves not much reduced in size — *E. peregrinus*
            6. Rays 0.6–1 mm wide; pappus double, the inner whorl of bristles surrounded by short scales or bristles — *E. speciosus*
            6. Rays 1.5–4 mm wide; pappus a single whorl of bristles — *E. peregrinus*
        5. Plants 4–18 cm tall; stems scapose or with much-reduced cauline leaves
            7. Involucre woolly-villous with shiny, spreading, multicellular hairs
                8. Hairs of involucre purplish black, or at least with dark purple cross-walls
                    9. Rays less than 7 mm long and 1 mm wide — *E. humilis*
                    9. Rays more than 7 mm long and 1 mm wide — *E. melanocephalus*
                8. Hairs of involucre white or clear, the bases or lowermost cross-walls sometimes light purple or reddish
                    10. Hairs of stem stiff, appressed to ascending
                        11. Basal leaves linear; involucre usually lacking glands — *E. ochroleucus*
                        11. Basal leaves oblanceolate to obovate; involucre glandular, at least near the tips of the bracts
                            12. Stems decumbent and often purplish at the base; leaves gradually narrowed to the petiole, the blades usually 3-nerved — *E. eatonii*
                            12. Stems erect to ascending and greenish at the base; leaves somewhat abruptly narrowed to the petiole, the blades 1-nerved — *E. tener*
                    10. Hairs of stem soft, spreading
                        13. Plants caulescent; basal leaves spatulate; stems with some glandular hairs — *E. simplex*
                        13. Plants subscapose to scapose; basal leaves oblanceolate, stems lacking glandular hairs
                            14. Plants densely woolly-villous — *E. lanatus*

14. Plants glabrate, puberulent, or finely pubescent, the stems sometimes loosely villous
    15. Ray flowers 40–65; pappus of 20–30 bristles — *E. goodrichii*
    15. Ray flowers 15–35; pappus of 15–20 bristles — *E. rydbergii*
7. Involucre variously pubescent or glandular, but not woolly-villous with shiny, spreading, multicellular hairs
  16. Plants scapose or nearly so, the cauline leaves lacking or greatly reduced and bractlike
    17. Caudex with slender branches, stout taproot lacking; basal leaves mostly 4 cm long or more, glabrous — *E. garrettii*
    17. Caudex simple or with short, compact branches; from a stout taproot; basal leaves mostly 2 cm long or less, finely pubescent to glabrate
      18. Pappus bristles 12 or fewer; disk corollas 2–3 mm long — *E. radicatus*
      18. Pappus bristles 14 or more; disk corollas 3–5 mm long — *E. rydbergii*
  16. Plants caulescent, the upper leaves often reduced in length
    19. Stems usually purple at base; plants fibrous-rooted, lacking a central underground axis
      20. Involucre glandular, with spreading pubescence; basal leaves oblanceolate — *E. ursinus*
      20. Involucre lacking glands, or nearly so, with appressed pubescence; basal leaves linear — *E. gracilis*
    19. Stems usually green at base; plants lacking fibrous roots, with a branched caudex and taproot or a central underground axis
      21. Hairs of stem spreading; basal leaves pubescent, the hairs sharp-curved — *E. caespitosus*
      21. Hairs of stem appressed, ascending, or lacking; basal leaves glabrous, or pubescent, the hairs straight or only slightly curved
        22. Basal leaves linear; involucre usually lacking glands — *E. ochroleucus*
        22. Basal leaves oblanceolate to obovate; involucre glandular, at least at the tips of the bracts
          23. Leaves glabrous or glabrate, often pubescent at the base, the tips rounded to retuse — *E. leiomerus*
          23. Leaves finely pubescent, the tips acute
            24. Stems decumbent at the base; leaves gradually narrowed to the petiole, the blades usually 3-nerved — *E. eatonii*
            24. Stems ascending to erect; leaves somewhat abruptly narrowed to the petiole, the blades 1-nerved — *E. tener*

Erigeron acris L.
Sp. Pl., 863, 1753.

Bitter Fleabane

Erigeron acre L.
Erigeron asteroides Andrz. ex Bess.
Erigeron debilis (Gray) Rydb.
Erigeron droebachensis O.F. Muell.
Erigeron elatus (Hook.) Greene
Erigeron elongatus Ledeb.
Erigeron jucundus Greene
Erigeron kamtschaticus DC.
Erigeron lapiluteus A. Nels.
Erigeron nivalis Nutt.
Erigeron podolicum Bess.
Erigeron politus Fries
Erigeron yellowstonensis A. Nels.
Trimorpha acris (L.) Nesom

Biennial or short-lived perennial herb from a short, simple to branched caudex; stems 1 to a few, erect to decumbent, simple, glabrous to hirsute and somewhat glandular. Basal leaves oblanceolate, acute, the margins entire to remotely serrulate; cauline leaves oblanceolate to linear, reduced upward. Heads solitary to numerous in an upward-arching, corymbose raceme; involucre glandular, hirsute, or both, the bracts acuminate to attenuate, subequal, green to purplish. Ray flowers pistillate, numerous, pink or white, the inner series eligulate and narrowly tubular; disk flowers perfect, yellow. Akenes 2-nerved, pubescent; pappus of slender, barbellate, white to reddish bristles exceeding both the disk and the eligulate ray flowers.

Rocky meadows at timberline in the Absaroka, Beartooth, Medicine Bow, Teton, Uinta, and Wind River Ranges. Circumboreal; south in western North America to California, Utah, and Colorado.

Timberline specimens of short stature and few to solitary heads are var. *debilis* Gray.

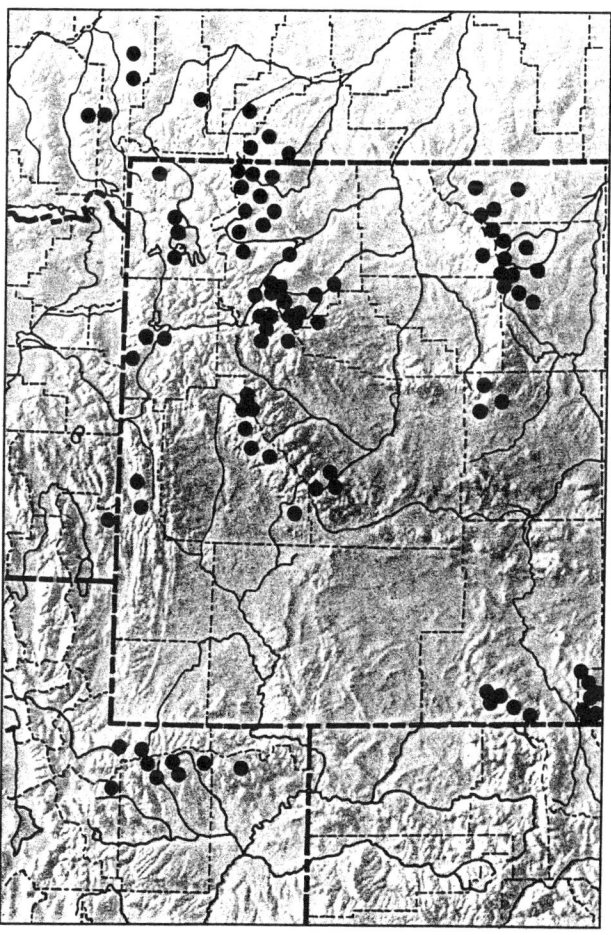

Erigeron caespitosus Nutt.
Trans. Am. Phil. Soc., II, 7:307, 1840.

Tufted Daisy

Diplopappus canescens Hook.
Diplopappus grandiflorus Hook.
Erigeron canescens (Hook.) T. & G.
Erigeron subcanescens Rydb.

Tufted, perennial herb from a stout, simple to branched caudex and taproot; stems 1 to several, often somewhat decumbent at the base, canescent or hirsute with short, spreading hairs. Basal leaves petiolate, the blades oblanceolate to spatulate, 3-nerved, entire, the tips rounded or obtuse; cauline leaves oblanceolate, oblong, or linear, becoming sessile and reduced upward. Heads solitary and terminal to few and axillary; involucre canescent and glandular, the bracts imbricate, thickened on the back. Ray flowers purplish, bluish, pink, or white; disk flowers yellow. Akenes 2-nerved, pubescent; pappus double, the outer series of scales.

Fellfields and timberline meadows of the Absaroka, Beartooth, Uinta, and Wasatch Ranges. Alaska and Yukon Territory south to Washington, Idaho, Arizona, New Mexico, and Nebraska.

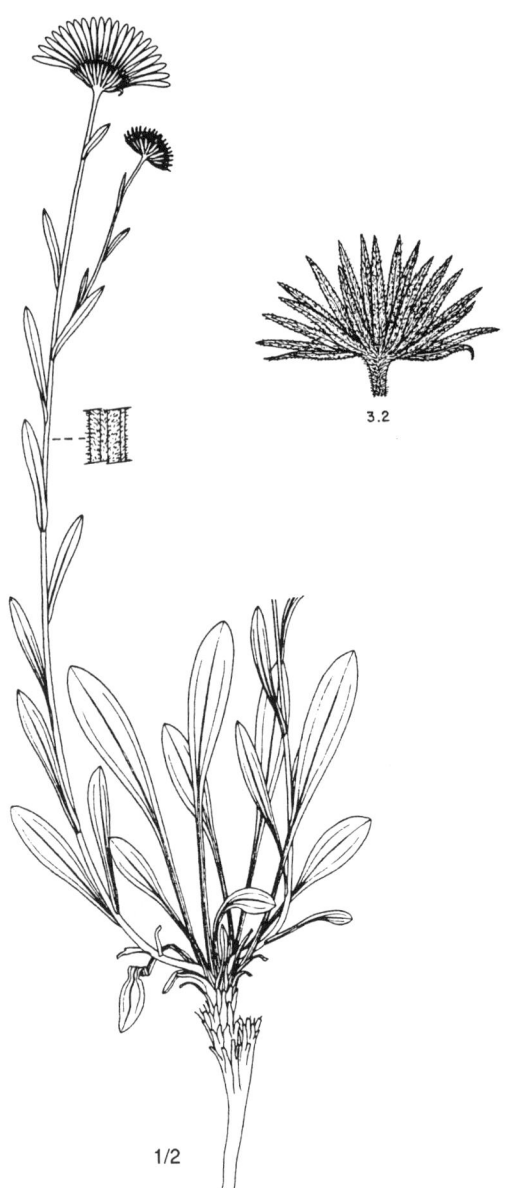
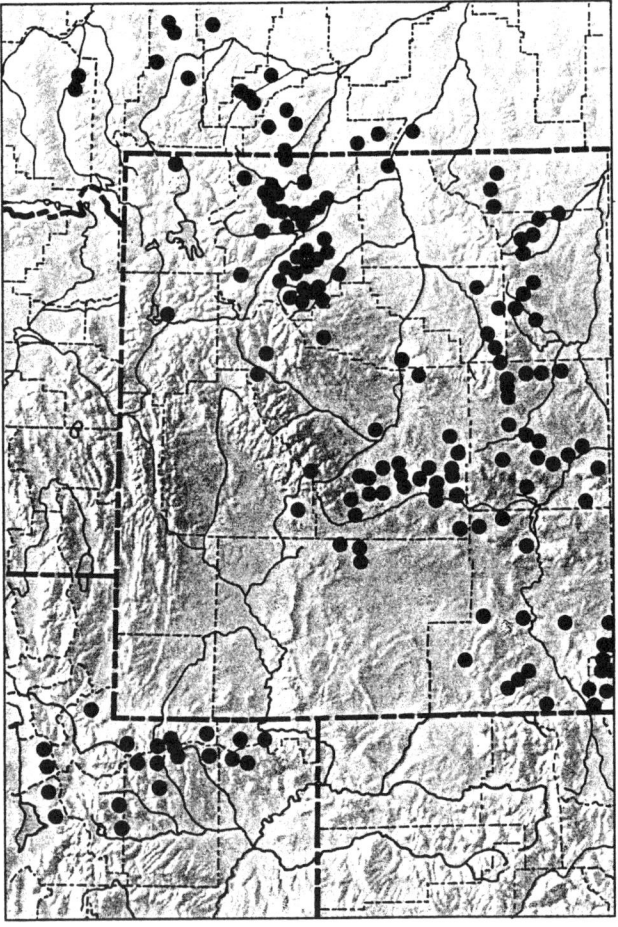

Erigeron compositus Pursh
Fl. Am. Sept. 2:535, 1814.

Fernleaf Daisy

Cineraria lewisii Rich.
Erigeron gormani Greene
Erigeron multifidus Rydb.
Erigeron pedatus Nutt.
Erigeron trifidus Hook.

Densely caespitose, subscapose, perennial herb from a simple to branched caudex and taproot; stems 1 to several, with a few reduced leaves. Basal leaves glandular and pubescent with spreading hairs, ternately dissected or trifid, the ultimate divisions linear to oblong; cauline leaves lobed to linear and entire, reduced upward. Heads solitary, terminal; involucres glandular and hirsute, the bracts subequal, purplish at the tips. Ray flowers lavender, bluish, pink, or white, sometimes reduced or wanting; disk flowers yellow. Akenes 2-nerved, pubescent; pappus single, of slender bristles.

Fellfields, ledges, talus, and scree in nearly all alpine ranges of the Middle Rocky Mountains. Not yet known from above timberline in the Medicine Bow Range. Alaska to Greenland; south in western North America to California, Arizona, Colorado, and South Dakota.

Two varieties occur in the alpine zone: var. *discoideus* Gray, with trifid leaves; and var. *glabratus* Macoun, with leaves 2–3 times ternate.

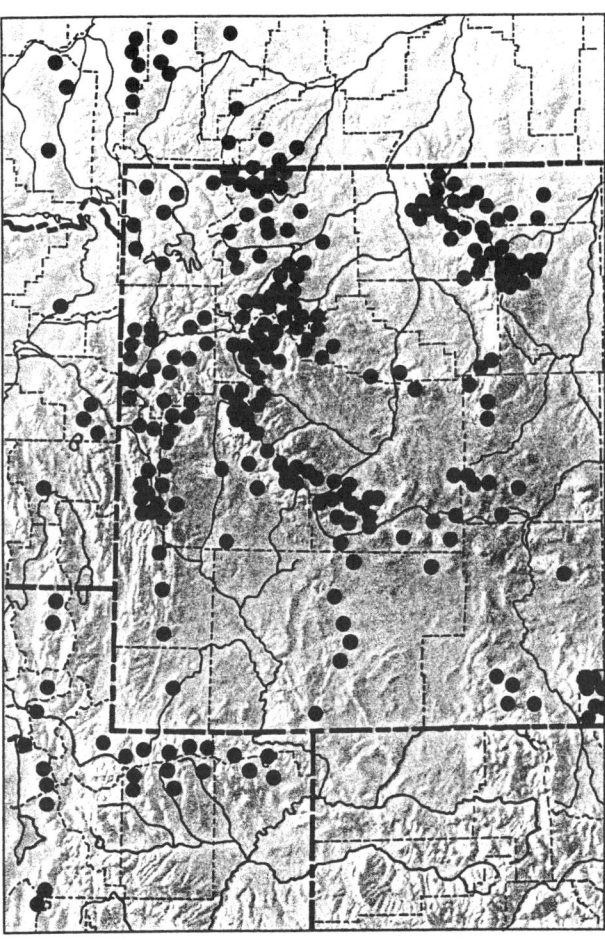

Erigeron eatonii Gray
Proc. Am. Acad. 16:91, 1880.

Eaton Daisy

Erigeron microlonchus Greene
Erigeron pacificus T.J. Howell
Erigeron plantagineus Greene
Erigeron robertianus Greene
Erigeron sonnei Greene

Perennial herb from a short, simple or branched caudex and taproot; stems strigose to strigose-hirsute with appressed to ascending hairs, decumbent, usually purplish at the base. Basal leaves petiolate, the blades oblanceolate or sometimes linear, 3-nerved, entire, the tips acute; cauline leaves narrowly oblanceolate to linear, reduced upward. Heads usually solitary and terminal, sometimes 2–4; involucres glandular and hirsute, the bracts imbricate, often purplish at the tips. Ray flowers mostly white, occasionally drying lavender or pink; disk flowers yellowish. Akenes 2–3-nerved, pubescent; pappus single, of slender bristles, or double with a few short, outer scales.

Fellfields near timberline in the Absaroka, Big Horn, Uinta, Wasatch, and Wind River Ranges. Washington and Oregon south to California, Utah, Arizona, and Colorado.

Our plants with imbricate, glandular, involucral bracts are the typical var. *eatonii*.

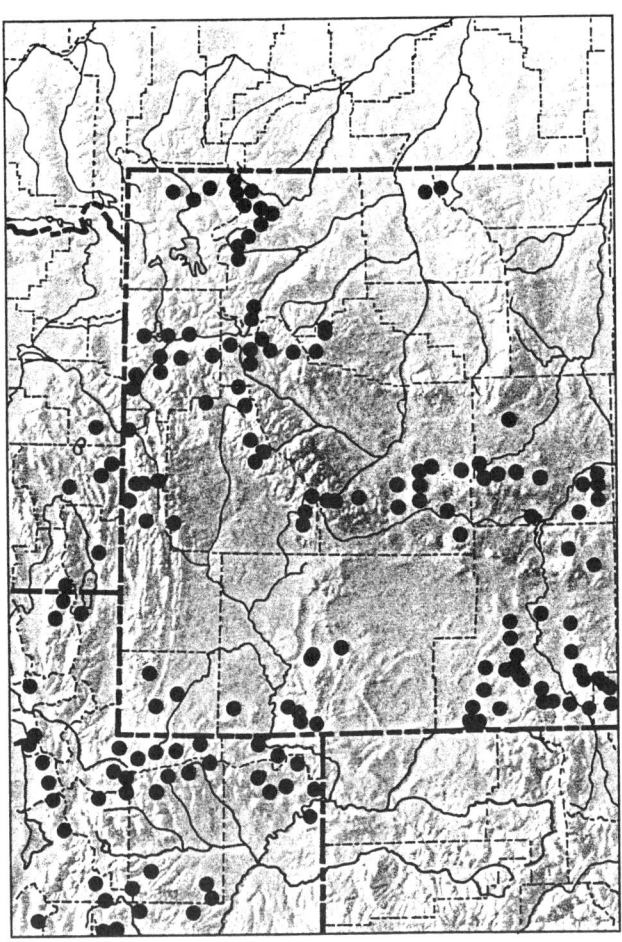

Erigeron flabellifolius Rydb.
Bull. Torr. Club 26:545, 1899.

Fanleaf Daisy

Low, caespitose, glandular perennial, less than 10 cm tall, from a slender taproot or branched rhizomatous caudex; stems solitary, simple. Basal leaves cuneate-flabelliform, ternate- to pentate-lobed, the lobes coarsely toothed; cauline leaves reduced upward, becoming entire and bractlike. Heads solitary, terminal, the involucre glandular. Ray flowers pink or white; disk flowers yellowish. Akenes with a pappus of simple bristles.

Scree and talus of the Absaroka and Beartooth Ranges. Southwestern Montana and northwestern Wyoming.

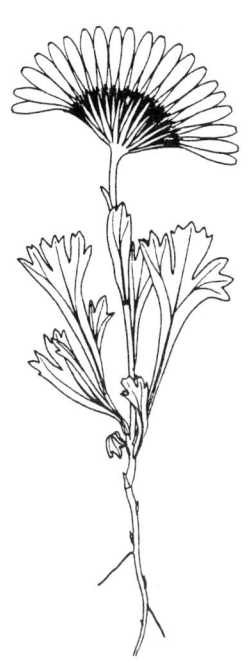

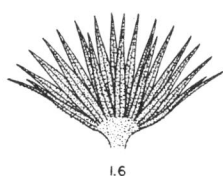

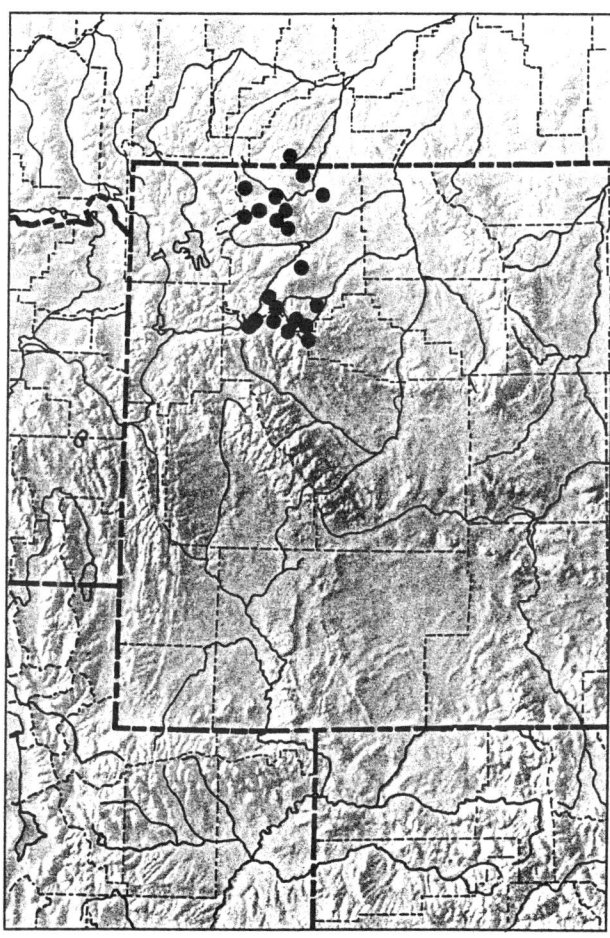

Erigeron garrettii A. Nels.
Man. Bot. Centr. Rocky Mtns., 526, 1909.

Garrett Daisy

Erigeron controversus Greene

Perennial, scapose to subscapose herb from a branched caudex; stems 1 to several, sparsely pubescent with appressed, stiff hairs. Basal leaves petiolate, the blades oblanceolate to spatulate, glabrous, obtuse or subacute; cauline leaves reduced, linear, or lacking. Heads solitary, terminal; involucres sparsely glandular, the bracts imbricate, linear-subulate, finely strigose. Ray flowers white to pink; disk flowers yellowish. Akenes 2-nerved, pubescent; pappus double, the outer series minute or lacking.

Ledges, cliffs, and rocky places near timberline in the Wasatch Range. Endemic to northern Utah.

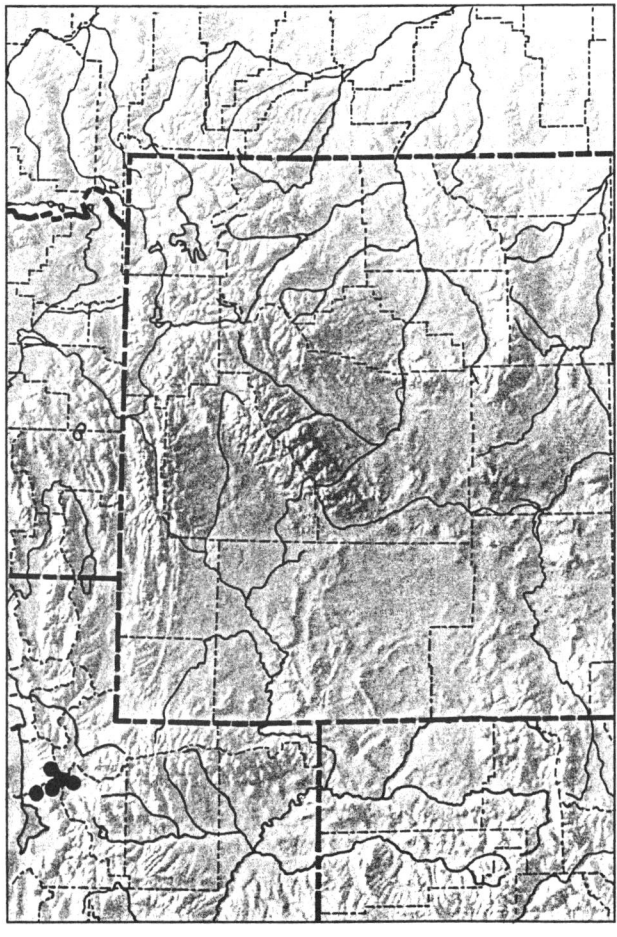

Erigeron goodrichii S.L. Welsh
Great Basin Nat. 43:366, 1983.

Goodrich Daisy

Perennial herbs from a simple to branched caudex and stout taproot; stems 1 to several, erect to decumbent. Basal leaves oblanceolate, finely pubescent, the tips obtuse; cauline leaves reduced upward, becoming linear and bractlike. Heads solitary, terminal; involucre villous to pilose with multicellular hairs, the bracts imbricate, attenuate, purplish and glandular near the tips. Ray flowers purple, pink, or white; disk flowers yellowish. Akenes 2-nerved, pubescent; pappus single, of slender bristles.

Fellfields and coarse talus of the Uinta Range. Endemic to northern Utah.

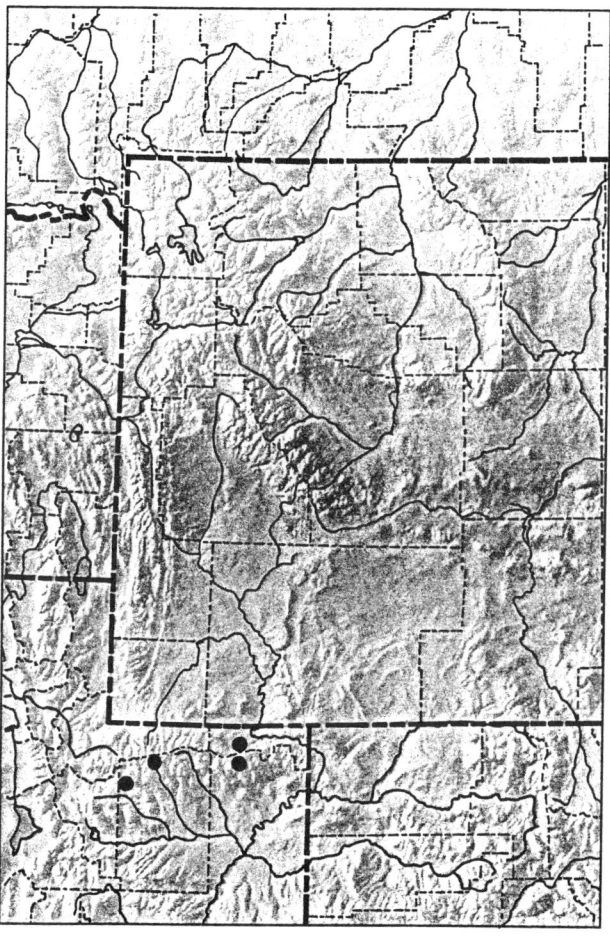

Erigeron gracilis Rydb.
Mem. N.Y. Bot. Gard. 1:404, 1900.

Slender Daisy

Erect, tufted, pubescent, perennial herb from a slender, branched caudex or rhizome with fibrous roots; stems 1 to several, decumbent and usually purplish at the base. Basal leaves narrowly oblanceolate to linear, entire; cauline leaves reduced upward, entire, lanceolate to linear. Heads solitary, terminal; involucre appressed-pubescent, lacking glands, the bracts oblong to lanceolate, acute. Ray flowers purple, pink, or bluish; disk flowers yellowish. Akenes 2-nerved, pubescent; pappus apparently double, the inner series of slender bristles, the outer of a few short scales.

Timberline meadows of the Absaroka and Beartooth Ranges. Southwestern Montana, western Wyoming, and eastern Idaho.

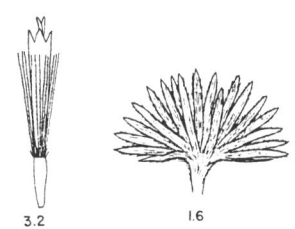
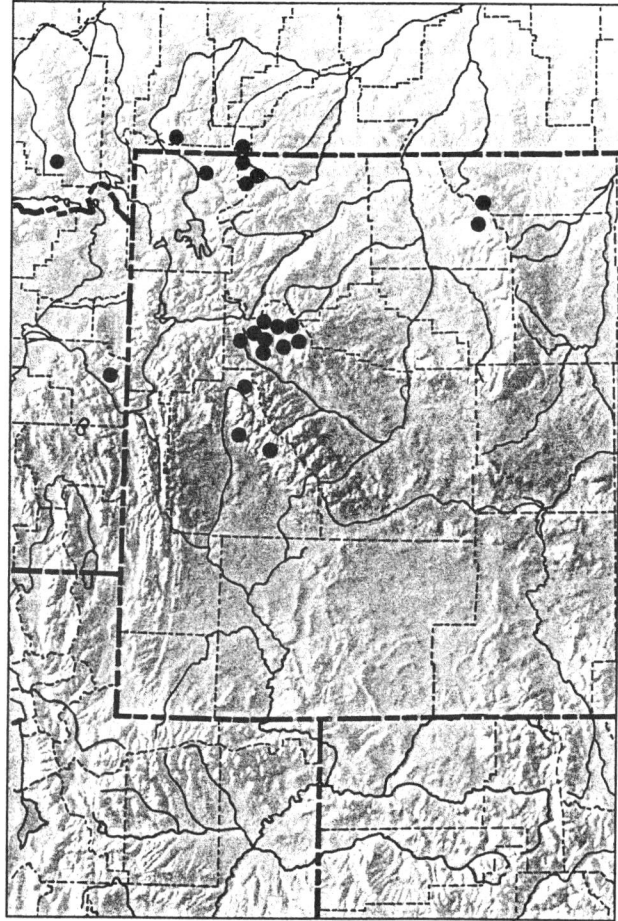

Erigeron humilis Grah.
Edinb. New Phil. J. 1828:175, 1828.

Low Daisy

Erigeron unalaschkensis (DC.) Vierh.

Slender, erect, tufted, villous to glabrate perennial from a taproot and a few fibrous roots; stems erect to somewhat decumbent, villous, the hairs of the upper portion commonly with purple cross-walls. Basal leaves petiolate, the blades oblanceolate to spatulate, entire, villous, becoming glabrate with age, the tips rounded; cauline leaves oblanceolate, lanceolate, or linear, acute, reduced upward. Heads solitary, terminal; involucre woolly-villous, the hairs blackish purple, or at least with dark purple cross-walls; bracts attenuate, subequal, dark purplish, at least on the tips. Ray flowers light purple to white; disk flowers yellowish. Akenes finely pubescent; pappus mostly single, of slender white or tawny bristles.

Fellfields, meadows, and moist rocky places of the Absaroka, Beartooth, Big Horn, Medicine Bow, and Wind River Ranges. Siberia; Alaska to Greenland; south in western North America to British Columbia, Montana, and Wyoming.

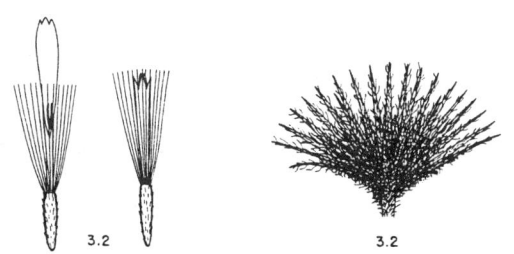

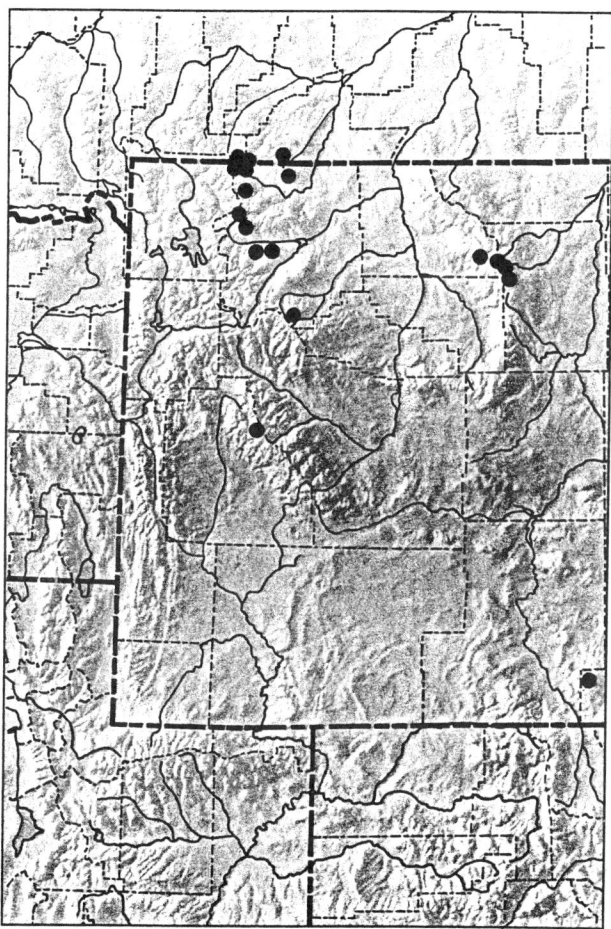

Erigeron lanatus Hook.
Fl. Bor.-Am. 2:17, 1834.

Woolly Daisy

Low, scapose, perennial herb, less than 6 cm tall, from a slender, branched caudex; stems solitary to few, woolly-villous, the long, tangled hairs becoming denser below the inflorescence. Leaves oblanceolate, acute, obtuse, or even 3-toothed at the apex, woolly-villous to pilose. Heads solitary, terminal; involucre woolly-villous with multicellular hairs, the cross-walls clear or sometimes pale purplish; bracts attenuate, greenish to purplish, the tips dark purple. Ray flowers white, bluish, or pinkish; disk flowers bright yellow. Pappus bristles mostly single, sometimes with short outer setae, the inner bristles white, slender, usually somewhat twisted.

Limestone talus on Gypsum Mountain of the northern Wind River Range. Alberta and British Columbia south to Wyoming and Colorado.

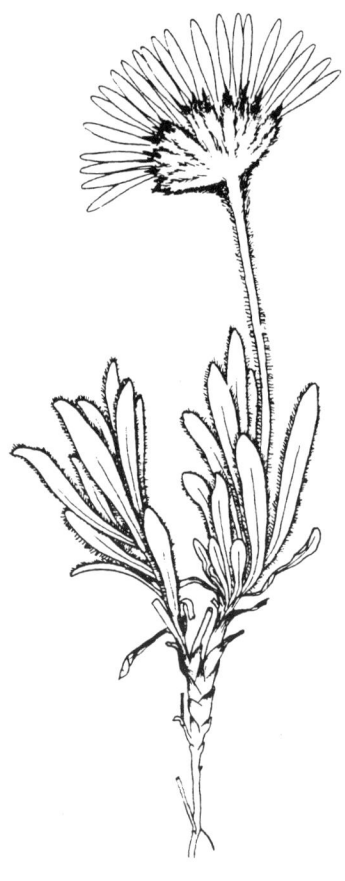

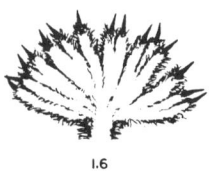

1.6

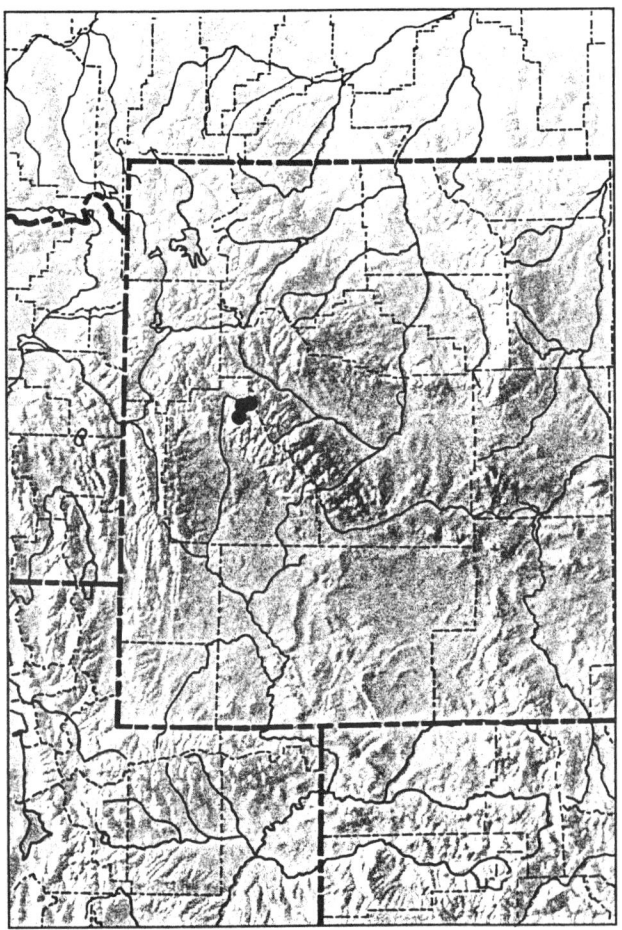

Erigeron leiomerus Gray
Syn. Fl. N. Am. 1:211, 1884.

Smooth Daisy

Erigeron minusculus Greene
Erigeron spathulifolius Rydb.

Low, subscapose, perennial herb less than 15 cm tall, from a slender, branched caudex or occasionally a taproot; stems spreading, glabrous or strigose. Basal leaves petiolate, glabrous to glabrate, the blades oblanceolate to spatulate, entire, the tips rounded to retuse; cauline leaves narrowly oblanceolate to linear, reduced upward. Heads solitary, terminal; involucre glandular, the bracts imbricate, purplish at least at the tips, acute. Ray flowers purple, bluish, or white; disk flowers yellow. Akenes 2-nerved, short-pubescent; pappus double, the inner series of slender bristles, the outer series of short, inconspicuous scales.

Fellfields, talus, and ledges of all western ranges. Not yet known from above timberline in the Beartooth, Big Horn, Hoback, and Medicine Bow Ranges. Wyoming and Idaho south to Utah, Colorado, and New Mexico.

4/5

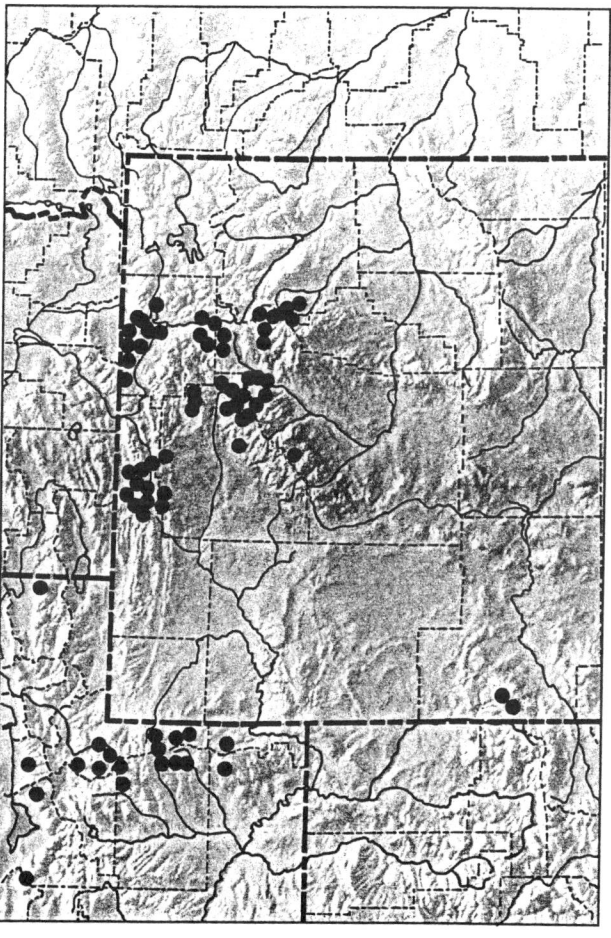

Erigeron melanocephalus (A. Nels.) A. Nels.
Bull. Torr. Club 26:246, 1899.

Blackhead Daisy

Somewhat tufted perennial herb from a simple to branched caudex; stems erect, less than 15 cm tall, villous with purple, multicellular hairs, at least below the head. Basal leaves petiolate, the blades oblanceolate to spatulate, villous to glabrate, the margins entire to ciliate, tips rounded to retuse; cauline leaves narrowly oblanceolate to linear, reduced upward. Heads solitary, terminal; involucres densely villous with multicellular hairs, the hairs with blackish or purple cross-walls, the bracts equal or subequal, attenuate, green or purplish, at least purple-tipped. Ray flowers white or pink; disk flowers yellow. Akenes 2-nerved, pubescent; pappus single, of slender, somewhat unequal bristles.

Fellfields and alp slopes of the Medicine Bow Mountains. Wyoming, Utah, Colorado, and New Mexico.

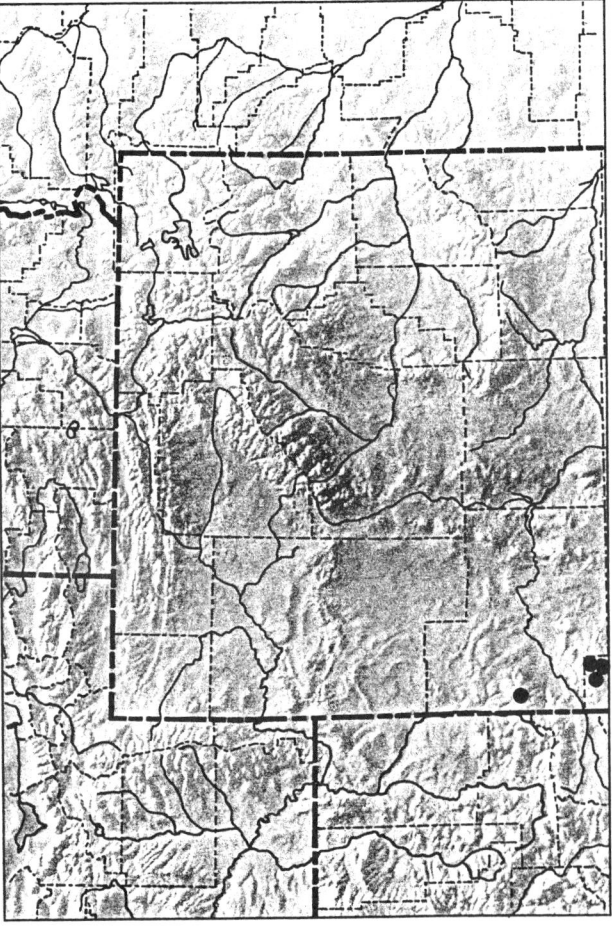

Erigeron ochroleucus Nutt.
Trans. Am. Phil. Soc., II, 7:311, 1840.

Pale Daisy

Erigeron laetevirens Rydb.
Erigeron macounii Greene
Erigeron montanensis Rydb.
Erigeron parryi Canby & Rose
Erigeron scribneri Canby ex Rydb.
Erigeron tweedyanus Canby & Rose
Wyomingia tweedyana (Canby & Rose) A. Nels.

Tufted, perennial herb from a simple or short-branched caudex and taproot; stems 1 to a few, strigose, or sometimes with loose, spreading hairs. Basal leaves linear, sometimes narrowly oblanceolate, entire, strigose to glabrate, the bases membranous and often purplish; cauline leaves linear, entire, much reduced upward. Heads mostly solitary, terminal, occasionally more than 1; involucre hirsute-villous, the hairs sometimes with the lower cross-walls reddish or purplish, the bracts subequal, often with purple tips. Ray flowers purple, blue, or white; disk flowers yellow. Akenes 2-nerved, pubescent; pappus double, the inner series of bristles, the outer series of a few short scales or thick bristles.

Fellfields, ledges, and scree of the Beartooth, Big Horn, Wind River, and Wyoming Ranges. British Columbia, Alberta, and Saskatchewan south to Wyoming and northwestern Nebraska.

Small, slender, subscapose alpine plants are var. *scribneri* (Canby) Cronq.

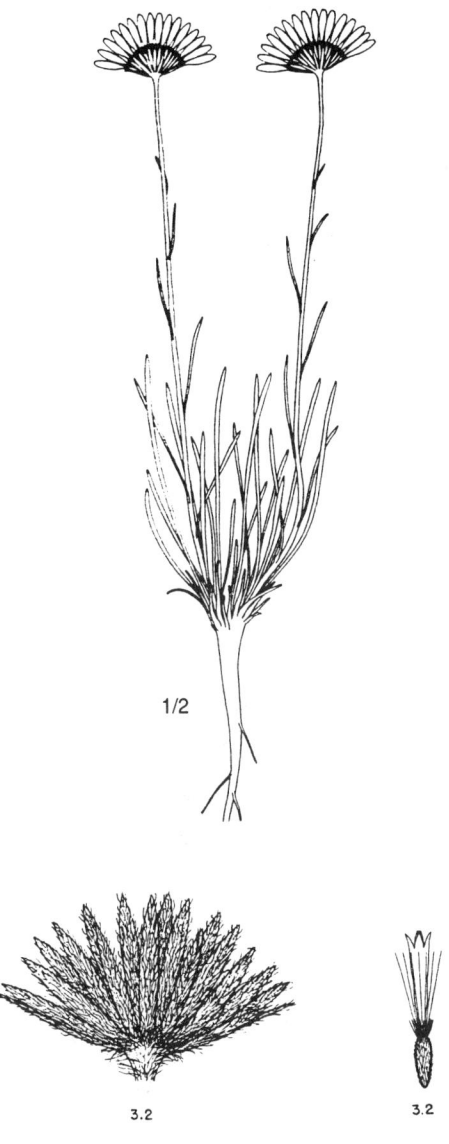

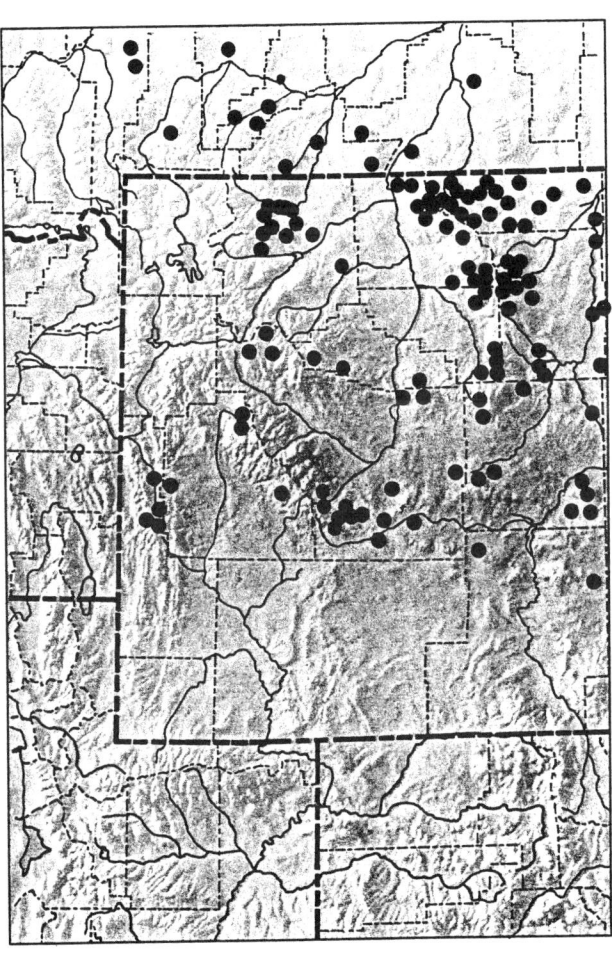

Erigeron peregrinus (Pursh) Greene
Pittonia 3:166, 1897.

Peregrine Daisy

Aster peregrinus Pursh
Aster glacialis Nutt.
Aster salsuginosus Rich.
Erigeron acutatus Greene
Erigeron angustifolius (Gray) Rydb.
Erigeron callianthemus Greene
Erigeron ciliolatus Greene
Erigeron glacialis (Nutt.) A. Nels.
Erigeron hesperocallis Greene
Erigeron loratus Greene
Erigeron membranaceus Greene
Erigeron obtusatus Greene
Erigeron regalis Greene
Erigeron salsuginosus (Rich.) Gray
Erigeron suksdorfii Greene
Erigeron thompsoni Blake
Erigeron unalaschkensis Less.

Erect, caulescent to subscapose, perennial herb from a short caudex or rhizome and fibrous roots; stems solitary, simple, glabrous to villous, at least below the head. Basal leaves petiolate, oblanceolate to obovate, glabrous or sometimes villous, the tips acute or obtuse; cauline leaves reduced upward, becoming sessile and ovate to lanceolate, somewhat clasping, glabrous or ciliate on the margins. Heads mostly solitary, sometimes 2 or more; involucre glandular, the bracts subequal, lanceolate to linear, attenuate, the tips sometimes reflexed. Ray flowers rose-purple, purple, bluish purple, or rarely white; disk flowers deep yellow. Akenes 4–7-nerved, usually 5-nerved and sparsely pubescent; pappus single, of long, slender bristles, or sometimes double with a few short outer bristles.

Meadows, willow thickets, alp slopes, and timberline ecotones in nearly all alpine ranges. Not yet known from above timberline in the Gros Ventre Range. Alaska and Yukon Territory south to California, Idaho, Utah, and New Mexico.

Plants with glandular involucral bracts, entire leaves, and rose-purple ray flowers are ssp. *callianthemus* (Greene) Cronq. Two varieties occur in the alpine zone:

1. Plants subscapose, usually less than 20 cm tall, the cauline leaves much reduced upward
   . . . var. *scaposus* (T. & G.) Cronq.
1. Plants caulescent, usually more than 30 cm tall, the cauline leaves not much reduced
   . . . var. *callianthemus*

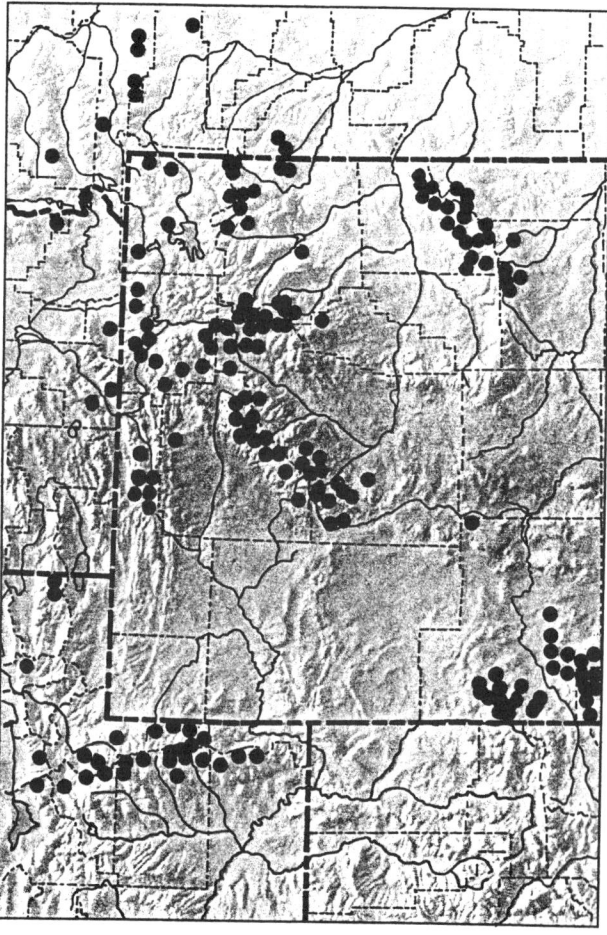

1/3

Erigeron pinnatisectus (Gray) A. Nels.
Bull. Torr. Club 26:246, 1899.

Pinnateleaf Daisy

Subscapose, perennial herb from a stout, simple to branched caudex; stems 1 to a few, glandular, and usually sparsely pubescent with spreading hairs. Basal leaves petiolate, pinnatifid, the ultimate segments linear, glabrous, the petioles usually pubescent; cauline leaves linear, entire, few and reduced upward. Heads solitary, terminal; involucre glandular, hirsute, the bracts subequal, often purplish. Ray flowers purple or blue; disk flowers yellow. Akenes 2-nerved, pubescent; pappus single of slender, subequal bristles.

Fellfields of the Medicine Bow Mountains. Wyoming, Colorado, and New Mexico.

Erigeron radicatus Hook.
Fl. Bor.-Am. 2:17, 1834.

Bigroot Daisy

Low, scapose to subscapose perennial herb, less than 6 cm tall, from a short-branched, compact caudex and stout taproot; stems 1 to several, finely hirsute, at least above. Basal leaves oblanceolate, glabrous to puberulent, the margins entire, ciliate; cauline leaves lacking or, if present, linear, finely pubescent, and reduced upward. Heads solitary, terminal; involucre viscid and short-villous, the bracts greenish, lanceolate to linear. Ray flowers white; disk flowers yellowish. Pappus double, the inner series of slender bristles, the outer series of a few short scales or bristles.

Rocky places in the Beartooth, Wind River, and Wyoming Ranges. Alberta and Saskatchewan south to northwestern Wyoming.

1.6

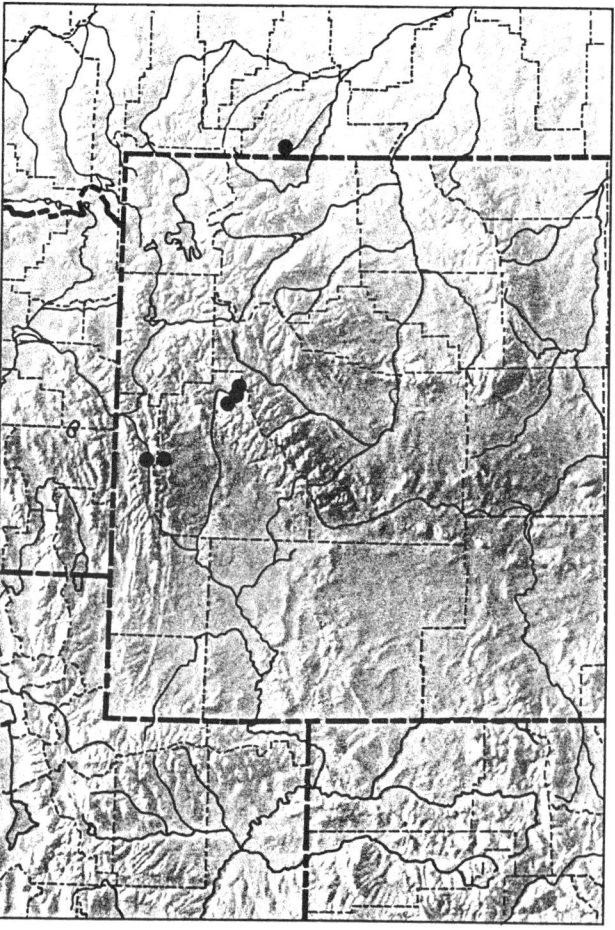

Erigeron rydbergii Cronq.
Brittonia 6:191, 1947.

Rydberg Daisy

Low, scapose to subscapose, perennial herb, less than 8 cm tall, from a simple to branched caudex; stems 1 to several, villous. Basal leaves oblanceolate, glabrous to puberulent, the margins entire, often ciliate; cauline leaves linear, reduced upward. Heads solitary, terminal; involucre somewhat viscid, puberulent, the bracts linear to oblong. Ray flowers lavender or white; disk flowers yellowish. Pappus double, the outer series of slender bristles, the inner series of a few short scales.

Rocky places in the northern alpine ranges. Not yet known from above timberline in the Teton Range. Montana, Idaho, and Wyoming.

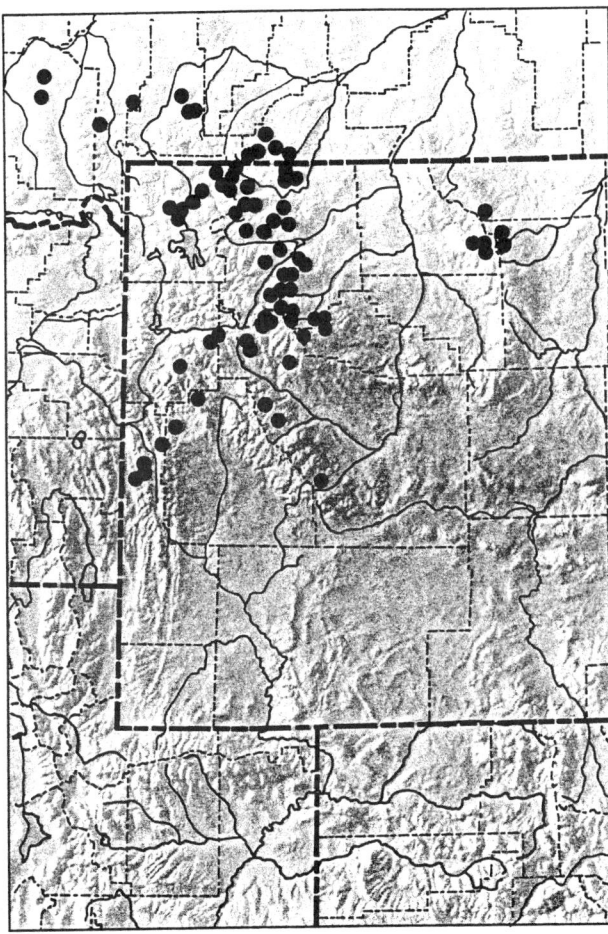

Erigeron simplex Greene
Fl. Franc., 387, 1897.

Oneflower Daisy

Erigeron leucotrichus Rydb.

Slender, somewhat tufted, perennial herb from a simple or branched caudex; stems 1 to a few, villous, with some glandular hairs. Basal leaves petiolate, oblanceolate to spatulate, glabrous to villous or somewhat hirsute, entire, the tips mostly obtuse; cauline leaves sessile, oblanceolate, oblong, or linear, reduced upward. Heads solitary, terminal; involucre woolly-villous with multicellular hairs having clear cross-walls, sometimes with reddish or purplish cross-walls near the base, the bracts sometimes with reflexed tips. Ray flowers lavender, bluish, pink, or white; disk flowers yellowish. Akenes 2-nerved, pubescent; pappus double, the inner series of long, slender bristles, the outer series of shorter scales or bristles.

Fellfields, alp slopes, and ledges in most of the alpine ranges. Not yet known from above timberline in the Hoback, Salt River, and Wasatch Ranges. Oregon south and east to Nevada, Montana, Arizona, and New Mexico.

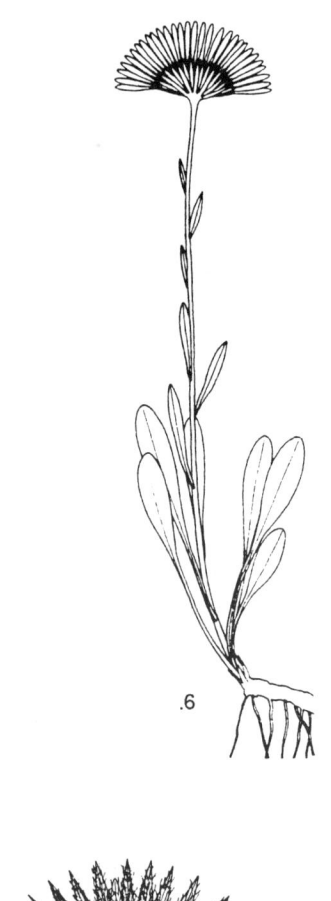
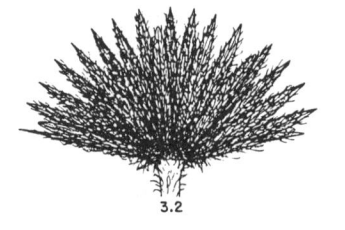 
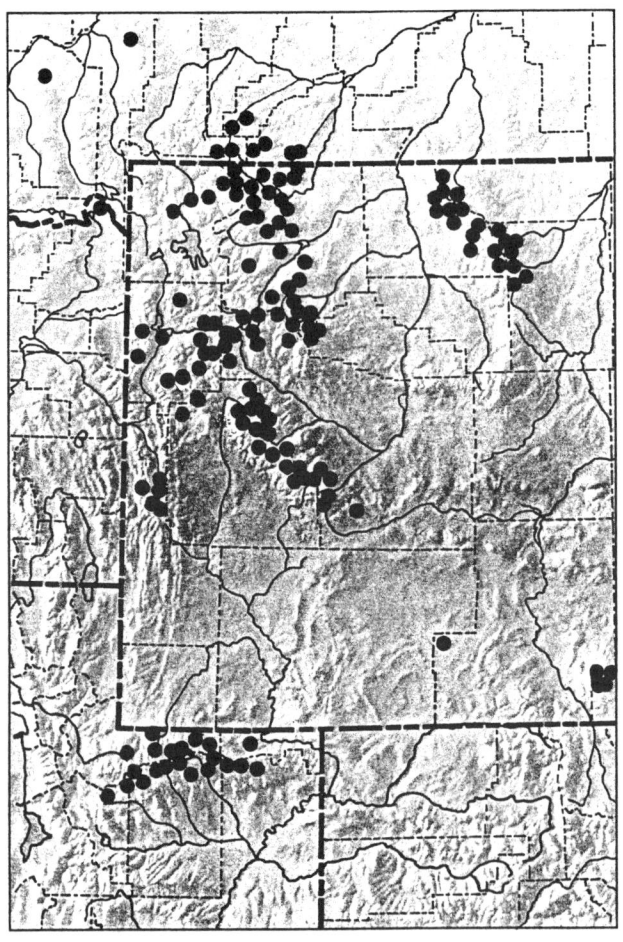

Erigeron speciosus (Lindl.) DC.
Prodromus 5:284, 1836.

Oregon Daisy

Stenactis speciosa Lindl.
Erigeron eucephaloides Greene
Erigeron grandiflorus Nutt.
Erigeron leiophyllus Greene
Erigeron macranthus Nutt.
Erigeron subtrinervis Rydb.
Erigeron uintahensis Cronq.

Erect, caulescent, perennial herb from a woody, usually branched caudex and fibrous roots; stems solitary to several, simple or branched, glabrate to glandular, at least above. Basal leaves petiolate, oblanceolate to obovate, early deciduous; cauline leaves sessile, lanceolate to oval, entire, ciliate, glabrous, glandular or spreading-pubescent, petiolate below, becoming sessile and reduced upward. Heads 1 to many; involucre pubescent to glandular, the bracts subequal, lanceolate, attenuate, the tips often purplish. Ray flowers purple, bluish purple, pinkish purple, or rarely white; disk flowers yellow. Akenes 2–4-nerved, mostly 2-nerved, pubescent; pappus double, the inner whorl of bristles, the outer of very small scales or bristles.

Meadows, timberline margins, and edges of krummholz in the Beartooth, Uinta, Wasatch, and Wind River Ranges. British Columbia and Alberta south to California, Arizona, New Mexico, and Nebraska.

Three varieties occur at and above timberline:
1. Upper leaves glandular
    . . . var. *uintahensis* (Cronq.) S.L. Welsh
1. Upper leaves glabrous
   2. Involucral bracts lacking long-spreading hairs; upper leaves ovate
    . . . var. *macranthus* (Nutt.) Cronq.
   2. Involucral bracts commonly with long-spreading hairs; upper leaves lanceolate
    . . . var. *speciosus*

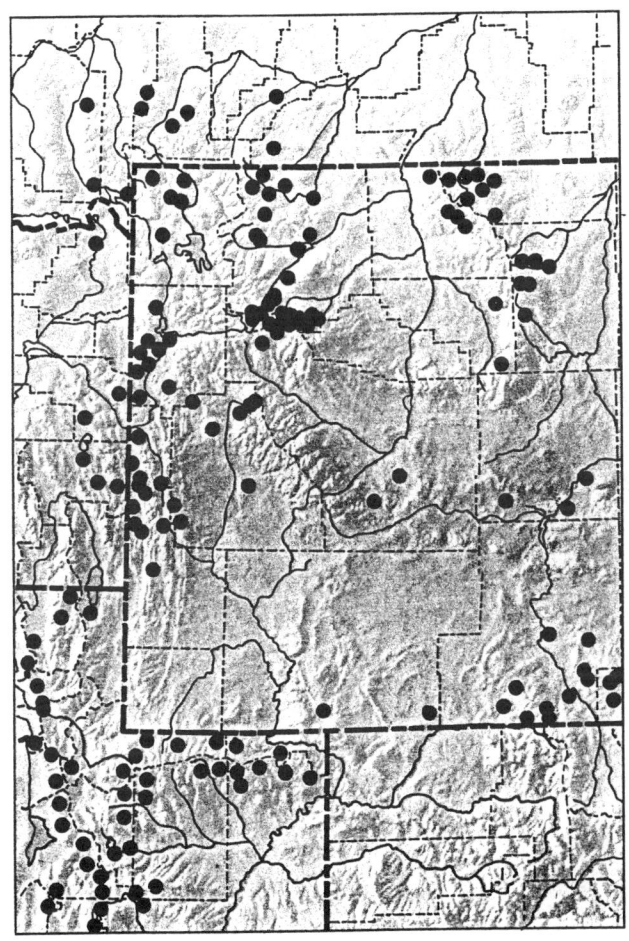

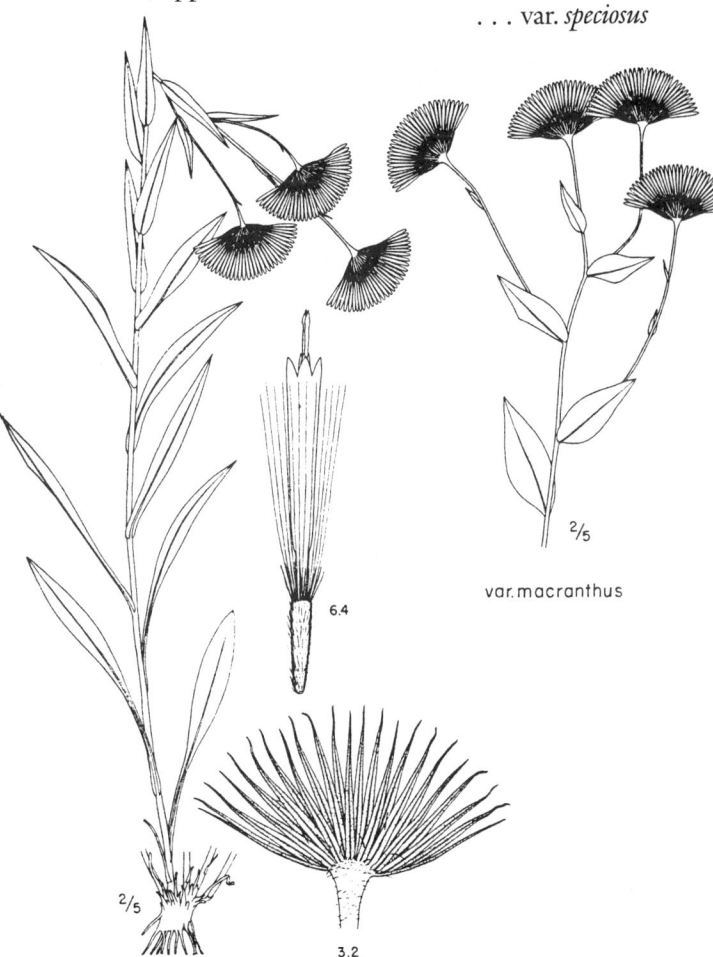

var. macranthus

var. speciosus

Erigeron tener (Gray) Gray
Proc. Am. Acad. 16:91, 1880.

Thin Daisy

Tufted, perennial herb from a simple to branched caudex and taproot; stems 1 to a few, slender, erect to ascending. Basal leaves petiolate, the blades oblanceolate to obovate, strigose, entire, the tips acute or nearly so; cauline leaves sessile, oblong to linear, reduced upward. Heads solitary, terminal; involucre glandular, often with spreading, multicellular hairs, the bracts imbricate, oblong to linear, greenish or purplish. Ray flowers mostly purple or bluish, sometimes white; disk flowers yellowish. Akenes 2-nerved, sparsely pubescent; pappus double, the inner series of long bristles, the outer series lacking or of short bristles.

Known from above timberline only in the Salt River Range. Oregon to California, Nevada, Utah, and Wyoming.

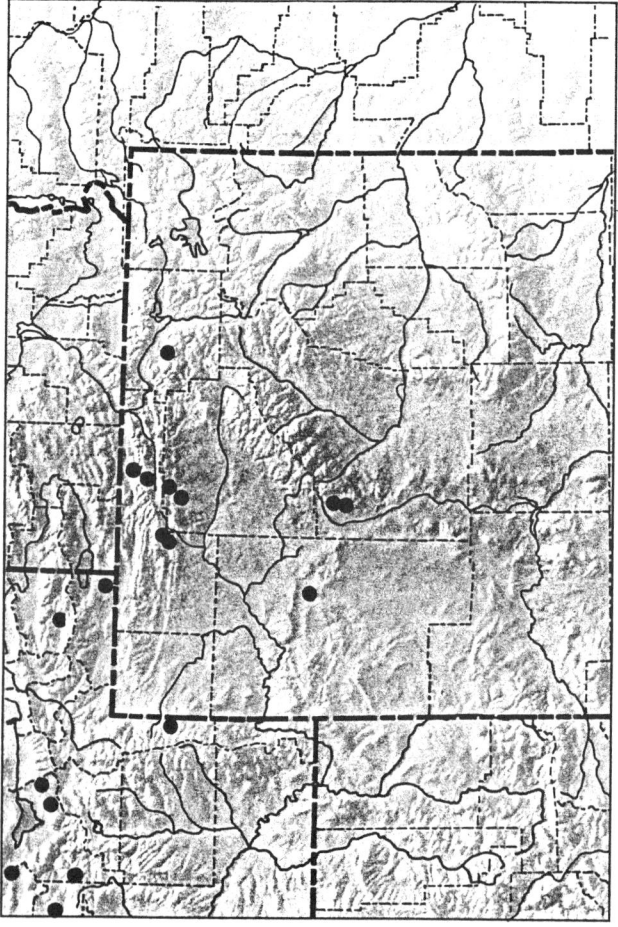

Erigeron ursinus D.C. Eat. in S. Wats.
Bot. King Exped., 148, 1871.

Bear River Daisy

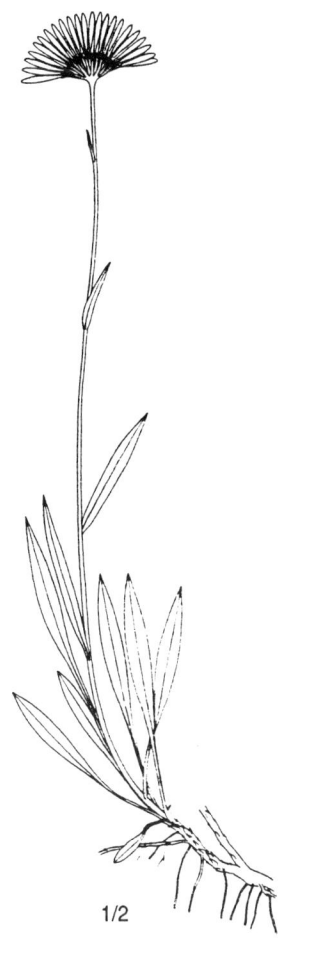

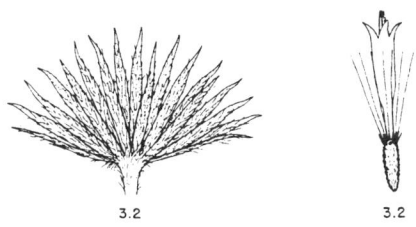

Slender, perennial herb from a branched rhizome and densely clustered fibrous roots; stems ascending-decumbent, simple, usually purplish at the base. Basal leaves petiolate, the blades oblanceolate to elliptic, entire, glabrous, acute, the margins ciliate; cauline leaves oblong to lanceolate, sessile, reduced upward. Heads solitary, terminal; involucre glandular and pubescent with spreading hairs, the bracts subequal, greenish or purplish, often reflexed at the tips. Ray flowers purple or bluish; disk flowers yellow. Akenes 2-nerved, pubescent; pappus double, the inner series of long, slender bristles, the outer series of short bristles or scales.

Fellfields and timberline meadows of nearly all alpine ranges in the Middle Rocky Mountains. Not yet known from above timberline in the Gros Ventre Range. Montana south to Idaho, Nevada, Arizona, and Colorado.

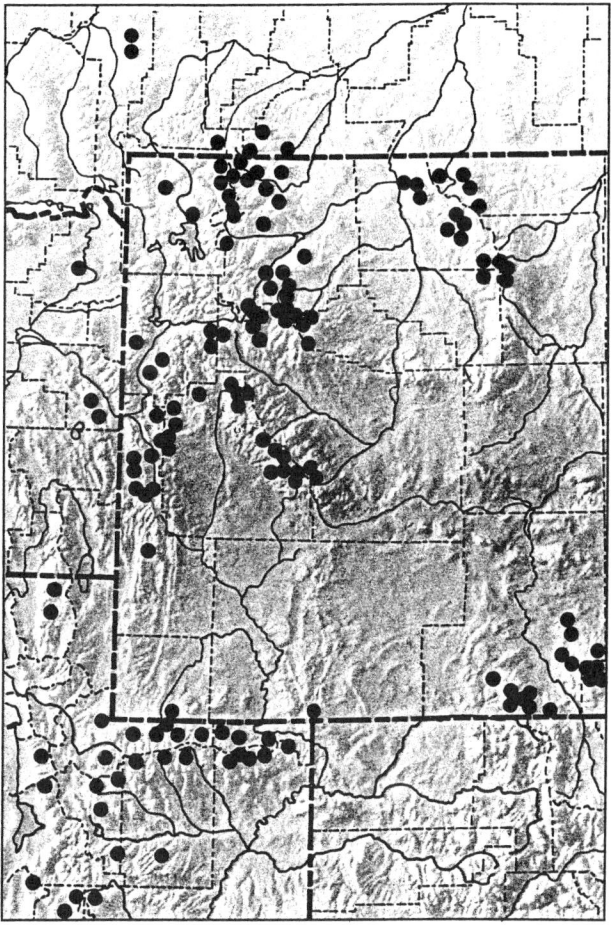

Eriophyllum Lag.
Eriophyllum, Woollyleaf

Our plants are perennial herbs or subshrubs with alternate, entire to ternately or pinnately cleft leaves. Heads typically radiate, the rays yellow, pistillate, and fertile, occasionally white; involucre of 1 or 2 series of erect bracts; receptacle flat or nearly so, naked. Disk flowers yellowish, perfect, and mostly fertile. Style branches flattened. Anthers sagittate at the base. Akenes slender, 4-angled, the pappus of chaffy or hyaline scales, occasionally lacking.

Eriophyllum lanatum (Pursh) Forbes
Hort. Woburn., 183, 1833.

Woolly Eriophyllum

Actinella lanata Pursh
Actinea lanata (Pursh) Steud.
Bahia achillaeoides DC.
Bahia gracilis Hook. & Arn.
Bahia integrifolia (Hook.) DC.
Bahia lanata (Pursh) DC.
Eriophyllum achillaeoides (DC.) Greene
Eriophyllum caespitosum Dougl. ex Lindl.
Eriophyllum gracile (Hook. & Arn.) Gray
Eriophyllum harfordii Rydb.
Eriophyllum integrifolium (Hook.) Greene
Eriophyllum leucophyllum Rydb.
Eriophyllum monoense Rydb.
Eriophyllum multiflorum (Nutt.) Rydb.
Eriophyllum nevadense Gand.
Eriophyllum pedunculatum A. Heller
Eriophyllum ternatum Greene
Eriophyllum trichocarpum Rydb.
Eriophyllum watsonii Gray
Helenium lanatum (Pursh) Spreng.
Trichophyllum integrifolium Hook.
Trichophyllum lanatum (Pursh) Nutt.
Trichophyllum multiflorum Nutt.

Tomentose, multistemmed perennial from a branched, often woody caudex. Leaves alternate, the upper ones sometimes opposite, entire and oblanceolate to ternately or pinnately cleft. Heads 1 or 2 on each stem, the receptacle naked; ray flowers and disk flowers yellow. Akenes glabrous to pubescent or glandular; pappus commonly of hyaline scales, occasionally lacking.

In the alpine zone known only from the Beartooth, Teton, and northwestern Wind River Ranges. British Columbia to Montana and south to California, Nevada, Utah, and Wyoming.

Mountain plants are var. *integrifolium* (Hook.) Smiley, which is less than 20 cm tall and has smaller heads than the typical var. *lanatum* and entire to ternately lobed leaves.

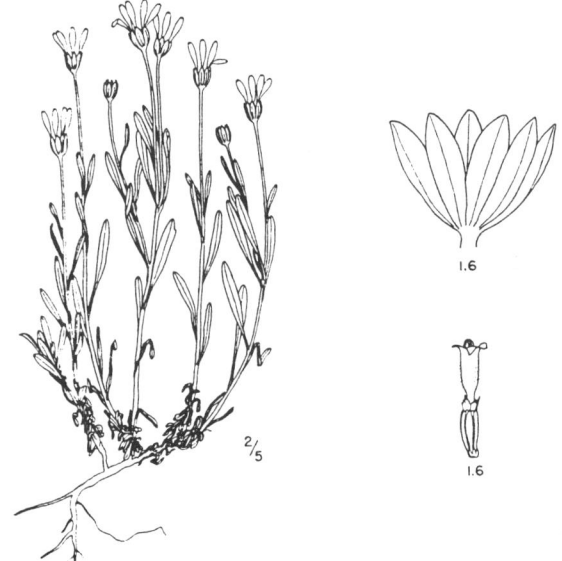

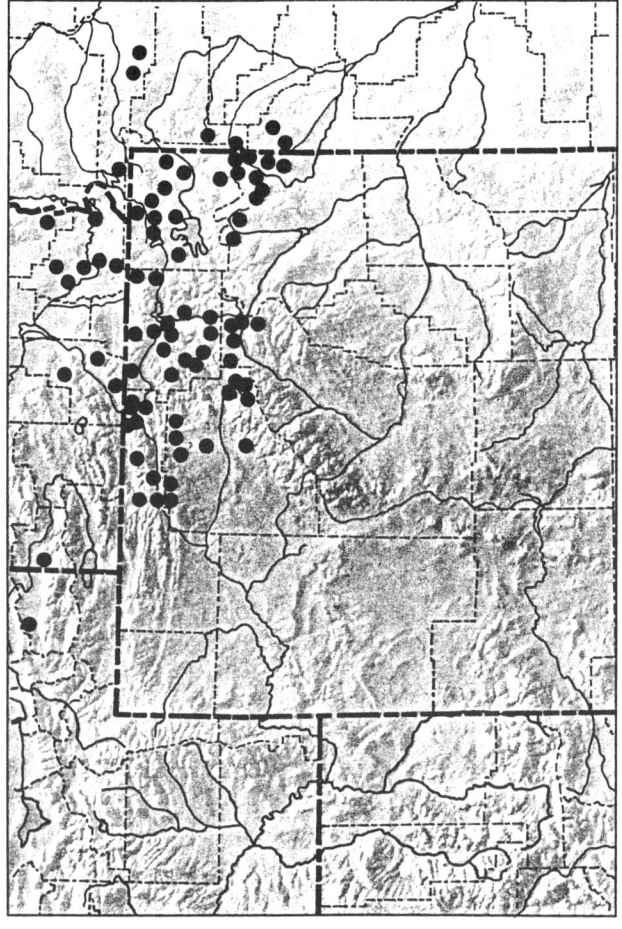

Gutierrezia Lag.
Snakeweed, Matchbrush

Annual or perennial herbs or shrubs with alternate, narrow, entire, and often punctate leaves. Heads numerous, few-flowered, in terminal corymbs. Involucre campanulate to turbinate, the bracts imbricate, scarious-margined, with green tips; receptacle flat, naked, often pitted (alveolate). Rays yellow, fewer than 10, pistillate, fertile. Disk flowers yellow, perfect, sometimes staminate or sterile. Anthers united. Style branches flattened. Pappus of scales or awns, often united at the base. Akenes terete, tapered toward the base, usually with several nerves.

Gutierrezia sarothrae (Pursh) Britt. & Rusby
N.Y. Acad. Sci. Trans. 7:10, 1887.

Broom Snakeweed

Solidago sarothrae Pursh
Brachyachyris euthamiae (Nutt.) Spreng.
Brachyris euthamiae Nutt.
Gutierrezia divaricata (Nutt.) T. & G.
Gutierrezia diversifolia Greene
Gutierrezia euthamiae (Nutt.) T. & G.
Gutierrezia fasciculata Greene
Gutierrezia filifolia Greene
Gutierrezia ionensis Lunell
Gutierrezia juncea Greene
Gutierrezia lepidota Greene
Gutierrezia linearis Rydb.
Gutierrezia longifolia Greene
Gutierrezia myriocephala A. Nels.
Gutierrezia scoparia Rydb.
Gutierrezia tenuis Greene
Xanthocephlum sarothrae (Pursh) Shinners

Shrub or subshrub, mostly less than 5 dm tall, from a woody caudex; branches numerous, mostly scabrous-puberulent, sometimes glabrous. Leaves alternate, linear, glabrous to puberulent or glandular-punctate, the margins entire to scabrous, often forming secondary fascicles in the lower axils. Heads numerous in terminal, corymbose inflorescences; involucre turbinate, glandular-resinous, the bracts imbricate, oblong, greenish-thickened at the acute tip. Receptacle naked. Ray flowers yellow, pistillate; disk flowers yellow, perfect. Akenes 5-angled, pubescent; pappus of short scales.

Common at lower elevations, this plains and foothills species is known from limestone soils of the Gallatin Formation above timberline in the northern Wind River Range. Alberta, Saskatchewan, and Manitoba south to California, Mexico, Texas, Oklahoma, and Kansas.

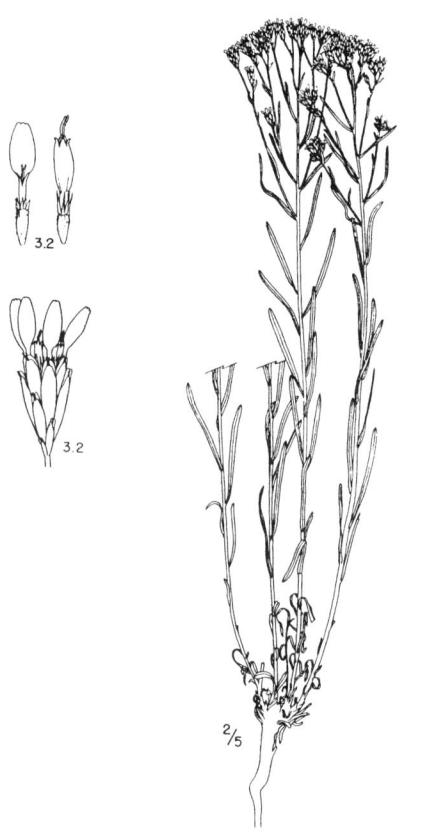
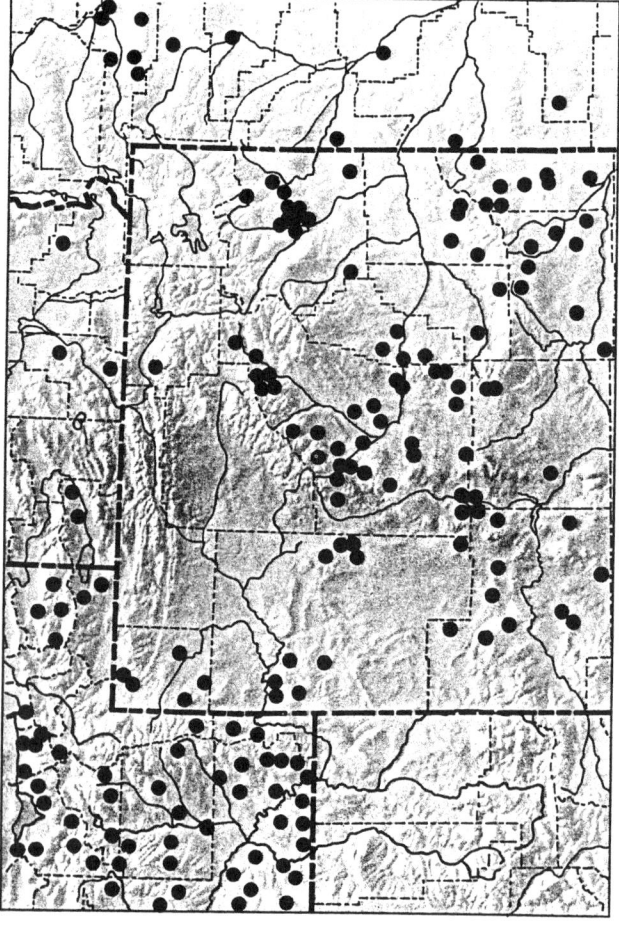

Haplopappus Cass.
Goldenweed

Annual or perennial herbs or shrubs from taproots, with alternate, entire or lobed leaves. Heads usually radiate, solitary or numerous, the involucral bracts usually imbricate, varying from green and somewhat leafy to wholly chartaceous, the receptacle flat, naked, somewhat alveolate. Ray flowers yellow, pistillate or neutral. Disk flowers usually perfect and yellow. Akenes cylindric or turbinate, angled or smooth, the pappus of numerous white or tawny, capillary bristles.

KEY TO THE SPECIES OF *HAPLOPAPPUS*

1. Plants shrubby
    2. Branchlets tomentose; ray flowers lacking — *H. macronema*
    2. Branchlets glabrous to glandular; ray flowers present — *H. suffruticosus*
1. Plants herbaceous
    3. Basal tuft of leaves lacking; cauline leaves little reduced upward
        4. Plants glandular-pubescent, 6–15 cm tall — *H. lyallii*
        4. Plants glabrate to villous, but not glandular, 1–6 cm tall — *H. pygmaeus*
    3. Basal tuft of leaves present; cauline leaves much reduced or wanting
        5. Leaves entire; rays 15 or fewer; plants less than 15 cm tall, mat forming from a much-branched caudex — *H. acaulis*
        5. Leaves shallowly toothed; rays 25 or more; plants 10–30 cm tall, with few to solitary stems from a simple or only slightly branched caudex — *H. uniflorus*

Haplopappus acaulis (Nutt.) Gray
Proc. Am. Acad. 7:353, 1868.

Stemless Goldenweed

Chrysopsis acaulis Nutt.
Aplopappus falcatus (Rydb.) S.F. Blake
Aplopappus nelsonii S.F. Blake
Aster caespitosus (Nutt.) Kuntze
Chrysopsis caespitosa Nutt.
Hoorebekia acaulis (Nutt.) M.E. Jones
Stenotus acaulis (Nutt.) Nutt.
Stenotus caespitosus (Nutt.) Nutt.
Stenotus falcatus Rydb.
Stenotus latifolius A. Nels.
Stenotus rudis A. Nels.
Stenotus scaber A. Nels.

Densely caespitose, subscapose, perennial herb, often forming mats, from a branched caudex and slender to stout taproot; stems few to several, scabrous. Basal leaves oblanceolate, glabrous to scabrous, often resinous-glandular, erect, usually prominently 3-nerved, entire, the tips acute to mucronate; cauline leaves few, narrowly oblong to linear, strongly reduced upward. Heads solitary, terminal; involucre hemispheric to campanulate, the bracts somewhat imbricate in 3 series, lanceolate to ovate, green on the midribs and at the acute or acuminate tips. Ray flowers yellow; disk flowers yellow. Akenes pubescent, sometimes glabrous; pappus white to tan.

Rocky ridges, talus, and scree of the western ranges. Not yet known from above timberline in the Big Horn, Medicine Bow, Uinta, and Wyoming Ranges. Saskatchewan south to California, Idaho, Utah, and Colorado.

Alpine plants with reduced cauline leaves and basal leaves that are scabrous to resinous-glandular are var. *acaulis*.

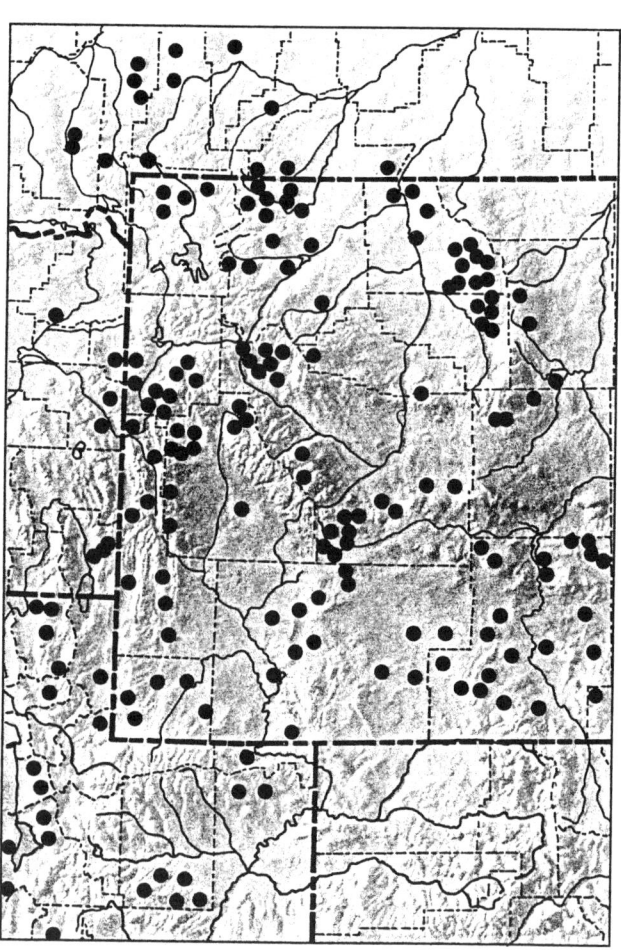

Haplopappus lyallii Gray
Proc. Acad. Phila. 1863:64, 1864.

Lyall Goldenweed

Aster jamesii Kuntze
Hoorebekia lyallii (Gray) Piper
Pyrrocoma lyallii (Gray) Rydb.
Stenotus lyallii (Gray) T.J. Howell
Tonestus lyallii (Gray) A. Nels.

Low, perennial herb, less than 15 cm tall, from a branched caudex and taproot; stems few to several, leafy, erect to decumbent, glandular-puberulent. Leaves sessile or short-petiolate, oblanceolate to spatulate, occasionally obovate, erect, entire, glandular-puberulent, somewhat reduced upward. Heads solitary, terminal, broadly campanulate to hemispheric; involucre glandular-puberulent, the bracts subequal, lanceolate, acute, the inner ones chartaceous, the outer bracts green and foliaceous, sometimes purplish or purple tipped. Ray flowers yellow; disk flowers yellow. Akenes pubescent or glabrous; pappus white.

Gravelly stream banks and lakeshores, talus, and scree of the Absaroka, Gros Ventre, and Wind River Ranges. British Columbia and Alberta south to Oregon, Nevada, Idaho, and Colorado.

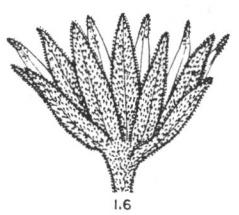

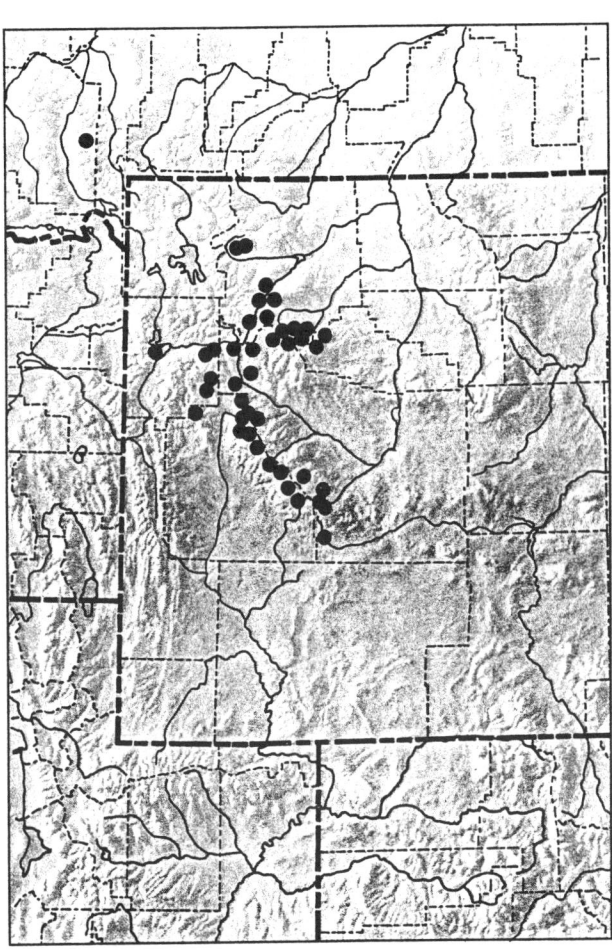

Haplopappus macronema Gray
Proc. Am. Acad. 6:542, 1865.

Cobwebby Goldenweed

Aster macronema (Gray) Kuntze
Bigelovia macronema (Gray) M.E. Jones
Ericameria discoidea (Nutt.) G.L. Nesom
Haplopappus discoideus (Nutt.) Hall & Hall
Macronema discoidea Nutt.
Macronema lineare Rydb.
Macronema obtusum Rydb.

Shrub or subshrub with basal branching, the branches leafy, white-tomentose to glandular-puberulent in the inflorescences. Leaves sessile, oblong to oblanceolate, glandular-puberulent to canescent-tomentose, the margins entire or somewhat undulate, tips obtuse to acute or mucronate. Heads single or few at the ends of branches, campanulate; involucre glandular-puberulent or scabrous, the bracts subequal, not imbricate, obtuse to acute or acuminate, the inner ones chartaceous, the outer bracts green, foliaceous, usually larger than the inner. Ray flowers lacking; disk flowers yellowish. Akenes villous; pappus whitish to tan.

Talus, scree, rocky ridges and ledges of the Absaroka and Wasatch Ranges. Oregon south and east to California, Idaho, Wyoming, Utah, and Colorado.

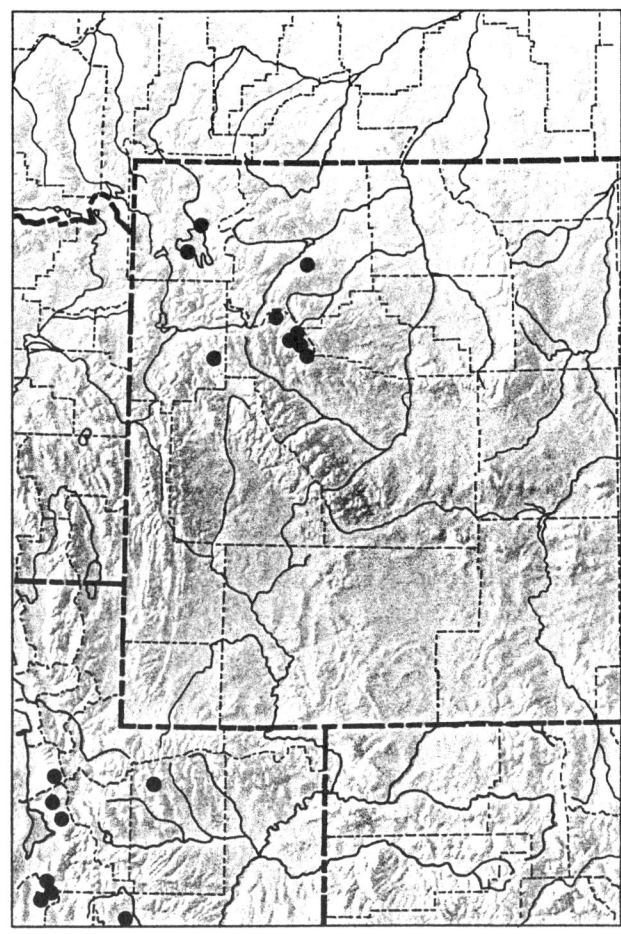

Haplopappus pygmaeus (T. & G.) Gray
Am. J. Sci., II, 33:239, 1862.

Pygmy Goldenweed

Stenotus pygmaeus T. & G.
Aster stenotus Kuntze
Macronema pygmaeum (T. & G.) Greene
Tonestus pygmaeus (T. & G.) A. Nels.

Low, tufted, perennial herb, less than 6 cm tall, from a branched caudex and stout taproot; stems few to several, puberulent. Basal leaves oblanceolate to spatulate, villous at least on the entire margins, the tips obtuse to broadly acute; cauline leaves well developed, scarcely reduced upward. Heads solitary, terminal, campanulate to hemispheric; involucre sparsely pubescent, the bracts subequal, hardly imbricate, the inner ones lanceolate, chartaceous, acuminate, the outer ones greenish or purplish, oblong. Ray flowers yellow; disk flowers yellow. Akenes sparsely pubescent; pappus whitish.

Rocky places in the Medicine Bow Mountains. Montana, Wyoming, Colorado, and New Mexico.

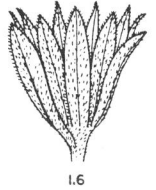

Haplopappus suffruticosus (Nutt.) Gray
Proc. Am. Acad. 6:542, 1865.

Shrubby Goldenweed

Macronema suffruticosum Nutt.
Aster suffruticosus (Nutt.) Kuntze
Ericameria suffruticosa (Nutt.) G.L. Nesom
Macronema grindelifolium Rydb.

Shrub or subshrub with brittle, glandular twigs. Leaves alternate, sessile or short-petiolate, oblanceolate or oblong, glandular, acute to acuminate, the margins undulate or curled. Heads solitary or few at the ends of branches, campanulate; involucre glandular, the bracts subequal, lanceolate to oblong, the outer ones often longer than the inner. Ray flowers yellow; disk flowers yellow. Akenes elongate; pappus whitish.

Scree and talus of the Beartooth, Gros Ventre, Hoback, Salt River, Teton, and Wyoming Ranges. Oregon south and east to Nevada, Idaho, Wyoming, and Montana.

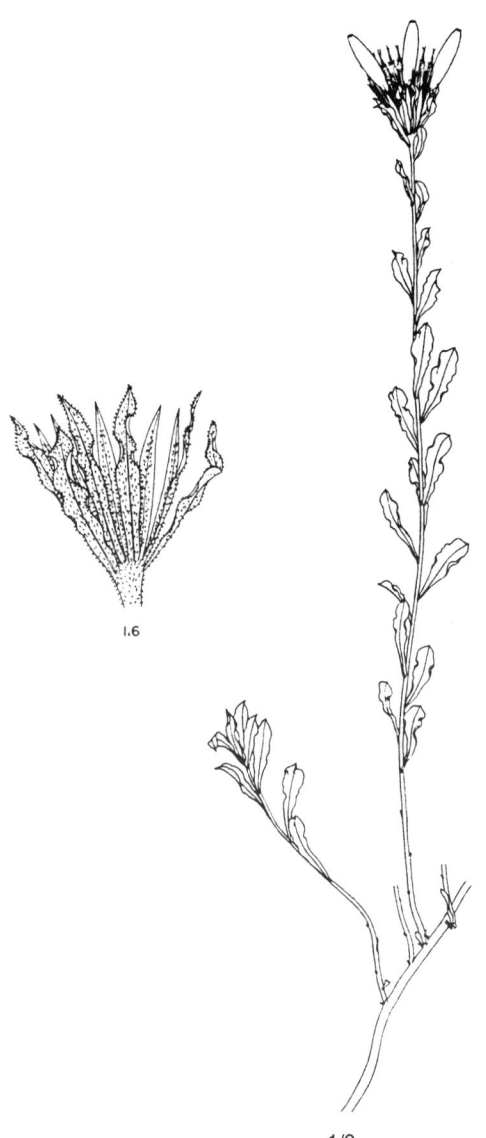

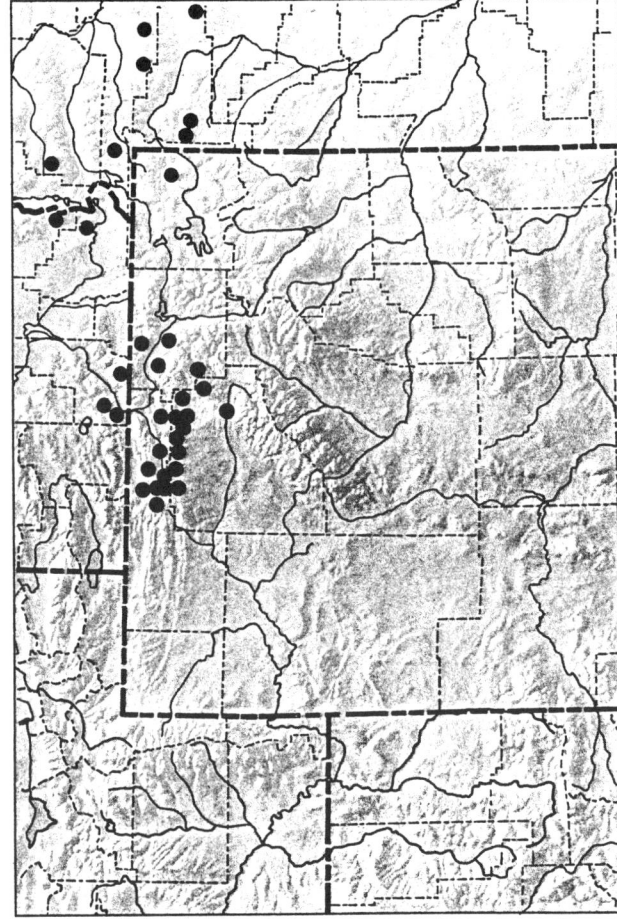

Haplopappus uniflorus (Hook.) T. & G.
Fl. N. Am. 2:241, 1842.

Singlehead Goldenweed

Donia uniflora Hook.
Aplopappus howellii Gray
Aplopappus inuloides (Nutt.) T. & G.
Aster uniflorus (Hook.) Kuntze
Haplopappus gossypinus (Greene) H.M. Hall
Homopappus inuloides Nutt.
Homopappus uniflorus (Hook.) Nutt.
Hoorebekia uniflora (Hook.) M.E. Jones
Pyrrocoma cheiranthifolia Greene
Pyrrocoma gossypina Greene
Pyrrocoma howellii (Gray) Greene
Pyrrocoma inuloides (Nutt.) Greene
Pyrrocoma linearis (Keck) Kartez & Gandhi
Pyrrocoma plantaginea Greene
Pyrrocoma sericea Greene
Pyrrocoma uniflora (Hook.) Greene

Tufted, perennial herb from a taproot, the caudex crowned with old, fibrous leaf bases; stems solitary or few, usually decumbent, glabrous to loosely tomentose. Basal leaves petiolate, the blades elliptic to oblanceolate, erect, glabrous to tomentose, the margins mostly toothed, sometimes entire; cauline leaves oblanceolate, oblong, or lanceolate, becoming reduced and sessile upward. Heads solitary and terminal, or occasionally 2–3 on long peduncles, hemispheric; involucre glabrous to finely tomentose, the bracts greenish throughout, oblong, acute, subequal, appressed, but not imbricate. Ray flowers yellow; disk flowers yellow. Akenes pubescent; pappus whitish.

Rocky meadows in the vicinity of timberline in the Absaroka and Beartooth Ranges. Alberta and Saskatchewan south to California, Nevada, Utah, and Colorado.

Middle Rocky Mountain plants, as described, and lacking narrow, linear leaves are var. *uniflorus*. They are very similar to *H. lanceolatus* (Hook.) T. & G. of lower elevations, which has several heads per plant and an involucre with chartaceous bracts green only at the tip, rather than greenish throughout, as in *H. uniflorus*. Some authors (Welsh et al., 1987) treat both as a single species, *H. lanceolatus*.

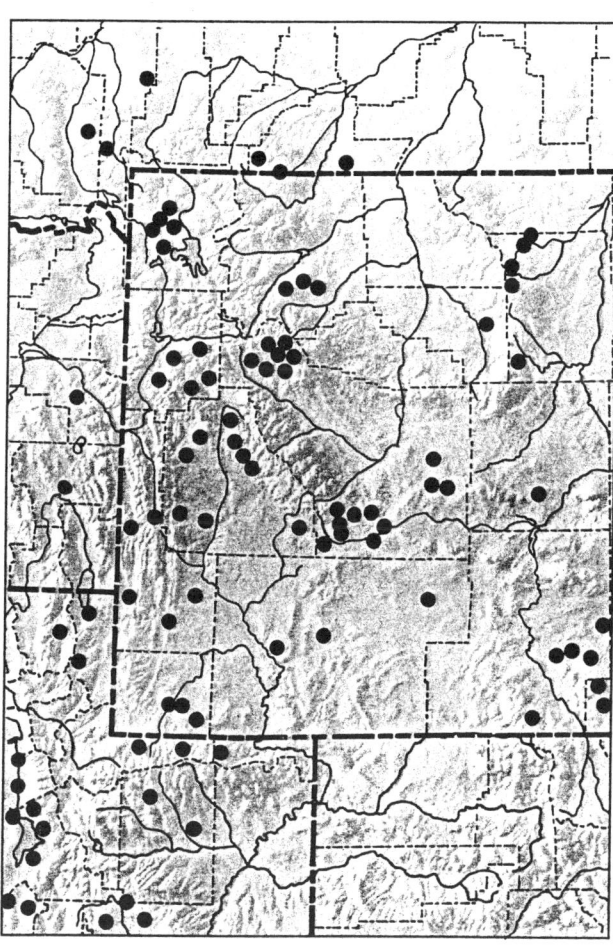

Heterotheca Cass.
Goldenaster

Annual, biennial, or perennial herbs from taproots, with alternate, simple, mostly entire leaves. Heads radiate, few to many, cymose, the involucral bracts narrow, imbricate in several series, the receptacle flat to somewhat convex, naked, pitted. Ray flowers golden yellow, pistillate, and fertile. Disk flowers yellow, numerous, perfect, and fertile. Style branches flattened, with elongate, pubescent appendages. Akenes compressed, pubescent; pappus mostly double, the inner series of long, whitish to tannish, capillary bristles; the outer series, when present, of short scales or coarse bristles.

Heterotheca villosa (Pursh) Shinners
Field & Lab. 19:71, 1951.

Hairy Goldenaster

Amellus villosus Pursh
Chrysopsis alpicola Rydb.
Chrysopsis amplifolia Rydb.
Chrysopsis angustifolia Rydb.
Chrysopsis arida A. Nels. ex Rydb.
Chrysopsis asprella Greene
Chrysopsis bakeri Greene
Chrysopsis ballardii Rydb.
Chrysopsis barbata Rydb.
Chrysopsis butleri Rydb.
Chrysopsis caudata Rydb.
Chrysopsis columbiana Greene
Chrysopsis compacta Greene
Chrysopsis cooperi A. Nels.
Chrysopsis depressa Rydb.
Chrysopsis foliosa Nutt.
Chrysopsis fulcrata Greene
Chrysopsis grandis Rydb.
Chrysopsis hirsuta Greene
Chrysopsis hirsutissima Greene
Chrysopsis hispida (Hook.) Nutt.
Chrysopsis horrida Rydb.
Chrysopsis imbricata A. Nels.
Chrysopsis mollis Nutt.
Chrysopsis pedunculata Greene
Chrysopsis pumila Greene
Chrysopsis resinolens A. Nels.
Chrysopsis villosa (Pursh) Nutt. ex DC.
Chrysopsis viscida (Gray) Greene
Chrysopsis wisconsinensis Shinners
Diplogon villosum (Pursh) Kuntze
Diplopappus hispidus Hook.
Diplopappus villosus (Pursh) Hook.
Heterotheca foliosa (Nutt.) Shinners
Heterotheca fulcrata (Greene) Shinners
Heterotheca horrida (Rydb.) V.L. Harms

Heterotheca pumila Greene
Heterotheca pumila Semple
Heterotheca viscida (Gray) V.L. Harms
Heterotheca wisconsinensis (Shinners) Shinners
Heterotheca zionensis Semple
Inula villosa (Pursh) Nutt.

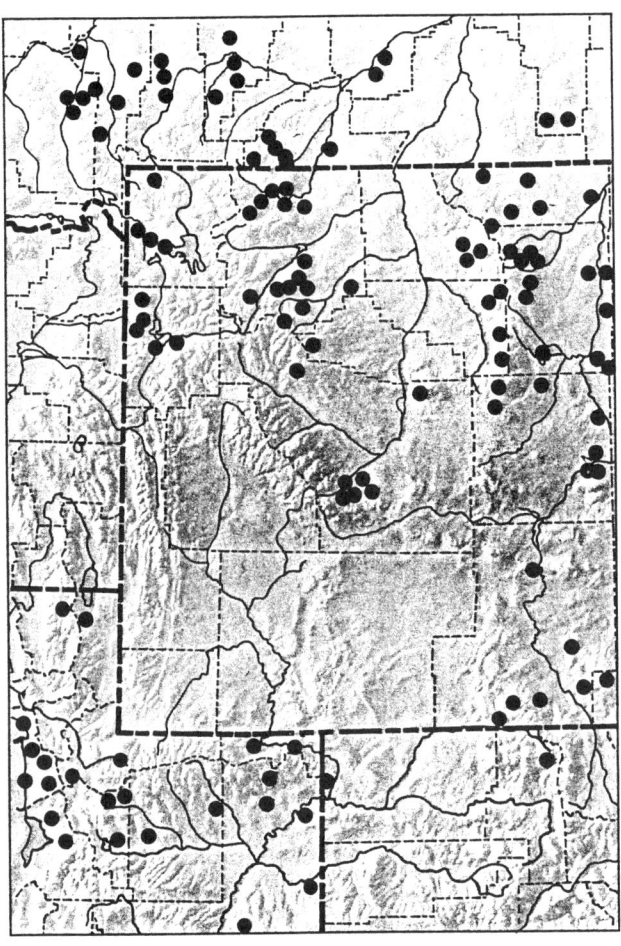

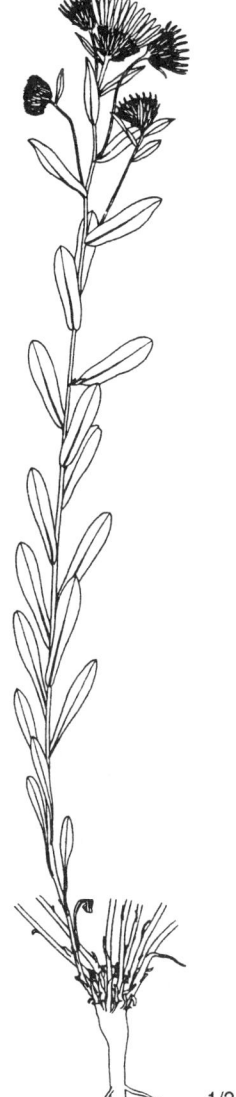

Strigose, multistemmed, perennial herb from a simple, often woody caudex and stout taproot. Leaves all cauline, alternate, sessile or short-petiolate, oblanceolate to somewhat elliptic, entire to denticulate, strigose (and often glandular), and grayish green; lower leaves often deciduous. Heads solitary, or several to many in a leafy corymb; involucre strigose, the bracts imbricate, lanceolate, attenuate, the margins thin-hyaline. Ray flowers golden yellow, pistillate; disk flowers perfect. Akenes pubescent, flattened; pappus tannish, the outer series much shorter than the inner.

Fellfields and rocky slopes at and above timberline in the Absaroka and Beartooth Ranges. British Columbia to Saskatchewan, south to California, Texas, Kansas, and Illinois.

The 2 varieties that are common in our region and might be expected as occasional above timberline in dry, rocky, exposed sites are var. *villosa,* with appressed hairs on the upper peduncles and leaves, usually lacking glands; and var. *hispida* (Hook.) Harms, with spreading hairs on the upper peduncles and leaves, often with glands.

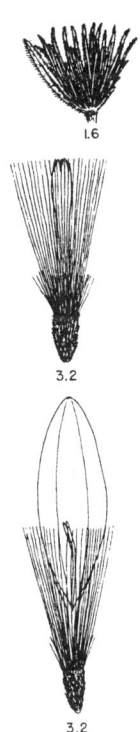

Hieracium L.
Hawkweed

Perennial herbs with milky juice, ours from short, stout rhizomes, with alternate and basal, entire to dentate leaves. Heads several to many, in open, terminal panicles or corymbs, the bracts of the involucre in 2–3 series, imbricate, the receptacle flat, usually naked. Flowers all ligulate, yellow or occasionally white. Akenes oblong, glabrous, 10-ribbed, somewhat contracted at the base, the apex truncate, not beaked, the pappus of brown or tan, or occasionally whitish, capillary bristles.

Hieracium gracile Hook.
Fl. Bor.-Am. 1:298, 1834.

Slender Hawkweed

Cholocrepis tristis (Willd. ex Spreng.) Löve & Löve
Hieracium hookeri Steud.
Hieracium triste Willd. ex Spreng.
Hieracium utahense Gand.
Pilosella gracillis (Hook.) F.W. Schultz & Schultz
Pilosella tristis (Willd. ex Spreng.) F.W. Schultz & Schultz

Slender, scapose or subscapose, perennial herb from a short, oblique caudex and fibrous roots; stems or scapes usually finely stellate-puberulent. Basal leaves petiolate, the blades oblanceolate to obovate, glabrous to puberulent, the margins entire to shallowly denticulate; cauline leaves much reduced, lanceolate, or lacking. Heads solitary or few to several and racemose; involucre stellate-pubescent, often with long, black, often gland-tipped hairs. Ray flowers yellow, often drying whitish. Akenes cylindric, striate; pappus tan.

Timberline meadows and lower-elevation fellfields of the Absaroka, Beartooth, Big Horn, Medicine Bow, Teton, Uinta, Wasatch, and Wind River Ranges. South America; Alaska and Mackenzie district of Northwest Territory south to California, Utah, and New Mexico.

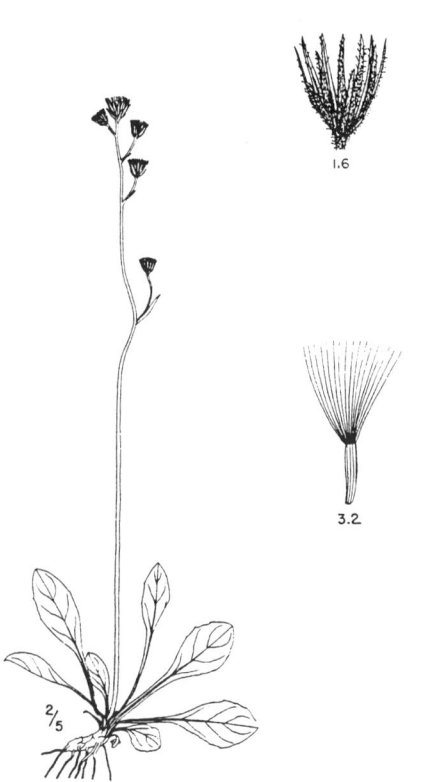
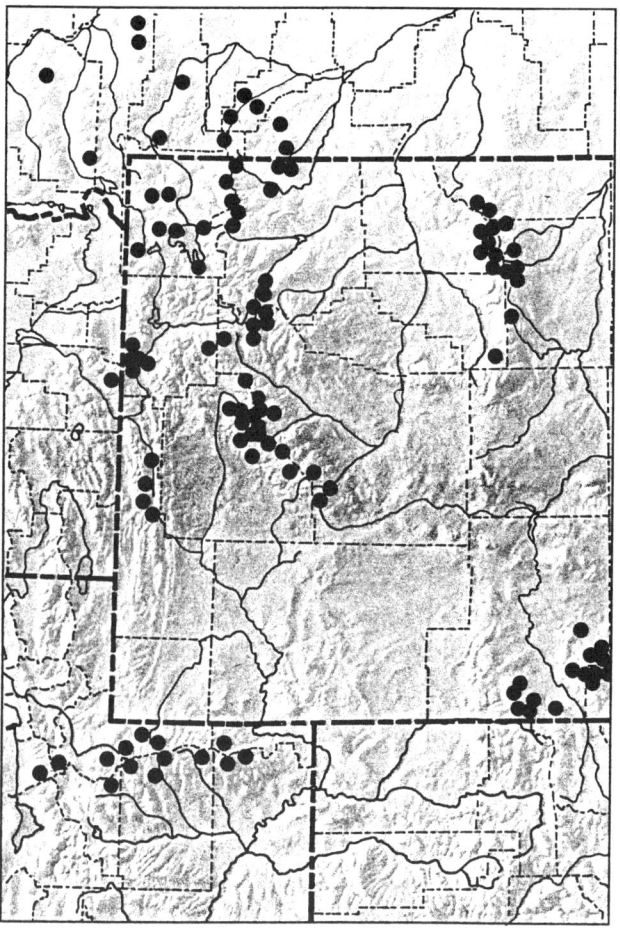

Hulsea T. & G.
Hulsea

Annual, biennial, or perennial herbs from stout, thick taproots, with viscid, aromatic, alternate, entire to pinnatifid leaves. Heads radiate, usually solitary, the involucral bracts narrow, in 2–3 series, the receptacle flat, with minute, horny teeth. Ray flowers yellow, pistillate, and fertile. Disk flowers yellow, glandular-viscid, perfect, and fertile. Style branches flattened, obtuse. Akenes linear, villous at least on the margins, the pappus of several hyaline scales.

Hulsea algida Gray
Proc. Am. Acad. 6:547, 1865.

Alpine Hulsea

Hulsea caespitosa A. Nels. & Kenn.
Hulsea carnosa Rydb.
Hulsea nevadensis Gand.

Somewhat tufted, glandular-puberulent, perennial herb from a simple or branched caudex and taproot; stems 1 to several. Basal leaves erect, short-petiolate, the blades oblanceolate to oblong, pinnately lobed or shallowly crenate-dentate; cauline leaves sessile, linear, reduced upward, with margins like the basal ones. Heads solitary, large, showy; involucre glandular, villous, the bracts subequal, lanceolate, attenuate. Ray flowers yellow; disk flowers yellowish. Akenes pubescent, cylindrical; pappus of short, lacerate scales.

Scree, talus, and rocky ledges of the Absaroka and Beartooth Ranges. Oregon, south and east to California, Nevada, Idaho, Wyoming, and Montana.

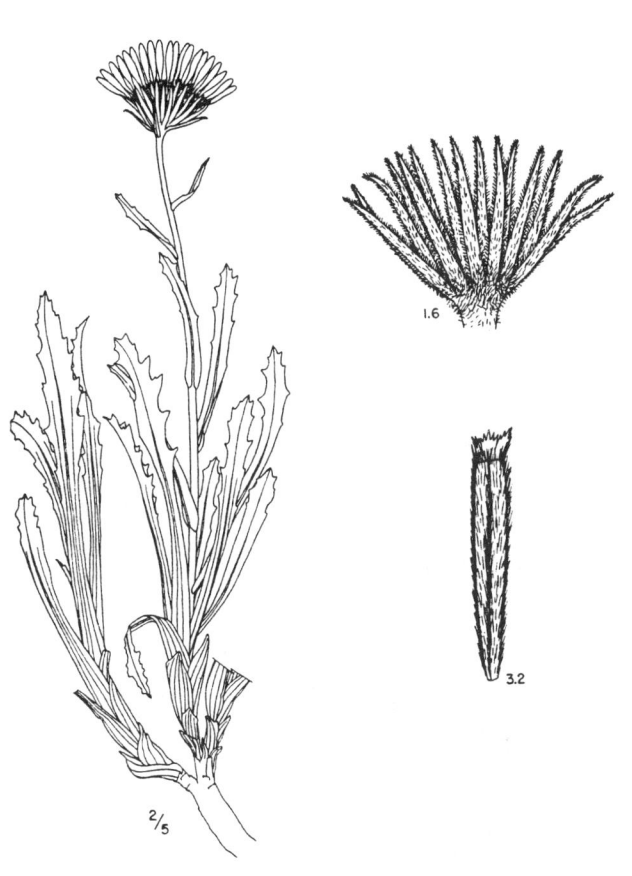

Hymenoxys Cass.
Hymenoxys

Annual or perennial, aromatic herbs from a stout taproot, with alternate or basal, entire to pinnatifid leaves. Heads radiate, solitary in ours, the involucral bracts in 2–3 series, the receptacle naked, hemispherical or convex. Ray flowers yellow, pistillate, and fertile, becoming reflexed with age. Disk flowers yellow, numerous, perfect, and fertile. Akenes turbinate, hairy, and pappus of 5–8 hyaline scales.

KEY TO THE SPECIES OF *HYMENOXYS*

1. Leaves entire, all basal — *H. acaulis*
1. Leaves pinnatifid, both basal and cauline — *H. grandiflora*

Hymenoxys acaulis (Pursh) K.L. Parker
Madroño 10:159, 1950.

Stemless Goldflower

Gaillardia acaulis Pursh
Actinea acaulis (Pursh) Spreng.
Actinea arizonica (Greene) A. Nels.
Actinea eradiata (A. Nels.) A. Nels.
Actinea epunctata (A. Nels.) A. Nels.
Actinea herbacea (Greene) B.L. Robins.
Actinea incana (A. Nels.) A. Nels.
Actinea simplex (A. Nels.) A. Nels.
Actinella acaulis (Pursh) Nutt.
Actinella argentea Gray
Actinella epunctata (A. Nels.) A. Nels.
Actinella eradiata (A. Nels.) A. Nels.
Actinella incana (A. Nels.) A. Nels.
Actinella lanata Nutt.
Actinella simplex (A. Nels.) A. Nels.
Cephalophora acaulis (Pursh) DC.
Hymenoxys argentea (Gray) K.L. Parker
Hymenoxys ivesiana (Greene) K.L. Parker
Picradenia acaulis (Pursh) Britt.
Ptilepida acaulis (Pursh) Britt.
Tetraneuris acaulis (Pursh) Greene
Tetraneuris arizonica Greene
Tetraneuris brevifolia Greene
Tetraneuris crandallii Rydb.
Tetraneuris epunctata A. Nels.
Tetraneuris eradiata A. Nels.
Tetraneuris herbacea Greene
Tetraneuris incana A. Nels.
Tetraneuris ivesiana Greene
Tetraneuris lanata (Nutt.) Greene
Tetraneuris lanigera Dan.
Tetraneuris pygmaea Woot. & Standl.
Tetraneuris septentrionalis Rydb.
Tetraneuris simplex A. Nels.
Tetraneuris trinervata Greene

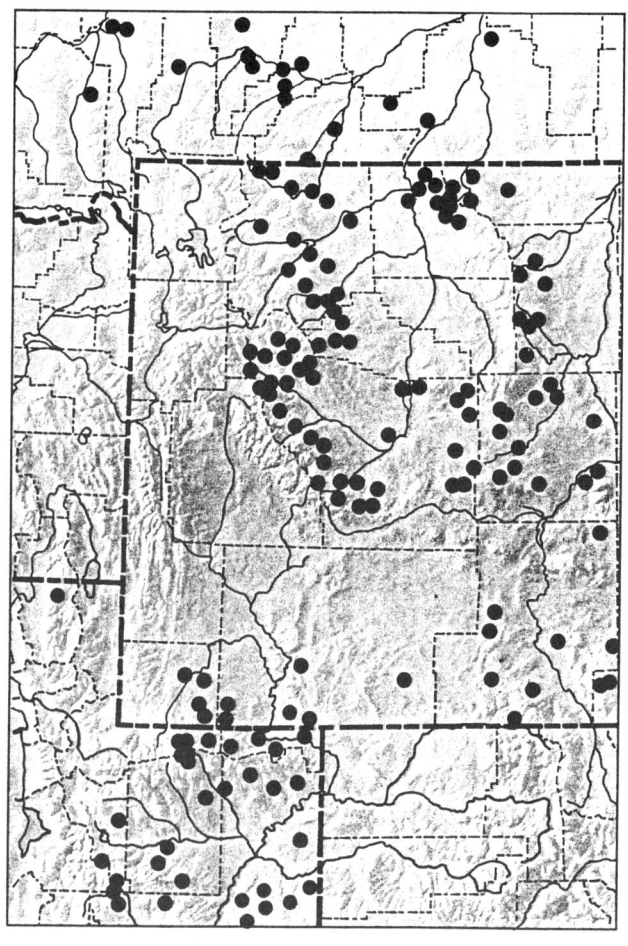

Scapose, villous to glabrous, perennial herb from a branched caudex and taproot, the caudex covered with old leaf bases. Leaves erect to ascending, punctate, linear to oblanceolate, the margins entire. Heads solitary on scapes exceeding the leaves; involucre villous, the bracts linear to oblong, distinct, subequal. Ray flowers bright yellow, showy, the corollas broadly 3-lobed or toothed, prominently nerved. Disk flowers yellow, short-toothed, pubescent. Akenes pubescent; the pappus of 5–7 acute or short-awned scales.

Rocky ledges and ridges, often on limestone, of the Big Horn, Uinta, and Wind River Ranges. Alberta and Saskatchewan south to California, Nevada, New Mexico, Texas, and Kansas.

The 2 alpine varieties are var. *acaulis,* which has punctate, glabrate to densely pubescent leaves; and var. *caespitosa* (A. Nels.) K.L. Parker, with leaves epunctate or with a few small glands, glabrous to sparsely pubescent.

Hymenoxys grandiflora (T. & G. ex Gray) K.L. Parker
Madroño 10:159, 1950.

Old-Man-of-the-Mountain

Actinella grandiflora T. & G. ex Gray
Actinea grandiflora (T. & G. ex Gray) Kuntze
Ptilepida grandiflora (T. & G. ex Gray) Rose
Rydbergia grandiflora (T. & G. ex Gray) Greene
Tetraneuris grandiflora (T. & G. ex Gray) K.L. Parker

Woolly-pubescent, perennial herb from a simple caudex and taproot; stems solitary or a few together, simple or branched below, densely villous. Basal leaves villous to glabrate, 1–2-ternately or pinnately compound, the ultimate segments linear, obtuse; petioles dilated at base; cauline leaves ternate to linear, entire, reduced upward. Heads large, showy, solitary, and terminal; involucre densely woolly, the bracts linear, subequal, distinct. Ray flowers yellow, the corollas 3-lobed or toothed; disk flowers yellow. Akenes pubescent; pappus of 5–8 slender, attenuate scales.

Fellfields and meadows, occasionally on talus, in most alpine ranges of the Middle Rocky Mountains. Not yet known from above timberline in the Beartooth, Hoback, Salt River, and Wyoming Ranges. Montana south to Idaho, Utah, New Mexico, and Colorado.

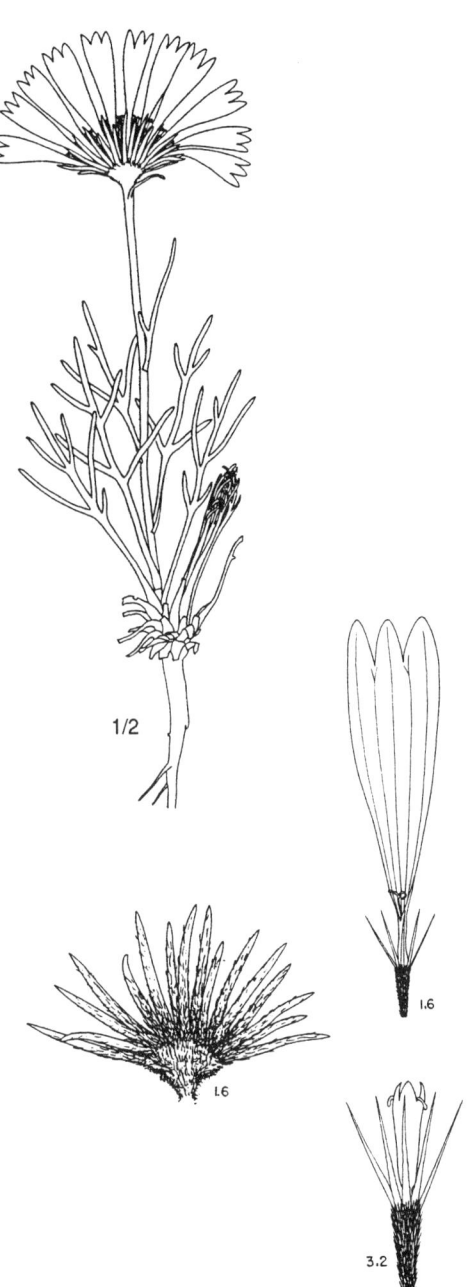

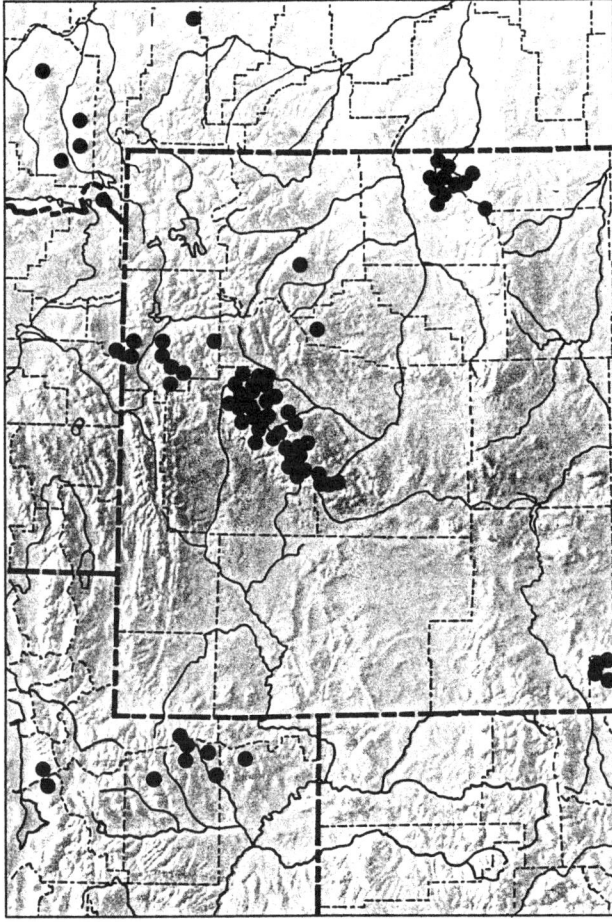

***Machaeranthera* Nees**
Tansyaster

Annual to perennial herbs from a taproot. Leaves alternate, simple, entire to spinulose-dentate, the apex spinulose. Heads radiate, rarely discoid, few to numerous in cymes or panicles, the involucral bracts in several series, linear to lanceolate, chartaceous at the base, with green, spreading, attenuate tips; receptacle naked, flat. Ray flowers blue, purple, or lavender, pistillate, fertile. Disk flowers yellowish, perfect, fertile. Akenes turbinate to oblong, glabrous to pubescent, usually several-nerved; pappus of whitish or brownish, unequal, capillary bristles.

KEY TO THE SPECIES OF *MACHAERANTHERA*

1. Plants 10–40 cm tall; basal leaves often withering by flowering time; heads several to many, the ray flowers bluish purple or pinkish purple *M. canescens*
1. Plants less than 10 cm tall; basal leaves usually persistent; heads solitary, the ray flowers white, fading to pink *M. kingii*

Machaeranthera canescens (Pursh) Gray
Pl. Wright. 1:89, 1852.

Hoary Tansyaster

Aster canescens Pursh
Aster attenuatus (T.J. Howell) Frye & Rigg
Aster glossophyllus Piper
Aster inornatus Greene
Aster leiodes S.F. Blake
Aster leucanthemifolius Greene
Aster rubricaulis Rydb.
Dieteria canescens (Pursh) Nutt.
Dieteria divaricata Nutt.
Dieteria incana (Lindl.) T. & G.
Dieteria pulverulenta Nutt.
Dieteria sessiliflora Nutt.
Dieteria viscosa Nutt.
Diplopappus incanus Lindl.
Machaeranthera angustifolia Rydb.
Machaeranthera asteroides (Torr.) Greene
Machaeranthera attenuata T.J. Howell
Machaeranthera divaricata (Nutt.) Greene
Machaeranthera glabella Greene ex Rydb.
Machaeranthera incana (Lindl.) Greene
Machaeranthera inornata Greene
Machaeranthera laetevirens Greene
Machaeranthera latifolia A. Nels.
Machaeranthera leptophylla Rydb.
Machaeranthera leucanthemifolia (Greene) Greene
Machaeranthera linearis Rydb.
Machaeranthera magna A. Nels.
Machaeranthera montana Greene
Machaeranthera paniculata A. Nels.
Machaeranthera pinosa Elmer
Machaeranthera pulverulenta (Nutt.) Greene
Machaeranthera ramosa A. Nels.
Machaeranthera scoparia Greene
Machaeranthera sessiliflora (Nutt.) Greene
Machaeranthera spinulosa Greene
Machaeranthera subalpina Greene
Machaeranthera superba A. Nels.
Machaeranthera verna A. Nels.
Machaeranthera viscosa (Nutt.) Greene

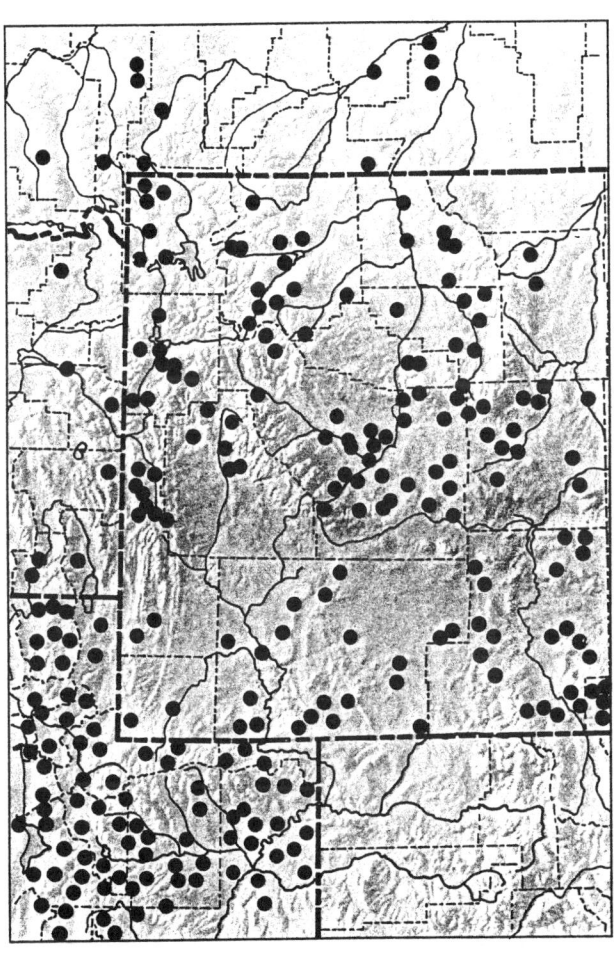

Biennial or short-lived perennial herb from a taproot; stems several to many, erect to spreading, simple or branched, glabrous to grayish-pubescent or glandular. Basal leaves oblanceolate to spatulate, toothed, the teeth and apex spinulose-tipped, often short-lived and withering at anthesis; cauline leaves linear to oblanceolate, glabrous to puberulent or glandular, entire to toothed, the teeth and apex spinulose-tipped. Heads several to many, in terminal panicles or corymbs; involucre canescent or glandular, the bracts imbricate, linear to oblong, each chartaceous at the base, the tip green, attenuate, and often spreading or recurved. Ray flowers bluish purple or pinkish purple; disk flowers yellowish. Akenes pubescent; pappus bristles whitish, numerous, somewhat unequal.

Rocky slopes of the Uinta and Wasatch Ranges. British Columbia to Saskatchewan and south to California, Arizona, and New Mexico.

Alpine plants that are biennial to perennial with involucral bracts 1–2 mm wide and abruptly attenuate at the apex are var. *latifolia* (A. Nels.) S.L. Welsh.

Machaeranthera kingii (D.C. Eat.) Cronq. & Keck
Brittonia 9:239, 1957.

King Tansyaster

Aster kingii D.C. Eat.
Tonestrus kingii (D.C. Eat.) G.L. Nesom

Low perennial, to 10 cm tall, from a branched caudex and taproot; stems single to few, erect to spreading, simple, stipitate-glandular, at least above. Basal leaves persistent, petiolate, the blades oblanceolate to spatulate, glabrous or glandular, entire or toothed, the teeth and apex spinulose-tipped; cauline leaves oblanceolate, becoming sessile and reduced upward. Heads solitary, or rarely 2–3; involucre stipitate-glandular, the bracts imbricate, linear to oblong, chartaceous at the base, the tips green and reflexed, the margins often purplish. Ray flowers white, fading to pink; disk flowers yellowish.

Scree, talus, and rocky ledges in the Wasatch Range. Endemic to north-central Utah.

Entire-leaved plants with short-stipitate glands are var. *kingii*. Near the southern limits of the range, var. *barnebyana* S.L. Welsh and Goodrich, with toothed leaves and long-stipitate glands, has been recognized. Following Watson (1978) this taxon has also been treated as *Aster kingii* D.C. Eaton by Welsh, et al. (1987, 1993), primarily on the basis of the 2n=18 chromosome number which suggests a stronger affinity to the genus *Aster* than to *Machaeranthera* with a base number of 2n=8 or 10.

Nothocalais (Gray) Greene
Nothocalais

Scapose, perennial herbs with milky juice, from taproots. Leaves entire or wavy-margined. Heads solitary and terminal on a naked peduncle, the bracts of the involucre mostly in 2 series, subequal in length, appressed, the receptacle naked. Flowers all ligulate, yellow, sometimes drying purplish. Akenes oblong to fusiform, 10-ribbed, lacking a beak; pappus of persistent, white, narrowly tapered scales.

Nothocalais nigrescens (Henders.) A. Heller
Muhlenbergia 1:8, 1900.

Black Nothocalais

Microseris nigrescens Henderson

Scapose, perennial herb from a taproot; scapes mostly villous, at least below the inflorescence. Leaves all basal, linear to lanceolate, glabrous to glabrate, entire, the margins ciliate. Heads solitary, large, showy; involucral bracts appressed to ascending, subequal, lanceolate to ovate, acuminate, dotted with purple or purplish black. Ray flowers yellow, often drying purplish. Akenes cylindric, slightly constricted near the apex, beakless; pappus white, of slender, narrow, attenuate scales which appear as bristles without magnification.

Meadows and fellfields of the lower alpine zone near timberline in the Beartooth and Wind River Ranges. Montana south to Idaho and Wyoming.

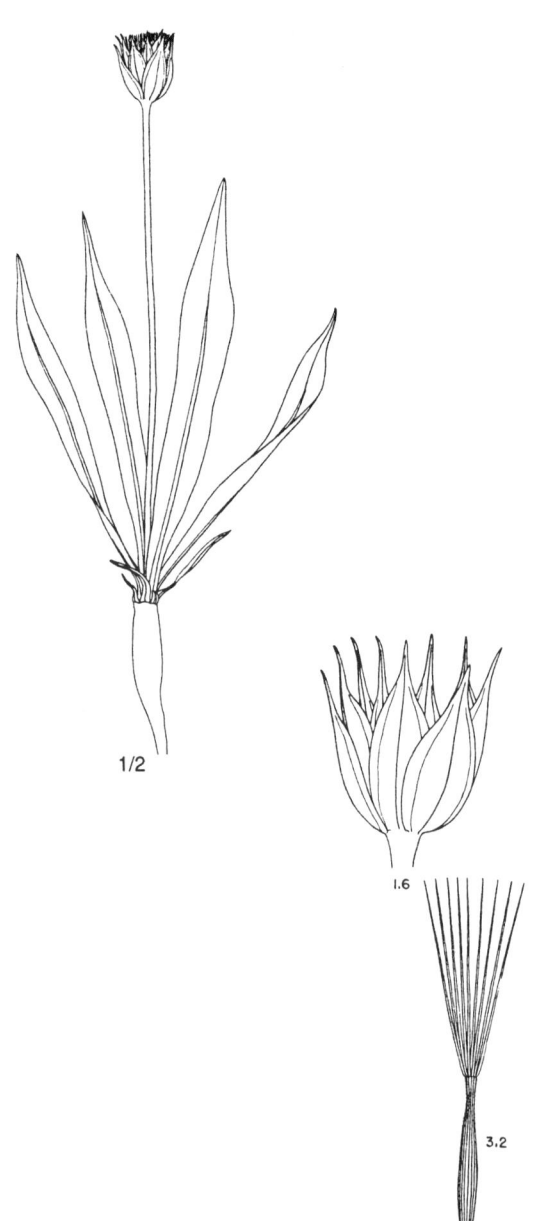

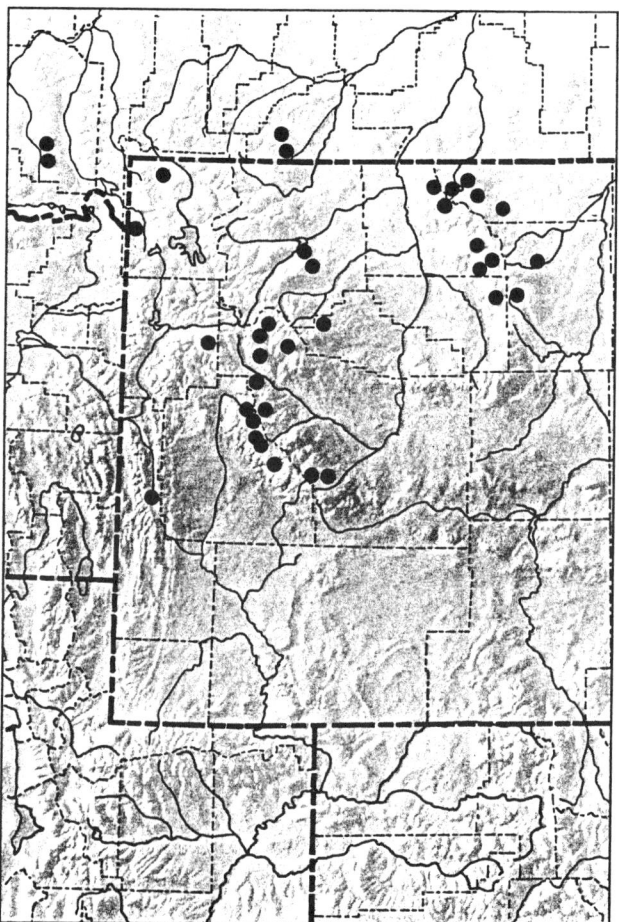

Petradoria Greene
Rock Goldenrod

Perennial herbs from a stout, often scaly taproot and branched, woody caudex; stems several to many, herbaceous or somewhat woody at the base, with alternate or basal, entire leaves. Heads radiate, in dense terminal cymes; involucres cylindric, resinous, the bracts in 4–5 series and mostly vertical ranks; receptacle naked. Ray flowers few, fertile, yellow, often recurved, sometimes lacking. Disk flowers few, sterile, yellow. Akenes compressed-turbinate, glabrous; pappus of brownish, capillary bristles.

Petradoria pumila (Nutt.) Greene
Erythea 3:13, 1895.

Rock Goldenrod

Chrysoma pumila Nutt.
Aster pumilus (Nutt.) Kuntze
Petradoria graminea Woot. & Standl.
Solidago graminea (Woot. & Standl.) S.F. Blake
Solidago petradoria S.F. Blake
Solidago pumila (Nutt.) T. & G.

Glabrous or glabrate, perennial herb from a stout taproot and branched, woody caudex; stems several to many, erect to ascending, and densely tufted. Basal leaves petiolate, lanceolate to oblanceolate, or elliptic, entire; cauline leaves alternate, sessile, becoming linear or oblong and reduced upward, often glandular-punctate. Heads several to many in compact cymes; involucre cylindrical, resinous, the bracts more or less keeled, in vertical ranks, yellowish, the tips brownish or greenish. Ray flowers 1–3, pistillate and fertile, yellow, often recurved, sometimes lacking; disk flowers 1–5, staminate, yellow. Akenes turbinate-compressed, glabrous, 5–10-nerved; pappus of brownish, capillary bristles.

Scree, talus, and other rocky places in the Wasatch Range. California, Idaho, and Wyoming south to Arizona and New Mexico.

Middle Rocky Mountain alpine plants are var. *pumila*.

2/3

Saussurea DC.
Saussurea

Perennial herbs with alternate, entire, dentate to sinuate-dentate, or pinnatifid leaves. Heads solitary, racemose, or corymbose, discoid; involucre imbricate, commonly in 2–4 series; receptacles flat, with numerous chaffy bracts. Flowers tubular, perfect, the corollas blue or purple, expanded at the throat. Anthers caudate; style with a ring of hairs below the branches. Akenes glabrous, nerved; pappus in 2 series, the inner plumose, united at the base; the outer of short, rigid, nonplumose bristles.

Saussurea weberi Hult.
Svensk. Bot. Tidsk. 53:202, 1959.

Weber Saussurea

Low, compact, perennial herb with thick stems and alternate, entire, arachnoid-floccose leaves. Upper leaves sessile or nearly so, the lower ones petiolate, with scattered yellow, resinous glands on the lower side. Heads tightly clustered; involucres of 2–3 series of ovate, blunt, imbricate bracts with dark margins and tips. Disk corollas purple. Akenes glabrous; pappus of inner united, plumose bristles and shorter outer, nonplumose bristles.

Locally common on rocky, exposed, limestone slopes, talus, and ridges in the Gros Ventre and northern Wind River Ranges. Montana, Wyoming, and Colorado.

Lipschitz (1979) has reduced *S. weberi* to a synonym of *S. densa*. Weber and Wittmann (1992) comment that "Lipschitz erred in synonymizing this (*S. weberi*) with *S. densa* Rydberg." I agree; both Colorado and Wyoming specimens that I've examined are short, compact plants with distinctly broad, ovate bracts that are mostly obtuse. For now it seems best to consider this an endemic with a distribution pattern centered on the Middle Rocky Mountains.

Senecio L.
Groundsel

Annual or perennial herbs with alternate, entire, toothed, or lobed leaves. Heads radiate, solitary to numerous, the involucral bracts essentially uniseriate, but often with a few smaller bractlets at the base, the receptacle flat or concave, naked. Ray flowers yellow, pistillate, and fertile. Disk flowers yellow or orange, perfect, and fertile. Akenes terete or slightly flattened, 5–10-nerved, the pappus of numerous white bristles.

KEY TO THE SPECIES OF *SENECIO*

1. Plants densely white-tomentose; pubescence of 2 types, the coarse, flattened hairs overlain by long, shaggy, white hairs; disk flowers orange *S. fuscatus*
1. Plants glabrous to tomentose; pubescence uniform if present; disk flowers yellow
    2. Basal tuft of leaves lacking; cauline leaves well developed, gradually or not at all reduced upward
        3. Plants dwarf, mostly less than 10 cm tall, freely branched, from a branched caudex; leaves flabellate *S. fremontii*
        3. Plants erect, more than 20 cm tall, unbranched, from fibrous roots; leaves lanceolate, elliptic, or triangular
            4. Leaf bases cuneate, the blades lanceolate or elliptic *S. serra*
            4. Leaf bases truncate, the blades triangular *S. triangularis*
    2. Basal tuft of leaves present, at least the basal leaves well developed; cauline leaves progressively reduced upward
        5. Plants glabrous, at least by flowering time, occasionally sparsely pubescent at base of stem and in leaf axils
            6. Heads nodding, mostly 3–4 cm broad and 17–25 mm high *S. amplectens*
            6. Heads erect or nearly so, mostly 1–3 cm broad and 7–15 mm high
                7. Largest leaves 10–20 cm long, acute, the margins entire to denticulate; bracts black-tipped *S. crassulus*
                7. Largest leaves 4–8 cm long, obtuse, at least some of the margins pinnatifid, lyrate, or crenate; bracts not black-tipped, sometimes slightly darker at the ends
                    8. Heads solitary or occasionally 2 or 3; basal leaves mostly 1–1.8 cm broad *S. cymbalarioides*
                    8. Heads several to many; basal leaves mostly 2–3.4 cm broad
                        9. Cauline leaves toothed or lobed less than half the distance to the midrib, the bases clasping; inflorescence compact, the peduncles less than 3 cm long *S. dimorphophyllus*
                        9. Cauline leaves lobed more than half the distance to the midrib, the bases not clasping; inflorescence open, the peduncles more than 3 cm long *S. streptanthifolius*
        5. Plants pubescent, usually in both the inflorescence and on lower leaf surfaces
            10. Heads 20–60; involucre mostly less than 6 mm wide, the bracts mostly 8 or fewer *S. atratus*
            10. Heads usually fewer than 20; involucre mostly more than 7 mm wide, the bracts mostly 13 or more

11. Plants with fibrous roots, lacking a rhizome or elongate caudex; pubescence loose, crisp-villous *S. integerrimus*
11. Plants from a rhizome or short, horizontal, or ascending caudex; pubescence fine, often appressed and dense
    12. Plants less than 15 cm tall; leaves less than 4 cm long and 15 mm wide
        13. Plants glabrate to sparsely tomentulose *S. werneriifolius*
        13. Plants white-tomentose or silvery-canescent *S. canus*
    12. Plants more than 15 cm tall; leaves more than 5 cm long and 15 mm wide
        14. Plants from a branched caudex; leaves entire, usually canescent; involucral bracts not black-tipped *S. canus*
        14. Plants from a short, horizontal rhizome; leaves toothed, greenish, with sparse pubescence; involucral bracts black-tipped
            15. Involucral bracts usually about 13, conspicuously black-tipped; ray flowers 7–15 mm long *S. lugens*
            15. Involucral bracts usually about 21, with inconspicuous black tips; ray flowers 6–10 mm long *S. sphaerocephalus*

Senecio amplectens Gray
Am. J. Sci., II, 33:240, 1862.

Alpine Groundsel

Ligularia amplectens (Gray) W.A. Weber
Ligularia holmii (Greene) W.A. Weber
Senecio holmii Greene
Senecio lactucinus Greene
Senecio seridophyllus Greene

Glabrous to sparsely tomentose, perennial herb from a short rhizome and fibrous roots; stems solitary, simple, tomentose when young, becoming glabrous or glabrate with age. Lower leaves petiolate, the blades obovate to elliptic, dentate; upper leaves becoming sessile, denticulate or entire. Heads 1–5, nodding, corymbose; involucre glabrous or glabrate, broadly campanulate to hemispheric, the bracts linear, acute to acuminate, brownish or purplish. Ray flowers yellow; disk flowers yellowish. Akenes glabrous; pappus white.

Meadows, fellfields, and rocky ledges of most alpine ranges in the Middle Rocky Mountains. Not yet known from above timberline in the Big Horn, Medicine Bow, Salt River, and Uinta Ranges. Montana south to Nevada, Utah, and Colorado.

Alpine plants less than 20 cm tall with glabrate herbage and the involucral bracts often purplish are var. *holmii* (Greene) Harrington.

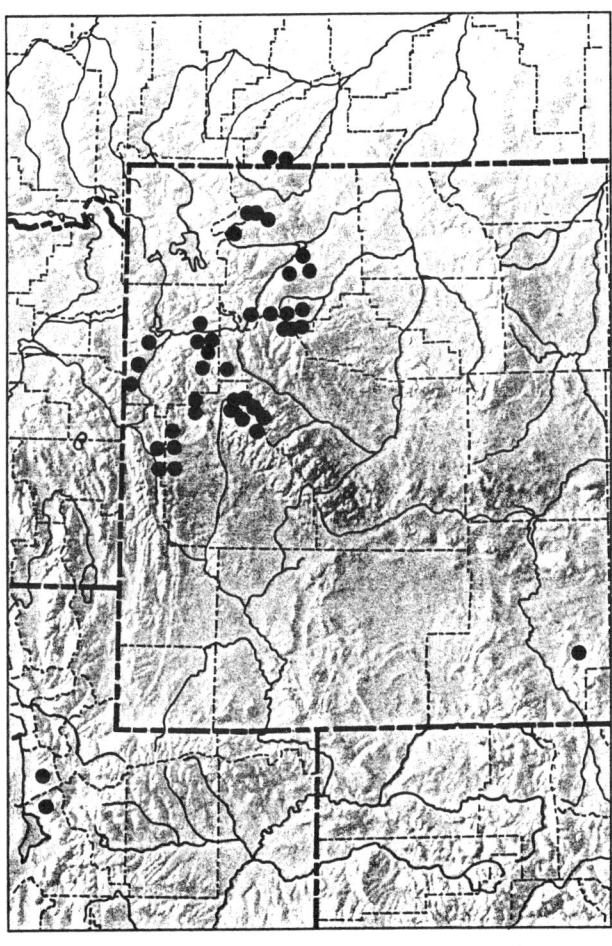

Senecio atratus Greene
Pittonia 3:105, 1896.

Black Groundsel

Senecio milleflorus Greene

Tomentose to glabrate, perennial herb from a simple to branched caudex and short, stout rhizome; stems 1 to several, erect or ascending. Basal leaves petiolate, the blades oblanceolate, oblong, elliptic, or obovate, dentate or sometimes nearly entire; cauline leaves reduced upward, becoming sessile, lanceolate. Heads numerous in a closed to somewhat open, corymbose inflorescence; involucre cylindrical to campanulate, the bracts few (commonly 8), greenish, usually tomentose at the base, the tips dark (often blackish), the margins somewhat scarious. Ray flowers 3–5, yellow; disk flowers yellow. Akenes glabrous; pappus white.

Rocky slopes and fellfields in the lower alpine zone of the Salt River, Uinta, and Wasatch Ranges. Wyoming, Utah, Colorado, and New Mexico.

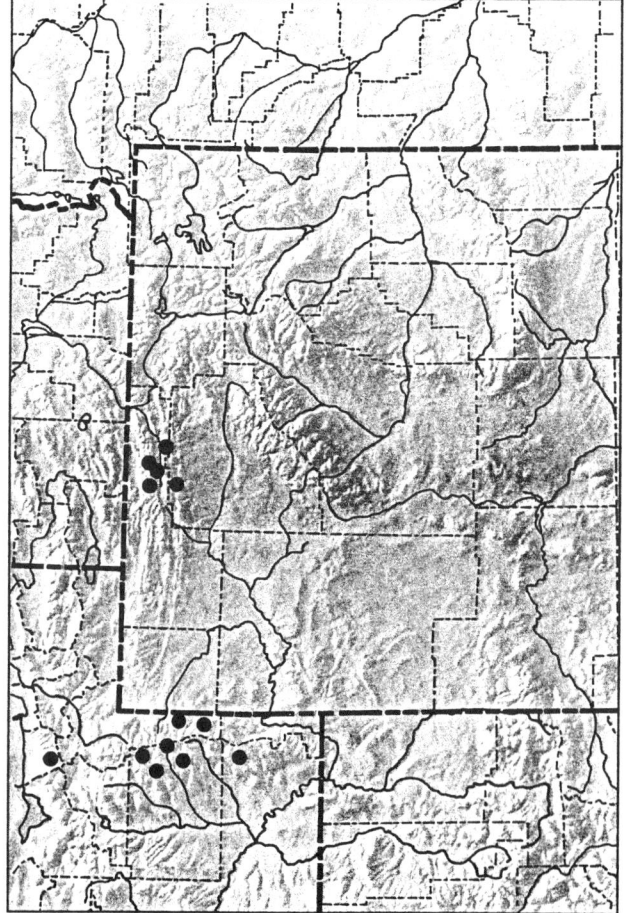

Senecio canus Hook.
Fl. Bor.-Am. 1:333, 1834.

Woolly Groundsel

Packera cana (Hook.) W.A. Weber & A. Löve
Senecio convallium Greenm.
Senecio hallii Britt.
Senecio harbourii Rydb.
Senecio howellii Greene
Senecio kernensis Greenm.
Senecio laramiensis A. Nels.
Senecio oreopolus Greenm.
Senecio purshianus Nutt.

Tufted, white-tomentose, perennial herb from a slender, branched caudex, or rarely with a short taproot; stems 1 to several, erect or ascending. Basal leaves petiolate, the blades oblanceolate to obovate, entire to dentate, the tips generally obtuse; cauline leaves strongly reduced upward, becoming sessile, lanceolate, entire, and bractlike at the base of the inflorescence. Heads several, corymbose; involucre campanulate, glabrous to tomentose, the bracts lanceolate, often with dark-attenuate tips. Ray flowers yellow; disk flowers yellow. Akenes glabrous; pappus whitish.

Fellfields and other rocky places in nearly all alpine ranges of the Middle Rocky Mountains. Not yet known from above timberline in the Salt River Range. British Columbia to Saskatchewan, south to California, Nevada, Colorado, and Kansas.

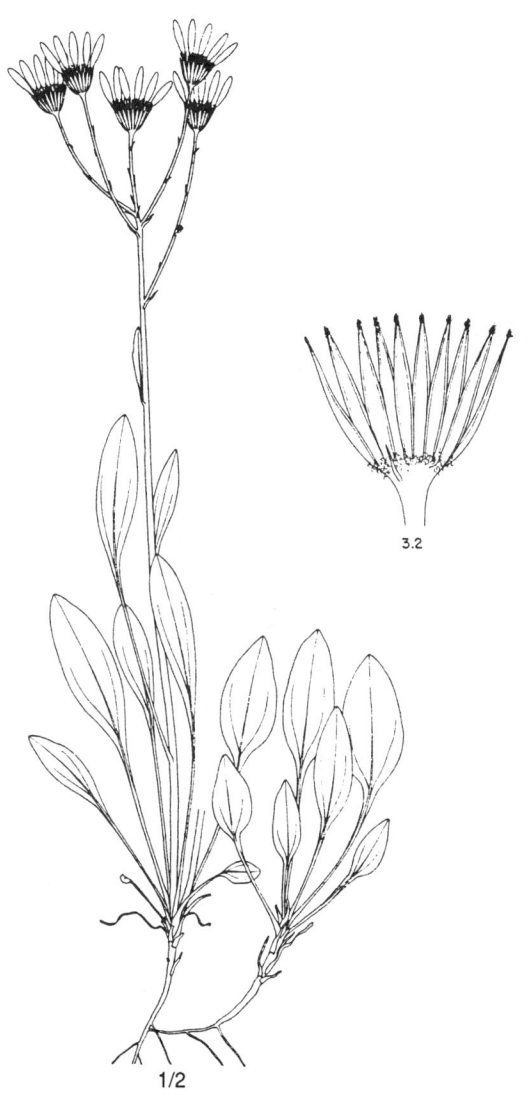

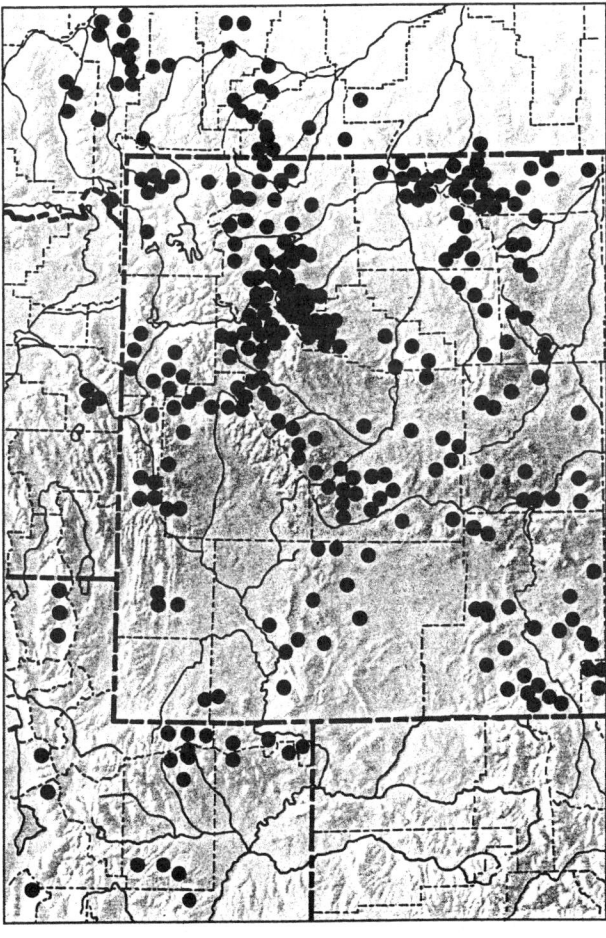

Senecio crassulus Gray
Proc. Am. Acad. 19:54, 1883.

Thickleaf Groundsel

Senecio lapathifolius Greene
Senecio semiamplexicaulis Rydb.

Glabrous, somewhat tufted, perennial herb from a short, erect, simple or branched caudex and fibrous roots; stems 1 to several, erect. Basal leaves petiolate, the blades oblanceolate to obovate or elliptic, entire to denticulate, acute; cauline leaves reduced upward, becoming sessile, lanceolate, and somewhat clasping. Heads several, corymbose; involucre campanulate, the bracts linear to oblong, greenish or brownish and black-tipped, the margins scarious. Ray flowers yellow; disk flowers yellow. Akenes glabrous; pappus whitish.

Meadows and stream banks of most alpine ranges of the Middle Rocky Mountains. Not yet known from above timberline in the Big Horn and Gros Ventre Ranges. Oregon to South Dakota, south to Idaho, Utah, and New Mexico.

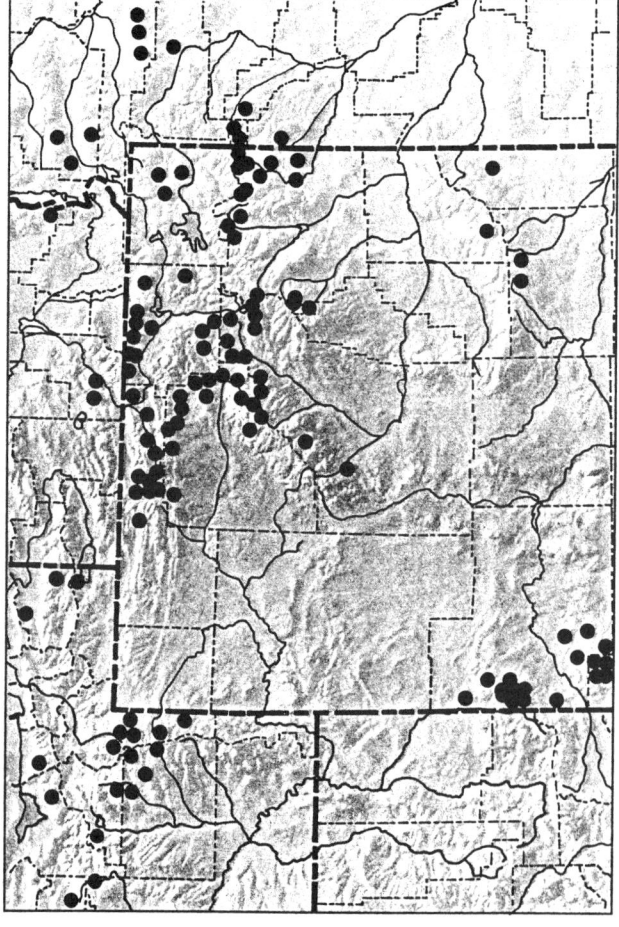

Senecio cymbalarioides Buek
Gen. Sp. Syn. Cand. 2:VI, 1840.

Cleftleaf Groundsel

Packera cymbalarioides (Buek) W.A. Weber & A. Löve
Senecio moresbiensis (Calder & Taylor) Dougl. & Ruyle-Dougl.
Senecio ovinus Greene
Senecio subnudus DC.

Glabrous, perennial herb from a slender rhizome; stems erect, 1 to a few, often floccose-tomentose in leaf axils when young. Basal leaves petiolate, the blades mostly obovate, entire to crenate or dentate; cauline leaves reduced upward, becoming sessile and pinnately lobed or parted. Heads mostly solitary and terminal, sometimes 2 or 3; involucre campanulate, glabrous or basally tomentose, the bracts lanceolate. Ray flowers yellow; disk flowers yellow. Akenes glabrous; pappus whitish.

Moist meadows and stream banks of the Absaroka, Beartooth, Big Horn, and Wind River Ranges. Alaska and Yukon Territory south to California, Nevada, Idaho, Colorado, and New Mexico.

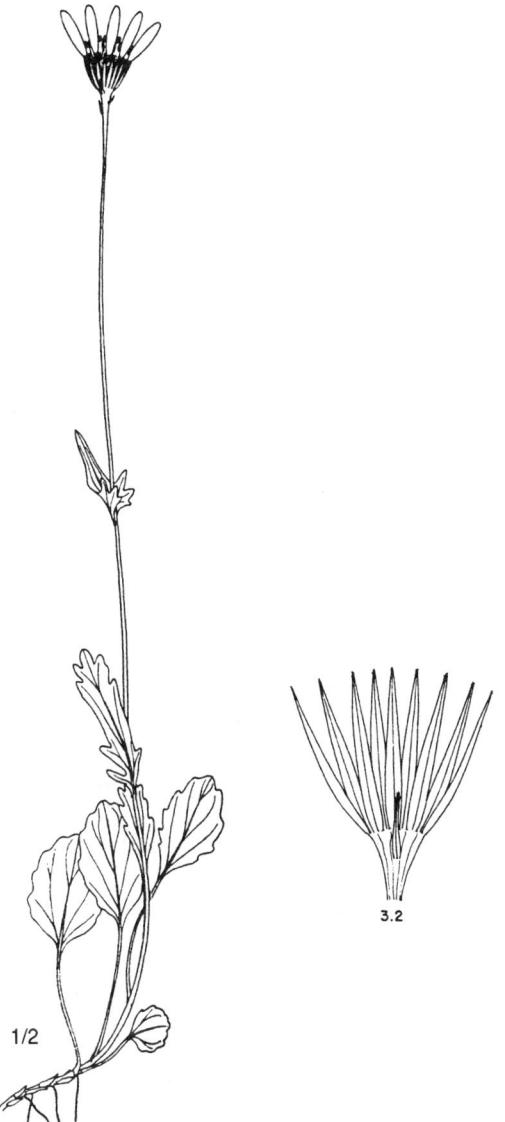

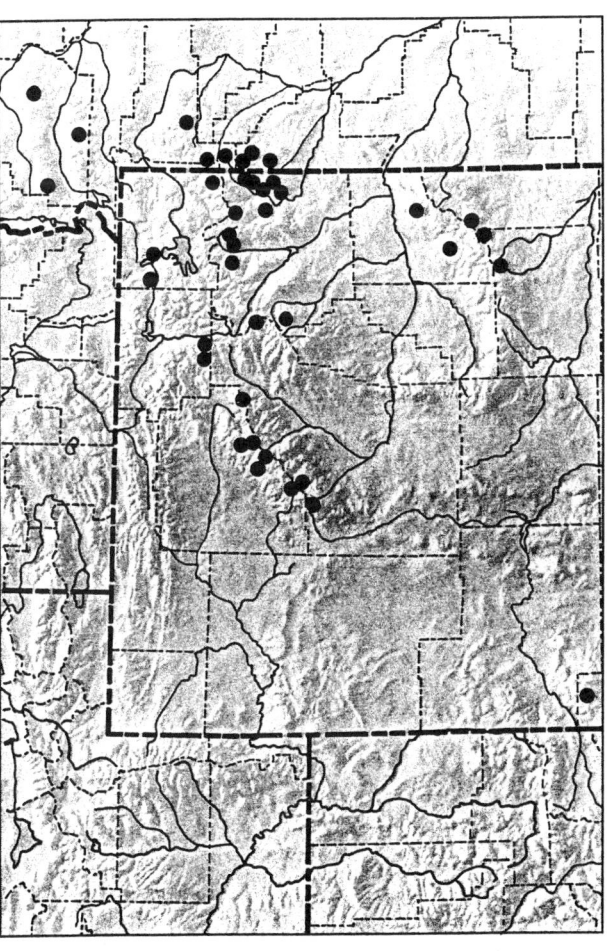

Senecio dimorphophyllus Greene
Pittonia 4:109, 1900.

Twoleaf Groundsel

Packera dimorphophyllus (Greene) W.A. Weber & A. Löve
Senecio heterodoxus Greene ex Rydb.

Glabrous or glabrate, perennial herb from short rhizomes; stems 1 to several, erect, simple or branched. Basal leaves petiolate, the blades ovate, obovate, or subreniform, entire to crenulate, obtuse, often cordate at the base; cauline leaves reduced upward, becoming sessile, pinnately lobed, and clasping at the base. Heads few to many, corymbose; involucre campanulate, glabrous, or sometimes sparsely tomentulose, the bracts lanceolate, attenuate, sometimes reddish on the tips. Ray flowers yellow; disk flowers yellow. Akenes glabrous; pappus white.

Moist rocky places of the Medicine Bow and northwestern Wyoming Ranges. Not yet known from above timberline in the Beartooth, Salt River, Teton, Uinta, Wasatch, and Wyoming Ranges. Wyoming, Utah, and Colorado.

Alpine plants with lax, branched stems 10–25 cm tall and ovate to subreniform blades on the basal leaves are var. *paysonii* T.M. Barkley.

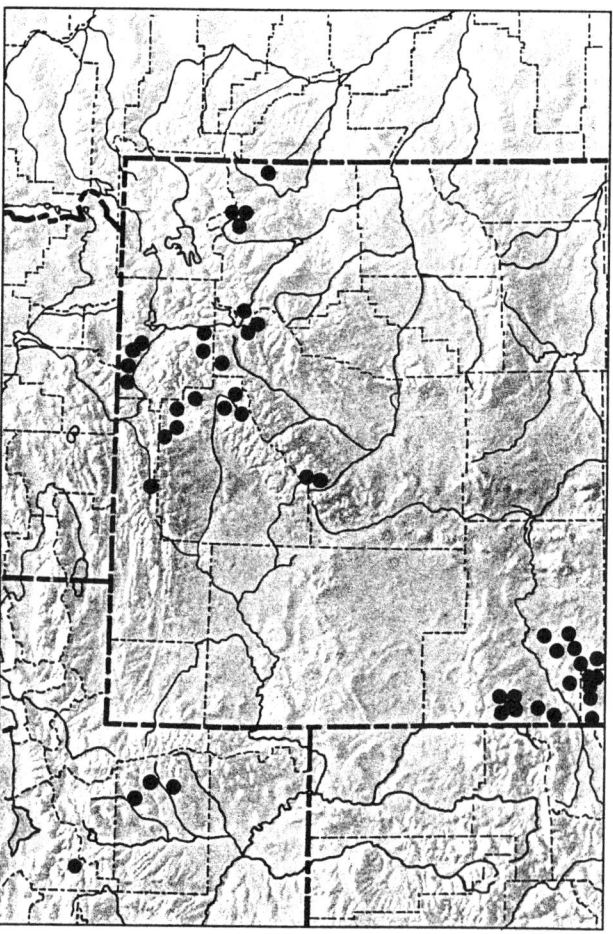

Senecio fremontii T. & G.
Fl. N. Am. 2:445, 1843.

Fremont Groundsel

Senecio blitoides Greene
Senecio carthamoides Greene
Senecio ductoris Piper
Senecio invenustus Greene
Senecio occidentalis Greene
Senecio occidentalis Rydb.

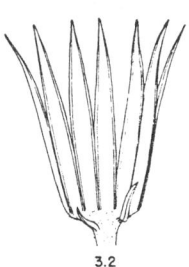

Glabrous, caespitose, perennial herb from a multi-branched caudex and slender taproot; stems several to many, decumbent. Leaves short-petiolate or sessile, oblanceolate, obovate, or spatulate, crenulate-dentate or dentate, rounded, the upper leaves subentire, the lowermost often reduced and dentate-lobed. Heads few to several, corymbose; involucre campanulate, glabrous to finely pubescent, the bracts lanceolate to oblong, attenuate, scarious on the margins. Ray flowers bright yellow; disk flowers yellow. Akenes glabrous to pubescent; pappus white.

Scree, talus, felsenmeer, and late snow beds in all alpine ranges of the Middle Rocky Mountains. British Columbia and Alberta south to California, Utah, and Colorado.

Two varieties occur in the alpine zone of the Middle Rocky Mountains: var. *fremontii* is mostly less than 20 cm tall, with leaf bases scarcely clasping; and var. *blitoides* (Greene) Cronq. is mostly more than 30 cm tall, with leaf bases clasping.

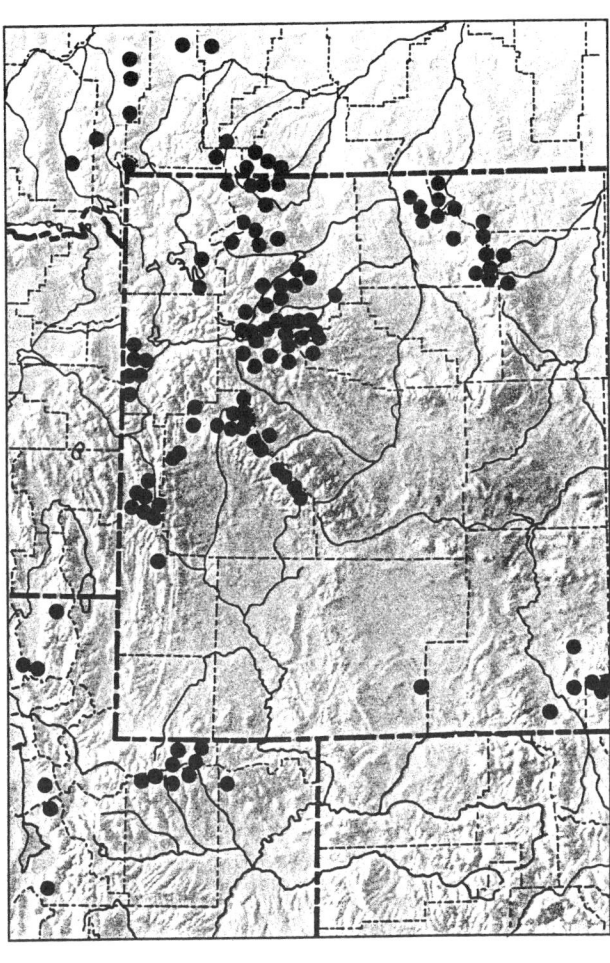

Senecio fuscatus Hayek
Allg. Bot. Zeit. 23:4, 1917.

Dusky Groundsel

Senecio bivestitus Cronq.
Senecio denalii A. Nels.
Senecio lindstroemii (Ostenf.) Pors.
Senecio tundricola Tolm.
Tephroseris fuscata (Hayek) Holub
Tephroseris lindstroemii (Ostenf.) Löve & Löve

Arachnoid-tomentose or villous, perennial herb from a short, erect caudex and fibrous roots; stems mostly solitary, simple. Basal leaves tufted, short-petiolate or sessile, oblanceolate to elliptic, acute or somewhat obtuse, entire, shallow-undulate, or irregularly toothed; cauline leaves reduced upward, sessile, lanceolate, the uppermost somewhat acuminate. Heads large, showy, solitary or few; involucre tomentose or villous, the bracts lanceolate, attenuate, often reddish purple. Ray flowers orange; disk flowers dark orange. Akenes pubescent.

Fellfields of the Absaroka and Beartooth Ranges. Circumpolar; south in widely scattered sites to northwestern Wyoming.

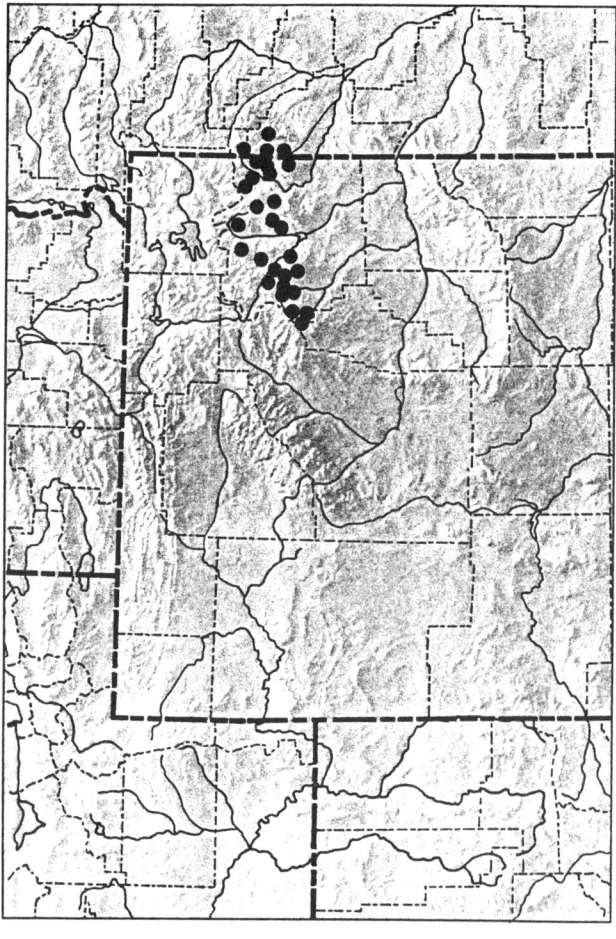

Senecio integerrimus Nutt.
Gen. Pl. 2:165, 1818.

Lambstongue Groundsel

Senecio arachnoideus Rydb.
Senecio atriapiculatus Rydb.
Senecio caulanthifolius Davy
Senecio columbianus Greene
Senecio condensatus Greene
Senecio cordatus Nutt.
Senecio dispar A. Nels.
Senecio exaltatus Nutt.
Senecio flintii Rydb.
Senecio foliosus Rydb.
Senecio fondinarum Greene
Senecio hookeri T. & G.
Senecio leibergii Greene
Senecio majus A. Heller
Senecio mendocinensis Gray
Senecio mesadenia Greene
Senecio ochraceus Piper
Senecio perplexus A. Nels.
Senecio scribneri Rydb.
Senecio sonnei Greene
Senecio solitarius Rydb.
Senecio vaseyi Greenm.
Senecio whippleanus Gray

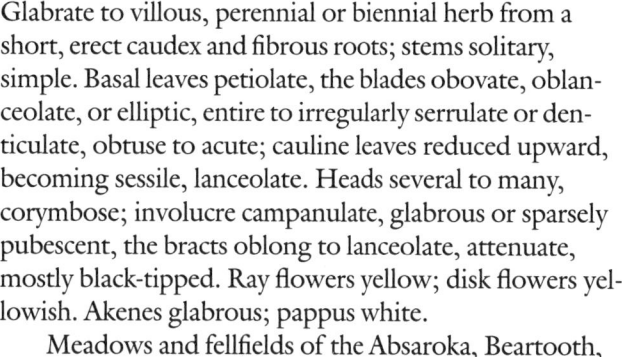

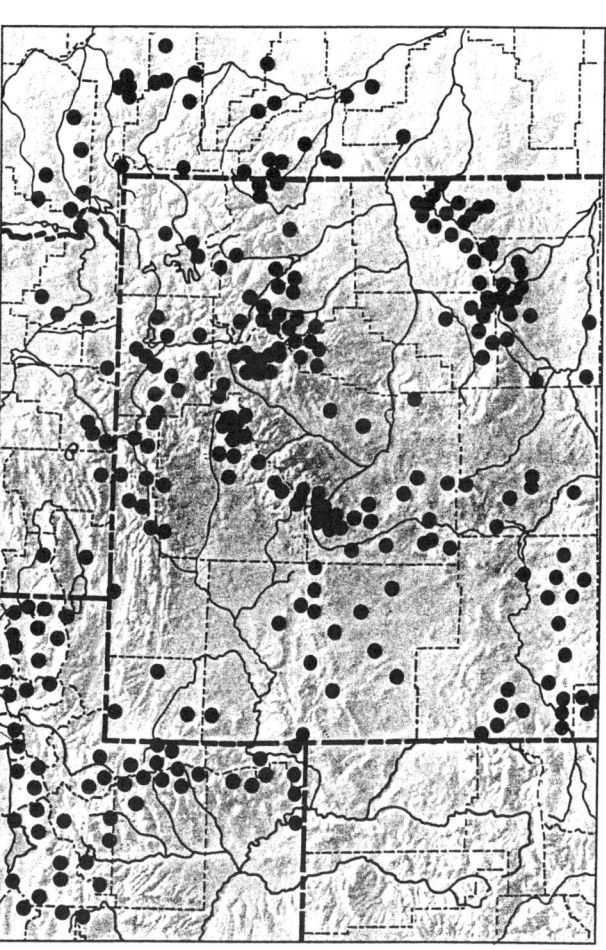

Glabrate to villous, perennial or biennial herb from a short, erect caudex and fibrous roots; stems solitary, simple. Basal leaves petiolate, the blades obovate, oblanceolate, or elliptic, entire to irregularly serrulate or denticulate, obtuse to acute; cauline leaves reduced upward, becoming sessile, lanceolate. Heads several to many, corymbose; involucre campanulate, glabrous or sparsely pubescent, the bracts oblong to lanceolate, attenuate, mostly black-tipped. Ray flowers yellow; disk flowers yellowish. Akenes glabrous; pappus white.

Meadows and fellfields of the Absaroka, Beartooth, Big Horn, Salt River, Teton, Uinta, Wasatch, and Wind River Ranges. British Columbia to Saskatchewan, south to California, Utah, Colorado, Iowa, and Minnesota.

Middle Rocky Mountain plants with black-tipped involucral bracts are var. *exaltatus* (Nutt.) Cronq.

Senecio lugens Richards.
Bot. App. Frankl. J., 748, 1823.

Blacktip Groundsel

Senecio glaucescens Rydb.
Senecio imbricatus Greene

Slender, glabrescent, perennial herb from a short, stout rhizome and fibrous roots; stems usually solitary, simple. Basal leaves petiolate, the blades oblanceolate, obovate, or elliptic, denticulate to nearly entire; cauline leaves few, reduced upward, becoming sessile, lanceolate, entire. Heads few to several, in a compact, corymbose cyme; involucre campanulate, loosely arachnoid to glabrous, the bracts lanceolate to oblanceolate, black-tipped, usually pubescent at the apex. Ray flowers yellow; disk flowers yellow. Akenes glabrous; pappus white.

Meadows and fellfields of the Absaroka, Beartooth, Big Horn, Gros Ventre, and Wind River Ranges. Alaska, Yukon, and Mackenzie district of Northwest Territories south to Washington, Montana, and northwestern Wyoming.

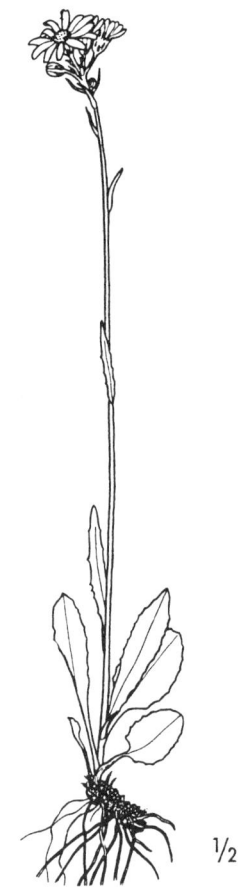

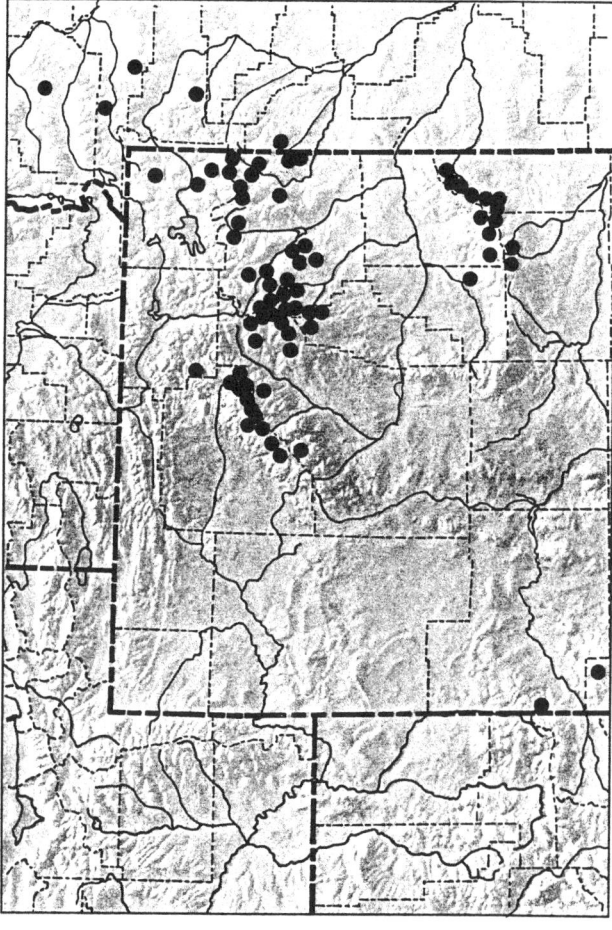

Senecio serra Hook.
Fl. Bor.-Am. 1:333, 1834.

Butterweed Groundsel

Senecio admirabilis Greene
Senecio andinus Nutt.
Senecio lanceolatus T. & G.
Senecio solidago Rydb.

Tall, glabrous to puberulent, perennial herb, to 1.5 m tall, from fibrous roots; rhizomes present or not; stems simple, branching in the inflorescence, single or clustered, erect to ascending. Leaves equally distributed, short petiolate to subsessile, the blades lanceolate or oblanceolate to elliptic, acute to acuminate, serrate to nearly entire, glabrous to glabrate, cuneate at base, somewhat reduced upward. Heads many to numerous in flat-topped, corymbose cymes; involucre campanulate, glabrous or glabrate, the bracts lanceolate to oblong, attenuate, scarious-margined, with a short tuft of hairs at the apex; often black-tipped. Ray flowers few, yellow; disk flowers yellow. Akenes glabrous, or nearly so; pappus white.

Sheltered stream banks and moist, rocky places near timberline in the Absaroka, Beartooth, and Salt River Ranges. Washington south and east to California, Utah, Colorado, Wyoming, and Montana.

Middle Rocky Mountain plants with involucral bracts more than 6 mm long and disk flowers 15 or more are var. *admirabilis* (Greene) A. Nels.

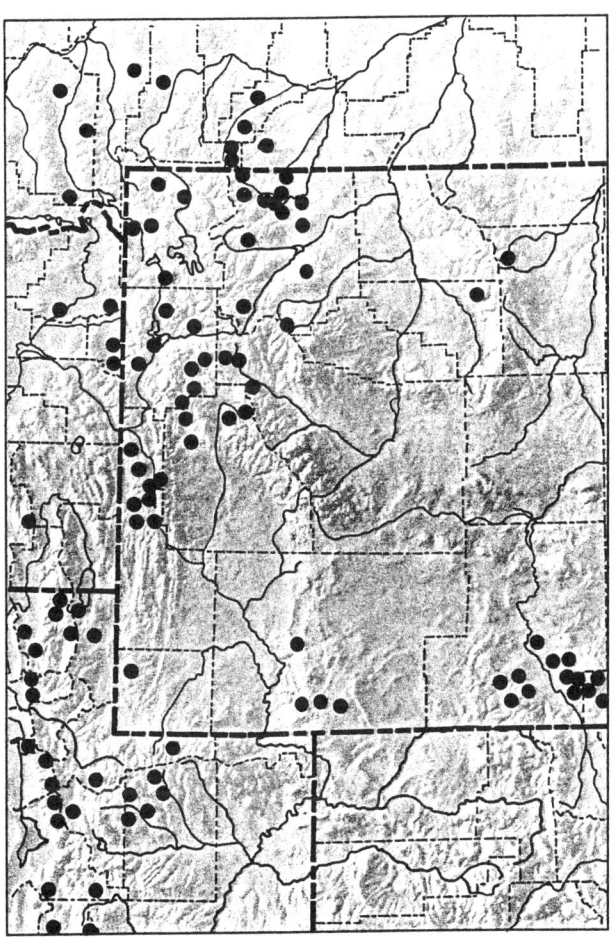

Senecio sphaerocephalus Greene
Pittonia 3:106, 1896.

Roundhead Groundsel

Senecio altus Rydb.
Senecio latus Rydb.
Senecio oreganus T.J. Howell

Glabrate, perennial herb from a short, horizontal rhizome and fibrous roots; stems erect or ascending, solitary, simple. Basal leaves petiolate, the blades oblanceolate to obovate or elliptic, entire or denticulate, obtuse; cauline leaves few, strongly reduced upward, becoming sessile, lanceolate, entire, and bracteate. Heads few to many, corymbose; involucre campanulate, glabrous or pubescent at the base, the bracts lanceolate, inconspicuously black-tipped, pubescent at the apex. Ray flowers yellow; disk flowers yellow. Akenes pubescent; pappus white.

Timberline meadows of the Absaroka and Beartooth Ranges. Oregon to Montana, south to Idaho, Utah, and Colorado.

Senecio streptanthifolius Greene
Erythea 3:23, 1895.

Twistedleaf Groundsel

Packera oodes (Rydb.) W.A. Weber
Packera streptanthifolia (Greene) W.A. Weber & A. Löve
Senecio acutidens Rydb.
Senecio adamsi T.J. Howell
Senecio aquariensis Greenm.
Senecio chapacensis Greene
Senecio cognatus Greene
Senecio cymbalarioides Nutt.
Senecio dileptiifolius Greene
Senecio farriae Greenm.
Senecio fraternus Piper
Senecio fulgens Rydb.
Senecio jonesii Rydb.
Senecio laetiflorus Greene
Senecio leonardi Rydb.
Senecio longipetiolatus Rydb.
Senecio oodes Rydb.
Senecio pammelii Greenm.
Senecio platylobus Rydb.
Senecio rubricaulis Greene
Senecio rydbergii A. Nels.
Senecio subcuneatus Rydb.
Senecio suksdorfii Greenm.
Senecio wardii Greene
Senecio willingii Greenm.

Glabrous or glabrate, perennial herb from a caudex or short rhizome and fibrous roots; stems solitary or few, erect, simple. Basal leaves petiolate, the blades elliptic to suborbicular, entire or dentate; cauline leaves reduced upward, becoming sessile, pinnately parted, scarcely clasping at the base. Heads few to many, corymbose; involucre campanulate, the bracts lanceolate to oblong, glabrous, with a short tuft of hairs at the apex. Ray flowers bright yellow; disk flowers yellow. Akenes glabrous; pappus white.

Moist, rocky slopes of the Absaroka, Teton, Uinta, Wasatch, Wind River, and Wyoming Ranges. Yukon Territory and Mackenzie district of Northwest Territories south to California, Utah, and New Mexico.

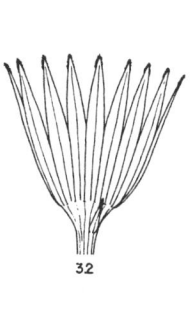
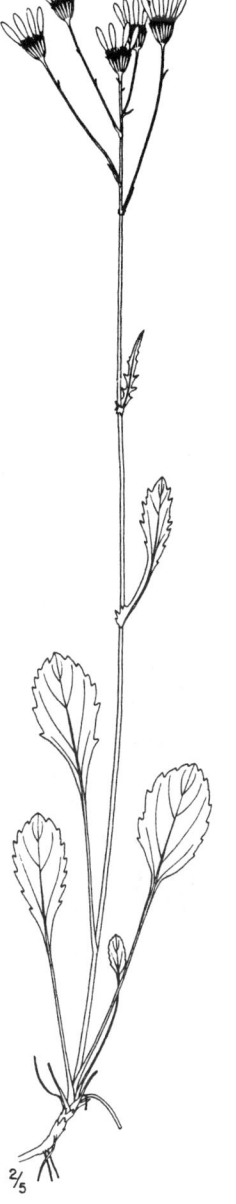
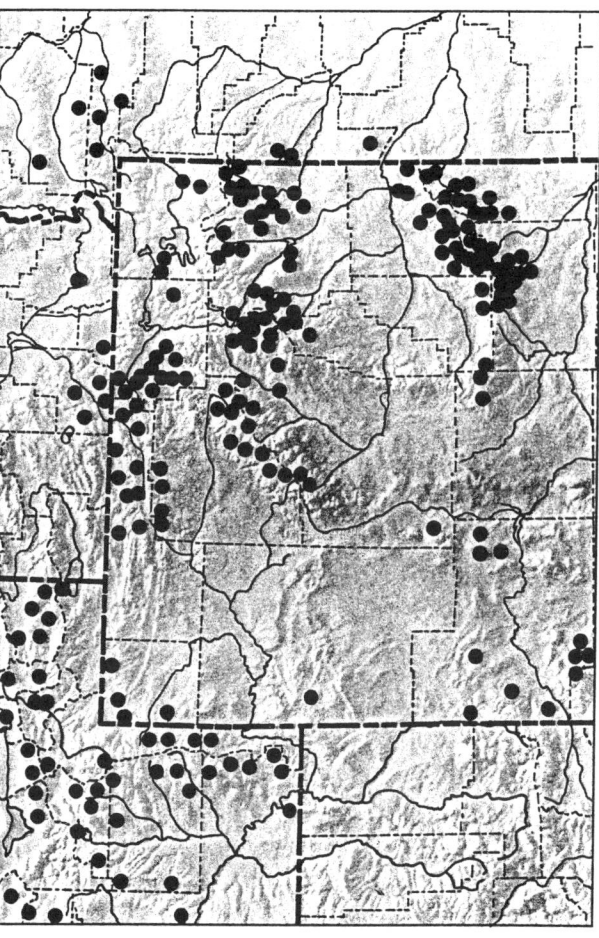

Senecio triangularis Hook.
Fl. Bor.-Am. 1:332, 1834.

Arrowleaf Groundsel

Senecio gibbonsii Greene
Senecio longidentatus DC.
Senecio prionophyllus Greene
Senecio saliens Rydb.
Senecio subvestitus T.J. Howell
Senecio trigonophyllus Greene
Senecio variifolius Rydb.

Tall, coarse, perennial herb, to 1.5 m tall, from a well-developed rhizome and fibrous roots; stems several to many, erect to ascending, glabrous, puberulent, or floccose near the nodes. Leaves equally distributed, petiolate, the blades triangular or triangular-hastate, acute, sharply serrate-dentate, glabrous, mostly truncate at the base, somewhat reduced upward. Heads several to many, in flat-topped, corymbose cymes; involucre campanulate, glabrous, the bracts oblong, attenuate, scarious-margined, with a short tuft of hairs at the apex. Ray flowers bright yellow; disk flowers yellow. Akenes glabrous; pappus white.

Stream banks and moist, rocky places near timberline in the Absaroka, Beartooth, Teton, Uinta, Wasatch, and Wind River Ranges. Alaska, Yukon Territory, and Mackenzie district of Northwest Territories south to California, Idaho, Utah, and New Mexico.

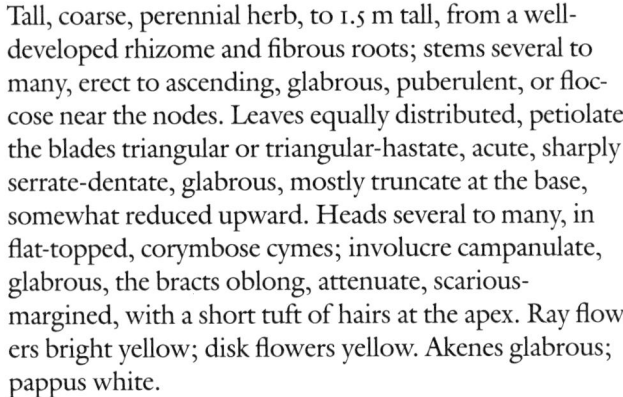

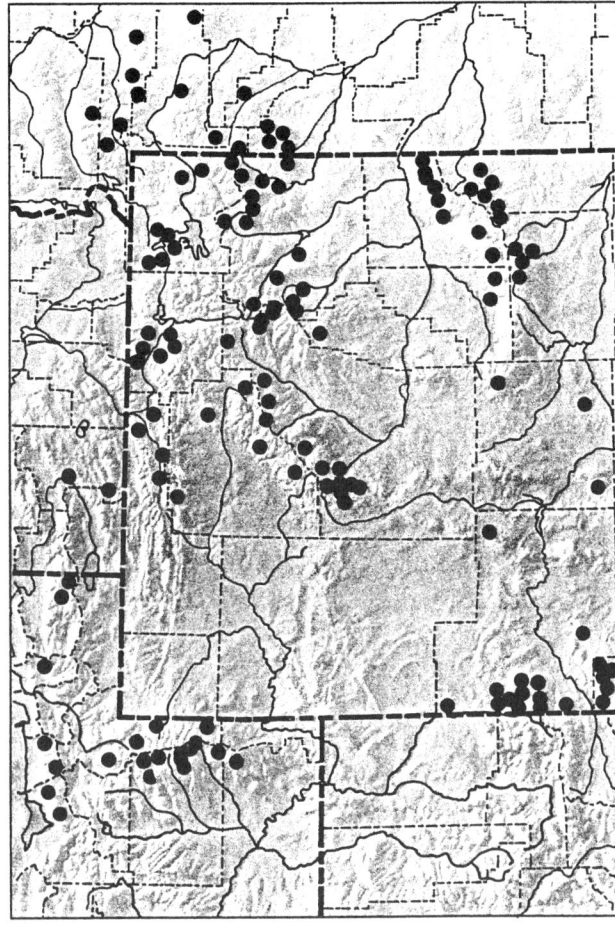

Senecio werneriifolius (Gray) Gray
Proc. Am. Acad. 19:54, 1883.

Mountain Groundsel

Packera werneriifolia (Gray) W.A. Weber & A. Löve
Senecio alpicola Rydb.
Senecio molinarius Greenm.
Senecio muirii Greenm.
Senecio pentodontus Greene
Senecio perennans A. Nels.
Senecio petraeus Klatt
Senecio petrocallis Greene
Senecio petrophilus Greene
Senecio saxosus Klatt
Senecio scaposus A. Nels.
Senecio speculicola J.T. Howell
Senecio turbinatus Rydb.

Low, tufted, subscapose, perennial herb, less than 20 cm tall, from a branched caudex; stems few to several, erect or ascending, tomentose, becoming glabrate with age. Basal leaves petiolate, the blades oblanceolate, elliptic, or spatulate, entire or denticulate near the tip, the margins slightly revolute; cauline leaves few, much reduced, sessile, lanceolate, the upper ones bracteate. Heads solitary or 2–6 in an open cyme; involucre campanulate, tomentose at the base, the bracts lanceolate, often purplish. Ray flowers yellow or lacking; disk flowers yellow. Akenes glabrous; pappus white.

Talus, scree, and other rocky places in the Absaroka, Beartooth, Gros Ventre, Teton, Uinta, Wasatch, and Wind River Ranges. Montana south to California, Nevada, Arizona, and Colorado.

Dorn (1992) recognized 2 varieties, both of which may occur in the alpine zone:

1. Plants usually more than 10 cm tall, erect; leaf blades oblanceolate . . . var. *werneriifolius*
1. Plants less than 10 cm tall, compact; leaf blades obovate to orbicular . . . var. *alpinus* (Gray) Dorn

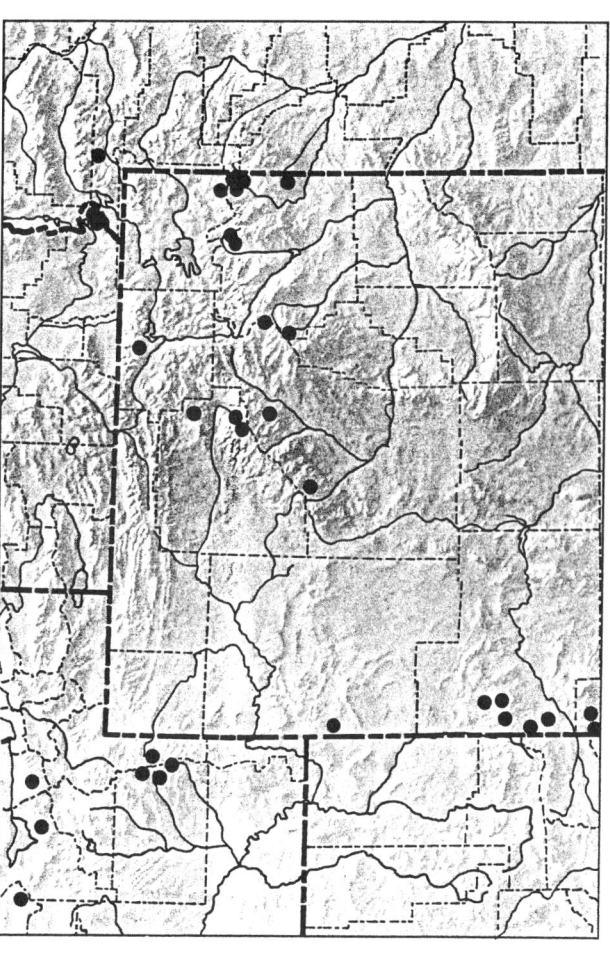

Solidago L.
Goldenrod

Perennial herbs from fibrous roots and a rhizome or caudex, with simple, alternate, entire or variously toothed leaves. Heads small, radiate in ours, in racemes or corymbs, the involucral bracts imbricate, in several series, usually with chartaceous bases and green tips, the receptacle flat, naked. Ray flowers yellow, pistillate, and fertile. Disk flowers yellow, perfect, and fertile. Akenes terete or angular, glabrous or hairy, 5–10-nerved, the pappus of white, capillary bristles.

KEY TO THE SPECIES OF *SOLIDAGO*

1. Stems glabrous; ray flowers 5–10 — *S. simplex*
1. Stems pubescent; ray flowers 10–20
    2. Petioles of lower leaves spreading-ciliate; stems villous with multicellular hairs — *S. multiradiata*
    2. Petioles of lower leaves with stiff, appressed hairs; stems puberulent with short, incurved hairs — *S. parryi*

Solidago multiradiata Ait.
Hort. Kew. 3:218, 1789.

Northern Goldenrod

Aster multiradiatus (Ait.) Kuntze
Solidago ciliosa Greene
Solidago corymbosa Nutt.
Solidago cusickii Piper
Solidago dilatata A. Nels.
Solidago laevicaulis Rydb.
Solidago scopulorum (Gray) A. Nels.

Somewhat tufted, perennial herb from a slender, branched rhizome; stems few to several, pubescent at least above, decumbent and purplish at the base. Basal leaves petiolate, the petioles ciliate-margined; blades oblanceolate or spatulate, glabrous, entire or serrate, acute to obtuse. Cauline leaves reduced upward, becoming sessile, lanceolate, entire. Heads few to many, racemose or corymbose; involucre campanulate, the bracts linear to lanceolate, ciliate, acute or acuminate, with a prominent midvein. Ray flowers linear to elliptic, 3-toothed, yellow, fewer than the disk flowers; disk flowers yellow. Akenes pubescent; pappus white.

Fellfields and meadows in all alpine ranges of the Middle Rocky Mountains. Siberia; Alaska south to California, Utah, and New Mexico, east to Quebec.

Middle Rocky Mountain plants with ciliate-margined petioles and bracts not sharply acute are var. *scopulorum* Gray.

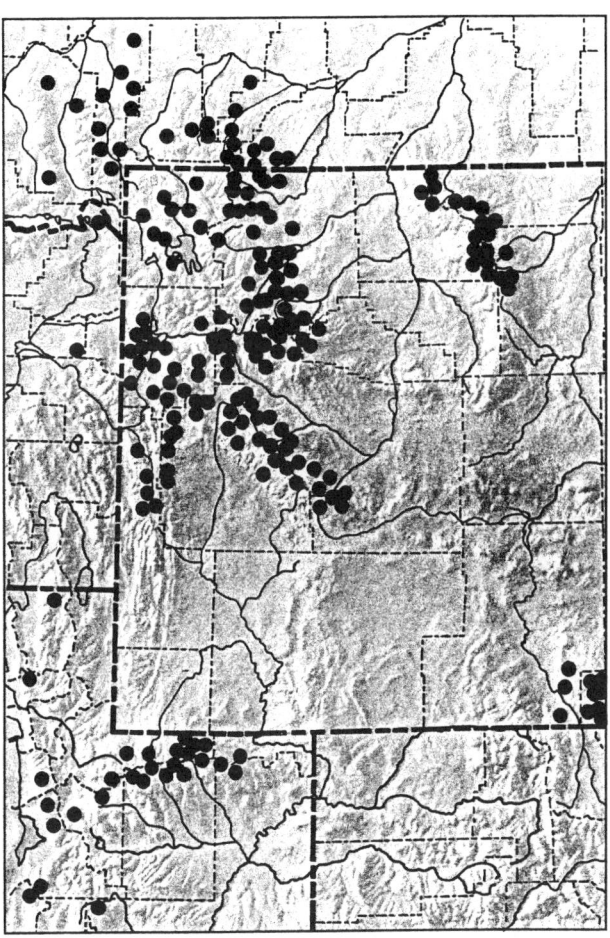

Solidago parryi (Gray) Greene
Erythea 2:57, 1894.

Parry Goldenrod

Aplopappus parryi Gray
Aster minor Kuntze
Oreochrysum parryi (Gray) Rydb.

Perennial herb from slender rhizomes; stems erect or ascending, pubescent at least above, often glabrous below. Basal leaves petiolate, the petioles scabrous or strigose; blades oblanceolate to elliptic, entire, glabrous or puberulent, obtuse. Cauline leaves reduced upward, becoming sessile and somewhat clasping. Heads few to many, cymose; involucre campanulate to hemispheric, the bracts ovate to lanceolate, ciliate, scarious-margined to chartaceous. Ray flowers yellow; disk flowers yellow. Akenes glabrate; pappus white.

Meadows, fellfields, and talus of the Medicine Bow, Uinta, and Wasatch Ranges. Wyoming, Utah, Colorado, Arizona, and New Mexico.

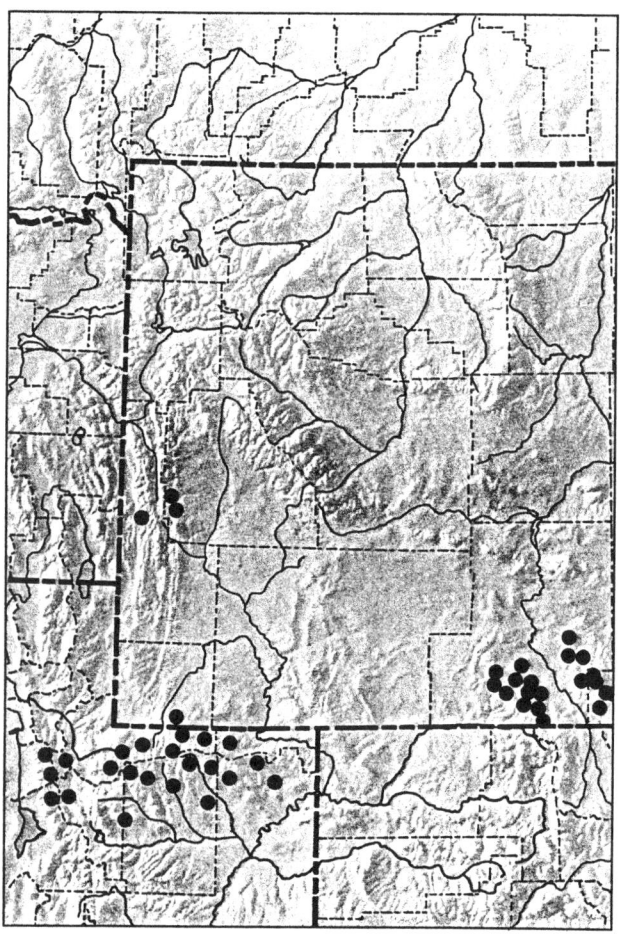

Solidago simplex H.B.K.
Nov. Gen. Sp. 4:81, 1818.

Alpine Goldenrod

Aster candollei Kuntze
Homopappus spathulatus (DC.) Nutt.
Solidago bellidifolia Greene
Solidago confertifolia DC.
Solidago decumbens Greene
Solidago gillmanii (Gray) Steele
Solidago glutinosa Nutt.
Solidago hesperius T.J. Howell
Solidago neomexicana (Gray) Woot. & Standl.
Solidago oreophila Rydb.
Solidago racemosa Green
Solidago randii (Porter) Britt.
Solidago spathulata DC.
Solidago spiciformis T. & G.
Solidago vespertina Piper
Solidago yukonensis Gand.

Mostly tufted, glabrous to puberulent, perennial herb from a stout caudex or short rhizome; stems erect or ascending, often decumbent and purplish at the base. Basal leaves petiolate, the blades oblanceolate or spatulate, entire to serrate near tip, rounded in alpine specimens; cauline leaves strongly reduced upward, becoming sessile, oblanceolate or oblong, entire, and somewhat clasping. Heads few to several, racemose; involucre campanulate, glabrous, the bracts oblong, imbricate, obtuse, with a prominent midvein. Ray flowers yellow, the corollas ovate to elliptic, 3-toothed; disk flowers yellow. Akenes pubescent; pappus white.

Fellfields, rocky slopes, and ridges of the Absaroka, Beartooth, Medicine Bow, Salt River, Uinta, and Wind River Ranges. Alaska, Yukon Territory, and Mackenzie district of Northwest Territories south to California, Utah, Arizona, and New Mexico, and east to Quebec.

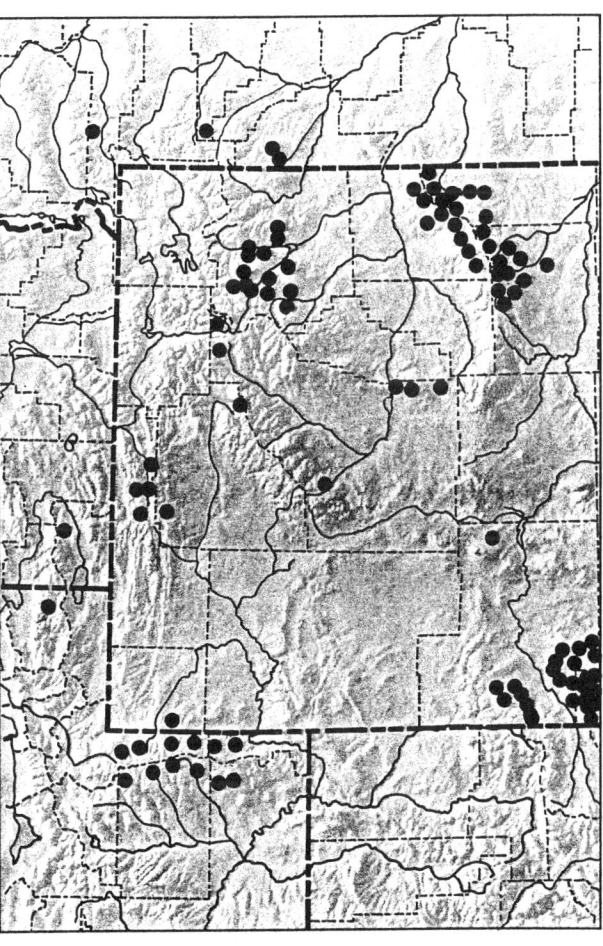

Sphaeromeria Nutt.
False Sagebrush

Shrubs or subshrubs with alternate, sometimes mostly basal, entire to pinnately or palmately lobed, glandular-punctate leaves. Heads several to many in terminal, capitate or corymbose inflorescences, the involucral bracts subequal or imbricate in 2–3 series; receptacle conical, naked. Flowers all tubular, yellow, perfect, the outer ones usually pistillate. Akenes slender, somewhat compressed, glabrous to glandular, 5–10-nerved; pappus lacking, or nearly so.

Sphaeromeria diversifolia (D.C. Eat.) Rydb.
N. Am. Fl. 34:242, 1916.

False Sagebrush

Tanacetum diversifolium D.C. Eat.

Low, dense subshrub, less than 40 cm tall; twigs of the year green, glabrous. Leaves alternate, sessile, simple, entire, and linear to pinnately lobed, glandular-punctate with small, scattered, dolabriform hairs. Heads few to several, in mostly flat-topped corymbs. Involucre campanulate, the bracts imbricate, yellowish, obtuse, the outer ones oblong and somewhat reduced, the inner ones broadly ovate with scarious margins. Disk corollas yellow, campanulate, glandular-dotted, the outer ones pistillate, the inner ones perfect. Akenes slender, compressed-terete, glandular; pappus lacking.

Scree and talus slopes of the Wasatch Range. Endemic to Nevada and Utah.

Taraxacum G.H. Weber ex Wiggers
Dandelion

Scapose, perennial herbs with milky juice, from taproots, the leaves pinnatifid, toothed, or rarely entire. Heads solitary and terminal on a hollow peduncle, the bracts of the involucre in 2 series, the outer usually shorter than the inner ones and often reflexed, the receptacle naked. Flowers all ligulate, yellow. Akenes oblong or fusiform, 4–5-ribbed, tapering into a spinulose beak bearing a pappus of persistent, white, capillary bristles.

KEY TO THE SPECIES OF *TARAXACUM*

1. Akenes red, reddish brown, or reddish purple at maturity; leaves entire or shallowly toothed .... *T. eriophorum*
1. Akenes blackish, brown, or olive at maturity; leaves deeply toothed to pinnatifid
    2. Akenes dark brown to blackish; plants mostly slender, 3–15 cm tall; involucre 6–12 mm high, the inner bracts ovate .... *T. lyratum*
    2. Akenes olive to brown; plants mostly coarse, 8–30 cm tall; involucre 12–22 mm high, the inner bracts lanceolate
        3. Outer involucral bracts appressed to somewhat spreading; inner involucral bracts usually corniculate .... *T. ceratophorum*
        3. Outer involucral bracts reflexed; inner involucral bracts not corniculate .... *T. officinale*

Taraxacum ceratophorum (Ledeb.) DC.
Prodromus 7:146, 1838.

Tundra Dandelion

Leontodon ceratophorus Ledeb.
Leontodon dumetorum (Greene) Rydb.
Leontodon leiospermum (Rydb.) Rydb.
Leontodon monticola (Nutt.) Rydb.
Taraxacum ambigens Fern.
Taraxacum amphiphron Böcher
Taraxacum arctogenum Dahlst.
Taraxacum brachyceras Dahlst.
Taraxacum carneocoloratum A. Nels.
Taraxacum carthamopsis Pors.
Taraxacum dumetorum Greene
Taraxacum eurylepium Dahlst.
Taraxacum hyperboreum Dahlst.
Taraxacum ingratum Hagl.
Taraxacum lacerum Greene
Taraxacum lapponicum Kihlm. ex Hand.-Maz.
Taraxacum latispinulosum M.P. Christens.
Taraxacum laurentianum Fern.
Taraxacum leiospermum Rydb.
Taraxacum longii Fern.
Taraxacum malteanum Dahlst.
Taraxacum maurolepium Hagl.
Taraxacum mitratum Hagl.
Taraxacum montanum Nutt.
Taraxacum multisimum Hagl.
Taraxacum naevosum Dahlst.
Taraxacum ovinum Rydb.
Taraxacum paucisquamosum M.E. Peck
Taraxacum pellianum Pors.
Taraxacum pseudonorvegicum Dahlst.
Taraxacum purpuridens Dahlst.
Taraxacum torngatense Fern.
Taraxacum trigonolobum Dahlst.
Taraxacum umbrinum Dahlst.

Scapose, perennial herb from a simple, or sometimes branched, caudex and taproot; scape to 30 cm tall, somewhat villous, at least below the head. Leaves lanceolate to oblanceolate, nearly entire to sinuate-dentate or pinnatifid, glabrous to sparsely pubescent, the petioles broadly winged. Heads solitary, large, showy; involucral bracts appressed to ascending, the outer ones ovate to lanceolate and much shorter than the narrowly lanceolate, corniculate inner bracts. Ray flowers yellow. Akenes brownish or olive, the beak 2–4 times as long as the body; pappus white.

Meadows and fellfields of most alpine ranges of the Middle Rocky Mountains. Not yet known from above timberline in the Teton, Wasatch, and Wyoming Ranges. Circumboreal; south in western North America to California, Utah, and New Mexico.

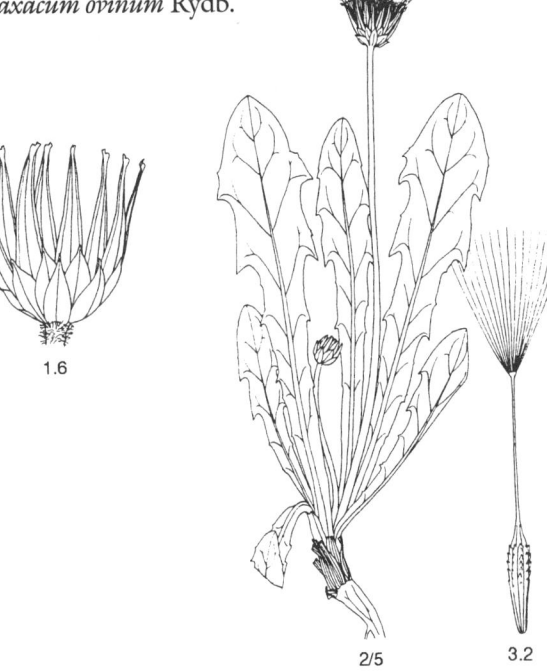

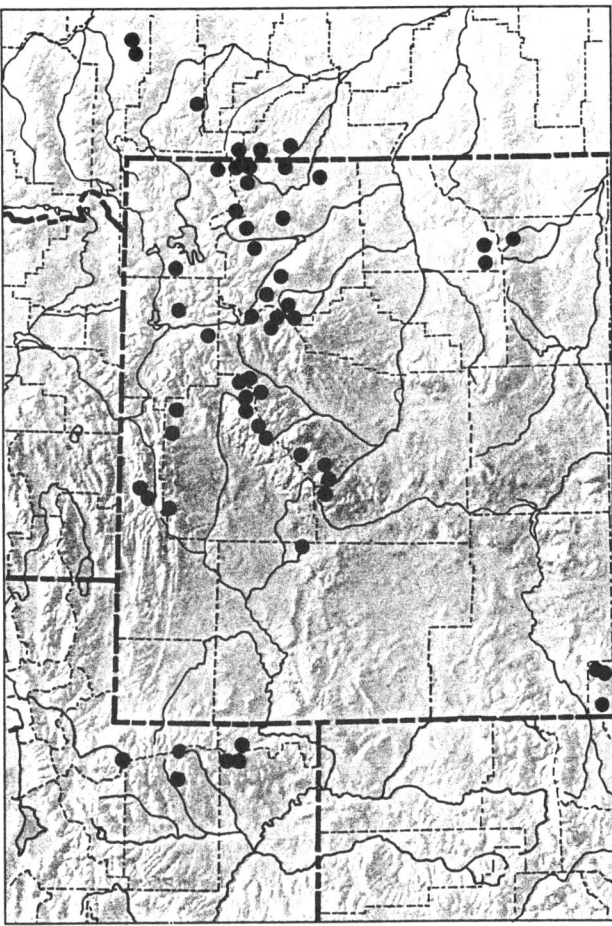

Taraxacum eriophorum Rydb.
Mem. N.Y. Bot. Gard. 1:454, 1900.

Rocky Mountain Dandelion

Leontodon ammophilum (A. Nels. ex Greene) Rydb.
Leontodon angustifolium (Greene) Rydb.
Leontodon eriophorum (Rydb.) Rydb.
Taraxacum ammophilum A. Nels. ex Greene
Taraxacum angustifolium Greene
Taraxacum olympicum G.N. Jones

Scapose, perennial herb from a taproot; scape to 20 cm tall, glabrous to sparsely pubescent. Leaves lanceolate to oblanceolate, entire to rather shallowly sinuate-dentate, glabrous or sparsely pilose. Heads solitary, showy; involucral bracts appressed to ascending, in 2 distinct series, the outer ones ovate and much shorter than the broadly lanceolate, scarious, noncorniculate inner bracts. Ray flowers yellow. Akenes red, reddish brown, or reddish purple, sharply quadrangular, the beak 2–4 times as long as the body; pappus white.

Timberline meadows of the Absaroka Range. Alaska to Washington, Wyoming, and Colorado.

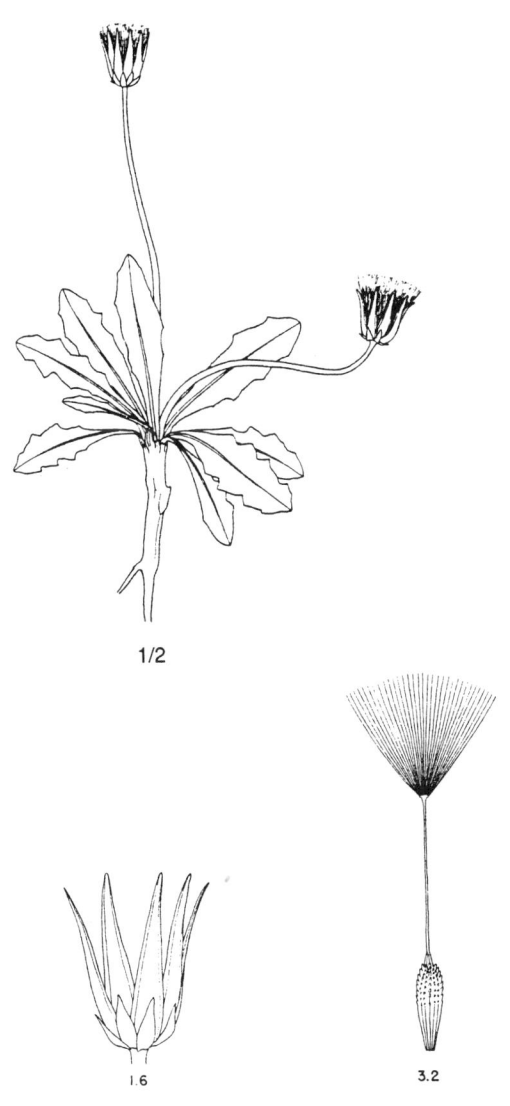

Taraxacum lyratum (Ledeb.) DC.
Prodromus 7:148, 1838.

Alpine Dandelion

Leontodon lyratus Ledeb.
Leontodon rupestre (Greene) Rydb.
Leontodon scopulorum (Gray) Rydb.
Taraxacum kamtschaticum Dahlst.
Taraxacum rupestre Greene
Taraxacum scopulorum (Gray) Rydb.
Taraxacum sibiricum Dahlst.

Low, glabrous, scapose, perennial herb from a taproot; scape less than 15 cm tall. Leaves lanceolate to oblanceolate, subentire to pinnatifid. Heads solitary, small; involucral bracts noncorniculate, blackish green, in 2 series, the outer ones ovate, appressed, and shorter than the broadly lanceolate, oblong, or mostly ovate inner bracts. Ray flowers yellow. Akenes dark brown to blackish, the beak about the same length as the body; pappus white.

Fellfields, talus, scree, and rocky ledges of most alpine ranges at high elevations in the Middle Rocky Mountains. Not yet known from above timberline in the Hoback, Salt River, and Wasatch Ranges. Siberia; Alaska and Yukon Territory south to Nevada, Arizona, and Colorado.

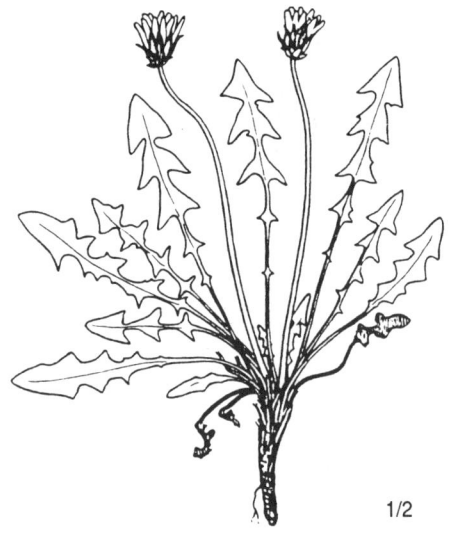

1/2

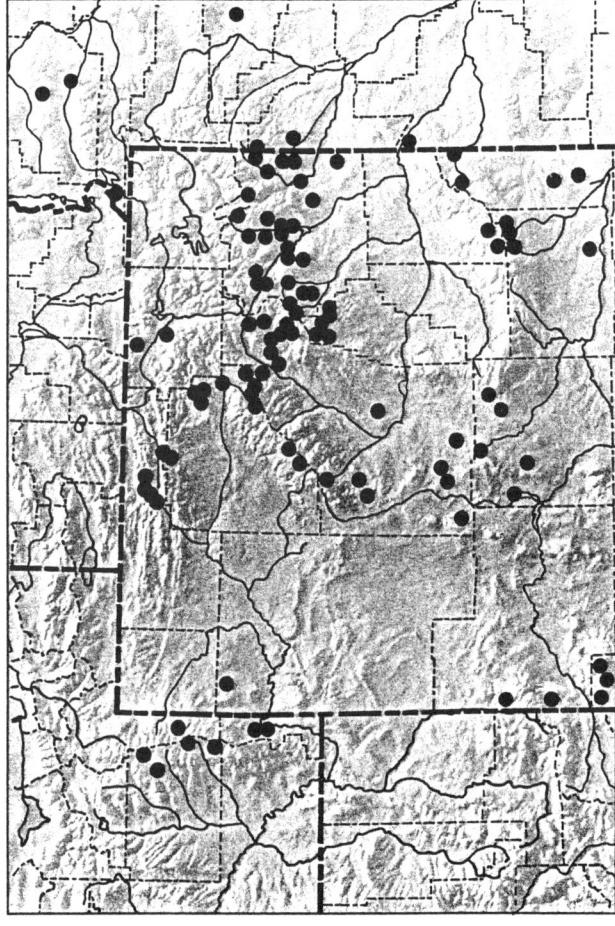

Taraxacum officinale G.H. Weber ex Wiggers
Prim. Pl. Holst., 56, 1780.

Common Dandelion

Leontodon latiloba (DC.) Britt.
Leontodon mexicanum (DC.) Rydb.
Leontodon taraxacum L. in part
Leontodon vulgare Lam.
Taraxacum atroglaucum M.P. Christens.
Taraxacum campylodes Hagl.
Taraxacum croceum auctt., non Dahlst.
Taraxacum curvidens M.P. Christens.
Taraxacum cyclocentrum M.P. Christens.
Taraxacum dahlstedtii Lindb. f.
Taraxacum davidssonii M.P. Christens.
Taraxacum devians Dahlst.
Taraxacum dilutisquameum M.P. Christens.
Taraxacum firmum Dahlst.
Taraxacum islandiciforme Dahlst.
Taraxacum kok-saghyz auctt., non Rodin
Taraxacum latilobum DC.
Taraxacum mexicanum DC.
Taraxacum palustre (Lyons) Symons in part
Taraxacum plentiflorum M.P. Christens.
Taraxacum retroflexum Lindb. f.
Taraxacum rhodolepis Dahlst.
Taraxacum taraxacum (L.) Karst.
Taraxacum turforsum (Schultz-Bip.) Soest
Taraxacum undulatum Lindb. f. & Marklund
Taraxacum vagans Hagl.
Taraxacum xanthostigma Lindb. f.
Taraxacum vulgare (Lam.) Schrank

Scapose, glabrous to villous, perennial herb from a simple or branched caudex and an often stout taproot; scape less than 30 cm tall in alpine plants, glabrous or villous, at least above. Leaves oblanceolate, pinnatifid or pinnately lobed, the terminal lobe broader and larger than the lateral ones, the blade tapering to a broad petiolar base. Heads solitary, large, showy; involucral bracts in 2 series, the outer ones broadly lanceolate and early-reflexed, the inner ones lanceolate, attenuate, noncorniculate, erect to ascending or somewhat reflexed with age. Ray flowers yellow. Akenes gray-brown to olive-brown, the beak 2–4 times as long as the body; pappus white.

Trailsides and disturbed habitats near trails and roads in most of the alpine ranges of the Middle Rocky Mountains. Not yet known from above timberline in the Gros Ventre, Hoback, and Salt River Ranges. Circumboreal; transcontinental in Canada and the United States.

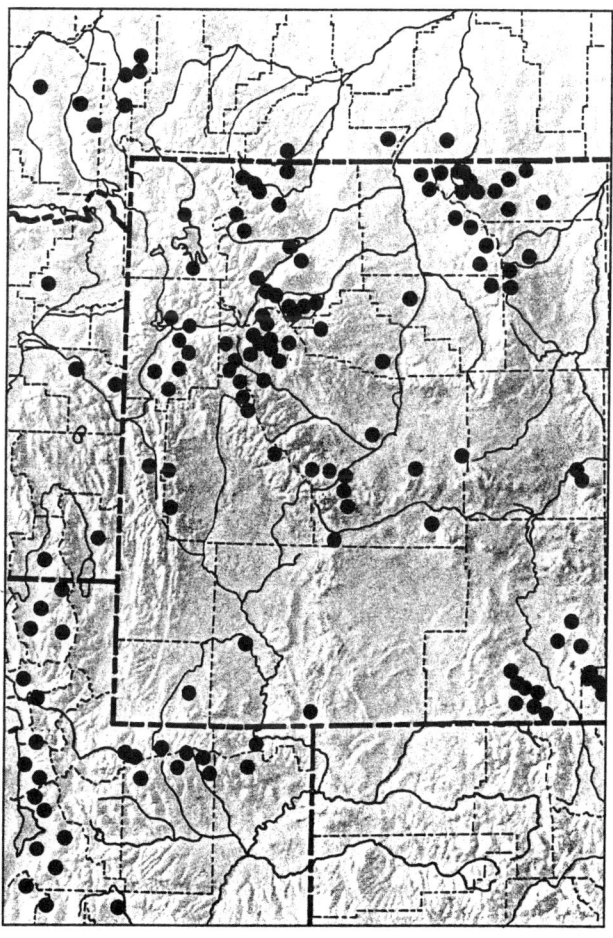

Townsendia Hook.
Townsendia

Annual, biennial, or perennial herbs from taproots, often tufted, with alternate or basal, entire leaves. Heads radiate, solitary or few, sessile or peduncled, the involucral bracts lanceolate, with scarious margins, appressed, somewhat imbricate, the receptacle flat, naked. Ray flowers white, violet, or purple, pistillate and fertile. Disk flowers yellow, numerous, perfect. Akenes flattened, glabrous, or usually pubescent, the pappus of the disk flowers of a single series formed of rigid, scabrous bristles, that of the ray flowers similar or of small scales.

KEY TO THE SPECIES OF *TOWNSENDIA*

1. Plants 1- to several-stemmed, annual, biennial, or short-lived perennial, from a taproot with an unbranched caudex, more than 8 cm tall — *T. parryi*
1. Plants caespitose, perennial, from a taproot with a branched caudex, less than 6 cm tall
    2. Plants woolly-villous; akenes with a readily deciduous pappus
        3. Plants small, less than 3 cm tall; heads less than 15 mm wide; rays lavender, pink, or purplish — *T. spathulata*
        3. Plants larger, up to 6 cm tall; heads more than 20 mm wide; rays white or lavender — *T. condensata*
    2. Plants not woolly-villous; akenes with a persistent pappus
        4. Involucral bracts linear to narrowly lanceolate, in 5–7 series; leaves linear to narrowly oblanceolate — *T. leptotes*
        4. Involucral bracts broadly lanceolate, obovate, or ovate, in 2–5 series; leaves oblanceolate to spatulate — *T. alpigena*

Townsendia alpigena Piper
Bull. Torr. Club 27:394, 1900.

Alpine Townsendia

Townsendia dejecta A. Nels.
Townsendia minima Eastw.
Townsendia montana M.E. Jones

Tufted, pulvinate, glabrate to strigose, perennial herb from a branched caudex and taproot. Leaves oblanceolate to spatulate, entire. Heads solitary, sessile or on short scapes up to 5 cm long; involucre glabrous or strigose, the bracts oblong to lanceolate, obtuse to somewhat acute, often purplish, the margins scarious. Ray flowers white, lavender, pink, or blue; disk flowers yellow. Akenes glabrous to sparsely pubescent with long hairs; scales of the pappus similar in ray and disk flowers.

Fellfields, talus, and rocky ridges of most alpine ranges of the Middle Rocky Mountains. Not yet known from above timberline in the Salt River Range. Montana south to Idaho, Utah, and northwestern Wyoming.

Alpine plants with slightly pedunculate heads, blue or white rays, and thin, flat, obtuse to subacute leaves are var. *alpigena*.

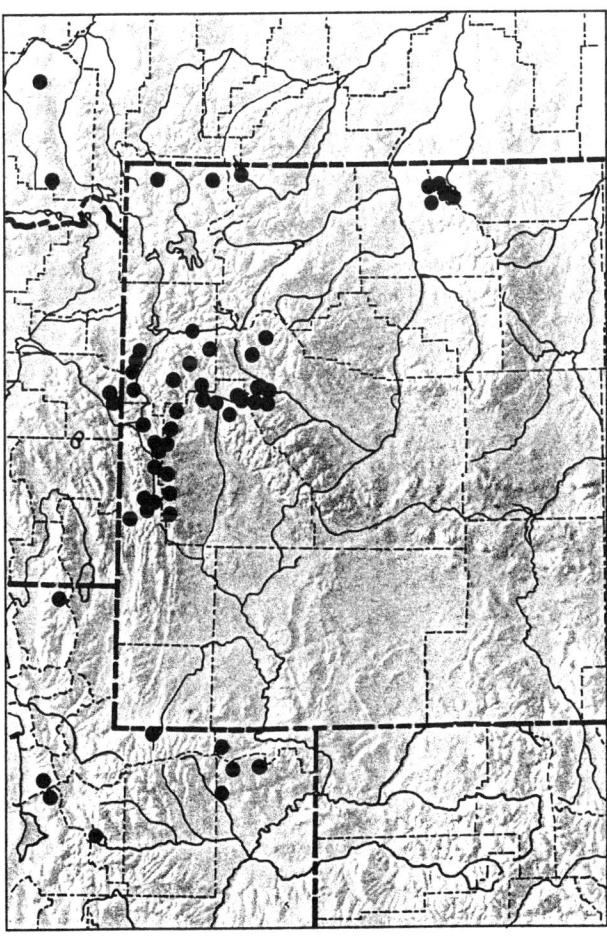

Townsendia condensata Parry ex Gray
Proc. Am. Acad. 16:83, 1880.

Cushion Townsendia

Townsendia anomala Heiser

Low, pulvinate, mostly woolly-villous, perennial herb less than 5 cm tall, from a branched caudex and taproot. Leaves spatulate, villous, entire. Heads solitary, sessile or on very short stems or scapes. Involucre villous, the bracts linear to lanceolate, acuminate, often purplish, the margins scarious, ciliate. Ray flowers white, pink, or lavender; disk flowers yellow. Akenes pubescent; pappus of ray and disk flowers similar, readily deciduous.

Fellfields of the Absaroka and Wind River Ranges. Alberta south to Montana, Wyoming, and Utah.

The typical var. *condensata* with mostly 1 or 2 inflorescences 17–40 mm wide occurs in both the Absaroka and Wind River Ranges. The recently described var. *anomala* (Heiser) Dorn, with 3 or more inflorescences 10–17 mm wide, is known only from the Absaroka Range.

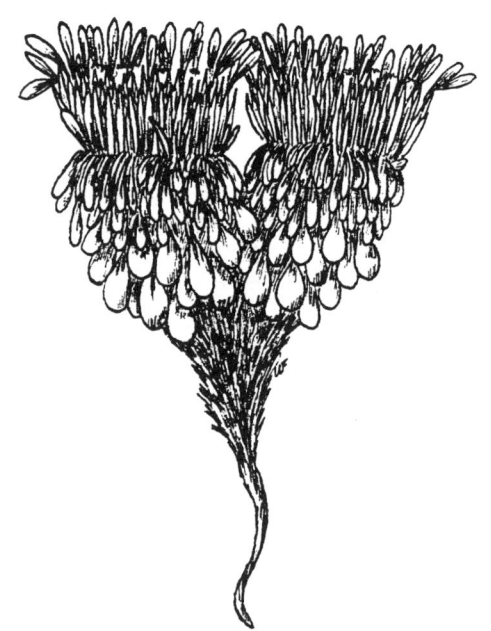

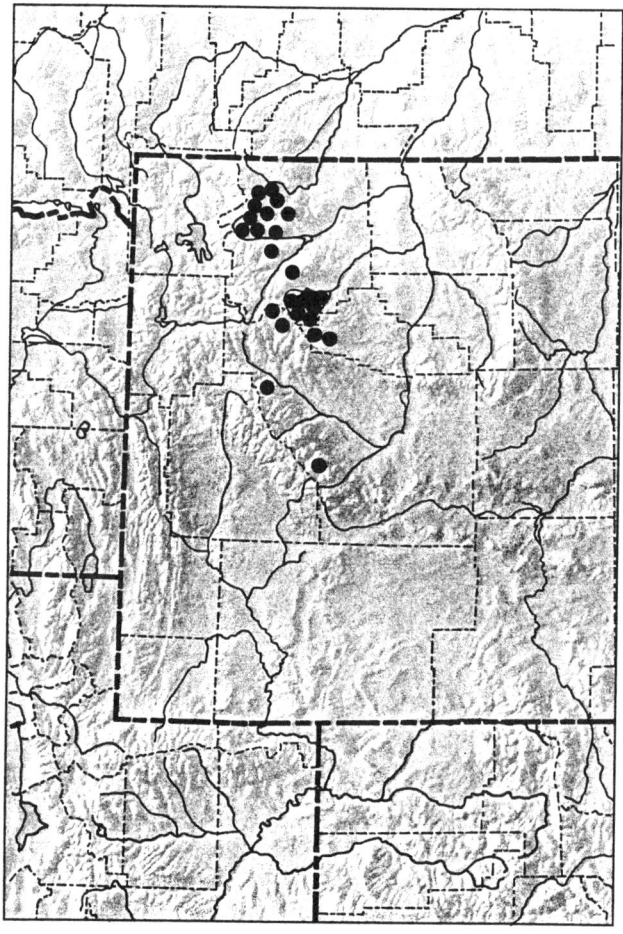

Townsendia leptotes (Gray) Osterh.
Muhlenbergia 4:69, 1908.

Slender Townsendia

Caespitose, strigose, perennial herb from a branched caudex and taproot. Leaves linear to narrowly oblanceolate, appressed-hirsute or strigose, entire. Heads solitary, sessile or rarely short-pedunculate; involucre glabrate, the bracts linear to lanceolate, often purplish, the margins scarious, ciliate. Ray flowers white or whitish, pink, or lavender; disk flowers yellow. Akenes pubescent, the hairs distinctly glochidiate; pappus of bristlelike scales and similar in ray and disk flowers.

Rocky ridges of the Absaroka, Gros Ventre, and Wind River Ranges. Montana, Idaho, Wyoming, Utah, and Colorado.

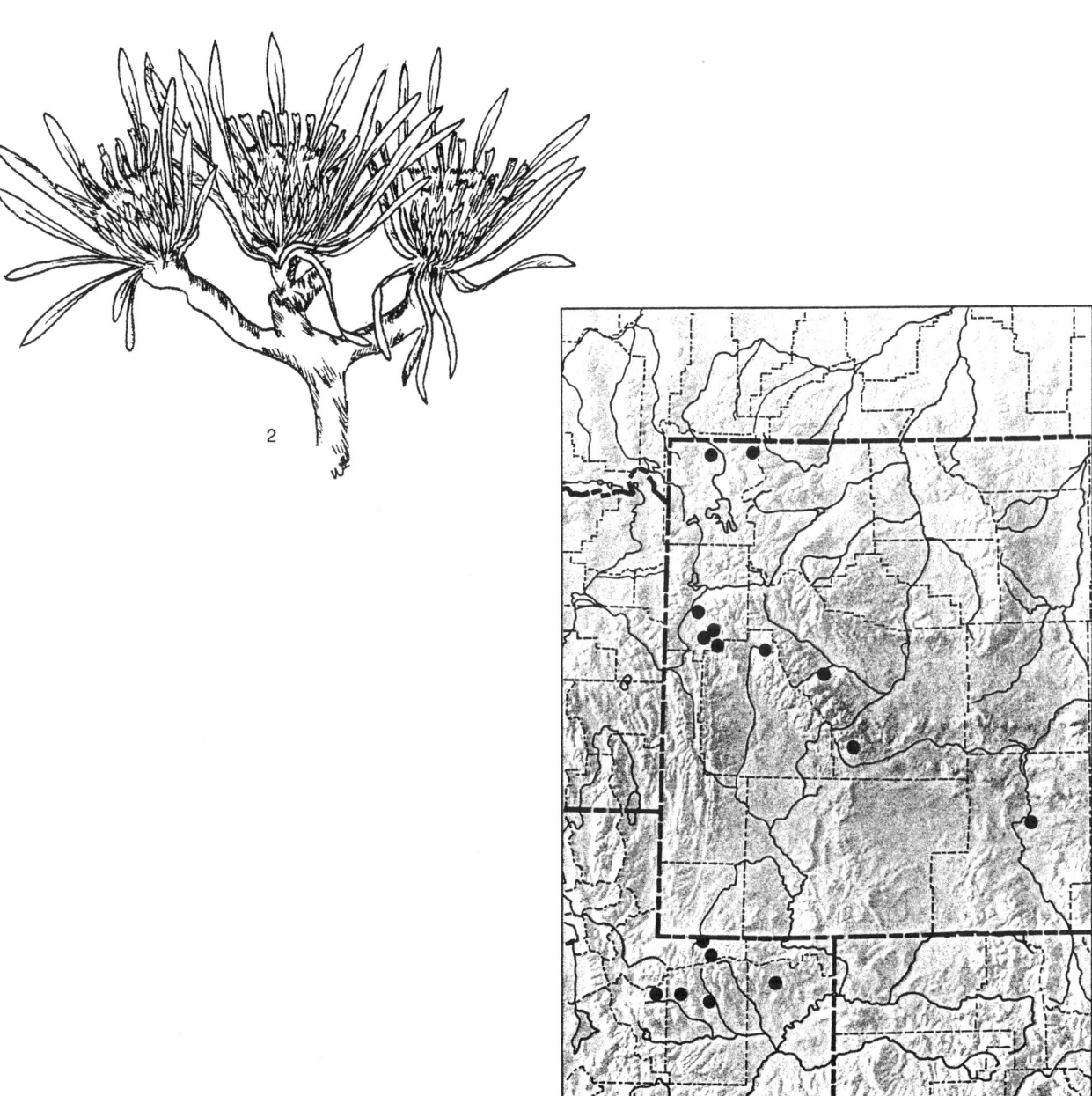

Townsendia parryi D.C. Eat. in Parry
Am. Nat. 8:212, 1874.

Parry Townsendia

Townsendia alpina (Gray) Rydb.

Annual, biennial, or perennial herb from a simple caudex and taproot; stems 1 to a few, subglabrate to strigose. Basal leaves crowded in a tuft, petiolate, oblanceolate to obovate, entire, usually glabrous above, strigose beneath; cauline leaves reduced upward and narrowly oblanceolate to lanceolate. Heads solitary and terminal, large, showy; involucre glabrate, the bracts lanceolate and acuminate, the margins scarious, ciliate. Ray flowers purple, lavender, or blue; disk flowers dark yellow. Akenes short-pubescent; pappus of long bristles and similar in ray and disk flowers.

  Fellfields and meadows of the Absaroka, Beartooth, Big Horn, Wind River, and Wyoming Ranges. Alberta south to Oregon, Idaho, and Wyoming.

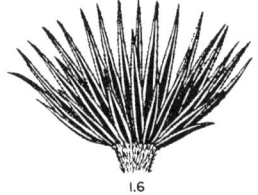

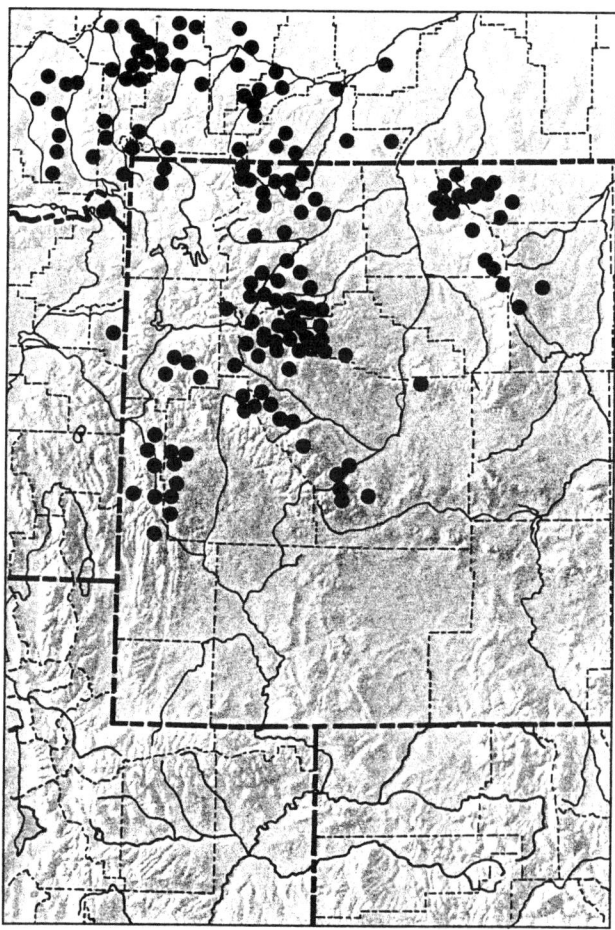

Townsendia spathulata Nutt.
Trans. Am. Phil. Soc., II, 7:305, 1840.

Spoonleaf Townsendia

Low, pulvinate, woolly-villous, perennial herb, less than 3 cm tall, from a branched caudex and taproot. Leaves obovate to spatulate, entire. Heads solitary, sessile or nearly so; involucre pubescent, the bracts linear to narrowly lanceolate, attenuate to acuminate. Ray flowers purple or lavender; disk flowers yellow. Pappus of ray and disk flowers similar.

Rocky slopes of the Absaroka Range. Montana, Idaho, and Wyoming.

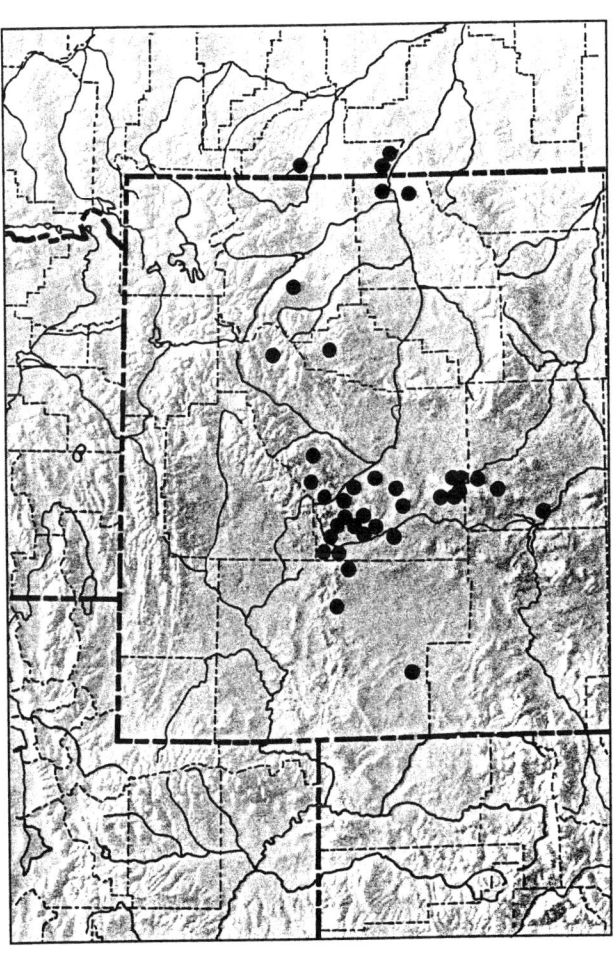

Tragopogon L.
Goatsbeard

Annual, biennial, or perennial herbs with milky juice, from thick taproots. Leaves alternate, entire, linear, nerved, with clasping bases. Heads large, solitary at the ends of branches; involucral bracts uniseriate, subequal to equal; receptacles flat, naked. Flowers ligulate, the corollas yellow or purple. Akenes linear, muricate, ribbed, terete or angled, constricted into a long, narrow beak, the outer akenes beakless; pappus of plumose bristles connate at the base.

Tragopogon dubius Scop.
Fl. Carn. 2:95, 1772.

Goatsbeard

Tragopogon major Jacq.

Erect, biennial or short-lived perennial from a taproot; stems mostly solitary, simple, glabrous to floccose at the nodes. Leaves alternate, linear or narrowly lanceolate, attenuate, loosely tomentose, at least when young, to glabrate. Heads solitary on hollow peduncles enlarged below the heads. Involucral bracts equal, lanceolate, attenuate, exceeding the yellow, ligulate flowers. Akenes slender, long-beaked; pappus plumose, whitish.

A common weed at lower elevations throughout the Middle Rocky Mountains, *T. dubius* has been collected from rocky slopes above timberline in the Salt River Range near Corral Creek Lake. Eurasia; introduced and established over much of temperate North America.

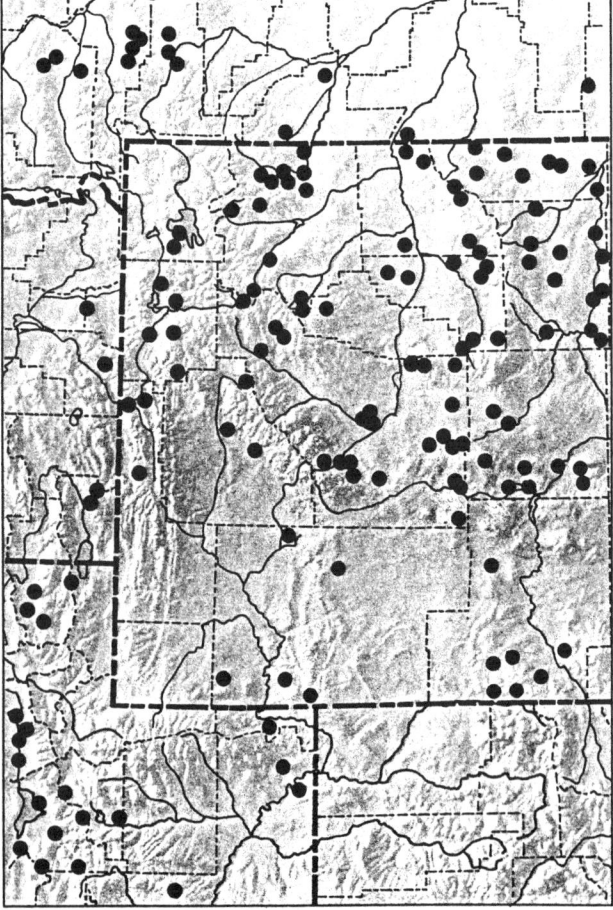

Betulaceae
Birch Family

Monoecious trees or shrubs with simple, alternate, stipulate leaves. Catkins precocious, the staminate catkins pendulous, loosely flowered, the flowers commonly 1–3 in the axil of each bract. Pistillate catkins erect or drooping, spicate or capitate, the calyx adnate to the 2-celled ovary, or lacking, the styles 2. Calyx 2–4-parted or lacking. Stamens 2–10. Fruit a nut or nutlet, winged, subtended or enclosed by a persistent involucre.

Betula L.
Birch

Our plants are low shrubs with smooth, resinous-aromatic bark and horizontal, elongate lenticels, the leaves petiolate, serrate, or crenate, and deciduous. Staminate catkins cylindric, elongate, the flowers 2–3 per cluster, the calyx 2–4-lobed, the stamens 2, each bifid at the tip. Pistillate catkins ovoid to oblong, the bracts 3-lobed, each subtending 2–3 flowers; calyx lacking; styles 2, short, persistent. Fruit a flat, oval to obovate, winged nutlet.

Betula nana L.
Sp. Pl. 2:983, 1753.

Bog Birch

Betula exilis Suk.
Betula hallii T.J. Howell
Betula crenata Rydb. ex Butler
Betula glandulifera (Regel) Butler
Betula glandulosa Michx.
Betula michauxii Sarg.
Betula terra-novae Fern.

Matted to erect, many-branched shrub with puberulent to pubescent, resinous, prominently warty twigs; bark grayish to somewhat reddish, not peeling. Leaves petiolate, the blades leathery, glandular at least below, broadly oval to orbicular, the margins serrate to crenate, cuneate to rounded at the base; petioles glabrous or puberulent. Pistillate catkins erect, 1–2 cm long; bracts puberulent, the lateral lobes divergent, shorter than the middle lobe; staminate catkins about 2 cm long, drooping. Nutlets narrowly winged.

Common in the upper montane and subalpine zones throughout most of the Middle Rocky Mountains. This species is only occasional in wet, boggy sites and margins of meadows in the lower alpine zone of the Beartooth, Medicine Bow, Uinta, Wasatch, and Wind River Ranges. Alaska to Greenland and south to California, Nevada, New Mexico, New York, and Maine.

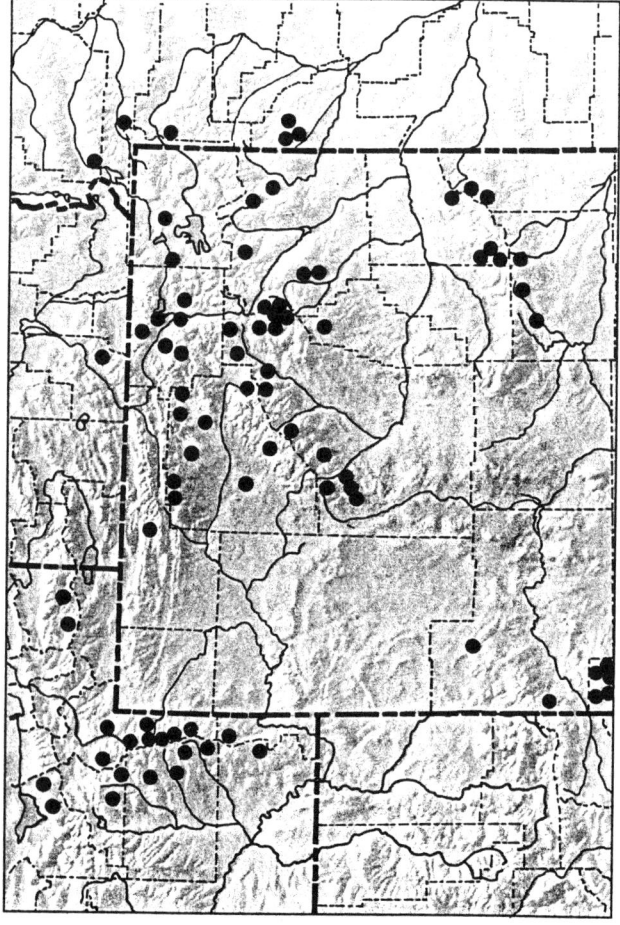

Boraginaceae
Borage Family

Ours are mostly rough-hairy herbs with exstipulate, usually alternate, simple leaves. Flowers hypogynous, 5-merous, regular, usually in coiled cymes. Calyx 5-parted. Corolla 5-lobed, salverform, campanulate, rotate, or funnelform, often bearing 5 appendages in the throat, these forming a small corona. Stamens 5, inserted on the tube of the corolla alternate with the lobes. Pistil 1, of 2 united carpels, the ovary usually 4-lobed, with 1 ovule in each lobe. Style single, originating from the base of the ovary between the 4 lobes, the stigmas 1 or 2. Fruit commonly 4 1-seeded nutlets.

KEY TO THE GENERA OF BORAGINACEAE

1. Corolla tubular, funnelform, or campanulate, the lobes erect; appendages lacking or poorly developed; nutlets with the dorsomarginal ridge incomplete .... *Mertensia*
1. Corolla salverform or rotate, the lobes spreading; appendages well developed, nearly closing the throat of the corolla; nutlets with the dorsomarginal ridge complete around the base or across the back
   2. Plants densely caespitose, less than 5 cm tall, forming cushionlike mats; nutlets attached obliquely .... *Eritrichum*
   2. Plants erect, more than 8 cm tall; nutlets attached basally
      3. Nutlets with barbed prickles along the margins; pedicels recurved in fruit .... *Hackelia*
      3. Nutlets lacking barbed prickles; pedicels not recurved in front
         4. Flowers blue .... *Myosotis*
         4. Flowers white to yellow .... *Cryptantha*

Cryptantha Lehm. ex G. Don
Miner's Candle

Annual, biennial, or perennial herbs, strigose or hirsute with coarse hairs that are often swollen basally (pustulate). Leaves basal and cauline, entire, linear to spatulate, alternate at least above. Flowers white to yellow, in scorpioid spikes or racemes, often cymose. Calyx of 5 lobes, becoming accrescent in fruit and exceeding the nutlets. Corolla salverform to funnelform; fornices well developed, often yellow; tube exceeding the calyx, the limb spreading. Stamens 5, included, the filaments short or lacking, attached below the middle of the corolla tube. Nutlets 4, or fewer by abortion, smooth, wrinkled, or tuberculate, attached laterally to the elongate gynobase.

Cryptantha celosioides (Eastw.) Pays.
Ann. Missouri Bot. Gard. 14:299, 1927.

Northern Miner's Candle

Oreocarya celosioides Eastw.
Cryptantha bradburiana Pays.
Cryptantha confusa Rydb.
Cryptantha hypsophila Johnst.
Cryptantha interrupta (Greene) Pays.
Cryptantha macounii (Eastw.) Pays.
Cryptantha nubigena (Greene) Pays.
Cryptantha sheldonii (Brand) Pays.
Cryptantha sobolifera Pays.
Cryptantha spiculifera (Piper) Pays.
Cryptantha subretusa Johnst.
Cynoglossum glomeratum Pursh
Oreocarya affinis Greene
Oreocarya cilio-hirsuta A. Nels. & Macbr.
Oreocarya glomerata (Pursh) Greene
Oreocarya interrupta Greene
Oreocarya macounii Eastw.
Oreocarya nubigena Greene
Oreocarya perennis (A. Nels.) Rydb.
Oreocarya sheldonii Brand
Oreocarya spiculifera Piper
Oreocarya subretusa (Johnst.) Abrams

Biennial or perennial, strigose herb from a taproot; caudex simple or branched; stems 1 to several, erect. Basal leaves broadly spatulate, tufted, tomentose, pustulate above and below; cauline leaves oblanceolate and reduced upward. Flowers white, the throat yellow, in paniculate spikes, the inflorescence often elongate at maturity. Calyx lobes hispid. Nutlets 4, ovate, rugose-tuberculate.

Rocky places in the lower alpine zone of the Beartooth and Wind River Ranges. British Columbia to Alberta and south to Oregon, Idaho, Wyoming, and Nebraska.

Alpine members of the *Cryptantha celosioides* complex with the ventral surfaces of the nutlets smooth, or nearly so, have been recognized by many authors as *C. nubigena* (Greene) Pays.

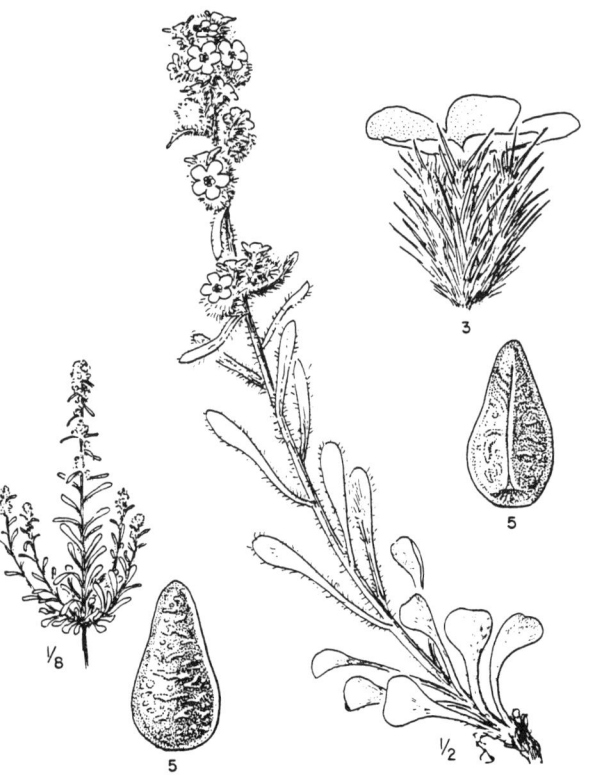

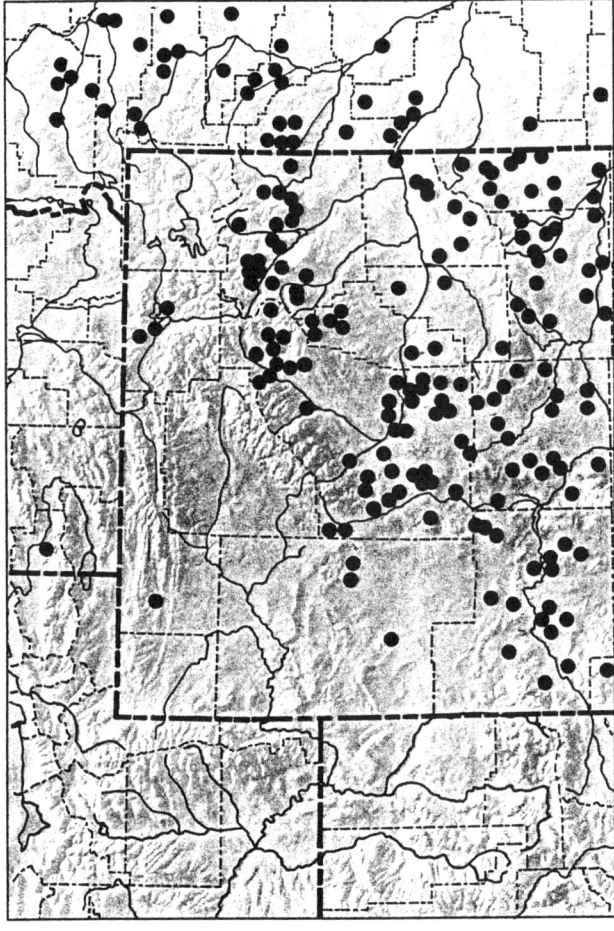

Eritrichum Schrad. ex Gaudin
Alpine Forget-Me-Not

Low, caespitose perennials with entire, villous leaves densely crowded on short branches. Flowers blue, in few-flowered, racemose cymes; the corolla salverform, 5-lobed, bearing 5 appendages in the throat. Stamens 5, included. Nutlets divergent, attached obliquely to the gynobase, with entire or toothed margins, the dorsomarginal ridge complete.

Eritrichum nanum (Vill.) Schrad. ex Gaudin
Comment. Soc. Regiae Sci. Gött. 4:186, 1820.

Alpine Forget-Me-Not

Myosotis nana Vill.
Eritrichum aretioides (Cham.) A. DC.
Eritrichum argenteum Wight.
Eritrichum chamissonis DC.
Eritrichum elongatum (Rydb.) Wight.

Densely caespitose to pulvinate, silvery-villous perennial, less than 5 cm tall. Leaves oblong to narrowly ovate, loosely villous, the surface visible beneath the hairs, which are tufted at the leaf apex. Flowers blue, with yellow throats. Nutlets 1–4, glabrous, the dorsal surface ridged.

Common throughout the alpine areas of the Middle Rocky Mountains on stony, exposed slopes and dry meadows. Not yet known from above timberline in the Hoback, Salt River, and Wasatch Ranges. Circumboreal; south in the Rocky Mountains to Montana, Oregon, and New Mexico.

Middle Rocky Mountain plants are var. *elongatum* (Rydb.) Cronq. and differ from the typical European var. *nanum* in being more densely pubescent. The closely related *E. howardii* (Gray) Rydb. is known from southern Montana and northern Wyoming and might be expected as occasional in the lower alpine zone. It is distinguished from *E. nanum* by its dense leaf pubescence nearly obscuring the surface and lack of a tuft of hairs at the leaf tip. This is one of the earliest species to bloom in the alpine zone and is very fragrant.

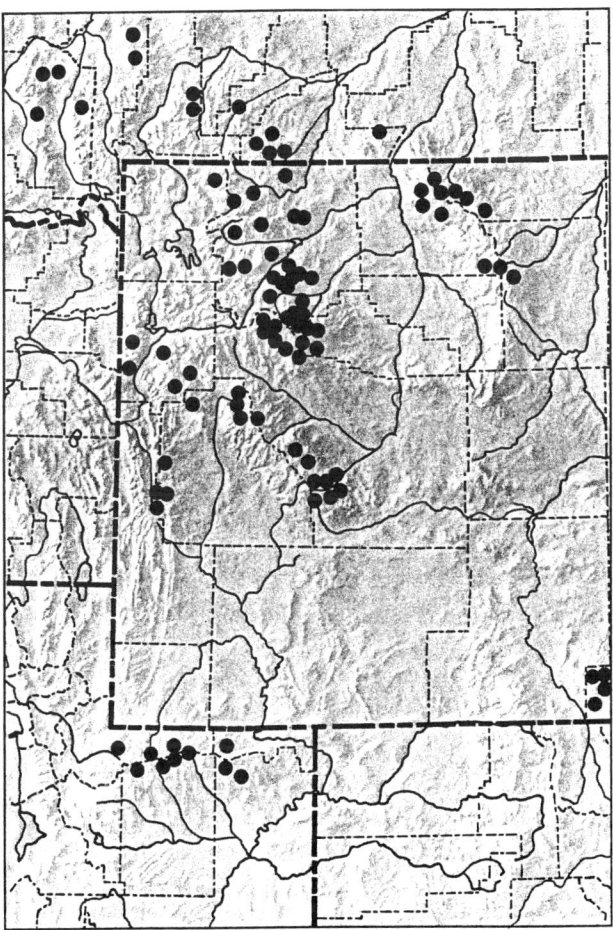

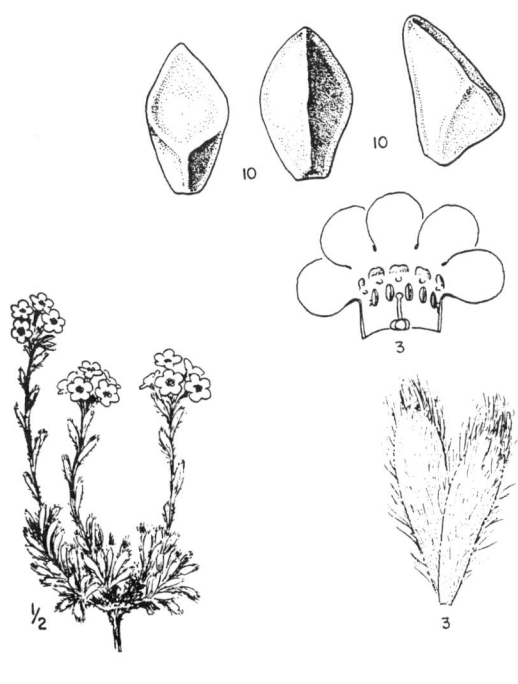

Hackelia Opiz
Tickweed

Biennial or perennial herbs with 1 to several stems from a taproot. Leaves alternate, entire, linear to spatulate, hirsute to strigose. Flowers blue or white, regular, in scorpioid racemes, the bracts minute or lacking; pedicels reflexed in fruit. Calyx lobes 5, distinct or nearly so; corolla rotate to salverform, barely exceeding the calyx; fornices well developed and conspicuous. Stamens 5, attached above the middle of the corolla tube. Ovary 4-lobed, the style slender, surpassed by the nutlets in fruit. Nutlets 4, attached medially to the broad gynobase, the dorsal surface with 2 marginal rows of glochidiate prickles.

KEY TO THE SPECIES OF *HACKELIA*

1. Corolla white, usually blue-veined and often light blue near the center   *H. patens*
1. Corolla blue, yellow near the center
    2. Plants mostly single-stemmed biennials from a simple caudex; corolla limb mostly 3–6 mm wide; dorsal face of nutlet lacking inframarginal prickles   *H. floribunda*
    2. Plants mostly multistemmed perennials from a branched caudex; corolla limb mostly 7–11 mm wide; dorsal face of nutlet with several inframarginal prickles   *H. micrantha*

Hackelia floribunda (Lehm.) I.M. Johnston
Contr. Gray Herb. 68:46, 1923.

Western Tickweed

Echinospermum floribundum Lehm.
Hackelia leptophylla (Rydb.) I.M. Johnston
Lappula floribunda (Lehm.) Greene

Biennial or weakly perennial herb with 1 to a few stout, strigose to hirsute stems, to 5 dm tall, from a taproot. Basal leaves petiolate, soon withering; lower cauline leaves oblanceolate, petiolate; upper cauline leaves reduced, becoming linear and sessile. Flowers in terminal 1-sided racemes; corolla blue with a yellow throat. Nutlets ovate, the prickles restricted to the margins, lacking inframarginal prickles on the dorsal face.

Common and widespread at lower elevations, this species can be expected in the lower alpine zone near timberline in any of the western ranges. Known from timberline margins of the Absaroka and Wyoming Ranges. British Columbia to Saskatchewan, south to California, Mexico, Colorado, and Nebraska.

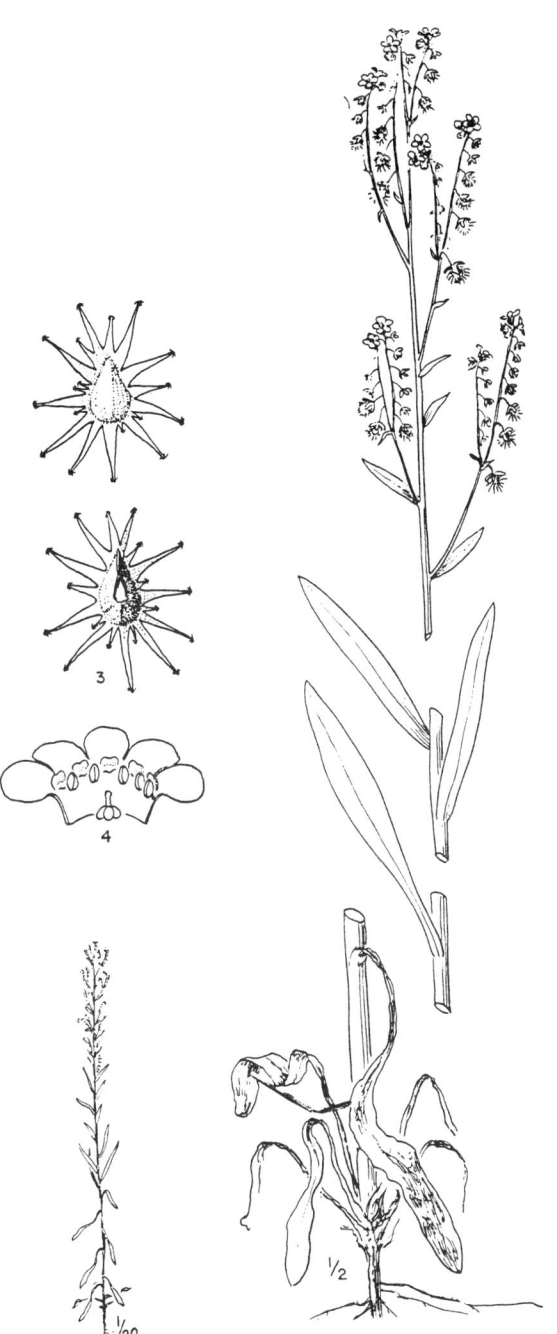

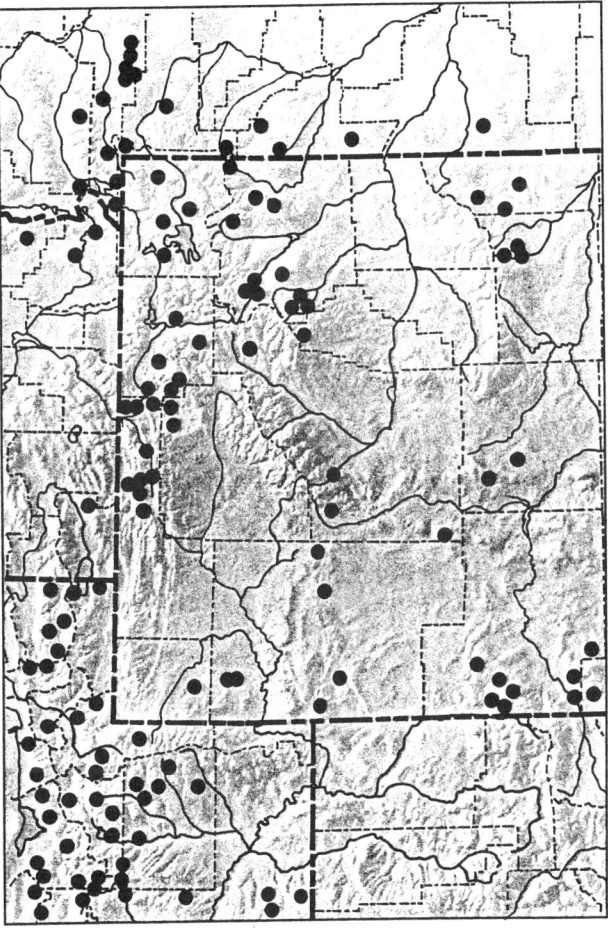

Hackelia micrantha (Eastw.) Gentry
Madroño 21:490, 1972.

Jessica Tickweed

Hackelia jessicae (McGregor) Brand
Lappula jessicae McGregor

Perennial herb from a taproot and branched caudex; stems several, strigose to hirsute, to 5 dm tall. Basal leaves petiolate, elliptic to oblanceolate, often persistent; cauline leaves reduced upward, from oblanceolate and petiolate below to lanceolate and sessile above. Flowers in short, few-flowered, terminal racemes; corolla blue with a yellow or white throat. Nutlets ovate, prickles both marginal and inframarginal, the marginal prickles larger and more numerous than the inframarginal ones.

Slopes and meadows near timberline. This species has been collected from the alpine zone on sedimentary meadow soils of the Beartooth, Salt River and Wyoming Ranges. British Columbia and Alberta south to Wyoming, Utah, and California.

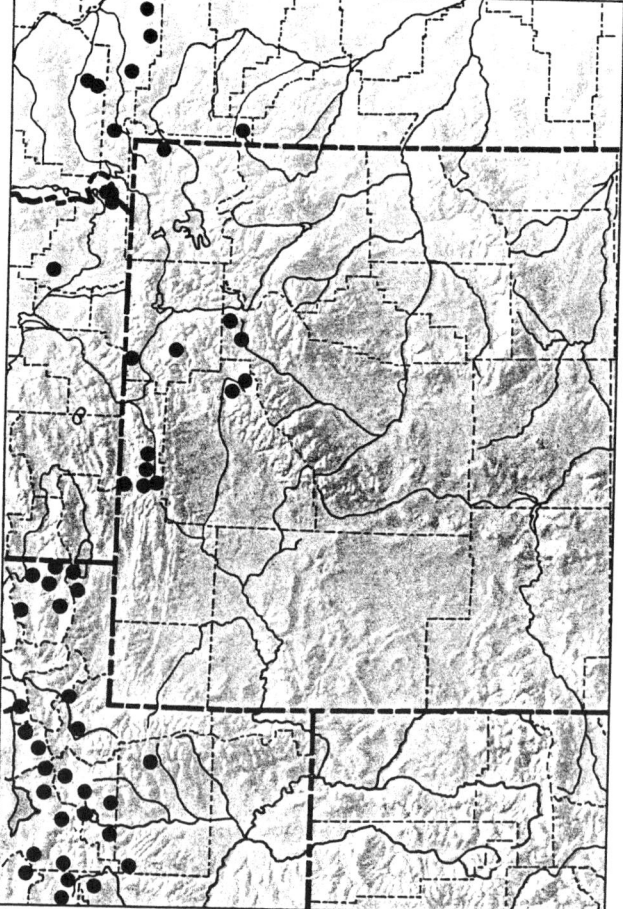

Hackelia patens (Nutt.) I.M. Johnston
Contr. Arnold Arb. 16:194, 1935.

Spreading Tickweed

Rochelia patens Nutt.
Echinospermum subdecumbens Parry
Lappula caerulescens Rydb.
Lappula subdecumbens (Parry) A. Nels.

Perennial herb from a taproot, with 1 to several strigose stems, the hairs retrose below. Basal leaves somewhat tufted, petiolate, oblanceolate or elliptic, strigose; cauline leaves oblanceolate or lanceolate, sessile, reduced upward. Flowers in terminal, paniculate clusters; corolla white with blue veins and a bluish or yellowish center. Nutlets ovate, marginal prickles much larger than the inframarginal ones.

Subalpine slopes and meadows, and in the lower alpine zone of the Salt River and Wyoming Ranges. Montana, Idaho, Wyoming, Utah, and Nevada.

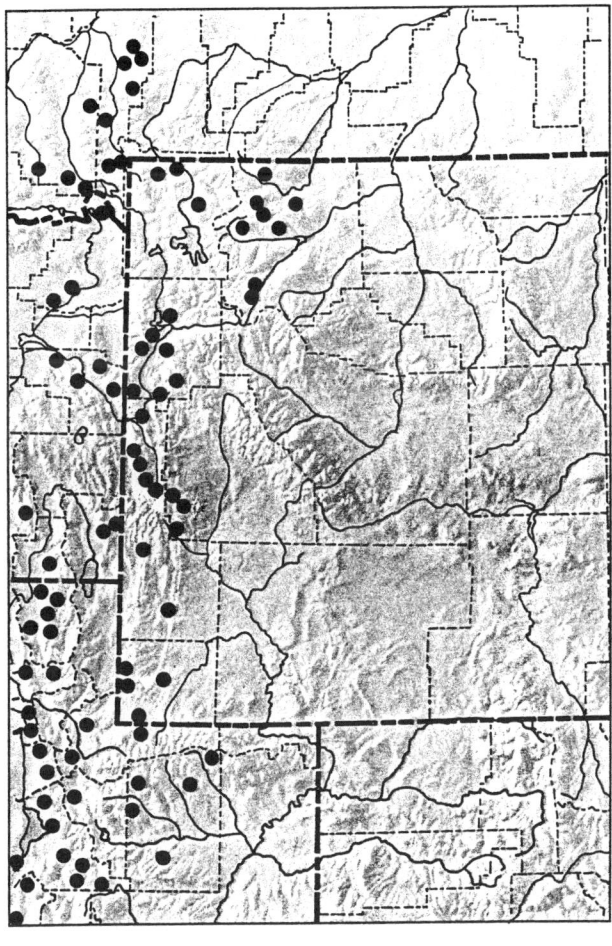

Mertensia Roth
Bluebells

Glabrous or pubescent, erect, perennial herbs with alternate, entire leaves. Flowers blue at maturity, usually pink in the bud, in short cymes or racemes. Calyx generally 5-cleft to below the middle. Corolla tubular, campanulate, or funnelform, shallowly 5-lobed, the fornices lacking or poorly developed. Stamens 5, inserted on the corolla tube at or below the level of the appendages. Nutlets 4, attached laterally to the gynobase, erect, the dorsomarginal ridge incomplete, wrinkled when dry.

KEY TO THE SPECIES OF *MERTENSIA*

1. Cauline leaves with evident lateral veins; stems usually more than 3 dm tall
    2. Calyx lobes mostly acute, 4 mm or more in length; limb of the corolla longer than the tube — *M. arizonica*
    2. Calyx lobes mostly obtuse, 3 mm or less in length; limb of the corolla equaling or shorter than the tube — *M. ciliata*
1. Cauline leaves without evident lateral veins; stems usually less than 3 dm tall
    3. Filaments attached in the corolla tube, the anthers not projecting beyond the throat — *M. alpina*
    3. Filaments attached near the throat of the corolla tube, the anthers projecting beyond the throat
        4. Corolla tube nearly the same length as the limb, bearing a ring of hairs within — *M. viridis*
        4. Corolla tube up to 2 times as long as the limb, glabrous within — *M. oblongifolia*

Mertensia alpina (Torr.) G. Don
Gen. Hist. Dichl. Pl. 4:372, 1838.

Alpine Bluebells

Pulmonaria alpina Torr.
Cerinthodes alpinum (Torr.) Kuntze
Mertensia obtusiloba Rydb.
Mertensia tweedyi Torr.

Multistemmed, perennial herb, less than 2 dm tall, from a branched caudex. Leaves petiolate, the blades lanceolate to oblanceolate, glabrous below, strigose above. Flowers dark blue; calyx cleft nearly to the base; corolla tube glabrous within; filaments attached in the corolla tube, the anthers not projecting beyond the throat; fornices evident. Nutlets rugose or papillose.

Widespread in moist meadows and fellfields throughout most of the northern Middle Rocky Mountain ranges. Not known from the western ranges. Southwestern Montana and eastern Idaho to Colorado and northern New Mexico.

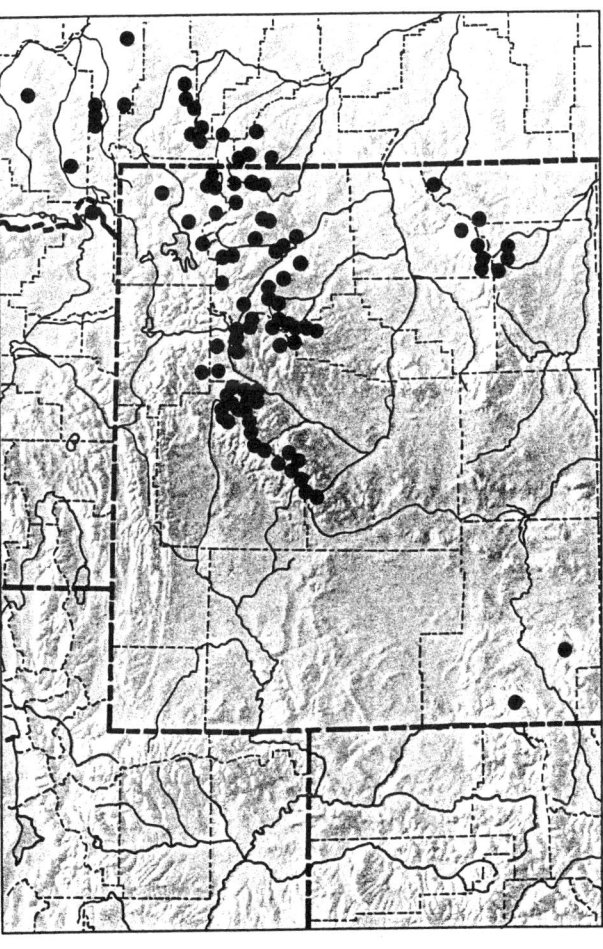

Mertensia arizonica Greene
Pittonia 3:197, 1897.

Arizona Bluebells

Mertensia leonardii Rydb.
Mertensia sampsonii Tidestr.

Erect, perennial herb, usually more than 3 dm tall, from a stout, branched caudex covered with old leaf bases; stems 1 to numerous, glabrous, to 60 cm in the alpine zone. Basal leaves long-petiolate, the blades lanceolate to elliptic or ovate, glabrous, the margins ciliate, often withering by flowering time; cauline leaves with prominent lateral veins, elliptic to oblanceolate, acute, the lower petiolate, the upper often reduced and sessile. Flowers nodding in axillary, cymose inflorescences. Calyx cleft nearly to the base, the lobes lanceolate, ciliate, mostly acute at the apex; corolla blue or occasionally white, with a ring of hairs within; filaments attached at the throat of the corolla; fornices evident. Nutlets rugose or papillose.

Stream banks, rocky ledges, and moist places among rocks in the Uinta and Wasatch Ranges. Wyoming, Utah, and Colorado.

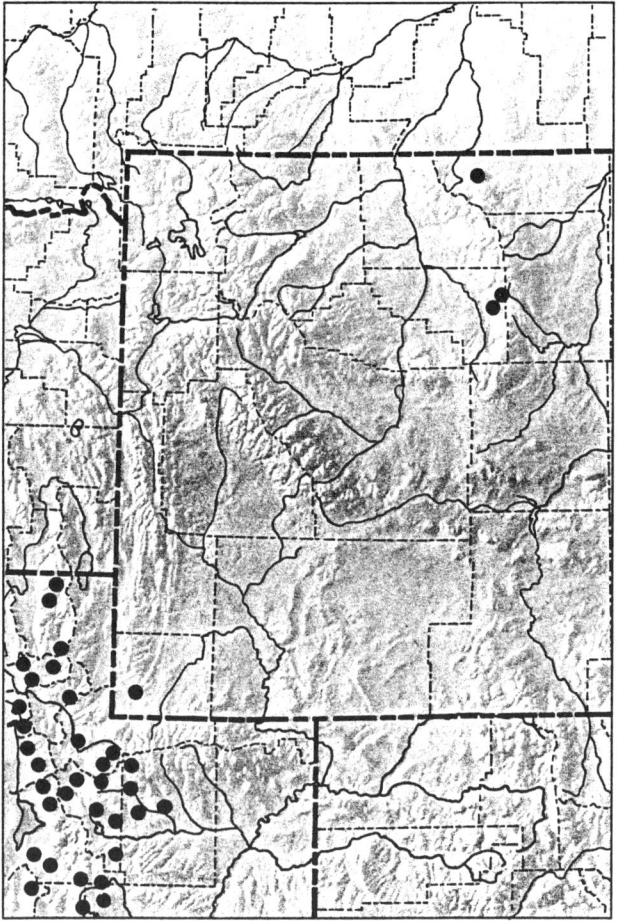

Mertensia ciliata (James ex Torr.) G. Don
Gen. Hist. Dichl. Pl. 4:372, 1838.

Mountain Bluebells

Pulmonaria ciliata James ex Torr.
Mertensia incongruens Macbr. & Pays.
Mertensia pallida Rydb.
Mertensia picta Rydb.
Mertensia polyphylla Greene
Mertensia punctata Greene
Mertensia subpubescens Rydb.

Erect, perennial herb, usually more than 3 dm tall; stems several to numerous, glabrous, to 6 dm in the alpine zone, from a fleshy, branched caudex covered with old leaf bases. Basal leaves long-petiolate, the blades oblong or elliptic to ovate, the margins ciliate; cauline leaves lanceolate to ovate, acute, glaucous, the lower ones petiolate with prominent lateral veins, the upper ones often reduced and lacking prominent veins. Flowers nodding in axillary, cymose inflorescences. Calyx cleft nearly to the base, the lobes lanceolate, ciliate, mostly obtuse at the apex; corolla blue or occasionally white, usually lacking a ring of hairs within; filaments attached at the throat of the corolla; fornices evident. Nutlets rugose or papillose.

Moist places among rocks and ledges, and protected, often partially shaded stream banks in nearly all alpine ranges. Not yet known from above timberline in the Hoback Range. South Dakota to Oregon, south to California, New Mexico, and Colorado.

Two varieties occur in the Middle Rocky Mountains: the most common, found throughout the region, is var. *ciliata,* which has glabrous leaves with ciliate margins; var. *subpubescens* (Rydb.) Macbr. & Payson, with leaves pubescent on the upper surface, has been seen from the southern Wind River Range, the Tetons, and the Beartooth Plateau.

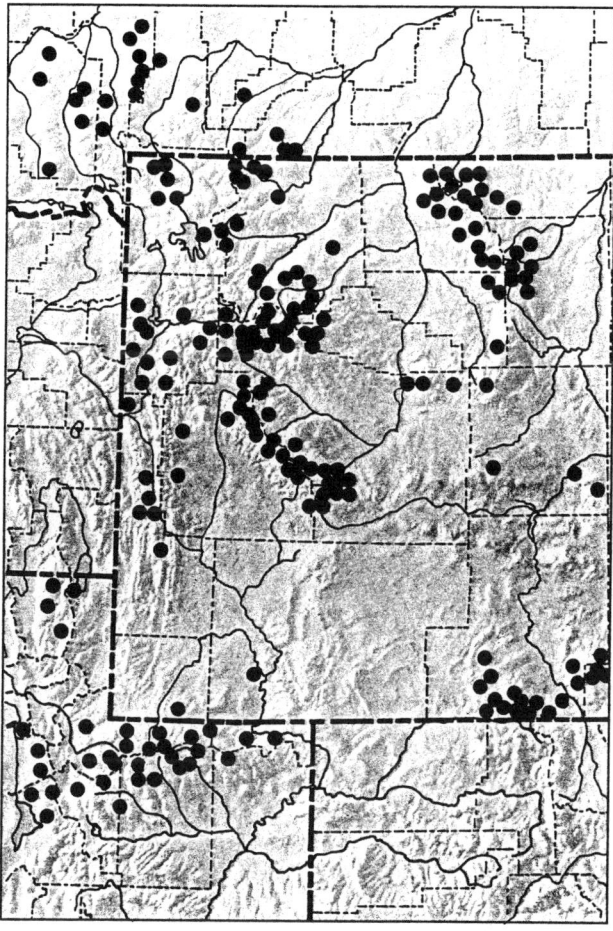

Mertensia oblongifolia (Nutt.) G. Don
Gen. Hist. Dichl. Pl. 4:372, 1838.

Oblongleaf Bluebells

Pulmonaria oblongifolia Nutt.
Mertensia coronata A. Nels.
Mertensia explicata Macbr.
Mertensia foliosa A. Nels.
Mertensia intermedia Rydb.
Mertensia nevadensis A. Nels.
Mertensia nutans T.J. Howell
Mertensia oreophila L.O. Williams
Mertensia praecox Smiley
Mertensia pubescens Piper
Mertensia stenoloba Greene
Mertensia tubiflora Rydb.

Perennial herb with 1 to several stems from a woody rootstock, often with a branched caudex. Basal leaves petiolate, the blades elliptic, oblanceolate, or narrowly ovate; cauline leaves petiolate below, sessile above, lanceolate to elliptic. Flowers blue, in a terminal inflorescence. Calyx cleft nearly to the base; corolla with tube longer than the limb, glabrous within; filaments expanded, conspicuous, attached at the throat of the corolla; fornices conspicuous. Nutlets rugose or papillose.

Scree, talus, fellfields, and other rocky places in the Beartooth, Big Horn, and Wind River Ranges. Washington to South Dakota, south to Nevada, Utah, Colorado, and Nebraska.

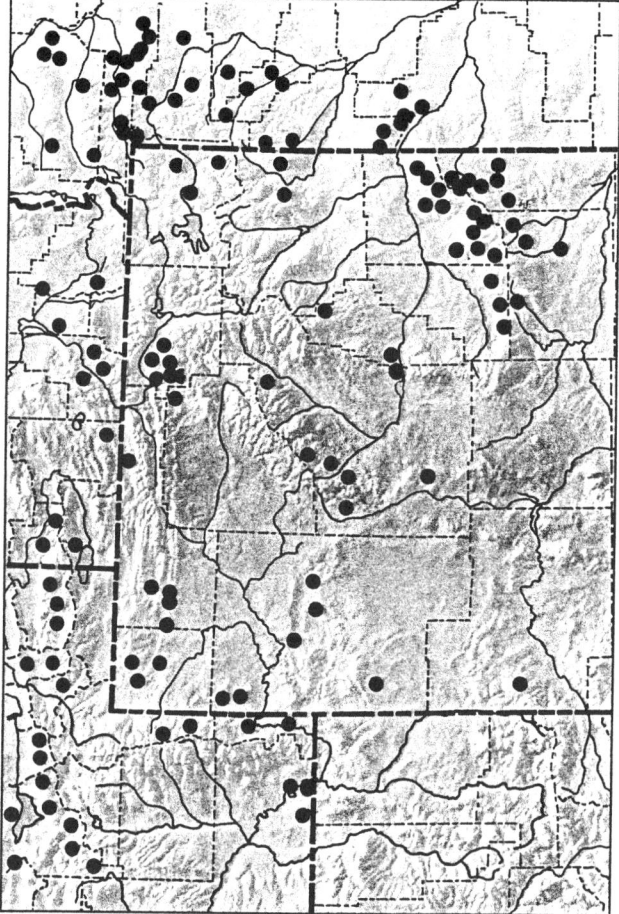

Mertensia viridis (A. Nels.) A. Nels.
Bull. Torr. Club 26:244, 1899.

Greenleaf Bluebells

Mertensia amoena A. Nels.
Mertensia bakeri Greene
Mertensia cana Rydb.
Mertensia canescens Rydb.
Mertensia coriacea A. Nels.
Mertensia cusickii Piper
Mertensia cynoglossoides Greene
Mertensia lateriflora Greene
Mertensia lineariloba Rydb.
Mertensia muricata Greene
Mertensia myosotifolia A. Heller
Mertensia nivalis (S. Wats.) Rydb.
Mertensia ovata Rydb.
Mertensia parryi Rydb.
Mertensia perplexa Rydb.

Perennial herb with few to several stems from a woody rootstock, often with a branched caudex. Basal leaves petiolate, oblanceolate to broadly elliptic, strigose above, usually glabrous below; cauline leaves sessile, lanceolate to ovate, strigose above, glabrous below. Flowers blue, in terminal cymes. Calyx cleft nearly to the base; corolla somewhat longer than the limb, with a ring of internal hairs below the middle within; filaments conspicuous, attached at the throat of the corolla; fornices conspicuous. Nutlets rugose or papillose.

Dry meadows and open slopes in the Absaroka, Big Horn, Medicine Bow, Uinta, Wasatch, and Wind River Ranges. Montana to Oregon, south to California, Nevada, and New Mexico.

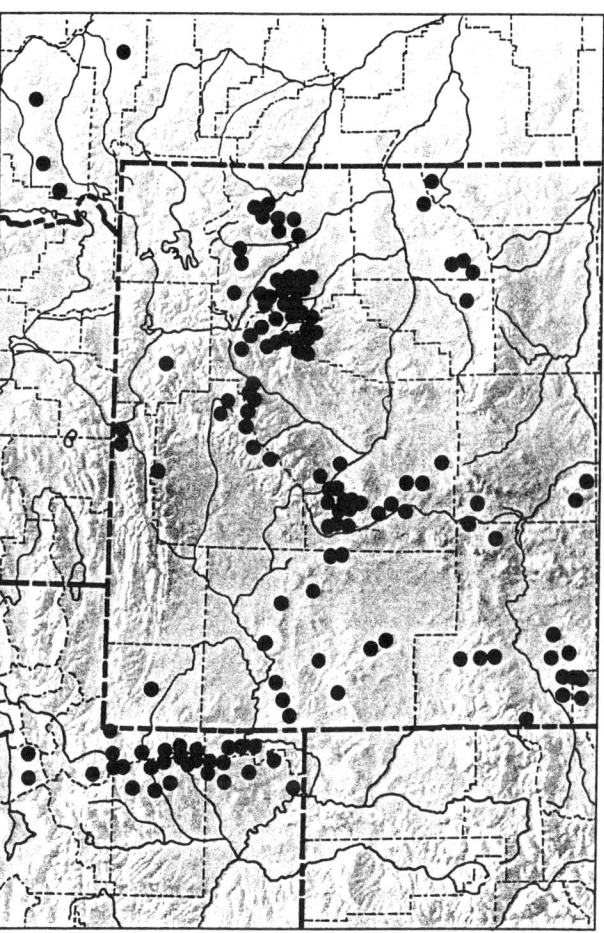

Myosotis L.
Forget-Me-Not

Annual or perennial herbs, usually low and somewhat hirsute, with alternate, entire leaves. Flowers blue, in slender, ebracteate racemes. Calyx 5-lobed, the lobes erect or spreading, pubescent at the base with some hooked or gland-tipped hairs. Corolla salverform, with an abruptly spreading, 5-lobed limb, the throat nearly closed by 5 well-developed fornices opposite the lobes. Stamens 5, included. Nutlets 4, attached basally to the gynobase, the dorsomarginal ridge complete.

Myosotis alpestris F.W. Schmidt
Fl. Boem. 3:26, 1794.

Mountain Forget-Me-Not

Multistemmed, hirsute to strigose, perennial or biennial from a short, branched caudex; roots fibrous. Leaves hirsute, the basal leaves oblanceolate to spatulate, petiolate; cauline leaves sessile, oblong to elliptic, gradually reduced upward. Flowers blue, rarely white, with a yellow throat, in open, terminal cymes. Calyx hirsute or strigose, cleft more than halfway to the base; corolla tube shorter than the calyx, the limb flat, spreading. Nutlets ovate, black, longer than the style.

Common in meadows and on well-vegetated slopes of the lower alpine zone, often near krummholz, in most western and northern ranges of the Middle Rocky Mountains. Eurasia; Alaska and Alberta south to Oregon, Idaho, Wyoming, Colorado, and South Dakota.

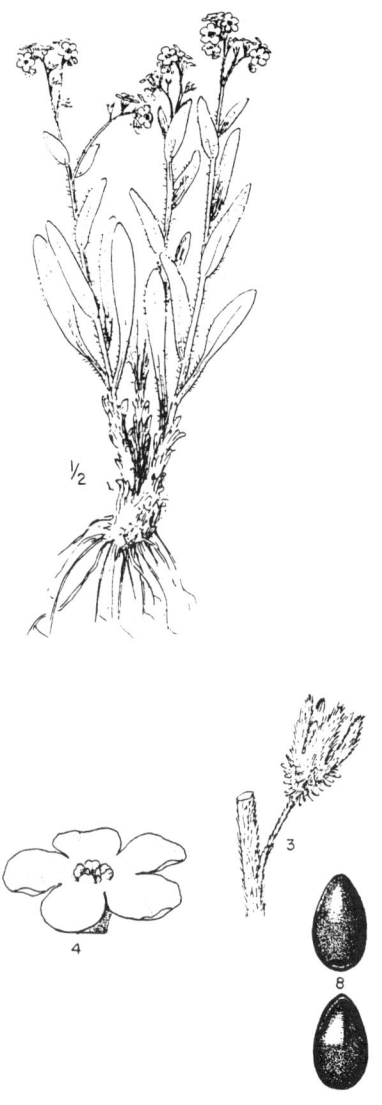

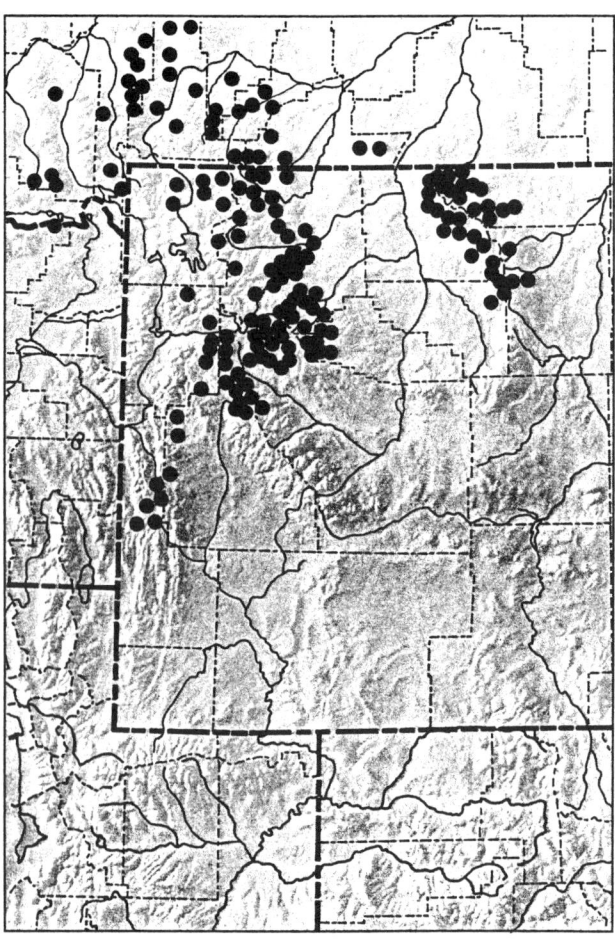

Brassicaceae
(Cruciferae)
Mustard Family

Annual, biennial, or perennial herbs, rarely woody at the base, with alternate or basal, simple or compound leaves. Flowers perfect, regular, hypogynous, usually in racemes. Sepals 4, often concave or saccate at the base. Petals 4, rarely 2 or lacking, usually long-clawed. Stamens 6, tetradynamous. Ovary 2-celled, the locules separated by a thin septum. Style present or lacking, often persistent, the stigma slightly lobed to discoid. Ovules single to many, in 1 or 2 rows in each locule. Fruit a silique or silicle, dehiscent by 2 valves, rarely indehiscent.

KEY TO THE GENERA OF BRASSICACEAE

1. Mature fruit (silicle) less than 4 times longer than broad
    2. Silicle terete or globose
        3. Pubescence lacking or entirely of simple hairs — *Rorippa*
        3. Pubescence, at least in part, of stellate or branched hairs, with or without additional simple hairs
            4. Silicles didymous, appearing as twin bladders — *Physaria*
            4. Silicles not didymous — *Lesquerella*
    2. Silicle strongly flattened
        5. Silicle flattened parallel to the broad septum — *Draba*
        5. Silicle flattened at a right angle to the narrow septum — *Thlaspi*
1. Mature fruit (silique) more than 4 times longer than broad
    6. Silique definitely flattened
        7. Petals large, more than 15 mm long; pubescence entirely of glandular hairs, or lacking and plants scapose — *Parrya*
        7. Petals small, less than 10 (to 12) mm long; pubescence of simple, branched, or stellate hairs, if lacking, plants caulescent — *Arabis*
    6. Silique terete or 4-angled
        8. Flowers bright yellow; siliques 4-angled — *Erysimum*
        8. Flowers white or cream; siliques terete
            9. Pubescence, if any, of simple hairs; leaves usually dentate to pinnatifid — *Rorippa*
            9. Pubescence of at least some branched or stellate hairs
                10. Petals cream; siliques torulose — *Descurainia*
                10. Petals white; siliques lacking evident constrictions between the seeds — *Smelowskia*

Arabis L.
Rockcress

Erect biennial or perennial herbs, sometimes woody at the base, glabrous or usually pubescent, with simple, forked, or stellate hairs. Leaves entire or toothed, the basal leaves petiolate, the cauline leaves usually sessile. Flowers small, white to purple, or sometimes cream. Stamens 6. Siliques sessile, mostly linear, flattened parallel to the septum. Seeds numerous, in 1 or 2 rows in each locule, orbicular to oval, often with a membranous wing.

KEY TO THE SPECIES OF *ARABIS*

1. Mature siliques spreading or reflexed
   2. Racemes mostly with 10 or fewer flowers, becoming secund with age; siliques spreading nearly at right angles to the rachis; fruiting pedicles less than 5 mm long ... *A. lemmonii*
   2. Racemes mostly with 15 or more flowers, never secund; siliques reflexed to pendulous; fruiting pedicles more than 5 mm long ... *A. holboellii*
1. Mature siliques erect to ascending
   3. Basal leaves pubescent with dendritic hairs
      4. Plant from a slender taproot; stems solitary or rarely 2; seeds oblong, wingless, in 2 rows ... *A. williamsii*
      4. Plant from a branched caudex; stems few to several; seeds orbicular, narrowly winged, in 1 or sometimes 2 rows
         5. Fruits less than 1.5 mm wide, divergent; stems slender, less than 1 mm thick at midlength; basal leaves less than 4 mm wide, often with a few teeth ... *A. microphylla*
         5. Fruits 2–3 mm wide, erect; stems stout, more than 1 mm thick at midlength; basal leaves often more than 4 mm wide, entire ... *A. lyallii*
   3. Basal leaves glabrous or pubescent with simple, forked, or dolabriform hairs
      6. Flowers white; cauline leaves cuneate at the base; outer sepals gibbous to saccate; seeds in 1 row ... *A. nuttallii*
      6. Flowers purple to lavender, rarely white; cauline leaves auriculate; outer sepals neither gibbous nor saccate; seeds in 1 or 2 rows
         7. Seeds broadly winged on one side only, in 2 rows, the rows sometimes imperfect; caudex usually simple; stems often more than 2 dm tall ... *A. drummondii*
         7. Seeds narrowly winged, in 1 row; caudex usually branched; stems less than 2 dm tall ... *A. lyallii*

Arabis drummondii Gray
Proc. Am. Acad. 6:187, 1866.

Drummond Rockcress

Arabis albertina Greene
Arabis brachycarpa (T. & G.) Britt.
Arabis confinis S. Wats.
Arabis connexa Greene
Arabis oxyphylla Greene
Arabis philonipha A. Nels.
Boechera drummondii (Gray) Löve & Löve
Erysimum drummondii (Gray) Kuntze
Streptanthus angustifolius Nutt. in T. & G.
Turritis drummondii (Gray) Lunell
Turritis stricta Grah.

Biennial or short-lived perennial from a simple or sometimes branched caudex. Stems 1 to several, erect, entirely glabrous or sometimes appressed-pubescent near the base. Basal leaves rosulate, oblanceolate to elliptic, petiolate, glabrous to pubescent; cauline leaves lanceolate, sessile, auriculate, acute, glabrous or nearly so. Flowers in loose, few-flowered racemes; pedicels glabrous. Sepals glabrous, linear-oblong, acute, often somewhat gibbous at the base. Petals pink, purplish, or white, and spatulate. Siliques erect, crowded, glabrous. Seeds in 2 rows, oblong, winged on one side.

Meadows and fellfields of most Middle Rocky Mountain ranges. Not yet known from above timberline in the Gros Ventre and Medicine Bow Ranges. British Columbia to Labrador, south to California, New Mexico, Nebraska, and Minnesota.

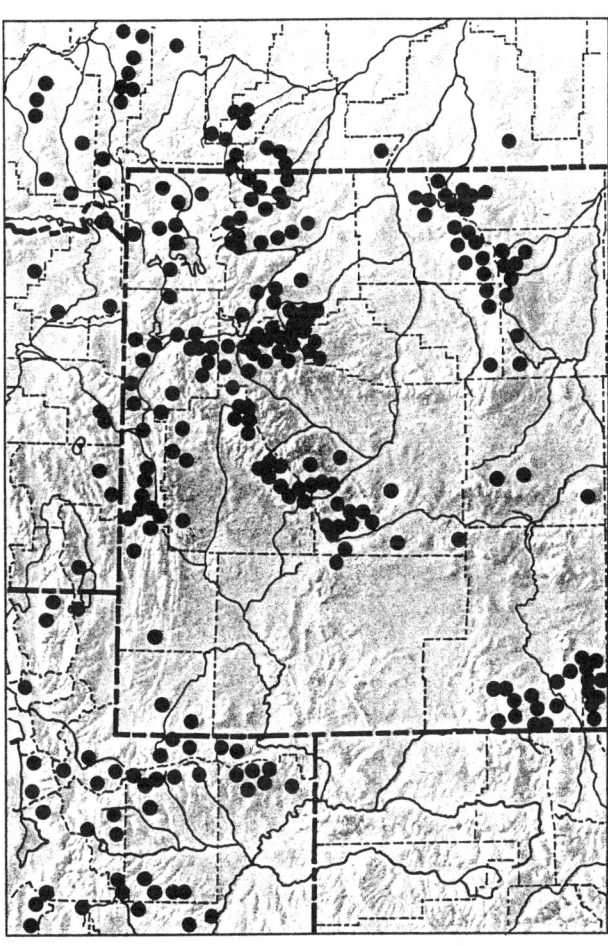

Arabis holboellii Hornem.
Fl. Dan. 11(32):5, Pl. 1879, 1827.

Holboell Rockcress

Arabis acutina Greene
Arabis bourgovii Rydb.
Arabis brevisiliqua Rydb.
Arabis caduca A. Nels.
Arabis collinsii Fern.
Arabis consanguinea Greene
Arabis dacotica Greene
Arabis divaricarpa A. Nels.
Arabis exilis A. Nels.
Arabis kochii Blank.
Arabis lignifera A. Nels.
Arabis lignipes A. Nels.
Arabis mcdougalii Rydb.
Arabis oblanceolata Rydb.
Arabis pendulocarpa A. Nels.
Arabis pinetorum Tidestr.
Arabis polyantha Greene
Arabis pratincola Greene in Fedde
Arabis retrofracta Grah.
Arabis rhodanthus Greene
Arabis secunda T.J. Howell
Arabis spatifolia Rydb.
Arabis stokesiae Rydb.
Arabis tenuis Greene
Boechera collinsii (Fern.) Löve & Löve
Boechera fendleri (S. Wats.) W.A. Weber
Boechera holboellii (Hornem.) Löve & Löve
Boechera retrofracta (Grah.) Löve & Löve
Boechera tenuis (Böcher) Löve & Löve
Erysimum holboellii (Hornem.) Kuntze
Sisymbrium pauciflorum Nutt. in T. & G.
Streptanthus virgatus Nutt. ex T. & G.
Turritis brachycarpa T. & G.
Turritis retrofracta (Grah.) Hook.

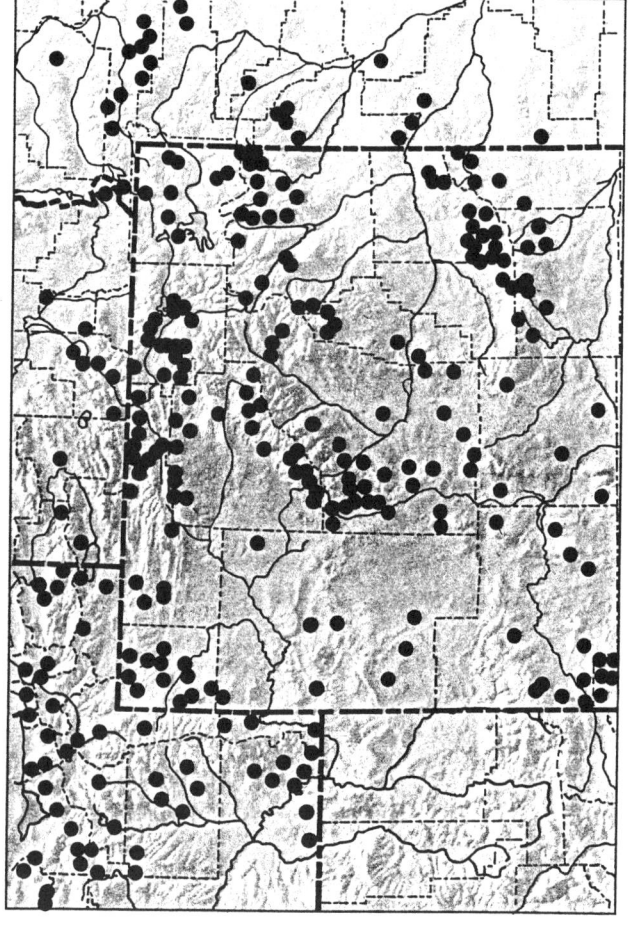

Biennial or perennial from a simple or branched caudex. Stems erect, 1 to several, simple or branched above, pubescent or sometimes glabrous above, the stellate or dendritic hairs appressed or spreading. Basal leaves rosulate, short-petiolate, oblanceolate, entire to somewhat dentate, acute to obtuse, pubescent with stellate, dendritic, simple or forked hairs; cauline leaves lanceolate to oblong, entire, the bases auriculate or tapered. Flowers in loose, many-flowered racemes. Sepals oblong, glabrous to pubescent. Petals pink, purplish, or white, spatulate. Siliques pendent, glabrous, straight to slightly curved, nerved below the middle. Seeds in 1 or 2 rows, the rows sometimes irregular, orbicular, narrowly winged.

Meadows and fellfields near timberline in the Absaroka, Teton, Wasatch, and Wind River Ranges. British Columbia to Quebec, south to California, New Mexico, Nebraska, and Minnesota.

Middle Rocky Mountain plants with pedicels curved-descending in fruit, siliques somewhat curved inward, and basal leaves with a few long, coarse, simple or forked hairs are var. *pinetorum* (Tidestr.) Rollins.

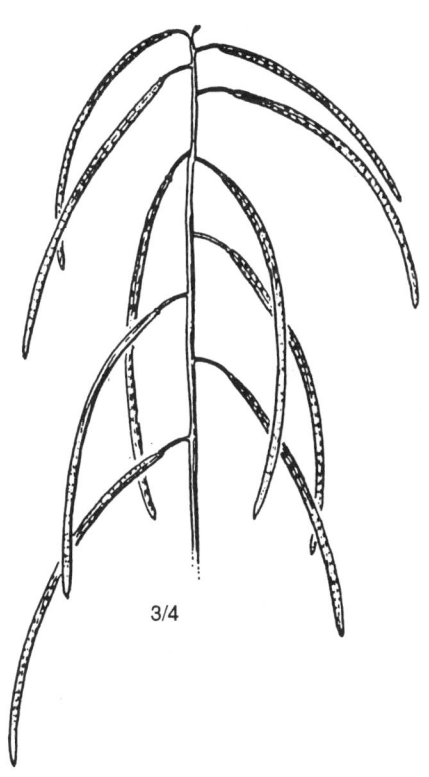

Arabis lemmonii S. Wats.
Proc. Am. Acad. 22:467, 1887.

Lemmon Rockcress

Arabis bracteolata Greene
Arabis depauperata A. Nels. & Kenn.
Arabis drepanoloba Greene
Arabis egglestonii Rydb.
Arabis kennedyi Greene
Arabis latifolia (S. Wats.) Piper
Arabis oreocallis Greene
Arabis polyclada Greene
Arabis semisepulta Greene
Boechera lemmonii (S. Wats.) W.A. Weber

Perennial from a branched caudex. Stems few to several, slender, erect, simple, pubescent at least below, with simple, or dendritic hairs. Basal leaves rosulate, oblanceolate, short-petiolate, entire to shallowly toothed, densely pubescent with dendritic hairs; cauline leaves sessile, oblong to lanceolate, auriculate, glabrous, or the lower ones pubescent. Flowers in loose racemes. Sepals oblong-lanceolate, glabrous to pubescent, often purplish. Petals purple or pinkish purple, spatulate. Siliques reflexed to horizontal-spreading, glabrous, straight or slightly curved, 1-nerved at the base. Seeds in 1 row, narrowly winged.

Fellfields, ledges, talus, and scree of the western Middle Rocky Mountain ranges. Alaska and Yukon Territory south to California, Idaho, Utah, and Colorado.

Alpine varieties may be distinguished as follows:
1. Plants more than 2 dm tall; stems few: siliques more than 2 mm wide
    . . . var. *drepanoloba* (Greene) Rollins
1. Plants less than 2 dm tall; stems several; siliques less than 2 mm wide
    . . . var. *lemmonii*

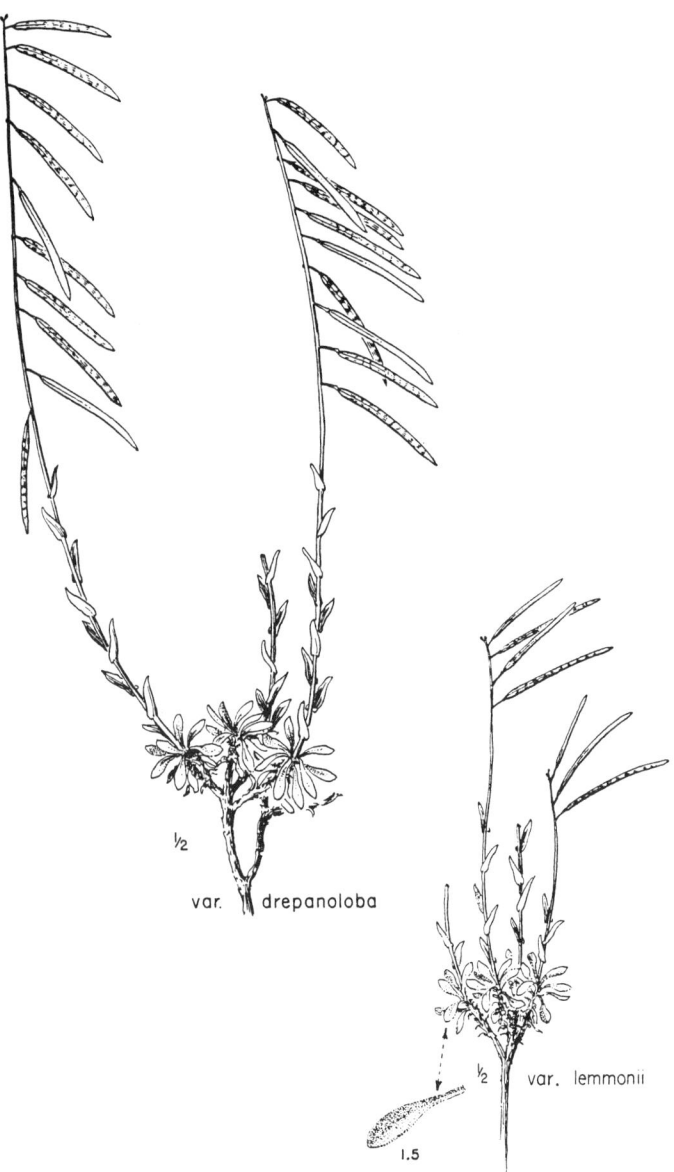
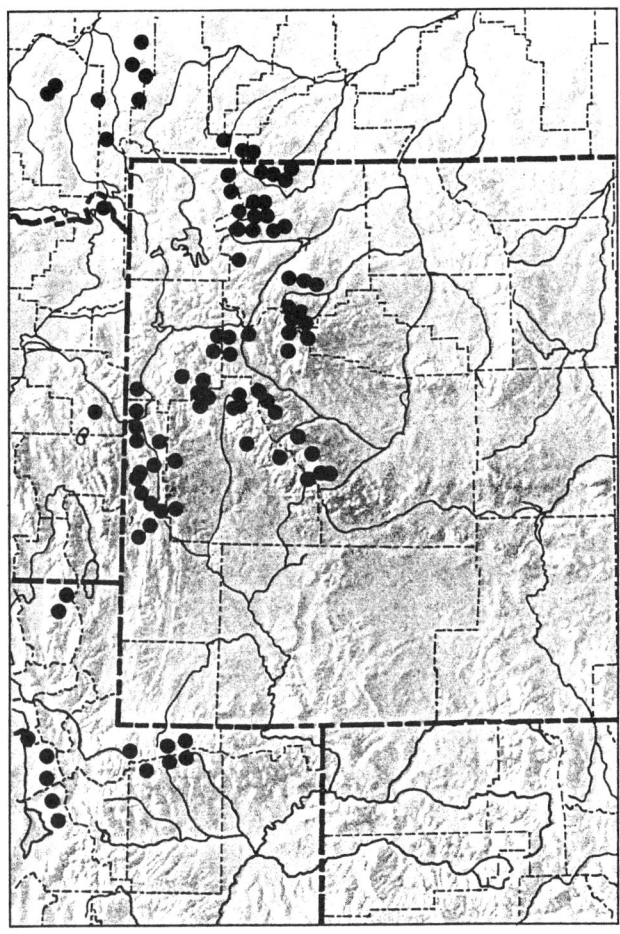

Arabis lyallii S. Wats.
Proc. Am. Acad. 11:122, 1875.

Lyall Rockcress

Arabis armerifolia Greene
Arabis densa Greene
Arabis multiceps Greene
Arabis nubigena Macbr. & Pays.
Arabis oreophila Rydb.
Arabis paupercula Greene

Caespitose perennial from a branched caudex. Stems few to several, glabrous, sometimes pubescent at the base with simple or forked hairs. Basal leaves rosulate, elliptic-oblanceolate, entire, acute, short-petiolate, glabrous or pubescent with small, branched hairs; cauline leaves oblong to ovate, sessile, auriculate, mostly glabrous. Flowers few in a short raceme. Sepals oblong-lanceolate, glabrous to pubescent, often purplish. Petals purple to rose-purple, spatulate. Siliques erect to spreading, glabrous, straight, 1-nerved. Seeds in 1 row, narrowly winged.

Fellfields and dry meadows of the western Middle Rocky Mountain ranges. British Columbia and Yukon Territory south to California, Nevada, Utah, and Wyoming.

Middle Rocky Mountain alpine plants are var. *lyallii*.

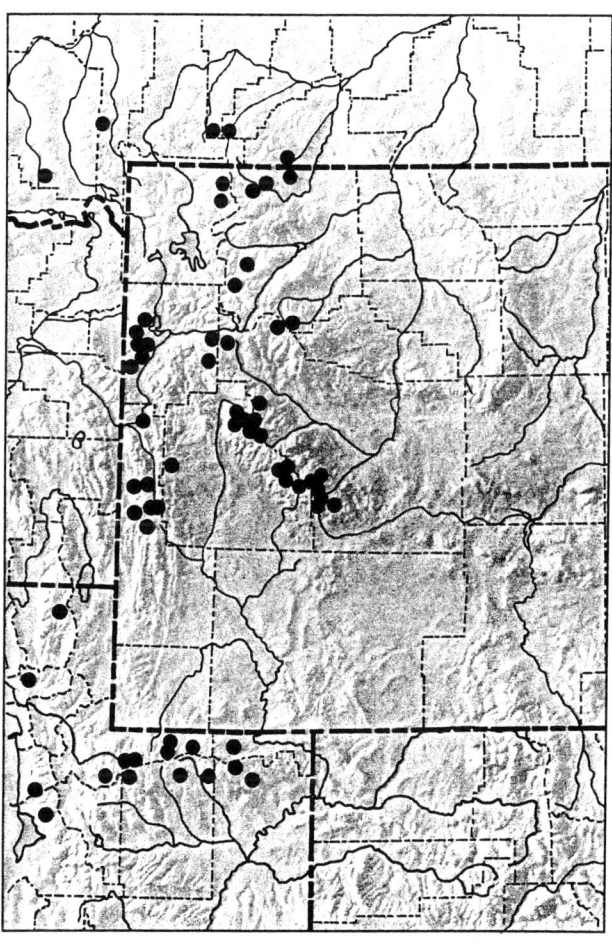

Arabis microphylla Nutt. in T. & G. *Fl. N. Am.* 1:82, 1838.

Small-Leaf Rockcress

Arabis densicaulis A. Nels.
Arabis macounii S. Wats.
Arabis tenuicula Greene

Tufted perennial from a branched caudex. Stems erect or decumbent at the base, few to many, simple, glabrous or more commonly pubescent, with simple or forked hairs at least below. Basal leaves rosulate, short-petiolate, oblanceolate, entire or with a few teeth, acute, pubescent with dendritic hairs; cauline leaves usually remote, sessile, auriculate-clasping, entire to shallowly toothed. Flowers in dense, several- to many-flowered racemes. Sepals mostly glabrous, oblong-lanceolate. Petals pink, lavender, or purplish, spatulate. Siliques erect to ascending, glabrous, straight or slightly curved, 1-nerved below the middle. Seeds in 1 row, orbicular, narrowly winged.

Rocky slopes, scree, and fellfields in the northern Wind River Range. British Columbia south to Oregon, Idaho, Utah, and Wyoming.

Alpine plants as described are var. *microphylla*.

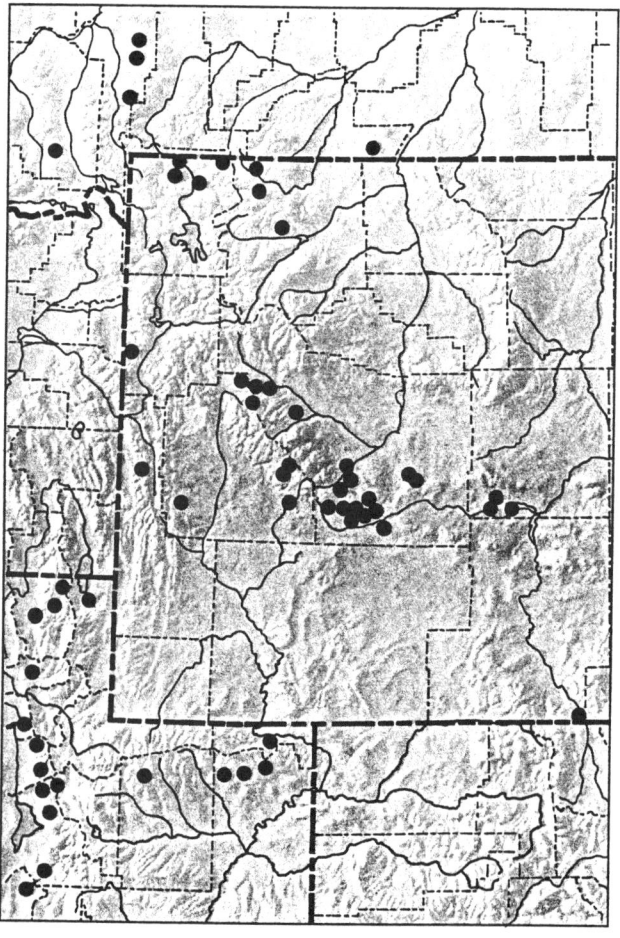

Arabis nuttallii Robins. in Gray
Syn. Fl. 1(1):160, 1895.

Nuttall Rockcress

Arabis bridgeri M.E. Jones
Arabis macella Piper
Arabis spathulata Nutt. in T. & G.
Erysimum nuttallii (Robins.) Kuntze

Perennial from a simple or few-branched caudex. Stems slender, few to several, glabrous above, hirsute below with simple or forked hairs. Basal leaves rosulate, oblanceolate to obovate, entire, glabrous to hirsute; cauline leaves few, lanceolate to obovate, glabrous or hirsute, sessile, tapering to the base. Flowers many in short, dense racemes. Sepals oblong-lanceolate, glabrous to sparsely hirsute, the outer pair saccate at the base. Petals white, occasionally lavender, and spatulate. Siliques erect to ascending, glabrous, straight, 1-nerved. Seeds in 1 row, wingless.

Meadows, willow thickets, and moist stream banks in the lower alpine zone of the Beartooth, Big Horn, Gros Ventre, Hoback, Salt River, Teton, and Wyoming Ranges. Alberta south to Washington, Idaho, Utah, and Wyoming.

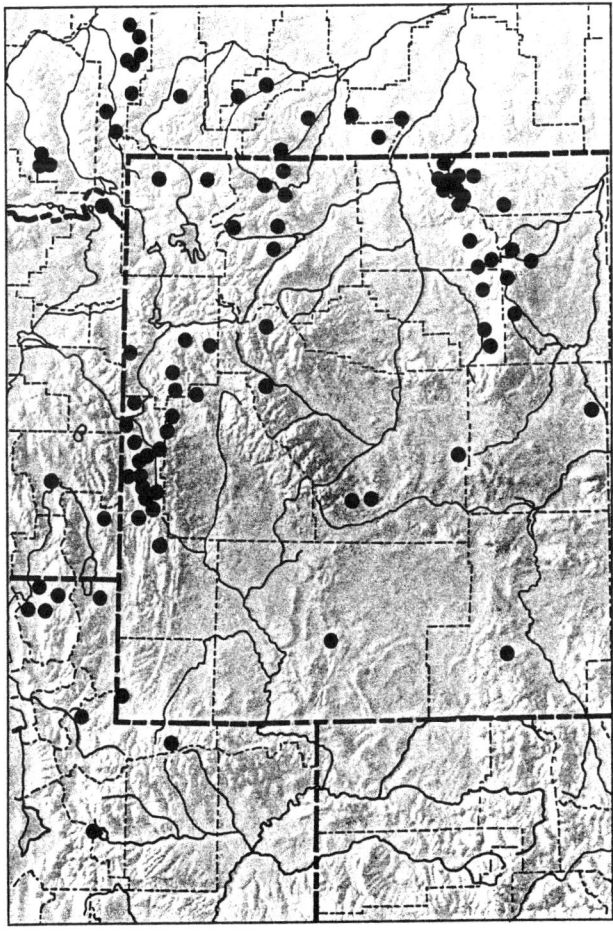

Arabis williamsii Rollins
Syst. Bot. 6:62, 1981.

Williams Rockcress

Biennial or perennial from a taproot or few-branched caudex. Stems erect, simple, usually solitary, sometimes 2 or 3, glabrous. Basal leaves weakly rosulate, erect or ascending, oblanceolate, entire, acute, pubescent with dendritic hairs; cauline leaves remote, sessile, oblong, entire, auricles lacking or only poorly developed in the uppermost leaves. Flowers in loose, elongate racemes. Petals pink to purplish, spatulate. Siliques erect, glabrous, straight or nearly so, 1-nerved below the middle. Seeds in 2 rows, oblong, wingless.

Rocky slopes in the lower alpine zone in the Absaroka and Wind River Ranges, mostly on limestone. Apparently endemic to Wyoming mountain ranges.

Alpine plants are var. *saximontana* (Rollins) Rollins.

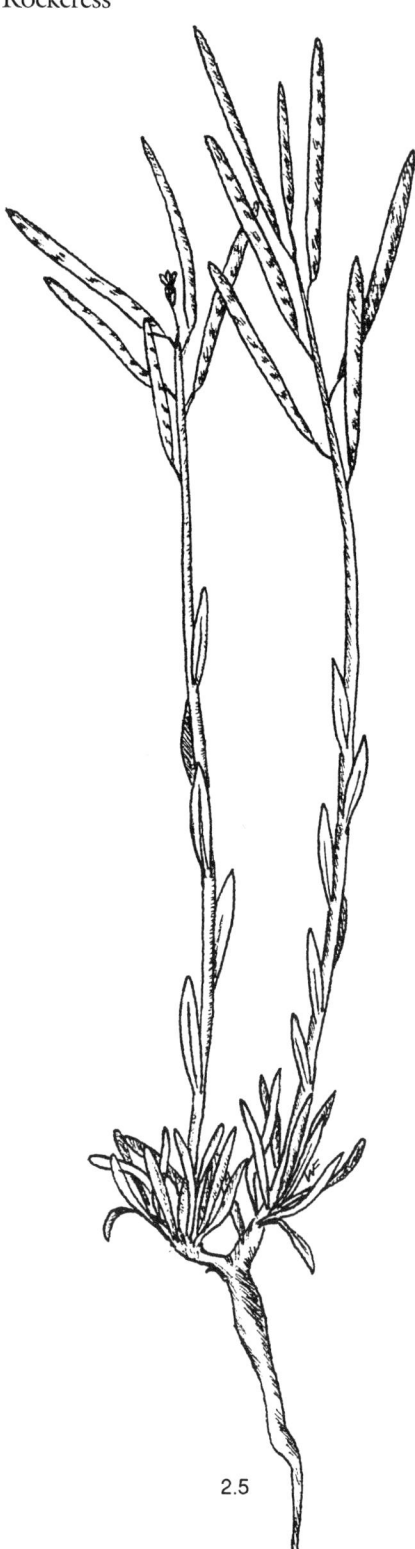

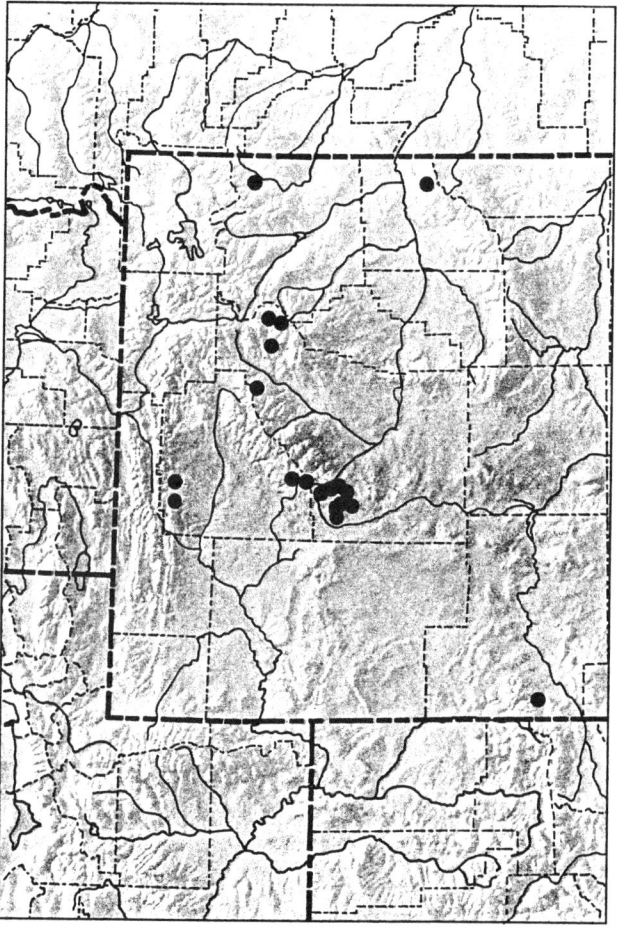

Descurainia Webb & Berth.
Tansymustard

Annual or biennial herbs with stellate pubescence, sometimes with simple or glandular hairs. Stems simple or branched, the leaves 1–3 times pinnately compound, often finely dissected and usually reduced upward; basal rosette present during early growth stages, usually lacking at anthesis. Flowers small, pale yellow, in terminal racemes. Sepals ovate, erect to spreading. Petals clawed, obovate, obtuse. Stamens 6, yellow. Style short, inconspicuous, the stigma entire. Siliques linear to clavate, terete or nearly so, the valves 1-nerved. Seeds uniseriate or biseriate.

KEY TO THE SPECIES OF *DESCURAINIA*

1. Stems branched from base, decumbent; fruiting pedicels appressed to rachis of inflorescence ..... *D. torulosa*
1. Stems single or branched well above the middle, erect; fruiting pedicels spreading ..... *D. incana*

Descurainia incana (Bernh. ex Fisch. & C.A. Mey.) Dorn
Vasc. Pl. Wyo., 296, 1988.

Mountain Tansy Mustard

Sisymbrium incanum Bernh. ex Fisch. & C.A. Mey.
Descurainia incisa (Engelm.) Britt.
Descurainia richardsonii (Sweet) Schulz
Descurainia rydbergii Schulz
Descurainia serrata (Greene) Schulz
Hesperis incisa (Engelm.) Kuntze
Sisymbrium hartwegianum Rydb.
Sisymbrium incisum Engelm. in Gray
Sisymbrium procerum (Greene) K. Schum.
Sisymbrium richardsonianum Rydb.
Sisymbrium richardsonii Sweet
Sisymbrium viscosum (Rydb.) Blank.
Sophia brevipes (Nutt.) Rydb.
Sophia californica Rydb.
Sophia hartwegiana (Fourn.) Greene
Sophia incisa (Engelm.) Greene
Sophia leptophylla Rydb.
Sophia procera Greene
Sophia purpurascens Rydb.
Sophia ramosa Rydb.
Sophia richardsonii (Sweet) Rydb.
Sophia serrata Greene
Sophia sonnei (Robins.) Greene
Sophia viscosa Rydb.

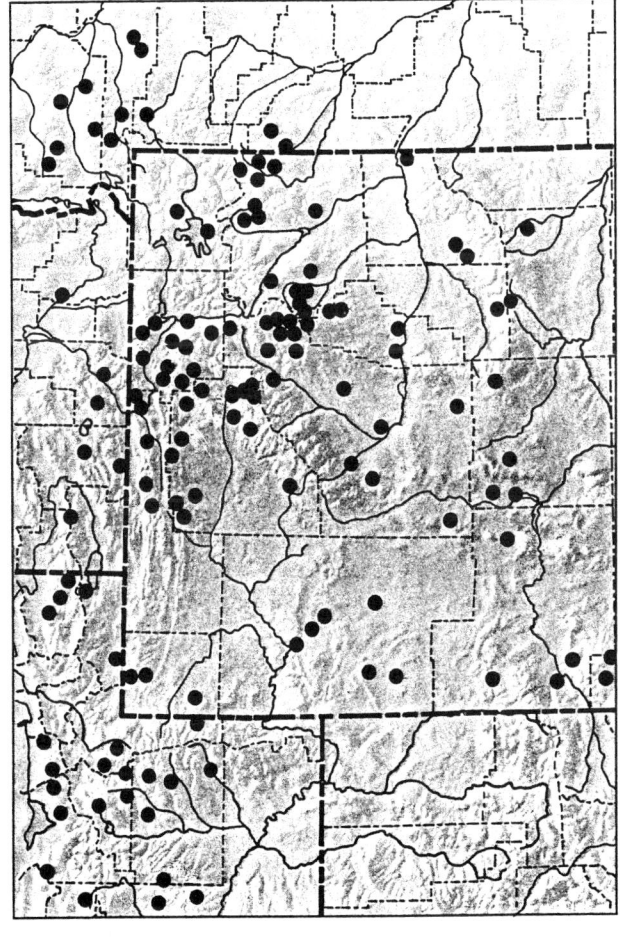

Biennial, mostly canescent herb with stems simple below and branched above, sometimes glandular. Leaves mostly cauline, pinnate above to bipinnate below, the segments lanceolate to linear. Flowers pale to bright yellow, in dense racemes. Petals small, barely exceeding the sepals. Siliques narrowly linear, straight to arcuate, ascending to appressed. Seeds uniseriate.

This is a widespread species in the Middle Rocky Mountains, common on the plains, roadsides, and other disturbed areas. It occurs above timberline in the Absaroka, Beartooth, Salt River, and Wind River Ranges. Midwestern and western North America from Yukon Territory to northern Mexico.

The two varieties that occur in the vicinity of timberline in the Middle Rocky Mountains are distinguished as follows:

1. Inflorescence glandular-pubescent; siliques ascending to widely spreading . . . var. *viscosa* (Rydb.) Dorn
1. Inflorescence glabrous to pubescent, but not glandular; siliques erect, appressed
 . . . var. *macrosperma* (O.E. Schulz) Dorn

Braya humilis (C.A. Mey.) Robins., Dwarf Braya, will key here. The first record from the Middle Rocky Mountains has recently been collected from limestone on Arrow Mountain in the northern Wind River Range. It is easily distinguished from *D. incana* as a small perennial, less than 15 cm tall, with mostly basal, entire, spatulate leaves.

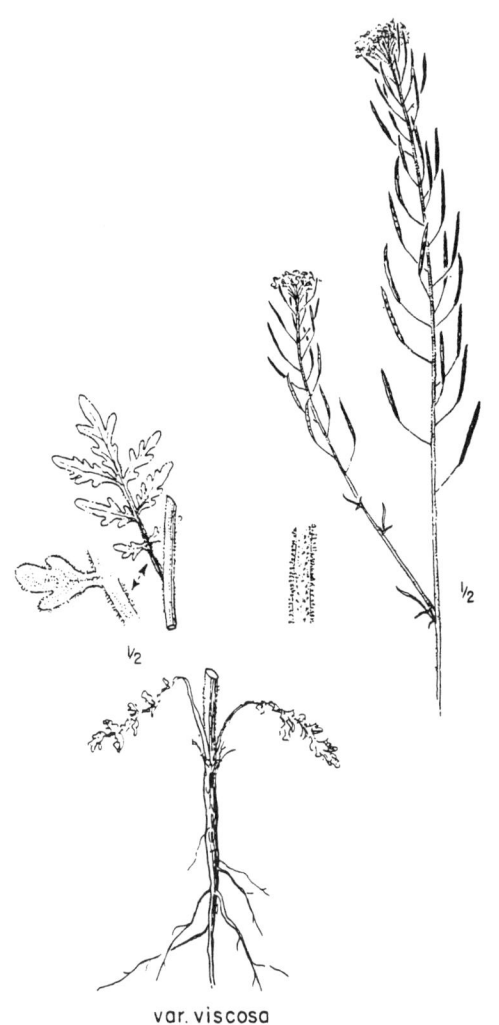

var. viscosa

var. macrosperma

Descurainia torulosa Rollins
J. Arnold Arb. 64:499, 1983.

Wind River Tansymustard

Biennial, or possibly perennial herb with several to many simple or branched, decumbent stems arising from a taproot with a simple crown. Basal leaves densely clustered, at least when young, pinnately lobed, short-petiolate. Cauline leaves few, pinnately lobed. Flowers cream-colored to yellowish, in racemes. Petals small, spatulate, exceeding the hyaline-margined sepals. Siliques terete, linear to slightly arcuate, torulose, erect and appressed to the rachis of the inflorescence, or ascending. Seeds uniseriate.

Scree and talus in the lower alpine zone of the Absaroka Range. Known only from the vicinity of the type locality (Scott, 761) near Brooks Lake, Togwotee Pass, and the head of Pickett Creek further north in the Absaroka Range. This taxon is apparently endemic to northwestern Wyoming.

Draba L.
Draba, Whitlow Grass

Low, annual or perennial herbs with simple or branching, leafy or scapose flowering stems. Pubescence of simple, branched, or stellate hairs, or a mixture. Leaves entire or dentate. Sepals erect or ascending, blunt. Petals yellow or white, entire, emarginate or bifid, narrowed below into a claw longer than the sepals. Silicle flattened parallel to the septum, rarely as much as 5 times as long as wide, sometimes twisted. Seeds several to numerous, in 2 rows in each locule.

KEY TO THE SPECIES OF *DRABA*

1. Plant annual, from a slender taproot
   2. Cauline leaves 0–2, the basal leaves 1–4 mm broad, 6–25 mm long   *D. crassifolia*
   2. Cauline leaves 1–8, the basal leaves 2–8 mm broad, 10–40 mm long
      3. Upper stem pubescent   *D. praealta*
      3. Upper stem glabrous   *D. albertina*
1. Plant biennial or perennial, from a taproot or branched caudex
   4. Plant caulescent, usually not caespitose or matted, the stems with 2 or more leaves well above the base
      5. Leaves glabrous or merely ciliate on the margins   *D. crassa*
      5. Leaves pubescent on one or both sides.
         6. Leaves with at least some pectinately branched hairs   *D. glabella*
         6. Leaves with simple to stellate hairs, but none pectinately branched
            7. Flowering stems glabrous above, sometimes sparsely pubescent below; inflorescence and silicles glabrous   *D. albertina*
            7. Flowering stems pubescent throughout; inflorescence and silicles pubescent
               8. Styles inconspicuous, less than 0.2 mm long or lacking   *D. praealta*
               8. Styles conspicuous, 0.2–1 mm long
                  9. Petals yellow; leaves with some simple hairs; flowering stems 1.5–2 mm thick   *D. aurea*
                  9. Petals white; leaves often with all branched hairs; flowering stems less than 1 mm thick   *D. breweri*
   4. Plant caespitose or mat forming, scapose or rarely with 1 or 2 reduced stem leaves near the base
      10. Basal leaves with numerous doubly pectinate hairs
         11. Basal leaves mostly less than 1.5 mm wide, with at least some doubly pectinate and sessile-appressed hairs; sepals and pedicels glabrous   *D. oligosperma*
         11. Basal leaves mostly more than 1.6 mm wide, with simple, branched-stellate, and stalked-spreading pectinate hairs; sepals and pedicels pubescent   *D. incerta*
      10. Basal leaves glabrous or variously pubescent, but only rarely with a few doubly pectinate hairs
         12. Leaves glabrous, at least on the upper surface, the margins ciliate
            13. Plant biennial, from a slender taproot, usually lacking a branched caudex covered by old leaf bases   *D. crassifolia*
            13. Plant perennial, mostly from a branched caudex densely covered by old leaf bases
               14. Petals white; styles 0.15 mm or less in length   *D. fladnizensis*
               14. Petals yellow, sometimes whitish; styles 0.2 mm or more in length   *D. densifolia*
         12. Leaves pubescent on upper surface
            15. Petals white, sometimes drying pale yellow; styles less than 0.5 mm long

16. Leaves ciliate, pubescent at least near the base, with simple or forked hairs, stellate hairs lacking ..... *D. fladnizensis*
16. Leaves with stellate hairs, sometimes combined with simple or forked hairs
   17. Silicles linear, oblong or lanceolate; basal leaves with simple, forked, and stellate hairs; stellate hairs with 9 or more rays, the individual hairs crowded and not easily distinguished ..... *D. lonchocarpa*
   17. Silicles obovate; basal leaves with simple, forked, cruciform, and stellate hairs; stellate hairs with 8 or fewer rays, the individual hairs loose and spreading ..... *D. porsildii*
15. Petals bright or at least distinctly yellow; styles more than 0.5 mm long
   18. Basal leaves linear to narrowly spatulate, 0.3–2 mm wide, prominently ciliate, with tangled, long-branched hairs below ..... *D. paysonii*
   18. Basal leaves ovate, oblanceolate, or obovate, 2–5 mm wide, not prominently ciliate, densely pubescent with simple, branched, and mostly 4-rayed hairs below ..... *D. ventosa*

Draba albertina Greene
Pittonia 4:312, 1901.

Alberta Draba

Draba deflexa Greene
Draba nitida Greene
Draba stenoloba Ledeb. (in part)

Biennial or perennial from a simple caudex; stem slender, erect, with simple to bifid hairs, often glabrous above. Basal leaves rosulate, obovate to lanceolate, entire to denticulate; cauline leaves mostly 1–3, sessile, lanceolate to ovate or elliptic, entire to denticulate; leaves with simple or bifid hairs above and bifid, trifid, or cruciform hairs below, rarely glabrous. Flowers in open racemes. Sepals pubescent with simple hairs or glabrous. Petals yellow or cream, often fading to white, spatulate to emarginate. Silicles glabrous, linear-elliptic to nearly ovate; styles less than 0.1 mm long, sometimes lacking.

Moist meadows and slopes of the Absaroka, Beartooth, Medicine Bow, Salt River, Uinta, Wasatch, and Wind River Ranges. Yukon Territory and western Mackenzie district of Northwest Territories, south to California and Arizona.

Draba stenoloba var. *nana* (Schulz) C.L. Hitchc. has been treated by various authors as a synonym of *D. albertina*. However, as Rollins (1993) pointed out, *D. albertina* differs in chromosome number (n=12) and in having leaf surfaces with simple or bifid hairs and stems with simple hairs. *Draba stenoloba* has a chromosome number of n=20, upper leaf surfaces with mostly 3- to many-branched hairs, and stems with cross-shaped or dendritic hairs. Based on these differences *D. stenoloba* does not occur in the Middle Rocky Mountains.

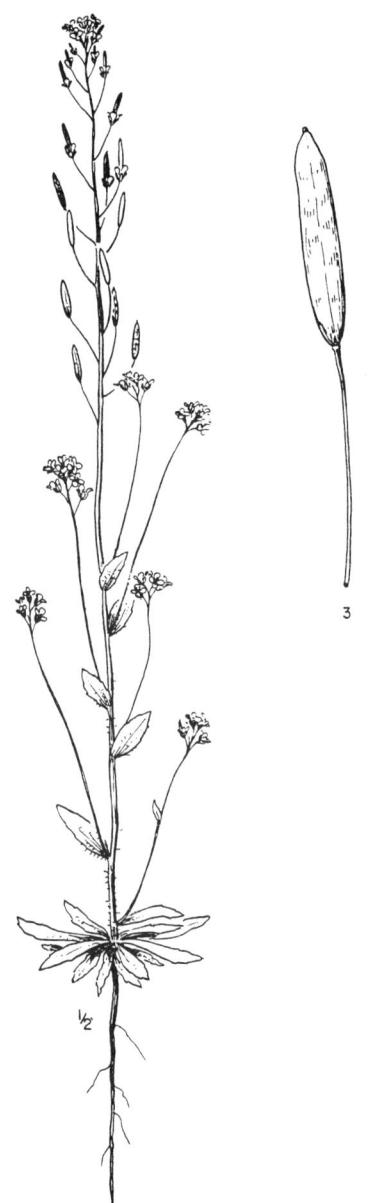

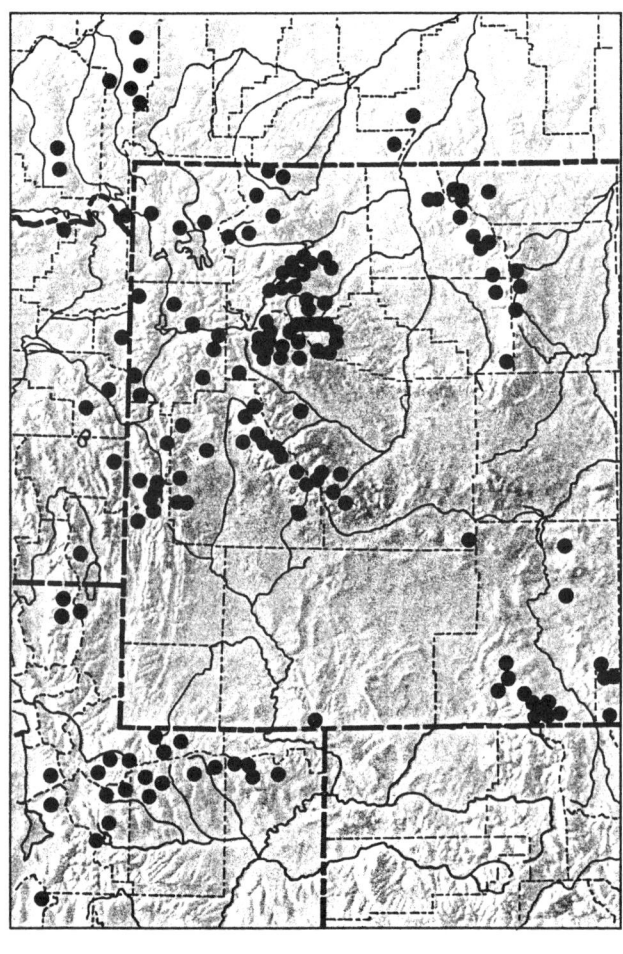

Draba aurea Vahl ex Hornem.
Fors. Dansk Oekon. Pl. 2:599, 1806.

Golden Draba

Draba aureiformis Rydb.
Draba bakeri Greene
Draba decumbens Rydb.
Draba luteola Greene
Draba mccallae Rydb.
Draba minganensis (Vict.) Fern.
Draba neomexicana Greene
Draba surculifera A. Nels.
Draba uber A. Nels.

Erect, rosulate, short-lived perennial from a simple to branched caudex; stems 1 to several, sometimes decumbent at the base, pubescent with simple and branched, often cruciform hairs. Basal leaves oblanceolate, petiolate, pubescent with mostly cruciform hairs on both surfaces, the margins entire to denticulate; cauline leaves reduced upward, sessile or nearly so, oblanceolate to ovate, entire to denticulate, canescent. Flowers in elongate, simple, or occasionally branched racemes, the lower ones often bracteate. Sepals greenish, pubescent with simple to branched hairs. Petals yellow, spatulate, rounded to emarginate. Silicles lanceolate, flat or twisted, pubescent or rarely glabrous; styles 0.3–1.5 mm long.

Rocky slopes, fellfields, and meadows of most alpine ranges. Not yet known from above timberline in the Big Horn, Gros Ventre, and Hoback Ranges. Greenland to Alaska, south in the Rocky Mountains to Idaho, Arizona, and New Mexico.

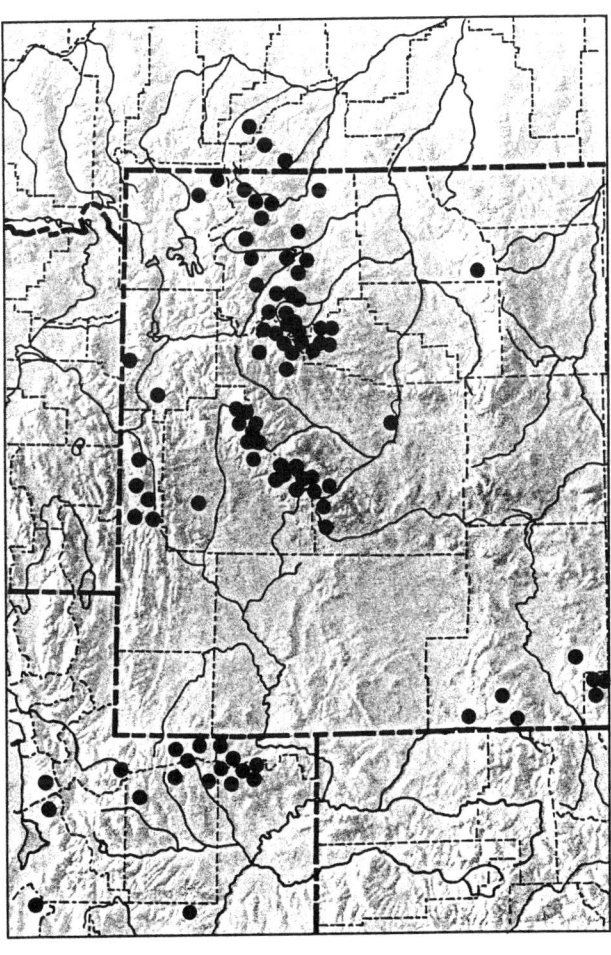

Draba breweri S. Wats.
Proc. Am. Acad. 23:260, 1888.

White Draba

Draba cana Rydb.
Draba stylaris J. Gay ex W.D.J. Koch
Draba valida Goodding

Rosulate perennial from a simple to branched caudex; stems several, simple or branched, pubescent with stellate, branched, and simple hairs. Basal leaves oblanceolate, entire to denticulate, canescent with stellate to branched hairs on both surfaces; cauline leaves several, somewhat reduced upward, lanceolate to ovate, denticulate, stellate- or branched-pubescent, sometimes with bifid hairs. Flowers several to many in racemes, the lower ones bracteate. Sepals pilose with simple and forked hairs. Petals white, spatulate, emarginate. Silicles lanceolate to oblong, somewhat twisted, pubescent with simple to stellate hairs; styles (0.2–) 0.4–0.8 (–1) mm long.

Rocky slopes, talus, scree, and dry meadows of most of the northern and western ranges. Not yet known from above timberline in the Hoback, Salt River, and Teton Ranges. Alaska and Yukon Territory south to Nevada, Utah, and Colorado.

Mulligan (1971) pointed out that *D. lanceolata* Royle, which has been used in various treatments and is treated as a synonym of *D. cana* by many authors, does not apply to North American plants with stellate pubescence on the fruits. Asiatic *D. lanceolata* has simple or forked hairs, but not branched stellate hairs as in North America. Rollins (1993) further clarified the nomenclatural situation by using the name *D. breweri*, which has priority over *D. cana*. Middle Rocky Mountain alpine plants thus become *Draba breweri* var. *cana* (Rydb.) Rollins.

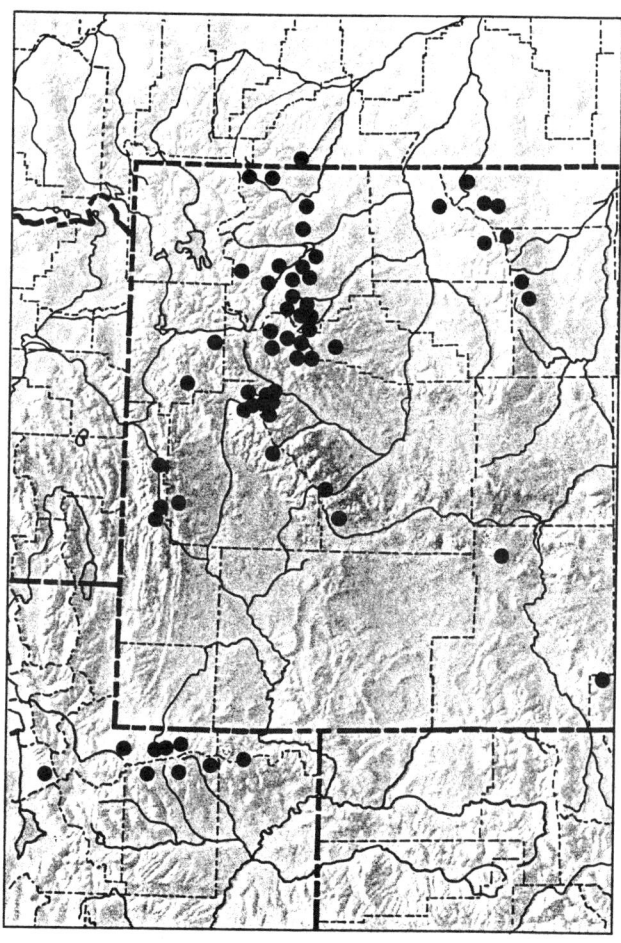

Draba crassa Rydb.
Mem. N.Y. Bot. Gard. 1:182, 1900.

Fleshy Draba

Draba chrysantha S. Wats.

Perennial from a thick, usually branched caudex covered by persistent leaf bases; stems several, often decumbent at the base, pubescent with simple and forked hairs. Basal leaves rosulate, oblanceolate, entire, glabrous, the margins sometimes ciliate; cauline leaves somewhat reduced upward, ovate or obovate, entire or nearly so, usually glabrous. Flowers in short, dense racemes, the pedicels soft-villous. Sepals pilose. Petals yellow, obovate, rounded. Silicles lanceolate to ovate, glabrous, flat or twisted, the margins undulate; styles 0.7–1 mm.

Fellfields, cliffs, talus, and scree of the Absaroka, Gros Ventre, Teton, Uinta, and Wind River Ranges. Southern Montana, Wyoming, Utah, and Colorado.

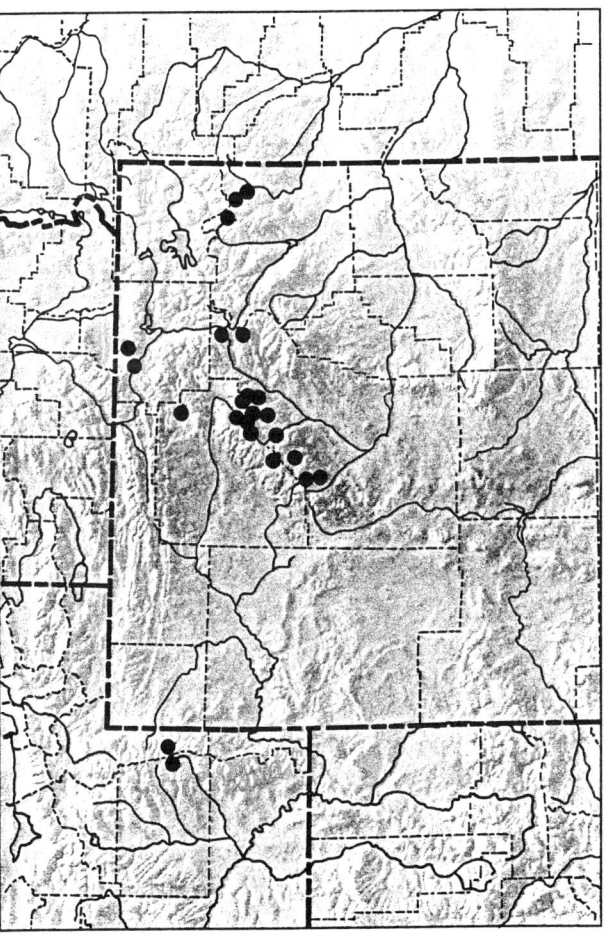

Draba crassifolia Grah.
Edinb. New Phil. J. 7:182, 1829.

Thickleaf Draba

Draba parryi Rydb.

Rosulate annual, biennial, or perennial from a simple or short-branched caudex; stems 1 to several, simple, glabrous above and sparsely pubescent with simple and forked hairs below. Basal leaves narrowly spatulate to oblanceolate, glabrous, occasionally with simple or forked hairs, the margins often ciliate; cauline leaves lacking, sometimes 1 or 2 near the stem base, these glabrous, reduced. Flowers in open racemes, the pedicels glabrous. Sepals glabrous. Petals spatulate, emarginate, yellow fading to white. Silicles elliptic to lanceolate, flat, glabrous; styles lacking.

Rocky slopes, talus, scree, snow beds, and dry rocky meadows in nearly all ranges of the Middle Rocky Mountains. Not yet known from above timberline in the Hoback Range. Alaska east to Greenland and Scandinavia, and south in western North America to California, Arizona, and Colorado.

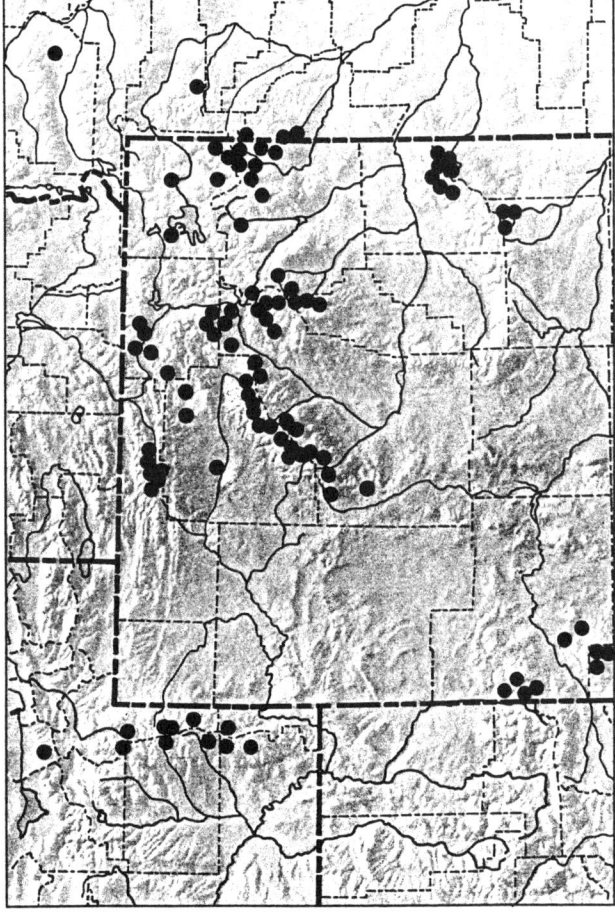

Draba densifolia Nutt. in T. & G.
Fl. N. Am. 1:104, 1838.

Nuttall Draba

Draba apiculata C.L. Hitchc.
Draba caeruleomontana Pays. & St. John
Draba globosa Pays.
Draba mulfordae Pays.
Draba nelsonii Macbr. & Pays.
Draba pectinata (S. Wats.) Rydb.
Draba sphaerula Macbr. & Pays.

Scapose perennial, often forming mats, from a simple to branched caudex. Scapes glabrous to pubescent with simple and bifid hairs. Leaves imbricate, oblong to oblanceolate, the margins ciliate with stiff, coarse, simple hairs, the upper surface usually glabrous, the lower surface often with a few forked, dendritic, or stellate hairs. Flowers in short, compact racemes. Sepals glabrous or pubescent with simple or forked hairs. Petals yellow or sometimes whitish, obovate, and rounded. Silicles ovate to elliptic, glabrous or stellate-pubescent; styles 0.2–1 mm long.

Talus, scree, rocky ridges, and slopes of the Absaroka, Beartooth, Gros Ventre, Teton, Uinta, Wasatch, and Wind River Ranges. British Columbia south to California, Utah, Idaho, and Wyoming.

Two varieties are recognized as follows:
1. Silicles and pedicels glabrous; styles less than 0.5 mm long . . . var. *apiculata* (C.L. Hitchc.) S.L. Welsh
1. Silicles and pedicels pubescent; styles 0.5–1.0 mm long . . . var. *densifolia*

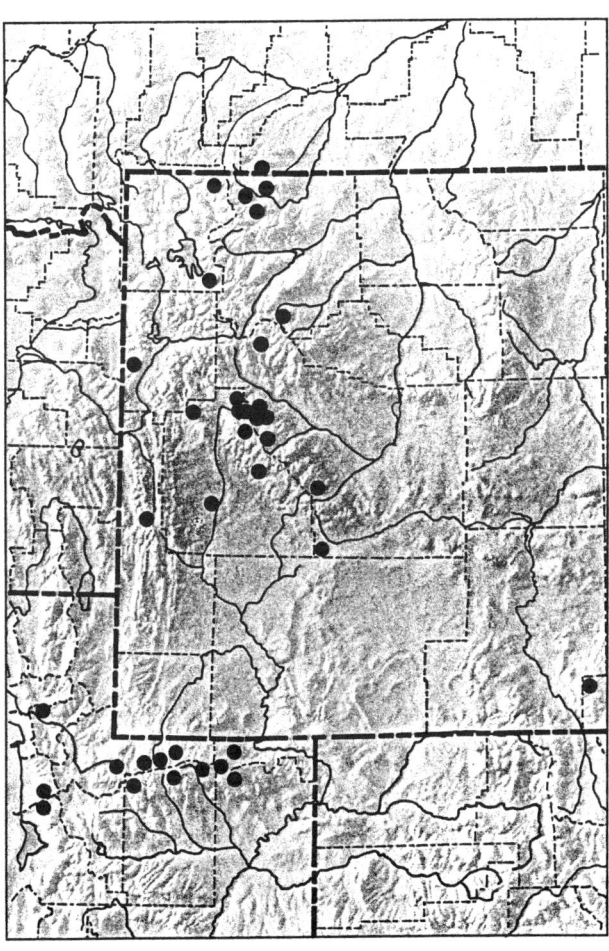

Draba fladnizensis Wulf. in Jacq.
Misc. Aust. Bot. 1:147, 1778.

Arctic Draba

Draba pattersonii Schulz
Draba tschuktschorum Trautv.

Rosulate perennial from a simple or branched caudex; stems 1 to several, glabrous or pubescent with simple to forked hairs. Basal leaves oblanceolate with prominent midribs, glabrous or pubescent with simple and forked hairs, the margins ciliate; cauline leaves lacking or 1–2, reduced. Flowers in short, congested, corymbose racemes, the pedicels glabrous. Sepals glabrous, sometimes pubescent with simple hairs. Petals white, spatulate, rounded to emarginate. Silicles oblong to ovate, glabrous; styles lacking.

Fellfields, talus, and scree at high elevations in the Absaroka, Beartooth, Big Horn, and Wind River Ranges. Circumpolar; south in western North America to Utah and Colorado.

Middle Rocky Mountain alpine plants are var. *pattersonii* (Schulz) Rollins.

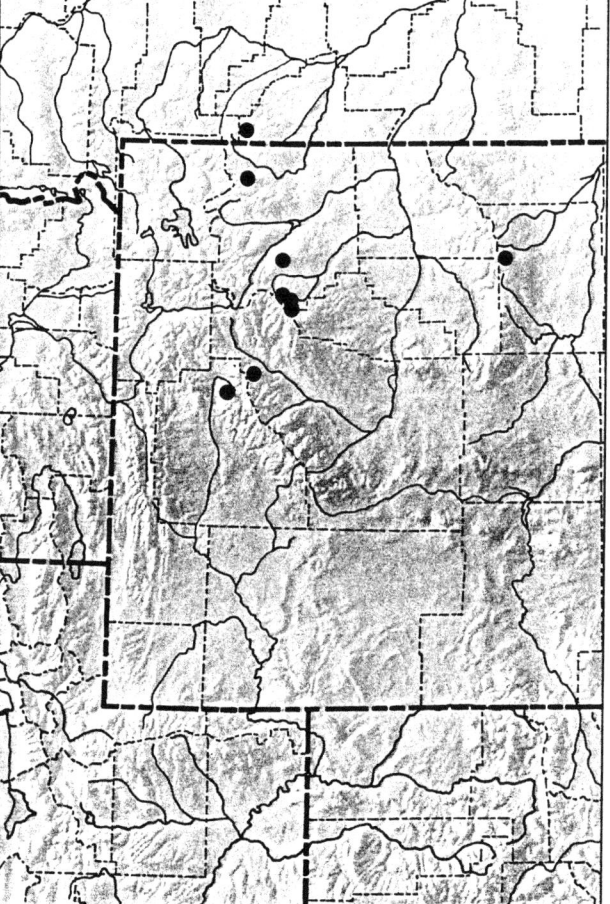

Draba glabella Pursh
Fl. Am. Sept. 2:434, 1814.

Smooth Draba

Draba canadensis Brunet
Draba daurica DC.
Draba henneana Schlecht.
Draba hirta L.
Draba juvenilis Kom.
Draba laurentiana Fern.
Draba megasperma Fern. & Knowlt.

Rosulate perennial from a simple caudex; stems 1 to several, simple to branched, stellate-pubescent, at least below. Basal leaves oblong to oblanceolate, stellate-pubescent, entire to dentate, the margins often ciliate; cauline leaves 2 or more, oblong to ovate, entire or denticulate. Flowers in short, corymbose racemes, the pedicels glabrous to pubescent with simple hairs. Sepals pubescent with simple to branched hairs, somewhat whitish-margined. Petals white, spatulate, emarginate. Silicles lanceolate, oblong, or narrowly ovate, glabrous or nearly so, flat to somewhat twisted; styles 0.2–0.5 mm long.

Rocky places in the Beartooth Range. Circumpolar; south in western North America to British Columbia, Alberta, and northwestern Wyoming.

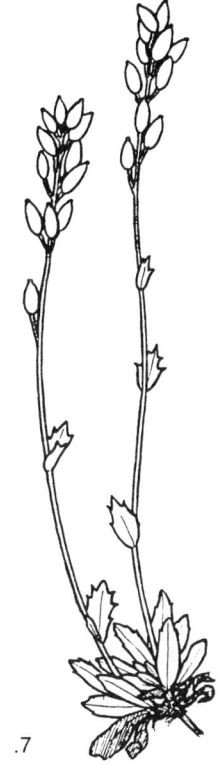

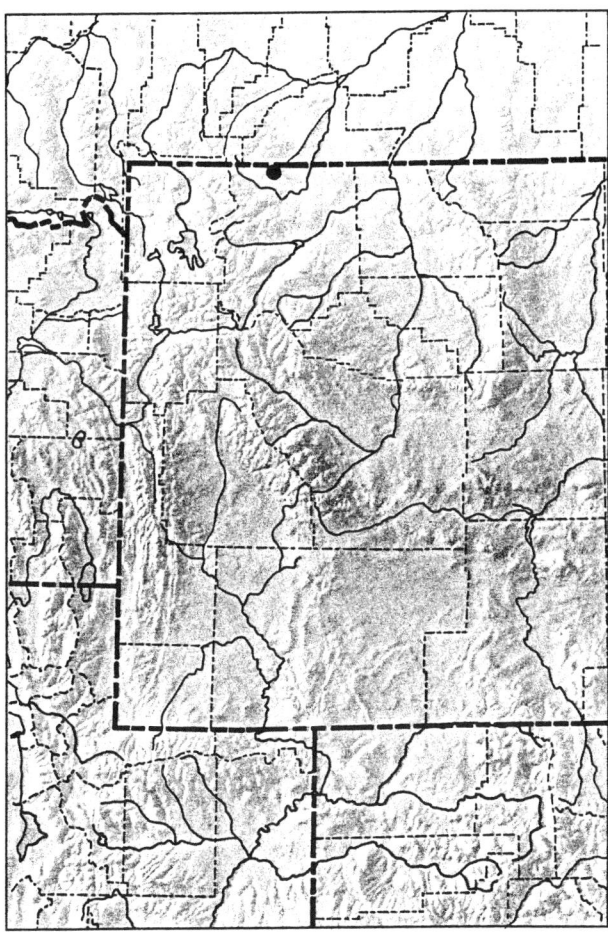

Draba incerta Pays.
Am. J. Bot. 4:261, 1917.

Yellowstone Draba

Draba laevicapsula Pays.
Draba peasei Fern.

Loosely caespitose, scapose perennial from a branched caudex; scapes several, pubescent with stellate, dendritic, and sometimes simple hairs, rarely with 1 reduced leaf. Basal leaves narrowly oblanceolate to linear, pubescent with branched to dendritic hairs, at least some of them doubly pectinate; margins entire, ciliate. Flowers in short, few-flowered racemes, the pedicels ascending, pubescent with simple or branched hairs. Sepals glabrous to pubescent with simple hairs. Petals spatulate, emarginate, yellow fading to cream or white. Silicles ovate to lanceolate, flat, pubescent with simple or branched hairs, sometimes glabrous; styles 0.4–1 mm long.

Fellfields of the Absaroka, Beartooth, Gros Ventre, Salt River, Uinta, and Wind River Ranges. Alaska and Yukon Territory south to Washington, Idaho, Utah, and Wyoming.

Draba incerta is closely related to *D. oligosperma*, from which it may be distinguished by its looser habit, wider, more spreading leaves, and spreading rather than appressed pubescence. Middle Rocky Mountain alpine plants are var. *incerta*.

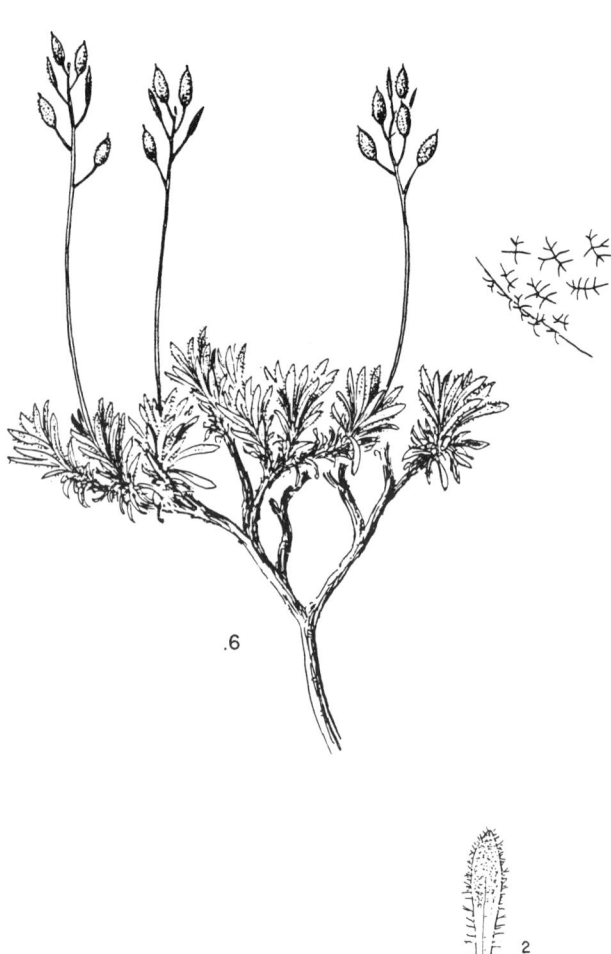

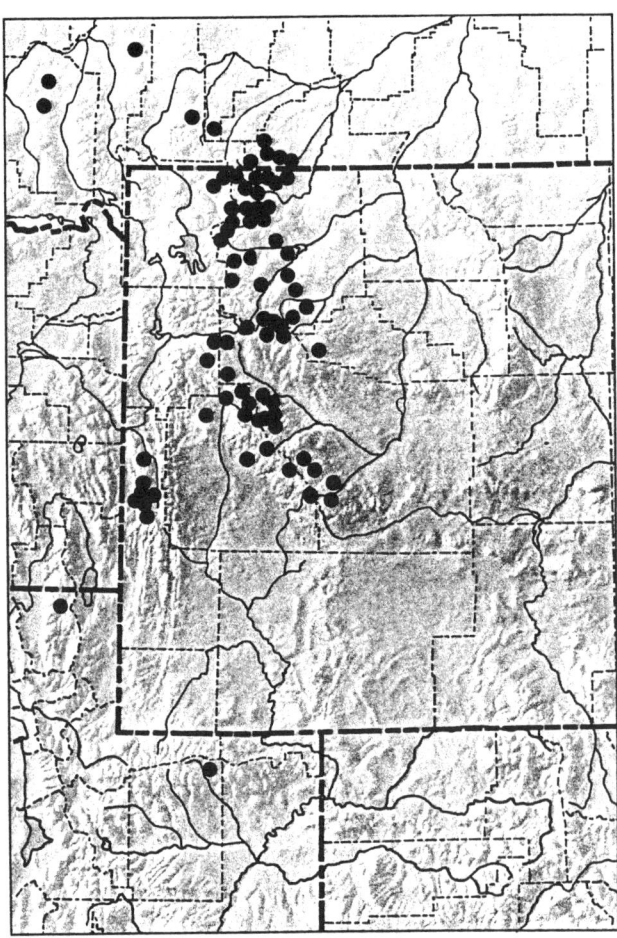

Draba lonchocarpa Rydb.
Mem. N.Y. Bot. Gard. 1:181, 1900.

Longfruit Draba

Caespitose perennial from a branched or, less frequently, simple caudex; stems glabrous to stellate-pubescent. Basal leaves oblanceolate to obovate, stellate-pubescent, the margins entire; cauline leaves lacking, or 1 or 2 reduced, the margins denticulate. Flowers in condensed to elongate racemes, the pedicels ascending to erect, glabrous to stellate-pubescent. Sepals glabrous to pubescent with simple, forked, or stellate hairs. Petals white, cuneate. Silicles elliptic, lanceolate, or narrowly ovate, flat or twisted, glabrous or stellate-pubescent; styles 0.2–0.5 mm long.

Fellfields, talus, ledges, and scree in most alpine ranges of the Middle Rocky Mountains. Not yet known from above timberline in the Hoback and Medicine Bow Ranges. Alaska; south in western North America to Washington, Nevada, Utah, and Colorado.

Alpine plants of the Middle Rocky Mountains are var. *lonchocarpa*.

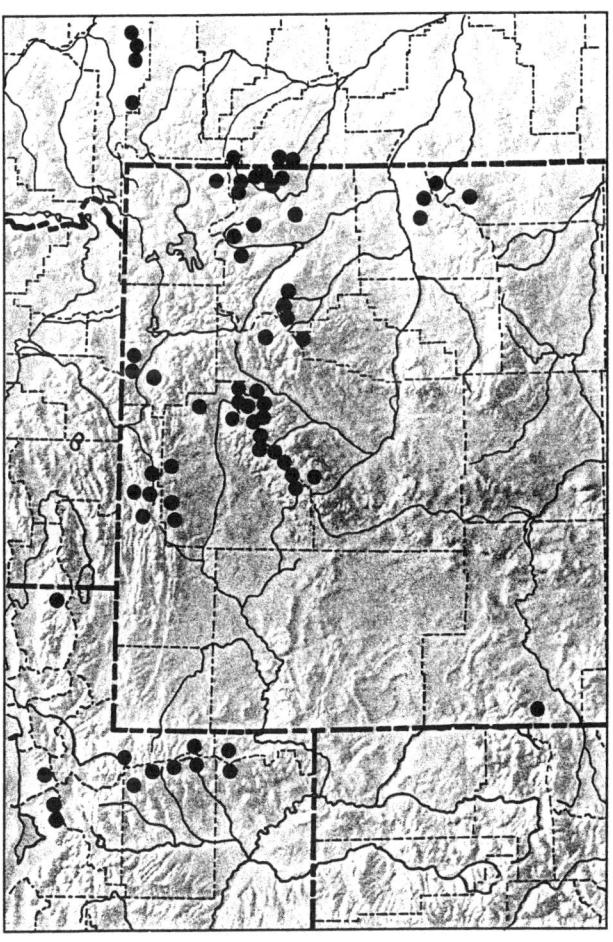

Draba oligosperma Hook.
Fl. Bor.-Am. 1:51, 1830.

Snowbank Draba

Draba andina (Nutt.) A. Nels.
Draba juniperina Dorn
Draba pectinipila Rollins
Draba saximontana A. Nels.
Draba subsessilis S. Wats.

Caespitose, scapose perennial from a stout, branched caudex; scapes several, pubescent with pectinate and stellate hairs to nearly glabrous. Leaves imbricate, linear to spatulate, at least the lower surfaces pubescent with appressed, sessile, doubly pectinate hairs, the margins ciliate with simple and pectinately branched hairs. Flowers in short racemes, somewhat elongate in fruit, the pedicels ascending, glabrous or pubescent. Sepals glabrous or pubescent with simple to branched hairs. Petals yellow or white, obovate, rounded to emarginate. Anthers indehiscent; pollen usually aborted. Silicles elliptic to ovate, flat, glabrous to pubescent with simple, forked, or doubly pectinate hairs; styles 0.1–1.5 mm long.

Fellfields, talus, scree, ledges, and other rocky places in all alpine ranges of the Middle Rocky Mountains. Alaska and Yukon Territory south to California, Idaho, Nevada, Utah, and Colorado.

Alpine material of this agamospermous species, 2n=64 (Mulligan, 1972), may be separated into 2 varieties.

1. Pubescence of silicles and scape doubly pectinate . . . var. *pectinipila* (Rollins) C.L. Hitchc.
1. Pubescence of silicles and scape, if present, of simple and forked hairs . . . var. *oligosperma*

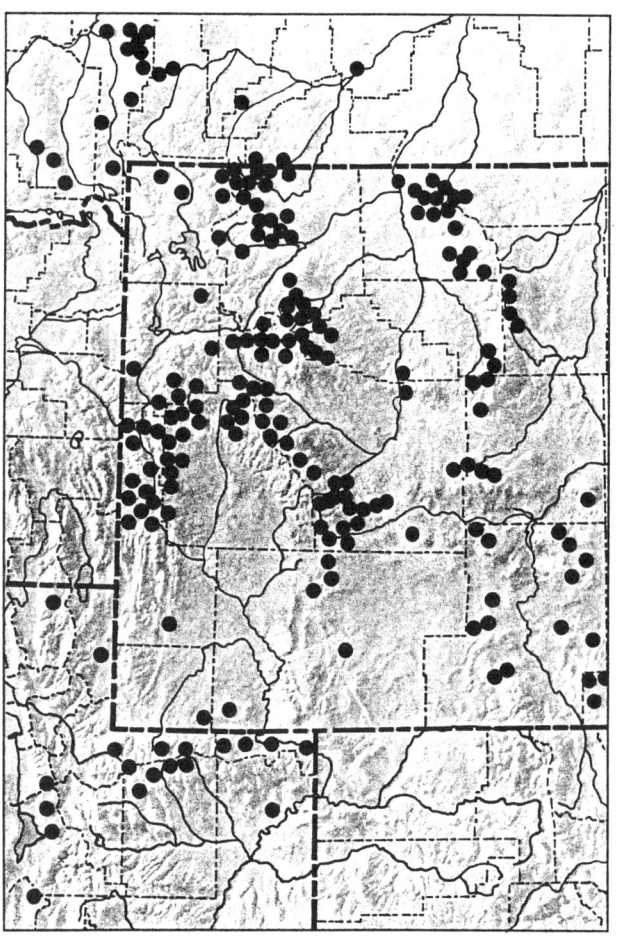

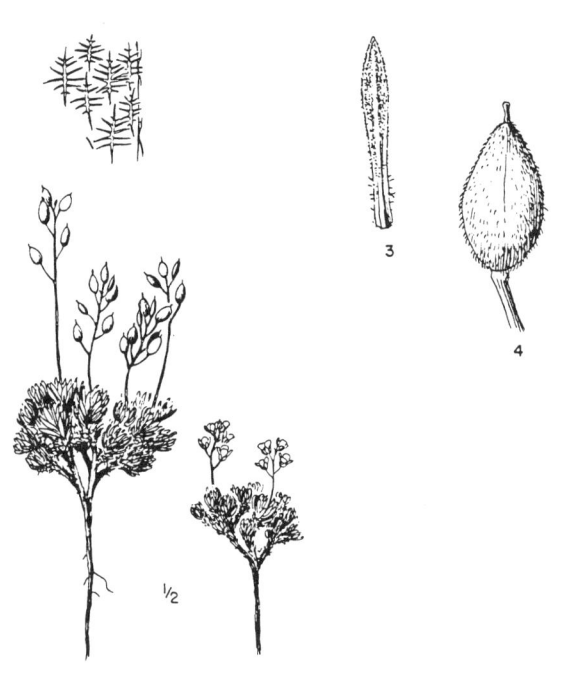

Draba paysonii Macbr.
Contr. Gray Herb. 56:52, 1918.

Payson Draba

Draba novolympica Pays. & St. John
Draba vestita Pays.

Caespitose, scapose perennial from a branched caudex covered with old leaf bases; scapes pubescent with simple and forked hairs. Leaves imbricate, linear to narrowly oblanceolate, the upper surfaces pubescent with simple and forked hairs, the lower with tangled, stalked, long-branched hairs; margins entire, prominently ciliate with simple and forked hairs. Flowers in short, somewhat congested racemes, the pedicels usually pubescent. Sepals pubescent with simple to branched hairs. Petals yellow, cuneate, and rounded. Silicles ovate, pubescent with simple and forked hairs; styles 0.5–1.8 mm long.

Fellfields and other rocky places in the Absaroka, Beartooth, Wind River, and Wyoming Ranges. British Columbia and Alberta south to California, Idaho, and Wyoming.

This agamospermous species, $2n = 42$ (Mulligan, 1972), can be separated into 2 varieties in the Middle Rocky Mountains as follows:

1. Silicles 5–8 mm long; styles 1–1.8 mm long
   . . . var. *paysonii*
1. Silicles 3–5 mm long; styles 0.5–0.7 mm long
   . . . var. *treleasii* (Schulz) C.L. Hitchc.

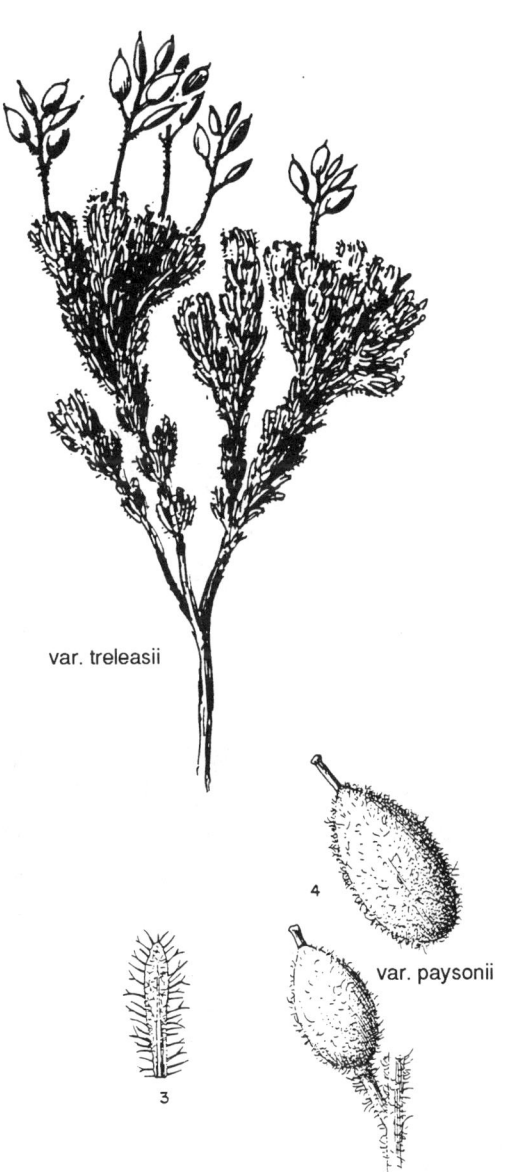

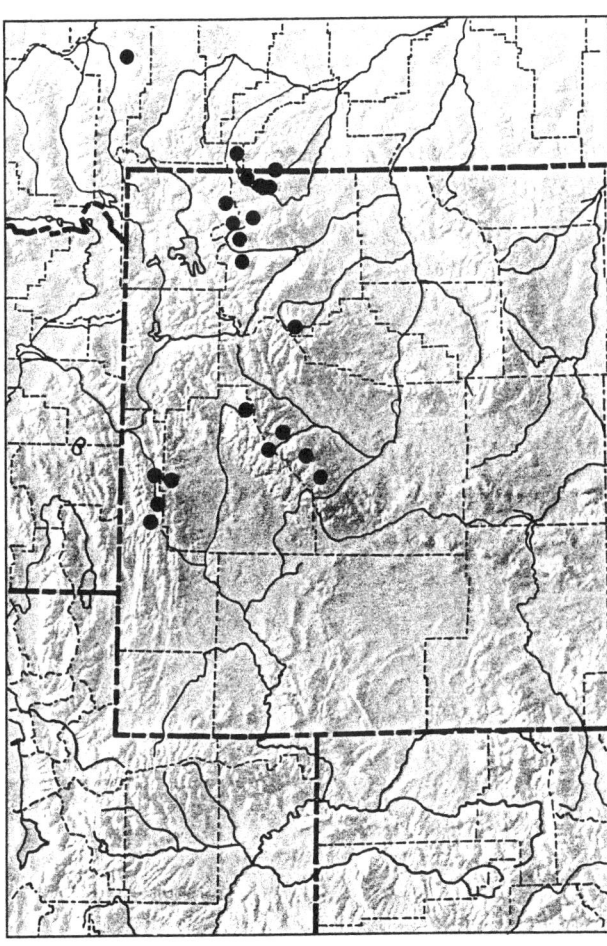

Draba porsildii Mulligan
Can. J. Bot. 52:1795, 1974.

Porsild Draba

Caespitose perennial from a simple or branched caudex; stems glabrous to very sparsely pubescent. Basal leaves oblanceolate, entire, with forked, cruciform, and stellate hairs; cauline leaves lacking, or sometimes 1 and reduced, with entire to minutely denticulate margins. Flowers in condensed to elongate racemes, the pedicels ascending to erect, shorter than the fruits, glabrous to sparsely pubescent. Sepals glabrous, ours with whitish margins and purplish midribs. Petals white, cuneate. Silicles obovate, flat, glabrous; styles about 0.25 mm long.

Fellfields, ledges, scree, and talus in the Absaroka, Beartooth, and Wind River Ranges. Yukon and Mackenzie district of Northwest Territories south to Wyoming and Colorado.

Middle Rocky Mountain alpine plants are var. *brevicula* (Rollins) Rollins.

2.5

fruit

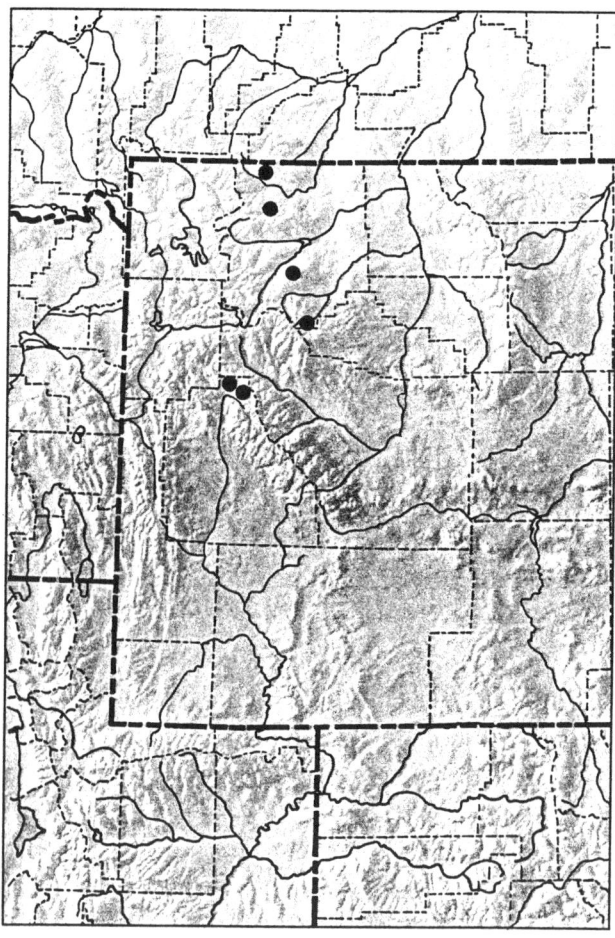

Draba praealta Greene
Pittonia 3:306, 1898.

Tall Draba

Draba cascadensis Pays. & St. John
Draba columbiana Rydb.
Draba dolichocarpa Schulz
Draba lapilutea A. Nels.
Draba yellowstonensis A. Nels.

Rosulate annual, biennial, or short-lived perennial from a simple or branched caudex; stems 1 to several, simple or branched, pubescent with simple, cruciform, and stellate hairs. Basal leaves oblanceolate, pubescent above and below with simple, forked, and stellate hairs, the margins entire to remotely denticulate; cauline leaves 1–8, lanceolate to ovate, the margins shallowly dentate. Flowers in elongate, open, usually bracteate racemes, the pedicels with forked hairs. Sepals pubescent with simple hairs. Petals white or yellowish, obovate, emarginate. Silicles lanceolate, elliptic, or oblong, pubescent with simple and forked hairs; styles very short, less than 0.2 mm long.

Meadows and rocky outcrops of the Beartooth and Wind River Ranges. Alaska, Yukon Territory, and Mackenzie district of Northwest Territories, south to Oregon, Idaho, and Wyoming.

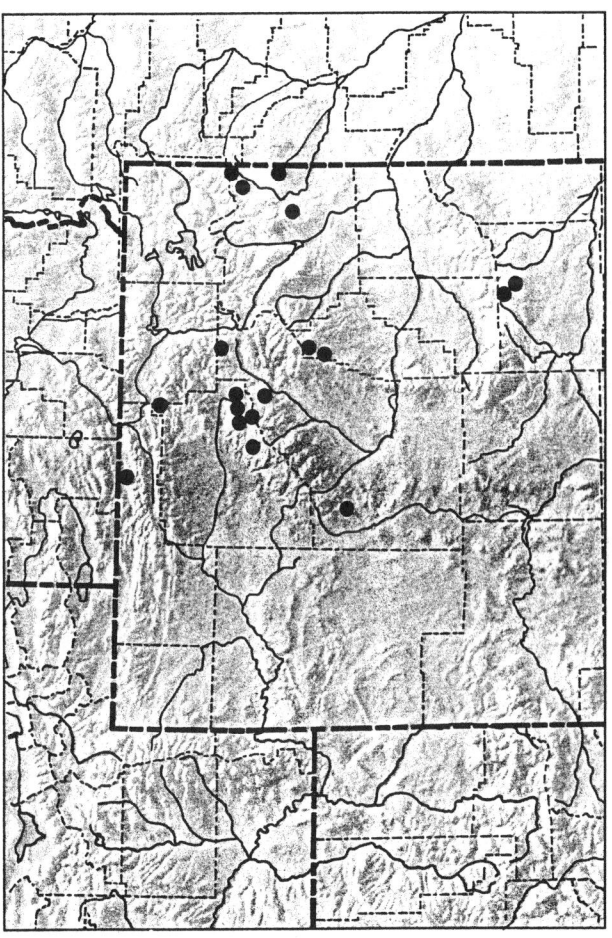

Draba ventosa Gray
Am. Nat. 8:212, 1874.

Wind River Draba

Caespitose, scapose perennial from a stout, much-branched caudex covered with old leaf bases; scapes pubescent with simple, forked, and stellate hairs. Leaves oblanceolate to obovate, pubescent with simple, forked, and branched to stellate hairs, the margins entire, not prominently ciliate. Flowers in short, compact racemes, the pedicels ascending, with simple to stellate hairs. Sepals pilose with mostly branched hairs. Petals yellow, obovate, and rounded. Silicles ovate to somewhat oval, flat, pubescent with simple and branched hairs; styles 0.5–1.2 mm long.

Felsenmeer, fellfields, talus, scree, and patterned ground of the Absaroka, Beartooth, Teton, Uinta, Wind River, and Wyoming Ranges. Alberta south to Utah and Wyoming.

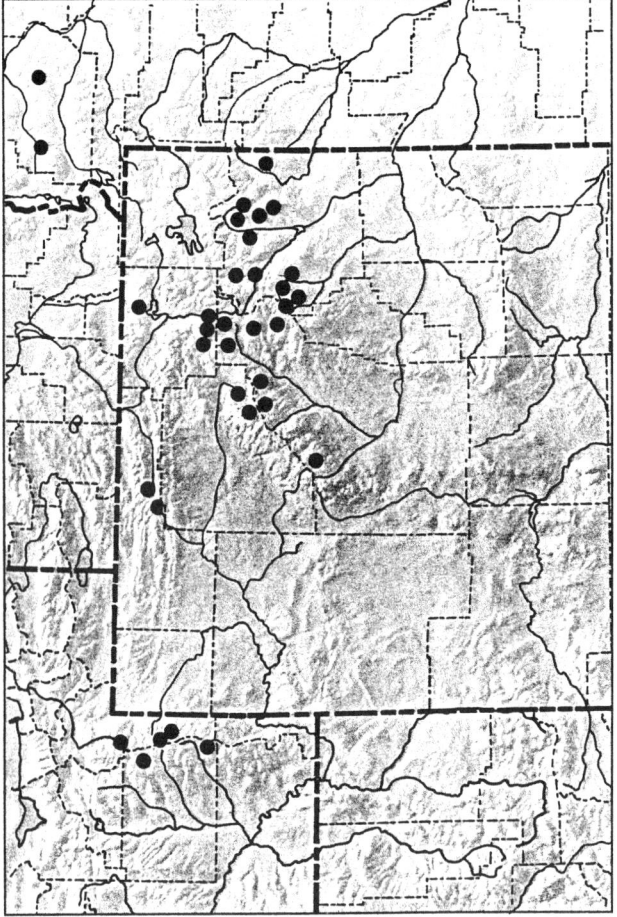

Erysimum L.
Wallflower

Annual, biennial, or perennial herbs with pubescence largely of 2-branched hairs, the leaves narrow, entire, dentate, or toothed. Flowers showy, ours are yellow. Sepals erect, often saccate at the base, the outer ones often minutely appendaged at the apex. Petals narrowed into a long claw. Silique elongate, more or less 4-angled, thinly to densely pubescent, strongly keeled by a prominent nerve on each valve. Seeds numerous, in 1 row in each locule.

Erysimum asperum (Nutt.) DC.
Prodromus 2:505, 1821.

Western Wallflower

Cheiranthus asper Nutt.
Cheiranthus alpestris (Cockll.) A. Heller
Cheiranthus angustatus Greene
Cheiranthus argillosus Greene
Cheiranthus aridus A. Nels.
Cheiranthus aridus Greene
Cheiranthus arkansanus (Nutt.) Greene
Cheiranthus asperrimus Greene
Cheiranthus bakeri Greene
Cheiranthus californicus (Greene) Greene
Cheiranthus capitatus Dougl. ex Hook.
Cheiranthus elatus (Nutt.) Greene
Cheiranthus nivalis Greene
Cheiranthus oblanceolatus (Rydb.) A. Heller
Cheiranthus pacificus Sheld.
Cheiranthus perennis (S. Wats. ex Cov.) Greene
Cheiranthus radicatus (Rydb.) A. Heller
Cheiranthus wheeleri (Rothr.) Greene
Cheirinia amoena (Greene) Rydb.
Cheirinia argillosa (Greene) Rydb.
Cheirinia arida (A. Nels.) Rydb.
Cheirinia arkansana (Nutt.) Moldenke
Cheirinia aspera (Nutt.) Britt.
Cheirinia asperrima (Greene) Rydb.
Cheirinia bakeri (Greene) Rydb.
Cheirinia brachycarpa Rydb.
Cheirinia cockerelliana (Daniels) Cockll.
Cheirinia desertorum Woot. & Standl.
Cheirinia elata (Nutt.) Rydb.
Cheirinia nevadensis (A. Heller) A. Heller
Cheirinia nivalis (Greene) Rydb.
Cheirinia oblanceolata (Rydb.) Rydb.
Cheirinia radicata (Rydb.) Rydb.
Cheirinia wheeleri (Rothr.) Rydb.
Erysimum alpestre (Cockll.) Rydb.
Erysimum amoenum (Greene) Rydb.
Erysimum angustatum Greene
Erysimum argillosum (Greene) Rydb.

Erysimum aridum (A. Nels.) A. Nels.
Erysimum arkansanum Nutt.
Erysimum asperrimum (Greene) Rydb.
Erysimum bakeri (Greene) Rydb.
Erysimum californicum Greene
Erysimum capitatum (Dougl. ex Hook.) Greene
Erysimum cockerellianum Daniels
Erysimum desertorum (Woot. & Standl.) Rossb.

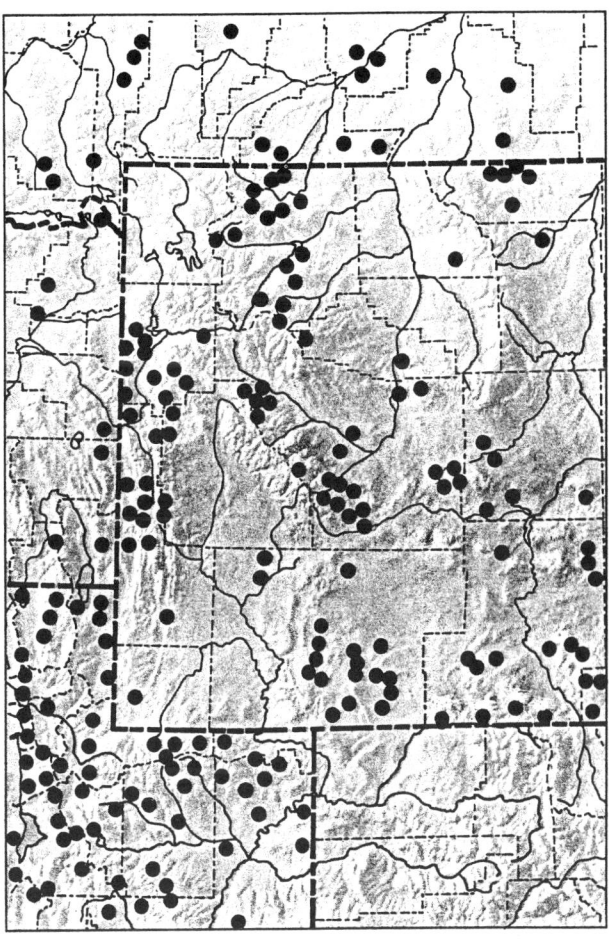

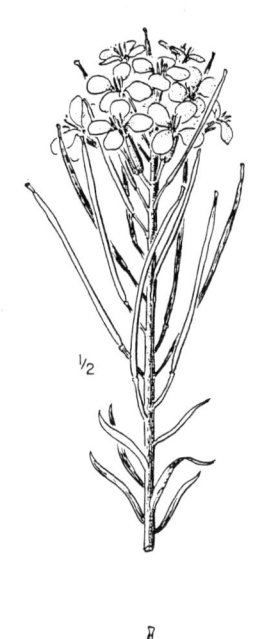

Erysimum elatum Nutt. in T. & G.
Erysimum moniliforme Eastw.
Erysimum nevadense A. Heller
Erysimum nivale (Greene) Rydb.
Erysimum oblanceolatum Rydb.
Erysimum perenne (S. Wats. ex Cov.) Abrams
Erysimum pumilum Nutt.
Erysimum radicatum Rydb.
Erysimum tilimi Gay
Erysimum wheeleri Rothr.

Biennial or short-lived perennial from a taproot and simple to branched caudex; stems erect, simple or occasionally branched, pubescent with appressed malpighian hairs. Basal leaves in a rosette, oblanceolate, with slender petioles, pubescent with malpighian hairs, the margins entire to dentate; cauline leaves lanceolate to linear, entire to toothed, the pubescence like the basal leaves. Flowers showy in dense, terminal racemes, the pedicels ascending to spreading. Sepals somewhat saccate at the base, sometimes purplish. Petals yellow, the blade about half as long as the slender claw. Siliques stout, linear, quadrangular to slightly flattened in cross section, ascending to spreading, 6–10 cm long; style forming a short beak. Seeds oblong, winged near the tip or wingless.

Fellfields and meadows of nearly all alpine ranges. Not yet known from above timberline in the Big Horn Range. British Columbia and Alberta south to California, New Mexico, Texas, Kansas, and Minnesota.

Erysimum asperum represents a complex group of populations adapted to a wide range of environmental conditions over an extensive geographic area. These have been variously treated at both the specific and infraspecific levels, based on their fruits, pubescence, stature, and leaf differences. Middle Rocky Mountain alpine plants of short stature that have been called *E. nivale* (Greene) Rydb., and most recently *E. capitatum* var. *purshii* (Durand) Rollins, are thus treated here as var. *asperum*.

Lesquerella S. Wats.
Bladderpod

Annual, biennial, or mostly perennial herbs with stellate or sometimes simple pubescence, the herbage often appearing more or less silvery. Stems few to many in ours, erect to spreading-decumbent. Leaves narrow to broad, entire to toothed, the basal ones usually forming a distinct rosette. Sepals erect, acute to obtuse, often hooded at the apex. Petals yellow, obovate to narrowly spatulate. Ovary subglobose, often substipitate, the style slender, persistent or rarely deciduous; stigma capitate. Silicles subglobose, ovate, short-oblong, or elliptic, sessile or short-stipitate, flattened to inflated. Seeds several in each locule, somewhat flattened, winged or wingless.

KEY TO THE SPECIES OF *LESQUERELLA*

1. Basal and cauline leaves similar in size and shape, the largest leaves seldom more than 3 mm broad, linear or narrowly oblanceolate; petals 5–6 mm long .... *L. alpina*
1. Basal and cauline leaves dissimilar in size and shape, the largest leaves usually more than 4 mm broad, ovate, obovate, or rhombic; petals 6–12 mm long
   2. Silicles flattened apically and along the margins .... *L. occidentalis*
   2. Silicles not flattened at the apex and along the margins
      3. Silicles obovoid to subglobose, not compressed, or somewhat compressed parallel to the septum; pedicels straight or curved .... *L. garrettii*
      3. Silicles elliptic to subglobose, compressed at right angles to the septum; pedicels strongly to weakly S-shaped
         4. Styles 2–4 mm long; siliques strongly compressed at right angles to the septum .... *L. paysonii*
         4. Styles 4–7 mm long; siliques only slightly compressed at right angles to the septum .... *L. utahensis*

Lesquerella alpina (Nutt. in T. & G.) S. Wats.
Proc. Am. Acad. 23:251, 1888.

Alpine Bladderpod

Vesicaria alpina Nutt. in T. & G.
Alyssum alpinum (Nutt. in T. & G.) Kuntze
Lesquerella condensata A. Nels.
Lesquerella curvipes A. Nels.
Lesquerella nodosa Greene
Lesquerella parvula Greene
Lesquerella spathulata Rydb.

Low, caespitose, densely stellate-pubescent perennial, mostly less than 10 cm tall, from a mostly simple caudex often covered with old leaf bases; stems several to many, erect to ascending. Basal leaves linear to oblanceolate, gradually to abruptly narrowed to the petiole, the margins entire; cauline leaves few, linear or narrowly oblanceolate like the basal ones. Flowers few, in short racemes, the pedicels straight, curved, or typically sigmoid. Sepals oblong to elliptic, stellate-pubescent, the outer pair hooded at the apex. Petals yellow, spatulate. Silicles sessile or nearly so, ovate, acute, the valves stellate-pubescent, somewhat inflated, becoming flattened at the apex; styles commonly 2–4 mm, more than half as long to as long as the silicles.

Fellfields, talus, and scree of the Beartooth, Big Horn, Gros Ventre, and Wind River Ranges. Alberta and Saskatchewan south to Idaho, Utah, Colorado, and Kansas.

Alpine plants of the Middle Rocky Mountains less than 10 cm tall, with flowers exceeding the narrowly oblanceolate basal leaves, are typical var. *alpina*.

var. alpina

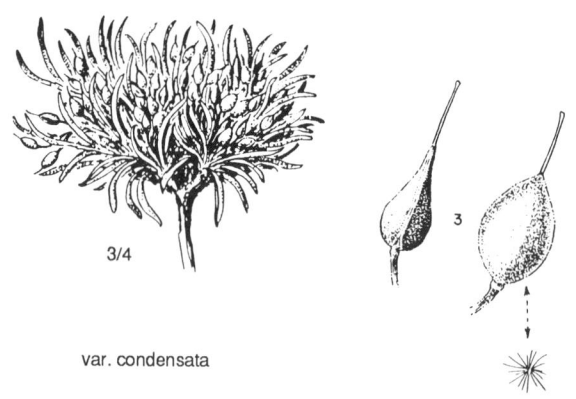
var. condensata

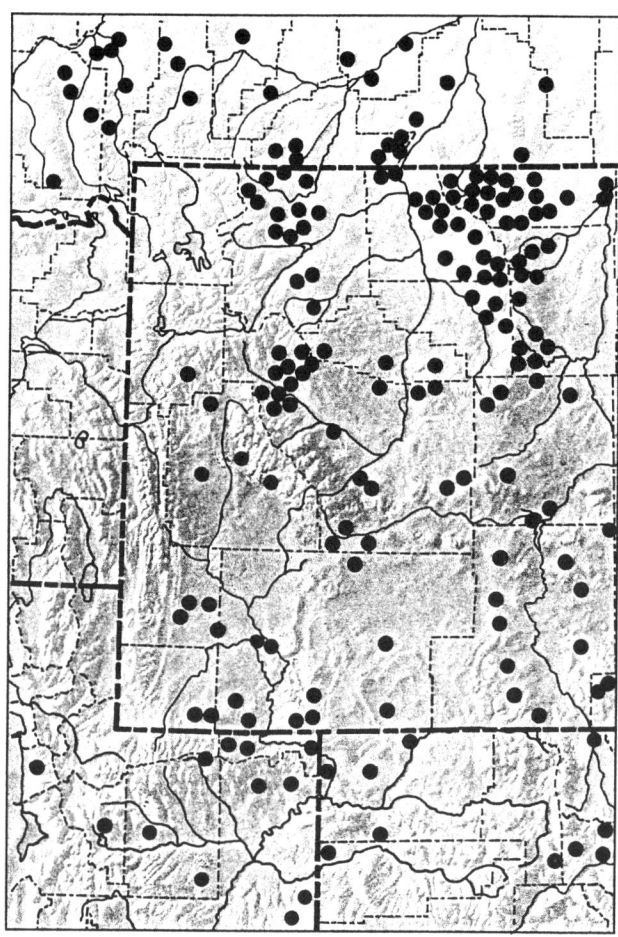

250 Brassicaceae

Lesquerella garrettii Pays.
Ann. Missouri Bot. Gard. 8:213, 1921.

Garrett Bladderpod

Caespitose, stellate-pubescent perennial from a simple to branched caudex covered with old leaf bases; stems ascending to spreading and decumbent. Basal leaves distinctly petiolate, the blades elliptic to ovate or obovate, entire or nearly so; cauline leaves obovate or oblanceolate, entire, sessile or short-petiolate, about half as wide as the basal ones. Flowers in short, loose, few-flowered racemes, the pedicels straight or curved, ascending. Sepals elliptic or linear, the outer pair hooded at the apex. Petals yellow, spatulate to obovate. Silicles subglobose or obovoid, short-stipitate, the valves densely stellate-pubescent; styles 4–7 mm, mostly longer than the silicles.

Talus and other rocky places in the Wasatch Range. Endemic to northern Utah.

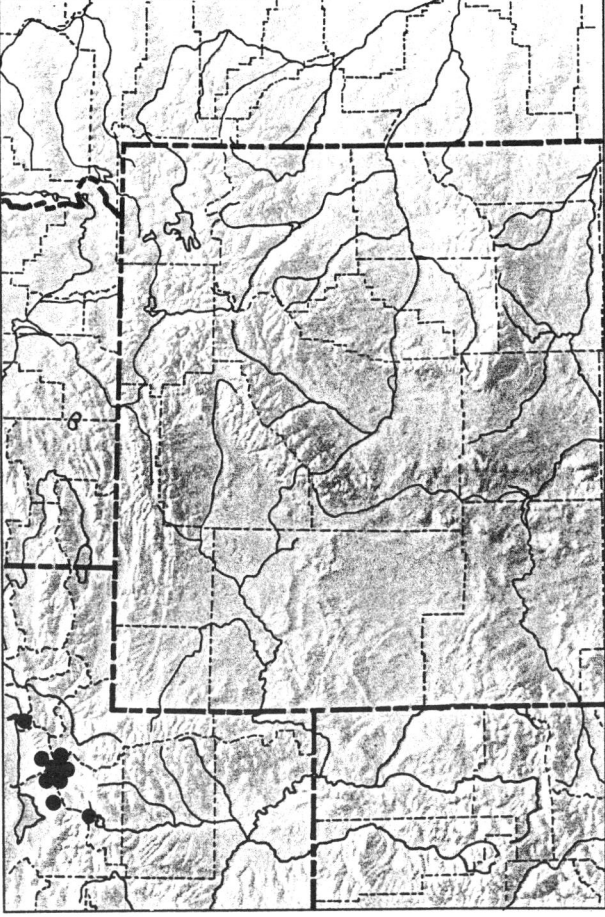

Lesquerella occidentalis (S. Wats.) S. Wats.
Proc. Am. Acad. 23:251, 1888.

Western Bladderpod

Vesicaria occidentalis S. Wats.
Lesquerella cusickii M.E. Jones
Lesquerella goodrichii Rollins

Caespitose, stellate-pubescent perennial from a slender to stout, simple or sometimes branched caudex covered with old leaf bases; stems simple, few to several, erect to decumbent-spreading. Basal leaves long-petiolate, the blades ovate to obovate, entire to undulate or toothed; cauline leaves short-petiolate below, becoming sessile upward, oblanceolate, entire or nearly so. Flowers in short, often loose racemes, the pedicels straight to curved, often sigmoid. Sepals oblong to elliptic, the outer pair hooded at the apex. Petals yellow, spatulate. Silicles ovate to elliptic, inflated to slightly compressed parallel to the septum, flattened at the apex and on the margins, the valves densely stellate-pubescent; styles 3–6 mm long, mostly shorter than the silicles.

Scree and other rocky places in the Wasatch Range. Washington south to California and east to Utah and Idaho.

Utah plants are var. *cinerascens* Maguire & Holmgren.

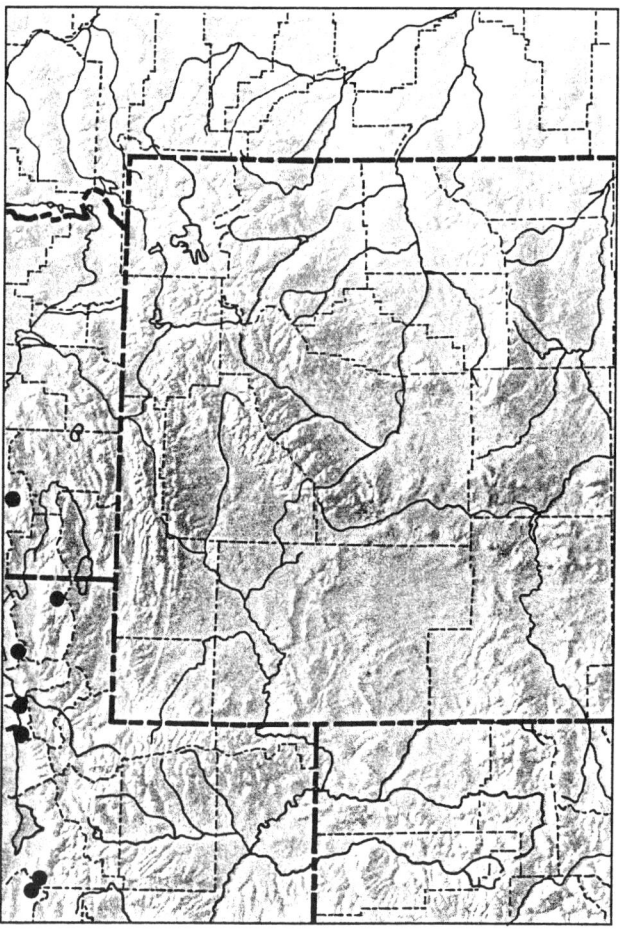

Lesquerella paysonii Rollins
Contr. Gray Herb. 171:44, 1950.

Payson Bladderpod

Caespitose, densely stellate-pubescent perennial from a simple caudex; stems slender, simple, decumbent. Basal leaves petiolate, the blades oblanceolate, rhombic, or ovate, the margins entire to undulate or shallowly lobed; cauline leaves petiolate, elliptic, much smaller than the basal ones, obtuse to acute. Flowers in short, compact racemes, the pedicels somewhat sigmoid, at least in fruit. Sepals oblong to elliptic, the outer pair hooded at the apex. Petals yellow, spatulate, sometimes with a purplish tint. Silicles elliptic, short-stipitate, strongly compressed at right angles to the septum, the valves densely stellate-pubescent; styles 2–4 mm long.

Talus, scree, and other rocky places of the Gros Ventre, Salt River, Teton, Wind River, and Wyoming Ranges. Endemic to western Wyoming and eastern Idaho.

Lesquerella fremontii Rollins & Shaw, Fremont Bladderpod, has been documented from limestones above timberline on Arrow Mountain in the northern Wind River Range. It is distinguished from *L. paysonii* by styles less than 2 mm long, petals less than 8 mm long, and fruits less strongly flattened.

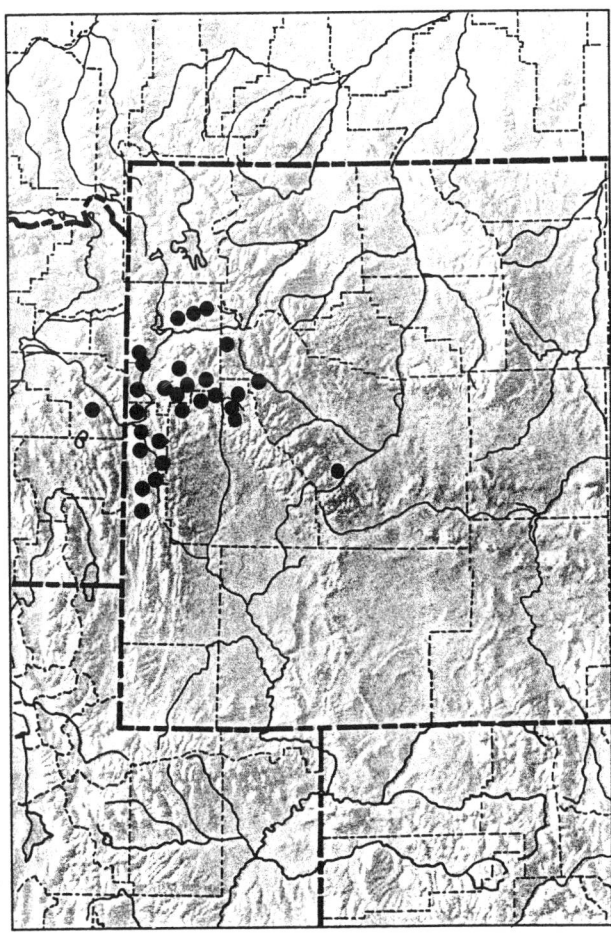

Lesquerella utahensis Rydb.
Bull. Torr. Club 30:252, 1903.

Utah Bladderpod

Caespitose, stellate-pubescent perennial from a simple or, rarely, once-branched caudex, often covered with old leaf bases; stems few to several, erect, ascending, or decumbent. Basal leaves petiolate, the blades ovate to broadly elliptic, entire or somewhat toothed, especially near the base; cauline leaves few, oblanceolate to elliptic, entire, sessile or short-petiolate. Flowers in loose racemes, the pedicels sigmoid or curved, ascending. Sepals linear to lanceolate, often pinkish, the outer pair hooded at the apex. Petals yellow, spatulate to cuneate. Silicles elliptic to subglobose, slightly compressed at right angles to the septum, sessile or substipitate, the valves sparsely stellate-pubescent; styles 4–7 mm, as long as or longer than the silicles.

Scree, talus, and rocky ledges of the Wasatch and Uinta Ranges. Endemic to northeastern Utah.

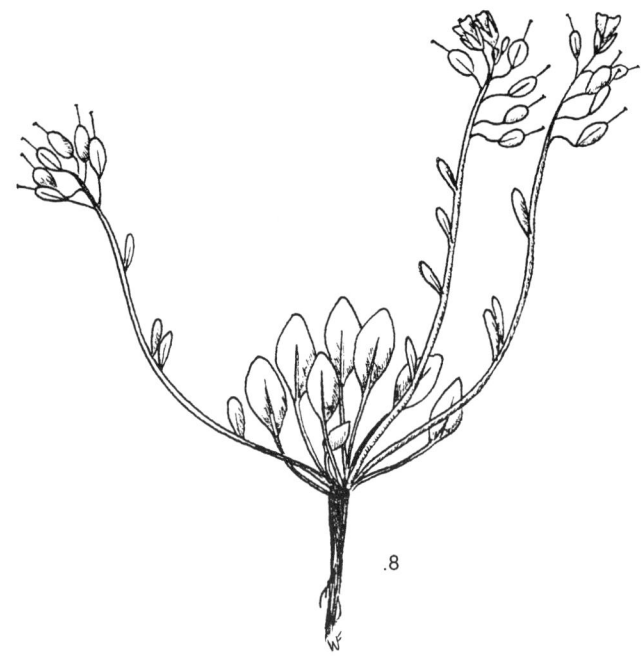

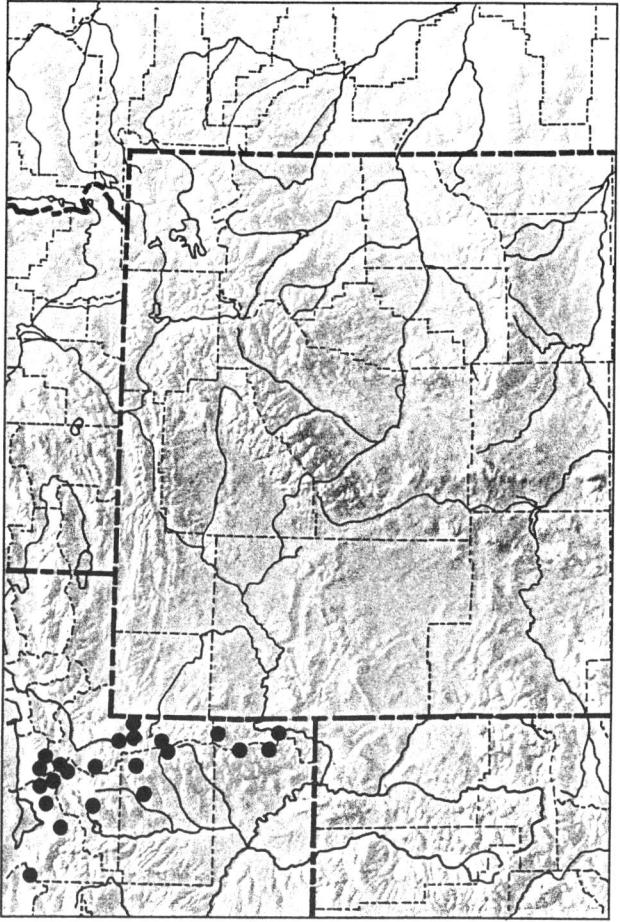

Parrya R. Br.
Parrya

Scapose perennials from a branching caudex and thick, fleshy taproot, the pubescence of simple, glandular hairs. Leaves in a basal tuft, spatulate to oblong, 2–6 cm long, coarsely toothed. Flowers showy, pink or purple, racemose, the petals with a broad blade and slender claw. Stamens 6. Stigma 2-lobed. Siliques oblong-linear, mostly 2–4 cm long, flattened parallel to the septum, the margins wavy, the valves nerved. Seeds several in each locule.

Parrya nudicaulis (L.) Regel
Bull. Soc. Imp. Nat. Mos. 34(3):176, 1861.

Smoothstem Parrya

Cardamine nudicaulis L.
Arabis nudicaulis (L.) DC.
Cardamine articulata Pursh
Neuroloma nudicaule (L.) DC.
Neuroloma rydbergii (Botsch.) Botsch.
Parrya macrocarpa R. Br.
Parrya platycarpa Rydb.
Parrya rydbergii Botsch.
Parrya turkestanica Korsh.

Scapose, glabrous to glandular-pubescent perennial from a simple or branched caudex or rhizome with marcescent leaf bases; scape single, glandular. Leaves petiolate, the blades elliptic, ovate, or oblanceolate, glabrous, the margins entire to dentate, glandular-ciliate. Flowers showy, in open racemes, the pedicels ascending. Sepals ovate, purplish with scarious margins, glabrous or glandular. Petals purple, lavender, or rarely white, obovate, emarginate to rounded. Siliques erect, oblong, glabrous to glandular, flattened parallel to the septum; style forming a distinct beak up to 3 mm long.

High altitude talus, fellfields, snow beds, and patterned ground in the Beartooth, Gros Ventre, Uinta, and Wind River Ranges. Circumpolar; south in western North America to Wyoming and northern Utah.

Middle Rocky Mountain plants with glands and dentate leaves are ssp. *nudicaulis*.

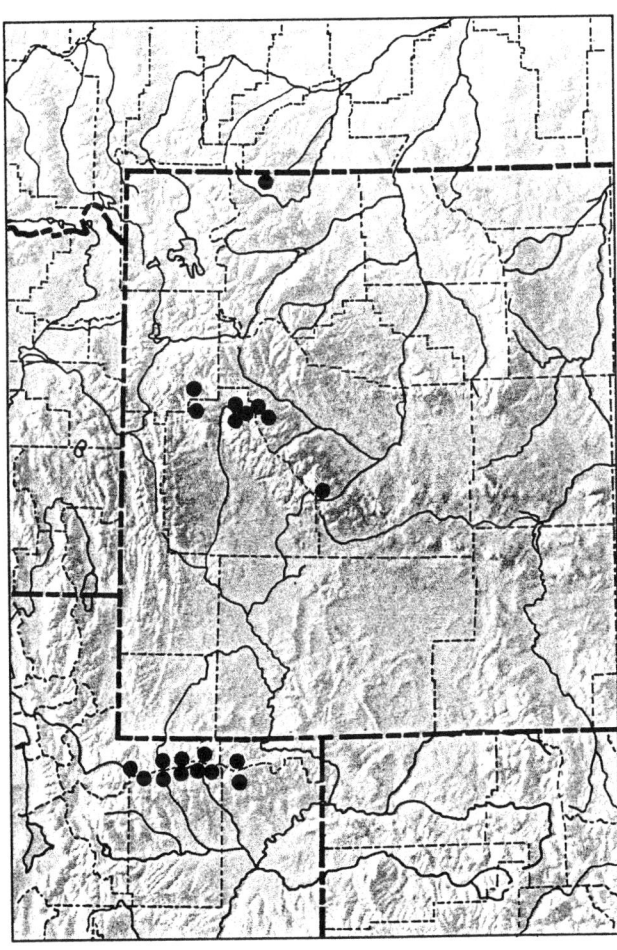

1/2

Physaria (Nutt. ex T. & G.) Gray
Twinpod

Perennial, caespitose herbs with stellate, silvery pubescence. Stems single or several from a taproot. Leaves simple, entire to toothed, mostly in a basal rosette; cauline leaves few, reduced upward. Flowers showy, yellow, in congested to elongate racemes. Sepals erect, pubescent, somewhat saccate at the tips. Petals spatulate, glabrous. Stamens 6. Silicles didymous, inflated, pubescent, more or less orbicular, often with a prominent apical cleft (sinus) between the two valves; styles persistent. Seeds biseriate, commonly 1–4 per locule.

KEY TO THE SPECIES OF *PHYSARIA*

1. Sinuses of mature fruits equal in depth; replum narrow, oblong to linear; seeds 1 or 2 per locule ..... *P. acutifolia*
1. Sinuses of mature fruits unequal; the upper much deeper than the lower; replum obovate; seeds mostly more than 2 per locule ..... *P. didymocarpa*

Physaria acutifolia Rydb.
Bull. Torr. Club 18:279, 1901.

Rydberg Twinpod

Physaria australis (Pays.) Rollins
Physaria repanda Rollins
Physaria stylosa Rollins

Caespitose perennials from a simple or branched caudex; stems simple, decumbent. Basal leaves entire, occasionally shallow-toothed, the blades obovate to orbicular, obtuse; cauline leaves entire, mostly oblanceolate. Flowers yellow, in congested racemes which elongate in fruit. Sepals linear to oblong; petals spatulate. Silicles erect, pubescent, the apical and basal sinuses narrow and equal in depth, the replum oblong, obtuse at the apex, the valves suborbicular to orbicular. Seeds 1–2 in each locule.

Typical of dry plains at lower elevations throughout the region, *P. acutifolia* occurs in the alpine zone of the Big Horn, Hoback, Uinta, and Wasatch Ranges. Wyoming and Idaho south to Utah, Colorado, and New Mexico.

Two weakly differentiated varieties occur in the alpine zone: var. *acutifolia,* with a simple caudex and basal leaves more than 12 mm wide; and var. *stylosa* (Rollins) S.L. Welsh, with a branched caudex and basal leaves less than 12 mm wide.

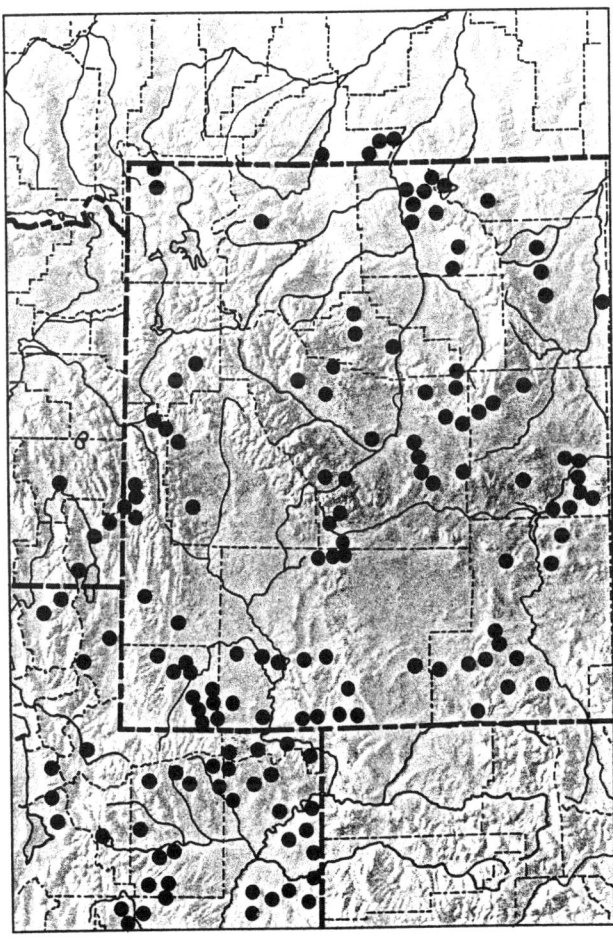

Physaria didymocarpa (Hook.) Gray
Gen. Illustr. 1:162, 1848.

Common twinpod

Vesicaria didymocarpa Hook.
Coulterina didymocarpa (Hook.) Kuntze
Physaria lanata (A. Nels.) Rydb.
Physaria macrantha Blank.

Caespitose, stellate-pubescent perennial from a stout taproot, often with a branched caudex; stems solitary or several, simple, decumbent. Basal leaves in dense rosettes, entire or shallowly toothed, the blades obovate to oblanceolate, mostly obtuse; cauline leaves few, reduced, oblanceolate. Flowers yellow, in short racemes, becoming elongate in fruit. Sepals lanceolate; petals spatulate. Silicles didymous, inflated, pubescent, the apical sinus with a deeper cleft than the shallowly cordate basal sinus; replum obovate, obtuse at the apex, the valves suborbicular. Seeds 2 or 3 in each locule.

Rocky places at and above timberline in the Absaroka and Big Horn Ranges. Alberta south to Washington, Idaho, Wyoming, and South Dakota.

Middle Rocky Mountain alpine plants with shallowly toothed basal leaves having appressed hairs are var. *didymocarpa*. The variety *lanata* A. Nels. occurs above timberline in the Big Horn Range of Wyoming. It is distinguished from typical *didymocarpa* by having basal leaves lanate, at least at the base, with spreading hairs.

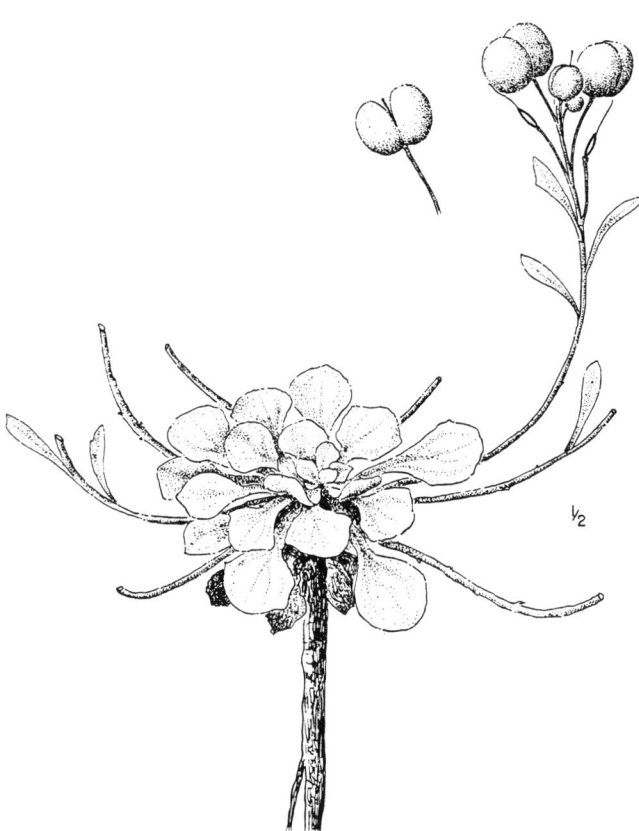

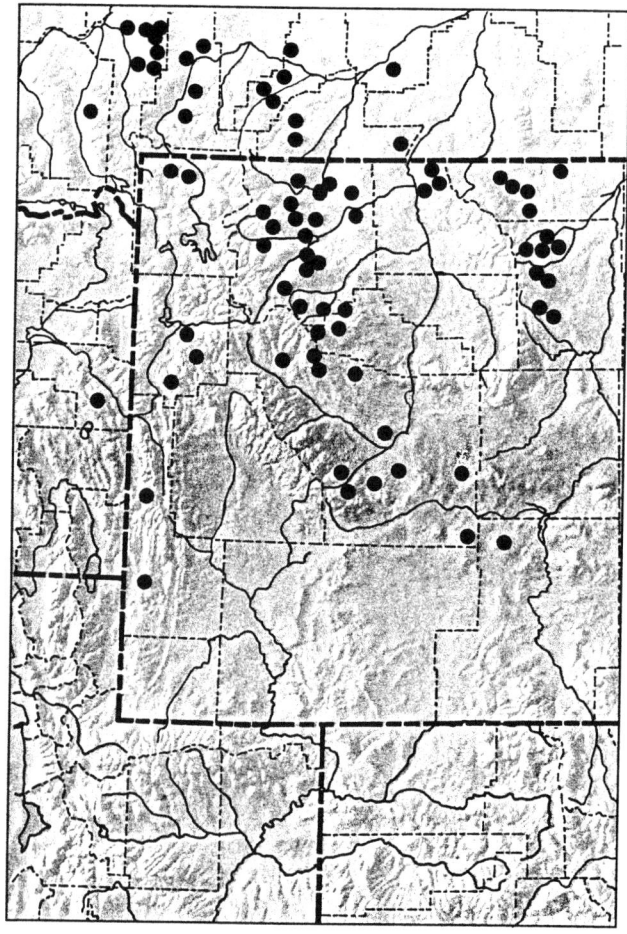

Rorippa Scop.
Yellowcress

Annual or perennial herbs of marshy or wet situations. Leaves dentate to pinnatifid, rarely entire, glabrous or the pubescence of simple hairs. Flowers yellow, mostly in terminal racemes. Sepals spreading. Petals spatulate, often scarcely exceeding the sepals. Short stamens flanked at the base by a pair of minute glands, these sometimes confluent into an annular gland; the long stamens separated by a short, conic gland. Ovary cylindric, the style very short, the stigma capitate. Siliques short, terete or nearly so, tipped with the short style and persistent stigma. Seeds very numerous, in 2 rows.

Rorippa curvipes Greene
Pittonia 3:97, 1896.

Obtuse Yellowcress

Nasturtium obtusum Nutt.
Nasturtium sphaerocarpum Gray
Radicula alpina (S. Wats.) Greene
Radicula curvipes (Greene) Greene
Radicula integra (Rydb.) A. Heller
Radicula obtusa (Nutt.) Greene
Radicula sphaerocarpa (Gray) Greene
Radicula underwoodii (Rydb.) A. Heller
Rorippa alpina (S. Wats.) Rydb.
Rorippa integra Rydb.
Rorippa obtusa (Nutt.) Britt.
Rorippa sphaerocarpa (Gray) Britt.
Rorippa underwoodii Rydb.

Glabrous annual or short-lived perennial from a taproot; stems 1 to several, prostrate-decumbent to erect, simple or branched from near the base. Leaves sessile to petiolate, the blades oblanceolate, obovate, or oblong; margins entire, crenate, serrate, or pinnatifid. Flowers minute, in elongate terminal and axillary racemes, the pedicels ascending to spreading or recurved in fruit. Sepals flat to slightly saccate at the base, deciduous after anthesis. Petals pale yellow, spatulate to oblong. Siliques oblong to lanceolate, straight to somewhat curved, acute to obtuse; style 0.5–1 mm long.

Stream banks and sandy, wet places in the lower alpine zone of the Absaroka, Beartooth, Medicine Bow, Uinta, Wasatch, and Wind River Ranges. British Columbia, Alberta, and Saskatchewan south to California, Arizona, New Mexico, Kansas, and South Dakota.

The low, spreading alpine plant, usually with a well-differentiated basal rosette of leaves, is var. *alpina* and may be distinguished from the other, more typically lower-elevation varieties, as follows:

1. Petals distinctly longer than the sepals
    . . . var. *alpina* (S. Wats.) Stuckey
1. Petals shorter than or subequal to the sepals
    2. Siliques less than 5 mm long; petals less than 1 mm long, shorter than the sepals . . . var. *curvipes*
    2. Siliques 3.5–8 mm long; petals more than 1 mm long, subequal to the sepals
        . . . var. *integra* (Rydb.) Stuckey

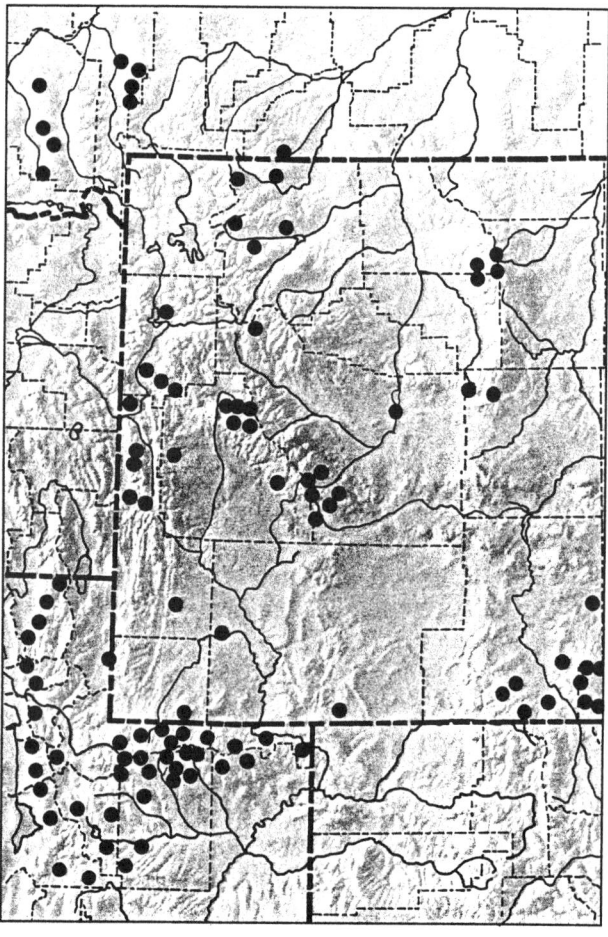

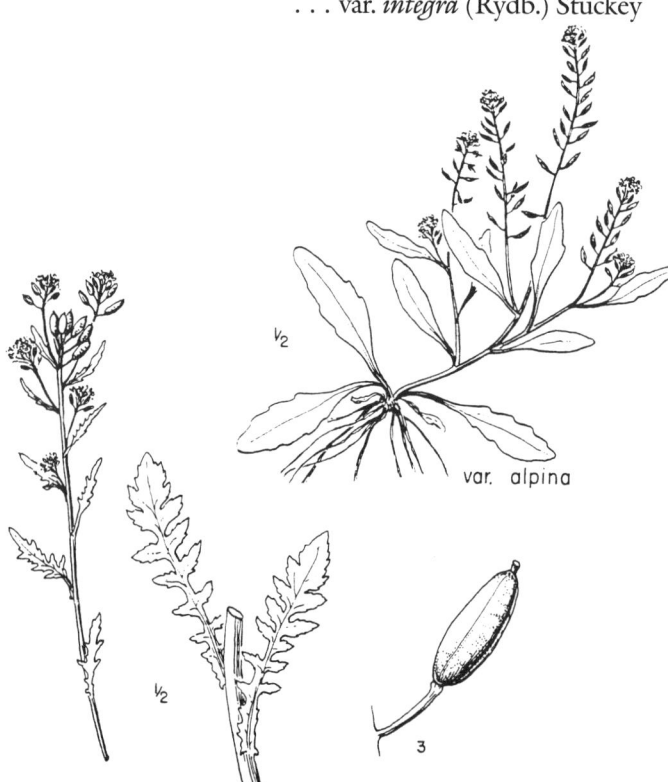

Smelowskia C.A. Mey.
Smelowskia

Tufted perennials from a branching caudex, the caudex covered with the remains of old leaf bases. Leaves chiefly basal, matted, pinnatifid, canescent with branched hairs. Flowers in short racemes, white or tinged with purple, the petals obovate, exserted, and spreading. Stamens 6. Siliques ovate to linear, sessile, subterete, the valves prominently nerved or keeled. Seeds several, in 1 row in each locule.

Smelowskia calycina (Steph. ex Willd.) C. A. Mey. in Ledeb. *Fl. Alt.* 3:170, 1831.

Alpine Smelowskia

Lepidium calycinum Steph. ex Willd.
Hutchinsia calycina (Steph. ex Willd.) Desv.
Smelowskia americana (Regel & Herder) Rydb.
Smelowskia lineariloba Rydb.
Smelowskia lobata Rydb.
Smelowskia porsildii (Drury & Rollins) Yurtsev

Caespitose perennial from a stout, branched caudex covered with old leaf bases; stems 1 to several, simple, pubescent with short, branched and long, simple or forked hairs. Basal leaves petiolate, pinnatifid to lobed or nearly entire, gray-pubescent with branched hairs; cauline leaves reduced upward and becoming sessile, pinnatifid, pubescent at the basal leaves. Flowers in congested terminal racemes, the pedicels ascending. Sepals oblong, pubescent, often purplish. Petals white to pink or rose, spatulate. Siliques erect, oblong, tapering at each end, terete or flattened parallel to the septum, glabrous, the valves keeled; style 1 mm long or less, forming a distinct beak.

Widespread and common in fellfields and other rocky places in the western alpine ranges. Asia and North America; south in western North America to Washington, Nevada, Utah, and Colorado.

Alpine plants of the Middle Rocky Mountains with pinnately lobed leaves and pedicels ascending in fruit are var. *americana* (Regel & Herder) Drury & Rollins.

See also note on p. 228.

Thlaspi L.
Pennycress

Erect, glabrous, annual or perennial herbs with simple, entire to dentate or lobed leaves, the cauline leaves auriculate-clasping at the base. Flowers small, white. Petals spatulate to obovate. Stamens 6. Silicles orbicular to cuneate, flattened at right angles to the septum, keeled or narrowly winged on the edges. Seeds 2 to several in each locule.

KEY TO THE SPECIES OF
THLASPI

1. Style less than 0.5 mm long; petals less than 4 mm long — *T. parviflorum*
1. Style more than 0.8 mm long; petals more than 4 mm long — *T. montanum*

Thlaspi montanum L.
Sp. Pl., 647, 1753.

Mountain Pennycress

Noccaea cochleariformis (DC.) Löve & Löve
Noccaea montana (L.) F.G. Mey.
Thlaspi australe A. Nels.
Thlaspi californicum S. Wats.
Thlaspi cochleariforme DC.
Thlaspi coloradense Rydb.
Thlaspi fendleri Gray
Thlaspi glaucum (A. Nels.) A. Nels.
Thlaspi hesperium (Pays.) G.N. Jones
Thlaspi nuttallii Rydb.
Thlaspi prolixum A. Nels.
Thlaspi purpurascens Rydb.
Thlaspi stipitatum A. Nels.

Rosulate, glabrous perennial from a simple or branched caudex; stems 1 to many, simple or branched. Basal leaves petiolate, the blades oblanceolate to obovate; margins entire to denticulate; cauline leaves reduced upward, sessile and auriculate, lanceolate, the margins entire. Flowers in terminal, congested racemes, becoming elongate in fruit, the pedicels slender, spreading to ascending. Sepals flattened at the base, greenish to purplish, 2–3 mm long. Petals white, spatulate, rounded, 4–6 mm long. Silicles obovate to obcordate, the margins slightly winged; styles prominent, 1–3 mm long.

Fellfields and slopes of the Gros Ventre, Hoback, Medicine Bow, Uinta, Wasatch, and Wyoming Ranges. Montana and Washington south to California, Arizona, New Mexico, and Texas.

Alpine plants of the Middle Rocky Mountains are var. *montanum*.

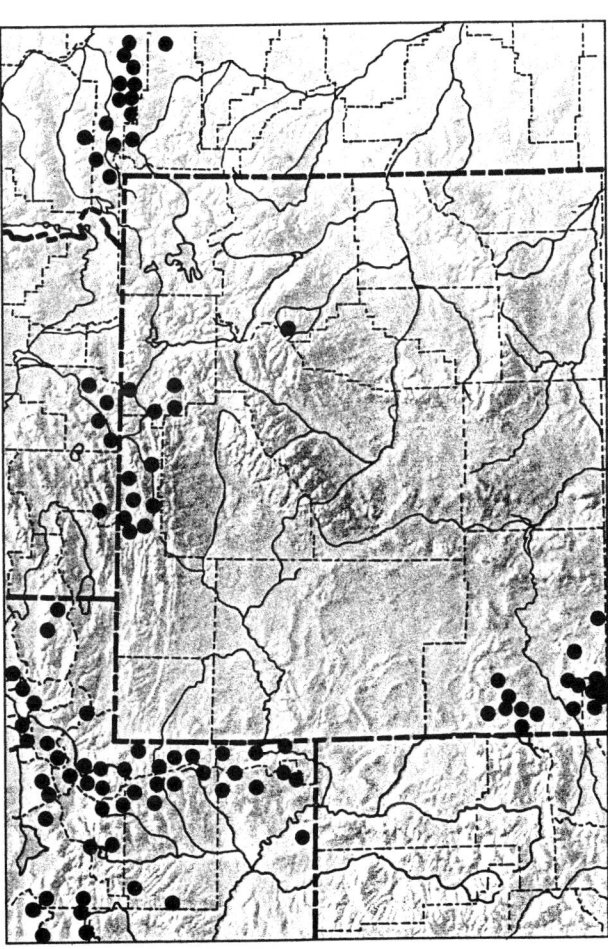

Thlaspi parviflorum A. Nels.
Bull. Torr. Club 27:265, 1900.

Smallflower Pennycress

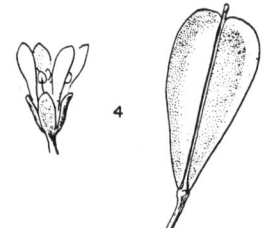

Glabrous perennial from a simple or branched caudex; stems 1 to many, mostly simple. Basal leaves in a rosette, petiolate, the blades oblanceolate to obovate; margins entire to denticulate; cauline leaves reduced upward, lanceolate, sessile and auriculate, the margins entire. Flowers in terminal, congested racemes, becoming elongate in fruit, the pedicels slender, spreading in fruit. Sepals flattened at the base, greenish, 1–1.5 mm long. Petals white, spatulate, rounded, 2.5–4 mm long. Silicles obovate, slightly emarginate at the tip; styles 0.5 mm or less in length.

Willow thickets, stream banks, and timberline meadows in the Absaroka, Beartooth, and Wind River Ranges. Endemic to eastern Idaho, western Wyoming, and southern Montana.

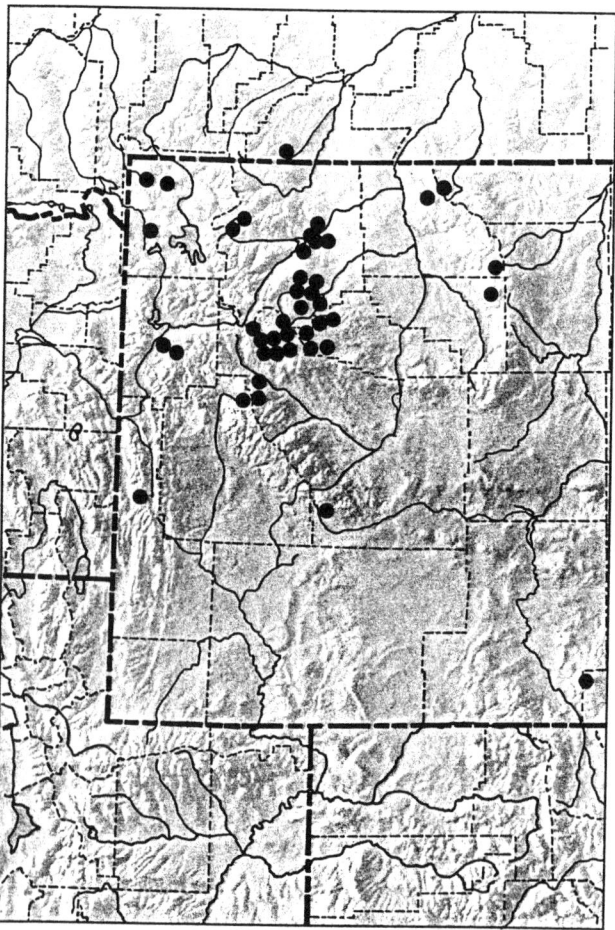

Callitrichaceae
Water Starwort Family

Aquatic, emergent, or terrestrial herbs with opposite, entire, linear to spatulate, exstipulate leaves. Flowers small and inconspicuous, 1–3 in the leaf axils, sessile or subsessile, usually imperfect and subtended by 2 hyaline bracts. Perianth lacking. Stamens 1. Pistil of 2 carpels, each carpel with 2 locules. Styles 2, distinct, slender, and elongate. Fruit a leathery schizocarp, dehiscing into 4 flattened, winged (in ours) mericarps.

Callitriche L.
Water Starwort

Submerged or emergent, annual or perennial herbs with slender stems. Leaves opposite, simple, exstipulate, the submerged ones linear; emergent leaves spatulate to obovate, clustered in terminal rosettes. Flowers 1–3 in leaf axils, polygamous; staminate flowers with a single stamen; pistillate flowers with a single, 2-carpellate pistil, the ovary 4-lobed; styles 2, distinct. Fruit a leathery, 4-lobed schizocarp.

Callitriche palustris L.
Sp. Pl., 2:969, 1753.

Water Starwort

Callitriche verna L.

Perennial, aquatic herbs from filamentous rhizomes, with slender, leafy stems. Submerged leaves linear, the apex retuse; emergent and floating leaves spatulate, the bases of both connate by a narrow wing. Flowers axillary. Fruits obovate to oval, narrowly winged, slightly retuse at the apex, reticulate in vertical lines.

The single record of this species in the alpine zone is from timberline at Echo Lake in the northern Wind River Range at 3156 m (10,350 ft). It can be expected elsewhere in relatively warm, shallow ponds and lakes in the lowest portions of the alpine zone or the alpine-subalpine ecotone.

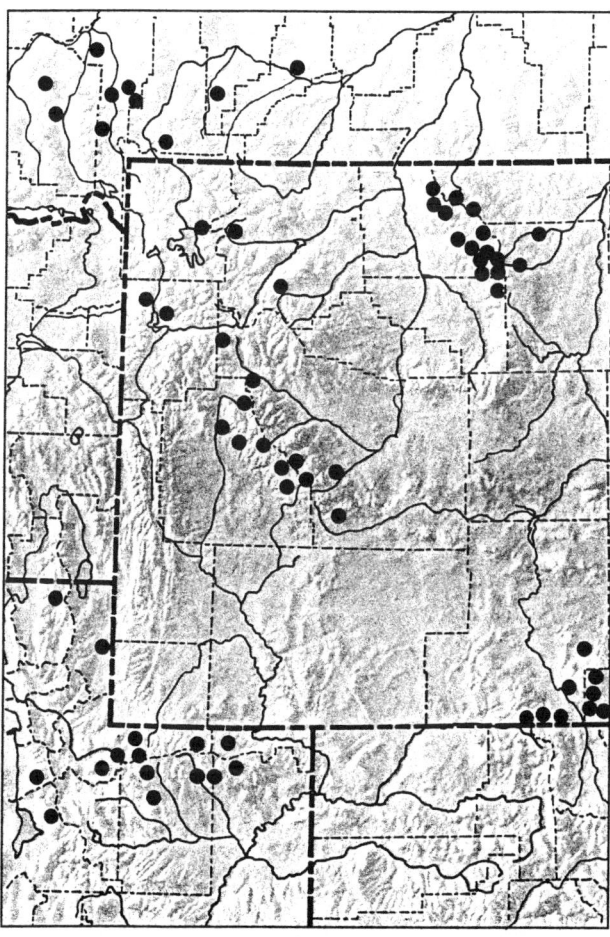

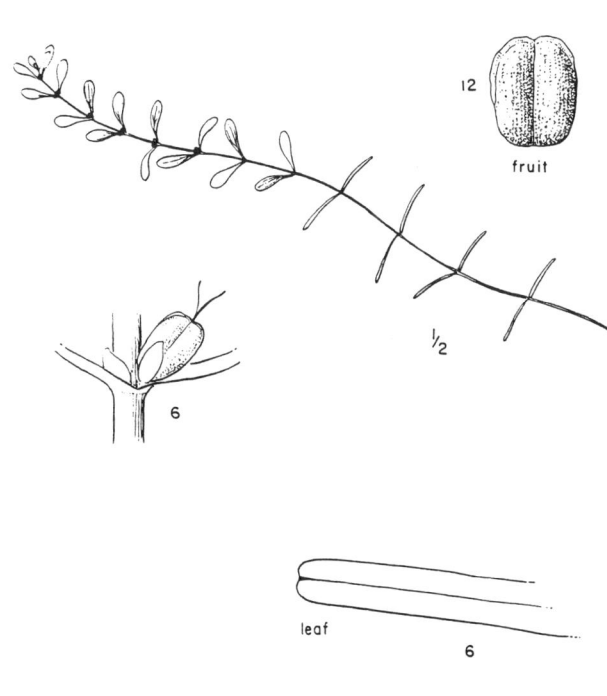

Campanulaceae
Bellflower Family

Annual or perennial herbs in ours, often with milky juice. Leaves alternate, simple, exstipulate. Flowers epigynous, perfect, 5-merous, regular or irregular, solitary or in racemes, spikes, or panicles. Calyx persistent, the corolla of united petals, campanulate, tubular, or bilabiate, usually 5-lobed. Stamens 5, alternate with the corolla lobes, the filaments dilated at the base. Pistil 1, of 3–5 united carpels, with a single, terminal style and commonly 3 stigmas. Fruit a many-seeded, poricidal capsule.

Campanula L.
Harebell

Perennial herbs with alternate, simple, exstipulate leaves. Flowers blue, solitary, racemose, or paniculate. Calyx 5-lobed, the lobes narrowly lanceolate. Corolla 5-lobed, campanulate. Stamens 5, distinct. Ovary 3–5-loculed, the style elongate; stigmas 3. Fruit a poricidal capsule with numerous seeds.

KEY TO THE SPECIES OF *CAMPANULA*

1. Corolla lobes narrowly lanceolate, about the same length as or slightly shorter than the tube; flowers erect or nearly so; capsules erect, opening by pores near the summit; plants usually less than 13 cm high .... *C. uniflora*
1. Corolla lobes deltoid, less than one-third the length of the tube; flowers nodding; capsules nodding, opening by pores near the base; plants usually more than 15 cm high .... *C. rotundifolia*

Campanula rotundifolia L.
Sp. Pl., 163, 1753.

Mountain Harebell

Campanula alaskana (Gray) Wight ex J.P. Anders.
Campanula dubia A. DC.
Campanula heterodoxa Bong.
Campanula intercedens Witasek
Campanula macdougalii Rydb.
Campanula petiolata A. DC.
Campanula sacajaweana M.E. Peck

Multistemmed, glabrous to finely hirsute perennial from a branched caudex or rhizome. Basal leaves entire to coarsely serrate, ovate to cordate, soon withering and lacking in flower; cauline leaves mostly entire, oblanceolate to linear, reduced upward. Flowers usually pendulous, blue, single to several in a racemose inflorescence. Calyx lobes entire, lanceolate to subulate, appressed or reflexed; corolla campanulate, the lobes less than one-half the length of the tube. Capsules nodding, opening by pores near the base.

Common at lower elevations and becoming alpine on dry, rocky slopes and ledges in most Middle Rocky Mountain ranges. Not yet known from above timberline in the Gros Ventre, Hoback, Salt River, Wasatch, and Wyoming Ranges. Cirumboreal; south in western America to California, New Mexico, Texas, and Nebraska.

Dwarf, 1-flowered alpine plants are var. *arctica* Lange.

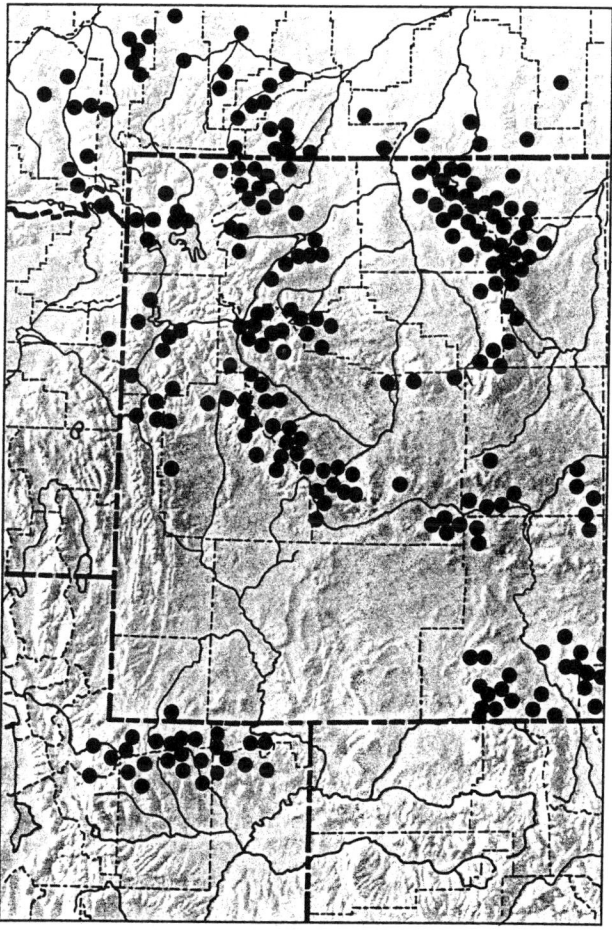

Campanula uniflora L.
Sp. Pl., 163, 1753.

Alpine Harebell

Perennial from a thick taproot and branched caudex. Stems 1 to several, often decumbent, glabrous or nearly so. Basal leaves petiolate, glabrous, oblanceolate to spatulate; cauline leaves sessile, lanceolate to linear, often reduced upward. Flowers blue, solitary, erect; calyx sparsely pubescent, the lobes lanceolate, erect; corolla narrowly campanulate, the lobes equaling or exceeding the tube. Capsule erect, opening by pores near the apex.

Occasional on slopes, fellfields, and meadows in the Absaroka, Beartooth, Big Horn, Medicine Bow, Uinta, and Wind River Ranges. Circumboreal; south in western North America to Wyoming and Colorado.

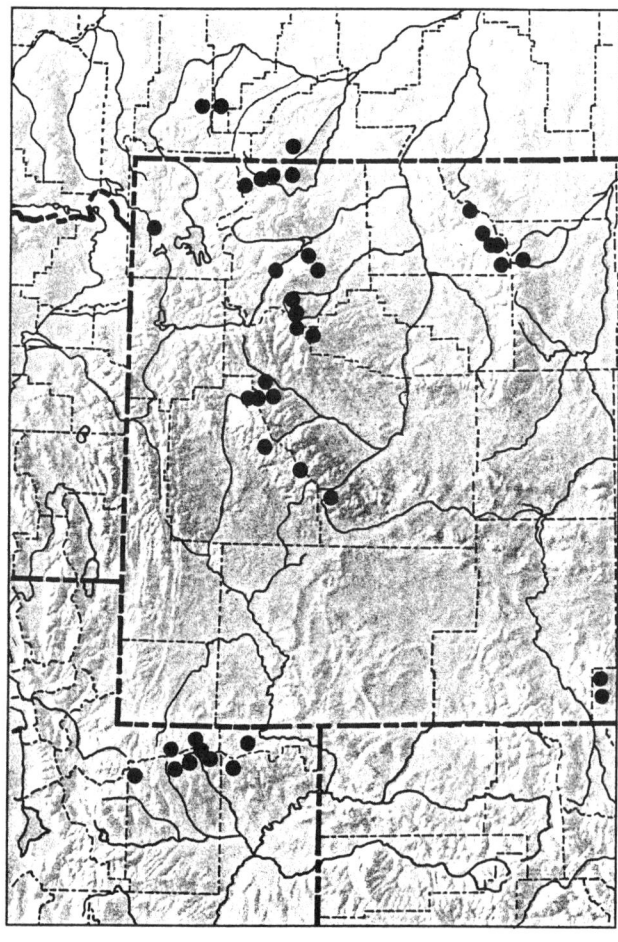

Caprifoliaceae
Honeysuckle Family

Mostly erect shrubs with opposite, simple or compound, usually exstipulate leaves. Flowers epigynous, usually 5-merous, perfect, regular or irregular, cymose. Calyx 3–5-lobed or 3–5-toothed. Corolla mostly 5-lobed and rotate, tubular, or funnelform, sometimes bilabiate. Stamens 5, epipetalous, usually as many as the corolla lobes and alternate with them, or rarely reduced to 4 and didynamous. Pistil 1, the ovary 1–5-celled, of 2–5 united carpels, the style elongate to obsolete, the stigmas capitate or 2–5-lobed. Fruit usually a berry, sometimes a drupe, or a capsule.

KEY TO THE GENERA OF CAPRIFOLIACEAE

1. Leaves pinnately compound; flowers in cymes; berries red, blue, or black — *Sambucus*
1. Leaves simple; flowers solitary or paired in leaf axils; berries white — *Symphoricarpos*

Sambucus L.
Elderberry

Shrubs, mostly 1–2 m high, with pithy stems and large, odd-pinnate leaves, the leaflets lanceolate to ovate, serrate. Flowers small, white, numerous, in large terminal cymes. Calyx lobes minute or obsolete. Corolla rotate-urceolate. Stamens 5. Ovary 3–5-celled, with 1 ovule in each cell, the style short, 3–5-lobed. Fruit a red, blue, or black berrylike drupe.

Sambucus racemosa L.
Sp. Pl., 270, 1753.

Red Elderberry

Sambucus callicarpa Greene
Sambucus melanocarpa Gray
Sambucus microbotrys Rydb.
Sambucus pubens Michx.

Shrub, to 1 m tall in the alpine zone; stems glabrous with age, often glaucous, with a white to dark brown pith. Leaves pinnate, 5–7-foliolate; leaflets elliptic to obovate, serrate, glabrous or glabrous to puberulent above and pubescent below, the base oblique. Flowers white to cream in a terminal, paniculiform, pyramidal to ovoid inflorescence. Fruit purplish black to red, rarely yellowish.

Coarse talus and rocky stream banks of the Teton, Uinta, and Wind River Ranges. Circumboreal; south in western North America to California, New Mexico, South Dakota, and Iowa.

North American plants are ssp. *pubens* (Michx.) House. In the Middle Rocky Mountains 2 varieties occur in the lower alpine zone: var. *melanocarpa* (Gray) McMinn, with purplish black fruits and glabrous to pubescent leaflets; and var. *microbotrys* (Rydb.) Kearney & Peebles, with red or orange fruits and glabrous leaves.

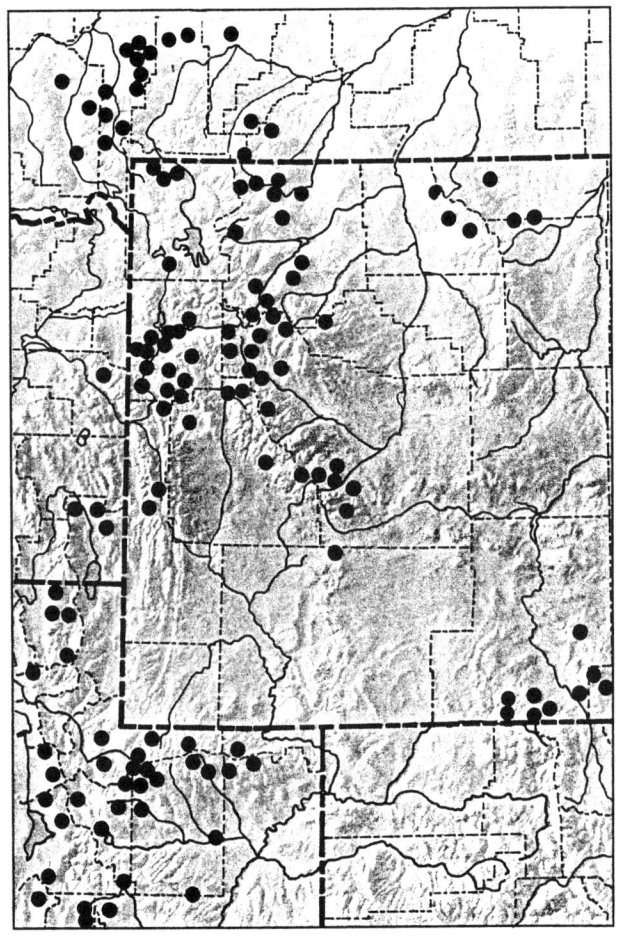

Symphoricarpos Duhamel
Snowberry

Erect to somewhat prostrate shrubs with shredding bark. Leaves deciduous, simple, opposite, short-petiolate, the margins entire to undulate, sometimes coarsely toothed or lobed; stipules lacking. Flowers white or pinkish, solitary in leaf axils or in terminal or axillary, racemose or spicate clusters. Calyx 4–5-lobed, the lobes persistent. Corolla regular to irregular, mostly 5-lobed, campanulate to funnelform or salverform. Stamens 4 or 5, included to exerted. Ovary with 4 locules, 2 abortive and 2 fertile; style 1, elongate, the stigma capitate or somewhat 2-lobed. Fruit a white or sometimes bluish or red, 2-seeded, berrylike drupe.

Symphoricarpos oreophilus Gray
J. Linn. Soc. Bot. 14:12, 1873.

Mountain Snowberry

Symphoricarpos tetonensis A. Nels.
Symphoricarpos utahensis Rydb.
Symphoricarpos vaccinioides Rydb.

Erect, much-branched shrub with glabrous to puberulent twigs; older branches with shredding bark. Leaves elliptic to ovate, short-petiolate, glabrous or glabrate, the margins entire, undulate-serrate, or lobed. Flowers solitary or paired in leaf axils, the upper often in short axillary or terminal racemes or spikes. Calyx glabrous, 4–5-lobed. Corolla white or pink, narrowly funnelform, symmetrical, mostly 5-lobed, the lobes much shorter than the tube; tube glabrous on the outside, glabrous to pubescent within. Stamens mostly 5, the anthers longer than the filaments. Style 1, glabrous, elongate, the stigma capitate. Drupes whitish, ovoid to ellipsoid; seeds 2.

This species occurs on rocky ledges and talus in the alpine zone of the Wasatch Range. British Columbia south to California, Mexico, and Texas.

Erect Middle Rocky Mountain plants with symmetrical corollas 6–10 mm long are var. *utahensis* (Rydb.) A. Nels.

Caryophyllaceae
Pink Family

Annual or perennial herbs with opposite, entire, simple, mostly exstipulate leaves. Stems erect or decumbent, usually swollen at the nodes. Flowers hypogynous, perfect, complete, or sometimes apetalous, 4–5-merous, solitary or cymose. Sepals distinct or united, persistent. Petals the same number as the sepals, or wanting, often with a spreading blade and a distinct claw. Stamens distinct, commonly twice as many as the petals. Pistil 1, of 2–5 united carpels, the ovary 1-celled, with free-central or basal placentation. Fruit a capsule opening by a lid, by valves, or by apical teeth.

KEY TO THE GENERA OF CARYOPHYLLACEAE

1. Sepals united, the petals clawed or lacking
   2. Sepals spinulose-tipped — *Paronychia*
   2. Sepals lacking spinulose tips
      3. Styles normally 3, rarely 4; capsules 3- or 6-toothed at the apex — *Silene*
      3. Styles normally 5, rarely 4; capsules 5- or 10-toothed at the apex — *Lychnis*
1. Sepals distinct or nearly so, the petals not clawed
   4. Leaves with thin, membranous stipules — *Spergularia*
   4. Leaves exstipulate
      5. Petals bifid or deeply emarginate; capsules opening by twice as many teeth as there are styles
         6. Styles mostly 3; capsules ovoid or oblong, dehiscent by valves — *Stellaria*
         6. Styles mostly 5; capsules cylindric, dehiscent by apical teeth — *Cerastium*
      5. Petals entire or shallowly emarginate, rarely lacking: capsules opening by as many or twice as many teeth as there are styles
         7. Styles usually 5; capsules dehiscent by 5 valves — *Sagina*
         7. Styles usually 3; capsules dehiscent by 3 or 6 valves
            8. Plants erect, tufted or loosely mat-forming: capsules apparently dehiscent by 6 valves — *Arenaria*
            8. Plants prostrate, densely mat-forming to pulvinate: capsules dehiscent by 3 valves — *Minuartia*

Arenaria L.
Sandwort

Erect, tufted, annual or perennial herbs with sessile, opposite, linear to subulate, exstipulate leaves. Flowers small, solitary or in open to contracted terminal cymes. Sepals 5, distinct, erect or ascending. Petals 5, white, entire to emarginate, rarely wanting. Stamens usually 10. Ovary 1-celled, the styles 3, rarely 2–5. Capsule globose to oblong, with few to many seeds, dehiscing by 6 valves.

KEY TO THE SPECIES OF ARENARIA

1. Inflorescence congested, capitate — *A. congesta*
1. Inflorescence open, cymose, elongate — *A. capillaris*

Arenaria capillaris Poir. in Lam.
Encyc. Meth. Suppl. 6:380, 1804.

Threadleaf Sandwort

Arenaria formosa (Fisch.) Regel
Arenaria nardifolia Ledeb.
Eremogone americana (Maguire) S. Ikonn.
Eremogone capillaris (Poir.) Fenzl

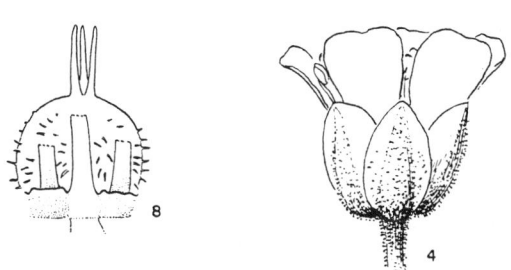

Caespitose, perennial herb from a branched, somewhat woody caudex; stems several, simple, glabrous below, glandular above. Leaves glabrous, linear, acuminate, erect, and usually curved, the margins scabrous. Flowers showy, terminal in open, bracteate cymes, the bracts lanceolate with scarious margins. Sepals 1-nerved, ovate, obtuse to broadly acute, with broad, scarious, often purplish margins. Petals white, erect to spreading, obovate, about twice as long as the sepals, entire to slightly emarginate. Stamens 10; styles 3. Capsule 6-valved, ovoid, longer than the sepals.

Reported from the Line Creek Plateau of the Beartooth Range by Lessica (1993). Eurasia; Alaska south to Oregon, Nevada, Idaho, and Montana.

Middle Rocky Mountain plants with glandular inflorescences are var. *americana* (Maguire) Davis.

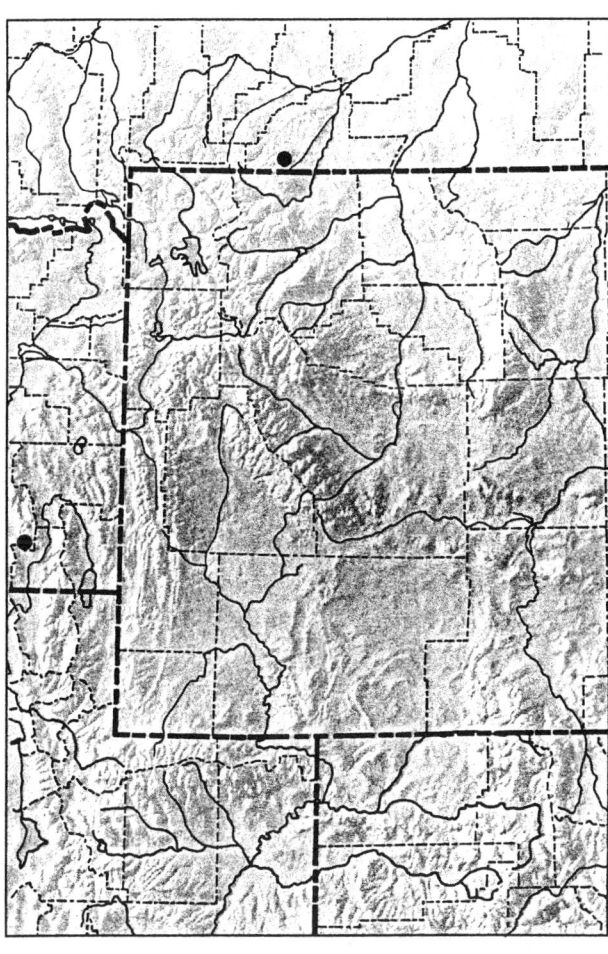

Arenaria congesta Nutt. in T. & G. *Fl. N. Am.* 1:178, 1838.

Ballhead Sandwort

Arenaria burkei T.J. Howell
Arenaria cephaloidea Rydb.
Arenaria glabrescens Piper
Arenaria lithophila (Rydb.) Rydb.
Arenaria subcongesta (S. Wats) Rydb.
Eremogone congesta (Nutt.) S. Ikonn.

Loosely caespitose, perennial herb from a branched, somewhat woody caudex; stems simple, mostly with 2–4 pairs of leaves. Leaves linear, glabrous, subulate, 1-nerved, erect or ascending. Flowers terminal in dense, capitate and bracteate cymes. Sepals ovate, obtuse, acute, or acuminate, scarious-margined, 1–3-nerved. Petals white, clawed, entire to slightly emarginate, about twice as long as the sepals. Stamens 10, alternately glandular at the base. Styles 3. Capsule 3-valved, each valve with 2 teeth, appearing 6-valved, equaling or exceeding the sepals.

Fellfields, rocky ledges, and talus in all alpine ranges of the Middle Rocky Mountains. Northwestern Canada south to California, Idaho, and Colorado.

The 2 alpine varieties may be distinguished as follows:

1. Inflorescence loose, the flowers with evident pedicels; sepals obtuse . . . var. *lithophila* Rydb.
1. Inflorescene dense, the flowers lacking evident pedicels; sepals acute . . . var. *congesta*.

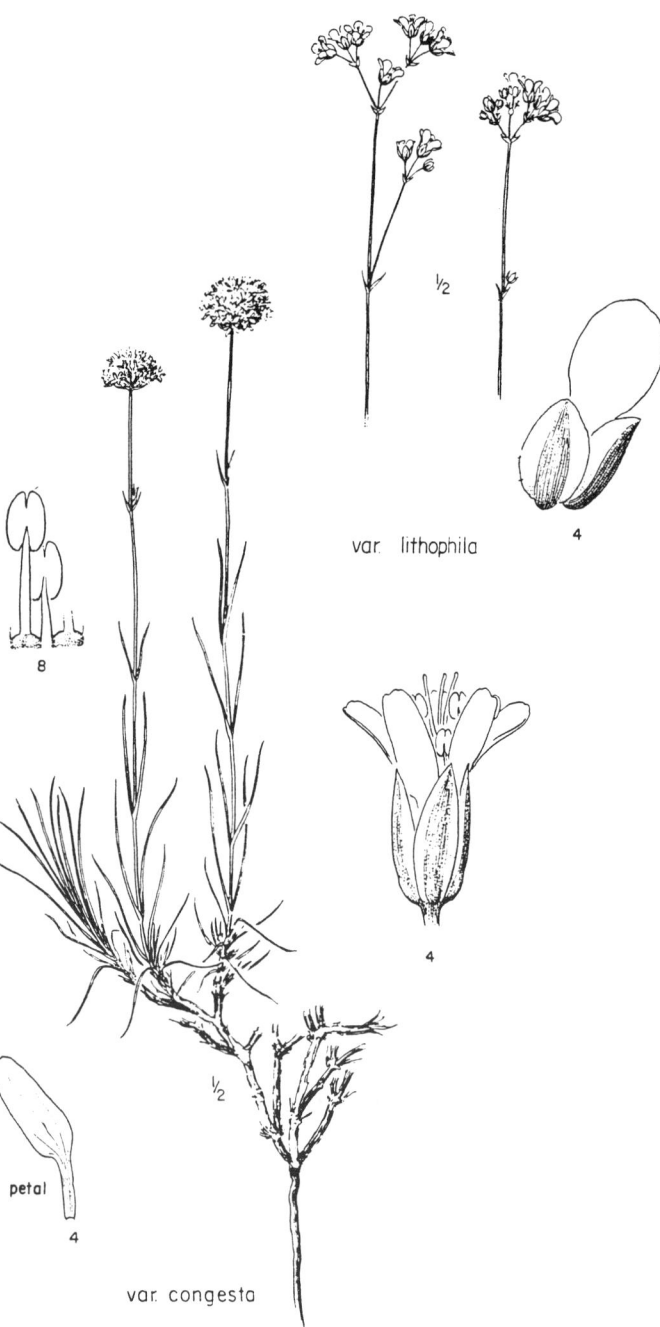

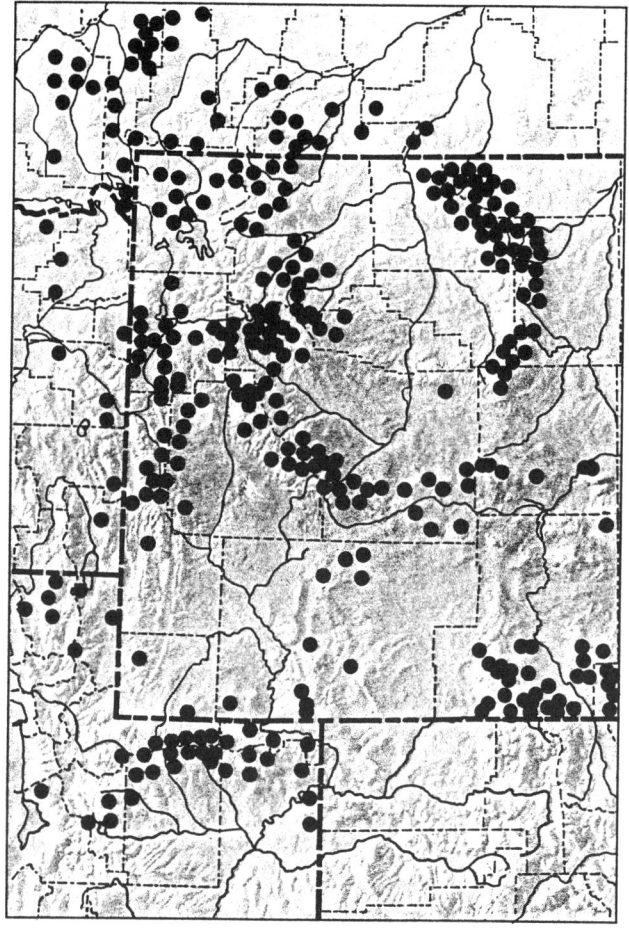

Cerastium L.
Chickweed

Mostly pubescent, annual or perennial herbs, with erect or prostrate stems and entire, opposite, sessile leaves. Flowers white, in terminal cymes. Sepals 5 or rarely 4, distinct. Petals the same number as the sepals, emarginate or cleft nearly to the base. Stamens 10, or sometimes 5. Ovary 1-celled; styles 5. Capsule cylindrical, straight or slightly incurved, dehiscent by twice as many apical teeth as there are styles.

KEY TO THE SPECIES OF *CERASTIUM*

1. Leaves linear to lanceolate, acute; bracts of inflorescence broadly scarious-margined; sterile shoots present in upper leaf axils ..... *C. arvense*
1. Leaves oblong, ovate, or oblanceolate, obtuse; bracts of inflorescence with a narrow scarious margin, or lacking; sterile shoots lacking in upper leaf axils ..... *C. beeringianum*

Cerastium arvense L.
Sp. Pl., 438, 1753.

Field Chickweed

Cerastium alsophilum Greene
Cerastium angustatum Greene
Cerastium bracteatum Raf.
Cerastium campestre Greene
Cerastium confertum Greene
Cerastium effusum Greene
Cerastium elongatum Pursh
Cerastium fuegianum (Hook. f.) A. Nels.
Cerastium graminifolium Rydb.
Cerastium latifolium Fenzl
Cerastium leibergii Rydb.
Cerastium nitidum Greene
Cerastium oblongifolium Torr.
Cerastium occidentale Greene
Cerastium oreophilum Greene
Cerastium patulum Greene
Cerastium scopulorum Greene
Cerastium sonnei Greene
Cerastium strictum L.
Cerastium subulatum Greene
Cerastium tenuifolium Pursh
Cerastium thermale Rydb.
Cerastium variabile Goodding
Cerastium velutinum Raf.
Cerastium vestitum Greene
Cerastium viride A. Heller
Cerastium villosum Muhl. ex Darl.

Caespitose, glabrous to glandular perennial from slender, branched rootstocks; stems erect, decumbent, or prostrate, with old, dried leaves at the base. Leaves linear, lanceolate or narrowly ovate, 1-nerved, acute to obtuse, at least the upper cauline leaf axils with secondary leaves or sterile shoots. Flowers in loose to compact, bracteate cymes, the bracts scarious-margined. Sepals lanceolate, glandular, acute, scarious-margined. Petals white, bifid, about twice as long as the sepals. Stamens 10. Styles 5. Capsule cylindric, equaling to twice as long as the sepals.

Fellfields, talus, and rocky slopes; known from all alpine ranges except the Hobacks. Widely distributed in North America and Eurasia.

This highly variable species is easily confused with *C. beeringianum* in alpine habitats. *Cerastium arvense* tends to have more acute leaves, scarious bracts in the inflorescence, and secondary leaves or sterile shoots in the upper leaf axils. Many alpine specimens appear to be intermediate between the two.

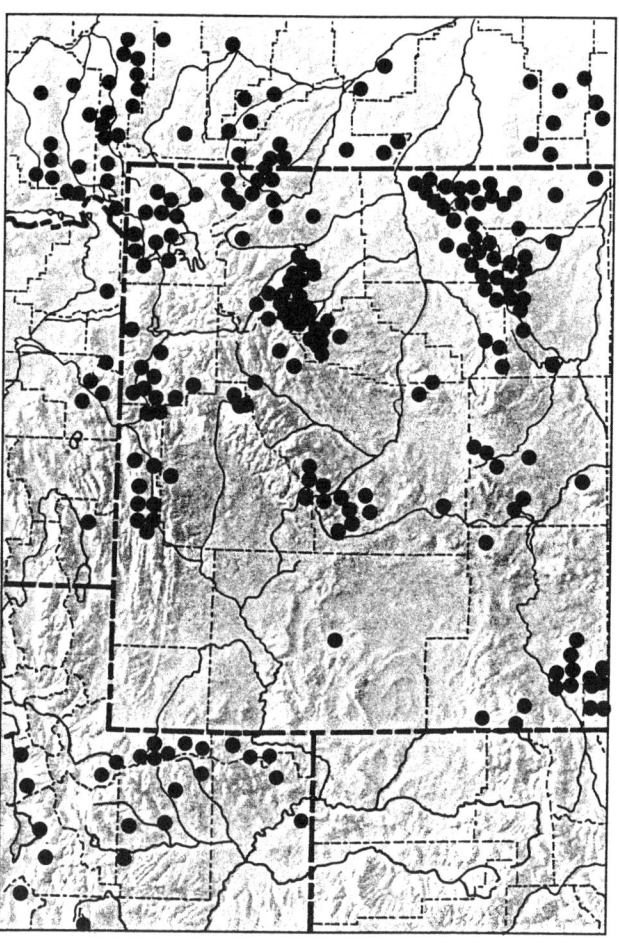

Cerastium beeringianum Cham. & Schlecht.
Linnaea 1:62, 1862.

Bering Chickweed

Cerastium bialynickii Tolm.
Cerastium buffumae A. Nels.
Cerastium earlei Rydb.
Cerastium pilosum Greene
Cerastium pulchellum Rydb.
Cerastium scammaniae Polunin
Cerastium terrae-novae Fern. & Wieg.

Caespitose, glandular-pubescent perennial from branched rootstocks; stems ascending to prostrate with mixed simple and glandular hairs. Leaves ovate to oblong-lanceolate, mostly obtuse; upper axils lacking secondary leaves or sterile shoots. Flowers in loose, bracteate cymes, occasionally solitary, the bracts lacking scarious margins. Sepals lanceolate to ovate, glandular, the inner scarious-margined and often purplish. Petals white, bifid, about twice as long as the sepals. Stamens 10. Styles 5. Capsule cylindrical, nearly twice as long as the calyx.

Fellfields, talus, and rocky slopes of most alpine ranges. Not yet known from above timberline in the Hoback and Medicine Bow Ranges. Siberia; transcontinental in North America, south in the West to California, Arizona, and New Mexico.

Middle Rocky Mountain plants with petals slightly longer than the sepals, glands of the pedicels very fine and short, retrorse hairs on lower internodes lacking or indistinct, and upper leaf surfaces glandular are ssp. *earlei* (Rydb.) Hult.

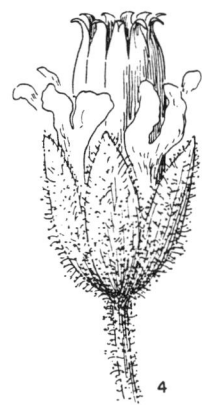

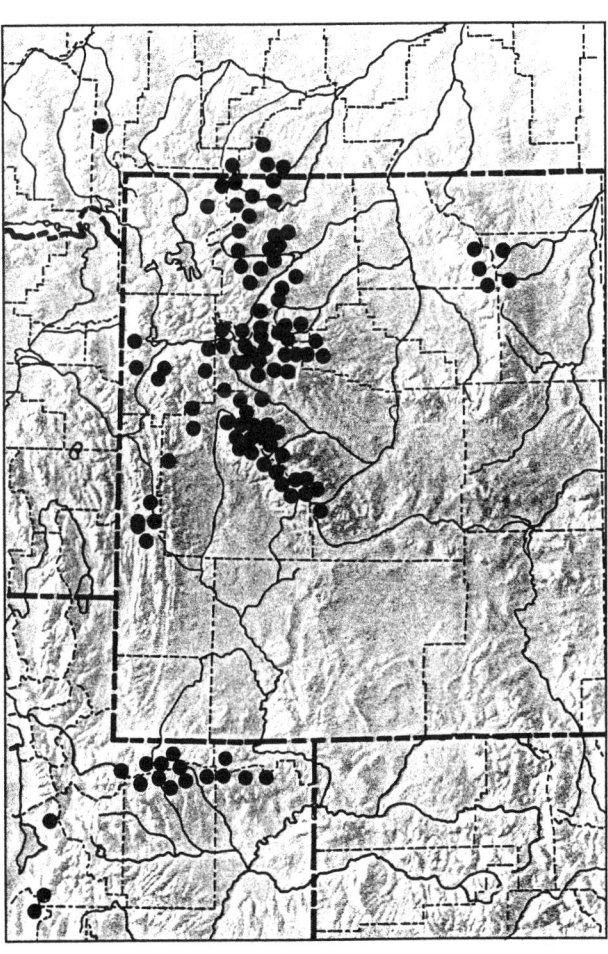

Lychnis L.
Campion

Perennial herbs with opposite, entire leaves. Flowers perfect or unisexual, in open cymes, or single and terminal. Calyx 5-lobed, 10-nerved, tubular, usually inflated in fruit. Corolla white, pink, purplish, or reddish, of 5 narrowly clawed petals, the blades entire or 2-cleft, usually with auricles at the junction of blade and claw, and with narrow to broad appendages on the inner side. Stamens 10, connate at the base. Styles 5, rarely 4. Capsule opening by 5 or 10 teeth.

KEY TO THE SPECIES OF *LYCHNIS*

1. Flowers usually 2 to several; calyx not inflated; plants mostly more than 20 cm tall .... *L. drummondii*
1. Flowers usually solitary; calyx inflated, at least in fruit; plants less than 18 cm tall .... *L. apetala*

Lychnis apetala L.
Sp. Pl., 437, 1753.

Alpine Campion

Agrostemma apetalum (L.) G. Don
Gastrolychnis apetala (L.) Tolm. & Kozh.
Gastrolychnis kingii (S. Wats.) W.A. Weber
Gastrolychnis uralensis Rupr.
Lychnis attenuata Farr.
Lychnis kingii S. Wats.
Lychnis montana S. Wats.
Melandrium apetalum (L.) Fenzl
Melandrium kingii (S. Wats.) Tolm.
Silene attenuata (Farr) Bocq.
Silene hitchguirei Bocq.
Silene kingii (S. Wats.) Bocq.
Silene uralensis (Rupr.) Bocq.
Silene wahlbergella Chowdhuri
Wahlbergella apetala (L.) Fries
Wahlbergella attenuata (Farr) Rydb.
Wahlbergella kingii (S. Wats.) Rydb.
Wahlbergella montana (S. Wats.) Rydb.

Erect, glandular-pubescent, perennial herb from a branched caudex and taproot. Basal leaves petiolate, linear-lanceolate to oblanceolate, pubescent, the hairs spreading to retrorse; cauline leaves sessile, narrower and shorter than the basal leaves, in 1–3 pairs. Flowers mostly solitary, erect to nodding when young, erect at fruiting. Calyx campanulate-urceolate, inflated, glandular-pubescent with purplish hairs, 10-nerved, the teeth acute. Petals pinkish or purplish, clawed, the blade bilobed with smaller lateral teeth or lobes. Stamens 10, included. Styles 5. Capsule 5-valved, the valves entire. Seeds wingless or wing-margined.

Fellfields, talus, and rocky slopes in the Absaroka, Beartooth, Big Horn, Uinta, and Wind River Ranges. Circumpolar; south in the Rocky Mountains to Colorado.

Two varieties occur in the alpine zone of the Middle Rocky Mountains:

1. Flowers nodding in bud; seeds winged; hairs of basal leaves spreading
   . . . var. *montana* (S. Wats.) C.L. Hitchc.
1. Flowers erect in bud; seeds wingless; hairs of basal leaves retrorse . . . var. *kingii* (S. Wats.) S.L. Welsh

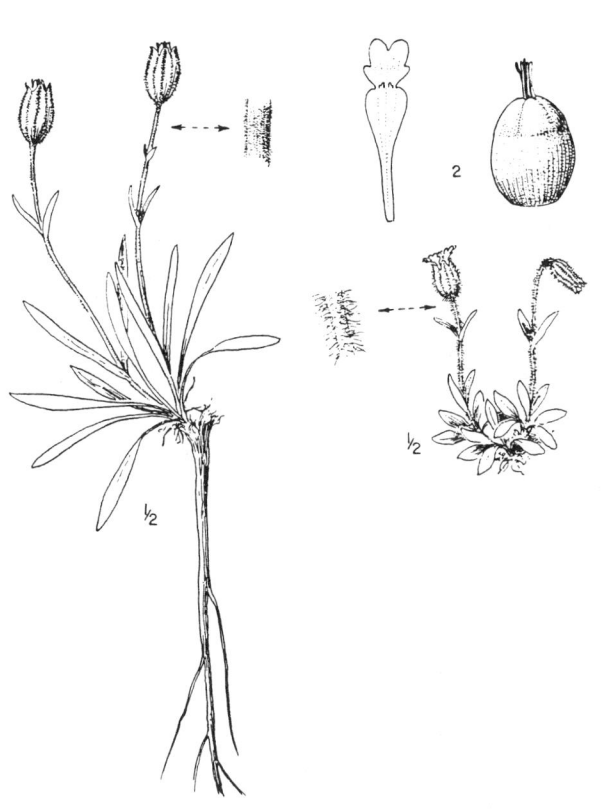
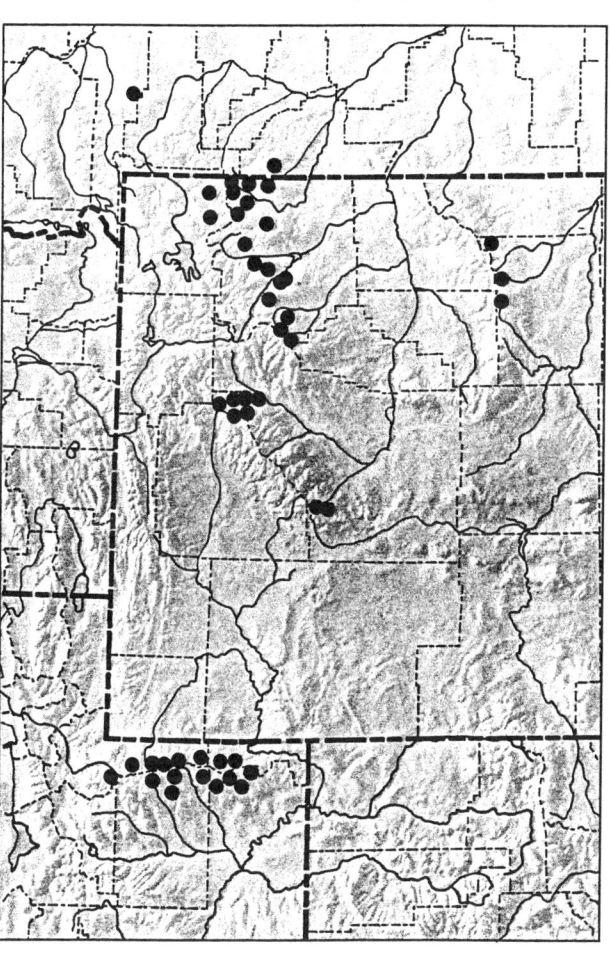

Lychnis drummondii (Hook.) S. Wats.
Bot. King. Exped. 37, 1871.

Drummond Campion

Silene drummondii Hook.
Gastrolychnis drummondii (Hook.) Löve & Löve
Lychnis pudica B. Boivin
Lychnis striata Rydb.
Melandrium drummondii (Hook.) Hult.
Wahlbergella drummondii (Hook.) Rydb.
Wahlbergella striata (Rydb.) Rydb.

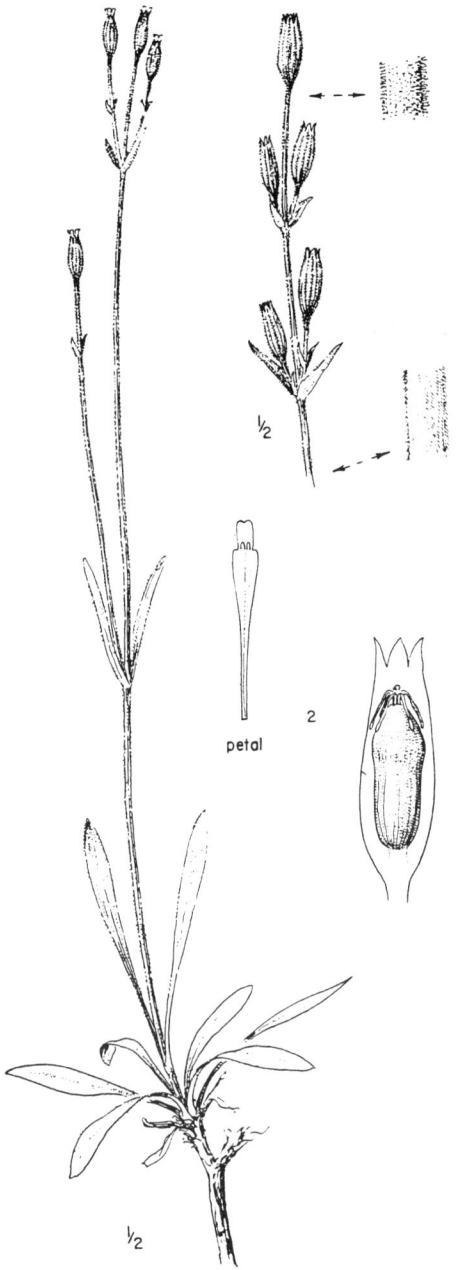

Erect, perennial herb from a branched caudex and stout taproot; stems 1 to several, retrorsely puberulent below, glandular above. Basal leaves tufted, elliptic to oblanceolate, petiolate; cauline leaves linear-lanceolate, sessile, reduced, in 2–4 pairs. Flowers 2 to several in a bracteate cyme, the pedicels erect. Calyx tubular, little inflated, glandular-puberulent, 10-nerved, the teeth triangular and acute. Petals white to pink, clawed, the claw wider than the blade; blade shallowly lobed to retuse. Stamens 10, included. Styles 5. Capsule 5-valved, the valves entire. Seed margins not winged.

Uncommon on fellfields of the Absaroka, Medicine Bow, Uinta, Wasatch, Wind River, and Wyoming Ranges. British Columbia and Alberta south to Nevada, Arizona, Colorado, Nebraska, and Minnesota.

Alpine plants with white petals are var. *drummondii*.

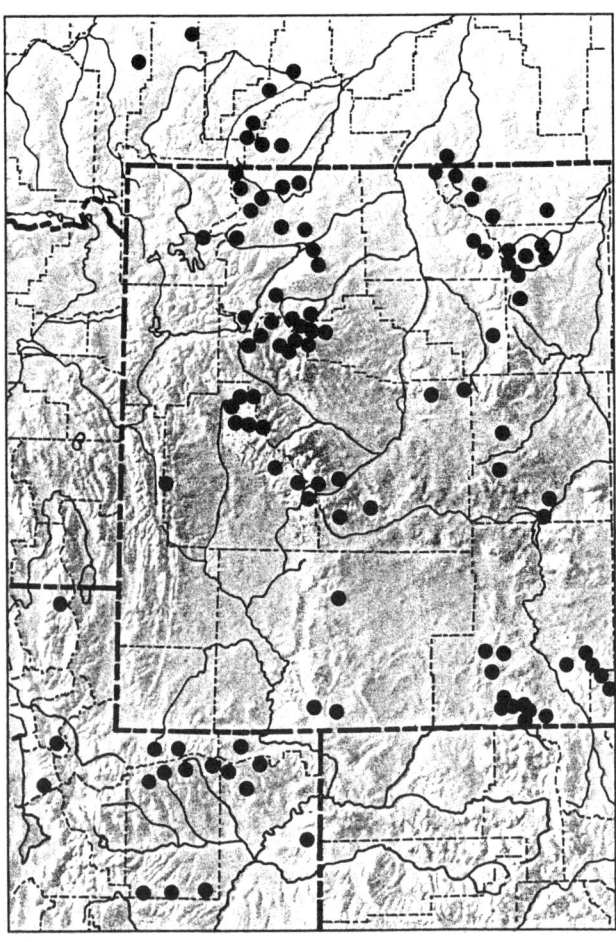

Minuartia L.
Sandwort

Low, densely to loosely mat-forming, annual or perennial herbs with opposite, sessile, linear to subulate, exstipulate leaves. Flowers small, or sometimes showy, solitary or in open to contracted, terminal cymes. Sepals 5, distinct, erect or ascending. Petals 5, white, entire to emarginate, or sometimes lacking. Stamens usually 10. Ovary 1-celled the styles 3, rarely 2–5. Capsule ovoid to oblong, with few to many seeds, dehiscing by 3 valves.

KEY TO THE SPECIES OF *MINUARTIA*

1. Sepals oblong, obtuse, conspicuously incurved at the apex . . . . . . *M. obtusiloba*
1. Sepals lanceolate, acute or acuminate at the apex, never incurved
    2. Plants glabrous; petals shorter than to longer than the sepals, or lacking
        3. Flowers showy; sepals 4–5 mm long; petals about 2 mm longer than the sepals . . . . . . *M. macrantha*
        3. Flowers not showy; sepals 2.5–4 mm long; petals lacking or not more than 1 mm longer than the sepals
            4. Plants usually less than 3 cm tall; stems mostly 1-flowered . . . . . . *M. austromontana*
            4. Plants usually more than 5 cm tall; stems mostly 2- to several-flowered
                5. Plants caespitose perennials forming small cushions; leaves acute . . . . . . *M. rubella*
                5. Plants erect annuals or weak perennials, not forming cushions; leaves subulate . . . . . . *M. dawsonensis*
    2. Plants glandular-pubescent; petals mostly longer than the sepals
        6. Leaves subulate, imbricate on flowering stems; sepals acuminate, 4–6 mm long; capsules shorter than the sepals . . . . . . *M. nuttallii*
        6. Leaves acute, mostly basal; sepals acute, 3–4 mm long; capsules equaling or exceeding the sepals . . . . . . *M. rubella*

Minuartia austromontana S.J. Wolf & Packer
Can. J. Bot. 57:1676, 1979.

Ross Sandwort

Alsine rossii (R. Br.) Fenzl
Alsinopsis rossii (R. Br.) Rydb.
Arenaria filiorum Maguire
Arenaria rossii R. Br.
Minuartia rolfii Nannf.
Minuartia rossii (R. Br.) Graebn.

Densely pulvinate, glabrous, perennial herb, less than 3 cm tall, from a taproot; stems erect to somewhat decumbent. Leaves crowded, linear, obtuse, 3-nerved, and strongly keeled, most with fascicles of secondary leaves in axils. Flowers 1, sometimes 2 per branch. Sepals glabrous, lanceolate, acute, 3-nerved, the middle nerve usually most prominent, often reddish or purplish. Petals white, spatulate, slightly shorter than to slightly longer than the sepals, or often lacking. Stamens 10. Styles 3. Capsule 3-valved, the valves entire, shorter than the sepals.

Fellfields in the Absaroka, Beartooth, Big Horn, Teton, and Wind River Ranges of northwestern Wyoming and southern Montana. Alberta to Oregon and south to Utah and Wyoming.

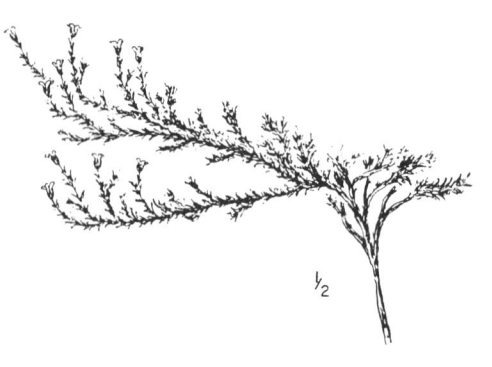

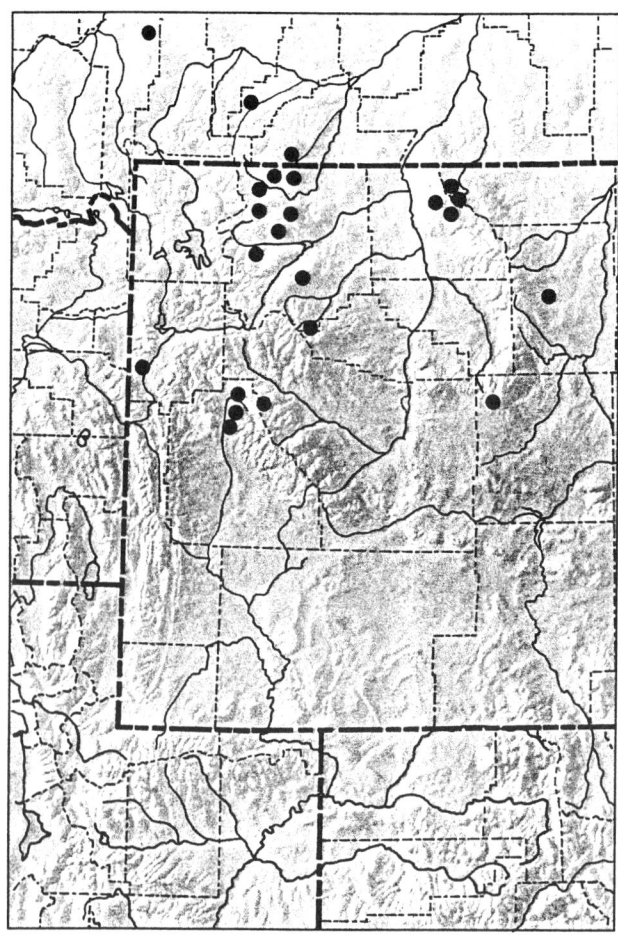

Minuartia dawsonensis (Britt.) House
Am. Midl. Nat. 7:132, 1921.

Rock Sandwort

Arenaria dawsonensis Britt.
Alsinanthe stricta (Sw.) Reichenb.
Alsine michauxii Fenzl
Alsine uliginosa (Murr.) Britt.
Alsinopsis dawsonensis (Britt.) Rydb.
Alsinopsis stricta (Michx.) Small
Alsinopsis tenella (Nutt.) A. Heller
Arenaria litorea Fern.
Arenaria macra A. Nels. & Macbr.
Arenaria stricta Michx.
Arenaria tenella Nutt.
Arenaria uliginosa (Murr.) Schleich.
Minuartia michauxii (Fenzl) Farw.
Minuartia stricta (Sw.) Hiern
Minuartia tenella (Nutt.) Mattf.
Minuopsis michauxii (Fenzl) W.A. Weber
Sabulina dawsonensis (Britt.) Rydb.
Sabulina stricta (Michx.) Small
Stellaria uliginosa Murr.

Glabrous to glandular-puberulent annual or weak perennial from a slender, branched caudex or taproot; stems erect or, more commonly, decumbent and rooting at the lower nodes. Leaves linear to linear-lanceolate, subulate, 3-nerved, often with secondary leaves or short branches in the axils. Flowers in open, 2- to several-flowered cymes, occasionally solitary. Sepals lanceolate, acute, 3-nerved, scarious-margined. Petals white, obovate, longer than the sepals. Stamens 10. Styles 3. Capsule 3-valved, ovoid, longer than the sepals.

Stream banks and other wet or damp, rocky places in the Absaroka and Wind River Ranges. Circumpolar; Alaska south and east over most of North America.

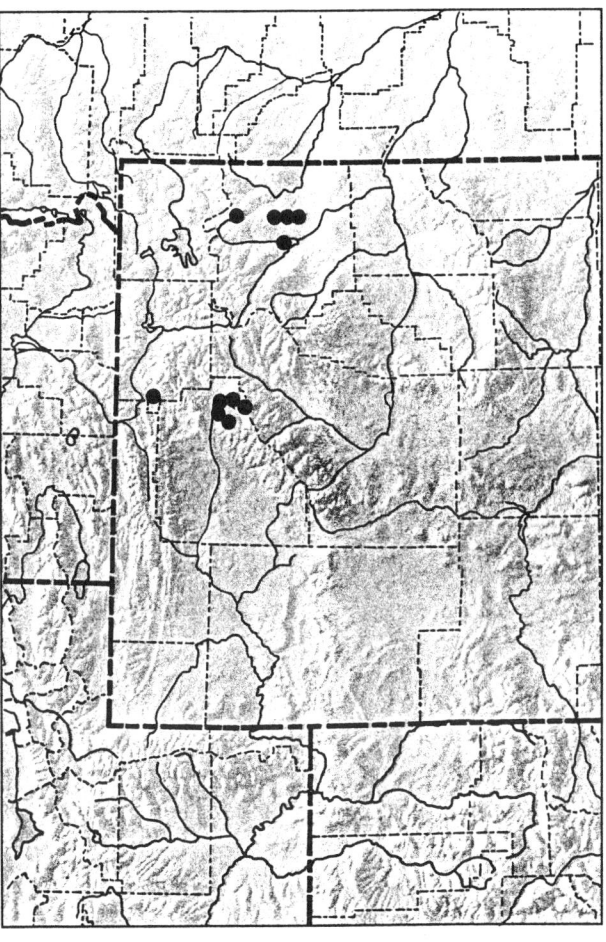

Minuartia macrantha (Rydb.) House
Am. Midl. Nat. 7:132, 1921.

Largeflower Sandwort

Alsinopsis macrantha Rydb.
Alsinanthe macrantha (Rydb.) W.A. Weber
Arenaria macrantha (Rydb.) A. Nels. ex Coult. & A. Nels.

Caespitose, glabrous, perennial herb; stems 4–10 cm tall, erect to spreading. Leaves narrowly linear, obtuse, 1-nerved. Flowers showy, solitary or 2–3 in open, terminal cymes. Sepals lanceolate, acute to acuminate, 3-nerved, scarious-margined. Petals white, spatulate, distinctly longer than the sepals. Stamens 10. Styles 3. Capsule 3-valved, the valves entire.

Rocky places, often on limestone, in the Gros Ventre and northern Wind River Ranges and near timberline in the Uinta Range. Endemic to Colorado, northern Utah, and Wyoming.

2.8

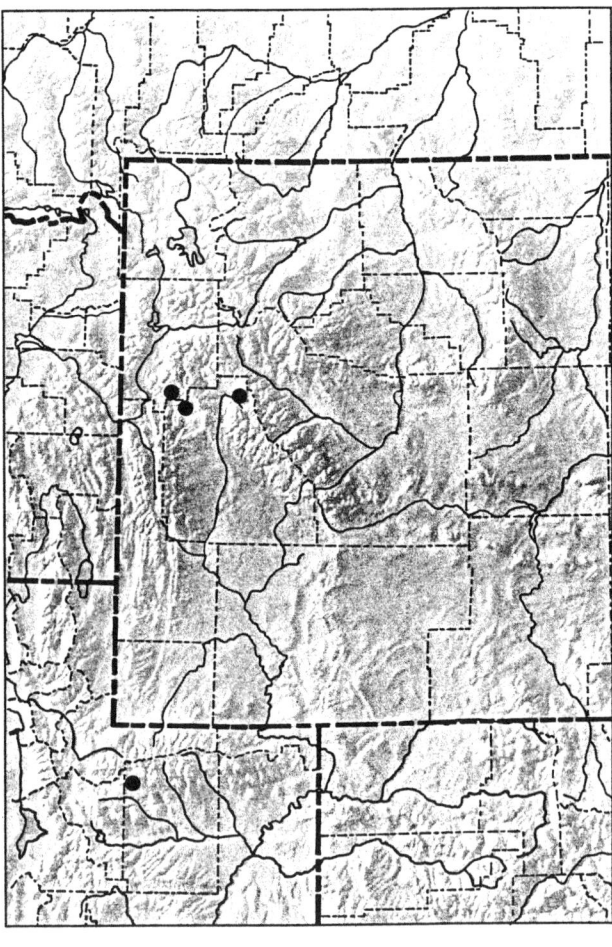

Minuartia nuttallii (Pax) Briq.
Ann. Cons. Jard. Bot. Geneve 13 & 14:385, 1911.

Nuttall Sandwort

Arenaria nuttallii Pax
Alsinopsis occidentalis A. Heller
Arenaria gregaria A. Heller
Arenaria pungens Nutt. in T. & G.
Minuartia pungens (Nutt. in T. & G.) Mattf.
Minuopsis nuttallii (Pax) W.A. Weber

Glandular-pubescent, caespitose, perennial herb from a branched caudex and taproot; stems decumbent. Leaves linear or lanceolate, subulate, 3-nerved, often imbricate and with secondary leaves in the axils, especially near the base of flowering stems. Flowers in loose, bracteate cymes. Sepals lanceolate to ovate, 1–3-nerved, acuminate. Petals white, ovate, entire, shorter than to twice as long as the sepals. Stamens 10, alternately glandular at the base. Styles 3. Capsule 3-valved, the valves entire, shorter than the sepals.

Fellfields and talus of most Middle Rocky Mountain ranges. Not yet known from above timberline in the Big Horn and Uinta Ranges. British Columbia south to California, Utah, and Wyoming.

Rocky Mountain plants that are moderately glandular to pilose with leaves that are erect-ascending and straight, not curved, are var. *nuttallii*.

Minuartia obtusiloba (Rydb.) House
Am. Midl. Nat. 7:132, 1921.

Arctic Sandwort

Alsinopsis obtusiloba Rydb.
Alsinanthe biflora (L.) Reichenb.
Alsine biflora (L.) Wahlenb.
Arenaria biflora (L.) S. Wats.
Arenaria obtusa Torr.
Arenaria obtusiloba (Rydb.) Fern.
Arenaria sajanensis Willd. ex Schlecht.
Lidia obtusiloba (Rydb.) Löve & Löve
Minuartia biflora (L.) Schinz & Thell.
Minuartia sajanensis (Willd. ex Schlecht.) House
Stellaria biflora L.

Densely caespitose, glandular-pubescent, perennial herb from a branched, sometimes woody caudex; stems decumbent, with old, dried leaves near the base. Leaves linear, obtuse to acute, glabrous to ciliate or glandular, the midvein forming a prominent keel. Flowers solitary or sometimes 2–3 on branches. Sepals erect to slightly spreading, 3-nerved, oblong, obtuse, cucullate at the apex. Petals erect, white, spatulate, shallowly emarginate, equal to or longer than the sepals. Stamens 10. Styles 3. Capsule 3-valved, the valves entire, nearly twice as long as the sepals.

Fellfields, ridges, rocky slopes, and talus. Known from all alpine ranges except the Wyoming Range. Eastern Canada; Alaska south in western North America to Oregon, Idaho, and New Mexico.

Dwarfed, high alpine specimens with sepals 2–3 mm long and petals equaling the sepals have been called *Minuartia biflora* (L.) Schinz & Thell.

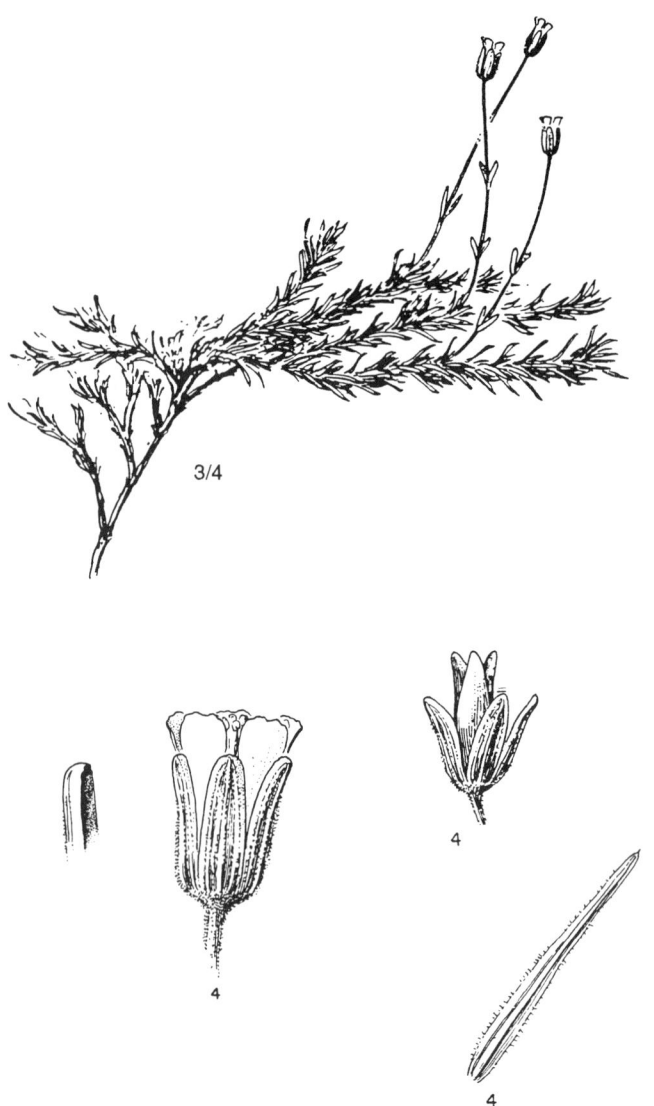

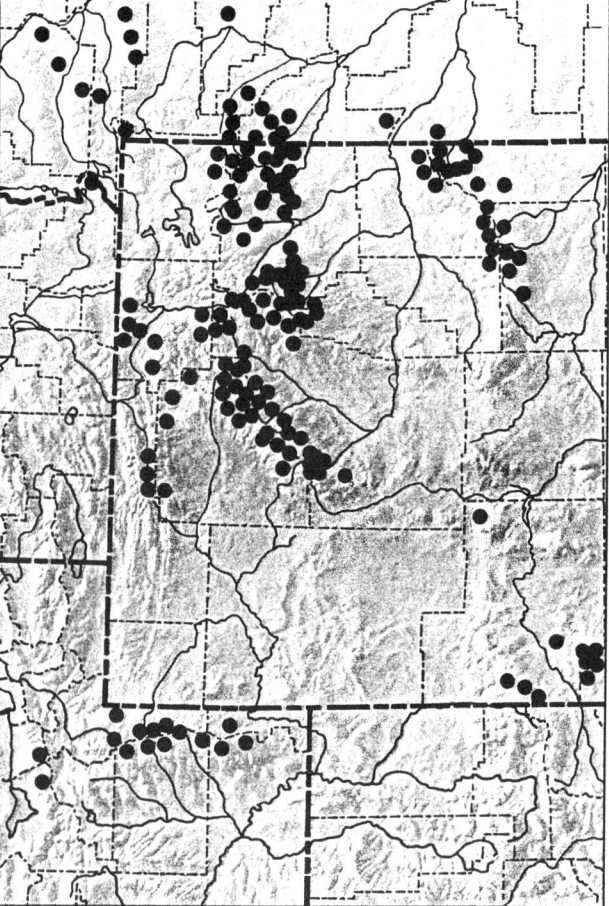

Minuartia rubella (Wahlenb.) Hiern
J. Bot. London 37:320, 1899

Red Sandwort

Alsine rubella Wahlenb.
Alsine hirta (Wormsk.) Hartm.
Alsinopsis propinqua (Richards.) Rydb.
Alsinopsis quadrivalvis (R. Br.) Rydb.
Arenaria aequicaulis (A. Nels.) A. Nels.
Arenaria propinqua Richards.
Arenaria quadrivalvis R. Br.
Arenaria rubella (Wahlenb.) Smith
Minuartia propinqua (Richards.) House
Minuartia quadrivalvis (R. Br.) House
Sabulina propinqua (Richards.) Rydb.
Tryphane rubella (Wahlenb.) Reichenb.

Caespitose, perennial herb from a slender taproot; stems decumbent to prostrate, glandular-pubescent, with old dried leaves at the base. Leaves linear to linear-lanceolate, acute to broadly acute, 3-nerved, often with secondary leaves in the axils. Flowers mostly in open cymes, occasionally solitary. Sepals lanceolate, acuminate, 3-nerved, scarious-margined. Petals white, obovate, mostly shorter than the sepals. Stamens 10. Styles 3. Capsule 3-valved, the valves entire, equal to or longer than the sepals.

Fellfields in most alpine ranges of the Middle Rocky Mountains. Not yet known from above timberline in the Gros Ventre and Hoback Ranges. Circumpolar; Alaska south to California, Nevada, and New Mexico.

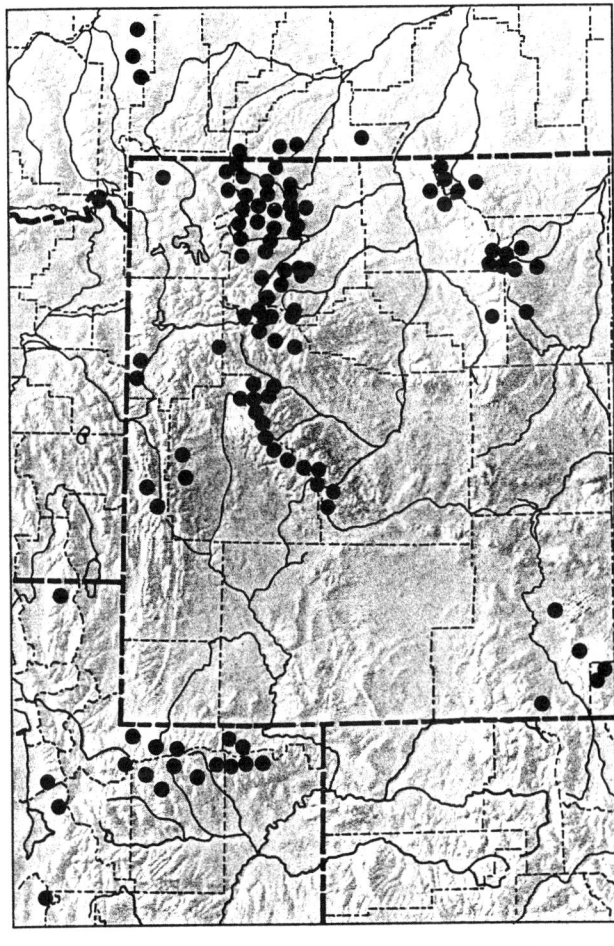

Paronychia Miller
Nailwort, Whitlow-wort

Ours are low, tufted, mat-forming, perennial herbs from a branched caudex. Leaves opposite, crowded, acute to subulate with scarious stipules. Flowers small, inconspicuous, single or clustered in few-flowered, terminal cymes. Sepals 5, cucullate, united at the base, awn-tipped. Petals lacking or minute. Stamens 5, opposite the sepals, sometimes alternating with staminodia. Ovary 1-celled, the style 2-parted near the apex. Fruit an ovoid, 1-seeded utricle.

Paronychia pulvinata Gray
Proc. Acad. Sci. Phil. 1863:58, 1864.

Rocky Mountain Nailwort

Low, densely caespitose, mat-forming perennial, from a branched, woody caudex. Leaves ovate, oblong, or elliptic, obtuse. Stipules ovate, obtuse, nearly as long as the leaves. Flowers bracteate, solitary in the leaf axils. Sepals oval, cuspidate. Petals lacking.

Known in the Middle Rocky Mountains only from the Medicine Bow and Uinta Ranges. Endemic to the high mountains of southern Wyoming, Colorado, and Utah.

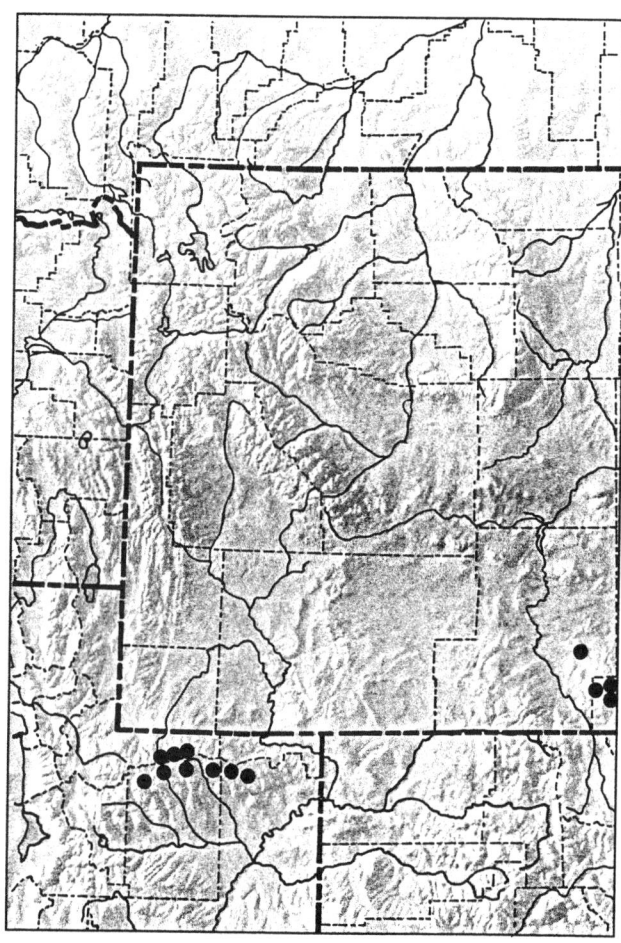

Sagina L.
Pearlwort

Annual or perennial herbs, often matted, with opposite, linear or subulate, basally connate leaves. Flowers white, perfect, long-pediceled, in terminal or axillary cymes. Sepals 4 or 5, distinct. Petals entire to retuse, the same number as the sepals, or lacking. Stamens and styles usually as many as the sepals. Capsule dehiscing by 4 or 5 valves.

Sagina saginoides (L.) Karst.
Deutsche. Fl. Pharm. Bot., 539, 1881.

Arctic Pearlwort

Spergula saginoides L.
Alsine linnaei (Presl) Krause
Alsinella saginoides (L.) Greene
Sagina linnaei Presl
Sagina micrantha (Bunge) Fern.
Spergella saginoides (L.) Reichenb.

Low, tufted, perennial herb, less than 6 cm tall, from a slender taproot; stems glabrous, decumbent, often rooting at lower nodes. Leaves linear, subulate. Flowers mostly solitary, the pedicels recurved at fruiting. Sepals oval, obtuse, glabrous, erect to slightly spreading. Petals white, entire, shorter than the sepals. Stamens 10. Styles 5. Capsule 5-valved, the valves entire, exceeding the sepals.

Moist tundra, nivation basins, and stream banks in the Absaroka, Beartooth, Big Horn, Medicine Bow, Teton, Uinta, Wasatch, and Wind River Ranges. Circumpolar; south in western North America to California and Mexico.

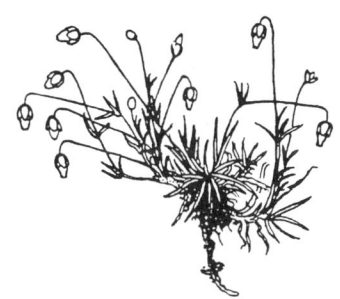

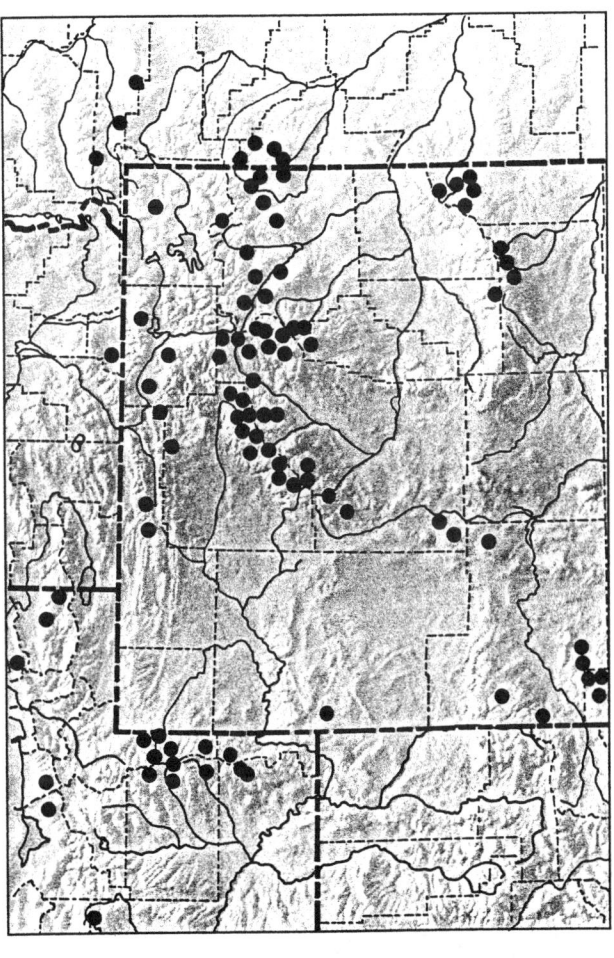

Silene L.
Catchfly, Campion

Erect or caespitose, annual or perennial herbs with exstipulate leaves. Flowers white or pink in ours, solitary or cymose. Calyx tubular, 5-toothed, usually 10-nerved. Petals 5, long-clawed, the blade bifid or sometimes 4-cleft, usually appendaged at the base. Stamens 10. Ovary 1-celled, rarely 2–4-celled, borne on a stipe (carpophore), the styles usually 3. Capsule dehiscent by 3 or 6 apical teeth.

KEY TO THE SPECIES OF *SILENE*

1. Plants densely caespitose, less than 5 cm tall; calyx glabrous — *S. acaulis*
1. Plants erect, more than 10 cm tall; calyx pubescent and often glandular
   2. Inflorescence an open, leafy cyme, the flowers less than 10 mm long; calyx less than 8 mm long; claw of petal about 5 mm long — *S. menziesii*
   2. Inflorescence a contracted cyme with reduced bracts, the flowers more than 12 mm long; calyx more than 9 mm long; claw of petal about 8–12 mm long
      3. Blades of petals 2-lobed; appendages linear to oblong, entire — *S. douglasii*
      3. Blades of petals 4-lobed; appendages obovate, erose — *S. parryi*

Silene acaulis (L.) Jacq.
Enum. Stirp. Vindob. 78, 242, 1762.

Moss Campion

Cucubalis acaulis L.
Silene exscapa Allioni

Low, densely pulvinate, perennial herb, less than 6 cm tall, from a branched caudex and stout, woody taproot. Stems short, crowded, with old dried leaves at the base. Leaves linear to lanceolate, glabrous, the margins glandular-ciliate. Flowers solitary on short peduncles or sessile, often imperfect. Calyx tubular to campanulate, glabrous, purplish, 10-nerved. Petals purple, rarely pinkish or whitish, clawed, the blade entire or bilobed. Stamens 10. Styles 3. Capsule elliptic, 3-loculed, exceeding the calyx.

Fellfields, rocky slopes, and ridges in all alpine ranges of the Middle Rocky Mountains except the Hoback Range. Circumpolar; south in western North America to Oregon, Nevada, Arizona, and New Mexico.

Rocky Mountain plants from northern Wyoming south to New Mexico and Arizona with a calyx 8–11 mm long and the blade of the petals oblong-oblanceolate, rounded to slightly emarginate are var. *subacaulescens* (F.N. Williams) Fern. & St. John.

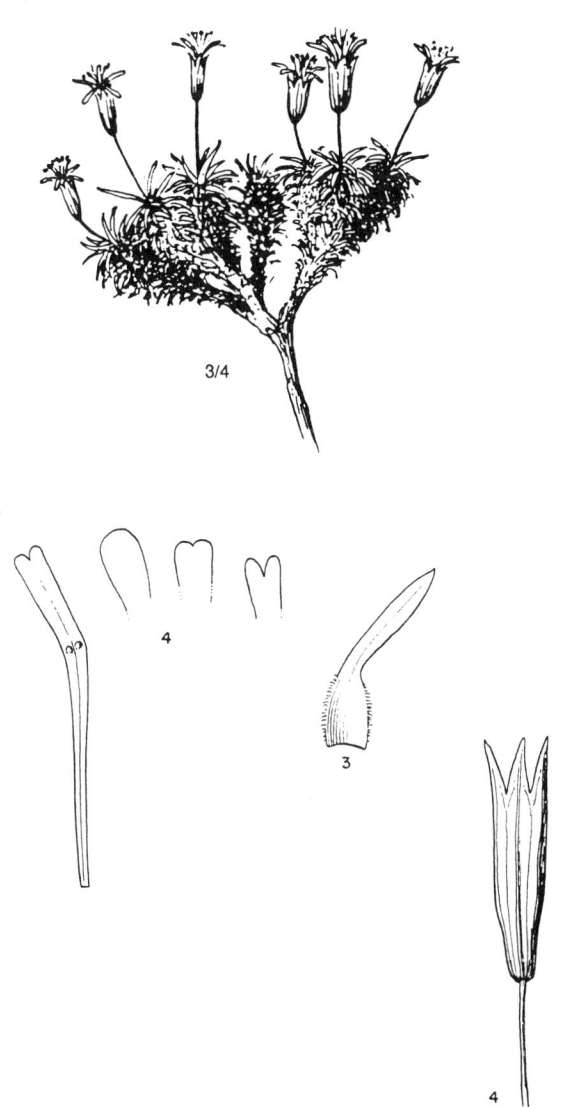

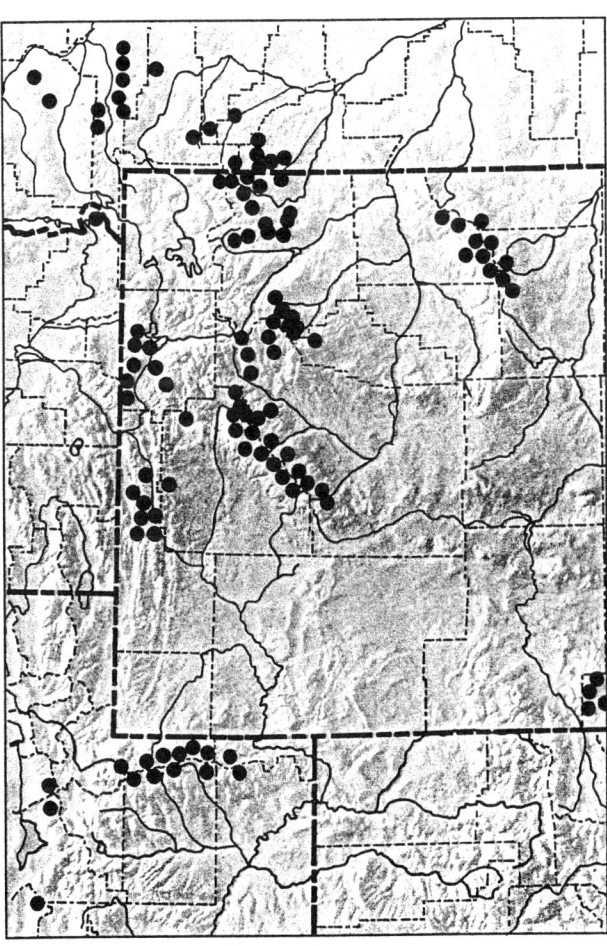

Silene douglasii Hook.
Fl. Bor.-Am. 1:88, 1830.

Douglas Campion

Silene dilatata Suksd.
Silene lyallii S. Wats.
Silene macrocalyx (Robins.) T.J. Howell
Silene monantha S. Wats.
Silene multicaulis Nutt.
Silene oraria M.E. Peck

Perennial from a taproot and branched caudex; stems simple, several to many, erect or decumbent-spreading, pubescent, the hairs mostly short-retrorse. Basal leaves reduced in size, early deciduous; cauline leaves opposite, reduced upward, petiolate, the blades oblanceolate to lanceolate or linear-lanceolate, glabrous to puberulent, rarely somewhat glandular, acute, the margins entire. Flowers terminal and solitary or 2 to several in short cymes. Calyx tubular, 10-nerved, becoming papery, inflated and campanulate in fruit, the teeth broadly ovate to nearly orbicular. Corolla white, greenish, or pinkish, the petals clawed with a bilobed blade; appendages 2, linear to oblong. Stamens 10. Stalk of the ovary (carpophore) puberulent; styles mostly 3. Capsule elliptic, 1-loculed; seeds brownish, papillate.

Talus, scree, and rocky ledges of the Wasatch Range. British Columbia south and east to California, Nevada, Utah, and Wyoming.

Middle Rocky Mountain plants that are densely pubescent, but mostly lacking glandular hairs are var. *douglasii*. At least some high altitude specimens with glandular hairs appear to be hybrids with *S. parryi*.

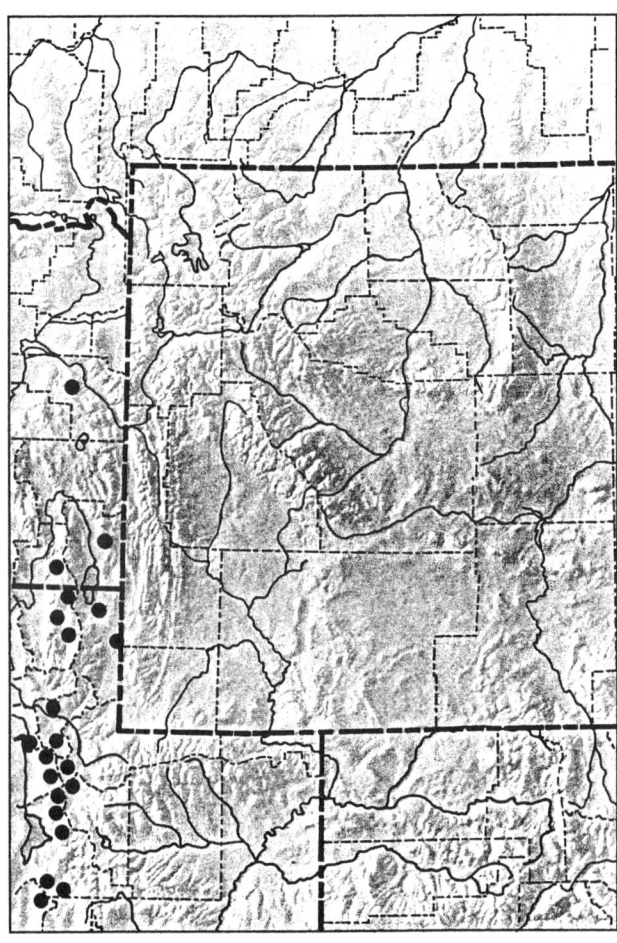

Silene menziesii Hook.
Fl. Bor.-Am. 1:90, 1830.

Menzies Catchfly

Anotites dorrii (Kell.) Greene
Anotites halophila Greene
Anotites latifolia Greene
Anotites macilenta Greene
Anotites menziesii (Hook.) Greene
Anotites nodosa Greene
Anotites picta Greene
Anotites tereticaulis Greene
Anotites viscosa Greene
Silene dorrii Kell.
Silene obovata Pors.
Silene stellarioides Nutt.
Silene williamsii Britt.

Erect to somewhat decumbent, perennial herb from slender rhizomes; stems simple or branched, pubescent and usually glandular, at least above. Leaves mostly cauline, lanceolate, elliptic, or oblanceolate, acute to acuminate, usually puberulent and ciliolate. Flowers imperfect, in terminal, open, leafy cymes. Calyx campanulate, glandular-pubescent, weakly 10-nerved, the teeth acute. Petals white, clawed, the blade bilobed, sometimes with small lateral teeth. Stamens 10. Styles 3. Capsule elliptic, 6-toothed, about as long as the calyx.

This lowland species occurs in the alpine zone of the Absaroka Range as var. *viscosa* (Greene) C.L. Hitchc. and Maguire. Alaska south to California and New Mexico.

Var. *menziesii*, distinguished from var. *viscosa* by the lack of glandular hairs on the lower stem, might also be expected in the vicinity of timberline.

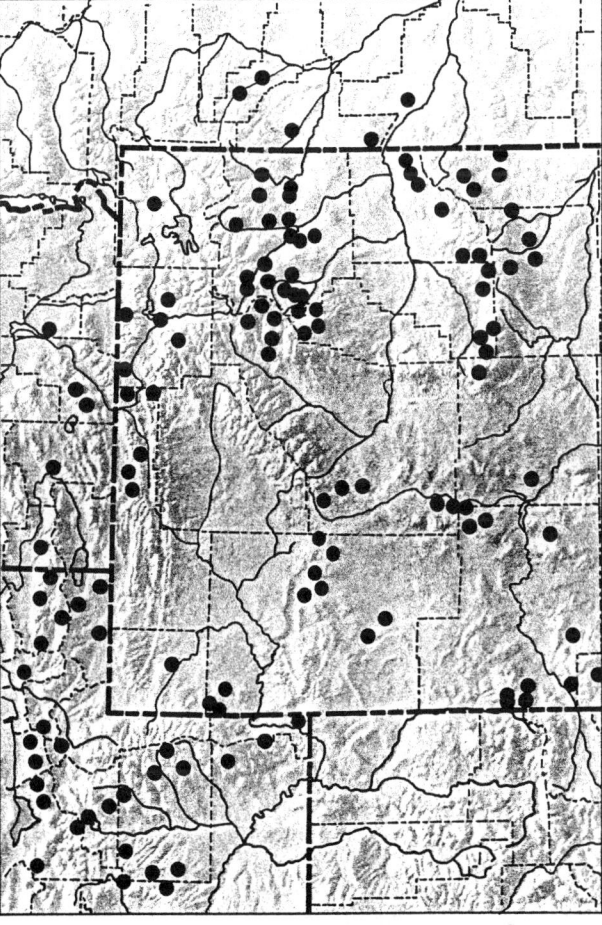

Silene parryi (S. Wats.) C.L. Hitchc. & Maguire
Univ. Wash. Pub. Biol. 13:36, 1947.

Parry Catchfly

Lychnis parryi S. Wats.
Lychnis elata S. Wats.
Silene macounii S. Wats.
Silene tetonensis E. Nels.
Silene tetragyna Suksd.
Wahlbergella parryi (S. Wats.) Rydb.

Erect, perennial herb from a branched caudex; stems 1 to several, simple, pubescent and glandular, at least above. Basal leaves linear to oblanceolate, petiolate, puberulent and often glandular; cauline leaves oblanceolate, in 2–4 pairs, reduced upward. Flowers in contracted cymes in the upper leaf axils. Calyx tubular-campanulate, glandular-pubescent, inflated in fruit, with 10 prominent nerves. Petals whitish or purplish, the claw erose above; blades nearly equally 4-lobed; appendages obovate, erose. Stamens 10, included or nearly so. Styles 3, usually exerted. Capsule elliptic, 1-celled.

Rocky slopes of most of the alpine ranges of the Middle Rocky Mountains. Not yet known from above timberline in the Big Horn and Wasatch Ranges. British Columbia and Alberta south to Idaho and Wyoming.

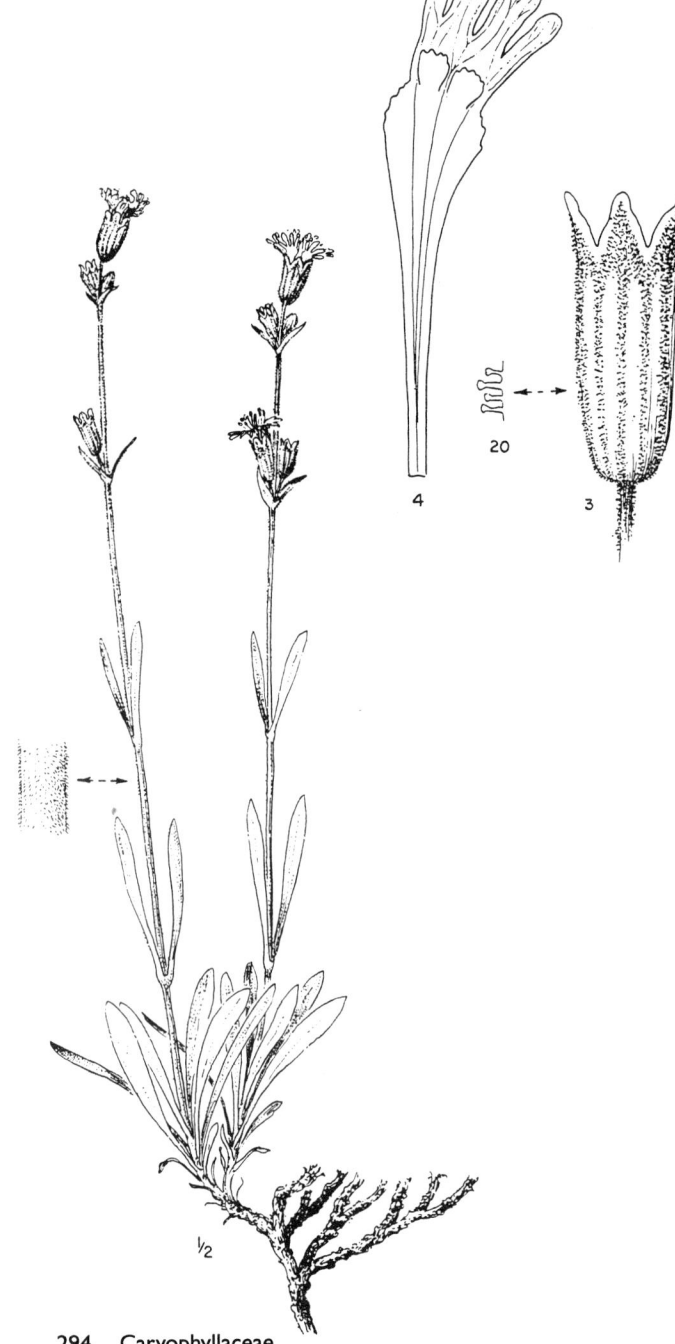
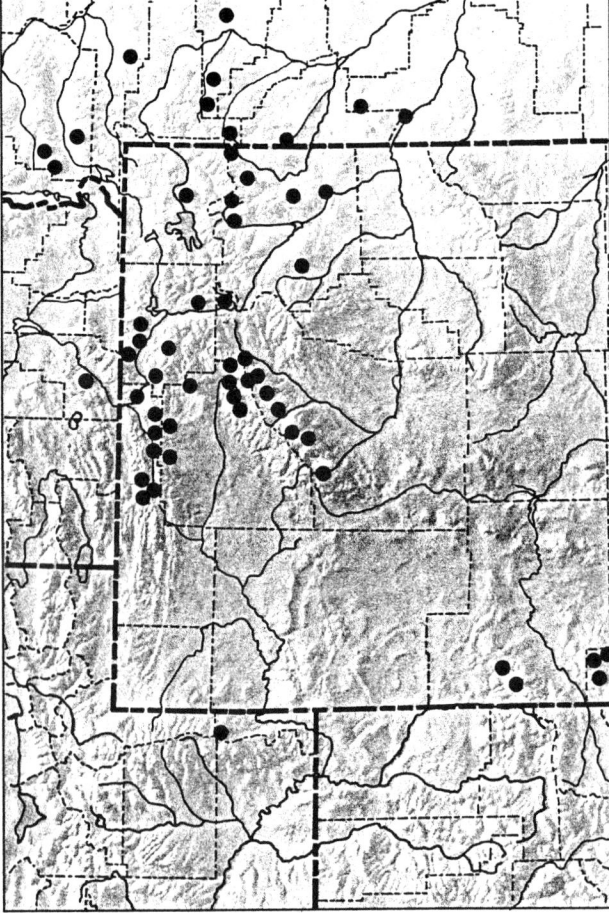

Spergularia (Pers.) J. & K. Presl
Sand Spurry

Low, annual or perennial herbs with opposite, linear, fleshy, stipulate leaves. Flowers white to pink, in open, terminal, leafy cymes. Sepals 5, distinct, scarious-margined. Petals usually 5 (rarely lacking), entire, shorter than or equaling the sepals. Stamens 2–10. Styles 3, distinct. Ovary 1-celled. Capsule 3-valved, dehiscent to base. Seeds smooth to papillate, sometimes with an entire to erose wing.

Spergularia rubra (L.) J. & K. Presl
Fl. Cechica, 94, 1819.

Red Sand Spurry

Arenaria rubra L.
Alsine rubra (L.) Crantz
Buda rubra (L.) Dumort
Corion rubrum (L.) N.E. Brown
Fasciculus ruber (L.) Dulac
Lepigonum rubrum (L.) Wahlb.
Melargyra rubra (L.) Raf.
Spergula rubra (L.) Dietr.
Stipularia rubra (L.) Haw.
Tissa rubra (L.) Britt.

Low, prostrate, spreading annual from a taproot. Leaves linear, glabrous, fascicled, mucronate. Stipules thin, membranous, conspicious, acuminate. Flowers pinkish, solitary in the leaf axis. Sepals lanceolate, glandular. Petals obtuse, shorter than the sepals. Stamens 6–10. Capsules barely exceeding the calyx. Seeds dark brown, minutely papillate.

A common annual weed in many habitats, *S. rubra* occurs at timberline in the Big Horn and Medicine Bow Ranges. Europe; South America; widely distributed in western North America.

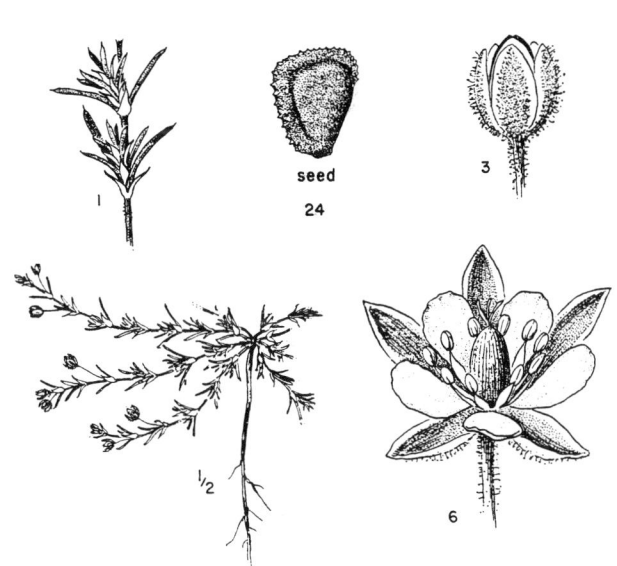
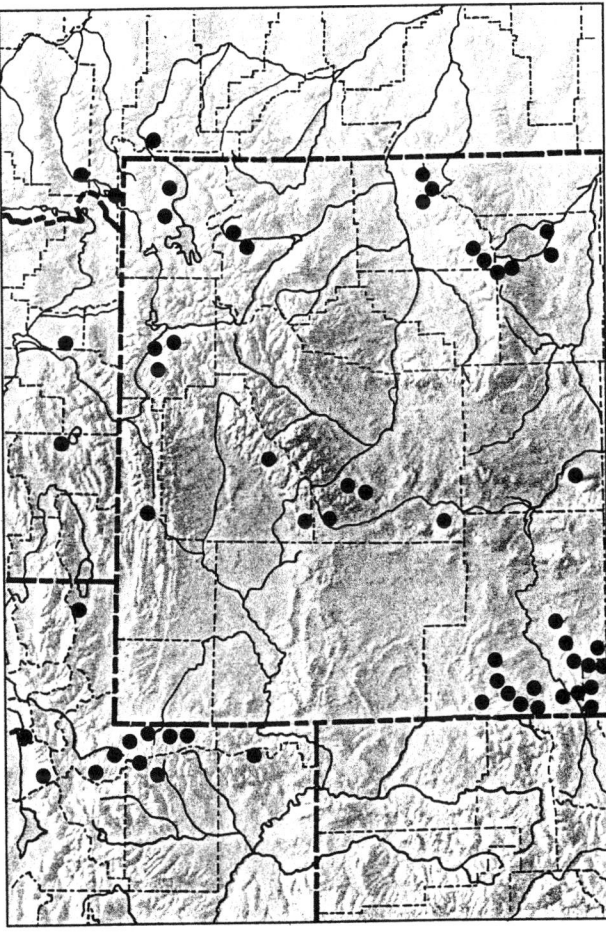

Stellaria L.
Starwort

Slender, annual or perennial herbs with weakly ascending stems, the leaves opposite, exstipulate. Flowers white, solitary or cymose. Sepals 5, distinct or nearly so. Petals commonly 5, bifid or parted nearly to the base. Stamens 10 or fewer. Ovary 1-celled, the styles usually 3. Capsule globose to oblong, dehiscent by twice as many valves as there are styles.

KEY TO THE SPECIES OF *STELLARIA*

1. Leaves about as long as broad; sepals obtuse; flowers solitary in leaf axils ... *S. obtusa*
1. Leaves distinctly longer than broad; sepals acute; flowers solitary or in terminal or axillary cymes
    2. Petals shorter than the sepals, or lacking; flowers in terminal, leafy or scarious-bracteate cymes
        3. Cymes scarious-bracteate, the bracts much smaller than the upper leaves; pedicels reflexed in age; leaves oblong-lanceolate ... *S. umbellata*
        3. Cymes leafy-bracteate, the bracts about the same size as the upper leaves; pedicels erect or ascending; leaves ovate to narrowly lanceolate ... *S. calycantha*
    2. Petals equaling or exceeding the sepals; flowers solitary, axillary, or cymose
        4. Bracts of the inflorescence scarious; leaves linear to linear-lanceolate; sepals 4–6 mm long ... *S. longipes*
        4. Bracts of the inflorescence green, foliaceous; leaves lanceolate to oblong-lanceolate; sepals 2–3 mm long ... *S. crassifolia*

Stellaria calycantha (Ledeb.) Bong.
Mem. Acad. St. Petersb., VI, 2:127, 1832.

Northern Starwort

Arenaria calycantha Ledeb.
Alsine borealis (Bigel.) Britt.
Alsine brachypetala (Bong.) T.J. Howell
Alsine oxyphylla (Robins.) A. Heller
Alsine simcoei T.J. Howell
Arenaria alpestris (Fries) Rydb.
Micropetalon lanceolatum (Michx.) Pers.
Spergulastrum lanceolatum Michx.
Stellaria alpestris Fries
Stellaria borealis Bigel.
Stellaria brachypetala Bong.
Stellaria longifolia Bong.
Stellaria oxyphylla Robins.
Stellaria simcoei (T.J. Howell) C.L. Hitchc.
Stellaria sitchana Steud.
Stellularia borealis (Bigel.) Kuntze

Slender, perennial herb from rhizomes; stems glabrous, erect to prostrate, simple or branched, angled. Leaves ovate to narrowly lanceolate, entire, sessile, ciliate at the base. Flowers axillary and in terminal leafy-bracteate cymes. Sepals 3-nerved, lanceolate, acute, scarious-margined. Petals white, shorter than the sepals, or lacking. Capsule ovoid, dark brown, longer than the sepals.

Meadows and timberline margins of the Absaroka, Beartooth, and Wind River Ranges. Circumpolar; south in western North America to California, Utah, and Colorado.

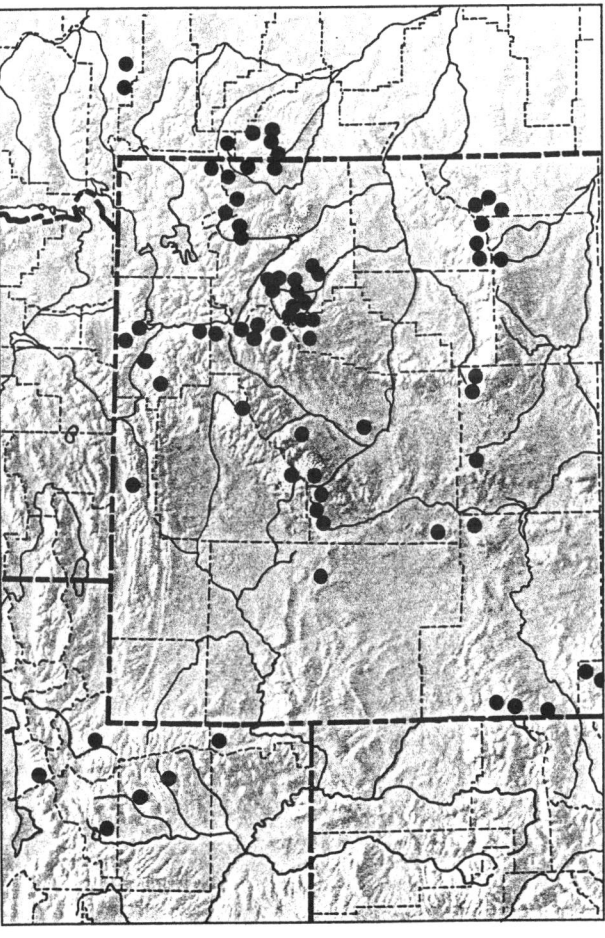

Stellaria crassifolia Ehrh.
Hannov. Bot. Mag. 8:116, 1784.

Thickleaf Starwort

Alsine crassifolia (Ehrh.) Britt.

Slender, perennial herb from rhizomes; stems branched, angled, weakly ascending, decumbent, or prostrate. Leaves sessile, lanceolate, elliptic, or oblong-lanceolate, acute to obtuse. Flowers solitary in leaf axils, often nodding; pedicels filiform. Sepals lanceolate, acute to acuminate, scarious-margined. Petals white, exceeding the sepals. Capsule oblong to ovoid, pale brown, longer than the sepals.

Wet meadows, alpine bogs, and snow beds of the Beartooth and Wind River Ranges. Circumboreal; south in western North America to Idaho and Colorado.

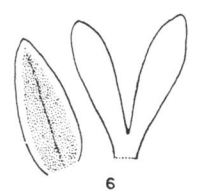

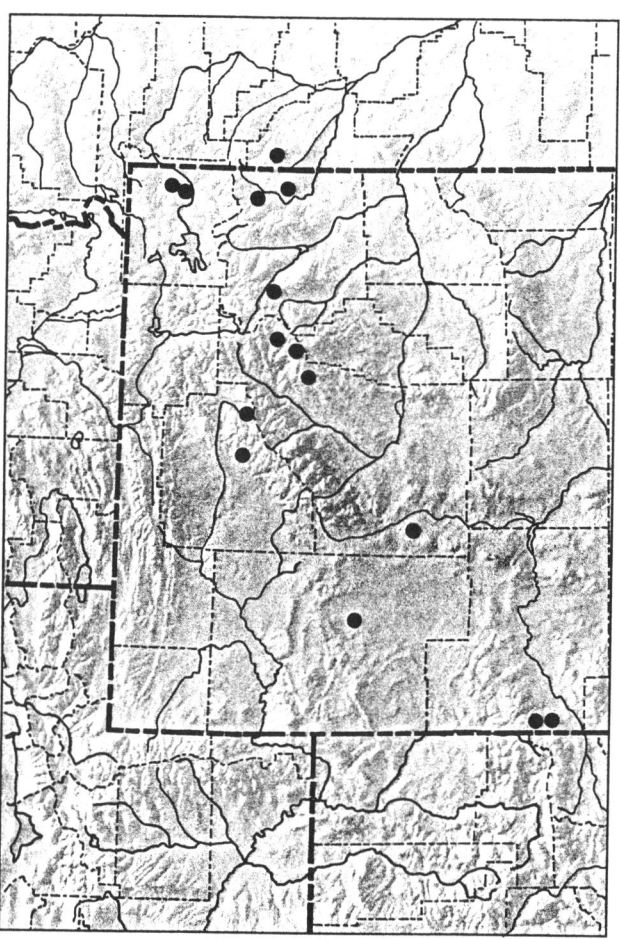

Stellaria longipes Goldie
Edinb. New Phil. J. 6:327, 1822.

Longstalk Starwort

Alsine laeta (Richards.) Rydb.
Alsine longipes (Goldie) Cov.
Alsine palmeri Rydb.
Alsine stricta (Richards.) Rydb.
Alsine strictiflora Rydb.
Alsine subvestita (Greene) Rydb.
Alsine validus Goodding
Stellaria arenicola Raup
Stellaria dulcis Gervais
Stellaria edwardsii R. Br.
Stellaria hultenii B. Boivin
Stellaria laeta Richards.
Stellaria laxmannii Fisch.
Stellaria monantha Hult.
Stellaria stricta Richards.
Stellaria strictiflora (Richards.) Macoun
Stellaria subvestita Greene
Stellularia longipes (Goldie) Macmill.

Mat-forming or tufted, perennial herb from rhizomes; stems erect to prostrate, angled, and glabrous, glabrate, or villous. Leaves sessile, linear-lanceolate to ovate, acute, glabrous and entire, often ciliate at the base, sometimes bluish green, glaucous. Flowers solitary and axillary or in leafy-bracteate cymes; bracts scarious-margined. Sepals lanceolate to ovate, glabrous, scarious-margined, 3-nerved, acute. Petals white, longer than the sepals. Capsule ovoid, dark brown to black or purplish, longer than the sepals.

Meadows and fellfields in most of the Middle Rocky Mountains. Not yet known from above timberline in the Hoback, Salt River, and Teton Ranges. Eurasia; transcontinental in Canada, south in western North America to California, New Mexico, Minnesota, and New York.

Alpine plants can be differentiated as follows:
1. Leaves blue-green, glaucous; flowers mostly single, terminal . . . var. *monantha* (Hult.) Welsh
1. Leaves yellow-green, not glaucous; flowers 1-several, solitary, axillary, or cymose . . . var. *longipes*

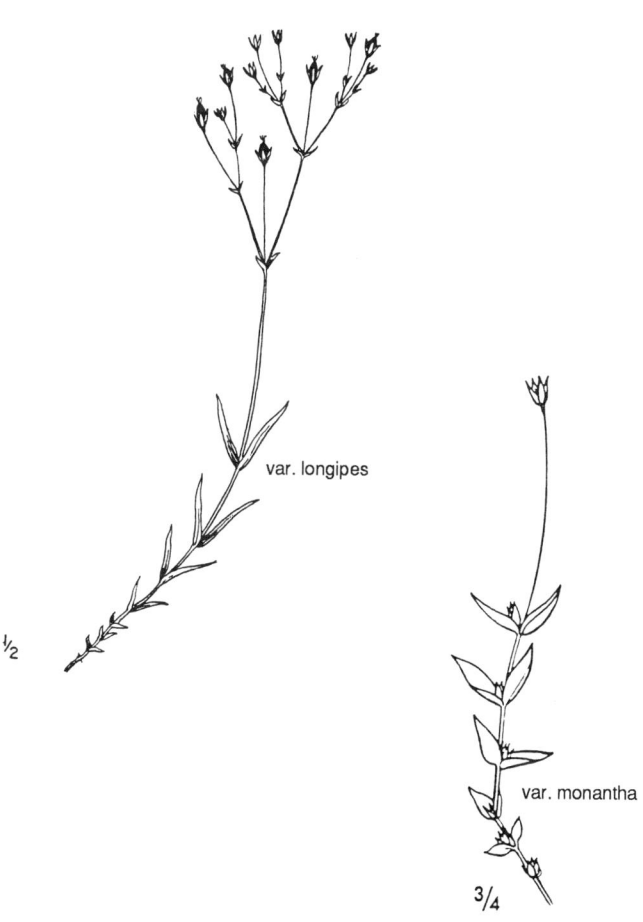

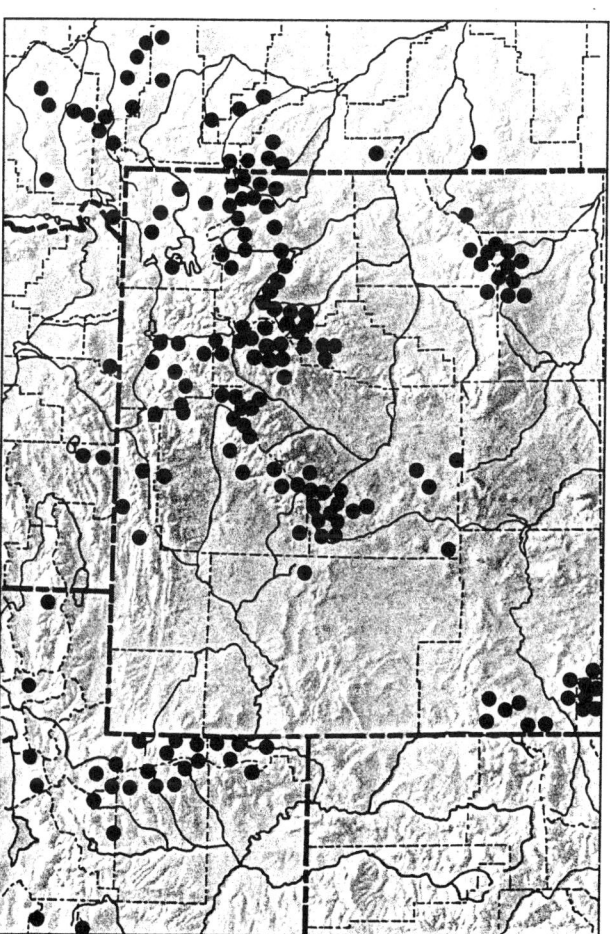

Stellaria obtusa Engelm.
Bot. Gaz. 7:5, 1882.

Blunt Starwort

Alsine obtusa (Engelm.) Rose
Alsine polygonoides Greene
Alsine viridula Piper
Alsine washingtoniana (B.L. Robins.) A. Heller
Stellaria viridula (Piper) St. John
Stellaria washingtoniana B.L. Robins

Low, mat-forming, perennial herb, less than 10 cm tall; stems decumbent or prostrate, glabrous, rooting at the lower nodes. Leaves ovate, acute, glabrous or ciliate at the base, the upper ones sessile, the lower ones often short-petiolate. Flowers solitary on filiform, glabrous pedicels in the leaf axils. Sepals ovate, obtuse. Petals lacking. Stamens 8–10. Styles 3–4. Capsule ovoid to globose, equaling or exceeding the sepals.

Stream banks and moist willow thickets of the Wasatch Range. British Columbia and Alberta south to California, Utah, and Colorado.

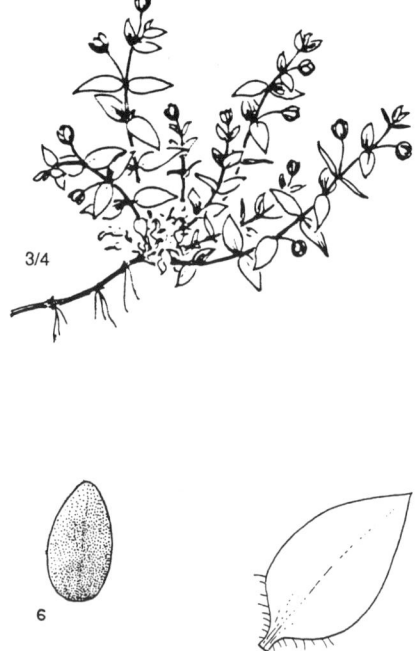

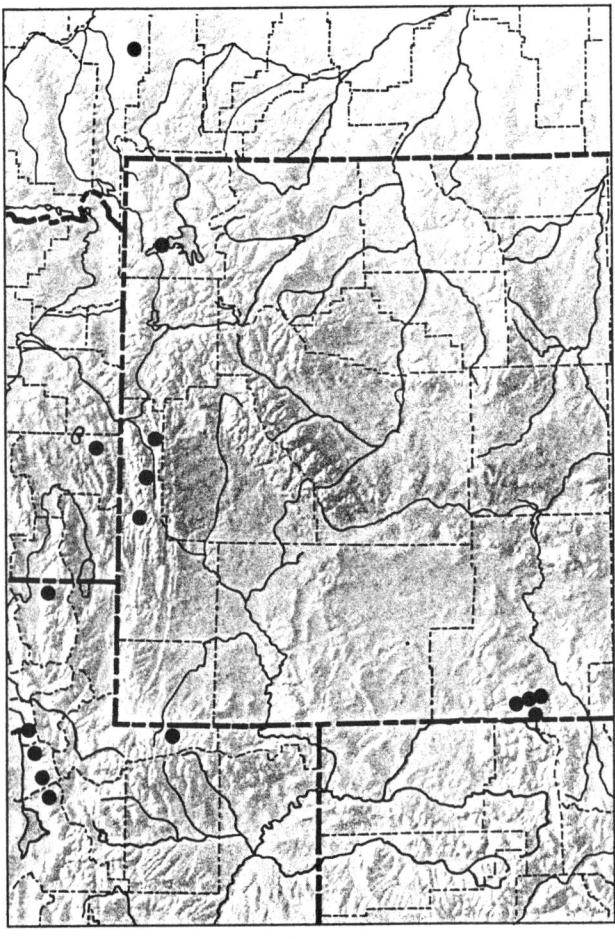

Stellaria umbellata Turcz. ex Kar. & Kir.
Bull. Soc. Imp. Nat. Mosc. 15:173, 1842.

Umbrella Starwort

Alsine baicalensis Cov.
Stellaria gonomischa B. Boivin
Stellaria weberi B. Boivin
Stellularia umbellata (Turcz. ex Kar. & Kir.) Kuntze

Slender, glabrous, perennial herb from rhizomes; stems erect or ascending, somewhat angled. Leaves sessile, elliptic to ovate, acute, sometimes ciliate at base. Flowers axillary and in scarious-bracteate, terminal, subumbellate cymes; pedicels capillary, somewhat reflexed with age. Sepals lanceolate to ovate, acute, scarious-margined. Petals minute or lacking. Capsule ovoid to oblong, pale, about twice as long as the sepals.

Moist meadows and slopes in most of the Middle Rocky Mountain alpine ranges. Not yet known from above timberline in the Hoback, Salt River, and Wyoming Ranges. Siberia; Alaska south to California, Idaho, and Colorado.

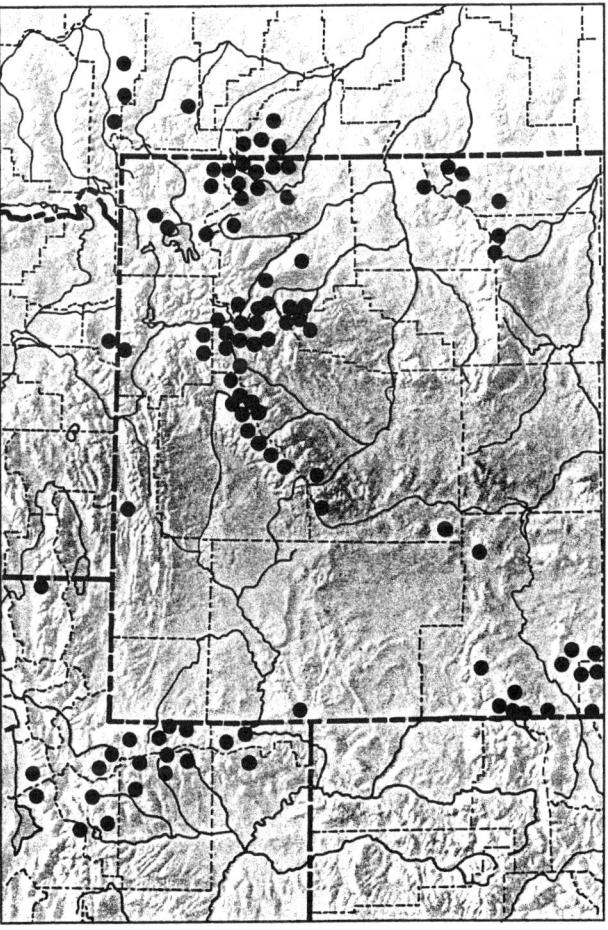

Celastraceae
Staff Tree Family

Trees, shrubs, or woody vines with simple, opposite or alternate, petiolate, deciduous or evergreen leaves; stipules lacking or small and early deciduous. Flowers small, perigynous, regular, 4- or 5-merous, perfect or unisexual, in axillary cymes, racemes, or panicles. Sepals united at the base; petals distinct, rarely lacking. Stamens 4 or 5. Pistil 1, with 2–5 locules. Style 1. Stigma 2–5-lobed. Fruit a capsule, the seeds usually enclosed by a fleshy aril.

Paxistima Raf.
Mountain Lover

Glabrous, prostrate, evergreen shrub. Leaves opposite, short-petiolate, serrulate, at least above the middle. Flowers perfect, regular, in axillary clusters. Sepals 4, petals 4, maroon to lavender. Stamens 4. Pistil 1, the ovary 2-celled. Capsule ovoid; seeds mostly 2.

Paxistima myrsinites (Pursh) Raf.
Sylva Tell., 42, 1838.

Mountain Lover

Ilex myrsinites Pursh
Myginda myrtifolia Nutt.
Oreophila myrtifolia (Nutt.) Nutt. in T. & G.
Paxistima macrophylla Farr
Paxistima myrsinites (Pursh) Wheeler
Paxistima schaefferi Farr

Low, mostly prostrate, glabrous, evergreen shrub with opposite, glossy leaves, the blades short-petiolate, elliptic to obovate, the margins serrate. Flowers inconspicuous in axillary cymes. Sepals 4. Petals 4, lavender, maroon, or dark red. Stamens 4. Pistil with 2 locules. Capsule 2-valved, ovoid; seeds brown with a white aril.

A species typical of forests and forest margins at lower elevations, *P. myrsinites* has been collected near timberline from the lower alpine zone of the Gros Ventre, Salt River, Uinta, and Wasatch Ranges. It might be expected as occasional in the lower alpine zone at other favorable locations.

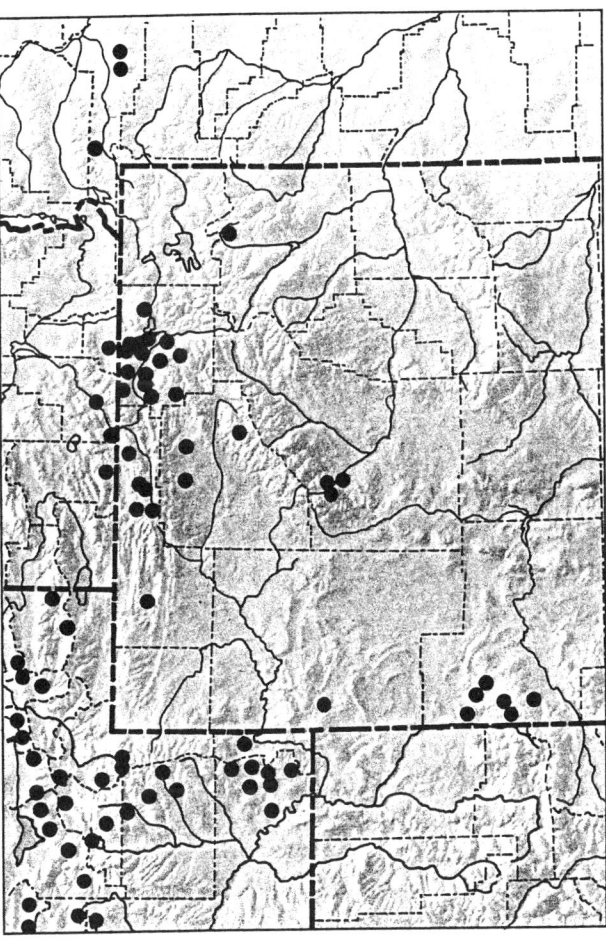

Chenopodiaceae
Goosefoot Family

Annual to perennial herbs or shrubs, often with succulent stems. Leaves alternate or opposite, exstipulate, glabrous or pubescent, often farinose with inflated or scalelike hairs. Flowers inconspicuous, apetalous, perfect or unisexual, glomerate in leaf axils, or in spikes, racemes, panicles. Calyx greenish, mostly 5-lobed or 5-parted, sometimes reduced to a single scale, sometimes lacking. Stamens as many as the calyx lobes and opposite them, or sometimes fewer. Pistil 1, superior, the ovary 1-celled, with 1 ovule. Styles mostly 1–3. Fruit a utricle, with a single seed.

KEY TO THE GENERA OF CHENOPODIACEAE

1. Sepals 3–5, partially enclosing the fruit — *Chenopodium*
1. Sepals 1–3, occasionally lacking, not enclosing the fruit — *Monolepis*

Chenopodium L.
Lamb's Quarter, Goosefoot

Annual to perennial, usually branched herbs, often farinose or glandular. Leaves alternate, usually petiolate, linear to deltoid, entire to toothed, lobed, or hastate. Flowers greenish, perfect, in glomerules, these in terminal or axillary spikes or panicles. Perianth fleshy, 5-lobed, often keeled and usually enclosing the fruit. Stamens mostly 5, opposite the perianth lobes. Styles usually lacking; stigmas 2–5. Fruit a utricle; seed 1, lenticular.

KEY TO THE SPECIES OF *CHENOPODIUM*

1. Fruits laterally flattened, the seeds vertical; leaves hastately lobed — *C. capitatum*
1. Fruits dorsally flattened, the seeds horizontal; leaves entire or nearly so — *C. fremontii*

Chenopodium capitatum (L.) Asch.
Fl. Brandenb. 1:572, 1864.

Strawberry Blite

Blitum capitatum L.
Blitum hastatum Rydb.
Chenopodium overi Aellen
Morocarpus capitatus Scop.

Glabrous, erect annual, with stems single or branched from the base. Leaves fleshy, with triangular-hastate blades, the margins entire, shallowly lobed, or sinuate-dentate; petioles of cauline leaves shorter than the blades. Flowers in terminal spikes and densely glomerate in the upper leaf axils. Calyx lobes 3–5, ovate, becoming red and fleshy at maturity, or remaining greenish or pinkish, but not fleshy. Stamens 3–4; stigmas 2. Fruits vertical, laterally flattened, the seed erect, oblong.

Scree slopes and rocky ledges in the Absaroka, Uinta, Wasatch, and Wyoming Ranges. Eurasia; transcontinental in Canada, south in the West to California, Nevada, and New Mexico.

Alpine plants with glomerules less than 5 mm in diameter and leaf margins entire or nearly so are var. *parvicapitatum* S.L. Welsh.

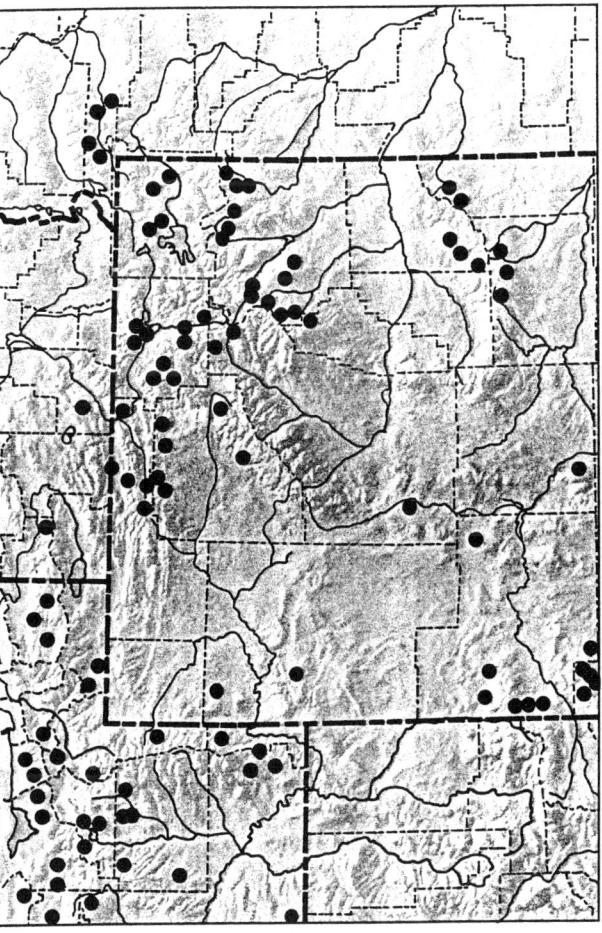

Chenopodium fremontii S. Wats.
Bot. King Exped., 287, 1871.

Fremont Goosefoot

Botrys fremontii (S. Wats.) Lunell
Chenopodium aridum A. Nels.
Chenopodium atrovirens Rydb.
Chenopodium hians Standl.
Chenopodium incanum (S. Wats.) A. Heller
Chenopodium incognitum H.A. Wahl.
Chenopodium watsonii A. Nels.
Chenopodium wolfii Rydb.

Annual from a slender taproot; stems simple or, more commonly, branched, glabrous to sparsely farinose. Leaves petiolate, the blades lanceolate to elliptic, entire or nearly so in ours, glabrous and green above, farinose and grayish green below, obtuse, becoming reduced upward. Inflorescence farinose, the flowers in terminal and lateral, often glomerulate spikes. Calyx lobes 5, ovate, farinose, keeled on the midnerve. Stamens 5; stigmas 2. Fruits horizontal, dorsally flattened; the seed horizontal, black, shiny, slightly rugulose to smooth.

Timberline and krummholz margins of the Wasatch and Wind River Ranges. British Columbia and Alberta south to Mexico, Texas, and Iowa.

Alpine plants with lanceolate to elliptic leaves and margins entire or nearly so are var. *atrovirens* (Rydb.) Fosberg.

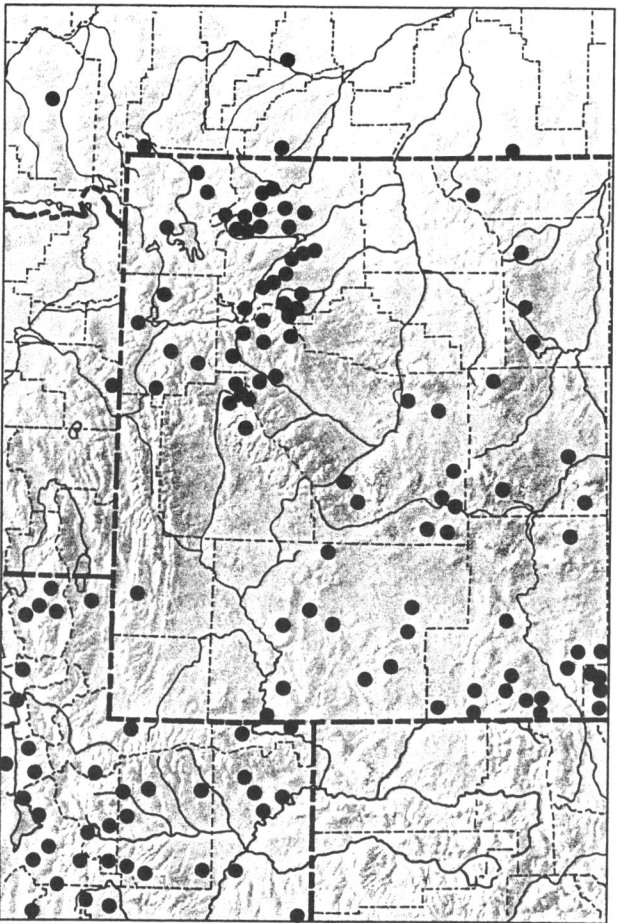

Monolepis Schrad.
Povertyweed

Annual herbs, mostly farinose, with branched, often spreading or prostrate stems. Leaves alternate, fleshy, sessile to petiolate, simple, entire to hastate or 3-lobed or 3-parted. Flowers greenish, perfect, and unisexual, in glomerules, these in terminal or axillary spikes or panicles. Perianth of 1–3 bractlike scales or lacking, the scales not enclosing the fruit. Stamens 1 or lacking; stigmas 2. Fruit a flattened utricle; seed 1, erect, lenticular.

Monolepis nuttalliana (Schult.) Greene
Fl. Franc., 168, 1891.

Nuttall Povertyweed

Blitum nuttallianum Schult. in R. & S.
Blitum chenopodioides Nutt.
Monolepis chenopodioides (Nutt.) Moq. in DC.

Prostrate to spreading, often purplish-tinged, annual herb from a slender taproot; stems simple or branched near the base, glabrate to farinose. Leaves fleshy, petiolate, the blades 3-parted to hastate, glabrate to farinose, reduced upward. Flowers greenish to reddish, in dense axillary glomerules, perfect or pistillate. Calyx lobes 1–3, oblong to oblanceolate, acute, or rarely lacking. Stamens 1, rarely 2, in perfect flowers, lacking in pistillate flowers; stigmas 2. Fruits vertical, laterally flattened; seed erect, dark brown to black.

Scree, talus, and rocky ledges in the lower alpine zone of the Wasatch Range. Alaska and Yukon Territory south in western North America to New Mexico and Texas, east to Manitoba and Missouri.

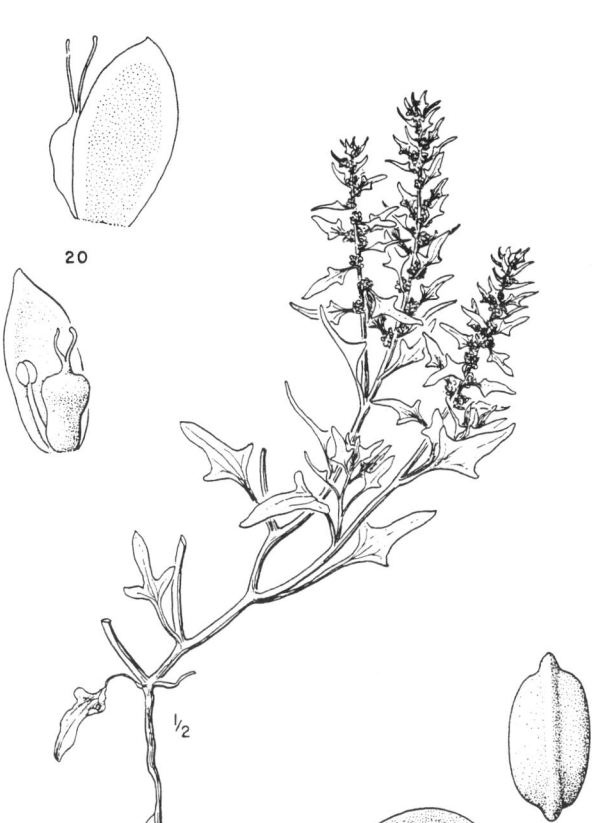

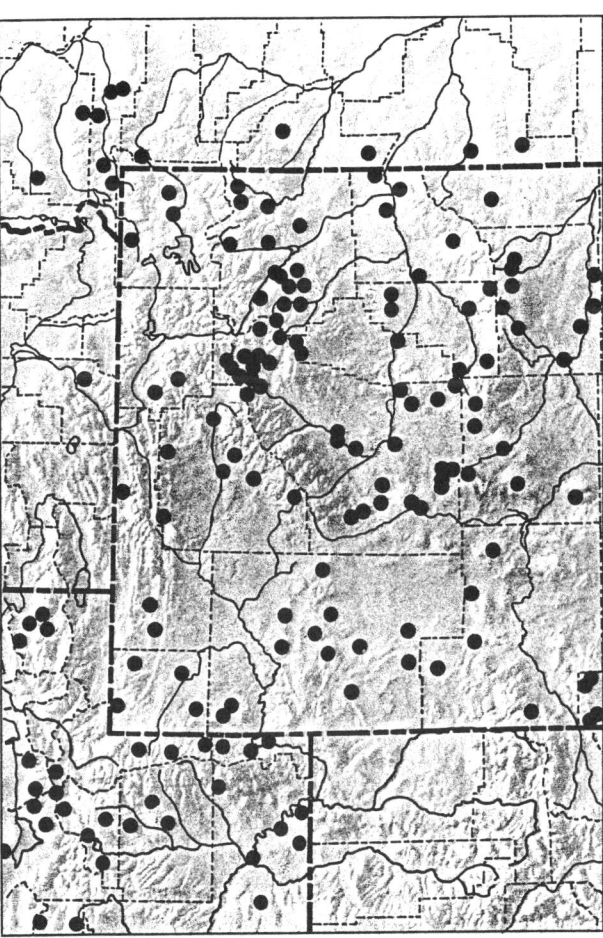

Crassulaceae
Orpine Family

Ours are succulent herbs with simple, alternate or opposite, fleshy, exstipulate leaves. Flowers hypogynous, cymose, regular, perfect, 4–5-merous. Sepals free or united. Petals the same number as the sepals, distinct or somewhat united. Stamens as many or twice as many as the petals, inserted on the corolla tube in gamopetalous flowers. Carpels 5, usually distinct, or somewhat united at the base, each carpel associated with a glandular appendage. Fruit a group of follicles, rarely a capsule.

Sedum L.
Stonecrop

Perennial herbs with alternate or whorled, simple, succulent, glabrous leaves. Flowers red, pink, or yellow, in terminal cymes. Sepals and petals 4 or 5, the sepals united at the base. Stamens 8 or 10, free or adnate to the petals below. Carpels as many as the petals, distinct. Fruit a group of 4 or 5 many-seeded follicles.

KEY TO THE SPECIES OF *SEDUM*

1. Flowers pink or red; leaves mostly cauline
    2. Petals light pink to rose, 6–9 mm long; leaves oblanceolate or narrower — *S. rhodanthum*
    2. Petals deep red or purplish, less than 5 mm long; leaves obovate to oblong — *S. rosea*
1. Flowers yellow; leaves mostly basal
    3. Cauline leaves mostly opposite; leaves broadly obovate — *S. debile*
    3. Cauline leaves mostly alternate; leaves lanceolate
        4. Leaves keeled, acuminate or subulate; bulbils present in upper leaf axils; follicles divaricate at maturity — *S. stenopetalum*
        4. Leaves not keeled, acute to obtuse; bulbils lacking in upper leaf axils; follicles erect at maturity — *S. lanceolatum*

Sedum debile S. Wats.
Bot. King Exped., 102, 1871.

Weakstem Stonecrop

Amerosedum debile (S. Wats.) Löve & Löve
Cotyledon debilis (S. Wats.) Fedde
Echeveria debilis (S. Wats.) A. Nels. & Macbr.
Gormania debilis (S. Wats.) Britt.

Tufted perennial from slender, branching rhizomes; stems prostrate, decumbent, or erect. Leaves opposite or nearly so, obovate to suborbicular, fleshy, sessile. Inflorescence cymose, of 3–14 mostly 5-merous flowers. Calyx lobes broadly lanceolate. Petals yellow, oblong, acuminate, somewhat connate at the base. Stamens 10, included. Pistils 5, erect, connate at the base; styles divergent, at least at maturity. Follicles slightly divergent.

Scree slopes and other rocky places in the alpine zone of the Gros Ventre, Salt River, Teton, Uinta, Wasatch, and Wyoming Ranges. Oregon, Idaho, and western Wyoming south to Utah and Nevada.

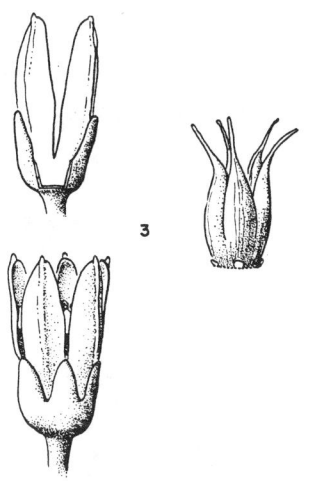
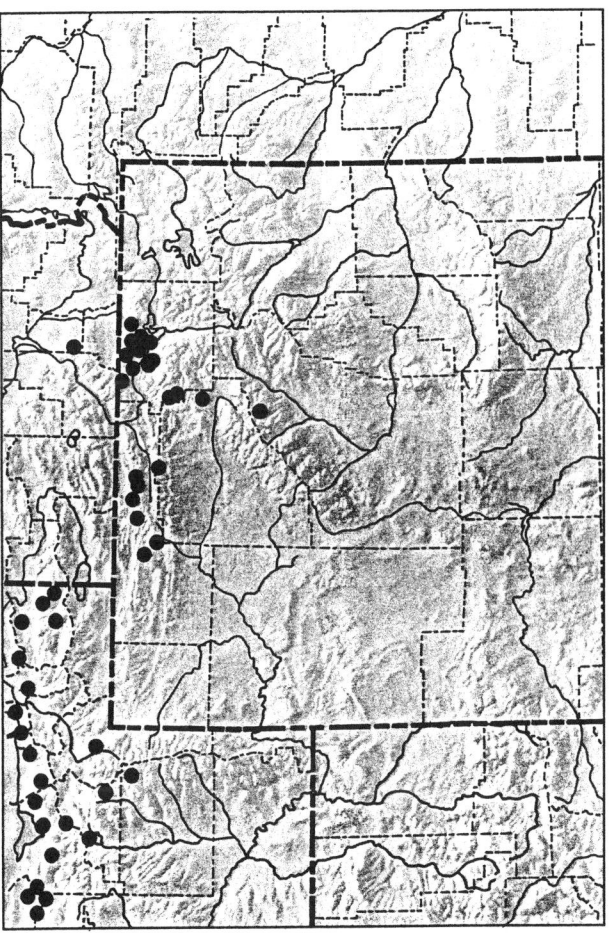

Sedum lanceolatum Torr.
Ann. Lyc. N.Y. 2:205, 1827.

Lanceleaf Stonecrop

Amerosedum lanceolatum (Torr.) Löve & Löve
Amerosedum nesioticum (G.N. Jones) Löve & Löve
Amerosedum subalpinum (Blank.) Löve & Löve
Sedum nesioticum G.N. Jones
Sedum rupicolum G.N. Jones
Sedum subalpinum Blank.

Perennial herb, often tufted, from slender, branching rhizomes; stems erect, mostly decumbent at the base. Leaves in sterile, basal rosettes or cauline and alternate, linear and terete in ours, sessile, succulent, lacking keels. Inflorescence a capitate to short-racemose cyme. Calyx lobes lanceolate, acute to acuminate. Petals yellow, lanceolate, acuminate, distinct, widely divergent and exceeding the stamens. Stamens 10. Pistils 5, erect, basally connate. Follicles erect, the stylar beaks divergent.

Rocky slopes, rock ledges, ridgetops, and dry meadows throughout the alpine areas of the Middle Rocky Mountains. Alaska south to California, New Mexico, Nebraska, and South Dakota.

Middle Rocky Mountain alpine plants are var. *lanceolatum*.

var. lanceolatum

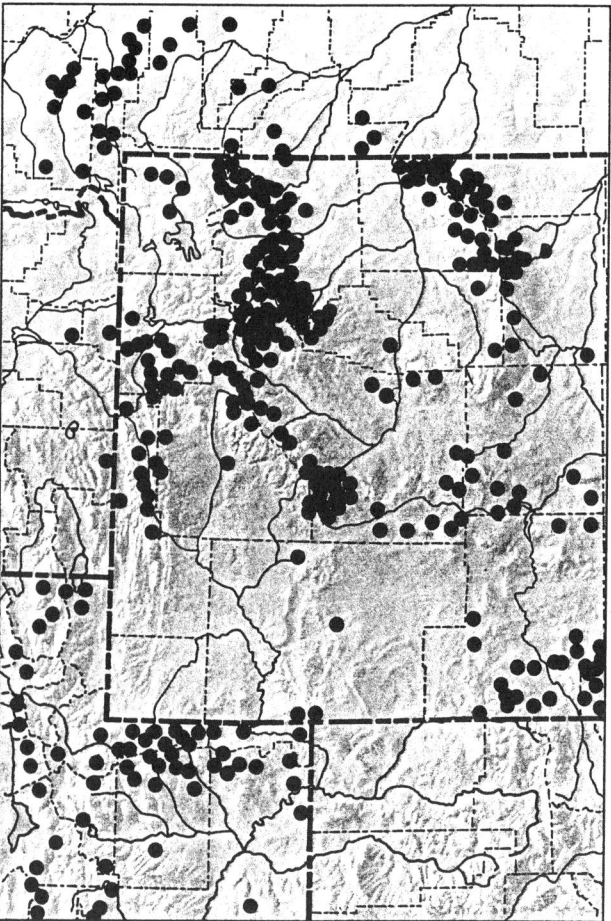

Sedum rhodanthum Gray
Am. J. Sci., II, 33:405, 1862.

Rosecrown

Clementsia rhodantha (Gray) Rose

Erect, glabrous, perennial herb from thick, scaly rhizomes; stems simple, usually several clustered together. Leaves alternate, sessile, succulent, oblong to oblanceolate, entire or sometimes finely toothed. Inflorescence a terminal, bracteate panicle that can be capitate or elongate. Calyx lobes narrowly lanceolate, acute. Petals pink, rose, or rarely white, lanceolate, acute, erect, about twice as long as the stamens. Stamens 10, adnate to the petals. Pistils 5, erect, distinct. Follicles erect, the stylar beaks short-divergent.

Stream banks and moist meadows in most of the alpine ranges of the Middle Rocky Mountains. Not yet known from above timberline in the Hoback, Salt River, Wasatch, and Wyoming Ranges. Montana south to Utah, Arizona, and Colorado.

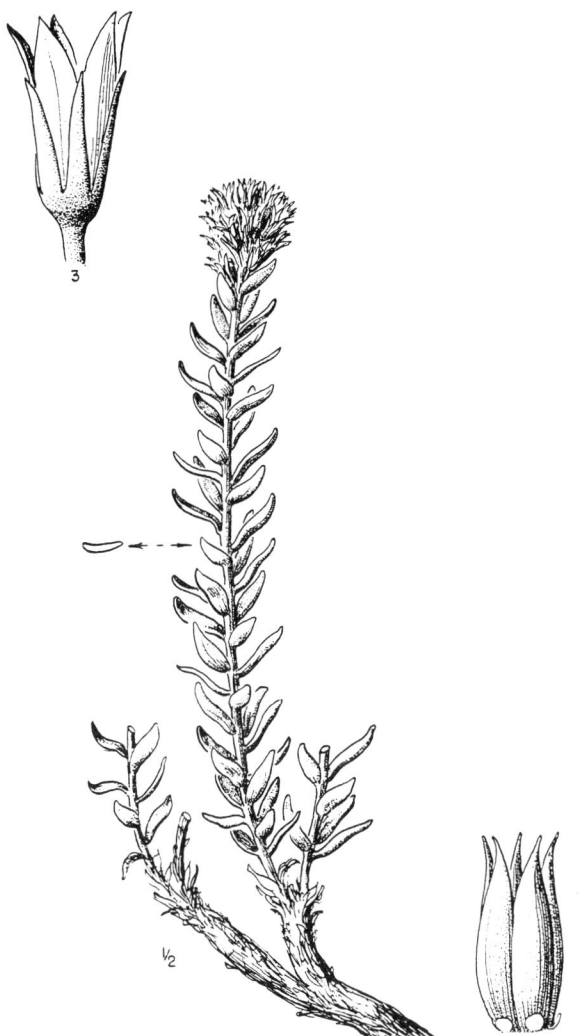

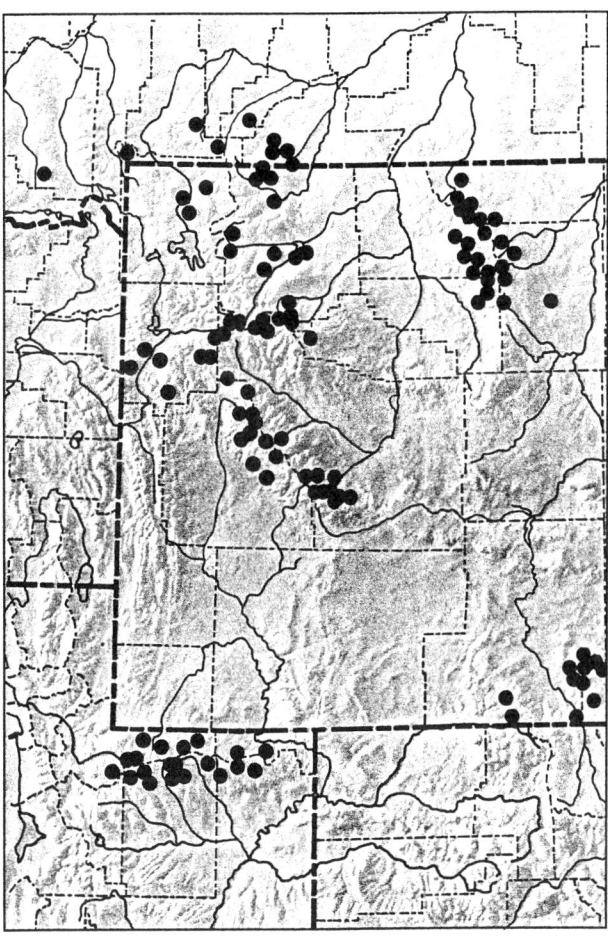

Sedum rosea (L.) Scop.
Fl. Carn., 2nd ed., 1:326, 1772.

King's Crown, Roseroot

Rhodiola rosea L.
Rhodiola alaskana Rose
Rhodiola atropurpurea (Turcz.) Trautv. & C.A. Mey.
Rhodiola integrifolia Raf.
Rhodiola polygama (Rydb.) Britt. & Rose
Rhodiola roanensis Britt.
Sedum alaskanum (Rose) Rose ex Hutchins.
Sedum atropurpureum Turcz.
Sedum frigidum Rydb.
Sedum integrifolium (Raf.) A. Nels. ex Coult. & A. Nels.
Sedum polygamum Rydb.
Sedum rhodiola DC.
Sedum rhodioloides Raf.
Tolmachevia integrifolia (Raf.) Löve & Löve

Low, dioecious, usually tufted perennial from a thick, scaly, branched, fleshy rhizome; stems erect, mostly less than 12 cm tall. Leaves alternate, glabrous and glaucous, flattened, sessile, obovate to oblong, the margins entire to serrulate. Flowers 4-merous, unisexual, in dense, flat-topped cymes. Sepals lanceolate, distinct or nearly so. Petals dark purple, distinct, linear to oblong, obtuse to acute. Stamens 8 or 10, equaling or exceeding the petals, lacking in pistillate flowers. Pistils 5, erect. Follicles erect, distinct, the stylar beaks short-divergent.

Wet meadows, bogs, fellfields, nivation basins, and scree slopes in the Absaroka, Beartooth, Big Horn, Medicine Bow, Teton, and Wind River Ranges. Circumboreal; in western North America south to California, Nevada, and Colorado.

Low-growing North American alpine plants with crowded, fleshy leaves are var. *integrifolium* (Raf.) Berger.

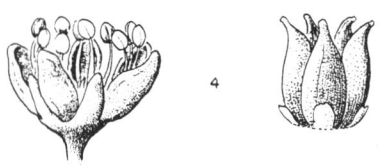
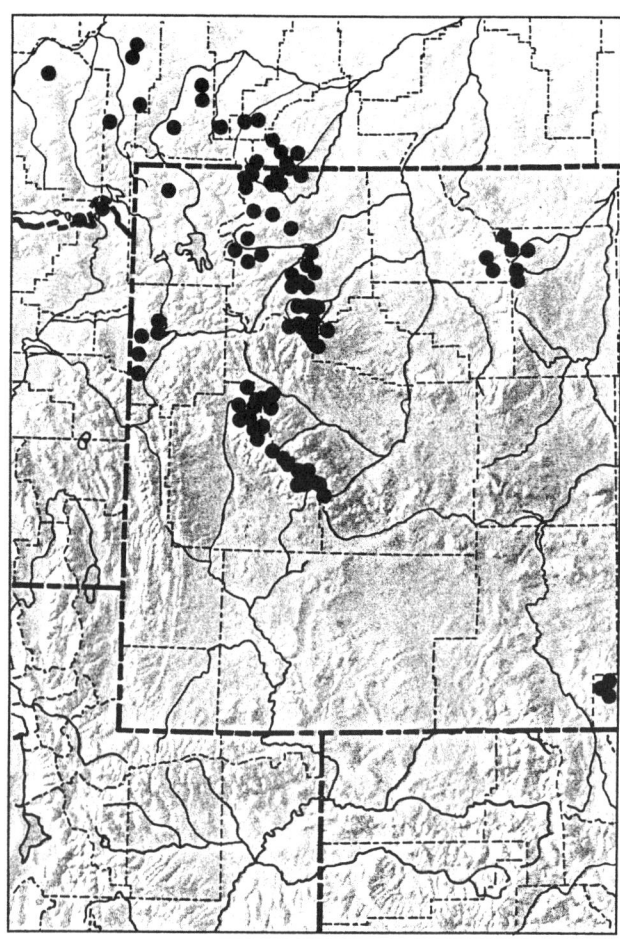

Sedum stenopetalum Pursh
Fl. Am. Sept., 324, 1814.

Narrowpetal Stonecrop

Amerosedum stenopetalum (Pursh) Löve & Löve
Sedum douglasii Hook.
Sedum uniflorum T.J. Howell

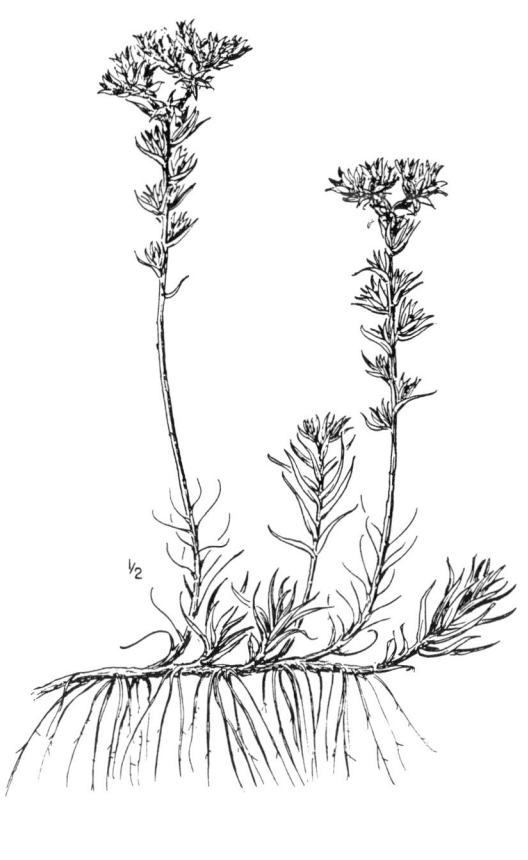

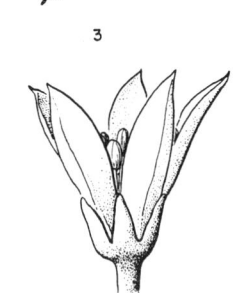

Tufted, glabrous perennial from rhizomes, with short, sterile shoots and erect flowering stems. Leaves alternate, sessile, linear, terete, keeled on the dorsal surface, acuminate or subulate, at least the middle ones deciduous by anthesis. Flowers 5-merous in a compact, branched cyme. Calyx lobes lanceolate. Petals yellow, lanceolate, acute to acuminate, somewhat spreading. Stamens 8–10, included. Pistils 5, erect. Follicles strongly divaricate.

Reported by Spence and Shaw (1981) from talus, cliffs, and rock faces of the Teton Range. Based on 2 non-alpine collections from the base of the Tetons and alpine occurrences in adjacent ranges of the Northern Rocky Mountains, I tentatively include this species as a member of the Middle Rocky Mountain alpine flora. British Columbia and Alberta south to Colorado and California.

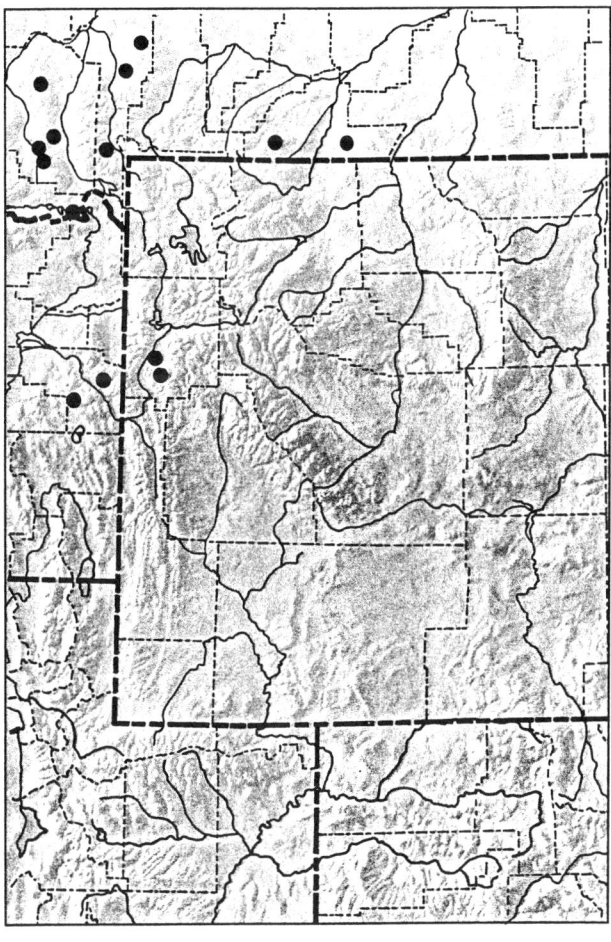

Cupressaceae
Cypress Family

Monoecious or dioecious, evergreen trees or shrubs; ours are shrubs with small, opposite or whorled, subulate or scalelike, crowded, resinous leaves. Cones small, terminal or axillary, the staminate cones with thin, peltate, soon deciduous scales, the pistillate cones berrylike in ours, of 1 to several fleshy, imbricate scales. Seeds angled, 1 or more in each cone.

Juniperus L.
Juniper

Monoecious or dioecious, evergreen, aromatic trees or (in the alpine zone) shrubs with decumbent or prostrate branches from taproots; bark shredded in thin, mostly longitudinal strips. Leaves scalelike to narrowly lanceolate, opposite or in whorls of 3, imbricate, decurrent on the branch. Pistillate cones ovoid, berrylike, of 2–8 fleshy, coalescent scales, glaucous, bluish or reddish brown. Seeds 1–6, wingless.

KEY TO THE SPECIES OF *JUNIPERUS*

1. Leaves linear, needlelike, spreading to ascending, in whorls of 3 at each node ... *J. communis*
1. Leaves scalelike, appressed, opposite ... *J. horizontalis*

Juniperus communis L.
Sp. Pl., 1040, 1753.

Common Juniper

Juniperus alpina (Sm.) S.F. Gray
Juniperus canadensis Lodd. ex Burgsd.
Juniperus depressa (Pursh) Raf.
Juniperus nana Willd.
Juniperus sibirica Burgsd.

Low, prostrate shrubs forming dense mats, the branches decumbent. Bark thin, papery, shredding, grayish to reddish brown. Leaves in whorls of 3, subulate or narrowly lanceolate, stiff and sharp-pointed, the margins entire. Staminate cones solitary, axillary, greenish. Ovulate cones fleshy, berrylike, blue, usually glaucous. Seeds 1–3 per cone.

Widespread on dry tundra and fellfields near timberline in most of the Middle Rocky Mountain ranges. Not yet known from above timberline in the Gros Ventre, Hoback, Salt River, and Wyoming Ranges. Eurasia; Alaska to Greenland, south to South Carolina, Georgia, Illinois, Nebraska, Colorado, New Mexico, Arizona, and California.

Densely matted alpine plants with all branches trailing are var. *montana* Ait.

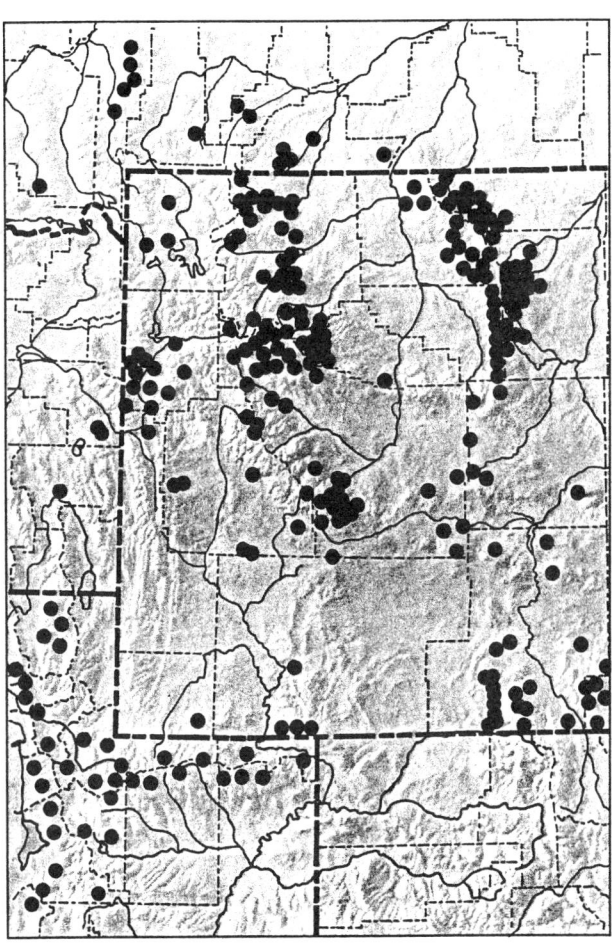

Juniperus horizontalis Moench
Meth., 699, 1794.

Creeping Juniper

Juniperus hudsonica Forbes
Juniperus prostrata Pers.
Juniperus repens Nutt.
Juniperus sabina L.
Sabina horizontalis (Moench) Rydb.
Sabina prostrata (Pers.) Antoine

Prostrate, matted, glabrous shrubs with trailing branches. Bark shredding, grayish to reddish brown. Leaves opposite, scalelike, imbricate, with a dorsal gland that is often depressed, apex acute to acuminate, the margins entire. Staminate cones yellowish to purplish, solitary at the ends of branches. Ovulate cones fleshy, dark blue, glaucous. Seeds 2–4 per cone.

Typical of plains and foothills, *J. horizontalis* occurs in meadows and fellfields of the Absaroka Range at 3140 m. Alaska to Nova Scotia, south to New York, Illinois, Iowa, Nebraska, Wyoming, and Colorado.

Middle Rocky Mountain plants are var. *horizontalis*.

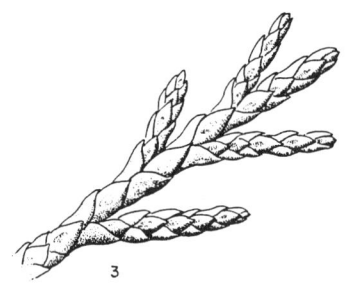

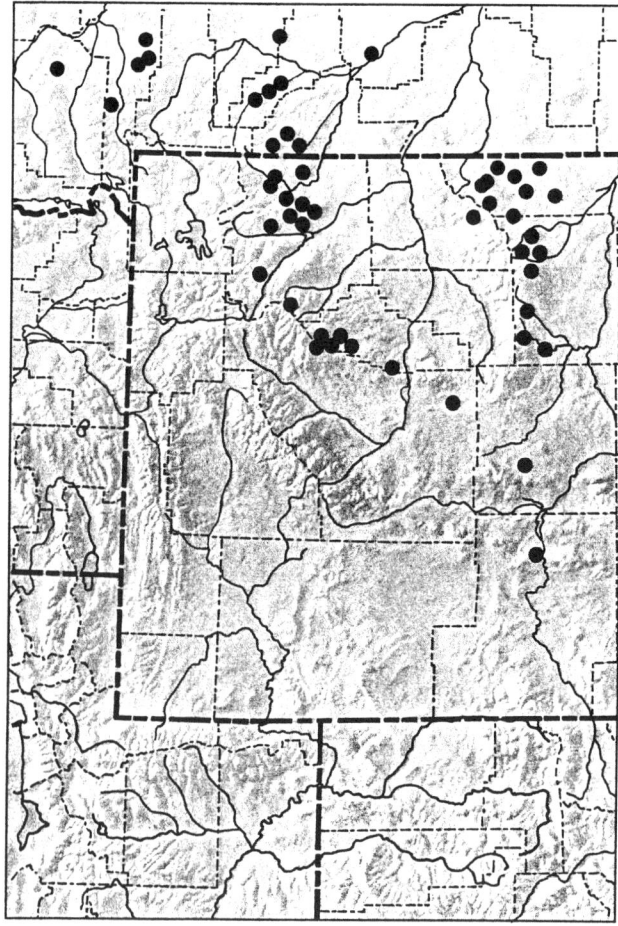

Cyperaceae
Sedge Family

Mostly perennial, grasslike or rushlike herbs with slender, usually triangular, jointless stems. Rhizomes usually present. Leaves, if present, 3-ranked, narrow, with closed sheaths, or reduced to bladeless sheaths. Flowers perfect or unisexual, in spikes or spikelets, each flower subtended by a chaffy scale, the scales 2-ranked or spirally imbricate, deciduous or persistent. In *Carex* and *Kobresia* the ovary is enveloped by an additional saclike scale, the perigynium. Perianth lacking or consisting of hypogynous scales or bristles. Stamens 1–3. Pistil 1, the style 2–3-cleft, the ovary with a single ovule. Fruit a lenticular or triangular akene.

KEY TO THE GENERA OF CYPERACEAE

1. Flowers perfect; perigynium lacking; perianth present
   2. Perianth of numerous long-exerted, cottonlike bristles — *Eriophorum*
   2. Perianth of 3 or 4 short bristles — *Eleocharis*
1. Flowers unisexual; perigynium present; perianth lacking
   3. Perigynium closed except for the apical opening — *Carex*
   3. Perigynium split down one side nearly to the base — *Kobresia*

Carex L.
Sedge

Perennial, rhizomatous, grasslike herbs with trigonous, solid culms. Leaves 3-ranked, filiform to linear, or lacking, with closed sheaths. Plants monoecious, or rarely dioecious, the flowers in spikes. Spikes 1 to many, distinct or aggregated, each spike either staminate, pistillate, androgynous, or gynaecandrous. Flowers unisexual, solitary in the axils of scales, the perianth lacking. Staminate flowers with 3 stamens, rarely 2. Pistillate flowers with 1 pistil surrounded by a saclike perigynium, the stigmas 2 or 3. Akenes trigonous or lenticular.

KEY TO THE SPECIES OF *CAREX*

1. Spike solitary
   2. Flowers all staminate or all pistillate on the same plant — *C. scirpoidea*
   2. Flowers both staminate and pistillate on the same plant
      3. Stigmas 2; akenes lenticular
         4. Plants lacking rhizomes; perigynia 3.5–4 mm long, stipitate to substipitate, tapering at the base — *C. nardina*
         4. Plants rhizomatous; perigynia 1.9–3.5 mm long, sessile, rounded at the base — *C. capitata*
      3. Stigmas 3; akenes trigonous
         5. Perigynia thin-walled, strongly inflated; rachilla well developed, more than half as long as the akene; pistillate scales persistent; plants rhizomatous
            6. Perigynia small, 2.5–4 mm long; spike slender, narrowly cylindrical; akene nearly filling the perigynium — *C. subnigricans*
            6. Perigynia large, 4–6.5 mm long; spike broadly ovate; akene much smaller than the perigynium — *C. breweri*
         5. Perigynia leathery, not inflated; rachilla obsolete, or if well developed, then plants lacking the above combination of characters
            7. Pistillate scales deciduous; perigynia stipitate, at least the lower ones often reflexed at maturity
               8. Plants rhizomatous; leaves flat, 1.5–2.8 mm wide; staminate flowers conspicuous; perigynia early reflexed — *C. nigricans*

8. Plants lacking rhizomes; leaves involute, 0.25–1.25 mm wide; staminate flowers few, inconspicuous; perigynia erect or ascending until full maturity ..... *C. pyrenaica*
 7. Pistillate scales persistent; perigynia neither stipitate nor reflexed at maturity
  9. Plants tufted, lacking rhizomes; leaves narrow, wiry, about 0.5 mm wide
   10. Perigynium finely pubescent, broadly ovate to orbicular ..... *C. filifolia*
   10. Perigynium glabrous, rarely with a few short hairs near the beak, ovate or obovate
    11. Perigynium finely striate, flattened to plano-convex, with thin margins, the beak not hyaline, not completely filled by the akene; scales about the same width as the perigynia, lacking hyaline margins ..... *C. nardina*
    11. Perigynium nerveless, turgid, lacking thin margins, the beak hyaline, completely filled by the akene; scales broader than the perigynia, with conspicuous hyaline margins ..... *C. elynoides*
  9. Plants rhizomatous, the stems single or a few together; leaves flat, 1–3 mm wide
   12. Perigynia 1–3 per spike, 5–6 mm long, beakless; pistillate scales short-awned ..... *C. geyeri*
   12. Perigynia more than 3 per spike, 3–4 mm long, beaked; pistillate scales obtuse to acute or cuspidate
    13. Leaves circinate at the tip, 1.5–3 mm wide; perigynia 5–15 per spike, the beak about 0.2 mm long ..... *C. rupestris*
    13. Leaves straight at the tip, 1–1.5 mm wide; perigynia 4–6 per spike, the beak 0.5–1 mm long ..... *C. obtusata*
1. Spikes more than 1
 14. Plants dioecious, rarely with some androgynous spikes ..... *C. douglasii*
 14. Plants monoecious, pistillate and staminate flowers in the same or separate spikes
  15. Stigmas mostly 2; akenes lenticular
   16. Terminal spike staminate, rarely androgynous or gynaecandrous; lateral spikes pedunculate, or if sessile then elongate, cylindric
    17. Perigynia golden yellow, fleshy at maturity; lowest bract long-sheathing ..... *C. aurea*
    17. Perigynia greenish, brownish, or purplish black at maturity; lowest bract sheathless or sometimes with a short sheath
     18. Lowest bract shorter than the inflorescence; pistillate scales with obsolete or slender midveins ..... *C. scopulorum*
     18. Lowest bract equaling or exceeding the inflorescence; pistillate scales with conspicuous midveins or with a broad, light-colored center
      19. Style continuous with the akene, persistent, flexuous and strongly bent against the akene;

    lowest spike long-pedunculate, widely spreading
    to drooping   *C. saxatilis*
  19. Style jointed with the akene, deciduous, straight
    or slightly bent; lowest spike sessile to short-
    pedunculate, erect   *C. aquatilis*
16. Terminal spike androgynous or gynaecandrous; lateral
  spikes sessile, short, oblong to ovoid
  20. Spikes (at least the terminal one) androgynous
    21. Plants tufted, often with multiple stems, lacking
      creeping rhizomes
      22. Perigynia broadest near the base; leaf sheaths with
        transverse ridges (cross-rugulose); perigynia with
        body tapering into the beak, 2.9–3.6 mm long   *C. neurophora*
      22. Perigynia broadest above the base and near the
        middle; leaf sheaths lacking transverse ridges;
        perigynia abruptly contracted into the beak,
        3.4–5 mm long   *C. hoodii*
    21. Plants from creeping rhizomes
      23. Rhizomes short-creeping, compact, the internodes
        very short; stems clustered on the rhizomes   *C. jonesii*
      23. Rhizomes long-creeping, slender, the internodes
        long; stems single or a few together
        24. Spikes in an ovoid-linear head, the lower ones
          distinct; perigynia 2.5–3 mm long, not stipitate   *C. stenophylla*
        24. Spikes in a globose head, crowded and
          indistinguishable; perigynia 3.25–4.5 mm long,
          stipitate   *C. maritima*
  20. Spikes (at least the terminal one) gynaecandrous
    25. Scales nearly the same length as the perigynia,
      concealing them above, or the tips of perigynia
      beaks sometimes barely visible
      26. Perigynia plump to the margins; akene about as
        large as the perigynial cavity
        27. Plants usually less than 20 cm tall; spikes mostly
          3 or fewer; perigynia elliptic to obovate, less
          than 3.4 mm long   *C. bipartita*
        27. Plants usually more than 25 cm tall; spikes
          mostly 5 or more; perigynia ovate to ovate-
          lanceolate, more than 3.5 mm long
          28. Inflorescence elongate, the spikes distinct,
            approximate; scales of the perigynia with
            broad hyaline margins   *C. praticola*
          28. Inflorescence ovate to nearly orbicular, the
            spikes dense; scales of the perigynia lacking
            hyaline margins   *C. macloviana*
      26. Perigynia flattened except where distended by the
        much smaller akene
        29. Inflorescence of loose spikes, ellipsoid; perigynia
          3–4 mm long, 0.8–1.2 mm wide, the beak poorly
          defined   *C. leporinella*
        29. Inflorescence of dense spikes, ovoid; perigynia
          3.5–5 mm long, 1.3–2 mm wide, the beak
          evident, 0.5–1 mm long   *C. phaeocephala*

25. Scales shorter and often narrower than the perigynia, the perigynia beaks and usually the margins evident and conspicuous
    30. Perigynia flattened, membranous, scalelike, thin, except where distended by the akene
        31. Perigynia 6–7 mm long, narrowly lanceolate   *C. ebenea*
        31. Perigynia 2.5–6 mm long, ovate
            32. Inflorescence pale greenish or straw-colored; perigynium with crinkled margins, the beak flat, margined and serrulate to the tip   *C. straminiformis*
            32. Inflorescence olive-green to dark brown; perigynium lacking crinkled margins, the beak terete, neither margined nor serrulate in the distal 0.5–1 mm of the tip   *C. macloviana*
    30. Perigynia planoconvex, plump to the margins, not thin and scalelike
        33. Rhizomes present, mostly short-creeping
            34. Perigynia spreading at maturity, lanceolate to narrowly ovate, 2.7–3.3 mm long, not white-punticulate, gradually tapered to a dark brown to black, oblique-cleft, entire beak   *C. illota*
            34. Perigynia ascending at maturity, ovate to broadly elliptic, 1.7–2.5 mm long, white-punticulate, contracted to a short, often serrulate beak   *C. praeceptorum*
        33. Rhizomes lacking, or at least inconspicuous; plants tufted from fibrous roots
            35. Inflorescence elongate, interrupted, the lower spikes distinct, approximate or remote
                36. Perigynia lanceolate to ovate, 2.7–3.5 mm long, widely spreading at maturity   *C. echinata*
                36. Perigynia ovate to elliptic, 1.8–2.6 (–3) mm long, ascending at maturity
                    37. Spikes with (9–) 15–30 perigynia, the perigynia with short beaks less than 0.25 mm long; leaves glaucous, 2–4 mm wide   *C. canescens*
                    37. Spikes with 5–10 perigynia, the perigynia with long beaks more than 0.5 mm long; leaves green, 1–2.5 mm wide   *C. brunnescens*
            35. Inflorescence ovoid to ellipsoid, compact, the spikes closely aggregated
                38. Perigynia more than 5.7 mm long   *C. petasata*
                38. Perigynia less than 5.1 mm long   *C. macloviana*
15. Stigmas mostly 3; akenes trigonous
    39. Perigynia hirsute to puberulent
        40. Plants tufted, lacking creeping rhizomes; bracts of inflorescence exceeding the terminal spike   *C. rossii*
        40. Plants from short-creeping rhizomes; inflorescence lacking bracts, or the single bract short, inconspicuous, less than half the length of the terminal spike   *C. scirpoidea*

39. Perygynia glabrous, sometimes serrulate on the margins, or rarely scabrous at the base of the beak
    41. Style continuous with the akene, indurate, persistent; perigynia usually more than 5 mm long, the beak 1–2 mm long (less in *C. saxatilis*)
        42. Lowest spike long-pedunculate, widely spreading to drooping at maturity; perigynia not inflated ..... *C. saxatilis*
        42. Lowest spike sessile to short-pedunculate, erect at maturity; perigynia inflated
            43. Perigynia ovate, spreading at maturity, the lower ones often reflexed ..... *C. rostrata*
            43. Perigynia lanceolate, ascending at maturity, never reflexed ..... *C. vesicaria*
    41. Style jointed with the akene, not indurate, at length withering and deciduous; perigynia usually less than 5 mm long, the beak less than 1 mm long
        44. Lowest bract of the inflorescence with a closed sheath 0.5–4.5 cm long
            45. Perigynia ovate, 2.2–3.2 mm long, trigonous, the margins entire ..... *C. capillaris*
            45. Perigynia lanceolate, 3–5 mm long, compressed, the margins finely serrulate, at least near the beak
                46. Leaves 1–3 mm wide; terminal spike gynaecandrous or pistillate; spikes drooping ..... *C. misandra*
                46. Leaves 3–9 mm wide; terminal spike androgynous or staminate; spikes erect ..... *C. luzulina*
        44. Lowest bract of the inflorescence sheathless, or with a closed sheath less than 0.3 cm long
            47. Terminal spike staminate
                48. Lower spikes spreading or nodding; perigynia strongly flattened, ovate; midvein of scales prominent, often excurrent ..... *C. podocarpa*
                48. Lower spikes erect-ascending; perigynia nearly round in cross section, little or not at all flattened, obovate; midvein of scales obscure ..... *C. raynoldsii*
            47. Terminal spike gynaecandrous, rarely pistillate or interspersed with staminate and pistillate flowers
                49. Spikes closely aggregated in a dense, terminal head; peduncle of lowest spike lacking or very short and not readily visible ..... *C. nova*
                49. Spikes approximate in a loose, elongate inflorescence; peduncle of lowest spike evident, often nearly as long as the spike
                    50. Perigynia small, 1.8–2.8 mm long, elliptic; scales shorter than the perigynia ..... *C. norvegica*
                    50. Perigynia large, 2.7–3.5 mm long, obovate; scales about the same length as the perigynia
                      51. Plants mostly more than 30 cm tall; peduncle of lowest spike about the same length as the spike; pistillate scales lanceolate, mostly lacking hyaline margins, narrower than the perigynia, the perigynial margins readily visible ..... *C. atrata*

51. Plants mostly less than 30 cm tall; peduncle of lowest spike less than half the length of the spike; pistillate scales broadly ovate, with narrow hyaline margins, as wide as the perigynia, the perigynial margins not visible .... *C. albonigra*

Carex albonigra Mack. in Rydb.
Fl. Rocky Mtns. 137, 1060, 1917.

Black-and-White Sedge

Tufted, somewhat aphyllopodic perennial from short rhizomes; culms erect, often purplish at the base, with persistent old leaves conspicuous. Leaves mostly basal; sheaths purplish, the lower ones lacking blades, the upper with well-developed, flat blades. Lowest bract shorter than the headlike inflorescence; spikes 2–4, usually 3, sessile or nearly so, the terminal one gynaecandrous, the lateral ones pistillate; pistillate scales wider than and subequal to or shorter than the perigynia, ovate, black to blackish purplish with white-hyaline margins; perigynia elliptic to obovate, sometimes nearly orbicular, granular-papillate, at least the apex purplish black; beak short, shallowly bidentate; stigmas 3; akenes trigonous, sessile to substipitate.

Dry meadows and fellfields of most Middle Rocky Mountain alpine ranges. Not yet known from above timberline in the Hoback, Salt River, Wasatch, and Wyoming Ranges. Alaska south to California, Arizona, and Colorado.

Carex albonigra is closely related to *C. atrata,* and the two are sometimes difficult to distinguish in the alpine zone. It seems to differ mostly in being shorter (usually less than 30 cm tall) and having the lowest spike nearly sessile and perigynia more uniformly papillate than *C. atrata*.

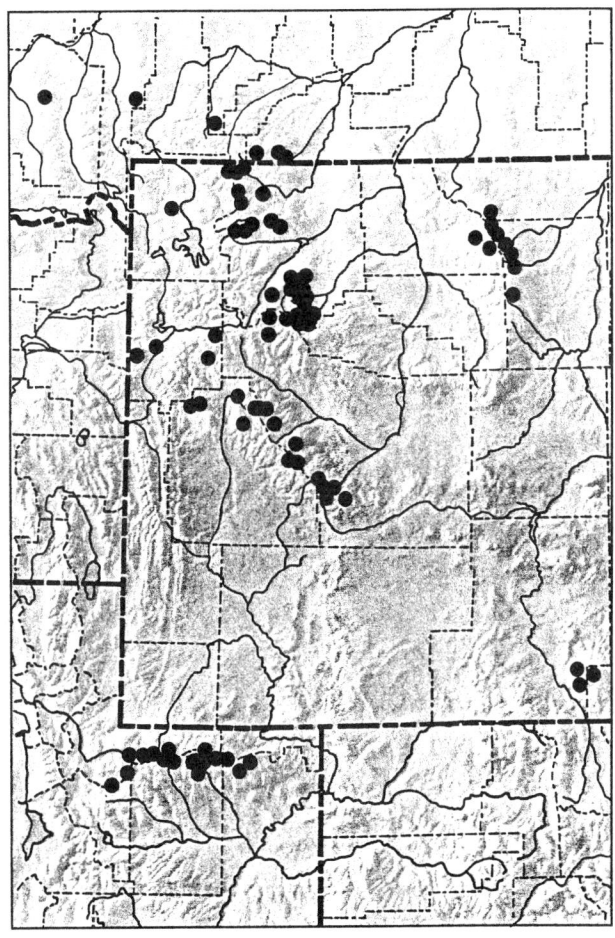

Carex aquatilis Wahlenb.
Kongl. Vetensk. Acad. Nya Handl. 24:165, 1803.

Water Sedge

Carex acutina L.H. Bailey
Carex acutinella Mack.
Carex altior Rydb.
Carex howellii Bailey
Carex interimus Maguire
Carex panda C.B. Clarke
Carex sitchensis Prescott ex Bong.
Carex stans Drej.
Carex substricta (Kükenth.) Mack. in Rydb.
Carex suksdorfii Kükenth. in Fedde
Carex variabilis L.H. Bailey
Neskiza aquatilis (Wahlenb.) Raf.

Perennial, phyllopodic herb with culms solitary to tufted, from long-creeping rhizomes. Leaves shorter than to about as long as the culms, mostly channeled at the base, becoming flat near the tip, often glaucous-green; sheaths reddish- or purplish-dotted. Lowest bract shorter than to longer than the inflorescence. Spikes 3–6, erect, sessile to pedunculate, the terminal one staminate; lateral spikes pistillate, the upper ones sometimes androgynous; pistillate scales ovate to oblong-ovate, brownish to purplish black, mostly with a pale greenish midrib, acute to acuminate, narrower and shorter than to exceeding the perigynia; perigynia obovate to elliptic, erect-ascending, flat, nerveless or nearly so, sessile or short-stipitate, the beak very short, entire or oblique; stigmas 2; akenes lenticular.

Stream banks, lakeshores, and wet meadows in the lower alpine zone of the Absaroka, Beartooth, Big Horn, Gros Ventre, Medicine Bow, Uinta, and Wind River Ranges. Circumboreal; south in western North America to California, Arizona, New Mexico, and Nebraska.

Short alpine plants with glaucous-green leaves and a single staminate spike have been called ssp. *stans* (Drej.) Hult. They do not seem to be consistently separable from the taller ssp. *altior* (Rydb.) Hult. or ssp. *aquatilis* of lower elevations.

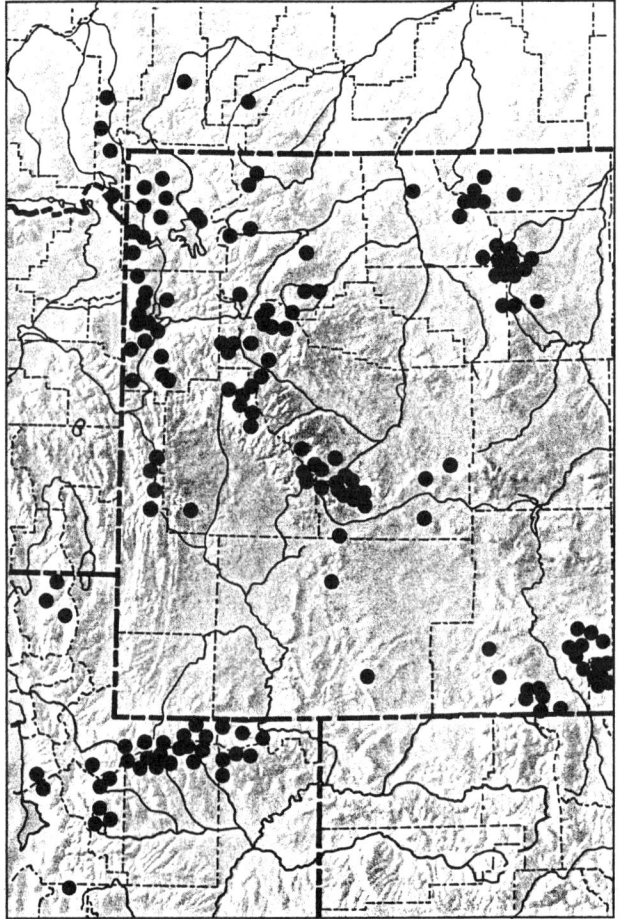

Carex atrata L.
Sp. Pl., 976, 1753.

Blackscale Sedge

Carex apoda Clokey
Carex atrosquama Mack.
Carex bella L.H. Bailey
Carex chalciolepis Holm
Carex epapillosa Mack. in Rydb.
Carex heteroneura W. Boott
Carex uncompahgre Kelso
Trasus atratus (L.) S.F. Gray

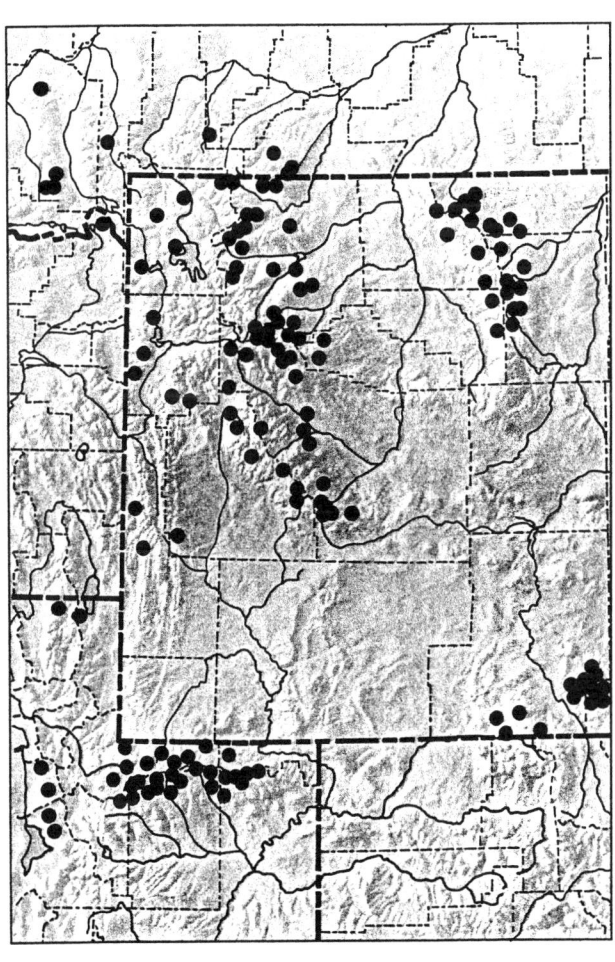

Tufted, aphyllopodic perennial from fibrous roots; rhizomes lacking or very short. Leaves glabrous, the blades flat, becoming channeled near the tip; sheaths brownish to purplish. Lowest bract leaflike, shorter than to about the same length as the inflorescence, sheathless or with a short sheath. Spikes mostly 3–5, the upper erect and sessile; terminal spike gynaecandrous; lateral spikes pistillate, the upper sometimes gynaecandrous; the lowermost pedunculate, often weakly divergent; pistillate scales blackish purple, narrower and shorter than to somewhat longer than the perygynia, lanceolate to oblong, acute; perigynia elliptic, obovate, or nearly orbicular, entirely olive-green or purplish with greenish margins, flat to somewhat inflated, glabrous, often papillose, abruptly contracted to a short, slender, purplish, entire to shallowly bidentate beak; stigmas 3; akenes trigonous, short-stipitate.

Moist meadows, fellfields, and rocky slopes in most Middle Rocky Mountain alpine ranges. At present not known from above timberline in the Big Horn, Hoback, Salt River, and Wyoming Ranges. Circumboreal; south in western North America to California, Arizona, and Colorado.

Four varieties occur in the alpine zone of the Middle Rocky Mountains and may be distinguished as follows:

1. Perigynia inflated at maturity, olive-green becoming yellowish brown
   . . . var. *atrosquama* (Mack.) Kelso
1. Perigynia flattened at maturity, but swollen by the akene, mostly greenish with tinges of red or purple
   2. Upper pistillate scales longer than the perigynia; perigynia granular and roughened, at least above, and the upper margins ciliate-serrulate
      . . . var. *chalciolepis* (Holm) Kuekenth.
   2. Upper pistillate scales shorter than the perigynia; perigynia not granular and roughened, the margins not ciliate-serrulate
      3. Lateral spikes gynaecandrous, the lower nodding . . . var. *discolor* L.H. Bailey
      3. Lateral spikes pistillate, the lower erect . . . var. *erecta* W. Boott

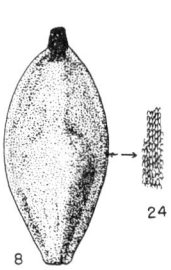 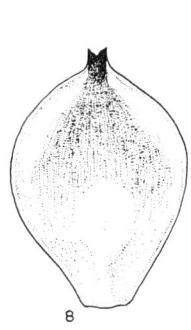 

Carex aurea Nutt.
Gen. Pl. 2:205, 1818.

Golden Sedge

Carex garberi Fern.
Carex hassei L.H. Bailey
Neskiza aurea (Nutt.) Raf.

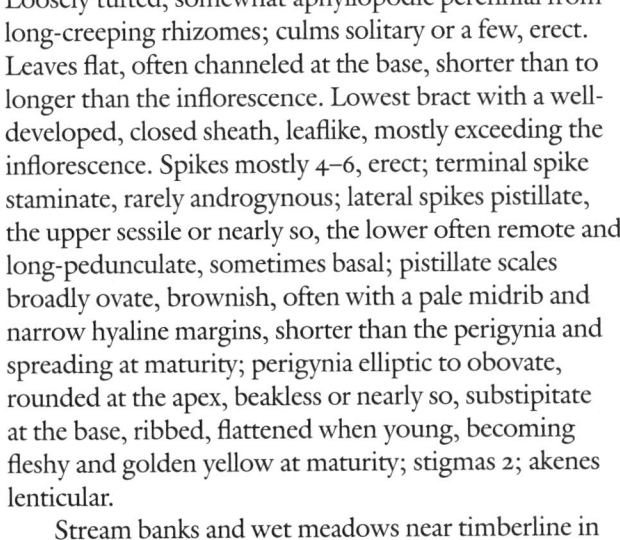

Loosely tufted, somewhat aphyllopodic perennial from long-creeping rhizomes; culms solitary or a few, erect. Leaves flat, often channeled at the base, shorter than to longer than the inflorescence. Lowest bract with a well-developed, closed sheath, leaflike, mostly exceeding the inflorescence. Spikes mostly 4–6, erect; terminal spike staminate, rarely androgynous; lateral spikes pistillate, the upper sessile or nearly so, the lower often remote and long-pedunculate, sometimes basal; pistillate scales broadly ovate, brownish, often with a pale midrib and narrow hyaline margins, shorter than the perigynia and spreading at maturity; perigynia elliptic to obovate, rounded at the apex, beakless or nearly so, substipitate at the base, ribbed, flattened when young, becoming fleshy and golden yellow at maturity; stigmas 2; akenes lenticular.

Stream banks and wet meadows near timberline in the Uinta and Wasatch Ranges. Alaska to Newfoundland; south in western North America to California, Utah, New Mexico, and Texas.

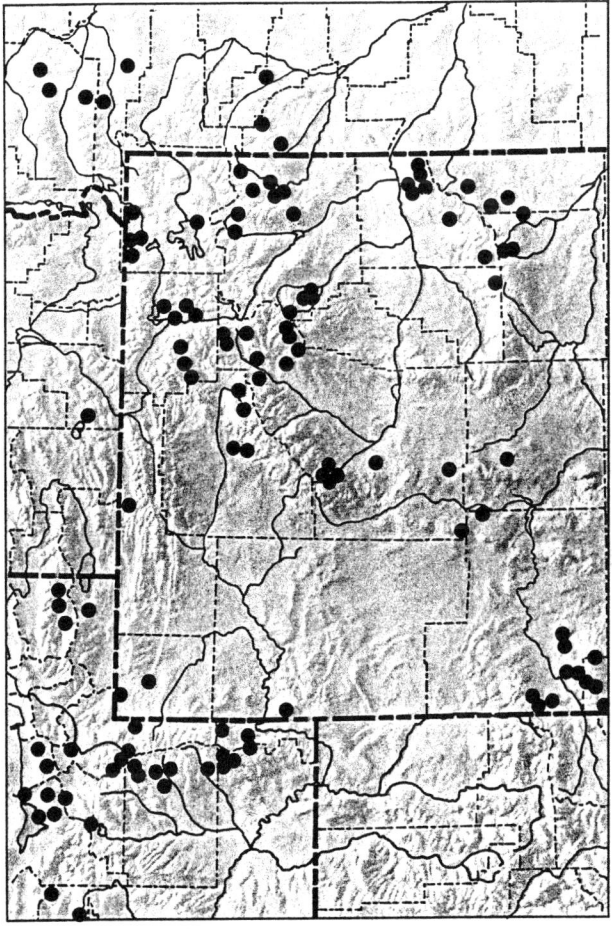

Carex bipartita Allioni
Fl. Pedem. 2:265, 1785.

Twotip Sedge

Carex lachenalii Schk.
Carex lagopina Wahlenb.
Carex tripartita Allioni

Somewhat aphyllopodic perennial from fibrous roots; rhizomes lacking or very short; culms solitary or a few, slender, erect, exceeding the leaves. Leaves basal, the blades flat. Lowest bract inconspicuous, attenuate, shorter than the first spike. Spikes mostly 2–4, sessile, gynaecandrous, approximate in an elongate head; pistillate scales oblong to ovate, brownish to reddish brown, with prominent hyaline margins and a pale midrib, about the same length and width as the perigynia, mostly concealing them; perigynia elliptic to obovate, often finely nerved, substipitate, mostly gradually tapered to a short beak; stigmas 2; akenes lenticular.

Wet meadows, nival basins, and patterned ground of the Absaroka, Beartooth, Big Horn, Uinta, and Wind River Ranges. Circumpolar; south in western North America to Utah and Colorado.

Plants at the southern margin of the range in Utah and Colorado have been called var. *austromontana* F.J. Herm. and are distinguished by having abruptly contracted beaks rather than gradually tapered ones, as in the typical var. *bipartita*.

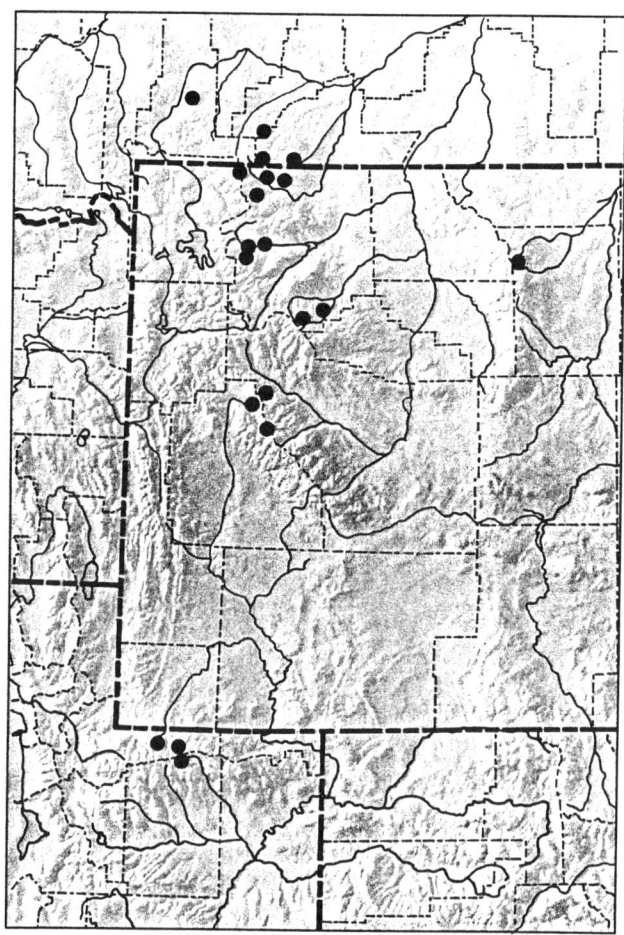

Carex breweri F. Boott
Illustr. Gen. Carex 4:142, pl. 455, 1867.

Brewer Sedge

Carex engelmannii L.H. Bailey
Carex paddoensis Suksd.

Phyllopodic perennial from creeping rhizomes; culms solitary or a few, slender, stiff, erect, equaling or exceeding the leaves. Leaves basal or nearly so, stiff, narrow, involute or channeled. Spike solitary, ebracteate, erect, androgynous, ovate; pistillate scales ovate, light brown to dark reddish brown, acute to acuminate, with narrow hyaline margins and 1–3 pale ribs, about the same length as or slightly shorter than the perigynia; perigynia ovate to elliptic, inflated, nerveless, substipitate, gradually or abruptly narrowed into a short beak; stigmas 3; akenes trigonous, much smaller than the perigynia.

Fellfields, talus, and other rocky places of the Absaroka, Beartooth, Gros Ventre, Teton, Uinta, Wasatch, and Wind River Ranges. British Columbia south and east to California, Utah, Colorado, and Wyoming.

Plants of the Middle Rocky Mountains with single-nerved pistillate scales and involute to deeply channeled leaves are var. *paddoensis* (Suksd.) Cronq.

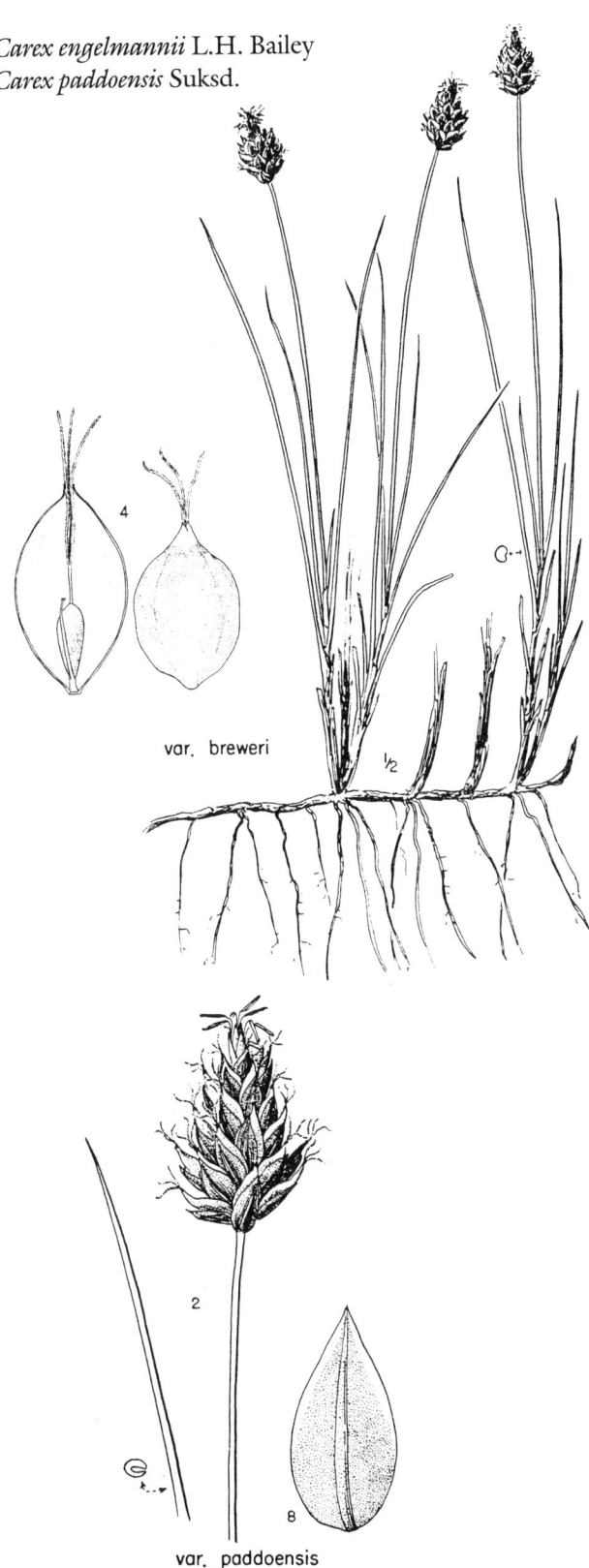

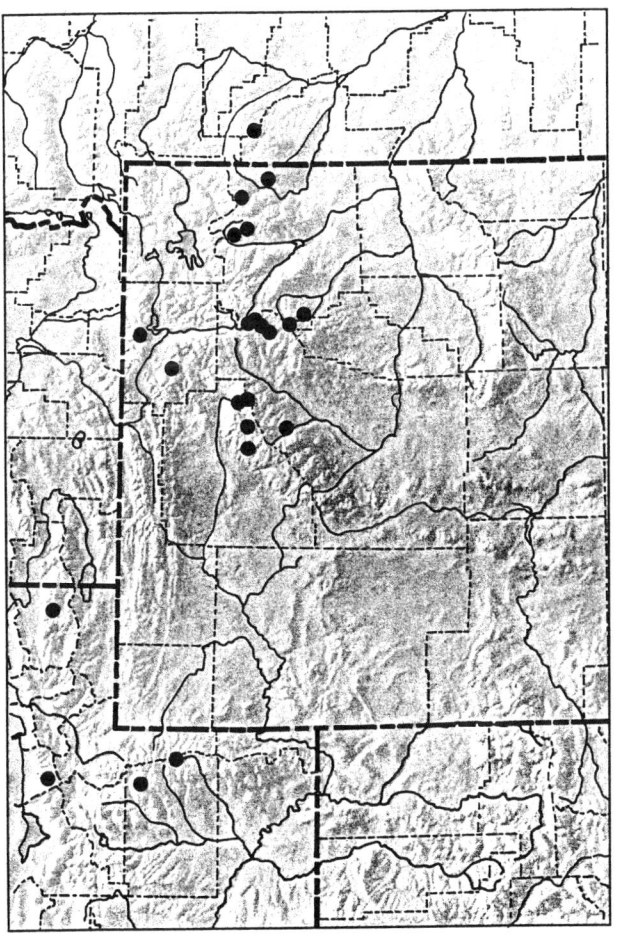

Carex brunnescens (Pers.) Poir. in Lam.
Encyc. Meth. Suppl. 3:286, 1813.

Brown Sedge

Carex vitilis Fries
Facolos brunnescens (Pers.) Raf.

Densely tufted, somewhat aphyllopodic perennial from thin, fibrous roots; rhizomes lacking or very short; culms few to several, slender, erect, exceeding the mostly basal leaves; blades flat, narrow. Lowest bract attenuate, reduced, and inconspicuous to longer than the lowest spike. Spikes 4–10, sessile, ovate, gynaecandrous, the upper approximate, becoming remote below; pistillate scales ovate, acute to somewhat obtuse, shorter than the perigynia, the margins hyaline-scarious with a pale greenish or brownish midrib; perigynia elliptic to ovate, greenish to brownish, finely nerved, substipitate, the apex gradually narrowed into a short, bidentate beak with serrulate margins; stigmas 2; akenes lenticular.

Wet meadows and fellfield seeps of the Big Horn, Medicine Bow, and Uinta Ranges. Circumboreal; south in western North America to Oregon, Idaho, Utah, and Colorado.

Middle Rocky Mountain plants are the weakly differentiated ssp. *brunnescens*.

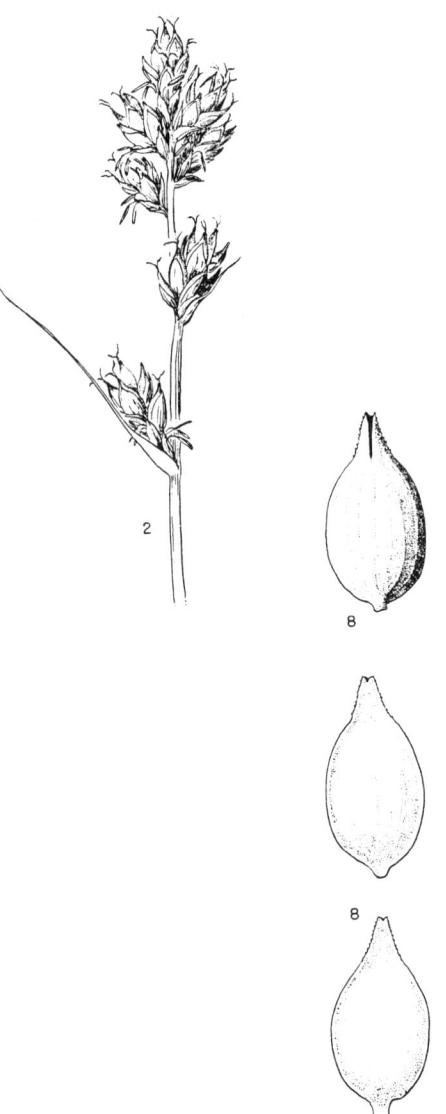
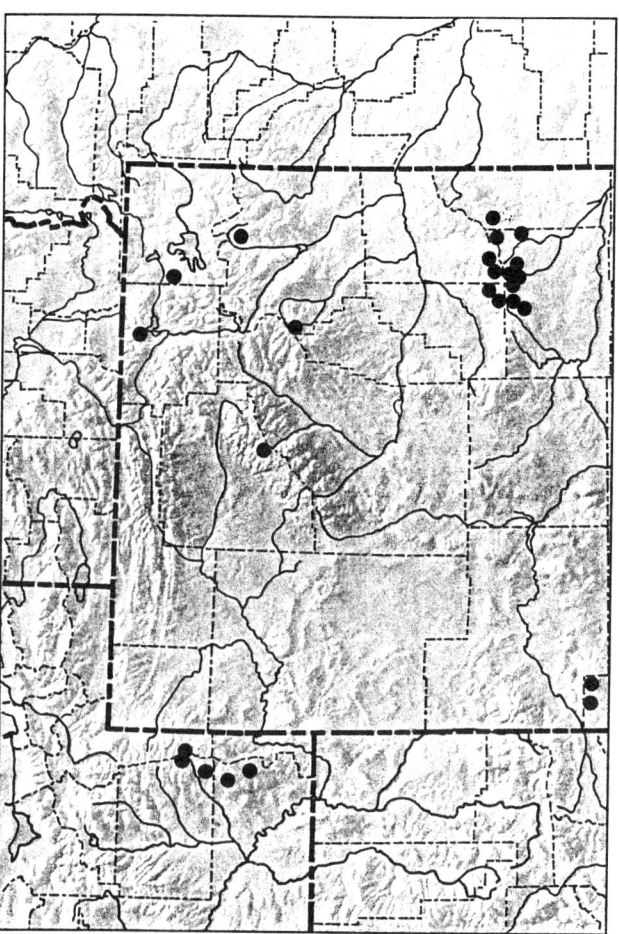

Carex canescens L.
Sp. Pl., 974, 1753.

Pale Sedge

Carex arctiformis Mack.
Carex curta Gooden
Carex lapponica O.F. Lang
Vignea canescens Reichenb.

Tufted, mostly phyllopodic perennial from fibrous roots and very short rhizomes, or rhizomes lacking; culms slender, erect to spreading, equaling or exceeding the leaves. Leaves mostly basal, the blades flat, glaucous, with hyaline sheaths. Lowest bract subulate, shorter than to longer than the lowest spike. Spikes 4–8, sessile, gynaecandrous, approximate above, becoming remote below; pistillate scales ovate, acute, with hyaline margins and a green midrib, shorter than the perigynia; perigynia planoconvex, ovate, greenish to brownish, finely nerved on both sides, the beak very short, with smooth to minutely serrulate margins; stigmas 2, akenes lenticular.

Wet meadows and other wet places near timberline in the lower alpine zone of the Beartooth and Uinta Ranges. Circumboreal; south in western North America to California, Arizona, and Colorado.

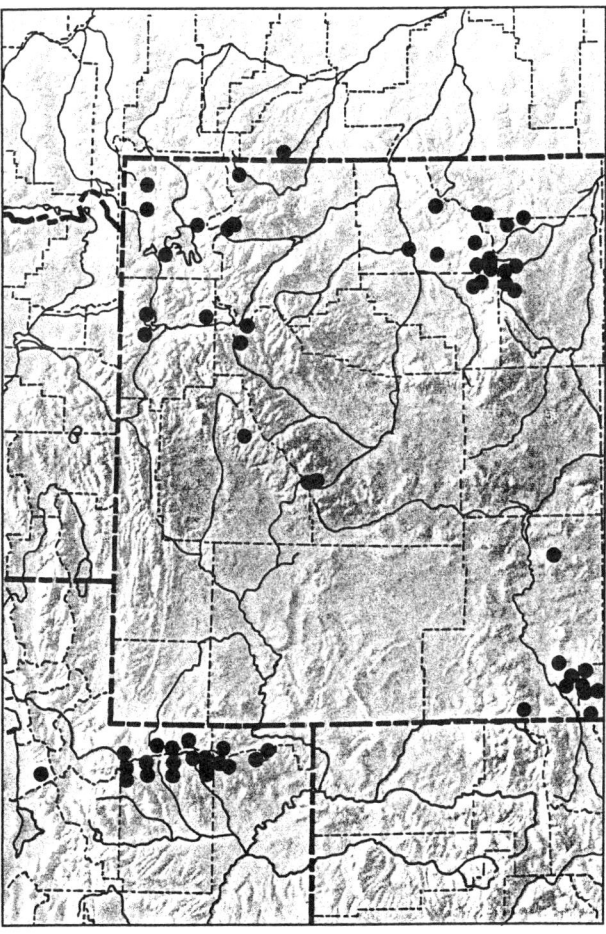

Carex capillaris L.
Sp. Pl., 977, 1753.

Hair Sedge

Carex chlorostachyas Stev.
Carex fuscidula Krecz.
Carex krausei Boeckl.
Trasus capillaris (L.) S.F. Gray

Loosely tufted, phyllopodic perennial from fibrous roots; rhizomes lacking or very short; culms slender, erect or decumbent. Leaves mostly basal, flat, thin, narrow, often channeled near the base. Lowest bract leaflike, with well-developed sheath, shorter than the inflorescence. Spikes 2–5, the terminal one small, slender, staminate; lateral spikes pistillate, ascending, spreading, or reflexed on slender, drooping peduncles; pistillate scales ovate to obovate, obtuse or acute, brown with hyaline margins and apex, wider and shorter than the perigynia; perigynia slightly inflated, elliptic, greenish brown, 2-ribbed, stipitate, the apex gradually tapering to a short beak; stigmas 3; akenes trigonous.

Wet stream banks, meadows, and rocky places of the Absaroka, Beartooth, Big Horn, Medicine Bow, Uinta, and Wind River Ranges. Circumpolar; south in western North America to British Columbia, Nevada, New Mexico, and Colorado.

Alpine plants of the Middle Rocky Mountains that are less than 20 cm tall, have narrow leaves less than 2 mm wide, and short inflorescences less than 6 cm long are var. *capillaris*.

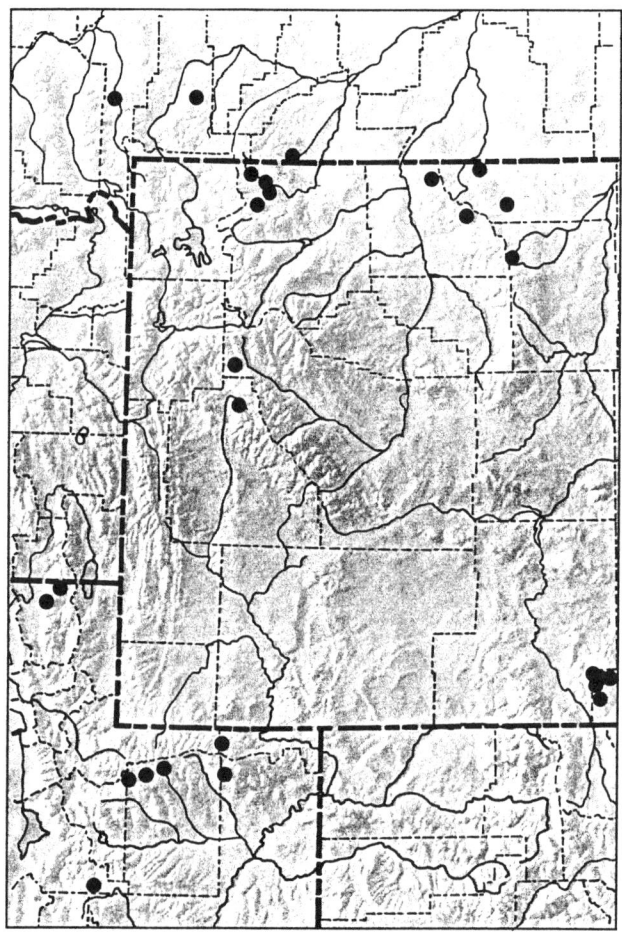

Carex capitata L.
Syst. Nat., 10th ed, 1261, 1759.

Capitate Sedge

Carex arctogena H. Smith
Diemisia capitata (L.) Raf.
Vignea capitata (L.) Reichenb. ex Moessl.

Tufted, aphyllopodic perennial from creeping, often ascending rhizomes with short internodes; culms solitary, slender, stiff, mostly longer than the leaves when mature. Leaves basal or nearly so, the blades narrow, involute; basal sheaths often purplish red. Spike solitary, ebracteate, broadly ovate to globose, androgynous; pistillate scales ovate to nearly orbicular, dark brown, with hyaline margins, shorter and narrower than the perigynia; perigynia ovate to broadly elliptic, spreading, sessile, rather abruptly contracted into a short, obliquely cleft beak; stigmas 2; akenes lenticular, the rachilla equaling or exceeding the akene in length.

Fellfields and seasonally wet meadows and alp slopes of the Absaroka, Beartooth, Big Horn, Uinta, and Wind River Ranges. Circumpolar; south in western North America to California, Nevada, Mexico, and Colorado.

Middle Rocky Mountain plants are ssp. *arctogena* (Sm.) Böcher.

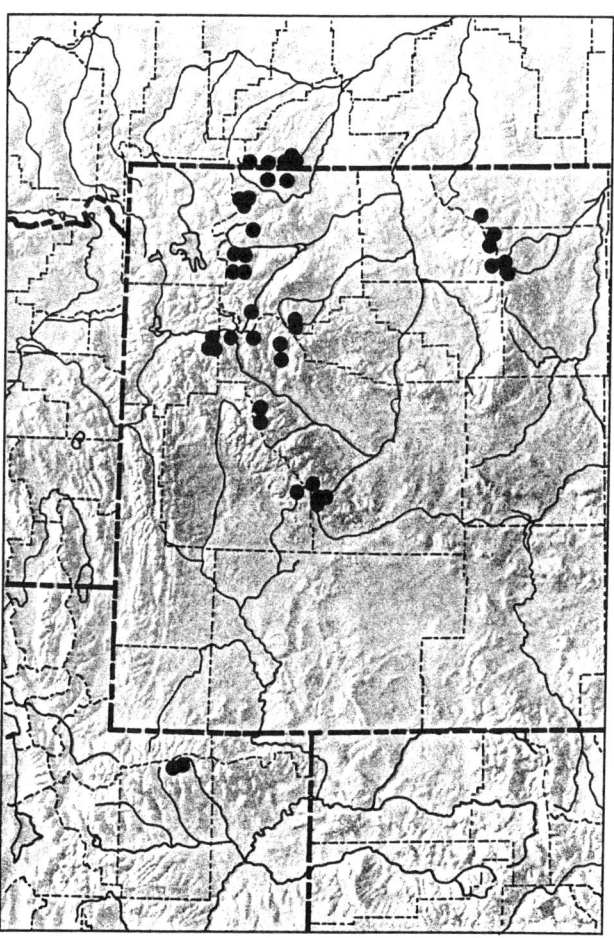

Carex douglasii F. Boott in Hook.
Fl. Bor.-Am. 2:213, 1839.

Douglas Sedge

Carex irrasa L.H. Bailey
Carex nuttallii Dewey

Slender, mostly dioecious, phyllopodic perennial from long-creeping rhizomes; culms solitary or a few together, equaling to exceeding the mostly basal leaves; blades narrow, flattened to involute. Lowest bract attenuate, shorter than the inflorescence, or lacking. Spikes several to numerous, sessile, congested, either all staminate or all pistillate, the staminate inflorescence elongate; pistillate inflorescence ovate, oblong, or suborbicular; pistillate scales lanceolate to ovate, acute to acuminate, with hyaline margins and a pale green midrib, wider and longer than the perigynia, concealing them above; perigynia brown or straw-colored, lanceolate to ovate, finely nerved, stipitate to substipitate, the apex narrowed to a long, serrulate, bidentate beak; stigmas 2, elongate and very conspicuous; akenes lenticular.

Common in dry or moist, often alkaline, situations at lower elevations, this species occurs at timberline in the Uinta Range. British Columbia to Manitoba and south to California, Utah, New Mexico, Colorado, and Iowa.

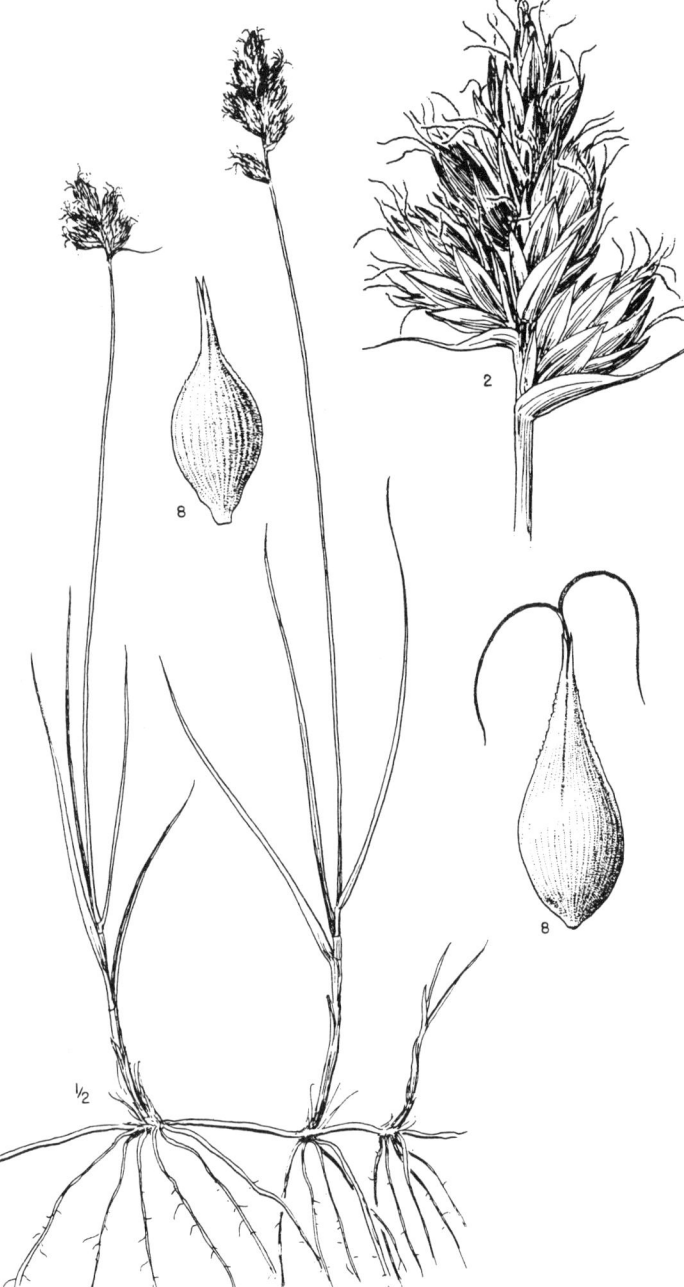
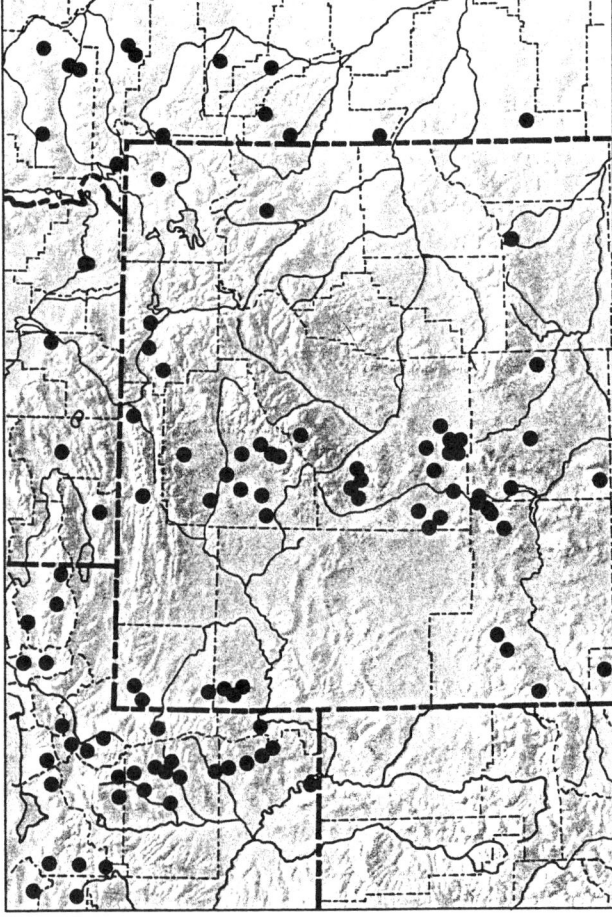

Carex ebenea Rydb.
Bull. Torr. Club 28:266, 1901.

Ebony Sedge

Densely tufted, aphyllopodic perennial from fibrous roots; rhizomes short or lacking; culms several to many, striate, scaberulous above. Leaves mostly basal, much shorter than the culms; blades flat. Lowest bract lacking or inconspicuous and shorter than the inflorescence. Spikes 5–10, gynaecandrous, sessile, in a dense ovoid to subglobose head; pistillate scales lanceolate to narrowly ovate, acute or obtuse, dark brownish or blackish, usually with narrow, hyaline margins, the midribs mostly lacking, shorter and mostly narrower than the perigynia; perigynia narrowly lanceolate, flattened, finely nerved, brownish to blackish, gradually narrowed into a long, slender, bidentate, finely serrulate beak (smooth at the apex), the base substipitate; stigmas 2; akenes lenticular.

Fellfields and meadows of the Medicine Bow, Uinta, and Wind River Ranges. Montana south to Arizona and New Mexico.

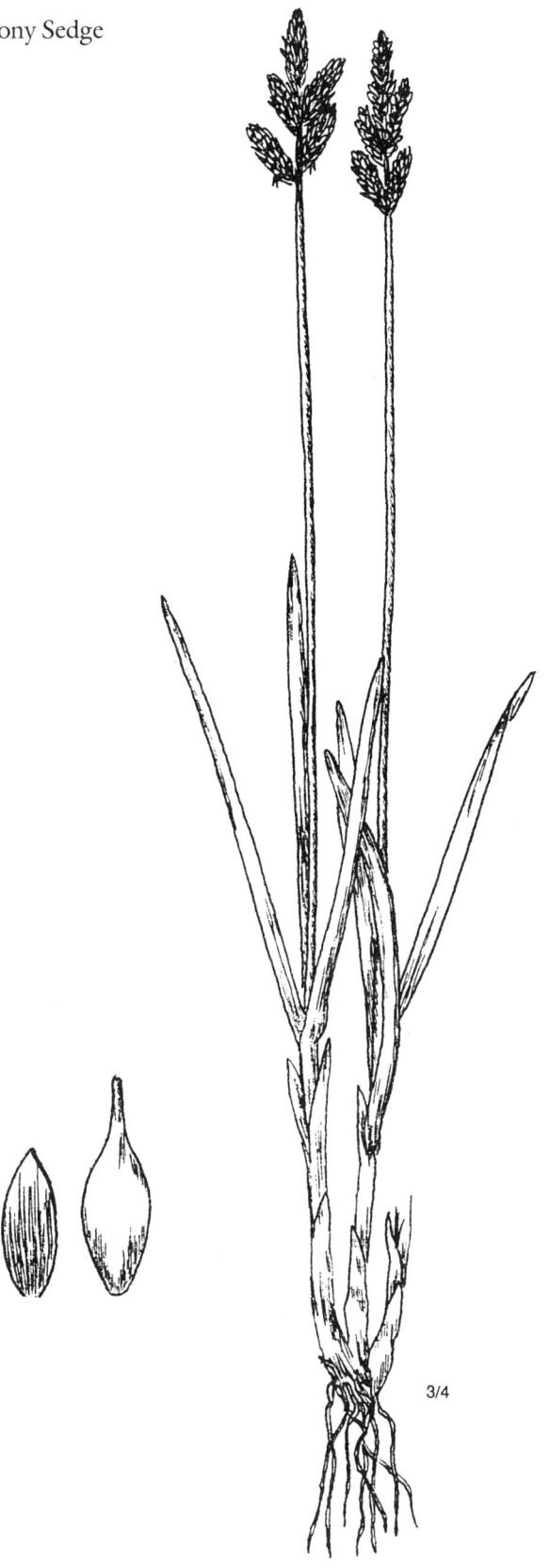

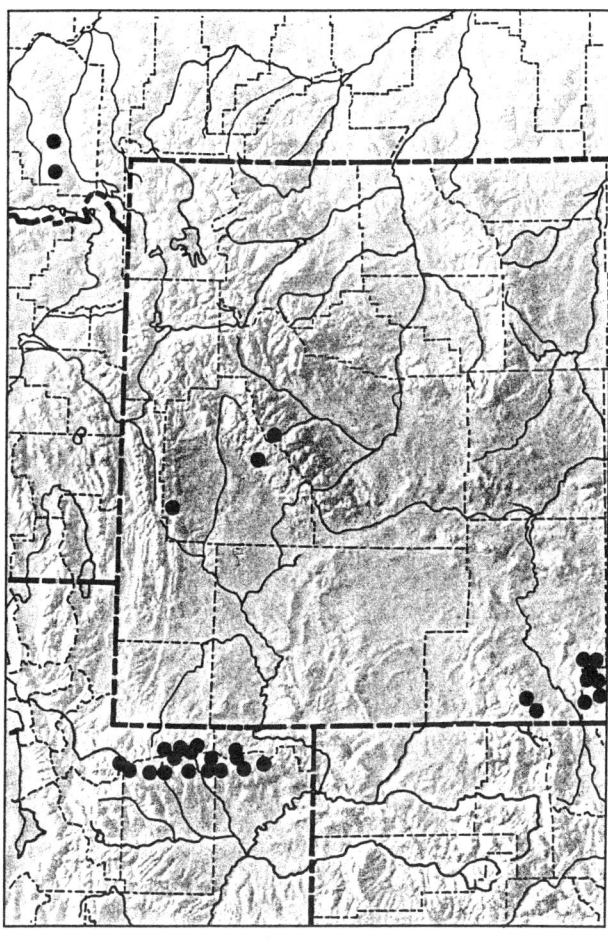

Carex echinata Murray
Prod. Stirp. Gott., 76, 1770.

Rough Sedge

Carex angustior Mack. in Rydb.
Carex cephalantha (L.H. Bailey) Bicknell
Carex hawaiiensis St. John
Carex josselynii (Fern.) Mack. ex Pease
Carex laricina Mack. ex Bright
Carex leersii Willd.
Carex muricata auctt. non L.
Carex ormantha (Fern.) Mack.
Carex phyllomanica W. Boott
Carex stellulata Gooden.
Carex sterilis Willd.
Carex svensonis Skottsberg
Caricina stellulata (Gooden.) St.-Lag. in Cariot
Vignea stellulata (Gooden.) Reichenb. ex Moessl.

Densely tufted, aphyllopodic perennial from fibrous roots; rhizomes lacking; culms erect, slender, scaberulous above. Leaves basal, mostly shorter than the culms; blades narrow, flat or sometimes channeled. Lowest bract ovate, awn-tipped, much shorter than the inflorescence. Spikes 2–6, sessile, approximate, gynaecandrous, the lateral ones occasionally pistillate; pistillate scales ovate, yellowish brown with a pale midrib, the margins hyaline, shorter and mostly narrower than the perigynia; perigynia widely spreading at maturity, lanceolate to ovate, planoconvex, nerved dorsally, gradually tapered to a serrulate, bidentate beak; stigmas 2; akenes lenticular.

Wet meadows near timberline in the Uinta Range. Circumboreal; south in western North America to California, Utah, and Colorado.

This species has been widely known as *C. muricata* L. According to Reznicek and Ball (1980) the use of this name for western North American plants is incorrect. Plants of the Middle Rocky Mountain alpine ecotone are thus *C. echinata* ssp. *echinata*.

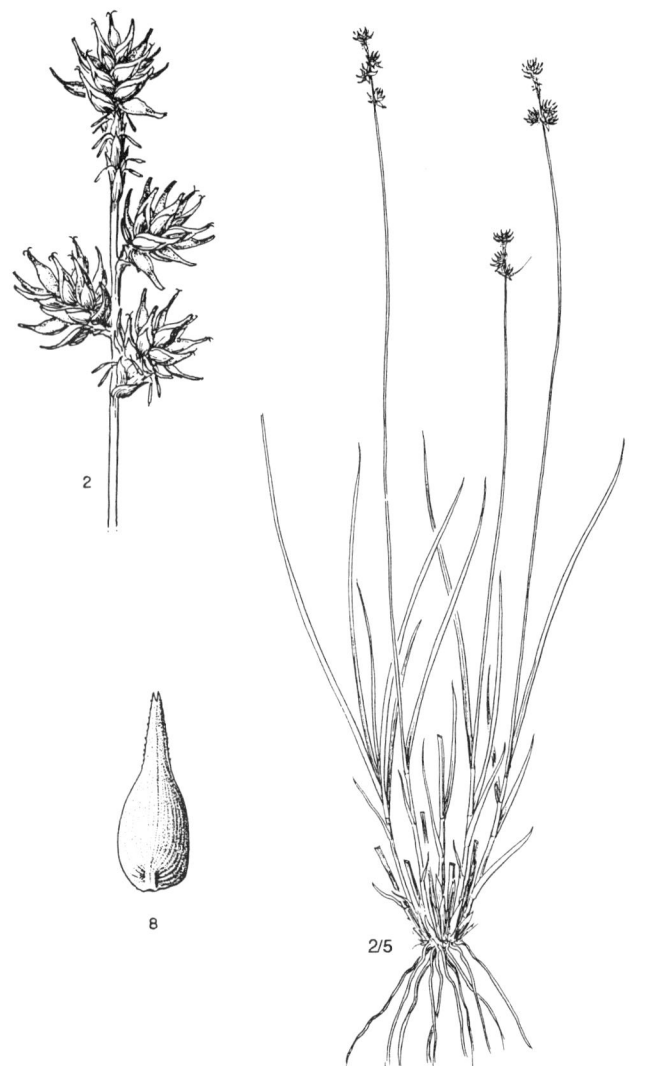
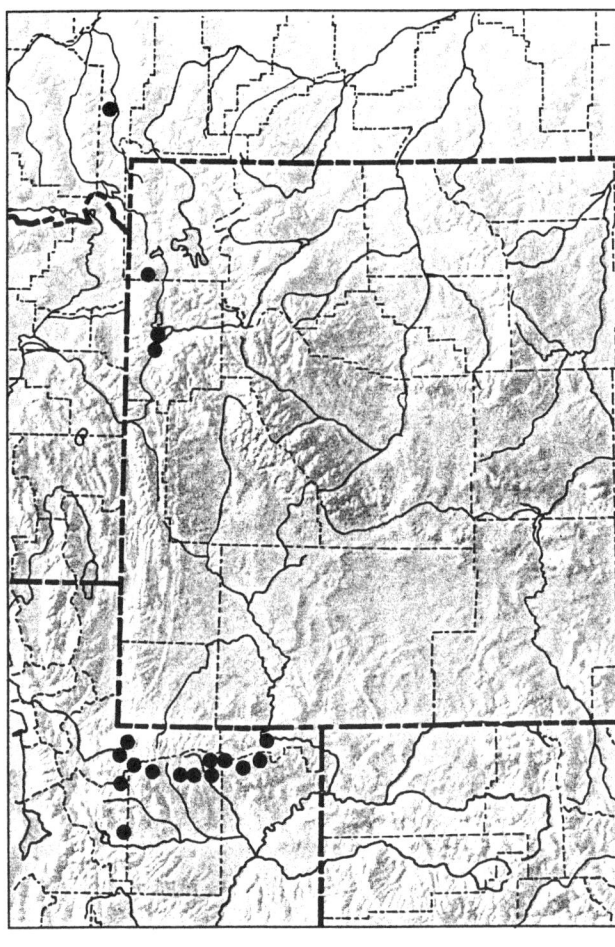

Carex elynoides Holm
Am. J. Sci., IV, 9:356, 1900.

Blackroot Sedge

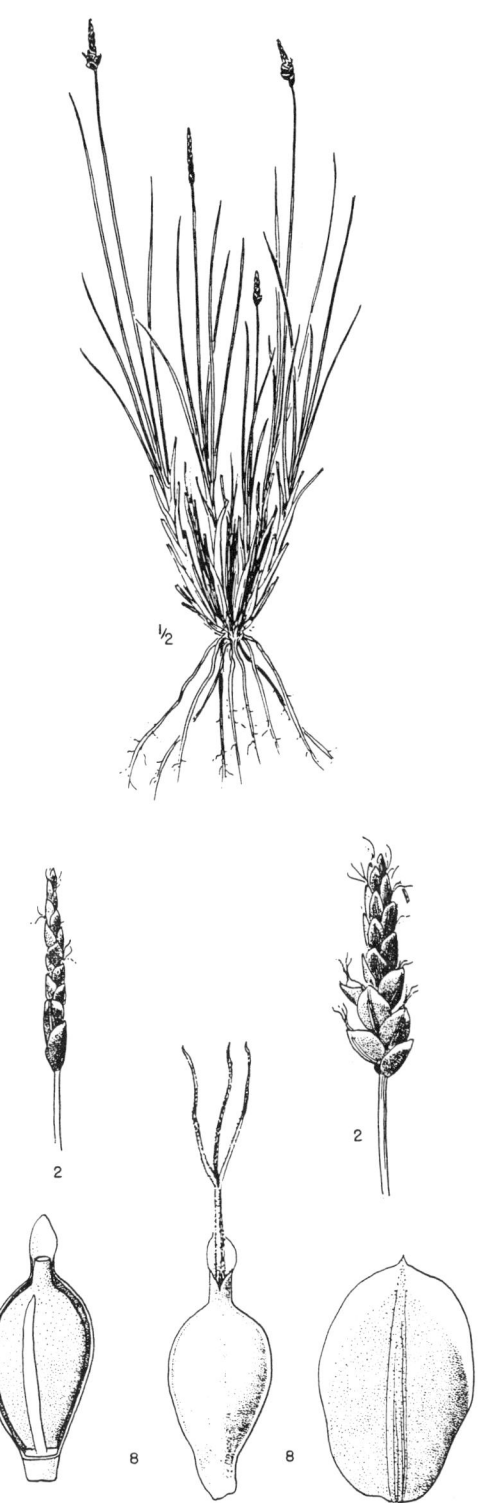

Densely tufted, mostly phyllopodic, fasciculate perennial from fibrous roots; rhizomes lacking; culms glabrous, slender to filiform, wiry. Leaves basal or nearly so, equaling or shorter than the culms; blades involute to terete, wiry. Spikes solitary, androgynous, narrowly cylindric, bractless; pistillate scales oblong to broadly obovate, brownish, the margins hyaline, wider and usually longer than the perigynia, mostly concealing them, the lower scales mucronate, becoming acute to obtuse upward; perigynia erect to ascending, obovate, greenish to brownish, glabrous or sometimes sparsely puberulent near the prominent short-cylindric, hyaline, oblique beak; rachilla about as long as the akene; stigmas 3; akenes trigonous.

Dry tundra, fellfields, and rocky slopes in all alpine ranges of the Middle Rocky Mountains. Montana south to Idaho, Utah, and Colorado.

Carex elynoides closely resembles *Kobresia bellardi,* which is distinguished by having the perigynium split down one side. Two other small similar sedges are *C. nardina,* which lacks hyaline margins on the pistillate scales; and *C. filifolia,* which has puberulent rather than glabrous or glabrate perigynia.

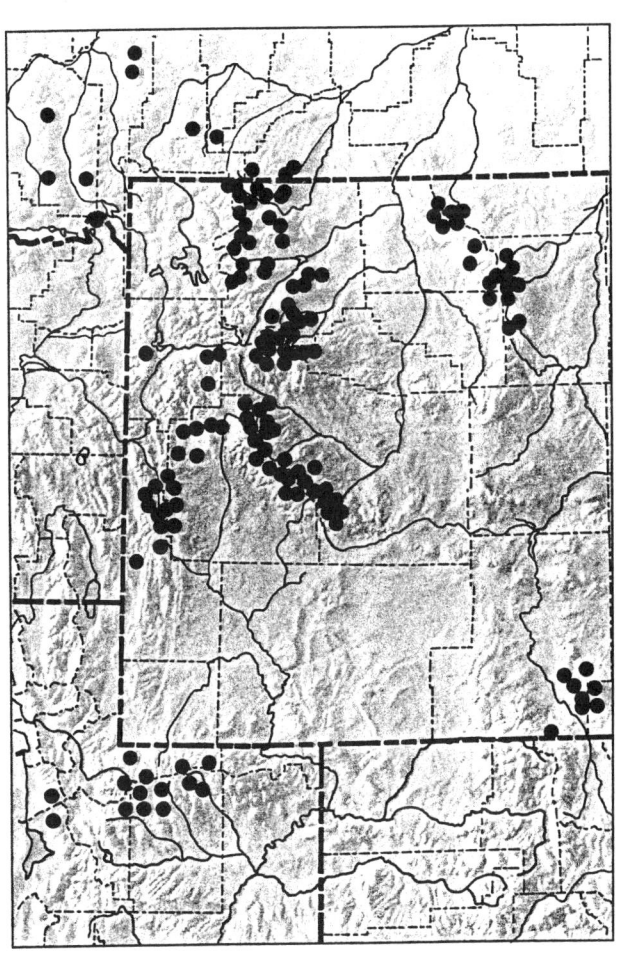

Carex filifolia Nutt.
Gen. Pl. 2:204, 1818.

Threadleaf Sedge

Carex elyniformis Porsild
Olotrema filifolia (Nutt.) Raf.
Uncinia filifolia (Nutt.) Nees in Wied-Neuwied

Caespitose, mostly phyllopodic, fasciculate, perennial herb from slender, fibrous roots; rhizomes lacking; culms several to many, slender, wiry. Leaves basal or nearly so, equaling or shorter than the culms; blades involute to terete, wiry. Spikes solitary, androgynous, cylindric, and bractless; pistillate scales broadly obovate, brownish, often with a pale midrib, the margins broadly hyaline, wider than the perigynia, the lower scales mucronate, the upper ones rounded or sometimes erose; perigynia erect to ascending, obovate to elliptic, light to dark brown at maturity, finely puberulent at least above the middle, beaked, the beak short, hyaline, oblique; rachilla about as long as the akene; stigmas 3; akenes trigonous.

Timberline in the Absaroka and Gros Ventre Ranges. Alaska and Yukon Territory south to Oregon, Nevada, Utah, New Mexico, Texas, and Nebraska.

Carex filifolia resembles *C. elynoides,* from which it may be distinguished by its finely puberulent perigynia, at least on the upper half.

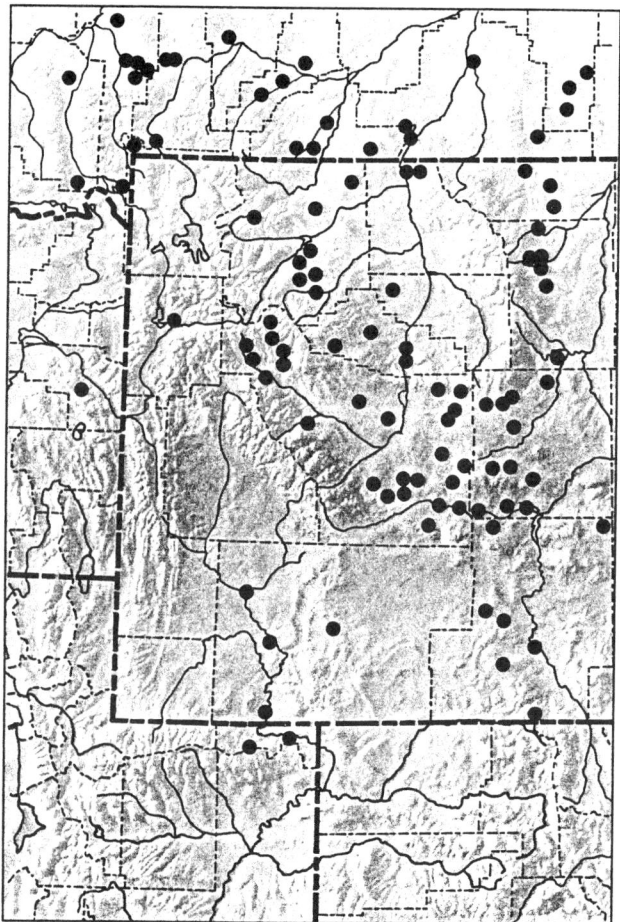

Carex geyeri F. Boott
Trans. Linn. Soc. 20:118, 1846.

Elk Sedge

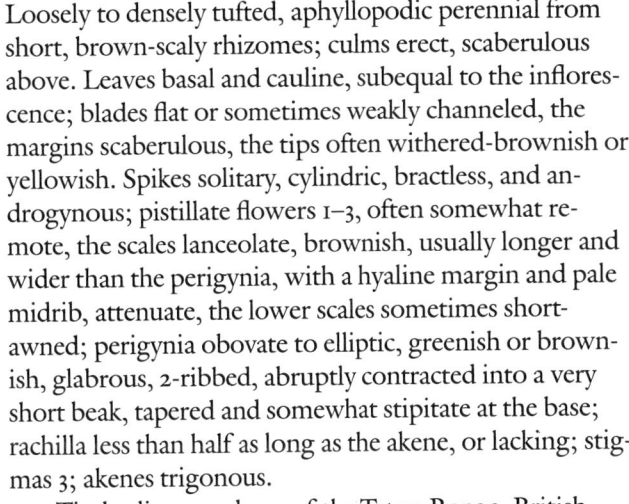

Loosely to densely tufted, aphyllopodic perennial from short, brown-scaly rhizomes; culms erect, scaberulous above. Leaves basal and cauline, subequal to the inflorescence; blades flat or sometimes weakly channeled, the margins scaberulous, the tips often withered-brownish or yellowish. Spikes solitary, cylindric, bractless, and androgynous; pistillate flowers 1–3, often somewhat remote, the scales lanceolate, brownish, usually longer and wider than the perigynia, with a hyaline margin and pale midrib, attenuate, the lower scales sometimes short-awned; perigynia obovate to elliptic, greenish or brownish, glabrous, 2-ribbed, abruptly contracted into a very short beak, tapered and somewhat stipitate at the base; rachilla less than half as long as the akene, or lacking; stigmas 3; akenes trigonous.

Timberline meadows of the Teton Range. British Columbia and Alberta south to California, Utah, and Colorado.

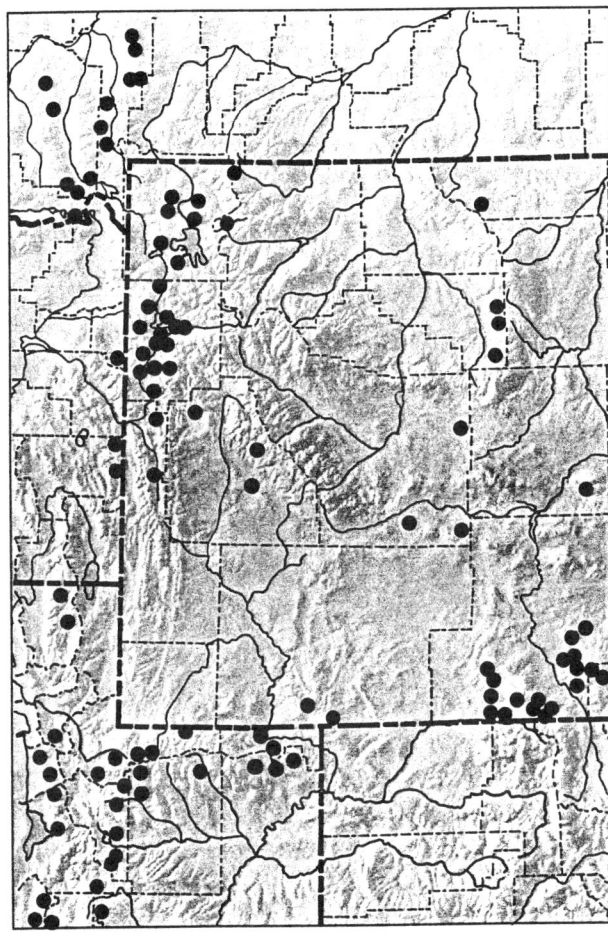

Carex hoodii F. Boott in Hook.
Fl. Bor.-Am. 2:211, 1839.

Hood Sedge

Densely caespitose, aphyllopodic perennial from fibrous roots; rhizomes lacking; culms erect, slender, scaberulous. Leaves basal and cauline, shorter than the culms; blades slender, flat or sometimes channeled. Lowest bract shorter than the inflorescence, or lacking. Spikes several to many, few-flowered, androgynous in a dense oblong to orbicular head; pistillate scales ovate, brown with a pale midrib, scarious-hyaline on the margins, equaling or shorter than the perigynia; perigynia ovate to elliptic, dark brown, with winged margins, nerveless or nearly so, the beak prominent, about one-third the length of the body, bidentate, serrulate on the margins; stigmas 2; akenes lenticular.

Meadows near timberline in the Absaroka, Salt River, and Wind River Ranges. British Columbia to Saskatchewan, south to California, Utah, Colorado, and South Dakota.

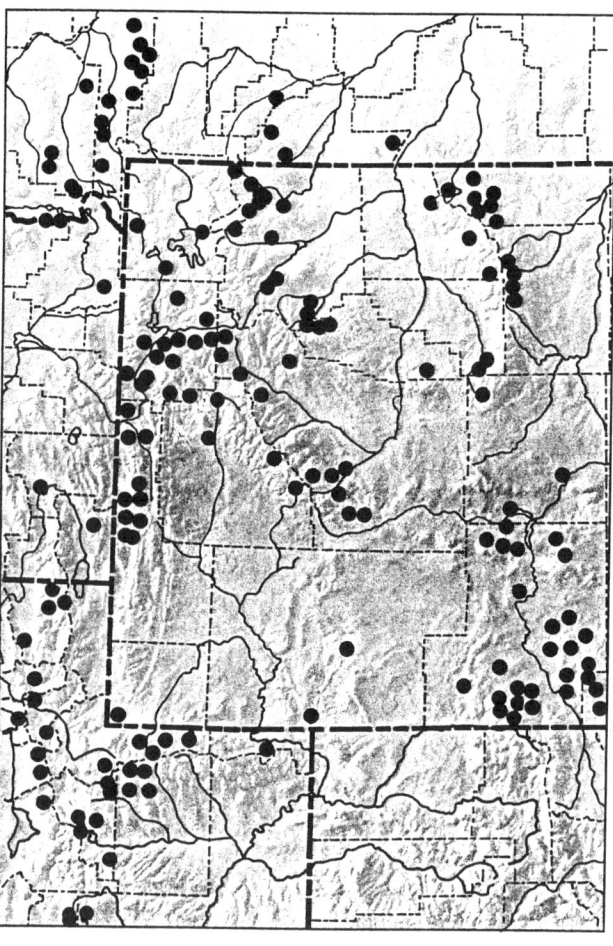

Carex illota L.H. Bailey
Mem. Torr. Club 1:15, 1889.

Sheep Sedge

Densely tufted, aphyllopodic perennial from short-creeping rhizomes; culms slender, erect, scaberulous. Leaves mostly basal, much shorter than the culms; blades flat or sometimes channeled. Lowest bract lacking or much shorter than the inflorescence. Spikes several, gynaecandrous, sessile, clustered in a dense, ovate to suborbicular head; pistillate scales shorter and mostly narrower than the perigynia, ovate, obtuse, dark brownish or greenish black, with narrow, hyaline margins, often with a pale midrib; perigynia entire, lanceolate to narrowly ovate, planoconvex, brownish to greenish above, yellowish brown below, obscurely nerved, gradually tapered to a dark brown to black, oblique-cleft beak; stigmas 2; akenes lenticular.

Stream banks, wet meadows, and willow thickets in all except the Gros Ventre, Hoback, Salt River, and Wyoming Ranges. British Columbia south to California, Utah, and Colorado.

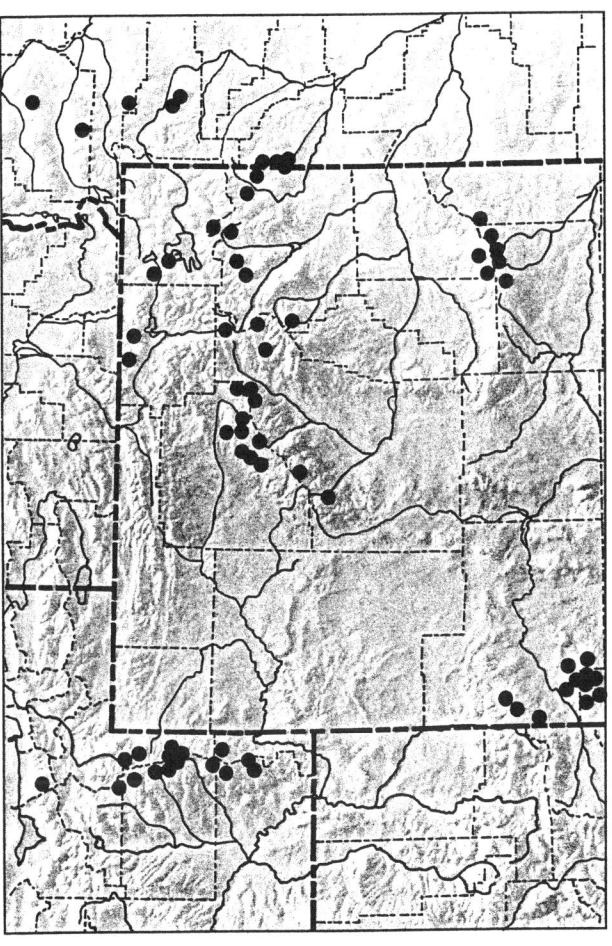

Carex jonesii L.H. Bailey
Mem. Torr. Club 1:16, 1889.

Jones Sedge

Loosely tufted, somewhat aphyllopodic perennial from creeping rhizomes; culms erect, slender, scaberulous below the inflorescence. Leaves mostly basal, much shorter than the culms; blades flat, stiff; sheaths not cross-rugulose. Lowest bract ovate, inconspicuous, much shorter than the inflorescence. Spikes few to several, androgynous, sessile, in a dense oblong to orbicular head; pistillate scales ovate, brownish to blackish, the margins hyaline, slightly shorter and wider than the perigynia, mostly concealing them; perigynia spreading, entire, lanceolate to narrowly ovate, dark brown, planoconvex, nerved on both surfaces, gradually tapered to a deeply bidentate beak, the base short-stipitate; stigmas 2; akenes lenticular.

Timberline meadows of the Wyoming Range. Washington south to California, Utah, and Colorado.

Carex jonesii closely resembles *C. neurophora*, but differs by the presence of creeping rhizomes and the lack of cross-rugulose leaf sheaths.

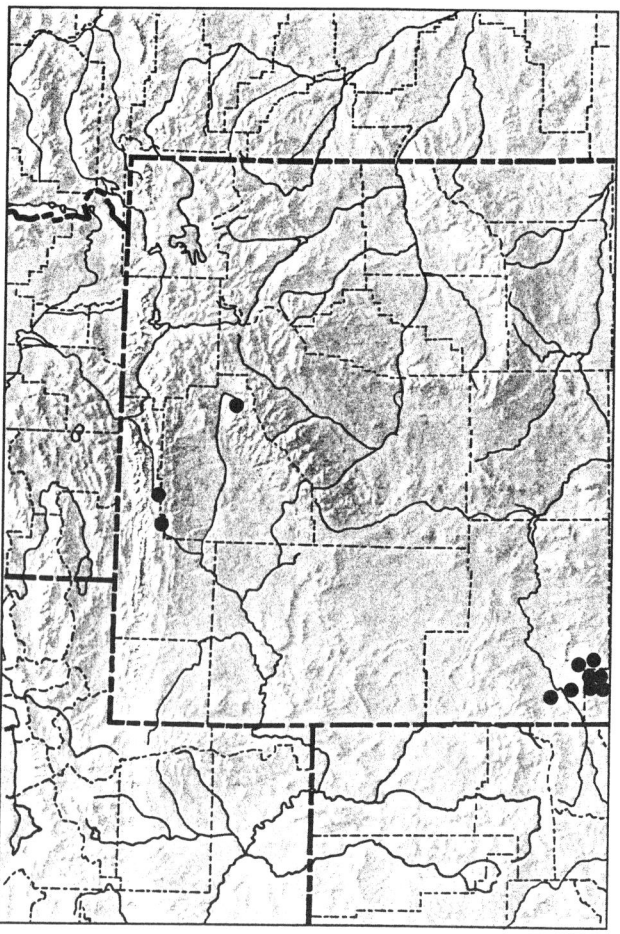

Carex leporinella Mack.
Bull. Torr. Club 43:605, 1917.

Sierra Hare Sedge

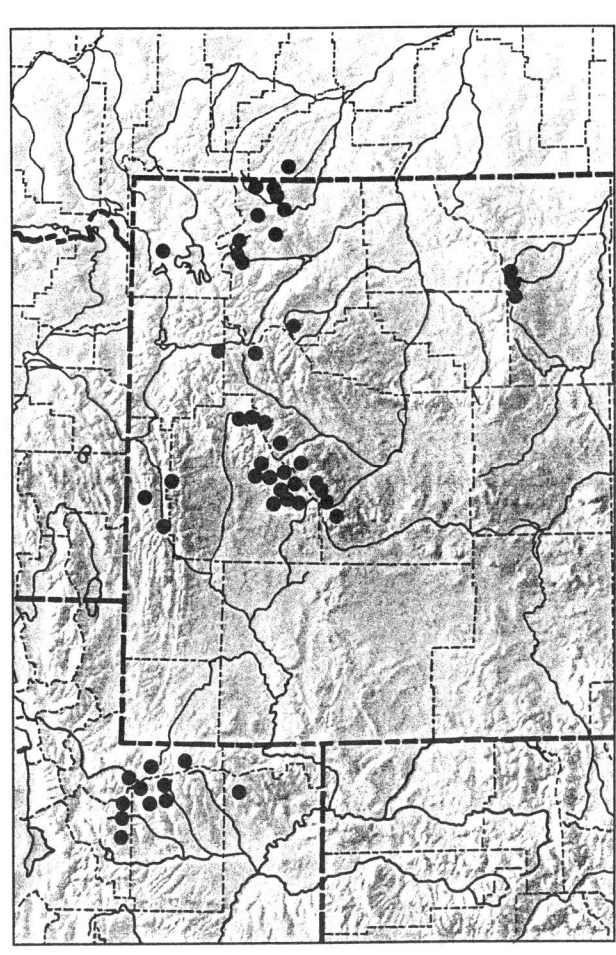

Densely tufted, aphyllopodic perennial from fibrous roots; rhizomes lacking; culms slender, stiffly erect, glabrous. Leaves mostly basal, mostly shorter than the culms; blades flat or channeled, sometimes involute. Lowest bract shorter than the inflorescence, sometimes lacking. Spikes few to several, gynaecandrous, loosely aggregated into an elongate, ovoid head, the lower spikes often remote; pistillate scales ovate, brown, often with a pale midrib, acute, with hyaline margins, as long and wide as the perigynia, mostly concealing them; perigynia planoconvex, lanceolate to narrowly elliptic, serrulate, conspicuously nerved above, faintly nerved below, gradually tapering to a shallowly bidentate beak; stigmas 2; akenes lenticular.

Fellfields and timberline meadows of the Absaroka, Beartooth, Big Horn, Salt River, Uinta, Wind River, and Wyoming Ranges. Washington south to California, Nevada, Utah, and Wyoming.

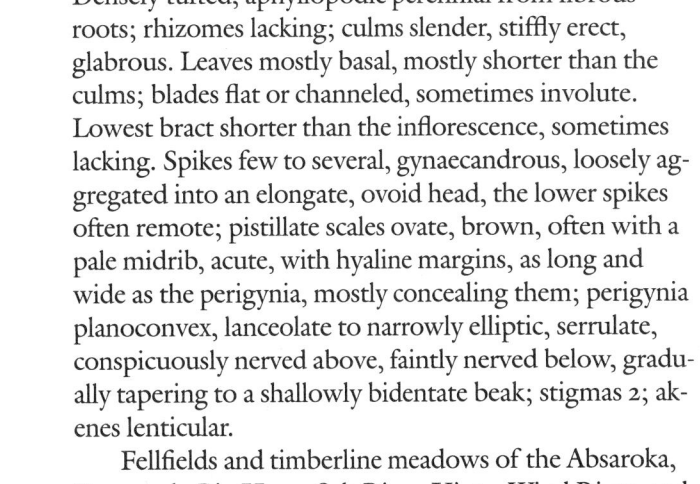

Carex luzulina Olney
Proc. Am. Acad. 7:395, 1868.

Woodrush Sedge

Carex ablata L.H. Bailey
Carex fissuricola Mack.

Loosely tufted, aphyllopodic perennial from fibrous roots; rhizomes lacking; culms solitary to few, erect. Leaves mostly basal, shorter than the culms; cauline leaves with closed sheaths 3–5 cm long; blades flat, broad, thick. Lowest bract with well-developed sheath and blade, shorter than the inflorescence. Spikes 3–5 in an elongate inflorescence; terminal spike staminate; lateral spikes pistillate, erect, the upper ones sessile, the lower pedunculate; pistillate scales ovate, acute to obtuse, dark brown with hyaline margins and a pale midrib, shorter and often narrower than the perigynia; perigynia lanceolate to narrowly ovate, flattened, faintly nerved, the margins ciliolate-serrulate, gradually tapering to a bidentate beak; stigmas 3; akenes trigonous.

Wet timberline meadows of the Teton and Wind River Ranges. British Columbia south and east to California, Montana, Wyoming, and Utah.

Middle Rocky Mountain plants are var. *atropurpurea* Dorn.

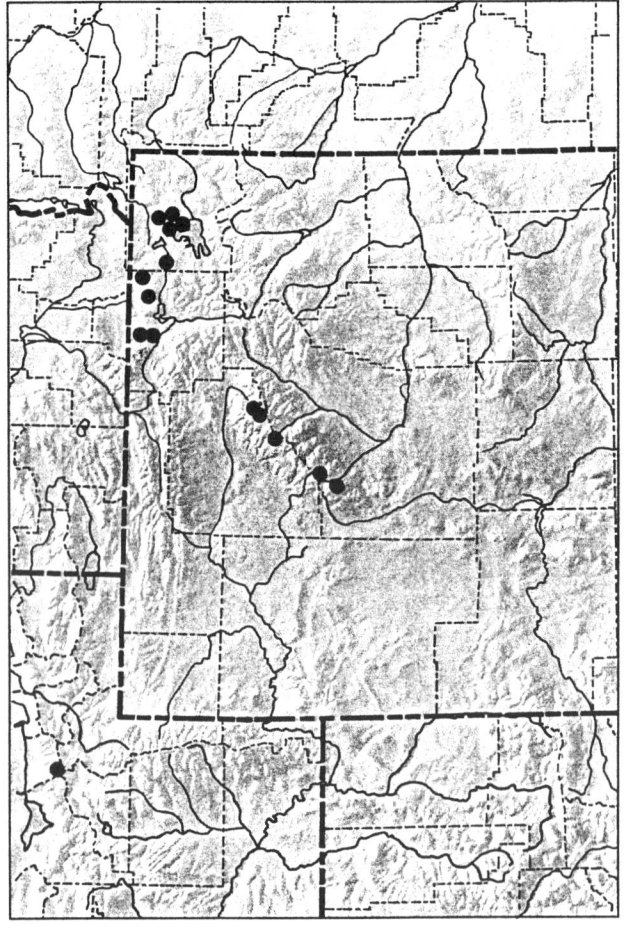

Carex macloviana D'Urv.
Mem. Soc. Linn. Paris 4:599, 1826.

Maclovian Sedge

Carex festiva Dewey
Carex festivella Mack.
Carex haydeniana Olney in King
Carex incondita F.J. Hermann
Carex microptera Mack.
Carex multimoda L.H. Bailey
Carex nubicola Mack.
Carex olympica Mack.
Carex pachystachya Cham. ex Steud.
Carex platylepis Mack.
Carex preslii Steud.
Carex pyrophila Gand.
Carex soperi Raup
Carex stenoptera Mack.
Carex subfusca W. Boott

Densely tufted, strongly to weakly aphyllopodic perennial from fibrous roots; rhizomes lacking, rarely present and very short; culms erect, slender, glabrous to scaberulous below the inflorescence. Leaves mostly basal, shorter than the culms; blades flat, firm. Lowest bract shorter than the inflorescence. Spikes several to many, sessile, gynaecandrous, tightly to loosely aggregated in an elongate to orbicular head; pistillate scales light brown to brownish black or copper-colored, lanceolate to ovate, acute to obtuse, often with a pale midrib, shorter and mostly narrower than the perigynia; perigynia lanceolate, elliptic, or ovate, greenish, brown, or coppery, nerved or nerveless, gradually narrowed to a bidentate beak; stigmas 2; akenes lenticular.

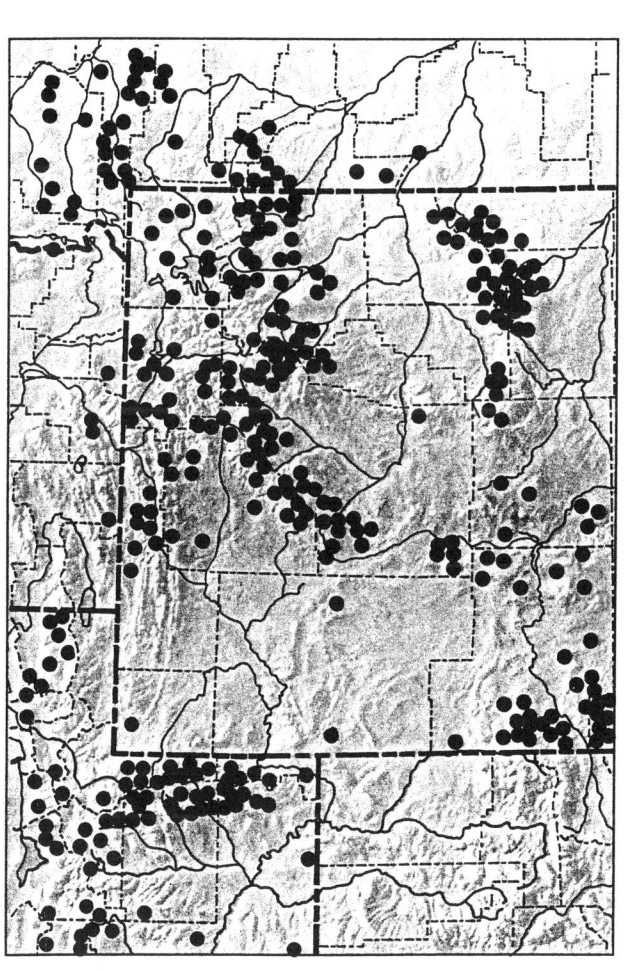

var. pachystachya

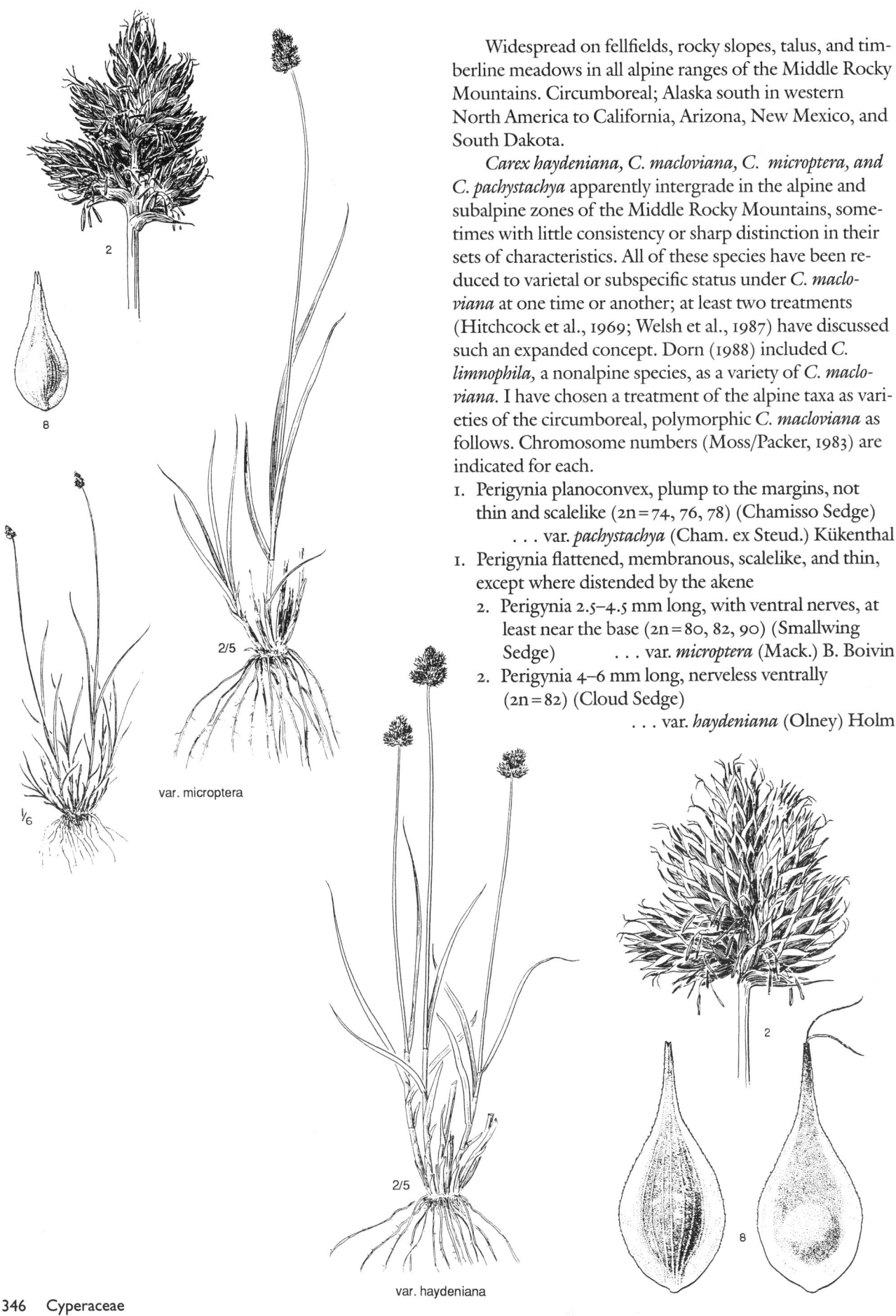

Widespread on fellfields, rocky slopes, talus, and timberline meadows in all alpine ranges of the Middle Rocky Mountains. Circumboreal; Alaska south in western North America to California, Arizona, New Mexico, and South Dakota.

Carex haydeniana, C. macloviana, C. microptera, and C. pachystachya apparently intergrade in the alpine and subalpine zones of the Middle Rocky Mountains, sometimes with little consistency or sharp distinction in their sets of characteristics. All of these species have been reduced to varietal or subspecific status under *C. macloviana* at one time or another; at least two treatments (Hitchcock et al., 1969; Welsh et al., 1987) have discussed such an expanded concept. Dorn (1988) included *C. limnophila,* a nonalpine species, as a variety of *C. macloviana*. I have chosen a treatment of the alpine taxa as varieties of the circumboreal, polymorphic *C. macloviana* as follows. Chromosome numbers (Moss/Packer, 1983) are indicated for each.

1. Perigynia planoconvex, plump to the margins, not thin and scalelike ($2n = 74, 76, 78$) (Chamisso Sedge)
   . . . var. *pachystachya* (Cham. ex Steud.) Kükenthal
1. Perigynia flattened, membranous, scalelike, and thin, except where distended by the akene
   2. Perigynia 2.5–4.5 mm long, with ventral nerves, at least near the base ($2n = 80, 82, 90$) (Smallwing Sedge) . . . var. *microptera* (Mack.) B. Boivin
   2. Perigynia 4–6 mm long, nerveless ventrally ($2n = 82$) (Cloud Sedge)
      . . . var. *haydeniana* (Olney) Holm

346   Cyperaceae

Carex maritima Gunn.
Fl. Norveg. 2:131, 1772.

Incurved Sedge

Carex amphilogos K. Koch
Carex arctica Deinb.
Carex banata Sm. in Rees
Carex bucculenta Krecz.
Carex camptotropa Krecz.
Carex danaensis Stacey
Carex hyalinolepis Steud.
Carex incurva Lightf.
Carex incurviformis Mack. in Rydb.
Carex jucunda Krecz.
Carex juncifolia All.
Carex orthocaula Krecz.
Carex psammogaea Steud.
Carex pseudofoetida Kükenth.
Carex psychroluta Krecz.
Carex setina (Christ.) Krecz.
Carex transmarina Krecz.
Caricina incurva (Lightf.) St.-Lag. in Cariot
Olotrema juncifolia (All.) Raf.
Vignea incurva (Lightf.) Reichenb.

2/3

Loosely tufted, somewhat aphyllopodic perennial from creeping rhizomes; culms slender, short, less than 4 cm tall. Leaves mostly basal, longer than the culms; blades thick, flat, becoming involute near the tips. Spikes few, few-flowered, androgynous, in a dense ovate to orbicular head; pistillate flowers few, the scales ovate, obtuse, brown, hyaline on the margins, shorter than the perigynia; perigynia ovate, dark brown, with a short beak about one-third the length of the body, stipitate at the base; stigmas 2; akenes lenticular.

Fellfields and other rocky places in the Absaroka and Wind River Ranges. Circumpolar; south in western North America to California, Montana, Wyoming, and Colorado.

This complex species is represented worldwide by a number of races which have been treated as separate species by some authors. Hulten (1962) regarded many of them as doubtful in their distinction. Accordingly, the disjunct Middle Rocky Mountain population of northwestern Wyoming and southwestern Montana is here treated as var. *incurviformis* (Mack.) B. Boivin.

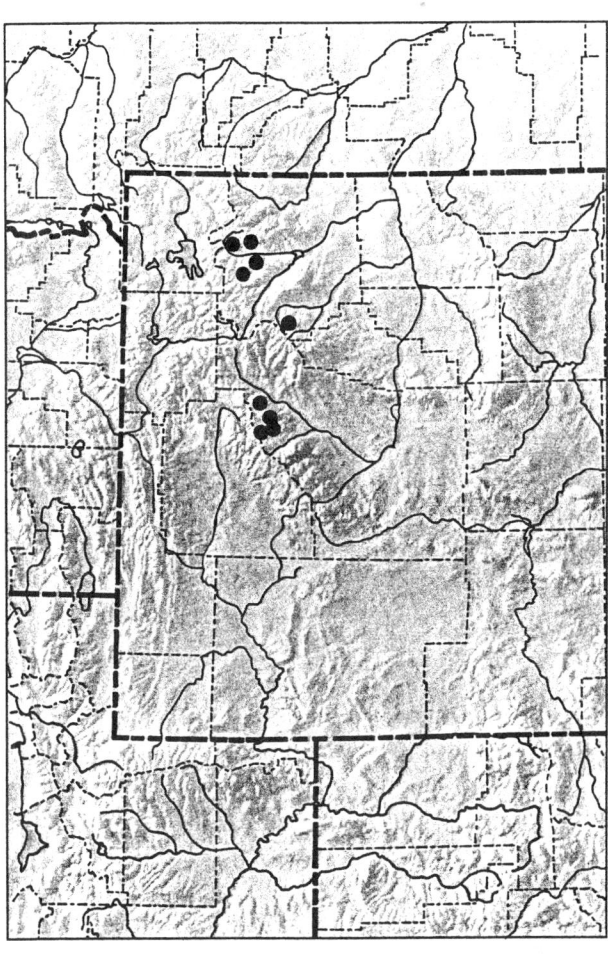

Carex misandra R. Br.
Chlor. Melv., 25, 1823.

Shortleaf Sedge

Loosely to densely tufted, phyllopodic perennial from fibrous roots; rhizomes lacking; culms solitary to few, slender, erect, nodding. Leaves basal, much shorter than the culms; blades relatively short, flat, thick, channeled below. Lowest bract with a closed sheath, shorter than the inflorescence. Spikes 2–4, nodding on long, slender peduncles, the terminal one gynaecandrous, the lateral ones pistillate; pistillate scales ovate, purplish black, the margins and tip hyaline, shorter and wider than the perigynia, sometimes with a pale midrib; perigynia flattened, lanceolate, purplish black above, light brownish or greenish below, gradually tapered to a bidentate, ciliate-serrulate beak, the base stipitate; stigmas 3; akenes trigonous.

Wet meadows, stream banks, and willow thickets of the Beartooth, Uinta, and Wind River Ranges. Circumboreal; south in western North America to Utah, Colorado, and Wyoming.

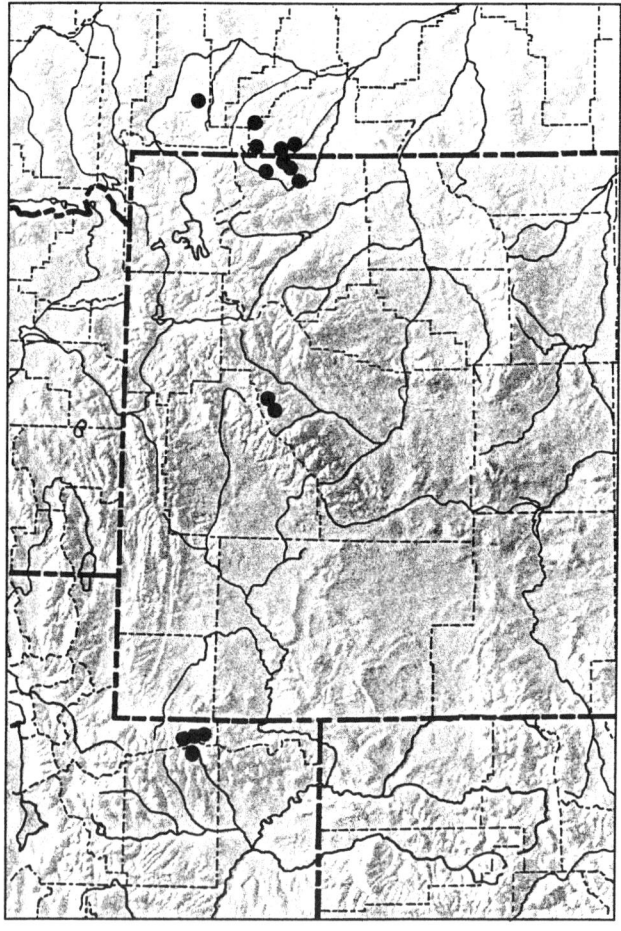

Carex nardina Fries
Nov. Fl. Suec. Mant. 2:55, 1839.

Hepburn Sedge

Carex hepburnii F. Boott in Hook.
Carex stantonensis M.E. Jones

Densely tufted, dwarf, mostly phyllopodic, fasciculate perennial, less than 15 cm tall, from fibrous roots; rhizomes lacking; culms few to several, short, slender, wiry, glabrous. Leaves mostly basal, equaling or longer than the culms; blades slender, folded or involute and nearly terete, wiry, the margins scaberulous. Bract lacking. Spikes solitary, erect, androgynous, cylindric; pistillate scales broadly ovate, brown, with hyaline margins and a pale midrib, shorter than but about as wide as the perigynia, the lower ones short-mucronate; perigynia elliptic, ovate, or obovate, planoconvex, and glabrous, the distal one-third serrulate on the margins, gradually or abruptly narrowed to a short, dark brown, bidentate, terete beak, stipitate to substipitate at the base; rachilla about as long as or slightly shorter than the akene; stigmas 2 or 3; akenes lenticular or trigonous.

Fellfields and seasonally dry slopes and meadows in all alpine Middle Rocky Mountain ranges except the Big Horn, Hoback, Salt River, and Wyoming Ranges. Circumboreal; south in western North America to Washington, Nevada, Utah, and Colorado.

Rocky Mountain plants with chromosome numbers of 2n = 68 (A. Löve et al., 1971) are ssp. *hepburnii* (F. Boott) Löve, Löve, and Kapoor. *Carex nardina* resembles *C. elynoides* but lacks the distinct hyaline margins on the pistillate scales of the latter.

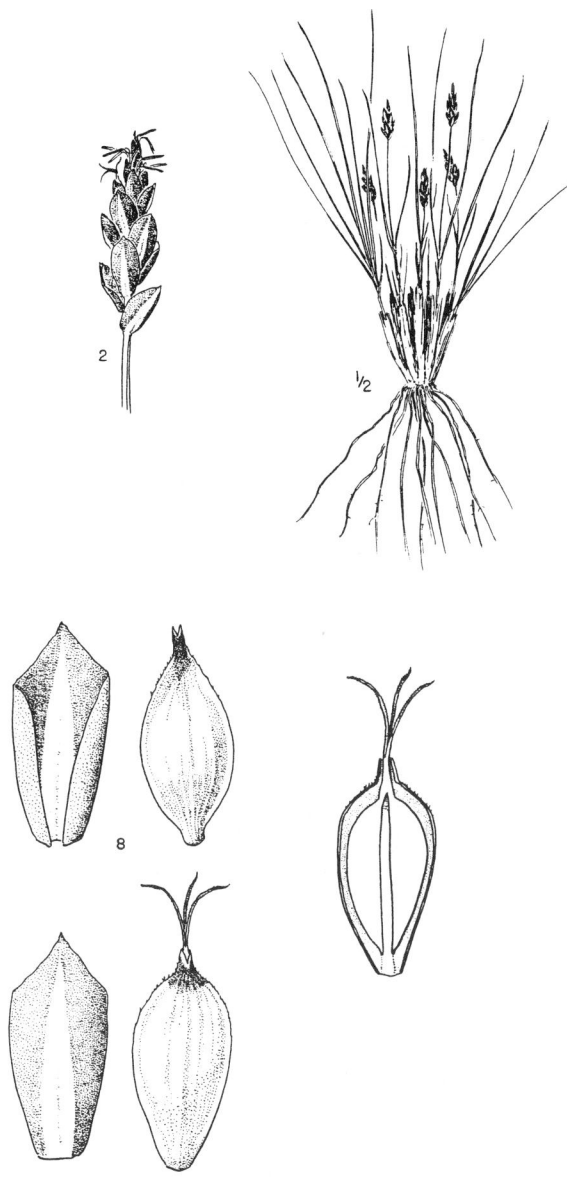

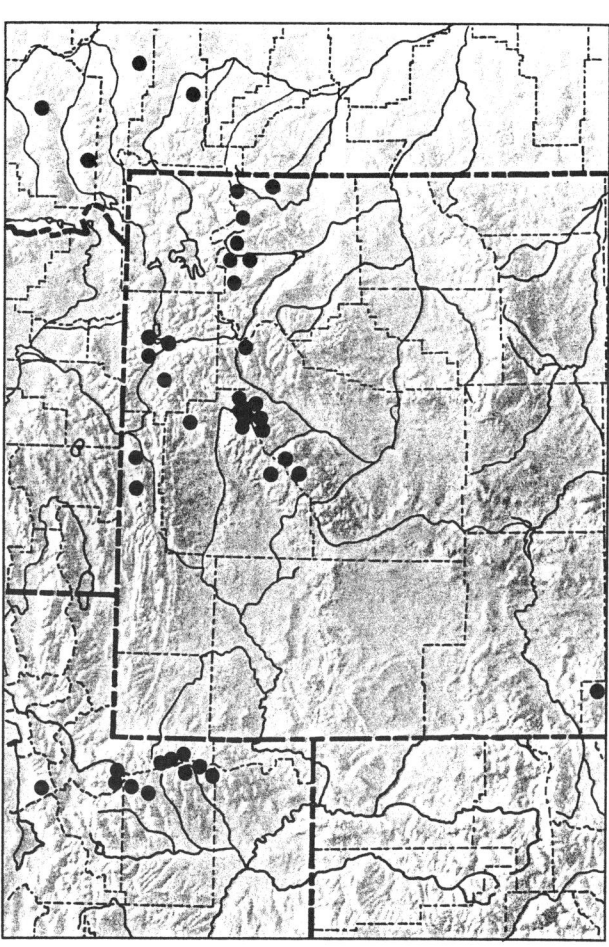

Carex neurophora Mack. in Abrams
Illustr. Fl. Pac. St. 1:298, 1923.

Alpine Nerved Sedge

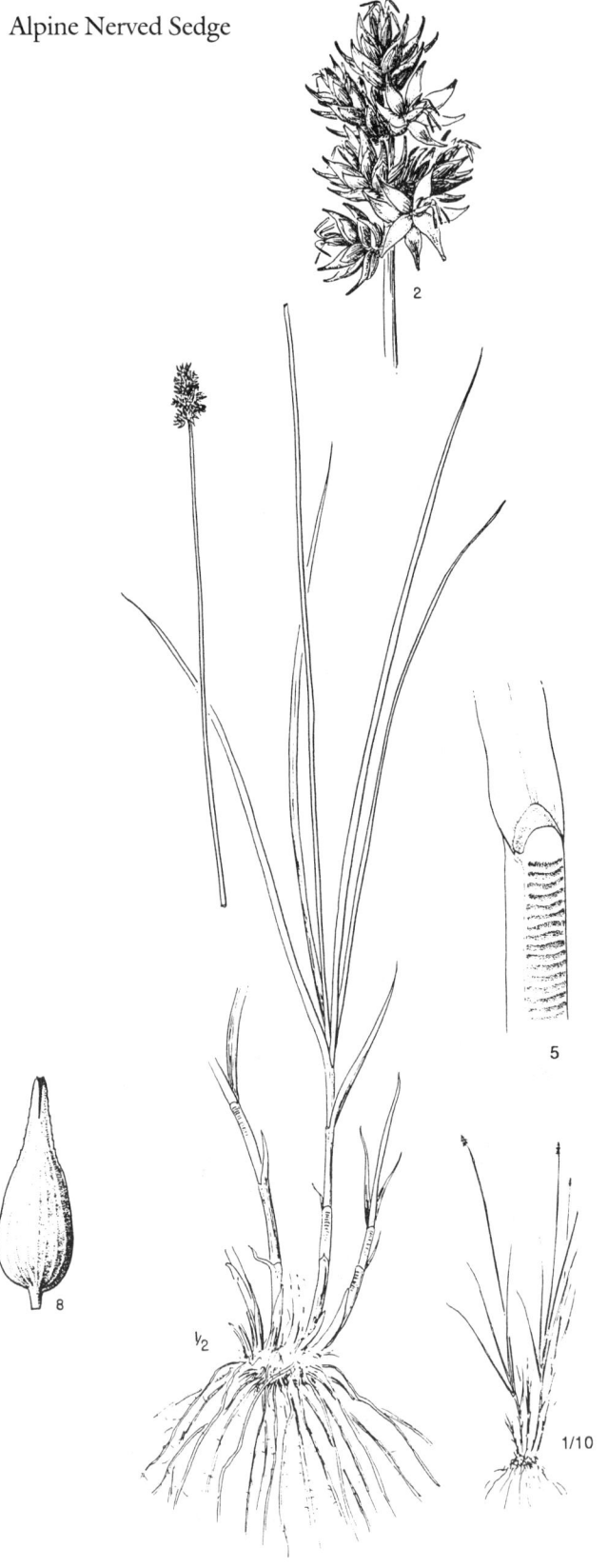

Loosely tufted, aphyllopodic perennial from fibrous roots; rhizomes usually very short; culms few to several, erect, stout. Leaves mostly basal, shorter than the culms; blades flat; sheaths cross-rugulose ventrally. Lower bract inconspicuous or lacking. Spikes 5–10, few-flowered, androgynous, sessile in an ovoid to subglobose head; pistillate scales ovate, brownish, the midrib pale green; perigynia lanceolate, planoconvex, light to dark brown, nerved on both surfaces, gradually tapered to a serrulate, bidentate beak, the base stipitate; stigmas 2; akenes lenticular.

Wet meadows and stream banks of the Wind River Range. Washington south and east to Nevada, Utah, Colorado, Montana, and Wyoming.

C. neurophora resembles *C. jonesii* but it differs by the lack of long rhizomes and the presence of cross-rugulose leaf sheaths.

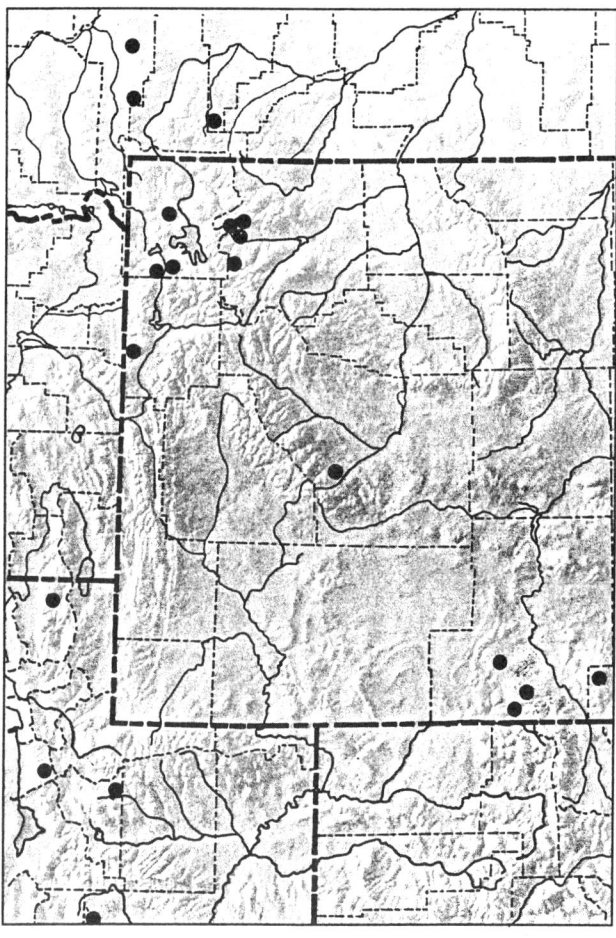

Carex nigricans C.A. Mey.
Mém. Acad. Imp. Sci. St. Pétersb., Sér. 6, Sci. Math.
1:211, pl. 7., 1831.

Black Alpine Sedge

Loosely tufted or occasionally solitary, aphyllopodic perennial from stout, short- to long-creeping rhizomes; culms erect, stiff. Leaves mostly basal, shorter than the culms; blades flat or channeled, stiff, glabrous with scaberulous margins. Lowest bract lacking. Spike solitary, androgynous (sometimes staminate or pistillate); staminate flowers erect to ascending; pistillate flowers widely spreading to reflexed, early deciduous, pistillate scales purplish black, ovate, much shorter than the perigynia, thin-margined, sometimes with a pale midrib, early deciduous; perigynia lanceolate to narrowly ovate, greenish to brownish, gradually tapered to the dark purplish, bidentate beak, stipitate; rachilla lacking; stigmas 3; akenes trigonous.

Fellfields, meadows, stream banks, and snow beds of all alpine ranges except the Big Horn, Hoback, Salt River, Wasatch, and Wyoming Ranges. Alaska south to California, Utah, and Colorado.

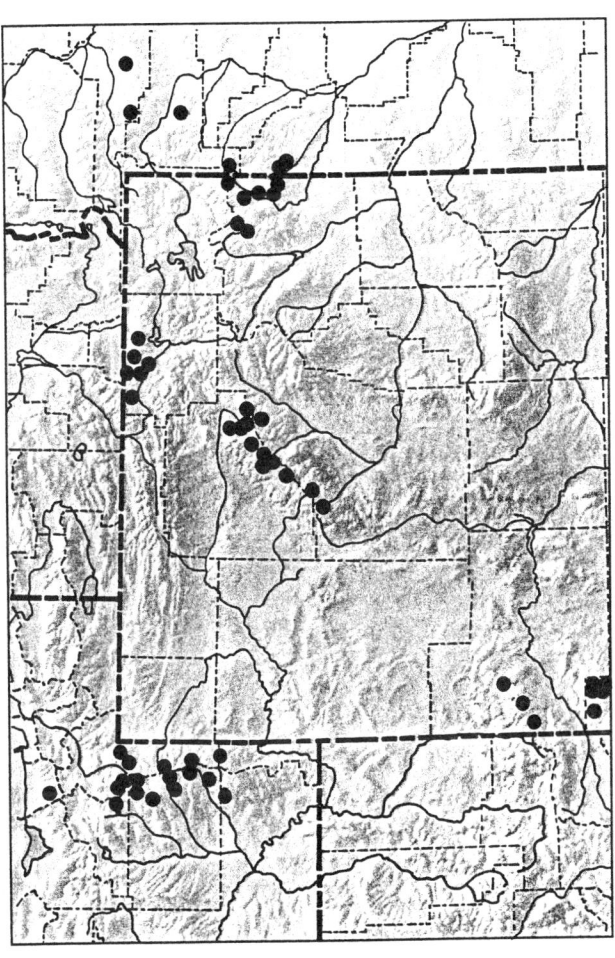

Carex norvegica Retz.
Prodr. Fl. Scand., 179, 1779.

Norway Sedge

Carex alpina Lilj.
Carex angarae Steud.
Carex halleri Gunn.
Carex media R. Br.
Carex stevenii Holm
Carex vahlii Schkuhr

Loosely or densely tufted, aphyllopodic perennial from short-creeping, slender rhizomes; culms erect, slender, glabrous, sometimes scaberulous below the inflorescence. Leaves mostly basal, much shorter than the culms; blades flat, the margins often involute, glabrous. Lowest bract sheathless, shorter than to slightly longer than the inflorescence. Spikes 2–5, mostly 3, oblong to ovoid, pedunculate, the terminal one gynaecandrous; lateral spikes pistillate, erect; pistillate scales purplish black, ovate, acute to obtuse, the margins hyaline, midrib inconspicuous or lacking, mostly shorter and wider than the perigynia; perigynia obovate to elliptic, greenish or brownish, rather abruptly contracted to a short, purplish black, bidentate beak, papillose, nerveless except for the well-developed marginal nerves, tapering to a short, broad, stipitate base; stigmas 3; akenes trigonous.

Stream banks and wet meadows, mostly near timberline, in all alpine ranges except the Hoback, Salt River, Teton, and Wyoming Ranges. Circumboreal; south in western North America to Idaho, Utah, and New Mexico.

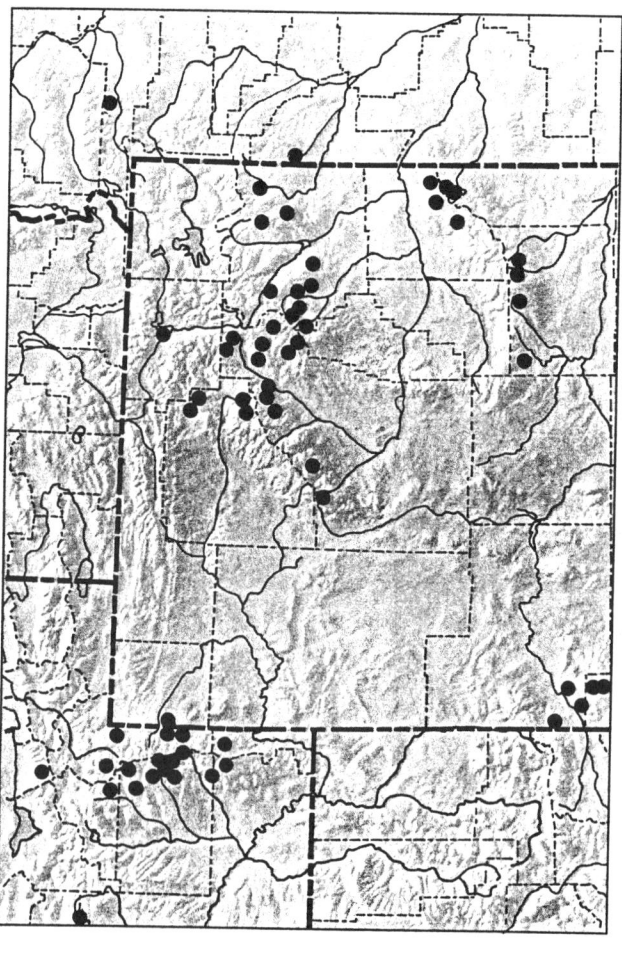

Carex nova L.H. Bailey
J. Bot. 26:322, 1888.

New Sedge

Carex elbertiana L. Kelso
Carex estesiana L. Kelso
Carex melanocephala Turcz.
Carex nelsonii Mack. in Rydb.
Carex pelocarpa Hermann

Loosely to densely tufted, somewhat aphyllopodic perennial from short-creeping, branched rhizomes; culms solitary to few, erect to somewhat nodding, often purplish at the base. Leaves mostly basal, shorter than the culms; blades flat, the margins somewhat revolute, the lower sheaths purplish. Lowest bract sheathless, shorter than to exceeding the inflorescence. Spikes mostly 3, sessile or short-pedunculate, ovoid, in a dense head, the terminal one gynaecandrous, lateral spikes pistillate; pistillate scales lanceolate to ovate, purplish black, hyaline margins lacking or very narrow, the midrib lacking or inconspicuous, equaling or shorter and narrower than the perigynia; perigynia greenish to brownish, sometimes entirely purplish at maturity, elliptic to orbicular, flattened or inflated, nerveless except for 2 marginal nerves, glabrous or papillose, abruptly narrowed to the purplish, bidentate beak, rounded to substipitate at the base; stigmas 3; akenes trigonous.

Fellfields and meadows of all alpine ranges except the Hoback and Salt River Ranges. Oregon and Montana south to Nevada, Utah, and New Mexico.

Carex nova is closely related to *C. albonigra* and *C. atrata* and is often difficult to distinguish from them, particularly in the case of immature specimens. Specimens with perigynia more ascending and inflated, and pistillate scales lacking hyaline margins have been called *C. nelsonii* Mack. in Rydb., but I can find no consistent separation between the groups of characters.

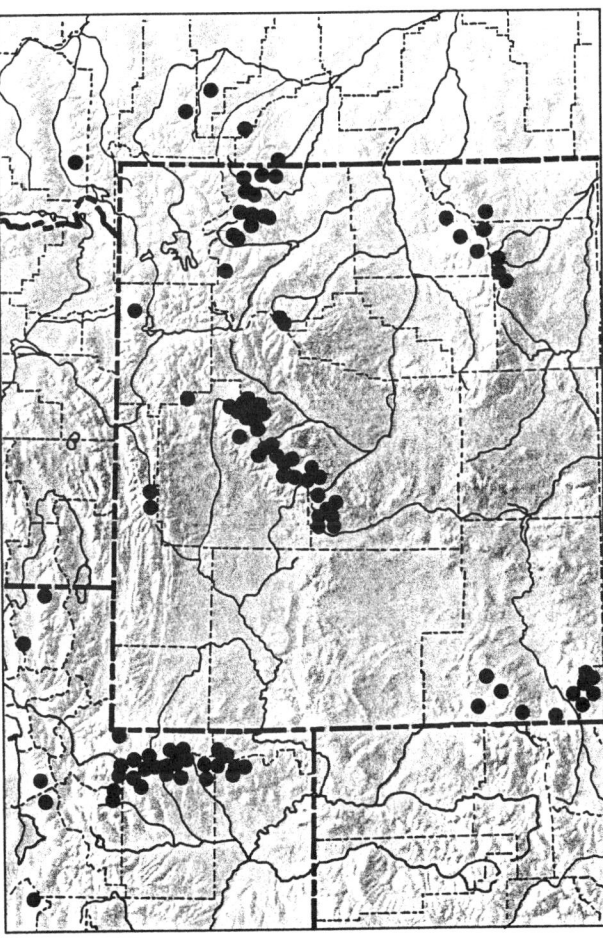

Carex obtusata Lilj.
Kongl. Vetensk. Acad. Nya Handl. 14:69, 1793.

Blunt Sedge

Genersichia obtusata (Lilj.) Heuffel

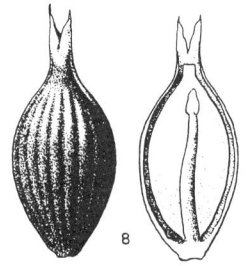

Slender, aphyllopodic perennial from long-creeping, purplish or brownish black, scaly rhizomes; culms mostly solitary or a few together. Leaves basal or nearly so, shorter than to nearly equaling the culms; blades stiff, flat or somewhat channeled. Lowest bract lacking. Spikes solitary, cylindrical, androgynous; the pistillate flowers few, spreading to ascending at maturity; pistillate scales ovate to elliptic, dark brown with hyaline margins, acute to awl-shaped, shorter than to about the same length as the perigynia and mostly narrower; perigynia ovate to elliptic, dark brown to blackish, finely to coarsely ribbed, gradually tapering to a short, bidentate or oblique, hyaline beak, the base stipitate; rachilla about the same length as the akene, often with a flattened, hyaline tip; stigmas 3; akenes trigonous.

Dry fellfields and rocky ridgetops of the Absaroka and Beartooth Ranges. Eurasia; Alaska south to New Mexico, South Dakota, and Minnesota.

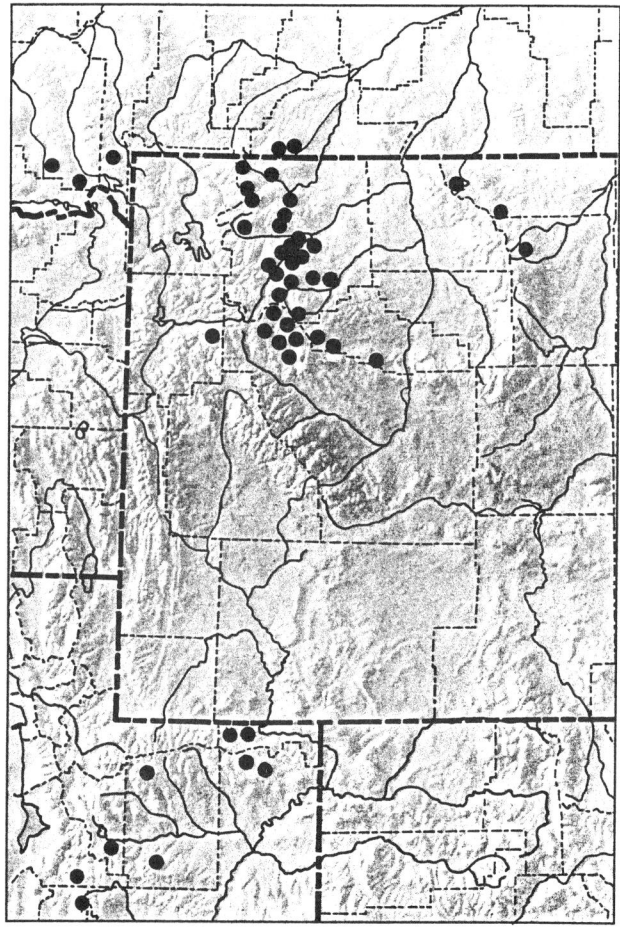

Carex petasata Dewey
Am. J. Sci. 29:246, 1836.

Liddon Sedge

Carex constanceana Stacey
Carex liddoni F. Boott in Hook.
Carex rufovariegata Boeck.

Tufted, aphyllopodic perennial from fibrous roots; rhizomes lacking or very short; culms slender, erect, glabrous. Leaves basal and cauline, shorter than the culms; blades flat or slightly channeled. Lowest bract lacking. Spikes 3–6, gynaecandrous, sessile in an erect, spikelike head; pistillate scales ovate, brownish, the margins hyaline, midrib pale, about the same width as or slightly narrower than the perigynia, but shorter so that the beaks are evident; perigynia lanceolate, planoconvex, finely nerved on both sides, winged near the serrulate margins, gradually tapered to a shallowly bidentate or oblique beak; stigmas 2; akenes lenticular.

Timberline margins of the Absaroka, Beartooth, and Wyoming Ranges. Alaska south to California, Arizona, and Colorado.

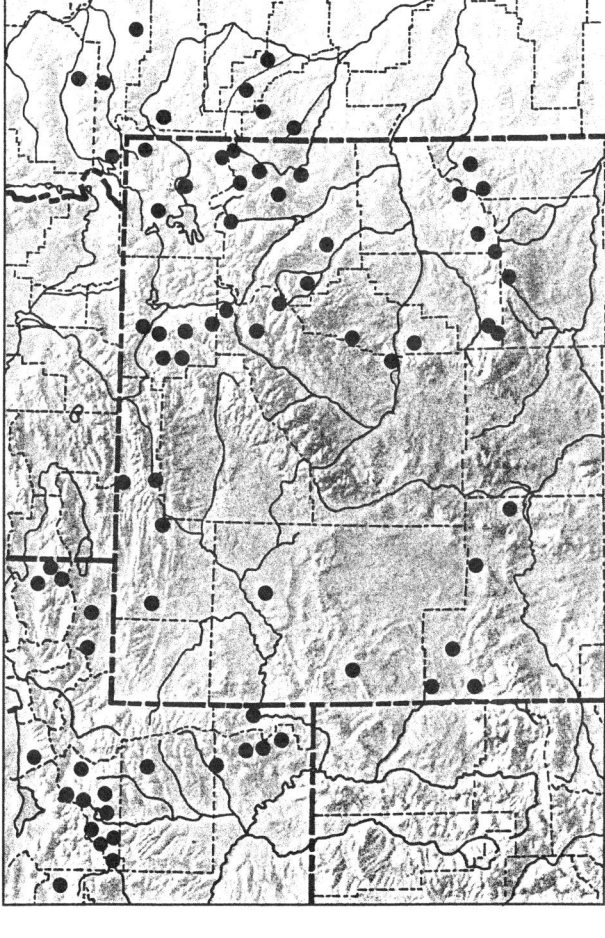

Carex phaeocephala Piper
Contr. U.S. Natl. Herb. 11:172, 1906.

Dunhead Sedge

Carex eastwoodiana Stacey

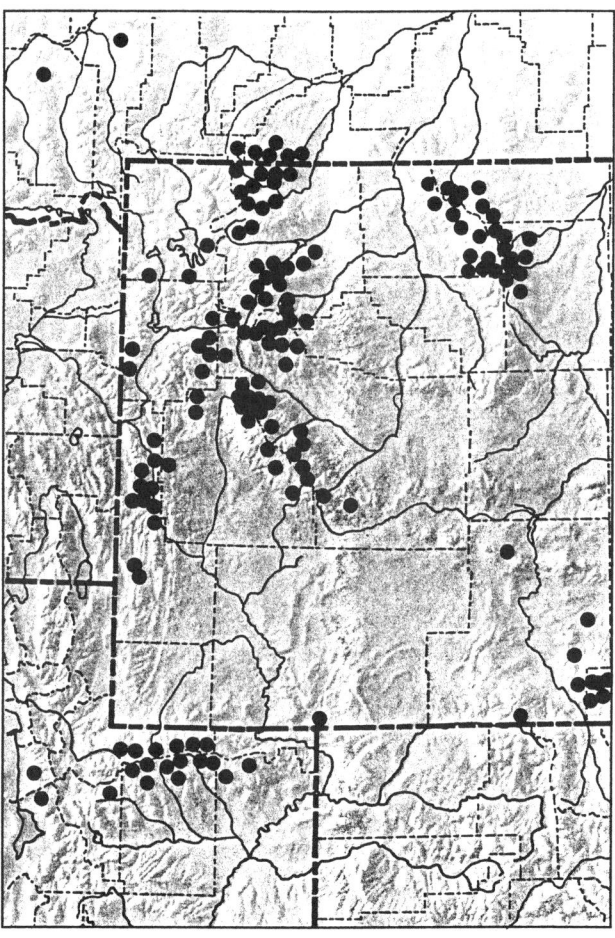

Densely tufted, somewhat aphyllopodic perennial from fibrous roots; rhizomes lacking; culms several to many, slender, erect, scaberulous above. Leaves basal, shorter than to nearly equaling the culms; blades stiff, flat, channeled or involute. Lowest bract short, inconspicuous or sometimes lacking. Spikes 2–7, sessile, gynaecandrous, aggregated in an ovoid head, the lower ones somewhat approximate; pistillate scales ovate, acute, brownish with hyaline margins and a pale midrib, about the same length and width as the perigynia, concealing them; perigynia lanceolate to elliptic, brownish, planoconvex, narrowly winged, finely nerved on the dorsal surface, sometimes nerved ventrally, the distal margins finely serrulate, gradually or abruptly tapered to a dark brown, bidentate to oblique beak; stigmas 2; akenes lenticular.

Fellfields, talus, scree, ledges, and rocky ridges of all alpine ranges; not yet collected above timberline in the Hoback Range. Alaska south to California, Idaho, Utah, and Colorado.

Carex phaeocephala is a common alpine species that closely resembles *C. petasata* but is distinguished by its ovoid head and pistillate scales concealing the perigynia.

Carex podocarpa R. Br. in Richards.
Bot. App. Frankl. J., 751, 1823.

Shortstalk Sedge

Carex behringensis C.B. Clarke
Carex montanensis L.H. Bailey
Carex nigella F. Boott in Hook.
Carex paysonis Clokey
Carex spectabilis Dewey
Carex tolmiei F. Boott in Hook.
Carex venustula Holm

Loosely tufted, aphyllopodic perennial from fibrous roots and short- to long-creeping, brown-scaly rhizomes; culms stout, erect, brown-scaly at base. Leaves basal and cauline, shorter than to nearly equaling the culms; blades flat, the margins often revolute. Lowest bract well developed, sheathless, shorter than to about as long as the inflorescence. Terminal spike staminate, erect, short-pedunculate; lateral spikes mostly pistillate, the upper one sometimes androgynous, sessile and erect-ascending above, becoming slender-pedunculate and spreading-reflexed below; pistillate scales ovate, purplish black, usually with a pale midrib that is mostly excurrent, forming a short awn or cuspidate tip; perigynia ovate to elliptic, greenish to partly purplish black, flattened, short-beaked, the beak slender, bidentate; stigmas 3; akenes trigonous.

Timberline meadows, willow thickets, and moist fellfields of all alpine ranges except the Hoback, Medicine Bow, Salt River, Wasatch, and Wyoming Ranges. Siberia; Alaska south to California, Nevada, Utah, and Wyoming.

Middle Rocky Mountain plants are var. *paysonis* (Clokey) B. Boivin.

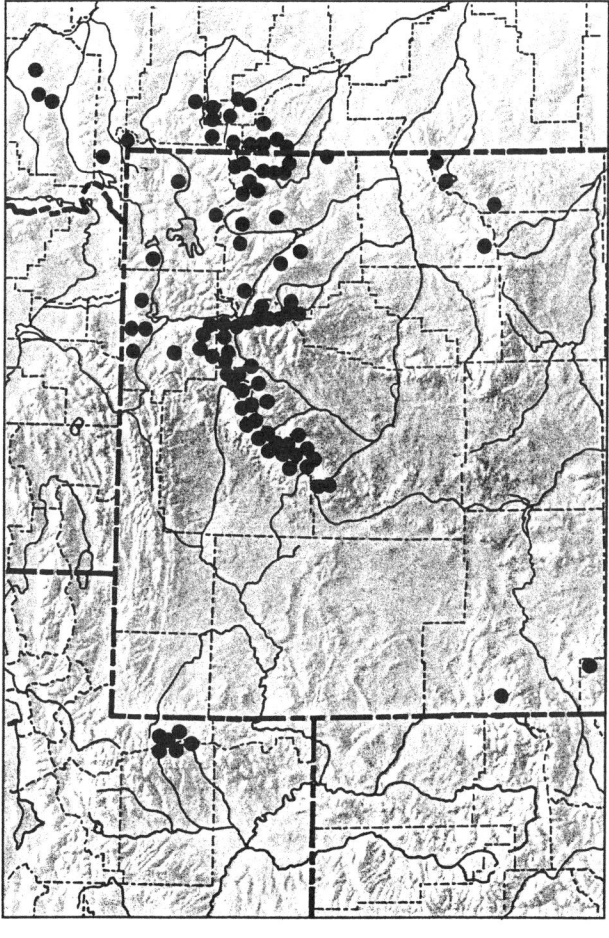

Carex praeceptorum Mack.
N. Am. Fl. 18:95, 1931.

Early Sedge

Loosely to densely tufted, aphyllopodic perennial from short-creeping rhizomes; culms 1 to several, erect, slender, often weakly decumbent. Leaves mostly basal, shorter than the culms; blades channeled. Lowest bract short, inconspicuous. Spikes 4–6, sessile, gynaecandrous, approximate or densely aggregated into an ovoid head; pistillate scales ovate, dark brown or chestnut, with hyaline margins and a pale green midrib, shorter than the perigynia; perigynia ovate to broadly elliptic, planoconvex, brownish or greenish, distinctly nerved on both sides and finely roughened on the surface, contracted into a short, often serrulate beak, substipitate at the base; stigmas 2; akenes lenticular.

Wet meadows, stream banks, and lake margins near timberline and in the lower alpine zone of the Absaroka, Beartooth, Big Horn, Medicine Bow, Uinta, Wasatch, and Wind River Ranges. Washington south and east to California, Nevada, and Colorado.

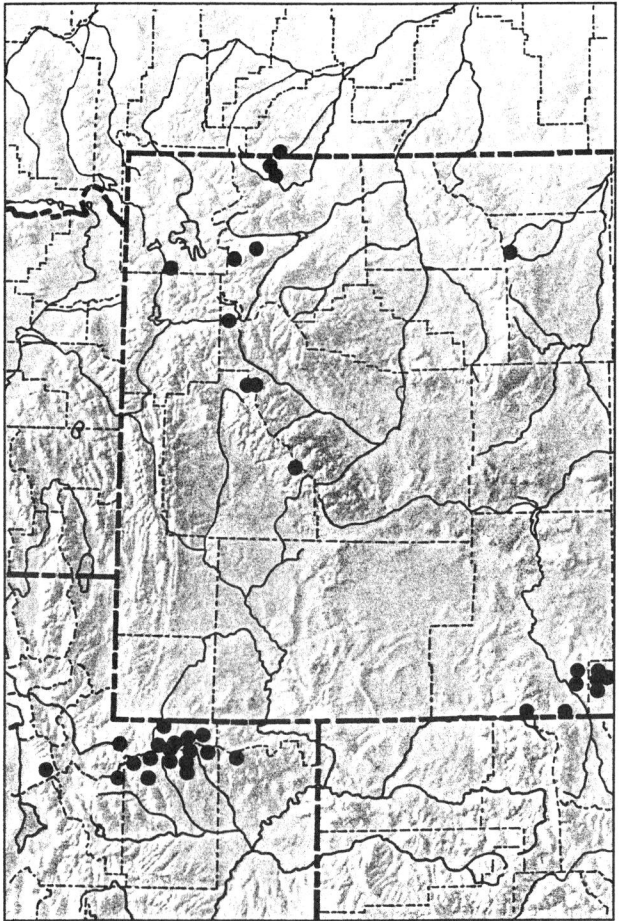

Carex praticola Rydb.
Mem. N.Y. Bot. Gard. 1:84, 1900.

Meadow sedge

Carex piperi Mack. ex Piper & Beattie
Carex pratensis Drej.

Loosely to densely tufted, aphyllopodic perennial from fibrous roots; rhizomes lacking or very short; culms erect, slender, somewhat flexuous or nodding. Leaves basal and, on the lower culms, shorter than the culms; blades flat, the sheaths hyaline ventrally. Lowest bract lacking, or if present, reduced and scalelike. Spikes 3–7, mostly 5 in alpine specimens; ovoid to oblong, gynaecandrous, sessile, distinct, approximate, the inflorescence nodding at maturity; pistillate scales ovate, acute, brownish, with a green midrib and conspicuous hyaline margins; perigynia ovate to ovate-lanceolate, greenish or brownish, gradually tapered to a long, serrulate, shallowly bidentate beak, finely nerved dorsally, usually nerveless ventrally; stigmas 2; akenes lenticular.

Wet meadows, stream banks, and pond margins near timberline and in the lower alpine zone of the Absaroka, Big Horn, and Wind River Ranges. Alaska to Greenland, south in western North America to California, Utah, Colorado, and North Dakota.

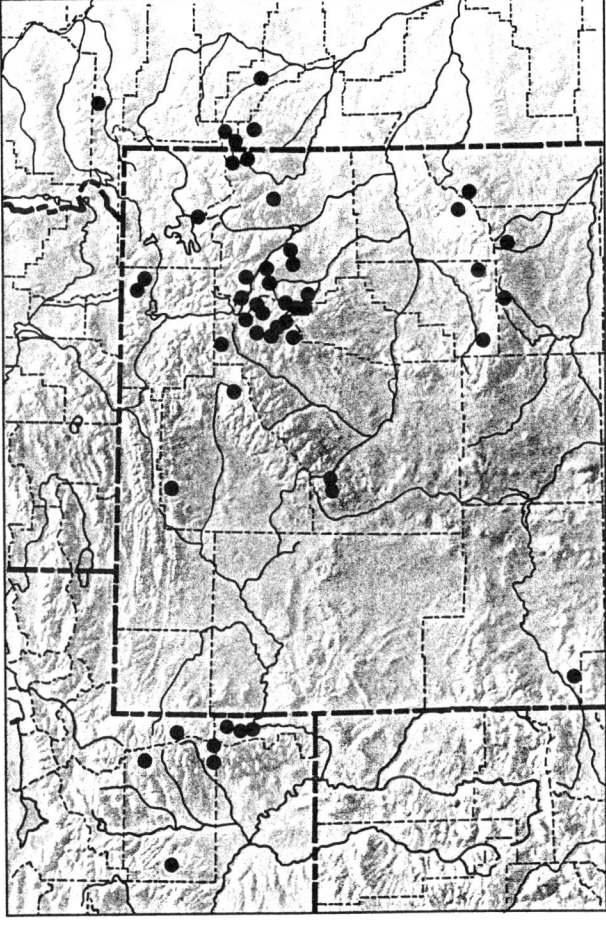

Carex pyrenaica Wahlenb.
Kongl. Vetensk. Acad. Nya Handl. 24:139, 1803.

Pyrenees Sedge

Callistachys pyrenaica (Wahlenb.) Heuffel
Carex crandallii Gand.
Carex jacob-peteri Hult.
Carex micropoda C.A. Mey.
Psyllophora pyrenaica (Wahlenb.) Schur

Densely tufted, aphyllopodic perennial from fibrous roots; rhizomes lacking; culms several to many, erect, slender, wiry. Leaves mostly basal, shorter than, equaling, or sometimes slightly exceeding the culms; blades very narrow, mostly involute, wiry. Lowest bract lacking. Spike solitary, erect, androgynous, oblong to narrowly elliptic; pistillate scales ovate, dark brown to chestnut, with hyaline margins and pale midribs, shorter than the perigynia, deciduous with age; perigynia lanceolate to narrowly ovate, tan to dark brown, nerveless, gradually tapering to a short, oblique beak, stipitate, spreading with age; stigmas 3; akenes trigonous.

Fellfields, meadows, and rocky ridges of most alpine ranges. Not yet known from above timberline in the Salt River, Wasatch, and Wyoming Ranges. Circumpolar; south in western North America to California, Utah, and Colorado.

Carex pyrenaica closely resembles *C. nigricans,* but is distinguished by its lack of rhizomes, more erect perigynia at maturity, and other characters mentioned in the key.

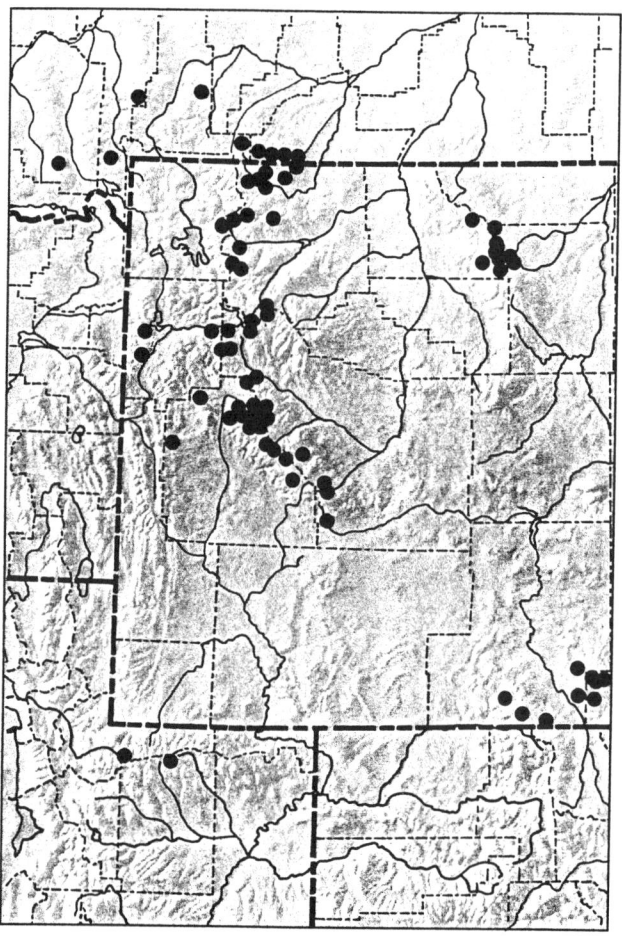

Carex raynoldsii Dewey
Am. J. Sci., II, 32:39, 1861.

Raynold Sedge

Carex lyallii F. Boott

Loosely tufted, mostly phyllopodic perennial from short-creeping rhizomes; culms few to several, erect, stout. Leaves basal and cauline, shorter than the culms; blades flat with revolute margins. Lowest bract conspicuous, shorter than to slightly exceeding the inflorescence. Terminal spike staminate, sessile, erect; lateral spikes pistillate, ascending, sessile above, becoming long-pedunculate and approximate below; pistillate scales ovate, acute, purplish black, the midrib obscure or lacking, sometimes with very narrow, hyaline margins, shorter and narrower than the perigynia; perigynia somewhat inflated, obovate or elliptic, light green, glabrous, finely nerved on both sides, abruptly contracted into a short, purplish black, entire to bidentate beak, substipitate; stigmas 3; akenes trigonous.

Timberline meadows and slopes of the Absaroka, Beartooth, Gros Ventre, Medicine Bow, Uinta, Wasatch, and Wind River Ranges. British Columbia and Alberta south to California, Nevada, Utah, and Colorado.

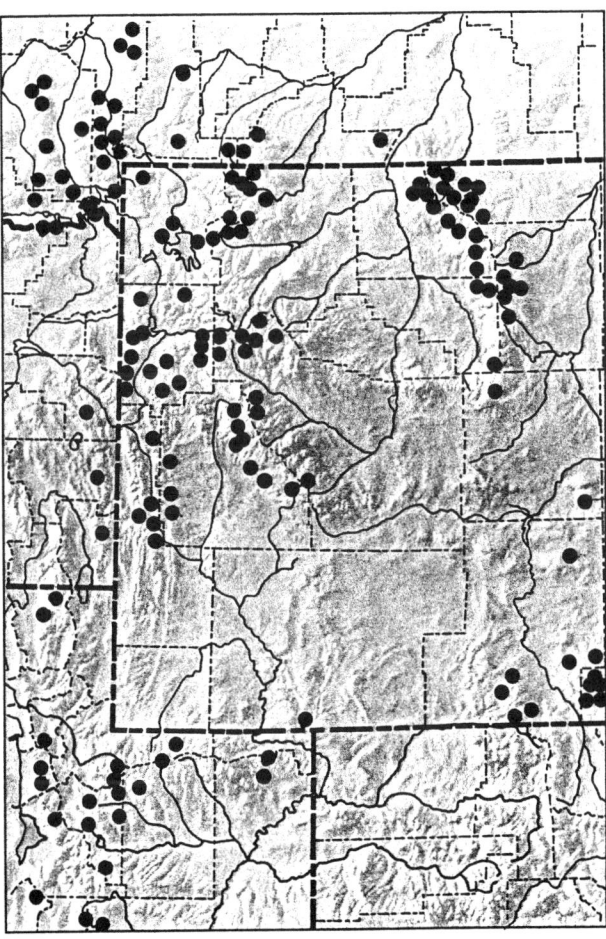

Carex rossii F. Boott in Hook.
Fl. Bor.-Am. 2:222, 1839.

Ross Sedge

Carex brevipes W. Boott in S. Wats.
Carex diversistylis Roach
Carex farwellii Mack.

Loosely to densely tufted, aphyllopodic perennial from fibrous roots; rhizomes lacking; culms few to several, slender, erect. Leaves basal and cauline, equaling or longer than the culms; blades thin, soft, flat, the margins somewhat revolute. Inflorescences both terminal and basal; terminal spike staminate, sessile or short-pedunculate; lateral spikes pistillate, few-flowered, sessile to pedunculate, the lowest bract exceeding the inflorescence, at least above; pistillate scales ovate, brownish, acute to somewhat awned, with hyaline margins and pale midribs, shorter than the perigynia; perigynia elliptic to ovate, puberulent, abruptly narrowed to a prominent bidentate beak, marginally nerved, stipitate; stigmas 3; akenes trigonous.

Timberline meadows and slopes of the Absaroka, Teton, Uinta, Wasatch, and Wind River Ranges. Alaska and Yukon Territory south to California, Arizona, and Colorado.

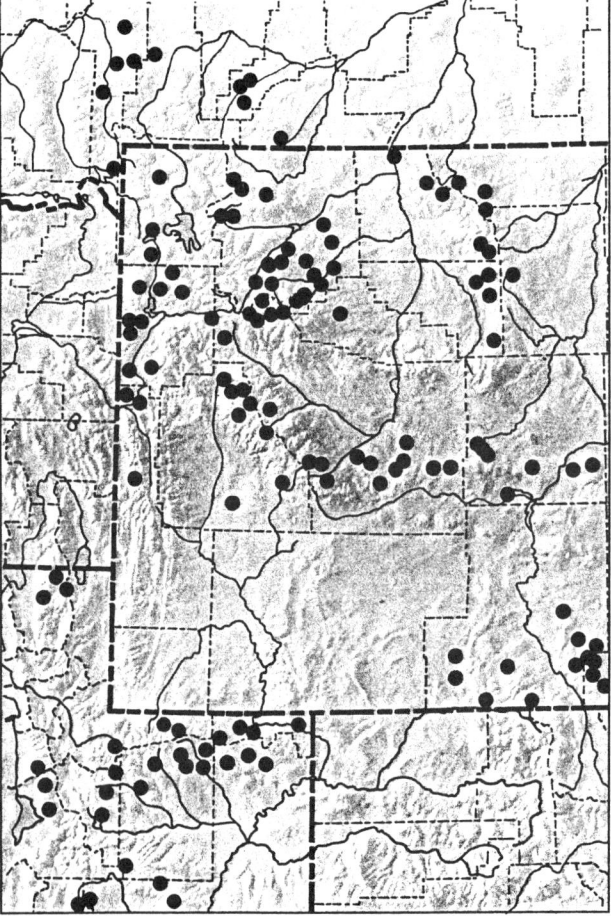

Carex rostrata Stokes ex With.
Brit. Pl. 2(2):1059, 1787.

Beaked Sedge

Carex inflata Huds.
Carex utriculata F. Boott in Hook.

Solitary or loosely tufted, somewhat aphyllopodic perennial from stout, scaly, creeping rhizomes; culms erect, stout, the bases thick and often spongy. Leaves cauline, glabrous, the blades flat, septate-nodulose, somewhat channeled near the base. Inflorescence of 5–7 cylindric, approximate to remote spikes, the upper staminate, the lower pistillate, sessile above, becoming short-pedunculate below; middle spikes sometimes androgynous; lowest bract exceeding the inflorescence; pistillate scales lanceolate to ovate, acuminate, light brown to chestnut, with a pale midrib, about the same length and narrower than the perigynia; perigynia inflated, ovate to subglobose, brownish to chestnut, glabrous, contracted into a prominent, bidentate beak, widely spreading at maturity; stigmas 3; akenes trigonous.

Wet places at timberline in the Absaroka, Beartooth, Big Horn, Medicine Bow, Uinta, Wasatch, and Wind River Ranges. Circumboreal; south in western North America to California, Arizona, New Mexico, and Nebraska.

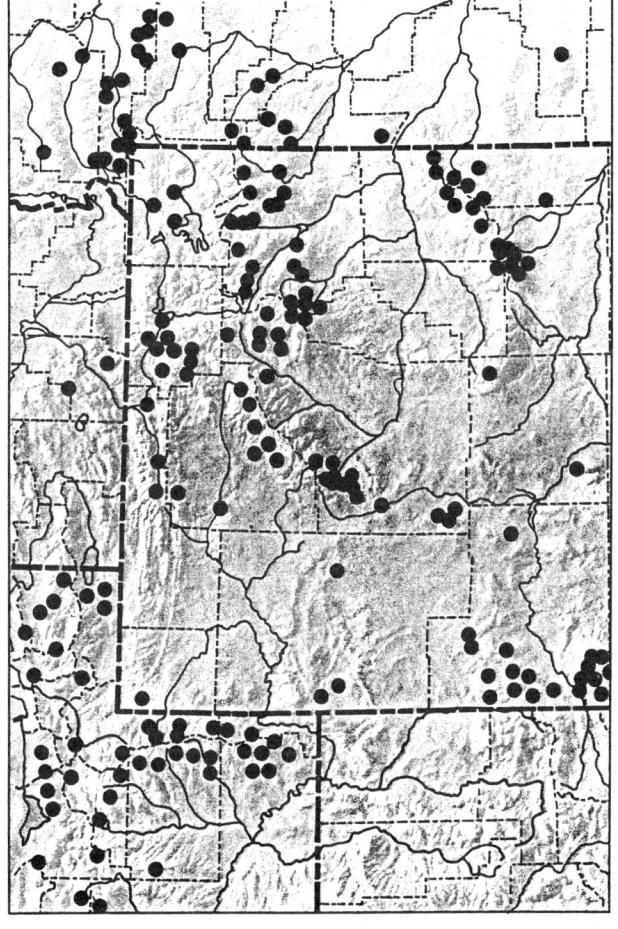

Carex rupestris Allioni
Fl. Pedem. 2:264, 1785.

Curly Sedge

Carex drummondiana Dewey
Caricinella rupestris (All.) St.-Lag. ex Cariot
Edritria rupestris (All.) Raf.

Solitary or loosely tufted, somewhat aphyllopodic perennial from slender, scaly, creeping rhizomes; culms solitary or a few together, stout, short, mostly less than 12 cm tall. Leaves basal, slightly shorter than to exceeding the culms; blades flat or channeled, tapering to a slender, circinate, often dried tip. Lowest bract lacking. Spike solitary, cylindric, androgynous; pistillate scales obovate to orbicular, obtuse to rounded, brownish to brownish purple, with hyaline margins and a pale midrib, about the same length as the perigynia and mostly concealing them; perigynia elliptic, greenish to brownish, glabrous, nerved, with a very short, truncate beak, short-stipitate; stigmas 3; akenes trigonous.

Fellfields and other rocky places in most alpine ranges of the Middle Rocky Mountains. Not yet seen from above timberline in the Gros Ventre, Hoback, Salt River, and Wasatch Ranges. Circumpolar; south in western North America to Utah, Colorado, and New Mexico.

Two weakly differentiated varieties occur in the alpine zone of the Middle Rocky Mountains. Typical var. *rupestris* has somewhat slender, flexuous spikes, and the lower perigynia fewer and more separated than var. *drummondiana* (Dewey) L.H. Bailey, which has stout, straight spikes and several lower perigynia that are strongly overlapping.

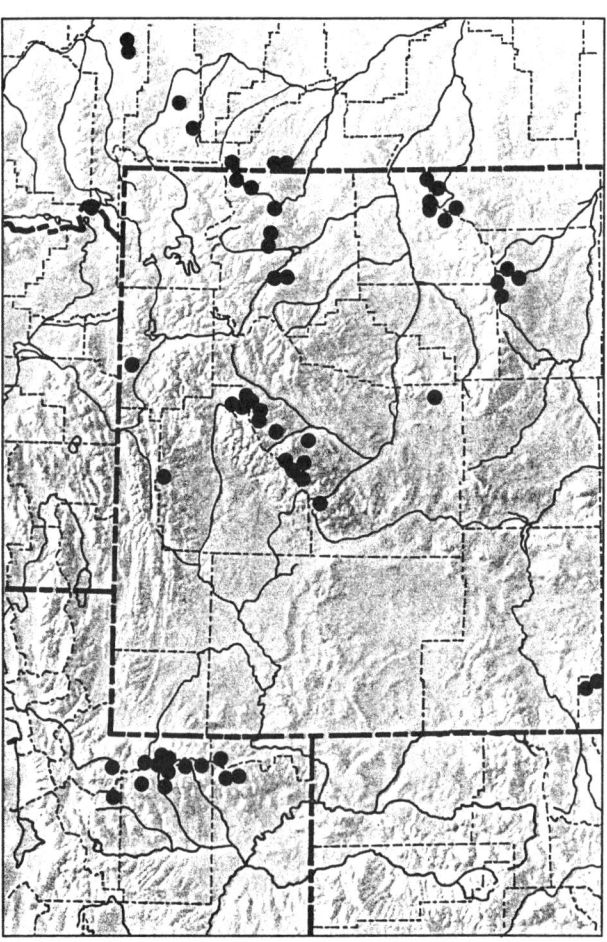

Carex saxatilis L.
Sp. Pl., 976, 1753.

Russet Sedge

Carex ambusta F. Boott
Carex miliaris Michx.
Carex physocarpa J. & K. Presl
Carex procerula Krecz.
Carex pulla Gooden.
Carex rhomalea (Fern.) Mack.

Solitary or loosely tufted, phyllopodic perennial from slender, creeping rhizomes; culms erect, reddish or purplish at the base. Leaves mostly basal, shorter than to about equaling the culms; blades flat with revolute margins, glabrous, somewhat septate-nodulose near the base. Lowest bract foliose, longer than the inflorescence. Upper spike staminate, erect; lower spikes pistillate, widely separate, nearly sessile and erect-ascending above, becoming long-pedunculate and spreading to drooping below; pistillate scales ovate, acute to acuminate, or sometimes erose, dark brown to purplish black, with hyaline margins and a pale midrib, narrower and shorter than the perigynia; perigynia ovate to suborbicular, brownish or purplish black, marginally nerved, obscurely nerved dorsally near the base, nerveless or nearly so ventrally, abruptly short-beaked, the beak truncate or shallowly bidentate, often purplish black; stigmas 2 or sometimes 3; akenes lenticular or sometimes trigonous, the style continuous with the akene and strongly bent or recurved against it.

Wet meadows and lake margins at timberline in the Big Horn, Uinta, and Wind River Ranges. Circumboreal; south in western North America to Washington, Nevada, Utah, and Colorado.

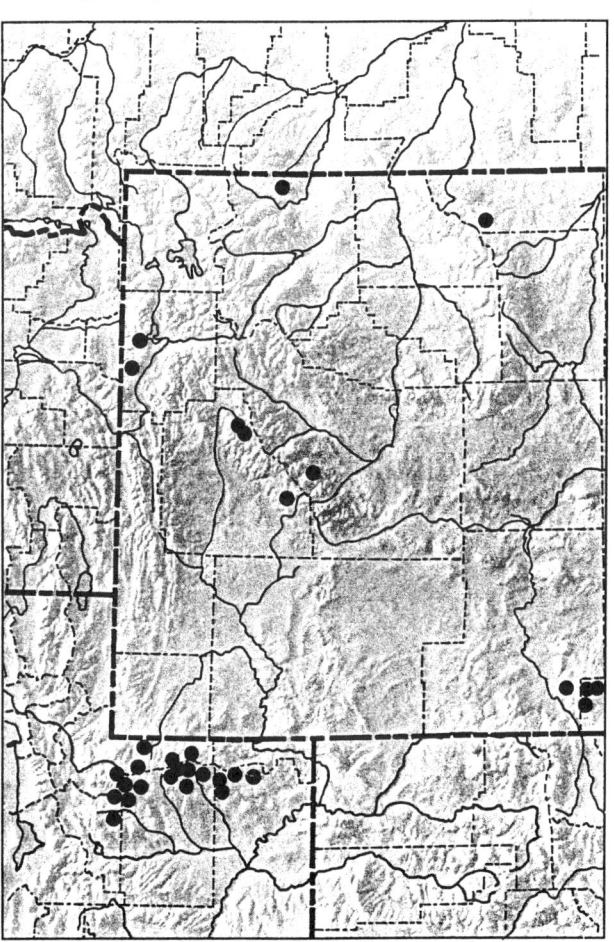

Carex scirpoidea Michx.
Fl. Bor.-Am. 2:171, 1803.

Singlespike Sedge

Carex athabascensis F.J. Herm.
Carex curatorum Stacey
Carex michauxii Schw.
Carex pseudoscirpoidea Rydb.
Carex scirpiformis Mack.
Carex scirpina Tuckerm.
Carex stenochlaena (Holm.) Mack.

Solitary or loosely tufted, mostly phyllopodic (in ours) perennial from stout, scaly, creeping rhizomes; culms single to a few together, erect, reddish or purplish at the base and covered by remains of old leaves. Leaves basal, shorter than to nearly equaling the culms; blades flat, somewhat channeled near the base. Lowest bract lacking or inconspicuous and much shorter than the inflorescence. Spike solitary, erect, cylindric, staminate or pistillate, rarely with an additional spike at the base; staminate scales similar to the pistillate scales; pistillate scales usually pubescent at least near the base, ovate, obtuse to acute, dark brown or purplish black, with hyaline margins and pale midrib, usually erose and ciliolate, at least near the apex, about the same length and width as the perigynia; perigynia pubescent, obovate, obscurely nerved, abruptly narrowed to a short, bidentate or oblique beak; stigmas 3, akenes trigonous.

Fellfields, dry meadows, and other well-drained sites in the Absaroka, Beartooth, Big Horn, Uinta, and Wind River Ranges. Transcontinental in Canada; south in the West to California, Arizona, and Colorado.

Alpine plants of the Middle Rocky Mountains are var. *pseudoscirpoidea* (Rydb.) Cronq.

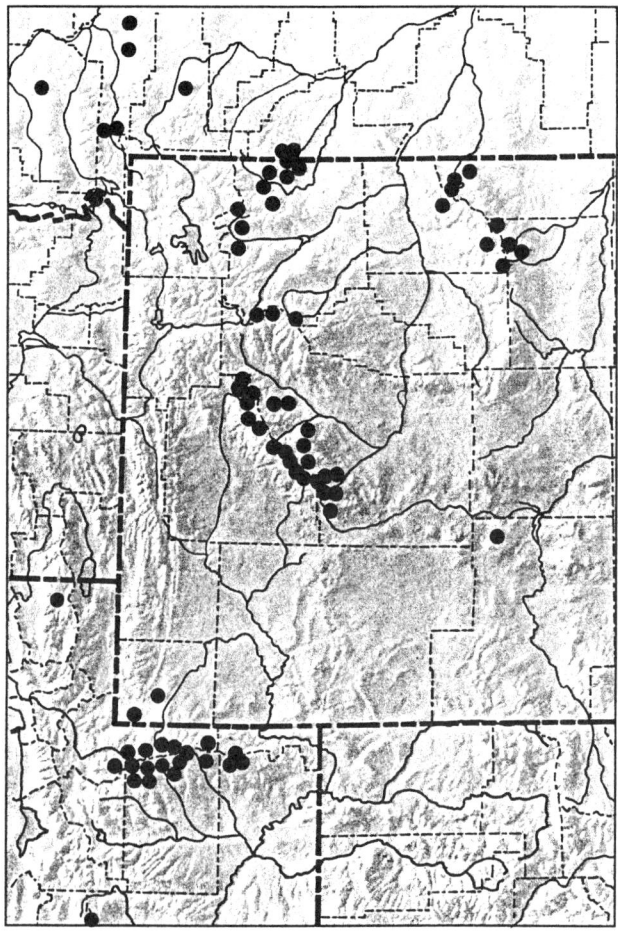

Carex scopulorum Holm
Am. J. Sci., IV, 14:422, 1902.

Rock Sedge

Carex accedens Holm
Carex campylocarpa Holm
Carex chimaphila Holm
Carex gymnoclada Holm
Carex miserabilis Mack.
Carex prionophylla Holm
Carex rigida Good.
Carex spreta L.H. Bailey

Solitary to loosely tufted, phyllopodic perennial from stout, scaly, brownish or purplish, branched, creeping rhizomes; culms stout, erect, often purplish at the base. Leaves basal, shorter than the culms; blades flat, with revolute margins. Lowest bract foliose, shorter than the inflorescence. Terminal spike staminate, sessile, erect, cylindric; lateral spikes pistillate, ovoid, erect-ascending, the upper spikes sessile, becoming pedunculate below, sometimes the middle ones androgynous, or the upper staminate; pistillate scales lanceolate, obovate, or ovate-elliptic, obtuse, purplish black, the margins narrowly hyaline, often with a pale, slender midrib, narrower and shorter than to longer than the perigynia; perigynia somewhat inflated, obovate to elliptic or suborbicular, pale greenish below and purplish black above, marginally nerved, abruptly narrowed to a very short, truncate or bidentulate beak; stigmas 2, rarely 3; akenes lenticular, rarely trigonous.

Wet meadows, stream banks, lake and pond margins, and snowmelt seeps in most alpine ranges; not yet reported from above timberline in the Hoback, Salt River, Teton, Wasatch, and Wyoming Ranges. Alaska south to California, Nevada, Utah, and Colorado.

Middle Rocky Mountain alpine plants are distinguished as follows:

1. Pistillate scales ovate-elliptic, shorter than the perigynia; perigynia spreading, the beak mostly 0.2–0.5 mm long . . . var. *scopulorum*
1. Pistillate scales lanceolate, longer than the perigynia; perigynia ascending, the beak 0.2 mm or less in length . . . var. *chimaphila* (Holm) Kükenth.

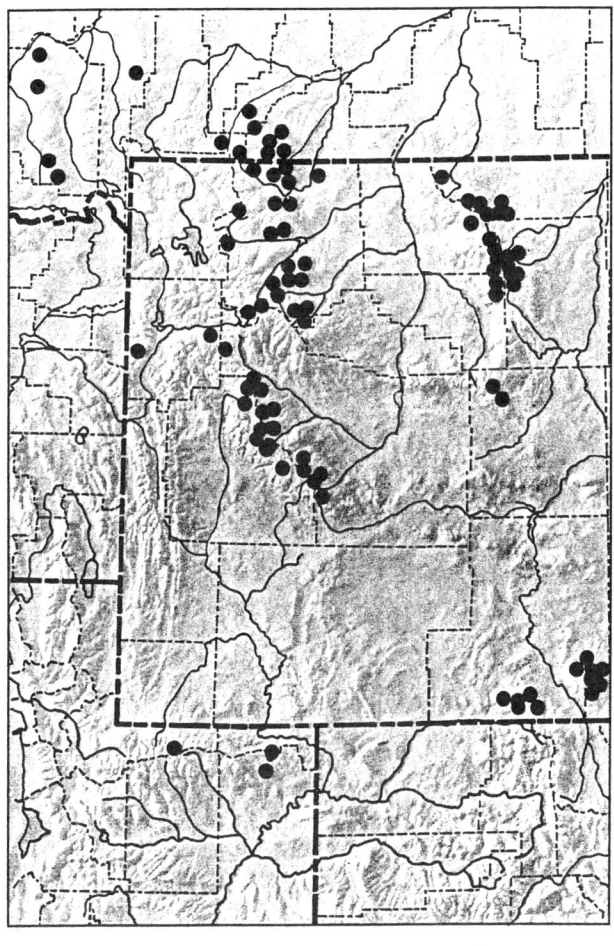

Carex stenophylla Wahlenb.
Kongl. Vetensk. Acad. Nya Handl. 24:142, 1803.

Needleleaf Sedge

Carex duriuscula C.A. Mey.
Carex eleocharis L.H. Bailey
Vignea stenophylla (Wahlenb.) Reichenb.

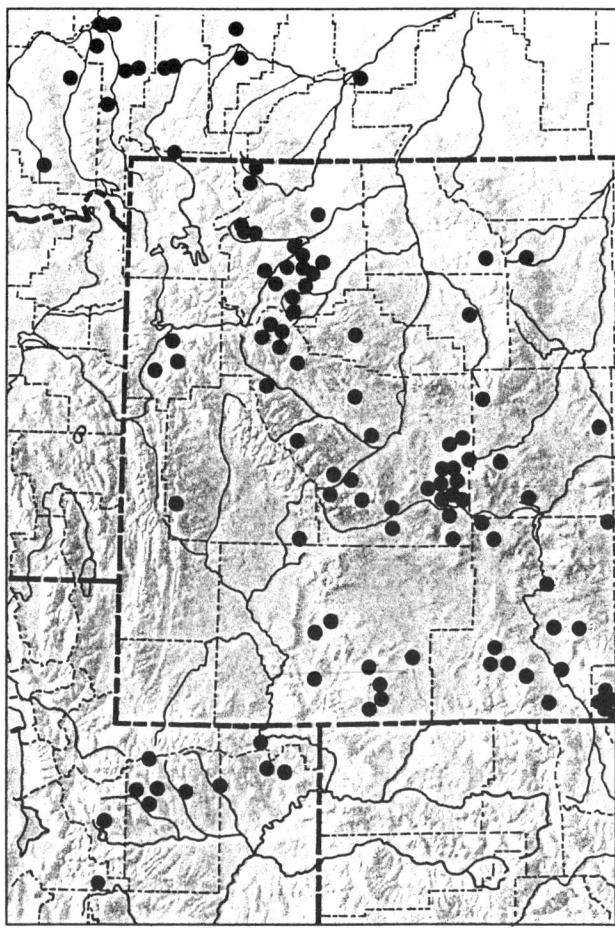

Solitary or loosely tufted, mostly phyllopodic perennial from slender, brown, scaly rhizomes and densely clustered fibrous roots; culms slender, erect, or sometimes decumbent. Leaves mostly basal, shorter than the culms; blades flat or channeled, the margins involute. Lowest bract lacking or inconspicuous. Spikes several, androgynous (rarely all staminate or all pistillate), sessile in a dense ovoid head, the lower spikes approximate; pistillate scales ovate to broadly elliptic, acute to short-acuminate, brown, with hyaline margins and light brown, often somewhat keeled midrib, wider and about the same length as the perigynia; perigynia ovate to elliptic, brownish to blackish brown, marginally nerved, gradually tapered to a short, oblique beak, short-stipitate; stigmas 2; akenes lenticular.

Timberline in the Absaroka and Uinta Ranges. Circumboreal; Alaska south to California, Arizona, and Colorado.

Carex straminiformis L.H. Bailey
Mem. Torr. Club 1:24, 1889.

Shasta Sedge

Densely tufted, mostly aphyllopodic perennial from fibrous roots; rhizomes lacking; culms few to several, stout, erect to curved. Leaves cauline, but clustered near the base, much shorter than the culms; blades flat or somewhat channeled, thick. Lowest bract lacking or inconspicuous. Spikes several, gynaecandrous, sessile in a dense ovate to oval head; pistillate scales ovate to lanceolate, acute, brownish with broad, hyaline margins and a greenish midrib, shorter and much narrower than the perigynia; perigynia thin and flattened, broadly ovate, pale green or straw-colored, the margins distinctively crinkled-membranous, abruptly tapered to a flattened, margined, serrulate, bidentate beak, the dorsal surface finely and many-nerved; stigmas 2; akenes lenticular.

Reported from the alpine zone in moist, rocky places (Arnow et al., 1980) and cirques (Welsh et al., 1987) of the Wasatch Range. Washington and California east to Montana, Idaho, and Utah.

Carex subnigricans Stacey
Leafl. West. Bot. 2:167, 1939.

Dark Alpine Sedge

Carex rachillis Maguire

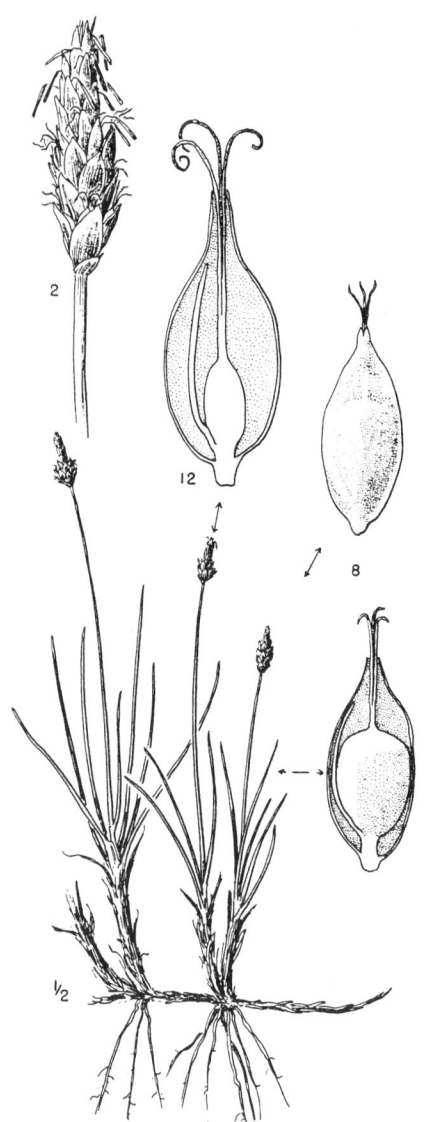

Loosely to densely tufted, somewhat aphyllopodic perennial from fibrous roots and short-creeping rhizomes; culms few to several, slender, stiff, erect. Leaves mostly basal, shorter than the culms; blades stiff, glabrous, channeled or involute. Lowest bract lacking or inconspicuous. Spike solitary, cylindric, erect, androgynous; pistillate scales lanceolate to ovate, acute to obtuse, light to dark brown, the margins narrowly hyaline, midrib very pale or lacking, about the same width and length as the perigynia; perigynia brownish, ovate to elliptic, glabrous, usually distended by the akene, gradually tapering to a short, oblique beak, the base stipitate; stigmas 3; akenes trigonous.

Timberline meadows of the Uinta and Wind River Ranges. Oregon south to California, Idaho, Nevada, and Utah.

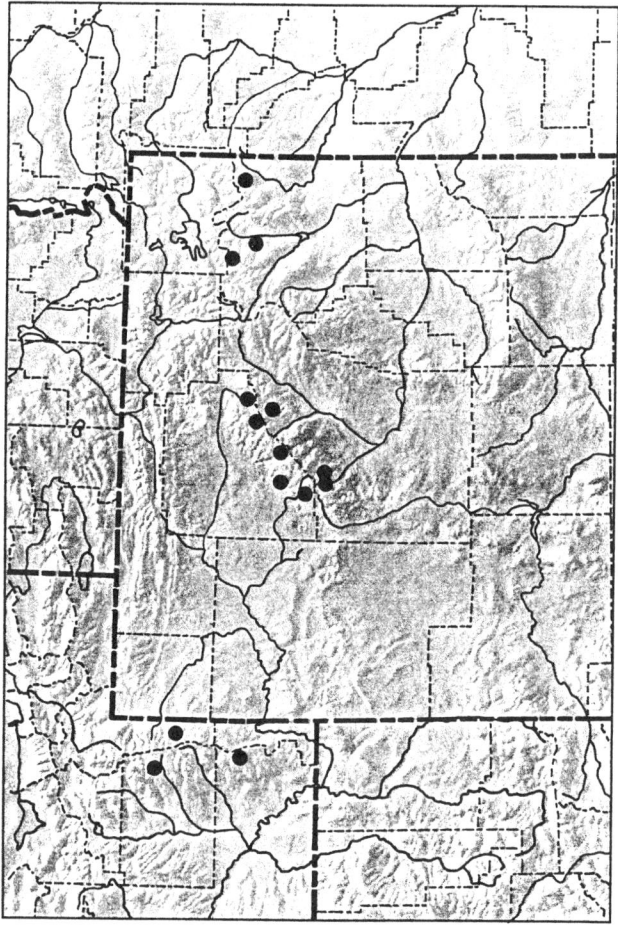

Carex vesicaria L.
Sp. Pl., 979, 1753.

Bladder Sedge

Carex exsiccata L.H. Bailey
Carex monile Tuckerm.
Carex raeana F. Boott
Trasus vesicarius (L.) S.F. Gray

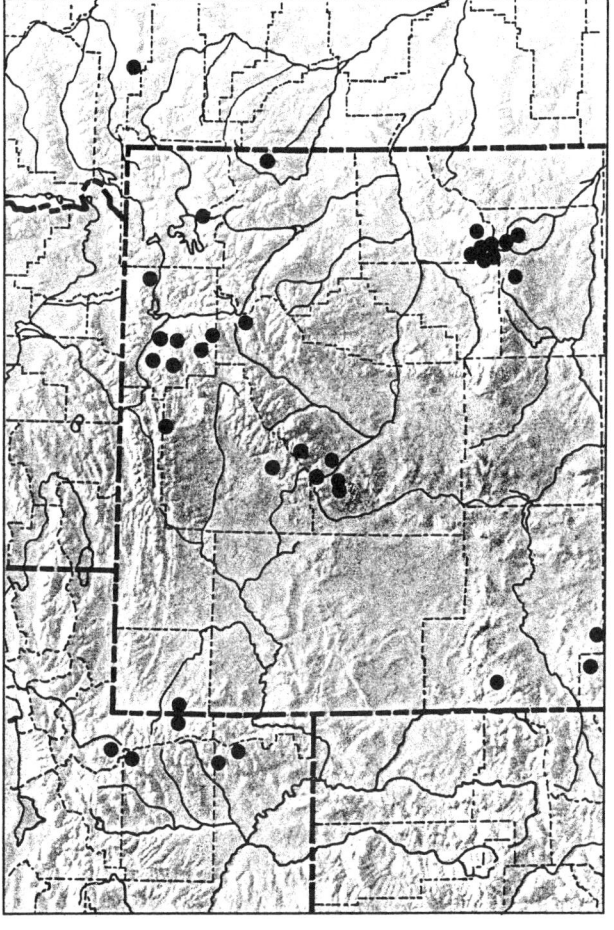

Loosely to densely tufted, aphyllopodic perennial from short-creeping, scaly, branched rhizomes; culms stout, erect, often purplish red at the base. Leaves cauline, shorter than the culms; blades flat with revolute margins, at least the lower sheaths septate-nodulose. Lowest bract conspicuous, foliose, exceeding the inflorescence. Spikes 3–7, erect, approximate to remote, the upper staminate, the lower pistillate, sessile, becoming short-pedunculate below; pistillate scales lanceolate to ovate, acute or short-acuminate, straw-colored to reddish brown, with narrow, hyaline margins and light tan midribs, shorter and narrower than the perigynia; perigynia lanceolate to narrowly ovate, glabrous, inflated, strongly nerved, greenish when young, becoming brownish at maturity, gradually tapered to a long, bidentate beak, substipitate; stigmas 3; akenes trigonous.

Wet meadows, stream banks, and lake and pond margins at timberline in the Big Horn, Uinta, and Wind River Ranges. Circumboreal; British Columbia to Newfoundland south to California, Arizona, New Mexico, Iowa, Missouri, Pennsylvania, and Delaware.

Middle Rocky Mountain plants are var. *vesicaria*.

Eleocharis R. Br.
Spikerush

Perennial, or less commonly annual, scapose herbs from fibrous roots or rhizomes, the culms usually simple and terete. Leaves reduced to bladeless sheaths. Inflorescence an erect, solitary, terminal, ebracteate spike. Flowers perfect, single in the axils of spirally imbricate or 2-ranked scales, the lower scales usually sterile. Perianth of 6–9 bristles, occasionally lacking. Stamens 2 or 3. Style 2–3-cleft, enlarged at the base and tuberculate in fruit. Akenes lenticular to trigonous, the tubercle either distinct or confluent with the body of the akene.

KEY TO THE SPECIES OF *ELEOCHARIS*

1. Tubercle of the akene forming a distinct cap, constricted at the base and differentiated from the body; culms less than 10 cm tall   *E. acicularis*
1. Tubercle of the akene not forming a distinct cap, neither constricted nor differentiated from the body; culms more than 10 cm tall   *E. quinqueflora*

Eleocharis acicularis (L.) R. & S.
Syst. Veg. 2:154, 1817.

Slender Spikerush

Scirpus acicularis L.
Clavula acicularis (L.) Dumort
Eleocharis reverchonii Svens.
Isolepis acicularis (L.) Schlecht.
Mariscus acicularis (L.) Moench
Scirpidium aciculare (L.) Nees

Small, tufted perennial less than 15 cm tall, from slender rhizomes; culms filiform, somewhat angled. Leaf sheaths purplish to pale, often scarious at the apex. Spikes flattened, 3–15-flowered, the scales lanceolate to ovate, with a green midrib and reddish to brownish, hyaline margins; perianth bristles commonly 3, equaling or shorter than the akene. Style branches 3; akenes weakly trigonous, longitudinally ribbed, with fine cross striations, yellowish to brownish; tubercle trigonous-conical, forming a distinct cap.

Although common at lower elevations, this species is rare in the alpine zone and is currently known only from wet tundra near receding snowfields in the Uinta and Wind River Ranges. Circumboreal; in North America from Greenland south to Florida and Mexico.

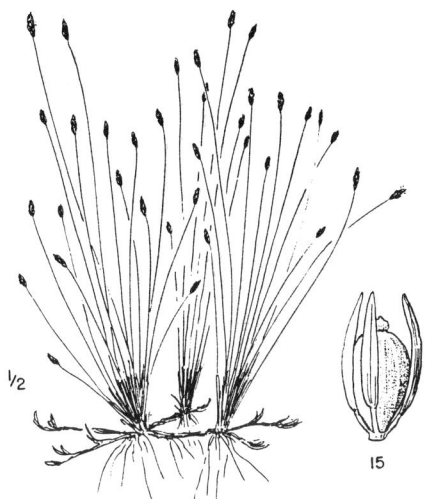
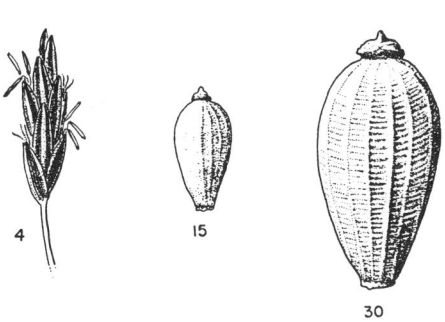
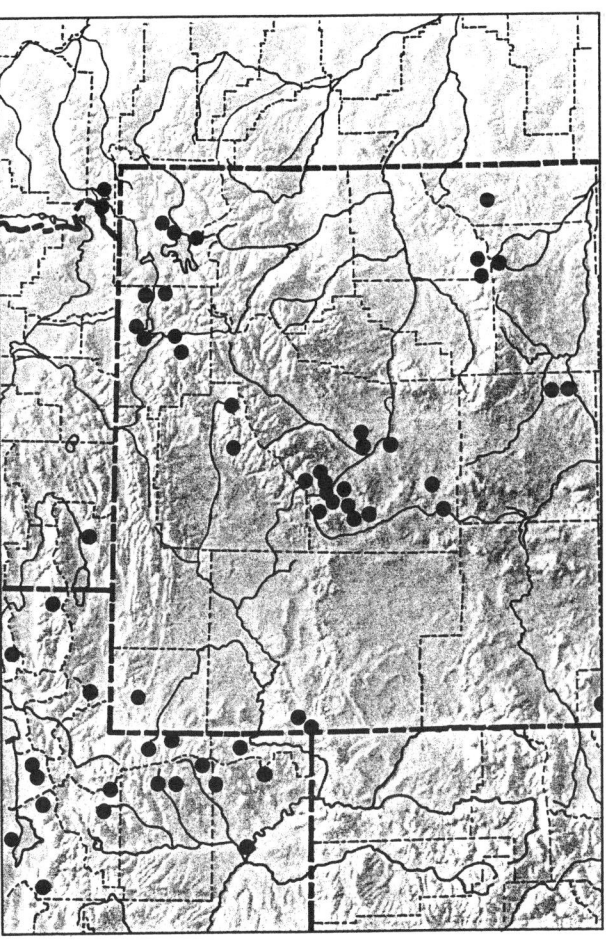

Eleocharis quinqueflora (F.X. Hartmann) O. Schwarz
Thüring. Bot. Ges. Mitt. 1:89, 1949.

Fiveflower Spikerush

Scirpus quinqueflorus F.X. Hartmann
Eleocharis bernardina Munz & Johnston
Eleocharis pauciflora (Lightf.) Link.
Eleocharis suksdorfiana Beauv.
Scirpus nanus Spreng.
Scirpus pauciflorus Lightf.

Tufted perennial from slender rhizomes and short, stout rootstocks with fibrous roots; culms several to many, erect, with 1 or 2 basal sheaths, the sheaths truncate, tannish yellow to reddish. Spikes ovate; mostly 2–9-flowered, the scales ovate, acute, dark brown to purplish, with hyaline margins, the midrib pale greenish or lacking; perianth bristles 2–6, equaling or exceeding the akene, sometimes reduced, rarely lacking. Style branches 3; akenes equally or unequally trigonous or rarely planoconvex, obovoid, olive-brown; tubercle triangular to lanceolate, not well differentiated from the body of the akene and not forming a distinct cap.

Stream banks and wet meadows at timberline in the Beartooth, Big Horn, Uinta, and Wasatch Ranges. Circumboreal; transcontinental in North America and south in the West to California, Utah, New Mexico, and Nebraska.

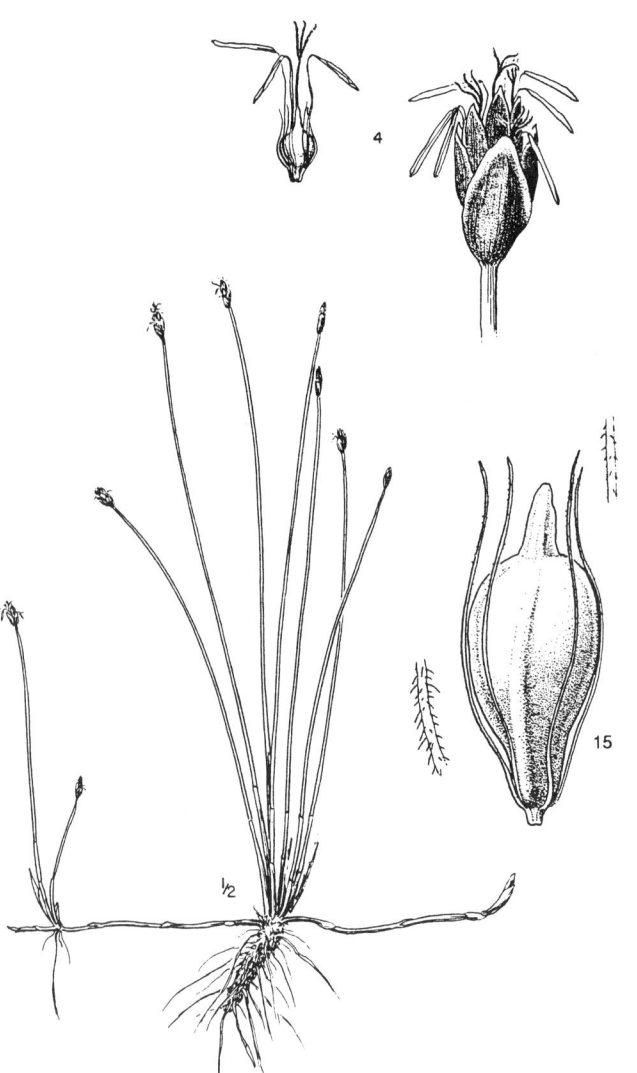

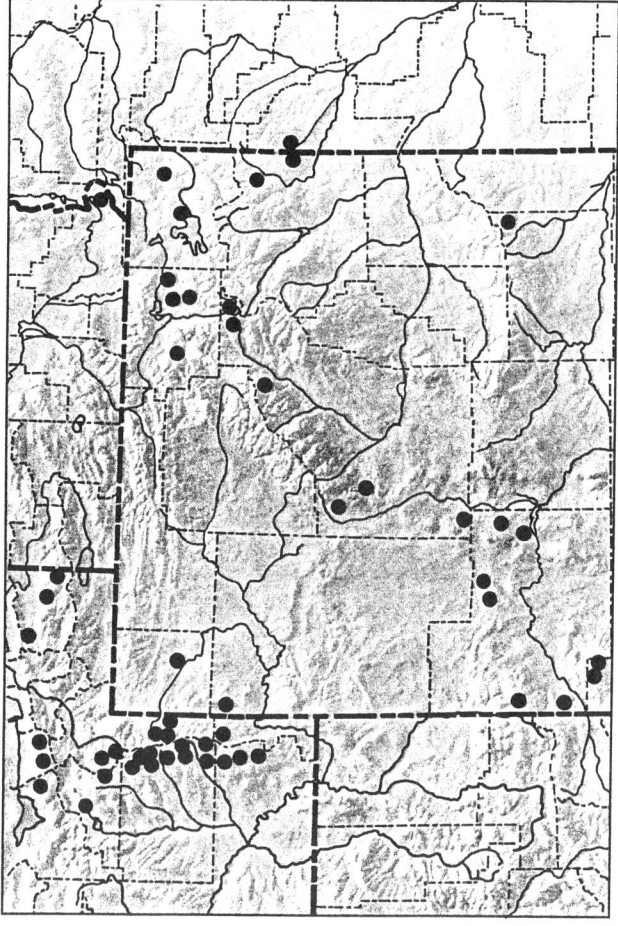

Eriophorum L.
Cottonsedge, Cotton Grass

Perennial herbs from fibrous roots, with or without rhizomes, the culms rounded or triangular. Leaves narrow, linear, or reduced to bladeless sheaths. Inflorescence with or without leafy bracts, the spikelets terminal, single or several, sessile or pedicellate. Flowers perfect, each subtended by a single scale; the perianth of many capillary bristles, these becoming elongate and cottonlike at maturity. Stamens 1–3. Ovary 1, the style 3-cleft. Akenes trigonous, oblong to obovate.

KEY TO THE SPECIES OF *ERIOPHORUM*

1. Spikelets mostly 2 or more, on slender pedicels, often drooping — *E. polystachion*
1. Spikelet solitary, sessile, erect
   2. Plants rhizomatous or stoloniferous, the culms not tufted; fertile scales of spikelet often with conspicuous pale margins — *E. scheuchzeri*
   2. Plants densely tufted, lacking rhizomes or stolons; fertile scales lacking pale margins — *E. callitrix*

Eriophorum callitrix Cham. ex C.A. Mey. *Mém Acad. Imp. Sci. St. Pétersb., Sér. 6, Sci. Math.* 1:203, 1830.

Tufted Cottonsedge

Densely tufted perennial from fibrous roots; rhizomes lacking; culms solitary to several, erect or decumbent at base. Leaves mostly basal, the blades linear; cauline leaves 1–2 or lacking, inserted below the middle of the culm; sheaths inflated. Spikelet solitary, erect, obovoid, terminal, not subtended by a leafy involucre, the involucral bracts scalelike, ovate, acute; fertile scales erect, lanceolate, blackish or greenish black, with translucent margins; perianth bristles shiny, bright white; anthers 1 mm or less in length.

Bogs, wet stream banks, and willow thickets of the Beartooth and Wind River Ranges. Circumpolar; south to Newfoundland, British Columbia, and Wyoming.

Middle Rocky Mountain plants with dark scales and white perianth bristles are var. *callitrix*.

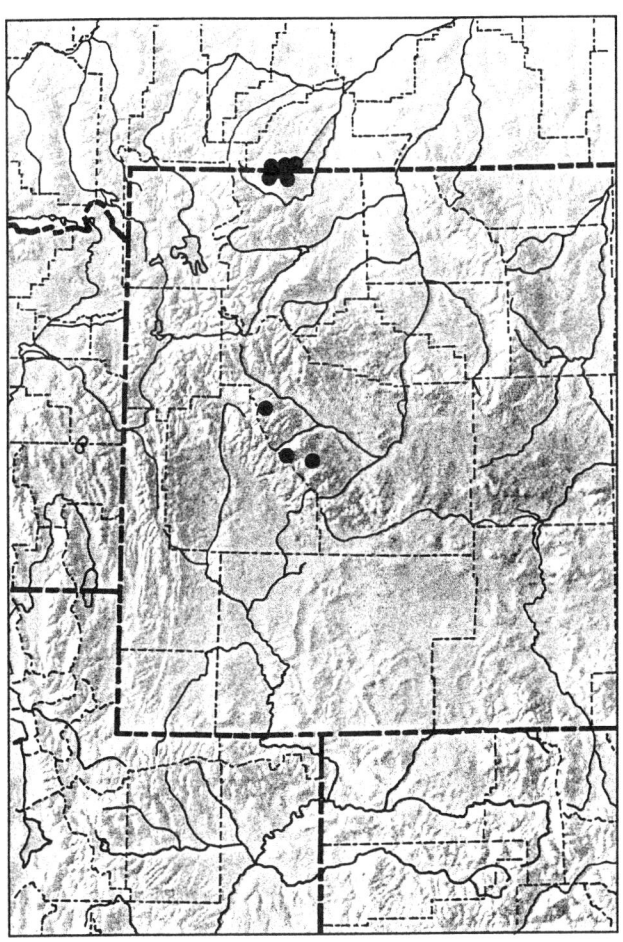

½

Eriophorum polystachion L.
Sp. Pl., 52, 1753.

Tall Cottonsedge

Eriophorum angustifolium Honck.
Eriophorum komarovii Vassiljev
Eriophorum ochreatum A. Nels.
Eriophorum subarcticum Vassiljev
Eriophorum triste (T. Fries) Hadac & A. Löve
Linagrostis polystachia (L.) Scop.
Scirpus angustifolius (Honck.) T. Koyama

Perennial from long-creeping rhizomes; culms erect, solitary or few together, obtusely angled. Leaves not forming a basal tuft, the blades flat below the middle and folded near the tip, scabrous at least on the margins; upper sheaths dark-banded at the summit. Spikelets ovoid or oblong, mostly 2–6 in an umbellate cyme, the peduncles mostly glabrous, drooping with age; involucral bracts unequal, the lowest ones foliaceous and nearly as long as the inflorescence; fertile scales lanceolate to ovate and greenish, purplish, or blackish, with hyaline margins and a light-colored midrib becoming obsolete near the tip; perianth bristles numerous, white; anthers 2.5–5 mm long.

Wet meadows and willow thickets near timberline in the Uinta Range. Circumboreal; south in western North America to Oregon, Utah, and New Mexico.

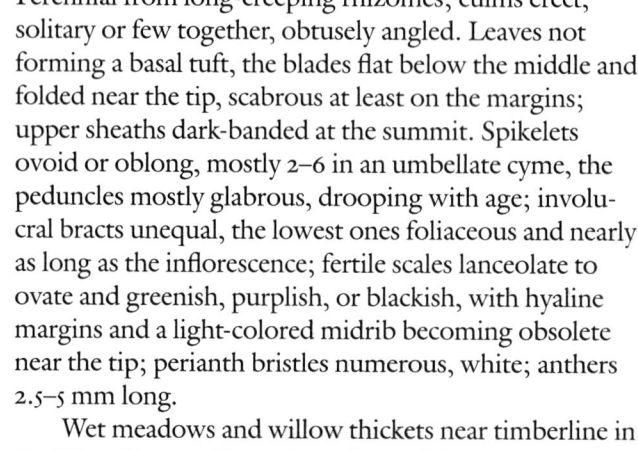
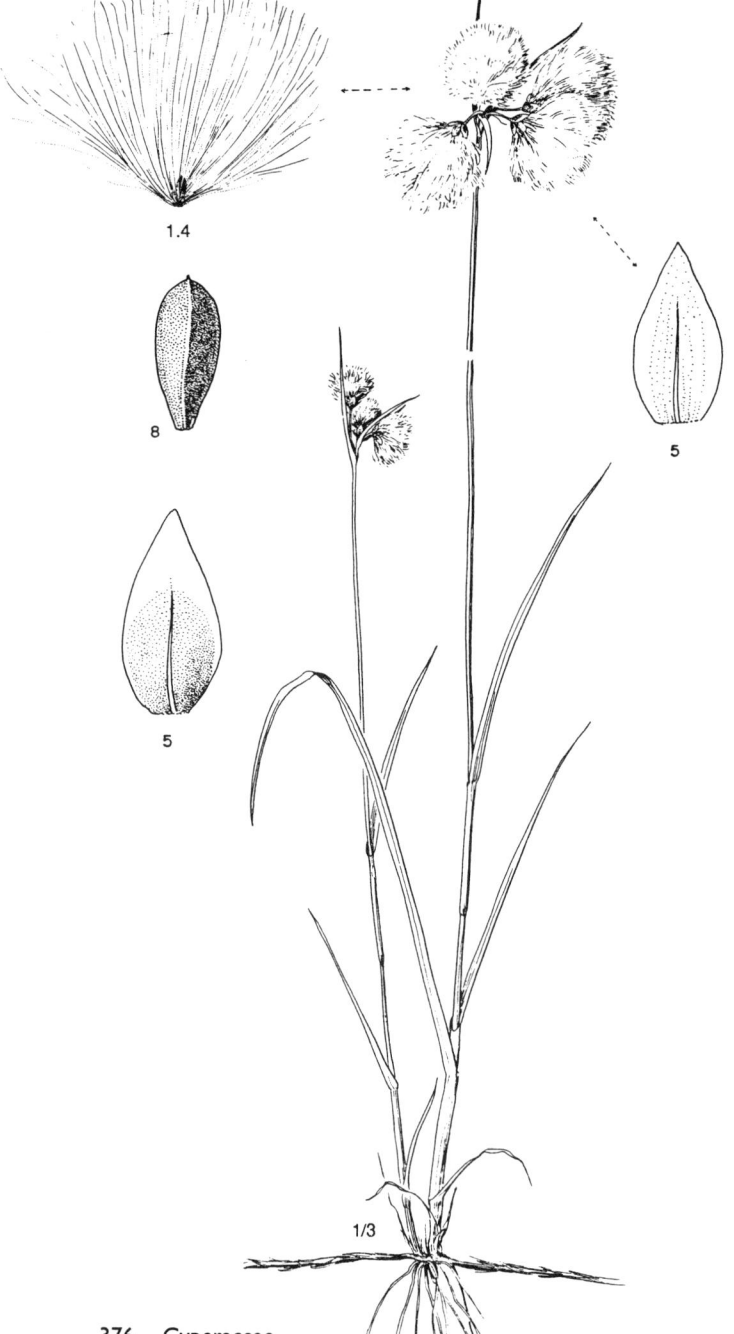
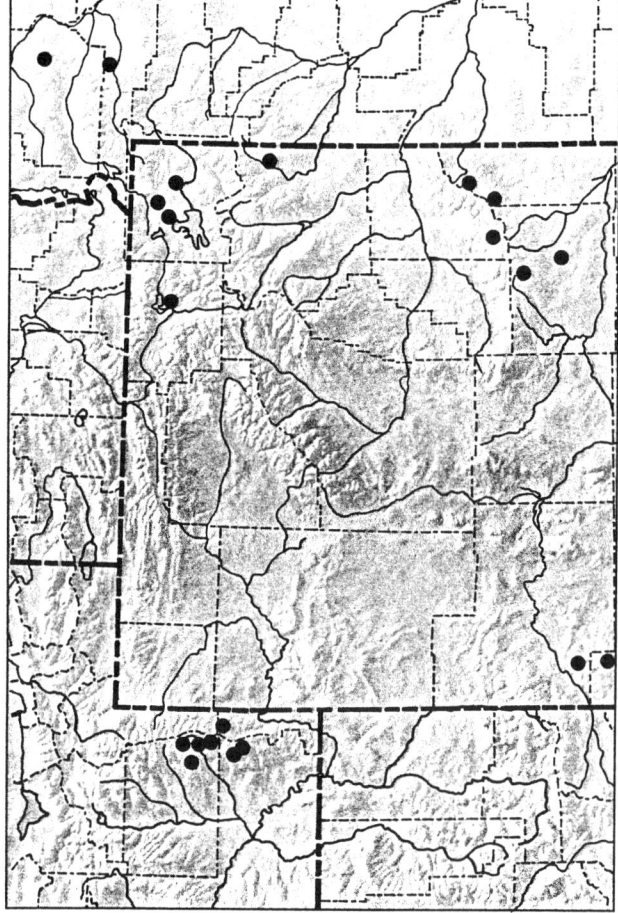

Eriophorum scheuchzeri Hoppe
Bot. Taschenb. 1800:104, 1800.

White Cottonsedge

Eriophorum capitatum Torr.
Eriophorum leucocephalum Boeck.

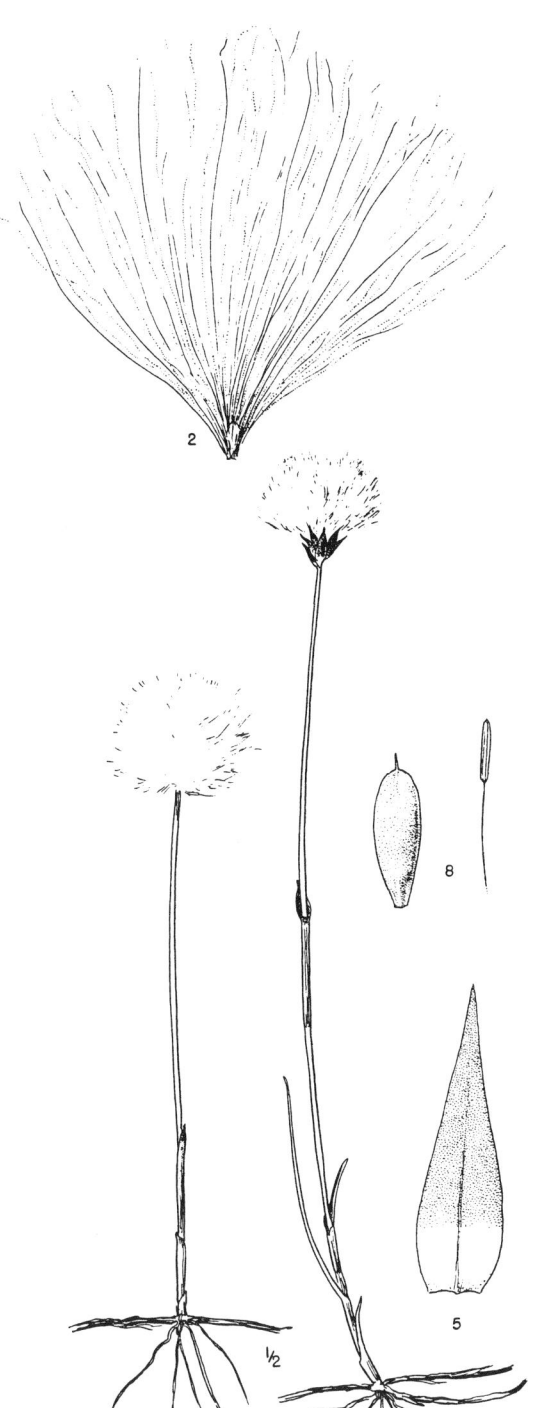

Perennial from long-creeping rhizomes; culms solitary, erect. Leaves mostly cauline, those near the base with short, filiform, channeled blades; sheaths well developed, inflated, the uppermost black-tipped. Spikelet solitary, subglobose, terminal; involucral bracts scalelike, blackish or purplish; fertile scales lanceolate to attenuate, grayish to blackish, with hyaline margins at least near the tips; perianth bristles numerous, white; anthers 0.5–1 mm long.

Bogs, wet stream banks, and willow thickets in the lower alpine zone of the Uinta and Wind River Ranges. Circumpolar; south in western North America to Utah and Colorado.

Slender plants of the Middle Rocky Mountains with filiform leaves have been called var. *tenuifolium* Ohwi.

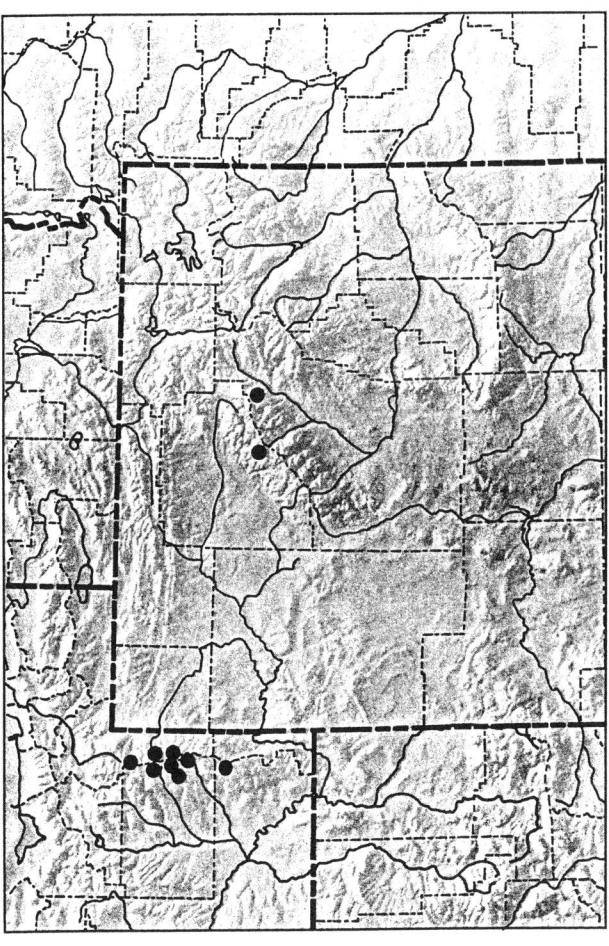

Kobresia Willd.
Alpinesedge, Kobresia

Low, tufted perennials with slender, erect culms, the leaves basally clustered. Flowers in terminal, solitary or compound, ebracteate spikes, usually in pairs, the staminate flowers above the pistillate. Perianth none. Stamens 3. Perigynium split down one side to the base, or nearly so. Stigmas 3, or rarely 2, exserted. Akenes trigonous, included in the perigynium or slightly exserted.

KEY TO THE SPECIES OF
KOBRESIA

1. Inflorescence 2–3 mm wide, linear — *K. myosuroides*
1. Inflorescence 3–7 mm wide, ovate to oblong — *K. schoenoides*

Kobresia myosuroides (Vill.) Fiori & Paol.
Fl. Analit. Ital. 1:125, 1896.

Mousetail Alpinesedge

Carex myosuroides Vill.
Carex affinis R. Brown
Carex bellardii All.
Carex hermaphroditica J.F. Gmel.
Carex scirpina (Willd.) Missb. & Krause in Sturm
Elyna bellardii (All.) Degl.
Elyna scirpina (Willd.) Pax in E. & P.
Elyna spicata Schrad.
Kobresia bellardii (All.) K. Koch
Kobresia scirpina Willd.
Scirpus bellardii (All.) Wahlenb.

Densely tufted perennial from fibrous roots; culms several to many, erect, with numerous brown, fibrillose sheaths at the base. Leaves basal, the blades erect, slender, flat or involute or channeled, mostly shorter than the culms. Spike solitary, narrow, linear, terminal; upper spikelet staminate, the lateral ones androgynous with 1 staminate and 1 pistillate flower, or occasionally pistillate; scales mostly less than 4 mm long, ovate to obovate, brownish in the center, the margins scarious; perigynia about 3.5 mm long, enclosing the akenes but with free margins; akenes brown, short-stipitate, with a persistent style base.

Fellfields and rocky ledges of the Absaroka, Beartooth, Big Horn, Uinta, and Wind River Ranges. Circumpolar; Alaska south to California, Idaho, Colorado, and New Mexico.

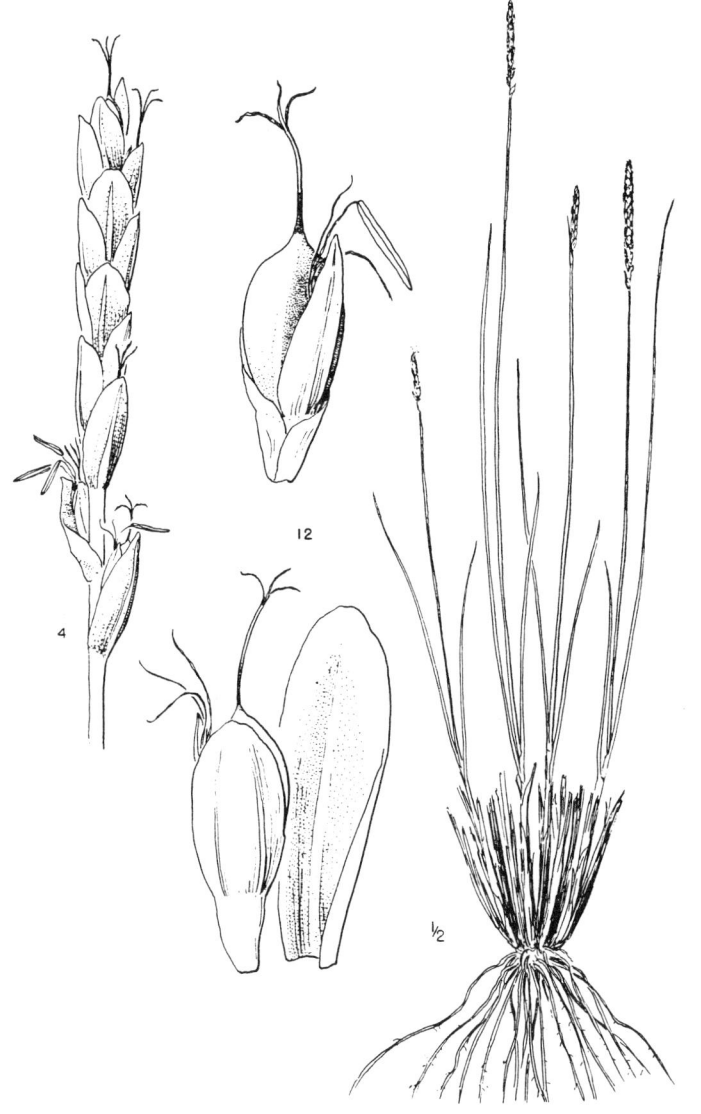
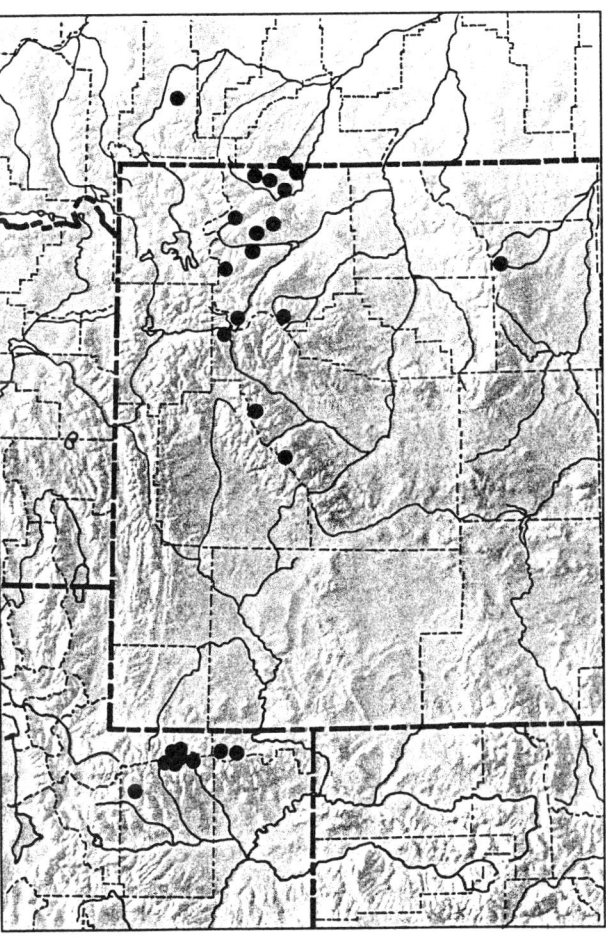

Kobresia schoenoides (C.A. Mey.) Steud.
Synops. Cyper., 246, 1855.

Northern Alpinesedge

Elyna schoenoides C.A. Mey.
Elyna sibirica Turcz. ex Ledeb.
Kobresia arctica Pors.
Kobresia hyperborea Pors.
Kobresia macrocarpa Clokey ex Mack.
Kobresia sibirica (Turcz. ex Ledeb.) Boeckl.

Densely tufted perennial from fibrous roots; culms few to several, erect, surrounded at the base by numerous brownish sheaths. Leaves basal, the blades erect, narrow, channeled, often exceeding the culms. Spike solitary, terminal, oblong to ovate; upper spikelet staminate, the lateral ones androgynous with 1 staminate and 1 pistillate flower; scales mostly more than 4 mm long, ovate, dark brown, the tips obtuse; perigynia about 5.5 mm long, enclosing the akenes but with free margins; akenes brown, short-stipitate, with a slender style.

Fellfields of the Absaroka and Beartooth Ranges. Circumpolar; Alaska south to Wyoming and Colorado.

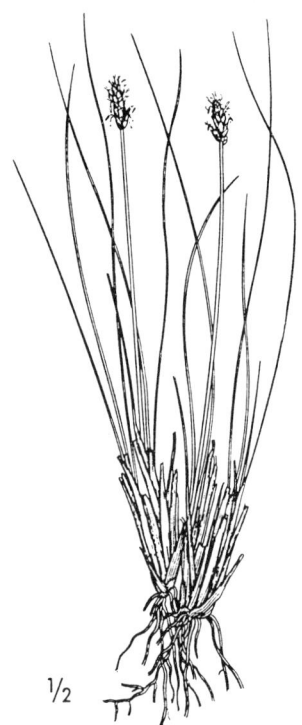

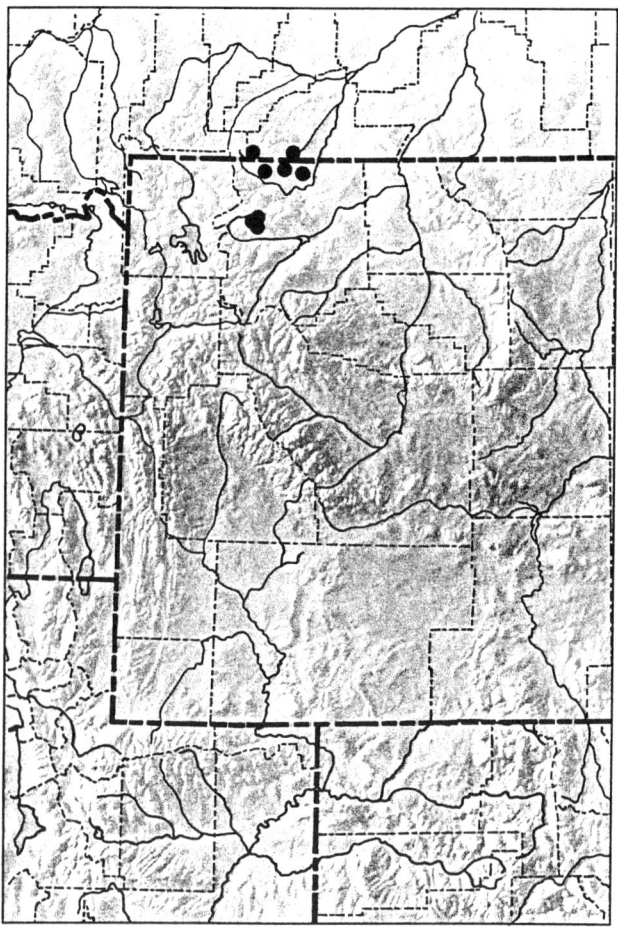

Ericaceae
Heath Family

Herbs or shrubs, sometimes saprophytic, with alternate or opposite, simple, often leathery leaves. Flowers hypogynous or epigynous, perfect, regular or nearly so, 4–5-merous. Petals nearly distinct or united, the corolla rotate, campanulate, or urceolate. Stamens the same number as the lobes of the corolla or twice as many, the anthers often appendaged. Pistil 1, of 4–5 carpels, with a single style and stigma. Fruit a capsule or berry.

KEY TO THE GENERA OF ERICACEAE

1. Petals distinct or nearly so; plants herbaceous, scapose or subscapose — *Pyrola*
1. Petals united, the corolla rotate, campanulate, or urceolate; plants often shrubby, often dwarfed or prostrate, caulescent
    2. Fruit a dry, brownish capsule
        3. Corolla rotate; leaves lanceolate to ovate, 2–3 times longer than wide — *Kalmia*
        3. Corolla campanulate; leaves linear, 5–8 times longer than wide — *Phyllodoce*
    2. Fruit a fleshy or leathery, red or purplish berry, or the calyx fleshy and the fruit berrylike
        4. Stems prostrate, herbaceous or slightly woody; fruit a capsule, enclosed by the fleshy calyx and appearing berrylike — *Gaultheria*
        4. Stems erect or somewhat prostrate, but definitely woody; fruit a berry
            5. Flowers epigynous; leaves thin, deciduous — *Vaccinium*
            5. Flowers hypogynous; leaves thick, evergreen — *Arctostaphylos*

Arctostaphylos Adans.
Bearberry

Erect or prostrate, evergreen shrubs with leathery, alternate, entire leaves. Flowers hypogynous, white or pink, in terminal racemes or panicles, or in ours, solitary, axillary, and nodding. Calyx 5-parted, the sepals nearly distinct. Corolla urceolate, 5-lobed, the lobes spreading or recurved. Stamens 10, included, the filaments dilated and hairy at the base, the anthers opening by terminal pores, each with 2 recurved dorsal appendages. Pistil of 5 united carpels, the ovary usually 5-celled, the style slender. Fruit red, berrylike, with 5 nutlets.

Arctostaphylos uva-ursi (L.) Spreng.
Syst. Veg. 2:287, 1825.

Kinnikinnick, Bearberry

Arbutus uva-ursi L.
Arctostaphylos adenotricha (Fern. & J.F. MacBr.) Löve, et al.
Arctostaphylos media Greene
Arctostaphylos officinalis Wimm. & Grab.
Arctostaphylos procumbens (Moench) Patze
Mairania uva-ursi (L.) Desv.
Uva-ursi buxifolia S.F. Gray
Uva-ursi procumbens Moench
Uva-ursi uva-ursi (L.) Britt.

Low, prostrate, mat-forming shrub with trailing, stolonlike branches. Bark reddish, exfoliating on older stems; younger stems puberulent to tomentose, sometimes stipitate-glandular. Leaves evergreen, coriaceous, petiolate, the blades obovate to spatulate, obtuse, entire, glabrous to puberulent. Flowers white to pink, pendent in terminal, few-flowered racemes. Sepals distinct, broadly ovate, acute. Corolla urceolate, 5-lobed, the lobes short-reflexed. Stamens 10, included; filaments pubescent; anthers subglobose. Ovary superior, glabrous. Fruit bright red, mealy, berrylike.

Occasional in the lower alpine zone near timberline in the Absaroka, Beartooth, Medicine Bow, Teton, Uinta, Wasatch, and Wind River Ranges. Eurasia; Alaska to Labrador, south to Virginia, Illinois, Nebraska, Colorado, New Mexico, and California.

Diploid plants, $2n=26$, of the lower alpine zone with stipitate glands intermixed with branchlet pubescence are ssp. *adenotricha* (Fern. & J.F. Macbr.) Calder & Taylor. Löve, et al. have proposed specific rank for such plants as *A. adenotricha* (Fern. & J.F. Macbr.) Löve, et al. Hexaploid plants, $2n=52$, of the montane zone with puberulent to tomentose branchlets are ssp. *coactilis* (Fern. & J.F. Macbr.) Löve, et al.

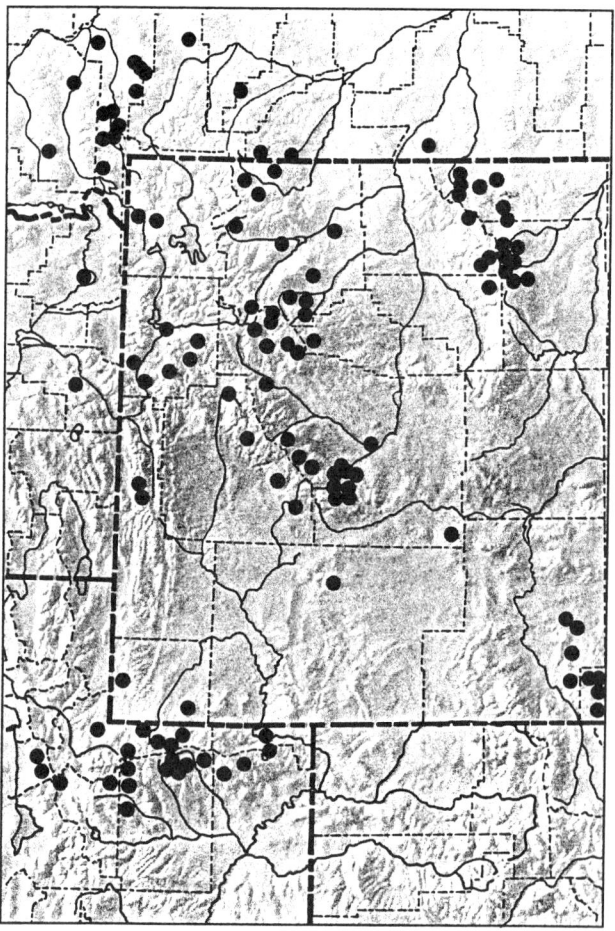

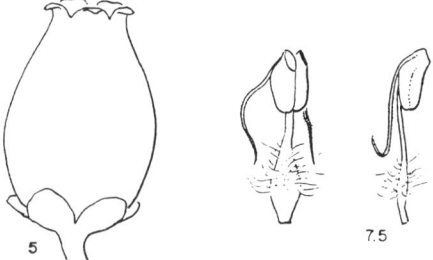

Gaultheria L.
Creeping Wintergreen

Low, prostrate to semierect shrubs with alternate, shiny, leathery, evergreen leaves. Flowers hypogynous, 4–5-merous, white or pink, axillary or in terminal racemes. Calyx usually deeply 5-lobed in ours. Corolla urceolate or campanulate. Stamens 8 or 10, included, the filaments dilated below, the anthers opening by terminal pores, sometimes awned. Pistil of 4 or 5 united carpels, the ovary 4–5-celled, the stigma entire or lobed. Fruit berrylike, the fleshy calyx surrounding a many-seeded capsule.

Gaultheria humifusa (Grah.) Rydb.
Mem. N.Y. Bot. Gard. 1:330, 1900.

Alpine Wintergreen

Vaccinium humifusum Grah.
Gaultheria myrsinites Hook.

Low, procumbent, evergreen shrub, often forming mats less than 5 cm tall. Stems prostrate, trailing, glabrous to puberulent or reddish-pilose. Leaves short-petiolate, the blades oval to broadly ovate, sometimes nearly orbicular, glabrous, obtuse or acute, entire to serrulate, the margins thickened. Flowers white to pinkish, solitary in the leaf axils. Calyx glabrous, nearly the same length as the campanulate, 5-lobed corolla. Stamens 8–10, included, lacking appendages. Fruit a capsule surrounded by the fleshy, red calyx, resembling a berry.

Rare to uncommon in moist meadows and alpine bogs near timberline in the Beartooth, Gros Ventre, Teton, and Uinta Ranges. British Columbia and Alberta south to Colorado, Idaho, and California.

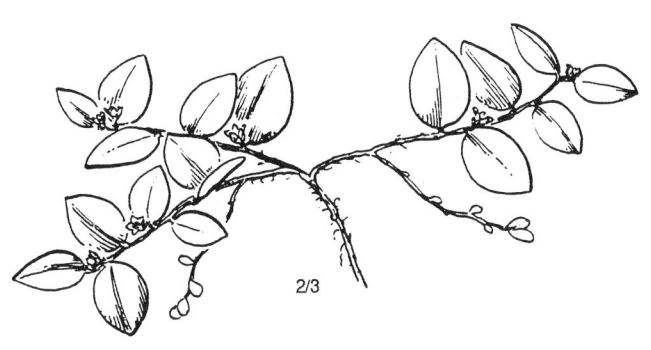
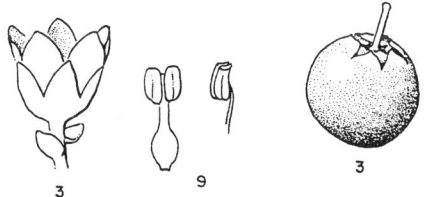
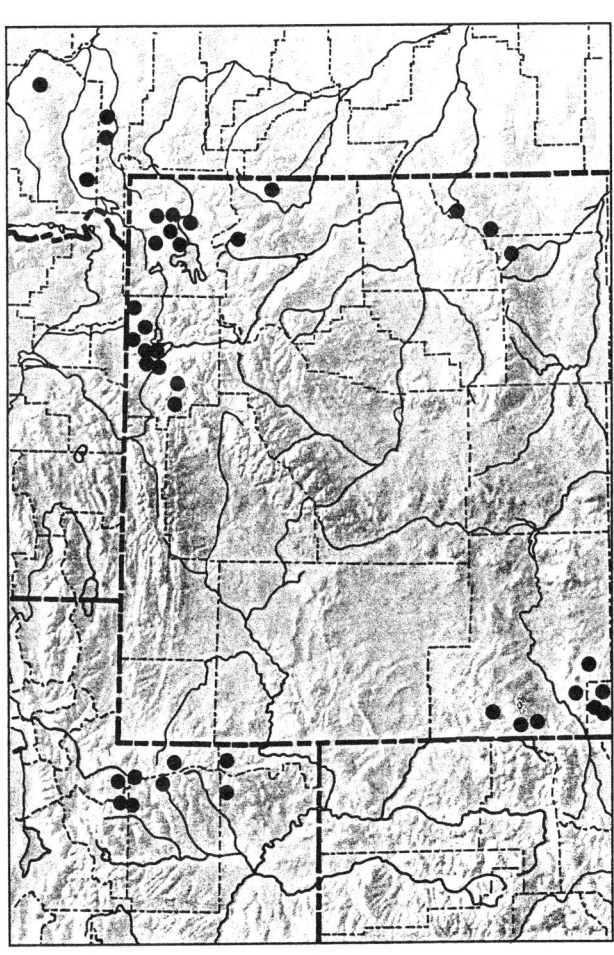

Kalmia L.
Bog Laurel

Low, branching, evergreen shrubs, 5–15 cm high, with opposite, entire, glabrous, leathery leaves. Flowers hypogynous, pink or rose, showy, in terminal or axillary corymbs. Calyx 5-lobed, leathery. Corolla 5-lobed, rotate, each lobe with 2 basal pouches. Stamens 10, opening by terminal pores, at first enclosed in the basal pouches of the corolla. Pistil of 5 united carpels, the style elongate, straight. Fruit a globose, many-seeded, septicidal capsule.

Kalmia microphylla (Hook.) A. Heller
Bull. Torr. Club 25:581, 1898.

Bog Laurel

Kalmia glauca L'Herit. ex Ait.
Kalmia occidentalis Small
Kalmia polifolia Wang.

Low, erect, evergreen shrub with short rhizomes. Leaves opposite, oval to linear-elliptic, coriaceous, glabrous above and whitish-pubescent beneath, sessile to short-petiolate, the margins revolute or flat. Flowers pink to rose, in terminal corymbs, the pedicels slender and red. Sepals ovate, concave, glabrous, the margins ciliolate. Corolla rotate, shallowly 5-lobed. Stamens 10, slightly exerted, the filaments pubescent. Fruit an ovoid to sub-globose capsule.

Wet meadows, bogs, nivation basins, and stream banks; particularly common on the shores of tarn lakes near timberline and in the lower alpine zone of most alpine ranges of the Middle Rocky Mountains. Not yet known from above timberline in the Hoback, Salt River, and Wyoming Ranges. Alaska, Yukon Territory, and Alberta south to California and Colorado.

Alpine plants less than 10 cm tall, with leaves less than 2 cm long and flowers less than 12 mm broad are var. *microphylla*.

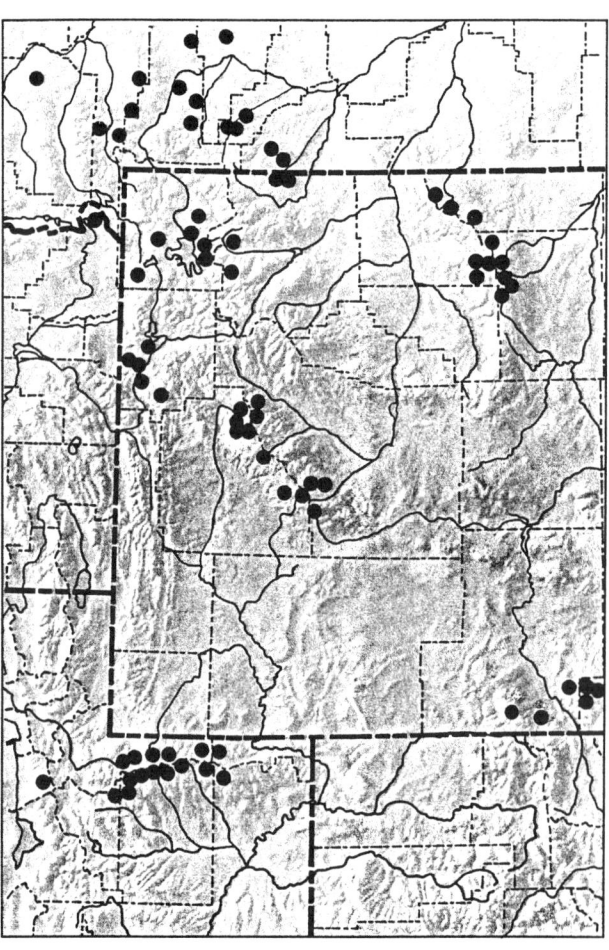

½

Phyllodoce Salisb.
Mountain Heath

Low, spreading, evergreen shrubs with small, alternate, linear, obtuse, crowded leaves. Flowers hypogynous, nodding, pink or cream in ours, in terminal, umbellate clusters. Calyx 5-parted, persistent. Corolla 5-lobed, urceolate to campanulate, with spreading lobes. Stamens 10, included, the anthers opening by terminal pores. Pistil of 5 united carpels, the stigma obscurely 5-lobed or capitate. Fruit a globose, septicidal capsule with numerous seeds.

KEY TO THE SPECIES OF *PHYLLODOCE*

1. Corolla pink or rose; sepals glandular, obtuse — *P. empetriformis*
1. Corolla cream-colored; sepals glandular, acute — *P. glanduliflora*

Phyllodoce empetriformis (Sw.) D. Don
Edinb. New Phil. J. 17:160, 1834.

Pink Mountain Heath

Menziesia empetriformis Sw.
Bryanthus empetriformis (Sw.) Gray
Menziesia grahamii Hook.
Phyllodoce grahamii (Hook.) Pursh

Low, densely branched, usually matted shrub with puberulent young stems, becoming glabrous when older. Leaves sessile or nearly so, evergreen, acicular, with revolute, glandular-serrulate margins. Flowers pink to rose, clustered in leaf axils near the ends of branches. Calyx lobes ovate, obtuse, eglandular, often reddish. Corolla campanulate, 5-lobed, the lobes strongly recurved. Stamens 10, included, the filaments glabrous; anthers reddish. Ovary glandular-pubescent; style exerted.

Locally common on wet alpine tundra, stream banks, lakeshores, and solifluction terraces of the Absaroka, Beartooth, Gros Ventre, Teton, and Wind River Ranges. Alaska south to Montana, Wyoming, Idaho, and California.

The 2 species of *Phyllodoce* tend to hybridize where they occur together in the Teton and Beartooth Ranges. These plants, sometimes called *P. intermedia* (Hook.) Camp, have a yellowish or pale reddish corolla, sepals acute or obtuse, and are moderately glandular.

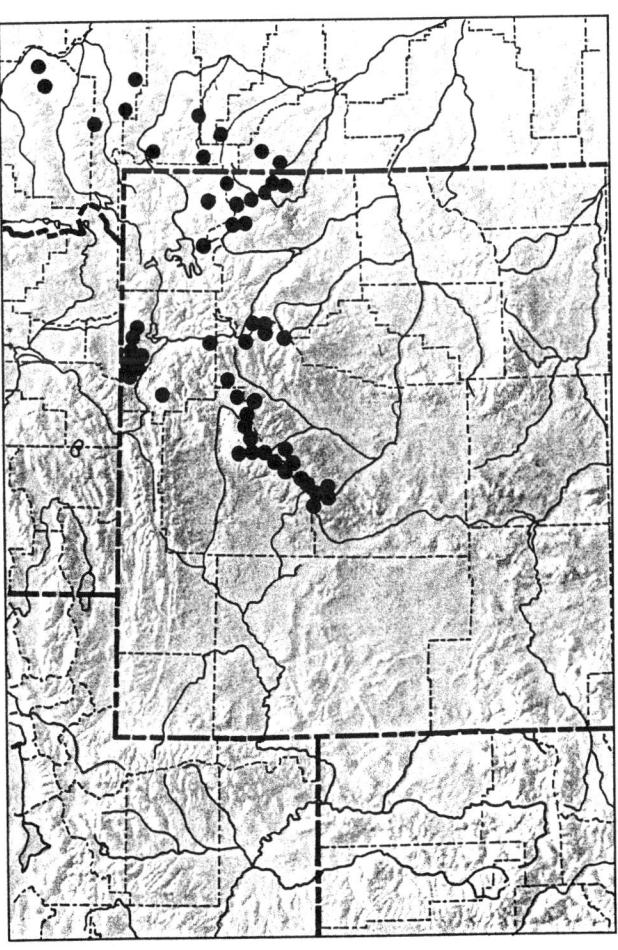

2/3

Phyllodoce glanduliflora (Hook.) Cov.
Mazama 1:196, 1897.

Cream Mountain Heath

Menziesia glanduliflora Hook.
Bryanthus glanduliflorus (Hook.) Gray

Low, usually matted shrub with glandular-puberulent young stems. Leaves sessile or nearly so, evergreen, acicular, with revolute margins, glandular-puberulent throughout. Flowers cream-colored, clustered in leaf axils near the ends of branches, the pedicels glandular-pubescent. Calyx lobes lanceolate, acute, glandular. Corolla urceolate, 5-lobed, the lobes spreading, the tube glandular. Stamens 10, included, the filaments pubescent. Ovary glandular, the style included.

Talus, scree, and rocky slopes of the Absaroka, Beartooth, and Teton Ranges. Alaska and Yukon Territory south to Wyoming, Idaho, and Oregon.

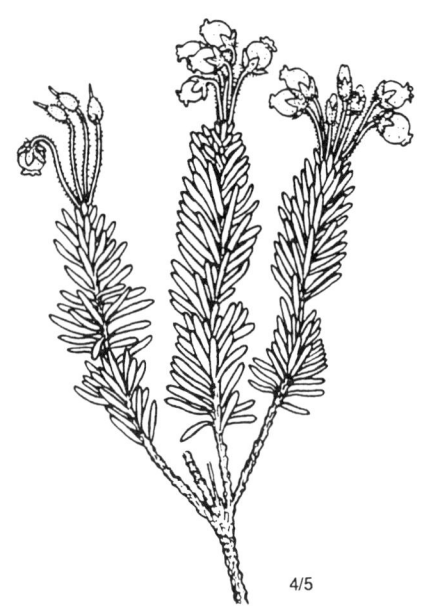

4/5

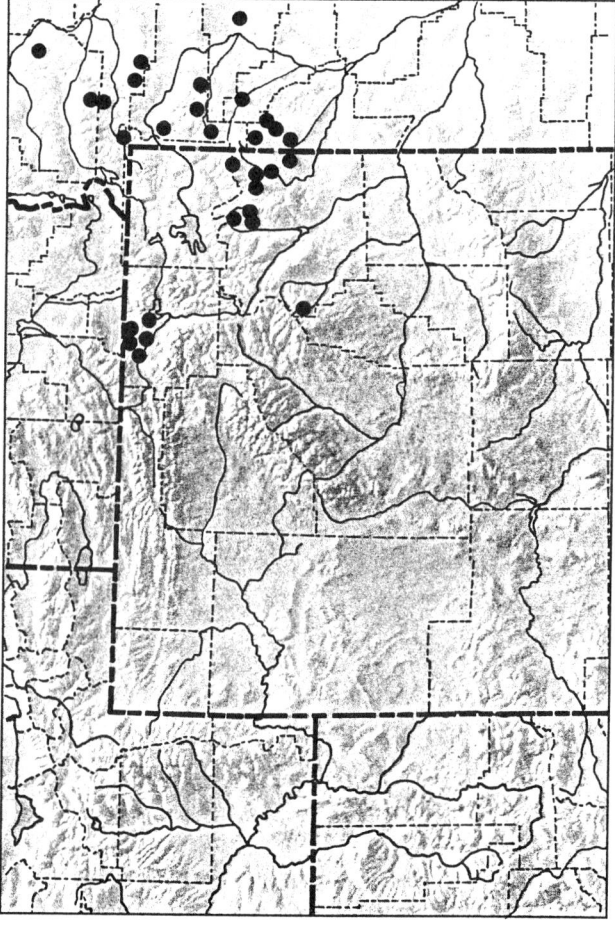

Pyrola L.
Pyrola, Wintergreen

Glabrous, scapose or subscapose, perennial herbs with ovate to orbicular, evergreen leaves. Scapes bracteate, bearing terminal racemes of nodding, hypogynous flowers. Calyx 5-parted. Petals distinct or slightly united, sometimes unequal, white, pink, or greenish. Stamens 10, the anthers opening by a pair of pores at the base (appearing apical by inversion). Pistil of 5 united carpels, the style straight or declined and curved upward, the stigma 5-lobed or 5-rayed. Capsule globose, many-seeded, dehiscent from the base upward.

KEY TO THE SPECIES OF PYROLA

1. Style straight; stigma peltate, broader than the style
   2. Raceme distinctly 1-sided; style exerted — *P. secunda*
   2. Raceme not 1-sided; style shorter than or equaling the petals — *P. minor*
1. Style curved; stigma the same diameter as the style — *P. picta*

Pyrola minor L.
Sp. Pl., 396, 1753.

Small Pyrola

Amelia minor (L.) Alef.
Braxilia minor (L.) House
Erxlebenia minor (L.) Rydb.
Pyrola conferta Fisch. ex Ledeb.

Scapose, perennial herbs from slender rhizomes. Leaves basal, petiolate, broadly ovate to orbicular, crenulate. Flowers nodding, in short, crowded racemes. Sepals broadly deltoid, pinkish. Petals pink to rose, orbicular. Stamens 10, included, the filaments thin, flattened. Style erect, straight, included or nearly so, the stigma peltate.

Pyrola minor is typical of coniferous forest but has been collected above timberline in the lower alpine zone of the Teton and Wind River Ranges. Eurasia; transcontinental in northern North America, south in the West to California and Colorado.

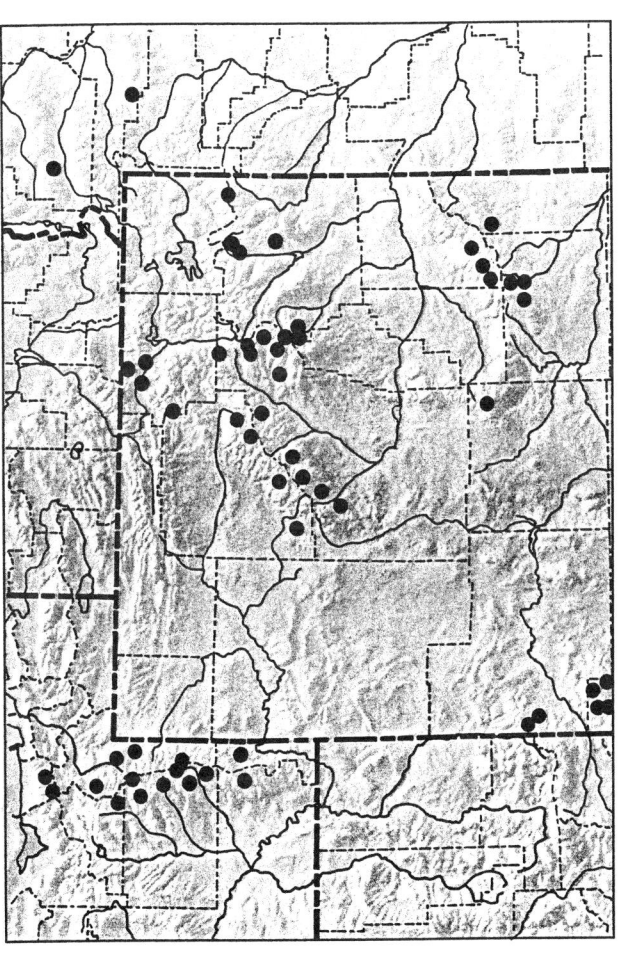

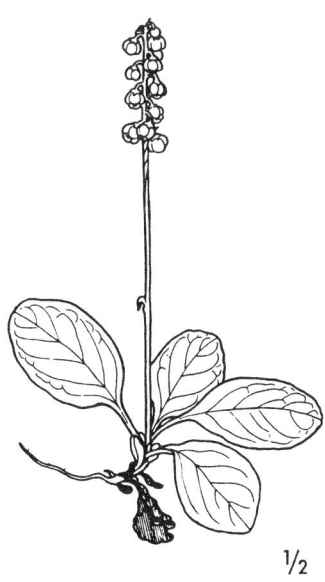

½

Pyrola picta Smith in Rees
Cycl. 29:*Pyrola*, No. 8, 1819.

Painted Pyrola

Pyrola aphylla Sm.
Pyrola blanda Andres
Pyrola conardiana Andres
Pyrola dentata Sm.
Pyrola pallida Greene
Pyrola paradoxa Andres
Pyrola septentrionalis Andres
Pyrola sparsifolia Suksdorf

Slender, perennial herb from rhizomes. Leaves forming basal rosettes, petiolate, the blades oblanceolate or spatulate to ovate, the margins entire to serrulate, the upper surfaces often whitish-mottled or whitish-veined. Flowers nodding in loose racemes, the pedicels spreading to somewhat reflexed. Sepals green to reddish, deltoid to ovate. Petals oval to obovate, cream or greenish. Stamens 10, included, the filaments slightly dilated at the base; anthers apiculate. Style exerted, strongly curved; stigma the same diameter as or slightly larger than the style.

Pyrola picta var. *dentata* (Sm.) Dorn is typical of coniferous forests in Teton County, Wyoming, but has been collected from above timberline in the lower alpine zone of the Teton Range. British Columbia to Montana, south to Wyoming, Idaho, and California.

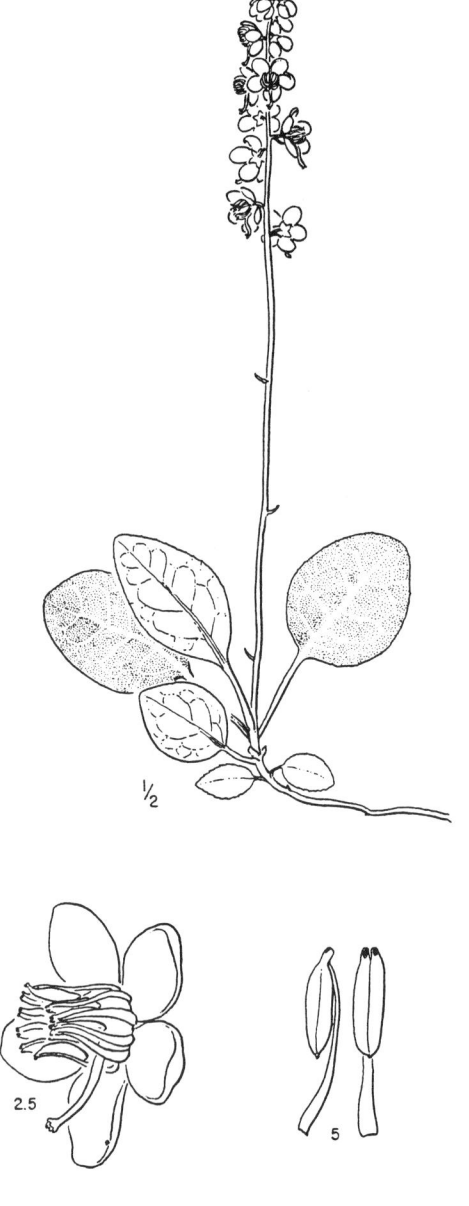

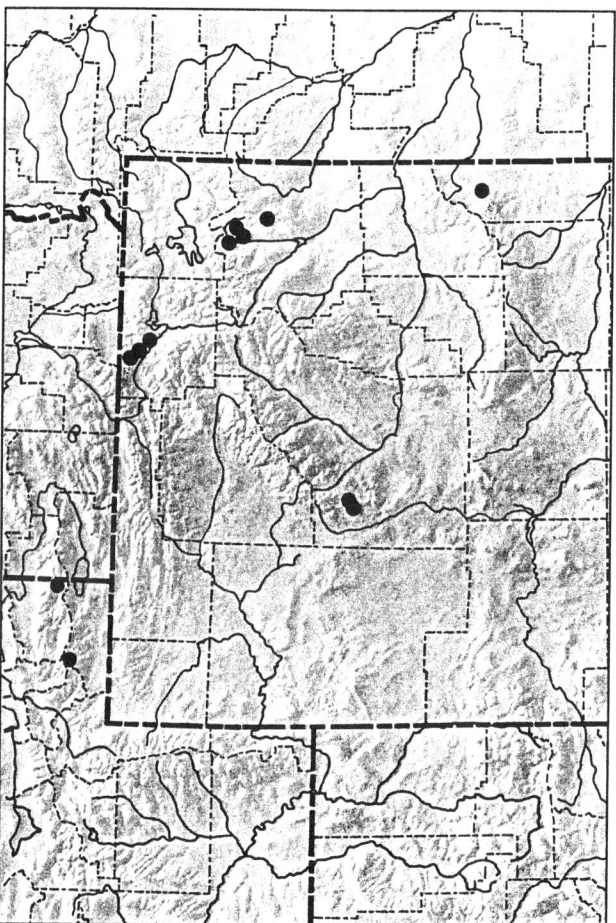

Pyrola secunda L.
Sp. Pl., 396, 1753.

One-Sided Wintergreen

Actinocyclus secundus (L.) Klotzsch
Orthilla secunda (L.) House
Ramischia elatior Rydb.
Ramischia secunda (L.) Garcke
Ramischia secundiflora Opiz

Perennial herb from creeping rhizomes, often forming mats; stems decumbent, often woody at the base. Leaves petiolate, the blades ovate to orbicular, the margins serrulate, denticulate, or nearly entire, acute to obtuse, dark green above, pale beneath. Flowers in terminal, secund racemes; pedicels reflexed. Sepals oval or elliptic, ciliolate. Petals greenish to white, each with 2 basal tubercles on the inside surface. Stamens 10, included, the filaments slender; anthers not apiculate, the pores terminal. Style straight, exerted; stigma peltate.

Forest margins at timberline in the Big Horn, Uinta, Wasatch, and Wind River Ranges. Circumboreal; south in western North America to California and Mexico.

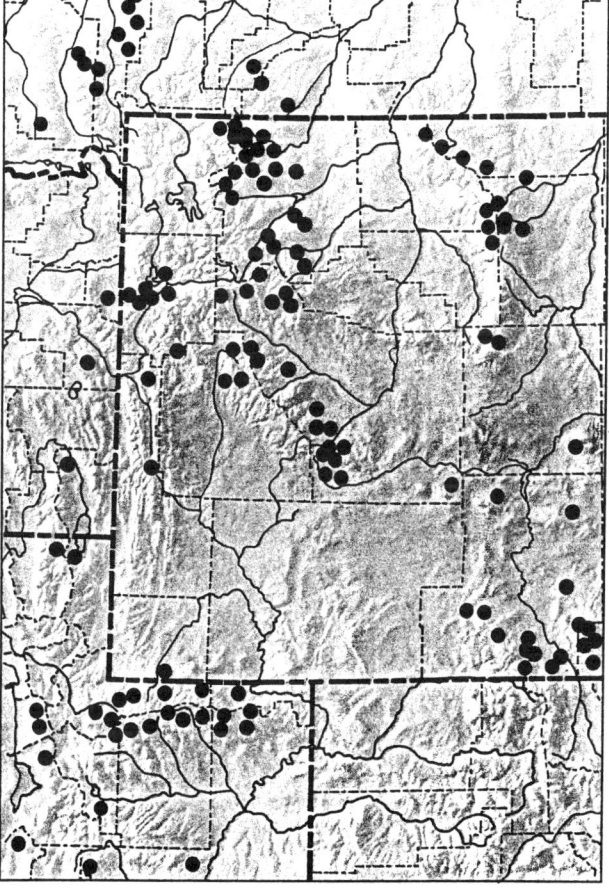

Vaccinium L.
Blueberry, Huckleberry

Low to moderately tall, perennial herbs or shrubs, with alternate, simple, entire or serrulate leaves. Flowers epigynous, solitary or racemose, the calyx toothed or lobed. Corolla pink or white, urceolate, shallowly to deeply 4–5-lobed, the lobes erect or spreading. Stamens twice as many as the corolla lobes, the anthers prolonged into tubes, sometimes awned. Pistil of 4–5 united carpels, ovary 4–5-celled or 8–10-celled by false partitions. Fruit a several-seeded, red or blue berry.

KEY TO THE SPECIES OF *VACCINIUM*

1. Branchlets green, sharply angled; leaves distinctly serrulate; flowers single in the leaf axils
    2. Branchlets glabrous; berries red — *V. scoparium*
    2. Branchlets pubescent; berries bluish to blue-black — *V. myrtillus*
1. Branchlets brownish, terete or subterete; leaves entire to remotely serrulate; flowers 1–4 in leaf axils
    3. Calyx shallowly lobed, the lobes rounded, deciduous in fruit; leaves serrulate above the middle; flowers solitary — *V. cespitosum*
    3. Calyx deeply lobed, the lobes triangular, persistent in fruit; leaves entire; flowers 2–4 — *V. uliginosum*

Vaccinium cespitosum Michx.
Fl. Bor.-Am. 1:234, 1803.

Dwarf Blueberry

Vaccinium arbuscula (Gray) Merriam
Vaccinium nivictum Camp
Vaccinium paludicola Camp

Low, somewhat prostrate shrub, to 3 dm, the branches terete to somewhat angled, glabrous to puberulent. Leaves sessile to short-petiolate, obovate to oblanceolate, acute to obtuse, mostly glabrous above and glandular below, the bases cuneate; margins serrulate, at least in the upper half. Flowers solitary in leaf axils. Calyx with short, broad lobes, the lobes deciduous in fruit. Corolla white to pink, narrowly urceolate, the lobes very short in relation to the tube. Stamens 10; filaments glabrous; anthers awned. Berries subglobose, blue, glaucous.

Streamsides, meadows, and protected slopes near timberline in the Wasatch and Uinta Ranges. Transcontinental in Canada; Alaska south to California, Utah, and Colorado.

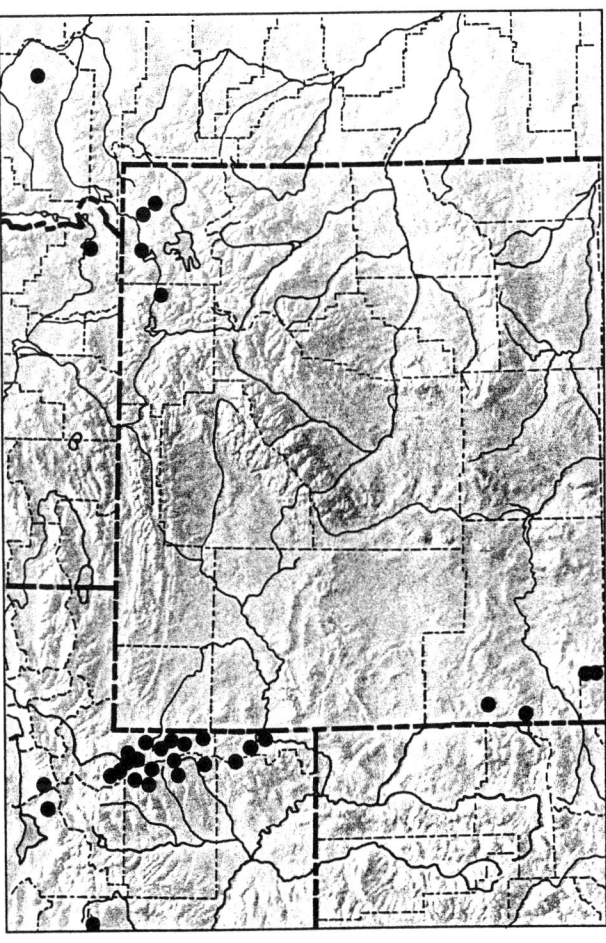

Vaccinium myrtillus L.
Sp. Pl., 349, 1753.

Low Blueberry

Vaccinium oreophilum Rydb.

Low shrub, to 3 dm, the branches greenish, widespreading, strongly angled, the grooves usually puberulent. Leaves short-petiolate, the blades ovate to lanceolate, acute to obtuse, glabrous, with prominent veins below, the bases rounded; margins serrulate. Flowers solitary in leaf axils. Calyx lobes shallow to obsolete. Corolla pink, broadly urceolate, the lobes short-reflexed. Stamens 10; filaments glabrous; anthers awned dorsally. Berries globose, bluish or blue-black.

Stream banks, meadows, and protected slopes near timberline in the Uinta Range. Eurasia; British Columbia and Alberta south to Washington, Utah, Arizona, and New Mexico.

Rocky Mountain plants that are geographically isolated from Eurasian plants are ssp. *oreophilum* (Rydb.) A. Löve et al.

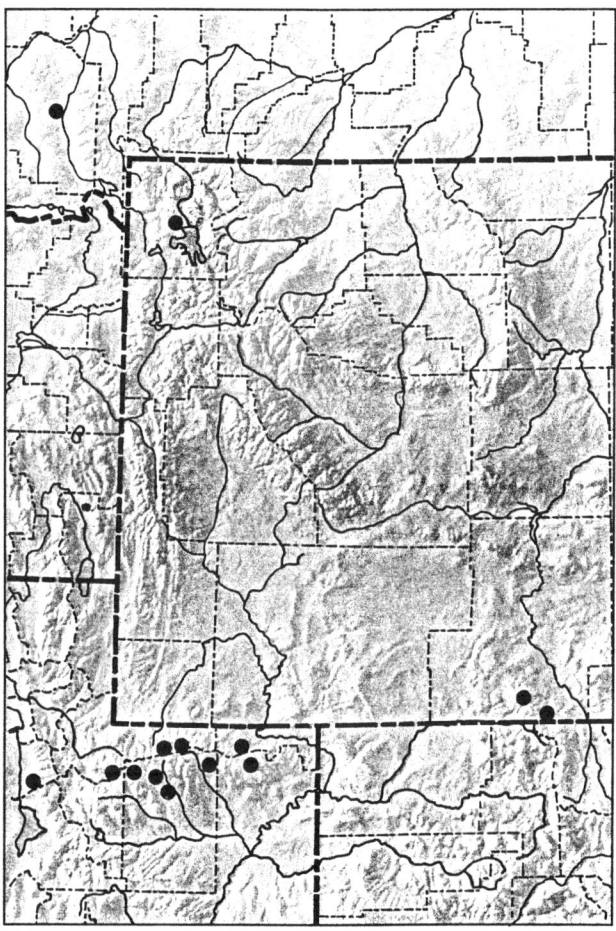

Vaccinium scoparium Leiberg ex Cov.
Mazama 1:196, 1897.

Grouseberry

Vaccinium erythrococcum Rydb.

Low, mat-forming shrub, much branched near the base, the branches yellow-green, angled, glabrous or sometimes finely puberulent. Leaves thin, lanceolate to ovate, acute, glabrous to puberulent, the margins serrulate. Flowers solitary in the leaf axils. Calyx shallowly lobed. Corolla pink or whitish, urceolate, the lobes very short, reflexed. Stamens 10; filaments glabrous; anthers awned. Berries globose, red, not glaucous.

Very common under stands of *Pinus albicaulis* or other conifers below timberline; occasional in the lower alpine zone of the Beartooth, Medicine Bow, Salt River, Teton, Uinta, Wasatch, and Wind River Ranges. British Columbia and Alberta south to South Dakota, Colorado, Idaho, and California.

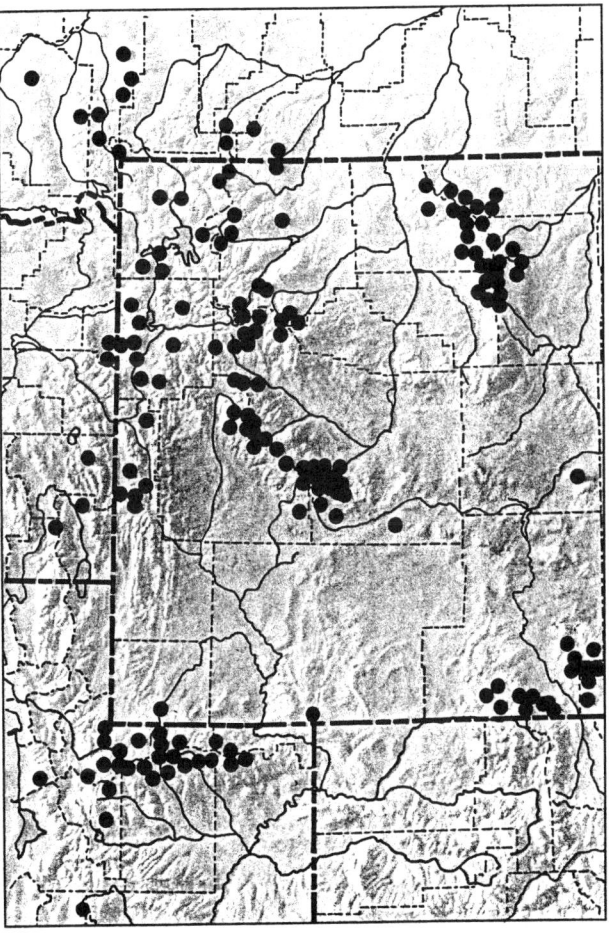

Vaccinium uliginosum L.
Sp. Pl. 1:350, 1753.

Bog Blueberry

Vaccinium occidentale Gray

Much-branched, often matted shrub, to 5 dm, the branches terete or nearly so, usually glabrous. Leaves narrowly elliptic or oblanceolate, entire, glaucous. Flowers in clusters of 2–4 in leaf axils. Calyx deeply lobed, the lobes ovate. Corolla pink, narrowly urceolate, the lobes very short in relation to the tube. Stamens 10; filaments glabrous; anthers awned. Berries globose, blue, glaucous.

Stream banks, snowmelt seeps, and lakeshores in the lower alpine zone of the Teton, Uinta, and Wind River Ranges. British Columbia south to California, Utah, and Wyoming. Middle Rocky Mountain plants are var. *occidentale* (Gray) H. Hara.

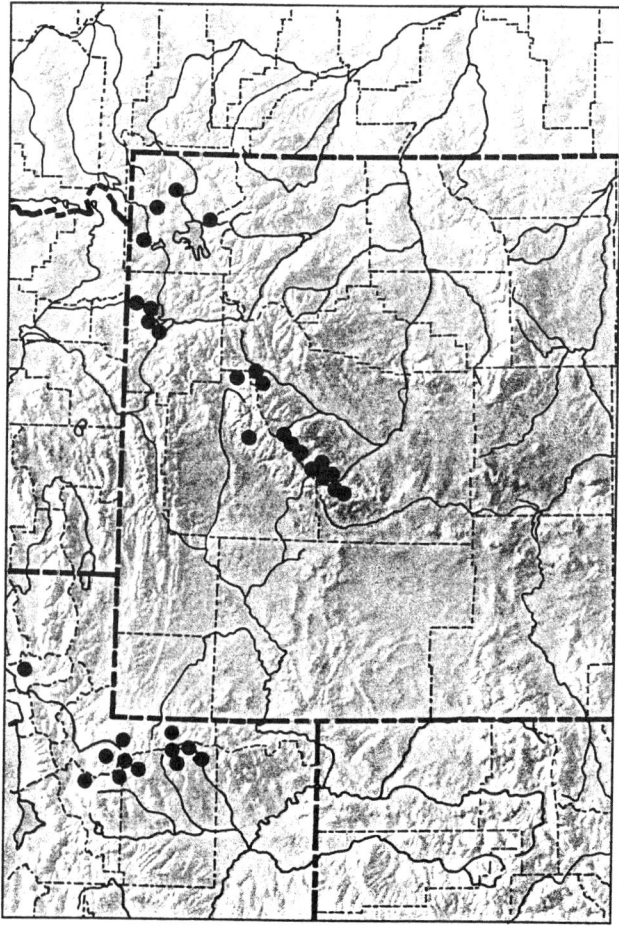

Fabaceae
(Leguminosae)
Pea Family

Herbs, shrubs, or trees with alternate, mostly compound leaves, the stipules present or lacking. Flowers perigynous, usually perfect, commonly irregular, 5-merous, the hypanthium often very short. Calyx usually tubular, of 5 united sepals. Corolla typically of 5 dissimilar petals, the median one (the banner or standard) larger than the others and enclosing them in bud, the 2 lateral ones (wings) on either side, and the 2 lower ones fused by the edges to form the keel. Stamens normally 10, monadelphous or diadelphous in ours. Pistil 1, of 1 carpel, the ovary sometimes 2-celled by the intrusion of sutures. Fruit a legume.

KEY TO THE GENERA OF FABACEAE

1. Leaves palmately compound or trifoliolate
   2. Stamens monadelphous, the anthers dimorphic; flowers in elongate racemes; leaves palmately 5–17-foliolate — *Lupinus*
   2. Stamens diadelphous, the anthers all alike; flowers in capitate spikes or racemes; leaves trifoliolate — *Trifolium*
1. Leaves pinnately compound
   3. Fruit a loment; keel surpassing the wings — *Hedysarum*
   3. Fruit a legume; keel mostly shorter than the wings
      4. Keel of the corolla blunt; plants usually caulescent — *Astragalus*
      4. Keel of the corolla with an abrupt, beaklike point; plants usually scapose — *Oxytropis*

Astragalus L.
Milkvetch

Our plants are perennial herbs with alternate, stipulate, odd-pinnate leaves. Flowers white, yellow, or purple, in spikes or racemes. Calyx 5-toothed. Corolla usually long and narrow, the narrow standard equaling or exceeding the wings and blunt keel. Stamens 10, diadelphous. Legumes few- to several-seeded, 1- or 2-celled by intrusion of the sutures, leathery or papery, sometimes inflated.

KEY TO THE SPECIES OF *ASTRAGALUS*

1. Leaflets spinulose-tipped; plants mat forming, the flowers and fruits very small — *A. kentrophyta*
1. Leaflets not spinulose-tipped; plants of various habits
   2. Pubescence of leaves mostly dolabriform
      3. Plants from thin, creeping rhizomes; fruits pendulous, 1-celled; fruiting pedicels 1.2–3 mm long; racemes 3–20-flowered, the flowers spreading — *A. miser*
      3. Plants from a thick taproot and branched caudex; fruits erect, 2-celled; fruiting pedicels 0.4–1 mm long; racemes 10–50-flowered, the flowers narrowly ascending — *A. adsurgens*
   2. Pubescence of leaves basifixed
      4. Wings of flowers bidentate at the apex, the lobes often unequal; stems clustered from a stout taproot — *A. australis*
      4. Wings of flowers entire; stems single or weakly clustered from adventitiously rooting, creeping caudex branches
         5. Fruits pendent, stipitate; racemes 7–15-flowered — *A. alpinus*
         5. Fruits erect, sessile or nearly so; racemes 2–3-flowered — *A. molybdenus*

Astragalus adsurgens Pall.
Sp. Astragal. 40, Tab. 31, 1800.

Standing Milkvetch

Astragalus chandonnetti Greene
Astragalus crandallii Gand.
Astragalus laxmanni Nutt.
Astragalus nitidus Dougl. ex Hook.
Astragalus striatus Nutt. in T. & G.
Astragalus sulphurescens Rydb.
Astragalus tananaicus Hult.
Astragalus viciifolius Hult.
Phaca adsurgens (Pall.) Piper
Tragacantha adsurgens (Pall.) Kuntze

Caespitose, perennial herb from a woody taproot and branched caudex; stems several, erect to ascending, strigose with dolabriform hairs. Leaves alternate, odd-pinnate; leaflets oblong to obovate, strigose; stipules deltoid to ovate, scarious, connate. Flowers numerous in congested terminal and axillary racemes. Calyx mixed black- and white-hairy, with 5 linear teeth. Petals purple to white; banner slightly reflexed; wings longer than the purple keel. Stamens 10, diadelphous. Legumes membranous, sessile or nearly so, erect, strigillose.

Timberline in the northern Absaroka Range. Alaska south to Washington, Idaho, New Mexico, Nebraska, Iowa, and Minnesota.

Middle Rocky Mountain plants are var. *robustior* Hook.

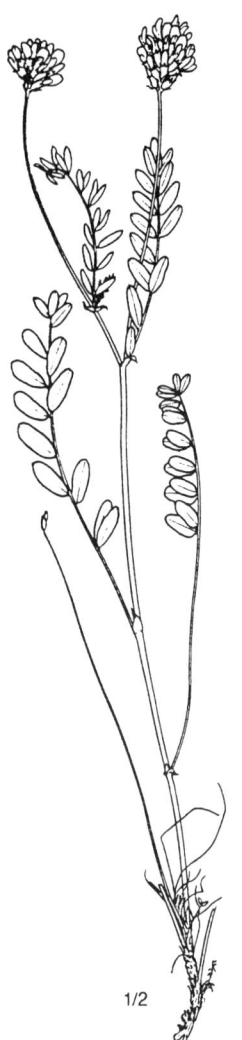

1/2

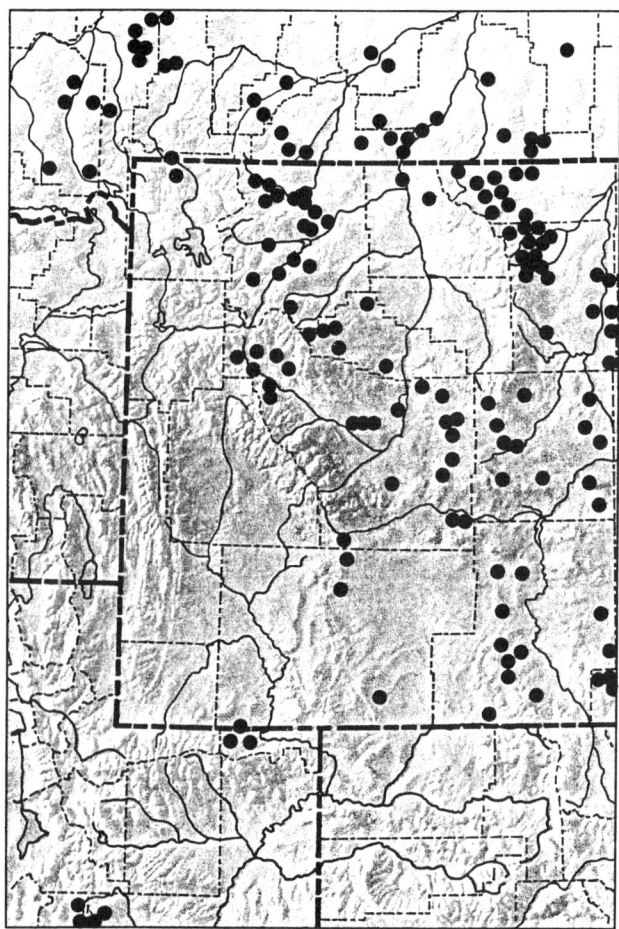

Astragalus alpinus L.
Sp. Pl., 760, 1753.

Alpine Milkvetch

Astragalus alpestris Bub.
Astragalus andinus M.E. Jones
Astragalus arcticus Bunge
Astragalus astragalinus (DC.) Sheld.
Astragalus astragalinus (Hook.) Löve & Löve
Astragalus brunetianus (Fern.) Rouss.
Astragalus giganteus Sheld.
Astragalus labradoricus DC.
Astragalus lapponicus (DC.) B. Schischk.
Astragalus pauciflorus Hook.
Astragalus phacinus Krause in Sturm.
Astragalus subpolaris Boriss. & B. Schischk.
Atelophragma alpinum (L.) Rydb.
Atelophragma labradoricum (DC.) Rydb.
Colutea astragalina (DC.) Poir.
Phaca alpina (L.) Piper
Phaca arctica (Bunge) Gand.
Phaca astragalina DC.
Phaca lapponica DC.
Phaca minima All.
Tium alpinum (L.) Rydb.
Tragacantha alpina (L.) Kuntze

Low, caulescent to subcaulescent, glabrate to strigillose, perennial herb, less than 20 cm tall, from creeping rhizomes; stems slender, spreading to ascending. Leaves alternate, odd-pinnate; the leaflets oblong to oval, obtuse or retuse, white-pubescent, at least below; stipules ovate to deltoid, at least the lower one connate. Flowers few to numerous, divergent, in open racemes. Calyx black-hirsute, the teeth lanceolate. Petals white to purplish; banner erect; wings clawed, shorter than the purple keel. Stamens 10, diadelphous. Legumes membranous, stipitate, pendulous, black-strigillose, the body ellipsoid-falcate.

Meadows and fellfields in most alpine ranges of the Middle Rocky Mountains. Not yet known from above timberline in the Hoback, Salt River, and Uinta Ranges. Circumpolar; south in the western United States to Oregon, Nevada, and New Mexico.

Western North American plants are var. *alpinus*.

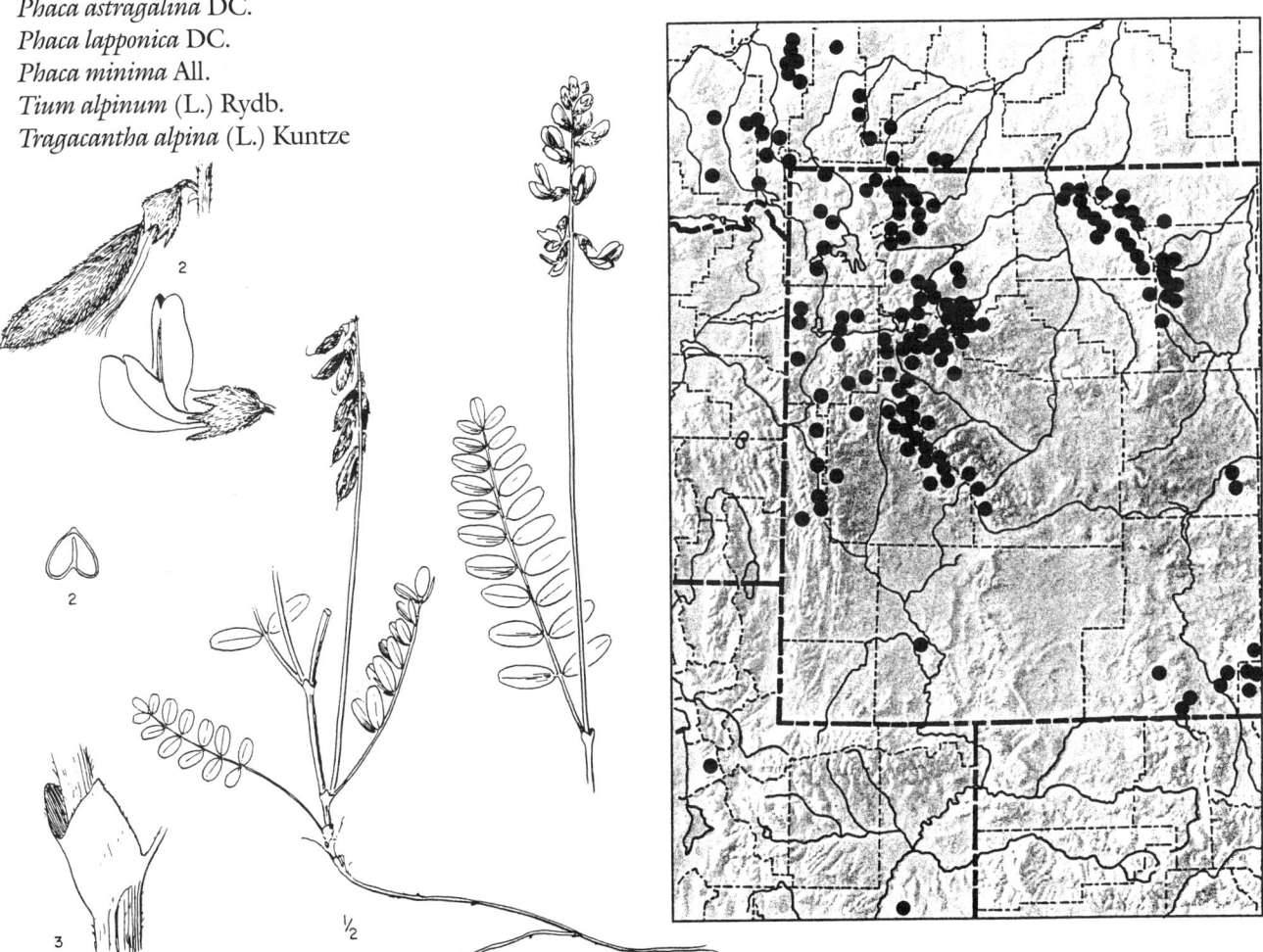

Astragalus australis (L.) Lam.
Fl. Franc. 2:637, 1778.

Indian Milkvetch

Phaca australis L.
Astragalus aboriginorum Richards.
Astragalus aboriginum Richards.
Astragalus forwoodii S. Wats.
Astragalus glabriusculus (Hook.) Gray
Astragalus lepagei Hult.
Astragalus linearis (Rydb.) Pors.
Astragalus richardsonii Sheld.
Astragalus scrupulicola Fern. & Weath.
Astragalus spatiosus (Sheld.) A. Heller
Astragalus vaginatus Hook.
Atelophragma aboriginorum (Richards.) Rydb.
Atelophragma forwoodii (S. Wats.) Rydb.
Atelophragma glabriuscula (Hook.) Rydb.
Atelophragma herriotii Rydb.
Atelophragma lineare Rydb.
Atelophragma wallowense Rydb.
Colutea australis (L.) Lam.
Homalobus aboriginorum (Richards.) Rydb.
Homalobus aboriginum (Richards.) Rydb.
Homalobus glabriusculus (Hook.) Rydb.
Homalobus spatiosus (Sheld.) A. Heller
Phaca aboriginorum (Richards.) Hook.
Phaca glabriuscula Hook.
Tragacantha aboriginorum (Richards.) Kuntze
Tragacantha glabriuscula (Hook.) Kuntze

Multistemmed, strigose to villous, perennial herb from a taproot and branched caudex. Leaves alternate, odd-pinnate; leaflets 7–15, elliptic to lanceolate, acute, sericeous to strigose or villous; stipules ovate to oblong, obtuse, connate below. Flowers several, in short racemes. Calyx black-hairy, with 5 linear teeth. Petals white or yellowish white, purple-veined; banner erect, obovate; wings bidentate; keel shorter than the wings, purple-tipped. Stamens 10, diadelphous. Legumes flattened-falcate, stipitate, pendulous, glabrous or pubescent.

Fellfields and scree of the Absaroka, Beartooth, Uinta, Wind River, and Wyoming Ranges. Circumpolar; transcontinental in Canada; Alaska south to Oregon, Nevada, Utah, New Mexico, and South Dakota.

Middle Rocky Mountain plants are var. *glabriusculus* (Hook.) Isely.

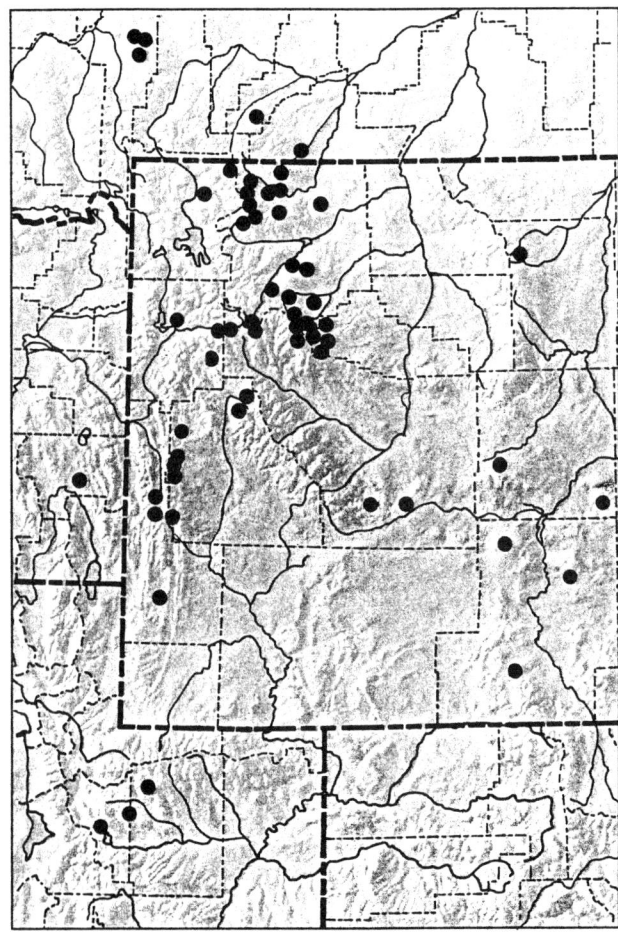

Astragalus kentrophyta Gray
Proc. Acad. Phila. 1863:30, 1863.

Prickly Milkvetch

Astragalus acauleatus A. Nels.
Astragalus centrophytus Clements
Astragalus impensus (Rydb.) Woot. & Standl.
Astragalus jessiae M.E. Peck
Astragalus montanus (Nutt. in T. & G.) M.E. Jones
Astragalus tegetarius S. Wats.
Astragalus viridis (Nutt. ex T. & G.) Sheld.
Homalobus aculeatus (A. Nels.) Rydb.
Homalobus montanus (Nutt. in T. & G.) Britt. in Britt. & Brown
Homalobus tegetarius (S. Wats.) Rydb.
Homalobus wolfii Rydb.
Kentrophyta aculeata (A. Nels.) Rydb.
Kentrophyta coloradoensis (M.E. Jones) Rydb.
Kentrophyta impensa (Sheld.) Rydb.
Kentrophyta minima Rydb.
Kentrophyta montana Nutt. in T. & G.
Kentrophyta rotunda (M.E. Jones) Rydb.
Kentrophyta tegetaria (S. Wats.) Rydb.
Kentrophyta ungulata (M.E. Jones) Rydb.
Kentrophyta viridis Nutt. ex T. & G.
Kentrophyta wolfii (Rydb.) Rydb.
Phaca viridis (Nutt. ex T. & G.) Piper
Tragacantha montana (Nutt. in T. & G.) Kuntze
Tragacantha tegetaria (S. Wats.) Kuntze

Low, prostrate, perennial herb, less than 5 cm tall; stems much branched, silvery-canescent, the hairs basifixed. Leaves alternate, pinnate; leaflets linear to lanceolate, spinulose-tipped; stipules membranous, connate, somewhat spinulose. Flowers solitary or in 2–3-flowered, axillary racemes. Calyx pubescent, the lobes linear to lanceolate. Petals purplish or whitish with a purple tinge; banner erect; wings clawed, longer than the purple keel. Stamens 10, diadelphous. Legumes sessile, compressed, ovoid to elliptic, grayish-strigillose.

Fellfields and scree in nearly all alpine ranges of the Middle Rocky Mountains. Not yet known from above timberline in the Medicine Bow Range. Oregon, Idaho, Montana, and South Dakota south to Utah, New Mexico, and Nebraska.

Mountain plants with basifixed hairs are var. *implexus* (Canby) Barneby.

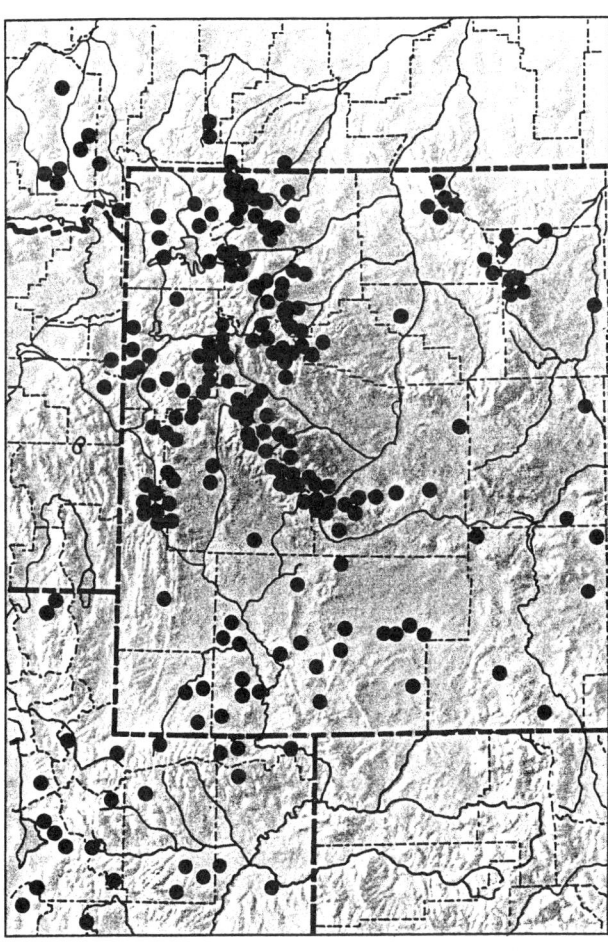

Astragalus miser Dougl. ex Hook.
Fl. Bor.-Am. 1:153, 1834.

Timber Milkvetch

Astragalus carltonii Macbr.
Astragalus decumbens (Nutt. ex T. & G.) Gray
Astragalus divergens Blank.
Astragalus garrettii Macbr.
Astragalus gracilis Nutt.
Astragalus griseopubescens Sheld.
Astragalus hylophilus (Rydb.) A. Nels.
Astragalus palliseri Gray
Astragalus rydbergii Macbr.
Astragalus serotinus Gray ex Cooper
Astragalus strigosus Coult. & Fisher
Homalobus camporum Rydb.
Homalobus decumbens Nutt. ex T. & G.
Homalobus decurrens Rydb.
Homalobus divergens (Blank.) Rydb.
Homalobus hitchcockii Rydb.
Homalobus humilis Rydb.
Homalobus hylophilus Rydb.
Homalobus microcarpus Rydb.
Homalobus miser (Dougl. ex Hook.) Rydb.
Homalobus oblongifolius Rydb.
Homalobus palliseri (Gray) Rydb.
Homalobus paucijugus Rydb.
Homalobus serotinus (Gray ex Cooper) Rydb.
Homalobus strigosus (Coult. & Fisher) Rydb.
Homalobus tenuifolius Nutt. in T. & G.
Phaca decumbens (Nutt. ex T. & G.) Piper
Phaca misera (Dougl. ex Hook.) Piper
Phaca parviflora Nutt. in T. & G.
Phaca serotina (Gray ex Cooper) Piper
Tium misera (Dougl. ex Hook.) Kuntze
Tium miserum (Dougl. ex Hook.) Rydb.
Tragacantha decumbens (Nutt. ex T. & G.) Kuntze
Tragacantha misera (Dougl. ex Hook.) Kuntze
Tragacantha serotina (Gray) Kuntze

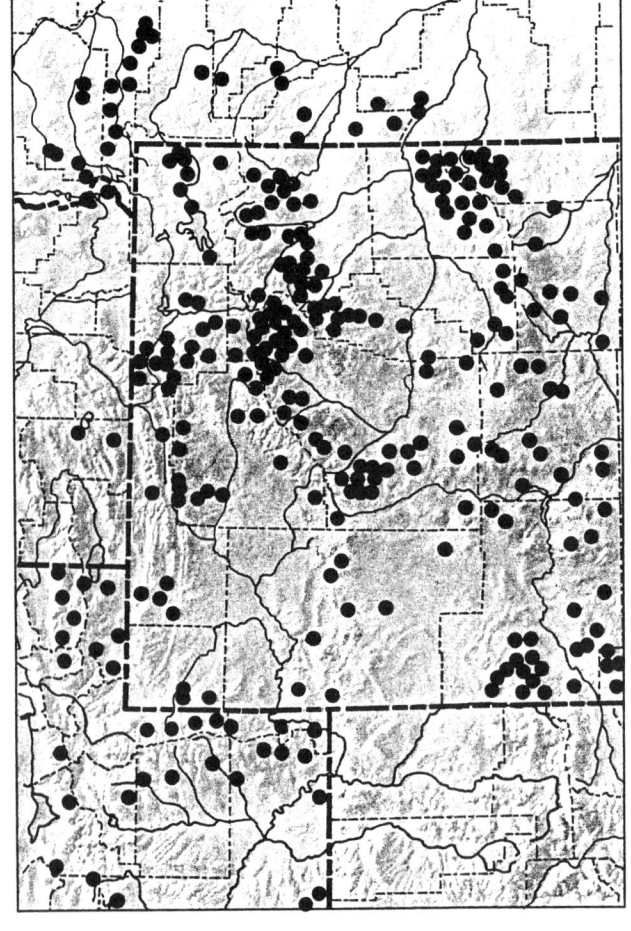

Tufted, perennial herb from a taproot and branched caudex; stems erect to ascending, often decumbent at the base, strigillose, the hairs basifixed to dolabriform. Leaves alternate, odd-pinnate; leaflets linear to lanceolate or oval, glabrous to strigose beneath; stipules lanceolate to deltoid, connate. Flowers few to numerous in short, loose, spreading racemes. Calyx pubescent, the teeth linear to lanceolate. Petals white to purplish; banner erect, purple-veined; wings purple-veined, about the same length as the purple keel. Stamens 10, diadelphous. Legumes laterally compressed, sessile to substipitate, linear-lanceolate, pendulous, strigillose to glabrous.

Rocky ridges, ledges, talus, and scree in the Absaroka, Beartooth, Uinta, Wasatch, Wind River, and Wyoming Ranges. British Columbia and Alberta south to Washington, Nevada, Utah, Colorado, and South Dakota.

There are 2 alpine varieties: var. *decumbens* (Nutt.) Cronq. which has dolabriform hairs on stems and leaflets; and var. *hylophilus* (Rydb.) Barneby, with basifixed hairs on the stems and leaflets.

Astragalus molybdenus Barneby
Leafl. West. Bot. 6:70, 1950.

Leadville Milkvetch

Astragalus plumbeus Barneby
Astragalus shultziorum Barneby

Low, slender, weakly rhizomatous, perennial herb from a branched caudex; stems prostrate to decumbent, pubescent, the hairs basifixed. Leaves alternate, odd-pinnate; leaflets lanceolate to ovate, subsessile, obtuse to acute, glabrate above, pubescent below; stipules membranous, ovate, connate. Flowers 1–3, ascending, in short, loose racemes. Calyx black-strigillose, the teeth linear to narrowly lanceolate, subulate. Petals whitish lavender; banner recurved, obovate, retuse; wings incurved, about the same length or slightly longer than the purple-tipped keel. Stamens 10, diadelphous. Legumes sessile, ascending, ovate, acuminate, glabrous to black-strigillose.

Scree slopes and rocky ridgetops on Madison limestone of the Gros Ventre, Salt River, Teton, and Wind River Ranges.

Middle Rocky Mountain alpine and subalpine plants are distinguished from typical *A. molybdenus* of the Southern Rocky Mountains as follows (modified from Barneby, 1981).

1. Leaflets of upper leaves 17–25; racemes 3–6-flowered; pod abruptly contracted distally into a short, deltate beak; ovules 6; seeds ± 4–4.5 mm long
 . . . var. *molybdenus*
1. Leaflets of upper leaves 9–15 (–17); racemes 1–3-flowered, mostly 2-flowered; pod attenuate distally into a narrowly lanceolate beak; ovules 8–9; seeds ± 2.5–2.7 mm long
 . . . var. *shultziorum* (Barneby) Scott, comb. nov.

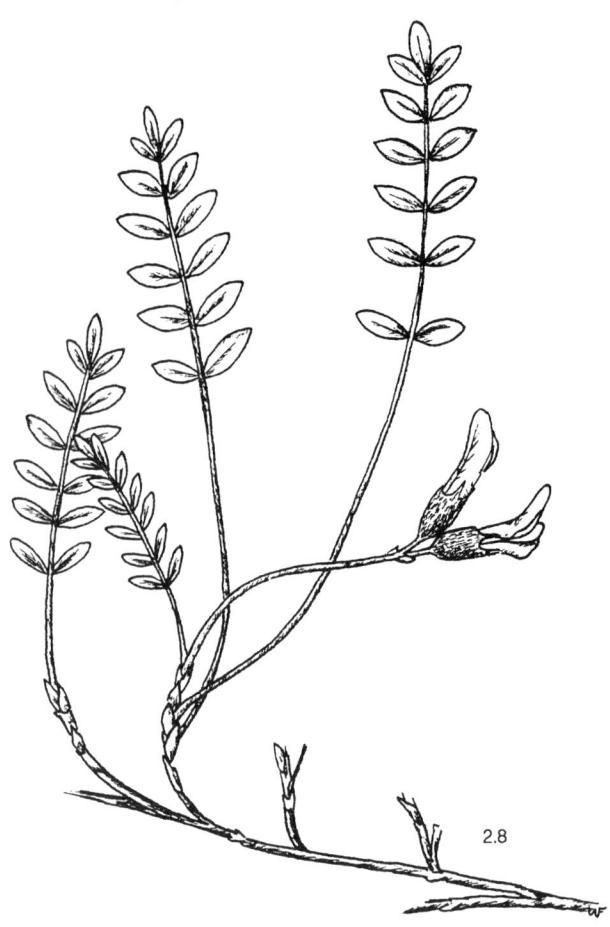

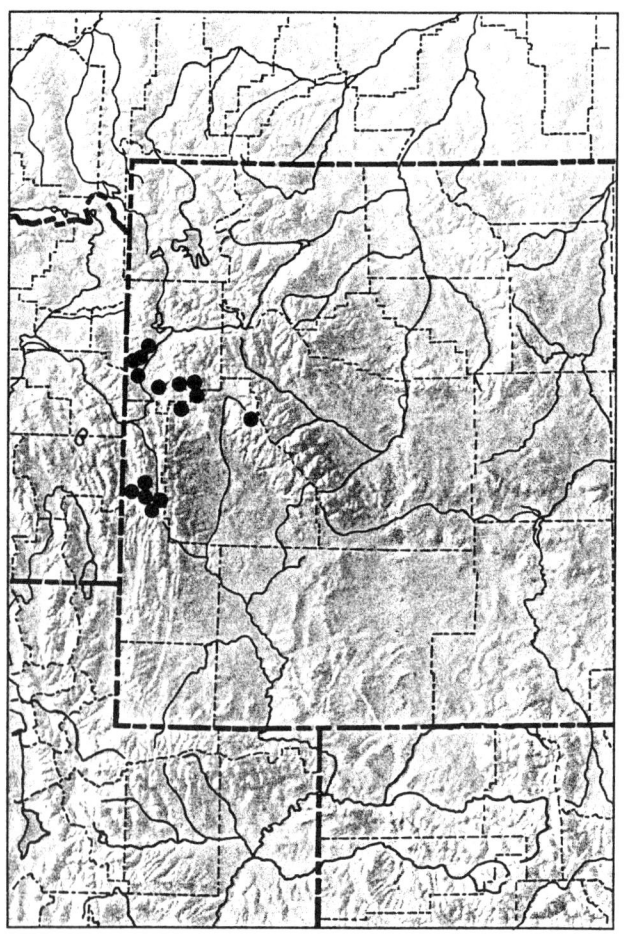

Hedysarum L.
Sweetvetch

Perennial herbs from woody taproots, occasionally frutescent at the base, with odd-pinnate, petiolate leaves, the leaflets sessile, glandular-dotted above, the stipules membranous. Flowers yellow, white, pink, or purple in ours, in axillary, peduncled racemes. Calyx bracteolate, campanulate, the lobes subequal and subulate. Keel obtuse, longer than the banner and the auriculate wings. Stamens 10, diadelphous. Fruit a loment, usually breaking transversely.

KEY TO THE SPECIES OF *HEDYSARUM*

1. Flowers yellow, sometimes white — *H. sulphurescens*
1. Flowers pink to purple
    2. Calyx lobes subequal, the upper lobes nearly the same size and shape as the lower 3; loment segments cross-corrugated — *H. boreale*
    2. Calyx lobes unequal, the upper lobes shorter and broader than the other 3; loment segments not cross-corrugated — *H. occidentale*

Hedysarum boreale Nutt.
Gen. Pl. 2:110, 1818.

Northern Sweetvetch

Hedysarum canescens Nutt. in T. & G.
Hedysarum carnosulum Greene
Hedysarum cinerascens Rydb.
Hedysarum gremiale Rollins
Hedysarum mackenzii Richards.
Hedysarum pabulare A. Nels.
Hedysarum utahense Rydb.

Erect, perennial herb from a stout, woody taproot and branched caudex; stems several to many, branched, glabrate to strigose. Leaves alternate, odd-pinnate; leaflets lanceolate to elliptic, usually glabrous above and strigose below, acute to obtuse; stipules brownish, ovate and connate below, becoming reduced and distinct upward. Flowers several to numerous, erect, spreading, or the lower ones somewhat reflexed, in short racemes. Calyx pubescent, the teeth linear, subulate, nearly equal in length. Petals purplish, rarely white; banner obovate, erect; wing with an auriculate basal lobe; keel truncate. Stamens 10, diadelphous. Loments pubescent, pendulous, the segments orbicular, transversely reticulate.

Rocky slopes and stream banks of the Salt River and Teton ranges. Alaska south to Oregon, Arizona, New Mexico, Oklahoma, Minnesota, and Newfoundland.

Middle Rocky Mountain plants with upper leaflet surfaces glabrous, or at least less pubescent than below, have been distinguished from var. *boreale* as var. *pabulare* (A. Nels.) Dorn.

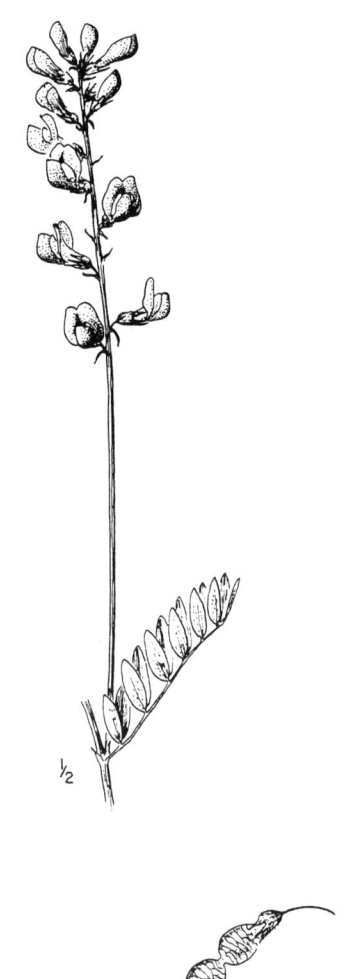
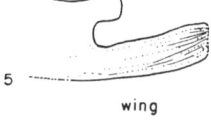
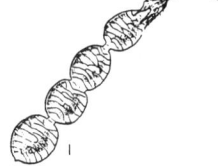
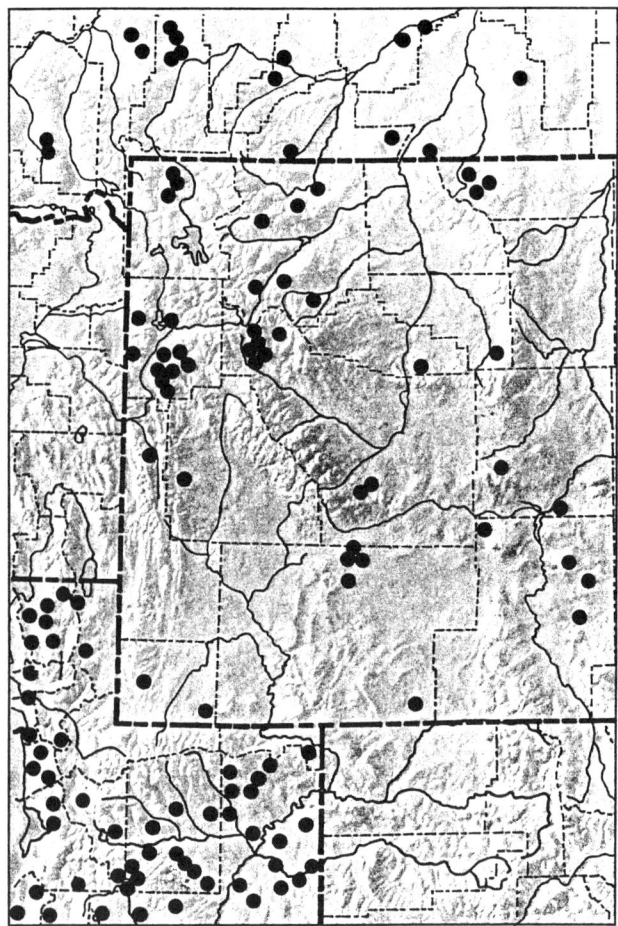

Hedysarum occidentale Greene
Pittonia 3:19, 1896.

Western Sweetvetch

Hedysarum lancifolium Rydb.
Hedysarum marginatum Greene
Hedysarum uintahense A. Nels.

Erect, perennial herb from a branched caudex and stout, woody taproot; stems several, branched, strigose to glabrate. Leaves alternate, odd-pinnate; leaflets ovate to oblong, glabrous above, sparsely strigose beneath; stipules brownish, membranous. Flowers pendent in a loose, elongate raceme. Calyx strigillose, the upper 2 teeth deltoid, the lower 3 teeth lanceolate and longer than the upper. Petals purple; banner obovate, emarginate; wing shorter than the keel, with a linear auricle, the auricles united. Stamens 10, diadelphous. Loments glabrous to strigillose, stipitate, pendulous, reticulations not transverse.

Scree, talus, and ledges of the Absaroka, Hoback, Salt River, Teton, Wind River, and Wyoming Ranges. Washington and Montana south to Idaho and Colorado.

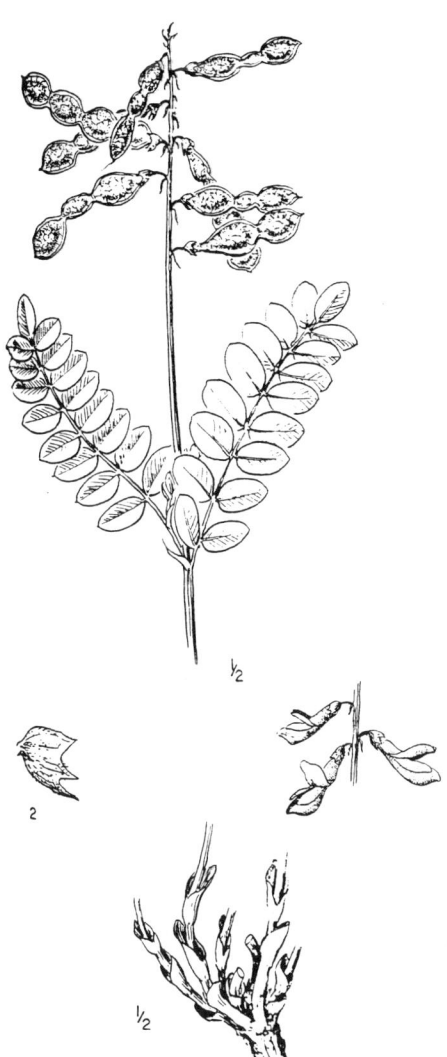

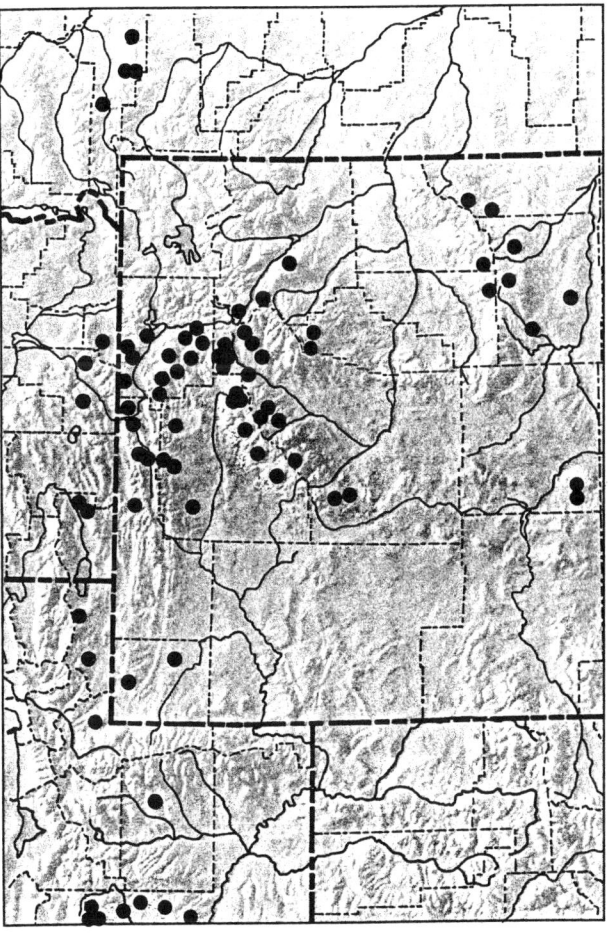

Hedysarum sulphurescens Rydb.
Bull. Torr. Club 24:251, 1897.

Sulfur Sweetvetch

Hedysarum albiflorum (Macoun) Fedtsch.
Hedysarum flavescens Coult. & Fisher

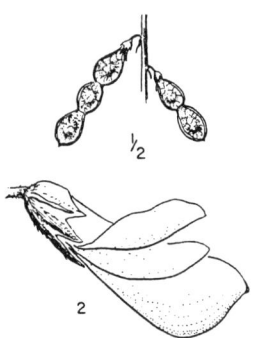

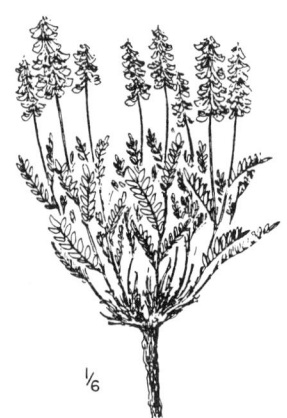

Tufted, multistemmed, strigillose, perennial herb from a taproot and branched caudex. Leaves alternate, odd-pinnate; leaflets ovate to oval, glabrate above, sparsely pubescent below; stipules brown, membranous, connate below, becoming distinct above. Flowers numerous, spreading to reflexed, in dense racemes. Calyx teeth unequal, the upper 2 teeth shorter and broader than the lower 3 teeth. Petals sulphur yellow to whitish; banner not reflexed; wings with a slender auricle, the auricles united. Stamens 10, diadelphous. Loments pendulous, the segments reticulate, glabrous, wing-margined.

Timberline meadows of the Absaroka, Beartooth, and Big Horn Ranges. British Columbia and Alberta south to Washington, Idaho, and Wyoming.

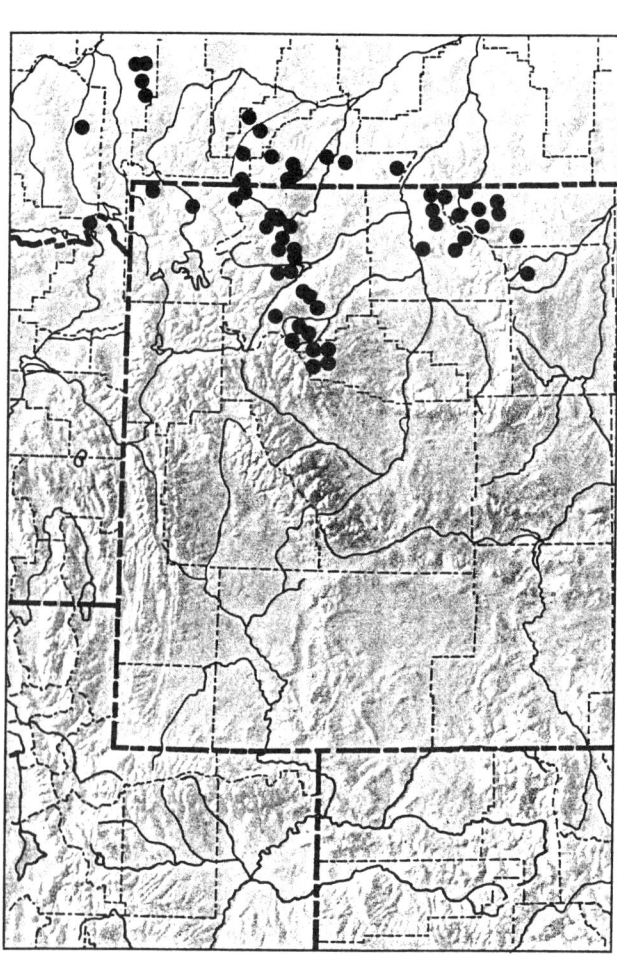

Lupinus L.
Lupine

Annual or perennial herbs with alternate, palmately compound, stipulate leaves, the leaflets 5–17, entire, oblanceolate to obovate. Flowers racemose, purple, bluish, or white in ours. Calyx bilabiate, the lips toothed or entire. Corolla glabrous or pubescent, the banner with a median groove and reflexed margins, the wings more or less enclosing the falcate keel. Stamens 10, monadelphous, the anthers dimorphic. Legumes flattened, 2- to many-seeded.

KEY TO THE SPECIES OF *LUPINUS*

1. Plants caespitose, the stems rarely more than 3 cm tall; leaves exceeding the inflorescence — *L. lepidus*
1. Plants erect, the stems more than 10 cm tall; leaves shorter than the inflorescence
    2. Banner pubescent on the back
        3. Pubescence of stem appressed or ascending — *L. argenteus*
        3. Pubescence of stem widely spreading or woolly — *L. leucophyllus*
    2. Banner glabrous on the back, occasionally with a few hairs
        4. Banner only slightly reflexed from the keel; leaflets often glabrous on the upper surface — *L. argenteus*
        4. Banner strongly reflexed from the keel; leaflets short-strigose on the upper surface — *L. wyethii*

Lupinus argenteus Pursh
Fl. Am. Sept., 468, 1814.

Silvery Lupine

Lupinus achilleaphilus C.P. Smith
Lupinus adscendens Rydb.
Lupinus alicanescens C.P. Smith
Lupinus aliumbellatus C.P. Smith
Lupinus alpestris A. Nels.
Lupinus alsophilus Greene
Lupinus alturasensis C.P. Smith
Lupinus capitis-amnicoli C.P. Smith
Lupinus cariciformis C.P. Smith
Lupinus charlestonensis C.P. Smith
Lupinus christianus C.P. Smith
Lupinus clarkensis C.P. Smith
Lupinus corymbosus A. Heller
Lupinus davisianus C.P. Smith
Lupinus decumbens Torr.
Lupinus depressus Rydb.
Lupinus edward-palmeri C.P. Smith
Lupinus equi-coeli C.P. Smith
Lupinus evermannii Rydb.
Lupinus floribundus Greene
Lupinus foliosus (T. & G.) Nutt. ex. Hook.
Lupinus fremontensis C.P. Smith
Lupinus garrettianus C.P. Smith
Lupinus hullianus C.P. Smith
Lupinus ingratus Greene
Lupinus jonesii Blank.
Lupinus lacuum-trinitatum C.P. Smith
Lupinus lanatocarinatus C.P. Smith
Lupinus laxiflorus Dougl. ex Lindl.
Lupinus laxus Rydb.
Lupinus leptostachyus Greene
Lupinus lucidulus Rydb.
Lupinus macounii Rydb.
Lupinus maculatus Rydb.
Lupinus minearanus C.P. Smith
Lupinus monticola Rydb.
Lupinus montis-cookii C.P. Smith
Lupinus montis-liberatatis C.P. Smith
Lupinus myrianthus Greene
Lupinus parviflorus Nutt. ex Hook. & Arn.
Lupinus perplexus C.P. Smith
Lupinus pulcherrimus Rydb.

Lupinus roseolus Rydb.
Lupinus rubricaulis Greene
Lupinus seclusus C.P. Smith
Lupinus serradentum C.P. Smith
Lupinus sitgreavsii S. Wats.
Lupinus sparhawkianus C.P. Smith
Lupinus spathulatus Rydb.
Lupinus stenophyllus Rydb.
Lupinus summae C.P. Smith
Lupinus tenellus Doug. ex G. Don
Lupinus tenellus Dougl. ex J.G. Agardh
Lupinus varneranus C.P. Smith

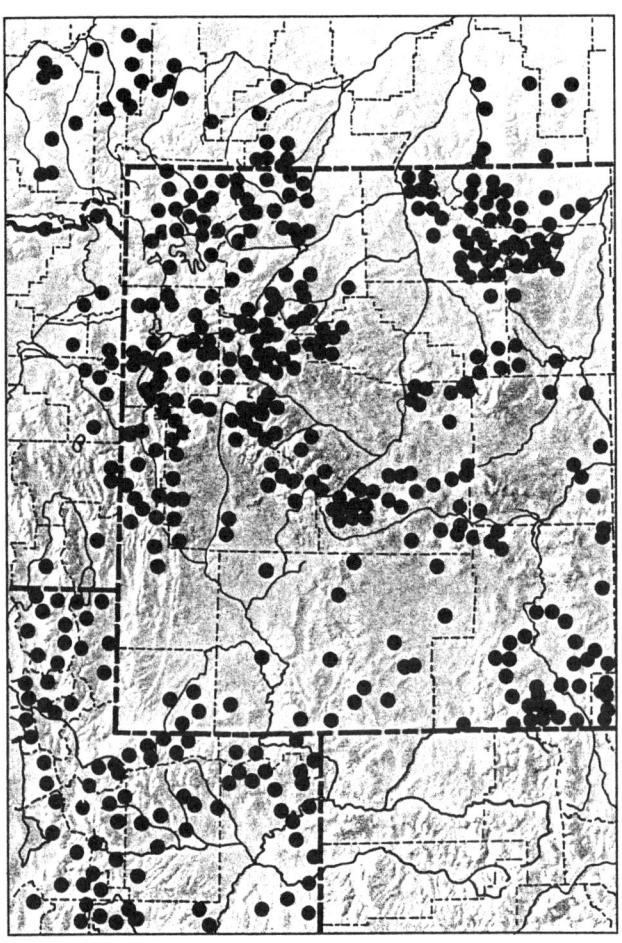

Erect, perennial herb from a stout, woody taproot and branched caudex; stems 1 to several, simple or branched, strigose to sericeous, at least below. Leaves alternate, palmate, long-petiolate, the petioles becoming shorter upward; leaflets oblanceolate to narrowly obovate, glabrate above, strigose to sericeous below; stipules partially fused to the petioles. Flowers numerous, spreading to ascending, in loose to compact racemes. Calyx pubescent, bilabiate, gibbous on one side, the upper lobe bidentate, the lower entire. Petals bluish purple to white; banner slightly reflexed, glabrous to pubescent on the back; wings glabrous; keel glabrous, ciliate. Stamens 10, monadelphous. Legumes elliptic to oval, pubescent.

Meadows, stream banks, and rocky slopes in all alpine ranges of the Middle Rocky Mountains. Montana south to Idaho, Utah, and Wyoming.

The two weakly differentiated varieties which occur in the alpine zone may be distinguished as follows:

1. Racemes with flowers congested; plants less than 2.5 dm tall . . . var. *depressus* (Rydb.) C.L. Hitchc.
1. Racemes with obvious spaces between the flowers; plants more than 2.5 dm tall . . . var. *argenteus*

var. argenteus

var. depressus

Lupinus

Lupinus lepidus Dougl. ex Lindl.
Bot. Reg. 14:Pl. 1149, 1828.

Dwarf Lupine

Lupinus abortivus Greene
Lupinus alcis-temporis C.P. Smith
Lupinus amnicoli-cervi C.P. Smith
Lupinus aridus Dougl. ex Lindl.
Lupinus brachypodus Piper
Lupinus caespitosus Nutt. ex T. & G.
Lupinus cusickii S. Wats.
Lupinus danaus Gray
Lupinus fruticulosus Greene
Lupinus hellerae A. Heller
Lupinus lenorensis C.P. Smith
Lupinus longivallis C.P. Smith
Lupinus lyallii Gray
Lupinus markleanus C.P. Smith
Lupinus minimus Dougl. ex Hook.
Lupinus minutifolius Eastw.
Lupinus paulinus Greene
Lupinus perditorum Greene
Lupinus piperi Robins.
Lupinus sinus-meyersi C.P. Smith
Lupinus torreyi Gray ex S. Wats.
Lupinus volutans Greene
Lupinus watsonii A. Heller

Low, caespitose, sericeous, perennial herb, usually less than 20 cm tall, from a woody taproot; stems shorter than petioles of the lower leaves. Leaves alternate, palmate, long-petiolate; leaflets oblanceolate, sericeous above and below. Flowers spreading to erect, in short, dense racemes that are usually exceeded by the lower leaves. Calyx sericeous, the upper lip bidentate, the lower entire. Petals bluish purple; banner reflexed, glabrous; wings glabrous, longer than the ciliate keel. Stamens 10, monadelphous. Legumes pubescent, elliptic, spreading to erect.

Rocky slopes and ridgetop meadows of the Absaroka, Big Horn, Gros Ventre, Wind River, and Wyoming Ranges. British Columbia and Alberta south to California, Idaho, and Colorado.

Middle Rocky Mountain alpine plants with inflorescences exceeded by the long-petiolate basal leaves are var. *caespitosus* (Nutt.) Detling.

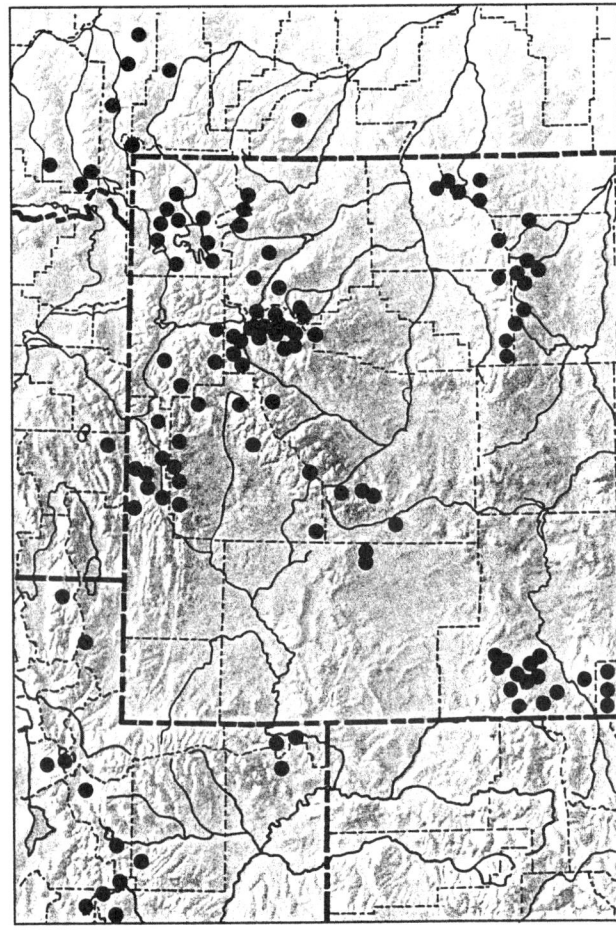

Lupinus leucophyllus Dougl. ex Lindl.
Bot. Reg. 13:Pl. 1124, 1828.

Velvet Lupine

Lupinus agropyrophilus C.P. Smith
Lupinus andersonianus C.P. Smith
Lupinus canescens T.J. Howell
Lupinus cyaneus Rydb.
Lupinus enodatus C.P. Smith
Lupinus erectus Henderson
Lupinus falsoerectus C.P. Smith
Lupinus forslingii C.P. Smith
Lupinus lysichitophilus C.P. Smith
Lupinus macrostachys Rydb.
Lupinus plumosus Dougl. ex Lindl.
Lupinus pureriae C.P. Smith
Lupinus retrorus Henderson
Lupinus salicisocius C.P. Smith
Lupinus tenuispicus A. Nels.

Erect, densely woolly-canescent, perennial herb from a woody taproot and branched caudex; stems 1 to several, simple or branched. Leaves alternate, palmate, the lower ones long-petiolate; leaflets oblanceolate, sericeous above and below. Flowers numerous in dense, spikelike racemes. Calyx sericeous, slightly gibbous, the upper lip bidentate, the lower entire. Petals white, lavender, or bluish; banner reflexed, pubescent on the back; wings glabrous, sometimes sparsely pubescent near the base, longer than the dorsally ciliate keel. Stamens 10, monadelphous. Legumes woolly-villous, the hairs rusty.

Known in the alpine zone only from rocky slopes of the Salt River Range. Washington and northern California south and east to Idaho, Montana, and Wyoming.

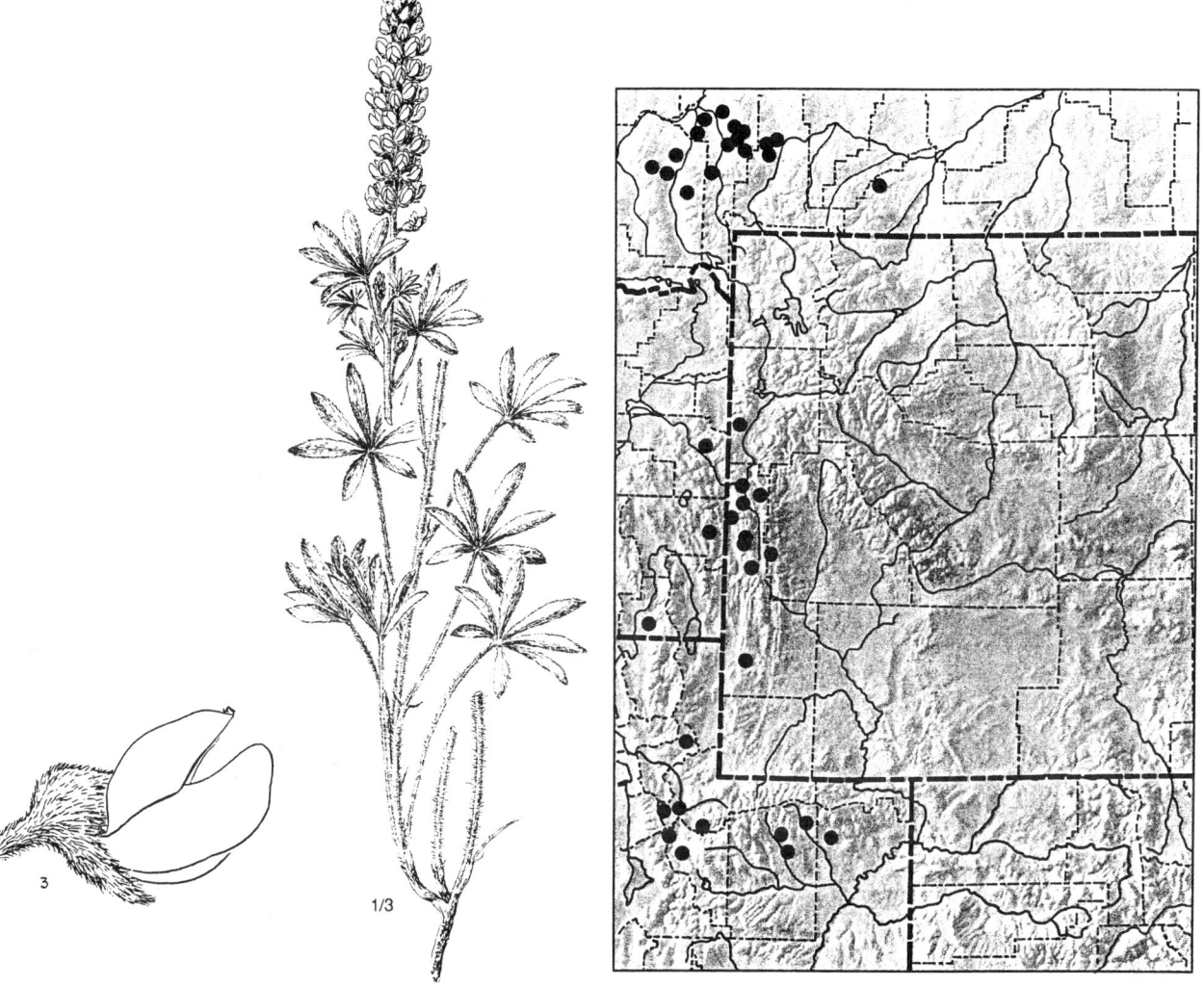

Lupinus wyethii S. Wats.
Proc. Am. Acad. 8:525, 1873.

Wyeth Lupine

Lupinus amniculi-salicis C.P. Smith
Lupinus candicans Rydb.
Lupinus comatus Rydb.
Lupinus diversalpicola C.P. Smith
Lupinus flavescens Rydb.
Lupinus humicola A. Nels.
Lupinus rydbergii Blank.

Erect, perennial herb from a woody taproot and short-branched caudex; stems 1 to several, simple, strigose. Leaves long-petiolate, at least below, alternate, palmate; leaflets oblanceolate, acuminate to apiculate, strigillose above and below. Flowers numerous, spreading-erect, in dense to somewhat open racemes, much exceeding the leaves. Calyx minutely hirsute, the upper lip bidentate, the lower entire. Petals purple to blue; banner reflexed, glabrous; wings glabrous, longer than the terminally ciliate keel. Stamens 10, monadelphous. Legumes sparse to densely pubescent.

Timberline margins of the southern Absaroka and northern Wind River Ranges. British Columbia and Alberta south to California, Idaho, and Colorado.

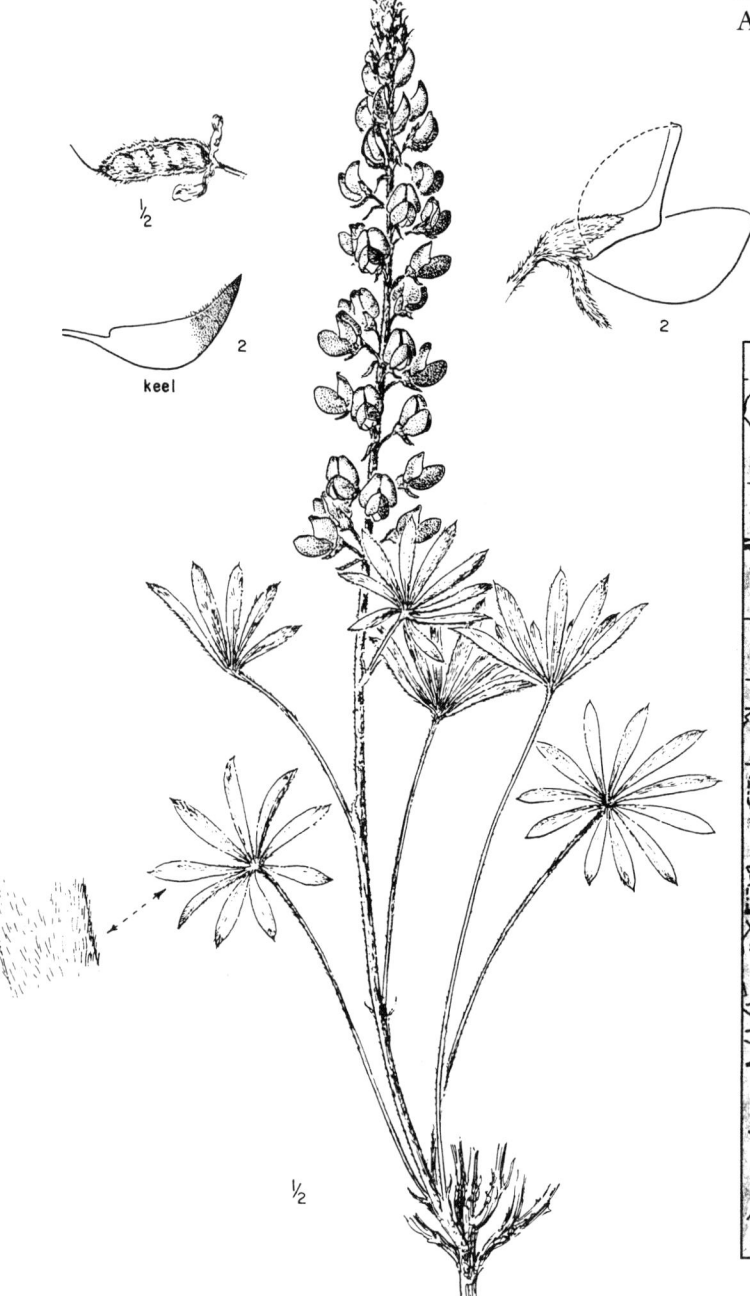

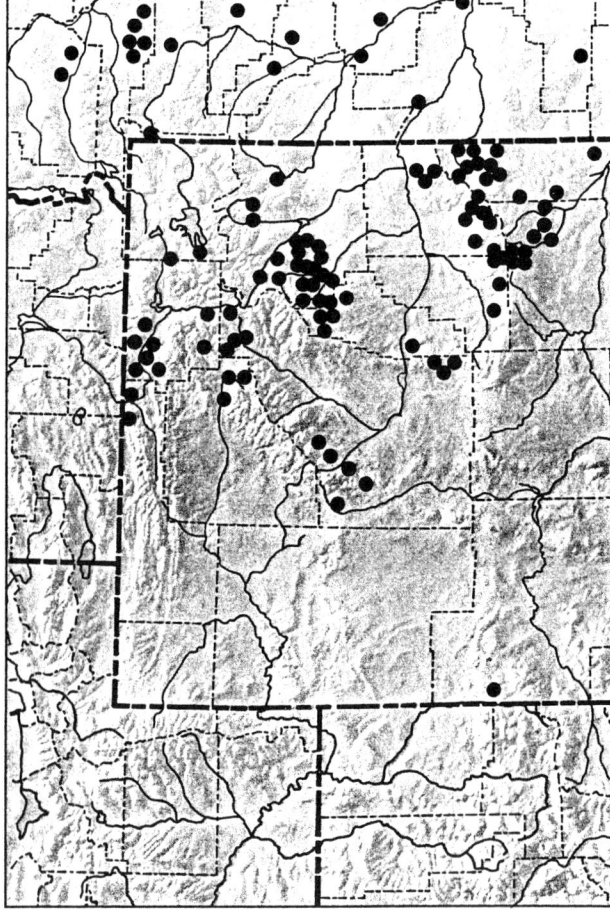

Oxytropis DC.
Locoweed

Perennial, usually scapose herbs with alternate, odd-pinnate, stipulate leaves, the stipules more or less adnate to the petiole. Flowers spicate or racemose, white, cream, or purple in ours. Calyx campanulate, 5-toothed, the teeth lanceolate, subequal. Corolla long and narrow, the keel prolonged into a prominent, ascending beak, the wings 2-lobed or emarginate, usually longer than the keel. Stamens 10, diadelphous. Legumes sessile or stipitate, 1-celled, or 2-celled by the intrusion of the ventral (upper) suture. Seeds several to many.

KEY TO THE SPECIES OF OXYTROPIS

1. Stipules only shortly adnate to the base of the petiole; legumes pendulous — *O. deflexa*
1. Stipules adnate to the base of the petiole for more than half their length; legumes erect or spreading
   2. Plants glandular-viscid — *O. viscida*
   2. Plants not glandular-viscid
      3. Flowers white or yellowish, keel sometimes purple-tinged
         4. Flowers mostly less than 17 mm long; legumes thin-walled, somewhat membranous — *O. campestris*
         4. Flowers more than 18 mm long; legumes thick-walled, often fleshy — *O. sericea*
      3. Flowers purple
         5. Racemes 1–3-flowered
            6. Flowers less than 1 cm long, the banner less than 1 cm long; legumes sessile — *O. parryi*
            6. Flowers more than 1 cm long, the banner more than 1.3 cm long; legumes stipitate — *O. podocarpa*
         5. Racemes 4–20-flowered
            7. Calyx inflated, densely villous, the hairs both black and white, bracts spreading-pilose on back; scapes short, less than 12 cm tall, equaling or scarcely exceeding the leaves; legumes not rupturing the calyx at maturity — *O. lagopus*
            7. Calyx not inflated, sparsely hispid or hirsute, the hairs white; bracts appressed-pilose on the back; scapes more than 12 cm tall, exceeding the leaves; legumes rupturing the calyx at maturity — *O. nana*

Oxytropis campestris (L.) DC.
Astragalogia, 59, 1802.

Plains Locoweed

Astragalus campestris L.
Aragallus albertinus Greene
Aragallus alpicola Rydb.
Aragallus campestris (L.) Greene
Aragallus cervinus Greene
Aragallus dispar (A. Nels.) K. Schum.
Aragallus gracilis A. Nels.
Aragallus luteolus Greene
Aragallus macounii Greene (in part)
Aragallus monticola (Gray) Greene
Aragallus villosus Rydb.
Astragalus albertinus (Greene) Tidestr.
Astragalus alpicola (Rydb.) Tidestr.
Astragalus grayanus Tidestr. in Tidestr. & Kitt.
Astragalus mazama (St. John) G.N. Jones
Astragalus rydbergianus (A. Nels.) Tidestr.
Astragalus sordidus Willd.
Astragalus varians Rydb.
Oxytropis alaskana A. Nels.
Oxytropis albertina (Greene) Rydb.
Oxytropis alpicola (Rydb.) M.E. Jones
Oxytropis cascadensis St. John
Oxytropis chartacea Fassett
Oxytropis columbiana St. John
Oxytropis cusickii Greenm.
Oxytropis dispar (A. Nels.) K. Schum.
Oxytropis gracilis (A. Nels.) K. Schum.
Oxytropis hyperborea Pors.
Oxytropis johannensis (Fern.) A. Heller
Oxytropis jordalii Pors.
Oxytropis luteola (Greene) Piper & Beattie
Oxytropis macounii (Greene) Tidestr. (in part)
Oxytropis mazama St. John
Oxytropis monticola Gray
Oxytropis okanoganea St. John
Oxytropis olympica St. John
Oxytropis paysoniana A. Nels.
Oxytropis rydbergii A. Nels.
Oxytropis terrae-novae Fern.
Oxytropis varians (Rydb.) K. Schum.
Oxytropis villosa (Rydb.) K. Schum.
Spiesia campestris (L.) Kuntze
Spiesia monticola (Gray) Kuntze

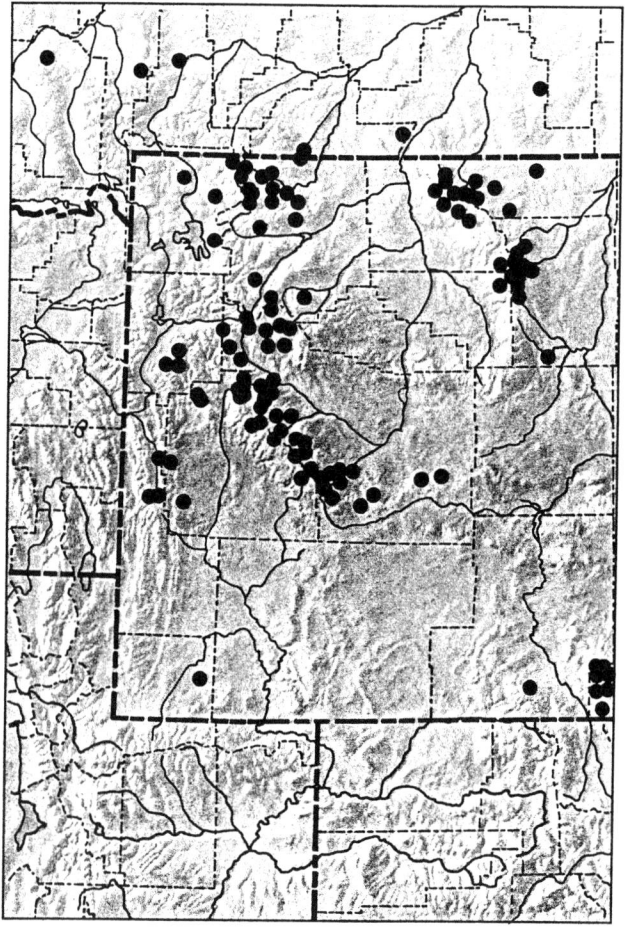

Scapose or short-caulescent, canescent to strigose, perennial herb from a stout taproot and branched caudex. Leaves odd-pinnate; leaflets lanceolate to elliptic, pilose; stipules glabrate to pilose, membranous, adnate for more than half their length, ciliate. Flowers spreading to erect, numerous in short, spikelike racemes. Calyx pilose, with mixed black, gray, or white hairs, the teeth lanceolate. Petals mostly whitish to pale yellow; banner erect, shallowly emarginate; wings narrow, shallow-emarginate, exceeding the often purplish keel. Stamens 10, diadelphous. Legumes mostly sessile or very short-stipitate, oblong to elliptic.

Fellfields, meadows, stream banks, and rocky slopes of the northern alpine ranges. Not yet known from above timberline in the Hoback, Uinta, and Wasatch Ranges. Circumboreal; Alaska south to Washington, Idaho, Colorado, and South Dakota.

Alpine plants are var. *cusickii* (Greenm.) Barneby, with glabrous stipules, leaflets less than 17, and scapes rarely more than 15 cm tall.

3/4

stipules

Oxytropis deflexa (Pall.) DC.
Astragalogia, 33, 1802.

Nodding Locoweed

Astragalus deflexus Pall.
Aragallus deflexa (Pall.) A. Heller
Aragallus foliolosus (Hook.) Macoun
Astragalus retroflexus Pall.
Oxytropis foliolosa Hook.
Oxytropis foliosa Hook. ex T. & G.
Oxytropis retrorsa Fern.

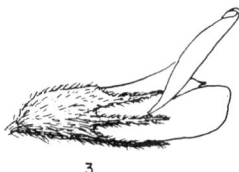

var. foliolosa

Scapose to subacaulescent, villous to pilose, perennial herb from a slender taproot and weakly branched caudex. Leaves odd-pinnate; leaflets lanceolate, ovate or elliptic, sparsely to densely villous; stipules lanceolate, pilose at least dorsally. Flowers few to numerous, in short to elongate racemes. Calyx black- and white-villous or strigose, the teeth linear to lanceolate. Petals purple; banner obovate, slightly emarginate, scarcely reflexed; wings exceeding the short-beaked keel. Stamens 10, diadelphous. Legumes stipitate, ellipsoid, reflexed to pendulous, strigose with mixed white and black or grayish hairs.

Fellfields, talus, scree, and rocky slopes of the Absaroka, Gros Ventre, Teton, Uinta, and Wind River Ranges. Circumboreal; south in western North America to California, New Mexico, and North Dakota.

Acaulescent alpine plants of northwestern Wyoming having bluish purple flowers about 10 mm long, in 7–10-flowered racemes, with the banner about twice as long as wide are var. *foliolosa* (Hook.) Barneby.

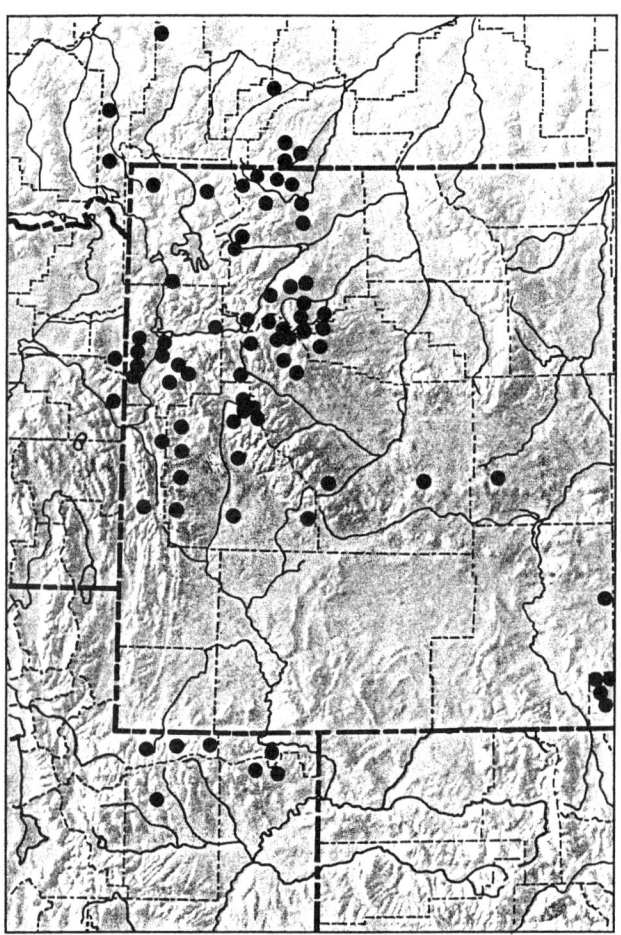

Oxytropis lagopus Nutt.
J. Acad. Phila. 7:17, 1834.

Haresfoot Locoweed

Aragallus atropurpureus Rydb.
Aragallus blankenshipii A. Nels.
Aragallus lagopus (Nutt.) Greene
Astragalus blankenshipii (A. Nels.) Tidestr.
Astragalus lagopus (Nutt.) Tidestr.
Oxytropis argentata Pursh
Oxytropis blankenshipii (A. Nels.) K. Schum.
Spiesia lagopus (Nutt.) Kuntze

Low, silky-pilose, caespitose perennial, less than 10 cm tall, from a taproot and short-branched caudex. Leaves odd-pinnate; leaflets lanceolate to elliptic; stipules membranous, pilose or villous on the back. Flowers in dense, spikelike, often subcapitate racemes, the peduncles villous, about the same length as the leaves. Calyx densely villous with white and black hairs, the teeth linear-lanceolate. Petals purplish; banner emarginate; wings exceeding the slender-beaked keel. Stamens 10, diadelphous. Legumes erect, short-stipitate, ovoid to oblong, silky-pilose, prominently beaked.

Scree slopes and other rocky places in the Absaroka, Beartooth, Big Horn, and Wyoming Ranges. Alberta, Montana, Idaho, and Wyoming.

Mountain plants with mostly 9–11 leaflets and the calyx early inflated in fruit are var. *lagopus*.

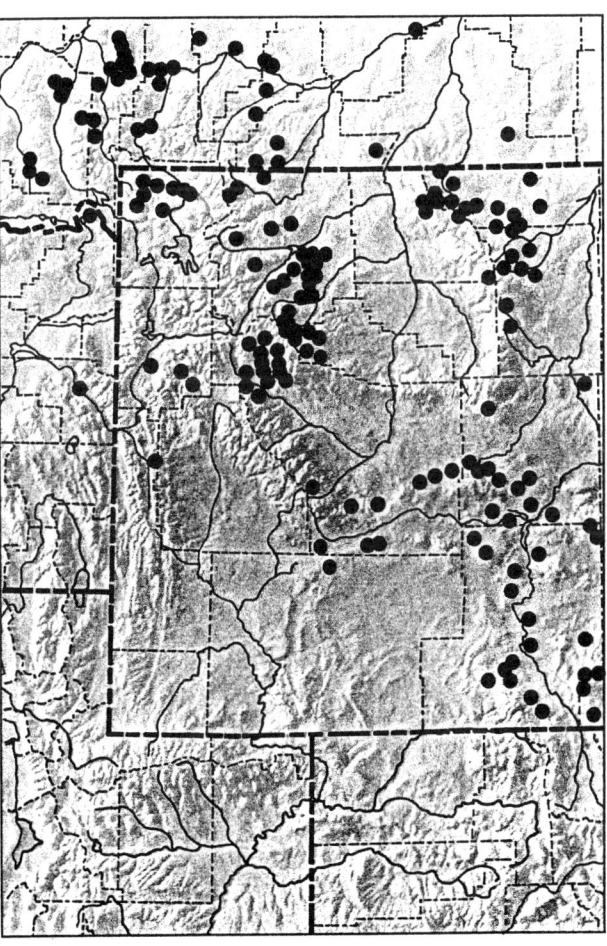

Oxytropis nana Nutt. in T. & G.
Fl. N. Am. 1:340, 1838.

Dwarf Locoweed

Aragallus argophyllus Rydb.
Aragallus besseyi Rydb.
Aragallus collinus A. Nels.
Aragallus nanus (Nutt.) Greene
Aragallus ventosus Greene
Oxytropis besseyi (Rydb.) Blank.
Oxytropis collina (A. Nels.) K. Schum.
Oxytropis lunelliana A. Nels.
Oxytropis obnapiformis C.L. Porter

Caespitose, canescent to silvery-pubescent, acaulescent or nearly so, perennial herb from a stout taproot and branched caudex. Leaves odd-pinnate; leaflets lanceolate to elliptic, pilose on both upper and lower surfaces; stipules thin, membranous, adnate to the petioles for more than half their length, ciliate. Flowers usually numerous, in dense racemes that elongate in fruit. Calyx hispid to strigose, the teeth narrowly lanceolate. Petals purple to reddish purple; banner erect, ovate, emarginate; wings narrow, emarginate, usually shorter than the keel. Stamens 10, diadelphous. Legumes short-stipitate, ovate to oblong, inflated, beaked, sericeous to villous, often purple-mottled.

Rocky, gravelly, and sandy areas, typically from lower elevations but has been collected above timberline in the Beartooth Range. Montana and Idaho south to Wyoming, Colorado, and Nebraska.

Middle Rocky Mountain plants of higher elevations are var. *besseyi* (Rydb.) Isely.

var. besseyi

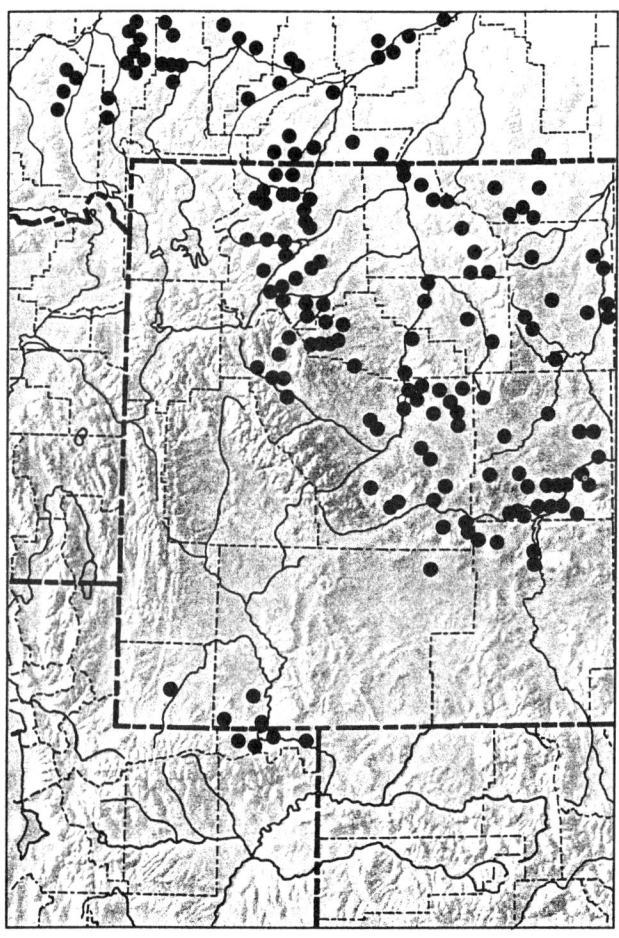

Oxytropis parryi Gray
Proc. Am. Acad. 20:4, 1884.

Parry Locoweed

Aragallus parryi (Gray) Greene
Astragalus parryanus (Gray) Tidestr.
Spiesia parryi (Gray) Kuntze

Low, pulvinate, gray-strigose perennial, less than 10 cm tall, from a taproot and branched caudex. Leaves odd-pinnate; leaflets oblong to lanceolate, gray-strigose; stipules membranous, adnate to the petioles. Flowers 1–3, in short racemes, the peduncles about the same length as the leaves. Calyx villous, at least some of the hairs black, the teeth lanceolate. Petals purple, the banner somewhat retuse and only slightly reflexed; wings longer than the short-beaked keel. Stamens 10, diadelphous. Legumes sessile, black-pubescent, erect to ascending, oblong, short-beaked.

Rocky ridges and scree slopes of the Absaroka, Beartooth, Salt River, Uinta, and Wind River Ranges. Idaho and Wyoming south to California, Nevada, and New Mexico.

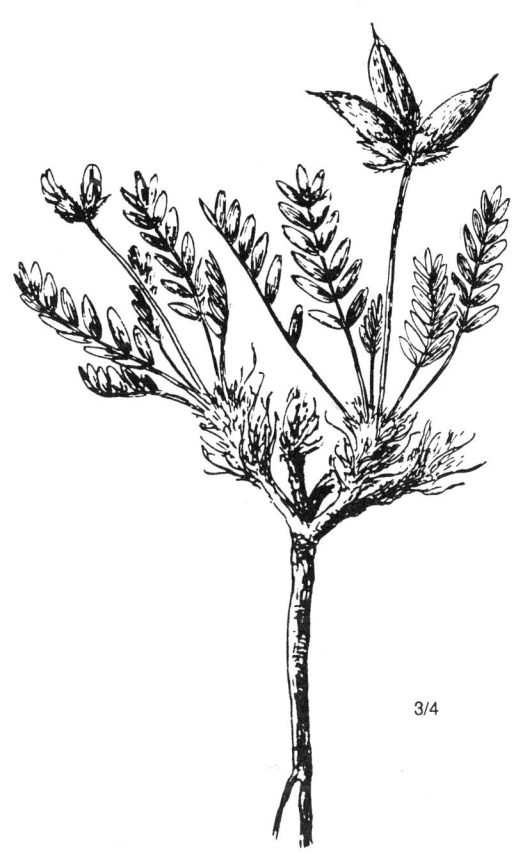

3/4

3

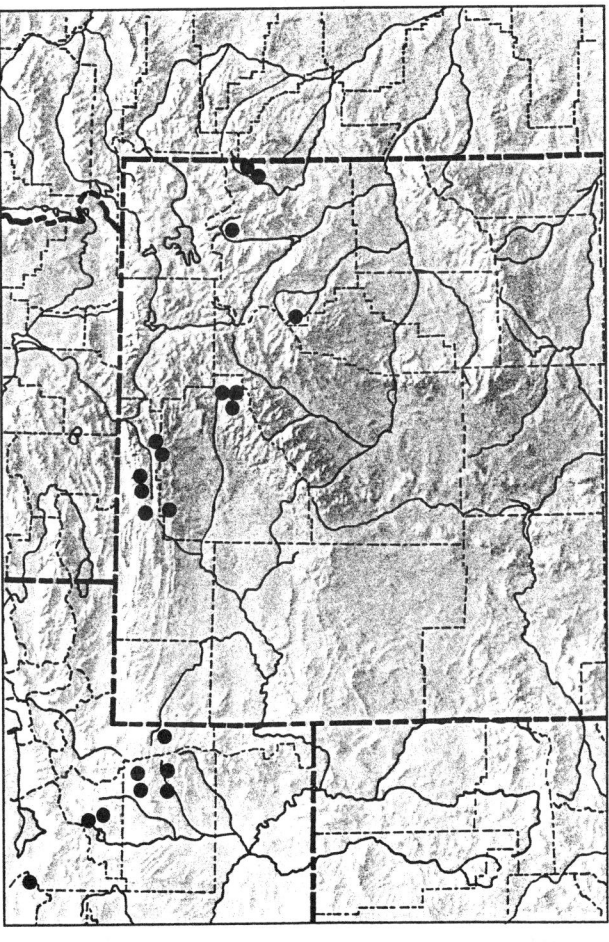

Oxytropis podocarpa Gray
Proc. Am. Acad. 6:234, 1864.

Alpine Locoweed

Aragallus hallii (Bunge) Rydb.
Aragallus inflatus (Hook.) A. Nels.
Aragallus podocarpus (Gray) A. Nels.
Oxytropis hallii Bunge
Spiesia inflata (Hook.) Britt.
Spiesia podocarpa (Gray) Kuntze

Caespitose, acaulescent perennial from a taproot and branched caudex. Leaves odd-pinnate; leaflets linear to oblong, silky-villous; stipules membranous, glabrate, ciliate on the margins, adnate to the petiole. Flowers 1–3, in short racemes; peduncles about the same length as the leaves. Calyx villous, the teeth lanceolate. Petals purple to bluish purple; banner shallowly emarginate, only slightly reflexed; wings longer than the keel. Stamens 10, diadelphous. Legumes stipitate, erect, elliptic to ovoid, becoming inflated with a prominent beak.

Scree slopes of the Absaroka and central Wind River Ranges. Alberta south to Wyoming and Colorado.

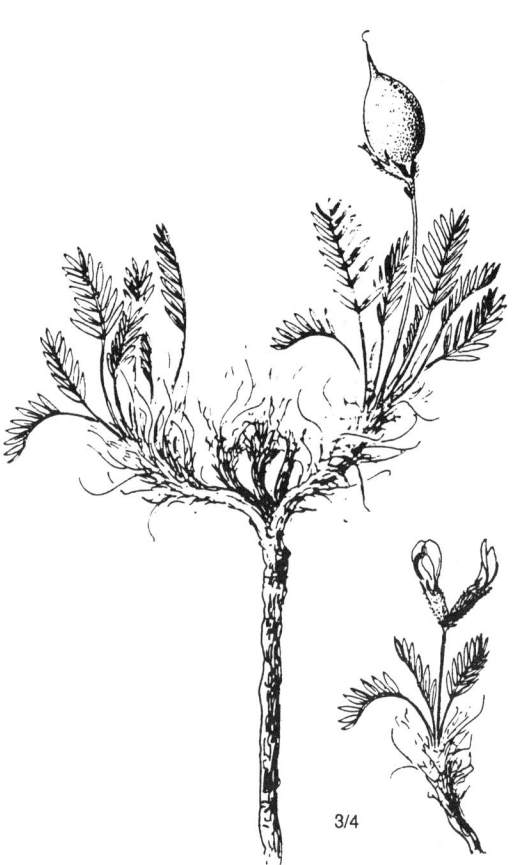

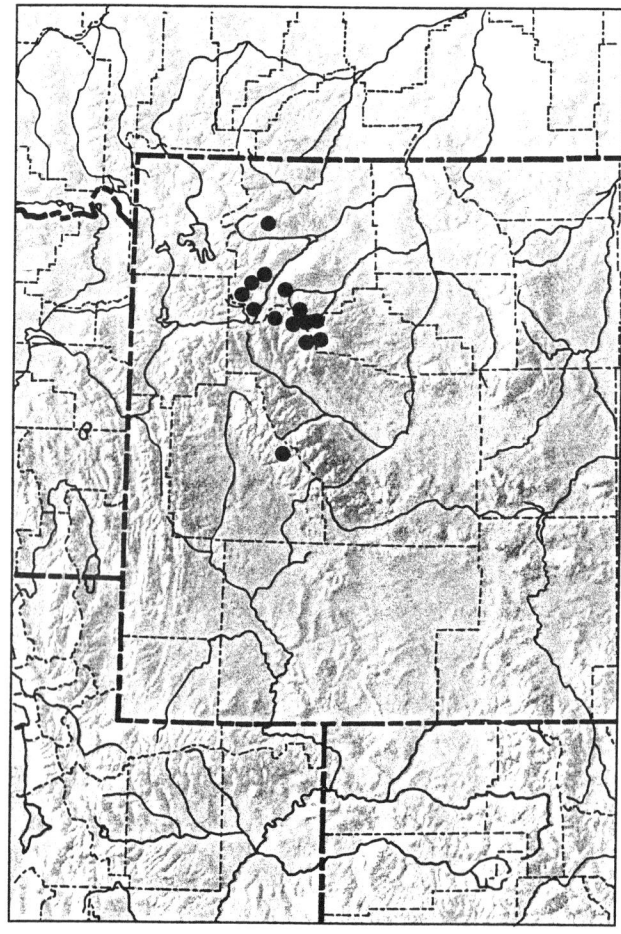

Oxytropis sericea Nutt. in T. & G.
Fl. N. Am. 1:339, 1838.

Silky Locoweed

Aragallus albiflorus A. Nels.
Aragallus macounii Greene (in part)
Aragallus majusculus Greene
Aragallus melanopontus Greene
Aragallus pinetorum A. Heller
Aragallus saximontanus A. Nels.
Aragallus sericea (Nutt.) Greene
Aragallus spicatus (Hook.) Rydb.
Astragalus albiflorus Gand.
Astragalus saximontanus (A. Nels.) Tidestr.
Oxytropis albiflora (Gand.) K. Schum.
Oxytropis condensatus (A. Nels.) A. Nels.
Oxytropis macounii (Greene) Rydb. (in part)
Oxytropis pinetorum (A. Heller) K. Schum.
Oxytropis saximontanus A. Nels.
Oxytropis spicata (Hook.) Standl.

Acaulescent, caespitose, silvery-pilose perennial from a stout taproot and short-branched caudex. Leaves odd-pinnate; leaflets oblong to lanceolate, opposite or subopposite; stipules membranous, pilose dorsally, adnate to the petiole for more than half their length. Flowers mostly 10–30, in a loose raceme. Calyx strigose with both black and white hairs, the teeth lanceolate. Petals yellowish or white, sometimes purple-tipped; banner deeply emarginate, erect or nearly so; wings emarginate, exceeding the short-beaked keel. Stamens 10, diadelphous. Legumes sessile, erect, oblong, silky-strigose, with a prominent beak.

Rocky places in the Absaroka, Beartooth, and Hoback Ranges. Alaska south to Utah, New Mexico, Oklahoma, and Kansas.

Mountain plants of northwestern Wyoming with corollas more yellow than white and lacking a purple-tipped keel are var. *spicata* (Hook.) Barneby.

var. spicata

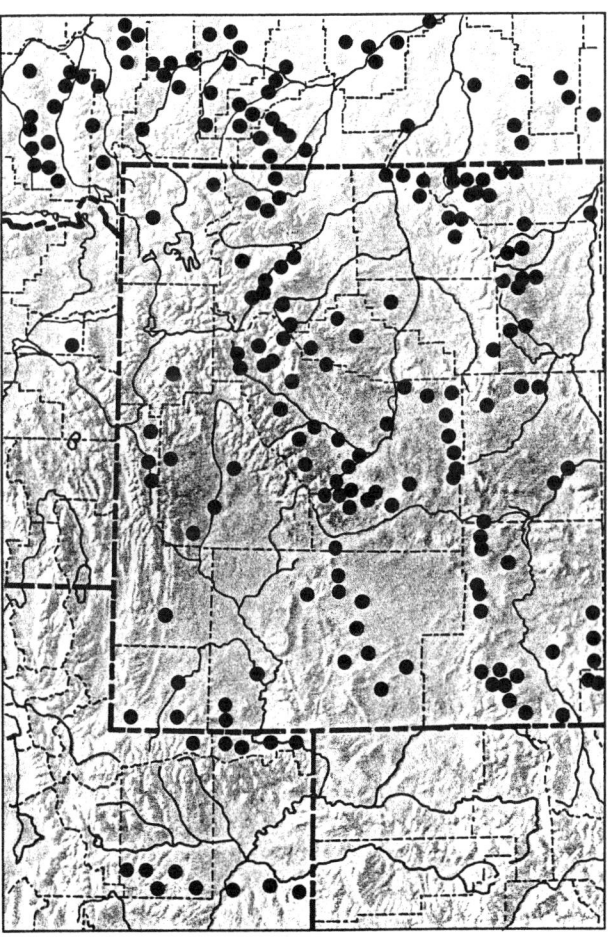

Oxytropis viscida Nutt. in T. & G.
Fl. N. Am. 1:341, 1838.

Sticky Locoweed

Aragallus hudsonicus Greene
Aragallus viscidulus Rydb.
Aragallus viscidus (Nutt.) Tidestr.
Astragalus viscidus (Nutt.) Tidestr.
Oxytropis gaspensis Fern. & Kelsey
Oxytropis glutinosa Pors.
Oxytropis hudsonica (Greene) Fern.
Oxytropis ixodes Butters & Abbe
Oxytropis sheldonensis Pors.
Oxytropis verruculosa Pors.
Oxytropis viscidula (Rydb.) Tidestr.
Spiesia viscida (Nutt.) Kuntze

Acaulescent, tufted, glandular perennial from a taproot and branched caudex. Leaves odd-pinnate; leaflets lanceolate to narrowly elliptic, pubescent and glandular; stipules membranous, pilose to villous dorsally, ciliate, adnate to petioles for more than half their length. Flowers 3–30, in capitate or elongate racemes. Calyx villous with dark gray or black hairs, glandular, the teeth linear-lanceolate. Petals purple to cream; banner slightly reflexed; wings emarginate, longer than the short-beaked keel. Stamens 10, diadelphous. Legumes oblong-lanceolate, pubescent with black hairs, prominently beaked.

Rocky slopes of the Absaroka, Beartooth, Gros Ventre, and Wind River Ranges. Alaska, Yukon Territory, and Mackenzie district of Northwest Territories, south to California and Colorado.

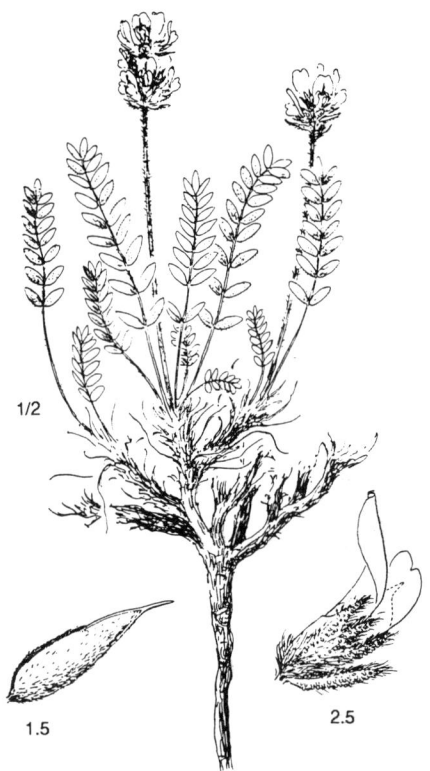

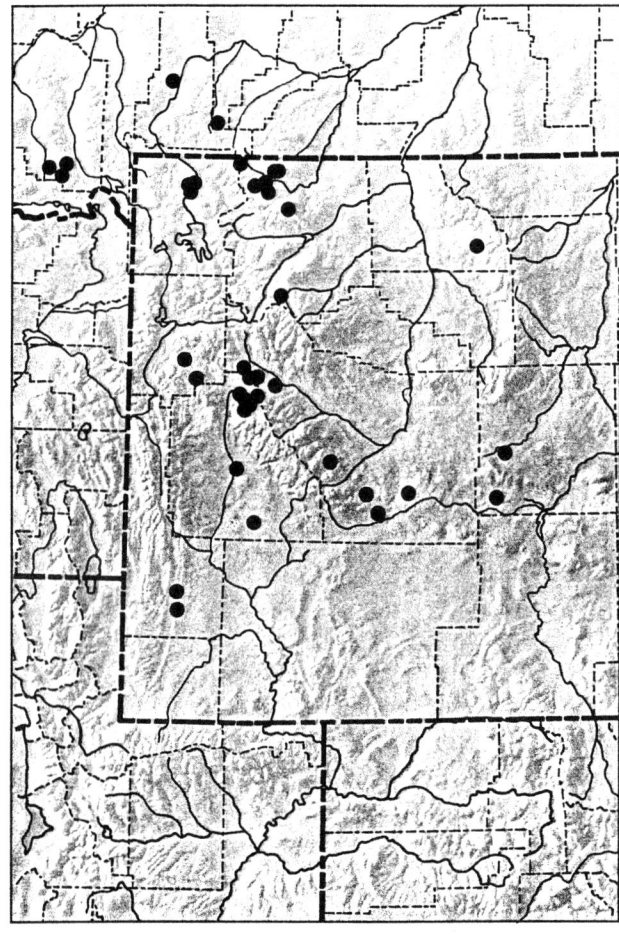

Trifolium L.
Clover

Annual or perennial herbs commonly with palmately 3-foliolate leaves, the leaflets subentire or denticulate. Flowers reddish or purple, in 1- to many-flowered, capitate spikes or racemes. Calyx 5-cleft, the lobes subequal. Corolla sometimes withering-persistent, the banner straight, usually appressed to the wings and keel, the wings longer than the keel. Stamens 10, diadelphous. Legumes sessile or stipitate, little longer than the usually persistent calyx. Seeds 1–6.

KEY TO THE SPECIES OF *TRIFOLIUM*

1. Heads 1–3-flowered; plants 2–5 cm tall ... *T. nanum*
1. Heads with 5 or more flowers; plants usually more than 5 cm tall
    2. Inflorescence subtended by an involucre; flowers erect
        3. Calyx pubescent; heads mostly 1–1.8 cm wide ... *T. dasyphyllum*
        3. Calyx glabrous; heads mostly 2–3.4 cm wide ... *T. parryi*
    2. Inflorescence lacking an involucre; flowers reflexed
        4. Plants rhizomatous; heads with more than 20 flowers; leaves mostly cauline; leaflets elliptic to oblanceolate ... *T. longipes*
        4. Plants from taproots; heads with less than 20 flowers; leaves mostly basal; leaflets broadly ovate to obovate ... *T. haydenii*

Trifolium dasyphyllum T. & G.
Fl. N. Am. 1:315, 1838.

Alpine Clover

Trifolium anemophilum Greene
Trifolium bracteolatum Rydb.
Trifolium lividum Rydb.
Trifolium scariosum A. Nels.
Trifolium stenolobum Rydb.
Trifolium uintense Rydb.

Acaulescent, pubescent to glabrate perennial from a taproot and branched, woody caudex covered with old stipules. Leaflets linear to oblanceolate, mostly entire, acute, strigose, at least below. Flowers 5–30, erect to spreading, in dense, globose heads, the peduncles exceeding the leaves; involucral bracts distinct or slightly connate, lanceolate, usually subulate. Calyx strigose, the lobes subequal, subulate. Petals purple to pinkish, the banner sometimes ochroleucous and purple-tipped. Legume 1–5-seeded.

Meadows, fellfields, and rocky slopes of the Absaroka, Beartooth, Medicine Bow, Uinta, and Wind River Ranges. Southern Montana south to Utah and Colorado.

Alpine plants with banner and wings ochroleucous, keel purplish violet, and leaflets acute are ssp. *dasyphyllum*. Those with flowers uniformly purple-violet and leaflets obtuse are ssp. *uintense* (Rydb.) J.M. Gillett.

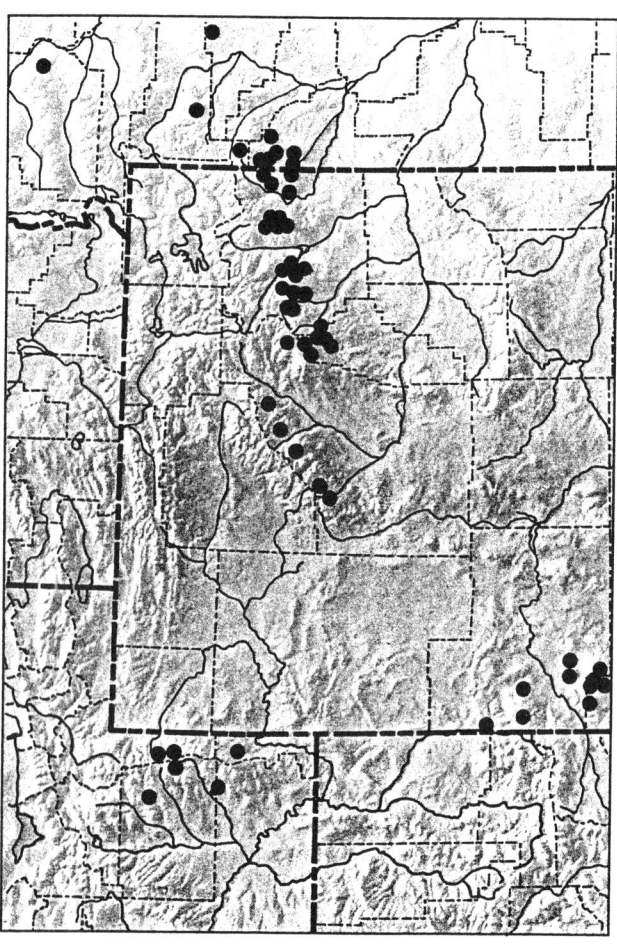

Trifolium haydenii Porter
Ann. Rep. U.S.G.S. Mont. 1871:480, 1872.

Hayden Clover

Low, pulvinate perennial, less than 10 cm tall, from a taproot and much-branched caudex covered with old stipules. Leaflets obovate to suborbicular, denticulate, the veins conspicuous. Flowers 5–20, reflexed in a loose, non-involucrate head. Calyx glabrous, the teeth lanceolate, subequal. Petals ochroleucous, often tinged with pink or purple, the banner and wings acute. Legume 1–2-seeded.

Known only from fellfields of the Absaroka and Beartooth Ranges. Endemic to southern Montana and northwestern Wyoming.

Alpine plants are var. *haydenii*.

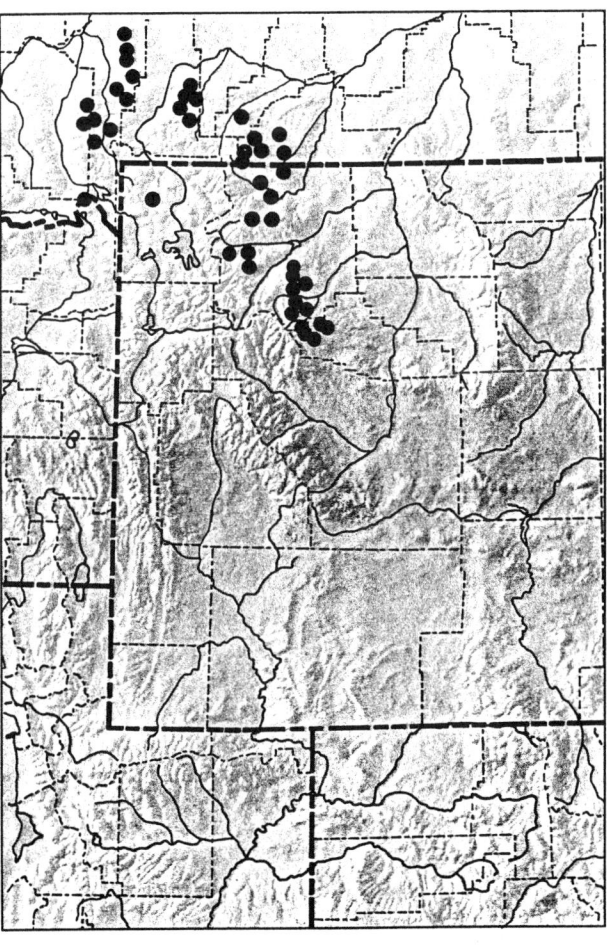

Trifolium longipes Nutt. in T. & G.
Fl. N. Am. 1:314, 1838.

Longstalk Clover

Trifolium atrorubens (Greene) House
Trifolium brachypus Blank.
Trifolium caurinum Piper
Trifolium confusum Rydb.
Trifolium covillei House
Trifolium elmeri Greene
Trifolium hansenii Greene
Trifolium multipedunculatum Kennedy
Trifolium oreganum T.J. Howell
Trifolium pedunculatum Rydb.
Trifolium rusbyi Greene
Trifolium rydbergii Greene
Trifolium shastense House

Caulescent, moderately to densely pubescent perennial from a slender taproot, often rhizomatous or stoloniferous, the stems usually decumbent at the base. Leaflets elliptic to lanceolate or oblanceolate, acute or obtuse, glabrous above, strigose below, the margins serrulate. Flowers more than 20, in dense, terminal, globose, non-involucrate heads, the lower ones reflexed with age. Calyx glabrous to strigose, the teeth narrowly lanceolate, subulate, subequal. Petals ochroleucous to pink or purple, the banner and wings acute. Legumes mostly 3–4-seeded.

Meadows and timberline margins of the Absaroka, Salt River, and Wind River Ranges. British Columbia south to California, Utah, and Colorado.

Middle Rocky Mountain plants with rhizomes and reflexed flowers are var. *reflexum* A. Nels.

var. reflexum

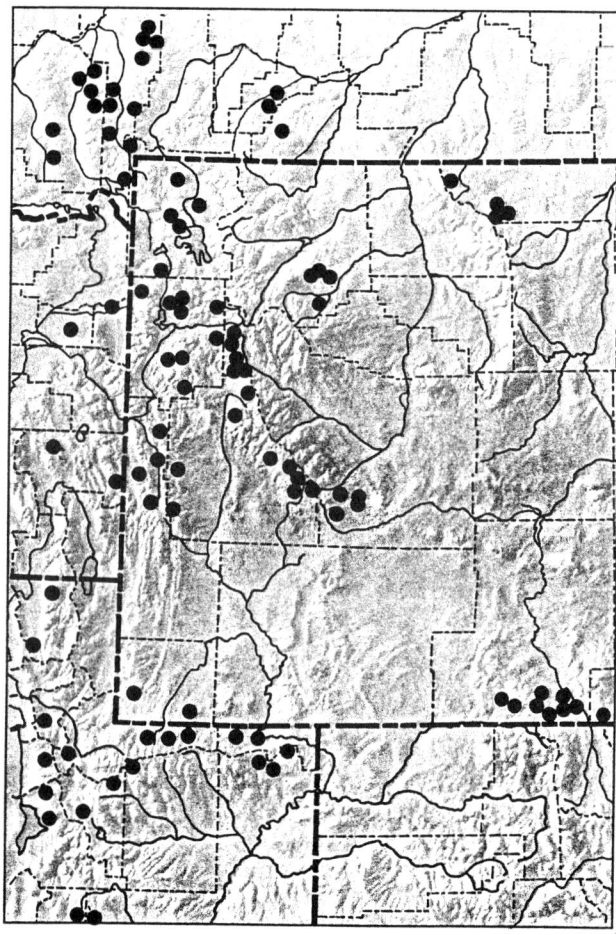

Trifolium nanum Torr.
Ann. Lyc. N.Y. 1:35, 1824.

Dwarf Clover

Low, densely pulvinate, acaulescent, glabrous perennial, mostly less than 5 cm tall, from a taproot and much-branched caudex. Leaflets oblanceolate to obovate, acute, the margins serrulate to entire. Flowers 1–3 in a pedunculate head about the same length as the leaves; involucre small, inconspicuous, of 1–4 distinct or slightly connate bracts. Calyx glabrous, the teeth lanceolate, subequal. Petals purple to reddish purple, turning brown on drying. Legumes 1–5-seeded.

Forming mats on fellfields, talus, scree, and rocky slopes in the Absaroka, Beartooth, Uinta, and Wind River Ranges. Southwestern Montana south to Utah, Colorado, and New Mexico.

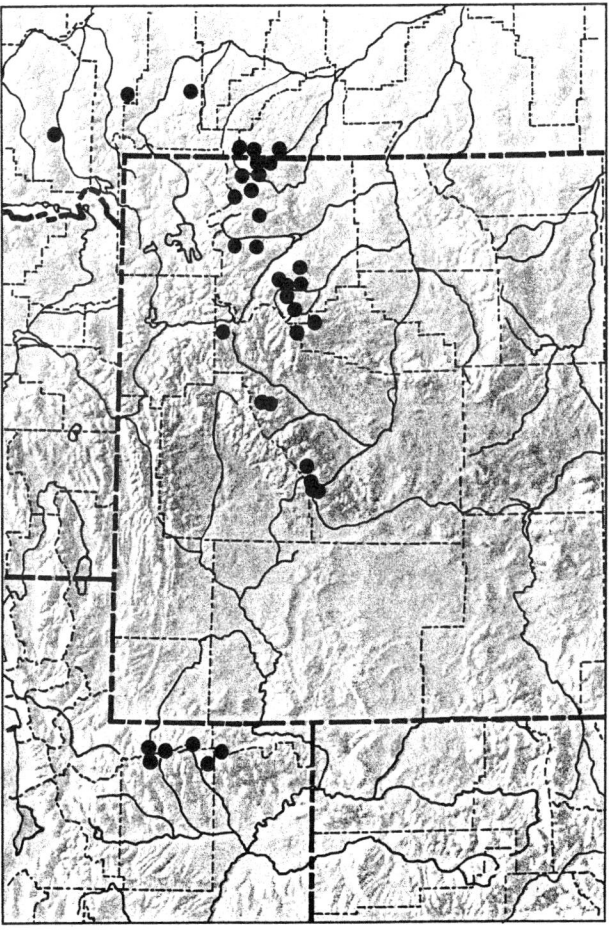

Trifolium parryi Gray
Am. J. Sci. 83:409, 1862.

Parry Clover

Trifolium inaequale Rydb.
Trifolium montanense Rydb.
Trifolium salictorum Greene ex Rydb.

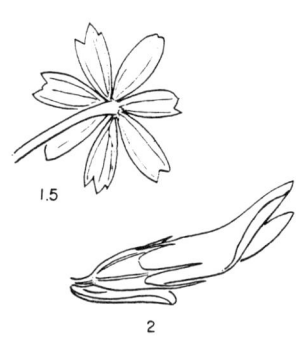

Caespitose perennial from a taproot and short-branched caudex. Leaflets elliptic to obovate, glabrous, serrulate to entire, acute or nearly so. Flowers 6–30, erect in globose heads, the peduncles exceeding the leaves; involucral bracts distinct, scarious, entire, acute to obtuse. Calyx glabrous, scarious, the teeth broadly lanceolate, subulate, unequal. Petals purple to reddish purple, drying brown. Legumes mostly 3–4-seeded.

Meadows and fellfields of the Absaroka, Beartooth, Medicine Bow, Uinta, and Wind River Ranges. Montana and Idaho south to Utah and New Mexico.

Two weakly differentiated subspecies occur in the Middle Rocky Mountains. According to Gillett (1965) they are recognized as follows:

1. Involucral bracts acute; inflorescence 2.1–2.9 cm long; flowers 1.4–2.2 cm long . . . ssp. *parryi*
1. Involucral bracts obtuse; inflorescence 1.4–2.4 cm long; flowers 1.2–1.7 cm long
. . . ssp. *montanense* (Rydb.) J. M. Gillett

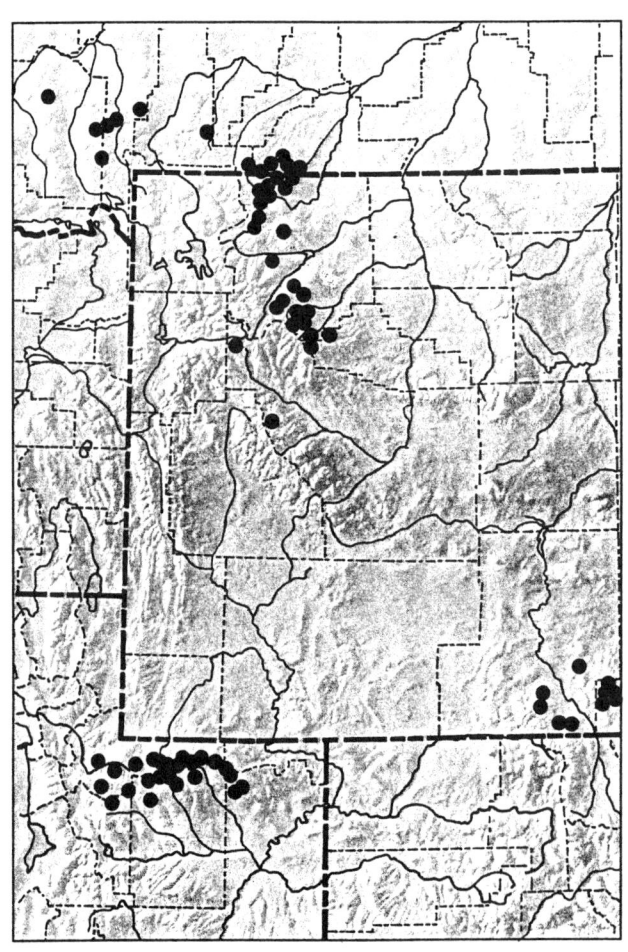

Fumariaceae
Fumitory Family

Annual or perennial, glabrous herbs with alternate or basal, exstipulate, variously compound to dissected leaves. Flowers hypogynous, perfect, irregular, in bracteate racemes or panicles. Sepals 2, small and bractlike. Petals 4, in an outer and inner series, 1 or 2 in the outer series spurred or saccate, the inner pair lacking spurs but connate at the tip. Stamens 6, diadelphous in 2 equal groups opposite the outer petals, the filaments of each set often united. Pistil 1; carpels 2; ovary with 1 locule; style 1; stigma usually 2-lobed. Fruit a 2-valved capsule.

Dicentra Bernh.
Bleeding Heart

Scapose or caulescent, glaucous, perennial herbs from fleshy, fascicled roots or rhizomes. Leaves ternately or pinnately dissected, long-petiolate. Flowers perfect, irregular, solitary or in racemes or panicles. Sepals 2, deciduous. Petals 4, white, pink, or purplish; the outer pair saccate or spurred with spreading to recurved tips, the inner pair connate toward the tip. Stamens 6, diadelphous, in 2 groups. Style slender; stigma bilobed. Fruit an elongate, more or less elliptic, several-seeded capsule.

Dicentra uniflora Kell.
Proc. Calif. Acad. 4:141, 1873.

Steershead

Bicuculla uniflora (Kell.) T.J. Howell
Capnorchis uniflora (Kell.) Kuntze
Corniveum uniflorum (Kell.) Nieuwl.
Diclytra uniflora (Kell.) Greene

Small, scapose perennial herb from a fascicle of fleshy, fusiform roots. Leaves petiolate, 1–2-ternate, the ultimate divisions oblanceolate, about equaling the scapes. Flowers single on each scape. Petals white to pink, the outer ones strongly recurved and spreading, the inner straight, connate. Stamens in 2 sets. Pistil 1; style slender, deciduous. Capsule elliptic.

Meadows and slopes along the western margin of the Middle Rocky Mountains. Collected at timberline in the Salt River Range and reported from the alpine zone in Yellowstone Park by McDougall and Bagley (1956), this species might be expected in the lower alpine zone of any of the western ranges. Washington to Wyoming and south to California, Utah, and Colorado.

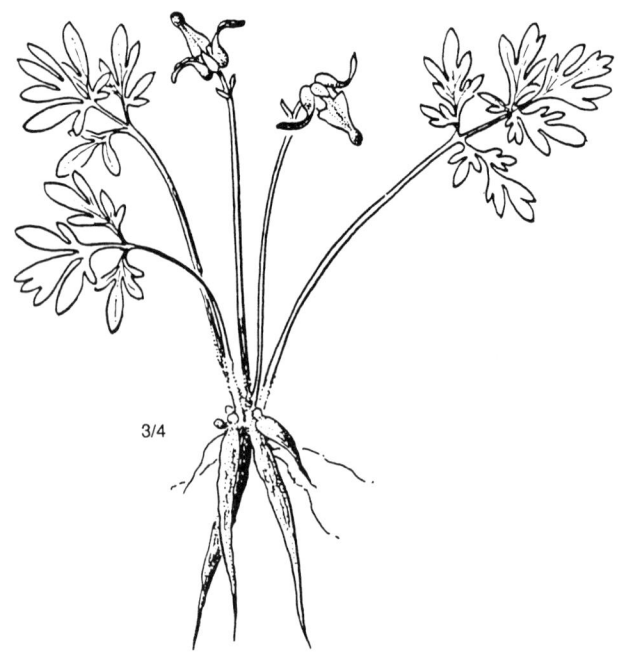

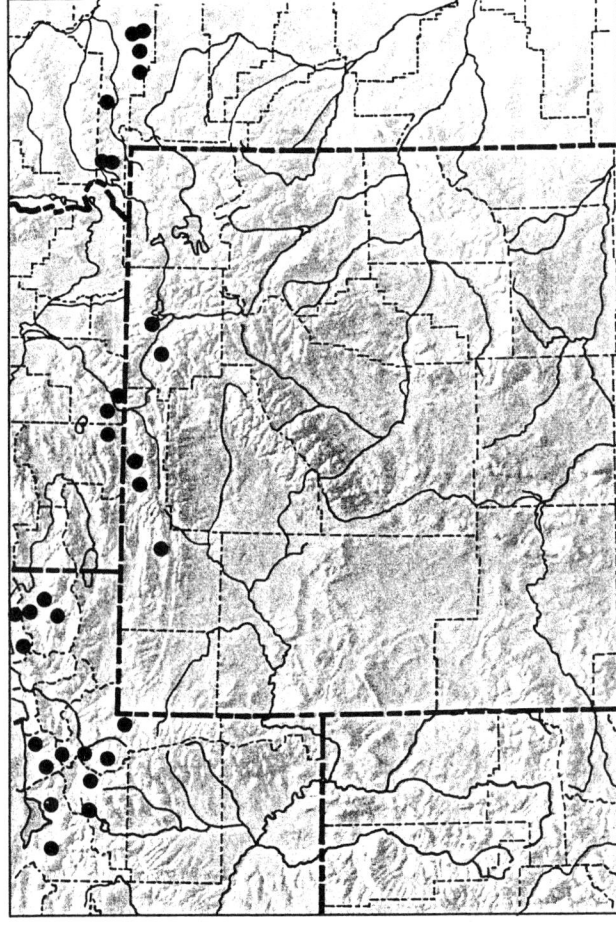

Gentianaceae
Gentian Family

Annual or perennial herbs, usually glabrous, the leaves simple, sessile, entire, exstipulate, opposite or whorled, or rarely alternate. Flowers often showy, terminal or axillary, hypogynous, perfect, regular, 4–5-merous, the corolla sympetalous, salverform, campanulate or tubular, or rotate. Stamens as many as the lobes of the corolla, alternate with them and inserted on the corolla tube. Pistil 1, of 2 united carpels, the ovary 1-celled, with 2 parietal placentae, the style single or none, the stigma single or, commonly, 2-lobed. Fruit a 2-valved, septicidal capsule with numerous seeds.

KEY TO THE GENERA OF GENTIANACEAE

1. Corolla tubular to campanulate, the lobes seldom as long as the tube, without glands or nectiferous pits at the base
    2. Corolla lobes fringed or erose; flowers 4-merous — *Gentianopsis*
    2. Corolla lobes not fringed; flowers mostly 5-merous
        3. Corolla plicate, with folds between the lobes — *Gentiana*
        3. Corolla not plicate, folds lacking — *Gentianella*
1. Corolla rotate, with 1 or 2 conspicuous pits or fringed glands at the base, the lobes usually twice as long as the tube
    4. Petals 4; style evident, at least 2 mm long; leaves whorled — *Frasera*
    4. Petals 5; style inconspicuous, scarcely 1 mm long; leaves opposite — *Swertia*

Frasera Walt.
Elkweed, Green Gentian

Tall, glabrous to puberulent, perennial herbs from a simple or branched caudex, the leaves opposite or whorled. Flowers 4-merous, rotate, in the upper leaf axils. Calyx cleft nearly to the base. Corolla greenish, deeply cleft, the lobes purple-spotted in ours, bearing a single or double, fringed gland at the base, sometimes with a broad appendage fused to the base of the lobe. Stamens inserted on the base of the corolla, the filaments distinct and alternate with small scalelike processes, or often united at the base. Ovary 1-celled; the style slender, persistent; the stigma 2-lobed or nearly entire. Capsule septicidal, leathery, the seeds often narrowly winged.

Frasera speciosa Dougl. ex Griseb. in Hook.
Fl. Bor.-Am. 2:66, 1838.

Elkweed

Frasera angustifolia (Rydb.) Rydb.
Frasera macrophylla Greene
Frasera scabra (M.E. Jones) Rydb.
Frasera stenosepala (Rydb.) Rydb.
Swertia radiata (Kell.) Kuntze
Tessaranthium angustifolium (Rydb.) Rydb.
Tessaranthium macrophyllum (Greene) Rydb.
Tesseranthium radiatum Kell.
Tesseranthium scabrum (M.E. Jones) Rydb.
Tesseranthium speciosum (Dougl.) Rydb.
Tesseranthium stenosepalum (Rydb.) Rydb.

Biennial or perennial herb with a single, erect, stout stem, to 0.5 m in the alpine zone, from a taproot. Leaves whorled, mostly in fours, somewhat glaucous, entire, strongly veined, obovate to spatulate, gradually reduced upward and becoming lanceolate. Flowers 4-merous, in axils of upper leaves. Calyx deeply parted, the lobes linear to lanceolate. Corolla rotate, greenish spotted with purple, the lobes acute, often shorter than the calyx, each with 2 fringed, basal glands. Stamens erect to ascending. Style about half the length of the ovary. Capsule slightly flattened.

Alpine meadows in most of the western ranges; not yet known from above timberline in the Medicine Bow and Teton Ranges. Washington to South Dakota and south to New Mexico and California; Mexico.

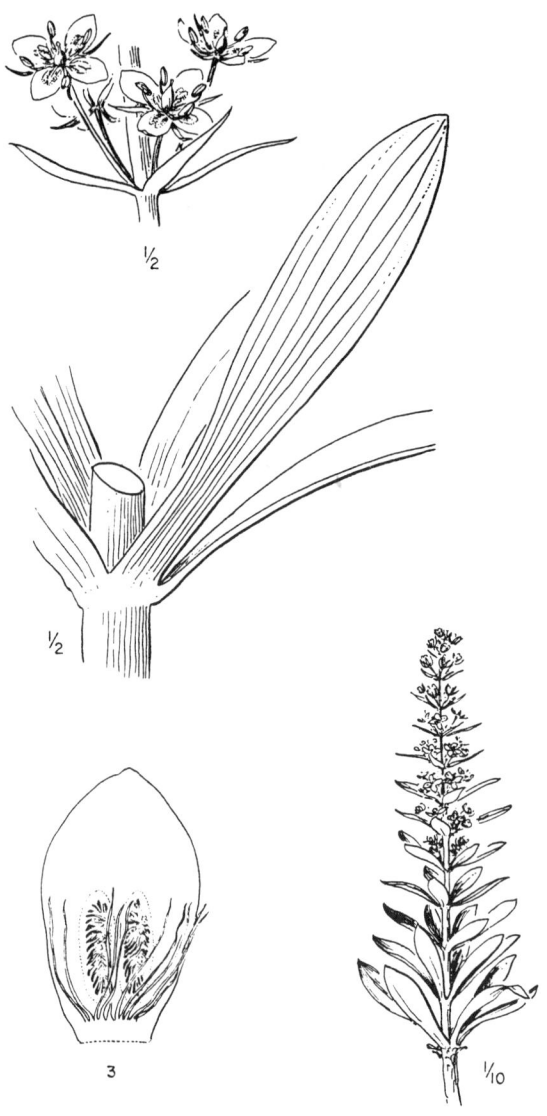

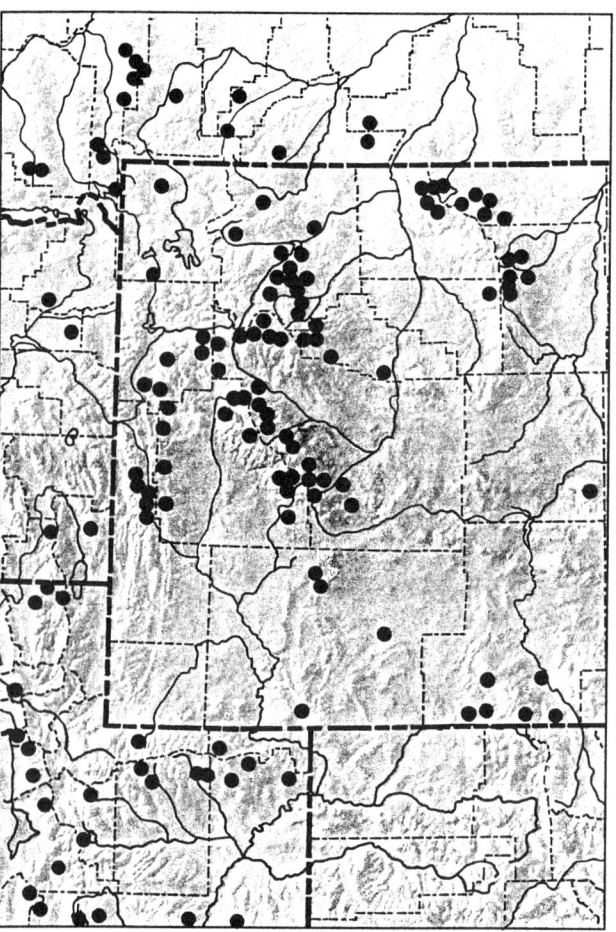

Gentiana L.
Gentian

Perennial or annual herbs, often with rhizomes; stems mostly simple, erect, glabrous or pubescent. Leaves opposite, sessile or petiolate. Flowers showy, solitary or in cymose clusters. Calyx 4–5-lobed, tubular, with a continuous internal membrane projecting above the lobes. Corolla 4–5-lobed and white, yellowish, greenish, blue, or purple, funnelform or campanulate, plicate in the sinuses, the folds sometimes toothed. Stamens 4–5, inserted on the corolla tube. Style short or lacking; stigmas 2. Capsules 1-loculed; valves 2; seeds numerous.

KEY TO THE SPECIES OF *GENTIANA*

1. Corolla greenish yellow, splotched with purple lines and dots; plaits of sinuses not toothed; leaves linear to narrowly oblanceolate ..... *G. algida*
1. Corolla deep blue, sometimes streaked with green; plaits of sinuses toothed; leaves lanceolate to ovate
    2. Plants annual or biennial, less than 10 cm tall; corollas less than 20 mm long ..... *G. prostrata*
    2. Plants perennial, often more than 10 cm tall; corollas more than 25 mm long
        3. Flowers usually single; leaves ovate or broadly ovate; stems glabrous ..... *G. calycosa*
        3. Flowers usually 2 or 3; leaves lanceolate; stems finely pubescent below nodes ..... *G. affinis*

Gentiana affinis Griseb. in Hook.
Fl. Bor.-Am. 2:56, 1838.

Rocky Mountain Pleated Gentian

Dasystephana affinis (Griseb.) Rydb.
Dasystephana bracteosa (Greene) Cockll.
Dasystephana forwoodii (Gray) Rydb.
Dasystephana interrupta (Greene) Rydb.
Dasystephana oregana (Engelm. ex Gray) Rydb.
Dasystephana parryi (Engelm.) Rydb.
Gentiana bigelovii Gray
Gentiana bracteosa Greene
Gentiana forwoodii Gray
Gentiana interrupta Greene
Gentiana menziesii Peck
Gentiana oregana Engelm. ex Gray
Gentiana parryi Engelm.
Gentiana remota Greene
Gentiana rusbyi Greene
Pneumonanthe affinis (Griseb.) W.A. Weber
Pneumonanthe bracteosa (Greene) Greene
Pneumonanthe forwoodii (Gray) Greene
Pneumonanthe parryi (Engelm.) Greene

Glabrous perennial from a stout caudex; stems 1 to several, erect or decumbent at the base. Leaves in pairs, the lower leaves reduced, with connate bases, the upper ones ovate to lanceolate, usually glandular-ciliolate near the base. Flowers 5-merous, crowded in racemose or capitate clusters in the upper leaf axils. Calyx tubular, the lobes linear to oblanceolate, usually unequal and about as long as the tube. Corolla funnelform, blue, sometimes greenish on the outside, the lobes ovate, acute or acuminate, plaited in the sinuses with cleft or entire appendages about one-third the length of the corolla lobes. Filaments partially adnate to the corolla tube. Ovary stipitate; style short, 2-cleft. Capsule long-stipitate; seeds winged.

Timberline meadows and rocky ridges in the lower alpine zone of the Absaroka and Beartooth Ranges. British Columbia and Alberta south to California and Arizona; Mexico.

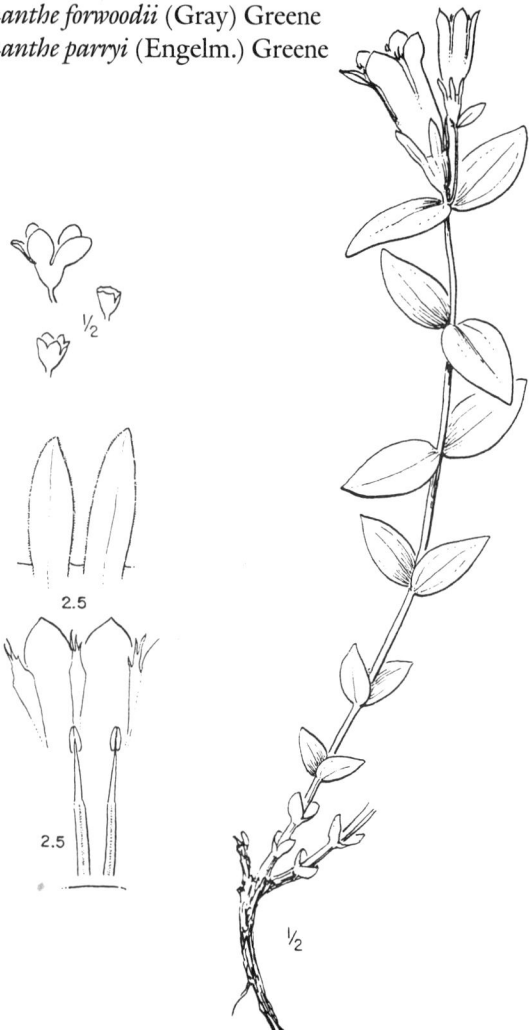

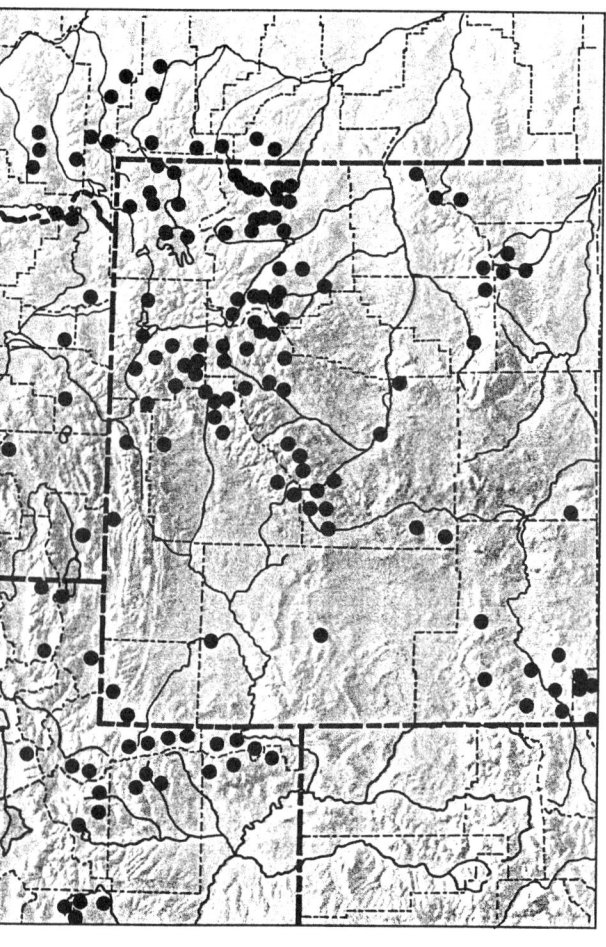

Gentiana algida Pall.
Fl. Ross. 1:107, 1789.

Arctic Gentian

Dasystephana romanzovii (Ledeb. ex Bunge) Rydb.
Gentiana frigida Haenke
Gentiana romanzovii Ledeb. ex Bunge
Gentianodes algida (Pall.) Löve & Löve
Gentianodes romanzovii (Ledeb. ex Bunge) Ledeb.

Low, tufted perennial from a short caudex, sometimes with short rhizomes; stems 1 to several, glabrous, with a basal rosette of linear-oblanceolate to oblanceolate leaves. Cauline leaves sessile, linear to oblong, somewhat connate at the base. Flowers 1–3, in terminal, bracteate clusters. Calyx funnelform, the lobes lanceolate, unequal, about one-half the length of the tube. Corolla greenish white or yellowish white and streaked with purple, funnelform, about twice as long as the calyx, the lobes triangular, acute to acuminate; sinuses strongly plicate. Filaments adnate below, the free portion narrowly triangular. Ovary stipitate; style 2-cleft. Capsule oblong-ovate.

Occasional in alpine meadows and fellfields of the Beartooth, Medicine Bow, Uinta, and Wind River Ranges. Siberia; Alaska south to Montana, Wyoming, and Colorado.

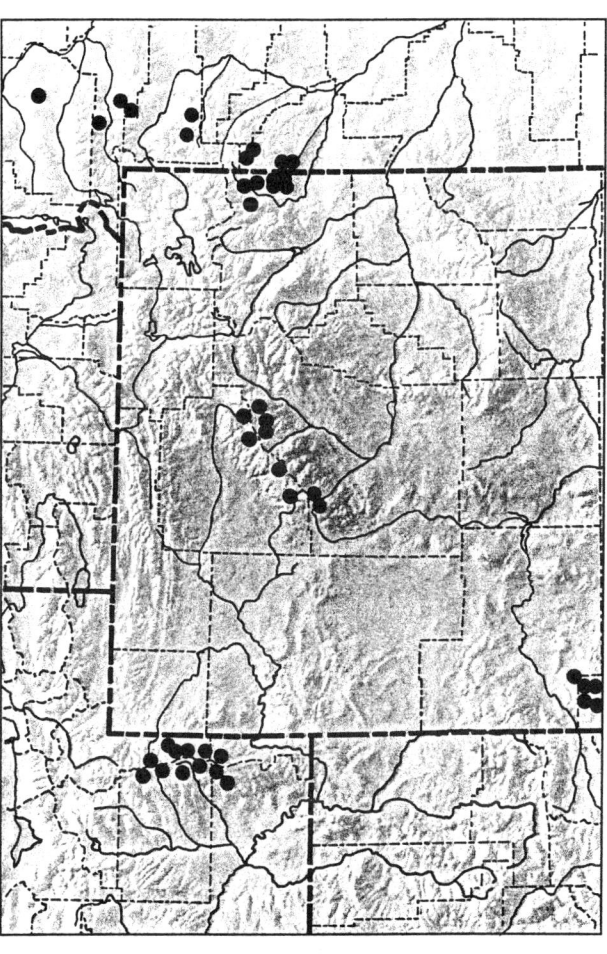

Gentiana calycosa Griseb. in Hook.
Fl. Bor.-Am. 2:56, 1838.

Explorer Gentian

Dasystephana calycosa (Griseb.) Rydb.
Dasystephana monticola (Rydb.) Rydb.
Dasystephana obtusiloba Rydb.
Gentiana cusickii Gand.
Gentiana gormani T.J. Howell
Gentiana idahoensis Gand.
Gentiana myrsinites Gand.
Gentiana saxicola English
Pneumonanthe calycosa (Griseb.) Greene

Perennial herb from a stout, often woody caudex. Stems 1 to several, often decumbent at the base. Leaves glabrous, ovate or broadly lanceolate, acute to obtuse, the lower distinctly connate-sheathing. Flowers mostly single, terminal, occasionally in bracteate clusters of 3. Calyx greenish or sometimes scarious, the lobes variable, from foliaceous-spreading to erect. Corolla funnelform, blue, sometimes streaked or spotted with green or yellow, the lobes ovate, acute; sinuses plicate, cleft into narrow, fringelike segments. Stamens adnate and wing-margined below, the free portion narrowly lanceolate. Ovary stipitate; style 2-cleft.

Meadows and fellfields of the Absaroka, Gros Ventre, Salt River, Teton, Uinta, and Wind River Ranges. British Columbia south to Wyoming and west to California.

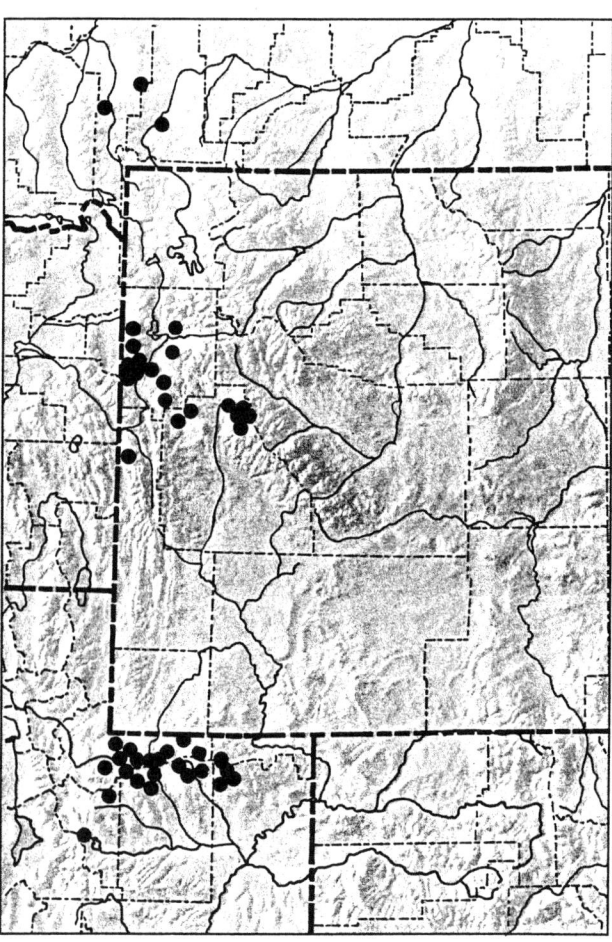

Gentiana prostrata Haenke ex Jacq.
Coll. Bot. 2:66, 1788.

Moss Gentian

Chondrophylla americana (Engelm.) A. Nels.
Chondrophylla aquatica (L.) W.A. Weber
Chondrophylla fremontii (Torr.) A. Nels.
Chondrophylla prostrata (Haenke ex Jacq.) Anderson
Ciminalis fremontii (Torr.) W.A. Weber
Ciminalis prostrata (Haenke ex Jacq.) Löve & Löve
Ericoilea prostrata (Haenke ex Jacq.) Borkh.
Gentiana aquatica L.
Gentiana fremontii Torr.
Gentiana humilis Gray

Annual or biennial herb from slender roots; stems simple or branched from the base, often decumbent-spreading. Leaves ovate to obovate, connate and sheathing. Flowers terminal, solitary. Calyx funnelform, the lobes shallow, subequal, and broadly triangular. Corolla funnelform, blue, 4- or 5-lobed, the lobes ovate to obovate, somewhat imbricate and contorted; sinuses plicate, the plaits usually shallowly toothed. Stamens included, inserted above the middle of the corolla tube; filaments slender. Ovary stipitate; style branched. Capsule linear-oblong, stipitate.

Wet meadows and alpine bogs of the Absaroka, Beartooth, Big Horn, Uinta, and Wind River Ranges. Alaska to Alberta, south to California, Nevada, and Colorado.

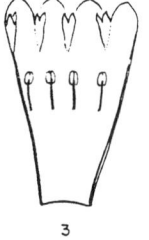
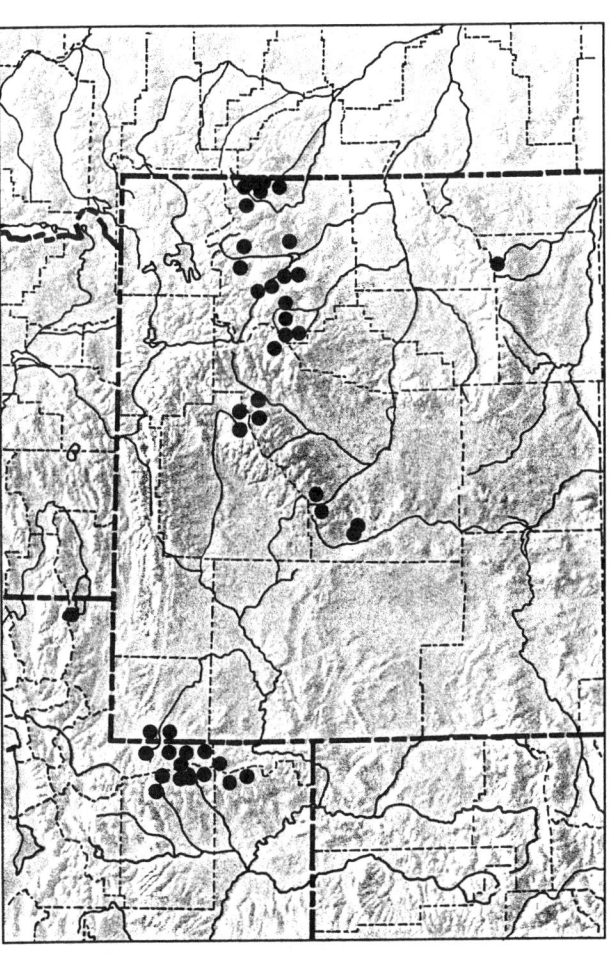

Gentianella Moench
Little Gentian

Ours are small, annual or biennial, mostly glabrous herbs from a taproot. Leaves opposite, sessile or nearly so. Flowers solitary to several, terminal, or on long peduncles from leaf axils. Calyx 5-lobed, funnelform, the lobes unequal in length. Corolla 5-lobed and blue, white, or yellowish green, salverform or tubular, usually appearing tubular, sinuses not plaited, the lobes fringed within. Stamens 5, inserted below or near the middle of the corolla tube. Style short or lacking; stigma 2-lobed. Capsules 1-loculed, sessile or short-stipitate; valves 2; seeds numerous.

KEY TO THE SPECIES OF *GENTIANELLA*

1. Fringe on inner edge of corolla lobe originating from 2 distinct scales; flowers on long, axillary peduncles; outer calyx lobes shorter and broader than the inner pair .... *G. tenella*
1. Fringe on inner edge of corolla lobe originating from a single scale; flowers in terminal clusters, lacking long peduncles; calyx lobes nearly similar in size and shape .... *G. amarella*

Gentianella amarella (L.) Börner
Fl. Deut. Volk, 543, 1912.

Annual Gentian

Gentiana amarella L.
Amarella acuta (Michx.) Raf.
Amarella amarella (L.) Cockll.
Amarella anisosepala (Greene) Greene
Amarella californica Greene
Amarella cobrensis Greene
Amarella conferta Greene
Amarella copelandi Greene
Amarella distegia (Greene) Greene
Amarella heterosepala (Engelm.) Greene
Amarella lemberti Greene
Amarella macounii Greene
Amarella plebeja (Ledeb. ex Spreng.) Greene
Amarella revoluta Greene
Amarella scopulorum Greene
Amarella strictiflora (Rydb.) Greene
Amarella tortuosa (M.E. Jones) Rydb.
Ericala acuta (Michx.) G. Don
Gentiana acuta Michx.
Gentiana anisosepala Greene
Gentiana distegia Greene
Gentiana heterosepala Engelm.
Gentiana plebeja Ledeb. ex Spreng.
Gentiana polyantha A. Nels.
Gentiana scopulorum (Greene) Tidestr.
Gentiana strictiflora (Rydb.) A. Nels.
Gentiana tortuosa M.E. Jones
Gentiana wrightii Gray
Gentianella acuta (Michx.) Hiit.
Gentianella clementis Rydb.
Gentianella heterosepala (Engelm.) Holub
Gentianella strictiflora (Rydb.) W.A. Weber
Gentianella tortuosa (M.E. Jones) J.M. Gillett

Glabrous, annual or biennial herb from a slender taproot. Stems single or branched, mostly from near the base. Basal leaves oblanceolate to spatulate; cauline leaves opposite, ovate to lanceolate, acute, not connate at the base. Flowers 4–5-merous, cymose, borne on slender pedicels, usually in the upper leaf axils, occasionally from near the base. Calyx tubular, often cleft nearly to the base, the lobes lanceolate, unequal. Corolla salverform and blue, purple, yellowish, or white, the lobes acute, erect or nearly so, with slender fimbriae at the base, the fimbriae originating from a single scale; sinuses lacking plaits or folds. Stamens included, inserted on the lower corolla tube. Ovary sessile or very shortly stipitate; style very short or lacking; stigmas 2. Capsule sessile or short-stipitate.

Moist meadows and willow bogs of the Absaroka, Beartooth, Big Horn, Gros Ventre, Medicine Bow, Teton, Uinta, Wasatch, and Wind River Ranges. Eurasia; Alaska south to Newfoundland, North Dakota, Colorado, and California; Mexico.

North American plants are ssp. *acuta* (Michx.) J.M. Gillett.

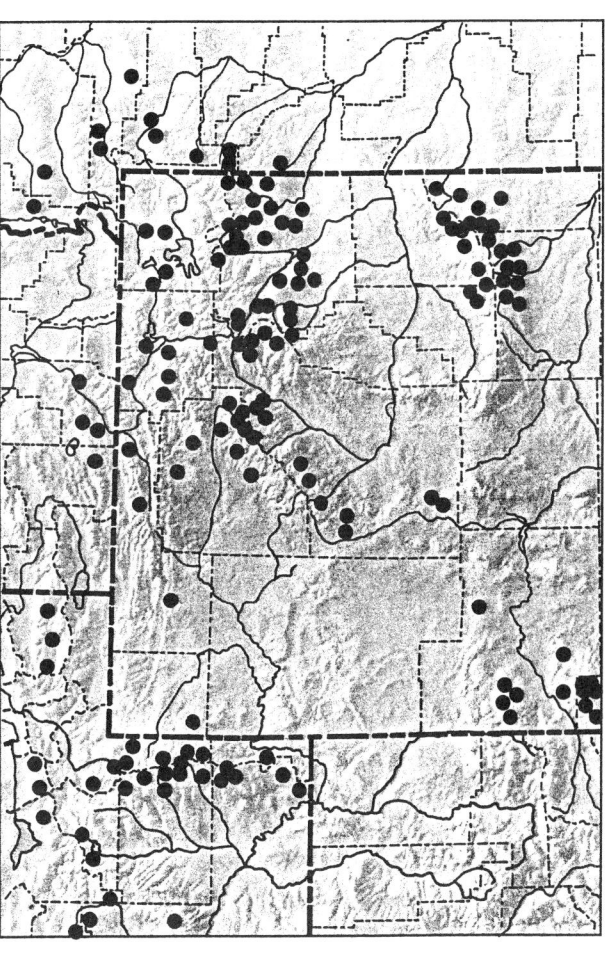

Gentianella tenella (Rottb.) Börner
Fl. Deut. Volk, 542, 1912.

Delicate Gentian

Gentiana tenella Rottb.
Amarella monantha (A. Nels.) Rydb.
Amarella tenella (Rottb.) Cockll.
Comastoma tenella (Rottb.) Toyokuni
Gentiana glacialis Thom.
Gentiana monantha A. Nels.
Lomatogonium tenellum (Rottb.) Löve & Löve

Glabrous, annual herb from a slender taproot. Stems single or, usually, branched from the base. Basal leaves oblanceolate, obtuse; cauline leaves oblanceolate to oblong. Flowers mostly 4-merous, single on peduncles longer than the stems. Calyx deeply lobed, the lobes distinct or nearly so, unequal, the outer 2 shorter than the inner ones. Corolla blue, purplish, or white, tubular to salverform, the lobes acute, with slender fimbriae at the base, the fimbriae originating from 2 distinct scales; sinuses lacking plaits or folds. Stamens included, inserted at or below the middle of the tube. Ovary sessile; style short.

Alpine meadows of the Absaroka, Beartooth, Medicine Bow, Uinta, and Wind River Ranges. Circumboreal; south in western North America to California, Arizona, and New Mexico.

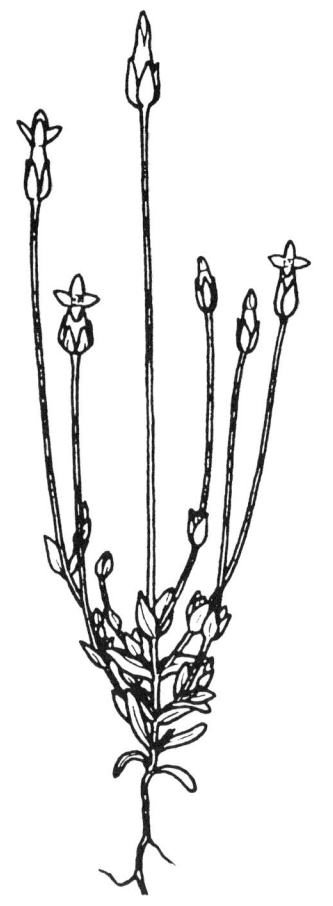
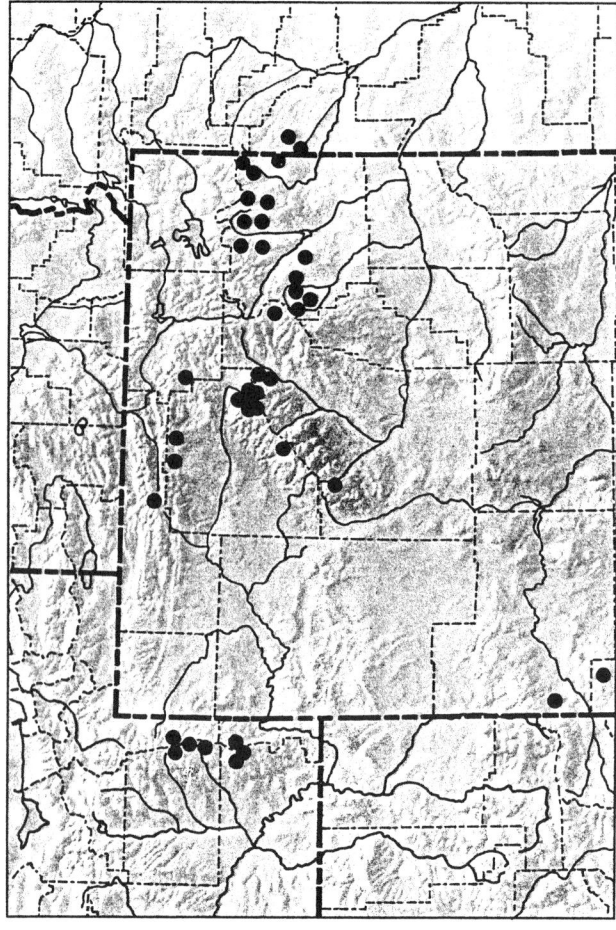

Gentianopsis Ma
Fringed Gentian

Annual to perennial, glabrous herbs. Leaves opposite, sessile or nearly so, often forming a loose basal cluster. Flowers showy, solitary to several in an axillary or terminal cyme. Calyx 4-lobed, the lobes with thin, scarious margins, 4-angled. Corolla 4-lobed, funnelform or salverform, the lobes fringed or toothed on the margins, lacking plaits or folds in the sinuses. Stamens inserted on the corolla tube above the middle. Style short; stigma 2-lobed. Capsules 1-loculed, elliptic, stipitate; valves 2; seeds numerous.

KEY TO THE SPECIES OF *GENTIANOPSIS*

1. Plants annual; flowers pedunculate, lacking leaflike bracts at their bases ... *G. detonsa*
1. Plants perennial; flowers not pedunculate, the bases subtended by 2 leaflike bracts ... *G. barbellata*

Gentianopsis barbellata (Engelm.) Iltis
SIDA Contr. Bot. 2:136, 1965.

Perennial Fringed Gentian

Gentiana barbellata Engelm.
Anthopogon barbellatas (Engelm.) Rydb.
Gentianella barbellata (Engelm.) J.M. Gillett

Glabrous, perennial herb from slender, rather fleshy roots. Stems mostly single. Leaves oblanceolate to linear, entire. Flowers solitary, sometimes several, in a terminal, bracteate cluster; bracts leaflike, 2 per flower. Calyx tubular, the lobes unequal, about as long as the tube. Corolla blue, funnelform, 4-lobed, the lobes nearly as long as the tube, fimbriate on the margins; sinuses not plaited. Filaments bearded below. Ovary and capsule stipitate.

Alpine slopes, meadows, and rocky summits of the Absaroka, Gros Ventre, Uinta, Wind River, and Wyoming Ranges. Wyoming to Utah, south to New Mexico and Arizona.

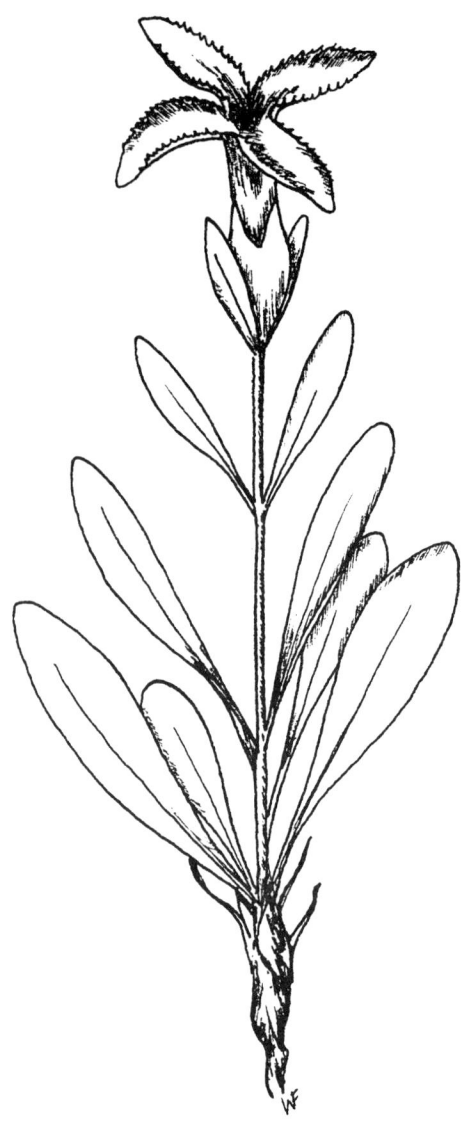

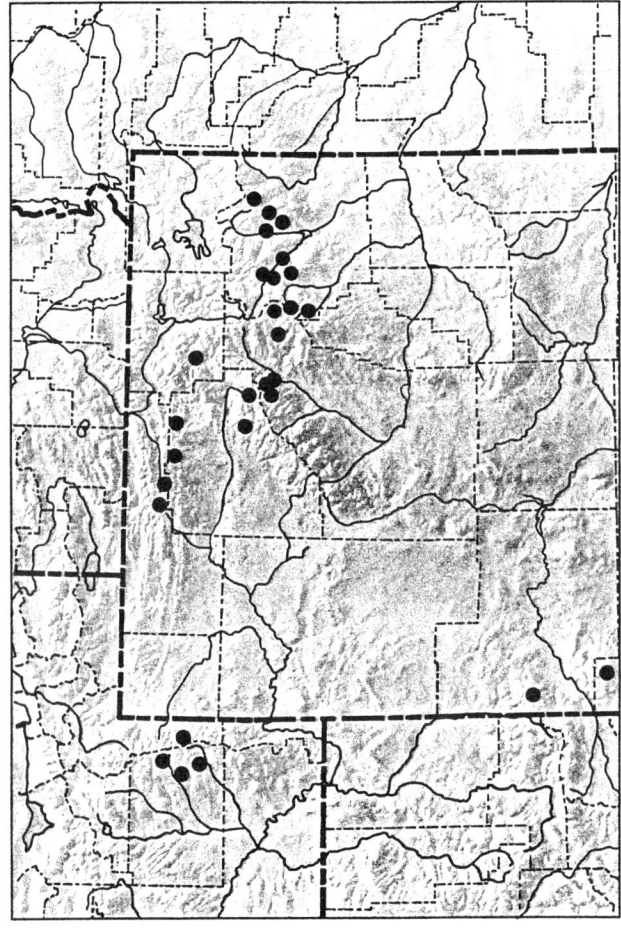

Gentianopsis detonsa (Rottb.) Ma
Acta Phytotax. Sinica 1:15, 19, 1951.

Rocky Mountain Fringed Gentian

Gentiana detonsa Rottb.
Anthopogon barbata (Froel.) Raf.
Anthopogon detonsa (Rottb.) Raf.
Anthopogon elegans (A. Nels.) Rydb.
Anthopogon macounii (Holm) Rydb.
Anthopogon thermalis (Kuntze) Rydb.
Gentiana barbata Froel.
Gentiana elegans A. Nels.
Gentiana macounii Holm.
Gentiana raupii Pors.
Gentiana richardsonii Pors.
Gentiana thermalis Kuntze
Gentianella detonsa (Rottb.) G. Don
Gentianopsis barbata (Froel.) Ma
Gentianopsis elegans (A. Nels.) Ma
Gentianopsis raupii (Pors.) Iltis
Gentianopsis thermalis (Kuntze) Iltis

Erect, glabrous annual from a slender taproot; stems several, branched from the base. Basal leaves oblanceolate to spatulate; cauline leaves sessile, lanceolate or oblong. Flowers 4-merous, single and terminal on long peduncles. Calyx tubular, the lobes acuminate, about as long as the tube. Corolla blue to bluish purple, the lobes rounded, erose on the margins; sinuses not plaited. Stamens included, adnate to the corolla tube below midlength. Ovary stipitate; style slender; stigma fringed.

Wet meadows of the lower alpine zone of the Absaroka, Beartooth, Gros Ventre, Medicine Bow, Uinta, Wasatch, and Wind River Ranges. Circumboreal; Alaska to Newfoundland, south to New York, South Dakota, Wyoming, Arizona, and California; Mexico.

Mountain plants of the Middle Rockies are var. *elegans* (A. Nels.) Holmgren.

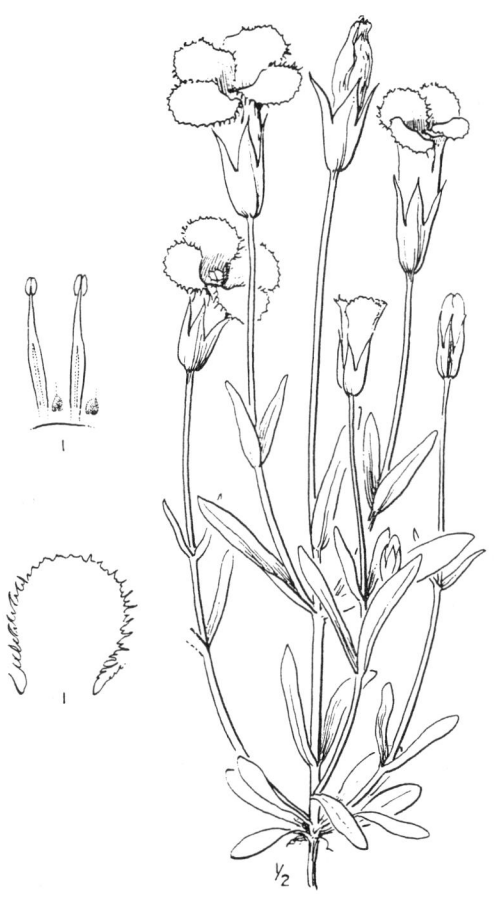

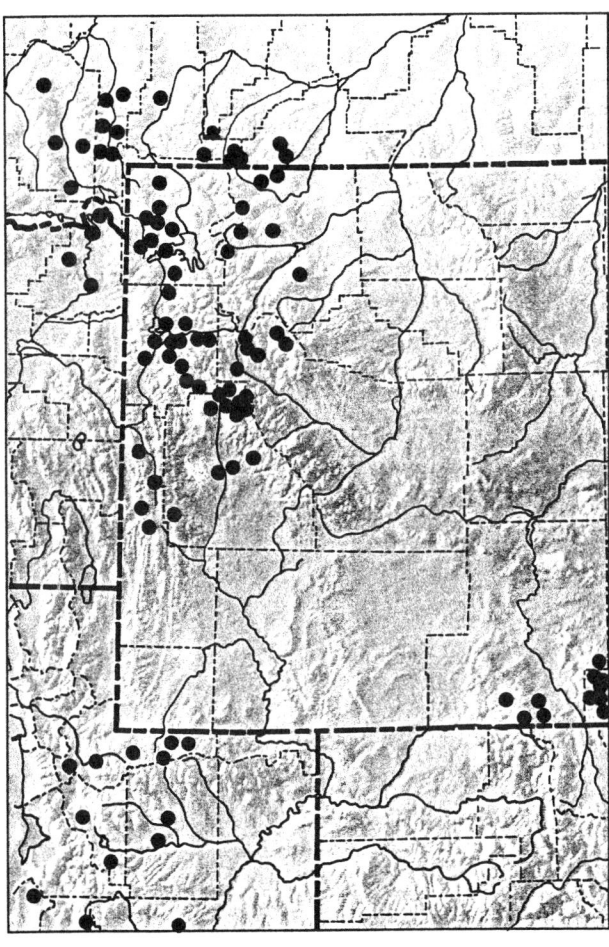

Swertia L.
Star Gentian

Slender, erect, perennial herbs with glabrous, opposite, entire leaves. Flowers 4–5-merous and blue, bluish purple, or rarely white, in panicles or racemes. Corolla rotate, deeply 5-lobed, each segment with a pair of basal glands. Ovary 1-celled, the style short or lacking, the stigma 2-lobed. Capsule ovate. Seeds winged in ours.

Swertia perennis L.
Sp. Pl. 1:226, 1753.

Star Gentian, Alpine Bog Swertia

Swertia congesta A. Nels.
Swertia fritillaria Rydb.
Swertia obtusa Ledeb.
Swertia occidentalis Greene
Swertia palustris A. Nels.
Swertia parallela Greene
Swertia scopulina Greene

Glabrous, perennial herb from a short, often woody rhizome; stem single, to 20 cm tall. Basal leaves petiolate, oblanceolate to obovate; cauline leaves sessile, opposite or occasionally alternate, ovate to lanceolate, reduced upward. Flowers 5-merous, in panicles or solitary in upper leaf axils. Calyx deeply 5-parted, the lobes narrowly lanceolate. Corolla rotate, dark bluish purple, often with darker streaks or spots, 5-parted nearly to the base, the lobes oblong to lanceolate, each with 2 fringed glands at the base. Stamens erect, free from the short corolla tube, or nearly so. Ovary sessile, the style very short. Capsule elliptic.

Wet meadows, stream banks, and nival basins in the Absaroka, Beartooth, Medicine Bow, Uinta, Wasatch, and Wind River Ranges. Circumboreal; Alaska south to California, Idaho, and Colorado.

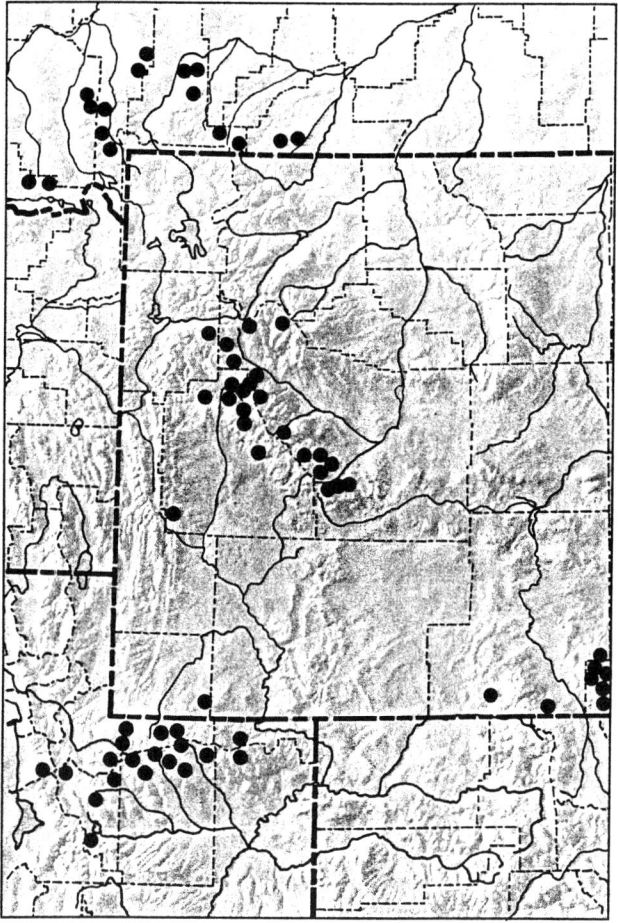

Geraniaceae
Geranium Family

Annual, biennial, or perennial herbs with alternate or opposite, simple or compound, stipulate leaves. Flowers perfect, regular, hypogynous, in cymes or umbels. Sepals 5, distinct, somewhat imbricate. Petals 5, distinct, white, pink, or rose in ours. Fertile stamens 10 or 5, the filaments connate at the base. Pistil 1, the ovary 5-lobed, of weakly united carpels; styles 5, the stigmas capitate or elongate. Fruit a beaked, elastic capsule, splitting at maturity.

Geranium L.
Wild Geranium

Annual or perennial herbs with alternate or opposite, simple, palmately lobed or divided, stipulate leaves. Flowers showy, white, rose, or pink, regular in ours, in pairs or cymes. Sepals 5, imbricate. Petals 5, distinct, imbricate, alternating with 5 glands. Stamens 10, in 2 sets of unequal length, the filaments connate at the base. Carpels 5. Capsules elastic, the stylar portions recurving from the top after seeds are shed.

KEY TO THE SPECIES OF *GERANIUM*

1. Petals white; inflorescence with mostly purplish glands — *G. richardsonii*
1. Petals pink or rose; inflorescence with mostly yellowish or whitish glands — *G. viscosissimum*

Geranium richardsonii Fisch. & Trautv.
Ind. Sem. Hort. Petrop. 4:37, 1837.

Richardson Geranium

Geranium albiflorum Hook.
Geranium gracilentum Greene
Geranium hookerianum Walp.
Geranium loloense St. John

Perennial, glabrous to glandular-pubescent herb with solitary or few, simple or branched, erect stems from a branched caudex and taproot, the caudex covered with old leaf bases. Basal leaves long-petiolate, the blades 5–7-palmately parted, the divisions coarsely toothed, strigose, the hairs covering the surface or sometimes restricted to the veins; cauline leaves resembling the basal ones, reduced upward. Flowers in terminal, few-flowered, glandular cymes, the glands mostly purple-tipped. Sepals narrowly elliptic, glandular, and apiculate. Petals white, with pink or purplish veins, pilose on the lower half of the inner surface. Stamens 10. Stylar column glandular-pubescent, prominently beaked. Valves of the capsule glandular-pubescent.

Timberline meadows, krummholz margins, and moist talus in the Beartooth, Uinta, and Wasatch Ranges. This species is common in the upper montane and subalpine zones throughout the Middle Rocky Mountains and should be expected in the lower alpine zone of all ranges. Yukon Territory south to California, Arizona, and New Mexico.

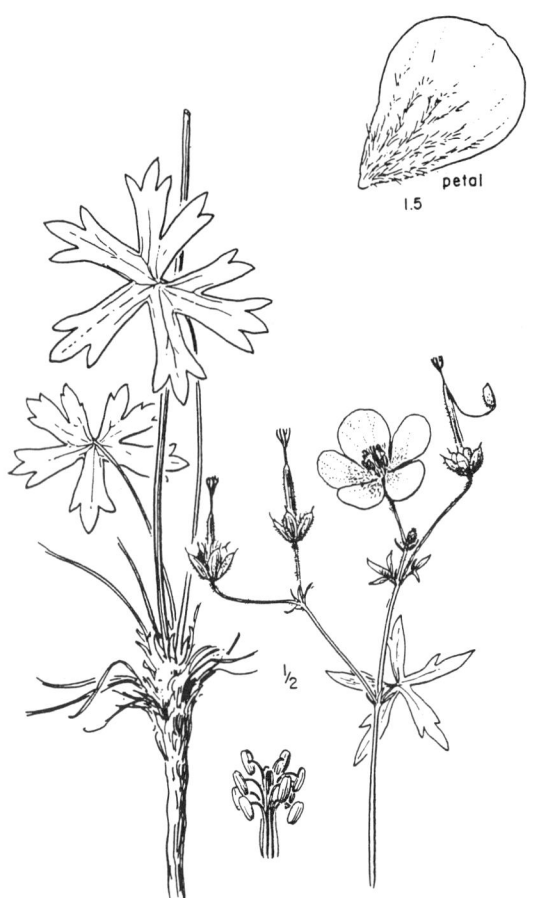
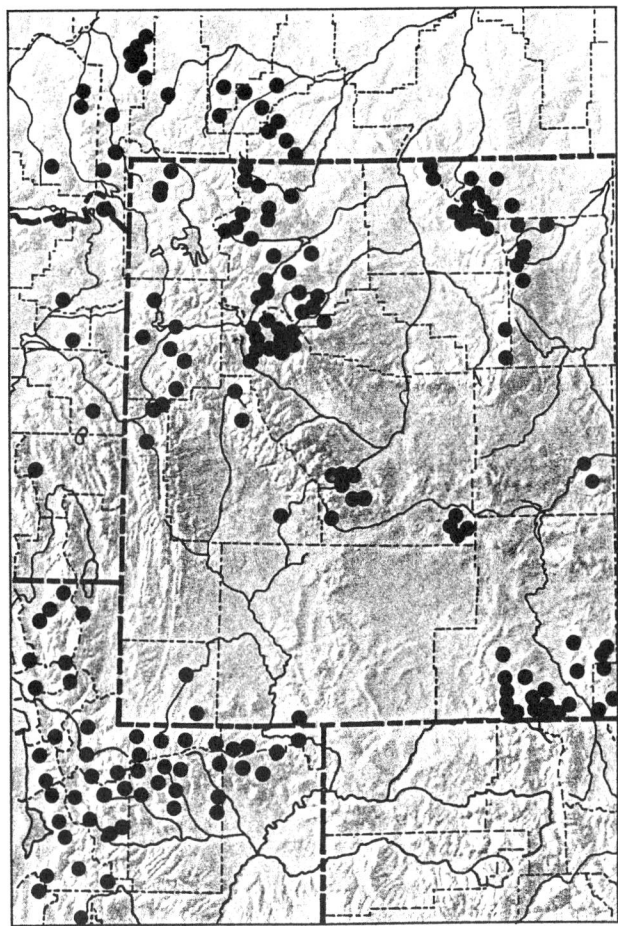

Geranium viscosissimum Fisch. & C.A. Mey. ex C.A. Mey. *Ind. Sem. Hort. Petrop.* 11: Suppl. 18, 1846.

Sticky Geranium

Geranium canum Rydb.
Geranium incisum Hanks.
Geranium nervosum Rydb.
Geranium strigosius St. John
Geranium strigosum Rydb.

Perennial, glandular-puberulent to pilose or viscid herb with solitary or clustered, simple or branched, erect stems, from a branched caudex and taproot, the caudex covered with old leaf bases. Basal leaves long-petiolate, the blades 5–7-palmately parted, the divisions coarsely toothed, strigillose to minutely hirsute, usually glandular; cauline leaves resembling the basal ones, reduced upward. Flowers in terminal, compact, glandular cymes, the glands mostly yellowish or whitish. Sepals oblong, glandular, apiculate. Petals pink to rose, with dark red or purple veins, pilose at the base of the inner surface. Stamens 10, the filaments pilose. Stylar column glandular, prominently beaked. Valves of the capsule glandular-pubescent.

Common in meadows and on mountain slopes at lower elevations throughout the Middle Rocky Mountains. *Geranium viscosissimum* occurs near timberline in all western ranges and has been collected from the lower alpine zone in the Absaroka, Salt River, and Wasatch Ranges. British Columbia to Alberta and Saskatchewan, south to South Dakota, Colorado, Utah, Nevada, and California.

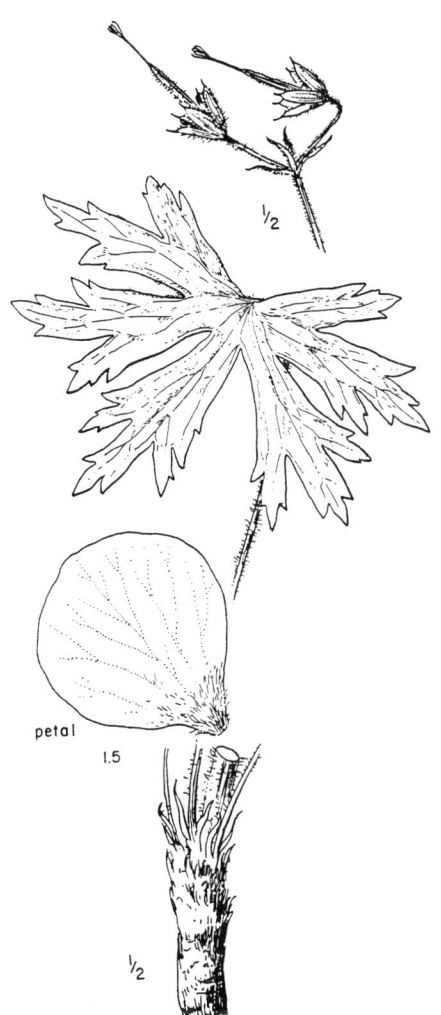

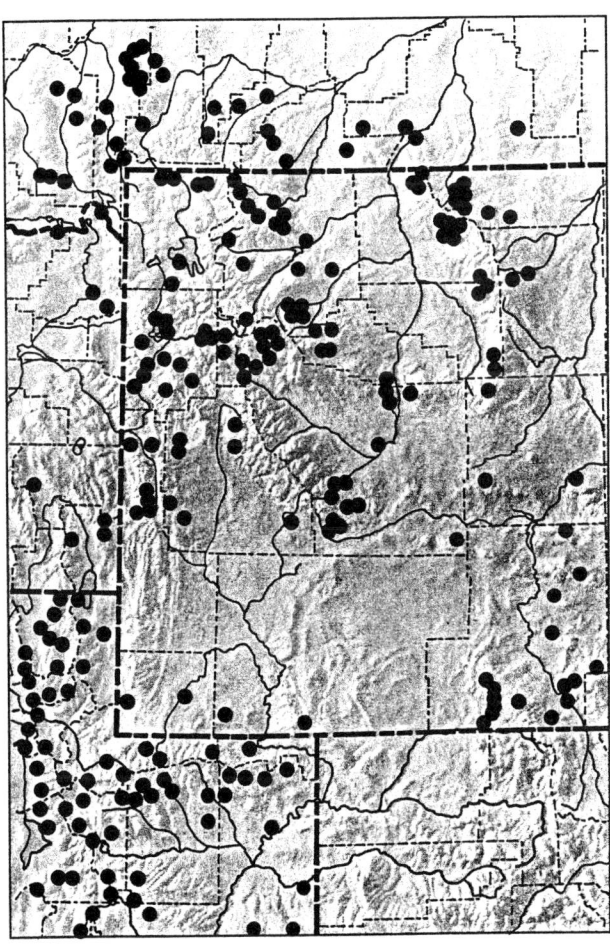

Grossulariaceae
Gooseberry Family

Shrubs with alternate, palmately veined, broadly ovate to reniform, often glandular-puberulent leaves, the stipules adnate to the petiole or lacking. Flowers epigynous, 5-merous, perfect, regular; the hypanthium rotate, tubular, or campanulate. Inflorescence racemose. Sepals 5, petaloid, spreading to sharply reflexed. Petals 5, usually erect, always smaller than the sepals. Stamens 5, alternate with the petals. Pistil 1, the ovary 1-celled with 2 parietal placentae; styles 2, connate or distinct, the stigmas capitate. Fruit a globose, many-seeded, glabrous to glandular-pubescent berry.

Ribes L.
Gooseberry, Currant

Shrubs with glabrous to glandular-pubescent branches, sometimes armed with prickles. Leaves alternate, palmately lobed, and veined; stipules adnate to the petioles or lacking. Flowers solitary or racemose. Calyx of 5 erect to spreading lobes; hypanthium rotate to tubular. Petals 5, erect to spreading, shorter than the sepals. Stamens 5, included to exerted, alternate with the petals. Pistil of 2 united carpels, the ovary inferior; styles 2, distinct or united. Fruit a red to blackish berry in ours.

KEY TO THE SPECIES OF *RIBES*

1. Stems unarmed with spines or prickles — *R. cereum*
1. Stems armed with nodal spines
   2. Hypanthium tubular; berries glabrous — *R. oxyacanthoides*
   2. Hypanthium rotate; berries glandular-pubescent
      3. Leaf blades glandular-pubescent; berries red — *R. montigenum*
      3. Leaf blades pubescent to glabrous, lacking glands; berries purplish black — *R. lacustre*

Ribes cereum Dougl.
Trans. Hort. Soc. Lond. 7:512, 1830.

Wax Currant

Cerophyllum inebrians (Lindl.) Spach
Ribes inebrians Lindl.
Ribes pumilum Nutt.
Ribes reniforme Nutt.
Ribes spathianum Koehne
Ribes viscidulum Bergen

Unarmed shrub with spreading branches, the younger ones glandular-pubescent; the older branches gray-brown to reddish brown, lacking pubescence. Leaves petiolate, the blades cordate to reniform, shallowly 3–5-lobed, the margins crenate to crenate-dentate, glabrous to glandular-pubescent on one or both surfaces. Inflorescence a 2–8-flowered, pendulous, glandular-pubescent raceme; bracts ovate to obovate in ours, longer than the jointed pedicels. Calyx and hypanthium pinkish, cylindrical, glandular-pubescent, the lobes ovate, spreading to reflexed. Petals white to pinkish, obovate, erect, less than half as long as the sepals. Anthers with a terminal cup-shaped gland, equaling or shorter than the filaments. Ovary partially fused to the hypanthium, glandular-pubescent; styles connate, glabrous to glandular-pubescent. Fruit a globose, red berry, glabrous to sparsely glandular.

Moist talus slopes, fellfields, and rocky stream banks in the lower alpine zone of the Hoback, Uinta, Wasatch, and Wind River Ranges. British Columbia south to California, Arizona, New Mexico, and Nebraska.

Middle Rocky Mountain plants are mostly var. *inebrians* (Lindl.) C.L. Hitchc., having entire, denticulate or 2–3-lobed, ovate to obovate bracts, and leaves and calyx glandular-pubescent. Typical var. *cereum* with 6-lobed bracts might also be expected near timberline.

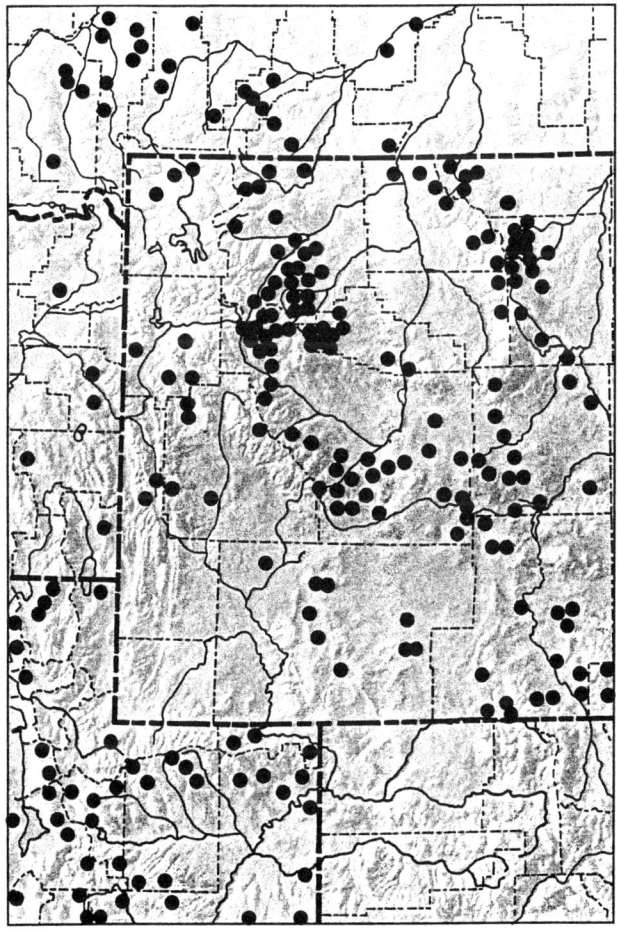

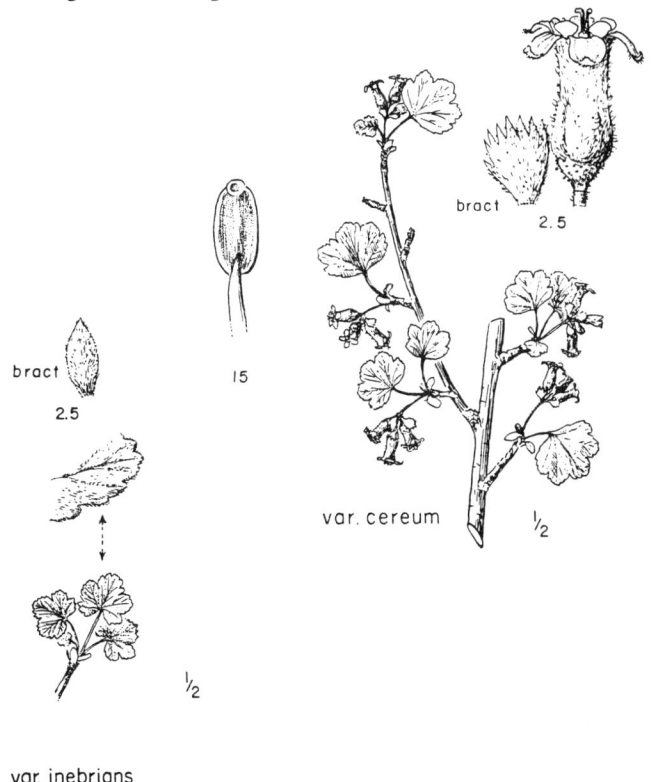

Ribes lacustre (Pers.) Poir. in Lam.
Encyc. Meth. Suppl. 2:856, 1812.

Swamp Gooseberry

Limnobotrya lacustris (Pers.) Rydb.
Limnobotrya parvula (Gray) Rydb.
Ribes echinatum Dougl.
Ribes grossularioides Michx. ex Steud.
Ribes parvulum (Gray) Rydb.

Armed shrub with spreading branches, the branches with nodal spines; internodes bristly with slender prickles; bark grayish. Leaves petiolate, the petioles stipitate-glandular, the blades cordate, 3–5-lobed or 3–5-parted, the margins crenate-dentate, glabrous on both surfaces or pubescent; glands lacking. Inflorescence a 5–15-flowered, spreading to pendulous, glandular-puberulent raceme; bracts ciliate-glandular. Hypanthium rotate; sepals ovate and yellowish, pinkish, or reddish. Petals pink, flabellate, about half as long as the sepals. Stamens subequal to the petals. Ovary reddish- to purplish-glandular. Fruit a globose, purplish black, glandular-bristly berry.

Stream banks and moist places around rocks near timberline in the Beartooth, Big Horn, and Wind River ranges. Alaska south to California, Utah, Colorado, and South Dakota, and east to Newfoundland.

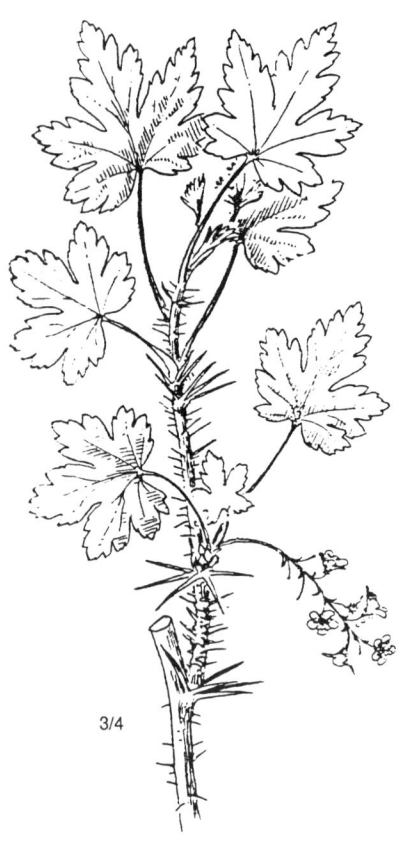

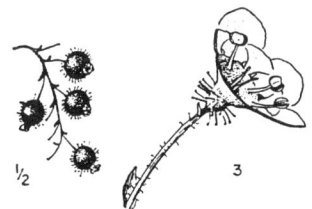

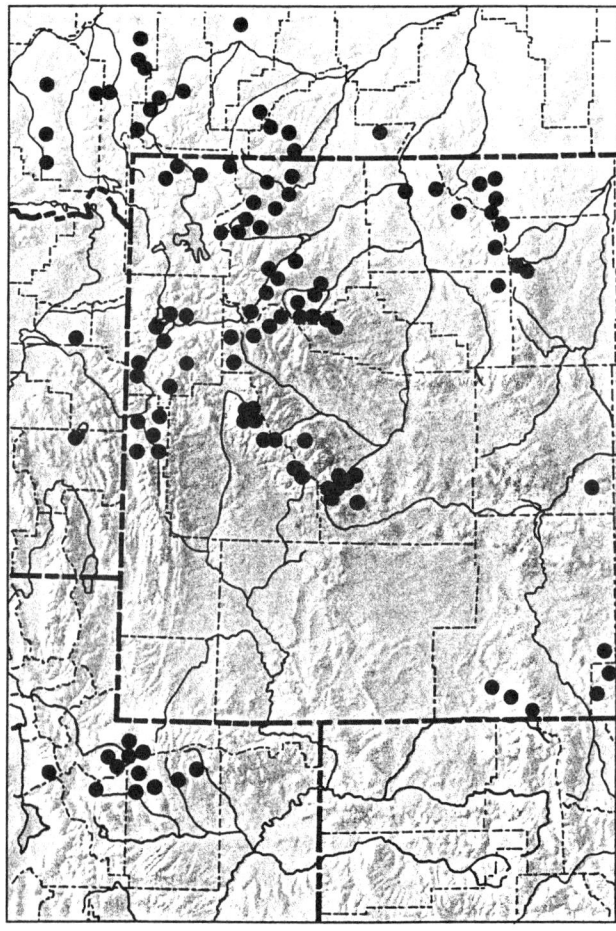

Ribes montigenum McClatchie
Erythea 5:38, 1897.

Mountain Gooseberry

Limnobotrya montigena (McClatchie) Rydb.
Ribes lentum (M.E. Jones) Cov. & Rose
Ribes molle T.J. Howell
Ribes nubigenum McClatchie

Shrub with spreading branches, the branches bristly on the internodes and with spreading or reflexed nodal spines; bark gray, smooth. Leaves petiolate, the blades cordate, 3–5-lobed or 3–5-cleft, the margins dentate or crenate-dentate, usually densely glandular-pubescent on both surfaces. Inflorescence a 4–8-flowered, pendulous, glandular-pubescent raceme; bracts shorter than the jointed pedicels. Hypanthium rotate; sepals obovate, yellowish or pink. Petals pink, flabellate, about half as long as the sepals. Stamens yellow, about as long as the petals. Ovary purplish-glandular. Fruit a globose, red-orange, glandular-bristly berry.

Common in rocky situations, fellfields, and on ledges near and above timberline in all alpine ranges of the Middle Rocky Mountains. British Columbia south to California and New Mexico.

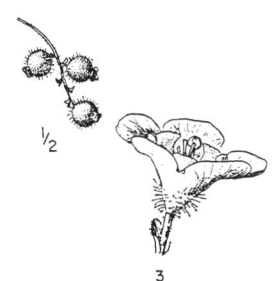

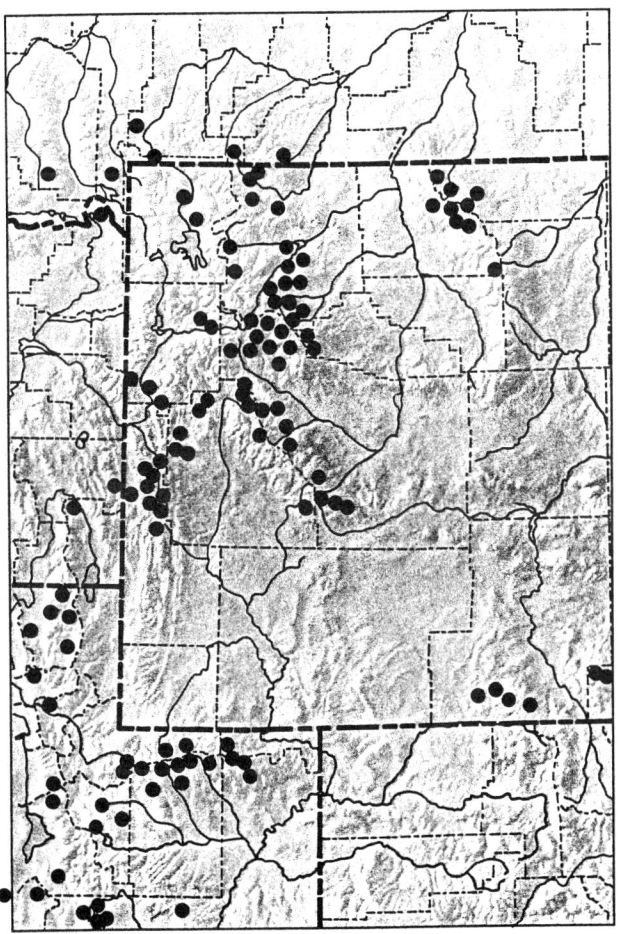

Ribes oxyacanthoides L.
Sp. Pl., 201, 1753.

Canada Gooseberry

Grossularia cognata Greene) Cov. & Britt.
Grossularia irrigua (Dougl.) Cov. & Britt.
Grossularia neglecta Berger
Grossularia nonscripta Berger
Grossularia oxyacanthoides (L.) Mill.
Grossularia setosa (Lindl.) Cov. & Britt.
Ribes camporum Blank.
Ribes cognatum Greene
Ribes hendersonii C.L. Hitchc.
Ribes irriguum Dougl.
Ribes leucoderme A. Heller
Ribes nonscripta (Berger) Standl.
Ribes saximontanum E. Nels.
Ribes saxosum Rydb.
Ribes setosum Lindl.

Shrub with ascending, bristly branches, puberulent when young; nodal spines commonly 1–3, spreading to reflexed, pale brown; bark of young branches yellowish, becoming gray with age. Leaves petiolate, the blades cordate to orbicular, mostly 5-lobed, the margins dentate to crenate-dentate, usually glandular-pubescent, at least below. Inflorescence a 1–3-flowered, axillary corymb or raceme; bracts about the same length as the nonjointed pedicels. Hypanthium glabrous, campanulate; sepals green to white, oblong, reflexed-spreading. Petals white to pink, oblanceolate, finely erose and rounded, erect. Stamens about as long as the petals; filaments glabrous, about twice as long as the anthers. Ovary inferior; styles connate for about half their length, pilose below the middle. Fruit a globose, glabrous, purplish blue berry.

Known in the alpine zone only from the Absaroka, Big Horn, and Wind River Ranges, although subalpine specimens approach timberline in the Beartooth Range. Transcontinental in Canada; south in the United States to Michigan, Nebraska, Wyoming, and Idaho.

Alpine plants with a hypanthium longer than the sepals are ssp. *setosum* (Lindl.) Sinnott.

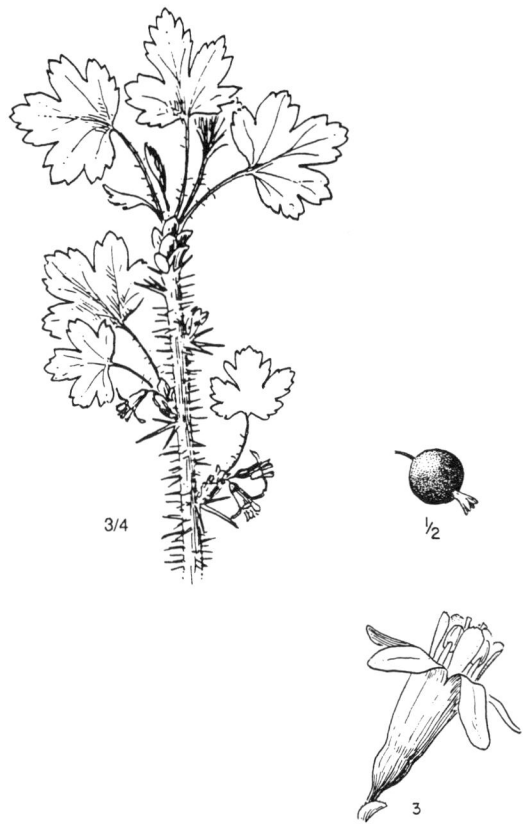

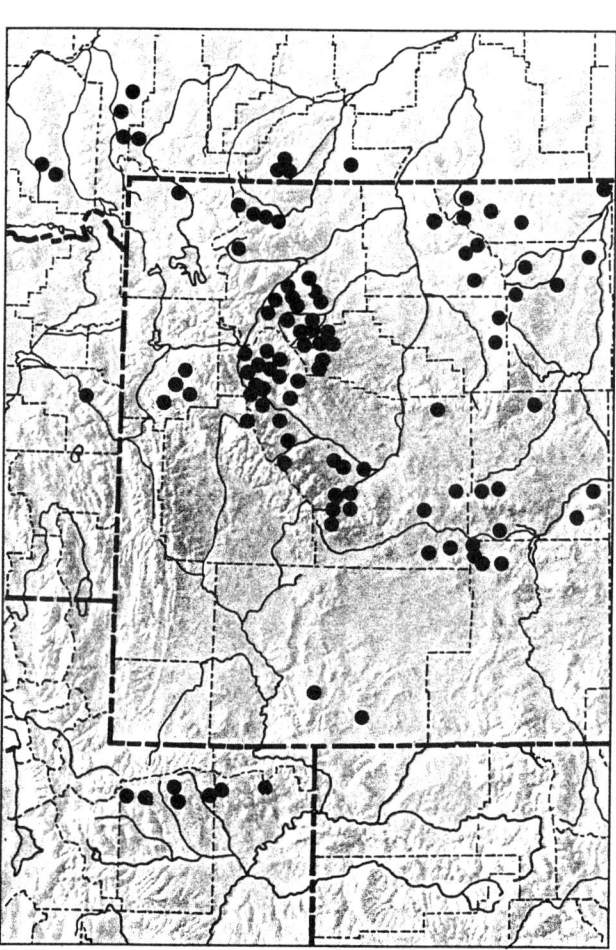

Hydrocharitaceae
Frogbit Family

Submersed, aquatic herbs. Leaves basal or cauline, the cauline leaves opposite or whorled, sessile, linear to oblong, the margins entire to serrulate. Flowers mostly imperfect, regular, in a 1- to many-flowered, sessile to pedunculate spathe. Sepals 3, green. Petals 3, white, or lacking. Stamens 2–9. Ovary 1, inferior, usually with 3 carpels; styles and stigma 3. Fruit many-seeded, indehiscent, capsulelike.

Elodea Michx.
Waterweed

Submerged, aquatic, perennial herbs with fibrous, nodal roots. Leaves cauline, sessile, linear to narrowly lanceolate, opposite or whorled, entire or serrulate. Flowers 1 to several in axillary, sessile to pedunculate, 2-lobed spathes; staminate flowers mostly solitary, sessile or with a slender, pedicel-like hypanthium, stamens 3–9, the filaments sometimes connate; pistillate flowers solitary with an elongate, slender, pedicel-like hypanthium, stigmas 3. Fruit cylindrical, indehiscent, with 1–5 seeds.

Elodea canadensis Michx.
Fl. Bor.-Am. 1:20, 1803.

Canada Waterweed

Anacharis canadensis (Michx.) Planch.
Anacharis planchonii (Casp.) Rydb.
Elodea brandegae St. John
Elodea ioensis Wylie
Elodea linearis (Rydb.) St. John
Elodea planchonii Casp.
Philotria canadensis (Michx.) Britt.
Philotria ioensis (Wylie) Wylie
Philotria linearis Rydb.

Slender-stemmed, submersed, often branched, perennial herb. Lower leaves opposite, the upper leaves in whorls of 3, crowded, linear to oblong, serrulate, obtuse. Staminate spathe pedunculate, with a single flower; sepals 3; petals 3, white; stamens 9, the inner 3 stamens elevated on a short tube. Pistillate spathe cylindrical, with a single flower and filiform hypanthium; sepals 3; petals 3, white; staminodia 3; stigmas 3, 2-cleft. Capsule 2–5-seeded.

Known only from lakes and ponds of the subalpine-alpine ecotone in the Snowy Range of the Medicine Bow Mountains. Transcontinental in Canada and the United States; south to Virginia, Alabama, Oklahoma, Nevada, and California.

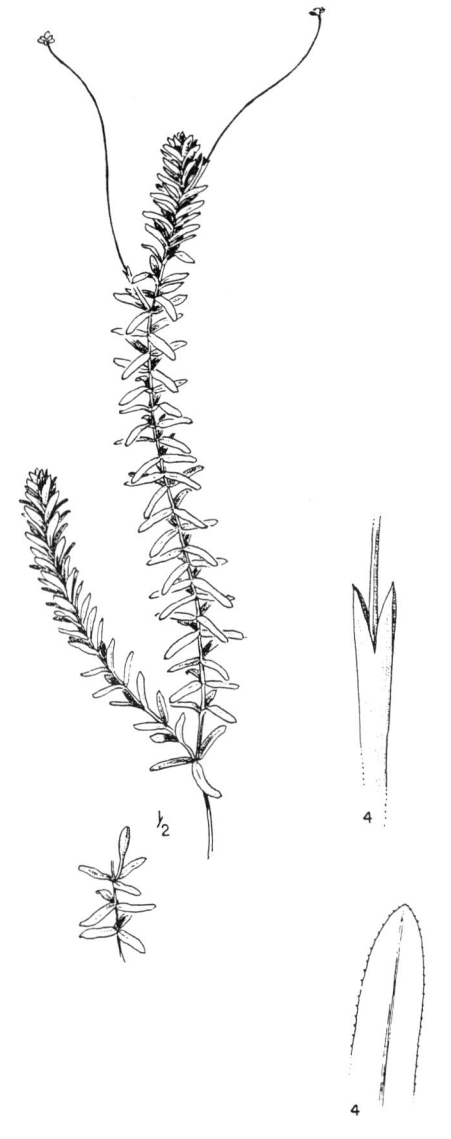

Hydrophyllaceae
Waterleaf Family

Annual or perennial herbs with usually alternate, simple or pinnate, exstipulate leaves. Flowers hypogynous, 5-merous, perfect, regular, in coiled or elongate cymes. Sepals 5, more or less united at the base. Petals 5, united, the lobes of the corolla often appendaged at the base. Stamens 5, inserted on the corolla tube, alternate with the lobes, usually well exserted. Pistil 1, of 2 united carpels, the ovary mostly 1-celled but sometimes 2-celled, the styles 2, distinct or partly united, the stigmas capitate. Fruit a loculicidal capsule, or sometimes berrylike.

Phacelia Juss.
Scorpionweed, Phacelia

Annual or perennial, mostly hirsute herbs from taproots, with alternate, entire to pinnatifid leaves. Flowers whitish to purple in ours, in scorpioid inflorescences. Calyx deeply cleft, the lobes equal or unequal. Corolla tubular to nearly rotate, usually with vertical folds or plaits in the tube. Stamens 5, usually long-exserted. Ovary 1-celled or nearly 2-celled by fusion of the placentae, the style 2-cleft. Capsule 2-valved, 2- to many-seeded.

KEY TO THE SPECIES OF *PHACELIA*

1. Leaves entire; stems suberect to prostrate; flowers white to lavender — *P. hastata*
1. Leaves pinnatifid; stems erect; flowers dark purple — *P. sericea*

Phacelia hastata Dougl. ex Lehm.
Stirp. Pug. 2:20, 1830.

Hastate Phacelia

Phacelia alpina Rydb.
Phacelia frigida Greene
Phacelia leucophylla Torr.
Phacelia magellanica (Lam.) Cov.
Phacelia nervosa Rydb.

Single- or multistemmed perennial from a stout taproot and branched caudex; stems erect to decumbent. Leaves strigose to hispid, the basal ones tufted, entire, lanceolate, petiolate; cauline leaves petiolate to sessile, entire to hastate, reduced upward. Flowers white to lavender, in short, dense, terminal cymes. Calyx hispid; corolla campanulate; stamens exerted, nearly twice the length of the corolla, the filaments villous near the middle. Ovules 4, the style exerted.

Talus and scree slopes in all alpine ranges of the Middle Rocky Mountains except the Medicine Bow Range. British Columbia and Alberta south to California and Colorado.

Alpine plants of the Middle Rocky Mountains with whitish, lavender or purplish flowers are var. *alpina* (Rydb.) Cronq.

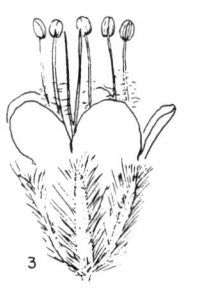

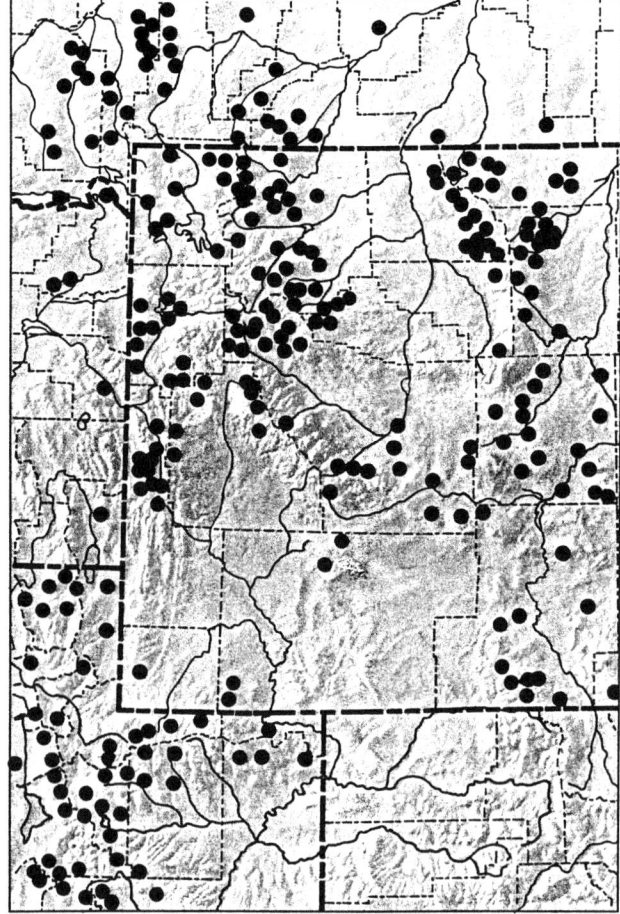

Phacelia sericea (Grah. ex Hook.) Gray
Am. J. Sci., II., 34:254, 1862.

Silky Phacelia

Eutoca sericea Grah. ex Hook.
Phacelia ciliosa (Rydb.) Rydb.

Multistemmed or occasionally single-stemmed perennial from a taproot and branched caudex covered with remnants of old leaf bases. Stems spreading, sericeous, usually with some spreading hairs. Basal leaves petiolate, pinnatifid, the lobes entire to cleft; cauline leaves short-petiolate, pinnatifid, reduced upward. Flowers dark purple, in a dense, terminal panicle. Calyx with linear lobes longer than the tube. Corolla campanulate, pubescent. Stamens exerted, more than twice as long as the corolla; filaments glabrous. Style cleft nearly to the middle.

Fellfields, felsenmeer, rock ledges, and Neoglacial moraines in most of the alpine ranges of the Middle Rocky Mountains. Not yet known from above timberline in the Big Horn and Wasatch Ranges. Alberta and British Columbia south to California, Nevada, and Colorado.

Short, sericeous, alpine plants less than 30 cm tall with weakly ciliate petioles are var. *sericea*.

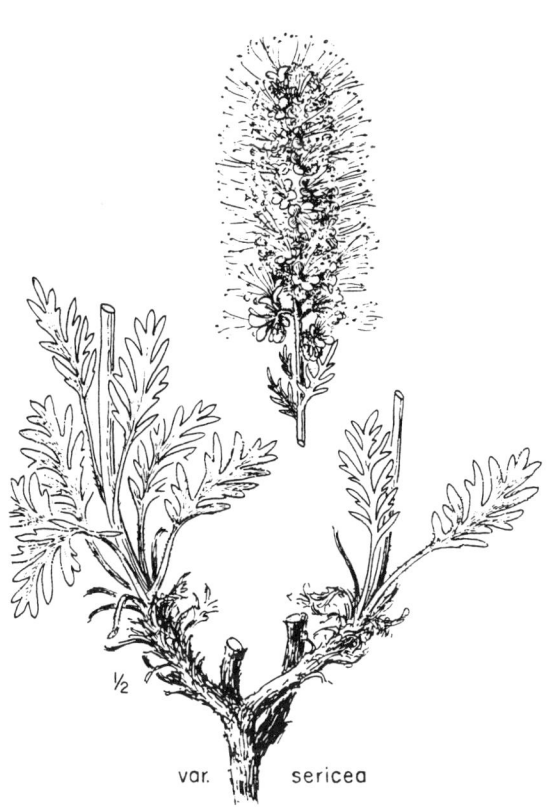

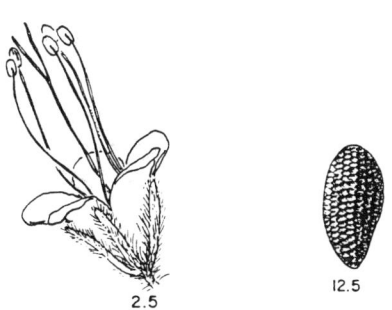

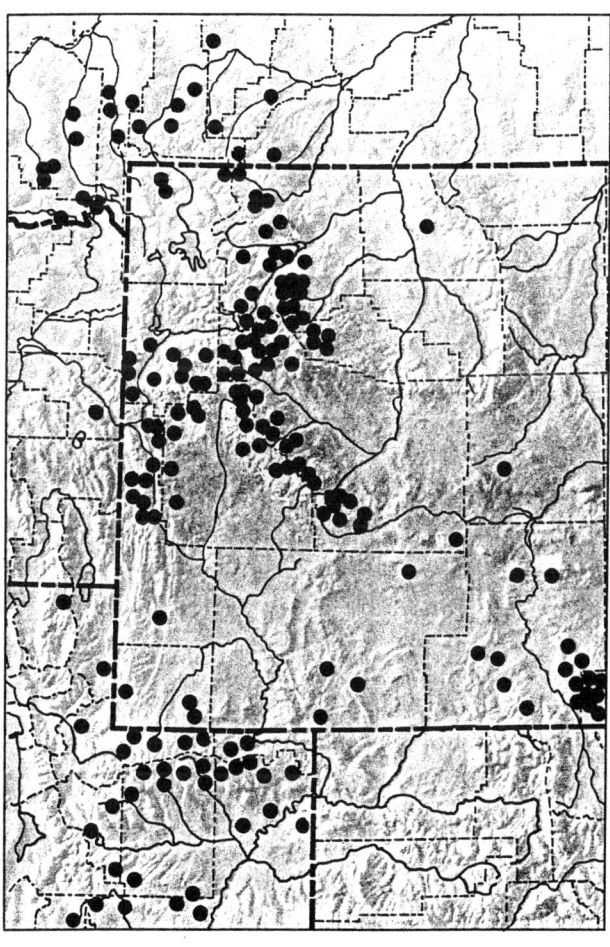

Hypericaceae
(Clusiaceae)
St. John's Wort Family

Annual or perennial, glabrous herbs or subshrubs with opposite, sessile, entire, exstipulate leaves, the blades punctate with translucent, black, or purplish glands. Flowers hypogynous, perfect, regular, solitary or in bracteate cymes. Sepals 4 or 5, free or partially connate. Petals 4 or 5, distinct, yellow in ours. Stamens few to numerous, free or connate basally into 3–5 fascicles. Pistil 1, the ovary with 3–5 carpels; styles 3–5. Fruit a capsule.

Hypericum L.
St. John's Wort

Our plants are glabrous herbs from horizontal rootstocks. Leaves opposite, simple, entire, exstipulate, glandular-punctate. Flowers perfect, cymose in ours. Sepals 5; petals 5, yellow. Stamens few to numerous, free or basally connate in groups. Pistil 1; styles 3–5, distinct or connate below. Fruit a septicidal capsule.

Hypericum scouleri Hook.
Fl. Bor.-Am 1:111, 1830.

Scouler St. John's Wort

Hypericum formosum H.B.K.
Hypericum nortonae M.E. Jones

Perennial herb from well-developed rhizomes and stolons; stems erect, simple to branched, at least above. Leaves opposite, sessile, ovate, oval, or obovate, obtuse, punctate with purplish black glands, at least on the margins. Flowers solitary in axils, or in few-flowered cymes. Sepals ovate, glandular-dotted, obtuse. Petals yellow, punctate on the margins. Stamens numerous, commonly in 3 groups. Styles slender. Capsule 3-lobed, with 3 locules.

Rocky slopes in the lower alpine zone of the Salt River and Teton Ranges. British Columbia to Alberta, south to California, Arizona, and Colorado.

Plants of the Salt River and Teton Ranges are ssp. *scouleri*.

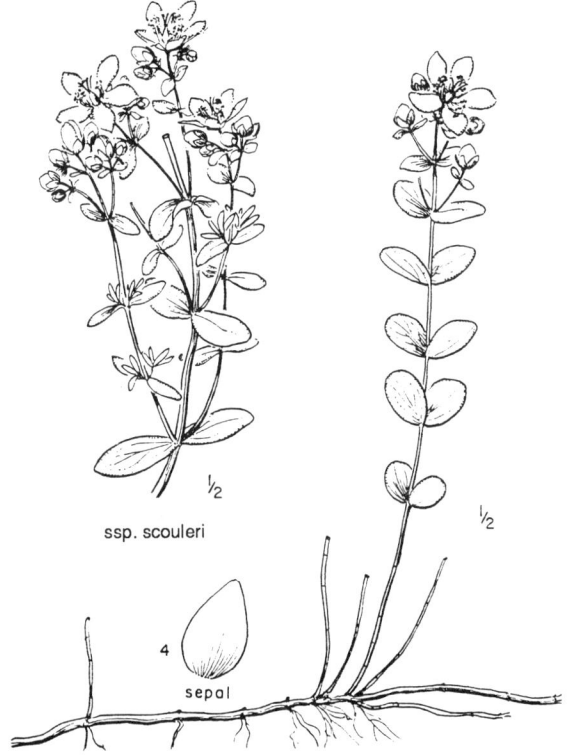

Isoetaceae
Quillwort Family

Perennial, aquatic or amphibious herbs from a short, cormlike stem; roots numerous, fibrous, branched; leaves crowded on the cormlike stem and linear, acicular, or subulate, with expanded bases. Sporangia contained in the leaf bases, the outer leaves bearing megasporangia, the inner ones with microsporangia; microspores minute, ovoid or oblong; megaspores larger, spherical, with various ridges and sculpturing.

Isoetes L.
Quillwort

Perennial, tufted, mostly aquatic herbs from short, thick, cormlike stems; roots numerous, fibrous. Leaves terete or subterete, linear, sheathing, and broadened at the base. Sporangia sessile, solitary, borne on the inner surfaces of the leaf bases, the microsporangia mostly on the inner leaves, megasporangia on the outer leaves.

Isoetes bolanderi Engelm.
Am. Nat. 8:214, 1874.

Bolander Quillwort

Calamaria bolanderi (Engelm.) Kuntze
Isoetes pygmaea Engelm.

Small perennial from a 2-lobed rootstock with acicular to subulate leaves, less than 10 cm long, erect to spreading, narrowly hyaline-winged near the broadened base; stomata few to numerous. Sporangium oval to orbicular, the velum incomplete, not covering the lower portion of the sporangium; megaspores white or blue-gray, with tubercles and short ridges; microspores finely spinulose.

Known only from shallow lake and pond shorelines at timberline in the northern Wind River Range. Subalpine records from the Medicine Bow Range are very near timberline. British Columbia to Montana, south to California, Arizona, and Colorado.

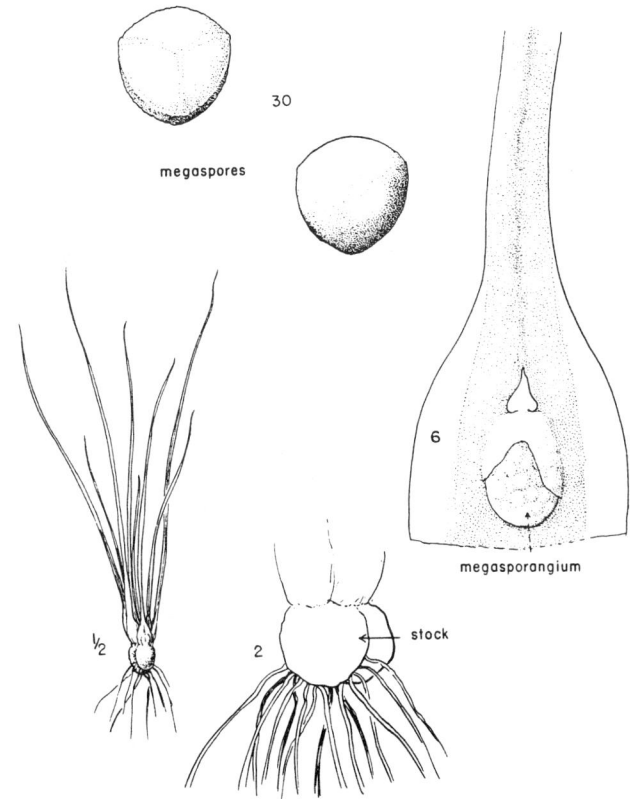

Juncaceae
Rush Family

Perennial or annual, tufted or, usually, rhizomatous herbs of moist situations. Leaves basal, less commonly cauline, sometimes reduced to sheaths; the blades, when present, flat or terete, linear or filiform. Flowers solitary or, usually, in panicles, corymbs, cymes, or heads, regular, hypogynous, perfect or unisexual. Perianth segments 6 and green, brown, or black, scalelike, in 2 series. Stamens 6 or 3, the pollen in tetrads. Pistil 1, of 3 united carpels, the ovary 3-celled or 1-celled, with 3 parietal placentae; stigmas 3. Fruit a loculicidal capsule with 3 to many seeds.

KEY TO THE GENERA OF JUNCACEAE

1. Plants glabrous; capsule many-seeded; leaf sheaths open — *Juncus*
1. Plants, at least the leaves and sheaths, pubescent; capsule 3-seeded; leaf sheaths closed — *Luzula*

Juncus L.
Rush

Perennial, less commonly annual, glabrous herbs, 2–8 dm tall. Leaves sometimes reduced to bladeless sheaths, the sheaths open; the blades, when present, more or less terete or channeled. Inflorescence a corymb, head, or panicle, appearing terminal or lateral. Flowers small, greenish or brownish, either single and basally subtended by 2 bractlets or, if 2 or more, without bractlets, but each in the axil of a bract. Perianth segments 6, scarious, equal. Stamens 6 or 3. Ovary 3-celled or 1-celled. Capsules many-seeded, the seeds minute.

KEY TO THE SPECIES OF *JUNCUS*

1. Inflorescence appearing lateral, the subtending bract erect, appearing as a short continuation of the stem; leaves mostly basal
   2. Uppermost leaf sheath bearing a well-developed blade 1.5 cm or more in length; capsules acute — *J. parryi*
   2. Uppermost leaf sheath bristle-tipped, the blade reduced to a mere rudiment 0.8 cm or less in length; capsules retuse — *J. drummondii*
1. Inflorescence obviously terminal, the bracts not strictly erect, more or less flat or channeled; leaves both basal and cauline
   3. Leaf blades laterally flattened so that one edge is toward the stem — *J. castaneus*
   3. Leaf blades terete and hollow, or deeply channeled or involute
      4. Flowers 5–7 mm long; leaf blades channeled or involute, often more than 3 mm wide; septae lacking; seeds with long appendages at both ends — *J. castaneus*
      4. Flowers 3–4.5 mm long; leaf blades terete or nearly so, mostly less than 2 mm wide; septae complete; seeds lacking appendages
         5. Plants less than 15 cm tall, caespitose, from fibrous roots; leaves basal
            6. Perianth segments dark purplish brown; capsules retuse; bract erect, longer than the inflorescence — *J. biglumis*
            6. Perianth segments pinkish to light chestnut; capsules mucronate; bract divergent, about the same length as the inflorescence — *J. triglumis*
         5. Plants more than 20 cm tall, rhizomatous or stoloniferous; leaves cauline
            7. Heads mostly 1 per stem; flowers 12 or more per head, the perianth dark purplish brown to nearly black; anthers much shorter than the filaments — *J. mertensianus*
            7. Heads mostly 5 to many per stem; flowers fewer than 12 per head, the perianth light to dark brown; anthers as long as or longer than the filaments — *J. nevadensis*

Juncus biglumis L.
Sp. Pl., 328, 1753.

Twoflower Rush

Low, tufted, perennial herb, less than 15 cm tall, from a very short rhizome; stems erect, channeled, with a single leaf. Leaves mostly basal; blades terete, septae complete. Inflorescence terminal, with 2 flowers; involucral bract single, erect, dark brown, lanceolate, longer than the inflorescence. Perianth segments oblong, obtuse, purplish brown to blackish. Stamens 6; anthers shorter than the filaments. Capsule obovate, retuse, longer than the perianth. Seeds appendaged at each end.

Wet patterned ground, nival basins, stream banks, and snowmelt drainages in the Absaroka, Beartooth, and Wind River Ranges. Circumpolar; Alaska south to Quebec, Colorado, and Wyoming.

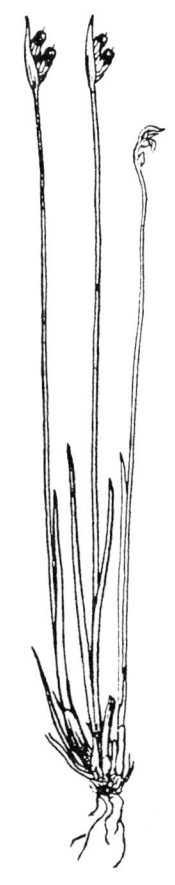

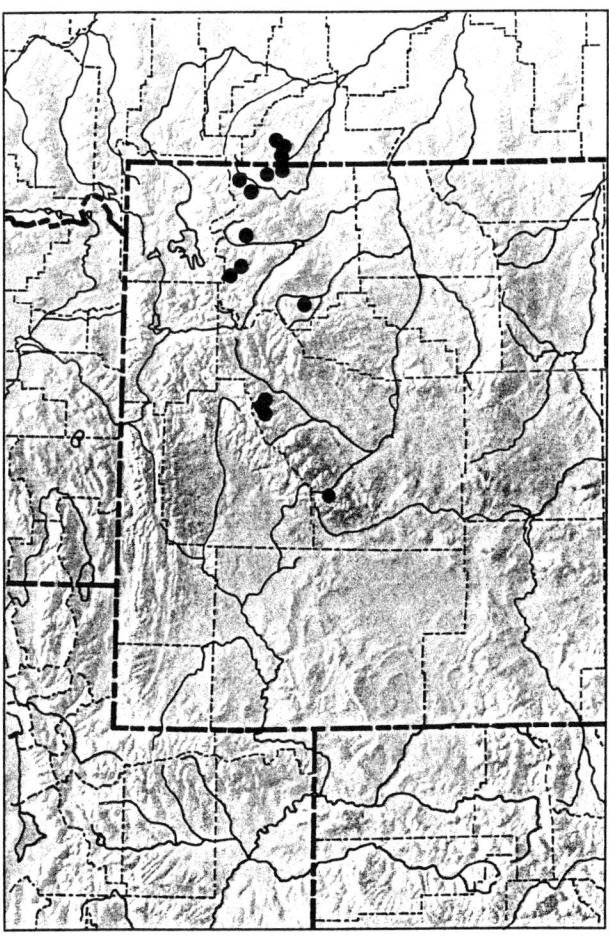

Juncus castaneus J.E. Smith
Fl. Brit. 1:383, 1800.

Chestnut Rush

Juncus leucochlamys Zing. ex Krecz.

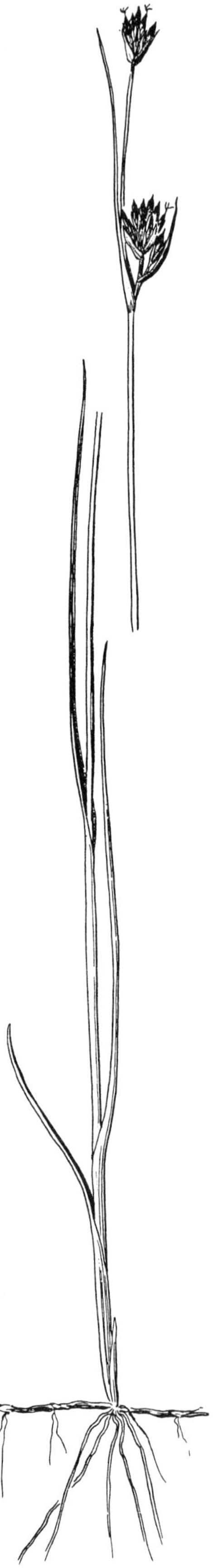

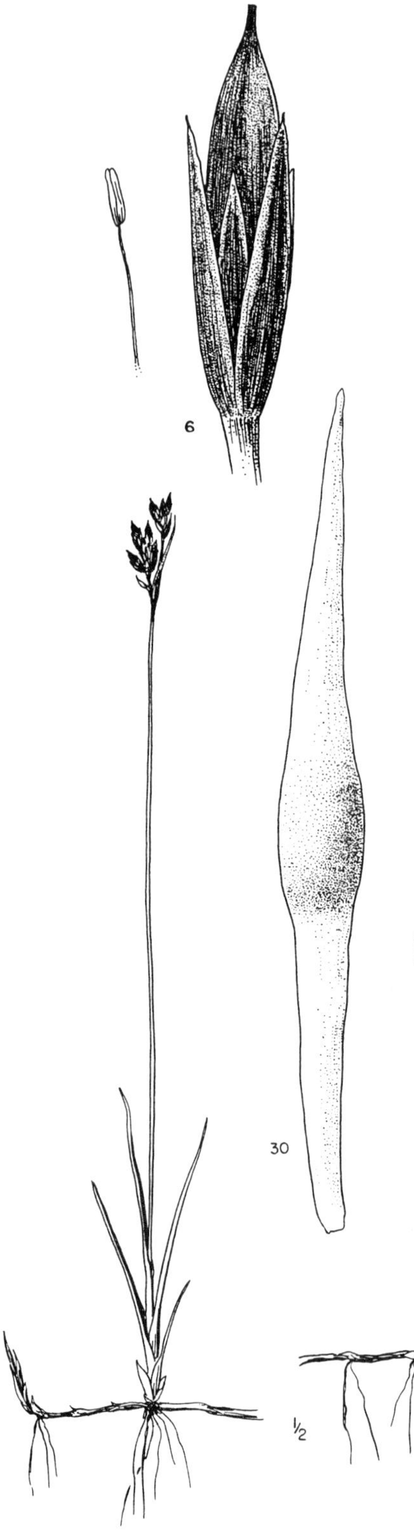

Erect, perennial herb from rhizomes or stolons; stems single, stiff, leafy. Leaves erect, mostly canaliculate, sometimes flattened, lacking septae. Inflorescence terminal, with 1–3 heads, the lower ones sessile or nearly so; flowers 4–9 per head; involucral bract single, erect, equaling or exceeding the inflorescence. Perianth chestnut brown, the outer segments acute and slightly longer than the obtuse inner segments. Stamens 6; anthers shorter than the filaments. Capsule chestnut brown to purplish, mucronate, longer than the perianth. Seeds appendaged at each end.

Wet meadows, bogs, and stream banks in the Absaroka, Beartooth, Medicine Bow, Uinta, and Wind River Ranges. Circumpolar; south in the Rocky Mountains to New Mexico.

Middle Rocky Mountain alpine plants are ssp. *castaneus*.

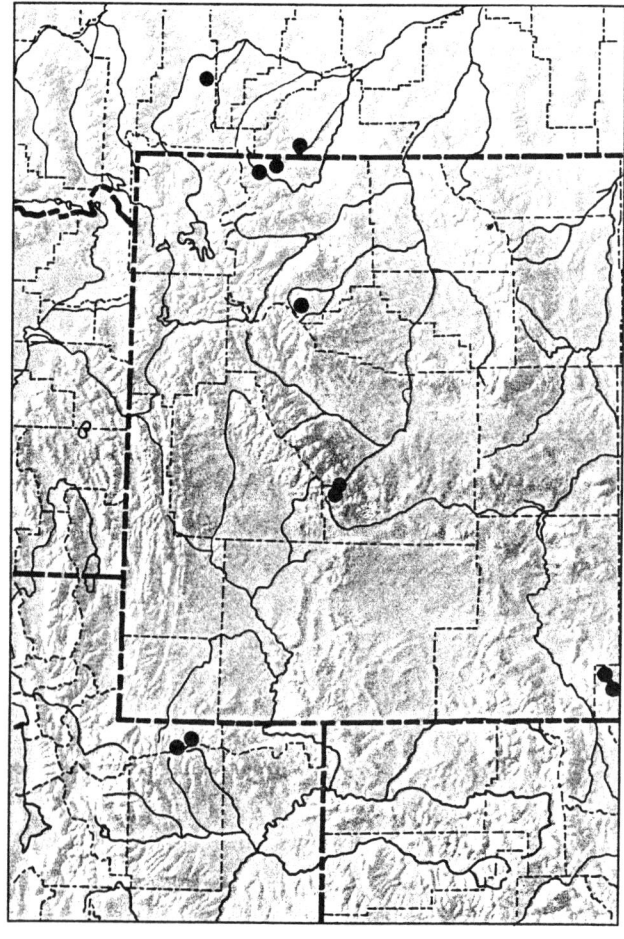

Juncus drummondii E. Mey. in Ledeb.
Fl. Ross. 4:235, 1853.

Drummond Rush

Juncus pauperculus Schwarz
Juncus subtriflorus (E. Mey.) Cov.

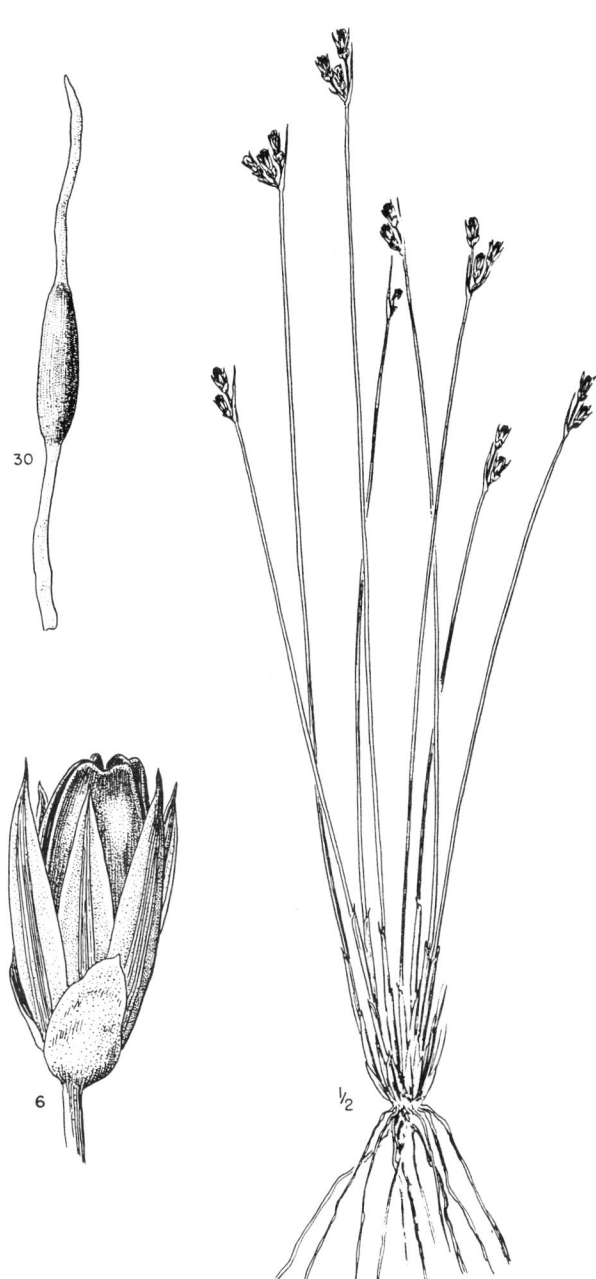

var. drummondii

Tufted, perennial herb from fibrous roots; stems terete, several to numerous. Leaves reduced to short, brownish, bristle-tipped sheaths; blades lacking. Inflorescence 1–5-flowered, appearing lateral; involucral bract terete, erect, equaling or slightly exceeding the inflorescence. Perianth segments brown, acute to acuminate, with a green to straw-colored midrib. Stamens 6; anthers equaling or shorter than the filaments. Capsule dark brown, oblong, retuse, equaling or exceeding the perianth. Seeds appendaged at each end.

Rocky slopes, talus, scree, and snow bed margins of most alpine ranges; not yet known from the Salt River and Wyoming Ranges. Alaska south to California, Arizona, and New Mexico.

Middle Rocky Mountain plants with capsules about the same length as the perianth are var. *drummondii*.

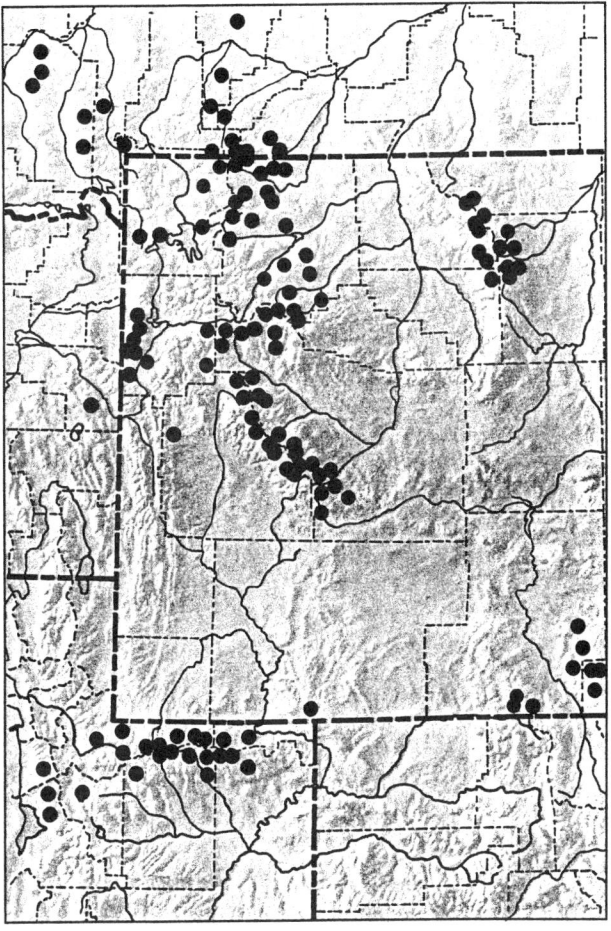

Juncus mertensianus Bong.
Mem. Acad. St. Petersb., VI, 2:167, 1832.

Subalpine Rush

Juncus slwookoorum S.B. Young

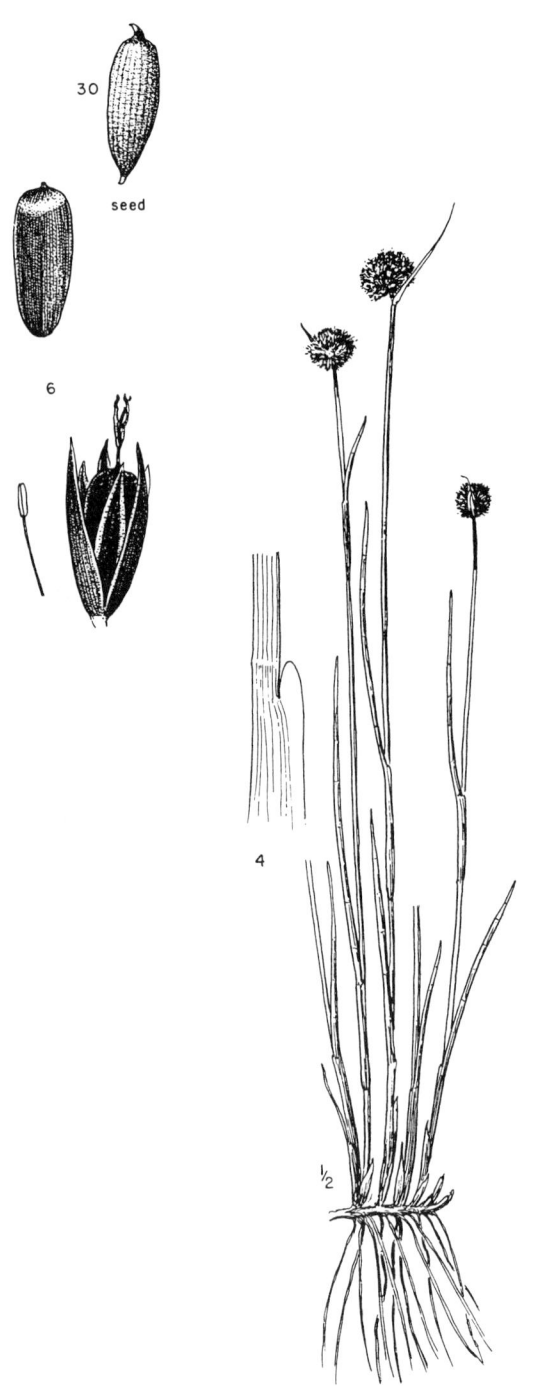

Densely tufted, perennial herb from brownish, scaly rhizomes; stems erect, leafy, somewhat flattened. Leaves both basal and cauline, the blades subterete; septae complete. Inflorescence a solitary, terminal, densely flowered head; involucral bract shorter than to exceeding the inflorescence. Perianth segments lanceolate, subulate, dark brownish black. Stamens 6; anthers much shorter than the filaments. Capsule oblong to obovate, retuse, about as long as the perianth. Seeds appendaged or not.

Wet meadows, stream banks, and lakeshores of most alpine ranges; not yet known from the Salt River and Wyoming Ranges. Alaska south to California, Arizona, and New Mexico.

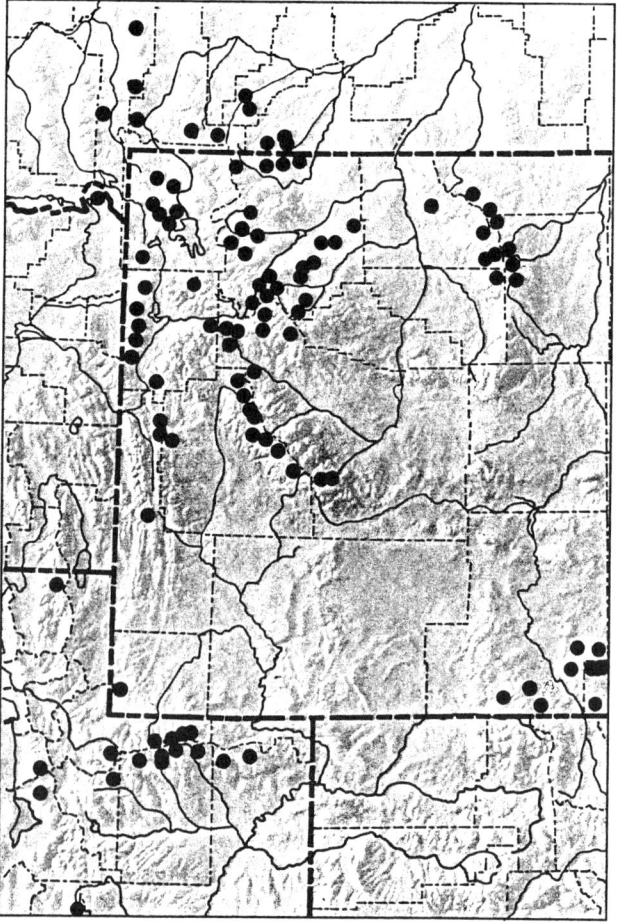

Juncus nevadensis S. Wats.
Proc. Am. Acad. 14:303, 1879.

Nevada Rush

Juncus badius Suksd.
Juncus columbianus Cov.
Juncus duranii Ewan
Juncus inventus Henders.
Juncus suksdorfii Rydb.
Juncus truncatus Rydb.

Erect, slender, perennial herb from short-creeping rhizomes; stems single or a few together, terete. Leaves mostly cauline, the sheaths auriculate and scarious-margined; blades terete, hollow, often becoming channeled above; septate, the septae complete. Inflorescence terminal, paniculate, of 4–15 few-flowered heads; involucral bracts 1–3, shorter than the inflorescence. Perianth segments lanceolate, light brown to dark brown, the outer acuminate and slightly longer than the acute inner whorl. Stamens 6; anthers much longer than the filaments. Capsule ovoid to oblong, rounded and abruptly contracted into a short beak, equaling or shorter than the perianth. Seeds apiculate at each end.

Pond margins at timberline in the Big Horn Range. British Columbia and Alberta south to California and New Mexico.

Middle Rocky Mountain plants in the vicinity of timberline with heads mostly 5 or fewer are var. *badius* (Suksd.) C.L. Hitchc.

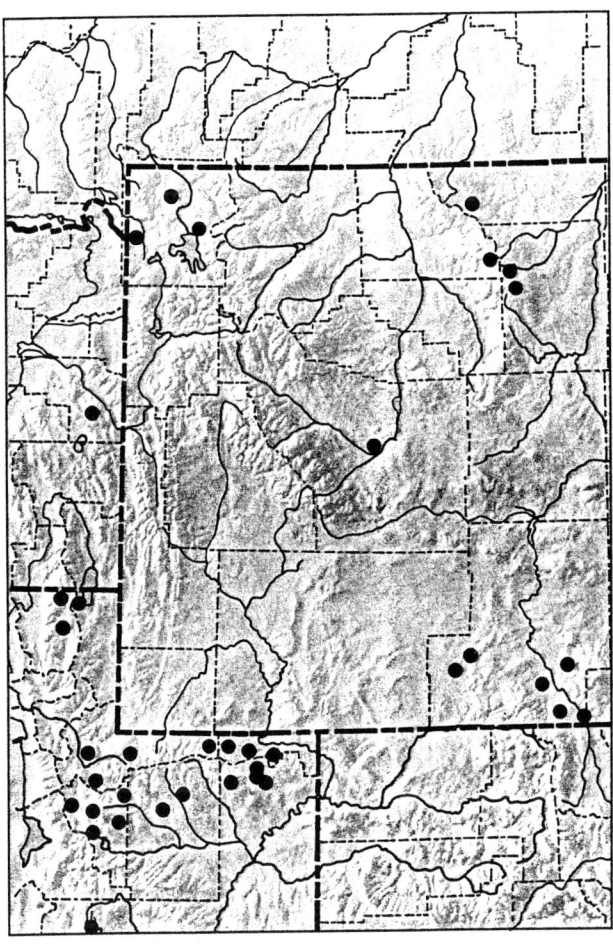

Juncus parryi Engelm.
Trans. Acad. Sci. St. Louis 2:446, 1866.

Parry Rush

Densely tufted, perennial herb from fibrous roots; stems several to numerous, erect, terete. Basal and lower cauline leaves represented by a brownish, bristle-tipped sheath; upper cauline leaves with an elongate, slender, terete blade; septae lacking. Inflorescence appearing lateral, 1–3-flowered; involucral bract terete, erect, exceeding the inflorescence. Perianth segments lanceolate, the outer ones acuminate and slightly longer than the acute inner whorl, scarious-margined. Stamens 6; anthers longer than the filaments. Capsule oblong, acute, longer than the perianth. Seeds appendaged at each end.

Rocky slopes, talus, scree, and snow bed margins in all alpine ranges. British Columbia and Alberta south to Idaho and Colorado.

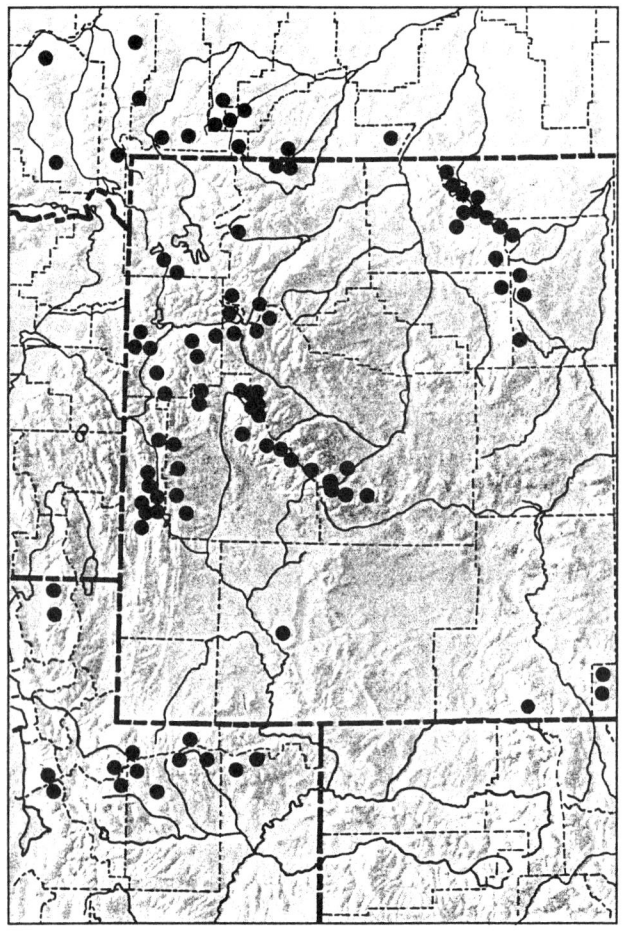

Juncus triglumis L.
Sp. Pl., 328, 1753.

Threeflower Rush

Juncus albescens (Lange) Fern.
Juncus schischkini Kryl. & Serg.

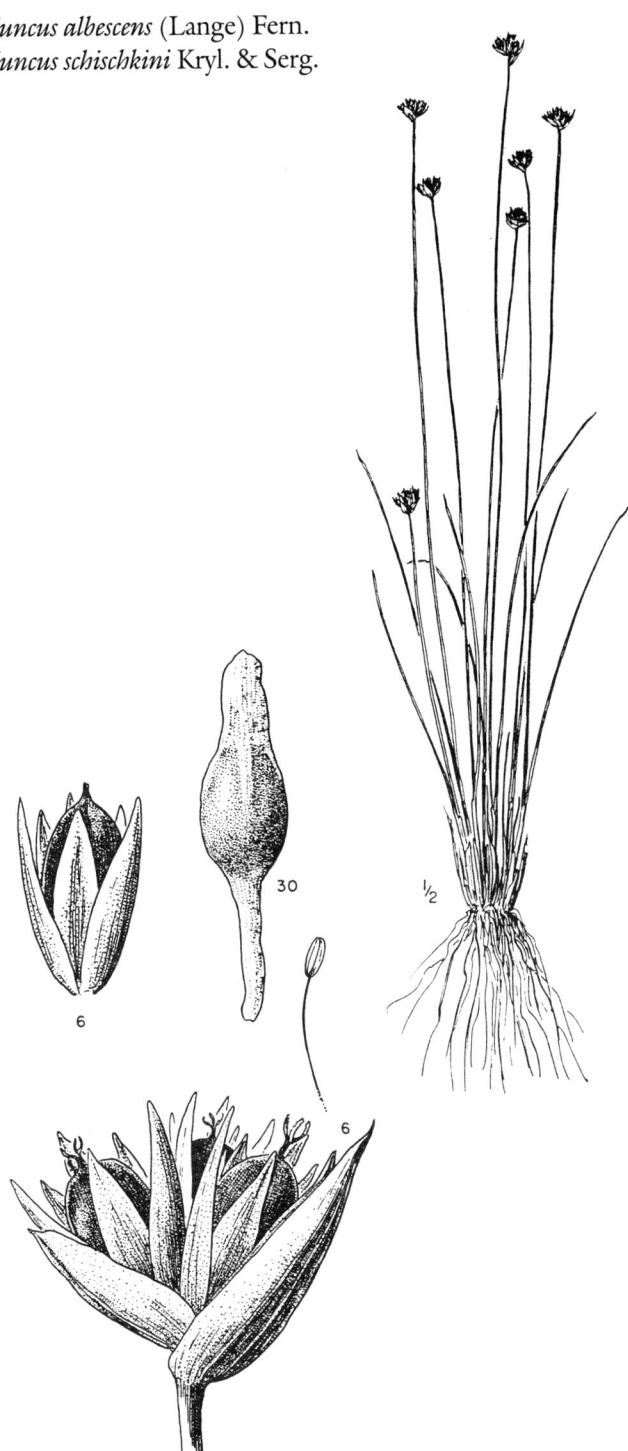

Low, tufted, perennial herb, less than 15 cm tall, from fibrous roots; stems terete, slender, stiff. Leaves basal or nearly so, the sheaths brownish, auriculate; blades terete, filiform; septae complete. Inflorescence terminal, the head 1–5-flowered; involucral bracts 1–3, equaling or shorter than the inflorescence. Perianth segments ovate-lanceolate, acute, reddish brown, becoming dark at maturity. Stamens 6; anthers shorter than the filaments. Capsule oblong, obtuse to mucronate, reddish to blackish brown, longer than the perianth. Seeds appendaged at each end.

Wet meadows and stream banks of the Beartooth, Medicine Bow, Uinta, and Wind River Ranges. Circumpolar; south in western North America to Utah and Colorado.

Rocky Mountain plants are var. *albescens* Lange.

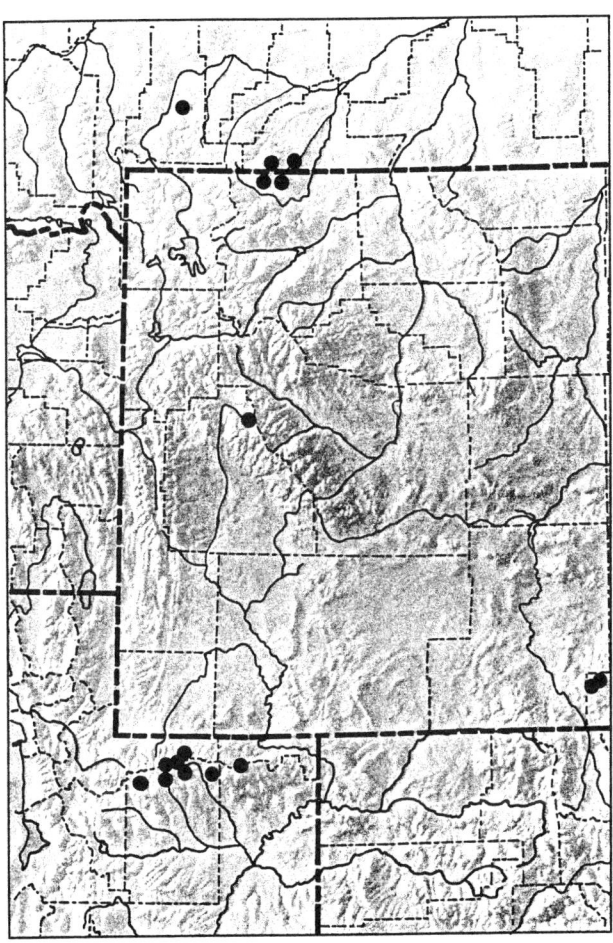

Luzula DC.
Woodrush

Low, tufted, perennial herbs with simple, erect culms, the leaves mostly basal, with closed sheaths, the blades flat or channeled, usually hairy on the margins. Inflorescence paniculate or corymbose, congested or open. Flowers bracteolate, the bracteoles usually toothed or lobed. Perianth segments 6. Stamens 6. Ovary 1-celled, with 3 often short-carunculate seeds.

KEY TO THE SPECIES OF *LUZULA*

1. Flowers crowded, subsessile, in spikelike glomerules — *L. spicata*
1. Flowers single, on slender pedicels, in a loose panicle
    2. Anthers much longer than the filaments; style about 1 mm long, forming a conspicuous beak in fruit; perianth mostly more than 3 mm long — *L. glabrata*
    2. Anthers shorter than the filaments; style 0.5 mm or less, not forming a beak; perianth mostly less than 2.5 mm long
        3. Cauline leaves 3 or more, the blades 4–10 mm wide; bracts of the inflorescence erose to entire — *L. parviflora*
        3. Cauline leaves 1–3, the blades 2–3 mm wide; bracts of the inflorescence lacerate and fimbriate — *L. wahlenbergii*

Luzula glabrata (Hoppe ex Rostk.) Desv.
J. Bot. 1:143, 1808.

Smooth Woodrush

Juncus glabratus Hoppe ex Rostk.
Juncodes glabratum (Hoppe ex Rostk.) Sheld.
Juncoides glabratum (Hoppe ex Rostk.) Piper
Juncoides majus Piper
Luzula hitchcockii Hämet-Ahti
Luzula piperi Hämet-Ahti

Erect, somewhat tufted, perennial herb with short rhizomes; stems with 2–3 leaves. Basal leaves linear-lanceolate, flat; cauline leaves lanceolate, acute, glabrous, mostly pilose at the base, reduced upward. Inflorescence an open, bracteate panicle with slender, nodding to erect pedicels; bracts and bracteoles lacerate-fimbriate; flowers single. Perianth segments lanceolate to ovate-lanceolate, acute, dark purplish brown to chestnut brown, with narrow, scarious margins. Anthers much longer than the filaments. Capsule broadly ovate with a short, prominent beak, purplish brown, equaling or exceeding the perianth. Seeds dark brown, lacking appendages at each end.

Krummholz margins and timberline meadows in the Beartooth Range. British Columbia and Alberta south to Oregon, Idaho, and northwestern Wyoming.

Middle Rocky Mountain plants are var. *hitchcockii* (Hämet-Ahti) Dorn.

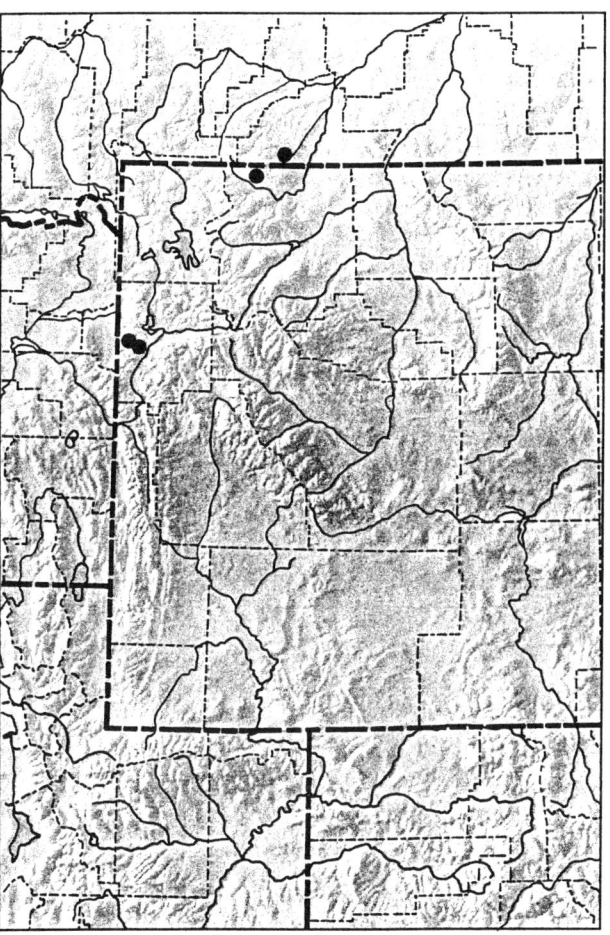

Luzula parviflora (Ehrh.) Desv.
J. Bot. 1:144, 1808.

Millet Woodrush

Juncus parviflorus Ehrh.
Juncodes parviflorum (Ehrh.) Sheld.
Juncoides parviflorum (Ehrh.) Cov.
Juncus melanocarpus Michx.
Luzula fastigiata E. Mey.
Luzula labradorica Steud.
Luzula melanocarpa (Michx.) Tolm.

Erect, somewhat tufted, perennial herb with short rhizomes; stems with 3–5 leaves, often decumbent at the base. Leaves linear-lanceolate, flat, glabrous, somewhat pilose near the sheath, the tips acuminate. Inflorescence an open, bracteate panicle with long, slender, nodding pedicels; bracts and bracteoles erose to entire; flowers single or sometimes 2 together. Perianth segments lanceolate, acute to acuminate, greenish to chestnut brown, scarious-margined. Anthers mostly shorter than the filaments. Capsule ovate, acute, greenish to dark brown, equaling or exceeding the perianth. Seeds brown, lacking appendages at each end.

Meadows, stream banks, and willow thickets of the Absaroka, Beartooth, Medicine Bow, Uinta, Wasatch, and Wind River Ranges. Circumpolar; south in western North America to California and New Mexico.

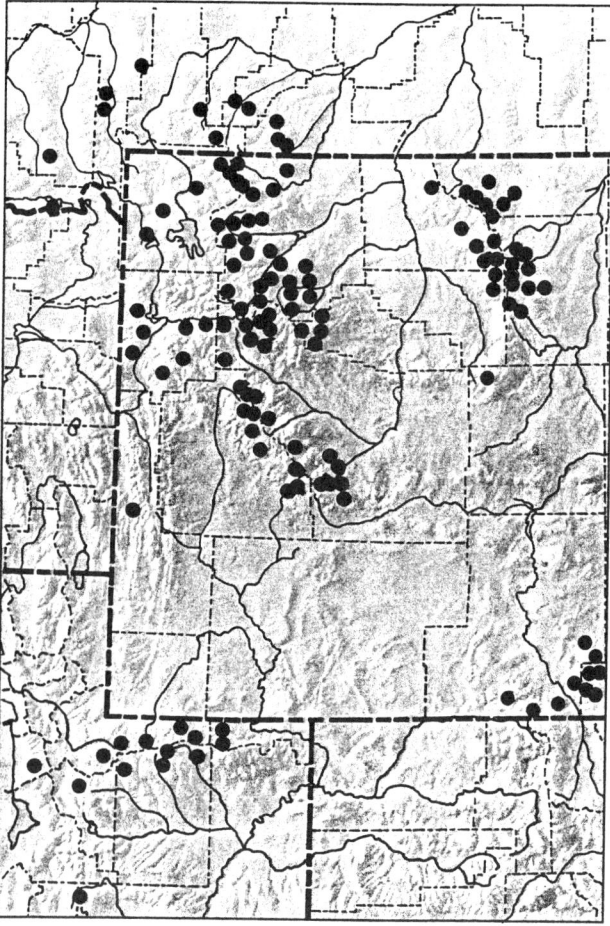

Luzula spicata (L.) DC. in Lam. & DC. *Fl. Franc.* 3:161, 1805.

Spike Woodrush

Juncus spicatus L.
Gymnodes spicata (L.) Fourr.
Juncodes spicatum (L.) Kuntze
Juncoides spicatum (L.) Cov.
Luciola spicata (L.) J.E. Smith
Luzula cusickii Gand.
Luzula orestera Sharsmith

Tufted, perennial herb from fibrous roots; stems simple, erect, slender. Leaves linear-lanceolate, flat, channeled or involute, acute, pilose at the base. Inflorescence a dense, spikelike panicle, nodding; involucral bract often involute, mostly shorter than the inflorescence; bracteoles fimbriate. Perianth segments lanceolate, acuminate, dark brown, the margins scarious. Anthers shorter than the filaments. Capsule ovate, acute, brownish black, equaling or shorter than the perianth. Seeds with a very short, rounded appendage at each end.

Fellfields, meadows, and alp slopes in nearly all alpine ranges. Not yet known from above timberline in the Hoback Range. Circumpolar; south in western North America to California, Arizona, and Colorado.

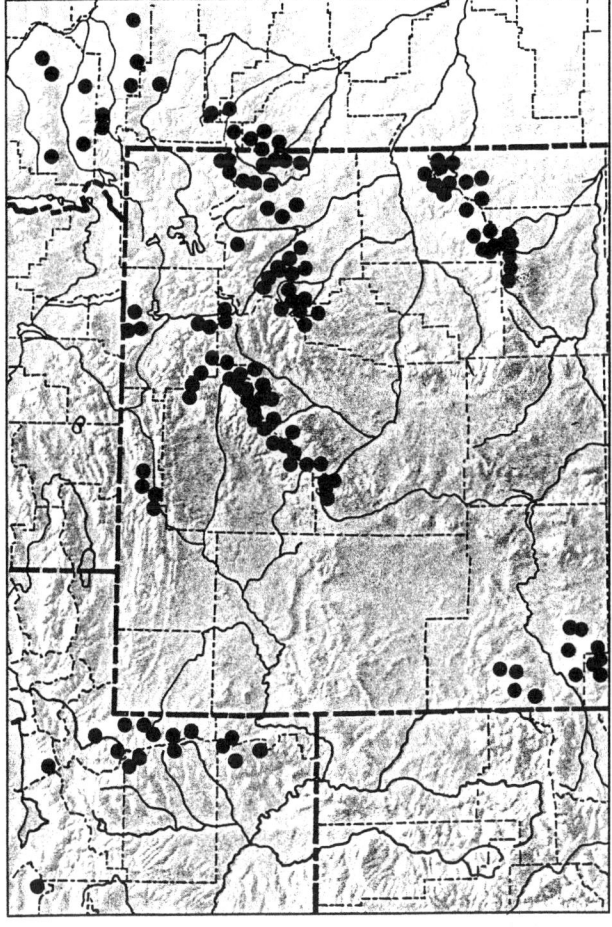

Luzula wahlenbergii Rupr.
Beitr. Pfl. Russ. Reich. 2:58, 1845.

Wahlenberg Woodrush

Juncoides piperi Cov.
Luzula piperi (Cov.) M.E. Jones

Erect, tufted, perennial herb from short rhizomes; stems with 2–3 leaves. Basal leaves linear-lanceolate, flat, glabrous, mostly pilose at the base and on the sheaths; cauline leaves reduced upward. Inflorescence a few-flowered, open, bracteate panicle, the pedicels slender, nodding; bracts and bracteoles lacerate-fimbriate; flowers single. Perianth segments ovate-lanceolate, acuminate or sometimes lacerate at the tip, dark brown or purplish brown, the margins somewhat scarious. Anthers equal to or shorter than the filaments. Capsule ovate, blackish brown, longer than the perianth. Seeds brown to yellow-brown, mucro-umbonate, but not appendaged at each end.

Wet meadows and bogs in the Beartooth and Teton Ranges. Circumpolar; south in western North America to Washington, Idaho, and Wyoming.

This taxon is very closely related to *L. parviflora* and is often difficult to separate from it.

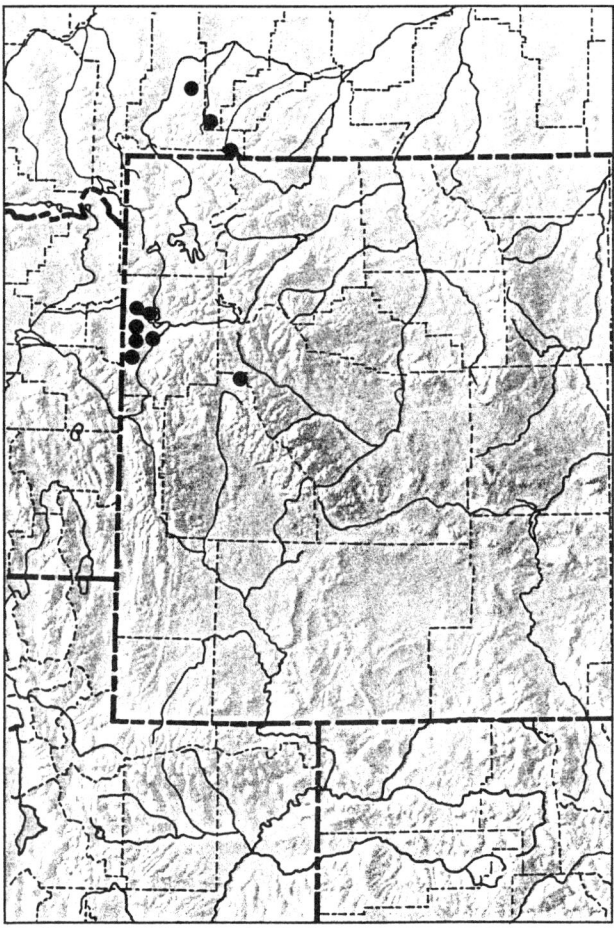

Lamiaceae
(Labiatae)
Mint Family

Annual to perennial herbs, sometimes woody, often aromatic, with mostly square stems and simple, opposite, exstipulate, entire to pinnatifid leaves. Flowers perfect, hypogynous, irregular, in small, few-flowered to dense axillary cymes, or terminal in spicate or racemose, verticillate clusters. Calyx regular or irregular, with 5 or 10 lobes, sometimes bilabiate. Corolla irregular, mostly bilabiate. Stamens 4 and didynamous, or 2 and epipetalous; staminodia often present. Pistil 1, of 2 carpels, the ovary mostly 4-cleft and 4-loculate; style basal between the 4 lobes, or sometimes terminal; stigma bifid, the branches unequal. Fruit a 1-seeded nutlet, usually in groups of 4 surrounded by the persistent calyx.

Monardella Benth.
Horsemint

Annual to perennial herbs with simple stems, sometimes branched near the base, forming dense clumps. Leaves opposite, simple, entire to serrate, sessile or petiolate. Flowers perfect, in a dense, terminal, verticillate head. Calyx tubular, 5-lobed, the lobes erect, linear to oblong, subequal. Corolla more or less bilabiate, the 5 lobes nearly subequal, oblong to linear. Stamens 4, exerted, didynamous; anther sacs divergent. Nutlets smooth, oval to oblong.

Monardella odoratissima Benth.
Lab. Gen. & Sp., 332, 1834.

Cloverhead Horsemint

Madronella discolor (Greene) Greene
Madronella glauca (Greene) Greene
Madronella nervosa (Greene) Greene
Madronella oblongifolia Rydb.
Madronella odoratissima (Benth.) Greene
Madronella parvifolia (Greene) Rydb.
Madronella sessilifolia Rydb.
Monardella discolor Greene
Monardella elegantula Gand.
Monardella glauca Greene
Monardella nervosa Greene
Monardella parvifolia Greene
Monardella purpurea T.J. Howell

Densely caespitose, strongly aromatic, perennial herb from a stout taproot and branched caudex; stems simple, erect or ascending, sometimes woody at the base. Leaves lanceolate, sometimes oblong to elliptic, entire, sessile to short-petiolate; upper leaves bracteate, forming a distinct purplish, ciliate-margined involucre below the inflorescence. Flowers in dense terminal heads. Calyx tubular to cylindrical, bilabiate or nearly so, nerved, the 5 lobes subequal. Corolla bilabiate, deep purple to lavender, the 5 lobes slender, oblong to linear, subequal, the tube exceeding the calyx. Stamens 4, didynamous, slightly exerted, the anther sacs parallel to divergent; filaments puberulent, at least below. Nutlets elliptic, brownish, glabrous.

Scree and talus slopes, particularly on limestone, at and above timberline in the Wasatch Range. Washington south and east to California, New Mexico, and Colorado.

Middle Rocky Mountain plants with elliptic or oblong floral bracts puberulent on the outside are var. *glauca* (Greene) St. John.

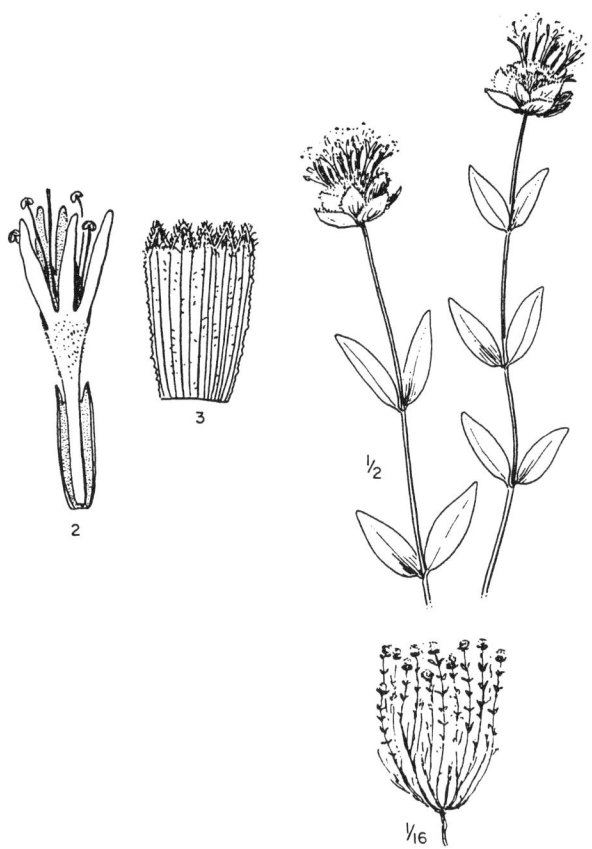
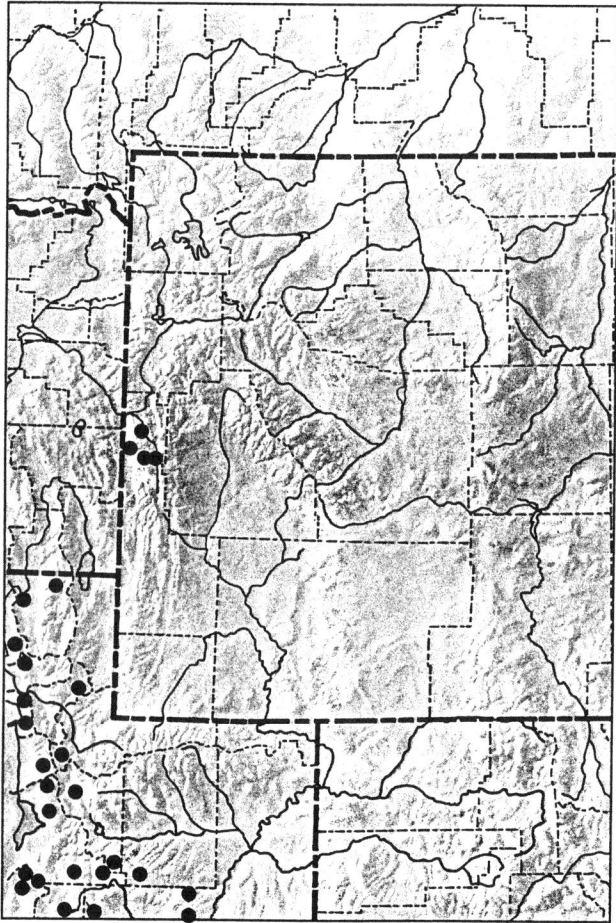

Liliaceae
Lily Family

Perennial, scapose or caulescent herbs, sometimes woody, from bulbs, corms, or rhizomes. Flowers usually in racemes, or sometimes solitary, regular, hypogynous, usually perfect, the 6 perianth segments in 2 whorls, usually petaloid. Stamens 6. Pistil 1, of 3 united carpels; stigmas 3, the ovary 3-celled, with axillary placentation. Fruit a berry or a loculicidal capsule.

KEY TO THE GENERA OF LILIACEAE

1. Perianth parts dissimilar, the sepals green and much narrower than the gland-bearing petals — *Calochortus*
1. Perianth parts similar, the sepals and petals lacking glands
    2. Flowers 1 or 2, white or yellow
        3. Perianth yellow, reflexed — *Erythronium*
        3. Perianth white with purple or green veins, not reflexed — *Lloydia*
    2. Flowers several to many, pink, purple, greenish, or white (if white, veins lacking)
        4. Inflorescence a raceme or panicle; styles 3
            5. Plants rhizomatous; leaves 2-ranked — *Tofieldia*
            5. Plants from bulbs; leaves not 2-ranked — *Zigadenus*
        4. Inflorescence an umbel; style 1 — *Allium*

Allium L.
Onion

Scapose or subscapose, perennial herbs with a strong odor and taste, from a tunicated, sometimes rhizomatous bulb. Leaves linear, flat or terete, sometimes hollow. Flowers in a terminal umbel subtended by 1–3 scarious bracts. Perianth segments 6 and white, pink, or purplish, distinct, similar, persistent. Stamens 6, often adnate to the base of the perianth, the filaments often flattened, the anthers short, introrse. Style 1, slender; the stigma entire or 3-lobed. Fruit an ovoid to obovoid, loculicidal capsule with 1–2 black seeds in each locule.

KEY TO THE SPECIES OF ALLIUM

1. Leaves hollow, terete; inflorescence subglobose — *A. schoenoprasum*
1. Leaves solid, flat or channeled; inflorescence umbellate
    2. Bulbs ovoid; rhizomes lacking; leaves much longer than the inflorescence — *A. brandegei*
    2. Bulbs elongate; rhizomes present or not, short; leaves much shorter than the inflorescence
        3. Inflorescence erect; perianth segments lanceolate, acuminate; stamens about half the length of the perianth; bulbs from a distinct rhizome — *A. brevistylum*
        3. Inflorescence nodding; perianth segments elliptic-ovate, obtuse; stamens exerted; bulbs from a short rhizome, or rhizomes lacking — *A. cernuum*

Allium brandegei S. Wats.
Proc. Am. Acad. 17:380, 1882.

Brandegee Onion

Allium diehlii (M.E. Jones) M.E. Jones
Allium minimum M.E. Jones

Perennial herb with bulb ovoid to globose, usually solitary; outer coat membranaceous, reticulate, the inner coat membranaceous, smooth, shiny, purple; rhizomes lacking. Leaves 2, linear, channeled, convex or nearly flat, much longer than the scape. Umbel several- to many-flowered, hemispherical to globose, erect to nodding, the pedicels straight or curved, shorter than the perianth; bracts of the inflorescence 2, ovate, acuminate. Perianth segments lanceolate, acute to acuminate, entire, whitish to pinkish. Stamens included, about three-fourths as long as the perianth. Ovary lacking a crest; style included; stigma capitate, entire. Capsule not crested, the seeds shiny, black.

Meadows and fellfields of the lower alpine zone near timberline in the Uinta, Wasatch, and Wind River Ranges. Oregon south to Nevada, Utah, and Colorado.

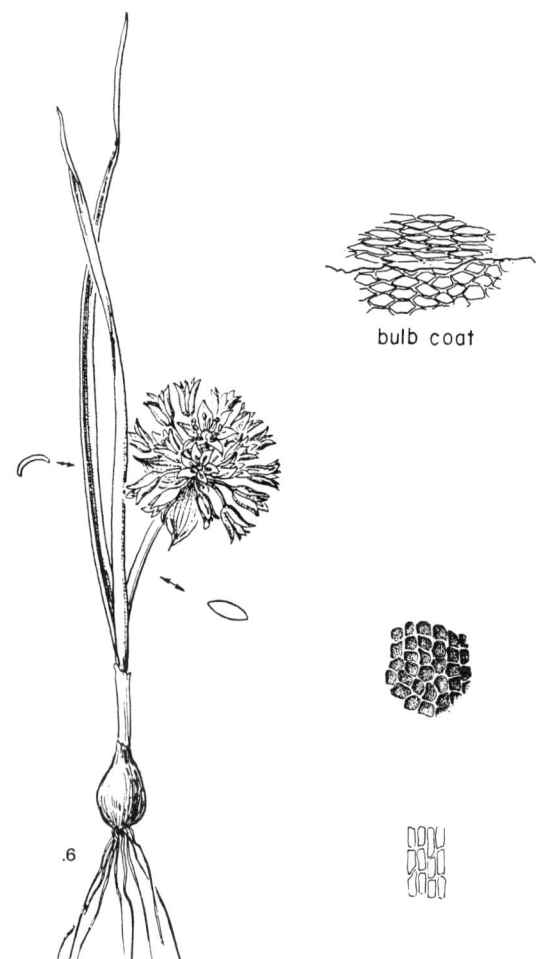
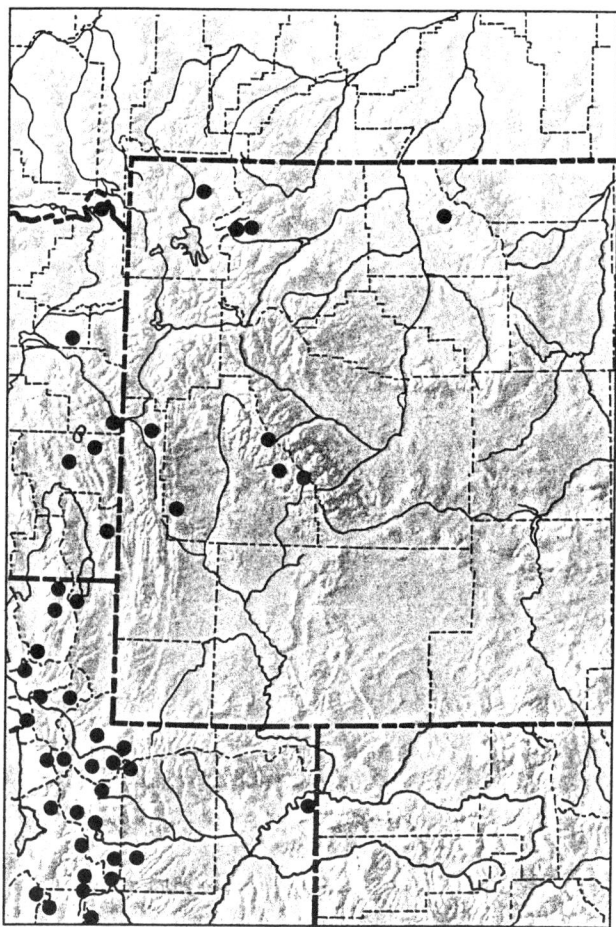

Allium brevistylum S. Wats.
Bot. King Exped., 350, 1871.

Shortstyle Onion

Perennial herb with bulb elongate, from a short rhizome, the outer coats finely striate, with parallel fibers. Leaves linear, flat, much shorter than the scape. Umbel 7–5-flowered, the pedicels about as long as the perianth, becoming elongate and curved in fruit; bracts of inflorescence 2, united at the base and usually offset to one side. Perianth segments lanceolate, acuminate, entire, subequal, pink. Stamens included, about half the length of the perianth, the filaments dilated. Ovary lacking a crest; style about half as long as the perianth; stigma shallowly 3-cleft. Capsule obcordate, the seeds dull black.

Wet meadows and stream banks in the Big Horn and Medicine Bow Ranges. Montana and Idaho south to Utah and Colorado.

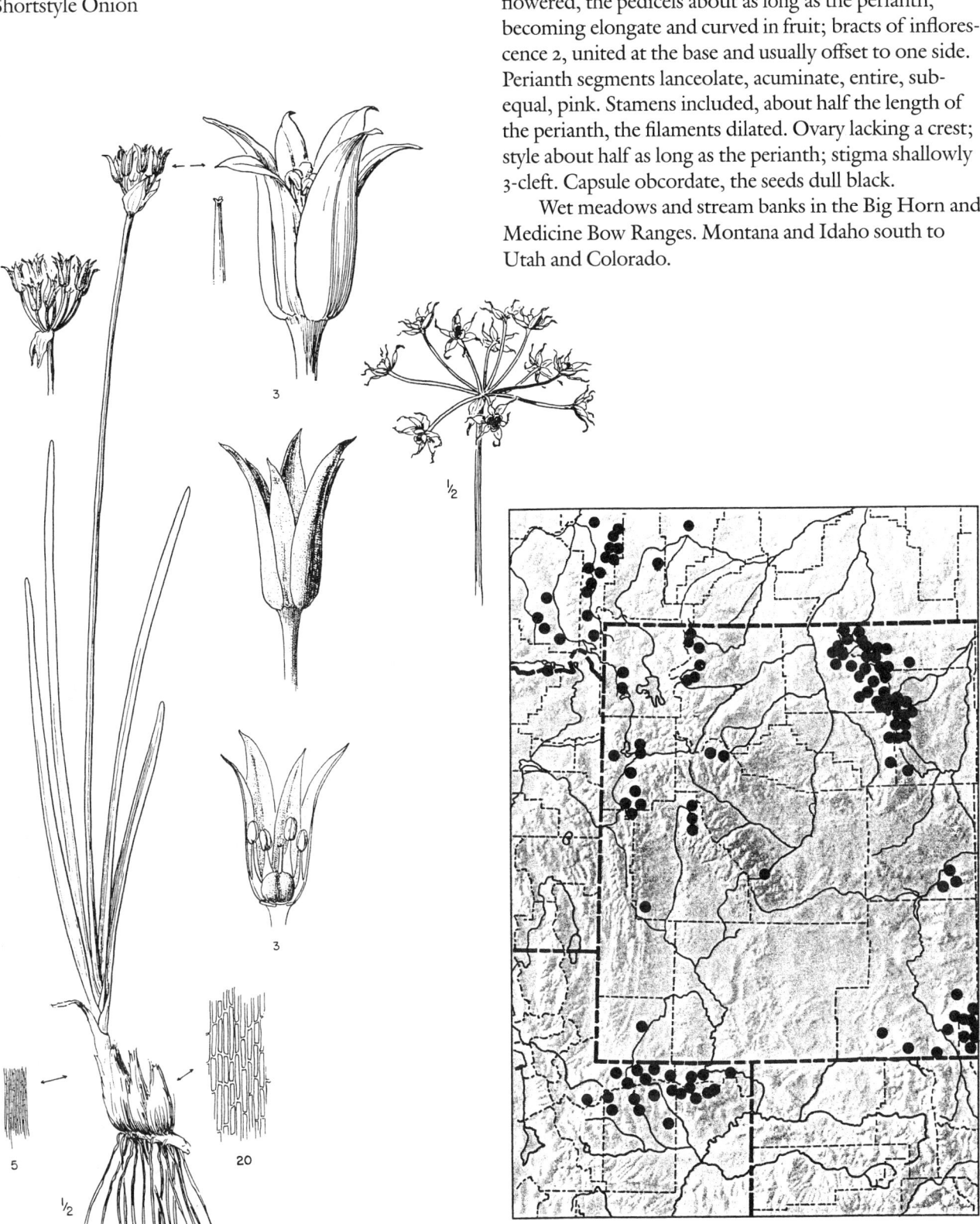

Allium cernuum Roth in Roem.
Arch. Bot. 1(3):40, 1798.

Nodding Onion

Allium alleghaniense Small
Allium natans Rydb.
Allium neomexicanum Rydb.
Allium recurvatum Rydb.

Perennial herb with bulb elongate, clustered from a short rhizome, the outer coat finely striate. Leaves linear, flat, channeled, or nearly terete, shorter than the scape. Umbel few- to many-flowered, nodding, the pedicels 2–3 times longer than the perianth, becoming elongate and bending upward in fruit; bracts of inflorescence 2, early deciduous. Perianth segments ovate to elliptic, obtuse to acute, entire, white or pink. Stamens exerted. Ovary strongly crested with 6 toothed processes; style exerted; stigma capitate, entire. Capsule crested, the seeds dull black.

Dry to seasonally moist tundra on the Beartooth Plateau. North America; south to Georgia and Mexico.

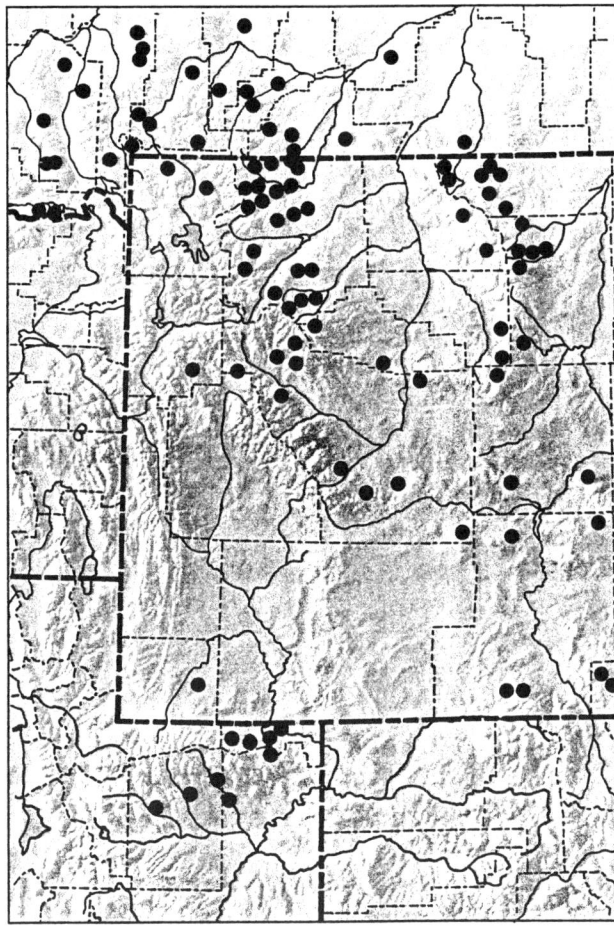

Allium schoenoprasum L.
Sp. Pl., 301, 1753.

Chive

Allium sibiricum L.

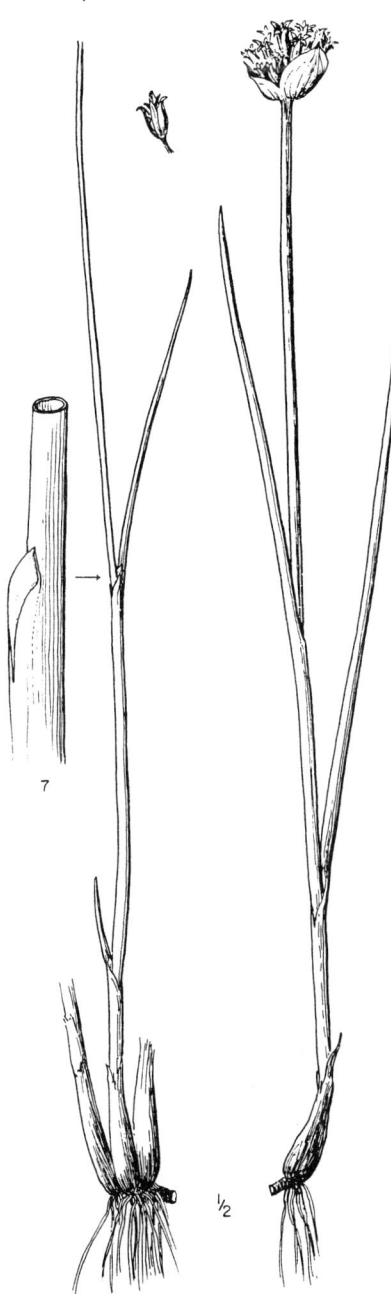

Perennial herb with bulb narrowly ovoid to elongate, from a short rhizome, the outer coat membranaceous, finely striate. Leaves linear, terete, hollow, usually 2 and shorter than the scape. Umbel several- to many-flowered, globose, the pedicels shorter than the perianth; bracts 2, ovate, distinct or nearly so. Perianth segments lanceolate to narrowly elliptic, entire, acute to acuminate, light to dark pink. Stamens included, the filaments dilated. Ovary lacking a crest; style included; stigma capitate, entire to shallowly lobed. Capsule not crested, the seeds shiny black.

Wet meadows, stream banks, and rocky sheetwash slopes of the Absaroka, Beartooth, Big Horn, Teton, and Wind River Ranges. Circumboreal; northern North America south to New York, Minnesota, Colorado, Idaho, and Washington.

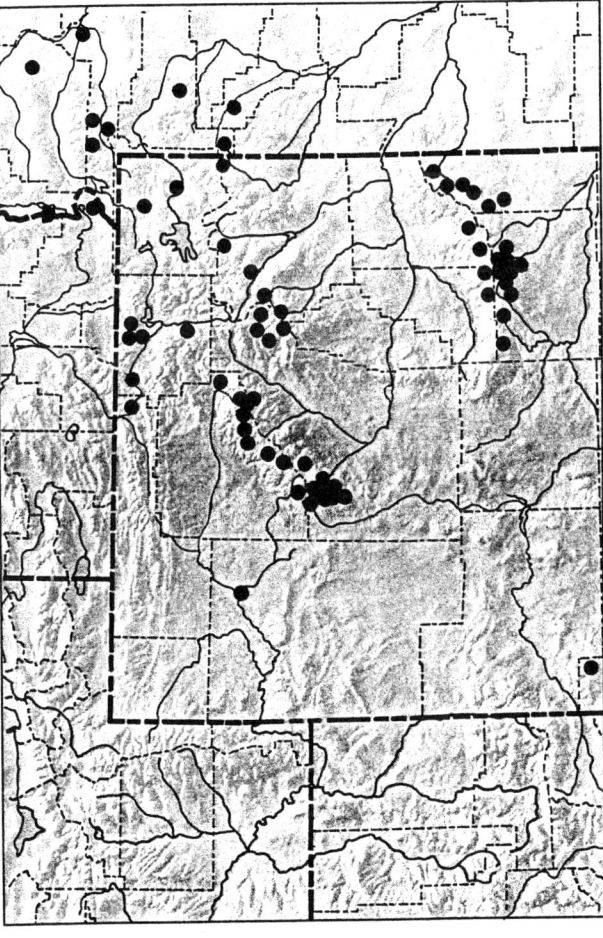

Calochortus Pursh
Sego Lily, Mariposa Lily

Slender, erect, perennial herbs from tunicated bulbs. Leaves few, linear, alternate or basal, reduced upward. Flowers showy, solitary or few in an umbel-like cluster. Perianth segments distinct; sepals green, narrowly lanceolate; petals broadly obovate, white or colored, with a large, fringed gland at the base. Stamens 6. Ovary triangular to 3-winged; style lacking, the stigma 3-lobed. Capsule septicidal, erect in ours, 3-angled or 3-winged.

Calochortus gunnisonii S. Wats.
Bot. King Exped., 348, 1871.

Gunnison Sego Lily

Perennial herb with bulb broadly ovate to globose, the outer coat finely striate. Leaves linear, channeled, reduced upward. Flowers erect, 1–3, in a subumbellate inflorescence. Sepals lanceolate, glabrous, greenish, with a transverse purple band and a purple spot below. Petals white, obovate, with a purple transverse band above the gland and a spot on the claw below it, densely bearded around the gland with branching, gland-tipped hairs. Anthers lanceolate, acute or sometimes acuminate, longer than the filaments. Ovary linear, 3-angled; stigma trifid. Capsule elliptic, 3-angled, erect.

Lower elevations up to timberline meadows in the Absaroka and Big Horn Ranges. The *Calochortus* sp. reported from Carter Mountain in the Absaroka Range (Thilenius & Brown, 1987) is probably this species. Montana and South Dakota south to Nebraska, New Mexico, and Arizona.

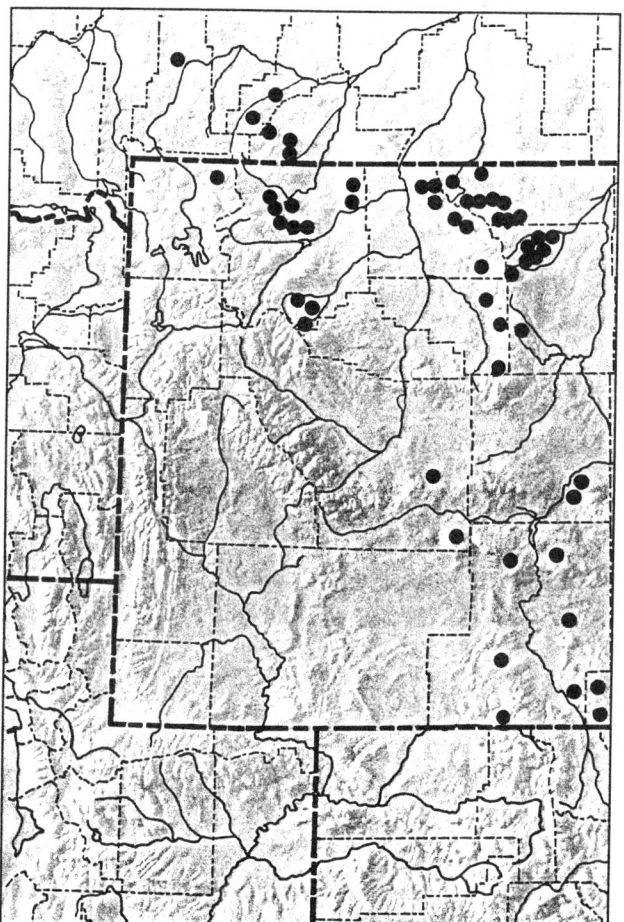

Erythronium L.
Glacier Lily

Perennial, scapose herbs from deep, elongate corms. Leaves linear, lanceolate, or elliptic, petiolate to sessile, paired, weakly sheathing at the base of the scape. Flowers showy, perfect, solitary or racemose, nodding, the perianth segments reflexed to spreading, distinct. Stamens 6. Style with a shallowly to deeply 3-lobed stigma. Fruit a 3-angled, many-seeded loculicidal capsule.

Erythronium grandiflorum Pursh
Fl. Am. Sept., 231, 1814.

Glacier Lily

Erythronium giganteum Lindl.
Erythronium idahoense St. John & G.N. Jones
Erythronium leptopetalum Rydb.
Erythronium nuttallianum Regel
Erythronium obtusatum Goodding
Erythronium pallidum G.N. Jones
Erythronium parviflorum (S. Wats.) Goodding
Erythronium utahense Rydb.

Scapose perennial with elongate corms. Leaves usually 2, oblong-elliptic, acute, gradually tapered to a short, broad petiole. Flowers 5 or less, racemose, showy, nodding, deep yellow in ours; perianth segments reflexed. Stamens 6; filaments broadened at the base; anthers mostly yellow in ours. Style slender; stigma 3-lobed, the lobes spreading. Capsule cylindric to club-shaped.

Open meadows and slopes of the lower alpine zone near timberline in the Absaroka, Beartooth, Medicine Bow, Teton, Uinta, and Wasatch Ranges. British Columbia to Montana, south to Colorado, Utah, Idaho, and California.

Our plants with stigmas exceeding 1 mm in length and perianth parts deep yellow are var. *grandiflorum*.

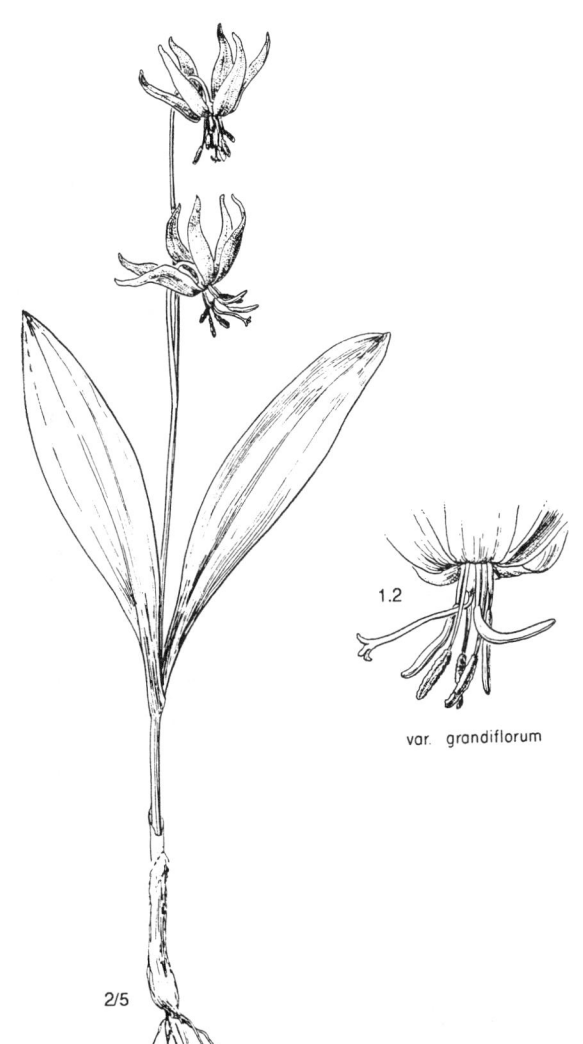

var. grandiflorum

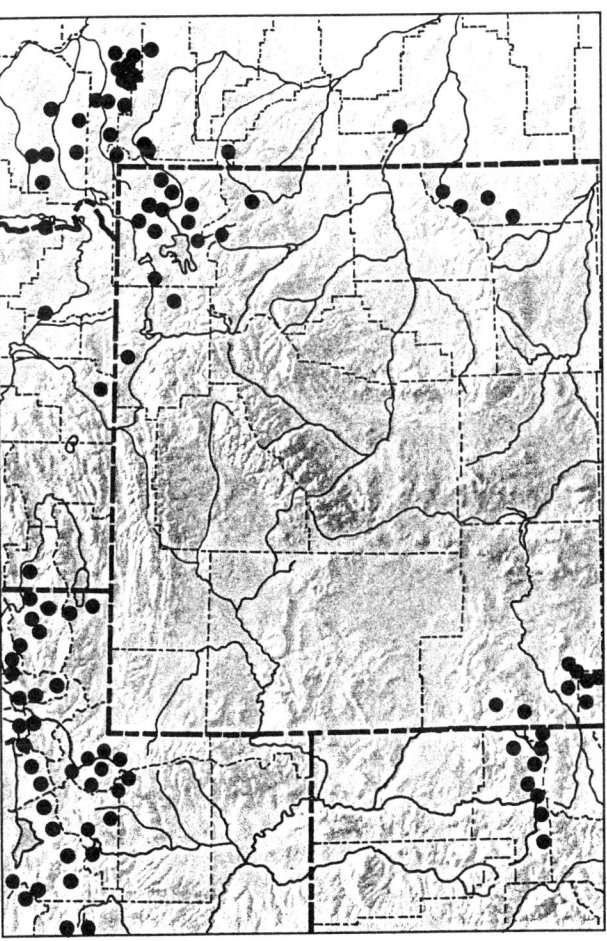

Lloydia Salisb. ex Reichenb.
Alp Lily

Dwarf perennials from tunicated bulbs arising from horizontal rhizomes. Stems 5–15 cm tall, the leaves alternate, linear. Flowers erect, solitary or less commonly 2–3, in a terminal raceme. Perianth segments 6, similar, distinct, white or creamy within, oblanceolate, obtuse or rounded at the apex, tinged with rose, and purple-veined or greenish-veined on the outer surface. Stamens 6, distinct, the filaments dilated below, the anthers basifixed. Style 1, the stigma 3-lobed. Fruit an ovoid or subglobose, erect, loculicidal, several-seeded capsule, dehiscent only at the apex.

Lloydia serotina (L.) Salisb. ex Reichenb.
Fl. Germ. Excurs., 102, 1830.

Alp Lily

Bulbocodium serotinum L.
Anthericum serotinum (L.) L.
Cronyxium serotinum (L.) Raf.
Gagea serotina (L.) Ker-Gawl
Rhabdocrinum serotinum (L.) Reichenb.

Small, slender, perennial herb from erect, fibrous-coated, bulblike rhizomes. Basal leaves 2, linear; cauline leaves 2–4, narrowly lanceolate to linear. Flowers solitary, occasionally 2. Perianth white, greenish white, or yellowish white, often purple-veined or purplish-tinged at the base, the segments oblanceolate to oblong, obtuse. Stamens about two-thirds as long as the perianth. Capsule obovoid.

Alp slopes, meadows, and occasionally in rock crevices throughout most of the alpine areas in the Middle Rocky Mountains. Not yet known from above timberline in the Hoback and Medicine Bow Ranges. Circumboreal; Alaska to Alberta and south to Oregon, Nevada, and New Mexico.

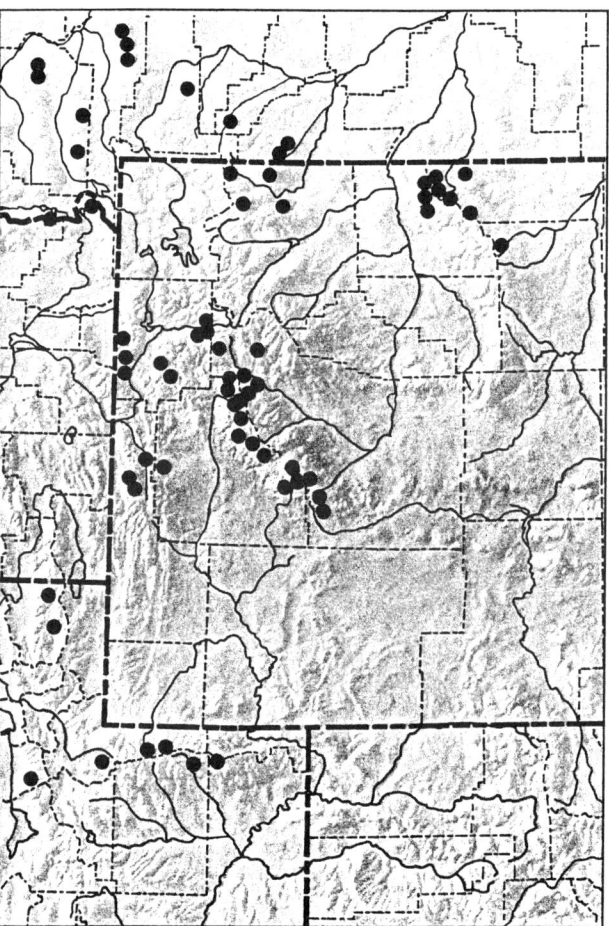

Tofieldia Huds.
Tofieldia

Small, slender, perennial herbs from short rhizomes. Leaves mostly basal, linear, equitant, conduplicate. Flowers small, whitish to greenish, in terminal, spikelike racemes. Perianth segments similar, distinct, oblong, spreading, subtended by 3 basally united bractlets. Stamens 6, the filaments narrow, lanceolate, flattened basally; anthers ovate to cordate. Styles 3, short, distinct, spreading to recurved. Ovary 3-lobed, at least at the summit. Fruit a 3-lobed, many-seeded, septicidal capsule.

Tofieldia glutinosa (Michx.) Pers.
Syn. Pl. 1:399, 1805.

Sticky Tofieldia

Narthecium glutinosum Michx.
Asphodeliris glutinosa (Michx.) Kuntze
Tofieldia intermedia Rydb.
Tofieldia occidentalis S. Wats.
Triantha glutinosa (Michx.) Baker
Trianthella glutinosa (Michx.) House

Single-stemmed, perennial herb from slender rhizomes; stem glandular-viscid above, glabrous to glandular below. Leaves linear, mostly basal. Raceme dense, subcapitate. Perianth white to greenish, the inner whorl slightly longer than the outer. Stamens about equaling the perianth segments. Styles 3, distinct. Capsules tipped by the recurved styles. Seeds with a slender, recurved, terminal appendage (funiculus).

Wet meadows and stream banks at timberline in the Gros Ventre and Teton Ranges. Alaska to the Atlantic Coast, south to North Carolina, the Appalachians, Wyoming, Idaho, and California.

Our plants with perianth segments 4 mm long or less, styles mostly less than 1 mm, seeds white or nearly so, and pubescence of long, slender hairs are var. *montana* (C.L. Hitchc.) Davis.

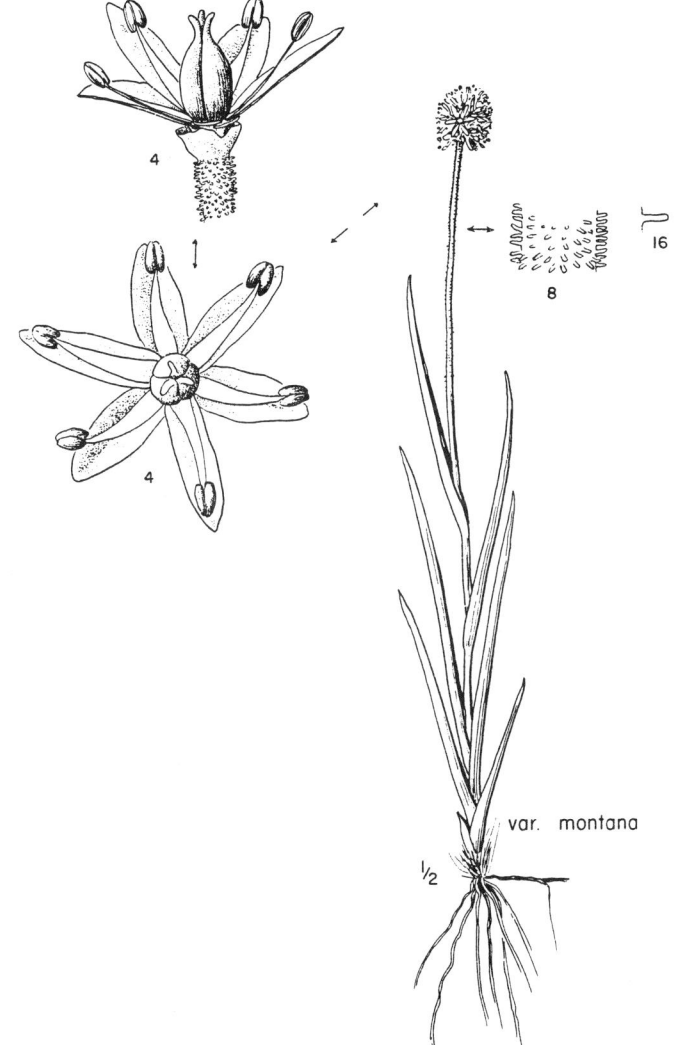

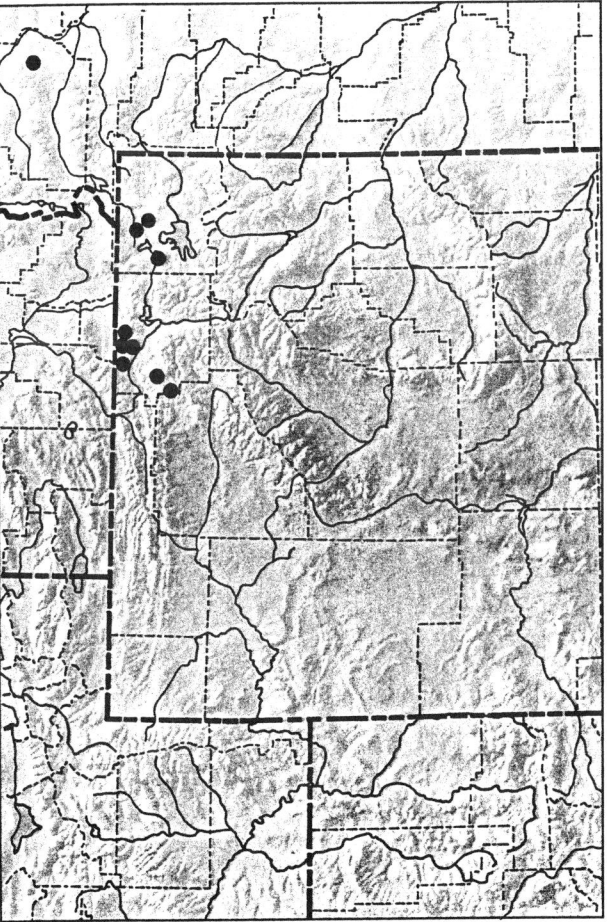

Zigadenus Michx.
Death Camas

Perennial herbs from tunicated bulbs, with erect stems and bearing linear, glabrous leaves from near the base. Flowers small and white, yellowish, or greenish, in a terminal, bracteate raceme or panicle. Perianth segments 6, widely spreading, similar, distinct or nearly so, each with a basal gland. Stamens 6, with slender filaments, the anthers extrorse. Ovary deeply 3-lobed, the styles 3, short, slender. Fruit a 3-lobed, loculicidal, many-seeded capsule.

Zigadenus elegans Pursh
Fl. Am. Sept., 241, 1814.

Wand Lily, Mountain Death Camas

Anticlea alpina (Blank.) A. Heller
Anticlea chlorantha (Richards.) Rydb.
Anticlea coloradensis (Rydb.) Rydb.
Anticlea elegans (Pursh) Rydb.
Anticlea longa (Greene) A. Heller
Zigadenus alpinus Blank.
Zigadenus chloranthus Richards.
Zigadenus coloradensis Rydb.
Zigadenus glaucus Nutt.
Zigadenus longus Greene
Zigadenus washakianus A. Nels.

Erect, perennial herb from an elongate-ovoid, fibrous-coated bulb. Leaves mostly basal, glaucous, linear, keeled at least below; cauline leaves linear-lanceolate, reduced upward. Inflorescence a 3- to several-flowered raceme, rarely a panicle. Flowers greenish white or yellowish white, the perianth segments spreading, ovate, oval, or obovate, obtuse, each with an obcordate, dark green, basal gland. Stamens erect, nearly as long as the perianth segments. Ovary partly inferior. Capsule ovoid to oblong; seeds tan.

Widely distributed in moist meadows and on alp slopes throughout all alpine ranges of the Middle Rocky Mountains. Alaska south to Saskatchewan, Minnesota, Missouri, Texas, New Mexico, Arizona, and Nevada.

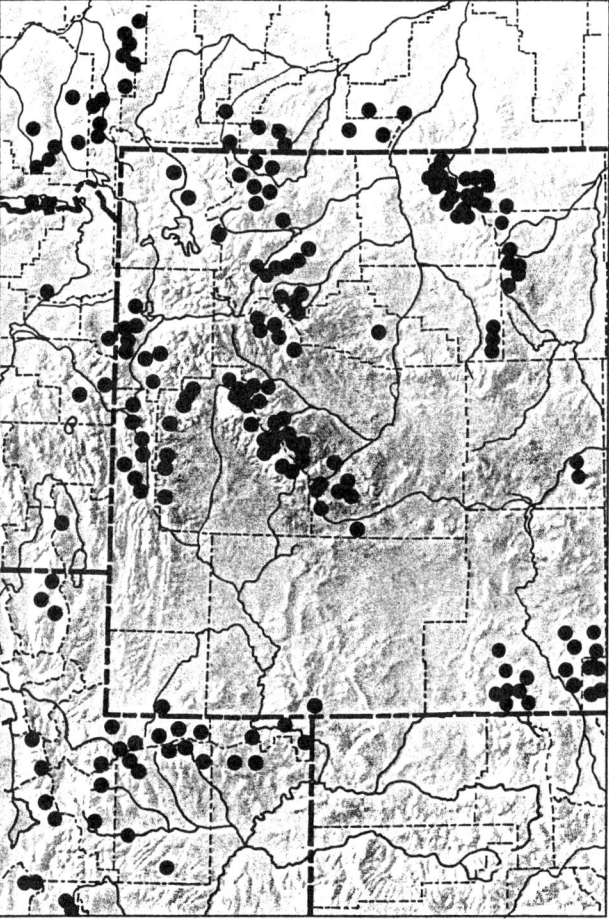

Linaceae
Flax Family

Annual to perennial, glabrous herbs, with alternate or opposite, simple, entire leaves. Flowers hypogynous, perfect, regular, in cymes or racemes. Sepals 5, imbricate, distinct or connate at the base, persistent. Petals 5, distinct, short-clawed, yellow or blue, fugacious. Stamens 5 and alternate with the petals, or 10; staminodia often present; filaments connate at the bases. Pistil 1, of 5 carpels; styles 5, distinct or connate at the bases; locules 5 or 10. Fruit a septicidal capsule.

Linum L.
Flax

Slender, annual or perennial herbs with glabrous stems. Leaves sessile, simple, acute, alternate in ours. Flowers regular, in racemes or panicles. Sepals 5, imbricate. Petals 5 and blue, white, yellow, or orange, early deciduous. Stamens 5, alternate with the petals, the bases connate. Pistil 1, of 5 united carpels; styles 5, distinct or connate at the bases. Capsule rounded, dehiscing into 5 (or 4) or 10 cells. Seeds flattened.

Linum perenne L.
Sp. Pl., 277, 1753.

Blue Flax

Adenolinum lewisii (Pursh) Löve & Löve
Linum lepagei B. Boivin
Linum lewisii Pursh

Multistemmed, glabrous, perennial herb from a branched, woody caudex. Leaves alternate, linear to linear-lanceolate, somewhat acute. Flowers in loose racemes or panicles, the pedicels slender, spreading. Sepals ovate, entire, acute or short-acuminate. Petals obovate, blue, fugacious. Stamens 5, alternating with staminodia; filaments connate at the base. Styles 5, distinct; stigmas capitate; carpels 5. Capsule 10-loculed.

In the alpine zone *L. perenne* occurs on scree, talus, rocky slopes, and ridges of the Absaroka, Beartooth, Hoback, Salt River, Teton, Wasatch, and northern Wind River Ranges. Eurasia; North America.

Native plants of western North America are var. *lewisii* (Pursh) Eat. & Wright. *Linum kingii* S. Wats. has been collected near timberline in the Wasatch Range and might be expected in the lower alpine zone in the same habitats as *L. perenne,* especially on limestone. It is distinguished by yellow petals, sepals glandular-ciliate, and capsules less than 4 mm long.

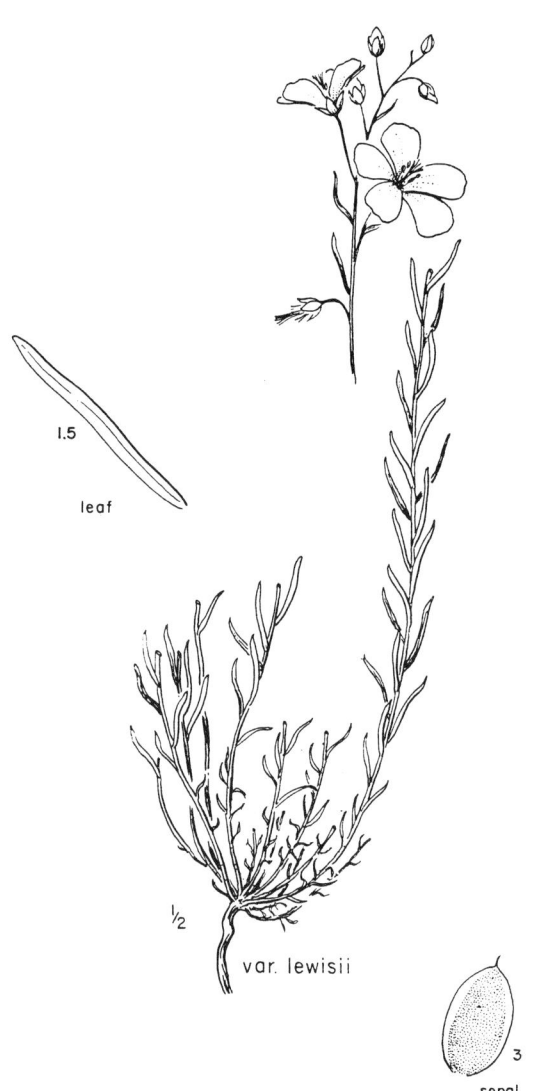

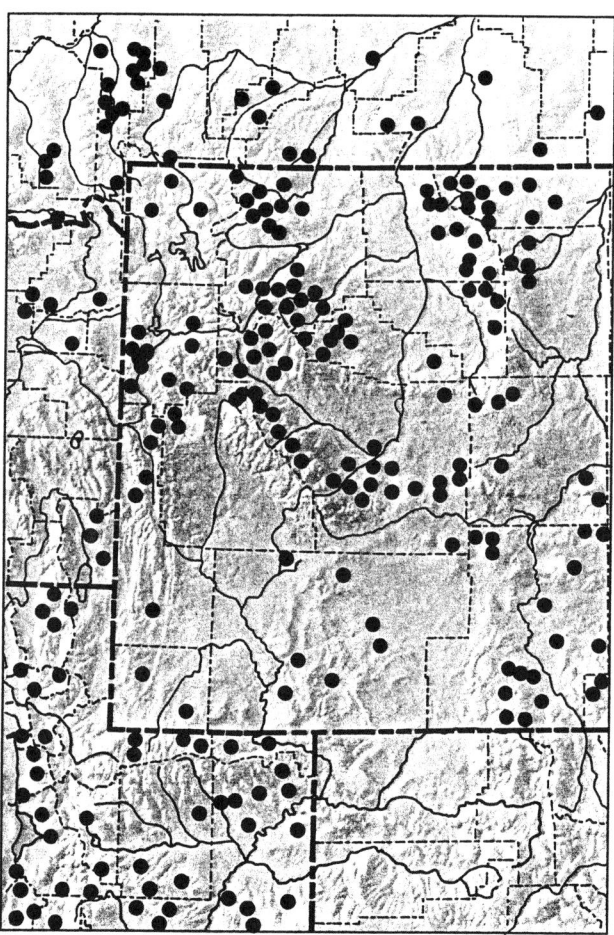

Lycopodiaceae
Clubmoss Family

Sporophytes low, evergreen, mosslike plants with erect or creeping stems. Leaves 1-nerved, densely imbricate, simple, linear to oblong. Sporangia orbicular to reniform, sessile in the axils of ordinary scale leaves, or in the axils of sporophylls forming terminal strobili. Spores numerous, uniform, smooth to variously sculptured. Gametophytes small, subterranean.

Lycopodium L.
Clubmoss

Low, evergreen, perennial herbs; stems decumbent to prostrate, branched, sometimes dichotomously, with adventitious roots. Leaves small, numerous, and crowded, lanceolate, 1-nerved, in 4–16 ranks. Fertile branches erect to ascending; sporangia ovoid to reniform, solitary in the axils of leaves, these often aggregated into terminal, clublike strobili; spores yellow. Gametophyte commonly subterranean and mycorrhizal.

Lycopodium selago L.
Sp. Pl., 1102, 1753.

Clubmoss

Huperzia selago (L.) Bernh. ex Mart. & Schrank.
Lycopodium appressum Desv.) Petrovic
Lycopodium arcticum Grossh.
Lycopodium porophilum Lloyd & Underw.
Plantanthus patens Beauv.
Plantanthus selago (L.) Beauv.
Selago vulgaris Schur
Urostachys selago (L.) Herter

Compact, tufted, dichotomously branched perennial from a decumbent base. Rhizomes slender, branched. Leaves dense, appressed to widely spreading, entire, triangular-lanceolate to linear or oblanceolate, acute to acuminate. Sporophylls shorter than the sterile leaves and produced in sections alternating with sterile sections. Sporangia reniform, pale greenish, the spores triangular.

Moist, rocky places. The single Middle Rocky Mountain alpine specimen was collected in 1932 at 11,000 ft from cliffs on the north side of the Grand Teton, Teton Range (L.O. Williams 1017). Transcontinental in northern Canada and south to Oregon, Colorado, and North Carolina.

Variation in this widespread, circumboreal species is discussed in A. Löve and Löve, 1965, under the name *Huperzia selago* (L.) Bernh.

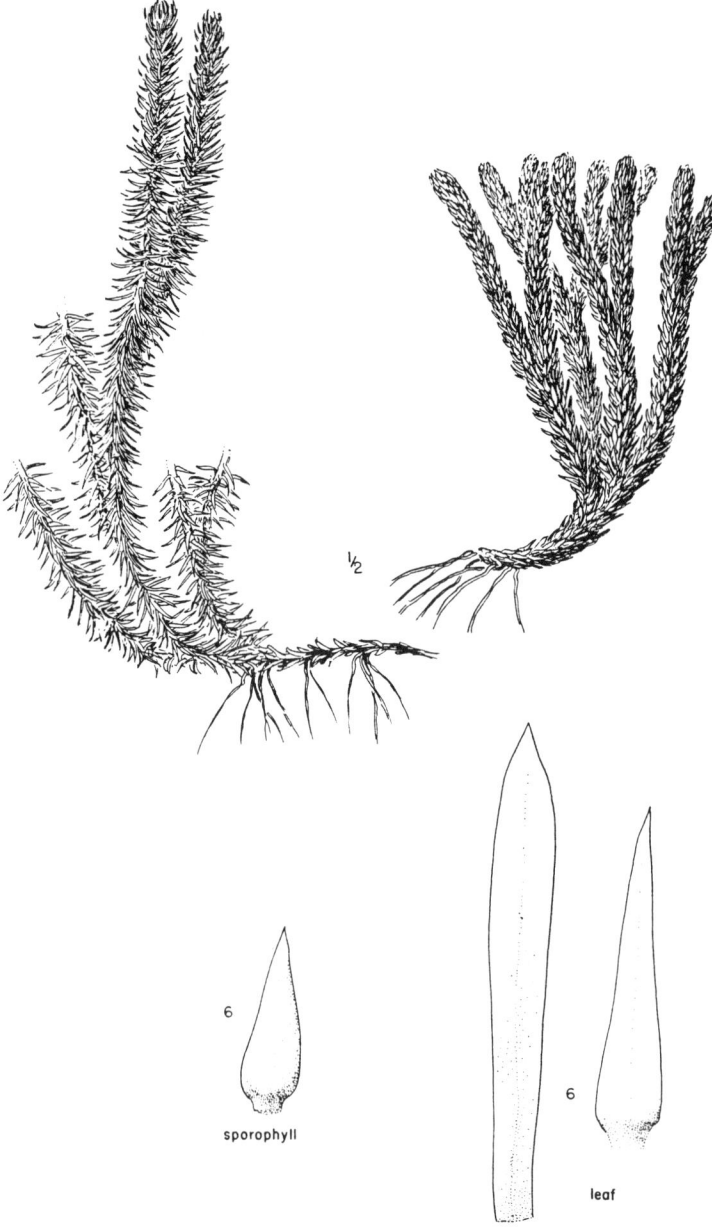

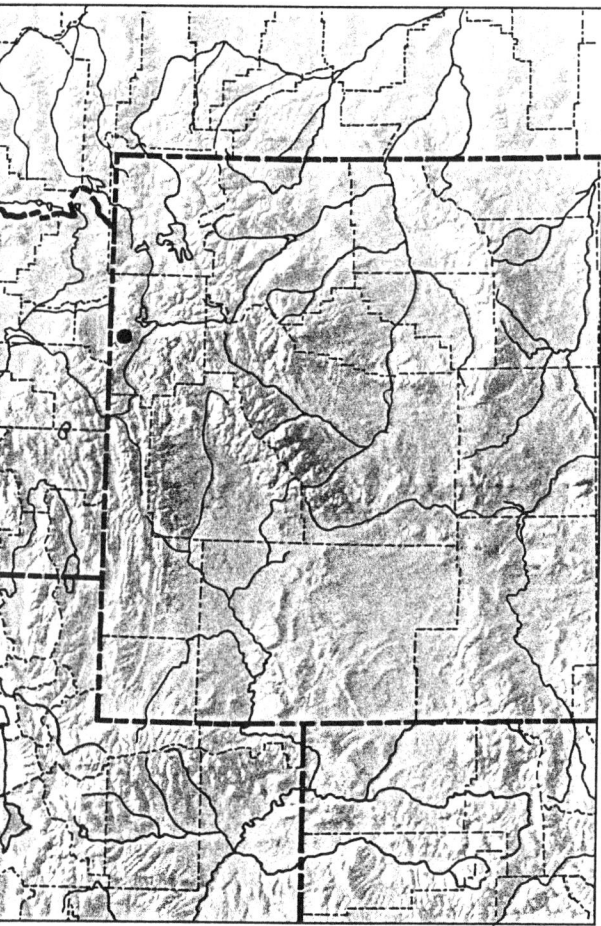

Onagraceae
Evening Primrose Family

Annual or perennial herbs with simple, alternate or opposite, exstipulate leaves. Flowers epigynous, often with a tubular, epigynous hypanthium, 4-merous, perfect, regular, solitary or in terminal spikes or racemes. Sepals 2 or 4, distinct or connate. Petals distinct, often clawed, rarely lacking. Stamens as many or twice as many as the petals. Ovary inferior, 2–5-celled, mostly 4-celled, the style single, the stigma capitate, discoid, or 4-lobed. Fruit a many-seeded capsule, rarely a berry, sometimes indehiscent and nutlike.

KEY TO THE GENERA OF ONAGRACEAE

1. Anthers basifixed; seeds with a tuft of hairs at one end — *Epilobium*
1. Anthers versatile; seeds lacking a tuft of hairs
   2. Petals less than 8 mm long — *Gayophytum*
   2. Petals more than 10 mm long — *Oenothera*

Epilobium L.
Willowherb, Fireweed

Annual or perennial herbs with mostly opposite, sessile or petiolate, entire or toothed, linear to ovate leaves. Flowers white to rose in ours, the hypanthium short or lacking. Inflorescence a raceme, or the flowers solitary in the upper leaf axils. Sepals 4. Petals 4, usually notched at the apex. Stamens 8, in 1 or 2 series, the anthers basifixed. Ovary 4-celled, the style single, the stigma entire and oblong or 4-lobed. Capsules 4-valved, fusiform or clavate, loculicidal. Seeds comose, with a prominent tuft of hairs at one end.

KEY TO THE SPECIES OF *EPILOBIUM*

1. Stigma 4-lobed; petals 8–25 mm long
   2. Plants 10–30 cm tall; racemes short, few-flowered; style glabrous — *E. latifolium*
   2. Plants more than 30 cm tall; racemes many-flowered; style pubescent at the base — *E. angustifolium*
1. Stigma entire, oblong; petals less than 8 mm long
   3. Turions present near the end of rhizomes — *E. halleanum*
   3. Turions lacking
      4. Leaves more than 3.5 cm long, serrulate; stems branched above the middle; seeds papillate, the papillae in rows — *E. ciliatum*
      4. Leaves less than 2.5 cm long, mostly entire; stems branched near the base; seeds glabrous or papillate, the papillae not in rows — *E. alpinum*

Epilobium alpinum L.
Sp. Pl., 348, 1753.

Alpine Willowherb

Epilobium anagallidifolium Lam.
Epilobium behringianum Hausskn.
Epilobium clavatum Trel.
Epilobium glareosum G.N. Jones
Epilobium hornemannii Reichenb.
Epilobium lactiflorum Hausskn.
Epilobium nutans Hornem.
Epilobium oregonense Hausskn.
Epilobium pulchrum Suksd.
Epilobium sertulatum Hausskn.
Epilobium treasianum Levl.

Caespitose, often reddish, stoloniferous perennial from rhizomes, lacking turions; stems 1 to many, often decumbent, simple or branched near the base, often decurrent-strigillose above. Leaves opposite, short-petiolate, and oblong, lanceolate, or ovate, entire to serrulate. Flowers mostly 1–4, racemose, nodding or erect, the pedicels to 20 mm; hypanthium short, to 2 mm long. Sepals lanceolate, about half as long as the petals. Petals reddish purple, pink, or rose, notched. Stigma entire. Capsule linear or clavate, reddish or purplish; seeds glabrous or papillate, the papillae not in rows, coma whitish.

Moist stream banks, meadows, solifluction terraces, fellfields, talus, and moraines of all alpine ranges.

Circumboreal; south in western North America to California, Idaho, and Colorado.

Many authors treat the *Epilobium alpinum* complex as 4 separate species. According to this concept, the 3 found in the Middle Rocky Mountain alpine zone would be distinguished as follows:

1. Capsules clavate to subclavate, 1.5–2 mm thick; leaves broadly ovate; seeds minutely papillate
   . . . *E. clavatum*
1. Capsules linear, about 1 mm thick; leaves laneolate to elliptic; seeds smooth or papillate
   2. Plants stoloniferous; stems mostly S-shaped, the cauline leaves entire . . . *E. alpinum*
   2. Plants lacking stolons; stems mostly straight, the cauline leaves denticulate . . . *E. hornemannii*

However, as treated here, the nearly continuous geographic variation in the *E. alpinum* complex is represented by two varieties:

1. Capsules linear, about 1 mm thick; leaves lanceolate to elliptic; seeds smooth or papillate . . . var. *alpinum*
1. Capsules clavate to subclavate, 1.5–2.0 mm thick; leaves broadly ovate: seeds minutely papillate
   . . . var. *clavatum* (Trel.) C.L. Hitchc.

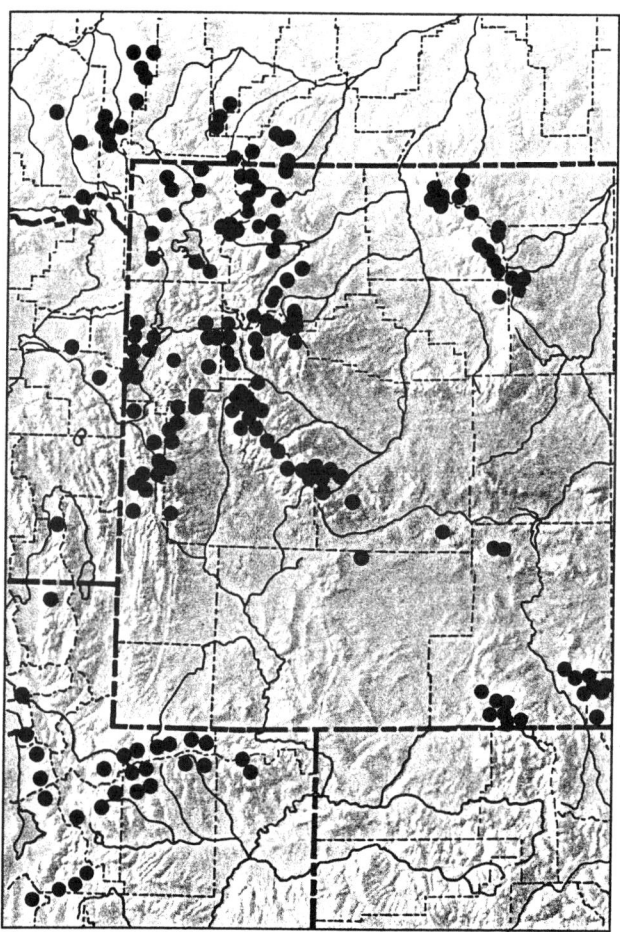

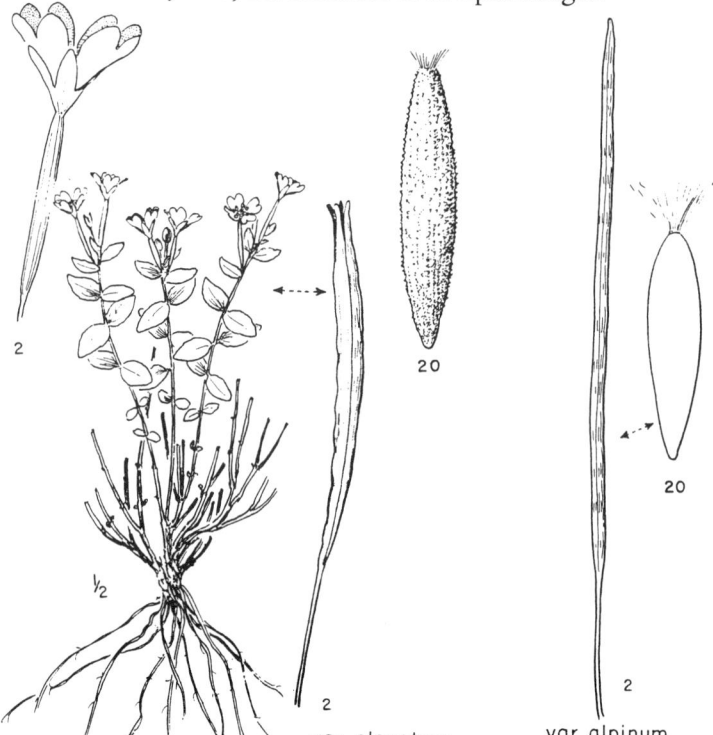

var. clavatum   var. alpinum

Epilobium angustifolium L.
Sp. Pl., 347, 1753.

Fireweed

Chamaenerion angustifolium (L.) Scop.
Chamaenerion exaltatum Rydb.
Chamaenerion spicatum (Lam.) S.F. Gray
Chamerion angustifolium (L.) Holub
Chamerion danielsii D. Löve
Chamerion platyphyllum (Daniels) Löve & Löve
Chamerion spicatum (Lam.) S.F. Gray
Epilobium spicatum Lam.

Stout, perennial herb from rhizomes; stems erect, mostly simple, glabrous below, pubescent above, at least in the inflorescence. Leaves alternate, sessile or nearly so, the blades lanceolate, entire, glabrous. Flowers in terminal, many-flowered, bracteate racemes; hypanthium very short, nearly lacking. Sepals narrowly lanceolate, spreading, acute, nearly as long as the petals. Petals purple, pinkish, or rose, and spreading, obovate, and clawed. Stamens 8, subequal; anthers purple, the filaments white. Style pubescent near the base; stigma 4-lobed. Capsule to 7 mm long, canescent; seeds reticulate, the coma whitish.

Willow thickets, stream banks, and margins of boulders in the lower alpine zone in most alpine ranges of the Middle Rocky Mountains. Not yet known from above timberline in the Gros Ventre, Hoback, Salt River, Teton, and Wyoming Ranges. Circumboreal; south in western North America to California, New Mexico, Colorado, and Nebraska.

Mountain plants in the vicinity of timberline are var. *angustifolium*.

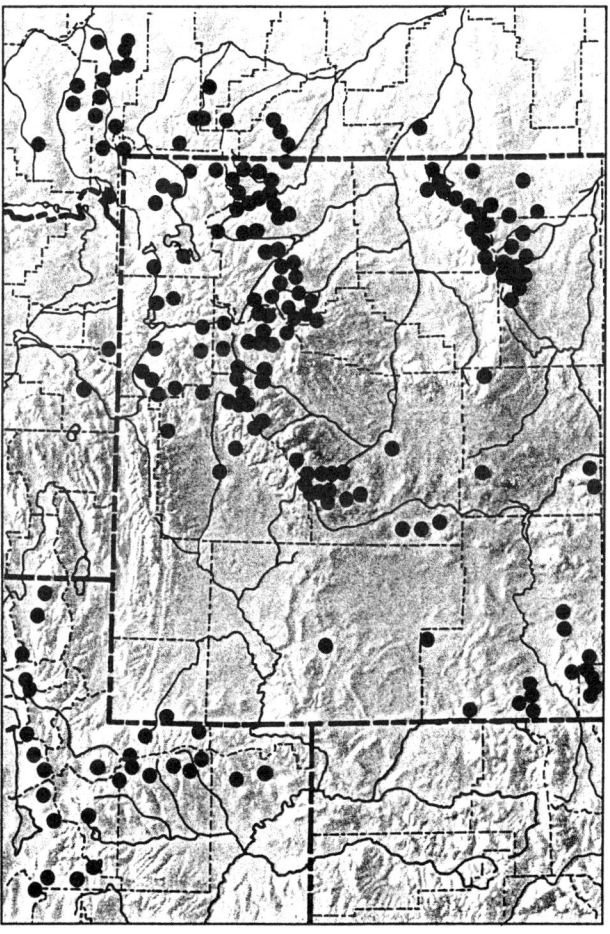

Epilobium ciliatum Raf.
Med. Repos. N.Y., II, 5:361, 1808.

Purpleleaf Willowherb

Epilobium adenocaulon Hausskn.
Epilobium affine Bong.
Epilobium americanum Hausskn.
Epilobium boreale Hausskn.
Epilobium brevistylum Barbey
Epilobium californicum Hausskn.
Epilobium cinerascens Piper
Epilobium delicatum Trel.
Epilobium ecomosum (Fassett) Fern.
Epilobium franciscanum Barbey
Epilobium glandulosum Lehm.
Epilobium griseum Suksd.
Epilobium macdougalii Rydb.
Epilobium occidentale Rydb.
Epilobium parishii Trel.
Epilobium perplexans (Trel.) A. Nels.
Epilobium praecox Suksd.
Epilobium ursinum Parish ex Trel.
Epilobium watsonii Barbey

Erect, perennial herb from rhizomes that produce rosettes of leaves at the tips; turions lacking; stems simple or branched, pubescent and often glandular above, decurrent-strigillose below. Leaves opposite, glabrate, the lower ones elliptic to obovate, petiolate; the upper ones lanceolate to ovate, short-petiolate or sessile. Inflorescence an erect, compound raceme. Hypanthium short, mostly less than 2 mm long. Sepals narrowly lanceolate, often glandular and reddish-tinged. Petals notched, purple or whitish. Anthers white; filaments purplish. Stigma entire or nearly so. Capsule slender, linear, reddish, pubescent, often glabrate with age; seeds papillate, ridged, the coma white.

Krummholz margins and stream banks in the lower alpine zone of the Absaroka Range. Transcontinental in North America, south to California, Colorado, Iowa, Tennessee, and North Carolina.

Middle Rocky Mountain alpine plants with lanceolate to ovate leaves and ebracteate, compound racemes are var. *ciliatum*.

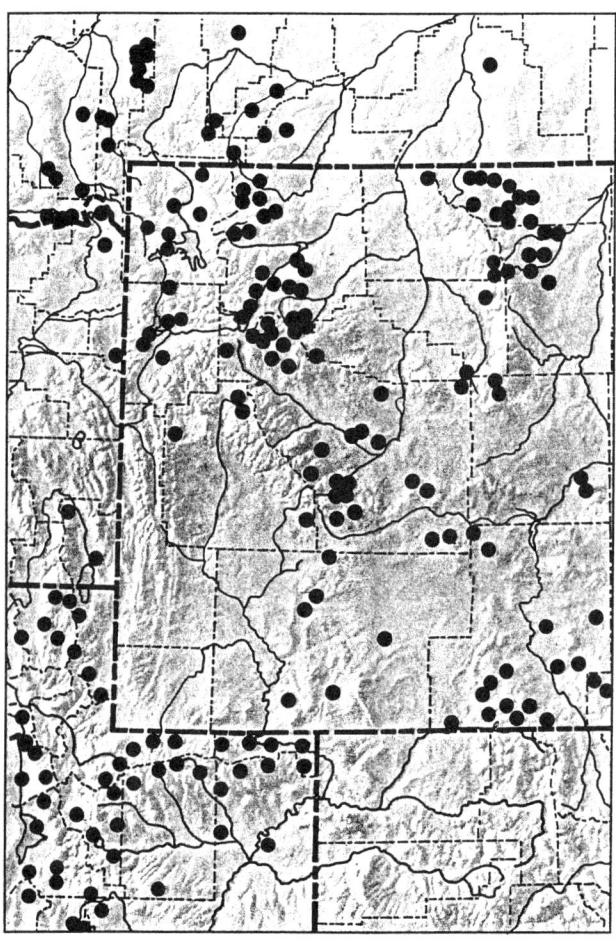

Epilobium halleanum Hausskn.
Mon. Epil., 261, 1884.

Hall's Willowherb

Epilobium pringleanum Hausskn.

Perennial herb from rhizomes; turions small, compact, globose; stems erect, simple or occasionally branched above the base, pubescent to glandular-puberulent above, with decurrent lines of pubescence below the petioles. Leaves opposite, acute, petiolate to nearly sessile, mostly glabrous, lanceolate to ovate, the margins entire to denticulate, becoming reduced upward. Inflorescence a few-flowered, bracteate raceme, nodding at first, becoming erect at anthesis. Hypanthium short, less than 2 mm long. Sepals lanceolate, green. Petals notched and pink, purple, or whitish. Anthers and filaments white to light yellow. Stigma entire. Capsule linear, slender, pubescent to nearly glabrous; seeds papillose, the coma whitish.

Meadows, stream banks, krummholz margins, and rocky ledges of the Absaroka, Teton, Uinta, Wasatch, and Wind River Ranges. Alaska south to California, Arizona, and Colorado.

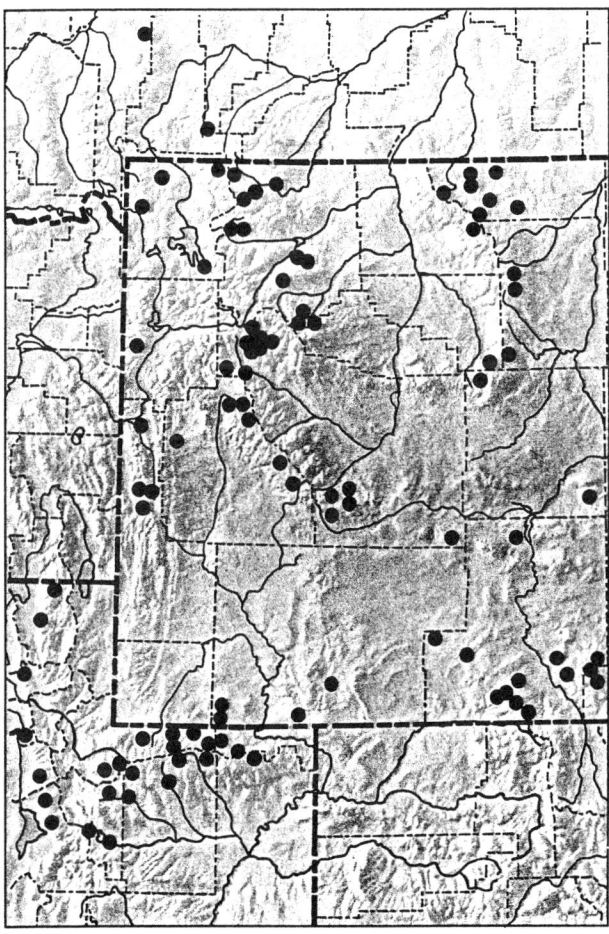

Epilobium latifolium L.
Sp. Pl., 347, 1753.

River Beauty

Chamaenerion latifolium (L.) Sweet
Chamaenerion subdentatum Rydb.
Chamerion latifolium (L.) Holub.
Chamerion subdentatum (Rydb.) Löve & Löve

Perennial herb from a stout, woody rootstock; stems simple to branched, erect, ascending, to somewhat decumbent, grayish-puberulent above, glabrous below. Leaves opposite or alternate, lanceolate to ovate, thick, glaucous, fleshy, entire or nearly so, subsessile, finely pubescent. Inflorescence a terminal, bracteate, few-flowered raceme. Hypanthium very short to lacking. Sepals narrowly lanceolate, purplish. Petals purple to pinkish, obovate, short-clawed, often purple-veined. Style glabrous, shorter than the stamens; stigma 4-lobed. Capsule clavate, canescent, the seeds with a whitish coma.

Braided streambeds, rocky places, ledges, and scree slopes of the Absaroka, Beartooth, Teton, Uinta, Wasatch, and Wind River Ranges. Eurasia; transcontinental in North America, south to California, Idaho, and Colorado.

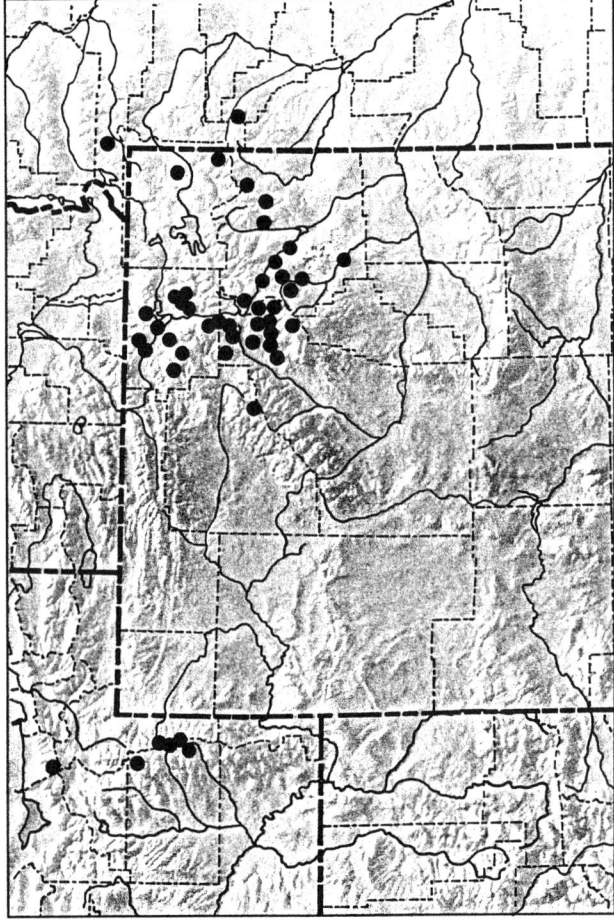

Gayophytum A. Juss.
Groundsmoke

Annual, caulescent herbs from taproots; stems erect to ascending, simple or branched. Leaves sessile or short-petiolate, alternate or opposite below, entire, linear to lanceolate; stipules lacking. Flowers small and inconspicuous in ours, the hypanthium lacking. Inflorescence a bracteate spike, raceme, or panicle, or the flowers solitary in the upper leaf axils. Sepals 4, reflexed at anthesis. Petals 4, white to yellowish or pinkish, obovate, entire to erose, sometimes with a short claw. Stamens 8, dimorphic, the anthers versatile. Ovary 2-celled, the stigma entire, capitate. Capsules 4-valved, linear or clavate. Seeds many, glabrous to puberulent, not comose.

KEY TO THE SPECIES OF
GAYOPHYTUM

1. Pedicels of flowers and fruits less than 2 mm long; capsules lacking constrictions between the seeds .... *G. racemosum*
1. Pedicels of flowers and fruits more than 3 mm long; capsules with constrictions between the seeds .... *G. diffusum*

Gayophytum diffusum T. & G.
Fl. N. Am. 1:513, 1840.

Bigflower Groundsmoke

Gayophytum eriospermum Cov.
Gayophytum intermedium Rydb.
Gayophytum lasiospermum Greene
Gayophytum nuttallii T. & G.

Annual from a slender taproot; stems erect to ascending, glabrous to densely pubescent, strongly branched, especially above. Leaves alternate, linear to lanceolate, sessile to petiolate, entire, glabrous to strigose, becoming reduced upward. Flowers in the upper leaf axils, short-pedicellate in ours, the pedicels erect to spreading. Sepals 4, divided nearly to the ovary, reflexed; petals white to pinkish, obovate, short-clawed. Capsule short-pedicellate, mostly terete, erect, spreading, or somewhat reflexed, glabrous to strigose, constricted between the seeds. Seeds glabrous to puberulent, in 1–2 rows per locule.

Scree, talus, and other rocky places in the lower alpine zone near timberline in the Wasatch Range. British Columbia south to California, Arizona, and New Mexico, east to South Dakota.

Alpine plants as described, with petals 1–3 mm long, are var. *strictipes* (Hook.) Dorn.

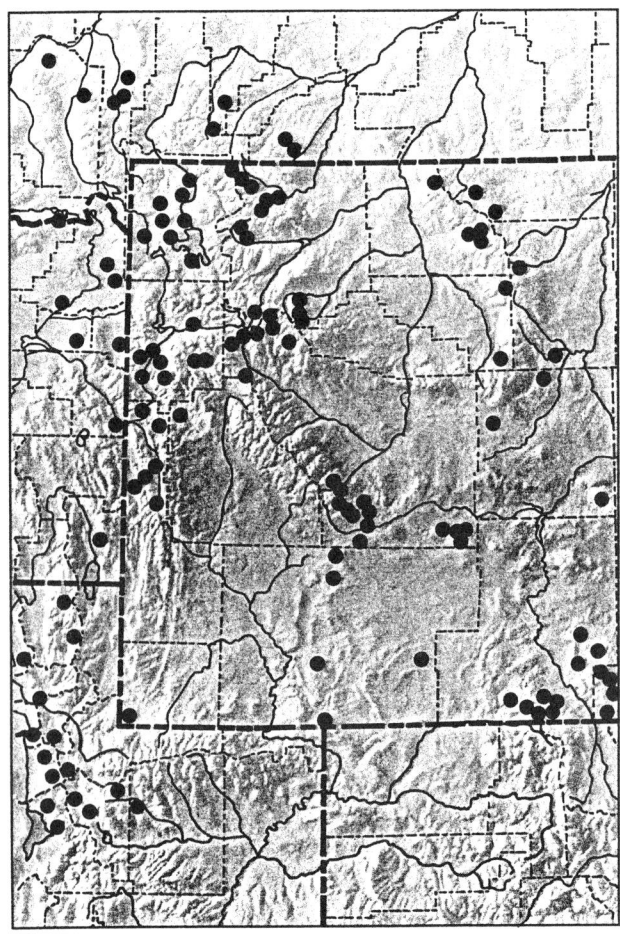

Gayophytum racemosum T. & G.
Fl. N. Am. 1:514, 1840.

Rocky Mountain Groundsmoke

Gayophytum caesium Nutt. ex T. & G.
Gayophytum decipiens Lewis & Szweykowski
Gayophytum helleri Rydb.

Annual from a slender taproot; stems erect to spreading, glabrous to puberulent, branched near the base, often unbranched above. Leaves alternate, linear to oblanceolate, sessile to petiolate, entire, glabrous to pubescent, little if at all reduced above. Flowers sessile to very short-pedicellate in the upper leaf axils. Sepals 4, divided nearly to the ovary, reflexed; petals white to pinkish, obovate, short-clawed. Capsule subsessile, flattened, erect to spreading, lacking constrictions between the seeds. Seeds glabrous to puberulent, in 2 rows per locule.

Rocky slopes and timberline margins of the Medicine Bow and Wasatch Ranges. Alberta south to California, Nevada, Utah, and Colorado.

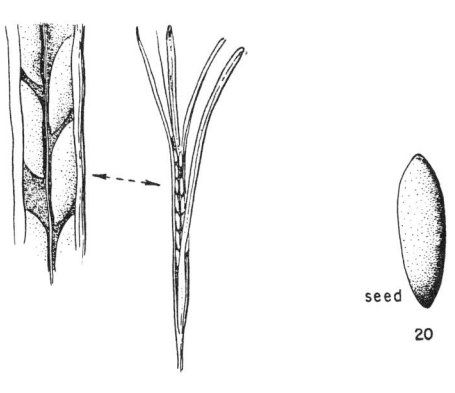

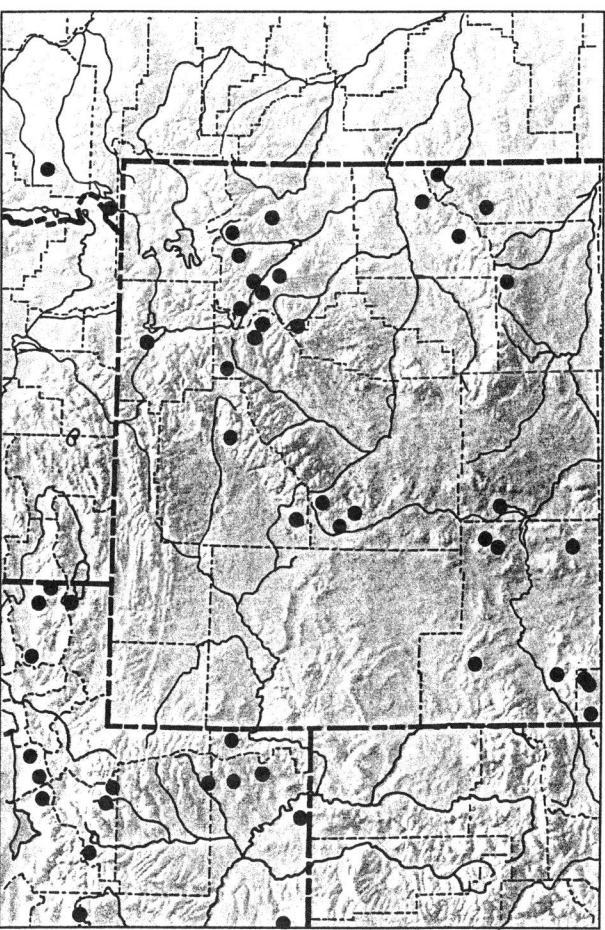

Oenothera L.
Evening Primrose

Annual to perennial, acaulescent to caulescent herbs. Leaves basal or alternate, entire to pinnatifid; stipules lacking. Flowers white or yellow, often fading to pink or rose, the hypanthium much longer than the ovary. Inflorescence a bracteate to ebracteate spike or raceme, or the flowers solitary in leaf axils. Sepals 4, reflexed at anthesis. Petals 4, large and conspicuous, entire to retuse or emarginate, lacking a claw. Stamens 8, subequal or dimorphic, the anthers versatile. Ovary 4-celled, the stigma capitate, discoid, or 4-lobed. Capsules loculicidal, 4-valved, straight to curved or coiled. Seeds numerous, not comose.

Oenothera cespitosa Nutt. in Fraser
Cat. no. 53, 1813.

Rock Rose

Oenothera idahoensis Mulford
Oenothera marginata Nutt. ex Hook. & Arn.
Oenothera montana Nutt. in T. & G.
Pachylophus cespitosus (Nutt.) Raimann
Pachylophus canescens Piper
Pachylophus crinitus Rydb.
Pachylophus cylindrocarpus A. Nels.
Pachylophus glabra A. Nels.
Pachylophus hirsutus Rydb.
Pachylophus macroglottis Rydb.
Pachylophus marginatus (Nutt. ex Hook. & Arn.) Rydb.
Pachylophus montanus (Nutt. in T. & G.) A. Nels.
Pachylophus psammophilus A. Nels. & Macbr.

Caespitose, acaulescent, villous to glabrate, perennial herb from a thick, woody taproot and simple or few-branched caudex. Leaves oblanceolate, petiolate, the blades sinuate-dentate to pinnatifid, or occasionally entire, glabrous, ciliate, or pubescent. Flowers solitary in leaf axils, sessile to pedicellate, erect (often somewhat drooping in bud); hypathium much longer than the petals, slender below, flared above, reddish, sparsely pubescent. Sepals lanceolate, reflexed at anthesis. Petals white to light pink, emarginate, spreading with age. Stamens subequal. Stigma of 4 linear lobes. Capsules sessile to pedicellate, oblong to ovoid, somewhat curved, 4-angled, tuberculate. Seeds dark brown, in 2 rows in each cell.

Common at lower elevations, *O. cespitosa* occurs at timberline on limestone scree and talus in the northern Wind River Range. Alberta and Saskatchewan, south in western North America to California, Idaho, Utah, New Mexico, and Nebraska.

Members of the northern Wind River Range population with short, somewhat curled hairs less than 0.5 mm long are var. *cespitosa*.

var. cespitosa

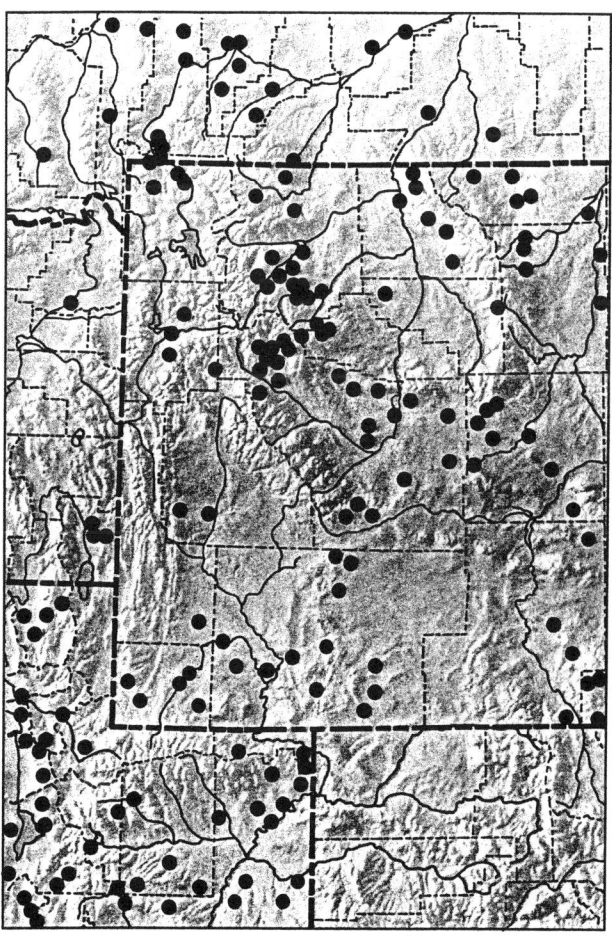

Ophioglossaceae
Adder's Tongue Family

Perennial, succulent herbs from a short, erect, subterranean stem; roots fleshy, mycorrhizal. Leaves simple or compound, of 2 kinds, a sterile blade and a fertile sporophyll united by a single petiole; bud partially or entirely enclosed by the enlarged petiole base. Sporangia in a terminal spike or panicle, numerous, homosporous, globose, bivalvate, the annulus lacking. Spores numerous, yellowish. Gametophytes subterranean, simple or branched, mycorrhizal.

Botrychium Sw.
Grape Fern

Slender, erect ferns from a short, erect rootstock; roots fleshy and clustered; common petiole often enlarged at the base. Fronds solitary, pinnately to ternately compound. Sporophyll paniculate or spikelike, the sporangia in lateral, globose, sessile or short-stalked clusters; spores yellowish. Gametophytes subterranean, saprophytic.

Botrychium lunaria (L.) Sw.
J. Bot. Schrad. 1800(2):110, 1801.

Moonwort

Osmunda lunaria L.
Botrychium minganense Vict.
Botrychium onondagense Underw.
Botrypus lunaria (L.) Rich.
Ophioglossum pinnatum Lam.
Osmunda lunata (L.) Salisb.

Glabrous, succulent, perennial herb with spreading, fleshy roots. Sterile blade sessile or nearly so, pinnate, the pinnae broadly lunate to reniform, overlapping, dichotomously veined. Fertile stalk (sporophyll) erect, racemose or paniculate, exceeding the sterile blade. Bud completely enclosed by the base of the common petiole.

Moist to wet meadows, ledges, willow thickets, and forest edges in the lower alpine zone of the Absaroka and Wind River Ranges. Circumboreal; Arctic Canada and Alaska south to California, Arizona, Colorado, Minnesota, Michigan, and New York.

Alpine plants that are smaller and more yellowish than typical *lunaria* are ssp. *occidentalis* Löve et al.

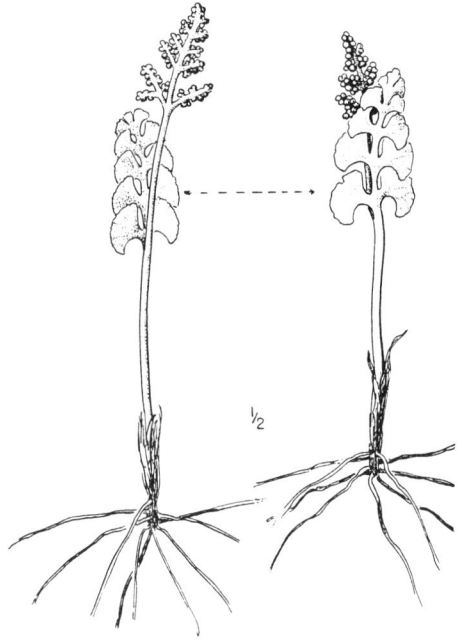

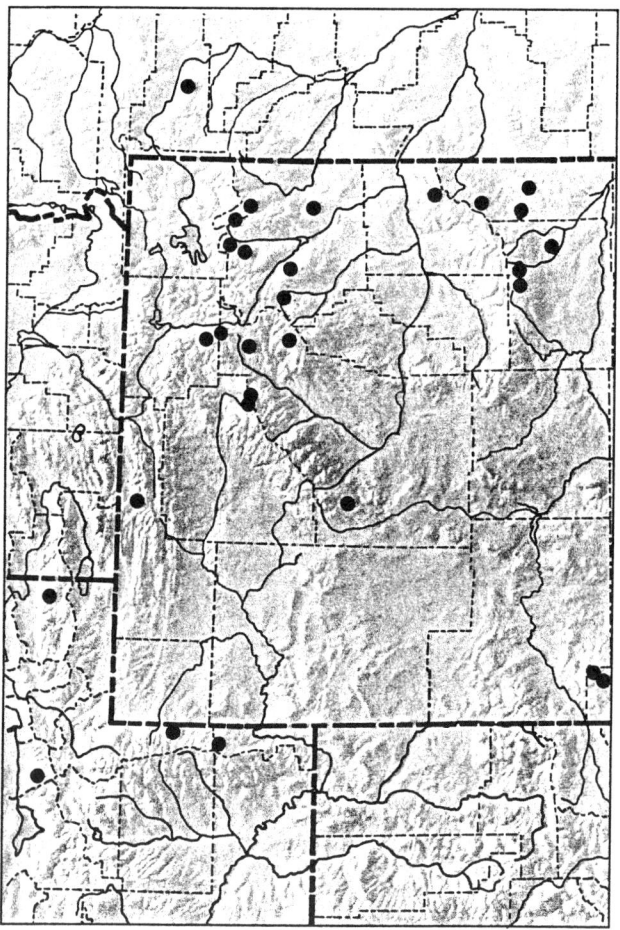

Orchidaceae
Orchid Family

Perennial herbs from corms, bulbs, or tuberous roots, often epiphytic, sometimes saprophytic. Leaves entire, sheathing, simple or sometimes reduced to scales. Flowers perfect, epigynous, irregular, solitary or in spikes or racemes. Sepals 3, usually petaloid. Petals 3, the lower petal (lip) unlike the lateral ones, sometimes with a spur at the base, often saccate. Stamens 1 or 2, united with the style and stigma into a column, the pollen often coherent in granular or waxy masses (pollinia). Pistil 1, of 3 united carpels, the ovary usually 1-celled with 3 parietal placentae. Style often prolonged into a beak (rostellum), the viscid stigmatic surface lying beneath the rostellum or in a cavity between the anther sacs. Fruit a 3-valved capsule with numerous minute seeds.

KEY TO THE GENERA OF ORCHIDACEAE

1. Lip with a distinct basal spur — *Habenaria*
1. Lip lacking a spur — *Spiranthes*

Habenaria Willd.
Bog Orchid

Glabrous herbs from tuberous roots, with 1 or more basal or cauline leaves and a terminal, bracteate spike or raceme of small, greenish or white flowers. Sepals and lateral petals similar in shape and color, the lip entire or 3-lobed, linear to obovate, prolonged at the base in a saccate or elongate spur. Stigmatic disks to which the pollinia are attached exposed, often widely separate.

Habenaria dilatata (Pursh) Hook.
Exot. Fl. 2:Pl. 95, 1825.

White Bog Orchid

Orchis dilatata Pursh
Habenaria borealis Cham.
Habenaria dilatatiformis Rydb.
Habenaria gracilis (Lindl.) S. Wats.
Habenaria graminifolia (Rydb.) Henry
Habenaria leptoceratitis (Rydb.) Henry
Habenaria leucostachys (Lindl.) S. Wats.
Limnorchis borealis (Cham.) Rydb.
Limnorchis dilatata (Pursh) Rydb.
Limnorchis dilatatiformis (Rydb.) Rydb.
Limnorchis foliosa Rydb.
Limnorchis gracilis (Lindl.) Rydb.
Limnorchis graminifolia Rydb.
Limnorchis leptoceratitis Rydb.
Limnorchis leucostachys (Lindl.) Rydb.
Platanthera dilatata (Pursh) Lindl. ex Beck
Platanthera gracilis Lindl.
Platanthera graminea Lindl.
Platanthera leucostachys Lindl.

Glabrous, perennial herb from a loose fascicle of slender, fleshy roots. Leaves cauline, oblong-elliptic to lanceolate, acute to obtuse, sheathing, reduced upward. Flowers white, yellowish, or greenish, in a lax to (more commonly) dense, spicate raceme. Upper sepal ovate to elliptic, hooded; lateral sepals elliptic-lanceolate, obtuse, usually falcate. Petals lanceolate, the upper two connivent with the upper sepal and forming a hood, the lip pendent, obtuse, dilated at the base; spur cylindrical to clavate, curved, shorter than to approximately twice as long as the lip. Column short, less than half the length of the upper petals, the base nearly surrounding the opening of the spur; pollinia 2 per pollen sac, the sacs only slightly separated by the connective.

Wet meadows and stream banks in the lower alpine zone of the Medicine Bow, Uinta, Wasatch, and Wind River Ranges. Alaska to Greenland and Labrador, south to New York, Michigan, South Dakota, New Mexico, Utah, and California.

Two varieties occur in the vicinity of timberline and the lower alpine zone: var. *leucostachys* (Lindl.) Ames has a spur 1.5–2 times as long as the lip; and var. *albiflora* (Cham.) Correll has a spur shorter than the lip.

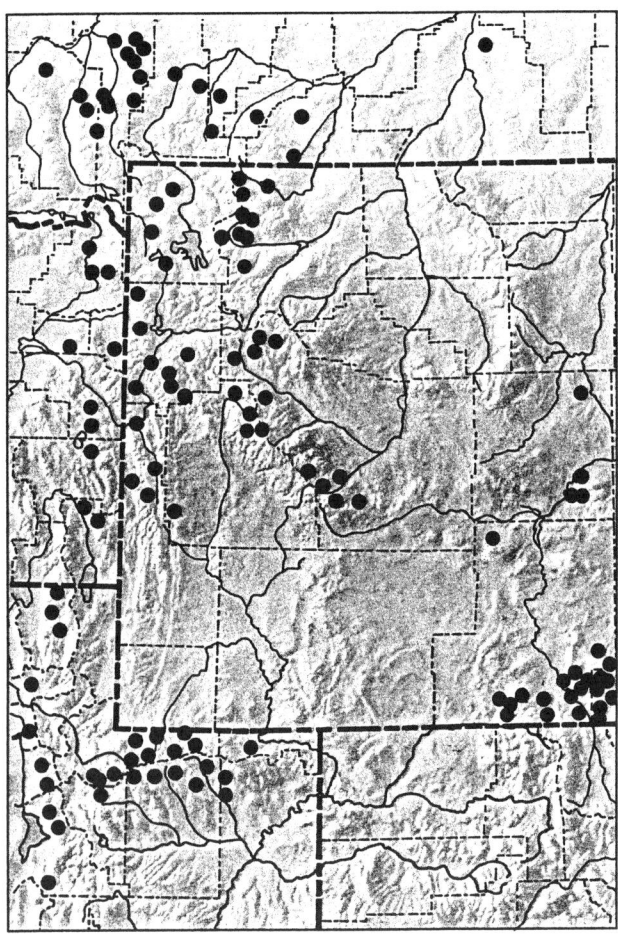

Spiranthes L.C. Rich.
Ladies' Tresses

Erect, perennial herbs with tuberous or fibrous roots. Leaves basal and cauline, sessile, entire. Flowers white to cream, in a terminal, spirally twisted spike. Sepals narrow, connivent, with 2 petals forming a hood enclosing the column and most of the lip, the lip oblong, sessile or short-clawed, usually short-tuberculate at the base. Column short, terete to clavate, with a single dorsal anther. Pollinia 2. Capsule erect, elliptic to ovoid.

Spiranthes romanzoffiana Cham.
Linnaea 3:32, 1828.

Hooded Ladies' Tresses

Gyrostachys porrifolia (Lindl.) Kuntze
Gyrostachys romanzowiana (Cham.) MacMill.
Gyrostachys stricta Rydb.
Ibidium porrifolium (Lindl.) Rydb.
Ibidium romanzoffianum (Cham.) House
Ibidium strictum (Rydb.) House
Orchiastrum porrifolium (Lindl.) Greene
Orchiastrum romanzoffianum (Cham.) Greene
Spiranthes porrifolia Lindl.
Spiranthes stricta (Rydb.) A. Nels. in Coult. & A. Nels.
Triorchis romanzoffiana (Cham.) Nieuwl.
Triorchis stricta (Rydb.) Nieuwl.

Slender, erect, caulescent, glabrate, perennial herb from fascicled, tuberous roots. Basal leaves sessile, linear to lanceolate, occasionally oblanceolate, the margins entire; cauline leaves sheathing, lanceolate, acute to acuminate, reduced upward. Flowers white or cream, in dense, terminal, spiral, spikelike racemes; floral bracts lanceolate, acuminate. Sepals and petals ovate-lanceolate, connivent, forming a somewhat falcate hood over the column and base of the lip; lower sepals with reflexed tips; lip deflexed, constricted above the middle and flared into a rounded, erose to crenulate tip. Column beaked; pollinia 2, each 2-cleft.

Wet meadows, stream banks, and seeps in the Uinta and Wasatch Ranges. Transcontinental in Canada; Alaska south in western North America to California, Arizona, New Mexico, and Nebraska.

Middle Rocky Mountain plants with the lip constricted below the tip and lacking prominent basal tuberosities are var. *romanzoffiana*.

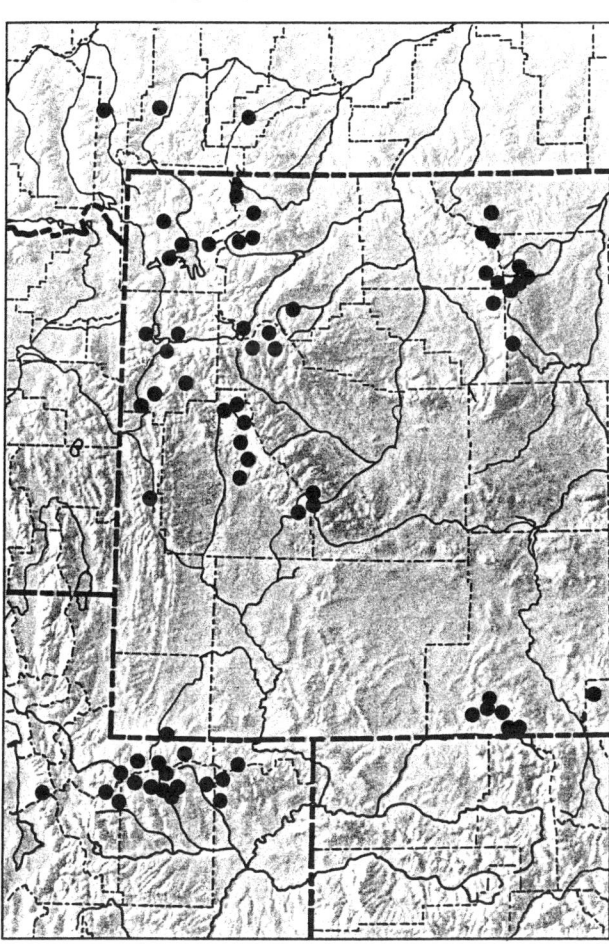

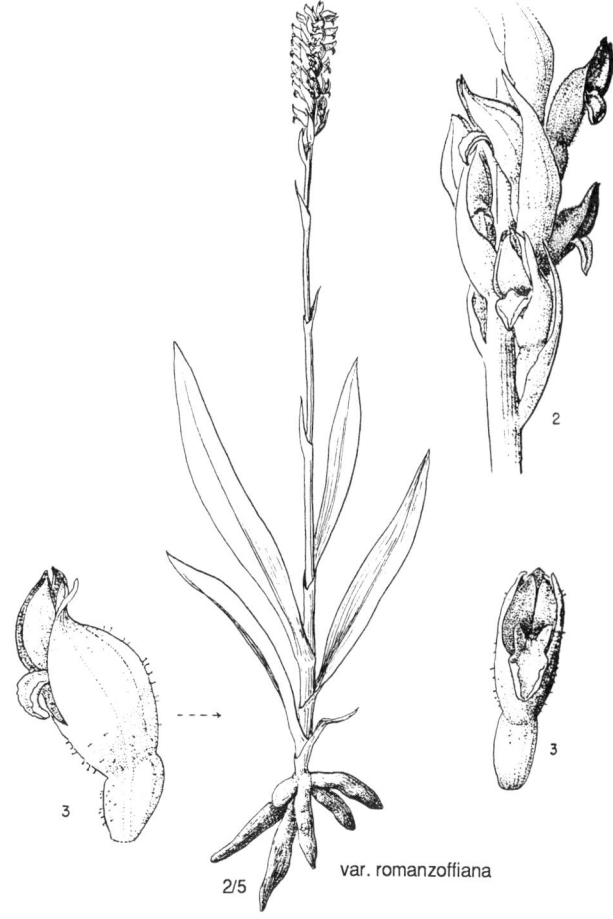

var. romanzoffiana

Papaveraceae
Poppy Family

Annual to perennial herbs with milky or colored sap. Leaves alternate or basal, simple, entire, lobed, or variously compound. Flowers mostly solitary and terminal, hypogynous, regular, perfect. Sepals 2 or 3, distinct, caducous. Petals mostly 4, distinct, imbricate in bud, fugacious. Stamens numerous, the filaments distinct. Pistil 1, with 2 to several carpels; ovary 1-celled with parietal placentation; stigmas the same number as the carpels. Fruit a capsule, dehiscent by pores or valves.

Papaver L.
Poppy

Tufted annual or perennial herbs from a taproot or fibrous roots, with milky or colored latex; stems erect or ascending, mostly simple. Leaves alternate or basal, pinnately lobed or dissected, mostly petiolate. Flowers perfect, solitary, showy, on long terminal or axillary peduncles, nodding in bud. Sepals 2, deciduous. Petals 4 and white, yellow, or red. Stamens numerous. Ovary with 4 to many carpels; stigmas united to form a discoid, peltate crown (stigmatic disk). Fruit a globose to nearly cylindrical capsule dehiscent by pores beneath the stigmatic disk; seeds minute, numerous.

Papaver kluanense D. Löve in Löve & Freedman
Bot. Notiser 109:178, 1956.

Arctic Poppy

Papaver alaskanum Hult.
Papaver alpinum L.
Papaver lapponicum (Tolm.) Nordh. in part
Papaver radicatum Rottb. in part

Caespitose perennial from a stout taproot, the caudex covered by persistent bases of old petioles. Scapes sparsely to densely hirsute. Leaves numerous, pinnately lobed or parted, petiolate, sparsely hirsute. Flowers solitary. Sepals 2, black-hirsute. Petals 4, yellow or whitish, obovate. Stamens numerous. Capsule obovoid, brown hispid.

Scree slopes, rocky ledges, and high mountain passes of the Absaroka, Beartooth, Big Horn, Uinta, and Wind River Ranges. Circumpolar; south in western North America to Colorado.

Papaver kluanense is a member of the arctic and northern alpine *P. radicatum* Rottb. complex and treated by some authors as that species, or as *P. lapponicum* (Tolm.) Nordh. ssp. *occidentale* (Lundstr.) Knaben. Following Löve (1969), who recognized it as cytologically separated (2n=42) from *P. radicatum* (2n=56), the name *P. kluanense* is used here to describe the high-altitude plants of the Middle Rocky Mountains. At 3818 m (12,600 ft), this is one of the highest-growing vascular plants in the Middle Rocky Mountains. It is rarely found below 3333 m (11,000 ft).

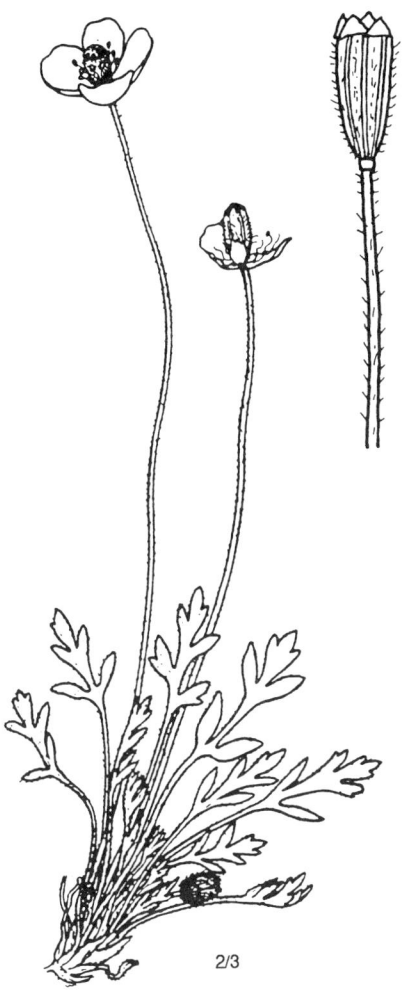

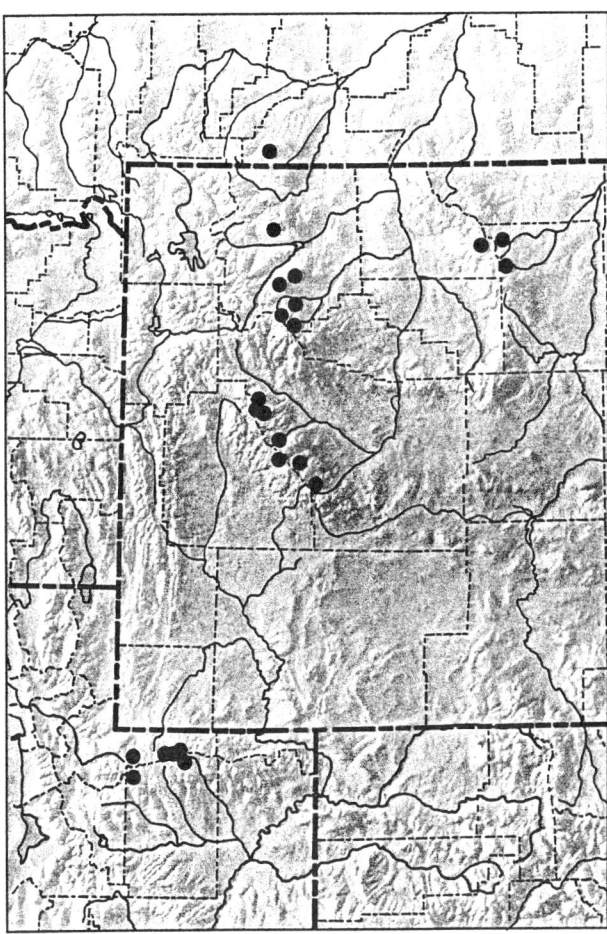

Pinaceae
Pine Family

Evergreen, resinous, monoecious trees with linear or needlelike leaves produced singly or in fascicles of 2–5. Staminate cones small, ovoid to oblong, early deciduous. Pistillate cones larger, of numerous woody or coriaceous scales subtended by included or exserted bracts, deciduous or persistent. Ovules naked on the scales of the pistillate cones. Stigmas none. Seeds 2 on each scale, winged or wingless.

KEY TO THE GENERA OF PINACEAE

1. Leaves in fascicles of 2–5; cones with thick, woody scales — *Pinus*
1. Leaves single; cones with thin, papery or leathery scales
   2. Leaves square in cross section; cones pendent — *Picea*
   2. Leaves flat; cones pendent or not
      3. Cones pendent, 3-pronged bracts present between cone scales; buds sharp-pointed — *Pseudotsuga*
      3. Cones erect, 3-pronged bracts lacking; buds blunt — *Abies*

Abies Mill.
Fir

Trees with narrow, pyramidal crowns and smooth, resinous bark. Leaves sessile, linear, flat, blunt at the apex, appearing 2-ranked because of twisting near the base. Buds resinous, blunt. Pistillate cones erect, oblong, the scales tightly imbricate, deciduous on the persistent cone axis. Seeds terminally winged.

Abies lasiocarpa (Hook.) Nutt.
N. Am. Sylva 3:138, 1849.

Subalpine Fir

Pinus lasiocarpa Hook.
Abies arizonica Merriam
Abies bifolia Murray
Abies subalpina Engelm. in Ward

Small tree with a narrow, pyramidal crown, or a low, prostrate shrub. Bark gray, smooth, with prominent resin blisters when young; older bark gray to gray-brown, fissured to platy; young twigs reddish-puberulent. Leaves single, blue-green, obtuse or notched at the end, ribbed beneath, erect, the lower on a branch twisted so that all appear to project upward. Staminate cones bluish. Ovulate cones erect, dark purple, oblong, the scales cuneate at the base, broadly obovate. Bracts lacerate, less than half as long as the scales. Seeds about 6 mm long, winged.

Typical of the subalpine zone and common at timberline throughout the region, *A. lasiocarpa* occurs in the alpine zone as krummholz or elfinwood islands. Individuals are often flagged on the upwind side by blowing ice crystals. Alaska and Yukon Territory south to California, Arizona, and New Mexico.

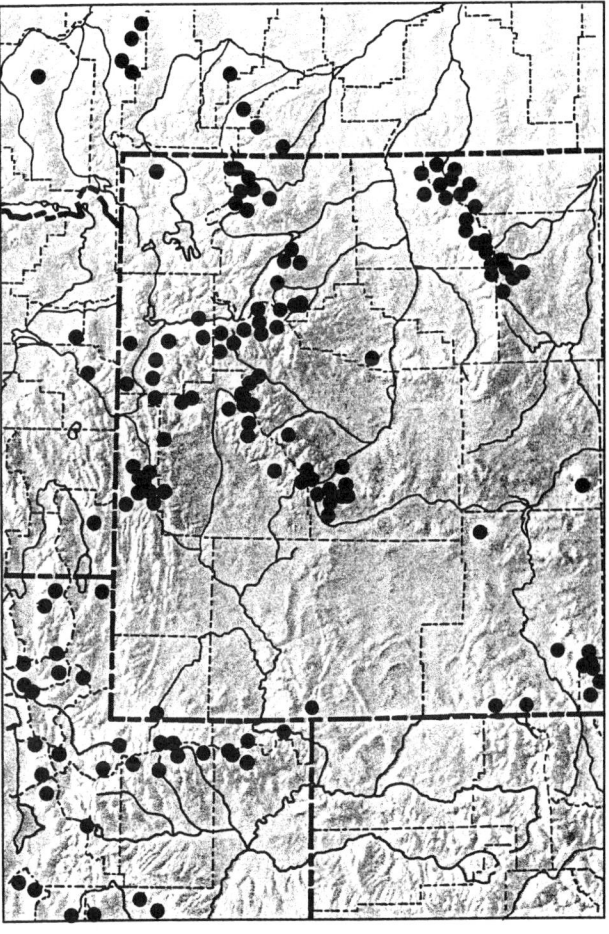

Picea A. Dietr.
Spruce

Trees with spire-shaped crowns and scaly bark. Leaves spirally arranged, linear, the blades 4-angled in cross section, sharply acute, and jointed at the base; on falling, leaving the persistent sterigmata on the twigs. Pistillate cones pendent, oblong, the scales loosely imbricate, thin-coriaceous. Bracts included, the entire cone falling at maturity. Seeds terminally winged.

Picea engelmannii Parry ex Engelm.
Trans. Acad. Sci. St. Louis 2:212, 1863.

Engelmann Spruce

Abies engelmannii Parry
Picea columbiana Lemmon

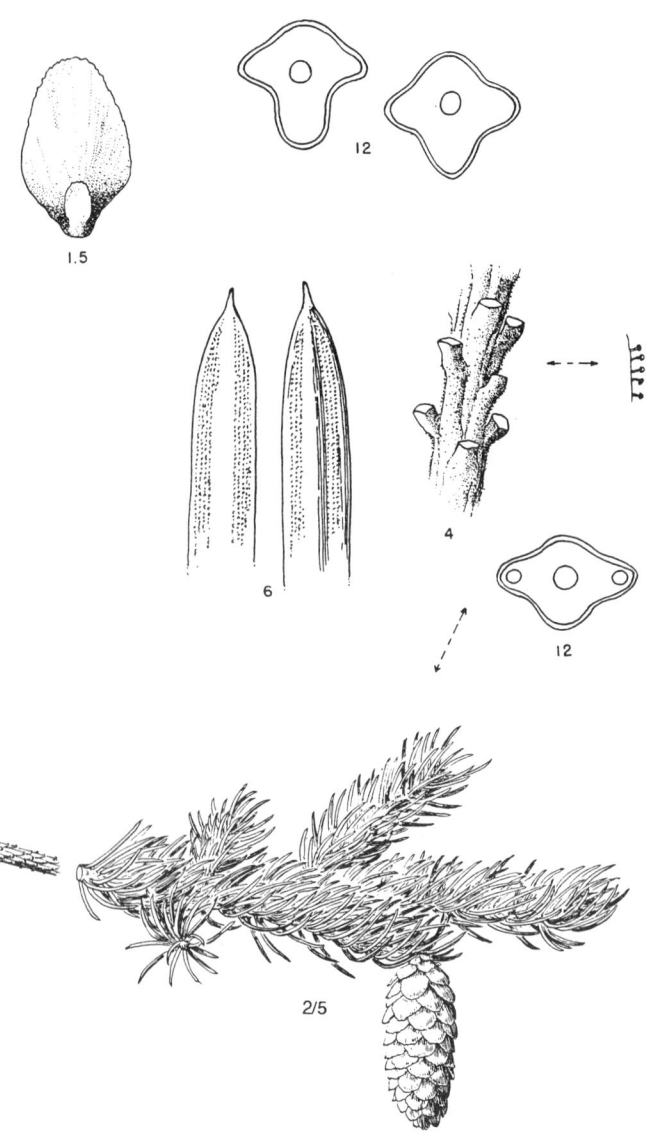

Tree with a pyramidal crown, or a low, prostrate shrub. Bark reddish brown, scaly; young twigs finely pubescent, rarely glabrous. Leaves single, blue-green, glaucous when young, quadrangular, acute, spreading in all directions from the branches. Staminate cones yellow at maturity, often purple when young. Ovulate cones pendent, brownish, the scales broadly ovate, erose, concealing the much shorter, oblong bracts. Seeds about 3 mm long, winged.

Common in the upper montane and subalpine zones, and found at timberline throughout the region. In the alpine zone *P. engelmannii* occurs as krummholz and elfinwood islands. Yukon Territory, Alberta, and British Columbia south to Colorado, New Mexico, Arizona, and California.

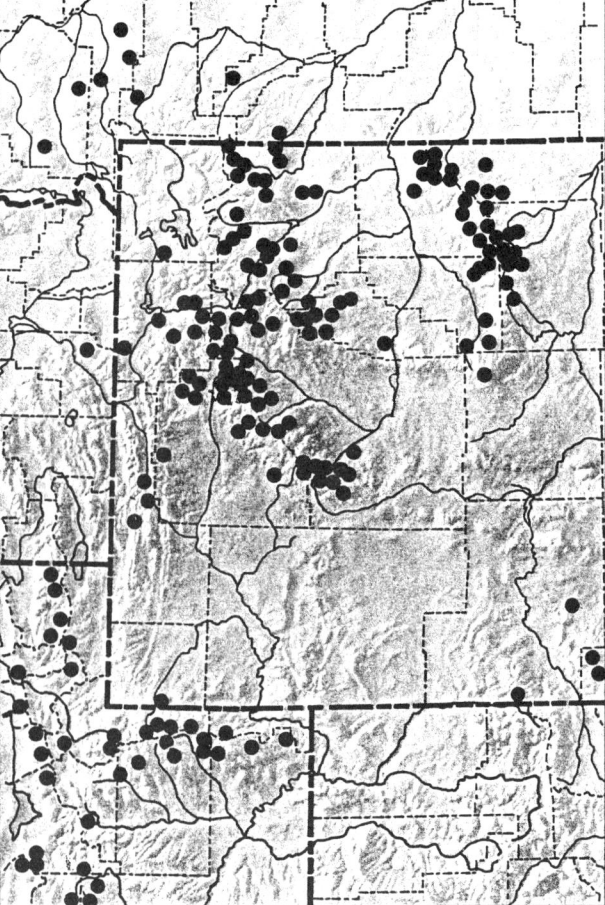

Pinus L.
Pine

Trees, or shrubs in the alpine zone, with needlelike leaves in fascicles of 2–5 in ours, enclosed at the base by a chaffy sheath, or rarely single. Pistillate cones large, woody, with thick scales, maturing in 2 years. Seeds winged or, in ours, wingless.

KEY TO THE SPECIES OF *PINUS*

1. Leaves in fascicles of 5
    2. Cones reddish purple, usually less than 8 cm long, the scales thickened toward the tip — *P. albicaulis*
    2. Cones light brown to olive, usually more than 8 cm long, the scales thinned toward the tip — *P. flexilis*
1. Leaves in fascicles of 2 — *P. contorta*

Pinus albicaulis Engelm.
Trans. Acad. Sci. St. Louis 2:209, 1868.

Whitebark Pine

Alpinus albicaulis (Engelm.) Rydb.
Pinus cembroides Newberry

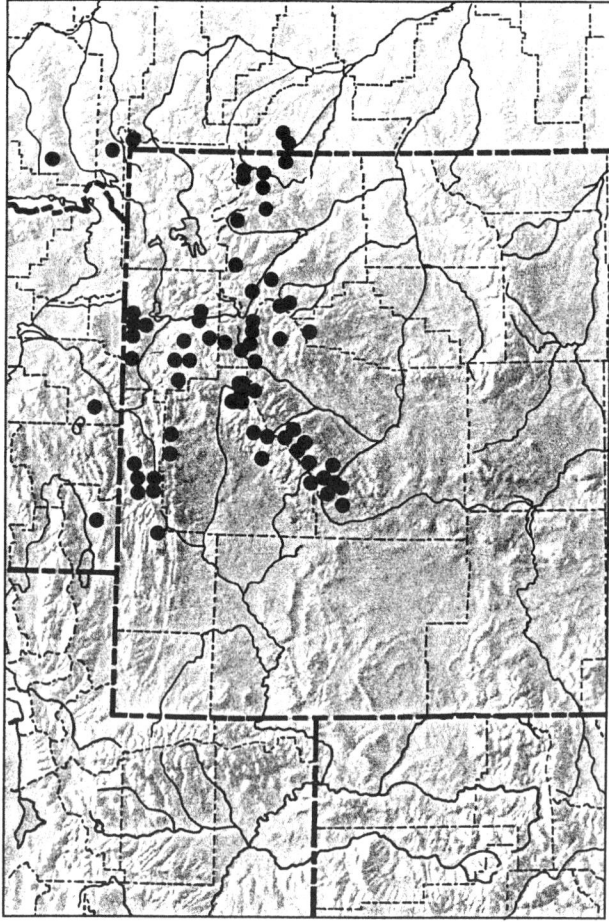

Low, spreading tree or prostrate shrub. Bark smooth, silver-gray when young; older bark grayish brown, scaly; young twigs pubescent. Leaves in fascicles of 5, dark yellow-green, stiff, slightly curved, clustered near the ends of branches. Staminate cones red. Ovulate cones reddish purple, ovoid, resinous, the scales densely imbricate, broad and thickened at the apex, unarmed. Seeds large, 8–12 mm long, wingless when shed.

Almost entirely restricted to the subalpine zone, *P. albicaulis* often forms extensive pure stands near timberline. In the alpine zone of the northwestern mountain ranges it forms occasional krummholz islands and aerodynamically sculpted forms downwind from felsenmeer and fellfield boulders. British Columbia and Alberta south to Wyoming, Idaho, Nevada, and California.

Pinus albicaulis closely resembles *P. flexilis* vegetatively, and the two are very difficult to distinguish without their cones. In the northern portion of the Middle Rockies *P. flexilis* occurs most commonly from the foothills to the upper montane zone. Most 5-needle pines at timberline are *P. albicaulis*. In the southern portion of the Middle Rockies all 5-needle pines at timberline are *P. flexilis*. *Pinus longaeva* D.K. Bailey, another 5-needle pine, has been collected north and west of the southern tip of the Wasatch Range, but to my knowledge is still not known from the Middle Rocky Mountains. It is distinguished from *P. flexilis* by generally shorter needles and cones, and cone scales with incurved prickles 4–6 mm long.

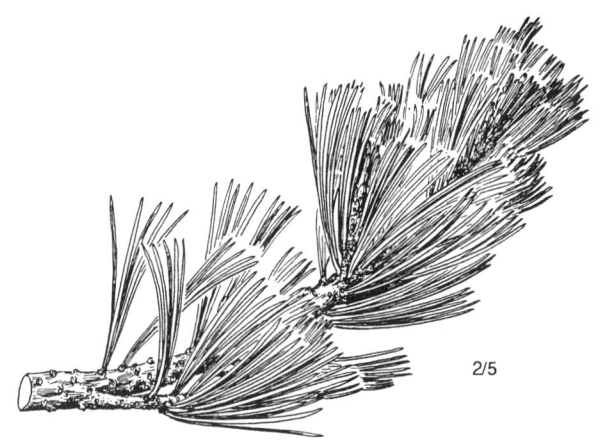

Pinus contorta Dougl. ex Loud.
Arb. Frut. Brit. 4:2292, 1838.

Lodgepole Pine

Pinus murrayana Grev. & Balf.

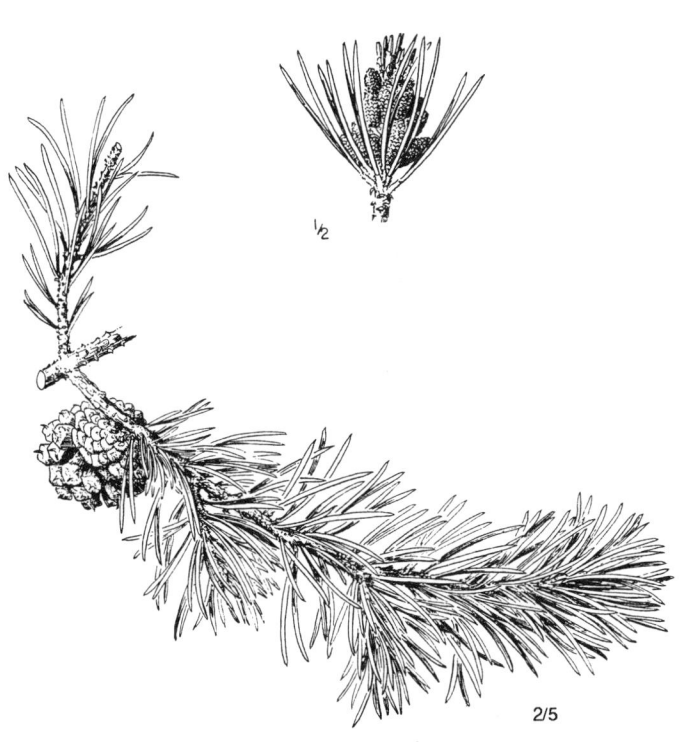

Slender tree with an oblong to pyramidal crown. Bark orange-brown to grayish, rough, scaly. Leaves in fascicles of 2, yellow-green, usually twisted, with persistent membranous sheaths. Staminate cones orange to red, yellow at maturity. Ovulate cones short-ovoid, asymmetrical, and slightly curved, often persistent and remaining closed for several years; scales narrow, thickened at the end and spine-tipped. Seed with an oblong wing.

Lodgepole pine is the most common conifer at middle elevations in the Middle Rocky Mountains and is occasional near timberline in the subalpine zone. It has been collected at timberline and in the lower alpine zone of the Medicine Bow and Wind River Ranges and may be expected in similar sites in the other ranges. Alaska and Yukon Territory south through the western American cordillera to Colorado, Utah, Nevada, and northern Baja California.

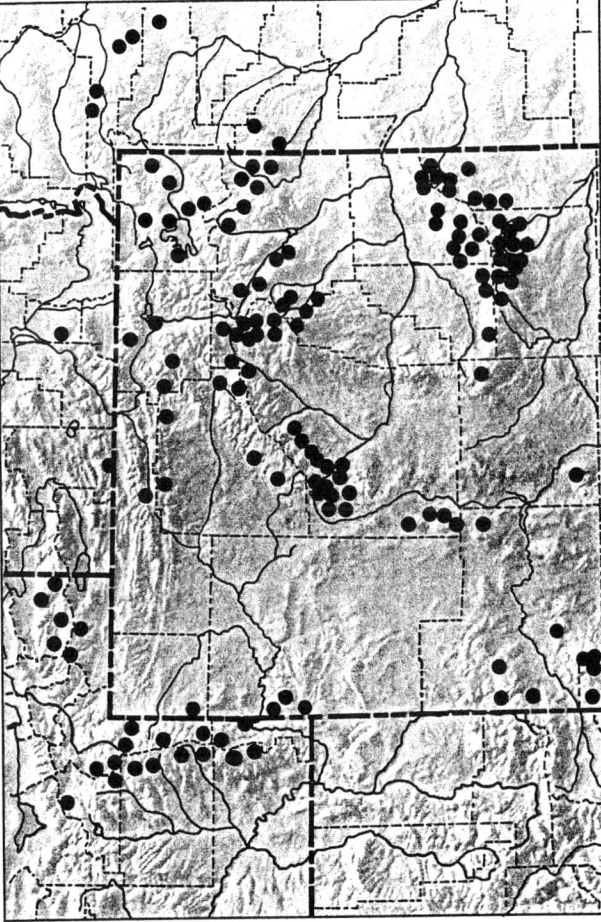

Pinus flexilis James
Rep. Long's Exped. Rocky Mtns. 2:27, 34, 1823.

Limber Pine

Alpinus flexilis (James) Rydb.

Low, spreading tree with a stout trunk, often shrublike in exposed habitats. Bark smooth, silver-gray when young, thick, cracked, and platy, dark brown to blackish when older; young twigs finely pubescent. Leaves in fascicles of 5, rigid, slightly curved, crowded near the ends of the branches. Staminate cones reddish. Ovulate cones ovoid to cylindrical, light brown to olive, falling entire from tree; scales thinned toward the apex, unarmed. Seeds 7–12 mm, dark brown, wingless when shed.

Exposed slopes, ridges, ledges, and rocky places. Occasional at timberline in the Absaroka, Big Horn, Uinta, Wasatch, and Wind River Ranges. British Columbia and Alberta south to California, Arizona, New Mexico, Nebraska, and the Dakotas.

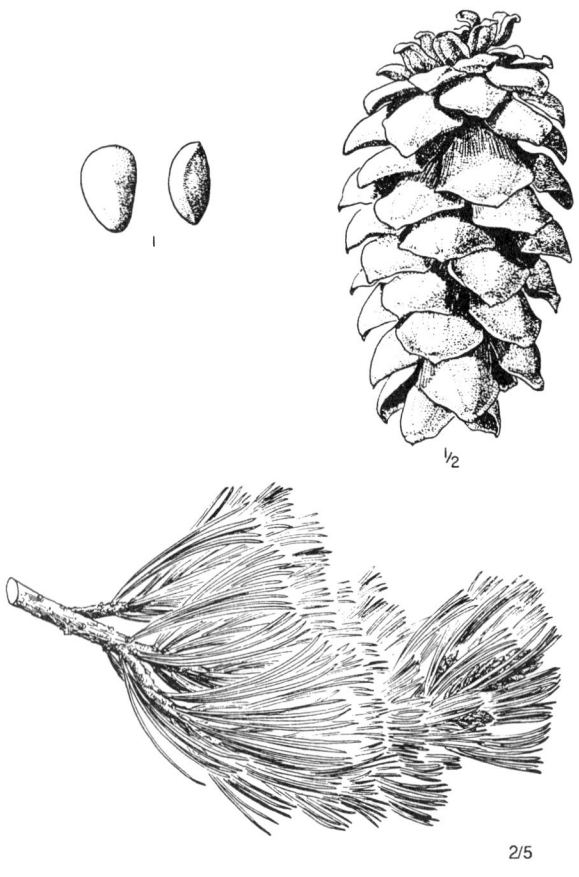
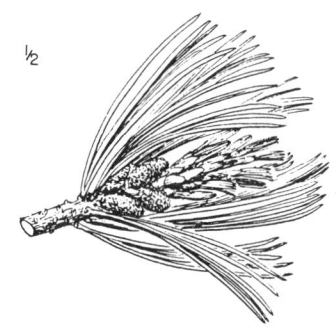
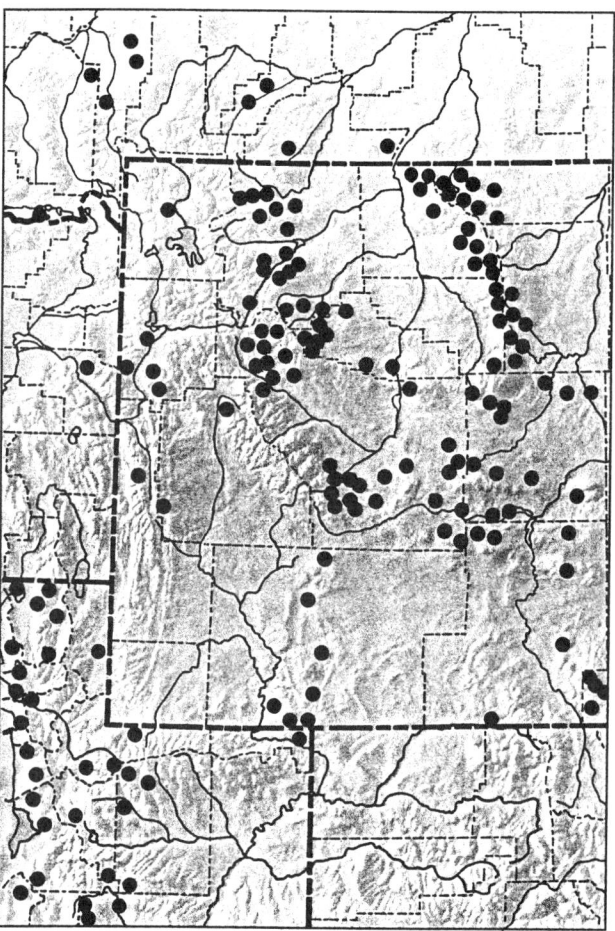

Pseudotsuga Carr.
Douglas Fir

Evergreen trees with broad crowns and platy bark. Leaves flat, single, short-petiolate, spirally arranged. Buds nonresinous, conical, and sharp-pointed. Pistillate cones pendent, oblong to ovoid, maturing in one season, the scales thin, rounded; bracts subtending the scales 3-lobed, conspicuously exerted. Seeds oblong, winged.

Pseudotsuga menziesii (Mirbel) Franco
Bol. Soc. Brot., 2, 24:74, 1950.

Douglas Fir

Abies menziesii Mirbel
Abies douglasii Hort. ex Loud.
Abies douglasii Lindl.
Abies lindleyana (Roezl) Murray
Abies mucronata Raf.
Abies taxifolia (Lamb.) Poir. in Lam.
Abies trigona Raf.
Pinus douglasii (Hort. ex Loud.) Sabine ex D. Don in Lamb.
Pinus taxifolia Lamb.
Pseudotsuga caesia (Schwerin) Flous
Pseudotsuga flahaulti Flous
Pseudotsuga glauca (Beissn.) Mayr
Pseudotsuga globulosa Flous
Pseudotsuga lindleyana (Roezl) Carr.
Pseudotsuga merrillii Flous
Pseudotsuga mucronata (Raf.) Sudw.
Pseudotsuga rehderi Flous
Pseudotsuga taxifolia (Lamb.) Britt. ex Sudw.
Pseudotsuga vancouverensis Flous
Tsuga lindleyana Roezl

Stout tree with a broad, pyramidal crown, becoming flattened with age; bark thick, brown, platy, with deep fissures between the plates. Branches slender, drooping or erect, pubescent near the tips. Leaves flat, linear, acute or obtuse, short-petiolate, spreading equally around the branch or sometimes turning upward. Staminate cones yellowish to reddish. Ovulate cones ovoid, pendent, maturing in one season and early deciduous; scales reddish brown, rounded, thin; bracts 3-lobed, conspicuously exerted. Seeds winged, the wing up to twice as long as the seed.

Occasional at timberline in the Big Horn, Wasatch, and northern Wind River Ranges on sedimentary substrates, particularly limestones. British Columbia and Alberta south to California, Mexico, Texas, and Colorado.

Middle Rocky Mountain trees with bracts mostly spreading or reflexed are var. *glauca* (Beissn.) Franco.

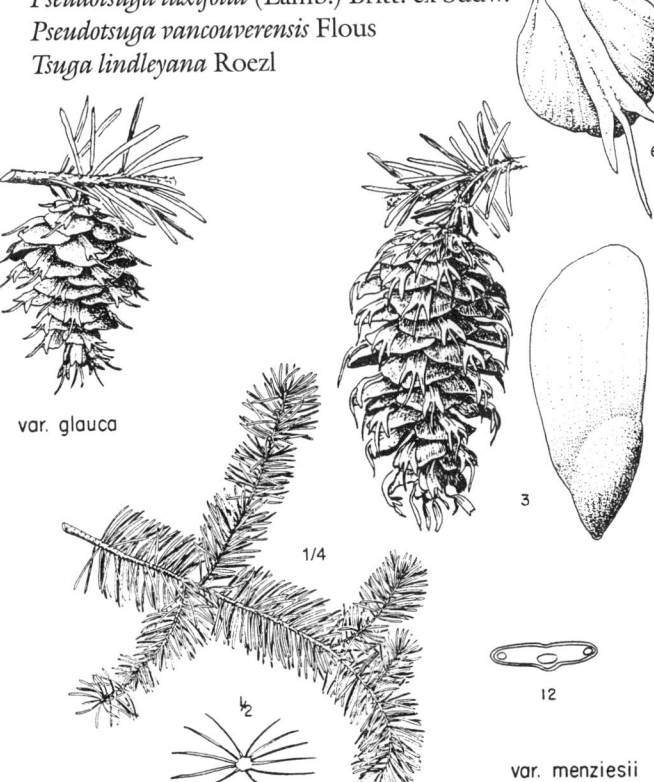
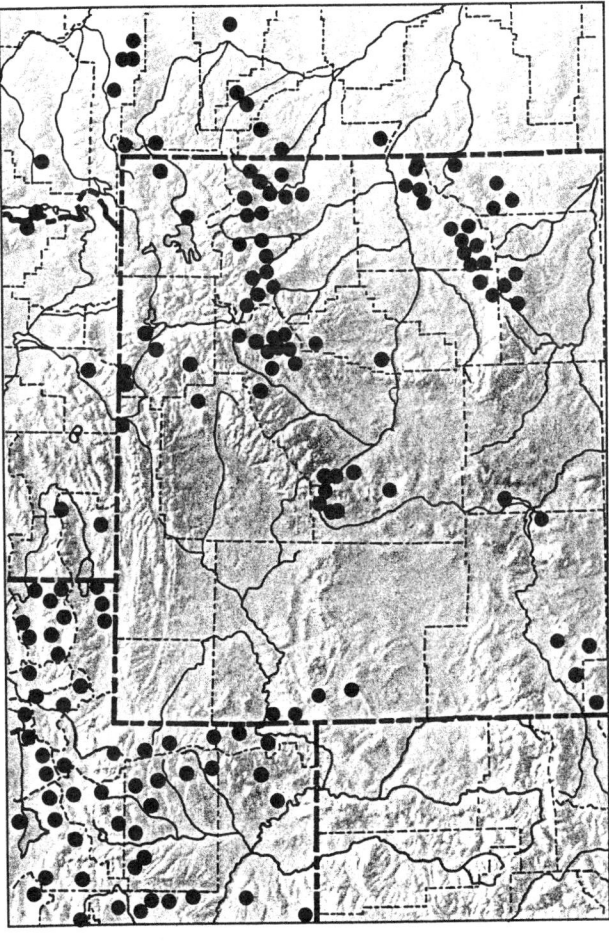

Plantaginaceae
Plantain Family

Annual or perennial herbs with simple, mostly basal leaves, the blades often ribbed. Flowers hypogynous, perfect or unisexual, regular, in terminal, bracteate spikes. Sepals 4, persistent. Petals 4, sympetalous, scarious, the corolla tubular to campanulate. Stamens 2 or 4, distinct, epipetalous, included or exerted, alternate with the corolla lobes. Pistil 1; ovary 2-carpellate in ours; style 1; stigma 1. Fruit a circumscissile capsule or an akene.

Plantago L.
Plantain

Annual to perennial, mostly acaulescent herbs from fibrous roots or taproots. Leaves simple, linear, oblanceolate, or ovate. Flowers small, greenish, perfect or unisexual, in dense, cylindrical, bracteate spikes. Sepals 4, connate at the base; petals 4, scarious. Stamens 4, rarely 2, inserted on the corolla. Ovary 2-loculed; style 1, filiform; stigma elongate, pubescent. Fruit a circumscissile capsule.

Plantago tweedyi Gray
Syn. Fl., 2nd ed., 2(1), 1886.

Tweedy Plantain

Glabrous to pubescent, perennial herb from a short, fibrous taproot. Leaves basal, oblanceolate to elliptic, entire or nearly so, gradually tapering to the petiole. Flowers perfect, in dense, bracteate spikes, the scapes barely exceeding the leaves. Calyx short, inconspicuous. Corolla lobes spreading to reflexed. Stamens 4. Capsule ovoid, circumscissile.

Moist timberline meadows and slopes. Known from the alpine zone only in the Beartooth, Medicine Bow, Uinta, and Wasatch Ranges. Idaho south to Utah, New Mexico, and Colorado.

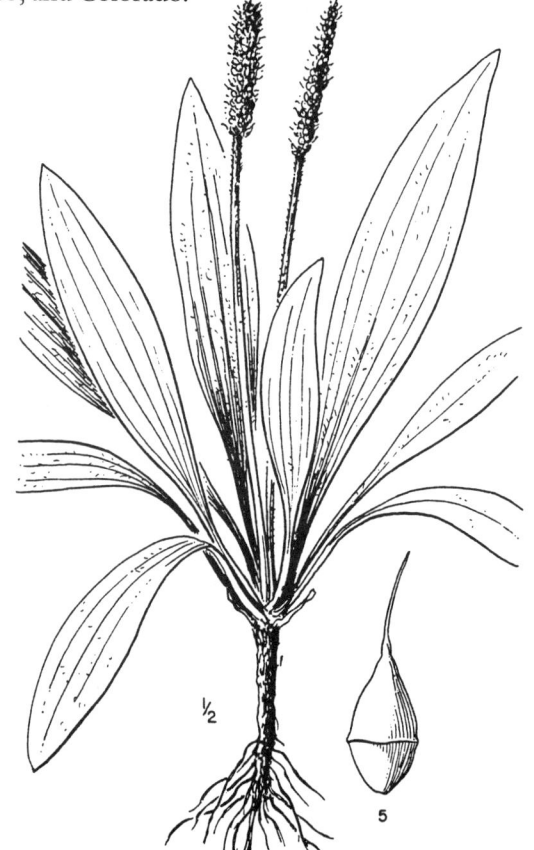

Poaceae
Grass Family

Annual or perennial herbs, in ours. Stems often branched at the base, in perennials forming flowering stems (culms) and sterile shoots (innovations). Culms round in cross section or somewhat flattened, usually hollow at the internodes and solid at the nodes. Leaves alternate and 2-ranked, consisting of a blade, a sheath, and a ligule; blades linear, entire, the veins parallel; the sheath encircling the stem at the base of the blade, the edges free or united; the ligule at the junction of the blade and sheath, commonly membranous. Inflorescence a spike, raceme, or panicle of spikelets, or sometimes of a single spikelet. Flowers (florets) small, perfect or unisexual, enclosed between 2 bracts, the lemma and the palea; perianth much reduced and consisting of 2 or 3 minute scales (lodicules). Florets sessile on the rachilla and bearing at the base 2 empty bracts (glumes), the florets and glumes, together forming a spikelet. Articulation above the glumes and between the florets or below the glumes, and the spikelets falling entire. Stamens 1–6, commonly 3. Pistil 1, usually with 2 plumose styles, the ovary with a single ovule. Fruit a caryopsis (grain), free or enclosed by the palea and lemma at maturity.

KEY TO THE GENERA OF POACEAE

1. Inflorescence a single, symmetrical spike (Hordeae)
   2. Spikelets widely divergent from the rachis, 2-ranked; internodes very short, mostly less than 2 mm long — *Agropyron*
   2. Spikelets ascending or appressed to the rachis, not 2-ranked; internodes mostly more than 3 mm long — *Elymus*
1. Inflorescence a raceme or panicle, sometimes narrow and spikelike, but never a true spike
   3. Spikelets with 1 perfect terminal floret and 2 sterile florets below it (Phalarideae) — *Hierochloe*
   3. Spikelets with 1 or more perfect florets, the sterile florets, if any, above the fertile floret
      4. Floret single in each spikelet (Agrostideae)
         5. Lemma indurate, terminally awned, and persistent around the mature fruit; nerves obscure — *Stipa*
         5. Lemma not indurate, awned or awnless; nerves evident
            6. Inflorescence a single, cylindric, spikelike raceme
               7. Articulation below the glumes; lemmas dorsally awned from below the middle — *Alopecurus*
               7. Articulation above the glumes; lemmas awnless — *Phleum*
            6. Inflorescence an open or contracted panicle
               8. Glumes longer than the lemmas
                  9. Callus of lemma with a tuft of hairs one-fourth or more as long as the lemma; palea evident — *Calamagrostis*
                  9. Callus of lemma without a tuft of hairs; palea minute or lacking — *Agrostis*
               8. Glumes shorter than the lemma or one of them lacking — *Phippsia*
      4. Florets 2 or more each spikelet
         10. Glumes shorter than the first lemma in the spikelet; awns, if present usually straight and terminal (Festuceae)
            11. Spikelets 15 mm long or more; lemmas awned from between the teeth of a minutely bidentate apex, or awnless — *Bromus*
            11. Spikelets less than 15 mm long; lemmas awned from the tip or awnless

12. Lemmas rounded on the back, slender-pointed, often awned   *Festuca*
12. Lemmas keeled at least at the summit, acute or blunt-pointed, awnless
    13. Plants dioecious; leaves tapering to a slender point, the sheaths open to the base   *Leucopoa*
    13. Plants with perfect flowers; leaves usually boat-shaped at the tip, the sheaths closed, at least near the base   *Poa*
10. Glumes equaling or exceeding the first lemma in the spikelet; awns of lemmas, if present, often dorsal and twisted or arising from between 2 apical teeth (Aveneae)
    14. Lemmas awned from between 2 apical teeth   *Danthonia*
    14. Lemmas awned from the back or awnless
        15. Lemmas convex on the back, not keeled
            16. Glumes 7 mm long or less; rachilla prolonged above the terminal floret   *Deschampsia*
            16. Glumes 10 mm long or more; rachilla not prolonged above the terminal floret   *Helictotrichon*
        15. Lemmas keeled on the back
            17. Callus and slender rachilla joint bearded; lemmas usually awned   *Trisetum*
            17. Callus and very short rachilla joint glabrous or minutely pubescent; lemmas usually awnless   *Koeleria*

Agropyron Gaertn.
Wheatgrass

Perennials, tufted in ours, from fibrous roots; culms erect to ascending. Leaves flat or involute; ligule membranous. Inflorescence a single symmetrical spike, the rachis continuous; internodes much shorter than the spikelets. Spikelets numerous, 2-ranked and widely divergent, 2–10-flowered, sessile, solitary at each joint of the rachis, flattened, strongly overlapping. Articulation above the glumes and between the florets. Glumes shorter to longer than the lowermost lemma, unequal, lanceolate, acute to awned. Lemmas keeled, mostly awned from the tip, 5-nerved. Palea well developed, bidentate, subequal to the lemma.

Agropyron cristatum (L.) Gaertn.
Nov. Comm. Petrop. 14:540, 1770.

Crested Wheatgrass

Bromus cristatus L.
Agropyron cristatiforme Sarkar
Agropyron desertorum (Fisch. ex Link) Schult.
Agropyron fragile (Roth) Candargy
Agropyron pectiniforme R. & S.
Agropyron sibiricum (Willd.) Beauv.
Avena cristata (L.) R. & S.
Costia cristata (L.) Willk.
Triticum cristatum (L.) Schreb.
Triticum desertorum Fisch. ex Link
Triticum pectinatum Bieb.
Triticum sibiricum Willd.
Zeia cristata (L.) Lunell

Tufted, perennial herb from fibrous roots; rhizomes lacking; culms several to many, erect to ascending. Leaves both basal and cauline, the blades flat, glabrous to pubescent; sheaths auriculate, glabrous, the lower ones sometimes pubescent; ligules erose. Spike oblong, ovate, or cylindric, dense, the rachis puberulent; spikelets 2–10-flowered, 2-ranked, and divergent from the rachis. Glumes oblong-lanceolate, unequal or subequal, awned or not, 1–3-nerved, keeled, the keel often puberulent; lemma ovate-lanceolate, glabrous to puberulent, keeled, the midnerve narrowed to a short awn; palea evident, keeled, minutely bidentate at the apex; anthers 2.5–4 mm long.

Scree, talus, and rocky places in the lower alpine zone of the Wasatch Range. Eurasia; introduced and widespread in western North America.

Alpine specimens from the Wasatch Range having a cylindric spike, spikelets 2–3-flowered, and awns of glumes 1.5–4 mm long are var. *desertorum* (Fisch. & Link) Dorn.

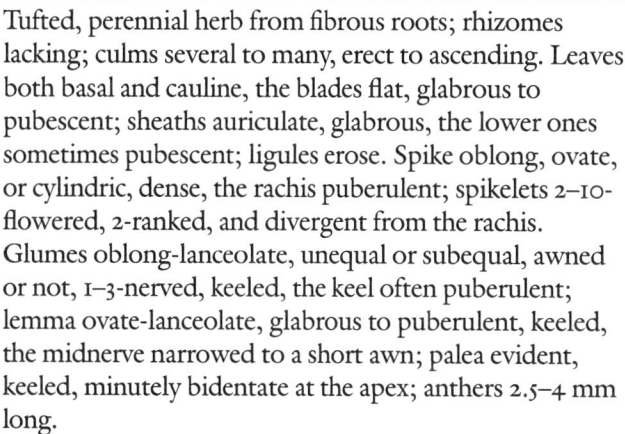

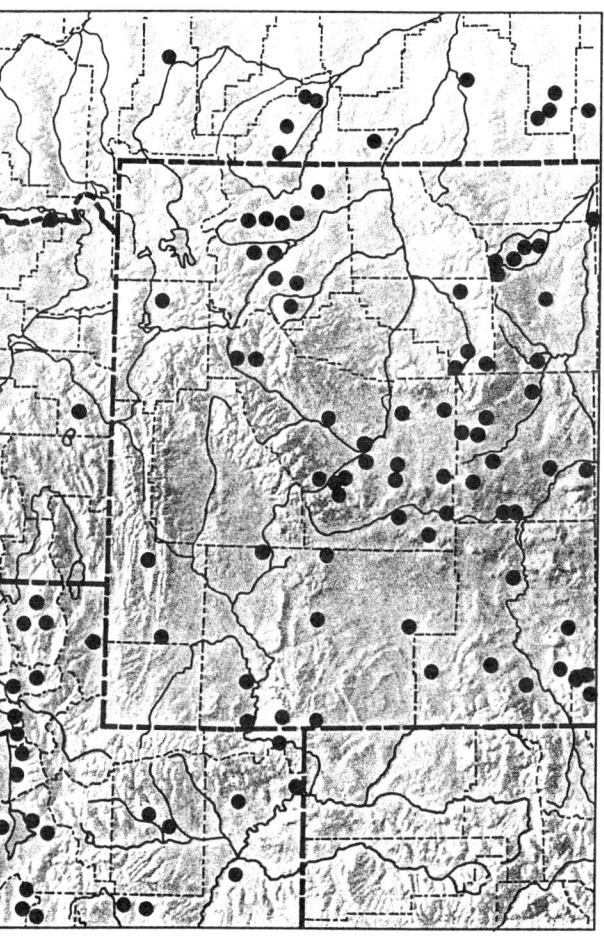

Agrostis L.
Bentgrass

Perennial or annual herbs mostly without rhizomes, usually caespitose. Leaf blades flat or revolute. Inflorescence an open or contracted panicle. Spikelets 1-flowered, the articulation above the glumes. Rachilla not usually prolonged beyond the palea. Glumes sub-equal, acute or acuminate, usually scabrous on the keel. Lemmas obtuse, usually shorter than the glumes, awnless or dorsally awned. Palea shorter than the lemma and 2-nerved, much reduced and nerveless, or obsolete.

KEY TO THE SPECIES OF *AGROSTIS*

1. Palea evident, more than half as long as the lemma
    2. Plants rhizomatous; panicle open; rachilla prolonged beyond the palea .... *A. thurberiana*
    2. Plants tufted, nonrhizomatous; panicle compressed; rachilla not prolonged beyond the palea .... *A. humilis*
1. Palea lacking or poorly developed and less than half as long as the lemma
    3. Lemmas awned, the awn 2–3 mm long .... *A. mertensii*
    3. Lemmas awnless or rarely awned, the awn less than 1.5 mm long
        4. Panicle contracted .... *A. variabilis*
        4. Panicle open or diffuse
            5. Branches of panicle branching below the middle; ligules 1–2 mm long; leaves 0.5–1.5 mm wide .... *A. idahoensis*
            5. Branches of panicle branching above the middle, often near the end; ligules 2–3 mm long; leaves 1–3 mm wide .... *A. scabra*

Agrostis humilis Vasey
Bull. Torr. Club 10:21, 1883.

Alpine Bentgrass

Podagrostis humilis (Vasey) Björkman

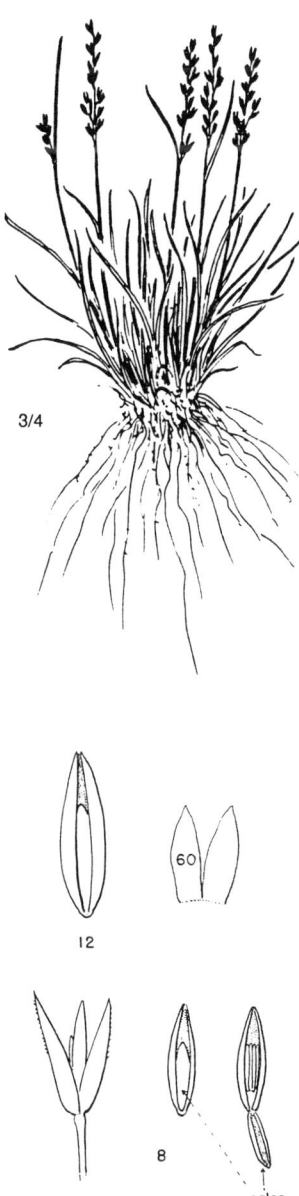

Tufted, glabrous to glabrate perennial from fibrous roots; rhizomes lacking. Leaves mostly basal, filiform, the blades flat or involute, glabrous below, puberulent above; sheaths glabrous, ligules obtuse. Panicle purplish, narrow, the branches erect to somewhat spreading; glumes subequal, lanceolate, acuminate, the keels scabrous at least near the apex; lemmas shorter than the glumes, purplish, awnless; palea two-thirds to three-fourths as long as the lemma; anthers 0.6–0.8 mm long.

Meadows, stream banks, fellfields, and patterned ground in most alpine ranges of the Middle Rocky Mountains. Not yet known from above timberline in the Big Horn, Hoback, Salt River, and Wyoming Ranges. British Columbia and Alberta south to Oregon, Nevada, Utah, and Colorado.

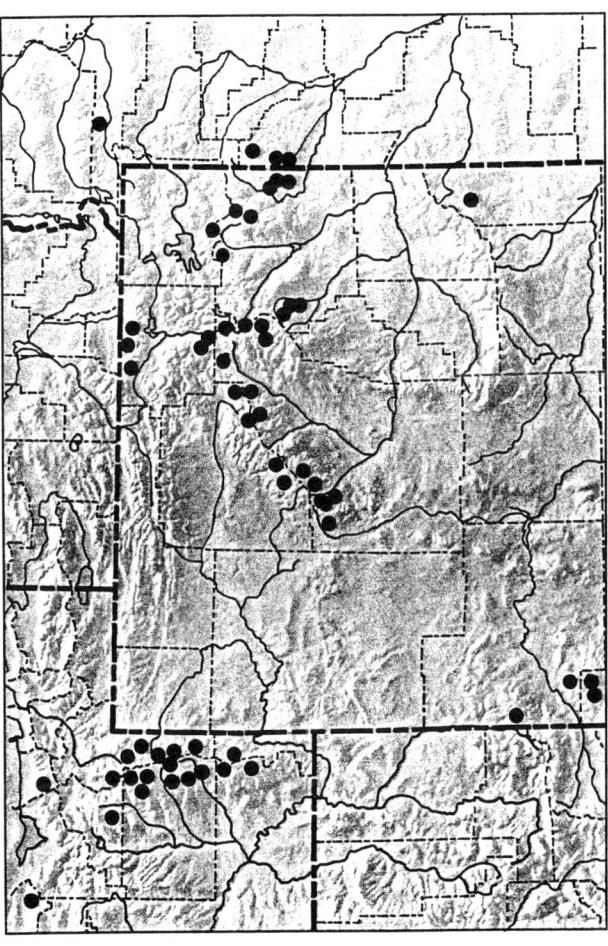

Agrostis idahoensis Nash
Bull. Torr. Club 24:42, 1897.

Idaho Bentgrass

Agrostis clavata auctt., non Trin.
Agrostis filicumis M.E. Jones
Agrostis tenuiculmis Nash in Rydb.
Agrostis tenuis Vasey

Erect, tufted perennial from fibrous roots; rhizomes lacking. Leaves mostly basal, the blades flat to involute, glabrous below, scabrous above; sheaths glabrous; ligules obtuse, erose to lacerate, those of the upper leaves becoming acute. Panicle open or diffuse, the branches capillary, ascending or spreading, rebranched below the middle; glumes equal or nearly so, purplish, acute to acuminate, scabrous on the keel; lemmas shorter than the glumes, glabrous, awnless; palea minute to lacking; anthers mostly less than 0.3 mm long.

Timberline meadows of the Absaroka, Big Horn, Gros Ventre, and Medicine Bow Ranges. Washington south and east to California, New Mexico, Colorado, Wyoming, and Montana.

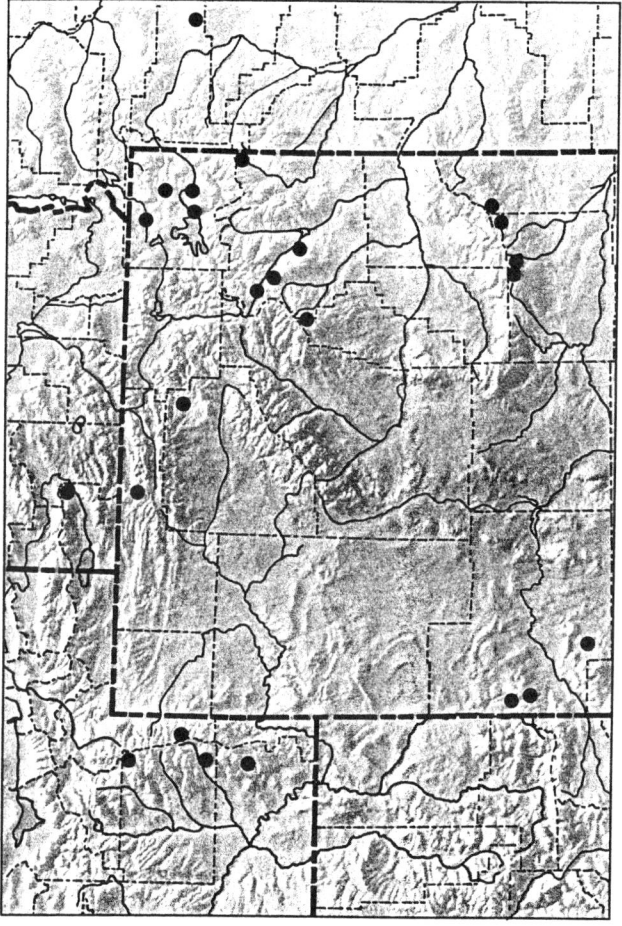

Agrostis mertensii Trin.
Linnaea 10:302, 1836.

Arctic Bentgrass

Agrostis bakeri Rydb.
Agrostis borealis Hartm.
Agrostis rupestris auctt., non All.

Low, tufted perennial, less than 4 dm tall, from fibrous roots; rhizomes lacking. Leaves mostly basal, the blades flat to somewhat involute, glabrous or nearly so; sheaths glabrous to scaberulous; ligule truncate, erose. Panicle narrowly pyramidal, the branches erect to ascending, branching near the middle; glumes subequal, acute to acuminate, scabrous on the keel, purplish; lemma nearly as long as the glumes, acute, bearing a bent, exserted awn 2–3 mm long from above the middle; palea minute to lacking; anthers 0.5–1 mm long.

Fellfields of the Absaroka and Medicine Bow Ranges. Circumpolar; south in western North America to Washington, Colorado, and Wyoming.

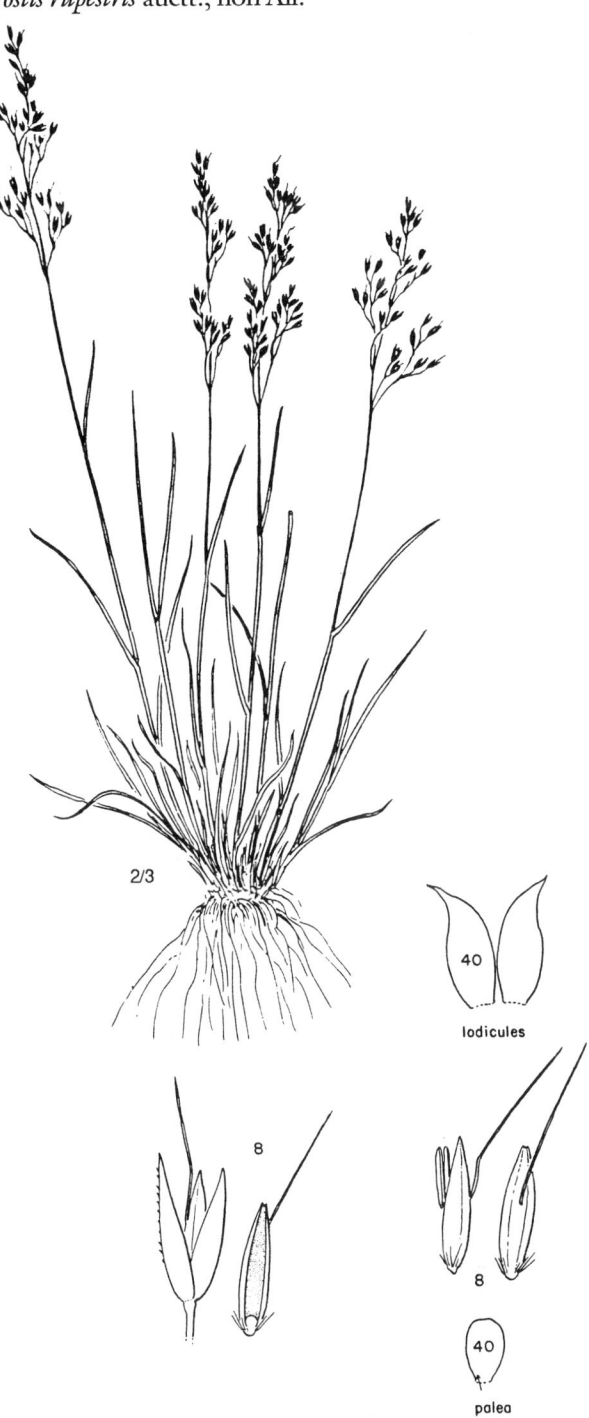

Agrostis scabra Willd.
Sp. Pl. 1:370, 1797.

Ticklegrass

Agrostis geminata Trin.
Agrostis hiemalis auctt., non (Walt.) B.S.P.
Agrostis michauxii Trin.
Agrostis nootkaensis Trin.
Agrostis nutkaensis Kunth
Agrostis scabriuscula Buckl.
Trichodium album Presl
Trichodium scabrum (Willd.) Muhl.
Vilfa scabra (Willd.) Beauv.

Tufted perennial from fibrous roots; rhizomes usually lacking. Leaves mostly basal, the blades flat to involute, scabrous, sometimes puberulent above; sheaths glabrous; ligules obtuse to acute, hyaline, somewhat lacerate to erose. Panicle open and diffuse, broadly elliptic to pyramidal, the branches slender, flexuous, spreading, rebranched only above the middle; glumes subequal to unequal, acute to acuminate, scabrous on the keel; lemmas two-thirds to three-fourths as long as the glumes, awnless or rarely with a short awn; palea lacking; anthers 0.3–0.6 mm long.

Timberline meadows in the Absaroka, Beartooth, Big Horn, Medicine Bow, Teton, Uinta, Wasatch, and Wind River Ranges. Siberia; transcontinental in North America, south in the West to California, New Mexico, and Texas.

Two varieties occur in the Middle Rocky Mountains: var. *scabra* has awnless lemmas; and var. *geminata* (Trin.) Swallen has short-awned lemmas.

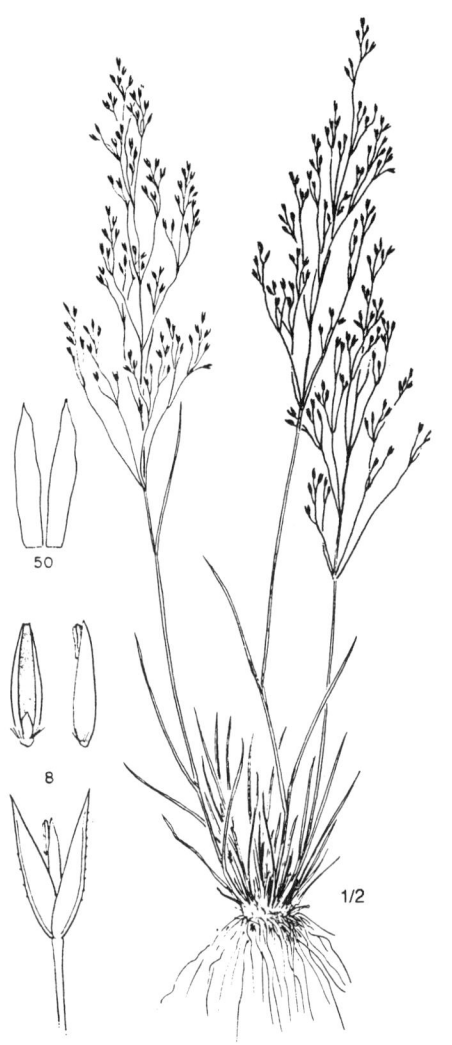

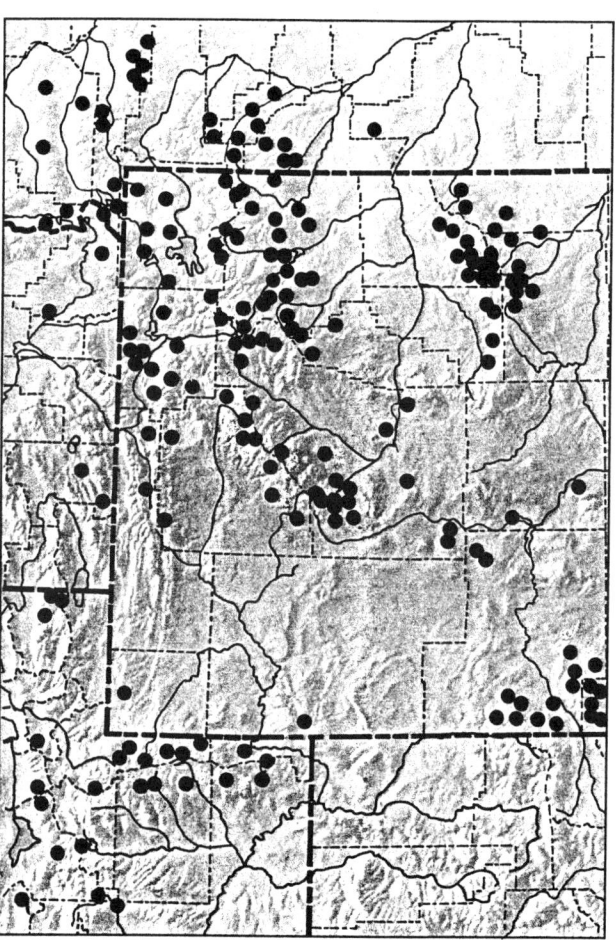

Agrostis thurberiana A.S. Hitchc.
U.S. Pl. Ind. Bull. 68:23, 1905.

Thurber Bentgrass

Agrostis atrata Rydb.
Podagrostis thurberiana (A.S. Hitchc.) Hult.

Loosely tufted perennial from short rhizomes; stems erect, often decumbent at the base. Leaves both basal and cauline, often somewhat crowded at the base, the blades flat, lax, glabrous or scaberulous; sheaths glabrous; ligules obtuse, entire to lacerate. Panicles narrow, open, oblong, the capillary branches erect or ascending; glumes subequal, acute, greenish to purplish, scabrous on the keel; lemmas nearly as long as the glumes, awnless, greenish or hyaline; palea two-thirds to three-fourths as long as the lemma, the rachilla often prolonged as a short bristle; anthers 0.5–0.6 mm long.

Wet meadows, moist willow thickets, and stream banks near timberline in most alpine ranges of the Middle Rocky Mountains. Not yet known from above timberline in the Gros Ventre, Hoback, Salt River, and Wyoming Ranges. British Columbia and Alberta south to California, Idaho, Utah, and Colorado.

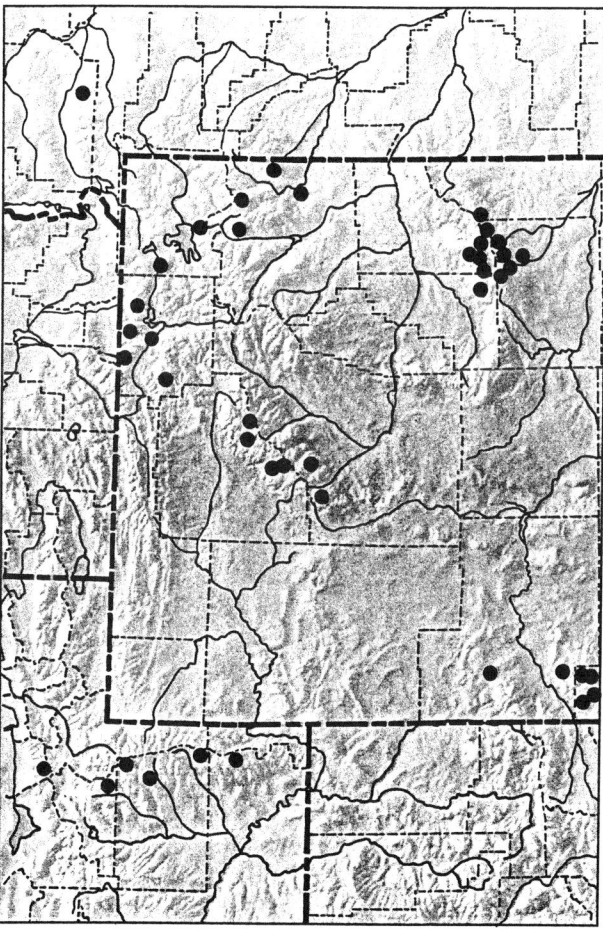

Agrostis variabilis Rydb.
Mem. N.Y. Bot. Gard. 1:32, 1900.

Mountain Bentgrass

Agrostis rossae auctt., non Vasey
Agrostis varians Trin.

Densely tufted perennial from fibrous roots; rhizomes lacking. Stems erect, glabrous. Leaves mostly basal, the blades flat to somewhat involute or folded, glabrous to scabrous or puberulent; sheaths glabrous to scabrous; ligules obtuse, lacerate. Panicles narrow, linear, the branches erect to somewhat spreading; glumes equal or nearly so, purple, acute to acuminate, scabrous on the keel; lemmas shorter than the glumes, usually purple-tinged, awnless or rarely awned with a short awn from below the middle; palea minute or lacking; anthers 0.3–0.6 mm long.

Fellfields, talus, and alp slopes of the Absaroka, Beartooth, Medicine Bow, Salt River, Uinta, and Wind River Ranges. British Columbia and Alberta south to California, Utah, and Colorado.

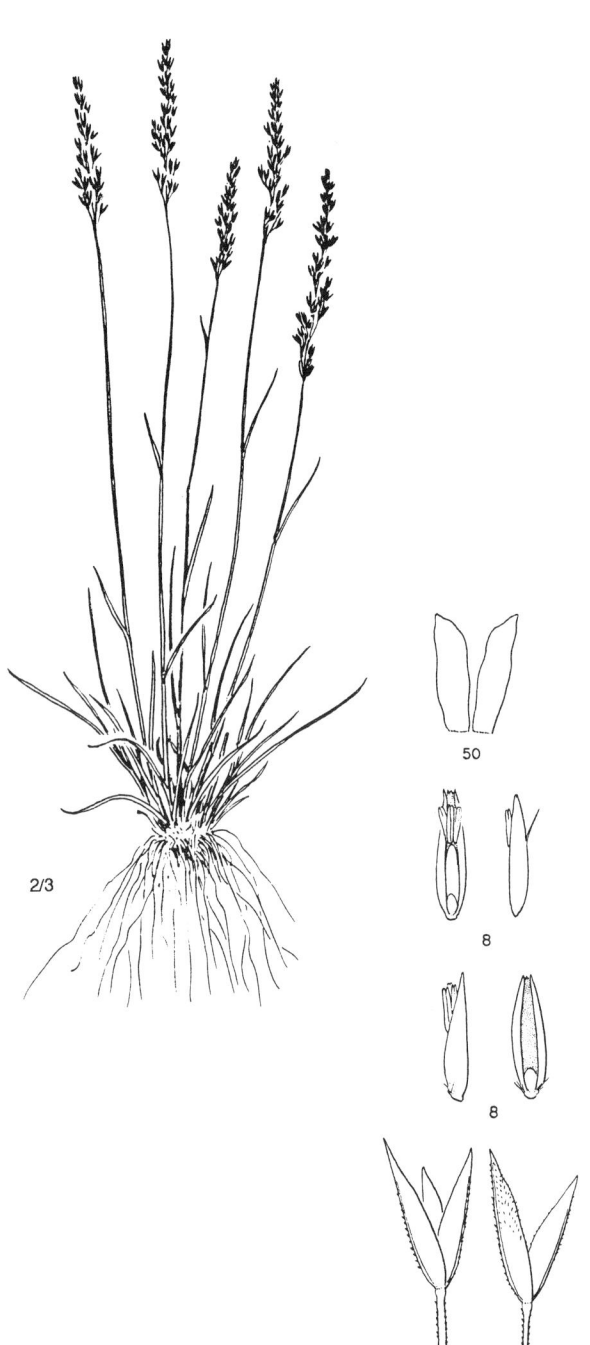

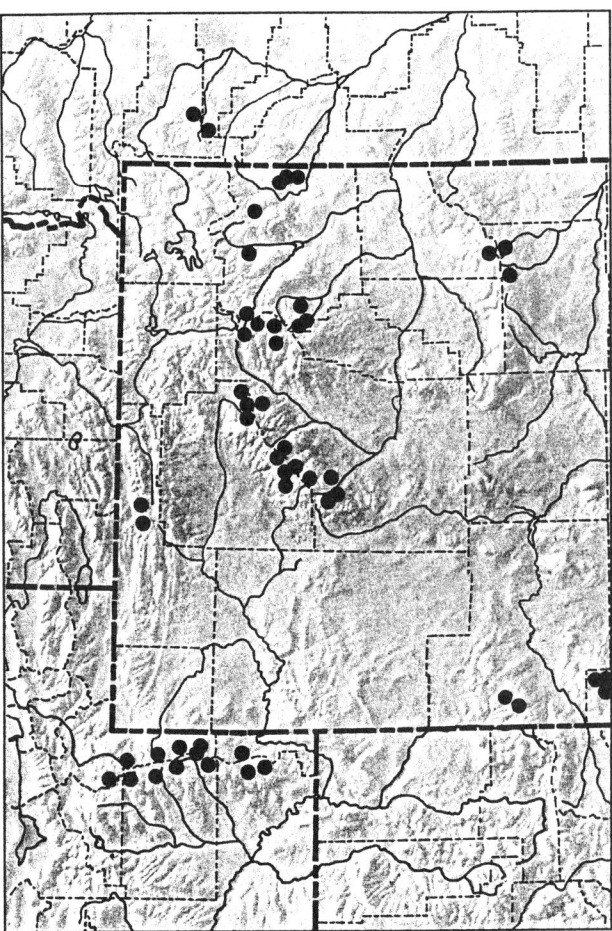

Alopecurus L.
Foxtail

Low to moderately tall perennials, or rarely annuals, with flat leaves and soft, spikelike panicles or racemes. Spikelets 1-flowered, laterally flattened, disarticulating below the glumes. Glumes subequal, ciliate on the keel. Lemma about as long as the glumes, bearing a dorsal awn from below the middle. Palea lacking or minute.

KEY TO THE SPECIES OF
ALOPECURUS

1. Glumes silky-villous to lanate over the entire surface; awns of lemmas much exceeding the glumes; panicle broadly cylindrical to ovoid, less than 3 cm long, about 1 cm thick ... *A. alpinus*
1. Glumes silky-villous only on the nerves and keel; awns of lemmas only slightly longer than the glumes; panicle narrowly cylindrical, more than 3 cm long, less than 1 cm thick ... *A. aequalis*

Alopecurus aequalis Sobol.
Fl. Petrop., 16, 1799.

Shortawn Foxtail

Alopecurus aristulatus Michx.
Alopecurus caespitosus Trin.
Alopecurus fulvus J.E. Smith in Sowerby
Alopecurus macounii Merr.
Tozzettia fulva (J.E. Smith in Sowerby) Lunell

Loosely tufted perennial, the culms single to few, erect, or decumbent at the base and rooting at the lower nodes. Leaf blades flat, glabrous to scabrous; sheaths glabrous; ligules lanceolate, the margins lacerate. Panicle spikelike, cylindrical, mostly 3–7 cm long, 3–7 mm wide; spikelets flattened, obovate; glumes obtuse, silky-villous on the nerves and keel; lemmas lanceolate to elliptic, about as long as the glumes, awned at or below the middle, the awn straight or curved, included or slightly exerted; anthers 0.5–1 mm long.

Stream banks and moist willow thickets at timberline in the Uinta, Wasatch, and Wind River Ranges. Circumboreal; south in western North America to northern Mexico.

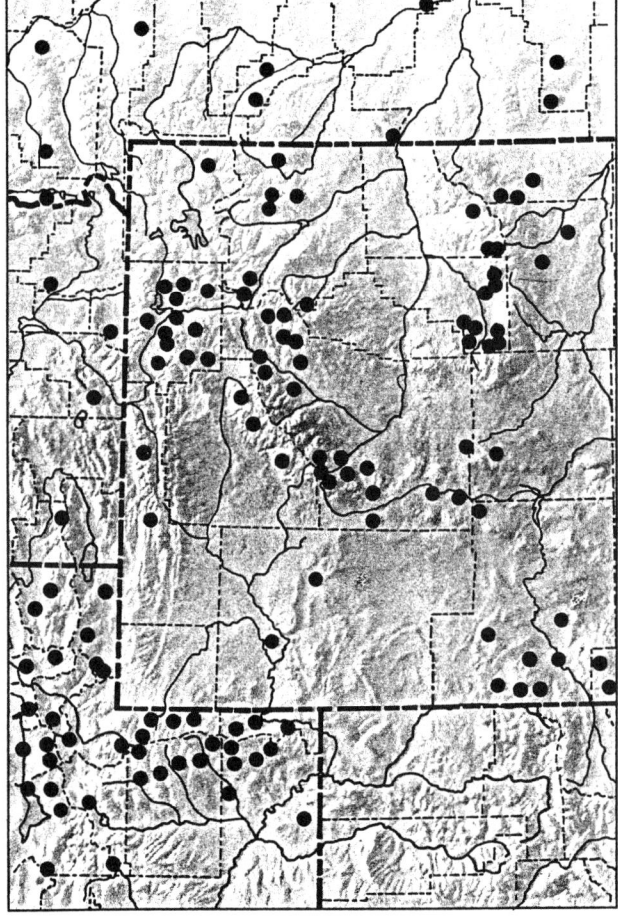

Alopecurus alpinus J.E. Smith in Sowerby
Eng. Bot., 1126, 1803.

Alpine Foxtail

Alopecurus beeringianus Gand.
Alopecurus borealis Trin.
Alopecurus glaucus Less.
Alopecurus occidentalis Scribn. & Tweedy
Alopecurus stejnegeri Vasey

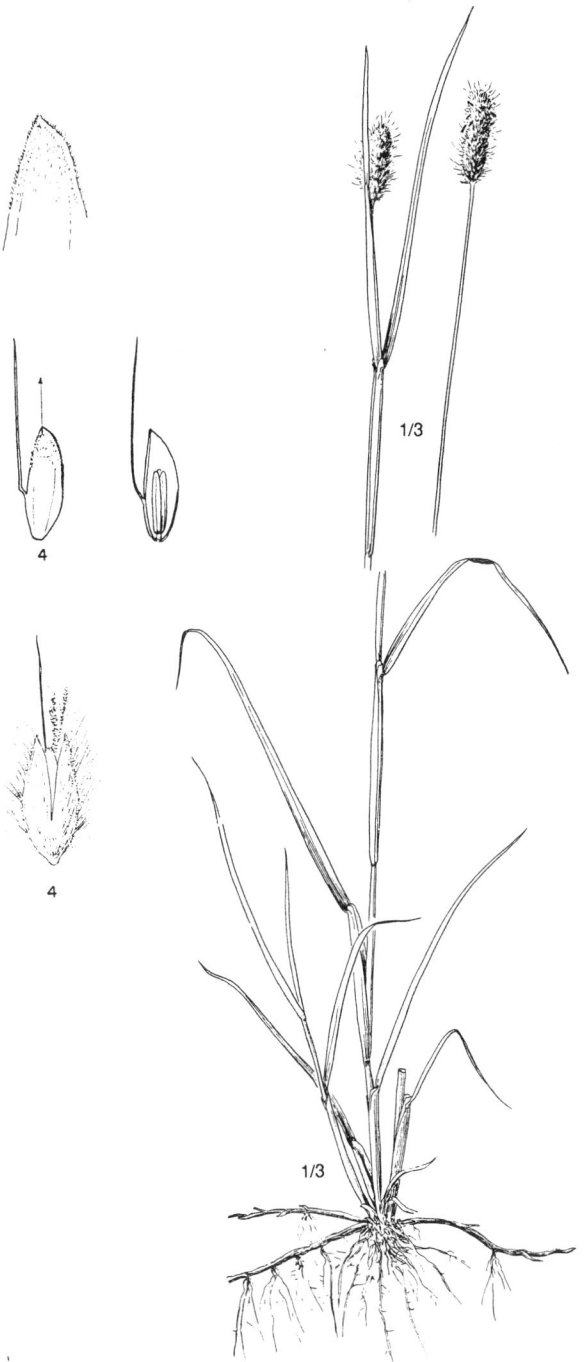

Perennial from slender, creeping rhizomes; culms solitary to several, erect or, often, decumbent at the base. Leaf blades flat, glabrous or scabrous; sheaths inflated, glabrous; ligules truncate, lacerate, and finely erose. Panicles spikelike, ovoid or oblong, mostly less than 3 cm long, about 1 cm wide or more; spikelets flattened, ovate to elliptic; glumes purplish, acute, silky-villous to lanate over their entire surface; lemmas elliptic, shorter than the glumes, awned below the middle, the awn straight or geniculate and twisted below, distinctly longer than the glumes; anthers 2–2.8 mm long.

Stream banks and moist willow thickets of the Uinta and Wind River Ranges. Circumpolar; south in western North America to Idaho, Utah, and Colorado.

Middle Rocky Mountain plants more than 3 dm tall and with prominent exerted awns on the lemmas are ssp. *glaucus* (Less.) Hult. Askell Löve et al. (1971) showed ssp. *glaucus* to be a hexadecaploid, 2n = 112.

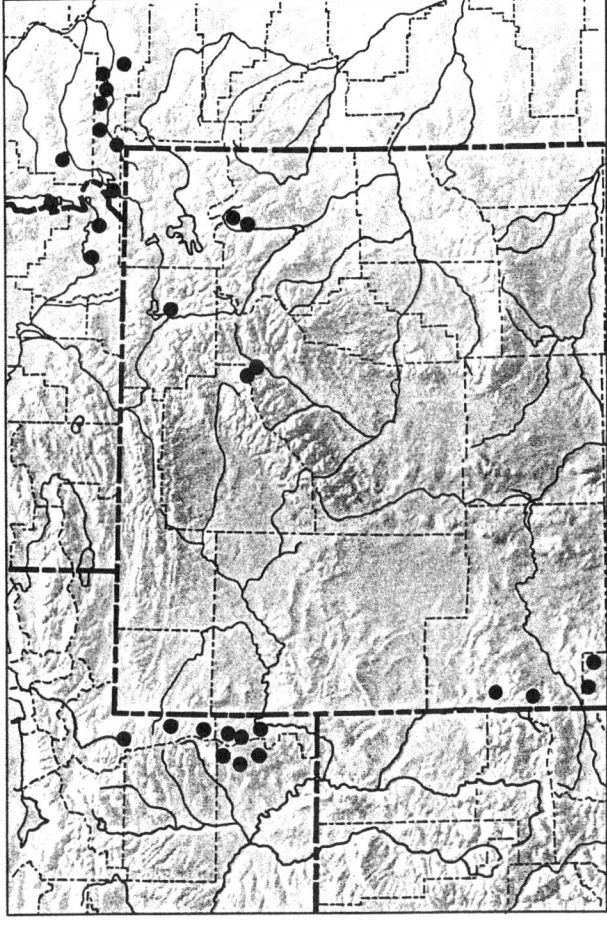

Bromus L.
Bromegrass

Low to moderately tall annual or perennial herbs with closed sheaths and flat blades. Spikelets rather large, usually more than 15 mm long, several-flowered, rounded to strongly flattened, in open or contracted panicles. Articulation above the glumes and between the florets, the glumes unequal, acute, and shorter than the first lemma. Lemmas convex on the back or keeled, 5–9-nerved, awned from between the 2 apical teeth or awnless. Paleas shorter than the lemmas, ciliate on the keels, at length adhering to the grain.

KEY TO THE SPECIES OF BROMUS

1. Plants rhizomatous — *B. inermis*
1. Plants lacking rhizomes
    2. Lemmas compressed, with distinct keels; spikelets strongly flattened; branches of inflorescence erect to ascending — *B. carinatus*
    2. Lemmas rounded on back; spikelets only slightly flattened, if at all; branches of inflorescence drooping
        3. Lemmas evenly pubescent; first glume 3-nerved — *B. anomalus*
        3. Lemmas unevenly pubescent, at least the tip glabrous; first glume 1-nerved — *B. ciliatus*

Bromus anomalus Rupr. ex Fourn.
Mex. Pl. 2:126, 1886.

Nodding Brome

Bromopsis anomala (Rupr. ex Fourn.) Holub
Bromopsis porteri (Coult.) Holub
Bromus frondosus (Shear) Woot. & Standl.
Bromus porteri (Coult.) Nash
Bromus scabratus Scribn.
Zerna anomala (Rupr. ex Fourn.) Henr.

Loosely tufted perennial from fibrous roots; rhizomes lacking; culms single or few, pubescent to puberulent at the nodes. Leaf blades flat or somewhat involute, stiff, scabrous; sheaths glabrous to sparsely pilose, lacking auricles, shorter than the internodes; ligules finely erose. Panicle open, the slender branches nodding; spikelets few, slightly compressed; glumes puberulent, the first 3-nerved, lanceolate, the second 3- or 5-nerved, elliptic; lemmas pilose on the back, rounded to slightly compressed, lanceolate, acute to shallowly bifid, the awn 2–4 mm long; anthers 2–3 mm long.

Meadows and fellfields of the Absaroka, Beartooth, Uinta, and Wasatch Ranges. British Columbia to Saskatchewan, south to California, Arizona, New Mexico, and Texas.

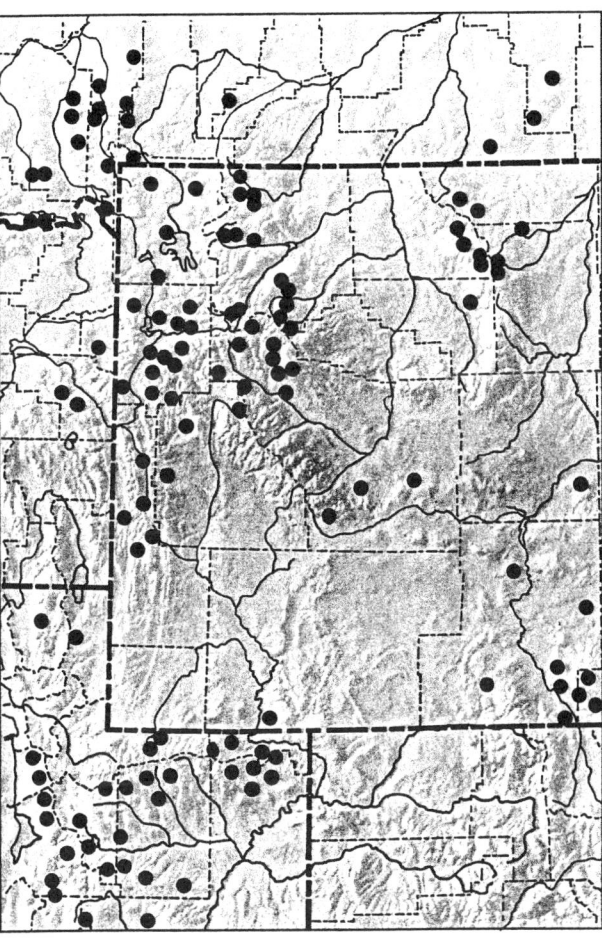

Bromus carinatus Hook. & Arn.
Bot. Beechey Voy., 403, 1840.

Mountain Brome

Bromus breviaristatus (Hook.) Buckl.
Bromus flodmannii Rydb.
Bromus hookerianus Thurb.
Bromus latior (Shear) Rydb.
Bromus marginatus Nees ex Steud.
Bromus maritimus (Piper) A.S. Hitchc. in Jeps.
Bromus multiflorus Scribn. in A. Nels
Bromus oregonus Nutt. ex Hook.
Bromus paniculatus (Shear) Rydb.
Bromus parviflorus Nutt. ex Gray
Bromus pauciflorus Nutt. ex Shear
Bromus polyanthus Scribn. ex Shear
Bromus subvelutinus Shear
Bromus virens Buckl.
Ceratochloa breviaristata Hook.
Ceratochloa carinata (Hook. & Arn.) Tutin
Ceratochloa grandiflora Hook.
Ceratochloa marginata (Nees ex Steud.) Jackson
Forasaccus breviaristatus (Hook.) Lunell
Forasaccus marginatus (Nees ex Steud.) Lunell

Erect, perennial herb, often loosely tufted, from fibrous roots; rhizomes lacking; culms solitary to a few, glabrous to pubescent. Leaf blades lax, flat or somewhat involute, glabrous to scabrous or pilose; sheaths pilose or occasionally glabrous; ligules erose. Panicles narrow, elliptic, compact or open, the branches erect to ascending, rarely deflexed; spikelets strongly flattened; glumes glabrous, scabrous, or puberulent, acuminate, the first 3-nerved, the second mostly 5-nerved and much longer than the first; lemmas lanceolate, keeled, glabrous, scabrous, or pubescent, bidentate, the awns mostly 3–7 mm long; anthers 1–8 mm long.

Timberline meadows of the Salt River and Teton Ranges. British Columbia to Saskatchewan, south to California, New Mexico, and Texas; Mexico.

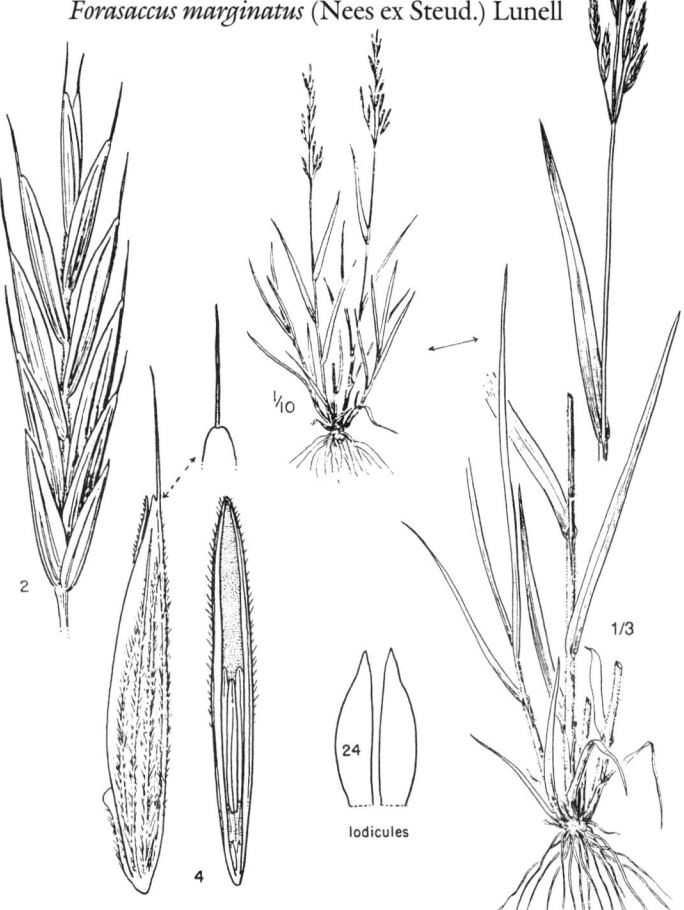

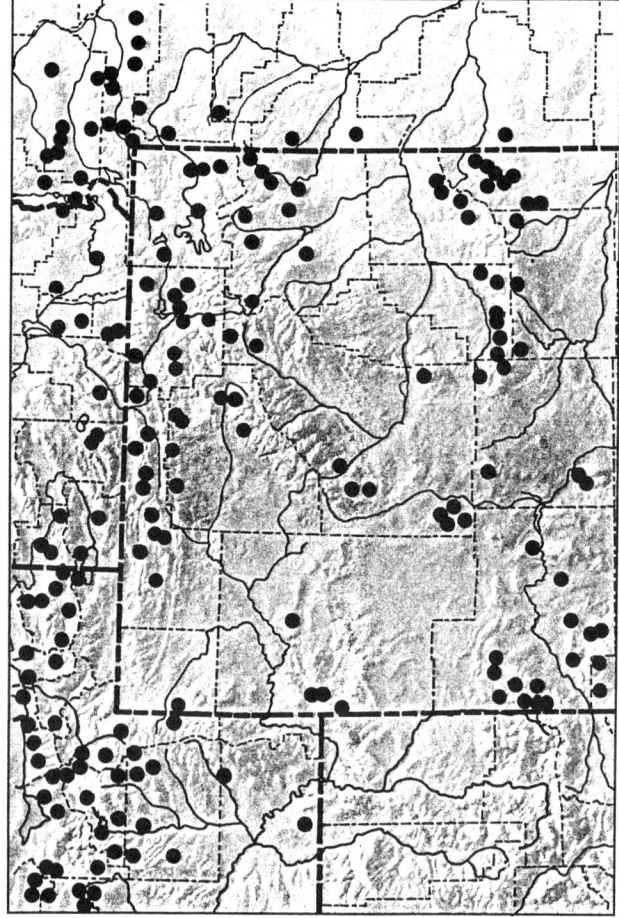

Bromus ciliatus L.
Sp. Pl., 76, 1753.

Fringed Brome

Bromopsis ciliata (L.) Holub
Bromus canadensis Michx.
Bromus dudleyi Fern.
Bromus richardsonii Link
Forasaccus ciliatus (L.) Lunell
Zerna ciliata (L.) Henr.
Zerna richardsonii (Link) Nevski

Loosely tufted perennial from fibrous roots; rhizomes lacking; culms erect, single to a few, glabrous or pubescent at nodes. Leaf blades flat, lax, glabrous to scabrous or sparsely pilose; sheaths glabrous to pilose; ligules finely erose, pilose-ciliate on the inner margins. Panicle pyramidal, loose, the slender branches nodding; spikelets mostly 7–9-flowered, slightly compressed; glumes glabrous to scabrous, the first lanceolate, 1-nerved (basally 3-nerved), the second elliptic, 3-nerved; lemmas elliptic, rounded on the back, 3–5-nerved, pubescent to nearly glabrous with some hairs along the margins, entire or shallowly bifid, the awn straight, 3–5 mm long; anthers 1.5–3 mm long.

Timberline meadows in the Absaroka, Beartooth, Uinta, and Wasatch Ranges. Siberia; transcontinental in Canada; south in the West to California, Arizona, New Mexico, Texas, and Nebraska; Mexico.

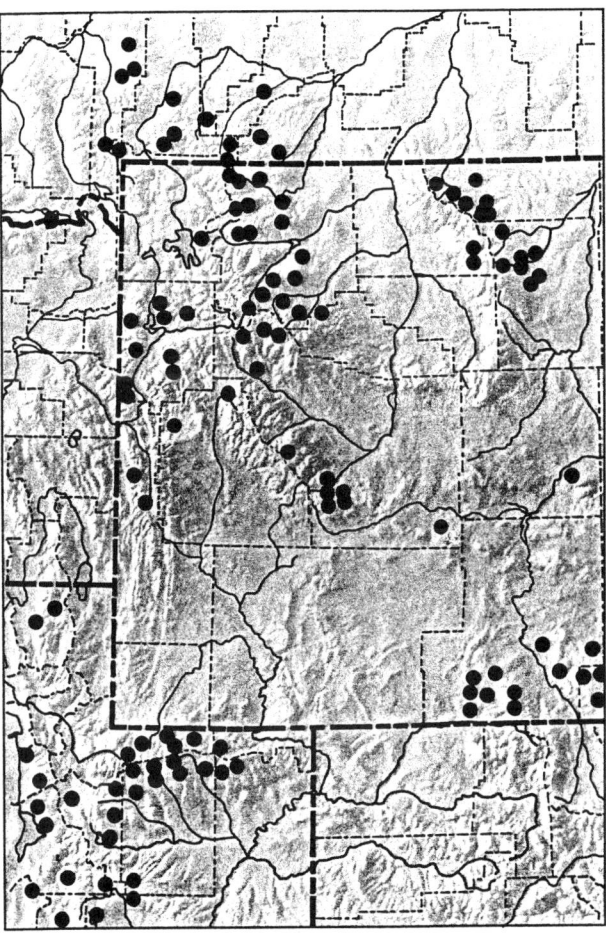

Bromus inermis Leyss.
Fl. Hal., 16, 1761.

Smooth Brome

Bromopsis dicksonii (Mitchell & Wilton) Löve & Löve
Bromopsis inermis (Leyss.) Holub
Bromopsis pumpelliana (Scribn.) Holub
Bromus arcticus Shear in Scribn. & Merr.
Bromus pumpellianus Scribn.
Festuca inermis (Leyss.) DC. & Lam.
Forasaccus inermis (Leyss.) Lunell
Forasaccus pumpellianus (Scribn.) Lunell
Schedonorus inermis (Leyss.) Beauv.
Zerna arctica (Shear) Tsvel.
Zerna inermis (Leyss.) Lindm.
Zerna pumpelliana (Scribn.) Tsvel.

Perennial from long-creeping rhizomes; culms solitary or loosely tufted, glabrous, erect or decumbent at the base. Leaf blades flat, glabrous to scabrous or pubescent; sheaths glabrous to pubescent; ligules finely erose. Panicle elliptic to oblong, compact to somewhat open, the branches erect or ascending, sometimes spreading with age; spikelets narrow, 5–10-flowered, somewhat compressed; glumes lanceolate, glabrous, the first usually 1-nerved, the second usually 3-nerved; lemmas lanceolate, mucronate or awned with short, straight awns less than 2 mm long, glabrous or pubescent; anthers 3–6 mm long.

Fellfields and timberline meadows of the Absaroka, Beartooth, Big Horn, Uinta, Wasatch, and Wind River Ranges. Circumboreal; Alaska south in western North America to California, Arizona, and New Mexico.

Native mountain plants with purplish spikelets and pubescent lemmas are ssp. *pumpellianus* (Scribn.) Wagnon. The 2 varieties may be recognized as follows:
1. Lemmas pubescent only on lower half and veins; glumes glabrous . . . var. *pumpellianus*
1. Lemmas pubescent to tip; glumes pubescent . . . var. *tweedyi* (Scribn.) C.L. Hitchc.

var. pumpellianus

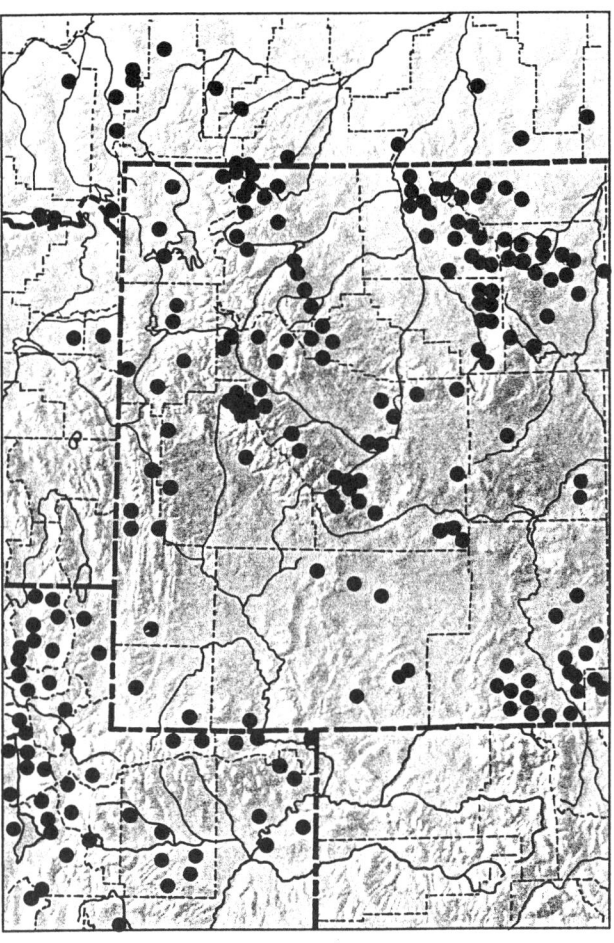

Calamagrostis Adans.
Reedgrass

Moderately tall perennials with flat or involute blades, usually from creeping rhizomes. Inflorescence an open or contracted, spikelike panicle. Spikelets small, 1-flowered, the articulation above the glumes, the rachilla prolonged beyond the palea as a short, usually hairy bristle. Glumes subequal, acute or acuminate. Lemma shorter than the glumes, rather thin, and bearing a tuft of hairs from the callus, dorsally awned, the awn straight or geniculate. Palea evident, shorter than the lemma.

KEY TO THE SPECIES OF *CALAMAGROSTIS*

1. Awn of lemma geniculate, twisted below the bend; panicle contracted
   2. Awn exceeding the glumes by 1–2 mm .... *C. purpurascens*
   2. Awn shorter than the glumes .... *C. montanensis*
1. Awn of lemma straight or slightly bent, not twisted below; panicle open or contracted
   3. Panicle mostly open, the branches spreading; leaf blades flat; glumes attenuate .... *C. canadensis*
   3. Panicle contracted, the branches appressed to ascending; leaf blades involute; glumes abruptly acute
      4. Plants from fibrous roots; callus hairs less than half as long as the lemma .... *C. scopulorum*
      4. Plants from rhizomes; callus hairs two-thirds as long to as long as the lemma .... *C. stricta*

Calamagrostis canadensis (Michx.) Beauv.
Essai Nouv. Agrostogr. 15, 152, 1812.

Bluejoint Reedgrass

Arundo canadensis Michx.
Arundo langsdorfii Link.
Calamagrostis alaskana Kearn.
Calamagrostis angustifolia Kom.
Calamagrostis anomala Suksd.
Calamagrostis atropurpurea Nash
Calamagrostis blanda Beal
Calamagrostis dubia (Scribn. & Tweedy) Scribn. in Vasey
Calamagrostis lactea Beal
Calamagrostis langsdorfii (Link.) Trin.
Calamagrostis macouniana (Vasey) Vasey
Calamagrostis michauxii Trin. ex Steud.
Calamagrostis nubila Louis-Marie
Calamagrostis oregonensis Buckl.
Calamagrostis pallida Vasey & Scribn. ex Vasey
Calamagrostis purpurea (Trin.) Trin.
Calamagrostis scabra (Kunth) Presl
Calamagrostis scribneri Beal
Deyeuxia canadensis (Michx.) Munro ex Hook.
Deyeuxia dubia Scribn. & Tweedy
Deyeuxia lactea (Beal) Beal
Deyeuxia macouniana Vasey
Deyeuxia scabra Kunth

Perennial from long-creeping rhizomes, the culms stout, glabrous, erect, solitary or tufted. Leaf blades flat or somewhat involute, glabrous to scaberulous; sheaths glabrous to scabrous; ligules membranous, lacerate to erose. Panicle open and broad with spreading branches to somewhat dense and contracted with erect-ascending branches; spikelets mostly 1-flowered; glumes lanceolate, acute to acuminate, scabrous at least on the keel, equal or slightly unequal, purplish in alpine specimens; lemmas glabrous to scaberulous, membranous, the apex 2–4-toothed, awned from just below to just above the middle, the awn slender, straight, fragile; callus hairs abundant, about as long as the lemma; anthers mostly 1–1.7 mm long.

Moist meadows and slopes in the lower alpine zone of the Beartooth, Teton, Uinta, Wasatch, and Wind River Ranges. Circumboreal; Alaska to Greenland, south to all but the southernmost United States.

Alpine plants with acute to acuminate glumes mostly more than 3 mm long are var. *canadensis*.

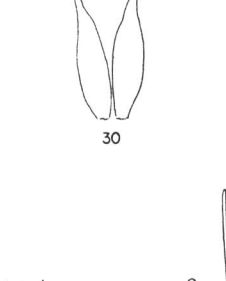
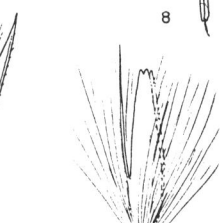
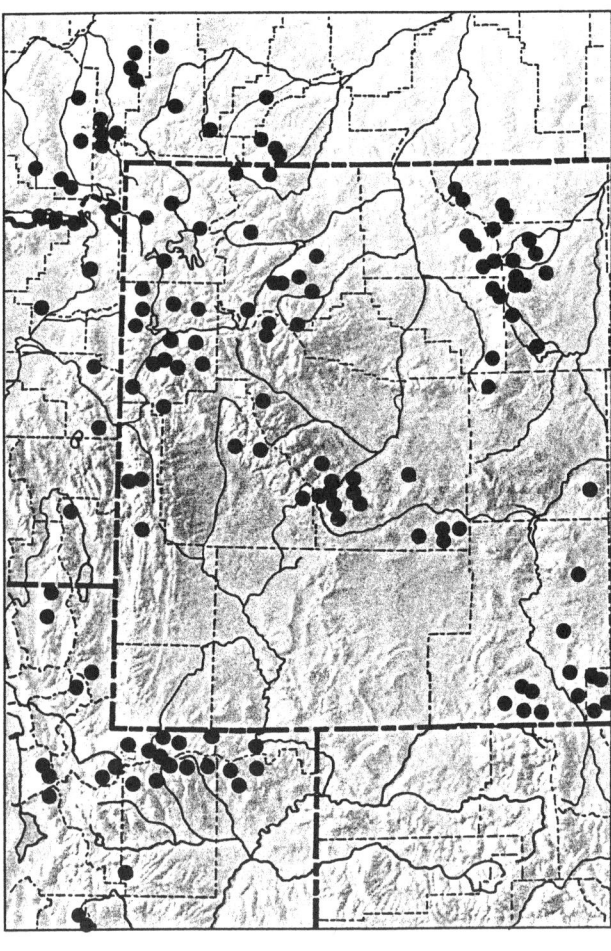

Calamagrostis montanensis (Scribn.) Scribn. in Vasey
Contr. U.S. Natl. Herb. 3:82, 1892.

Plains Reedgrass

Deyeuxia montanensis Scribn.

Perennial from creeping, slender rhizomes; culms erect, single to loosely tufted, scabrous below the panicle. Leaf blades involute, stiff, erect, scabrous on the ventral surface; sheaths glabrous to scaberulous; ligules mostly lacerate and ciliolate, acute to obtuse. Panicle dense, erect, contracted, and spikelike, the branches short, erect to ascending; glumes equal or nearly so, acuminate, sharply keeled, scabrous, at least on the keels; lemmas shorter than to nearly equaling the glumes, 4-toothed at the apex, awned from near the base, the awn geniculate, twisted below, shorter than to about equaling the glumes and exerted sidewise; callus hairs about half as long as the lemma; anthers 1.4–2 mm long.

Occasional in timberline meadows of the Absaroka, Beartooth, and Medicine Bow Ranges. British Columbia to Manitoba and south to Colorado, South Dakota, and Minnesota.

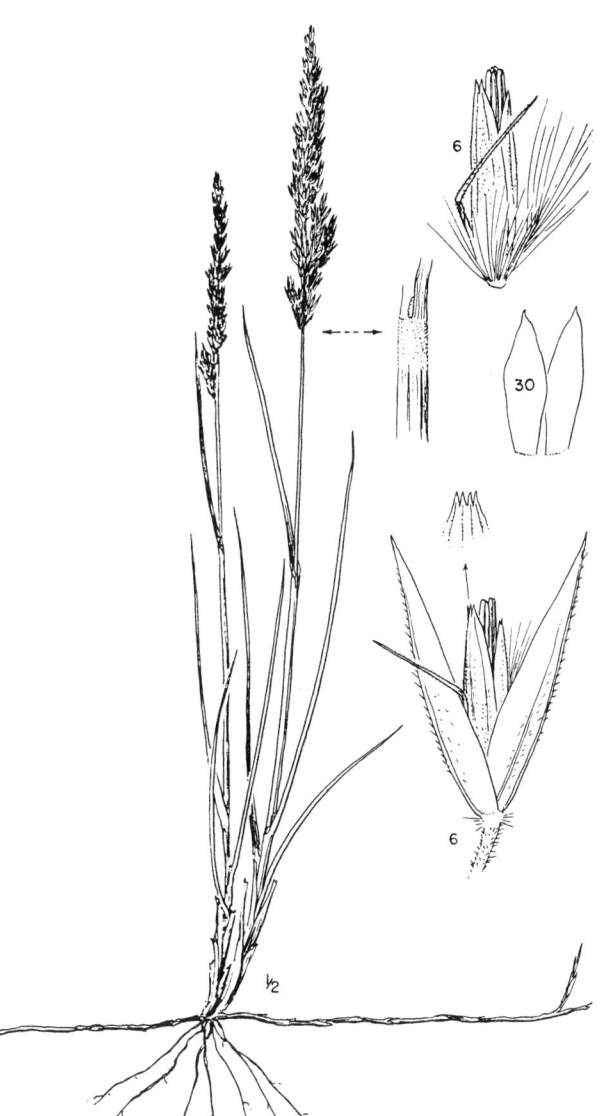

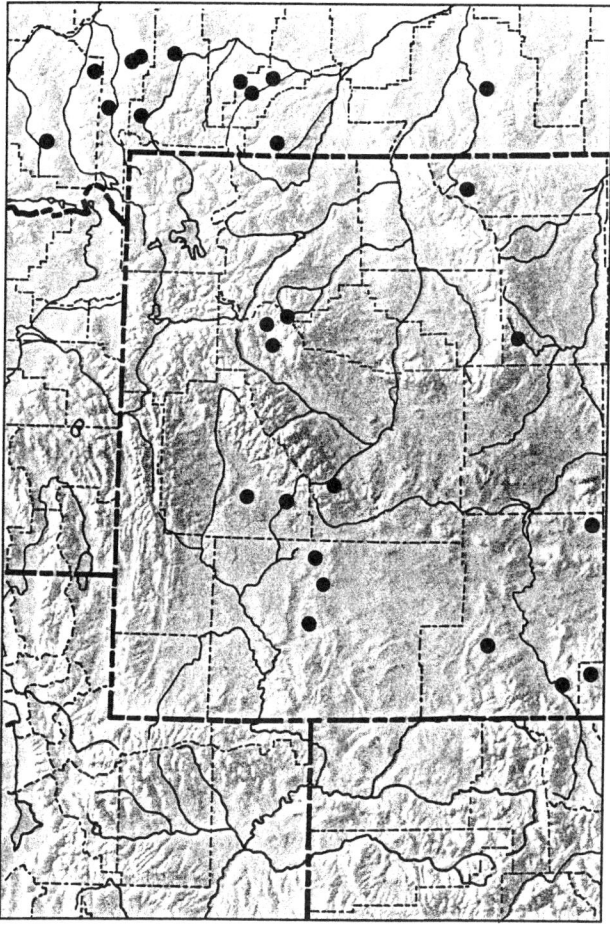

Calamagrostis purpurascens R. Br. in Richards. *Bot. App. Frankl. J.*, 731, 1823.

Purple Reedgrass

Arundo purpurascens (R. Br.) Shult.
Calamagrostis arctica Vasey
Calamagrostis laricina (Louis-Marie) Lalonde
Calamagrostis lepageana Louis-Marie
Calamagrostis maltei (Polunin) Löve & Löve
Calamagrostis poluninii Sørenson
Calamagrostis vaseyi Beal
Calamagrostis yukonensis Nash
Deschampsia congestiformis Booth
Deyeuxia purpurascens (R. Br.) Kunth

Tufted perennial from short rhizomes, or rhizomes sometimes lacking; culms erect, glabrous to puberulent near the base. Leaf blades stiff, erect, flat to involute, glabrous to scabrous; sheaths usually scabrous, sometimes glabrous; ligules obtuse to somewhat truncate, erose. Panicle dense, narrow, spikelike; spikelets mostly 1-flowered; glumes purplish, lanceolate, acute to acuminate, subequal, glabrous to scaberulous; lemmas glabrous to scaberulous, the apex with 4 bristlelike teeth, awned from near the base, the awn geniculate, twisted below, stout, exceeding the glumes, often by 2 mm; callus hairs less than half as long as the lemma; anthers 1.5–2.8 mm long.

Occasional in rocky places, ledges, talus, and fellfields of most alpine ranges in the Middle Rocky Mountains. Not yet known from above timberline in the Gros Ventre, Hoback, and Salt River Ranges. Circumboreal; south in western North America to California, Utah, Colorado, South Dakota, and Minnesota.

Middle Rocky Mountain plants are ssp. *purpurascens*.

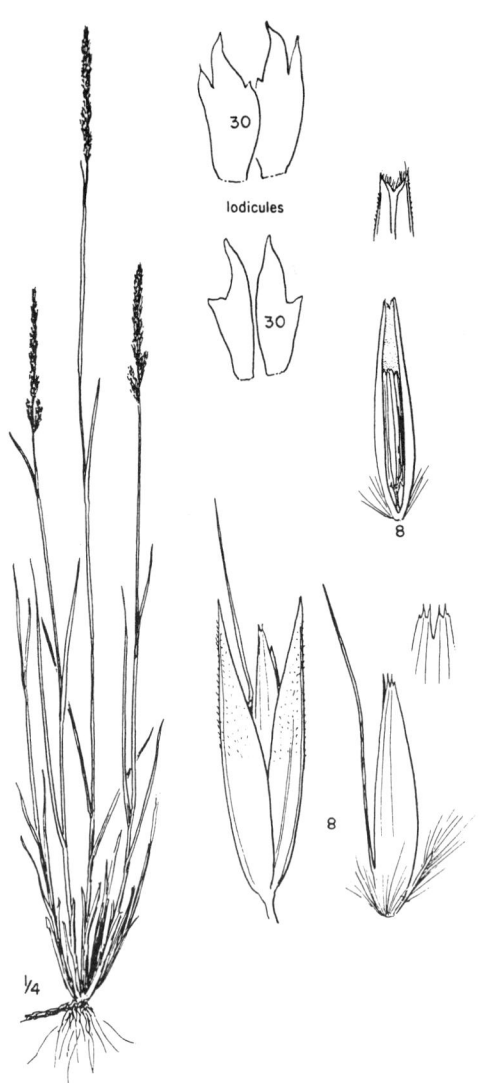

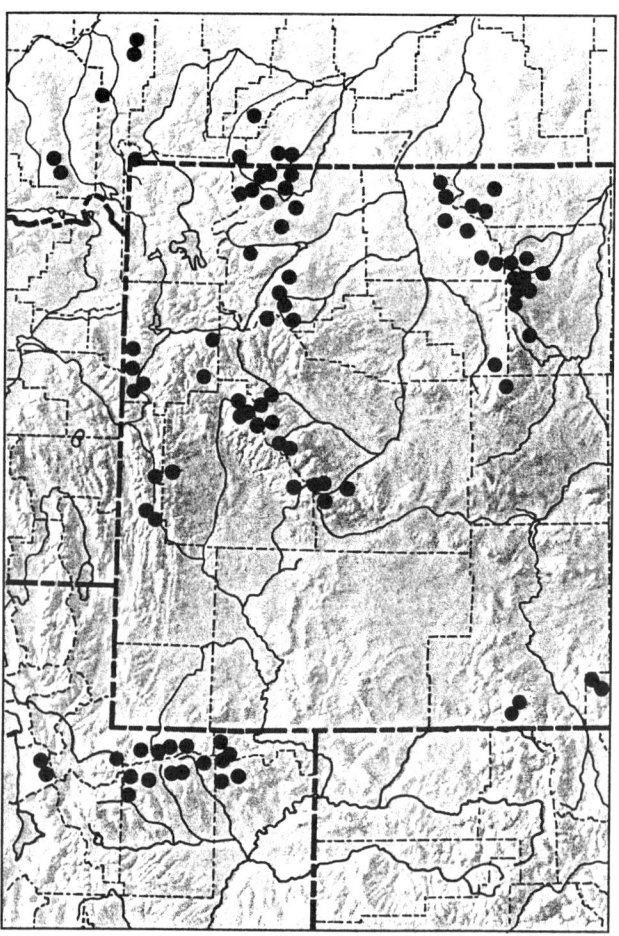

Calamagrostis scopulorum M.E. Jones
Proc. Calif. Acad. Sci., II, 5:722, 1895.

Jones Reedgrass

Tufted perennial, sometimes with short rhizomes; culms erect, glabrous. Leaf blades erect, stiff, involute to flat, scabrous ventrally; sheaths glabrous or scaberulous; ligules obtuse, erose to lacerate. Panicle narrow, spikelike, the branches erect to ascending; spikelets mostly 1-flowered; glumes mostly purplish, lanceolate, acute to acuminate, glabrous, the keel scaberulous; lemmas glabrous to scaberulous, with 4 narrow teeth at the apex, awned from the middle or below the middle, the awn straight or bent, slender, about as long as the lemma; callus hairs about one-third to half the length of the lemma; anthers 2–3 mm long.

Timberline meadows and krummholz margins of the Uinta and Wasatch Ranges. Montana south to Utah, Arizona, and New Mexico.

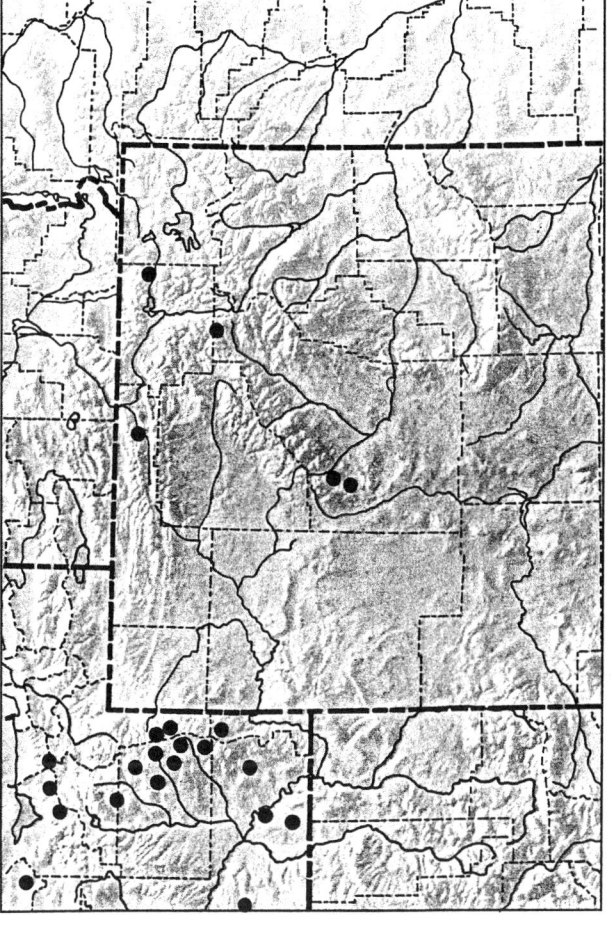

Calamagrostis stricta (Timm) Koel.
Descr. Gram., 105, 1802.

Northern Reedgrass

Arundo stricta Timm
Arundo neglecta Ehrh.
Calamagrostis americana (Vasey in Macoun) Scribn.
Calamagrostis borealis Laestad.
Calamagrostis californica Kearney
Calamagrostis chordorrhiza Pors.
Calamagrostis elongata (Kearney) Rydb.
Calamagrostis expansa Rickett & Gilly
Calamagrostis fernaldii Louis-Marie
Calamagrostis hyperborea Lange
Calamagrostis inexpansa Gray
Calamagrostis labradorica Kearney
Calamagrostis lacustris (Kearney) Nash
Calamagrostis laxiflora Kearney
Calamagrostis lucida Scribn.
Calamagrostis micrantha Kearney
Calamagrostis neglecta (Ehrh.) Gaertn., Mey. & Scherb.
Calamagrostis robusta (Vasey in Wheeler) Vasey
Calamagrostis wyomingensis Gand.
Deyeuxia americana (Vasey in Macoun) Lunell
Deyeuxia neglecta (Ehrh.) Kunth

Perennial from slender, creeping rhizomes; culms solitary or rarely 2 or more together, erect, scabrous, at least below the panicle. Leaf blades involute, sometimes flat, ascending, scabrous at least on the ventral surfaces; sheaths glabrous or scaberulous; ligules obtuse, lacerate or sometimes erose. Panicle congested, spikelike, often interrupted, the branches erect to ascending; spikelets mostly 1-flowered; glumes subequal, lanceolate to elliptic, acute or sometimes acuminate, glabrous or scaberulous; lemmas glabrous to scabrous, minutely 4-toothed at the apex, awned from just below the middle, the awn straight, slender, equaling or slightly exceeding the lemma; callus hairs two-thirds to as long as the lemma; anthers 1.2–2 mm long.

Timberline meadows and stream banks of the Beartooth and Wind River Ranges. Circumboreal; south in western North America to California, Arizona, New Mexico, and Nebraska.

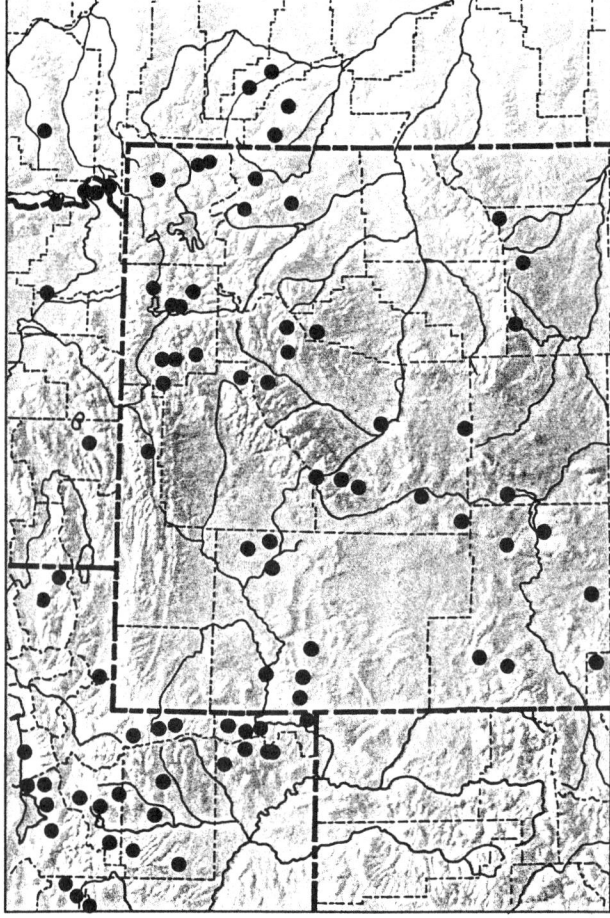

Danthonia DC.
Oatgrass

Tufted, low or moderately tall perennials with narrow, often involute, flat or folded leaves. Inflorescence an open or contracted, spikelike, often purplish panicle or raceme, or sometimes a single spikelet. Spikelets several-flowered, the articulation above the glumes and between the florets. Rachilla not prolonged above the upper floret. Cleistogamous spikelets produced in the lower sheaths. Glumes subequal in length, broad, acute, usually exceeding the uppermost floret. Lemmas shorter than the glumes, rounded on the back, bidentate at the apex, the teeth often elongate, bearing a flat, twisted, geniculate awn between the teeth. Palea papery, shorter than the lemma.

Danthonia intermedia Vasey
Bull. Torr. Club 10:52, 1883.

Timber Oatgrass

Danthonia canadensis Baum & Findlay
Danthonia cusickii (Williams) A.S. Hitchc.
Merathrepta intermedia (Vasey) Piper
Pentameris intermedia (Vasey) A. Nels. & Macbr.
Trisetum williamsii Louis-Marie

Densely tufted perennial from fibrous roots; culms erect, glabrous. Leaf blades flat or mostly involute, erect, glabrous or sparsely pilose above; sheaths glabrous, with tufts of hairs at the throat and on the collar; ligules forming a fringe of hairs. Panicle narrow, racemose, often interrupted and secund, the branches short, erect to ascending; spikelets 3–8-flowered; glumes lanceolate, acute, subequal; lemmas glabrous on the back, pilose on the margins, acuminate-toothed, the middle tooth forming a terminal awn; awn flattened, twisted at the base, mostly divergent, up to 10 mm long; callus pilose-bearded; anthers up to 4 mm long.

Dry meadows, fellfields, and margins of krummholz in most alpine ranges of the Middle Rocky Mountains. Not yet known from above timberline in the Gros Ventre, Hoback, and Wyoming Ranges. Alaska to Newfoundland, south in western North America to California, New Mexico, South Dakota, and Michigan.

Deschampsia Beauv.
Hairgrass

Perennials or less commonly annuals, with flat, folded, or somewhat involute blades. Inflorescence a narrow or open panicle of pale or purplish 2-flowered spikelets. Rachilla hairy, prolonged beyond the upper floret. Articulation above the glumes; the glumes shining, membranous, subequal. Lemmas membranous, rounded on the back, usually shorter than the glumes, truncate, 2–4-toothed at the apex, bearded at the base, bearing a straight or geniculate, dorsal awn from the middle or below it.

KEY TO THE SPECIES OF *DESCHAMPSIA*

1. Panicle narrow, the branches appressed to the rachis; blades filiform or capillary, about 1 mm wide — *D. elongata*
1. Panicle open, the branches spreading to ascending; blades thin and flat, usually 1.5–6 mm wide
    2. Glumes much exceeding the terminal floret; lemmas awned from near the middle — *D. atropurpurea*
    2. Glumes shorter than, equaling, or slightly exceeding the terminal floret; lemmas awned from near the base — *D. cespitosa*

Deschampsia atropurpurea (Wahlenb.) Scheele
Flora 27:56, 1844.

Mountain Hairgrass

Aira atropurpurea Wahlenb.
Aira latifolia Hook.
Avena atropurpurea (Wahlenb.) Link
Deschampsia hookeriana Scribn.
Deschampsia latifolia (Hook.) Vasey
Deschampsia pacifica Tatew. & Ohwi
Holcus atropurpureus (Wahlenb.) Wahlenb.
Vahlodea atropurpurea (Wahlenb.) Fries
Vahlodea flexulosa (Honda) Ohwi
Vahlodea latifolia (Hook.) Hult.

Loosely tufted perennial from fibrous roots; culms erect, glabrous, often purplish at the base. Leaves mostly basal, the blades flat or sometimes involute, scabrous or glabrous, acute to nearly acuminate; sheaths glabrous or scabrous; ligules erose, obtuse, puberulent. Panicle open, elongate to broadly pyramidal, the branches wide-spreading to drooping; spikelets purplish, mostly 2-flowered; glumes lanceolate to ovate, keeled or somewhat rounded on the back, acute to acuminate, subequal, the first 1-nerved, the second 3-nerved; lemmas glabrous to scabrous, erose-toothed at the apex; awned from about the middle, awn straight to geniculate, usually twisted near the base; callus hairs short, about half as long as the lemma; anthers 0.8–2 mm.

Occasional in moist meadows of the Beartooth and Teton Ranges. Circumboreal; south in western North America to California, Wyoming, and Colorado.

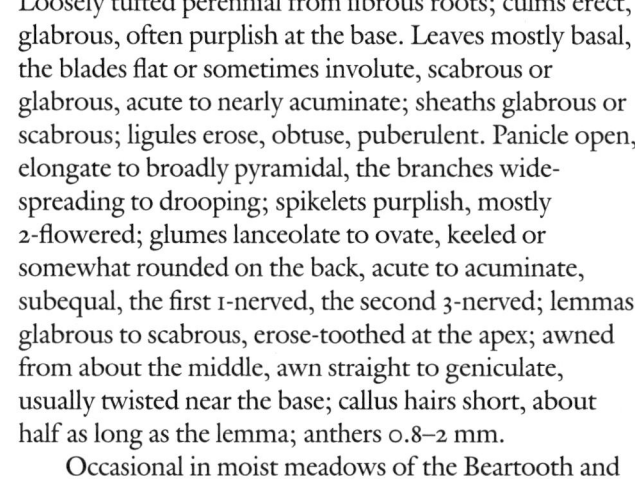

Deschampsia cespitosa (L.) Beauv.
Essai Nouv. Agrostogr. 19, 149, 160, 1812.

Tufted Hairgrass

Aira cespitosa L.
Agrostis caespitosa (L.) Salisb.
Aira alpicola (Rydb.) Rydb.
Aira ambigua Michx.
Aira holciformis (Presl) Steud.
Aira pungens (Rydb.) Rydb.
Aira sukatschewii Popl.
Avena caespitosa (L.) Kuntze
Campella caespitosa (L.) Link
Deschampsia alpicola Rydb.
Deschampsia alpina (L.) R. & S.
Deschampsia ambigua (Michx.) Beauv. ex Jackson
Deschampsia beringensis Hult.
Deschampsia bottonia (Wahl.) Trin.
Deschampsia brevifolia R. Br.
Deschampsia confinis (Vasey) Rydb.
Deschampsia curtifolia Scribn.
Deschampsia glauca Hartm.
Deschampsia holciformis Presl
Deschampsia hudsonica Abbe
Deschampsia komarovii Vassiljev
Deschampsia paramushirensis Honda
Deschampsia pumila (Ledeb.) Ostenf.
Deschampsia pungens Rydb.
Deschampsia sukatschewii (Popl.) Roshev.
Podionapus caespitosus (L.) Dulac

Caespitose perennial from fibrous roots; culms erect, glabrous. Leaves mostly basal and often matted, the blades folded to involute, sometimes flat, typically scabrous; sheaths scabrous to glabrous, somewhat keeled; ligules entire, membranous, acute to obtuse. Panicle contracted and elongate to open and pyramidal, the branches spreading, often drooping with age; spikelets purplish, mostly 2-flowered; glumes lanceolate to ovate, mostly acute, the first glume 1-nerved, the second 3-nerved; lemmas glabrous, membranous, erose-toothed at apex, awned from near the base, the awn straight or slightly geniculate, usually exerted beyond the glumes; callus hairs short, about one-third the length of the lemmas; anthers 0.7–1 mm long.

Common in moist meadows, along streams, and on alp slopes in nearly all alpine ranges. Not yet known from above timberline in the Wyoming Range. Circumboreal; south in western North America to Mexico.

Middle Rocky Mountain plants are distinguished from northern alpine and arctic populations as ssp. *cespitosa* var. *cespitosa*.

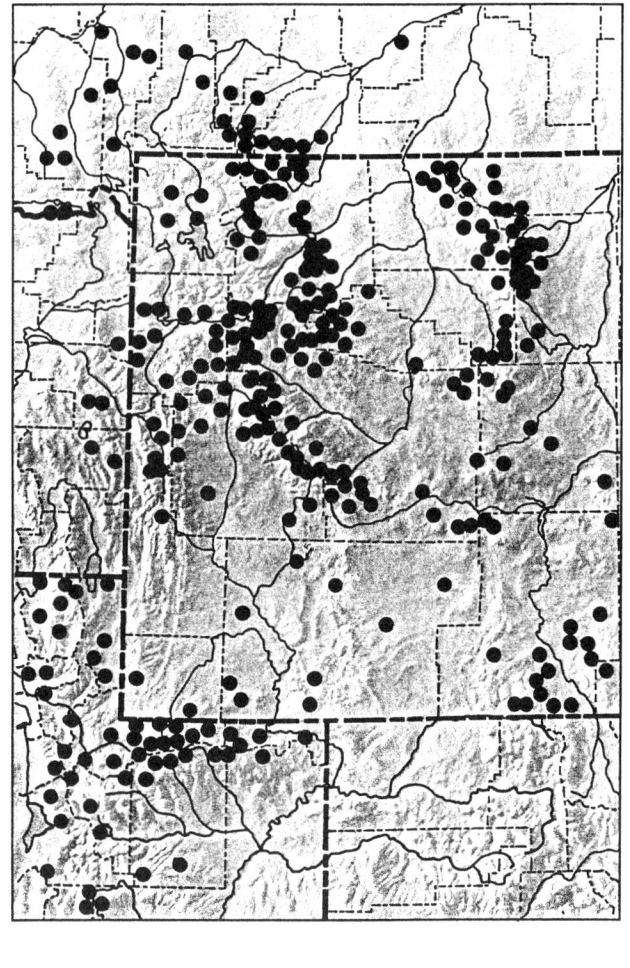

Deschampsia elongata (Hook.) Munro ex Benth.
Pl. Hartweg., 342, 1857.

Slender Hairgrass

Aira elongata Hook.
Aira vaseyana Rydb.
Deschampsia ciliata (Vasey ex Beal) Rydb.

Densely tufted perennial from thin, fibrous roots. Leaves mostly basal and filiform, the blades flat, folded, or involute, glabrous to scaberulous; sheaths mostly glabrous; ligules acute, entire, often splitting with age. Panicle narrow, erect, spikelike or racemose, the branches erect to ascending; spikelets mostly 2-flowered, often purplish; glumes lanceolate, subequal, acute, 3-nerved; lemmas glabrous, erose-toothed, shiny, often purplish, awned from just below the middle, the awn straight or nearly so; callus hairs about half the length of the lemmas; anthers mostly 0.5–0.7 mm long.

Timberline meadows, krummholz margins, and talus slopes of the Uinta and Wasatch Ranges. Alaska south to Mexico, in most of the western United States.

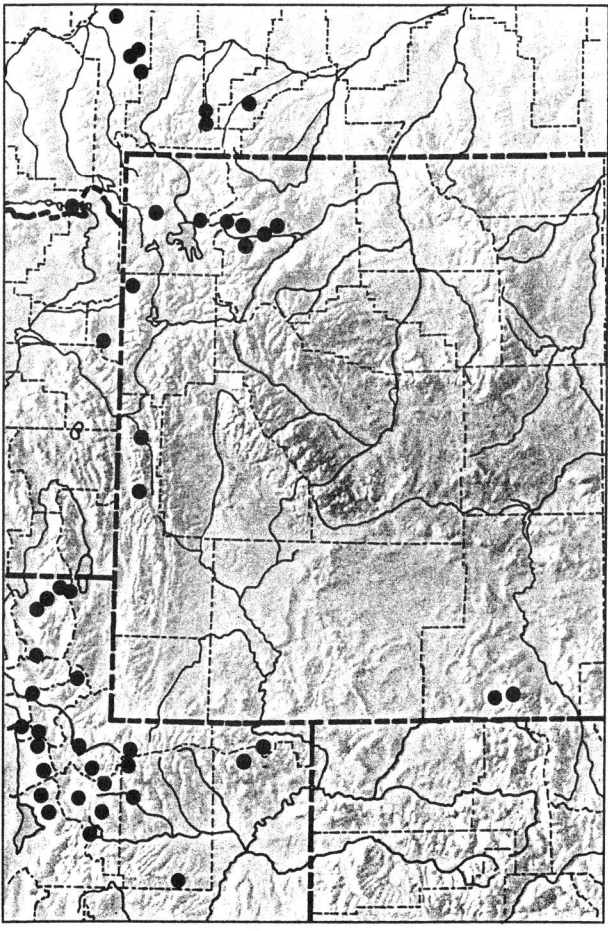

Elymus L.
Wildrye, Wheatgrass

Perennials, often tufted, less commonly annuals, from fibrous roots or rhizomes; culms erect, spreading, or somewhat geniculate. Leaves flat or involute; ligule membranous. Inflorescence a single, symmetrical, slender or bristly spike, sometimes disarticulating at maturity; spike greenish or purplish. Spikelets numerous, 2- to several-flowered, sessile, solitary to several at each joint of the rachis, subterete to flattened, usually overlapping. Articulation above the glumes and between the florets, or in the rachis. Glumes shorter than to longer than the lowermost lemma, equal to unequal, narrow and acute to awned, or nearly linear and awnlike, 1–5-nerved or nerves lacking; awns, if present, short, straight, and ascending to flexuous and divergent. Lemmas rounded on the back, acute or awned from the tip, 5–7-nerved. Palea well developed, shorter than or nearly as long as the lemma.

KEY TO THE SPECIES OF *ELYMUS*

1. Spikelets mostly 2 (1–3) at each joint of the rachis
   2. Awns straight, erect or ascending, less than 2 cm long; rachis not disarticulating at maturity .... *E. glaucus*
   2. Awns flexuous, divergent, 2 cm or more long; rachis disarticulating at maturity .... *E. elymoides*
1. Spikelets solitary at each joint of the rachis
   3. Plants rhizomatous
      4. Glumes lanceolate, broadest near the base, mostly 3–5-nerved, acuminate to short-awned .... *E. smithii*
      4. Glumes oblanceolate, broadest at or above the middle, mostly 5–7-nerved, acute to acuminate .... *E. lanceolatus*
   3. Plants lacking rhizomes
      5. Stems spreading- to ascending-decumbent; rachis disarticulating at maturity .... *E. scribneri*
      5. Stems erect, sometimes slightly decumbent at the base; rachis not disarticulating at maturity
         6. Spikelets shorter than or slightly exceeding the internode length; lemmas awned, the awns divergent .... *E. spicatus*
         6. Spikelets about twice as long as the internode length; lemmas awnless or awned, the awns straight .... *E. trachycaulus*

Elymus elymoides (Raf.) Swezey
Nebr. Pl. Doane Coll., 15, 1891.

Squirreltail

Sitanion elymoides Raf.
Aegilops hystrix Nutt.
Elymus brevifolius (J.G. Smith) M.E. Jones
Elymus glaber (J.G. Smith) Davy
Elymus hystrix (Nutt.) M.E. Jones
Elymus longifolius (J.G. Smith) Gould
Elymus sitanion J.A. Shult.
Hordeum elymoides (Raf.) Schenk
Polyantherix hystrix (Nutt.) Nees
Sitanion albescens Elmer
Sitanion basalticola Piper
Sitanion breviaristatum J.G. Smith
Sitanion brevifolium J.G. Smith
Sitanion californicum J.G. Smith
Sitanion ciliatum Elmer
Sitanion cinereum J.G. Smith
Sitanion glabrum J.G. Smith
Sitanion horteoides Suksd.
Sitanion hystrix (Nutt.) J.G. Smith
Sitanion insulare J.G. Smith
Sitanion latifolium Piper
Sitanion longifolium J.G. Smith
Sitanion molle J.G. Smith
Sitanion montanum J.G. Smith
Sitanion pubiflorum J.G. Smith
Sitanion rigidum J.G. Smith
Sitanion strigosum J.G. Smith
Sitanion velutinum Piper

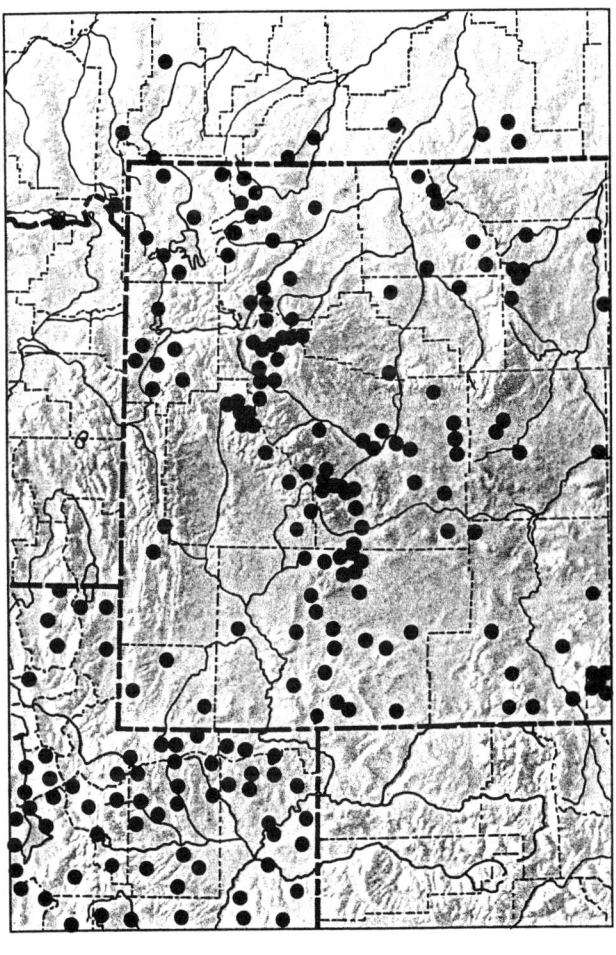

Tufted perennial from fibrous roots; culms erect to spreading, occasionally solitary, glabrous or sometimes pubescent. Leaf blades flat or involute, and glabrous, scabrous, or pubescent; sheaths glabrous to pubescent, open, the upper ones often inflated, auricles lacking or inconspicuous; ligules less than 1 mm long, membranous, ciliate. Spike erect to flexuous, often partially included in the upper sheath, the rachis disarticulating at maturity; spikelets mostly 2 per node, 2–6-flowered, the lowermost floret sometimes reduced to 1 or 2 glumes; glumes narrowly lanceolate to linear, tapering to a long (2–9 cm), scabrous, spreading to flexuous awn, occasionally cleft to form 2 or 3 awns; lemmas glabrous to puberulent, obscurely nerved, awned from the tip, the awn 2–9 cm long, scabrous, spreading to flexuous; palea nearly the same length as the lemma, often short-awned; anthers 1–2 mm long.

Fellfields, talus, scree, and other dry rocky places in the Absaroka, Big Horn, Teton, Uinta, Wasatch, Wind River, and Wyoming Ranges. British Columbia to California, Mexico, Texas, and South Dakota.

Two varieties of this complex, highly variable species may be recognized in our area:

1. Lowermost lemma reduced; glumes 5 or more per spikelet . . . var. *elymoides*
1. Lowermost lemma not reduced; glumes 4 per spikelet . . . var. *brevifolius* (J.G. Smith) Dorn.

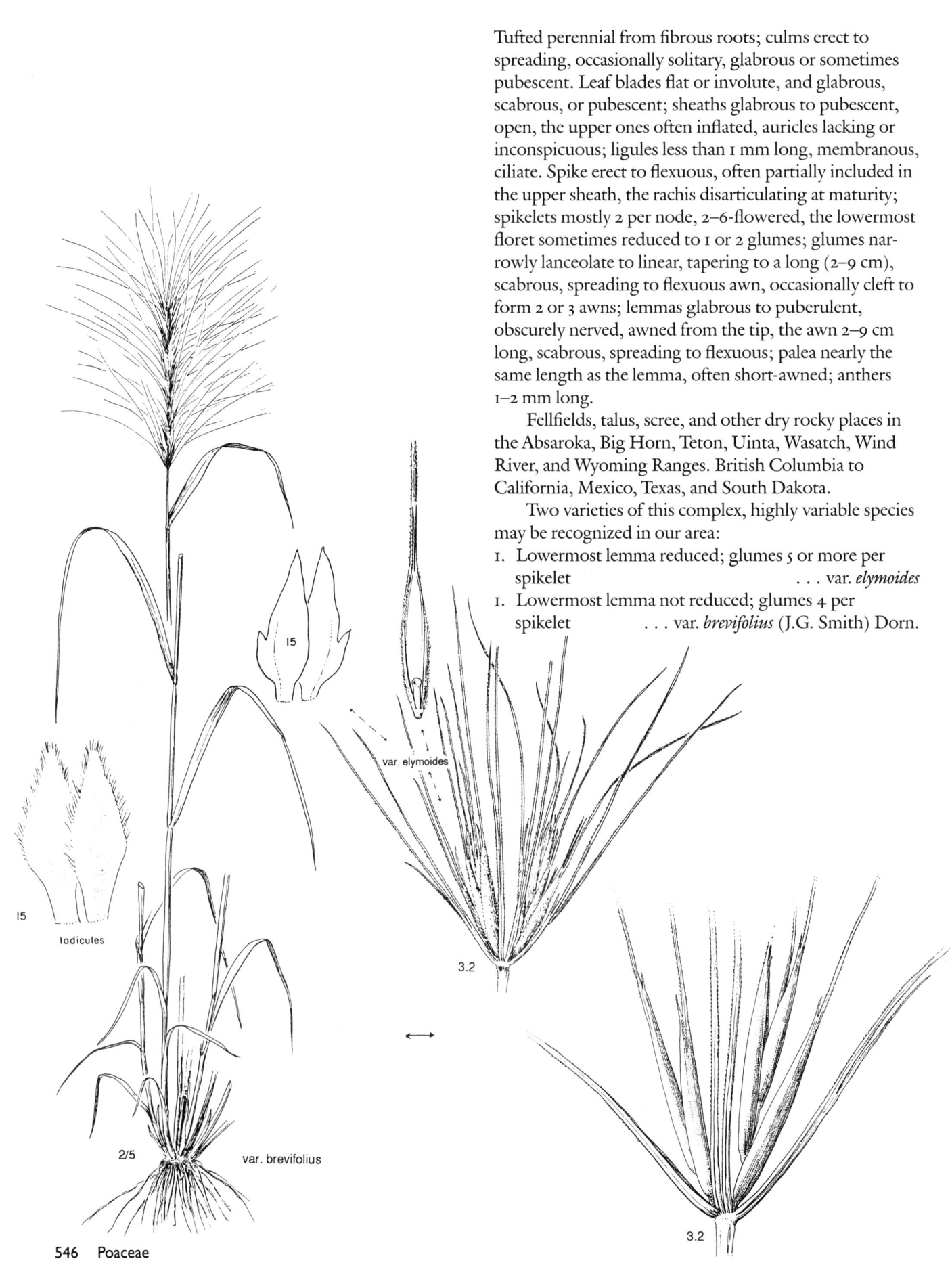

546 Poaceae

Elymus glaucus Buckl.
Proc. Acad. Phila. 1862:99, 1863.

Blue Wildrye

Clinelymus glaucus (Buckl.) Nevski
Elymus americanus Vasey & Scribn. ex Macoun
Elymus edentatus Suksd.
Elymus hispidulus Davy
Elymus howellii Scribn. & Merr.
Elymus marginalis Rydb.
Elymus nitidus Vasey
Elymus petersonii Rydb.
Elymus pubescens Davy in Jeps.
Elymus strigatus St. John
Elymus virescens Piper
Terrellia glauca (Buckl.) Lunell

Perennial from short rhizomes, or rhizomes lacking; culms single or more commonly in tufts, glabrous to pubescent. Leaf blades mostly flat, glabrous or scabrous, usually glaucous; sheaths glabrous, scabrous, or sparsely villous, auricles mostly well developed and clasping; ligules entire to erose-ciliate. Spike erect, the rachis glabrous to scabrous, continuous; spikelets 2–4-flowered, mostly 2 per node; glumes narrowly lanceolate or elliptic, acute to short-awned; lemmas glabrous or scabrous, rounded on the back, awned from the tip, the awn straight or occasionally spreading, 1–2 times as long as the body of the lemma; anthers 1.5–3 mm long.

Timberline margins of the Beartooth and Wind River Ranges. Alaska to Ontario, south to California, Mexico, New Mexico, Iowa, and Indiana.

Middle Rocky Mountain plants are the typical var. *glaucus*.

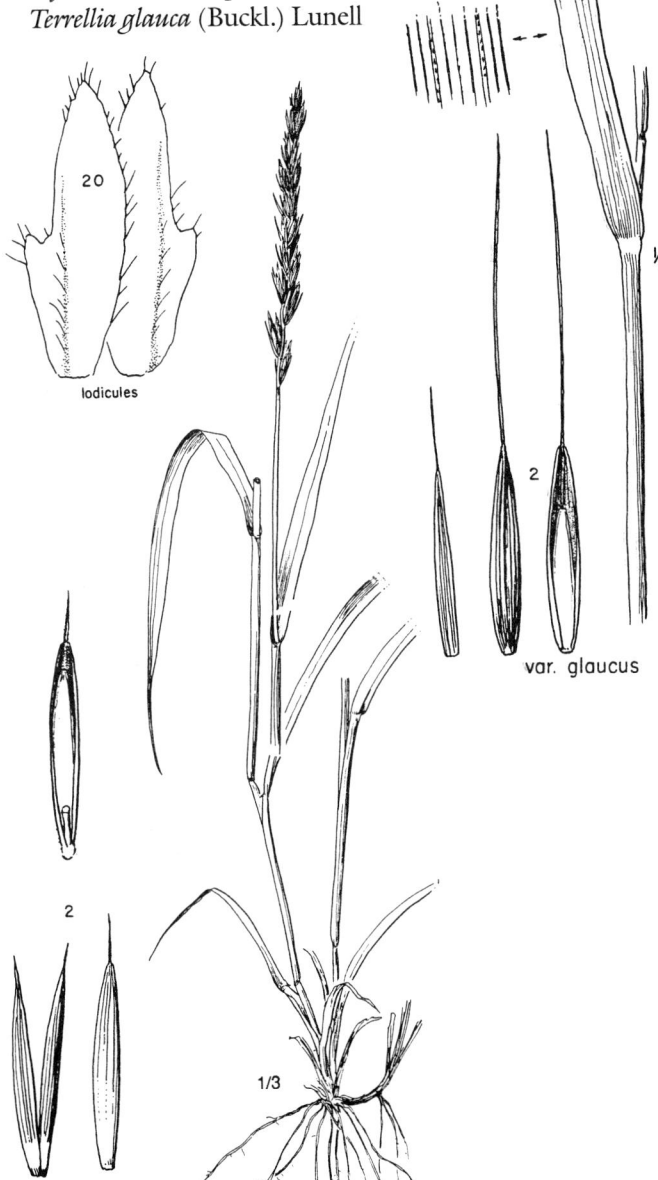
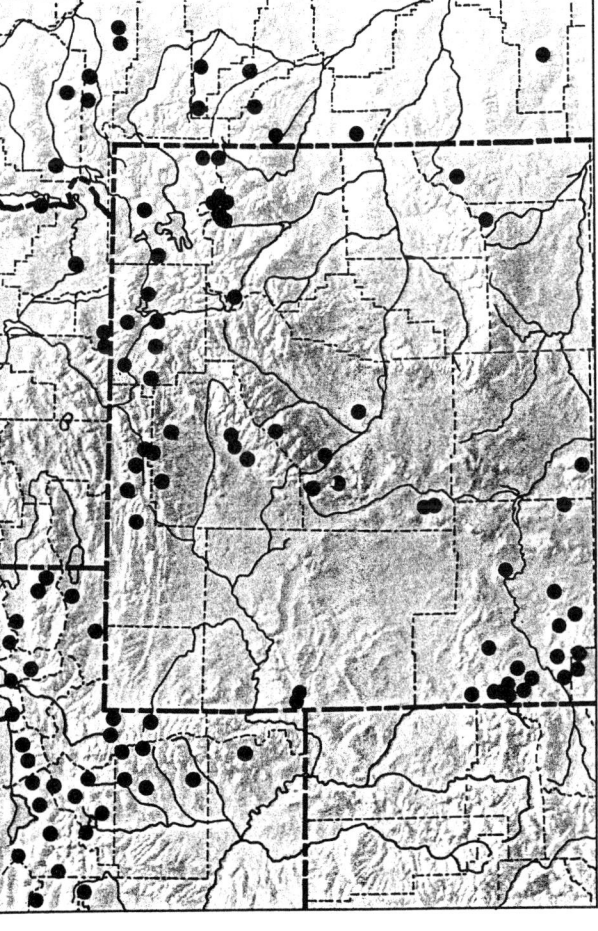

Elymus lanceolatus (Scribn. & Smith) Gould
Madroño 10:94, 1949.

Thickspike Wheatgrass

Agropyron lanceolatum Scribn. & Smith
Agropyron albicans Scribn. & Smith
Agropyron dasystachyum (Hook.) Scribn. & Smith
Agropyron elmeri Scribn.
Agropyron griffithsii Scribn. & Smith ex Piper
Agropyron psammophilum Gillett & Senn
Agropyron pseudorepens Scribn. & Smith (in part)
Agropyron riparium Scribn. & Smith
Agropyron subvillosum (Hook.) E. Nels.
Agropyron yukonense (Scribn. & Merr.) Dewey
Elymus albicans (Scribn. & Smith) A. Löve
Elymus griffithsii Scribn. & Smith ex Piper) A. Löve
Elymus riparius (Scribn. & Smith) Gould
Elymus rydbergii Gould
Elymus subvillosus (Hook.) Gould
Elytrigia dasystachya (Hook.) Löve & Löve
Elytrigia riparia (Scribn. & Smith) Beetle
Roegneria albicans (Scribn. & Smith) Beetle
Triticum dasystachyum (Hook.) Gray
Zeia albicans (Scribn. & Smith) Lunell
Zeia dasystachya (Hook.) Lunell
Zeia griffithsii (Scribn. & Smith ex Piper) Lunell
Zeia pseudorepens (Scribn. & Smith) Lunell (in part)
Zeia riparia (Scribn. & Smith) Lunell

Perennial from long-creeping rhizomes; culms erect to somewhat decumbent at the base, solitary or tufted, glaucous, glabrous. Leaf blades flat to involute, glabrous, sometimes scabrous above, usually stiff and ascending; sheaths glabrous to puberulent, auricles usually present, small; ligules erose-ciliate. Spike erect, the rachis continuous; spikelets mostly 3–8-flowered, solitary at each node of the rachis; glumes subequal, narrowly oblanceolate, the tips acute or acuminate, mostly pubescent, occasionally glabrous or scabrous; lemmas pubescent, obtuse to acute, or sometimes awn-tipped, rounded on the back; anthers 3–5 mm long.

Timberline margins of the Absaroka, Beartooth, and Wind River Ranges. Alaska south to California, Arizona, Colorado, Nebraska, and West Virginia.

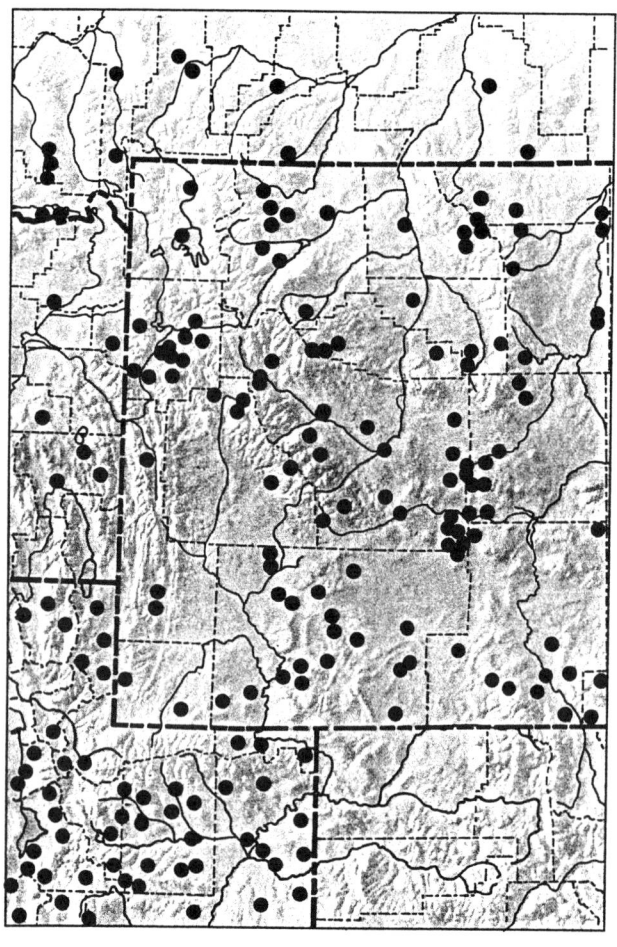

Elymus scribneri (Vasey) M.E. Jones
Contr. West. Bot. 14:20, 1912.

Scribner Wheatgrass

Agropyron scribneri Vasey
Agropyron bakeri E. Nels. (in part)
Sitanion marginatum Scribn. & Merr.

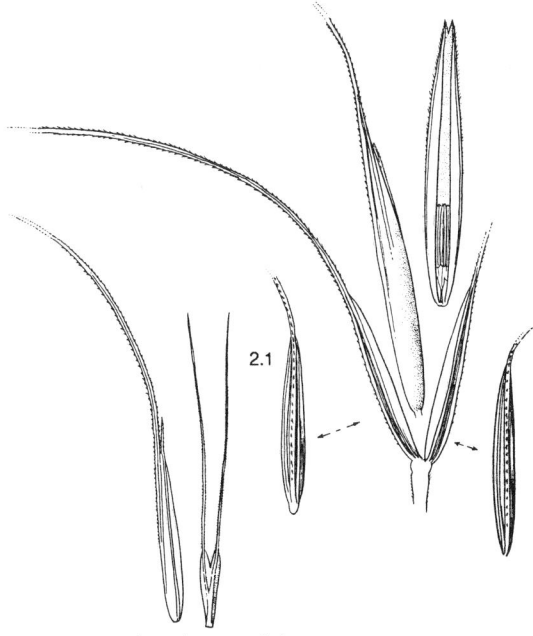

glumes from top spikelet

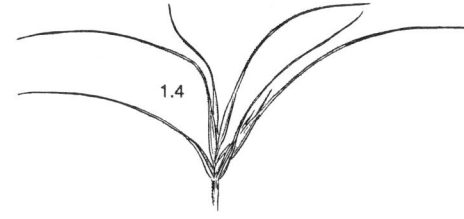

spikelet

Tufted perennial, lacking rhizomes; culms spreading to flexuous, sometimes nearly prostrate, decumbent at the base. Leaf blades flat to involute, glabrous to puberulent; sheaths glabrous, or more commonly pubescent, auricles small or lacking; ligules short-ciliolate to erose. Spike flexuous, often partially included, the rachis disarticulating at maturity; spikelets mostly solitary at each node, 3–5-flowered; glumes narrowly lanceolate to linear, awned, the awn scabrous, divergent; lemmas rounded on the back, glabrous to scaberulous, tapering to a divergent awn; anthers 1–2 mm long.

Fellfields, rocky meadows, ledges, and talus in all alpine ranges. British Columbia to Arizona and New Mexico.

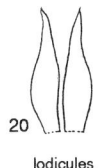

lodicules

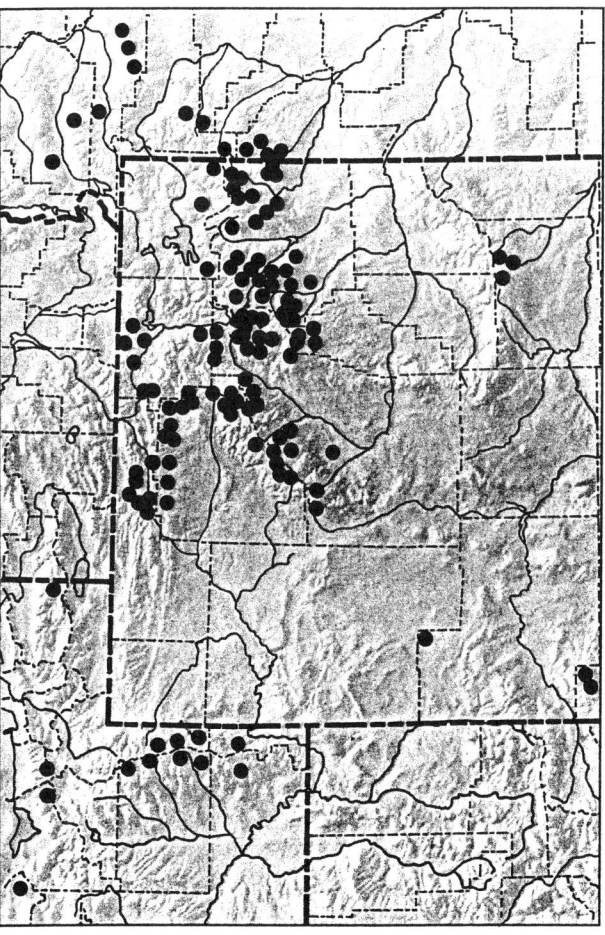

Elymus smithii (Rydb.) Gould
Madroño 9:127, 1947.

Western Wheatgrass

Agropyron smithii Rydb.
Agropyron molle (Scribn. & Smith) Rydb.
Agropyron occidentale (Scribn.) Scribn.
Agropyron palmeri (Scribn. & Smith) Rydb.
Elytrigia smithii (Rydb.) Nevski
Pascopyrum smithii (Rydb.) A. Löve
Zeia mollis (Scribn. & Smith) Lunell
Zeia occidentalis (Scribn.) Lunell
Zeia smithii (Rydb.) Lunell

Perennial from long-creeping rhizomes; culms erect, single or loosely tufted, glaucous. Leaves glaucous, the blades involute to flat, glabrous to short-villous; sheaths glabrous to puberulent, auricles conspicuous; ligules erose-ciliolate. Spike erect, the rachis not disarticulating; spikelets solitary at each node, mostly 5–8-flowered; glumes rigid, subequal, glabrous to puberulent, linear-lanceolate to lanceolate; lemmas lanceolate, glabrous to puberulent, mucronate to short-awned; paleas about the same length as the lemmas; anthers 3–5 mm long.

Occasional on fellfields and roadsides in the lower alpine zone of the Beartooth Range. Alaska south to Arizona, Texas, Kentucky, and Ontario.

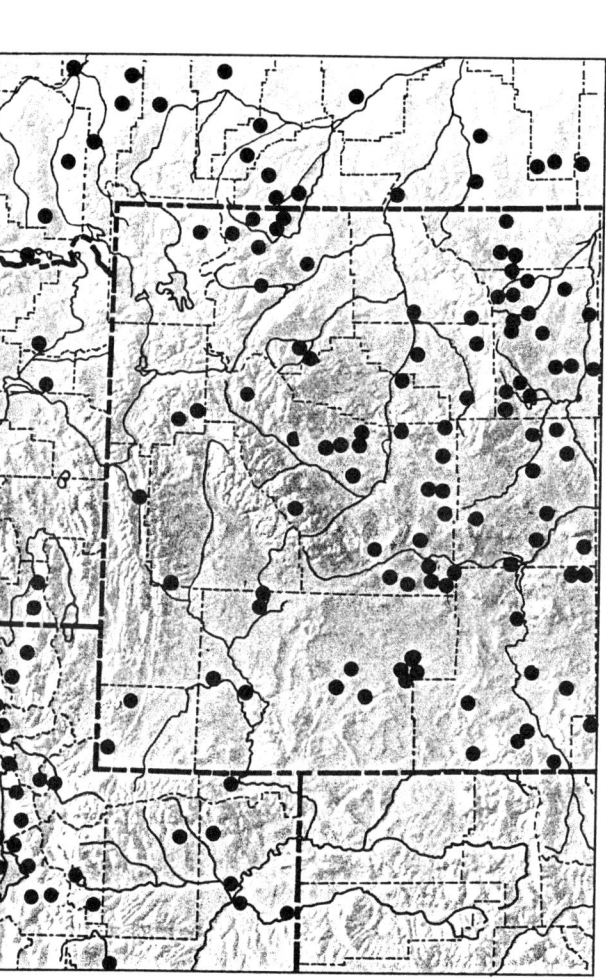

Elymus spicatus (Pursh) Gould
Madroño 9:125, 1947.

Bluebunch Wheatgrass

Festuca spicata Pursh
Agropyron divergens (Nees in Steud.) Nees ex Vasey
Agropyron inerme (Scribn. & Smith) Rydb.
Agropyron spicatum Pursh
Agropyron vaseyi Scribn. & Smith
Elytrigia spicata (Pursh) D. Dewey
Pseudoroegneria spicata (Pursh) A. Löve
Roegneria spicata (Pursh) Beetle
Schedonorus spicatus (Pursh) R. & S.
Triticum divergens Nees in Steud.
Zeia spicata (Pursh) Lunell

Tufted perennial, rhizomes lacking; culms glabrous, erect or nearly so. Leaf blades green or glaucous, involute or flat, glabrous to puberulent; sheaths glabrous to pubescent, often villous, auricles small, prominent; ligules erose-ciliolate. Spike erect, open, the rachis not disarticulating; spikelets solitary at the nodes, 5–8-flowered; glumes subequal, oblong to oblanceolate, obtuse, acute, or mucronate; lemmas elliptic, glabrous to scabrous, rounded on the back, acute or more commonly awned, the awn widely divergent; paleas about the same size and shape as the lemmas; anthers 4–6 mm long.

Timberline meadows of the Beartooth, Wind River, and Wyoming Ranges. Alaska south to California, Arizona, New Mexico, and Nebraska.

Middle Rocky Mountain plants are the typical var. *spicatus*.

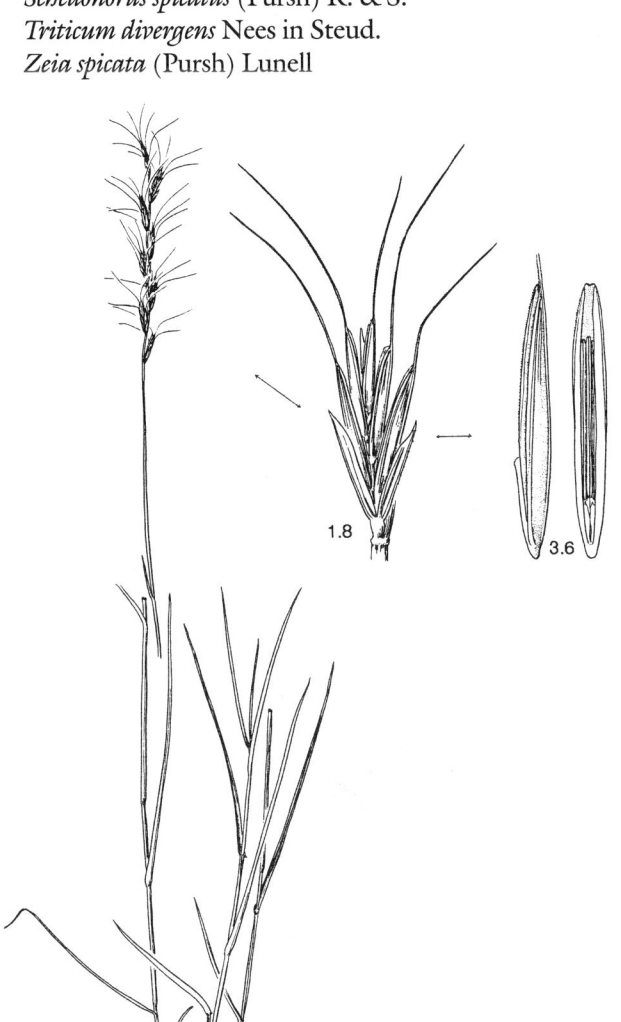

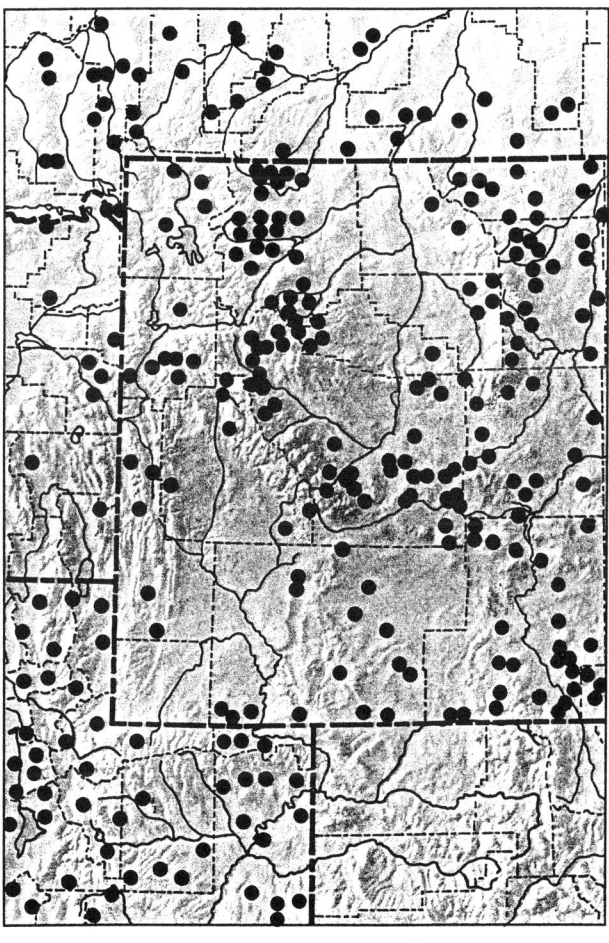

Elymus trachycaulus (Link) Gould ex Shinners
Rhodora 56:28, 1954.

Slender Wheatgrass

Triticum trachycaulum Link
Agropyron alaskanum Scribn. & Merr.
Agropyron andinum (Scribn. & Smith) Rydb.
Agropyron bakeri E. Nels. (in part)
Agropyron biflorum auctt.
Agropyron biflorum (Brign.) R. & S. ex Rydb.
Agropyron boreale (Turcz.) Drobov
Agropyron brevifolium Scribn.
Agropyron caninoides (Ram.) Beal
Agropyron caninum (L.) Beauv.
Agropyron gmelini (Griseb.) Scribn. & Smith
Agropyron latiglume (Scribn. & Smith) Rydb.
Agropyron novae-angliae Scribn. ex Brain., et al.
Agropyron pauciflorum (Schwein. in Keat.) A.S. Hitchc. ex Silveus
Agropyron pseudorepens Scribn. & Smith (in part)
Agropyron richardsonii (Schrad.) Schrad.
Agropyron subsecundum (Link) A.S. Hitchc.
Agropyron tenerum Vasey
Agropyron teslinense Pors. & Senn.
Agropyron trachycaulon (Link) Steud.
Agropyron trachycaulum (Link) Malte ex H.F. Lewis
Agropyron unilaterale Cassidy
Agropyron violaceum (Hornem.) Lange
Agropyron violescens (Ram.) Beal
Elymus pauciflorus (Schwein. in Keat.) Gould
Elymus sierrus Gould
Elymus subsecundus (Link) Löve & Löve
Roegneria borealis (Turcz.) Nevski
Roegneria canina (L.) Nevski
Roegneria latiglumis (Scribn. & Smith) Nevski
Roegneria pauciflora (Schwein. in Keat.) Hylander
Roegneria scandia Nevski
Roegneria trachycaula (Link) Nevski
Roegneria violaceum (Hornem.) Melderis
Roegneria virescens (Lange) Böcher
Triticum boreale Turcz.
Triticum caninum L.
Triticum pauciflorum Schwein. in Keat.
Triticum richardsonii Schrad.
Triticum subsecundum Link
Zeia canina (L.) Lunell
Zeia pseudorepens (Scribn. & Smith) Lunell (in part)
Zeia richardsonii (Schrad.) Lunell
Zeia tenera (Vasey) Lunell

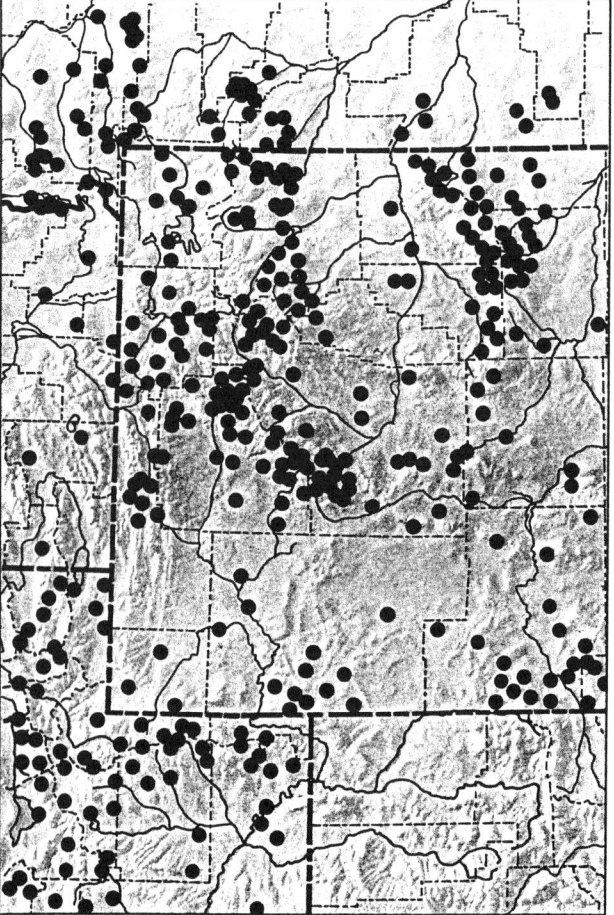

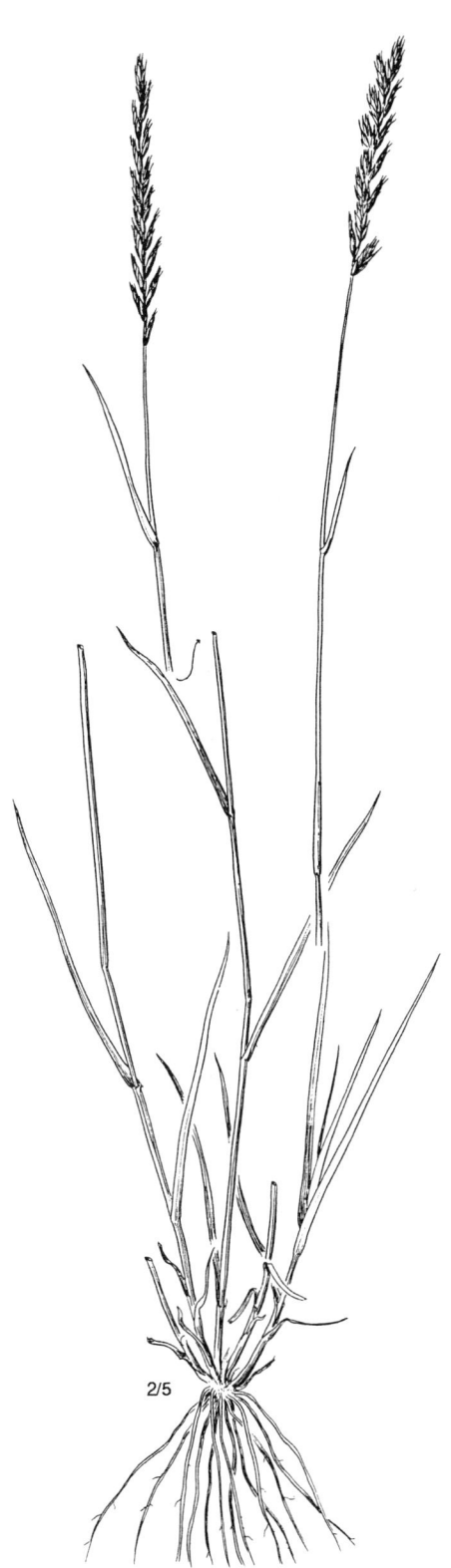

Tufted perennial, mostly lacking rhizomes; culms erect or decumbent at the base, glabrous to puberulent. Leaf blades ascending, flat to somewhat involute, and glabrous, scabrous, or pilose; sheaths glabrous to puberulent, occasionally pilose, auricles present or lacking; ligules entire to erose-ciliolate. Spike erect, open, the rachis not disarticulating; spikelets solitary at the nodes, 3–5-flowered; glumes subequal, oblong to oblanceolate, conspicuously 5–7-nerved, hyaline-margined, acute to short-awned; lemmas lanceolate, glabrous to scabrous, acuminate to long-awned, the awns straight or nearly so; anthers mostly 1–2 mm long.

Meadows and fellfields throughout the alpine zone of the Middle Rocky Mountains. Circumboreal; Alaska to Labrador and south to Mexico, Iowa, and New York.

Alpine plants are ssp. *trachycaulus*. There are 2 varieties:

1. Awns shorter than the body of the lemma, less than 1 cm long . . . var. *trachycaulus*
1. Awns longer than the body of the lemmas, 1–3 cm long or more . . . var. *andinus* (Scribn.& Smith) Dorn

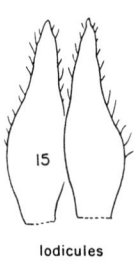

lodicules

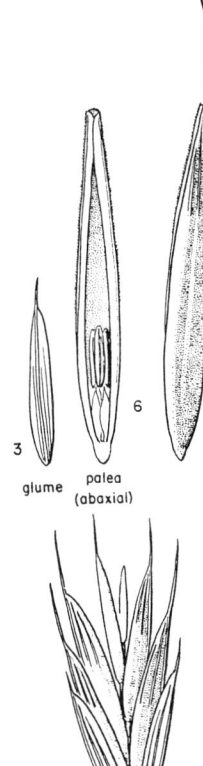

glume   palea (abaxial)

Festuca L.
Fescue

Erect annuals or perennials with flat, narrow to filiform blades, the spikelets in narrow or open panicles. Spikelets several-flowered, the articulation above the glumes and between the florets, the uppermost floret often reduced in size. Glumes narrow, acute, unequal, shorter than the first lemma. Lemmas rounded on the back, acute, awned or awnless, 5-nerved, the nerves often obscure. Palea about as long as the lemma.

KEY TO THE SPECIES OF *FESTUCA*

1. Culms tomentose to puberulent, at least below the panicle — *F. baffinensis*
1. Culms glabrous or scaberulous
    2. Plants short-rhizomatous, usually decumbent at the base; leaf bases reddish, fibrillose — *F. rubra*
    2. Plants lacking rhizomes, erect; leaf bases rarely reddish, not fibrillose — *F. ovina*

Festuca baffinensis Polunin
Bull. Natl. Mus. Can. 92:91, 1940.

Baffin Fescue

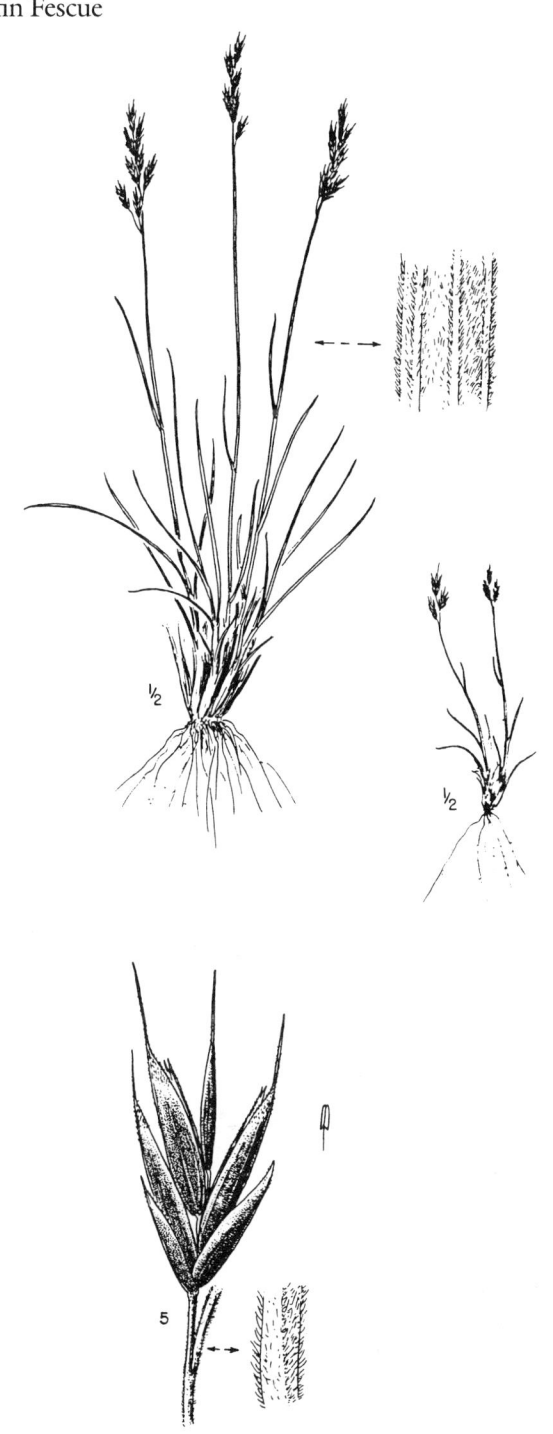

Low, tufted perennial from fibrous roots, rhizomes lacking; culms slender, less than 20 cm tall, puberulent or tomentose below the inflorescence. Leaves mostly basal, the blades filiform, flat to involute, glabrous to scabrous; sheaths glabrous to scaberulous, auricles lacking; ligules ciliolate. Panicle purplish, contracted; spikelets solitary at the nodes, 3–4-flowered; glumes lanceolate, glabrous to scabrous, acute or acuminate; lemmas lanceolate, glabrous to scabrous, short-awned, the awns slender, straight; anthers 0.3–0.5 mm long.

Fellfields, felsenmeer, talus, scree, and other rocky well-drained situations in the Absaroka, Beartooth, Gros Ventre, Wind River, and Wyoming Ranges. Alaska, the Canadian Arctic Archipelago, and Greenland south to Hudson Bay, Wyoming, and Colorado.

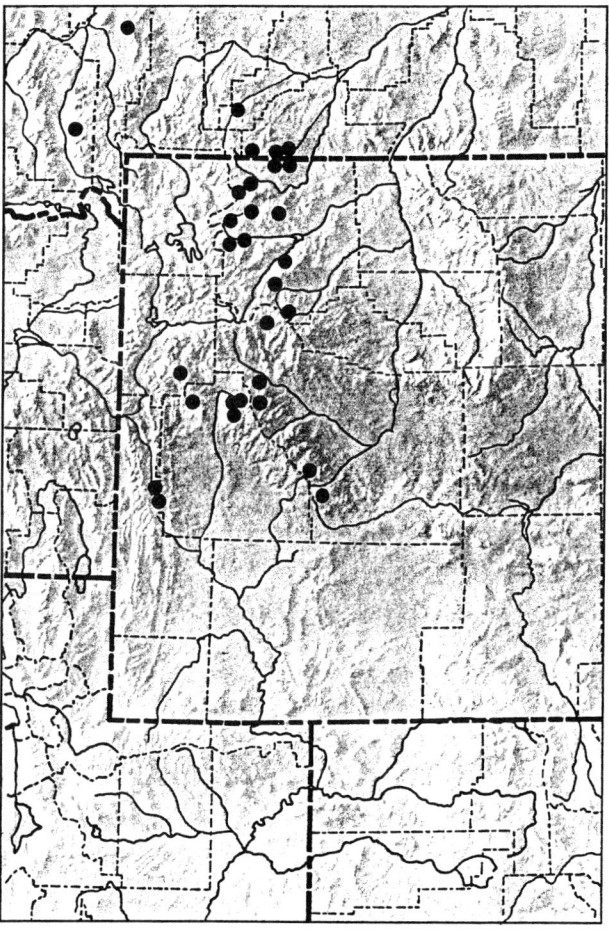

Festuca ovina L.
Sp. Pl., 73, 1753.

Sheep Fescue

Avena ovina (L.) Salisb.
Bromus ovinus (L.) Scop.
Festuca arizonica Vasey
Festuca brachyphylla J.A. Shult. ex J.A. & J.H. Shult.
Festuca brevifolia R. Br.
Festuca calligera (Piper) Rydb.
Festuca capillata Lam.
Festuca duriuscula L.
Festuca idahoensis Elmer
Festuca ingrata (Hack. ex Beal) Rydb.
Festuca minutiflora Rydb.
Festuca roemeri (Pavlick) Alexeev
Festuca saximontana Rydb.
Festuca supina Schur
Gnomonia ovina (L.) Lunell

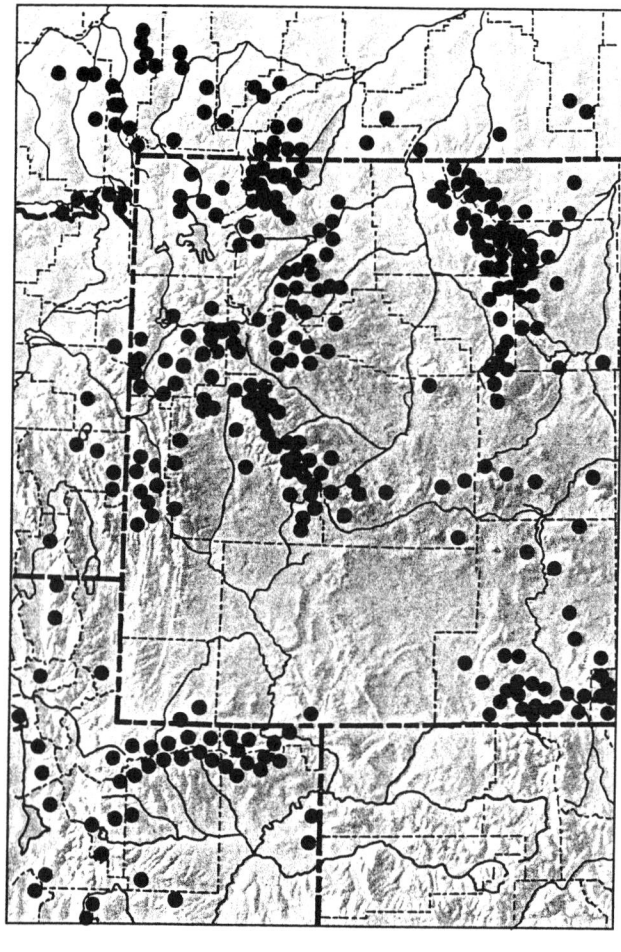

Densely tufted perennial from fibrous roots, rhizomes lacking; culms erect, glabrous. Leaves mostly basal, the blades filiform, folded-involute, glabrous to scabrous; sheaths glabrous to puberulent, auricles lacking; ligules ciliolate. Panicle narrow, contracted, often spikelike, erect, the branches erect-ascending, sometimes spreading at anthesis; spikelets mostly 3–7-flowered; glumes lanceolate, glabrous, acute; lemmas rounded to keeled, obscurely nerved, glabrous to scaberulous, the awns slender, erect; anthers 0.5–4 mm long.

Fellfields, alpine meadows, scree, and talus slopes in all alpine ranges. Circumboreal; south in western North America to California, New Mexico, and Colorado.

Three varieties occur at and above timberline in the Middle Rocky Mountains. They may be separated as follows:

1. Culms mostly more than 30 cm tall; panicles open, 7 cm or more long; anthers 2–4 mm long (*F. idahoensis*) . . . var. *ingrata* Hackel ex Beal
1. Culms mostly less than 30 cm tall; panicles contracted, less than 7 cm long; anthers 0.5–1.8 mm long
    2. Anthers more than 1 mm long; culms more than 20 cm tall . . . var. *rydbergii* St. Yves
    2. Anthers less than 1 mm long; culms less than 20 cm tall (*F. brachyphylla*) . . . var. *brevifolia* (R. Br.) S. Wats.

Common names for these infraspecific taxa are Idaho Fescue (var. *ingrata*), Rydberg Fescue (var. *rydbergii*), and Alpine Fescue (var. *brevifolia*).

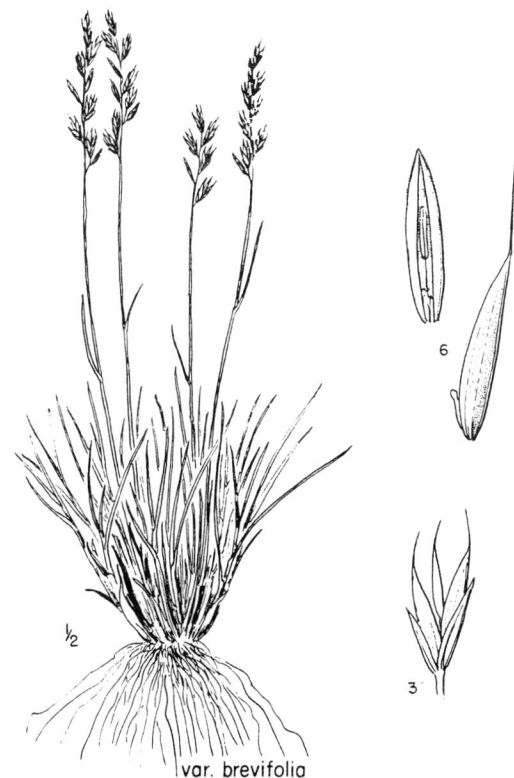

Festuca rubra L.
Sp. Pl., 74, 1753.

Red Fescue

Bromus secundus Presl
Festuca arenaria Osbeck
Festuca aucta Krecz. & Bobr.
Festuca densiuscula (Hack. ex Piper) Alexeev
Festuca earlei Rydb.
Festuca fallax Thuill.
Festuca heterophylla Lam.
Festuca kitaibeliana Schult.
Festuca lanuginosa (Mert. & Koch in Rohl) Scheele
Festuca multiflora Hoffman
Festuca oregana Vasey
Festuca prolifera (Piper) Fern.
Festuca pubescens Willd. ex Link
Festuca richardsonii Hook.
Festuca vallicola Rydb.

Perennial from short rhizomes; culms erect, sometimes decumbent at the base, solitary or several in loose tufts, glabrous or scabrous. Leaves mostly basal, the blades flat or, more commonly, folded-involute; sheaths glabrous to pubescent, the lower brown or reddish brown and fibrillose; auricles lacking; ligules ciliolate. Panicle erect, contracted, the branches ascending; spikelets mostly 4–7-flowered, often purplish; glumes unequal, lanceolate, acute; lemmas lanceolate, glabrous, scabrous or pilose, short-awned, the awn slender, about half as long as the body of the lemma; anthers 2–4 mm long.

Timberline meadows of the Absaroka and Wind River Ranges. Circumboreal; south throughout most of the eastern and western United States; in the West to California, New Mexico, and Texas.

Festuca rubra differs from *F. ovina* by having short rhizomes and reddish-strigillose leaf bases at least at the base of the plant. Two varieties of the mountains may be distinguished as follows: var *rubra* has lemmas that are glabrous to scabrous-ciliate on the back; var. *lanuginosa* Mert. & Koch in Rohl has lemmas that are pilose on the back.

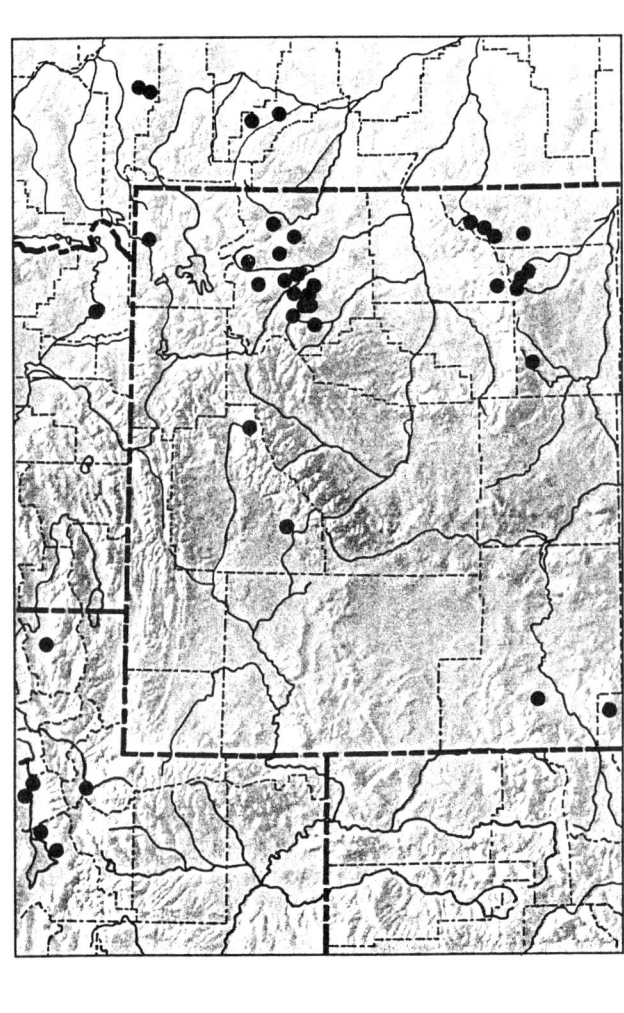

Helictotrichon Besser
Oatgrass

Tufted perennials with open panicles and flat or involute blades, the sheaths open. Spikelets bronze or purplish, less than 1.5 cm long, the rachilla bearded, not prolonged beyond the terminal floret. Articulation above the glumes and between the florets. Glumes thin, subequal, equaling or exceeding the first lemma in the spikelet. Lemmas rounded, shorter than the glumes, with a dorsal, twisted, geniculate awn. Palea shorter than the lemma.

KEY TO THE SPECIES OF *HELICTOTRICHON*

1. Plants 15–45 cm tall, the panicle 5–10 cm long; spikelets 3–6-flowered, the glumes shorter than the uppermost floret   *H. hookeri*
1. Plants 5–20 cm tall, the panicle 2–7 cm long; spikelets 2-flowered, the glumes equal to or longer than the uppermost floret   *H. mortonianum*

Helictotrichon hookeri (Scribn.) Henrard
Blumea 3:429, 1940.

Hooker Oatgrass

Avena hookeri Scribn.
Avena americana Scribn.
Avena versicolor Hook.
Avenochloa hookeri (Scribn.) Holub
Avenula hookeri (Scribn.) Holub

Tufted perennial from fibrous roots, rhizomes lacking; culms solitary to several, erect, glabrous. Leaves mostly basal, the blades flat or folded, glabrous to scabrous, the midrib prominent, the margins thickened, whitish; sheaths glabrous, keeled, open to the base; ligules membranaceous, lacerate to erose. Panicle erect, narrow, the branches erect, mostly with single spikelets; spikelets 3–6-flowered, disarticulating above the glumes and between the florets; glumes subequal, lanceolate, scarious, acute, shorter than the florets; lemmas brownish, lanceolate, scaberulous, the apex bifid, awned from a little above to a little below the middle, the awn twisted near the base, strongly geniculate; anthers 3–5 mm long.

Timberline meadows of the Absaroka and Beartooth Ranges. Alberta to Manitoba south to New Mexico, Colorado, North Dakota, and Minnesota.

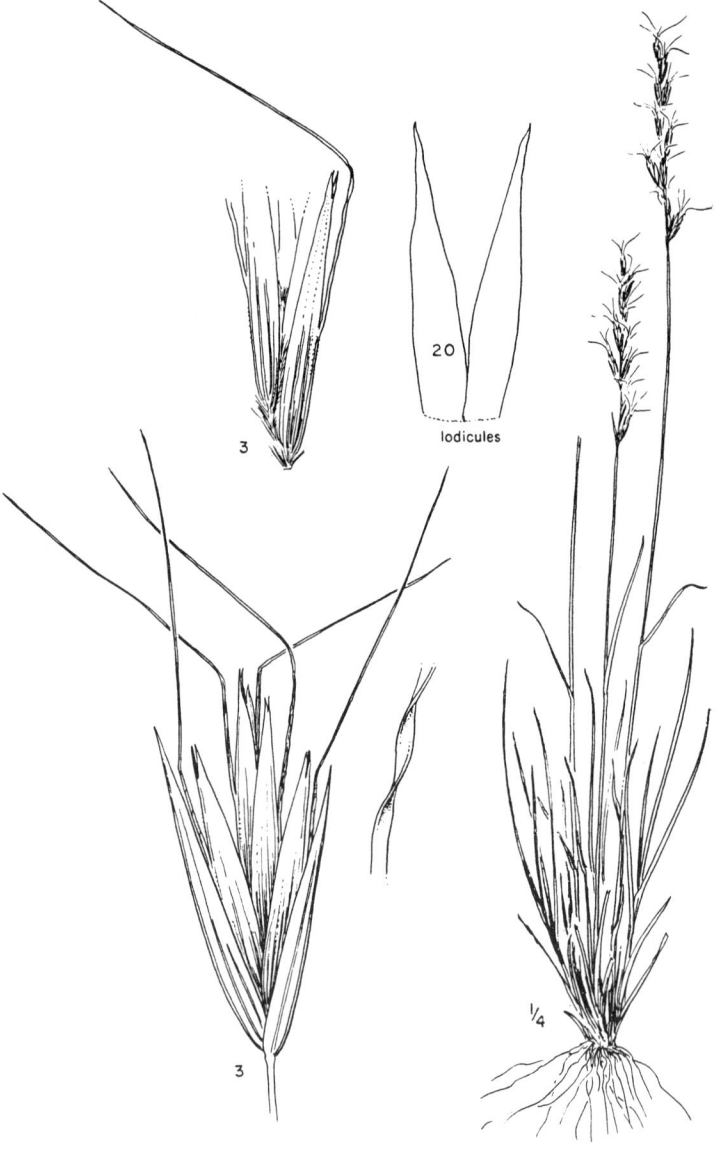

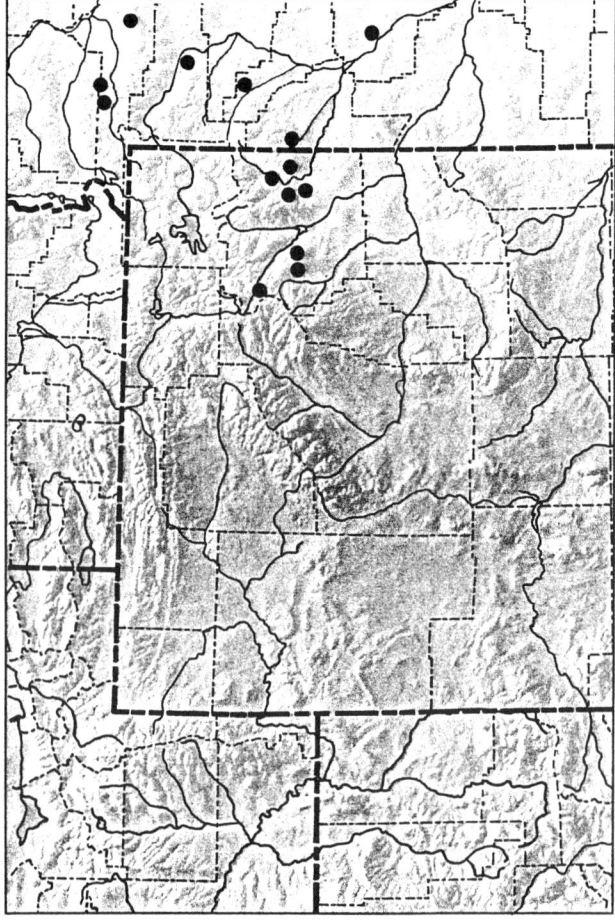

Helictotrichon mortonianum (Scribn.) Henrard
Blumea 3:429, 1940.

Alpine Oatgrass

Avena mortoniana Scribn.

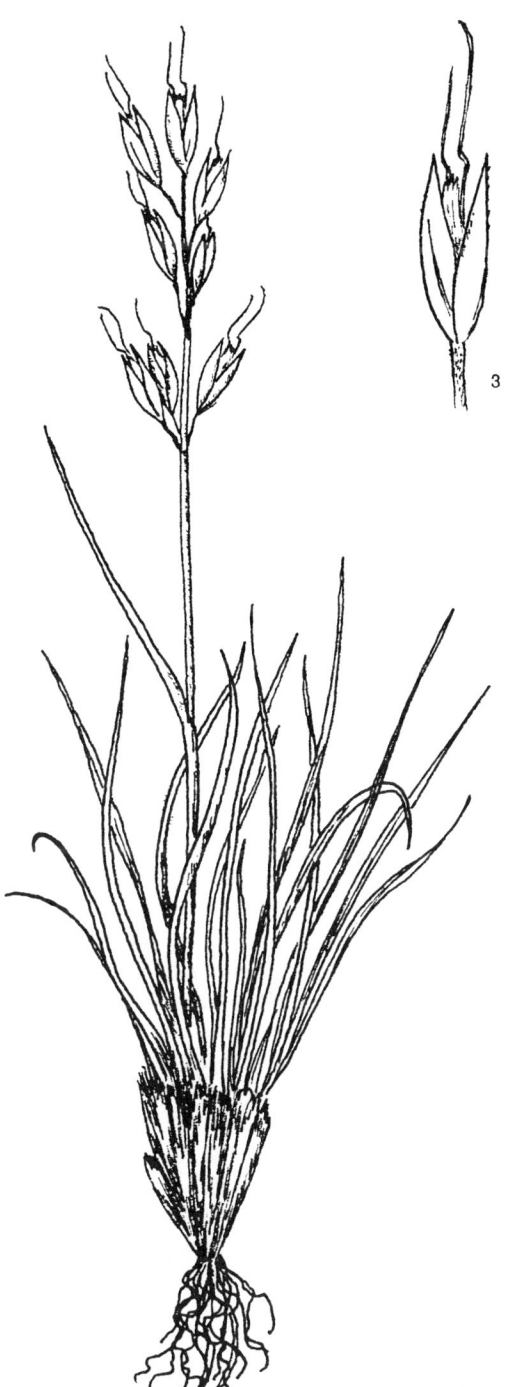

Tufted perennial from fibrous roots, rhizomes lacking; culms several, erect, glabrous. Leaf blades mostly involute, scaberulous; sheaths glabrous to puberulent, open to the base; ligules membranous. Panicle erect, narrow, compact to somewhat loose, the branches erect, each bearing a single spikelet; spikelets 2-flowered, the lower floret perfect, the upper usually sterile; glumes subequal, lanceolate, scabrous at least on the keel, acuminate, mostly longer than the florets; lemmas scaberulous, obscurely nerved, lanceolate, the apex minutely 4-toothed, awned from near the middle, the awn twisted near the base, strongly geniculate; anthers 1.5–2.5 mm long.

Meadows of the Uinta Range. Colorado and Utah south to New Mexico.

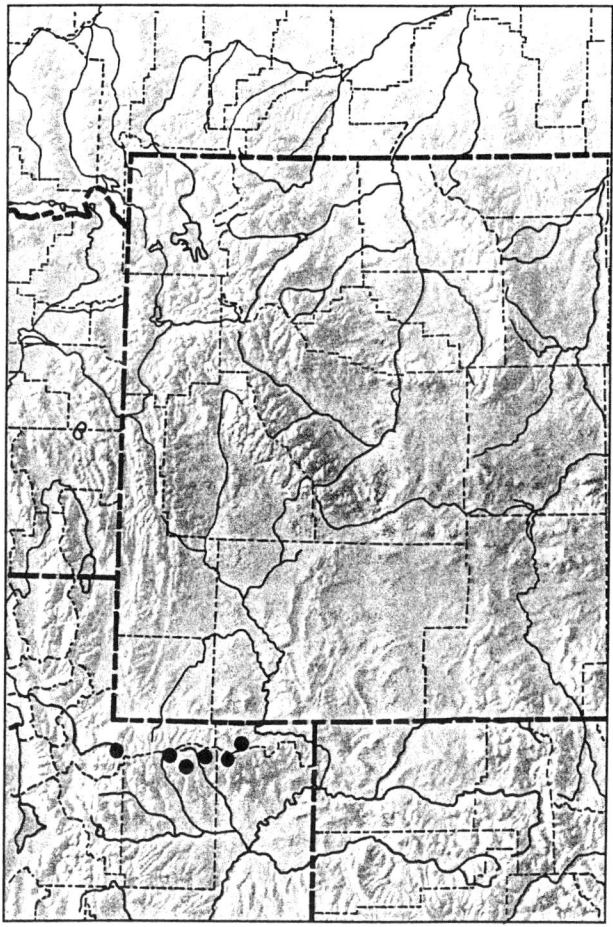

Hierochloe R. Br.
Sweetgrass

Erect perennials with fragrant herbage and small, open panicles of broad, bronze-colored spikelets. Spikelets with 1 perfect terminal floret and 2 inferior staminate florets that are attached to the fertile floret and fall with it. Glumes equal, broad and papery, 3-nerved. Staminate lemmas boat-shaped, ciliate-hairy, equaling the glumes, awnless or nearly so in ours. Fertile lemmas slightly indurate and similar to the sterile lemmas. Palea 3-nerved, rounded on the back.

Hierochloe odorata (L.) Beauv.
Essai Nouv. Agrostogr. 62, 164, 1812.

Sweetgrass

Holcus odoratus L.
Anthoxanthum nitens (Weber) Y. Schouten & Veldkamp
Avena odorata (L.) Koel.
Hierochloe arctica Presl
Hierochloe nashii (Bickn.) Kaczmarek
Savastana nashii Bickn.
Savastana odorata (L.) Scribn.
Torresia odorata (L.) A.S. Hitchc.

Perennial from long-creeping rhizomes; culms solitary or a few together, glabrous, often purplish at the base. Leaves mostly basal, reduced upward, the blades flat, glabrous to sparsely pubescent; sheaths glabrous to pubescent, open to the base, those of the basal leaves brownish to reddish; ligules membranous, acute to obtuse, erose. Panicle mostly open, pyramidal, the branches spreading to reflexed; spikelets bronze or purplish, compressed, ovate, 3-flowered, with 1 perfect terminal floret and 2 staminate florets below; glumes membranous, subequal, acute to acuminate, about the same length as or exceeding the florets; lemmas acute, pubescent, dimorphic, those of the staminate florets ovate, those of the perfect florets lanceolate; anthers 1–2.5 mm long, those of the staminate florets mostly longer than those of the perfect florets.

Timberline meadows and krummholz margins of the Big Horn, Uinta, and Wind River Ranges. Circumboreal; south in western North America to Nevada, Arizona, New Mexico, and Texas.

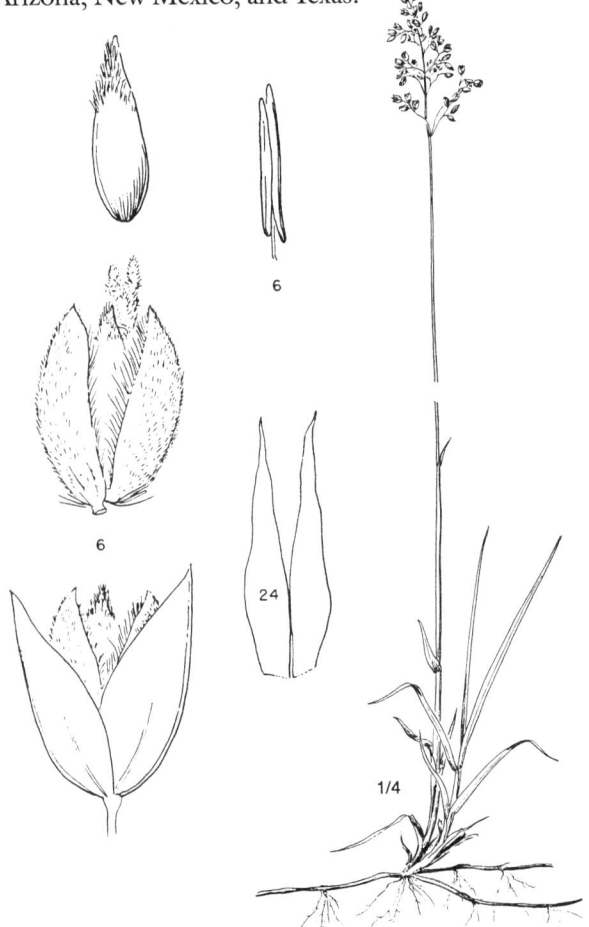

Koeleria Pers.
Junegrass

Tufted, erect perennials with narrow blades and shining, spikelike panicles. Spikelets 2–4-flowered, laterally flattened. Articulation above the glumes and between the florets, the rachilla glabrous, prolonged as a slender bristle or bearing a reduced floret at the tip. Glumes about equaling the first lemma, subequal, the first narrower and shorter than the second. Lemmas keeled, somewhat scarious, acute or short-awned, the awn, if present, originating from just below the apex. Palea thin, hyaline, equaling or slightly shorter than the lemma.

Koeleria macrantha (Ledeb.) Schult.
Mant. 2:345–346, 1824.

Junegrass

Aira macrantha Ledeb.
Aira gracilis (Pers.) Trin.
Airochloa gracilis (Pers.) Link
Koeleria cristata auctt., non Pers.
Koeleria elegantula Domin
Koeleria gracilis Pers.
Koeleria latifrons (Domin) Rydb.
Koeleria nitida Nutt.
Koeleria pyramidata auctt., non (Lam.) Beauv.
Koeleria robinsoniana Domin
Koeleria yukonensis Hult.

Tufted perennial from fibrous roots, rhizomes lacking; culms several to many, erect, glabrous to puberulent. Leaves mostly basal, the blades glabrous to scabrous or sometimes pubescent, flat or folded to involute; sheaths glabrous to scabrous or retrorse-hispid; ligules membranaceous, pubescent, ciliate, erose. Panicle narrow, cylindrical, spikelike, the spikelets flattened laterally, 2–4-flowered; glumes subequal, lanceolate to oblanceolate, glabrous or scabrous, acute to acuminate, or rarely awn-tipped, strongly keeled; lemmas lanceolate, scaberulous, acute to awn-tipped; anthers 1.1–2.5 mm long.

Dry timberline meadows, slopes, and fellfields of the Absaroka, Beartooth, Big Horn, Uinta, Wasatch, and Wind River Ranges. Circumboreal; south in western North America to northern Mexico and Texas.

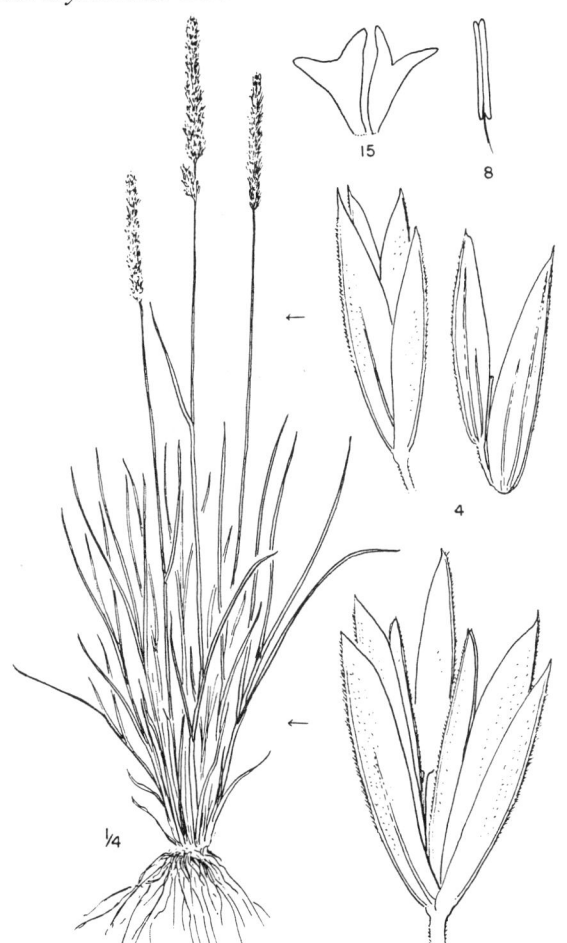
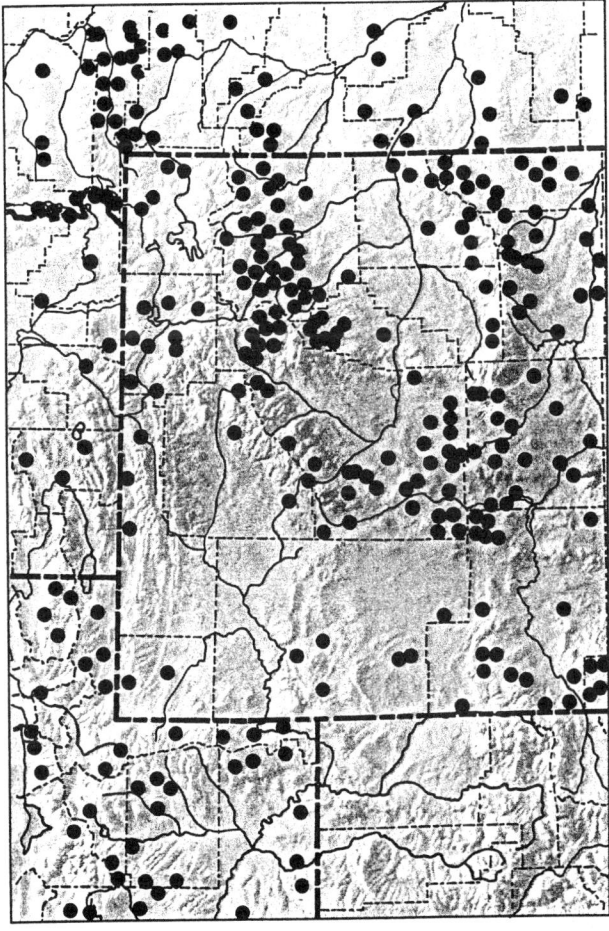

Leucopoa Griseb.
Spikegrass

Coarse, dioecious perennials producing short rhizomes and forming dense tufts. Blades flat or loosely involute, tapering to a slender point. Staminate and pistillate plants similar in appearance. Spikelets 3–5-flowered, in narrow panicles, the articulation above the glumes and between the florets. Glumes subequal in length, shorter than the lemmas. Lemmas more or less keeled, acute, awnless. Palea equaling the lemma, scabrous-ciliate on the keels.

Leucopoa kingii (S. Wats.) W.A. Weber
Univ. Colo. Stud. Ser. Biol. 23:2, 1966.

Spike Fescue

Poa kingii S. Wats.
Festuca confinis Vasey
Festuca kingii (S. Wats.) Cassidy
Festuca watsonii Nash in Britt.
Hesperochloa kingii (S. Wats.) Rydb.
Wasatchia kingii (S. Wats.) M.E. Jones

Densely tufted, dioecious perennial from fibrous roots; rhizomes lacking or very short; culms few to several, erect, glabrous. Leaves mostly basal, the blades stiff, glaucous, flat or somewhat involute, glabrous to scabrous; sheaths glabrous to scabrous, persistent, often forming dense basal clumps; ligules pubescent, erose-ciliolate. Panicle narrow, contracted, the branches erect-ascending; spikelets 3–5-flowered, the florets unisexual; glumes subequal, lanceolate to ovate, glabrous, mostly keeled on the back; lemmas lanceolate to ovate, acute or rarely awn-tipped, rounded on the back, glabrous or, more commonly, scabrous; anthers of staminate florets to 6 mm long.

Common on fellfields and in timberline meadows in all alpine ranges. Washington south to California and east to Montana, Nebraska, and Colorado.

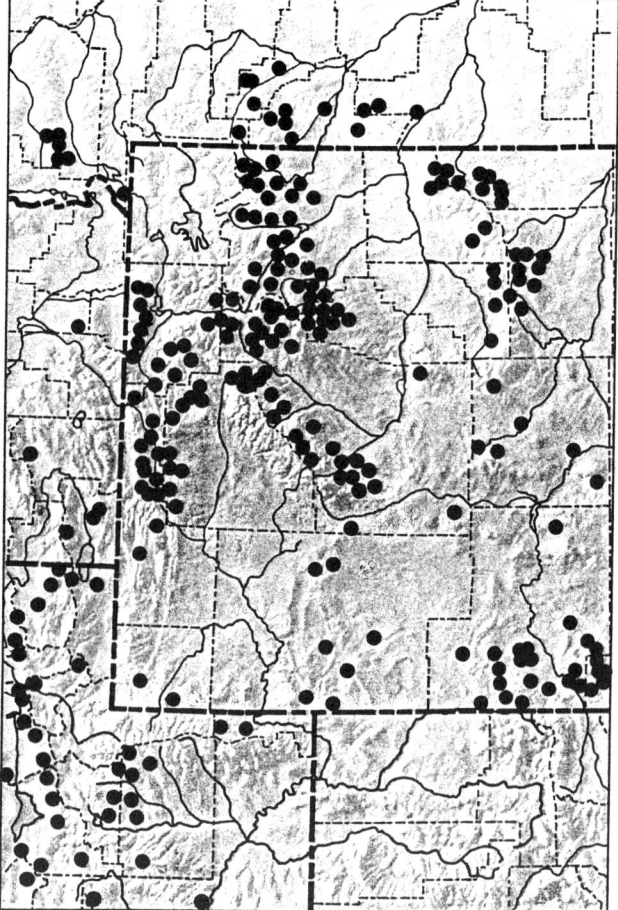

Phippsia (Trin.) R. Br.
Icegrass

Dwarf, tufted perennials with flat blades and narrow, few-flowered panicles of small, 1-flowered spikelets. Articulation above the glumes, the rachilla not prolonged beyond the palea. Glumes unequal, shorter than the lemma, the first glume sometimes wanting. Lemma 3-nerved, thin, membranous, somewhat keeled, abruptly acute. Palea dentate, somewhat shorter than the lemma.

Phippsia algida (Phipps) R. Br.
Chlor. Melv., 27, 1823.

Icegrass

Agrostis algida Phipps
Catabrosa algida (Phipps) T. Fries
Colpodium monandrum Trin.
Phippsia monandra (Trin.) Hook.
Poa algida (Phipps) Rupr.
Trichodium algidum (Phipps) R. & S.
Vilfa algida (Phipps) Trin.
Vilfa monandra Trin.

Low, tufted perennial, less than 10 cm tall, from fibrous roots; rhizomes lacking; culms several to many, glabrous, erect to decumbent. Leaves mostly basal, the blades flat, boat-shaped at the tip; sheaths glabrous; ligules membranous. Panicle narrow, contracted, the branches erect-ascending; spikelets 1-flowered, often purplish; glumes unequal, the first glume sometimes lacking, second glume erose at the apex; lemma glabrous or scaberulous, acute, lacking awns; paleas bifid, slightly shorter than the lemmas; anthers about 0.4 mm long.

Cold, wet, rocky places along streams and around receding snowbanks of the Beartooth and Wind River Ranges. Circumpolar; south in western North America to Wyoming and Colorado.

3/4

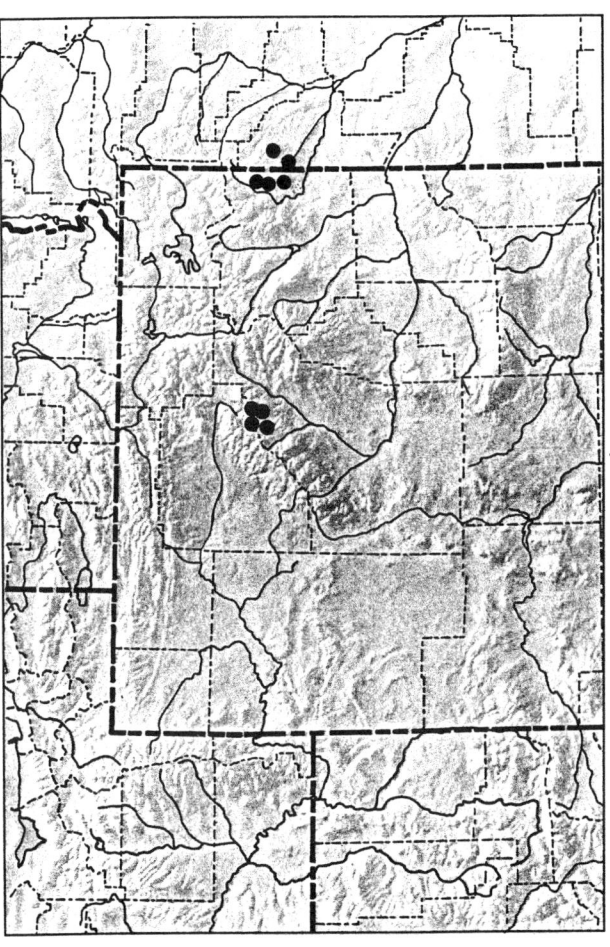

Phleum L.
Timothy

Caespitose perennials with erect culms and flat blades. Inflorescence a dense ovoid to cylindric, spikelike raceme or panicle. Spikelets 1-flowered, laterally compressed, the articulation above the glumes. Glumes equal, membranous, keeled, mucronate, or short-awned. Lemma shorter than the glumes, awnless, hyaline, broadly truncate, 3–5-nerved. Palea narrow, nearly as long as the lemma.

Phleum alpinum L.
Sp. Pl., 59, 1753.

Alpine Timothy

Phleum commutatum Gaudin
Phleum haenkeanum Presl
Plantinia alpina (L.) Bubani

Tufted perennial from a decumbent, rhizomelike base; culms several to many, not bulblike at the base. Leaf blades glabrous to scabrous, flat; sheaths glabrous, the middle and upper ones inflated, auricles small, rounded, or lacking; ligules truncate, entire or nearly so. Panicle short, compact, cylindrical to oblong, purplish; spikelets 1-flowered; glumes subequal, pubescent, awned, the keel ciliate with stout, spreading hairs; lemmas puberulent, the apex truncate, shallowly erose; anthers 1–2 mm long.

Meadows, stream banks, and mesic fellfields of nearly all alpine ranges. Not yet reported from the Salt River and Wyoming Ranges. Circumboreal; Alaska to Greenland, south to Maine and Michigan; mountainous areas of the western United States; arctic and alpine regions of the Southern Hemisphere.

North American plants with short-awned glumes and strongly inflated sheaths are var. *americanum* Fourn. *Phleum pratense* L., Common Timothy, is widespread in mountain meadows and is to be expected in sheltered sites near timberline, particularly along pack trails. It can be distinguished from *P. alpinum* as follows:

1. Panicle short, compact, less than 3 cm long; awns mostly more than 2 mm long; culms not bulbous at base; upper leaf sheaths inflated . . . *P. alpinum*
1. Panicle elongate, dense, more than 4 cm long; awns mostly less than 2 mm long; culms bulbous at base; upper leaf sheaths not inflated . . . *P. pratense*

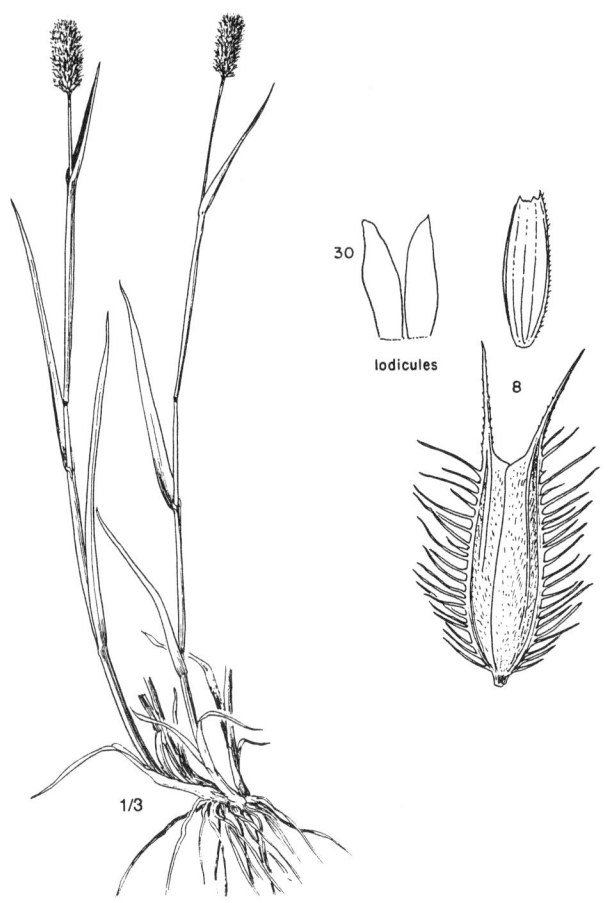

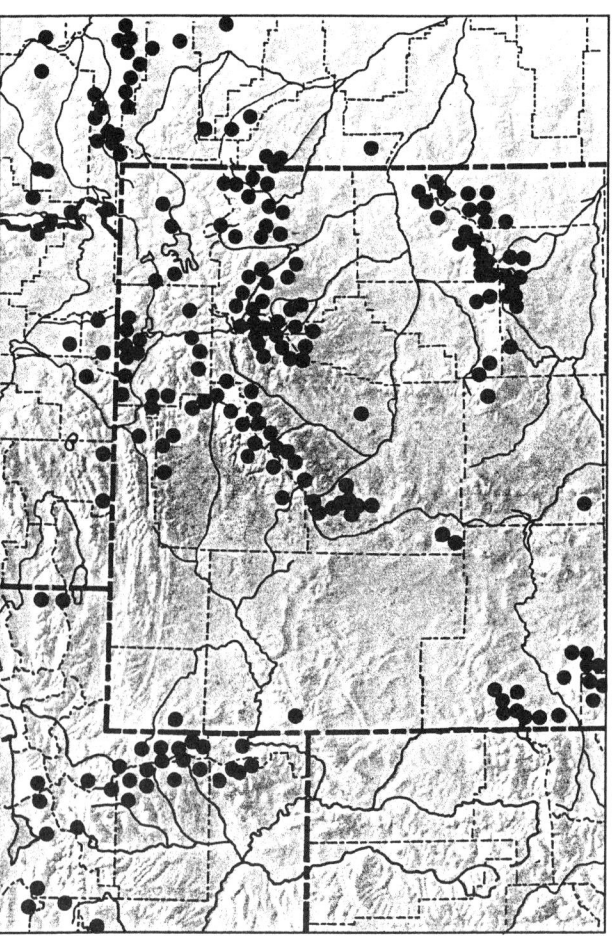

Poa L.
Bluegrass

Low to moderately tall perennials (rarely annuals in ours) with erect culms, the blades flat, folded or involute, ending in a boat-shaped apex. Spikelets 2- to several-flowered, in open or contracted panicles, the articulation above the glumes and between the florets. Glumes acute, keeled, somewhat unequal, shorter than the first lemma. Lemmas keeled to somewhat rounded, awnless, blunt, glabrous or pubescent, sometimes webbed at the base. Palea equaling the lemma or shorter.

KEY TO THE SPECIES OF *POA*

1. Plants annual, lacking rhizomes and remains of old stems — *P. annua*
1. Plants perennial, either rhizomes or old stem bases present
    2. Plants rhizomatous, the rhizomes long-creeping
        3. Lemmas webbed at the base
            4. Lemmas 4–5 mm long, pubescent between the keel and marginal nerves, at least on the lower half — *P. arctica*
            4. Lemmas less than 4 mm long, glabrous between the keel and marginal nerves — *P. pratensis*
        3. Lemmas not webbed
            5. Sheaths purplish, retrorsely pubescent; flowers unisexual, mostly pistillate (if perfect, then spikelets greenish) — *P. nervosa*
            5. Sheaths greenish or rarely purplish, glabrous or sparsely scabrous; flowers perfect; spikelets usually purplish — *P. arctica*
    2. Plants mostly tufted, sometimes decumbent at the base with very short rhizomes
        6. Lemmas webbed at the base
            7. Panicle open, with spreading to reflexed branches; glumes unequal in length; anthers less than 1 mm long
                8. Lower panicle branches reflexed at maturity; lemmas averaging 3 mm long; glumes nearly equal in length; palea keels with soft, wavy cilia — *P. reflexa*
                8. Lower panicle branches spreading to ascending; lemmas averaging 3.5 mm long; glumes unequal in length; palea keels with short, stiff cilia — *P. leptocoma*
            7. Panicle somewhat compact and narrow with ascending branches; glumes subequal in length; anthers 0.7–1.7 mm long
                9. Plants less than 20 cm tall; second glume 3.5–4.2 mm long; lemmas 3.5–4 mm long — *P. pattersonii*
                9. Plants often more than 25 cm tall; second glume 2.5–3.5 mm long; lemmas 3–3.5 mm long — *P. glauca*
        6. Lemmas not webbed
            10. Spikelets rounded; glumes and lemmas rounded on the back, the keels obscure — *P. secunda*
            10. Spikelets flattened; glumes and lemmas keeled
                11. Lemmas pubescent on the keel and marginal nerves, the internerves glabrous or with conspicuously shorter hairs
                    12. Flowers unisexual, mostly pistillate; leaf blades folded or involute, stiff — *P. fendleriana*
                    12. Flowers perfect; leaf blades flat, or folded and soft
                        13. Panicle about as broad as long; spikelets broadly ovate to subcordate — *P. alpina*
                        13. Panicle longer than broad; spikelets oblong or lanceolate

  14. Culms little if at all exceeding the crowded basal leaves; second glume more than 3.5 mm long, about equaling the first lemma   *P. pattersonii*
  14. Culms much exceeding the basal leaves; second glume less than 3.5 mm long, shorter than the first lemma   *P. glauca*
11. Lemmas glabrous, sometimes uniformly puberulent, at least on the lower half
  15. Culms less than 10 cm high; lemmas 2–3 mm long; spikelets 3–4 mm long, flowers mostly perfect   *P. lettermanii*
  15. Culms more than 15 cm high; lemmas 4–6 mm long; spikelets 5–8 mm long, flowers mostly pistillate   *P. fendleriana*

Poa alpina L.
Sp. Pl., 67, 1753.

Alpine Bluegrass

Poa vivipara (L.) Willd.

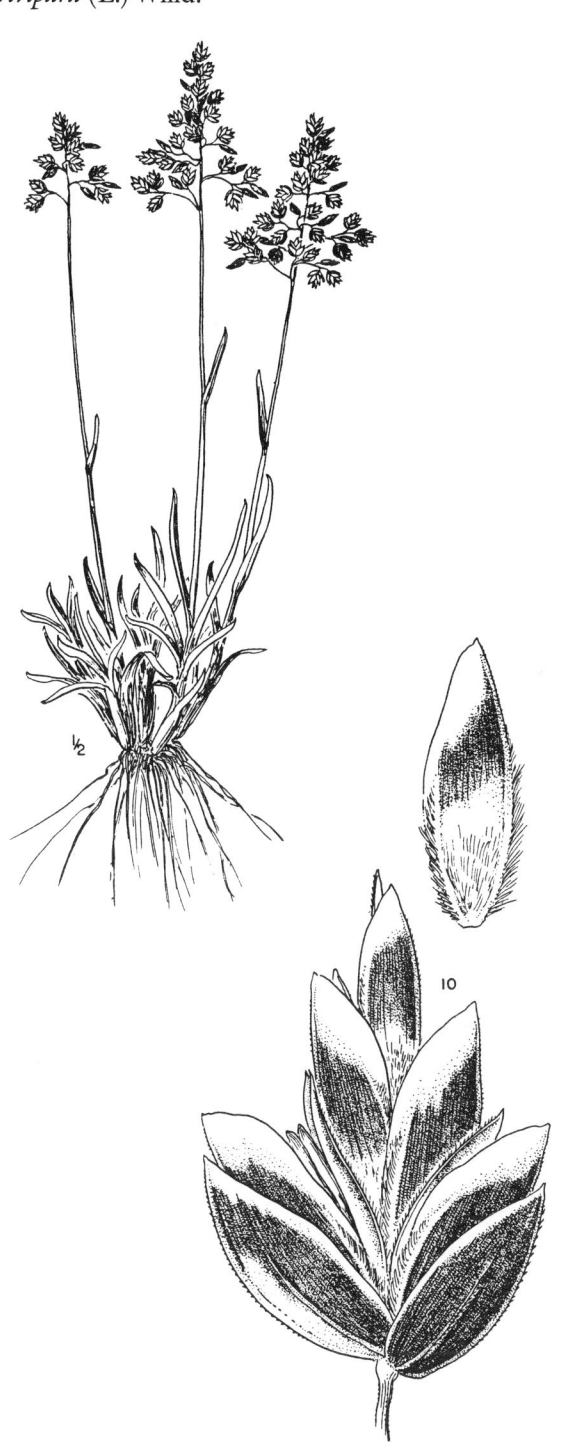

Tufted perennial from fibrous roots, rhizomes lacking; culms solitary to many, erect, glabrous. Leaves mostly basal, the blades flat to somewhat folded, glabrous to scabrous, abruptly tapered to a boat-shaped tip; sheaths glabrous to puberulent, open nearly to the base; ligules truncate to obtuse, lacerate. Panicle open to rather compact, usually pyramidal, the branches ascending to spreading; spikelets flattened, mostly purplish, broadly ovate to cordate, 3–6-flowered; glumes subequal, ovate, scabrous at least on the back, acute to nearly acuminate; lemmas lanceolate to ovate, acute, sparsely pubescent, villous on the keel and marginal nerves, not webbed at the base; palea about as long as the lemma, ciliate on the keel; anthers 1.2–2 mm long.

Common in meadows, fellfields, and wet rocky places in all alpine ranges. Circumpolar; arctic Canada and Alaska, south in the West to Oregon, Utah, Colorado, and Wyoming.

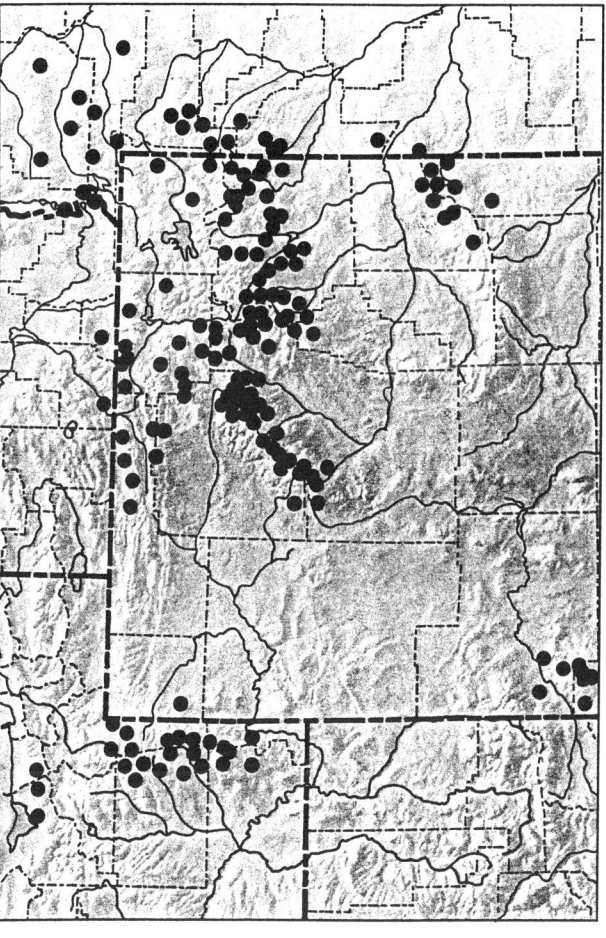

Poa annua L.
Sp. Pl., 68, 1753.

Annual Bluegrass

Tufted, glabrous annual or short-lived perennial from fibrous roots; culms several to many, glabrous, compressed-flattened, decumbent at the base and often rooting from lower nodes. Leaves mostly basal, the blades glabrous, flat or folded, somewhat scabrous on the margins; sheaths glabrous, open for about three-fourths to half their length; ligules glabrous, rounded, entire to erose. Panicle open, pyramidal, the branches 1 or 2 at each node, glabrous, spreading to ascending; spikelets compressed, 2–6-flowered; glumes unequal, lanceolate to oblanceolate, scarious-margined; lemmas elliptic, pubescent on the nerves and keel, at least near the base, acute, not webbed, but the hairs of the keel sometimes resembling a basal web; paleas puberulent on the keel; anthers 0.7–1.1 mm long.

Timberline margins of the Absaroka Range. Circumboreal; transcontinental in Canada; south throughout most of the United States.

This weedy species is widespread and locally common throughout the Northern Hemisphere and might be expected near timberline in all the alpine ranges of the Middle Rockies.

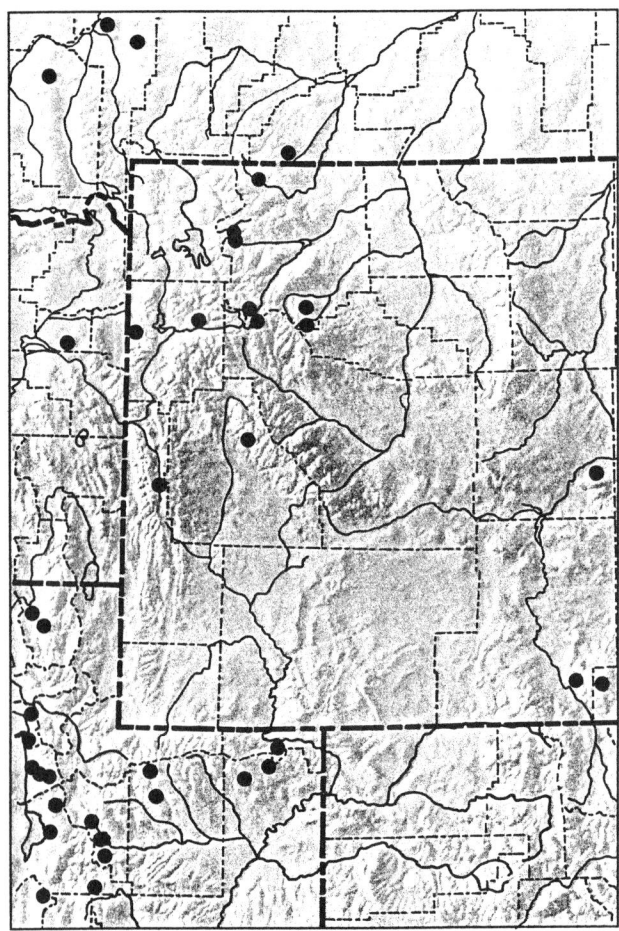

Poa arctica R. Br.
Sup. App. Parry's Voy., 288, 1823.

Arctic Bluegrass

Poa alpicola Nash in Rydb.
Poa aperta Scribn. & Merr.
Poa brintnellii Raup
Poa callichroa Rydb.
Poa cenisia All.
Poa grayana Vasey
Poa lanata Scribn. & Merr.
Poa laxa Haenke
Poa longiculmis Hult.
Poa longipila Nash in Rydb.
Poa phoenicea Rydb.
Poa tricholepis Rydb.
Poa tolmatchewii Roshev.
Poa williamsii Nash.

Tufted perennial from short, or occasionally long-creeping rhizomes; culms 1 to several, erect or slightly decumbent at the base, glabrous. Leaves mostly basal, the blades flat or folded, glabrous to sparsely scabrous, the tips abruptly acute; sheaths glabrous to scabrous, open to below midlength; ligules glabrous to minutely scabrous, rounded to acute, the margins erose. Panicle open and pyramidal to contracted and narrow, the branches mostly glabrous, ascending to spreading; spikelets purplish, elliptic, 3–7-flowered, the florets perfect; glumes oblong to lanceolate, keeled, acute to nearly acuminate; lemmas oblong to lanceolate, keeled, pubescent on nerves and keel, usually webbed at the base; paleas about the same length as the lemmas, pubescent on the keel; anthers 1.5–2.5 mm long.

Meadows and fellfields of most of the alpine ranges. Not yet known from above timberline in the Hoback and Wasatch Ranges. Circumboreal; Alaska south to Oregon, Nevada, and New Mexico.

Löve, Löve & Kapoor (1971) suggest ssp. *grayana* (Vasey) Löve et al. for larger Rocky Mountain plants with broad, flat leaves and spikelets larger than typical arctic and northern alpine ssp. *arctica*.

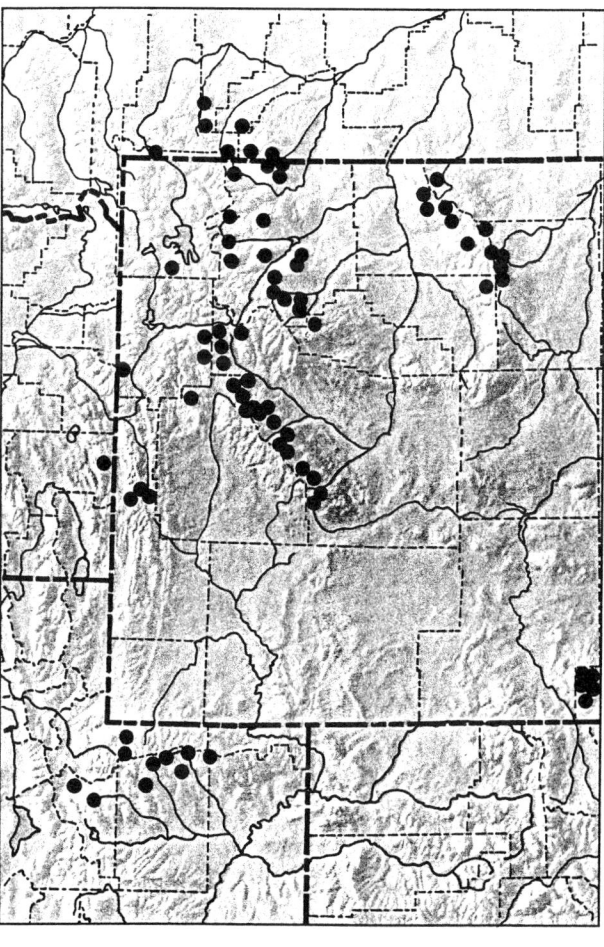

Poa fendleriana (Steud.) Vasey
Bull. U.S.D.A. Div. Bot. 13(2):Pl. 74, 1893.

Muttongrass

Eragrostis fendleriana Steud.
Atropis fendleriana (Steud.) Beal
Paneion longiligulum (Scribn. & Williams) Lunell
Panicularia fendleriana (Steud.) Kuntze
Poa albescens A.S. Hitchc.
Poa brevipaniculata Scribn. & Williams
Poa cottoni Piper
Poa cusickii Vasey
Poa eatonii S. Wats.
Poa epilis Scribn.
Poa filifolia Vasey
Poa idahoensis Beal
Poa longiligula Scribn. & Williams
Poa longipedunculata Scribn.
Poa nematophylla Rydb.
Poa paddensis Williams
Poa purpurascens Vasey
Poa scaberrima Rydb.
Poa scabrifolia A. Heller
Poa scabriuscula Williams
Poa spillmani Piper
Poa subaristida Scribn.
Poa subpurpurea Rydb.

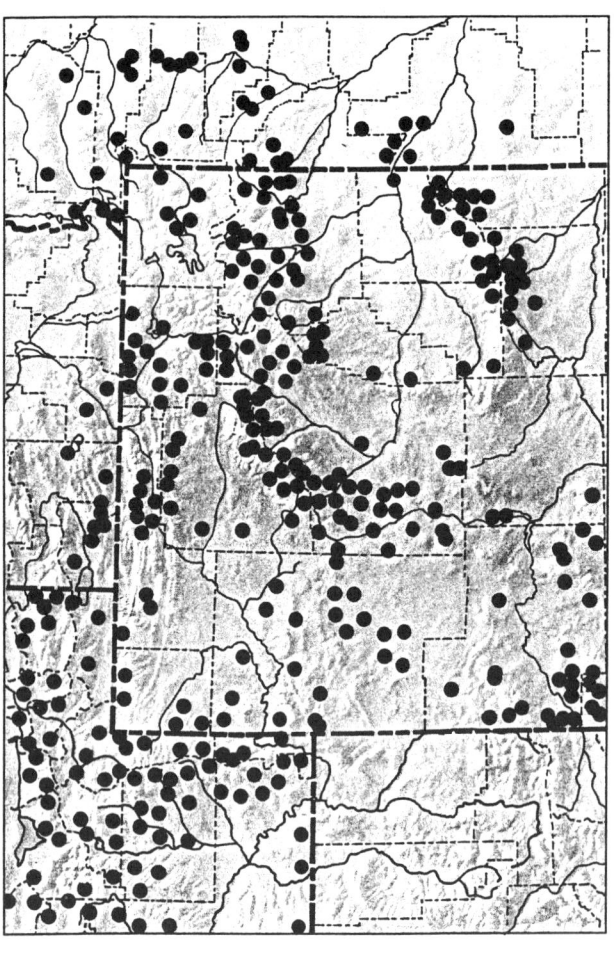

"Poa fendleriana"

Tufted, dioecious perennial from fibrous roots; rhizomes usually lacking, sometimes present and very short; culms erect, scabrous at least below the panicle. Leaves mostly basal, the blades folded or involute, the basal ones sometimes flat; sheaths compressed, open for half their length or more, glabrous to scabrous; ligules acute to truncate, glabrous to scaberulous, erose. Panicle narrow, compact, the branches 1–5 per node, erect to ascending; spikelets compressed, purplish, mostly 3–6-flowered, the florets mostly pistillate, sometimes staminate or perfect; glumes lanceolate to ovate, subequal, keeled, acute, glabrous to scaberulous or puberulent; lemmas keeled, pubescent to scabrous on the keel and marginal nerves, lacking webs, the margins membranous; anthers of staminate florets 2–3 mm long.

Fellfields, meadows, and rocky ledges in all alpine ranges of the Middle Rocky Mountains. Alaska south to California, New Mexico, Texas, and Nebraska.

Many authors treat *P. fendleriana, P. epilis,* and *P. cusickii* as separate species based on various combinations of pubescence of the lemmas and width of the leaf blades. I follow Welsh et al. (1987) and Arnow et al. (1980) in recognizing a single, polymorphic, dioecious (mostly pistillate) species for which the name *P. fendleriana* has priority.

"Poa cusickii"

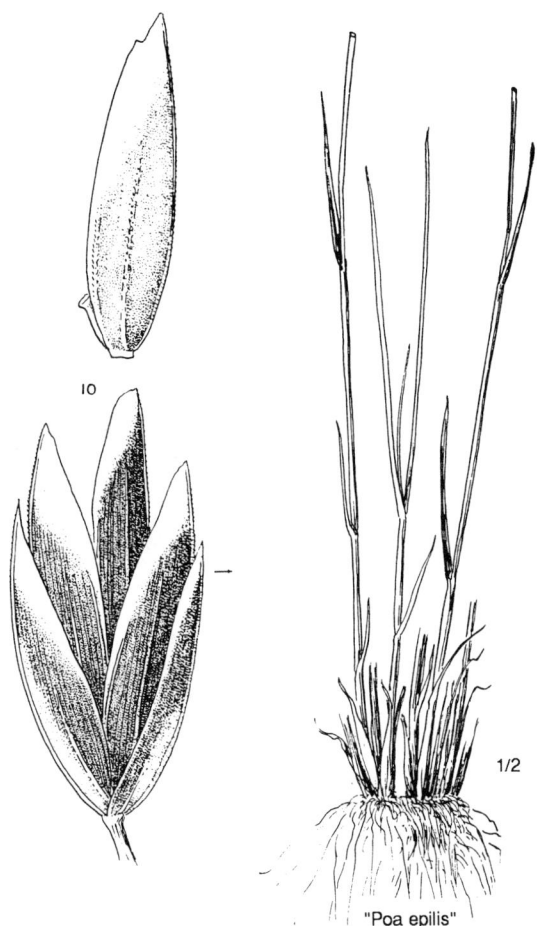

"Poa epilis"

Poa glauca Vahl
Fl. Dan. 6:3, 1790.

Greenland Bluegrass

Paneion interius (Rydb.) Lunell
Poa ammophila Pors.
Poa anadyrica Roshev.
Poa bryophila Trin.
Poa glaucantha Gaudin
Poa interior Rydb.
Poa rupestris Vasey
Poa rupicola Nash ex Rydb.
Poa scopulorum Butters & Abbe
Poa subtrivialis Rydb.

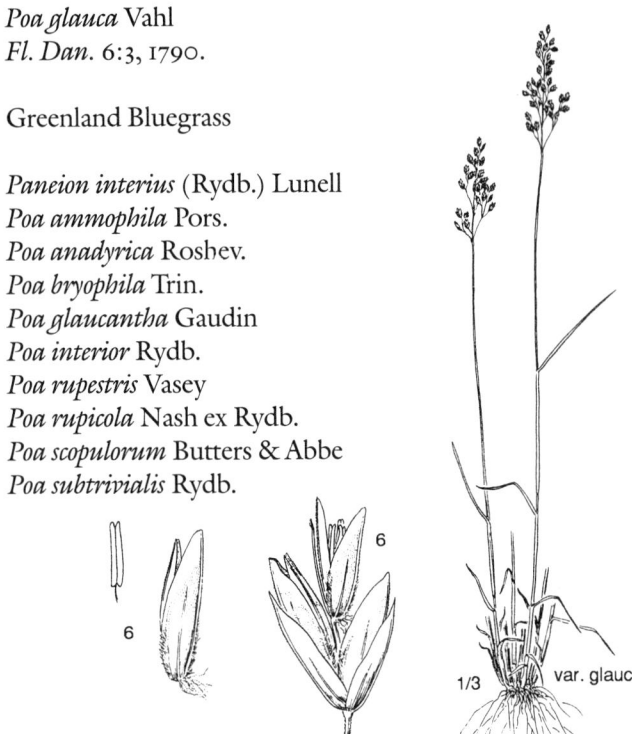

Tufted perennial from fibrous roots; rhizomes lacking; culms several to many, glabrous to scabrous, erect or sometimes decumbent and rooting at the base. Leaves mostly basal, the blades erect, short, and stiff, flat to folded or sometimes involute; sheaths rounded to somewhat compressed, glabrous to scabrous, open nearly to the base; ligules obtuse to truncate, scaberulous, ciliolate to erose. Panicle oblong to ovate, compact to somewhat open, the branches erect to ascending, mostly 2–3 per node, glabrous to scabrous; spikelets usually purplish, ovate or elliptic, 2–4-flowered, the florets perfect; glumes subequal, ovate to lanceolate, acute, keeled; lemmas ovate, acute, keeled, villous on the keel and marginal nerves, glabrous to pubescent between, lacking a web at the base or sparsely webbed; anthers mostly 1.2–1.6 mm long.

Fellfields and meadows of all alpine ranges of the Middle Rocky Mountains. Circumboreal; south in western North America to Arizona, New Mexico, and Nebraska.

Two varieties occur in the alpine zone and are recognized as follows:

1. Lemmas sparsely webbed at base, glabrous between keel and marginal nerves; branches of panicle scabrous throughout . . . var. *glauca*
1. Lemmas not webbed at base; pubescent to puberulent between keel and marginal nerves; branches of panicle glabrous or partially scabrous
. . . var. *rupicola* (Nash ex Rydb.) B. Boiv.

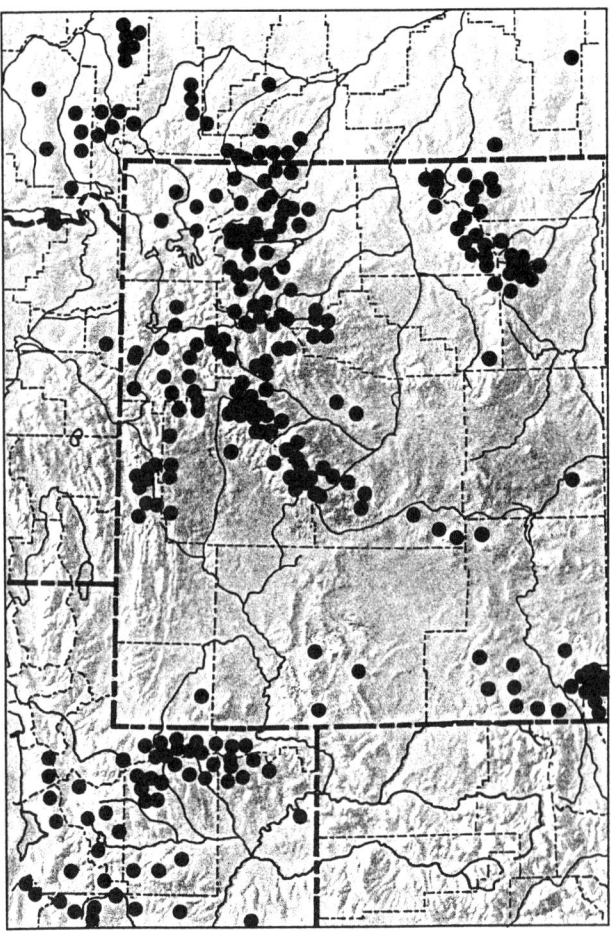

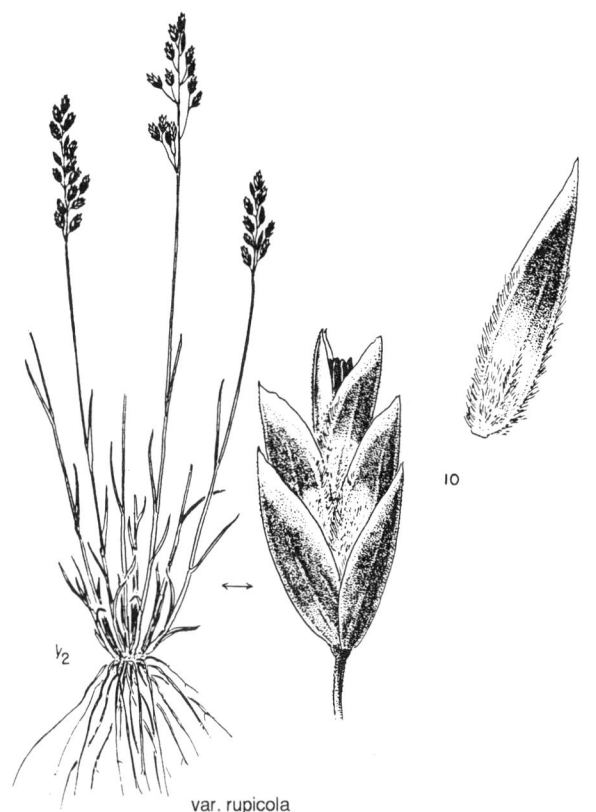

Poa leptocoma Trin.
Mem. Acad. St. Petersb., VI, 1:374, 1830.

Bog Bluegrass

Poa paucispicula Scribn. & Merr.

Loosely tufted perennial from slender, fibrous roots; culms few to several, glabrous, usually decumbent and rooting at the lower nodes. Leaf blades flat to folded, lax, glabrous or often scabrous, at least on the margins; sheaths glabrous to scabrous, terete or compressed, open for about half their length; ligules glabrous, obtuse to nearly truncate. Panicle open, nodding, the branches mostly 2 per node, slender, glabrous to scabrous, somewhat flexuous, spreading to reflexed at maturity; spikelets purplish, compressed, lanceolate to elliptic, 2–4-flowered, the florets perfect; glumes unequal, keeled, lanceolate to obovate, acuminate, scaberulous on the keels; lemmas keeled, membranous, glabrous, or pubescent on the keel and marginal nerves, glabrous between, acute, webbed at the base; anthers 0.3–0.9 mm long.

Wet meadows, seeps, nivation basins, and willow thickets of the Absaroka, Beartooth, Big Horn, Salt River, Uinta, Wasatch, and Wind River Ranges. Alaska south to California, Nevada, and New Mexico.

Middle Rocky Mountain plants more than 20 cm tall with scaberulous sheaths are var. *leptocoma*. *Poa leptocoma* is closely related to *P. reflexa,* from which it may be distinguished by its unequal glumes and longer lemmas. Some authors (Arnow et al. 1980) consider the two inseparable.

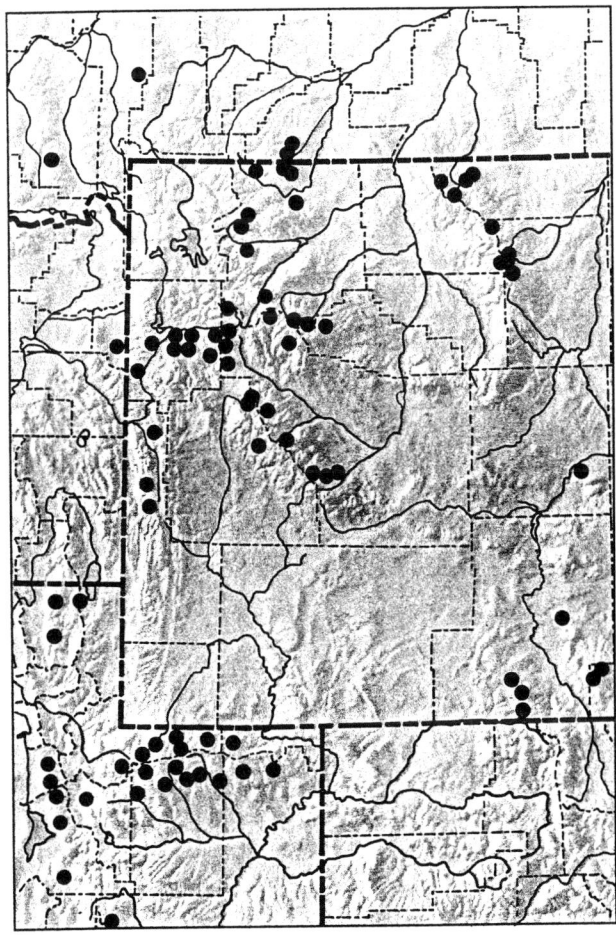

Poa lettermanii Vasey
Contr. U.S. Natl. Herb. 1:273, 1893.

Letterman Bluegrass

Atropis lettermani (Vasey) Beal
Poa brandegei Scribn. in Beal
Poa montevansii L. Kelso
Puccinellia lettermanii (Vasey) Ponert

Low, tufted perennial, less than 10 cm tall, from fibrous roots; rhizomes lacking; culms several to many, glabrous, erect or somewhat decumbent at the base. Leaf blades folded to flat, lax; sheaths glabrous to scaberulous, open nearly to the base; ligules glabrous, mostly truncate, erose. Panicle oblong, compact, the branches erect to ascending, glabrous or somewhat scaberulous; spikelets purple, elliptic, 2–4-flowered, the florets perfect; glumes unequal, glabrous, lanceolate to oblanceolate, keeled to somewhat rounded on the back, the tips membranous, acute; lemmas keeled, glabrous, the tips membranous, acute to somewhat rounded, not webbed at the base; anthers 0.3–0.8 mm long.

Fellfields of the Absaroka, Beartooth, Big Horn, Uinta, and Wind River Ranges. British Columbia to California, Nevada, Utah, and Colorado.

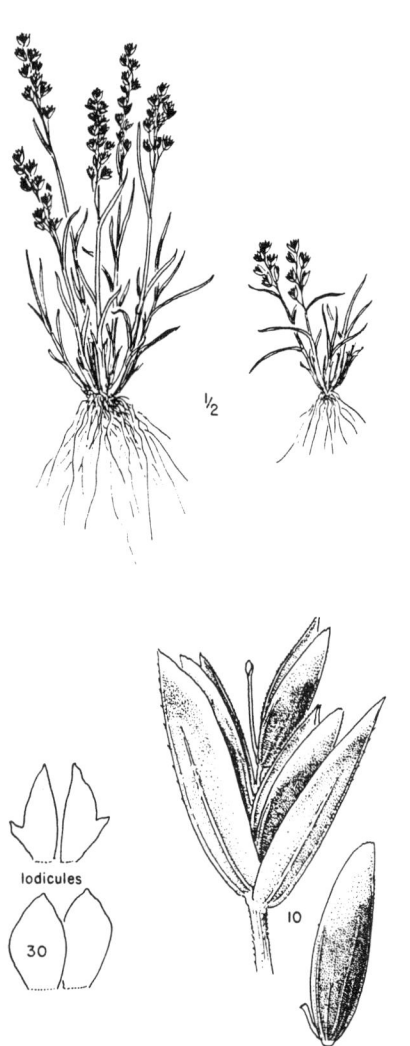

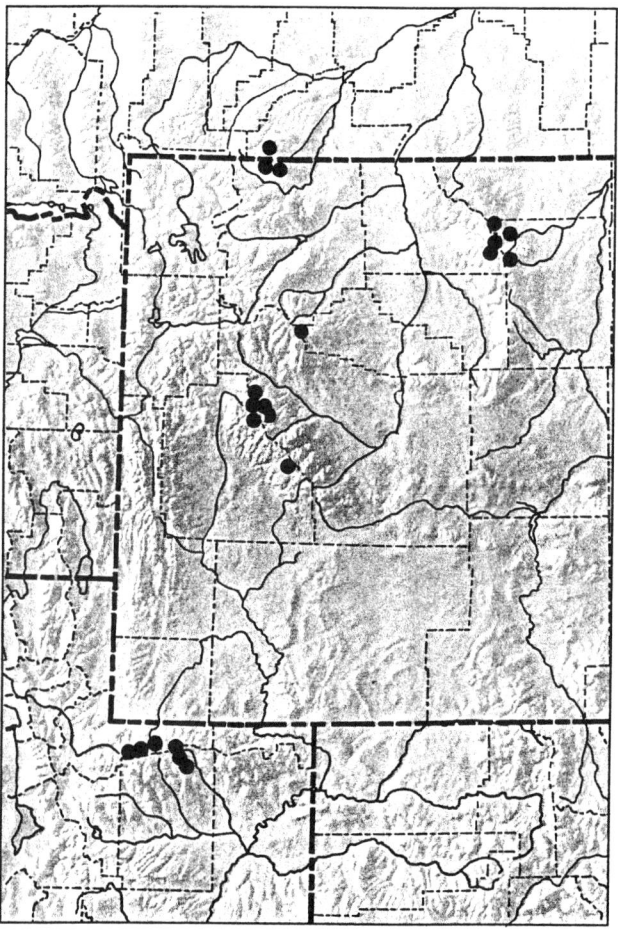

Poa nervosa (Hook.) Vasey
Bull. U.S.D.A. Div. Bot. 13(2):Pl. 81, 1893.

Wheeler Bluegrass

Festuca nervosa Hook.
Poa columbiensis Steud.
Poa curta Rydb.
Poa cuspidata Vasey
Poa olneyae Piper
Poa subreflexa Rydb.
Poa vaseyana Scribn. in Beal
Poa wheeleri Vasey in Rothr.

Loosely tufted, often dioecious perennial from rhizomes; culms solitary to several, erect to somewhat decumbent at the base. Leaves mostly basal, the blades flat or folded, glabrous to scabrous on the ventral surfaces; sheaths mostly retrorse-puberulent, rarely glabrous, closed for most of the length, often purplish below; ligules acute to truncate, puberulent to pubescent, ciliate or erose. Panicle compact to open and pyramidal, the branches erect to spreading or reflexed, mostly 1–3 per node; spikelets compressed, oblong to lanceolate, mostly 3–6-flowered, the florets perfect or pistillate; glumes unequal, lanceolate, keeled, glabrous or sometimes scabrous on the keels; lemmas keeled, strongly 5-nerved, glabrous or scaberulous on the back or on the keel and marginal nerves, acute to somewhat obtuse, not webbed at the base; anthers of perfect florets mostly 2.5–3 mm long.

Meadows, slopes, and fellfields in most alpine ranges; not yet reported from above timberline in the Beartooth, Gros Ventre, and Hoback Ranges. British Columbia and Alberta south to California, Nevada, and New Mexico.

Middle Rocky Mountain plants of the alpine zone that are functionally pistillate and lack long hairs on the sheath collars are var. *wheeleri* (Vasey in Rothr.) C.L. Hitchc.

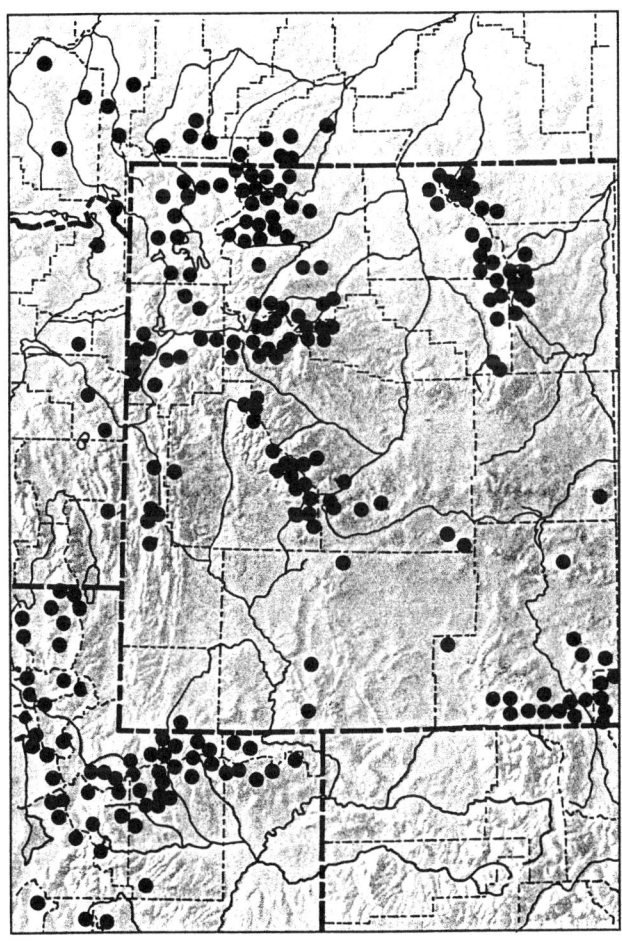

Poa pattersonii Vasey
Contr. U.S. Natl. Herb. 1:275, 1893.

Patterson Bluegrass

Low, tufted perennial, less than 20 cm tall, from fibrous roots; rhizomes lacking; culms several to many, erect or often decumbent at the base. Leaves mostly basal, lax, the blades flat or folded, scabrous; sheaths glabrous or occasionally puberulent, terete or compressed, open nearly to the base; ligules glabrous, scabrous, or puberulent, acute to obtuse or truncate, the margins lacerate. Panicle narrow, elongate, and usually compact, the branches glabrous to scabrous, erect to ascending; spikelets purplish, compressed, elliptic, 2–4-flowered, the florets perfect; glumes subequal, lanceolate to ovate, keeled, acute to nearly acuminate, the margins membranous; lemmas lanceolate, keeled, the margins membranous, pubescent on the keel and marginal nerves, glabrous to pubescent between, webbed or not at the base; anthers mostly 0.7–1.2 mm long.

Fellfields and rocky ledges in most Middle Rocky Mountain alpine ranges; not yet reported from the Hoback and Wasatch Ranges. British Columbia and Alberta south to Nevada, Utah, and Colorado.

Poa pattersonii is closely allied to *P. fendleriana, P. lettermanii,* and the arctic *P. abbreviata,* with which it is included by some authors as ssp. *pattersonii* (Vasey) Löve et al. (A. Löve et al., 1971).

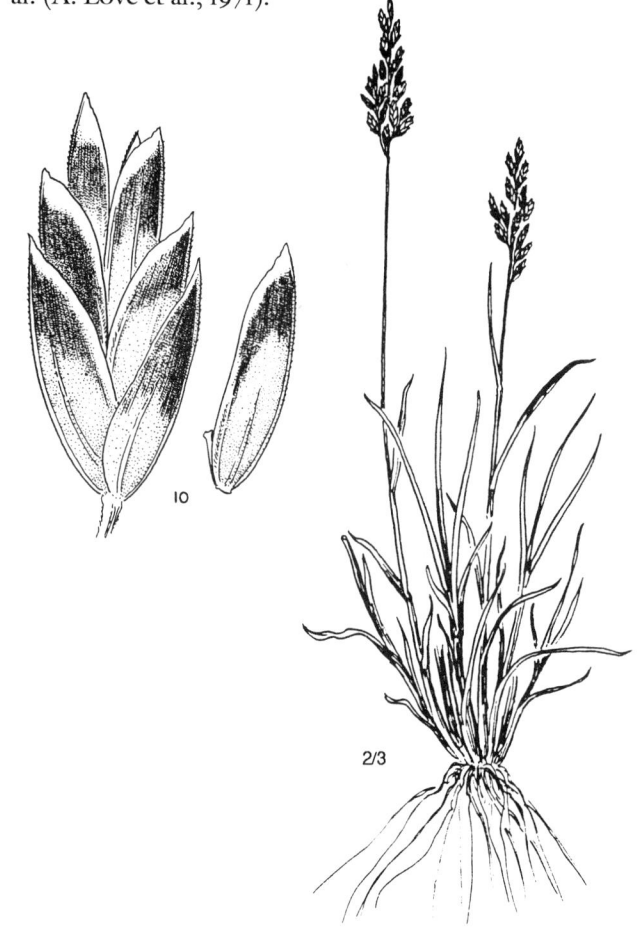

Poa pratensis L.
Sp. Pl., 67, 1753.

Kentucky Bluegrass

Paneion pratense (L.) Lunell
Poa agassizensis B. Boivin & D. Löve
Poa alpigena (Fries ex Blytt) Lindm. f.
Poa angustifolia L.
Poa peckii Chase

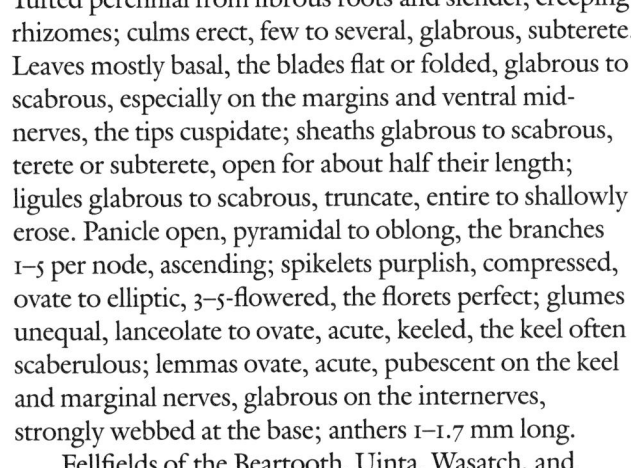

Tufted perennial from fibrous roots and slender, creeping rhizomes; culms erect, few to several, glabrous, subterete. Leaves mostly basal, the blades flat or folded, glabrous to scabrous, especially on the margins and ventral midnerves, the tips cuspidate; sheaths glabrous to scabrous, terete or subterete, open for about half their length; ligules glabrous to scabrous, truncate, entire to shallowly erose. Panicle open, pyramidal to oblong, the branches 1–5 per node, ascending; spikelets purplish, compressed, ovate to elliptic, 3–5-flowered, the florets perfect; glumes unequal, lanceolate to ovate, acute, keeled, the keel often scaberulous; lemmas ovate, acute, pubescent on the keel and marginal nerves, glabrous on the internerves, strongly webbed at the base; anthers 1–1.7 mm long.

Fellfields of the Beartooth, Uinta, Wasatch, and Wind River Ranges. Circumboreal; widespread in the United States except in the Southeast.

This species is widespread at lower elevations and probably more common in the alpine zone than the record suggests.

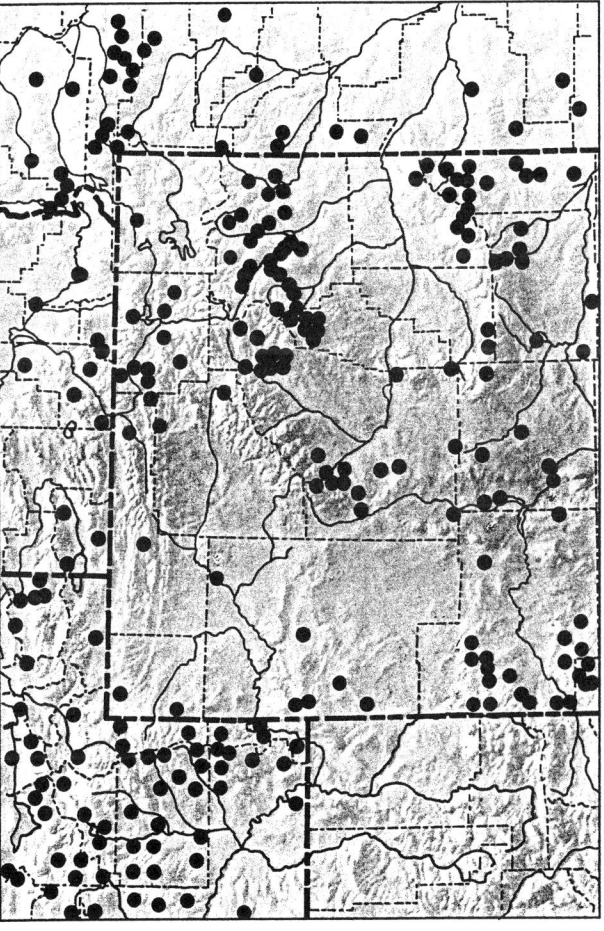

Poa reflexa Vasey & Scribn. ex Vasey
Contr. U.S. Natl. Herb. 1:276, 1893.

Nodding Bluegrass

Poa acuminata Scribn. in Beal
Poa pudica Rydb.

Tufted perennial from fibrous roots; rhizomes lacking; culms few to several, occasionally solitary, erect to somewhat decumbent at the base, glabrous. Leaves mostly basal, the blades flat or folded, glabrous; sheaths glabrous, open for about half their length; ligules glabrous, truncate to obtuse, strongly to shallowly erose. Panicle open, pyramidal, the branches 1–3 per node, spreading, at least the lower ones reflexed with age; spikelets purplish, compressed, elliptic to ovate, 2–5-flowered, the florets perfect; glumes subequal, lanceolate to obovate, keeled, acute, the keel often scaberulous; lemmas lanceolate to ovate, acute to obtuse, pubescent on the keel and marginal nerves, webbed at the base; anthers 0.2–0.8 mm long.

Stream banks, moist meadows, and willow thickets in most Middle Rocky Mountains alpine ranges; not yet reported from the Gros Ventre, Hoback, and Salt River Ranges. British Columbia south to Oregon, Nevada, Arizona, and Colorado.

See comments under *P. leptocoma*.

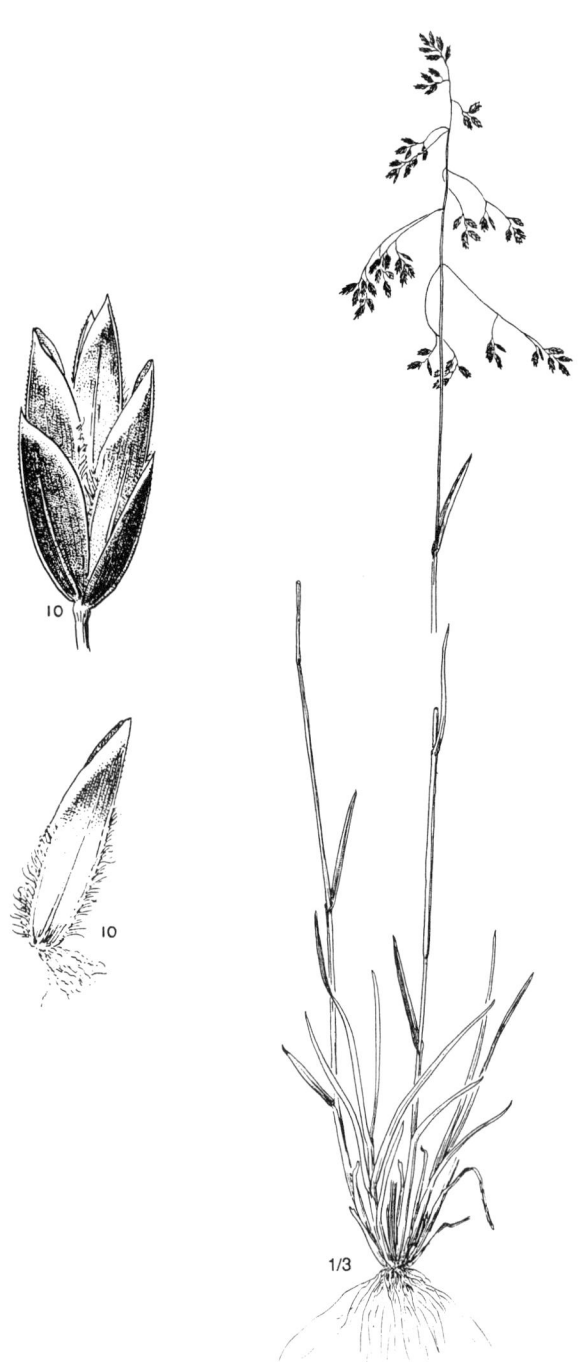

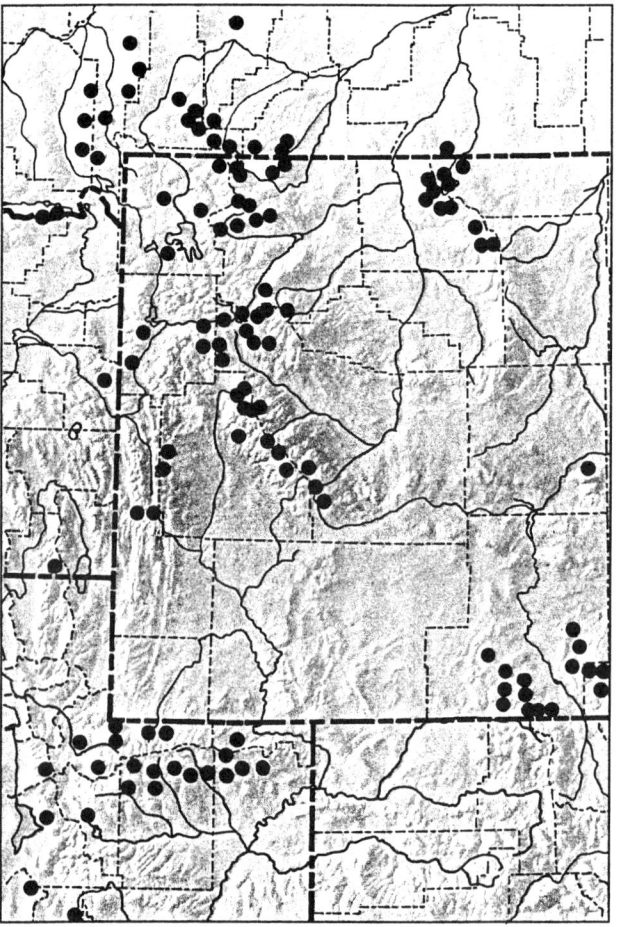

Poa secunda Presl
Rel. Haenk. 1:271, 1830.

Sandberg Bluegrass

Aira brevifolia Pursh
Airopsis brevifolia (Pursh) R. & S.
Atropis canbyi (Scribn.) Beal
Atropis laevis (Vasey) Beal
Atropis nevadensis (Vasey ex Scribn.) Beal
Atropis pauciflora Thurb. in S. Wats.
Atropis scabrella Thurb. in S. Wats.
Atropis tenuifolia (Buckl.) Thurb. in S. Wats.
Festuca patagonica Philippi
Festuca spaniantha Philippi
Glyceria canbyi Scribn.
Paneion sandbergii (Vasey) Lunell
Panicularia nuttalliana Kuntze
Panicularia scabrella (Thurb. in S. Wats) Kuntze
Panicularia thurberiana Kuntze
Poa acutiglumis Scribn.
Poa alcea Piper
Poa ampla Merr.
Poa andina Nutt. ex S. Wats.
Poa brachyglossa Piper
Poa buckleyana Nash.
Poa canbyi (Scribn.) T.J. Howell
Poa capillaris Scribn.
Poa confusa Rydb.
Poa englishii St. John & Hardin
Poa gracillima Vasey
Poa helleri Rydb.
Poa incurva Scribn. & Williams
Poa invaginata Scribn. & Williams
Poa juncifolia Scribn.
Poa laeviculmis Williams
Poa laevigata Scribn.
Poa laevis Vasey
Poa leckenbyi Scribn.
Poa limosa Scribn. & Williams
Poa lucida Vasey
Poa multnomae Piper
Poa nevadensis Vasey ex Scribn.
Poa nudata Scribn.
Poa orcuttiana Vasey
Poa pauciflora (Thurb. in S. Wats.) Benth. ex Vasey
Poa sandbergii Vasey
Poa saxatilis Scribn. & Williams
Poa scabrella (Thurb. in S. Wats.) Benth. ex Vasey
Poa tenerrima Scribn.
Poa tenuifolia Buckl.
Poa tenuifolia Nutt. ex S. Wats.
Poa thurberiana (Kuntze) Vasey
Poa truncata Rydb.
Poa wyomingensis Scribn. in Pammel
Puccinellia canbyi (Scribn.) Ponert
Puccinellia laevis (Vasey) Ponert
Puccinellia nevadensis (Vasey ex Scribn.) Ponert
Puccinellia scabrella (Thurb.) Ponert
Sporobolus bolanderi Vasey

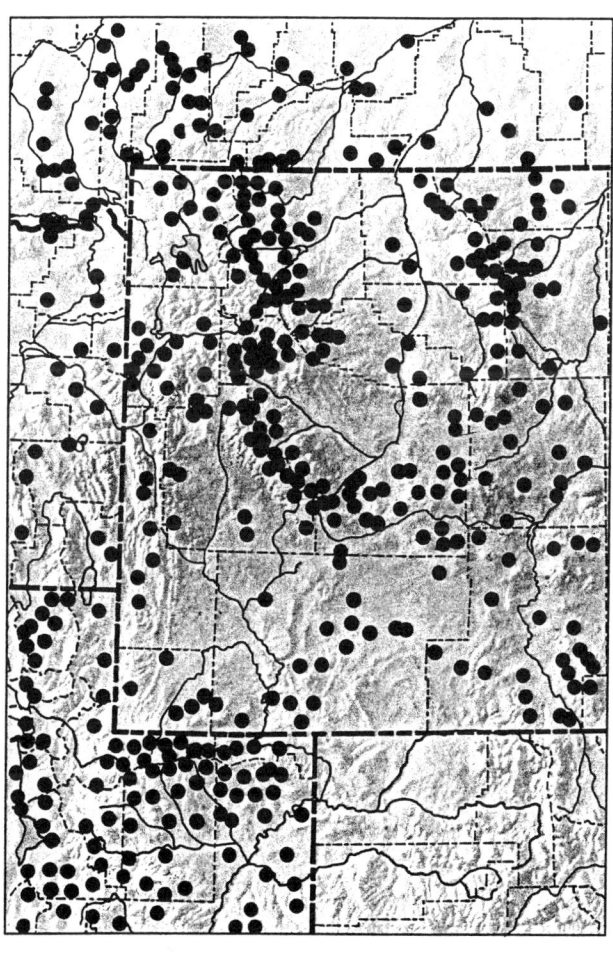

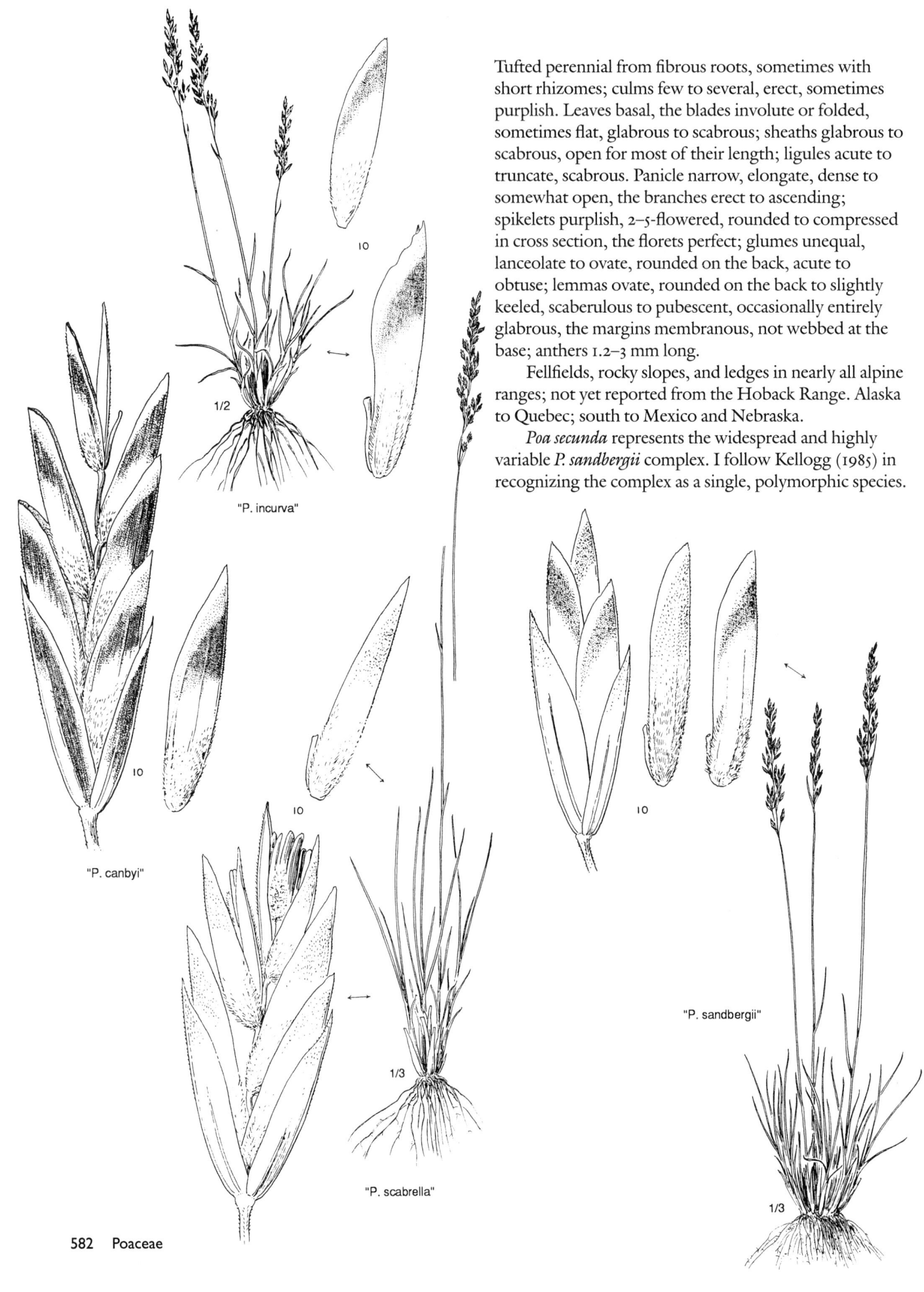

Tufted perennial from fibrous roots, sometimes with short rhizomes; culms few to several, erect, sometimes purplish. Leaves basal, the blades involute or folded, sometimes flat, glabrous to scabrous; sheaths glabrous to scabrous, open for most of their length; ligules acute to truncate, scabrous. Panicle narrow, elongate, dense to somewhat open, the branches erect to ascending; spikelets purplish, 2–5-flowered, rounded to compressed in cross section, the florets perfect; glumes unequal, lanceolate to ovate, rounded on the back, acute to obtuse; lemmas ovate, rounded on the back to slightly keeled, scaberulous to pubescent, occasionally entirely glabrous, the margins membranous, not webbed at the base; anthers 1.2–3 mm long.

Fellfields, rocky slopes, and ledges in nearly all alpine ranges; not yet reported from the Hoback Range. Alaska to Quebec; south to Mexico and Nebraska.

Poa secunda represents the widespread and highly variable *P. sandbergii* complex. I follow Kellogg (1985) in recognizing the complex as a single, polymorphic species.

Stipa L.
Needlegrass

Tufted perennials with narrow, usually convolute blades and open or contracted panicles. Spikelets 1-flowered, disarticulating above the glumes and leaving a bearded, sharp-pointed callus attached to the base of the floret. Glumes equal or nearly so, narrow, papery, acute to aristate. Lemma narrow, terete, strongly convolute and indurate around the palea, terminating in a prominent, persistent awn.

KEY TO THE SPECIES OF
STIPA

1. Palea about two-thirds as long as the lemma; callus straight, the tip blunt; awn glabrous, scabrous, or puberulent .... *S. lettermanii*
1. Palea about half as long as the lemma; callus curved, the tip acute; awn glabrous to plumose at the base .... *S. occidentalis*

Stipa lettermanii Vasey
Bull. Torr. Club 13:53, 1886.

Letterman Needlegrass

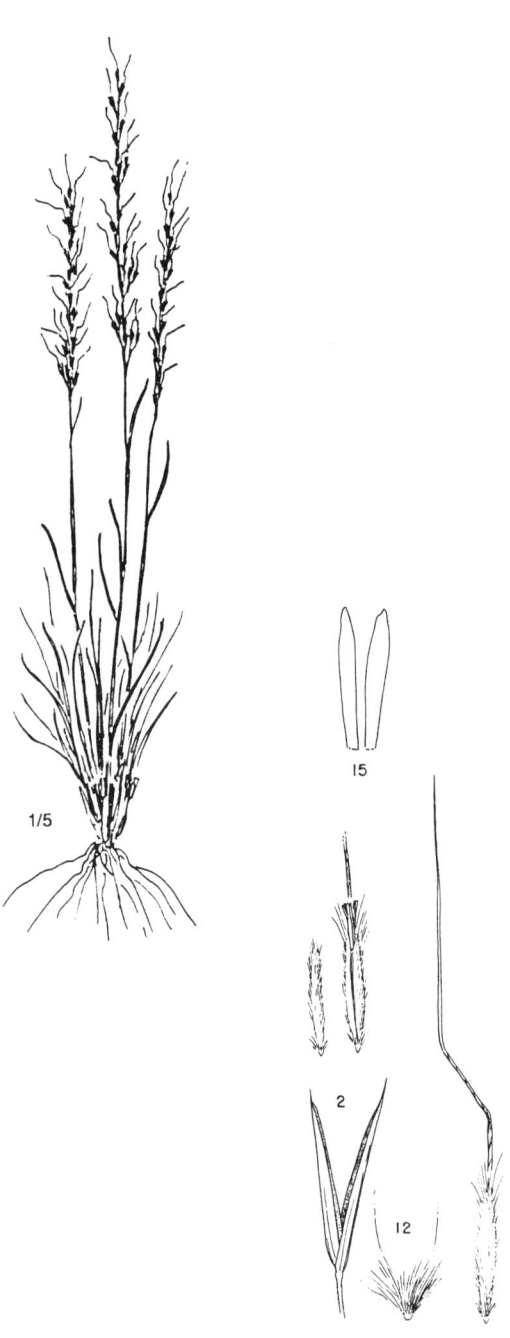

Medium-sized, tufted perennial, less than 4 dm tall, from fibrous roots; rhizomes lacking; culms few to several, glabrous to scaberulous. Leaves mostly basal, the blades very narrow, involute-filiform, glabrous to scabrous; sheaths glabrous or scaberulous; ligules truncate. Panicle narrow, elongate, the branches erect-ascending; spikelets greenish or purplish; glumes subequal, acuminate, 3-nerved, glabrous or scaberulous; lemmas pubescent, the awn slender, 1.5–2 cm long, glabrous to somewhat scabrous, mostly twice-geniculate, the 2 lower segments twisted; callus short, straight, with a blunt tip; palea slender, pubescent, about two-thirds as long as the lemma; anthers 1.7–2.5 mm long.

Timberline meadows of the Absaroka, Medicine Bow, Uinta, Wasatch, Wind River, and Wyoming Ranges. Washington south to California, Arizona, and New Mexico.

At lower elevations *S. lettermanii* can be distinguished from *S. occidentalis* on the basis of its shorter awns. At timberline and in the lower alpine zone this distinction does not appear to be a consistent difference; the awns of both species fall within the range of 15–22 mm.

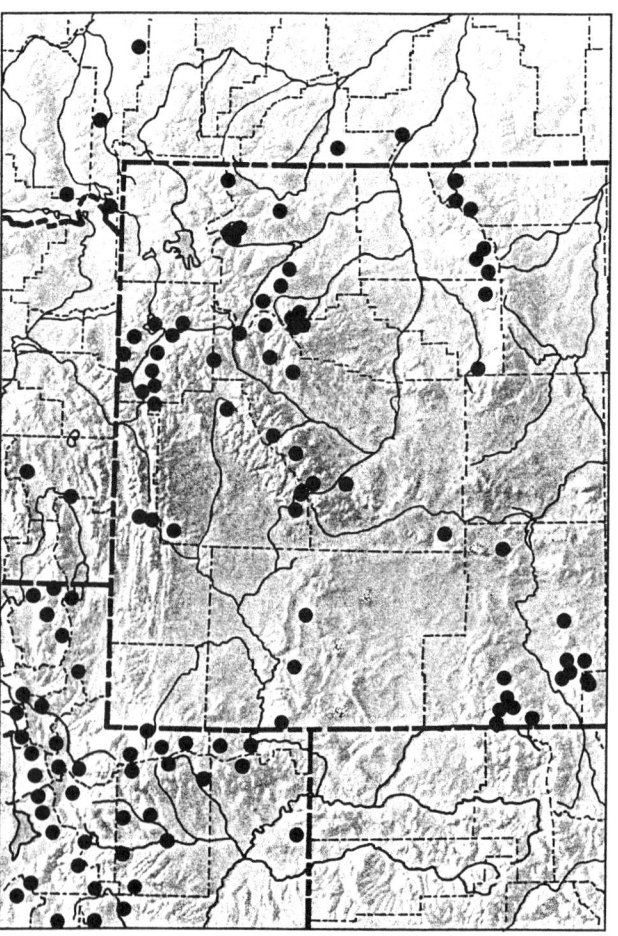

584 Poaceae

Stipa occidentalis Thurb. ex S. Wats.
Bot. King Exped. 5:380, 1871.

Western needlegrass

Stipa californica Merr. & Davy
Stipa columbiana Macoun
Stipa elmeri Piper & Brodie ex Scribn.
Stipa minor (Vasey) Scribn.
Stipa nelsonii Scribn.
Stipa oregonensis Scribn.
Stipa stricta Vasey
Stipa williamsii Scribn.

Densely tufted perennial from fibrous roots; rhizomes lacking; culms few to several, erect, glabrous to puberulent. Leaves both basal and cauline, the blades more or less flat to involute and filiform, glabrous to scaberulous; sheaths glabrous to pubescent; ligules truncate and entire, erose, or ciliate. Panicle long, narrow, the branches appressed-ascending, compact to somewhat loose; spikelets often purplish; glumes subequal, glabrous to scaberulous, obscurely 3-nerved; lemmas indurate, pubescent, the slender awn mostly twice-geniculate, glabrous to scabrous, often pubescent in the first 2 twisted segments; callus short, curved, the tip glabrous, acute to blunt; palea slender, pubescent, about half as long as the lemma; anthers mostly 2–3 mm long.

Timberline meadows and fellfields of the Absaroka, Uinta, Wasatch, and Wind River Ranges. Yukon Territory south to California, Mexico, Texas, and South Dakota.

Timberline plants with awns plumose on the first 2 segments are var. *occidentalis*. Those lacking plumose awns are var. *nelsonii* (Scribn.) C.L. Hitchc.

var. occidentalis

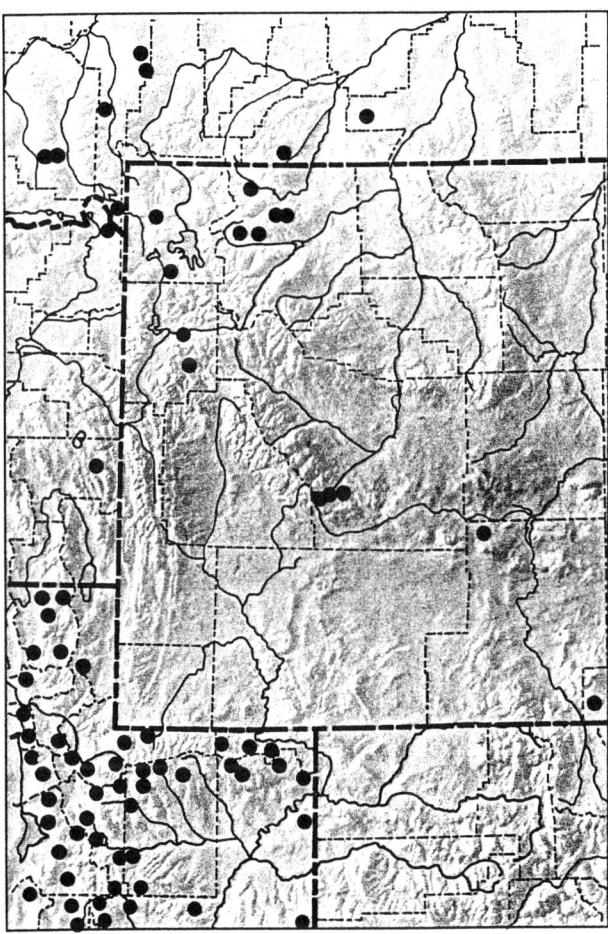

Trisetum Pers.
Trisetum

Tufted perennials (rarely annuals) with flat blades and usually dense, spikelike panicles. Spikelets 3–5-flowered, commonly 2-flowered, the rachilla prolonged beyond the upper floret. Callus and rachilla joint bearded. Articulation above the glumes and between the florets. Glumes subequal, acute, nearly as long as the spikelet. Lemmas keeled, 2-toothed at the apex, bearing a straight or geniculate awn from the back below the cleft apex. Palea usually shorter than the lemma.

KEY TO THE SPECIES OF *TRISETUM*

1. Lemmas awned, the awns exerted, geniculate, 3–6 mm long — *T. spicatum*
1. Lemmas awnless or with awns included, straight, less than 2 mm long — *T. wolfii*

Trisetum spicatum (L.) Richter
Pl. Eur. 1:59, 1890.

Spike Trisetum

Aira spicata L.
Aira subspicata L.
Avena airoides Koel.
Avena mollis Michx.
Avena spicata (L.) Fedtsch.
Avena subspicata (L.) Clairv.
Koeleria canescens Torr. ex Trin.
Koeleria spicata (L.) Reichenb. ex Willk. & Lange
Koeleria subspicata (L.) Reichenb.
Panicum spicatum (L.) Farw.
Rupestrina pubescens Prov.
Trisetaria airoides (Koel.) Baumg.
Trisetum airoides (Koel.) Beauv. ex R. & S.
Trisetum alaskanum Nash
Trisetum americanum Gand.
Trisetum congdoni Scribn. & Merr.
Trisetum majus (Vasey) Rydb.
Trisetum molle (Michx.) Kunth
Trisetum subspicatum (L.) Beauv.
Trisetum triflorum (Bigelow) Löve & Löve
Trisetum villosissimum (Lange) Louis-Marie

Tufted perennial from fibrous roots; rhizomes lacking; culms mostly several to many, erect, glabrous or puberulent. Leaves mostly basal, the blades flat or folded, glabrous to pubescent; sheaths mostly puberulent, sometimes glabrous; ligules usually pubescent, erose. Panicle narrow, dense, spikelike, the spikelets often purple or purplish, 2- or sometimes 3-flowered; glumes equal to unequal, lanceolate, glabrous, sometimes scabrous on the keels, acute to acuminate; lemmas lanceolate, glabrous to scaberulous, acute or mostly bifid at the apex, long-awned from above the middle, the awn geniculate, divergent at maturity; palea lanceolate, nearly as long as the lemma; anthers 0.7–1.5 mm long.

Fellfields, meadows, and ledges in all alpine ranges of the Middle Rocky Mountains. Circumboreal; south in western North America to Mexico; South America.

The *T. spicatum* complex is represented by considerable variation, which Hultén (1959) treated as 14 subspecies. There are 3 in the alpine zone of the Middle Rocky Mountains.

1. Culms glabrous below the panicle
   ... ssp. *majus* (Vasey) Hult.
1. Culms pubescent below the panicle
   2. Florets distinctly longer than the glumes; culms long-pilose ... ssp. *spicatum*
   2. Florets about the same length, or slightly exceeding the glumes; culms short-pubescent
   ... ssp. *molle* (Michx.) Hult.

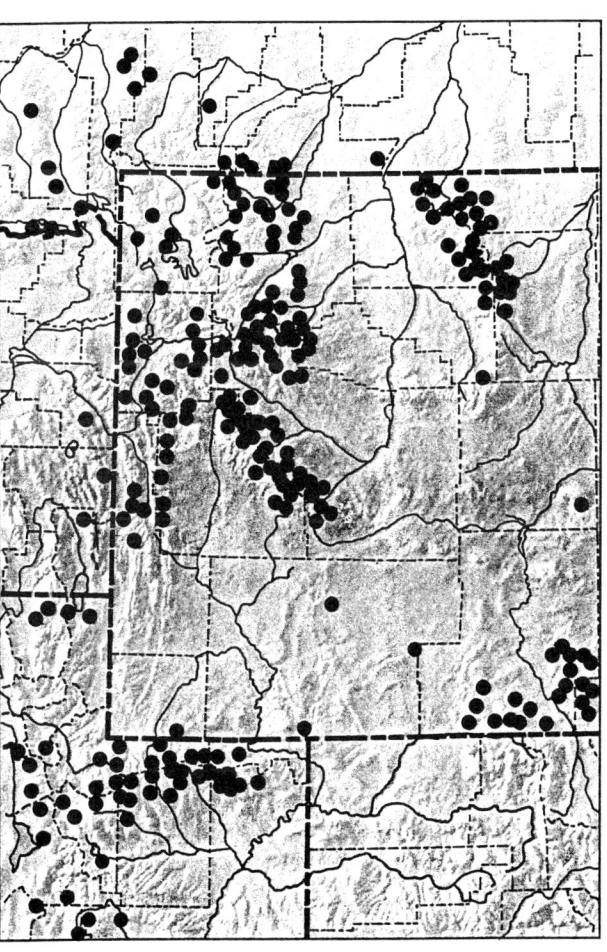

Trisetum wolfii Vasey
U.S.D.A. Monthly Rep. (Feb.–Mar.), 156, 1874.

Wolf Trisetum

Graphephorum brandegei (Scribn.) Rydb.
Graphephorum muticum (Boland. ex Thurb. in S. Wats.) A. Heller
Graphephorum wolfii (Vasey) Vasey ex Coult.
Trisetum brandegei Scribn.
Trisetum muticum (Boland. ex Thurb. in S. Wats.) Scribn.

Tufted perennial from fibrous roots; rhizomes lacking, or sometimes present and very short; culms few to several, erect. Leaves mostly basal, the blades flat, glabrous or scabrous, sometimes sparsely pilose; sheaths glabrous, scabrous, or pilose; ligules truncate, erose. Panicle narrow, spikelike, the branches erect to ascending; spikelets purplish, 2–3-flowered; glumes subequal, lanceolate, acute to nearly acuminate, the margins membranous; lemmas lanceolate, scaberulous, obtuse, awnless or with a short awn less than 2 mm long; palea oblong to lanceolate; anthers 0.6–1.5 mm long.

Timberline meadows of the Beartooth, Uinta, Wasatch, and Wind River Ranges; approaching timberline from below in the Absaroka Range. Alberta south to Washington, California, Utah, and New Mexico.

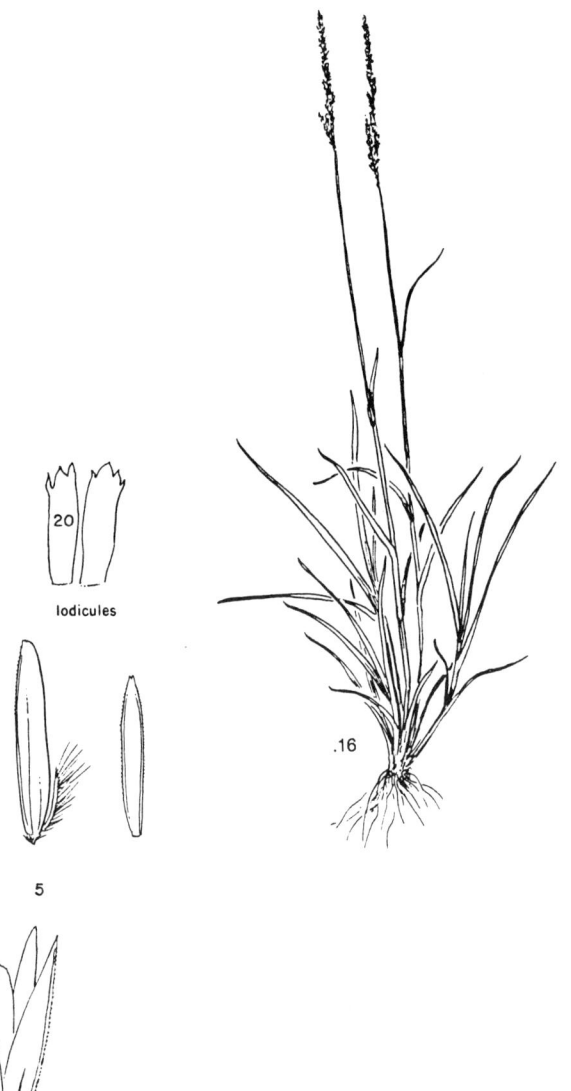

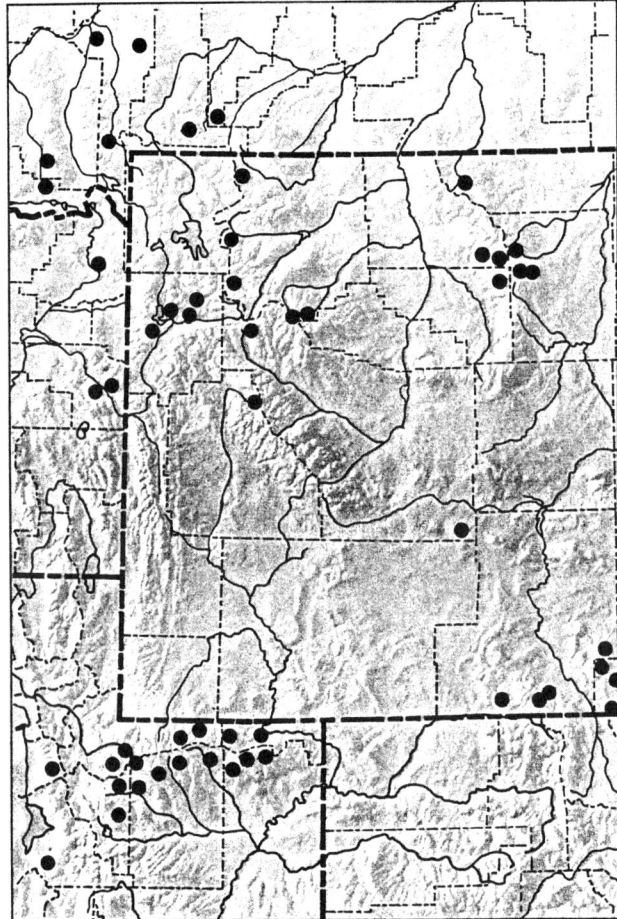

Polemoniaceae
Phlox Family

Annual or perennial herbs with alternate or opposite, exstipulate, simple or variously divided leaves. Flowers showy, solitary or in cymes, hypogynous, 5-merous. Calyx 5-lobed. Corolla 5-lobed and salverform, funnelform, campanulate, or rotate, convolute in the bud. Stamens inserted on the corolla tube, alternate with the lobes. Ovary 3-celled, with axillary placentation; style 1; stigmas 3. Fruit a loculicidal capsule.

KEY TO THE GENERA OF POLEMONIACEAE

1. Inflorescence subtended by a whorl of conspicuous, basally connate bracts; cotyledons persistent, green .... *Gymnosteris*
1. Inflorescence lacking a whorl of connate bracts; cotyledons deciduous
   2. Leaves simple and entire, or few-lobed, often crowded at the ends of prostrate stems
      3. Calyx tube uniformly textured, not ruptured by the developing capsule; leaves alternate .... *Collomia*
      3. Calyx tube hyaline between the costae, tending to split along the intercostal membranes; leaves opposite .... *Phlox*
   2. Leaves palmatifid, pinnatifid or pinnate in ours, either basal or distributed equally along the erect stems
      4. Leaves pinnately compound; calyx tube of uniform texture, remaining unbroken at maturity of capsule .... *Polemonium*
      4. Leaves palmatifid or pinnatifid; calyx tube with green costae separated by hyaline intervals (these sometimes very narrow) usually splitting along the intercostal membrane at maturity
         5. Leaves alternate, pinnatifid .... *Ipomopsis*
         5. Leaves opposite, palmatifid .... *Linanthus*

Collomia Nutt.
Collomia

Annual or perennial herbs with alternate, mostly entire leaves. Flowers pinkish or purplish in ours, in terminal, subcapitate, bracteate clusters. Calyx tube somewhat scarious, the lobes green. Corolla funnelform to salverform, the limb of short, spreading, obtuse lobes. Stamens unequally inserted on the corolla tube. Capsule obovoid, the seeds usually mucilaginous when wet.

KEY TO THE SPECIES OF *COLLOMIA*

1. Plants perennial, the several stems prostrate or decumbent .... *C. debilis*
1. Plants annual, the stem single, erect .... *C. linearis*

Collomia debilis (S. Wats.) Greene
Pittonia 1:127, 1887.

Alpine Collomia

Gilia debilis S. Wats.
Collomia larsenii (Gray) Pays.
Collomia hurdlei A. Nels.
Gilia howardii M.E. Jones
Gilia larseni Gray
Navarretia debilis (S. Wats.) Kuntze

Low, sprawling, perennial herb, less than 10 cm tall, from a taproot; stems decumbent, simple or branched, the tips erect. Leaves alternate, oblanceolate to spatulate, entire to tridentate or trifid. Flowers terminal, in bracteate clusters. Calyx tubular, the lobes nearly as long as the tube. Corolla lavender to white or yellowish, funnelform, the lobes much shorter than the tube. Stamens equally inserted below the expanded throat of the corolla tube, the anthers exerted.

Scree and talus slopes in the Absaroka, Big Horn, Gros Ventre, Salt River, Uinta, Wasatch, and Wyoming Ranges. Montana to Washington and south to Wyoming, Utah, and California.

Two varieties occur in the Middle Rocky Mountains: var. *ipomoea* Pays. with entire to tridentate leaves and deep rose pink flowers, and var. *debilis* with spatulate, entire, narrowly oblanceolate to trifid leaves and white to lavender flowers.

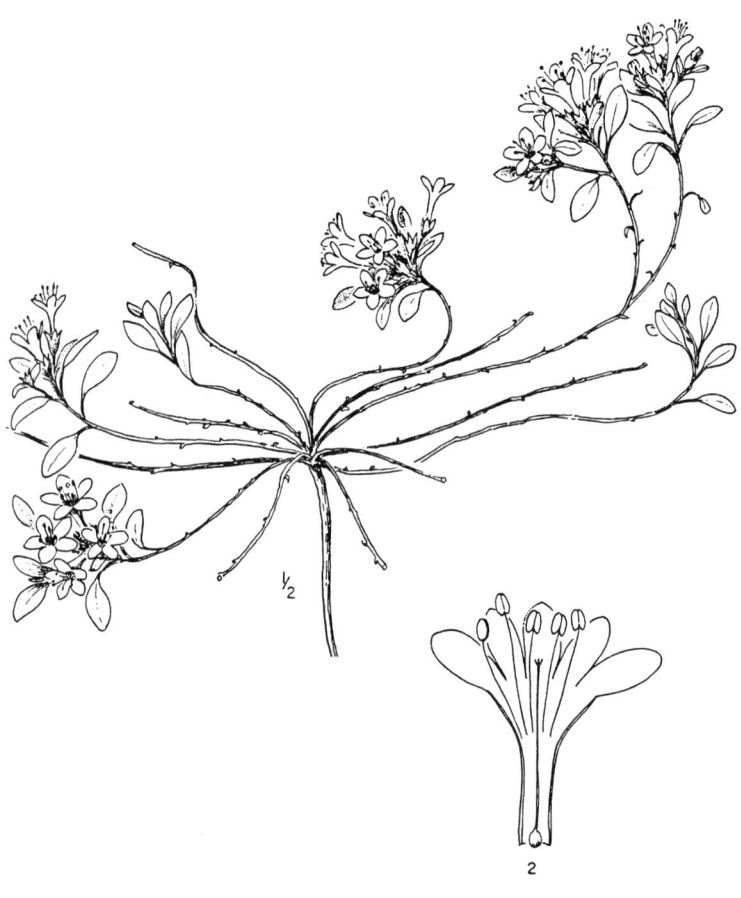
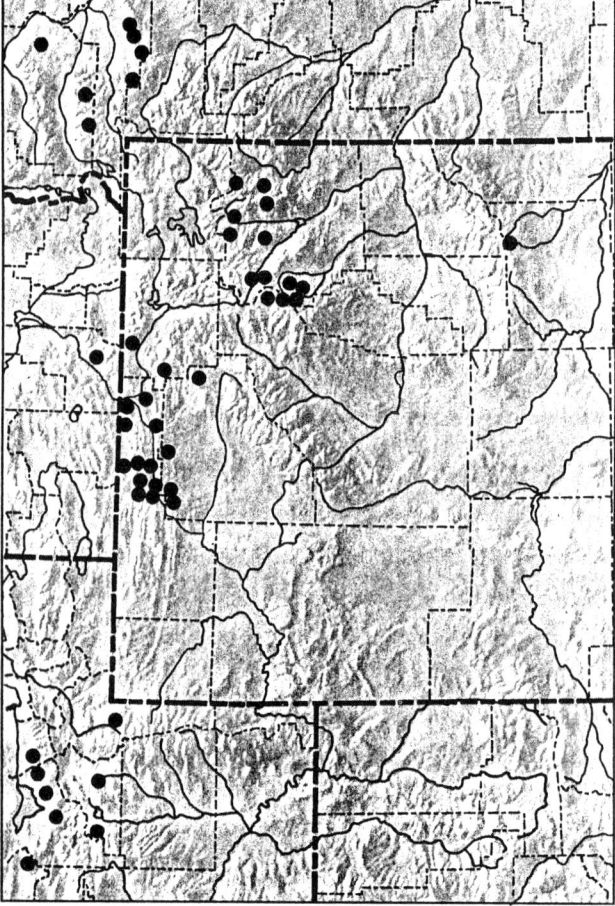

Collomia linearis Nutt.
Gen. Pl. 1:126, 1818.

Narrowleaf Collomia

Collomia parviflora Hook.
Gilia linearis (Nutt.) Gray
Navarretia linearis (Nutt.) Kuntze

Annual with erect, simple or branched, puberulent stems. Leaves alternate, sessile, lanceolate, linear, or elliptic, entire. Flowers terminal, in dense, glandular-pubescent, leafy-bracted, capitate to cymose clusters, the bracts lanceolate to ovate. Calyx tubular, glandular-pubescent, the lobes lanceolate, acute to subulate. Corolla pink, blue, or white, funnelform, about twice as long as the calyx. Stamens unequally inserted in the throat of the corolla tube, included.

Dry meadows and fellfields of the Absaroka and northern Wind River Ranges. Transcontinental in Canada; Alaska south to California, New Mexico, Nebraska, Missouri, and Wisconsin.

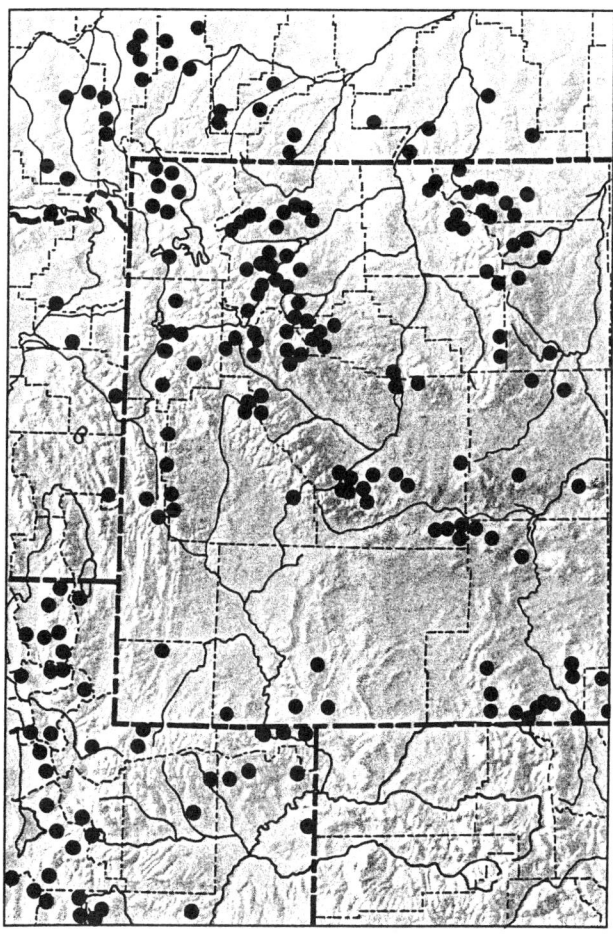

Gymnosteris Greene
Falsephlox

Very small, glabrous annuals with simple, leafless stems. Cotyledons green, persistent, connate. Inflorescence terminal, capitate, subtended by a conspicuous involucre of entire, basally connate bracts. Flowers sessile, the calyx usually scarious at the base, with green teeth. Corolla salverform and white, yellow, or pinkish. Stamens included, the filaments very short. Capsules dehiscent, not rupturing the calyx, or rupturing it irregularly. Seeds mucilaginous when wet.

Gymnosteris parvula A. Heller
Muhlenbergia 1:3, 1900.

Leafless Falsephlox

Gilia parvula Rydb.
Gymnosteris leibergii Brand
Gymnosteris rydbergii Tidestr.

Small annual, less than 5 cm tall; stems mostly simple, glabrous, with a terminal whorl of bracts and often a pair of connate cotyledons at the base; bracts lanceolate to ovate. Flowers single or in terminal clusters. Calyx scarious, inflated-tubular, the teeth herbaceous, ovate, acute. Corolla salverform, inconspicuous, and white, pinkish, or yellowish, the lobes shorter than the tube, oblong, obtuse to weakly acute. Stamens inserted equally in the corolla throat, the filaments very short to lacking.

Rare in dry alpine meadows of the Absaroka Range. Oregon to Wyoming and south to California and Colorado.

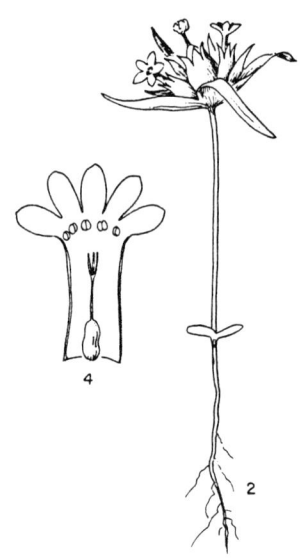

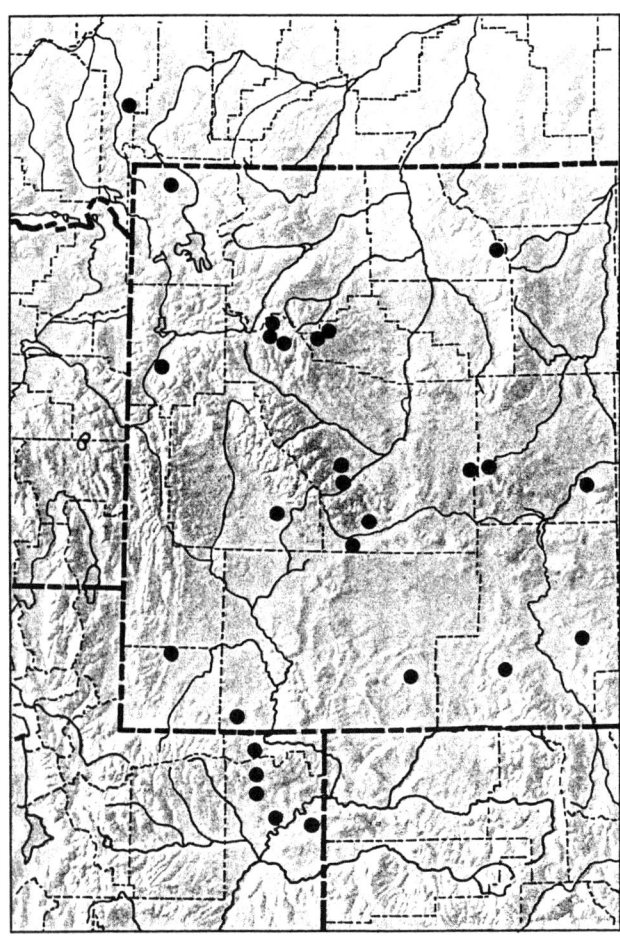

Ipomopsis Michx.
Ipomopsis, Gilia

Erect annual or perennial herbs from taproots, with simple, entire or pinnatifid, alternate leaves, the lower ones often opposite. Flowers white, pink, red, blue, or purple; in panicles, cymes, or capitate clusters; sometimes minute, but often showy. Calyx slightly accrescent, at length ruptured by the capsule. Corolla funnelform to salverform. Stamens equally or unequally inserted on the corolla tube. Capsule globose, ovoid or oblong, the seeds 1 to many in each locule, becoming mucilaginous when wet.

KEY TO THE SPECIES OF *IPOMOPSIS*

1. Inflorescence densely capitate or subcapitate, spicate; flowers small, the corolla tube mostly less than 12 mm long .... *I. spicata*
1. Inflorescence open, paniculate; flowers large, the corolla tube mostly more than 15 mm long .... *I. aggregata*

Ipomopsis aggregata (Pursh) V. Grant
El Aliso 3:360, 1956.

Fairy Trumpet

Cantua aggregata Pursh
Batanthes aggretata (Pursh) Raf.
Batanthes attenuata (Gray) Greene
Batanthes formosissima (Greene) Greene
Batanthes pulchella (Dougl. ex Hook.) Greene
Callisteris arizonica Greene
Callisteris attenuata (Gray) Greene
Callisteris formosissima Greene
Callisteris pulchella (Dougl. ex Hook.) Greene
Callisteris violacea Greene
Collomia aggregata (Pursh) Porter ex Rothr.
Gilia aggregata (Pursh) Spreng.
Gilia arizonica (Greene) Rydb.
Gilia attenuata (Gray) A. Nels.
Gilia bridgesii (Gray) Wherry
Gilia candida Rydb.
Gilia formosissima (Greene) Woot. & Standl.
Gilia pulchella Dougl. ex Hook.
Gilia tenuituba Rydb.
Gilia texana (Greene) Woot. & Standl.
Ipomopsis bridgesii (Gray) Wherry
Ipomopsis candida (Rydb.) W.A. Weber
Ipomopsis tenuituba (Rydb.) V. Grant
Ipomeria aggregata (Pursh) Nutt.

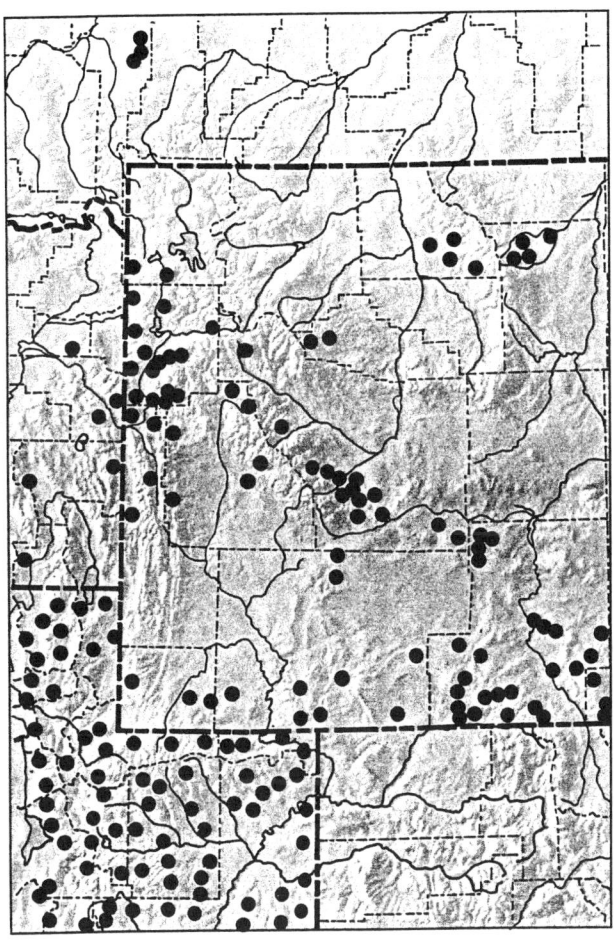

Erect, biennial or perennial herb from a taproot; stems 1 to several, stipitate-glandular, at least below the inflorescence, and pubescent with white, spreading hairs. Basal leaves pinnatifid, or sometimes 2-pinnatifid, persistent and forming an evident rosette, or sometimes early deciduous; cauline leaves pinnatifid, reduced upward. Inflorescence a bracteate panicle or raceme, the flowers showy, pedicellate. Calyx narrowly campanulate, the lobes attenuate; corolla salverform and red, salmon, pink, or white, the lobes acute to attenuate, spreading. Stamens equally to slightly unequally inserted above the middle of the corolla tube, included to short-exerted. Style elongate, included to short-exerted. Capsules with 1 to several seeds in each locule.

Slopes, meadows, and margins of krummholz in the vicinity of timberline in the Uinta, Wasatch, and Wind River Ranges. British Columbia south to California and Mexico, east to Montana and Wyoming.

White- or whitish-flowered plants of the subalpine and lower alpine zone in the southern portion of the Middle Rocky Mountains are var. *tenuituba* (Rydb.) Dorn. Because of the extensive introgression between populations of the *I. aggregata* complex (Grant & Wilken, 1988), I follow Dorn (1992) in treating the high mountain representatives as a variety rather than a distinct species.

Ipomopsis spicata (Nutt.) V. Grant
El Aliso 3:361, 1956.

Spike Ipomopsis

Gilia spicata Nutt.
Gilia cephaloidea Rydb.
Gilia globularis (Brand) W.A. Weber
Gilia tridactyla Rydb.

Erect perennial from a taproot; stems simple or branched from the base, arachnoid to tomentose. Basal leaves tufted, mostly pinnatifid; cauline leaves alternate, pinnatifid to trifid, the upper ones often entire. Flowers in a dense, terminal, capitate or spicate panicle. Calyx ovoid, the lobes cuspidate. Corolla white (brownish when dry), salverform, the lobes lanceolate, shorter than the tube. Stamens equally inserted at the sinuses of the corolla lobes, exerted, the filaments shorter than the anthers. Style included, shorter than the tube. Capsules with 1–2 seeds in each locule.

Seasonally dry ridges, meadows, and scree slopes in the Absaroka, Beartooth, Big Horn, and Wind River Ranges. Idaho to South Dakota, south to Kansas, New Mexico, and Utah.

Small, high-altitude plants usually less than 10 cm tall with a dense capitate or subcapitate inflorescence, pinnately 5-parted leaves, and corolla lobes more then 3 mm long are var. *orchidacea* (Brand) Dorn; those with twice-pinnately 7- or more-parted leaves and corolla lobes less than 3 mm long are var. *robruthiorum* (Wilken & Hartm.) Dorn.

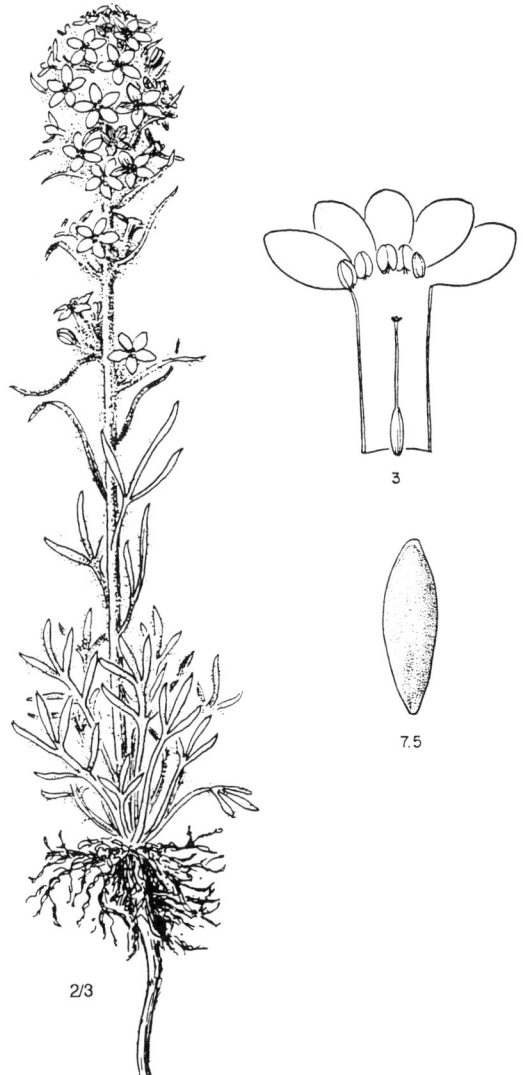

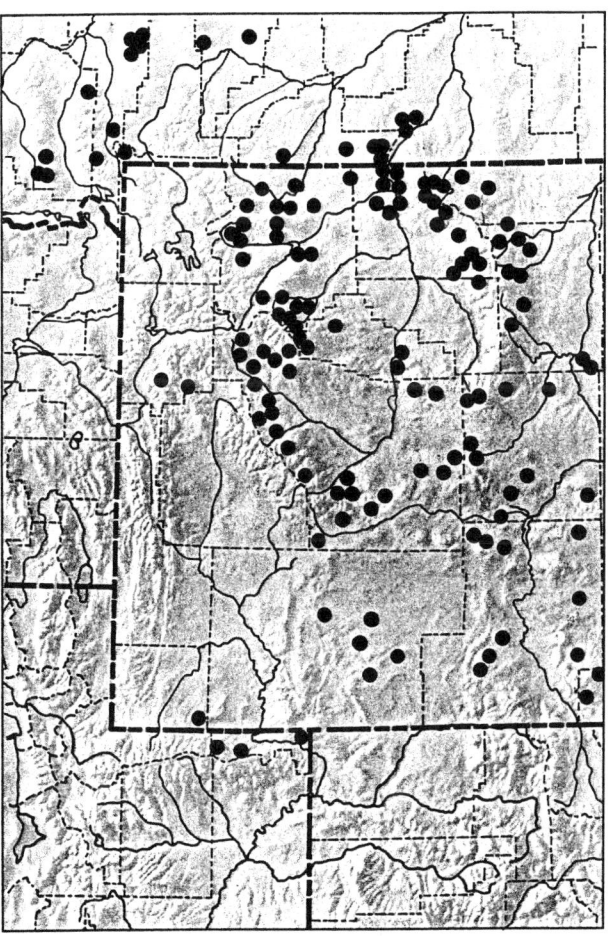

Linanthus Benth.
Flaxflower

Annual or, in ours, perennial herbs from a woody caudex. Leaves opposite, the blades palmately parted into narrow divisions. Flowers axillary or terminal in small, cymose inflorescences. Calyx campanulate, the tube scarious beneath the sinuses and between the green costae. Corolla salverform, funnelform, or campanulate, the lobes mostly shorter than the tube. Stamens about equal and inserted on the throat of the corolla. Capsules ovoid, the valves persistent after dehiscence.

Linanthus nuttallii (Gray) Greene ex Milliken
Univ. Calif. Pub. Bot. 2:54, 1904.

Nuttall Flaxflower

Gilia nuttallii Gray
Leptodactylon nuttallii (Gray) Rydb.
Linanthastrum nuttallii (Gray) Ewan
Navarretia nuttallii (Gray) Kuntze
Siphonella montana Nutt. ex Gray
Siphonella nuttallii (Gray) A. Heller
Siphonella parviflora Nutt. ex Gray

Perennial from a taproot, with a branched, woody caudex; stems simple or branched, pubescent at least above. Leaves opposite, palmately parted, appearing whorled. Flowers in compact, terminal, bracteate cymes. Calyx scarious beneath the sinuses. Corolla white or cream, the lobes obovate, the tube pubescent, equaling the calyx in length. Stamens equal, included or slightly exceeding the throat of the corolla. Mature capsules ovoid, rupturing the calyx; seeds 1–4 in each locule.

Typical of lower elevations, *L. nuttallii* occurs above timberline on rocky slopes, scree, and talus in the Gros Ventre, Teton, Uinta, and Wasatch Ranges. Washington to Wyoming, south to Colorado and California.

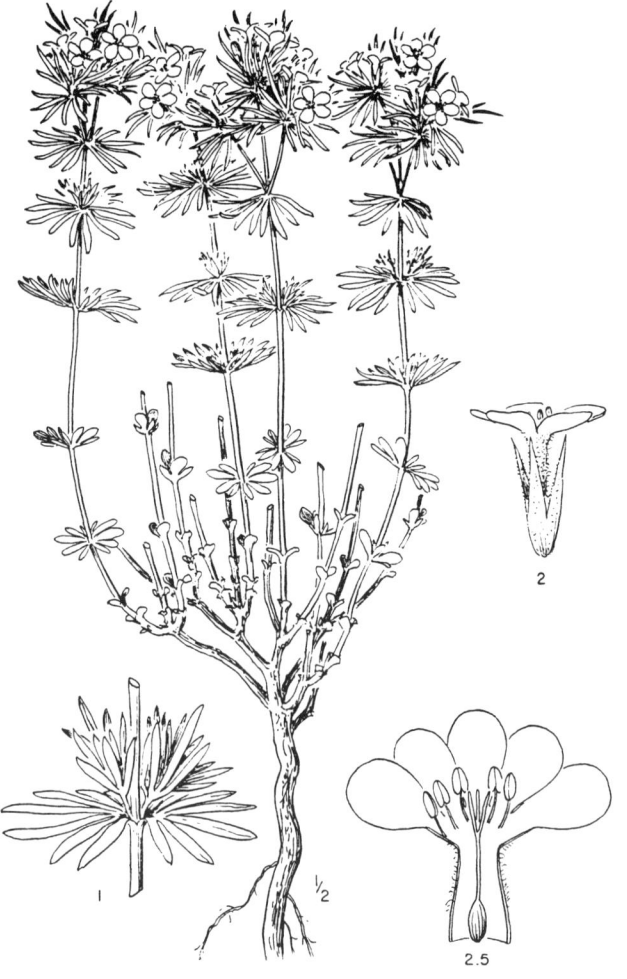
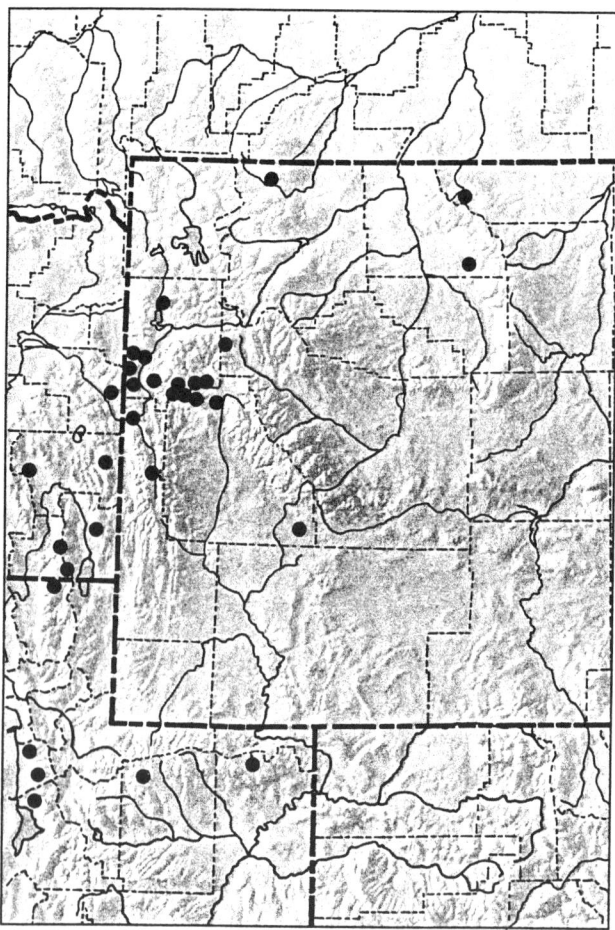

Phlox L.
Phlox

Ours are low, caespitose perennials with entire, sessile, opposite, often sharp-pointed leaves. Flowers white or bluish, solitary or cymose. Sepals 5, united below by scarious margins. Corolla 5-lobed, salverform. Stamens unequally inserted in the tube of the corolla, included. Ovary 3-celled, the slender style with 3 stigmas. Fruit an ovoid, few-seeded capsule enclosed by the calyx.

KEY TO THE SPECIES OF PHLOX

1. Principal leaves 10–20 mm long; styles mostly 5–9 mm long — *P. multiflora*
1. Principal leaves 4–10 mm long; styles 2–5 mm long
    2. Leaves mostly 0.5–1 mm wide at the middle, pungent, arachnoid on the margins; calyx arachnoid; plants forming dense cushions — *P. hoodii*
    2. Leaves mostly 1–2 mm wide at the middle, acute, ciliate on margins, at least near the base (rarely glabrous); calyx sparsely glandular-pubescent to glabrous; plants forming loose mats — *P. pulvinata*

Phlox hoodii Richards.
Bot. App. Frankl. J., 733, 1823.

Hood Phlox

Armeria canescens (T. & G.) Kuntze
Armeria hoodii (Rich) Kuntze
Fonna hoodii (Rich) Nieuwl. & Lunnell
Phlox canescens T. & G.
Phlox glabrata (E. Nels.) Brand
Phlox lanata Piper
Phlox scleranthifolia Rydb.

Pulvinate to caespitose, arachnoid perennial from a taproot. Leaves sessile, basally connate, linear, lanceolate, or subulate, the margins arachnoid, tips pungent. Flowers solitary, sessile to short-pedicellate, terminal. Calyx arachnoid, the lobes linear, subulate, about as long as the tube, midrib thickened and conspicuous. Corolla white to lavender, the throat often yellowish, the tube less than twice as long as the calyx, the lobes obovate to elliptical. Stamens unequally inserted in the throat of the corolla tube, included. Style about half the length of the corolla tube, included.

Exposed ridgetops and rocky slopes in the Absaroka, Big Horn, Hoback, Wasatch, Wind River, and Wyoming Ranges. Alaska and Yukon Territory south to Nebraska, New Mexico, and California.

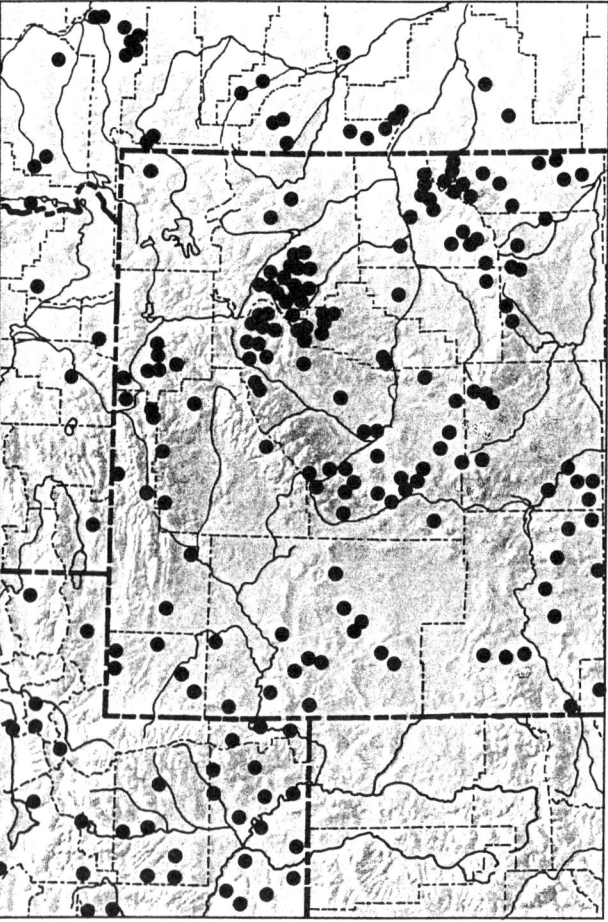

Phlox multiflora A. Nels.
Bull. Torr. Club 25:278, 1898.

Rocky Mountain Phlox

Phlox costata Rydb.
Phlox depressa (E. Nels.) Rydb.
Phlox patula A. Nels.

Caespitose, glabrous to pubescent perennial from a taproot; stems branched from the base, suberect. Leaves linear, scaberulous, apiculate, sessile and connate at the base. Flowers solitary, terminal, mostly pedicellate. Calyx glabrous or rarely pubescent, the lobes about as long as the tube, subulate, midrib thickened and conspicuous. Corolla white, occasionally bluish or pinkish, the tube about the same length as the calyx, the lobes obovate. Stamens unequally inserted in the throat of the corolla, included. Style about as long as the corolla tube, included.

Rocky slopes and dry meadows in the Absaroka, Beartooth, Big Horn, Gros Ventre, Wind River, and Wyoming Ranges. Montana and Idaho south to Nevada, Utah, and New Mexico.

Subspecies *depressa* (E. Nels.) Wherry can be distinguished from ssp. *multiflora* on the basis of its more compact habit and pedicels less than 8 mm long; typical ssp. *multiflora* is loosely caespitose, with pedicels up to 18 mm long.

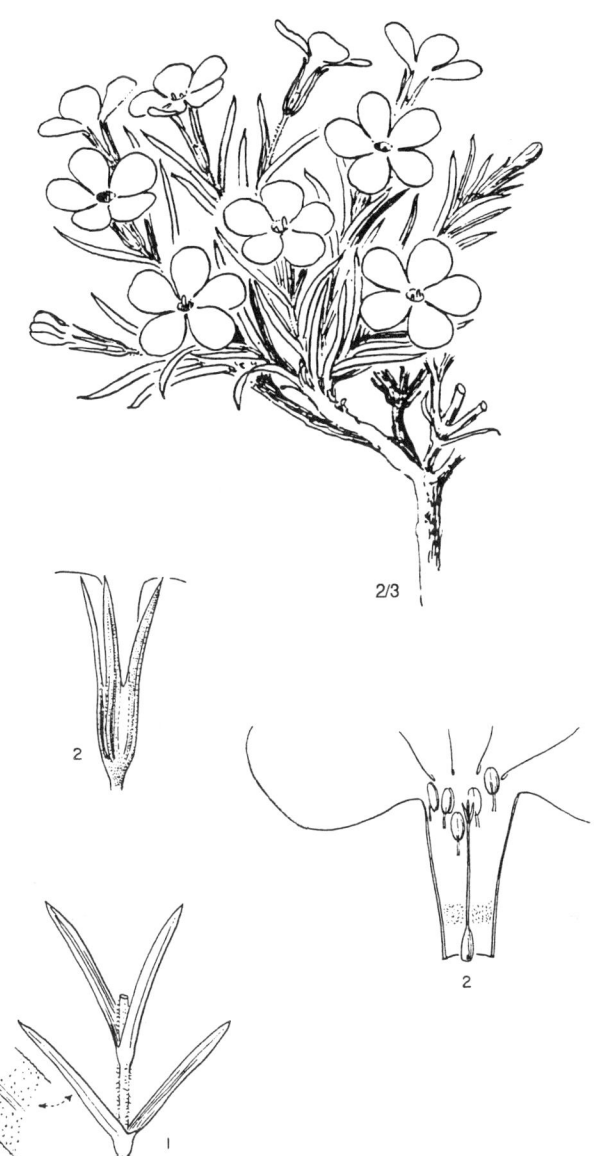
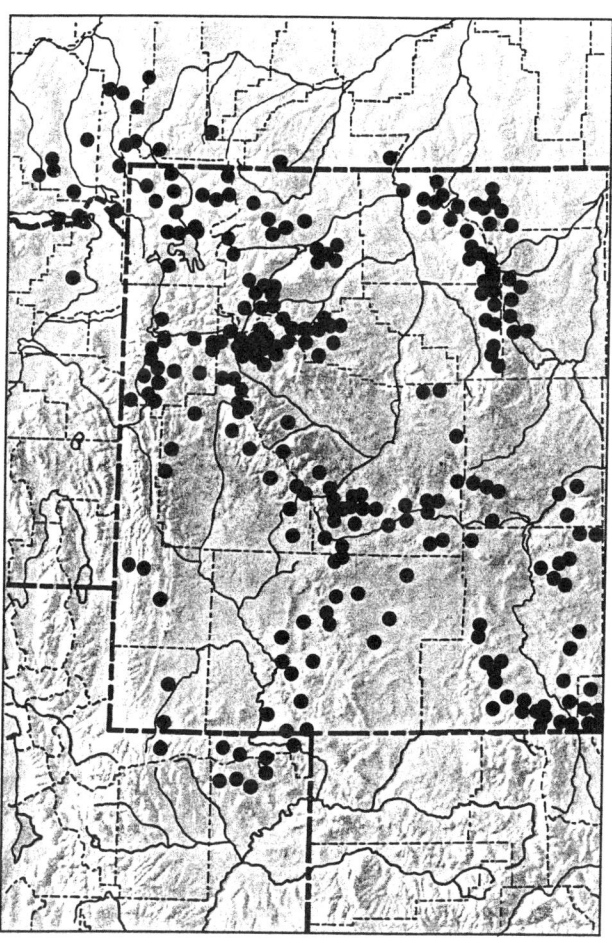

Phlox pulvinata (Wherry) Cronq.
Vas. Pl. Pac. NW 4:135, 1959.

Tufted Phlox

Phlox caespitosa Nutt.
Phlox condensata (Gray) E. Nels.

Pulvinate to caespitose perennial from a taproot; stems usually branched, erect to spreading, less than 6 cm long. Leaves sessile, linear to lanceolate, glabrous to pubescent or glandular, the margins ciliate at least near the base, apiculate. Flowers solitary, terminal, short-pedicellate. Calyx glabrous to sparsely glandular-pubescent, the lobes lanceolate, about as long as the tube, midrib obscure or lacking. Corolla white to bluish, the tube about twice as long as the calyx, the lobes obovate. Stamens unequally inserted in the throat of the corolla, included to slightly exserted. Style nearly as long as the corolla tube.

Common on exposed ridgetops and rocky slopes in most of the alpine ranges of the Middle Rocky Mountains. Not yet known from above timberline in the Big Horn, Hoback, Salt River, and Wyoming Ranges. Oregon to Montana and south to Nevada and New Mexico.

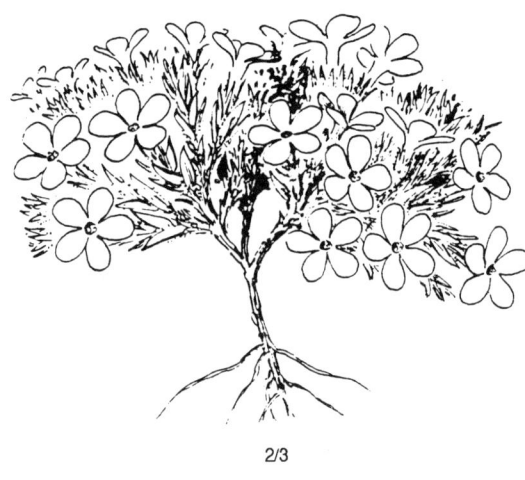

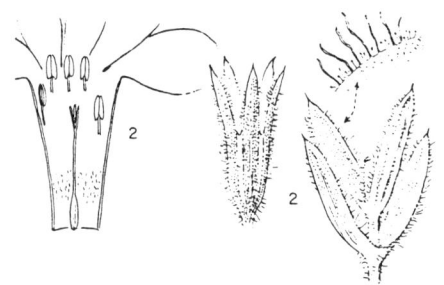

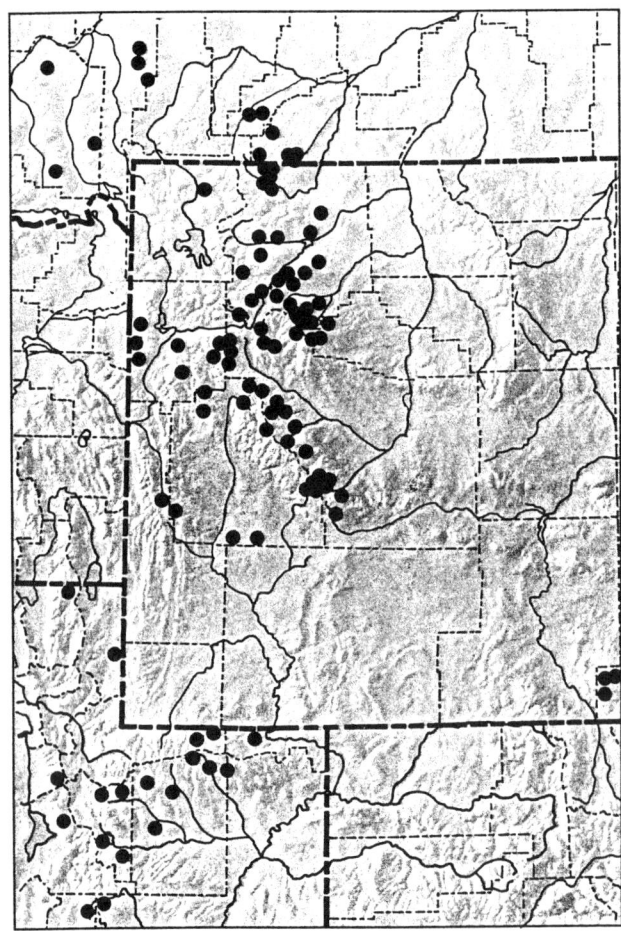

Polemonium L.
Jacob's Ladder

Low, erect, perennial herbs, the stems glandular-pubescent in ours, the leaves alternate, pinnate, often with verticillate leaflets. Flowers blue, in a cymose or subcapitate inflorescence. Calyx 5-lobed, campanulate, persistent. Corolla campanulate to funnelform, the lobes rounded. Stamens 5, included or exserted, usually pilose at the base. Ovary 3-celled, the style with 3 stigmas. Fruit a 3-valved, loculicidal capsule with black or brown, angled or winged seeds.

KEY TO THE SPECIES OF *POLEMONIUM*

1. Plants glabrate to sparsely glandular; leaflets 2-ranked, opposite or nearly so; corolla 6–11 mm long   *P. pulcherrimum*
1. Plants strongly glandular; leaflets verticillate or deeply 3–5-cleft and appearing verticillate; corolla 17–23 mm long   *P. viscosum*

Polemonium pulcherrimum Hook.
Curtis' Bot. Mag. 57:Pl. 2979, 1830.

Jacob's Ladder

Polemonium berryi Eastw.
Polemonium delicatum Rydb.
Polemonium fasciculatum Eastw.
Polemonium haydenii A. Nels.
Polemonium humile Lindl.
Polemonium lindleyi Wherry
Polemonium mexicanum Nutt.
Polemonium orbiculare Gand.
Polemonium oreades Gand.
Polemonium parvifolium Nutt. ex Rydb.
Polemonium pilosum (Greenm.) G.N. Jones
Polemonium rotatum Eastw.
Polemonium shastense Baker ex Eastw.

Multistemmed perennial from a taproot and branched caudex. Leaves pinnately compound, glabrate to sparsely viscid-pubescent, the leaflets ovate to elliptic; basal leaves clustered; cauline leaves reduced upward. Flowers in dense to somewhat lax, terminal cymes. Calyx campanulate, the lobes deltoid, acute, about as long as the tube. Corolla blue with a yellowish tube, campanulate or nearly rotate, the lobes obovate, obtuse, as long as to twice as long as the tube. Anthers whitish, inserted near the middle of the tube. Style and trifid stigma included or nearly so.

Occasional on rocky slopes and fellfields of the Absaroka, Beartooth, Uinta, Wasatch, and Wind River Ranges. Alaska south to California, Nevada, Utah, and Wyoming.

Middle Rocky Mountain plants are the typical var. *pulcherrimum*.

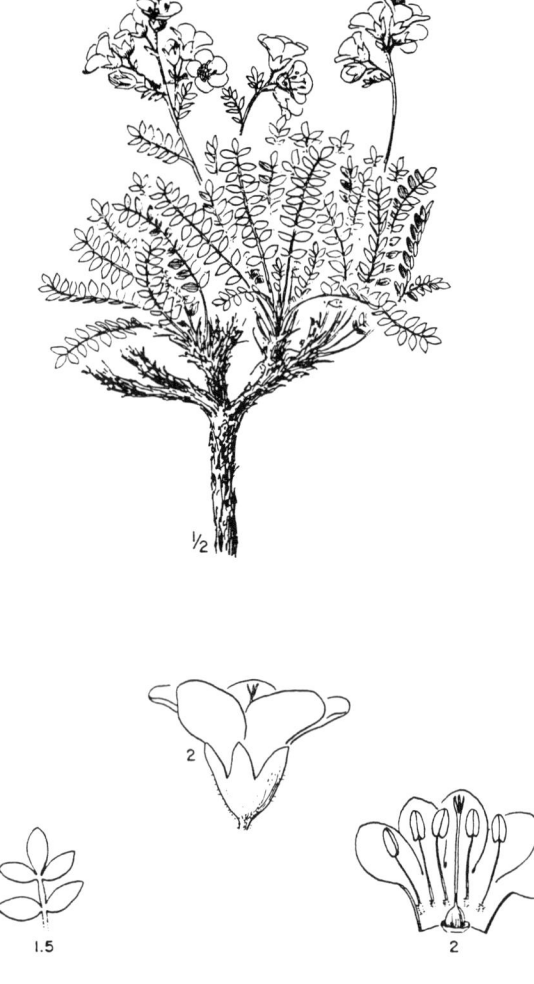

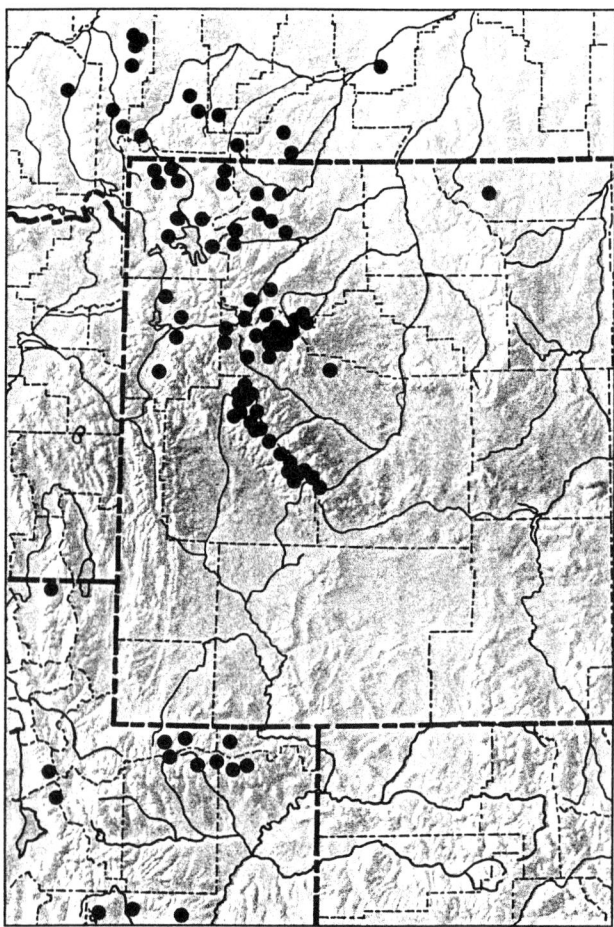

Polemonium viscosum Nutt.
J. Acad. Phila., II, 1:154, 1848.

Sky Pilot

Polemonium confertum Gray
Polemonium grayanum Rydb.
Polemonium speciosum Rydb.

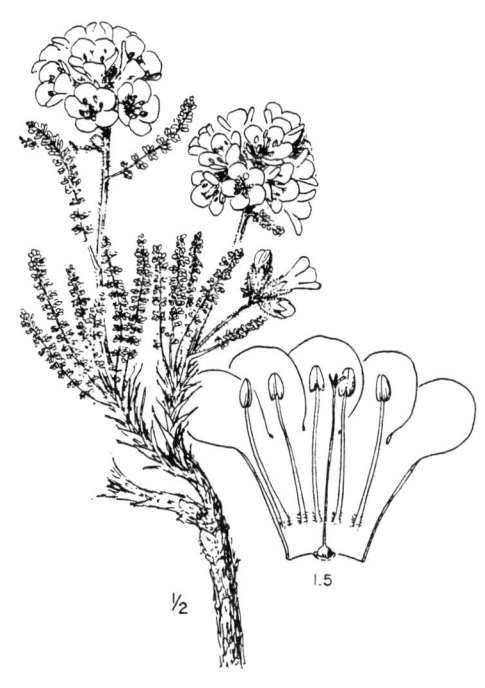

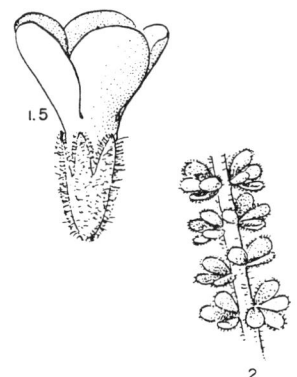

Strongly glandular-pubescent perennial from a taproot and branched caudex; stems single to several, spreading-erect. Leaves pinnate, mostly basal, the cauline ones reduced upward; leaflets 2–5-cleft to the base, the individual segments oblanceolate to obovate. Flowers in capitate or subcapitate cymes. Calyx tubular, glandular-puberulent, the lobes lanceolate, acute, about as long as the tube. Corolla blue, funnelform to campanulate, the lobes obovate, shorter than the tube. Anthers yellow, inserted below the middle of the tube. Style and trifid stigma included or nearly so.

Widespread and common in damp rocky places, talus slopes, meadows, and fellfields of all alpine ranges in the Middle Rocky Mountains. British Columbia and Alberta south to Oregon, Nevada, Arizona, and New Mexico.

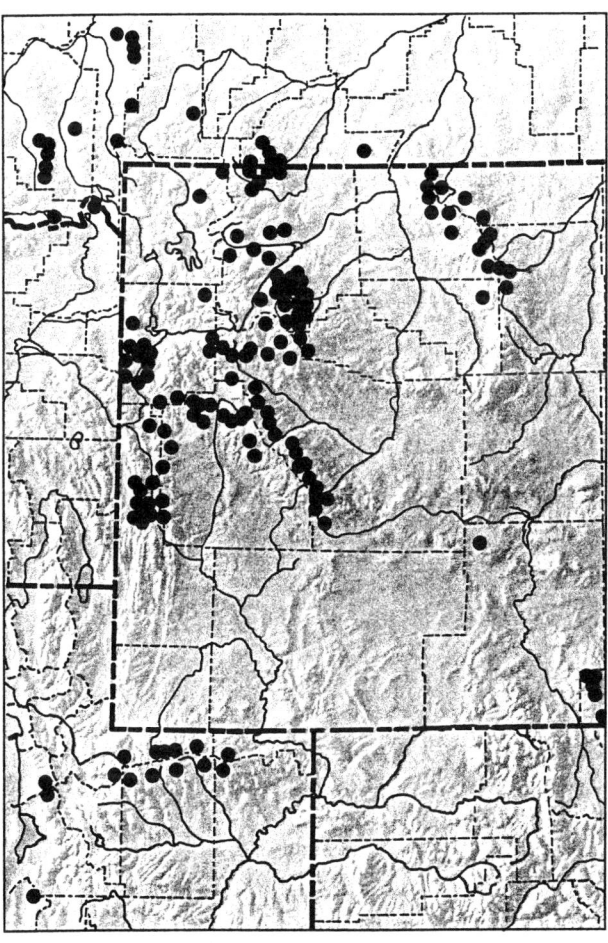

Polygonaceae
Buckwheat Family

Ours are herbs, often with conspicuously jointed stems, and with simple, alternate, entire, stipulate leaves, the stipules often forming sheaths above the swollen joints or the stems. Flowers small, mostly perfect, sometimes polygamous or dioecious, the perianth lobes 4–6, often persistent. Stamens usually 4–9, distinct or inserted at the base of the perianth, usually exserted. Pistil 1; the ovary 1-celled, with a single ovule; styles 2–3. Fruit a lenticular or trigonous akene.

KEY TO THE GENERA OF POLYGONACEAE

1. Dwarf annuals less than 2 cm tall, with a few sessile, entire leaves forming an involucre at the summit of the threadlike stem — *Koenigia*
1. Plants not as above in all respects, mostly 5 cm or more tall
   2. Stipular sheaths lacking; inflorescence subtended by an involucre of lanceolate to ovate bracts — *Eriogonum*
   2. Stipular sheaths present; inflorescence without an involucre
      3. Perianth with 3 outer and 3 inner segments — *Rumex*
      3. Perianth with 4 or 5 segments
         4. Perianth segments 4, the akenes broadly winged; leaf blades orbicular to reniform — *Oxyria*
         4. Perianth segments 5, the akenes not winged; leaf blades lanceolate to broadly ovate or cordate — *Polygonum*

Eriogonum Michx.
Wild Buckwheat

Annual or perennial herbs, the stems sometimes woody at the base, with mostly basal, exstipulate leaves. Flowers small, white, yellow, or pinkish; in terminal, involucrate umbels, cymes, or capitate clusters. Perianth 6-lobed, the segments equal or unequal. Stamens 9. Styles 3, with capitate stigmas. Akenes trigonous, sometimes winged.

KEY TO THE SPECIES OF *ERIOGONUM*

1. Plants less than 5 cm tall; inflorescence sessile or subsessile, rarely exceeding the leaves — *E. acaule*
1. Plants more than 6 cm tall; inflorescence on a peduncle exceeding the leaves
   2. Inflorescence umbellate, primary rays visible, the base subtended by large, verticillate, leaflike bracts
      3. Perianth pubescent externally; plants about 10 cm tall, tufted from a branched caudex — *E. flavum*
      3. Perianth glabrous externally; plants more than 10 cm tall, stoloniferous, forming large mats — *E. umbellatum*
   2. Inflorescence capitate or cymose, primary rays mostly lacking, the base lacking large, verticillate bracts
      4. Perianth segments oblong, nearly equal in size; flowers mostly yellow; leaves often revolute — *E. brevicaule*
      4. Perianth segments obovate, the outer ones about twice as broad as the inner; flowers mostly white, pink, or cream, leaves flat — *E. ovalifolium*

Eriogonum acaule Nutt.
Proc. Acad. Phila. 4:13, 1848.

Stemless Buckwheat

Branched, spreading perennial forming cushions or mats less than 6 cm tall. Leaves mostly basal, crowded, revolute, linear to oblong, white-tomentose. Inflorescence sessile or short-pedunculate, shorter than the leaves or nearly so. Involucre turbinate, the lobes erect. Perianth yellowish, pubescent. Akenes pubescent.

This plains species occurs above timberline in the Wyoming Range. Wyoming, Colorado, Idaho, and Nevada.

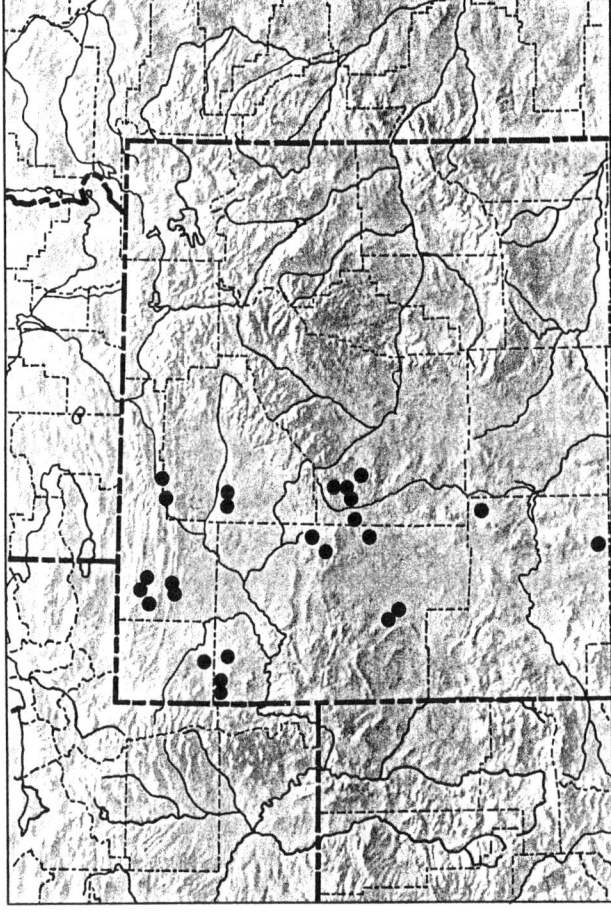

Eriogonum brevicaule Nutt.
Proc. Acad. Phila. 4:15, 1848.

Shortstem Buckwheat

Eriogonum campanulatum Nutt.
Eriogonum chrysocephalum Gray
Eriogonum desertorum (Maguire) R.J. Davis
Eriogonum ephedroides Reveal
Eriogonum grayi Reveal
Eriogonum lagopus Rydb.
Eriogonum loganum A. Nels.
Eriogonum medium Rydb.
Eriogonum micranthum Nutt.
Eriogonum nanum Reveal
Eriogonum nudicaule (Torr.) Small
Eriogonum orendense A. Nels.
Eriogonum viridulum Reveal
Eriogonum wasatchense M.E. Jones

Tufted to mat-forming perennial from a branched, woody caudex and taproot. Leaves all basal or on short stems, linear to oblanceolate, grayish-pubescent on one or both surfaces, petiolate, the margins undulate, shallowly toothed, or entire, involute or flat. Scapes or stems glabrous to glandular and floccose. Inflorescence cymose to capitate, exceeding the numerous basal leaves. Involucres campanulate to turbinate-conic, glabrous to floccose or tomentose, 5-lobed, the segments acute. Perianth yellow, cream, or white, glabrous, the lobes oblong to oval, nearly equal in size. Akenes glabrous.

Scree, talus, and rocky ridges of the Hoback, Uinta, Wasatch, and Wind River Ranges. Montana south to Idaho, Utah, Colorado, and Nebraska.

Two varieties occur at higher elevations in the Middle Rocky Mountains: var. *nanum* (Reveal) S.L. Welsh, with leaf margins undulate or broadly and shallowly toothed, and at least partially revolute; and var. *laxifolium* (T. & G.) Reveal, with leaf margins entire and revolute or not.

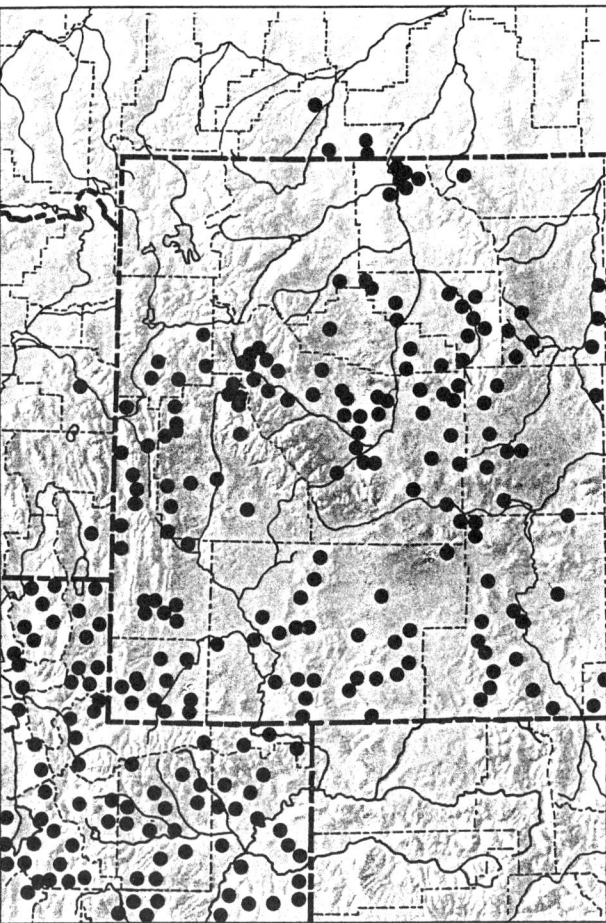

Eriogonum flavum Nutt. in Fraser
Cat. no. 34, 1813.

Yellow Buckwheat

Eriogonum chloranthum Greene
Eriogonum crassifolium Benth.
Eriogonum piperi Greene
Eriogonum polyphyllum Small ex Rydb.
Eriogonum sericeum Pursh

Caespitose, mat-forming perennial from a branched caudex and woody taproot. Leaves basal, clustered on the caudex branches, the blades narrowly oblong to spatulate, petiolate, silvery-tomentose below. Scapes tomentose to floccose, exceeding the leaves. Inflorescence an umbel subtended by several spreading, foliaceous, or occasionally reduced bracts. Involucres turbinate-campanulate, tomentose to pilose, the lobes shallow, erect or slightly spreading. Perianth yellow, sometimes reddish- or pinkish-tinged, pilose externally. Akenes villous above the middle.

Rocky places in the Absaroka, Beartooth, Salt River, and Wind River Ranges. Alaska; British Columbia and Alberta south to Oregon, Idaho, Wyoming, and Colorado.

Middle Rocky Mountain alpine plants having short stipes that are less than 1 mm long and often thicker than the pedicels are var. *flavum;* those with stipes more than 1 mm long and thinner than the pedicels are var. *piperi* (Greene) M.E. Jones.

var. flavum

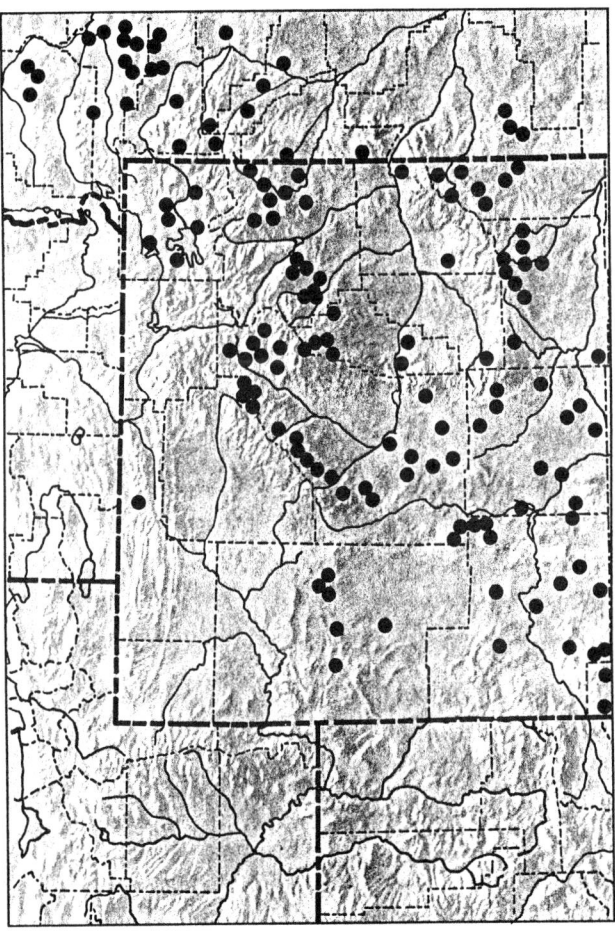

var. piperi

Eriogonum ovalifolium Nutt.
J. Acad. Phila. 7:50, 1834.

Cushion Buckwheat

Eriogonum davisianum S. Stokes
Eriogonum dichroanthum Gand.
Eriogonum nivale Canby
Eriogonum ochroleucum Small ex Rydb.
Eriogonum orthocaulon Small
Eriogonum purpureum (Nutt.) Benth.
Eriogonum rhodanthum A. Nels. & Kennedy
Eriogonum roseiflorum Gand.
Eriogonum rubidum Gand.
Eriogonum vineum Small
Eucycla ovalifolia (Nutt.) Nutt.
Eucycla purpurea Nutt.

Scapose, caespitose or sometimes pulvinate perennial, from a branched caudex and often woody taproot. Leaves clustered on the caudex branches, spatulate to petiolate with elliptic to ovate blades, white-tomentose. Scapes tomentose; inflorescence capitate, subtended by mostly 3 narrowly lanceolate bracts. Involucres campanulate, puberulent, 5-toothed; the teeth short, broad, erect or spreading. Perianth yellow, white, or pinkish, glabrous externally, not stipitate, the outer segments broadly obovate, about twice as broad as the narrowly spatulate inner ones. Akenes glabrous.

Talus and scree in most alpine ranges of the Middle Rocky Mountains. Not yet known from above timberline in the Big Horn and Medicine Bow Ranges. British Columbia and Alberta south to California and New Mexico.

Dwarf alpine plants of the Middle Rocky Mountains with spatulate leaves that are green or greenish above are var. *depressum* Blank. West of our area these intergrade with the var. *nivale* (Canby) M.E. Jones, which has orbicular, white-tomentose leaves.

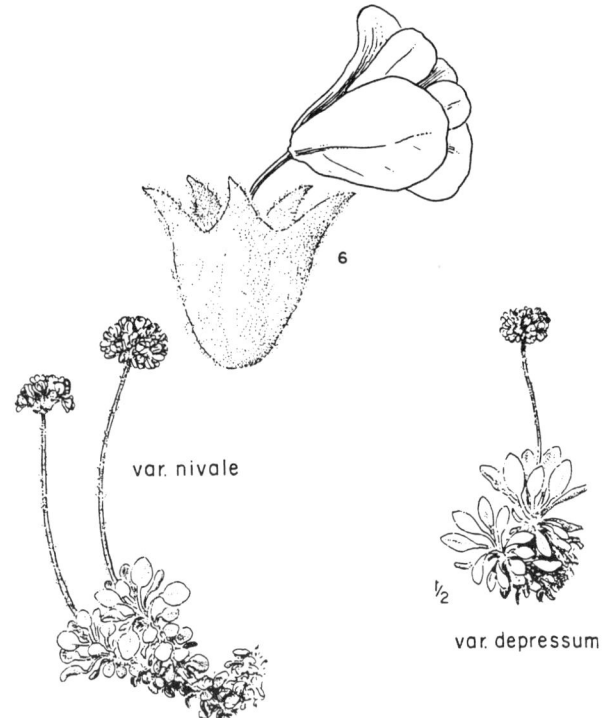

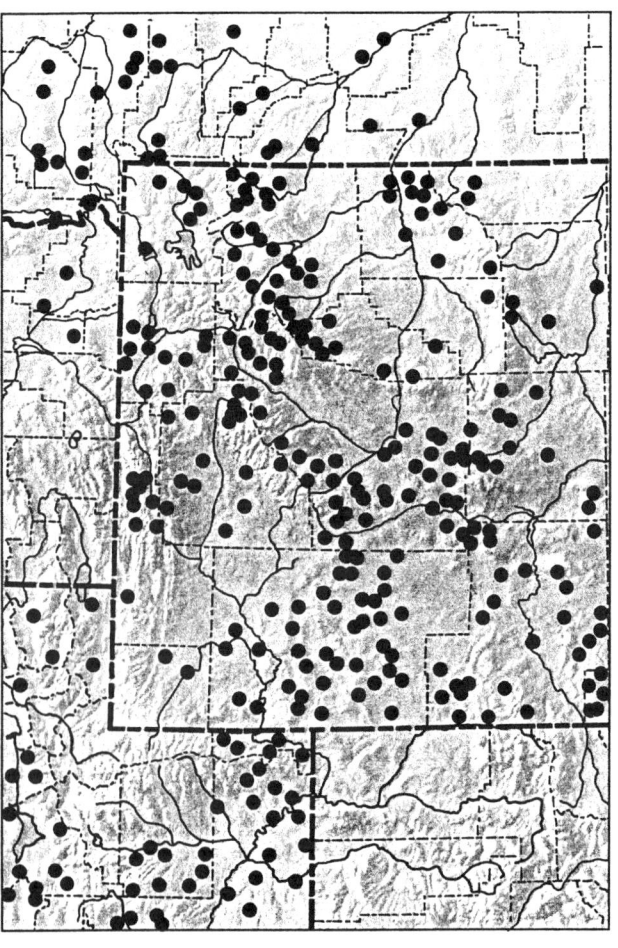

Eriogonum umbellatum Torr.
Ann. Lyc. N.Y. 2:241, 1828.

Sulfur Buckwheat

Eriogonum aridum Greene
Eriogonum biumbellatum Rydb.
Eriogonum cognatum Greene
Eriogonum covillei Small
Eriogonum croceum Small
Eriogonum ellipticum Nutt.
Eriogonum glaberrimum Gand.
Eriogonum hausknechtii Dammer
Eriogonum latum Small
Eriogonum montanum T.J. Howell
Eriogonum neglectum Greene
Eriogonum polyanthum Benth.
Eriogonum rydbergii Greene
Eriogonum stellatum Benth.
Eriogonum subalpinum Greene
Eriogonum tolmieanum Hook.
Eriogonum torreyanum Gray
Eriogonum umbelliferum Small

Scapose, mat-forming perennial from a much-branched, often woody caudex and taproot. Leaves usually basal, rarely cauline, the blades elliptic, spatulate to obovate, glabrous to sparsely pubescent above, tomentose below. Scapes tomentose or occasionally glabrous; inflorescence a simple or compound umbel subtended by a whorl of mostly oblanceolate, foliaceus bracts. Involucres turbinate to campanulate, tomentose, deeply lobed, the lobes spreading to reflexed. Perianth cream to yellowish, often pink- or purple-tinged, glabrous externally, stipitate, the segments spatulate. Akenes pubescent above the middle.

Slopes, ridges, moraines, talus, and scree of all alpine ranges of the Middle Rocky Mountains. British Columbia south to Oregon, Nevada, and Colorado.

Three varieties occur in the Middle Rocky Mountains:

1. Flowers cream-colored, often purple-tinged
 . . . var. *majus* Hook.
1. Flowers bright yellow
    2. Leaves tomentose beneath, glabrous to variously pubescent above . . . var. *umbellatum*
    2. Leaves glabrous or glabrate on both sides
 . . . var. *porteri* (Small) Stokes

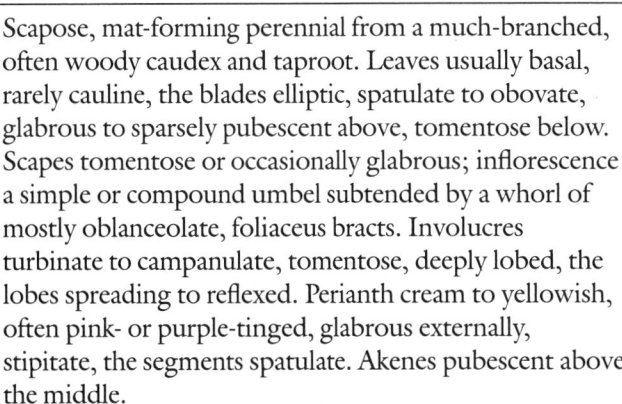

var. umbellatum

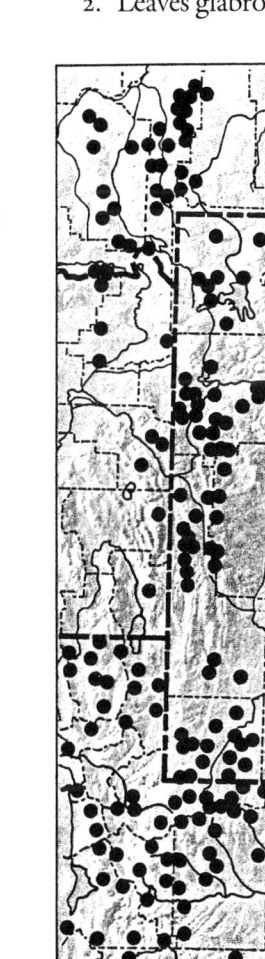

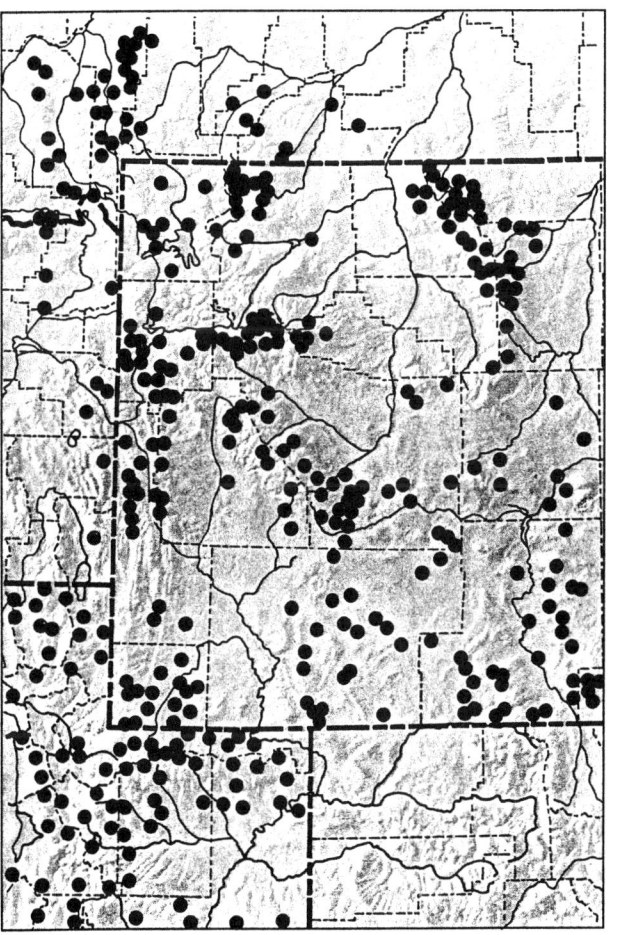

Koenigia L.
Koenigia

Low, dwarf annuals from fibrous roots, with slender, spreading stems; the leaves sessile, obovate or oblanceolate, sometimes lacking. Flowers greenish white, in terminal, involucrate clusters or single. Perianth 3-parted. Stamens 3. Akenes trigonous.

Koenigia islandica L.
Mant. 1:35, 1767.

Iceland Koenigia

Macounastrum islandicum (L.) Small

Tiny, filiform annual, less than 5 cm tall; stems glabrous, often reddish. Leaves opposite or nearly so, oblanceolate to obovate. Flowers 1 to a few, in axillary or terminal clusters. Perianth 3–4-lobed, the segments ovate, obtuse. Stamens 3. Styles 2. Akenes trigonous, dark brown.

Wet nival basins, patterned ground, stream banks, and lakeshores of the Beartooth and Wind River Ranges. Circumpolar; south in the Rocky Mountains to east-central Colorado.

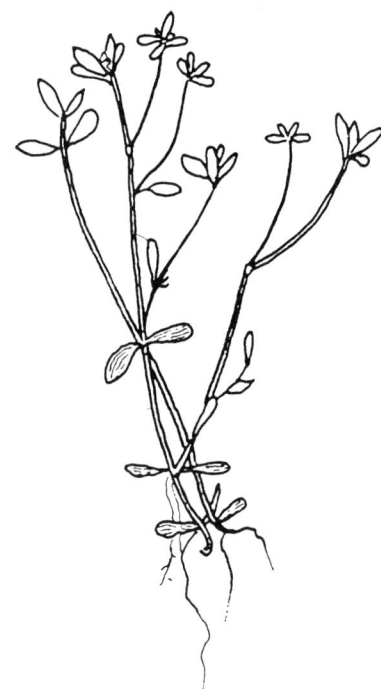

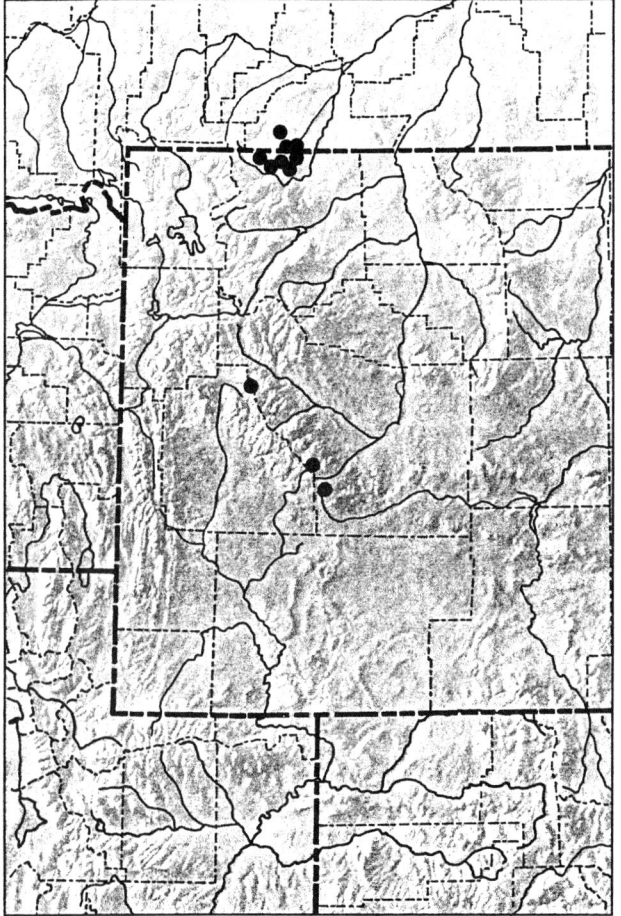

Oxyria Hill
Mountain Sorrel

Erect, glabrous, perennial herbs with sour juice, from thick, fleshy taproots; leaves long-petioled, entire, reniform to orbicular, mostly basal, with small sheathing stipules. Flowers small, greenish to reddish, clustered in terminal panicles; involucre none. Perianth of 4 segments, the outer pair larger than the inner pair. Stamens 6. Stigmas 2, tufted, persistent on the wings in fruit. Akenes lenticular, broadly winged.

Oxyria digyna (L.) Hill
Hort. Kew., 158, 1768.

Alpine Sorrel

Rumex digynus L.

Glabrous, perennial herb from a thick, fleshy taproot; stems simple or few-branched. Leaves long-petiolate, mostly basal, the blades cordate to reniform, margins undulate. Inflorescence a raceme or panicle. Perianth segments 4, reddish or greenish, the 2 outer becoming reflexed, the 2 inner ones keeled, erect. Stamens 6. Styles 2. Akenes orbicular, flat, broadly winged.

Rocky areas, ledges, talus, fellfields, glacial moraines, and outwash in all alpine ranges. Circumpolar; south in western North America to California, Arizona, and New Mexico.

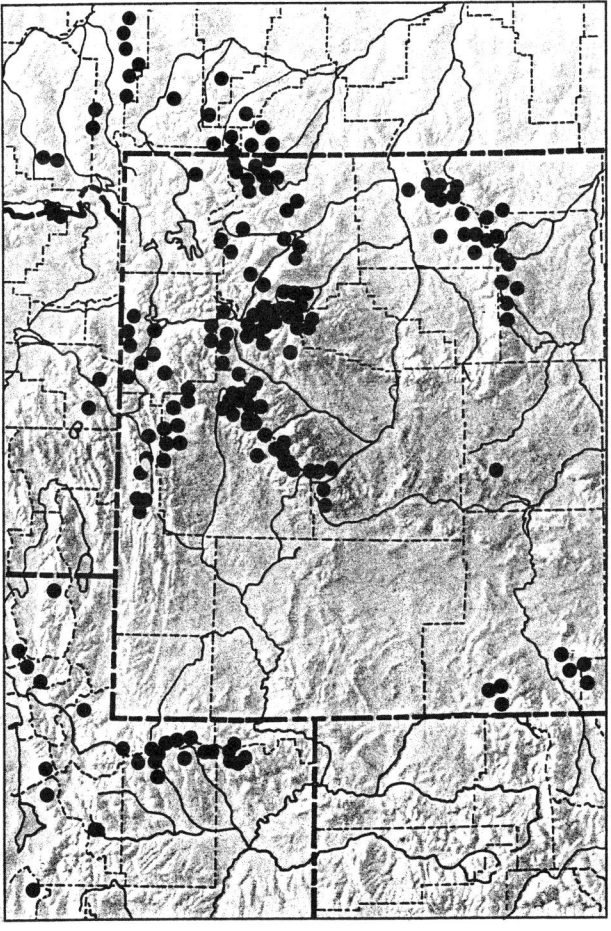

Polygonum L.
Knotweed

Annual or perennial, terrestrial or amphibious, sometimes aquatic herbs. Stems erect, prostrate or twining, usually swollen at the joints. Leaves subsessile or petiolate, with sheathing stipules. Flowers perfect, axillary or in spikelike terminal racemes, the perianth 4–6-parted, the outer segments often developing keels or conspicuous wings in fruit. Stamens 3–9, the filaments slender, often dilated at the base. Styles 2 or 3, distinct or partly united; the stigmas usually capitate. Akenes lenticular or trigonous, partly or wholly enclosed by the persistent perianth.

KEY TO THE SPECIES OF *POLYGONUM*

1. Plants perennial, from short rhizomes
   2. Racemes loose, narrowly cylindrical, 5–8 mm thick; flowers dimorphic, the lower ones reduced to ovoid bulbils ... *P. viviparum*
   2. Racemes dense, oblong, 10 mm or more thick; flowers all alike, the lower not reduced to bulbils ... *P. bistortoides*
1. Plants annual, from a taproot
   3. Pedicels of older flowers reflexed; inflorescence a loose, elongate raceme ... *P. douglasii*
   3. Pedicels of older flowers ascending; inflorescence a compact, short raceme, or flowers axillary
      4. Stems angled; leaves narrowly linear, about 1 mm broad; flowers in a crowded terminal cluster; akenes dark brown, striate ... *P. kelloggii*
      4. Stems terete; leaves lanceolate to obovate, more than 2 mm broad; flowers axillary; akenes black, smooth
         5. Leaves ovate or elliptic, less than twice as long as broad, scarcely reduced upward ... *P. minimum*
         5. Leaves lanceolate to narrowly oblong, more than twice as long as broad, reduced upward ... *P. sawatchense*

Polygonum bistortoides Pursh
Fl. Am. Sept., 271, 1814.

American Bistort

Bistorta bistortoides (Pursh) Small
Polygonum cephalophorum Greene
Polygonum glastifolium Greene
Polygonum linearifolium (S. Wats.) Greene
Polygonum vulcanicum Greene

Slender perennial from a starchy, horizontal to ascending rhizome with fibrous roots; stems simple, erect. Basal leaves long-petiolate, the blades elliptic, oblong, or oblanceolate, cuneate at the base, cauline leaves sessile, lanceolate, truncate or weakly cordate at base, reduced upward; stipular sheaths brownish, oblique at the apex, entire. Flowers in a crowded, solitary, terminal, spikelike raceme. Perianth white to rose, the segments 5, oblong, connate below. Stamens 8, exerted. Styles 3. Akenes trigonous, light brown, smooth, and shiny.

Alpine meadows and stream banks in nearly all ranges of the Middle Rocky Mountains; not yet reported from above timberline in the Hoback Range. British Columbia south to California and New Mexico.

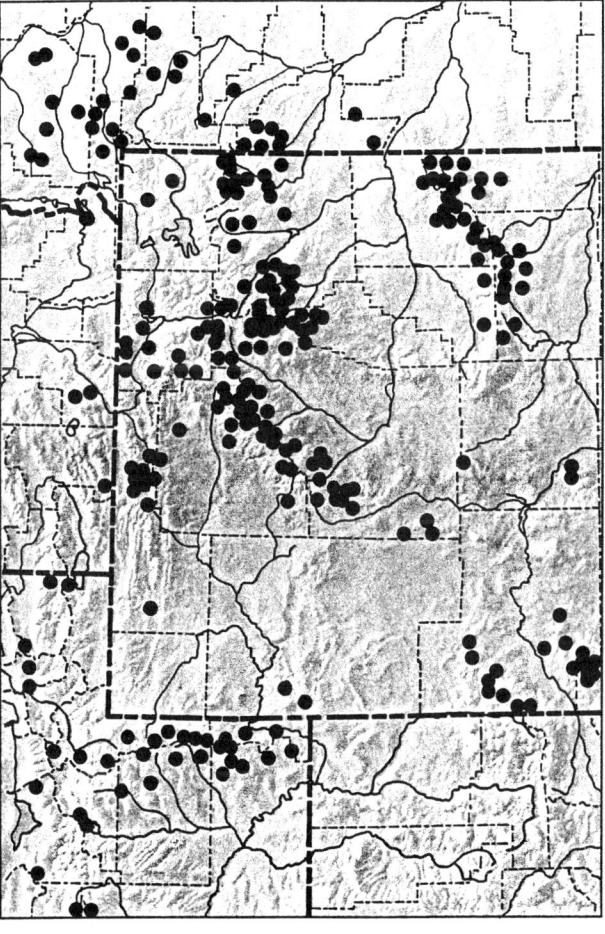

Polygonum douglasii Greene
Bull. Calif. Acad. Sci. 1:125, 1885.

Douglas Knotweed

Polygonum austinae Greene
Polygonum emaciatum A. Nels.
Polygonum engelmannii Greene
Polygonum microspermum (Engelm.) Small
Polygonum montanum Small
Polygonum pannosum Sharp

Slender annual with simple to branched, erect or ascending stems from a taproot. Leaves linear to oblong, sessile or short-petiolate, cuneate at the base, reduced upward; stipular sheaths short, lacerate. Flowers in loose terminal racemes, and sometimes upper leaf axils, the pedicels reflexed after anthesis. Perianth green with white or rose margins, the segments 5, oblanceolate, obtuse, distinct nearly to the base. Stamens 8. Styles 3. Akenes trigonous, black, smooth, and shiny.

Known from alpine tundra only in the Beartooth, Medicine Bow, Teton, and Wasatch Ranges. Nearly transcontinental in Canada; south in the western United States to New Mexico and Arizona.

Two varieties are to be expected as occasional in the lower alpine zone throughout the region: var *douglasii* has linear leaves less than 4 mm wide; var *latifolium* (Engelm.) Greene has oblong leaves 4 mm or more wide.

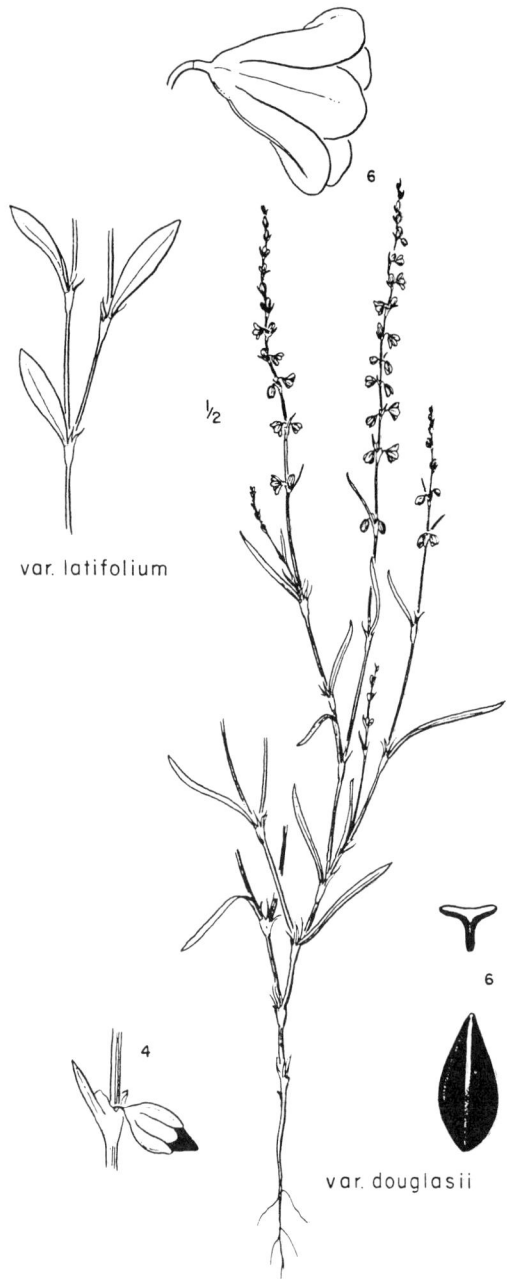

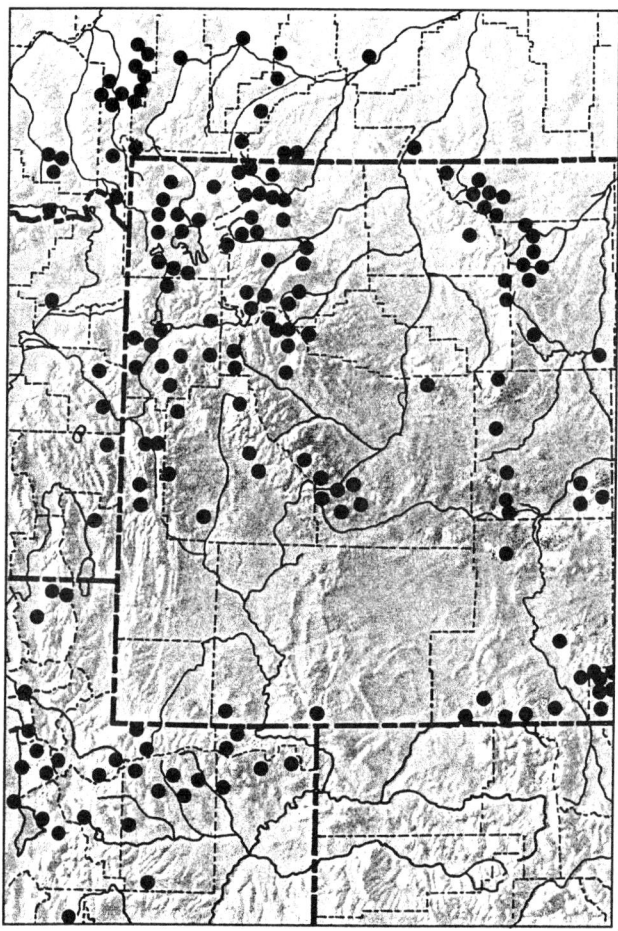

Polygonum kelloggii Greene
Fl. Franc., 134, 1891.

Kellogg Knotweed

Polygonum confertiflorum Nutt. ex Piper
Polygonum minutissimum Williams
Polygonum unifolium Small

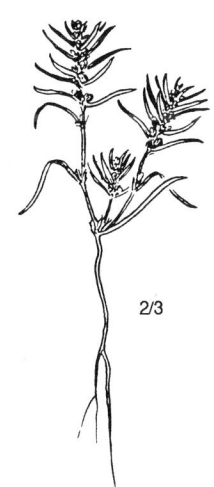

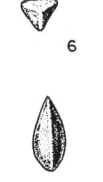

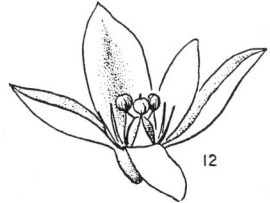

Small, glabrous annual, less than 6 cm tall in the alpine zone, from a slender taproot; stems angled, simple or branched from the base. Leaves linear to linear-lanceolate, acute, spreading to slightly reflexed. Flowers in a terminal, bracteate spike, and often in the lower leaf axils. Perianth green with white or pink margins, the segments 5, subequal, connate below. Anther-bearing stamens 3. Stigmas 3, the styles lacking or nearly so. Akenes trigonous, light brown, somewhat striate or at least granular.

Known from alpine tundra and willow thickets in the Beartooth, Big Horn, Medicine Bow, Uinta, Wasatch, and Wind River Ranges. British Columbia south to California, Arizona, and Colorado.

Middle Rocky Mountain alpine plants may be distinguished as follows:
1. Bracts of the upper inflorescence with very narrow white margins, the tips acute . . . var. *kelloggii*
1. Bracts of the upper inflorescence with prominent white margins, the tips obtuse
. . . var. *confertiflorum* (Nutt. ex Piper) Dorn

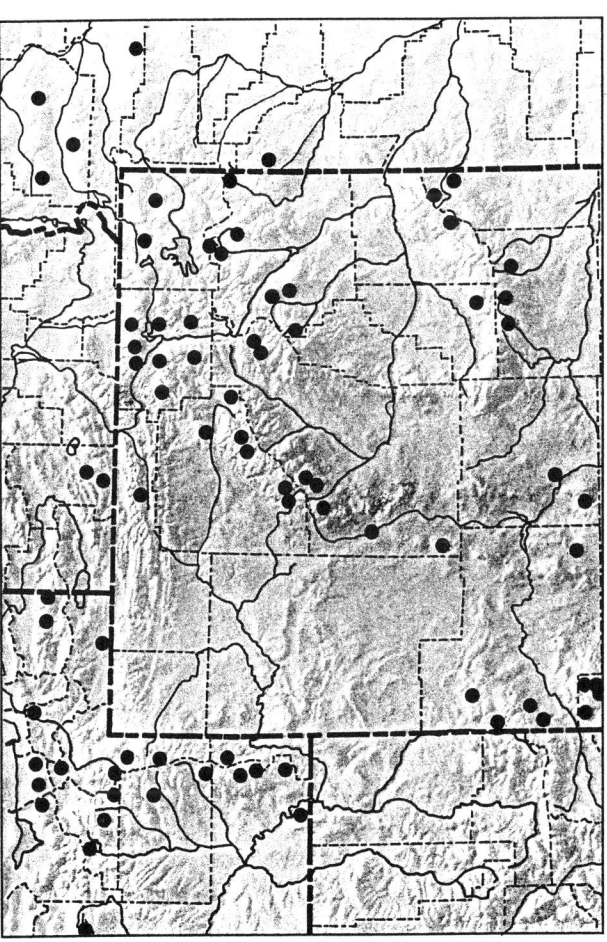

Polygonum minimum S. Wats.
Bot. King Exped., 315, 1871.

Broadleaf Knotweed

Annual with spreading-ascending stems branched from the base or occasionally simple, from a slender taproot. Leaves elliptic, ovate, or obovate; sessile or subsessile, scarcely reduced upward, often crowded at the end of branches; stipules lacerate. Flowers mostly 1–3 per axil. Perianth green with white or pink margins, the 5 segments connate at the base. Stigmas 3, styles lacking or nearly so. Akenes trigonous, black, smooth, and shiny.

Alpine meadows and rocky ledges of the lower alpine zone near timberline in the Absaroka, Uinta, Wasatch, and Wind River Ranges. British Columbia south to California, Idaho, Utah, and Colorado.

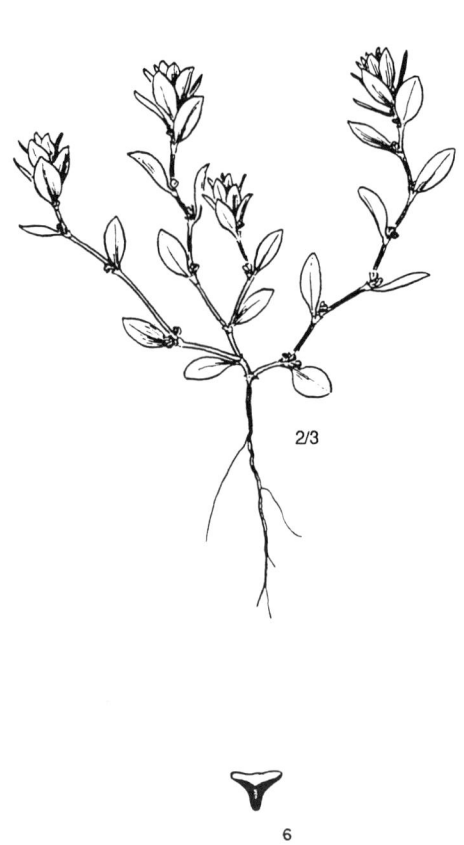

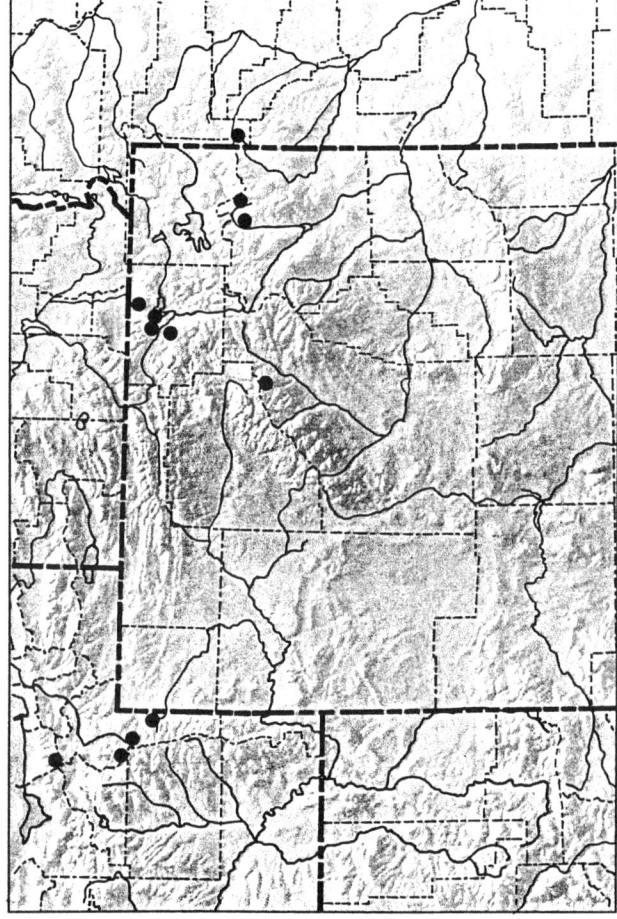

Polygonum sawatchense Small
Bull. Torr. Club 20:213, 1893.

Sawatch Knotweed

Slender, erect annual with ascending, somewhat 4-angled stems branched from the base or simple, from a taproot. Leaves oblanceolate, oblong, or elliptic; acute, sessile, somewhat revolute, reduced upward. Stipules lacerate. Flowers 1–4 per axil. Perianth green, often with lighter margins, the segments 5, connate below the middle. Stamens 6–8. Stigmas 3, the styles lacking or nearly so. Akenes trigonous, black, smooth, and shiny.

Known in the alpine zone from the Beartooth, Medicine Bow, and Salt River Ranges. Washington south to California, Colorado, and Nebraska.

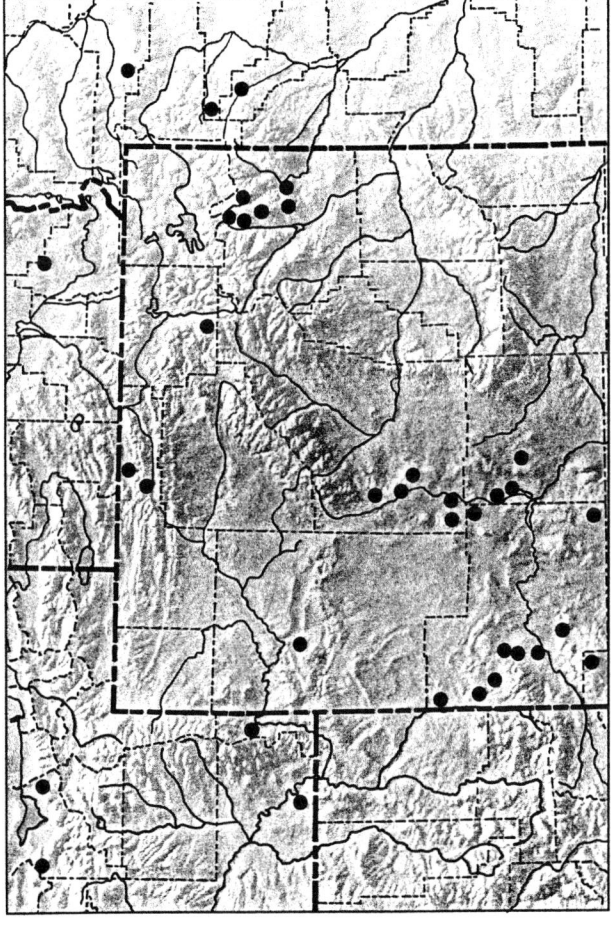

Polygonum viviparum L.
Sp. Pl., 360, 1753.

Alpine Bistort

Bistorta littoralis Greene
Bistorta macounii (Small) Greene
Bistorta ophioglossa Greene
Bistorta vivipara (L.) S.F. Gray
Polygonum fugax Small
Polygonum macounii Small ex Macoun

Slender perennial from a contorted, starchy, ascending to erect rhizome with fibrous roots; stems simple, erect. Leaves mostly basal, long-petiolate, the blades oblong to lanceolate, obtuse to acutish; cauline leaves linear to lanceolate, sessile above, somewhat revolute; stipules sheathing, brownish, oblique, entire. Inflorescence a terminal, spikelike raceme, the lower flowers replaced by purplish bulbils; upper flowers functionally imperfect. Perianth white to pink, the segments 5, ovate, connate at the base. Stamens 8, exserted. Styles 3. Akenes trigonous, dark brown, granular, dull.

Moist stream banks, meadows, and willow thickets in all but the Hoback, Salt River, and Wyoming Ranges. Circumpolar; south in western North America to New Mexico.

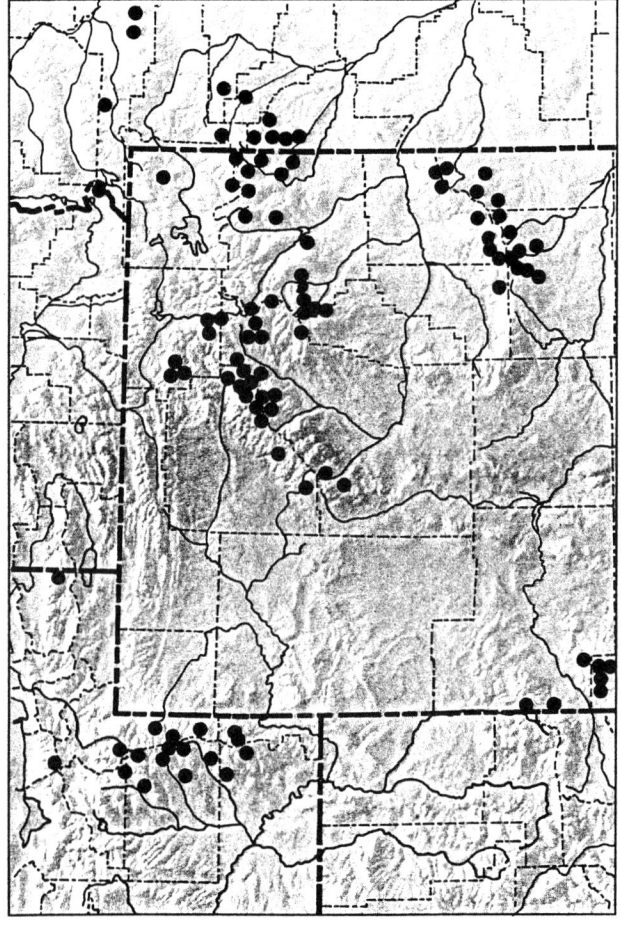

Rumex L.
Dock, Sorrel

Perennial herbs with leafy stems and simple, alternate, entire or wavy-margined leaves, the stipules sheathing, usually membranous. Flowers small, brownish or greenish, perfect or unisexual, in panicles. Perianth segments 6, the outer 3 segments not enlarged in fruit, the inner 3 segments often enlarged and usually winged. Stamens 6. Styles 3, the stigmas peltate. Akenes sharply triquetrous, usually smooth, brownish to black.

KEY TO THE SPECIES OF *RUMEX*

1. Flowers mostly unisexual; plants dioecious
    2. Leaves sagittate; outer perianth segments reflexed in fruit — *R. acetosa*
    2. Leaves elliptic to oblanceolate; outer perianth segments not reflexed in fruit — *R. paucifolius*
1. Flowers mostly perfect
    3. Plants from stout rhizomes — *R. densiflorus*
    3. Plants from taproots, lacking rhizomes — *R. salicifolius*

Rumex acetosa L.
Sp. Pl., 337, 1753.

Garden Sorrel

Acetosa alpestris (Jacq.) A. Löve
Lapathum alpestre Scop.
Rumex alpestris Jacq.
Rumex thyrsiflorus Fingerhuth

Dioecious perennial from a taproot, lacking rhizomes; stems simple, 1 to several, glabrous, somewhat succulent. Basal leaves long-petiolate, the blades sagittate; cauline leaves sessile, somewhat clasping. Inflorescence a panicle, the flowers imperfect. Perianth reddish, the outer segments strongly reflexed, lanceolate; inner segments erect, cordate, entire, with a small basal callosity. Akenes triquetrous, dark brown, shiny.

Meadows, snowmelt seeps, and stream banks in the Beartooth Range. Circumpolar; south in western North America to Oregon, Montana, and Wyoming.

Plants that just reach the Middle Rocky Mountains from the north and have ovate-triangular leaves, with somewhat divergent basal lobes, and broad, open sinuses are ssp. *alpestris* (Scop.) Löve.

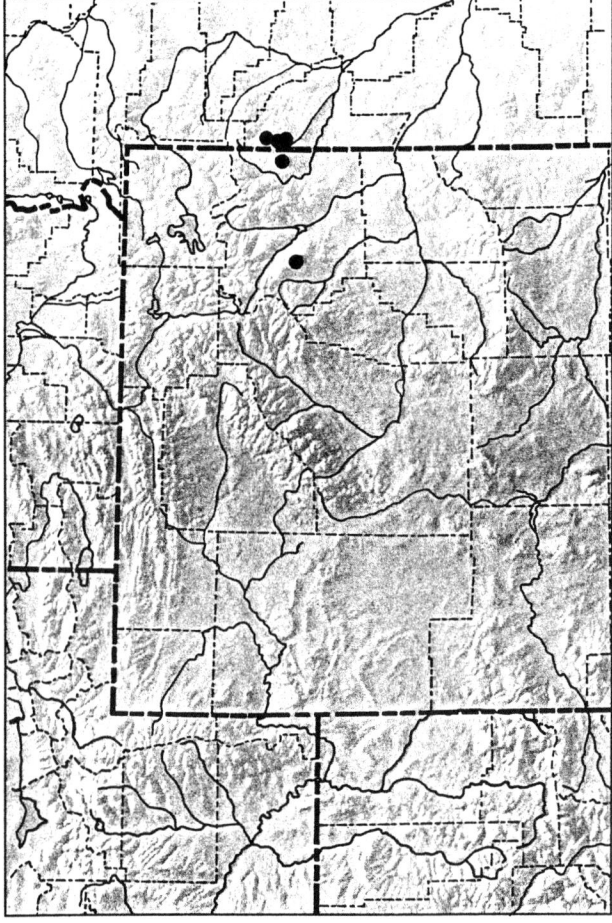

Rumex densiflorus Osterh.
Erythea 6:13, 1898.

Denseflower Dock

Rumex praecox Rydb.
Rumex pycnanthus Rech. f.

Stout, glabrous perennial from rhizomes; stems erect, grooved. Basal leaves petiolate, the blades elliptic to ovate; cauline leaves oblong to ovate. Flowers polygamous, in a dense panicle. Perianth reddish, the outer segments spreading, the inner ones erect, ovate, acute, denticulate at least near the base, callosities lacking. Akenes triquetrous.

Alpine tundra in the Medicine Bow Range. Endemic to Colorado, southern Wyoming, and southern Idaho.

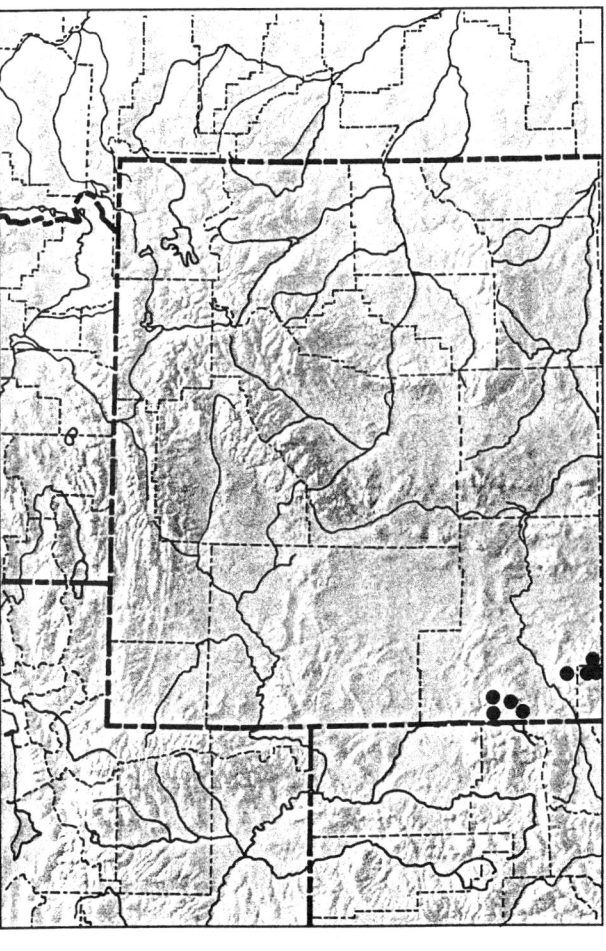

Rumex paucifolius Nutt.
J. Acad. Phila. 7:49, 1834.

Mountain Sorrel

Acetosa gracilescens (Rech. f.) A. Löve & Evenson
Acetosella paucifolius (Nutt.) A. Löve
Rumex geyeri (Meisn.) Trel.

Dioecious perennial from a taproot, lacking rhizomes; stems 1 to several, glabrous, erect, and simple. Leaves mostly basal, petiolate, the blades elliptic to obovate, cuneate at the base; cauline leaves lanceolate, sessile, reduced upward. Inflorescence a panicle of imperfect flowers, the branches erect. Perianth yellowish green or reddish; outer segments lanceolate, spreading, but not reflexed; inner segments cordate, entire, erect; callosities lacking. Akenes triquetrous, smooth.

Alpine meadows in the Absaroka, Beartooth, Big Horn, Salt River, Uinta, Wasatch, and Wind River Ranges. British Columbia and Alberta south to California and Colorado.

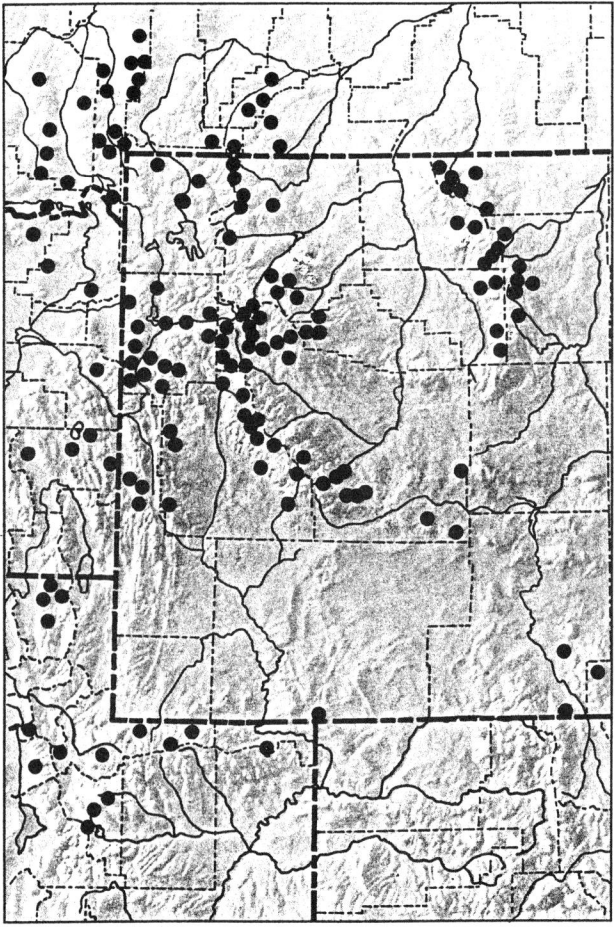

Rumex salicifolius Weinm.
Flora 4:28, 1821.

Beach Dock

Lapathum mexicanum (Meisn.) Nieuwl.
Rumex californicus Rech. f.
Rumex crassus Rech. f.
Rumex hesperius Greene
Rumex lacustris Greene
Rumex mexicanus Meisn.
Rumex quadrangulivalvis (Danser) Rech f.
Rumex transitorius Rech. f.
Rumex triangulivalvis (Danser) Rech f.
Rumex utahensis Rech. f.

Glabrous perennial from a taproot, lacking rhizomes; stems 1 to several, erect, branched above the base. Leaves petiolate near the base to subsessile or sessile above, narrowly elliptic, lanceolate, or oblanceolate, acute, entire or shallowly undulate. Inflorescence a leafy-bracteate panicle, the branches ascending; flowers mostly perfect. Perianth greenish to reddish; outer segments lanceolate, acute, spreading; inner segments ovate to deltoid, erect, entire or nearly so; callosities lacking. Akenes triquetrous, dark brown, smooth.

Meadows in the vicinity of timberline in the Absaroka, Medicine Bow, and Wasatch Ranges. Alaska south to California, Nevada, Utah, and Colorado.

Middle Rocky Mountain plants are ssp. *triangulivalvis* Danser.

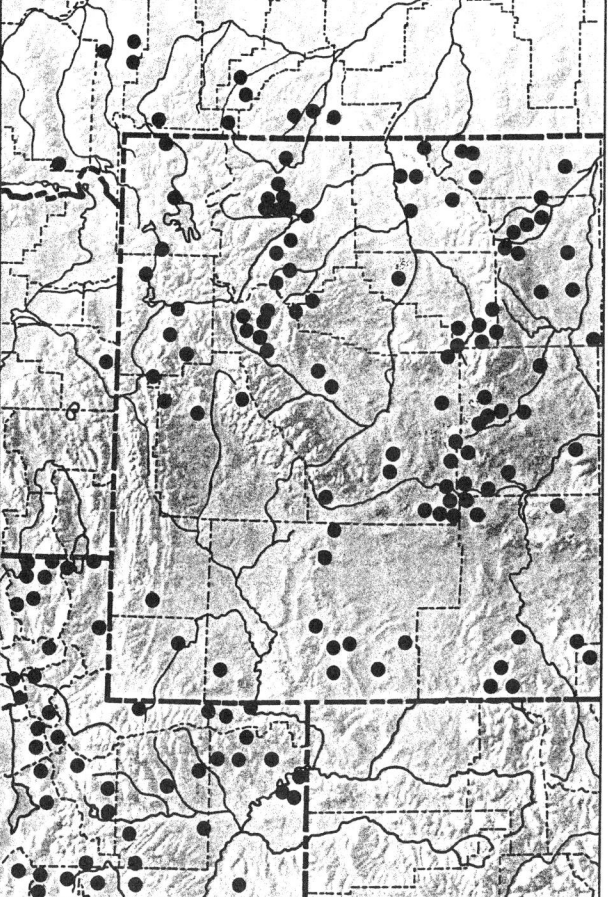

ssp. triangulivalvis

Portulacaceae
Purslane Family

Small, annual or perennial, succulent herbs from fleshy or fibrous roots, the leaves simple, entire, alternate or opposite, often basal. Flowers hypogynous, perfect, regular, mostly ephemeral. Sepals normally 2, distinct or united below, the lower sepal usually overlapping the upper. Petals 3–8, usually 5, entire to emarginate. Stamens commonly 5, opposite the petals. Pistil 1, of 2–3 carpels; the ovary 1-celled; the styles and stigmas 2–8, the styles often united below. Fruit a circumscissile, or 2–3-valved loculicidal, capsule with 2 to many black, shiny seeds.

KEY TO THE GENERA OF PORTULACACEAE

1. Petals 4; stamens 3; flowers in dense, umbellate clusters — *Spraguea*
1. Petals and stamens 5 or more; flowers solitary, racemose, or paniculate
   2. Leaves linear to narrowly oblanceolate in ours; capsules circumscissile; petals usually 6–8; stigmas frequently more than 3 — *Lewisia*
   2. Leaves lanceolate to obovate; capsules valvate; petals 5; stigmas 3 — *Claytonia*

Claytonia L.
Springbeauty

Low, glabrous, succulent, perennial herbs from fleshy taproots or corms, with 1 to several entire, basal leaves or, mostly, 2 exstipulate, cauline leaves. Flowers white or pink, in loose terminal racemes. Sepals 2, persistent, ovate. Petals 5, equal. Stamens 5, adnate to the petals. Ovary 1-celled, the style 3-cleft. Fruit a 3-valved capsule with 2–6 seeds.

KEY TO THE SPECIES OF *CLAYTONIA*

1. Basal leaves densely clustered, spatulate or obovate; plants from a thick, fleshy root — *C. megarhiza*
1. Basal leaves few or lacking, the 2 cauline leaves lanceolate to ovate; plants from a subglobose corm — *C. lanceolata*

Claytonia lanceolata Pursh
Fl. Am. Sept., 175, 1814.

Lanceleaf Springbeauty

Claytonia aurea A. Nels.
Claytonia chrysantha Greene
Claytonia flava A. Nels.
Claytonia multicaulis A. Nels.
Claytonia multiscapa Rydb.
Claytonia rosea Rydb.

Delicate, glabrous, perennial herb from a globose corm; stems 1 to several, glabrous. Basal leaves lanceolate to ovate, long-petioled, sometimes lacking at anthesis; cauline leaves 2, opposite, sessile, lanceolate to ovate. Flowers 3–15, in short, somewhat secund racemes; pedicels becoming recurved with age. Sepals 2, ovate to oval, obtuse, fleshy. Petals white to pink in ours, with pink or purple veins, obovate, rounded to emarginate at the apex. Stamens 5, opposite the petals and adnate to them at the base. Ovary with 3 styles. Capsule ovoid, the seeds black and shiny.

Moist meadows and edges of receding snowfields in most alpine ranges of the Middle Rocky Mountains. Not yet known from above timberline in the Hoback, Salt River, Uinta, and Wasatch Ranges. British Columbia and Alberta south to California and New Mexico.

Alpine plants are var. *lanceolata* and have ovate to ovate-lanceolate leaves less than 5 times as long as broad, and corms less than 2 cm broad.

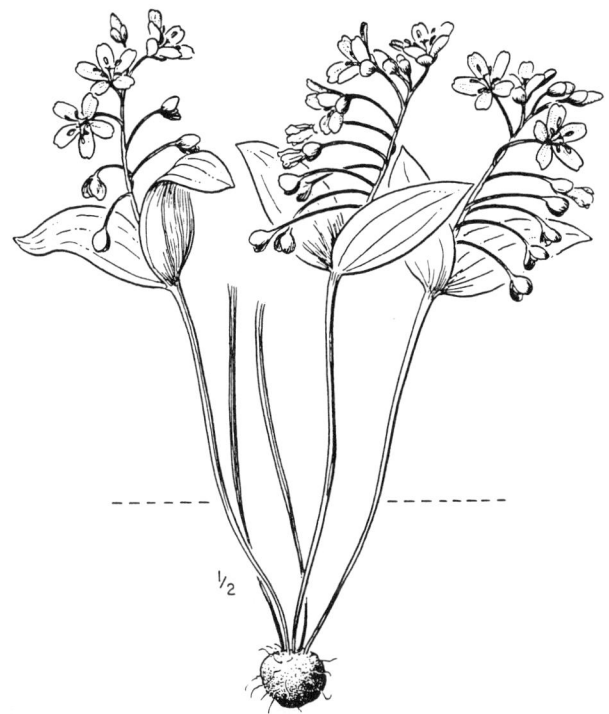

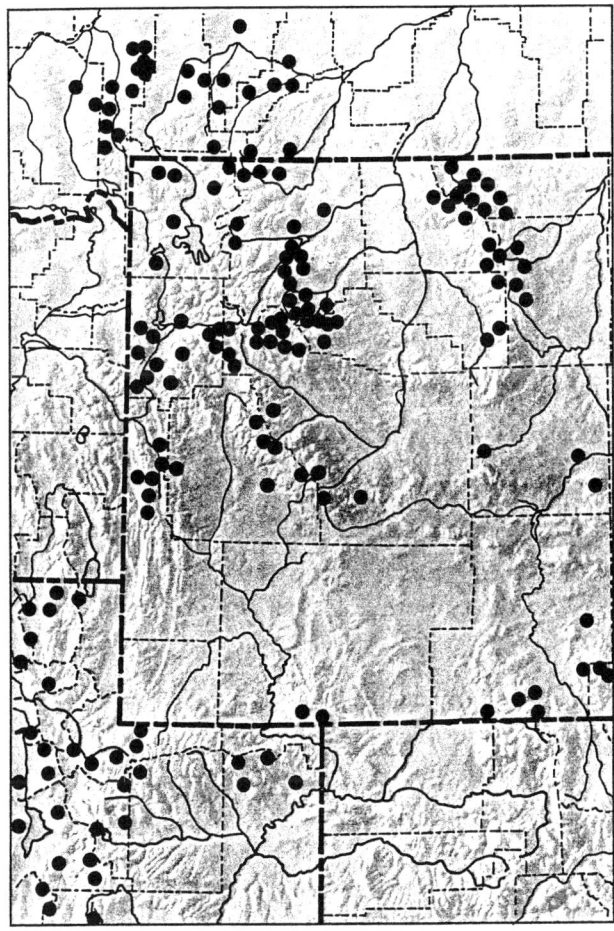

Claytonia megarhiza (Gray) Parry ex S. Wats.
Bibl. Ind. 1:118, 1878.

Alpine Springbeauty

Claytonia bellidifolia Rydb.
Claytonia nivalis English

Low, succulent, perennial herb from a thick, elongate, simple or branched taproot; stems 1 to several, short, about the same length as the basal rosette of leaves. Basal leaves fleshy, numerous, the blades spatulate, petioles broadly winged; cauline leaves 2, opposite, linear to oblanceolate. Flowers mostly 2–6, in a bracteate raceme. Sepals 2, ovate, obtuse to acute. Petals white to pink, obovate, shallowly emarginate, clawed. Stamens 5, adnate to the petals at the base. Ovary with 3 styles. Capsule ovoid, the seeds black, shiny.

Scree, talus, nivation basins, and late snow beds in all alpine ranges except the Hoback, Salt River, Wasatch, and Wyoming Ranges. Alberta south to Montana, Wyoming, Washington, Oregon, and New Mexico.

Alpine plants of the Middle Rocky Mountains are var. *megarhiza*, with white to pink flowers, basal leaves 1–1.5 cm broad, and broadly winged petioles.

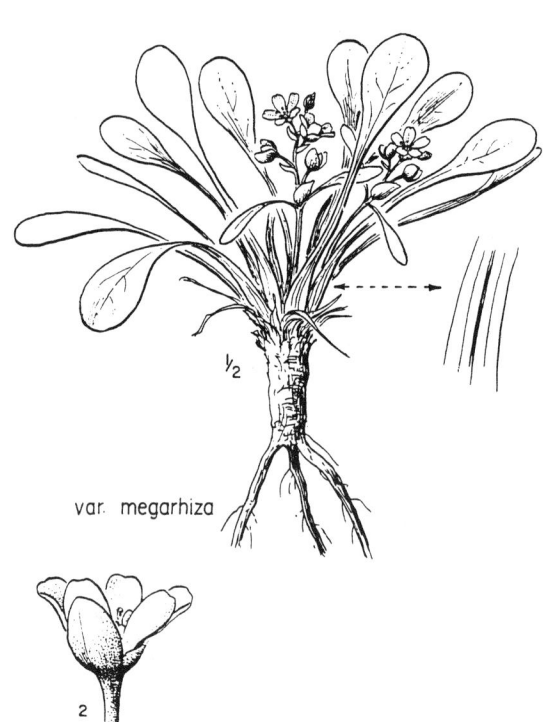

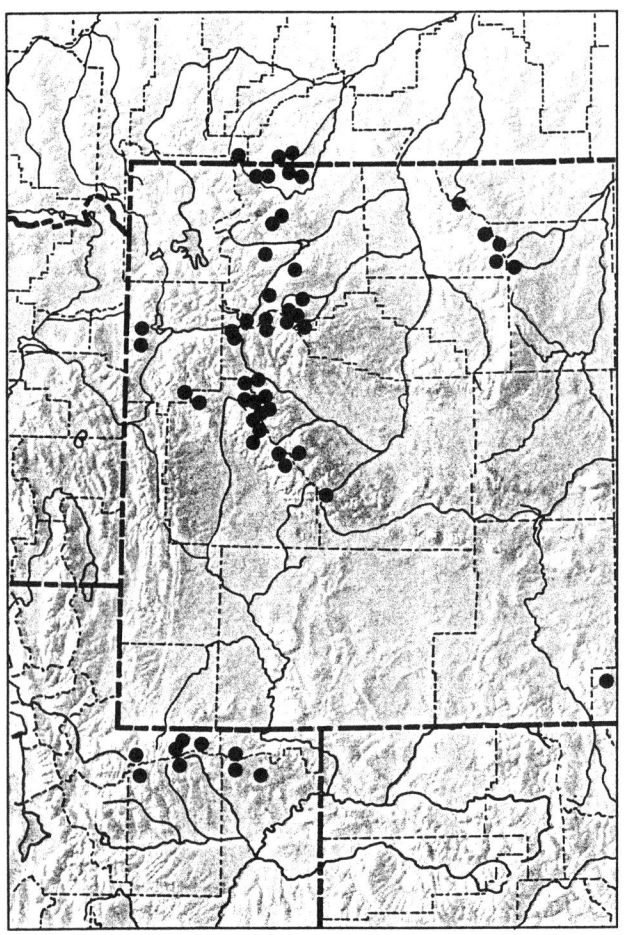

Lewisia Pursh
Bitterroot

Low, fleshy, perennial herbs from fleshy taproots or globose corms, the leaves mostly basal, spatulate to narrowly linear. Flowering scapes 1–3-flowered, usually shorter than or equaling the leaves, the flowers white or pink. Sepals 2–8. Petals 3–16. Ovary 1-celled, the style branches 3–8, slender. Fruit a globose to ovate capsule, circumscissile near the base and sometimes splitting upward by the valves.

KEY TO THE SPECIES OF *LEWISIA*

1. Plants from globose corms; basal leaves usually lacking — *L. triphylla*
1. Plants from elongate taproots; basal leaves present — *L. pygmaea*

Lewisia pygmaea (Gray) Robins. in Gray
Syn. Fl. 1(1):268, 1897.

Pygmy Bitterroot

Talinum pygmaeum Gray
Calandrinia grayi Britt.
Calandrinia nevadensis Gray
Calandrinia pygmaea (Gray) Gray
Claytonia grayana (Britt.) Kuntze
Lewisia aridorum (A. Heller) Clay
Lewisia exarticulata St. John
Lewisia glandulosa (Rydb.) Dempster
Lewisia minima (A. Nels.) A. Nels.
Lewisia nevadensis (Gray) Robins.
Oreobroma aridorum A. Heller
Oreobroma exarticulatum (St. John) Rydb.
Oreobroma grayi (Britt.) Rydb.
Oreobroma minima A. Nels.
Oreobroma nevadensis (Gray) T.J. Howell
Oreobroma pygmaea (Gray) T.J. Howell

Low, scapose, perennial herb from a thick, fleshy taproot. Leaves linear to oblanceolate, several to numerous, the petioles with membranous margins. Scapes 1-flowered, bracteate at or below the middle; the bracts 2, opposite, basally connate. Sepals 2, ovate, obtuse, glandular-serrulate, strongly veined and often reddish. Petals mostly 7, white to pink, oblanceolate. Stamens 4–9. Style branches 3–5. Capsule ovoid, the seeds dark brown, shiny.

Dry meadows, fellfields, and rocky places in most of the alpine ranges. Not yet known from above timberline in the Hoback and Salt River Ranges. Alaska south to California, Arizona, and New Mexico.

Middle Rocky Mountain alpine plants with glandular-serrulate, veined sepals and mostly linear leaves are var. *pygmaea*.

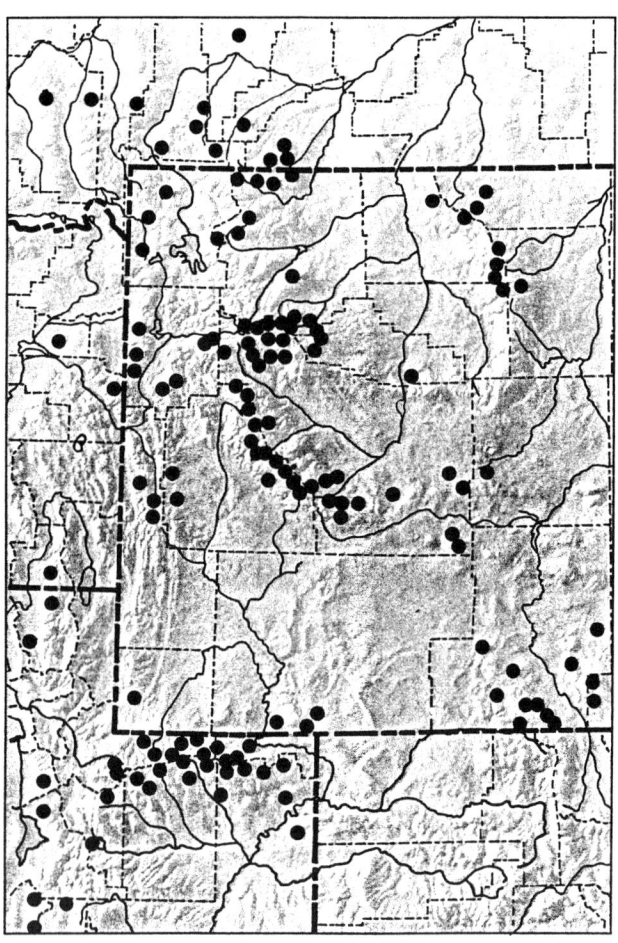

Lewisia triphylla (S. Wats.) Robins. in Gray
Syn. Fl. 1(1):269, 1897.

Threeleaf Bitterroot

Claytonia triphylla S. Wats.
Erocallis triphylla (S. Wats.) Rydb.
Oreobroma triphylla (S. Wats.) T.J. Howell

Slender, glabrous, perennial herb from a globose corm; stems 1 to several. Basal leaves linear, withering by flowering time; cauline leaves linear, usually in a whorl of 3. Flowers 2–15, in a bracteate, corymbose or paniculate inflorescence. Sepals 2, oval, entire, obtuse. Petals 5–8, white to pink, with pinkish veins. Stamens mostly 5. Style branches 3–5. Capsule ovoid, the seeds dark brown, shiny.

Meadows and moist rocky places in the lower alpine zone of the Gros Ventre, Teton, Uinta, and Wasatch Ranges. Washington to California and east to Montana, Wyoming, and Colorado.

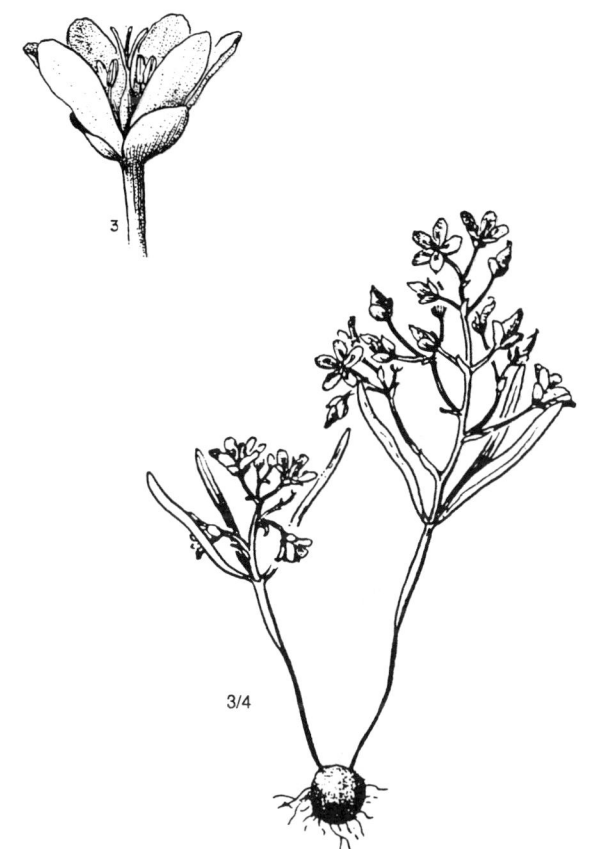

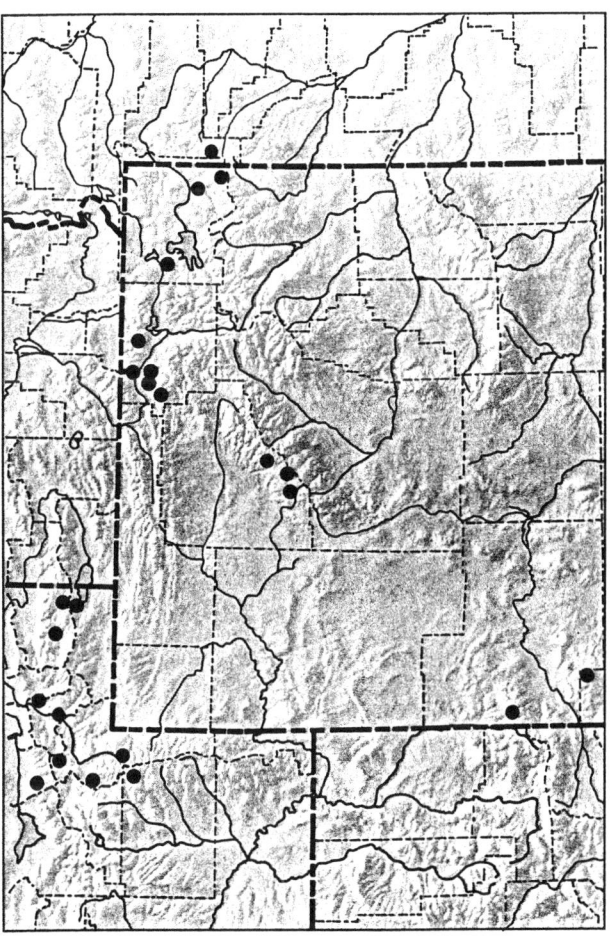

Spraguea Torr.
Pussypaws

Low, glabrous, perennial herbs with mostly basal, oblanceolate to obovate leaves. Flowers ephemeral, numerous, in dense, globose, umbellate clusters. Sepals 2, scarious-margined, subequal, much larger than the 4 petals. Stamens 3, opposite the larger petals. Style slender, the stigma 2-lobed. Fruit a 2-valved capsule.

Spraguea umbellata Torr. in Smith
Contr. Knowl. 6:4, 1853.

Umbellate Pussypaws

Calyptridium umbellatum (Torr.) Greene
Cistanthe umbellata (Torr.) Hershkovitz
Spraguea multiceps T.J. Howell

Annual to perennial herb from a simple to branched caudex. Leaves clustered basally, petiolate, the blades oblanceolate to spatulate, entire. Flowers in terminal, capitate heads on mostly leafless peduncles. Sepals 2, orbicular, white to pinkish. Petals 4, pink to white, obovate, shorter than the sepals. Stamens 3, exerted. Capsule ovoid, 1–2-seeded.

Scree, talus, and other rocky places in the Absaroka, Beartooth, Gros Ventre, Teton, Uinta, Wasatch, and Wind River Ranges. British Columbia to California and east to Montana, Wyoming, and Colorado.

Middle Rocky Mountain plants are var. *caudicifera* Gray.

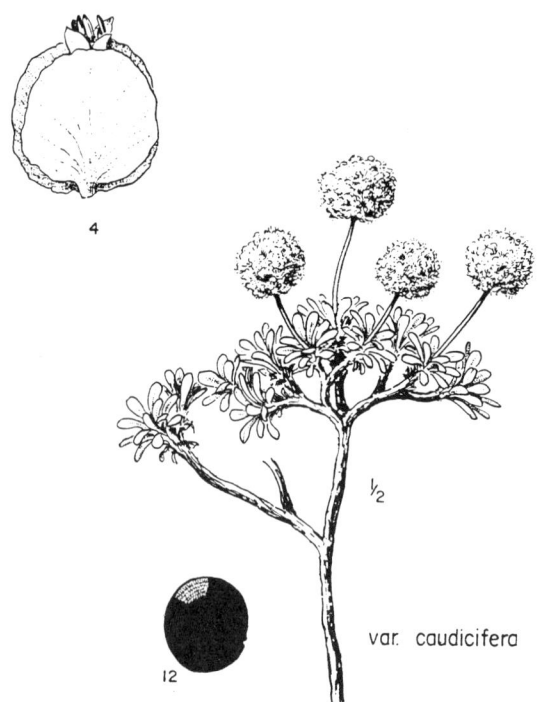

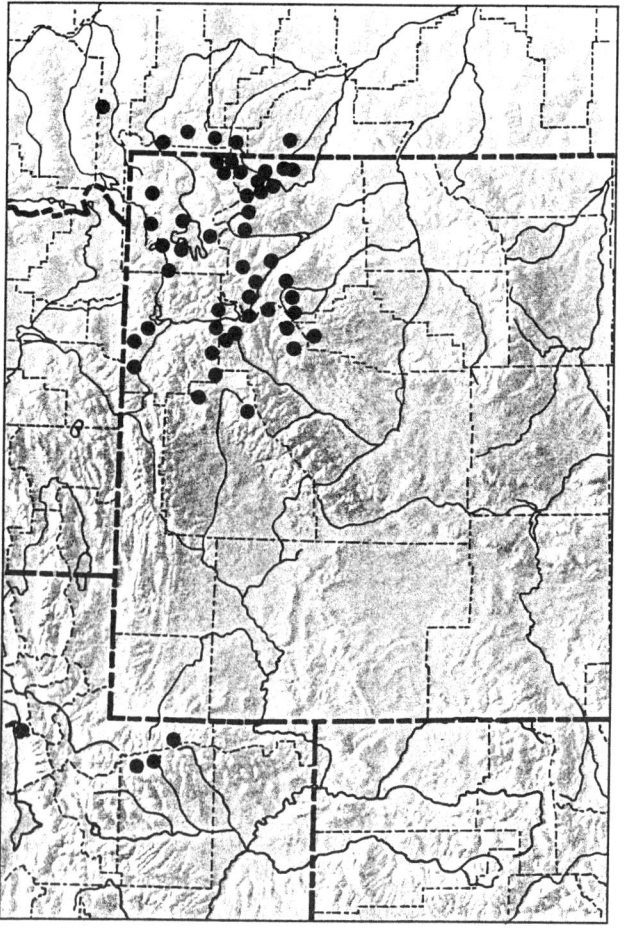

630 Portulacaceae

Primulaceae
Primrose Family

Scapose or sometimes caulescent, annual or perennial herbs with simple leaves. Flowers hypogynous, 5-merous, perfect, regular, complete in ours, axillary and solitary or, usually, in terminal racemes, umbels, or corymbs. Calyx shallowly to deeply lobed. Corolla salverform, funnelform, or rotate, shallowly lobed to deeply cleft, the lobes usually 5. Stamens 5, inserted on the corolla tube opposite the lobes. Pistil 1, the ovary 1-celled, of 5 united carpels, with free-central placentation; the style single; the stigma usually capitate. Fruit a capsule, dehiscing by 5 teeth or valves, or less commonly circumscissile.

KEY TO THE GENERA OF PRIMULACEAE

1. Corolla lobes reflexed, several times longer than the tube; stamens exserted, connivent — *Dodecatheon*
1. Corolla lobes erect or spreading, less than twice as long as the tube; stamens included, not connivent
   2. Flowers white, less than 5 mm long, the corolla tube shorter than the calyx — *Androsace*
   2. Flowers pink to reddish purple, more than 5 mm long, the corolla tube equaling or exceeding the calyx
      3. Plants caespitose, with small, narrow leaves 5–25 mm long — *Douglasia*
      3. Plants erect, the leaves 4–25 cm long — *Primula*

Androsace L.
Rockjasmine

Low, annual or perennial, scapose herbs with 1 to several flowering stems. Leaves clustered in a basal rosette. Flowers small, white, in involucrate umbels. Calyx turbinate to hemispheric, persistent, 5-lobed. Corolla 5-lobed, salverform or funnelform, the tube shorter than the calyx, sometimes constricted in the throat, the lobes spreading, shorter than the tube. Stamens 5, included. Ovary 1-celled, the style very short, the stigma capitate. Fruit a 5-valved capsule, more or less enclosed by the persistent calyx, the seeds few to many.

KEY TO THE SPECIES OF *ANDROSACE*

1. Plants pilose perennials; scapes 1 from each rosette; petals cream-colored — *A. chamaejasme*
1. Plants glabrous to sparsely pubescent annuals; scapes 1 to several from each rosette; petals white — *A. septentrionalis*

Androsace chamaejasme Wulfen in Jacq.
Coll. Bot. 1:194, 1787.

Dwarf Rockjasmine

Androsace albertina Rydb.
Androsace bungeana Schischk. & Bobr.
Androsace carinata Torr.
Androsace lehmanniana Spreng.
Drosace albertina Rydb.
Drosace carinata (Torr.) A. Nels.

Low, perennial herb, less than 8 cm tall, from prostrate stems with terminal rosettes of leaves; scapes single from each rosette, densely villous to pilose. Leaves glabrous above, pubescent below, oblanceolate, entire, ciliate. Flowers few to several, in compact, bracteate umbels; pedicels shorter than the flowers. Calyx lobes ovate to lanceolate, the teeth shorter than the tube. Corolla cream to whitish, with a yellow center, the tube about the same length as the calyx; capsule less than 3 mm long.

Rocky exposed ridges and slopes, mostly on limestone, in the Absaroka and Wind River Ranges. Circumboreal; Alaska, Yukon Territory and Mackenzie district of Northwest Territories south to Colorado and Utah.

Middle Rocky Mountain plants are var. *carinata* (Torr.) Knuth.

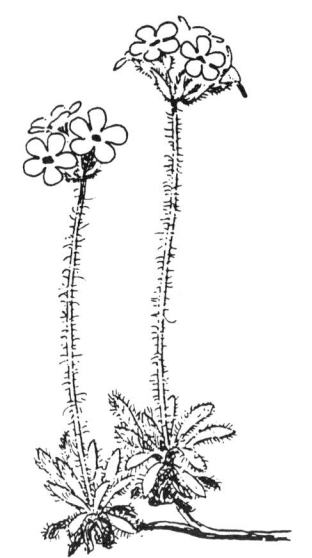
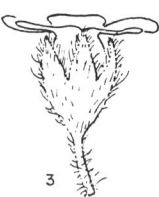

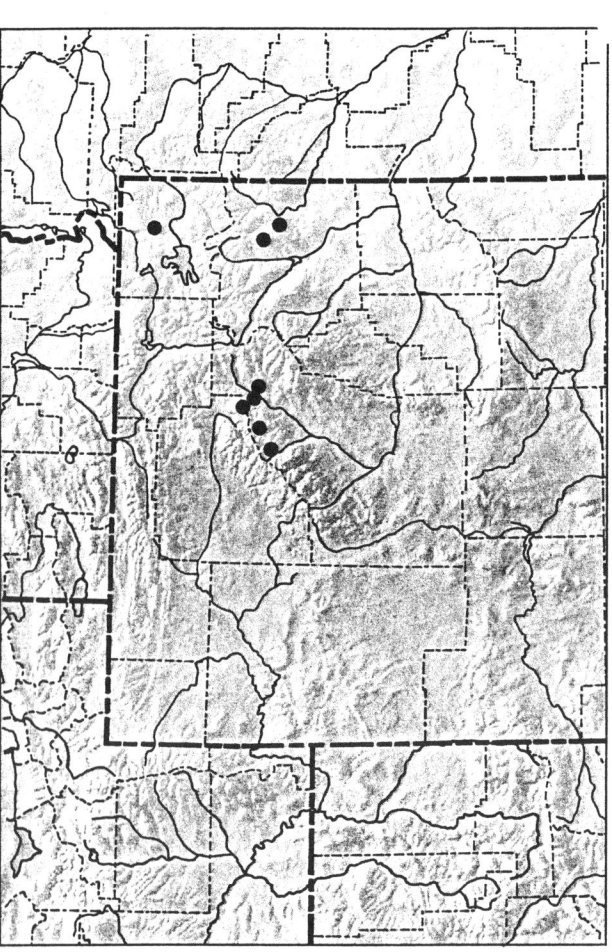

Androsace septentrionalis L.
Sp. Pl., 142, 1753.

Pygmyflower Rockjasmine

Amadea diffusa (Small) Lunell
Amadea puberulenta (Rydb.) Lunell
Androsace arguta Greene
Androsace diffusa Small
Androsace glandulosa Woot. & Standl.
Androsace gormani Greene
Androsace puberulenta Rydb.
Androsace subulifera Rydb.
Androsace subumbellata (A. Nels.) Small

Low, pubescent to nearly glabrous annual, less than 12 cm tall in the alpine zone, from a slender taproot. Leaves in a basal rosette, oblanceolate to linear, entire to denticulate, pubescent, the hairs simple, forked, or stellate. Scapes 1 to several, glabrate to pubescent or glandular. Flowers in involucrate umbels, the bracts linear to lanceolate; pedicels slender, much longer than the flowers. Calyx campanulate, 5-ridged, the lobes lanceolate to narrowly deltoid. Corolla white, shorter than or equaling the calyx. Capsule globose, about as long as the calyx.

Widespread in meadows, fellfields, nivation basins, and late snow beds throughout alpine areas of the Middle Rocky Mountains. Circumpolar; south in western North America to California, Arizona, and New Mexico.

Two varieties occur in the alpine areas of the Middle Rocky Mountains: the var. *subumbellata* A. Nels., having glabrous or sparsely pubescent scapes and pedicels, appears to be the more common; the var. *puberulenta* (Rydb.) Knuth, with densely puberulent scapes and pedicels, is known only from the Wind River Range.

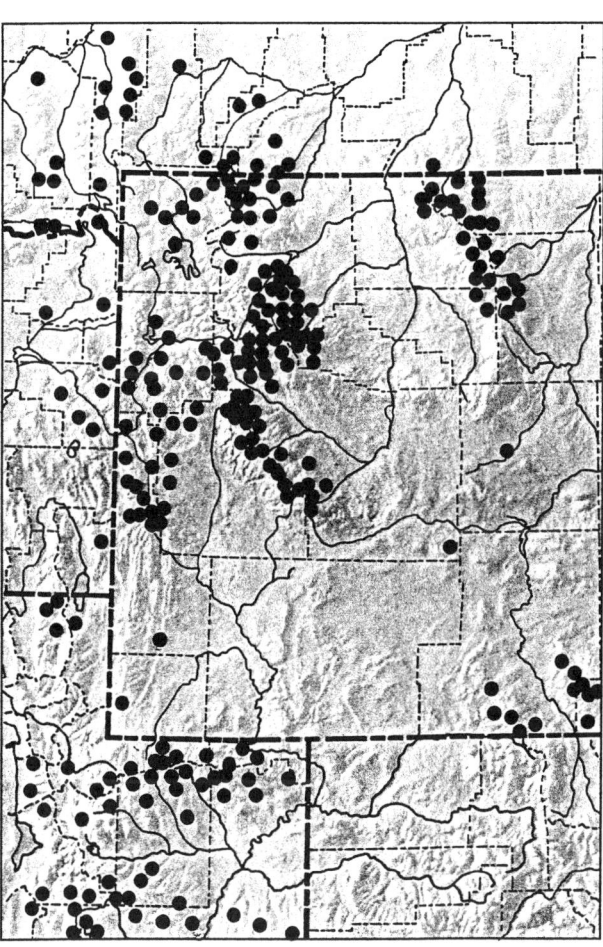

Dodecatheon L.
Shootingstar

Erect, scapose, perennial herbs, commonly glabrous, from fibrous or fleshy roots. Leaves petioled, simple, entire to dentate, in a basal rosette. Flowers showy, pink or purple, nodding on slender pedicels in a terminal, involucrate umbel. Calyx short-tubular, 5-parted, the lobes lanceolate. Corolla 5-cleft nearly to the base, the lobes reflexed at anthesis, the tube very short. Stamens 5, inserted on the corolla tube, the filaments short, distinct or more commonly united into a tube, the anthers connivent around the style. Ovary 1-celled, the style filiform, the stigma capitate. Fruit a valvate or operculate capsule with numerous seeds.

KEY TO THE SPECIES OF *DODECATHEON*

1. Corolla and calyx lobes 4; stigmas capitate — *D. alpinum*
1. Corolla and calyx lobes 5; stigmas not capitate
    2. Connectives of anthers transversely wrinkled; capsules operculate, the teeth truncate — *D. conjugens*
    2. Connectives of anthers smooth or longitudinally wrinkled; capsules valvate, the teeth acute — *D. pulchellum*

Dodecatheon alpinum (Gray) Greene
Erythea 3:39, 1895.

Alpine Shootingstar

Scapose, glabrous to sparsely gandular-pubescent, perennial herb from a short, erect rootstock and fibrous roots; scapes single, slender, glabrous to glandular-pubescent. Leaves gradually narrowed to the petioles, the blades oblanceolate, the tips rounded; margins entire or sinuate. Flowers single, or 2–9 in a terminal, involucrate umbel. Calyx glabrous, the lobes lanceolate. Corolla rose-pink to purple, the tube yellowish, the lobes reflexed. Filaments distinct or united into a tube; connectives purplish black, transversely wrinkled; anthers purplish black. Stigma capitate, much larger in diameter than the style. Capsule oblong-ovoid, valvate.

Alpine meadows and stream banks in the vicinity of timberline in the Uinta and Wasatch Ranges. Oregon south to California, Utah, and Arizona.

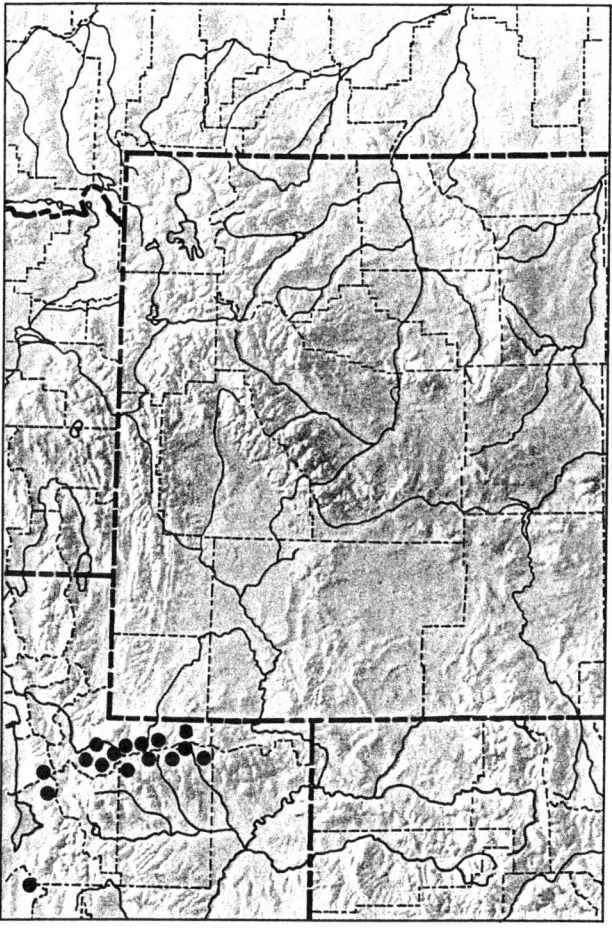

Dodecatheon conjugens Greene
Erythea 3:40, 1895.

Sailorcaps Shootingstar

Dodecatheon acuminatum Rydb.
Dodecatheon albidum Greene
Dodecatheon campestrum T.J. Howell
Dodecatheon cylindrocarpum Rydb.
Dodecatheon glastifolium Greene
Dodecatheon pubescens Rydb.
Dodecatheon pulchrum Rydb.
Dodecatheon viscidum Piper

Scapose, glabrous to somewhat pubescent perennial from fibrous roots, the scape single, slender. Leaves petiolate, the blades entire, lanceolate to oblanceolate or obovate. Flowers single, or 2–8 in terminal, involucrate umbels. Calyx glabrous, the lobes lanceolate. Corolla rose-pink to purple, rarely white, the tube yellowish, the lobes strongly reflexed. Filaments short, free or nearly so, yellowish or purplish; connectives red or purple, transversely wrinkled; anthers yellow to purple. Stigma the same diameter as the style. Capsule ovoid or nearly so, operculate.

Typical of montane meadows, *D. conjugens* occurs above timberline in the Absaroka and Teton Ranges and can be expected in other ranges near timberline. British Columbia and Alberta south to Wyoming and California.

Glabrous plants are var. *conjugens;* those with pubescence on the scape and leaves are var. *viscidum* (Piper) Mason ex St. John.

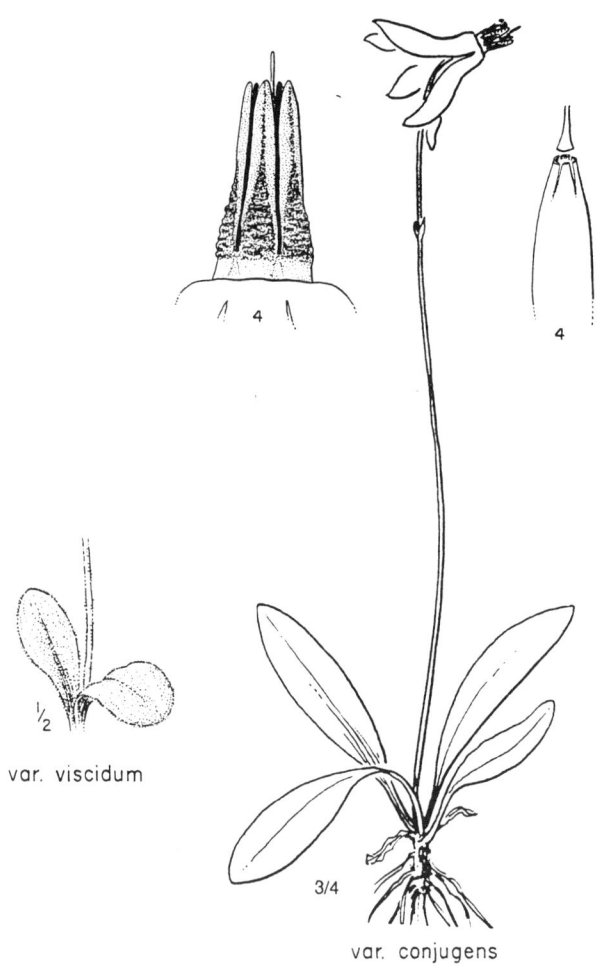

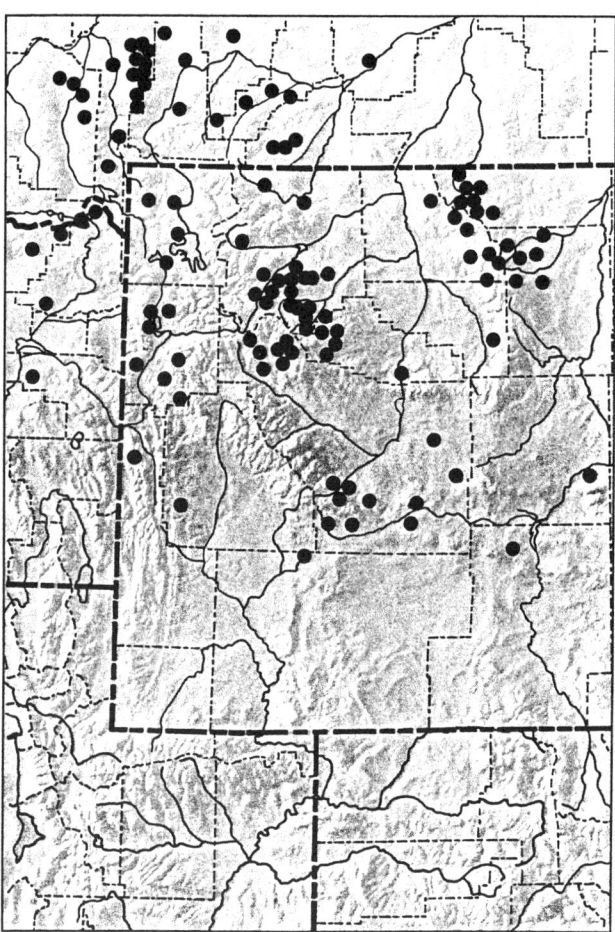

636 Primulaceae

Dodecatheon pulchellum (Raf.) Merr.
J. Arnold Arb. 29:212, 1948.

Darkthroat Shootingstar

Exinia pulchella Raf.
Dodecatheon amethystinum Fassett
Dodecatheon cusickii Greene
Dodecatheon meadia L.
Dodecatheon multiflorum Rydb.
Dodecatheon pauciflorum (Durand) Greene
Dodecatheon philoscia A. Nels.
Dodecatheon puberulentum A. Heller
Dodecatheon puberulum (Nutt.) Piper
Dodecatheon radicatum Greene
Dodecatheon salinum A. Nels.
Dodecatheon sinuatum Rydb.
Dodecatheon superbum Pennell & Stair
Dodecatheon thornense Lunell
Dodocatheon uniflorum Rydb.
Dodecatheon watsonii Tidestr.
Dodecatheon zionense Eastw.
Meadia cusickii (Greene) Kuntze
Meadia salina (A. Nels.) Lunell

Scapose, mostly glabrous perennial from a short, erect rootstock and fibrous roots; scapes single, glabrous or occasionally glandular-pubescent. Leaves petiolate, the blades mostly oblanceolate to obovate, usually entire. Flowers single, or 2–5 in terminal, involucrate umbels. Calyx glabrous, the lobes lanceolate to subulate, usually purple-spotted. Corolla rose pink to purple, rarely white, the tube yellowish, the lobes reflexed. Filaments united into a yellowish tube; connectives dark purple, smooth or longitudinally wrinkled; anthers yellow to dark purple. Stigma the same diameter as the style. Capsule nearly ovoid, glabrous to glandular-pubescent, valvate.

Alpine meadows and stream banks of the Absaroka, Beartooth, Big Horn, Teton, Uinta, Wasatch, and Wind River Ranges. Alaska to Alberta, North Dakota, and New York, south in the western United States to Colorado and Nebraska; Mexico.

Alpine plants with corolla lobes less than 12 mm long and leaves mostly less than 10 cm long are var. *pulchellum*.

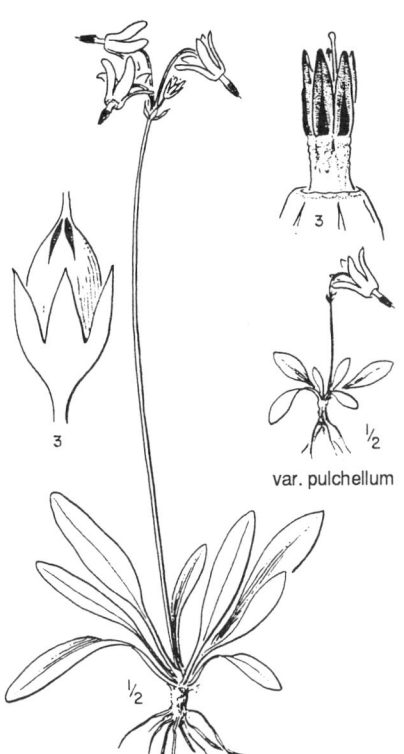

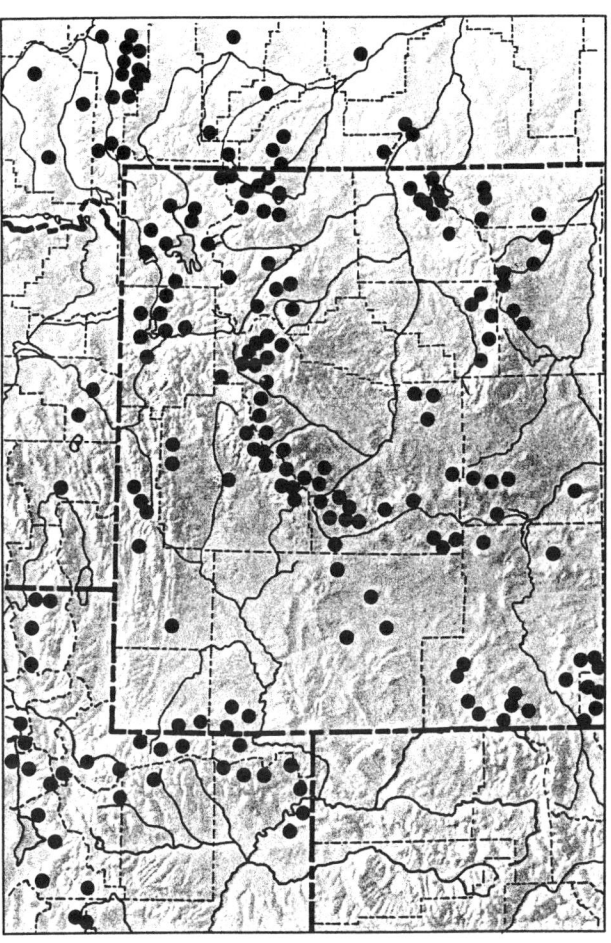

Douglasia Lindl.
Douglasia

Low, caespitose, perennial herbs, the peduncles and pedicels with very fine, forked or stellate pubescence, the stems prostrate or ascending, terminating in rosettes of small, entire or dentate leaves. Flowers red or pink, solitary, or several together in involucrate umbels. Calyx campanulate, 5-cleft to the middle, persistent. Corolla funnelform, the tube nearly equaling the calyx, the lobes spreading, the throat constricted and bearing 5 appendages alternate with the lobes. Stamens included, inserted on the upper third of the corolla tube, the filaments shorter than the anthers. Ovary 1-celled, the style filiform. Capsules subglobose, 5-valved from the apex, the seeds usually 2–3 by abortion.

Douglasia montana Gray
Proc. Am. Acad. 7:371, 1868.

Mountain Douglasia

Androsace uniflora Haussk.
Douglasia biflora A. Nels.
Gregoria montana (Gray) House
Primula montana (Gray) Derganc

Low, scapose, mat-forming perennial, less than 5 cm tall. Scapes pubescent with branched hairs, 1 to several from a basal rosette. Leaves linear-lanceolate, the margins scabrous-serrulate. Flowers single, rarely 2, terminal on the scape, sometimes with 1 or 2 bracts at the base. Calyx about as long as the corolla tube, 5-ridged, pubescent or glabrous, often reddish. Corolla dark pink, salverform, the lobes oblong, retuse. Capsules 1–5-seeded, equaling or slightly exceeding the calyx tube.

Alpine turf, fellfields, felsenmeer, and scree slopes of the Absaroka, Beartooth, Big Horn, Wind River, and Wyoming Ranges. British Columbia and Alberta south to Wyoming and Idaho.

3/4

2

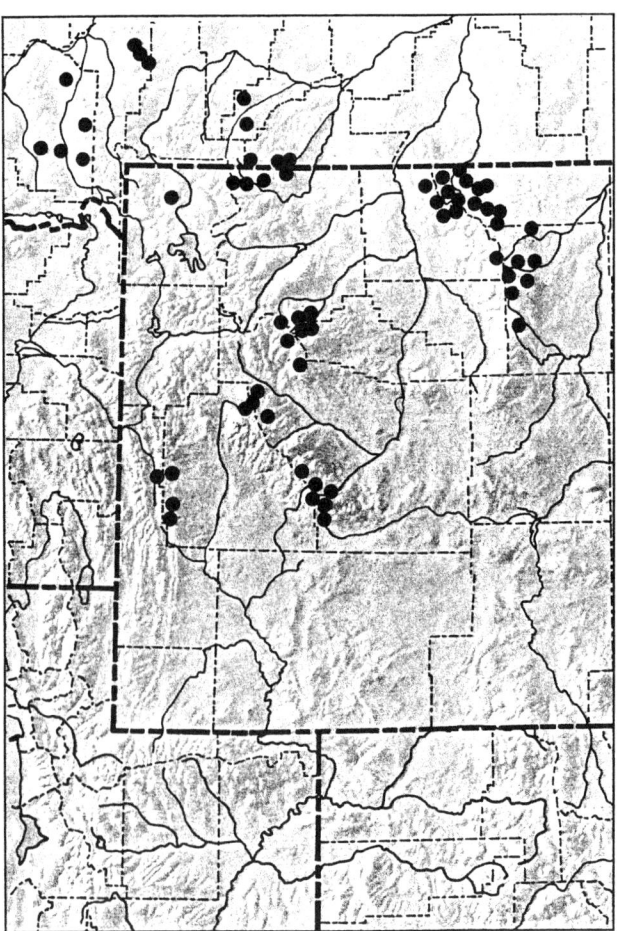

Primula L().
Primrose

Scapose, perennial herbs, the leaves in a basal rosette, simple, exstipulate, sometimes farinose beneath. Flowers red to pink or lavender, in an involucrate, terminal umbel. Calyx persistent, 5-lobed. Corolla salverform or funnelform, the tube longer than the calyx, the lobes emarginate. Stamens 5, included, inserted on the upper third of the corolla tube, the filaments short. Ovary 1-celled, the style filiform, the stigma capitate. Fruit a 5-valved, many-seeded capsule.

Primula parryi Gray
Am. J. Sci. 34:257, 1862.

Parry Primrose

Primula mucronata Greene

Scapose, fleshy, somewhat viscid perennial from fascicled roots. Leaves erect to ascending, oblanceolate, obtuse or acute, the margins entire to dentate. Flowers 3–10, in a terminal, involucrate umbel, the bracts lanceolate. Calyx campanulate, about equaling the corolla tube, the lobes lanceolate-acuminate. Corolla dark pink to reddish purple, with a yellow center, salverform, the lobes obovate, retuse. Capsule ovoid, about the same length as the calyx.

Fellfields, cirque walls, ledges, stream banks, and snow beds in all alpine areas of the Middle Rocky Mountains. Montana and Idaho south to Arizona and New Mexico.

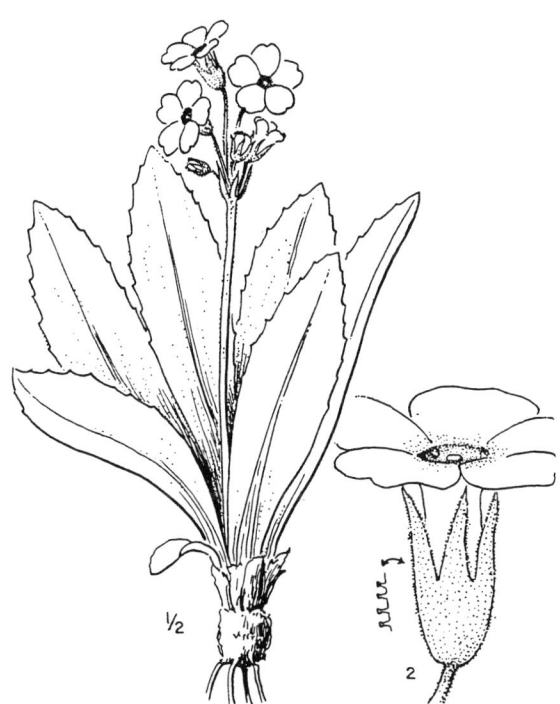

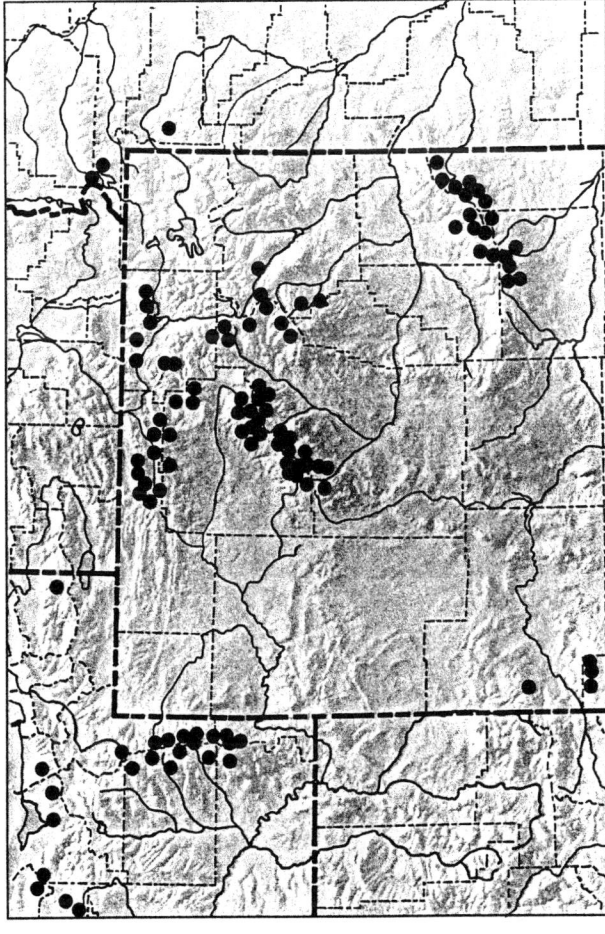

Ranunculaceae
Buttercup Family

Mostly perennial herbs, or sometimes shrubby climbers, with colorless, acrid juice and alternate, rarely opposite, simple or compound, exstipulate leaves. Flowers hypogynous, usually perfect, regular to irregular, solitary or in racemes or panicles. Sepals 3 to many, often petaloid and sometimes spurred. Petals none or indefinite, sometimes with a nectariferous claw at the base, rarely spurred. Stamens numerous and spirally arranged or cyclic. Fruit a berry, or a group of akenes or follicles.

KEY TO THE GENERA OF RANUNCULACEAE

1. Flowers irregular; sepals deep blue, the upper sepal spurred — *Delphinium*
1. Flowers regular; the sepals variously colored, the upper sepal not spurred
    2. Carpels 1-ovuled; fruit a group of akenes
        3. Cauline leaves (involucre) in whorls of 3; sepals large, showy, 7–30 mm long — *Anemone*
        3. Cauline leaves alternate, rarely opposite; sepals small, less than 7 mm long
            4. Petals present, yellow or white; plants with perfect flowers — *Ranunculus*
            4. Petals lacking, the sepals greenish; plants often dioecious — *Thalictrum*
    2. Carpels with several ovules; fruit a group of follicles or a berry
        5. Petals conspicuous, spurred at the base — *Aquilegia*
        5. Petals lacking or inconspicuous, never spurred at the base
            6. Petals small, linear; leaves palmately parted or cleft — *Trollius*
            6. Petals lacking; leaves entire or merely toothed — *Caltha*

Anemone L.
Anemone

Erect, perennial herbs with long-stalked basal leaves and 2–9 opposite or whorled cauline leaves (bracts) forming an involucre below the flower. Flowers white, cream, or purple, solitary or sometimes umbellate. Sepals petaloid, the petals wanting. Stamens numerous. Pistils numerous, the fruit a hemispherical or globose head of flattened, ribless akenes.

KEY TO THE SPECIES OF ANEMONE

1. Sepals 2.5–4 cm long; styles plumose, 2–4 cm long at maturity — *A. patens*
1. Sepals less than 2 cm long; styles rarely plumose, less than 1.5 cm long
    2. Basal leaves ternately compound, the ultimate divisions crenate; sepals greenish white, tinged with lavender at the base — *A. parviflora*
    2. Basal leaves ternately decompound, the ultimate divisions linear to lanceolate; sepals yellowish white, blue, purple, or reddish on the back
        3. Akenes and ovaries glabrous — *A. narcissiflora*
        3. Akenes and ovaries pubescent
            4. Styles filiform, straight, 1.5–4 mm long, yellowish in fruit — *A. lithophila*
            4. Styles thick, curved, 1–1.5 mm long, red or purple in fruit — *A. multifida*

Anemone lithophila Rydb.
Bull. Torr. Club 29:152, 1902.

Fellfield Anemone

Anemone drummondii S. Wats.

Erect, villous, perennial herb from a branched, usually woody caudex. Basal leaves long-petiolate; blades 3–4-ternate, the lobes linear-oblong; involucral leaves short-petiolate or sessile. Flowers usually single, the peduncle villous. Sepals mostly 6–9, ovate, white, villous and bluish on the outside. Filaments yellowish. Akenes villous, in a globose head, the mature styles filiform, straight, yellowish.

Rocky slopes and fellfields of the Absaroka, Beartooth and Wind River Ranges. Alberta and British Columbia south to Washington, Idaho, and northwestern Wyoming.

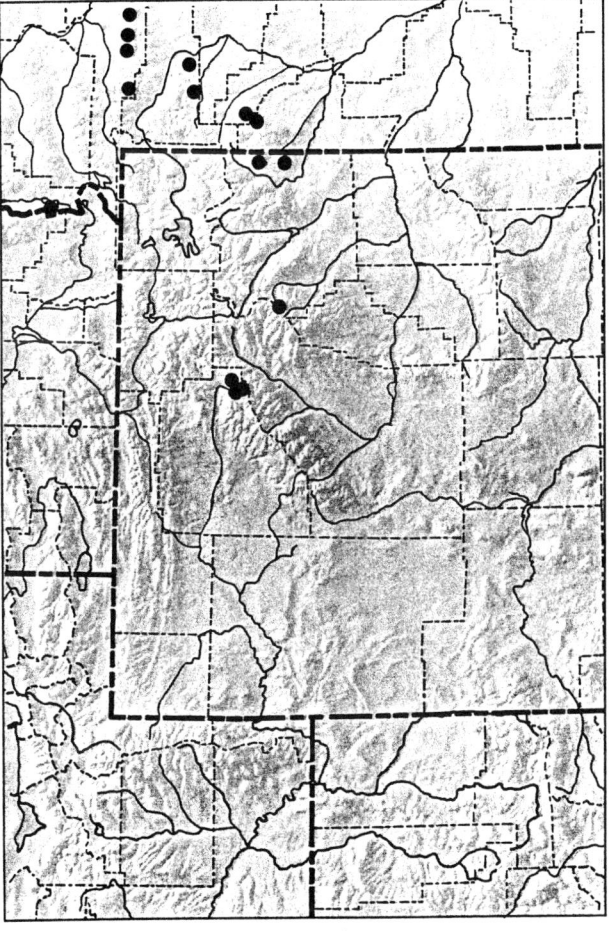

Anemone multifida Poir. in Lam.
Encyc. Meth. Bot. 1:364, 1810.

Cutleaf Anemone

Anemone globosa (T. & G.) Nutt. ex Pritz.
Anemone hudsoniana (DC.) Richards.
Anemone tetonensis Porter

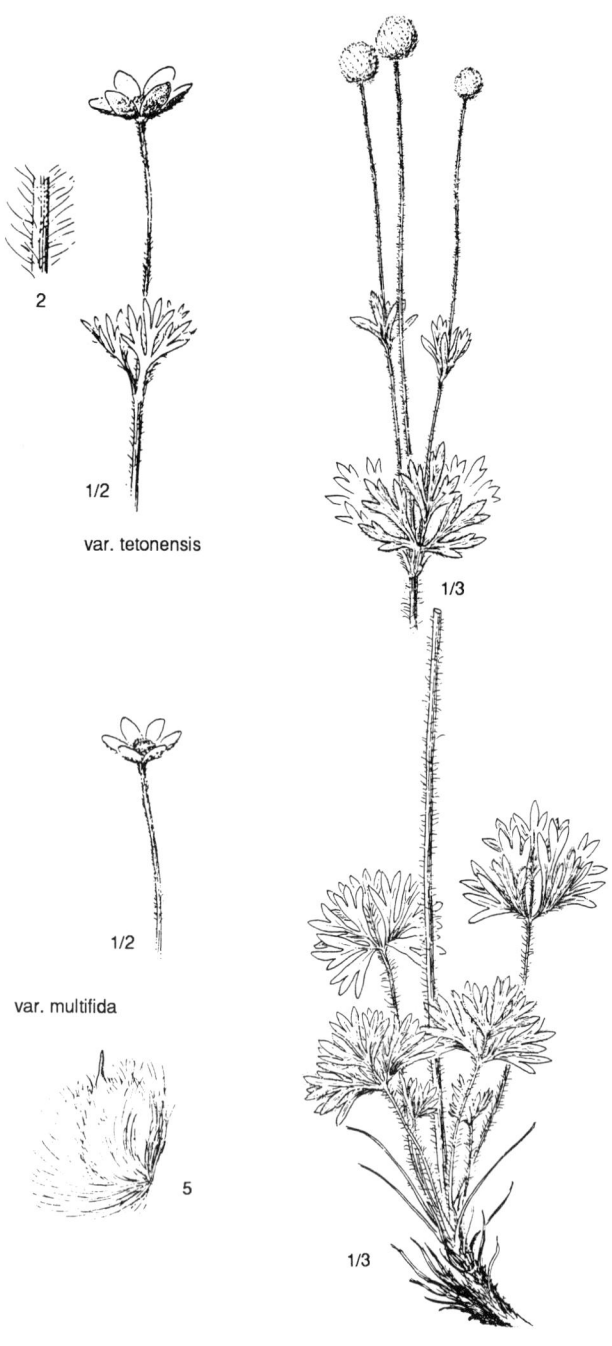

Erect perennial from a stout, usually branched caudex; stems and leaves villous; basal leaves petiolate; blades 2–3-ternate, the lobes linear to lanceolate; involucral leaves short-petiolate to sessile. Flowers 1–4, the peduncles villous. Sepals 5–9, ovate, white or yellow, villous, and often reddish, bluish, or purplish on the outside. Akenes villous, in a globose head, the styles thick, curved, reddish or purplish.

Meadows, rocky slopes, and fellfields in all alpine ranges of the Middle Rocky Mountains. Transcontinental in boreal North America; south in the West to California, New Mexico, and Colorado.

Two varieties occur in the alpine zone: var. *tetonensis* (Porter ex Britt.) C.L. Hitchc. has single flowers and ultimate leaf lobes less than 3 mm wide; var. *multifida* has 2–3 flowers and ultimate leaf lobes more than 3 mm wide.

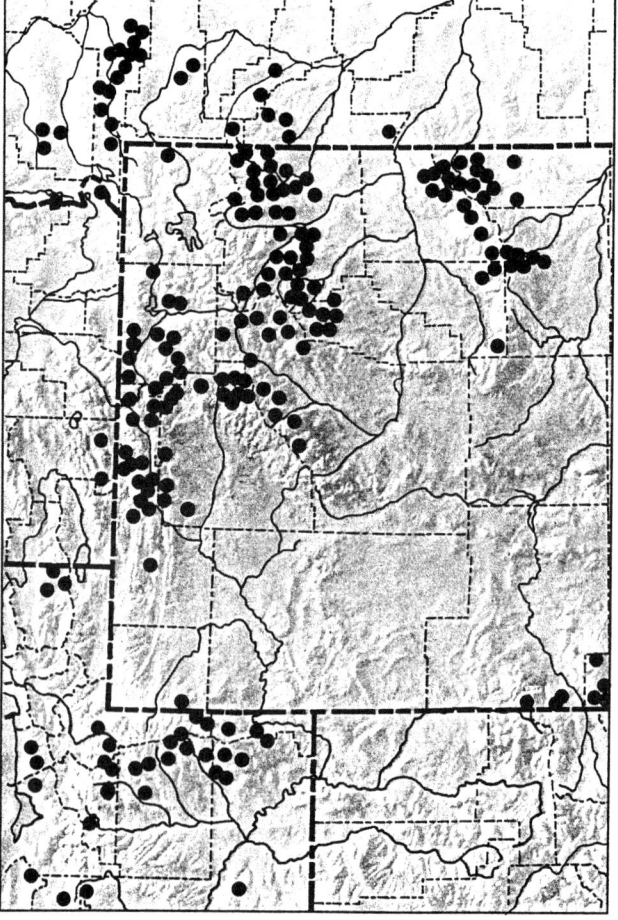

Anemone narcissiflora L.
Sp. Pl., 542, 1753.

Narcissus Anemone

Anemone zephyra A. Nels.

Erect, pilose perennial from a stout caudex. Basal leaves long-petiolate; blades 1–2-ternate, the ultimate lobes lanceolate; involucral leaves sessile. Flowers 1–3. Sepals 5–9, oval, white or yellow, often bluish on the outside. Akenes glabrous, in a globose head, the styles short and curved.

Fellfields of the Big Horn Range. Alaska south to Colorado and Wyoming.

Middle Rocky Mountain alpine plants are ssp. *zephyra* (A. Nels.) Löve and Löve.

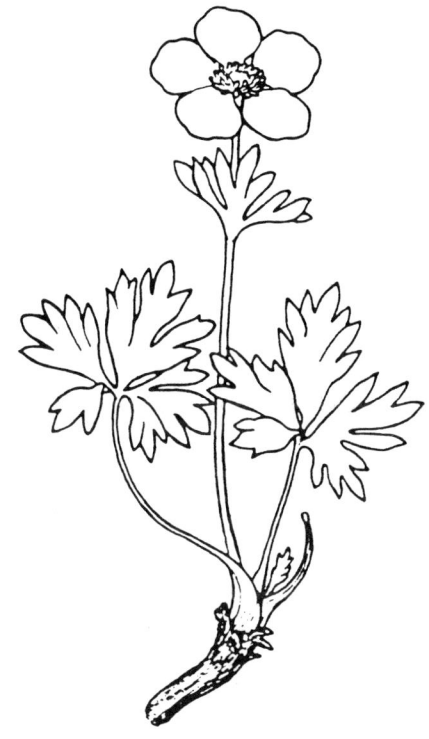

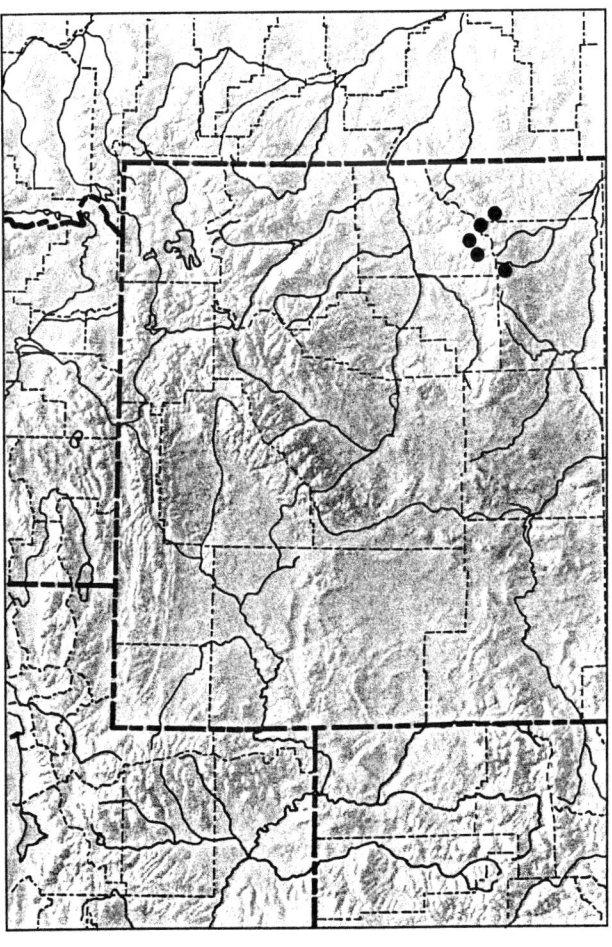

Anemone parviflora Michx.
Fl. Bor.-Am. 1:319, 1803.

Northern Anemone

Perennial, pubescent to nearly glabrous herb from a slender rhizome. Basal leaves long-petiolate, ternately compound; leaflets cuneate, distally crenate-dentate; involucral leaves sessile, 3-lobed. Flowers solitary, the peduncles pubescent. Sepals 5, oval, white, villous and often bluish purple on the outside. Akenes woolly-villous, in a globose head, the styles straight, glabrous.

Moist tundra slopes of the Absaroka, Beartooth, Big Horn, Medicine Bow, and Wind River Ranges. Transcontinental in boreal North America; south in the West to Oregon, Idaho, and Colorado.

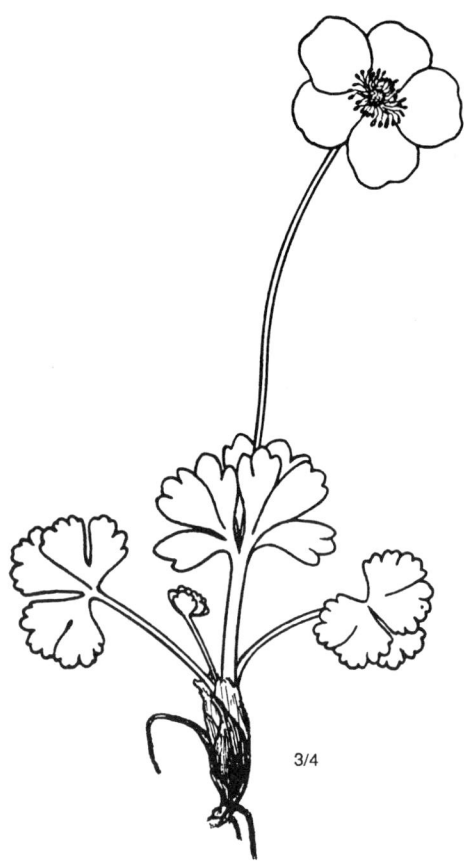

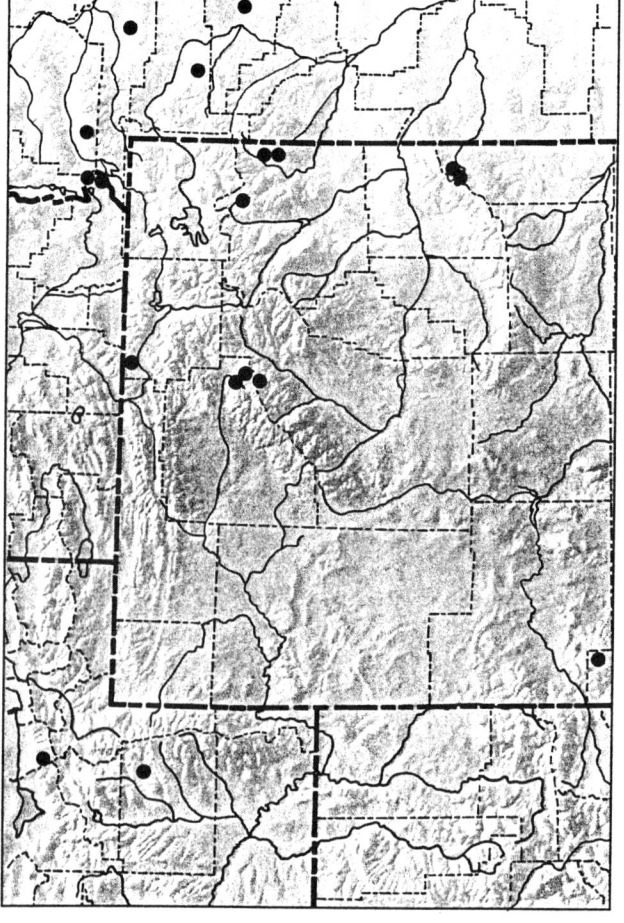

Anemone patens L.
Sp. Pl., 538, 1753.

Pasque Flower

Anemone hirsutissima (Pursh) MacMillan
Anemone ludoviciana Nutt.
Anemone multifida (Pritz.) Zamels
Anemone nuttalliana DC.
Anemone wolfgangiana Bess.
Pulsatilla hirsutissima (Pursh) Britt.
Pulsatilla ludoviciana (Nutt.) A. Heller
Pulsatilla nuttalliana (DC.) Spreng.
Pulsatilla patens (L.) P. Mill.

Erect, silky-villous perennial from a stout, branched or unbranched caudex. Basal leaves long-petiolate; blades ternately divided, the ultimate divisions linear; involucral leaves sessile, ternately divided. Flowers solitary, showy. Sepals 5–7, ovate, purple to bluish, villous on the outside. Stamens with glandlike staminodia. Akenes villous; style persistent, long-plumose.

Meadows and fellfields in the lower alpine zone of the Absaroka, Beartooth, Big Horn, Uinta, and Wind River Ranges. Circumboreal; Alaska south in the West through Washington to New Mexico, Texas, Colorado, and Nebraska.

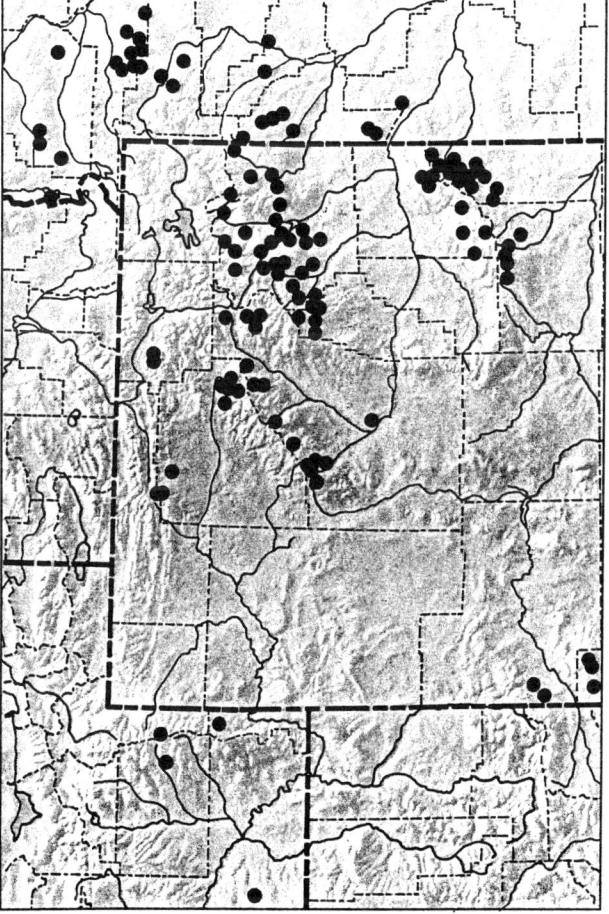

Aquilegia L.
Columbine

Perennial herbs with 2–3-ternately compound leaves, the leaflets lobed. Flowers large, yellow or blue in ours, solitary on the ends of branches. Sepals 5, petaloid. Petals 5, prolonged at the base into hollow spurs, the blades smaller than the sepals. Stamens numerous, the inner ones often sterile. Pistils 5, erect, with slender styles, forming a few erect, many-seeded follicles.

KEY TO THE SPECIES OF *AQUILEGIA*

1. Flowers yellow — *A. flavescens*
1. Flowers blue, purplish, lavender, or white
   2. Plants scapose, the flowers equaling or barely exceeding the leaves — *A. jonesii*
   2. Plants caulescent, the flowers much exceeding the leaves — *A. coerulea*

Aquilegia coerulea James
Rep. Long's Exped. Rocky Mtns. 2:15, 1823.

Colorado Columbine

Aquilegia leptocera Nutt.
Aquilegia oreophila Rydb.
Aquilegia piersoniana Williams
Aquilegia pinetorum Tidestr.
Aquilegia scopulorum Tidestr.

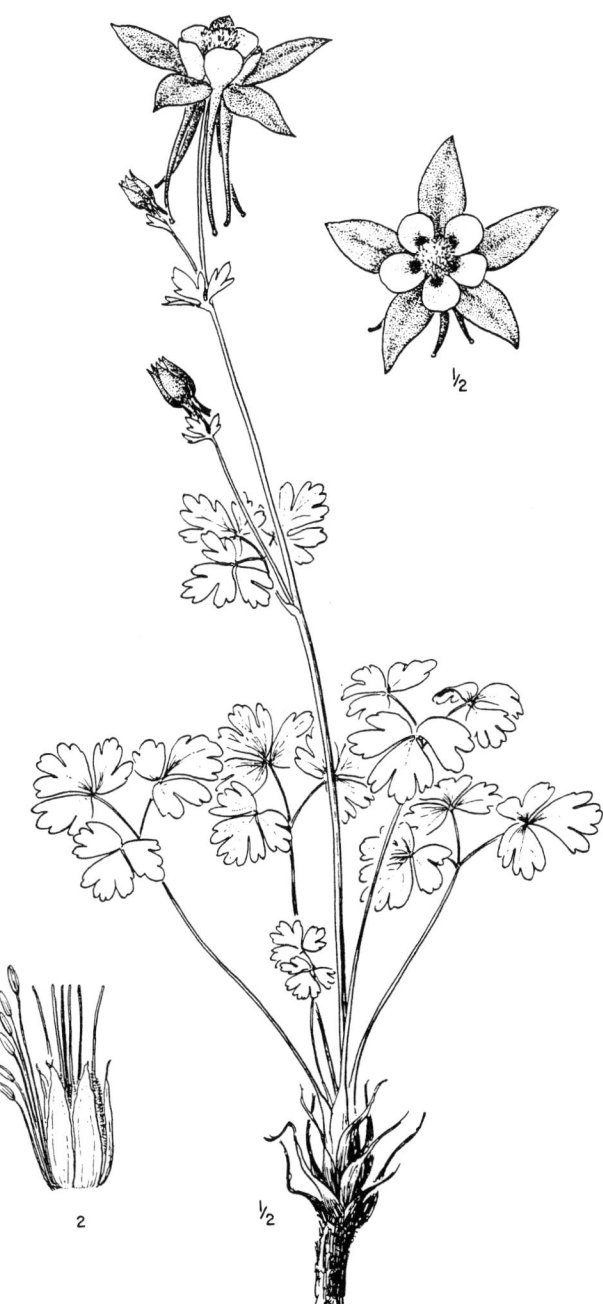

Erect, glabrous to glandular-puberulent, perennial herb from a simple or few-branched caudex. Basal leaves long-petiolate, 2-ternate; leaflets broadly cuneate, glaucous below, 2–3-cleft, the lobes rounded; cauline leaves reduced, short-petiolate, the leaflets similar to the basal ones. Flowers several, large, showy, erect. Sepals petaloid, white to bluish or lavender, spreading. Petals white or cream, the spurs straight or nearly so, about twice as long as the petals. Stamens included or nearly so, the inner ones modified into staminodia. Follicles 5–7, erect, pubescent.

Ledges, boulders, fellfields, and coarse talus slopes in most alpine ranges of the Middle Rocky Mountains. Not yet known from above timberline in the Big Horn, Hoback, Salt River, and Wyoming Ranges. Montana to Idaho and south to New Mexico and Arizona.

Three weakly differentiated and intergrading varieties can be distinguished in the Middle Rocky Mountains: var. *ochroleuca* Hook. has white or cream sepals; var. *calcarea* M.E. Jones has dark blue petals and sepals, and the ultimate leaf segments overlap; and var. *coerulea* has lavender petals and sepals, or at least the petals are lighter than the sepals, and the ultimate leaf segments do not overlap.

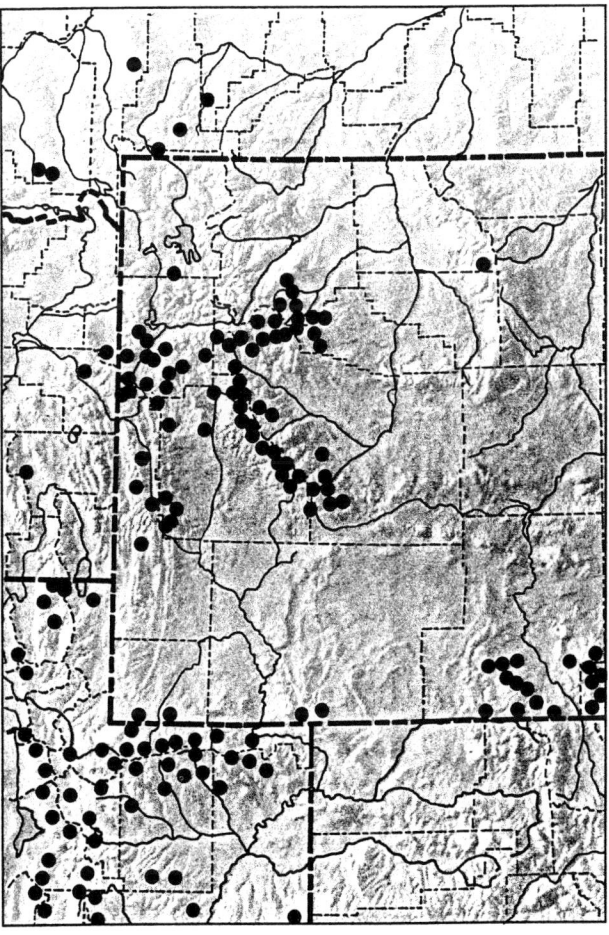

Aquilegia flavescens S. Wats.
Bot. King Exped., 10, 1871.

Yellow Columbine

Aquilegia rubicunda Tidestr.

Erect, perennial herb from a simple to few-branched caudex; stems glabrous, pilose, or glandular-pubescent. Basal leaves long-petiolate, 2–3-ternate; leaflets broadly cuneate, 2–3-cleft; cauline leaves reduced, short-petiolate, the leaflets similar to the basal ones. Flowers several, showy, nodding. Sepals yellow, spreading to reflexed. Petals yellow, the spurs curved, hooked at the apex, about the same length as to slightly longer than the petals. Stamens exerted, the inner ones modified into staminodia. Follicles 5, spreading, pilose to glandular-pubescent.

Talus and boulders near timberline in the lower alpine zone of the Absaroka, Beartooth, Teton, Uinta, and Wasatch Ranges. British Columbia and Alberta south to Washington, Utah, and Colorado.

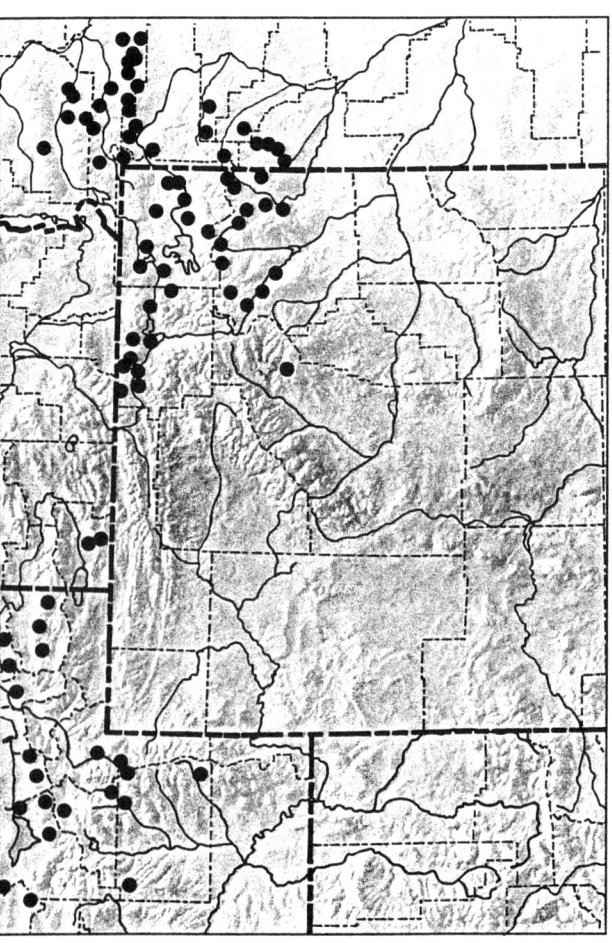

Aquilegia jonesii Parry
Am. Nat. 8:211, 1874.

Jones Columbine

Low, perennial herb, less than 12 cm tall, from a stout, often highly branched caudex. Leaves all basal, crowded, long-petiolate, biternate; leaflets crowded, orbicular, 3–4-cleft, finely pubescent and glaucous. Flowers single, showy, erect. Sepals deep blue, erect to ascending. Petals deep to light blue, the spurs straight or somewhat incurved, shorter than the sepals. Staminodia lacking. Follicles 5, erect, glabrous.

Rocky exposed ridgetops and scree on limestone, particularly Big Horn dolomite, of the Beartooth, Big Horn, and Wind River Ranges. Alberta south to northwestern Wyoming.

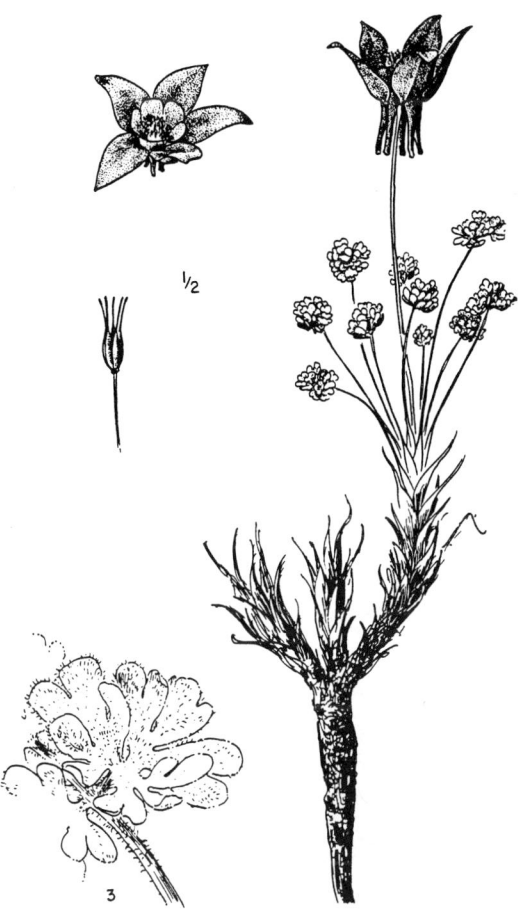

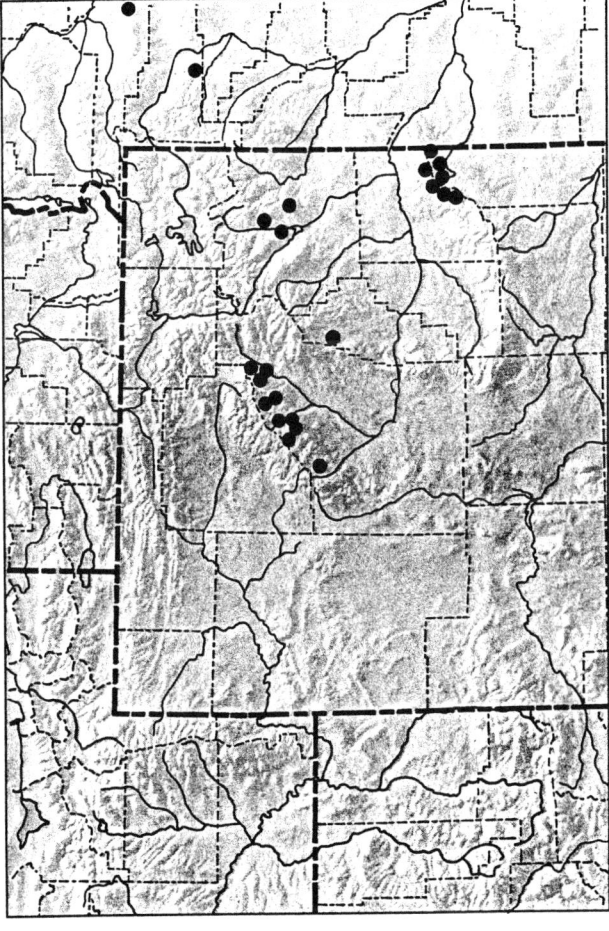

Caltha L.
Marsh Marigold

Low, glabrous, somewhat succulent, perennial herbs from fibrous roots, with cordate or reniform, mostly basal leaves. Flowers regular, showy, white to cream, in open racemes. Sepals 5–9, petaloid, early deciduous. Petals none. Stamens numerous. Pistils few to many, becoming several-seeded follicles.

Caltha leptosepala DC.
Prodromus 1:310, 1818.

Marsh Marigold

Caltha auriculata (Raf.) Merr.
Caltha biflora DC.
Caltha chelidonii Greene
Caltha howellii Huth
Caltha rotundifolia (Huth) Greene
Caltha uniflora Rydb.
Psychrophila auriculata Raf.
Psychrophila leptosepala (DC.) W.A. Weber

Scapose, glabrous, perennial herb from a short, erect caudex and fibrous to fleshy roots. Leaves petiolate, simple; blades oblong-oval, cordate at the base, the margins toothed to undulate. Flowers mostly 1, the sepals 6–12, oblong, white to cream, sometimes lavender or bluish on the outside. Petals lacking. Stamens numerous; filaments longer than the anthers, the inner ones often modified into staminodia. Follicles several to numerous, stipitate, beaked.

Wet meadows, nivation basins, and margins of receding snowfields in most of the alpine ranges. Not yet known from above timberline in the Hoback, Salt River, and Wyoming Ranges. Alaska south to Utah, Colorado, and New Mexico.

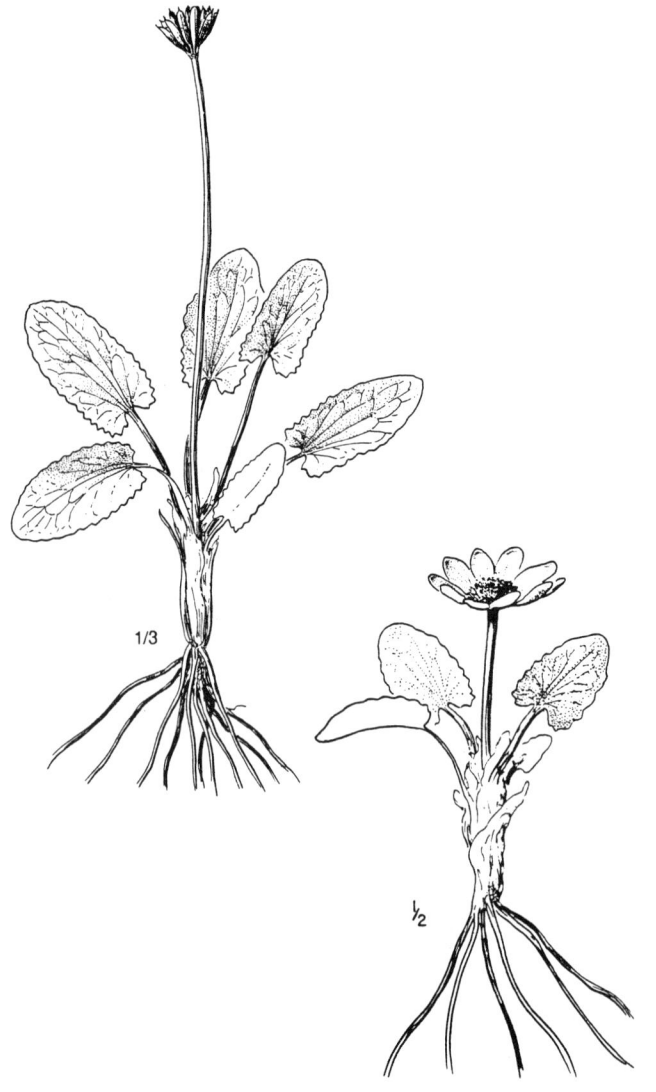

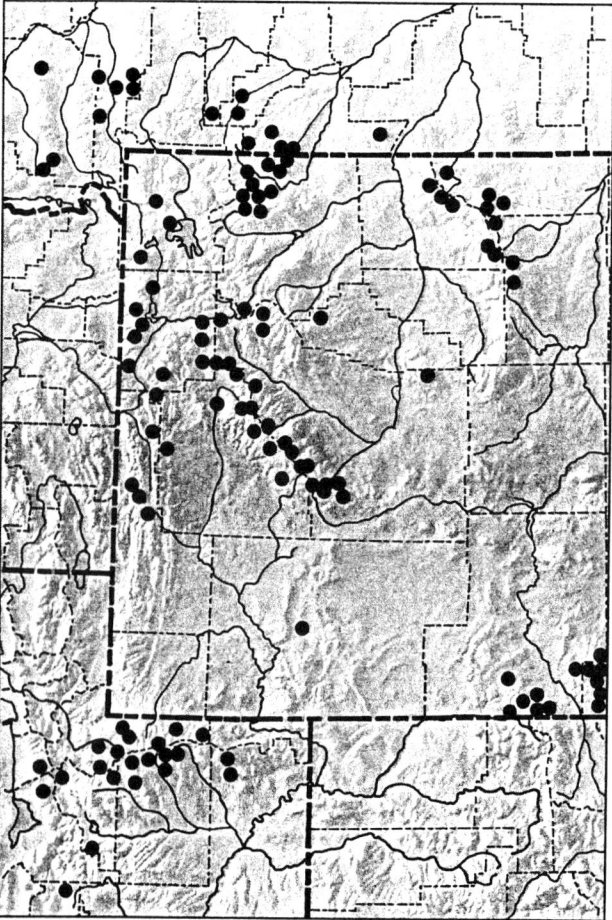

Delphinium L.
Larkspur

Erect, annual or perennial herbs from fibrous or tuberous, fascicled roots, with palmately lobed or divided leaves. Flowers irregular, blue or purple in ours, racemose or paniculate. Sepals 5, petaloid, the upper one prolonged from the base into a spur. Petals usually 4, the upper 2 spurred, the spurs enclosed by the spur of the upper sepal. Stamens numerous. Pistils 1–5, becoming erect to divergent, many-seeded follicles.

KEY TO THE SPECIES OF *DELPHINIUM*

1. Stems hollow; lobes of leaves toothed; sepals angled forward — *D. occidentale*
1. Stems solid; lobes of leaves dissected; sepals flared — *D. bicolor*

Delphinium bicolor Nutt.
J. Acad. Phila. 7:10, 1834.

Little Larkspur

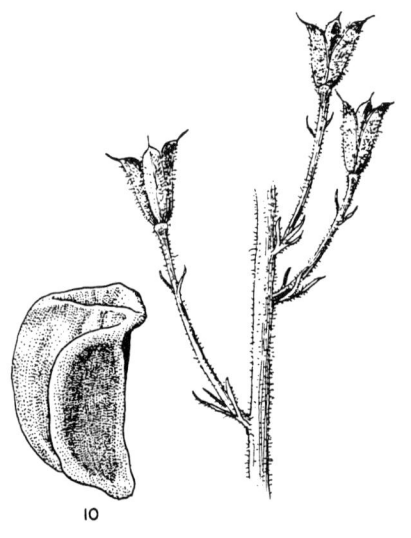

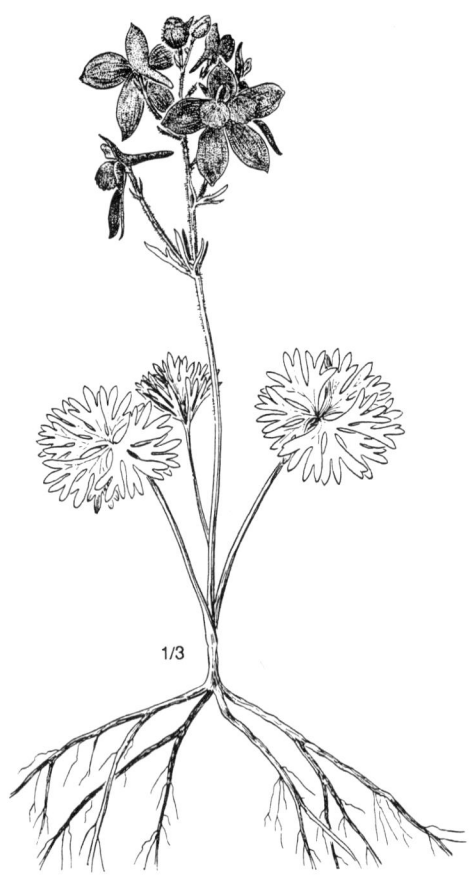

Erect, perennial herb, glabrate below, becoming pubescent above, from a branched, fibrous root system. Basal leaves long-petiolate; blades orbicular, palmately dissected, the leaflets 3–4-cleft. Inflorescence a short, few-flowered raceme. Sepals dark bluish purple, the dorsal blade smaller than the other four, the spur 10–20 mm long. Upper petals white, blue- or purple-veined; lower petals dark blue-purple, elliptic to orbicular, shallowly lobed. Follicles divergent, puberulent to subglabrate.

Meadows and scree slopes of the Absaroka, Beartooth, and Big Horn Ranges. Alberta and Saskatchewan south to Washington, Idaho, Wyoming, and South Dakota.

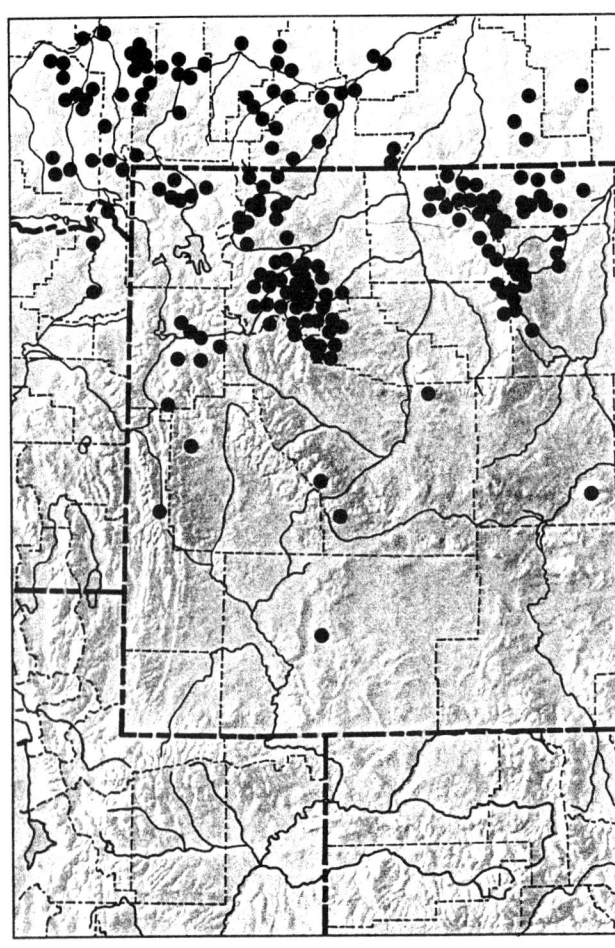

Delphinium occidentale (S. Wats.) S. Wats.
Bot. Calif. 2:428, 1880.

Duncecap Larkspur

Delphinium abietorum Tidestr.
Delphinium cucullatum A. Nels.
Delphinium multiflorum Rydb.
Delphinium reticulatum (A. Nels.) Rydb.
Delphinastrum occidentale (S. Wats.) Nieuwl.

Large, perennial herb, to 1 m tall in the lower alpine zone, from a stout, woody rootstock; stems hollow, glaucous below to hirsute or glandular, puberulent above. Leaves petiolate, palmately divided or cleft, the divisions cuneate and apically toothed. Inflorescence racemose or paniculate, the rachis glandular-puberulent. Sepals dark blue, purple, or white, the spur 10–12 mm long. Upper petals white or bluish; lower petals blue, bilobed. Follicles erect, glabrous to glandular-pubescent.

Meadows and slopes near timberline in the lower alpine zone of the northern Wind River Range. Montana to Oregon, south to Idaho and Colorado.

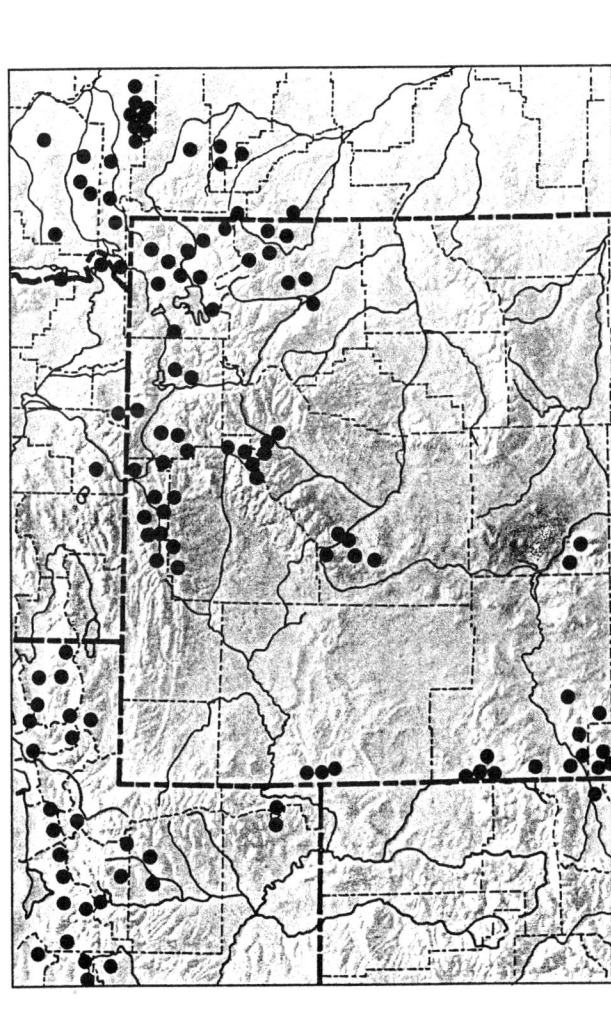

Ranunculus L.
Buttercup

Low, glabrous to pubescent, perennial herbs from fibrous or fascicled roots, with alternate, entire to lobed, parted or pinnately compound leaves. Flowers regular, yellow or white, solitary to cymose. Sepals 5, often deciduous. Petals usually 5, each with a nectiferous gland at the base. Stamens mostly numerous, occasionally few. Pistils several to many, forming a globose to oblong head of akenes, these usually beaked by the persistent styles.

KEY TO THE SPECIES OF *RANUNCULUS*

1. Plants aquatic, the stems floating or submerged
    2. Petals yellow; leaves 3–5-lobed — *R. hyperboreus*
    2. Petals white; leaves finely dissected — *R. aquatilis*
1. Plants terrestrial, often in wet places, the stems erect
    3. Leaves entire, at least in part
        4. Akenes glabrous; nectary scales entire to shallowly toothed; leaves all entire — *R. alismifolius*
        4. Akenes puberulent; nectary scales fimbriate; leaves both entire and lobed — *R. glaberrimus*
    3. Leaves all toothed, lobed, or dissected
        5. Stems less than 5 cm tall
            6. Petals less than 3 mm long; leaf blades less than 1 cm long; stylar beak straight or slightly curved — *R. pygmaeus*
            6. Petals more than 3 mm long; leaf blades 1–2 cm long; stylar beak strongly recurved — *R. gelidus*
        5. Stems more than 8 cm tall
            7. Petals less than 6 mm long, scarcely exceeding the sepals
                8. Basal leaves crenate; akenes mostly pubescent, the stylar beak 0.7–0.9 mm long, slightly recurved — *R. inamoenus*
                8. Basal leaves palmately lobed to parted; akenes glabrous, the stylar beak to 0.6 mm long, recurved
                    9. Mature akenes 1–1.5 mm long; petals broadly obovate — *R. verecundus*
                    9. Mature akenes 2–2.5 mm long; petals oblanceolate to obovate — *R. gelidus*
            7. Petals more than 7 mm long, conspicuously exceeding the sepals
                10. Plants pilose throughout; stylar beak sharply recurved — *R. pedatifidus*
                10. Plants glabrous or glabrate throughout, sometimes brown-pilose in the inflorescence; stylar beak straight to somewhat curved
                    11. Basal leaves finely dissected, the segments less than 2 mm wide; calyx glabrate to brown-pilose — *R. adoneus*
                    11. Basal leaves lobed to parted, the segments more than 2 mm wide; calyx glabrate — *R. eschscholtzii*

Ranunculus adoneus Gray
Proc. Acad. Phila. 15:56, 1864.

Alpine Buttercup

Glabrous, perennial herb from slender, fibrous roots. Stems 1 to several, erect, scapose. Leaves simple, petiolate, the blades orbicular to reniform, palmately divided, the divisions 2-lobed into narrowly linear segments. Flowers 1–3. Sepals green, sometimes purple-tinged, spreading-villous. Petals 5, yellow, broadly obovate, about twice as long as the sepals; nectary scale glabrous, triangular-truncate. Stamens numerous. Akenes oblong, glabrous, in an ovoid to cylindric cluster; stylar beaks curved.

Margins of snowbanks, wet meadows, and stream banks of the western portion of the Middle Rocky Mountains. Not yet known from above timberline in the Big Horn and Medicine Bow Ranges. Wyoming, Idaho, Utah, and Colorado.

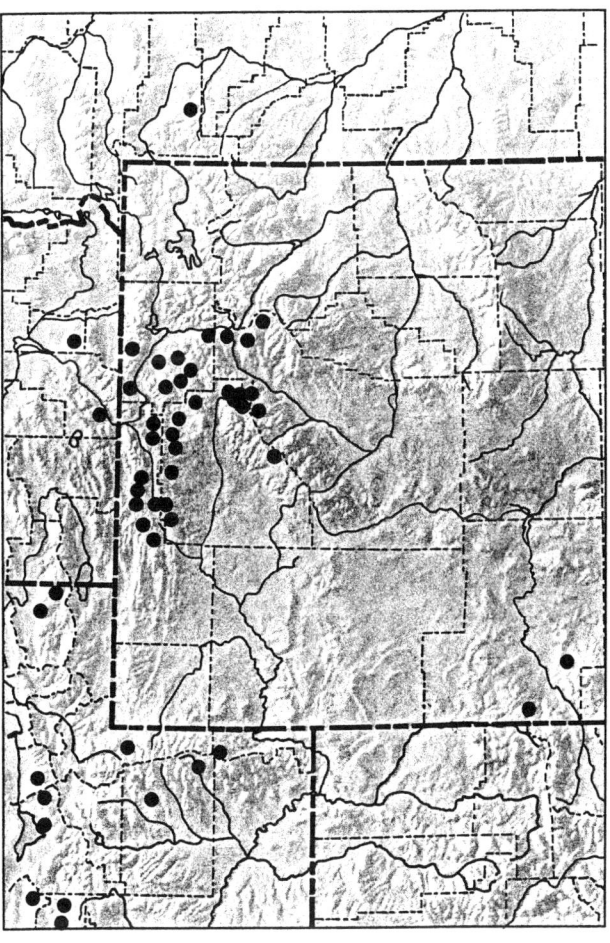

Ranunculus alismifolius Geyer ex Benth.
Pl. Hartweg., 295, 1848.

Plantainleaf Buttercup

Ranunculus alismellus Greene
Ranunculus calthaeflorus Greene
Ranunculus hartwegii Greene
Ranunculus lemmonii Gray
Ranunculus unguiculatus Greene

Glabrous to pilose, perennial herb from slender, fibrous roots; stems erect, branched, not rooting at the nodes. Basal leaves simple, petiolate, the blades lanceolate to narrowly ovate, entire to serrulate; cauline leaves somewhat reduced upward, alternate or opposite, sessile or short-petiolate, the margins entire to serrulate. Inflorescence corymbose. Sepals 5, spreading, pubescent. Petals 5, yellow, obovate; nectary scale oblong to ovate, forming a pocket. Stamens numerous. Akenes obovate, glabrous, in a subglobose head; stylar beak stout, thick, straight.

Stream banks and wet meadows of the Medicine Bow, Uinta, and Wasatch Ranges. British Columbia south to California, Nevada, and Colorado.

var. montanus

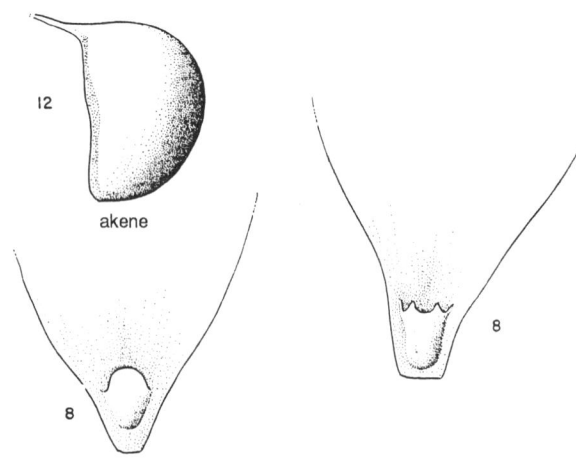

akene

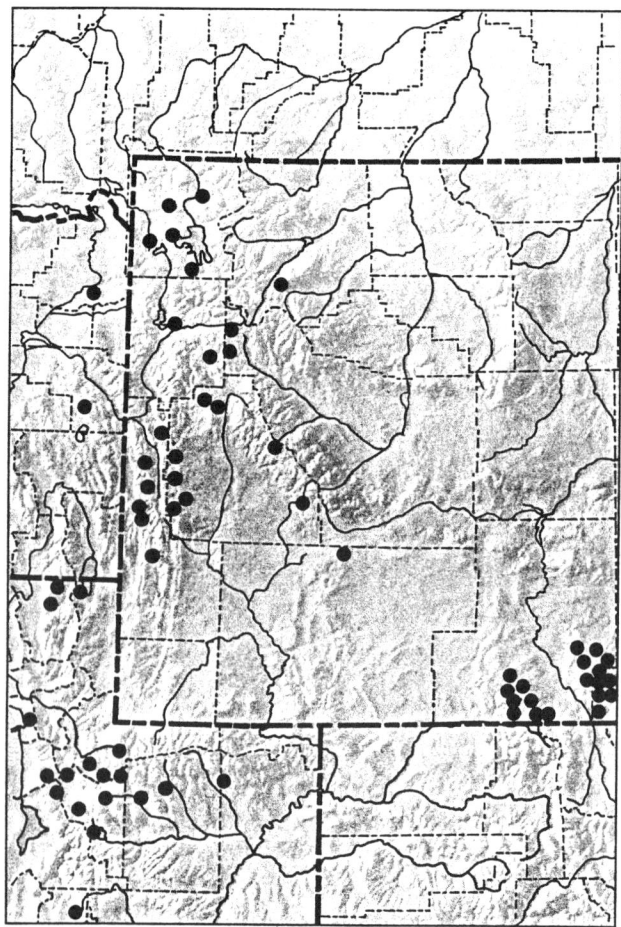

Ranunculus aquatilis L.
Sp. Pl., 556, 1753.

White Watercrowfoot

Batrachium aquatile (L.) Dumort.
Batrachium bakeri Greene
Batrachium confervoides (Fries) Fries
Batrachium drouetii (F.W. Schultz) Nym.
Batrachium flaccidum (Pers.) Rupr.
Batrachium porteri (Britt.) Rydb.
Batrachium trichophyllum (Chaix) F.W. Schultz
Ranunculus capillaceus Thuill.
Ranunculus porteri Britt.
Ranunculus trichophyllus Chaix

Aquatic, perennial herb; stems elongate, branched, glabrous to finely pubescent, often rooting at the nodes. Submerged leaves petiolate, the blades dissected into filiform segments; floating leaves 3-lobed to 3-parted, the segments toothed. Flowers axillary, the pedicels stout, ascending to recurved in fruit. Sepals light green, membranous, deciduous. Petals white, sometimes yellow at the base; nectary scale hemispheric, very small. Stamens numerous. Akenes obovate, glabrous to hirsute, transversely ridged, in a globose cluster; stylar beak short-apiculate.

Ponds and slow-moving streams throughout the Middle Rocky Mountains, and in the alpine-subalpine ecotone of the Absaroka Range. Circumboreal; western North America south to California and New Mexico.

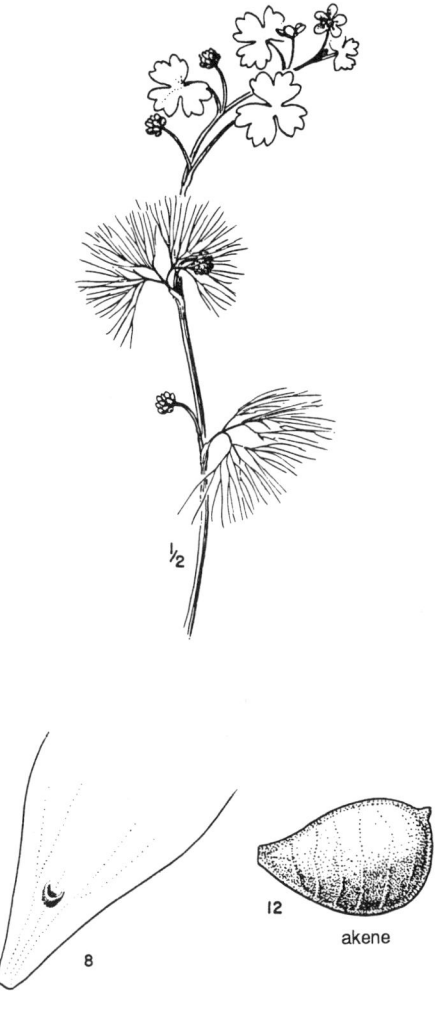

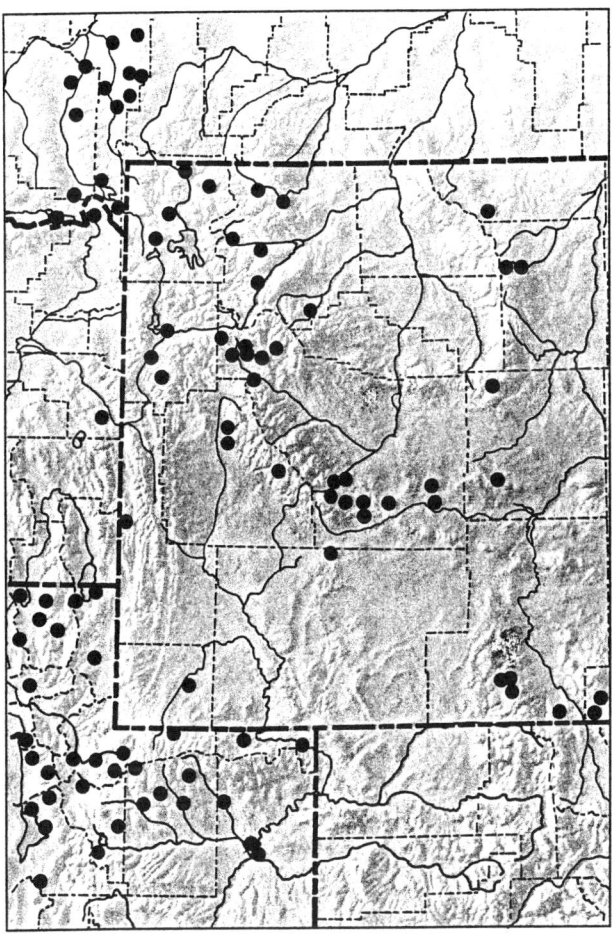

Ranunculus eschscholtzii Schlecht.
Anim. Ranunc. 2:16, pl. 1, 1820.

Snowbed Buttercup

Ranunculus eximus Greene
Ranunculus helleri Rydb.
Ranunculus ocreatus Greene
Ranunculus oxynotus Gray
Ranunculus saxicola Rydb.
Ranunculus suksdorfii Gray
Ranunculus trisectus Eastw.

Glabrous or pubescent, perennial herb from a short, stout caudex and slender, fibrous roots. Stems 1 to several, erect, not rooting at the nodes. Basal leaves petiolate, the blades reniform, 3–5-lobed or parted, often cordate at the base, ultimate segments lanceolate; cauline leaves sessile, 3–5-parted. Flowers 1–3. Sepals shorter than the petals, usually pilose and purplish beneath. Petals yellow, obovate; nectary scale small, deltoid, forming a pocket. Stamens numerous. Akenes oblong to obovate, glabrous or pubescent in a cylindric to oval head; stylar beak slender, prominent, straight or slightly curved.

Wet meadows, stream banks, and margins of snowfields in most alpine ranges. Not yet known from above timberline in the Hoback and Salt River Ranges. Siberia; Alaska south to New Mexico and Arizona.

Three varieties occur in the alpine zone of Middle Rocky Mountain ranges and can be distinguished as follows:

1. Ultimate basal leaf lobes usually sharply acute
    . . . var. *eximus* (Greene) Benson
1. Ultimate basal leaf lobes rounded or obtuse, not sharply acute
    2. Akenes pubescent; styles mostly more than 1 mm long; middle division of basal leaves usually twice-lobed into narrow or linear segments
        . . . var. *trisectus* (Eastw.) Benson
    2. Akenes glabrous; styles about 1 mm long; middle division of basal leaves not more than once-lobed
        . . . var. *eschscholtzii*

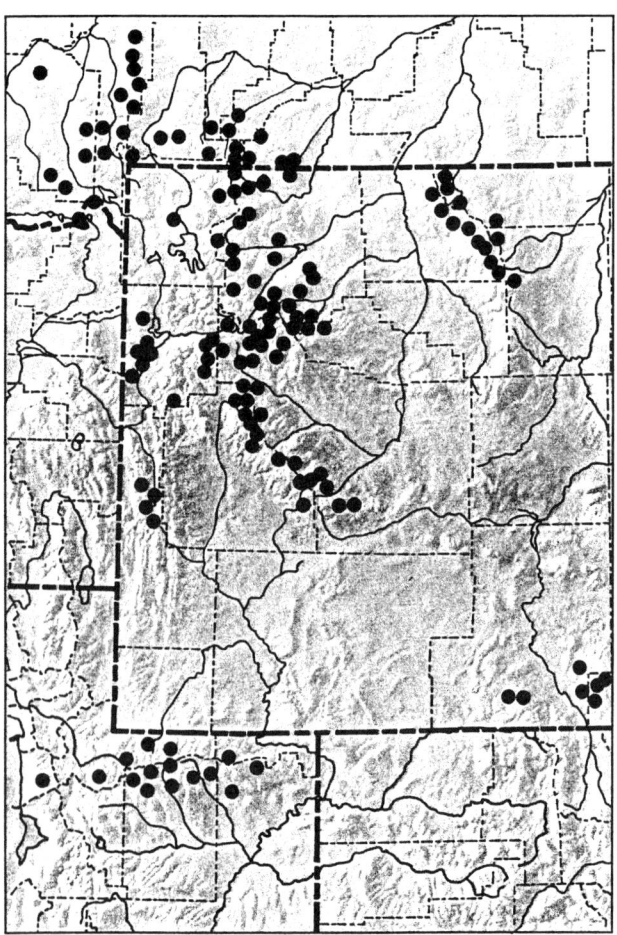

Ranunculus gelidus Kar. & Kir.
Bull. Soc. Nat. Mosc. 15:133, 1842.

Tundra Buttercup

Ranunculus drummondii Greene
Ranunculus grayi Britt.
Ranunculus hookeri Regel
Ranunculus pedatifidus Hook.
Ranunculus ramulosus M.E. Jones

Perennial herb from a short caudex and slender, fibrous roots; stems 1 to several, erect, glabrous below and usually pubescent above. Basal leaves petiolate, the blades cordate to reniform, mostly 3-parted, the lobes again cleft into oblong, obtuse segments; cauline leaves short-petiolate, 3–5-lobed. Flowers 1–3. Sepals spreading or reflexed, usually pubescent beneath. Petals yellow, up to twice as long as the sepals, short-clawed; nectary scale very small, forming a small pocket. Stamens numerous. Akenes broadly ovate, mostly glabrous, in a cylindric or ovoid head; stylar beak short, stout, strongly recurved.

Talus, scree, and wet fellfields of the Uinta Range. Siberia; Alaska south to Montana, Utah, and Colorado.

Rocky Mountain plants are ssp. *grayi* (Britt.) Hult.

Ranunculus glaberrimus Hook.
Fl. Bor.-Am. 1:12, 1829.

Sagebrush Buttercup

Ranunculus ellipticus Greene
Ranunculus reconditus A. Nels. & J.F. Macbr.
Ranunculus waldronii Lunell

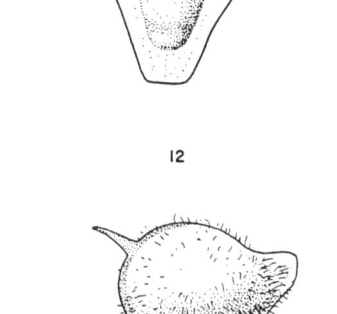

Low, tufted, glabrous to sparsely hirsute, perennial herb from thick, fleshy roots; stems 1 to several and prostrate, ascending, or erect. Basal leaves long-petiolate; blades simple, elliptic, ovate, or obovate, entire, undulate or 3-lobed, the base cuneate; cauline leaves short-petiolate to sessile, entire or 2–3-lobed. Flowers solitary or few. Sepals spreading, elliptic, glabrous to hirsute, often purplish. Petals 5, yellow, obovate, often turning white with age; nectary scale oblong, truncate or retuse, usually ciliate, forming a deep pocket. Stamens numerous. Akenes broadly obovate, glabrous to puberulent, in a globose head; stylar beak slender, prominent, straight to slightly curved.

An early ephemeral of meadows, stream banks, snowbank margins, and willow thickets of the Beartooth, Big Horn, and Wind River Ranges. British Columbia and Alberta south to California, New Mexico, and Nebraska.

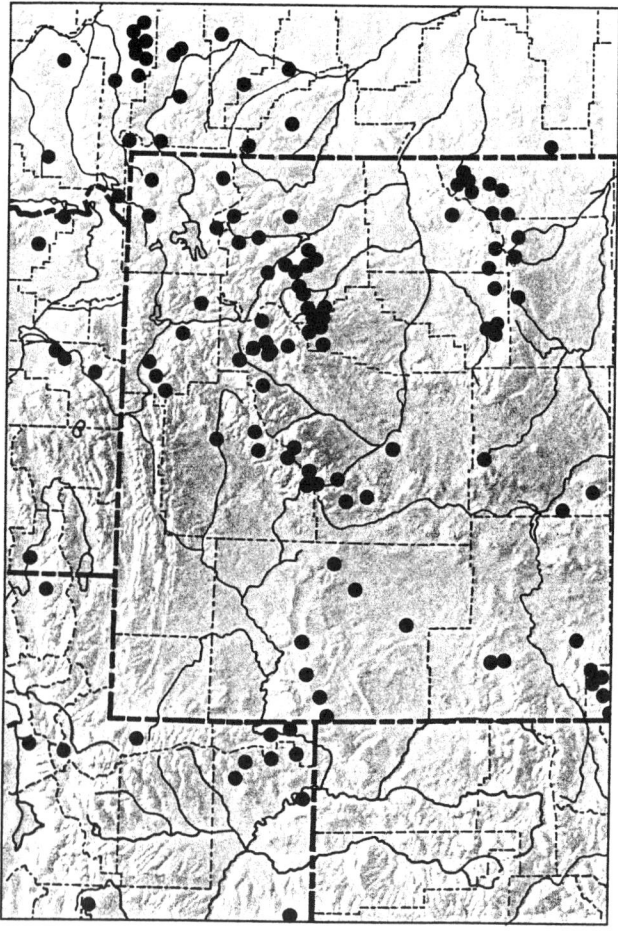

Ranunculus hyperboreus Rottb.
Skrift. Kongel. Dansk. Vidensk. Selsk. 10:458, 1700.

Arctic Buttercup

Ranunculus intertextus Greene
Ranunculus natans C.A. Mey. in Ledeb.

Prostrate, glabrous, perennial herb with weak, branched, horizontal stems rooting at the nodes; roots slender, fibrous. Leaves alternate, long-petiolate, blades cordate, 3-lobed, the 2 lateral lobes each 1-cleft. Flowers solitary in the leaf axils. Sepals spreading, about the same length as the petals. Petals 5, yellow, obovate, short-clawed; nectary scale forming a shallow pocket. Stamens numerous. Akenes obovate, glabrous, in an ovoid or globose head; stylar beak short, stout, straight.

Shallow meltwater ponds and small streams of the Absaroka, Beartooth, and Big Horn Ranges. Circumpolar; south in North America to Idaho and Colorado.

Middle Rocky Mountain plants are var. *hyperboreus*.

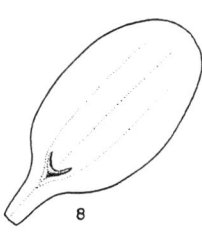

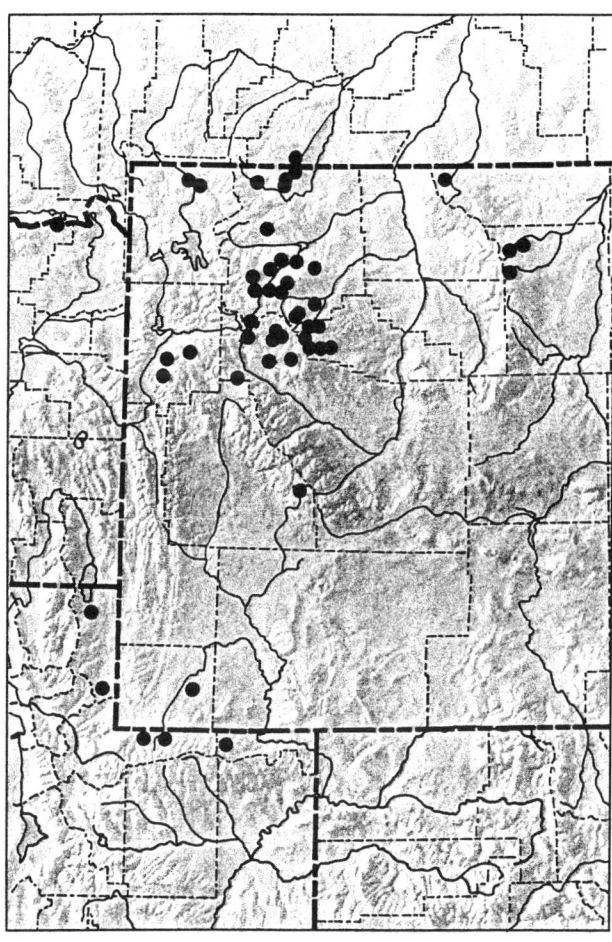

Ranunculus inamoenus Greene
Pittonia 3:91, 1896.

Unpleasant Buttercup

Ranunculus alpeophilus A. Nels.
Ranunculus arizonicus J.G. Lemmon ex Gray
Ranunculus micropetalus (Greene) Rydb.
Ranunculus utahensis Rydb.

Erect, pubescent or subglabrate, perennial herb from fibrous roots; stems 1 to several, simple or branched, hollow. Basal leaves long-petiolate; blades simple, orbicular, obovate, cuneate at the base, crenate to crenately 3–5-lobed; cauline leaves sessile, 3–7-parted. Flowers 1 to several. Sepals spreading or reflexed, obovate, about half the length of the petals. Petals 5, yellow, elliptic to obovate; nectary scale glabrous, forming a pocket. Stamens numerous. Akenes obovate, densely puberulent, sometimes glabrous, in a cylindric head; stylar beak slender, recurved.

Moist meadows and slopes of the montane and subalpine zones. Known from the timberline ecotone of the Absaroka, Medicine Bow, Uinta, Wasatch, and Wind River Ranges and can be expected elsewhere in the lower alpine zone. British Columbia and Alberta south to Washington, Utah, Arizona, and New Mexico.

Middle Rocky Mountain alpine plants may be distinguished as follows:
1. Plants pubescent; akenes puberulent
    . . . var. *inamoenus*
1. Plants subglabrate; akenes glabrous
    . . . var. *alpeophilus* (A. Nels.) Benson

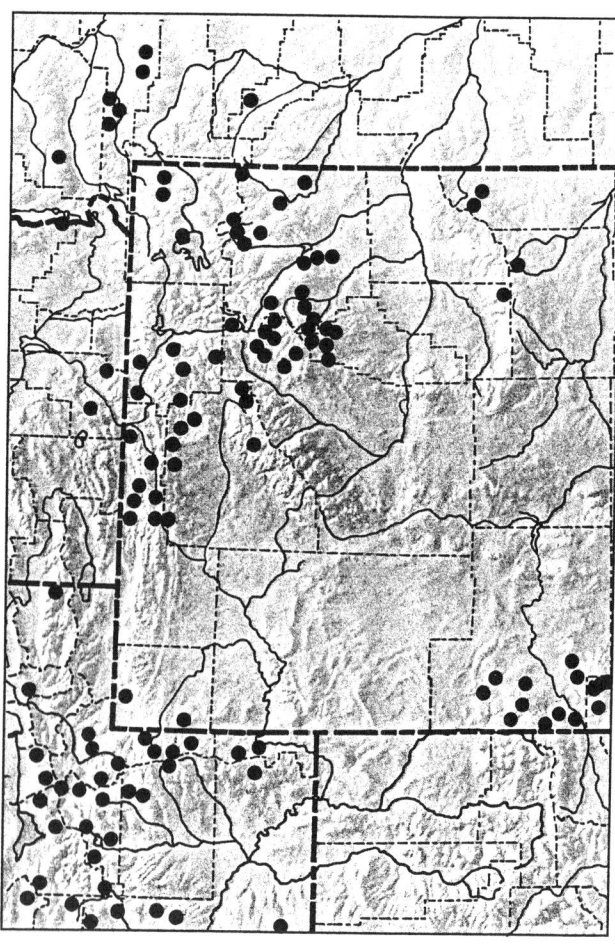

Ranunculus pedatifidus Smith in Rees
Cycl. 29: *Ranunculus*, 72, 1814.

Birdfoot Buttercup

Ranunculus affinis R. Br.
Ranunculus apetalus Farr

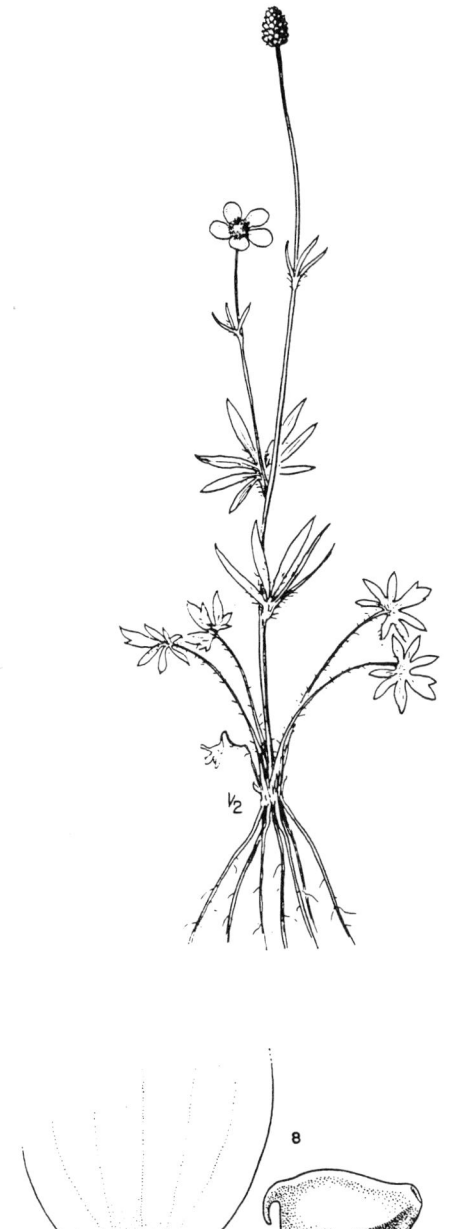

Erect, pilose, perennial herb with 1 to several stems from a short caudex and fibrous roots. Basal leaves long-petiolate; blades cordate, palmately 5–7-cleft; cauline leaves sessile or short-petiolate, 5–7-cleft, the ultimate divisions linear. Flowers 1 to a few. Sepals spreading, villous about half the length of the petals. Petals 5, yellow, obovate; nectary scale glabrous, forming a small pocket. Stamens numerous. Akenes obovate, puberulent to glabrate, in a cylindric or elliptic head; stylar beak slender, strongly recurved.

Meadows and fellfields of the Absaroka, Beartooth, Medicine Bow, and Uinta Ranges. Circumpolar; Alaska south to New Mexico and Arizona.

North American plants are ssp. *affinis* (R. Br.) Hult.

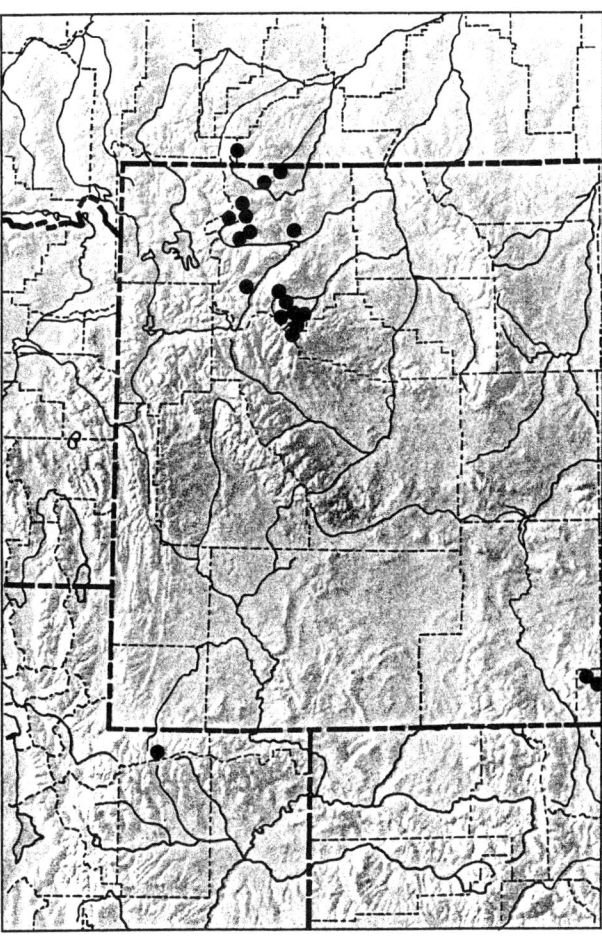

Ranunculus pygmaeus Wahlenb.
Fl. Lappl., 157, 1812.

Dwarf Buttercup

Ranunculus sabinii R. Br.

Dwarf, perennial herb, mostly less than 5 cm tall, from a stout caudex and fibrous roots; stems ascending to erect, glabrous or sparsely pubescent. Basal leaves petiolate; blade cordate to reniform, 3-lobed, the lateral lobes again 2-lobed, the middle lobe entire; cauline leaves reduced, sessile, 3-cleft, the lobes linear to narrowly lanceolate. Flowers solitary. Sepals 5, spreading, pubescent. Petals 5, yellow, short-clawed, about as long as the sepals; nectary scale truncate, forming a small pocket. Stamens numerous. Akenes obovate, glabrous, in a subglobose head; stylar beak short, slender, straight or slightly curved.

Margins of snowbanks, snowmelt seeps, stream banks, and patterned ground in the Absaroka, Beartooth, Big Horn, Medicine Bow, and Wind River Ranges. Circumpolar; south in western North America to Montana, Wyoming, and Colorado.

North American alpine plants with petals less than 5 mm long are ssp. *pygmaeus*.

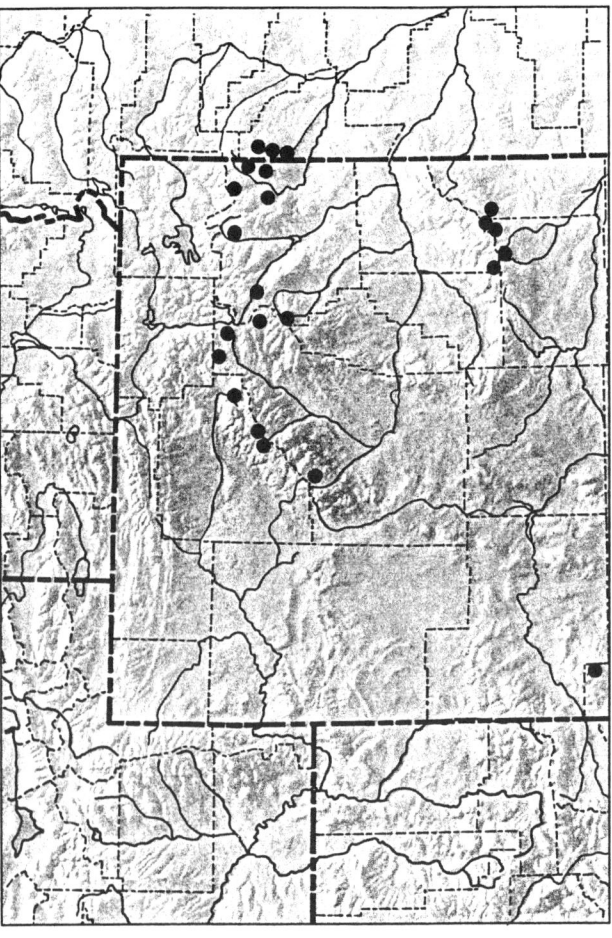

Ranunculus verecundus Robins. ex Piper
Contr. U.S. Natl. Herb. 11:274, 1906.

Timberline Buttercup

Glabrous or glabrate, perennial herb from a stout caudex and slender, fibrous roots; stems 1 to several, decumbent or ascending, glabrous below and usually puberulent to villous above. Basal leaves long-petiolate; blades cordate or reniform, palmately 3-cleft, the lobes again crenately lobed; cauline leaves sessile, 3–5-cleft into oblanceolate or linear, entire segments. Flowers 1–5. Sepals 5, spreading, pubescent, purplish, about the same length as the sepals. Petals 5, yellow, broadly obovate, often shallowly retuse; nectary scale glabrous, forming a very shallow pouch. Stamens numerous. Akenes obovate, glabrous, in a cylindric head; stylar beak stout, recurved.

Talus and scree in the Absaroka Range. British Columbia and Alberta south to Oregon, Idaho, and northwestern Wyoming.

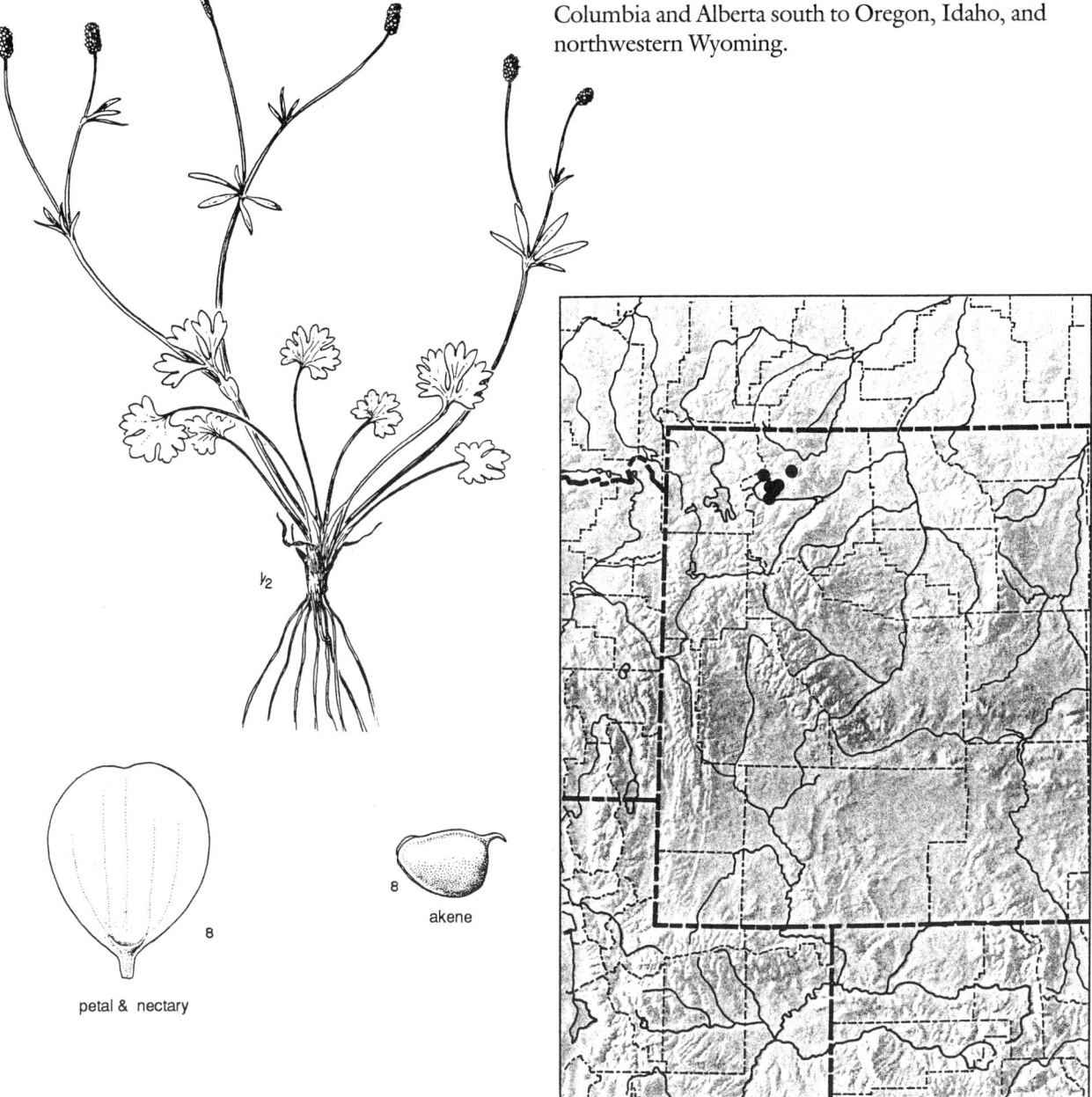

Thalictrum L.
Meadowrue

Erect, perennial herbs from thickened roots, with ternately compound leaves, the divisions and leaflets stalked. Flowers small, greenish white, and perfect, unisexual, or a mixture of the two, in panicles or racemes. Sepals 4–5, petaloid or greenish. Petals lacking. Stamens numerous, exserted, the filaments dilated toward the apex. Pistils few to several, in fruit forming a head of sessile or short-stipitate, ribbed or nerved akenes.

KEY TO THE SPECIES OF *THALICTRUM*

1. Plants scapose or subscapose; inflorescence a raceme; flowers perfect — *T. alpinum*
1. Plants caulescent; inflorescence a panicle; flowers unisexual, the plants dioecious
    2. Akenes more than twice as long as wide, spreading to reflexed; stigma usually purple, the style mostly 3–4.5 mm long — *T. occidentale*
    2. Akenes less than twice as long as wide, ascending to erect; stigma rarely purple, the style mostly 1.5–3 mm long
        3. Inflorescence leafy-bracteate; leaflets greenish beneath, lacking prominent veins; akenes strongly compressed — *T. fendleri*
        3. Inflorescence lacking leafy bracts; leaflets glaucous and prominently veined beneath; akenes terete to subterete — *T. venulosum*

Thalictrum alpinum L.
Sp. Pl., 545, 1753.

Alpine Meadowrue

Thalictrum leiophyllum Greene

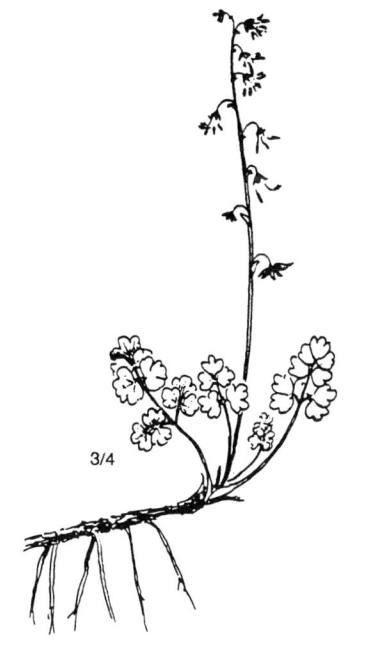

Glabrous to glandular, erect, perennial herb from a slender rhizome; stems mostly simple, somewhat decumbent at the base, scapose or with a single cauline leaf. Leaves basal or nearly so, 2–3 times ternate; leaflets obovate to orbicular, glaucous, 3-lobed, broadly cuneate at the base, the margins revolute. Flowers perfect, in a loose, bracteate raceme, the pedicels often recurved. Sepals 5, elliptic, reddish or purplish. Stamens 8–15, filaments purplish, anthers brownish, apiculate. Pistils 3–5. Akenes lanceolate, ovate or obovate, with prominent, thick ribs, the beak slender, recurved.

Meadows and fellfields in the Absaroka and Wind River Ranges. Circumpolar; south in western North America to Oregon, Utah, and New Mexico.

North America material is var. *hebetum* B. Boivin.

Thalictrum fendleri Engelm. ex Gray
Pl. Fendl., 5, 1849.

Fendler Meadowrue

Thalictrum polycarpum (Torr.) S. Wats.
Thalictrum stipitatum Rydb.

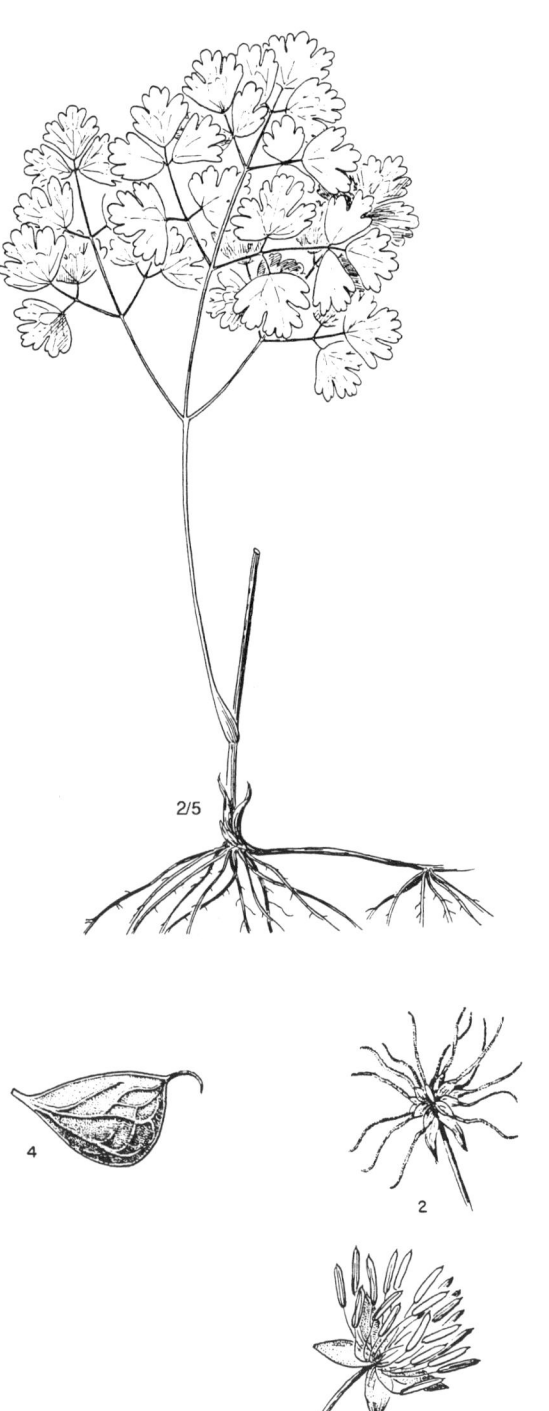

Slender, glabrous to glandular-pubescent, perennial herb from a rhizome; stems simple, usually purplish and branched above the middle. Leaves mostly cauline, 3–4-ternate; leaflets ovate to orbicular, 3-lobed and crenately 2–3-toothed, glabrous above and glabrous to glandular-puberulent beneath. Flowers unisexual, in an open, leafy-bracteate panicle. Sepals whitish to green. Stamens numerous, the anthers shorter than the filaments. Pistils 7–16. Akenes strongly compressed, obovate, prominently nerved, glabrous to glandular, the beak stout, recurved.

Timberline and krummholz margins in the lower alpine zone of the Beartooth and Wind River Ranges. Oregon to Wyoming and south to Baja California and New Mexico.

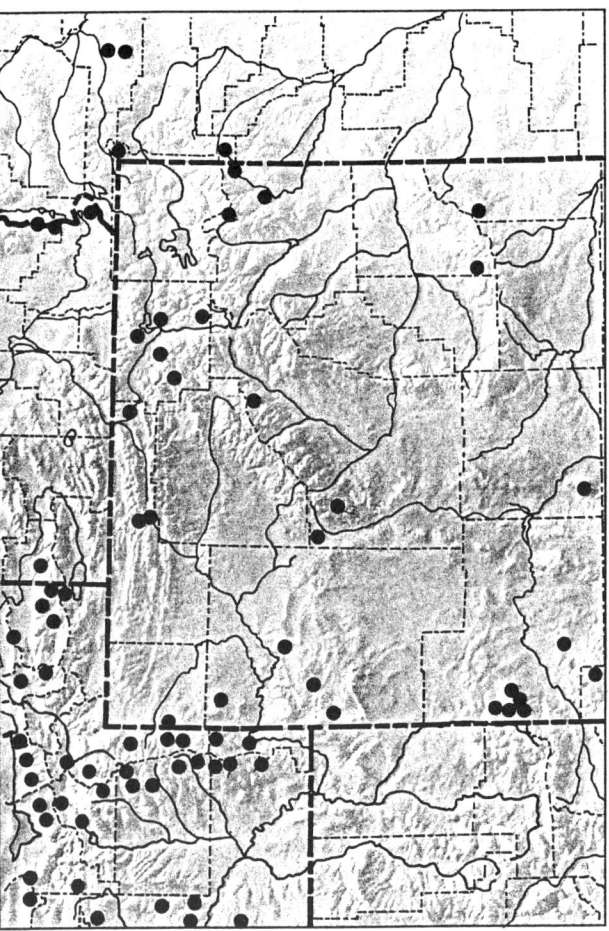

668 Ranunculaceae

Thalictrum occidentale Gray
Proc. Am. Acad. 8:372, 1872.

Western Meadowrue

Thalictrum breitungii B. Boivin
Thalictrum megacarpum Torr.
Thalictrum propinquum Greene
Thalictrum rainierense St. John

Glabrous to glandular-puberulent, erect, perennial herb from a rhizome; stems stout, simple. Leaves mostly cauline, 3–4-ternate; leaflets often glaucous, obovate to orbicular or cordate, 3-lobed, the lobes crenately 2–3-lobed or toothed. Flowers unisexual, in an open, leafy-bracteate panicle. Sepals whitish to green, or sometimes purplish. Stamens numerous, the anthers shorter than the filaments. Pistils 10–18. Akenes slightly compressed, elliptic, tapering equally at both ends, mostly glandular-puberulent, the beak straight to only slightly recurved.

Krummholz and timberline margins in the lower alpine zone of the Absaroka, Beartooth, and Wind River Ranges. British Columbia to Alberta, south to California, Idaho, and Wyoming.

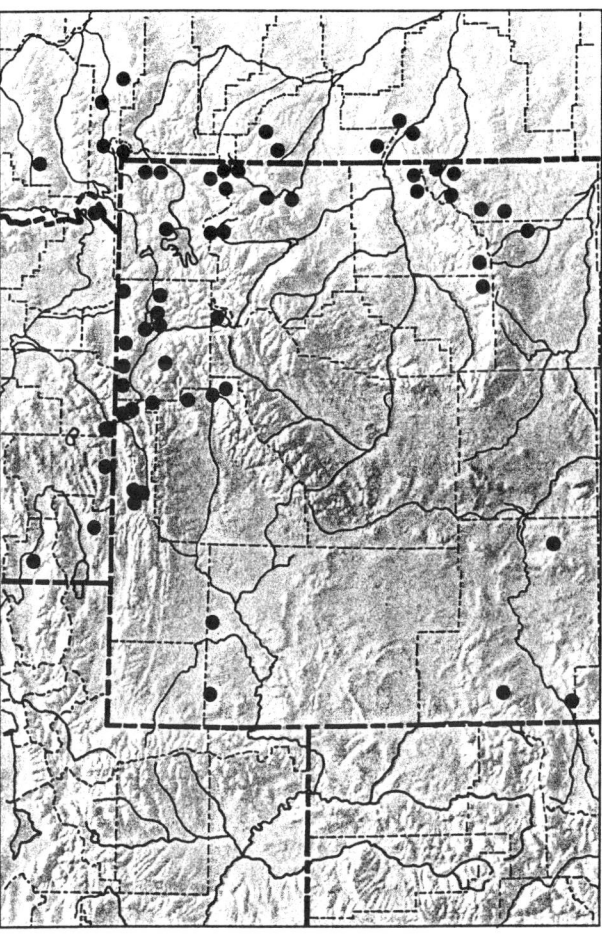

Thalictrum venulosum Trel.
Proc. Bost. Soc. Nat. Hist. 23:302, 1886.

Veiny Meadowrue

Thalictrum columbianum Rydb.
Thalictrum confine Fern.
Thalictrum fissum Greene
Thalictrum turneri B. Boivin

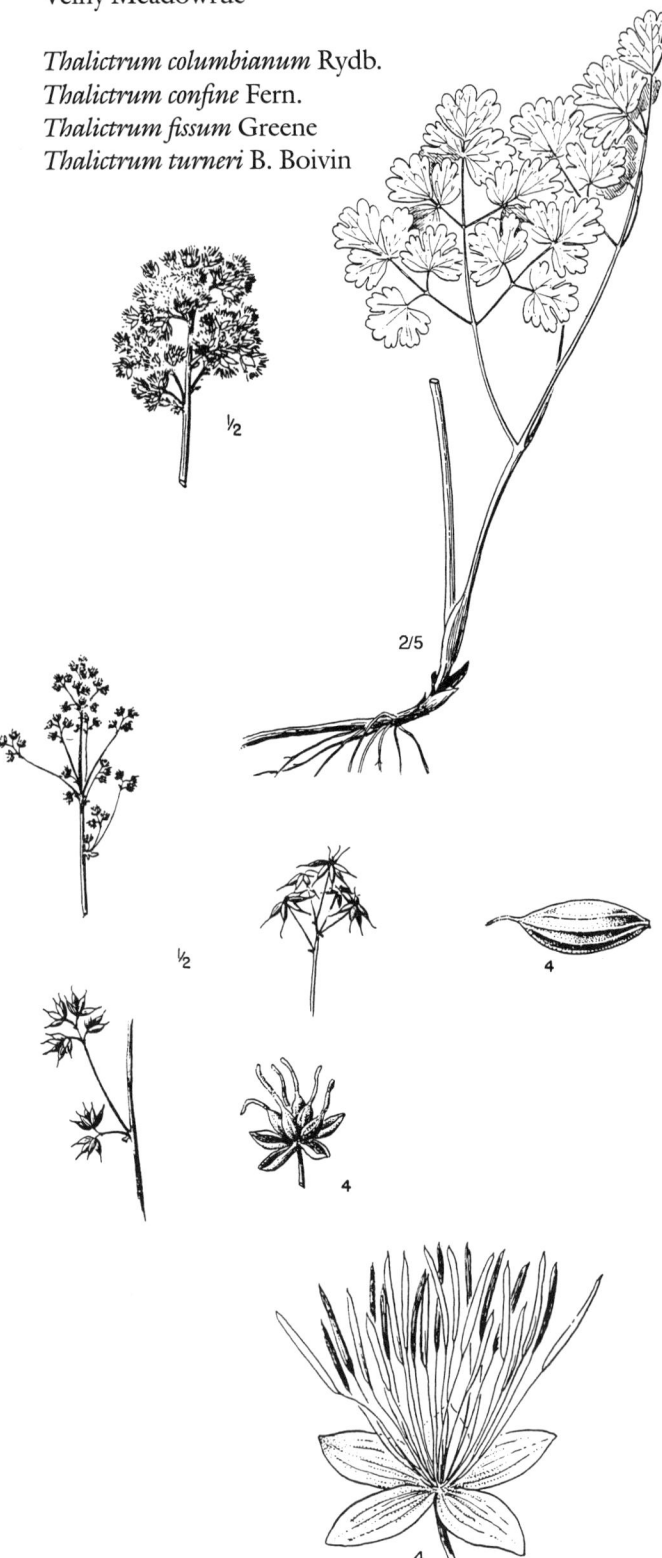

Slender, glabrous to sparsely glandular-puberulent, perennial herb from a rhizome; stems hollow, single or somewhat branched. Leaves long-petiolate, 2–4-ternate; leaflets ovate to orbicular, 3-lobed, the lobes crenately 2–3-toothed, prominently veined and glaucous beneath. Flowers unisexual, in a compact, narrow to open panicle. Sepals 4, spreading, greenish white to purplish. Stamens numerous, the anthers shorter than the filaments. Pistils 5–16. Akenes terete to subterete, obovate, ribbed, the beak slender, recurved.

Woods, thickets, and other shaded, moist habitats, mostly at lower elevations but occurring along timberline margins of the Absaroka Range. British Columbia and Alberta south in western North America to Oregon, New Mexico, and Nebraska.

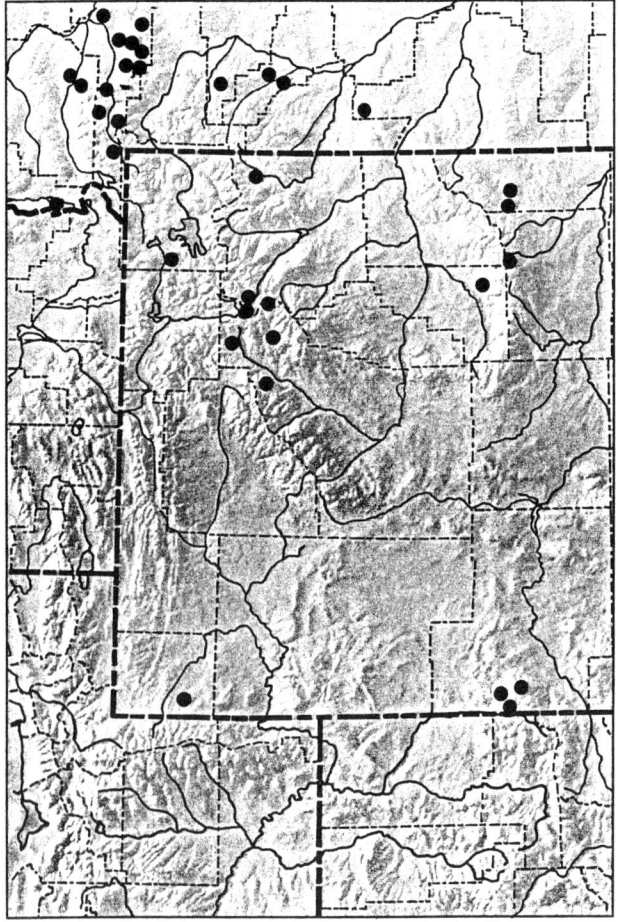

Trollius L.
Globeflower

Glabrous, perennial herbs with palmately lobed or parted leaves. Flowers solitary, terminal, pale yellow in ours. Sepals 5–15, petaloid. Petals 5–15, greenish, smaller than the sepals, with a minute nectiferous pit at the base. Stamens numerous. Pistils numerous, becoming several-seeded follicles in fruit.

Trollius laxus Salisb.
Trans. Linn. Soc. 8:303, 1807.

Globeflower

Trollius albiflorus (Gray) Rydb.
Trollius americanus DC.

Glabrous, perennial herb from a short caudex and fibrous roots; stems 1 to several, erect, 2–4-leaved. Basal leaves long-petiolate; blades cordate to reniform, palmately 5-cleft, the segments obovate, toothed; cauline leaves cleft like the basal ones, reduced upward, short-petiolate, becoming sessile. Flowers large, showy, terminal, solitary on each peduncle. Sepals 5–9, ovate to obovate, petaloid, cream to white. Petals reduced and inconspicuous or lacking. Stamens numerous. Pistils numerous. Follicles transversely ribbed, slender-beaked.

Stream banks, wet meadows, and receding snowbanks in the lower alpine zone of most Middle Rocky Mountain ranges. Not yet observed in the alpine zone of the Hoback, Salt River, Wasatch, and Wyoming Ranges. British Columbia and Alberta, south in western North America to Washington, Idaho, and Colorado.

Plants of western North America are var. *albiflorus* Gray.

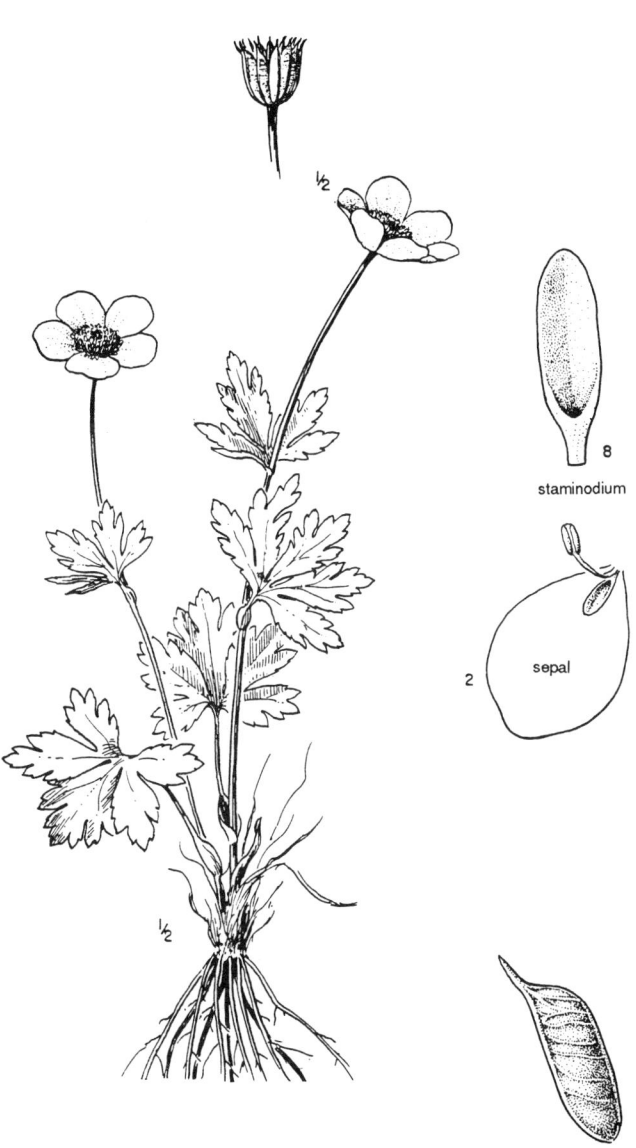

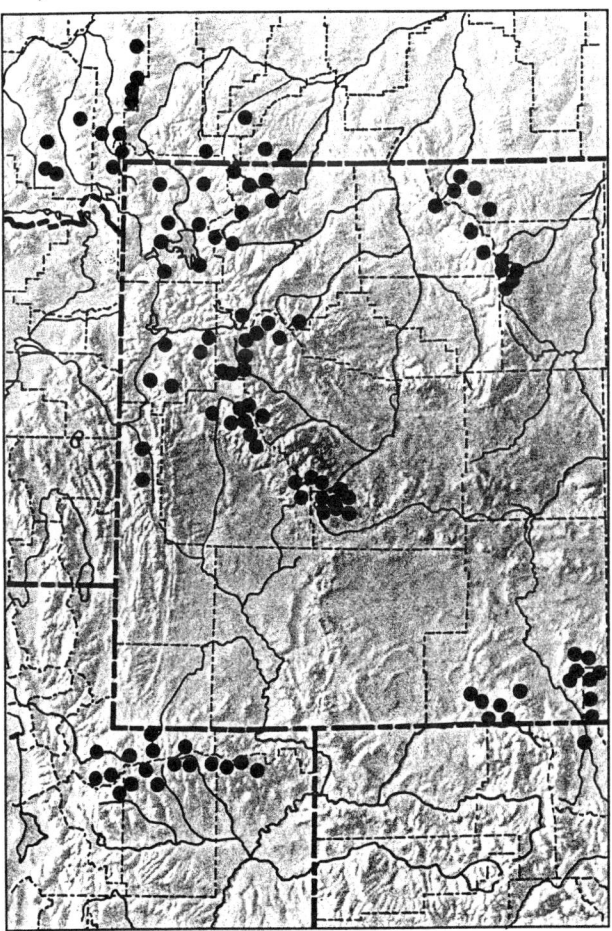

Rosaceae
Rose Family

Herbs, shrubs, or trees with alternate or occasionally opposite, simple or compound, stipulate leaves. Flowers perigynous, 5-merous, perfect, regular, the hypanthium cup-shaped or urceolate, free or adnate to the ovary. Inflorescence usually a raceme or cyme, or the flowers solitary. Sepals usually alternating with 5 bractlets. Petals occasionally lacking. Stamens distinct, numerous, in whorls of 5, usually 15 or more, rarely fewer. Carpels 1 to many, distinct or united, the styles distinct or connate. Fruit a pome, a drupe, or a group of akenes, follicles, or drupelets.

KEY TO THE GENERA OF ROSACEAE

1. Plants with woody stems; erect shrubs
   2. Stems armed with sharp prickles; fruit fleshy, an aggregate of drupelets — *Rubus*
   2. Stems not armed; fruit dry, an akene or a group of akenes
      3. Leaves simple, entire, the margins revolute; petals lacking; fruits with an elongate, plumose style — *Cercocarpus*
      3. Leaves toothed, lobed, or compound, the margins plane; petals present; fruits lacking an elongate, plumose style
         4. Flowers single and axillary, or 3–9 in cymes; petals yellow, more than 5 mm long — *Potentilla*
         4. Flowers numerous, in compound panicles; petals white, pale yellowish, or pink, less than 3 mm long — *Holodiscus*
1. Plants herbaceous, or if woody then the stems low, sprawling or depressed, and the plants mat-forming
   5. Leaves simple; plants forming dense mats
      6. Flowers in narrow racemes or panicles — *Petrophyton*
      6. Flowers solitary
         7. Leaves with crenate, often revolute margins, white-tomentose beneath; stipules present — *Dryas*
         7. Leaves entire, with flattened margins, lacking tomentum; stipules lacking — *Kelseya*
   5. Leaves compound or deeply pinnatifid; plants not forming dense mats
      8. Stamens 5; pistils 15 or fewer
         9. Leaves ternately compound, the leaflets 3-toothed at the apex — *Sibbaldia*
         9. Leaves pinnately compound, the leaflets dissected — *Ivesia*
      8. Stamens 10 or more; pistils numerous
         10. Receptacle enlarged, hemispheric, becoming fleshy in fruit; petals white; plants stoloniferous; leaves trifoliolate — *Fragaria*
         10. Receptacle neither enlarged nor becoming fleshy in fruit; petals mostly yellow, sometimes cream or white; plants stoloniferous or not; leaves pinnate, lyrate or trifoliolate
            11. Styles persistent or only the terminal segment deciduous, elongate and plumose in fruit; leaves lyrate — *Geum*
            11. Styles deciduous or short and inconspicuous in fruit, never plumose; leaves rarely lyrate — *Potentilla*

Cercocarpus H.B.K.
Mountain Mahogany

Erect shrubs or small trees with alternate, simple, evergreen, entire or toothed leaves; stipules adnate to the petioles. Flowers small, perfect, nonshowy, solitary or several, axillary. Sepals 5, deciduous, the hypanthium tubular. Petals lacking. Stamens 15–25, in 2–3 whorls on the limb of the hypanthium. Pistil 1, the style terminal. Fruit a terete, villous akene, the style persistent, elongate, and plumose.

Cercocarpus ledifolius Nutt. ex T. & G. *Fl. N. Am.* 1:427, 1840.

Curl-Leaf Mountain Mahogany

Cercocarpus hypoleucus Rydb.
Cercocarpus intricatus S. Wats.

Erect, evergreen shrub with short branches; bark grayish on young branches, becoming reddish and furrowed. Leaves short-petiolate, narrowly elliptic, lanceolate, or sometimes oblanceolate, entire, revolute, glabrous to glabrate and shiny above, densely white-tomentose beneath. Flowers sessile, solitary or 2–3 in the leaf axils. Calyx tomentose to villous, the sepals deciduous; hypanthium tubular, persistent. Petals lacking. Stamens 15–25, the anthers glabrous. Pistil 1, with 1 carpel. Akenes terete, villous; style elongate, plumose.

Rocky ridges, talus, and scree at timberline in the Wasatch Range. Washington and California, south and east to Arizona, Colorado, Wyoming, and Montana.

Plants that occur at or near timberline in the Wasatch Range with leaves more than 3 mm wide and margins only weakly revolute are var. *ledifolius*.

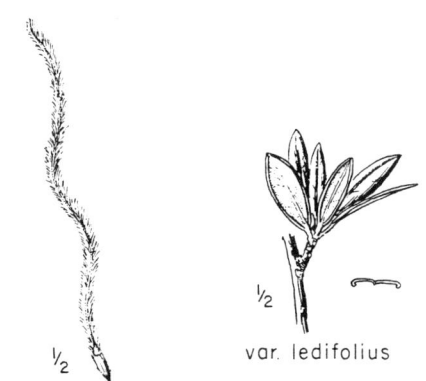

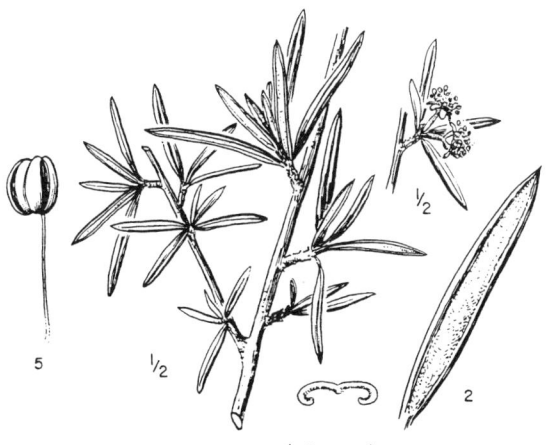

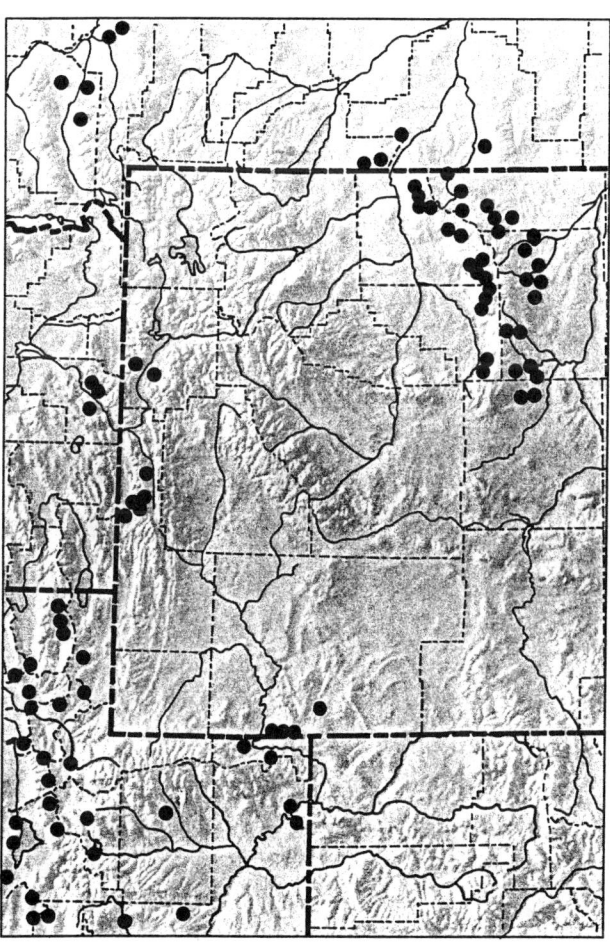

Dryas L.
Mountain Dryad

Low, prostrate shrubs, often rooting from the branches and forming large, loose mats. Leaves evergreen, alternate, white-tomentose beneath in ours, the margins crenate. Flowers showy, white or cream, on elongate, naked peduncles. Sepals 8–10, persistent. Petals 8–10, oval to obovate. Stamens numerous, inserted at the mouth of the hypanthium. Pistils numerous, the styles terminal and persistent. Fruit a group of akenes, the styles becoming elongate and plumose in fruit.

Dryas octopetala L.
Sp. Pl., 501, 1753.

Alpine Dryad

Dryas alaskensis Pors.
Dryas hookeriana Juz.
Dryas kamtschatica Juz.
Dryas octopetala Kuntze
Dryas punctata Juz.

Prostrate, mat-forming, perennial shrub with horizontal branches. Leaves oblong to ovate, the margins crenate and somewhat revolute, subcordate at the base, dark green, rugose and glabrous or glabrate above, white-tomentose beneath; veins with sessile to stalked, brownish glands. Flowers solitary; peduncles white-tomentose, with purplish black glands, often with a single reduced bract. Calyx white-tomentose, with purplish black glands, the lobes narrowly lanceolate. Petals white to cream, elliptic to obovate. Stamens numerous. Pistils numerous. Akenes with plumose styles.

Fellfields and exposed ridges in most of the Middle Rocky Mountain alpine ranges; not yet known from above timberline in the Hoback, Medicine Bow, Salt River, and Wasatch Ranges. Circumpolar; south in western North America to Oregon, Idaho, and Colorado.

Rocky Mountain plants are var. *hookeriana* (Juz.) Breit. Those with extremely narrow leaves (less than 5 mm wide) have been called var. *angustifolia* C.L. Hitchc.

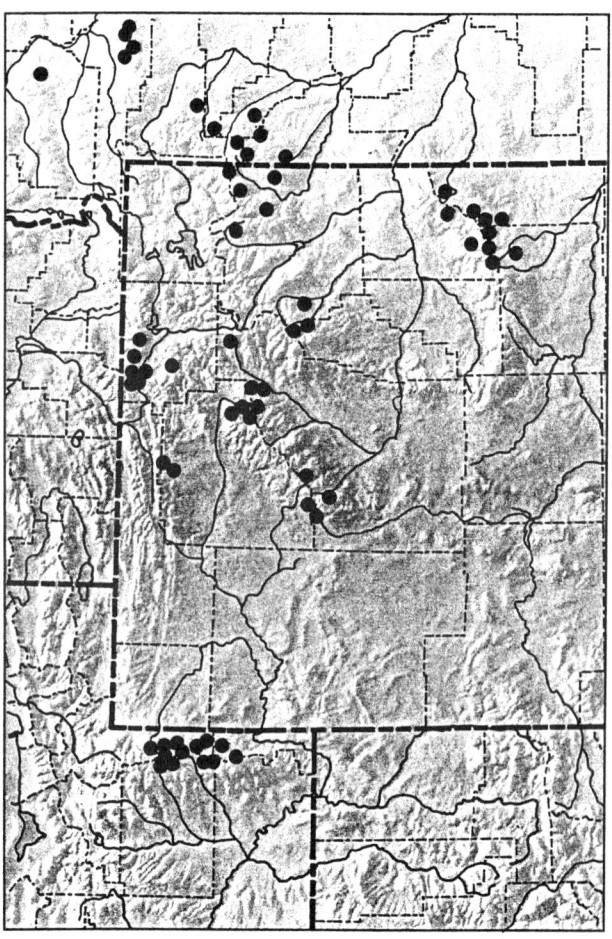

½

Fragaria L.
Strawberry

Perennial, stoloniferous herbs from short, scaly rootstocks. Leaves long-petiolate, mostly basal, trifoliolate, the leaflets serrate; stipules membranous, adnate to the base of the petiole. Flowers showy, perfect to unisexual, solitary to several in bracteate cymes or racemes. Sepals 5, ovate to lanceolate, alternating with large, foliaceous bracts of about the same size. Petals 5, white to pink, obovate, rounded, deciduous. Stamens 20–40. Pistils numerous on a hemispheric or conic receptacle that enlarges and becomes fleshy, juicy, and red in fruit; styles short, attached laterally on the ovaries. Akenes numerous, brownish, on the surface or embedded in pits on the surface of the receptacle.

Fragaria vesca L.
Sp. Pl., 494, 1753.

Wild Strawberry

Fragaria americana (Porter) Britt.
Fragaria australis (Rydb.) Rydb.
Fragaria bracteata A. Heller
Fragaria californica Cham. & Schlecht.
Fragaria canadensis Michx.
Fragaria crinita Rydb.
Fragaria glauca (S. Wats.) Rydb.
Fragaria grayana Vilm. ex J. Gay
Fragaria helleri Holz.
Fragaria latiuscula Greene
Fragaria multicipita Fern.
Fragaria ovalis (Lehm.) Rydb.
Fragaria pauciflora Rydb.
Fragaria platypetala Rydb.
Fragaria prolifica Baker & Rydb.
Fragaria pumila Rydb.
Fragaria retrorsa Greene
Fragaria sibbaldifolia Rydb.
Fragaria suksdorfii Rydb.
Fragaria truncata Rydb.
Fragaria virginiana Duchn.
Potentilla ovalis Lehm.
Potentilla vesca (L.) Scop.

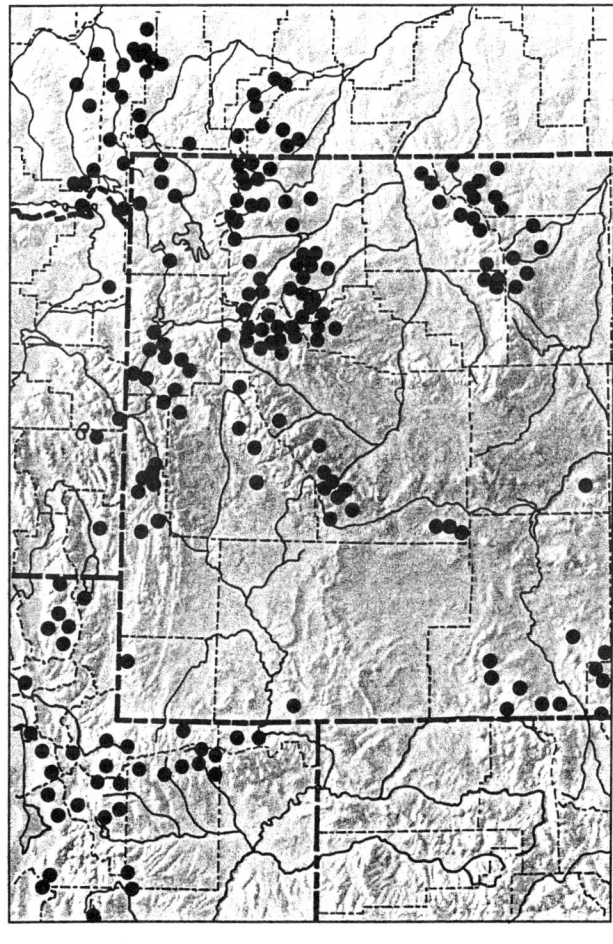

Pubescent, rosulate, stoloniferous herb from thick rootstocks; scapes shorter than to longer than the leaves. Leaves long-petiolate, trifoliolate, the leaflets short-petiolulate, narrowly to broadly obovate, glabrous and often glaucous on the upper surface, silky-villous on the lower surface, serrate; petioles glabrous, appressed-strigose, or spreading-villous. Flowers perfect to unisexual, few to many in bracteate cymes or racemes. Sepals linear-lanceolate, acuminate, longer and wider than the bracteoles. Petals white to pink, obovate to nearly orbicular, rounded. Fruit fleshy, the akenes sunken in pits or on the surface of the receptacle.

Typically of lower-elevation forest margins, meadows, and stream banks, this species occurs at timberline in the Absaroka, Beartooth, and Uinta Ranges, where it is often locally common on talus and scree. Alaska, Yukon Territory and Mackenzie district south in western North America to California, Utah, and New Mexico.

In the timberline specimens that I have examined I find no consistent correlation between the characters traditionally used to separate *F. virginiana* from *F. vesca:* the relative length of the terminal tooth on each leaflet relative to the adjacent teeth and the height of the inflorescence relative to the leaves. If, following Arnow et al. (1980), the two are treated as a single species, our alpine plants would be var. *bracteata* (A. Heller) Davis.

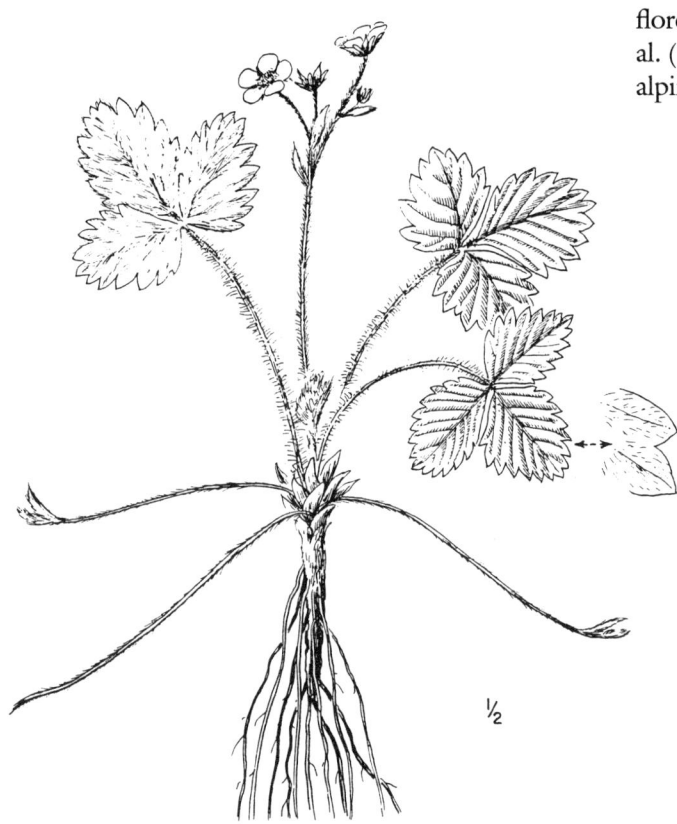

Geum L.
Avens

Rhizomatous, perennial herbs with mostly basal, pinnate or pinnatifid leaves. Cauline leaves 3-lobed or trifoliolate, the upper ones often simple. Inflorescence an open, bracteate cyme or corymb. Sepals 5, erect to reflexed, with 5 alternating bractlets. Petals 5 and yellow, cream, or reddish. Stamens numerous. Pistils numerous, the styles terminal, entire or jointed. Fruit a group of akenes, each tipped by the persistent, plumose style.

KEY TO THE SPECIES OF GEUM

1. Sepals reflexed in flower; styles geniculate and hooked at the tip — *G. macrophyllum*
1. Sepals erect in flower; styles straight
    2. Flowers rotate, erect; petals yellow, broadly obovate — *G. rossii*
    2. Flowers campanulate, nodding; petals pink, red, or cream, elliptic — *G. triflorum*

Geum macrophyllum Willd.
Enum. Pl. Hort. Berol., 557, 1809.

Largeleaf Avens

Geum perincisum Rydb.
Geum oregonense Rydb.

Erect, hirsute to hispid, perennial herb from a short caudex and horizontal rhizome. Basal leaves interrupted lyrate-pinnate, hirsute; leaflets crenate or dentate, sometimes sharply incised, the terminal segment much larger than the others. Cauline leaves 3–5-lobed, short-petiolate to nearly sessile, reduced upward. Flowers in asymmetrical cymes. Hypanthium short, saucer-shaped. Sepals 5, pubescent, triangular, reflexed. Petals 5, yellow. Stamens numerous. Pistils numerous. Akenes hirsute, the styles geniculate; lower persistent stylar segment glandular, hooked at the tip.

Stream banks and willow thickets of the timberline ecotone in the Absaroka and Wind River Ranges. Asia; Alaska south to Colorado, South Dakota, Michigan, and Vermont; Mexico.

Rocky Mountain plants are var. *perincisum* (Rydb.) Raup.

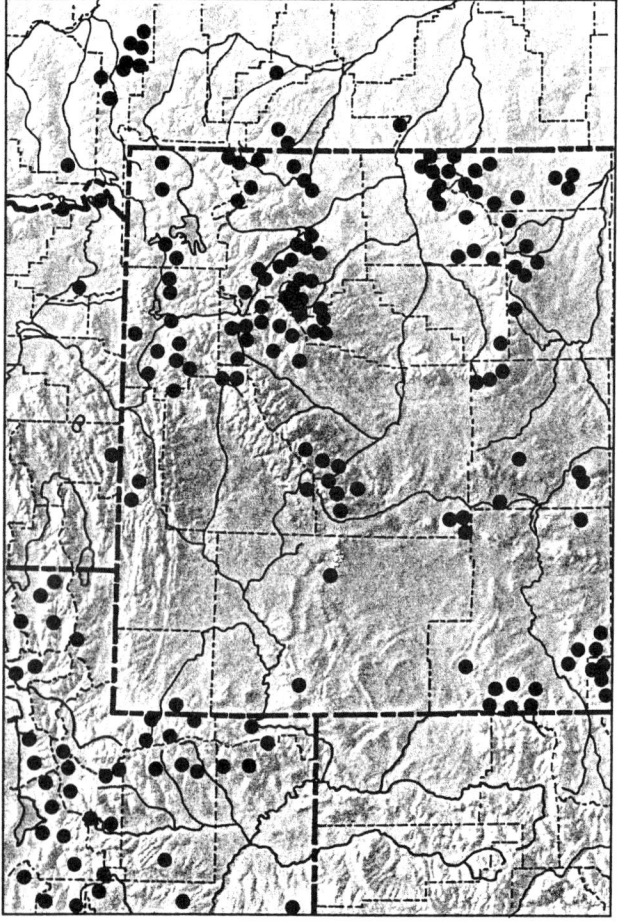

Geum rossii (R. Br.) Ser. in DC.
Prodromus 2:553, 1825.

Alpine Avens

Sieversia rossii R. Br.
Acomastylis depressa Greene
Acomastylis gracilipes (Piper) Greene
Acomastylis humilis (R. Br.) Rydb.
Acomastylis rossii (R. Br.) Greene
Acomastylis sericea (Greene) Greene
Acomastylis turbinata (Rydb.) Greene
Geum gracilipes (Piper) M.E. Peck
Geum sericeum Greene
Geum turbinatum Rydb.
Potentilla gracilipes Piper
Potentilla nivalis Torr.
Sieversia gracilipes (Piper) Greene
Sieversia humilis R. Br.
Sieversia sericea (Greene) Greene
Sieversia turbinata (Rydb.) Greene

Erect, perennial herb from a thick, woody caudex. Basal leaves crowded, interrupted pinnate to pinnatifid; leaflets oblanceolate to obovate, entire to coarsely 3–5-toothed, glabrate to silky-sericeous. Cauline leaves alternate, sessile, pinnatifid, reduced upward. Flowers 1–4; hypanthium turbinate. Sepals 5, triangular, often purplish. Petals 5, yellow, broadly obovate, sometimes retuse. Stamens numerous. Pistils numerous, rarely fewer than 10. Akenes pubescent, the styles glabrous.

Fellfields of nearly all Middle Rocky Mountain alpine ranges; not yet known from above timberline in the Hoback Range. Siberia; Alaska south to Oregon, Nevada, Arizona, and New Mexico.

Alpine plants of the Middle Rocky Mountains are var. *turbinatum* (Rydb.) C.L. Hitchc.

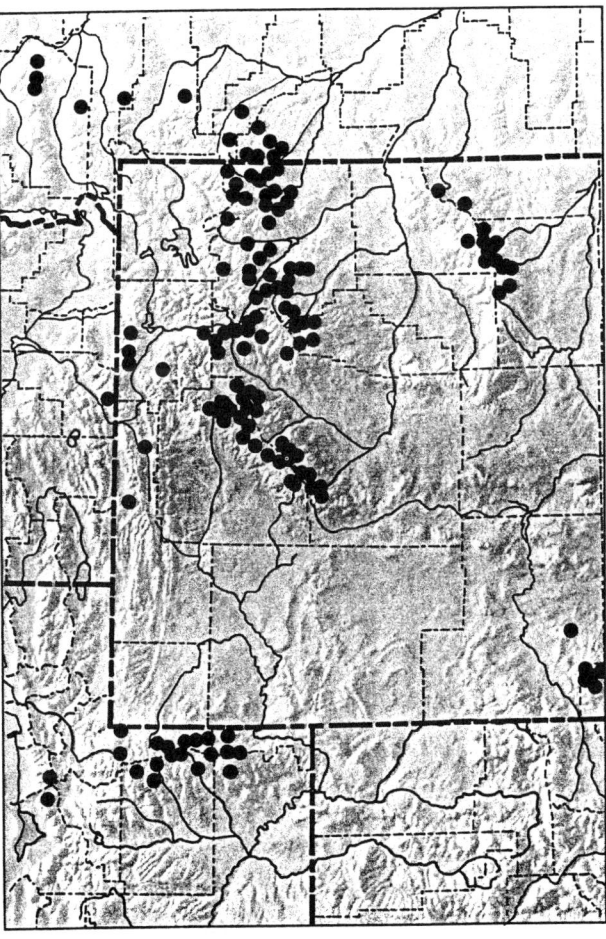

Geum triflorum Pursh
Fl. Am. Sept., 736, 1814.

Old Man's Whiskers

Erythrocoma affinis Greene
Erythrocoma campanulata Greene
Erythrocoma canescens Greene
Erythrocoma ciliata (Pursh) Greene
Erythrocoma flavula Greene
Erythrocoma grisea Greene
Erythrocoma triflora (Pursh) Greene
Geum campanulatum (Greene) G.N. Jones
Geum canescens (Greene) Munz
Geum ciliatum Pursh
Sieversia campanulata (Greene) Rydb.
Sieversia canescens (Greene) Rydb.
Sieversia ciliata (Pursh) G. Don
Sieversia flavula (Greene) Rydb.
Sieversia grisea (Greene) Rydb.
Sieversia rosea Grah.
Sieversia triflora (Pursh) R. Br.

Erect, perennial herb from a thick, scaly rhizome; stems usually reddish purple, pubescent. Basal leaves pubescent, interrupted lyrate-pinnate or pinnatifid; leaflets oblanceolate, irregularly cleft or lobed, the larger terminal segment 3-lobed, the lobes terminally toothed. Cauline leaves 2, opposite, subsessile, pinnatifid, much smaller than the basal leaves. Flowers 1–9, mostly 3, nodding, in cymes; hypanthium hemispheric, reddish purple. Sepals 5, reddish purple, triangular; bractlets 5, linear, mostly longer than the sepals, often reflexed. Petals 5, cream to pinkish, elliptic. Stamens numerous. Akenes pubescent, plumose, the terminal segment glabrous.

Meadows and moist slopes of the Absaroka, Beartooth, Big Horn, Uinta, Wasatch, and Wind River Ranges near timberline. Transcontinental; British Columbia and Alberta south to Utah, New Mexico, and Nebraska.

Middle Rocky Mountain plants are var. *ciliatum* (Pursh) Fassett.

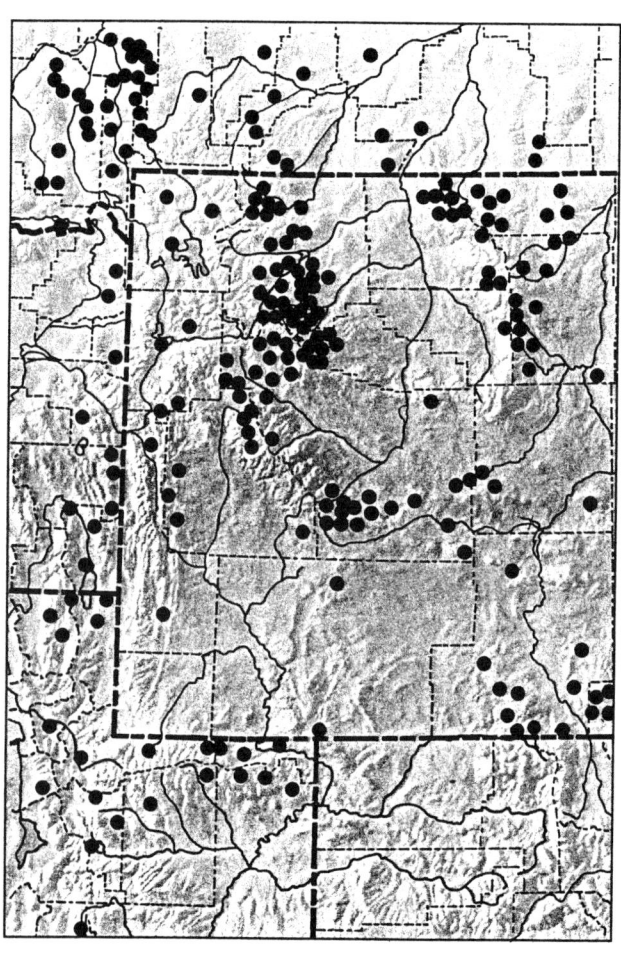

Holodiscus Maxim.
Holodiscus

Erect, unarmed, many-branched shrubs with simple, alternate, toothed to somewhat lobed, exstipulate leaves. Inflorescence a panicle. Sepals 5, erect, persistent, ebracteate; hypanthium shallow, rotate. Petals 5, white to yellowish or pink, elliptic to oval in ours. Stamens numerous. Pistils 5, distinct, pubescent or glandular, the styles terminal. Fruit a pubescent akene borne on a short stipe; style persistent.

Holodiscus dumosus (Nutt. in T. & G.) A. Heller
Cat. N. Am. Pl., 4, 1898.

Mountain Spray

Spiraea dumosa Nutt. in T. & G.
Holodiscus glabrescens (Greenm.) A. Heller ex Jepson
Holodiscus microphyllus Rydb.
Schizonotus dumosus (Nutt.) Koehne
Sericotheca concolor Rydb.
Sericotheca dumosa (Nutt.) Rydb.
Sericotheca glabrescens (Greenm.) Rydb.
Sericotheca microphylla (Rydb.) Rydb.

Low, many-branched shrub; bark of young branches reddish, smooth, becoming gray or grayish red and exfoliating with age. Leaves simple, short-petiolate, the blades elliptic, ovate, or obovate, cuneate at the base, toothed or lobed, glabrous to pubescent on one or both surfaces, often glandular. Inflorescence a dense, often compound panicle, sometimes reduced to a raceme. Flowers perfect; sepals villous, whitish to pinkish; petals oval and whitish, yellowish, or pinkish, early deciduous. Stamens included or nearly so. Pistils villous, often glandular; akenes flattened, villous, short-stipitate.

Talus and scree at timberline in the Uinta and Wasatch Ranges. Oregon south and east to Arizona, New Mexico, and Colorado.

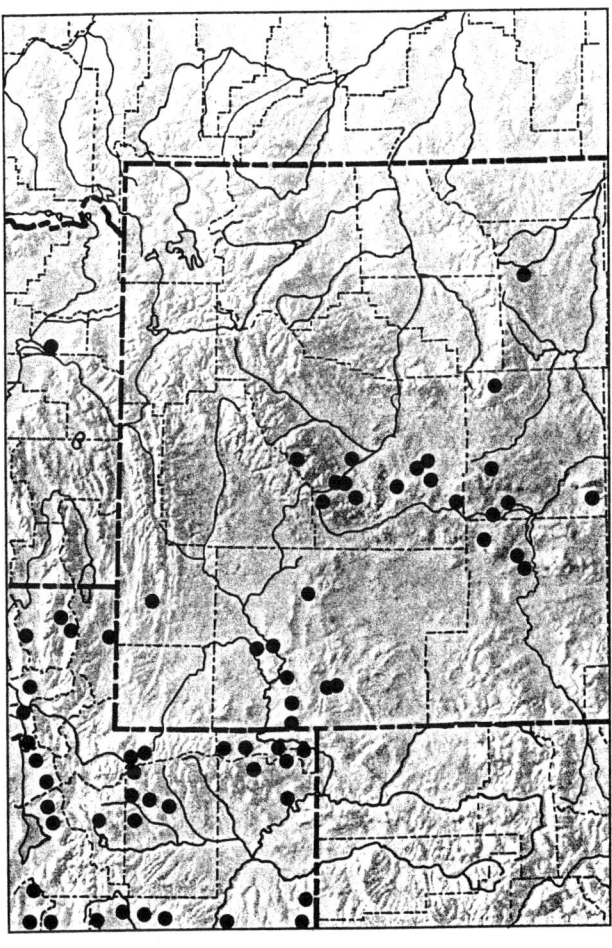

Ivesia T. & G.
Ivesia

Perennial herbs from taproots with pinnately compound, mostly basal leaves. Flowers perfect, complete, in terminal, usually dense cymes. Sepals 5. Petals 5 and white, purplish, or yellow, weakly clawed. Stamens 5–20, inserted on the hypanthium. Pistils mostly 2–6; styles straight, deciduous. Fruit an akene.

KEY TO THE SPECIES OF *IVESIA*

1. Flowers yellow; stems erect or ascending — *I. gordonii*
1. Flowers white; stems decumbent-spreading — *I. utahensis*

Ivesia gordonii (Hook.) T. & G. in Newberry
Pac. R.R. Rep. 6(3):72, 1857.

Gordon Ivesia

Horkelia gordonii Hook.
Ivesia alpicola Rydb. ex T.J. Howell
Potentilla gordonii (Hook.) Greene

Tufted, perennial, scapose to subscapose, glandular herb from a stout taproot and branched caudex; stems erect or ascending. Basal leaves pinnately compound, the leaflets deeply incised. Cauline leaves lacking or single and reduced-pinnate; stipules lobed. Flowers in dense, capitate cymes. Hypanthium campanulate or turbinate, the calyx lobes narrowly deltoid, erect. Petals yellow, spatulate, usually clawed, shorter than to nearly equaling the sepals. Stamens 5, the filaments filiform. Pistils 2–6, commonly 2–4, the styles subterminal. Fruit an akene.

Rocky exposed ridges, dry meadows, and fellfields of the Absaroka, Gros Ventre, Teton, Uinta, Wasatch, Wind River, and Wyoming Ranges. Washington to Montana, south to Colorado, Utah, and California.

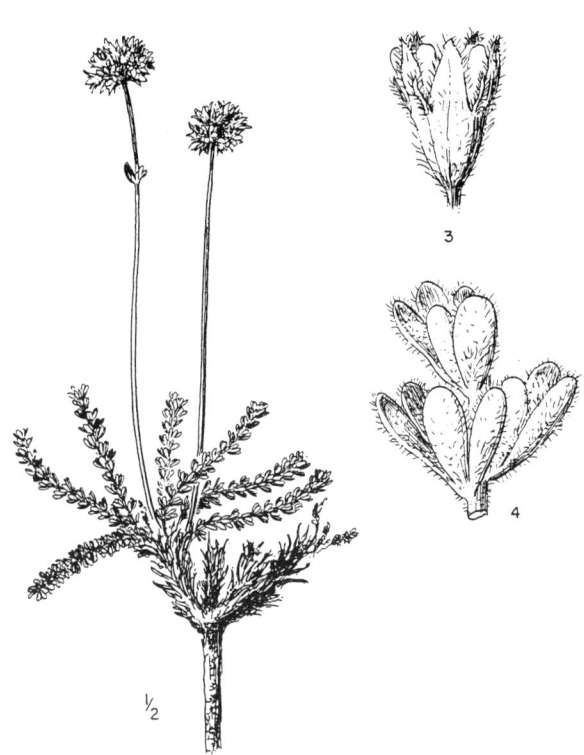

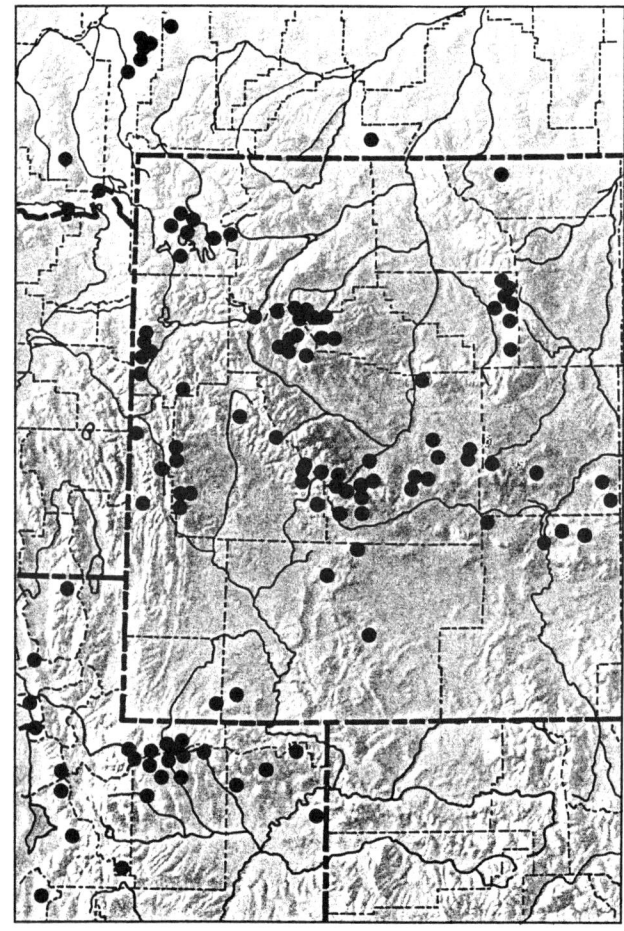

Ivesia utahensis S. Wats.
Proc. Am. Acad. Arts 17:371, 1882.

Utah Ivesia

Horkelia utahensis (S. Wats.) Rydb.
Potentilla utahensis (S. Wats.) Greene

Tufted, glandular, perennial herb from a taproot, the caudex covered with persistent leaf bases; stems decumbent and spreading. Basal leaves pinnate, the leaflets incised to the base; cauline leaves lacking or reduced-pinnate. Flowers in dense, capitate cymes. Hypanthium campanulate, the calyx lobes narrowly deltoid, erect. Petals white, spatulate. Stamens 5. Pistils 2, the styles lacking glands. Fruit an akene.

Talus and rocky ledges near timberline in the Uinta and Wasatch Ranges. Endemic to northern Utah.

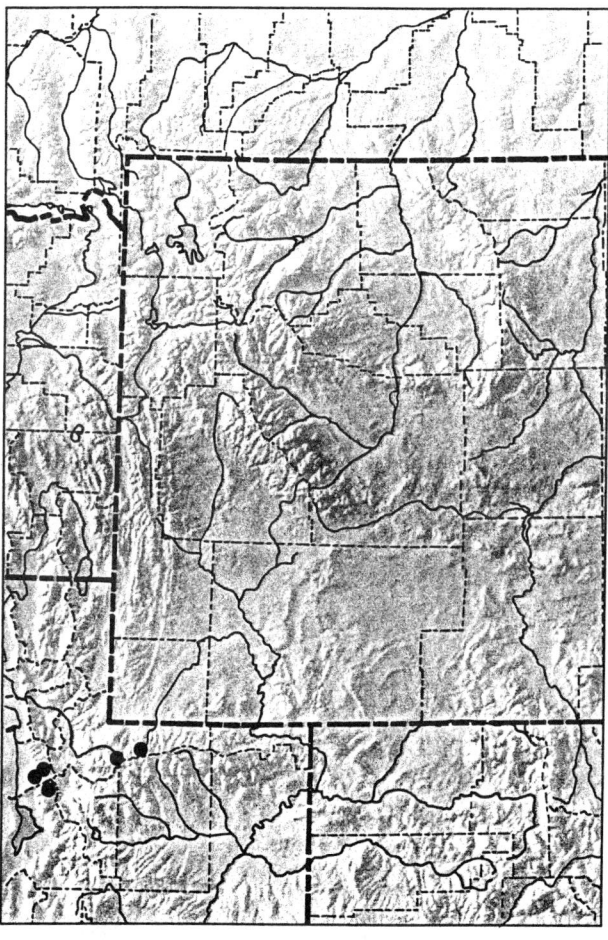

Kelseya (S. Wats.) Rydb.
Kelseya

Low, mat-forming, lithophilous shrubs with crowded, alternate, entire, exstipulate leaves. Flowers pink, terminal on the ends of short branches. Sepals 5; petals 5. Stamens mostly 10, inserted on the hypanthium. Pistils 3–5, usually 3. Fruit follicle-like (dehiscent on 2 sutures).

Kelseya uniflora (S. Wats.) Rydb.
Mem. N.Y. Bot. Gard. 1:207, 1900.

Oneflower Kelseya

Eriogynia uniflora S. Wats.
Luetkea uniflora (S. Wats.) Kuntze
Spiraea uniflora (S. Wats.) Piper

Densely caespitose, highly branched, low shrub with imbricate, oblanceolate, leathery, silky-pubescent (sericeous) leaves. Flowers pink or purplish, nearly sessile. Hypanthium campanulate, the calyx lobes lanceolate or nearly so. Petals elliptic to oblanceolate, exceeding the sepals. Stamens purplish, slightly exerted. Styles equaling or slightly exceeding the stamens. Follicles elliptical.

This species is a calciphile restricted to cracks and crevices in limestone rocks. In the alpine zone it is known only from the Big Horn Range. Endemic to Montana, Idaho, and Wyoming.

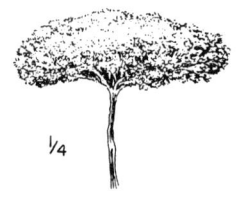

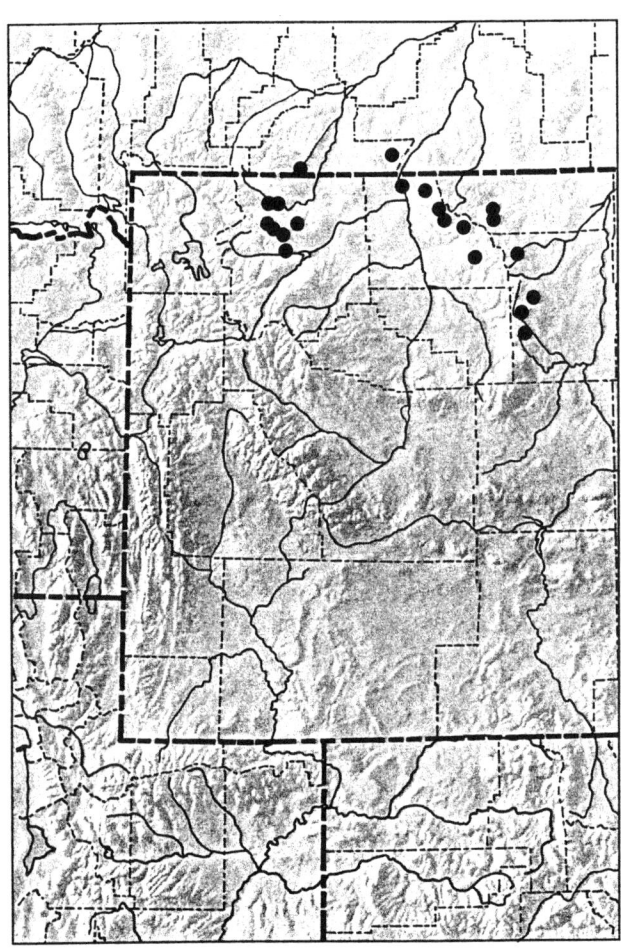

Petrophyton (Nutt.) Rydb.
Rockmat, Rockspiraea

Dense, mat-forming shrubs of rock surfaces. Leaves simple, entire, alternate, and spatulate. Flowers small, complete, in spikelike racemes or panicles. Sepals 5. Petals 5. Stamens numerous, usually 20. Pistils 5, distinct. Fruit a follicle.

Petrophyton caespitosum (Nutt.) Rydb.
Mem. N.Y. Bot. Gard. 1:206, 1900.

Tufted Rockmat

Spiraea caespitosa Nutt.
Eriogynia caespitosa (Nutt.) S. Wats.
Luetkea caespitosa (Nutt.) Kuntze

Low, caespitose, mat-forming perennial with densely imbricate, spatulate, sericeous leaves. Flowers small, numerous, in dense, spikelike racemes or panicles. Hypanthium sericeous, broadly campanulate, the calyx lobes erect, lanceolate. Petals white, spatulate to linear. Stamens 20, nearly twice as long as the petals. Pistils 5; styles sparsely pubescent, at least at the base. Follicles glabrous or pubescent.

Rock surfaces, usually limestone. Known from the alpine zone of the Gros Ventre, Teton, Wasatch, and Wyoming Ranges. Oregon to South Dakota, south to Colorado, Texas, New Mexico, Arizona, and California.

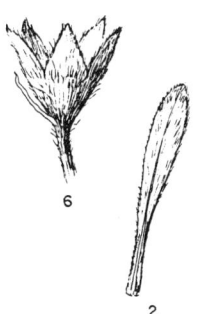
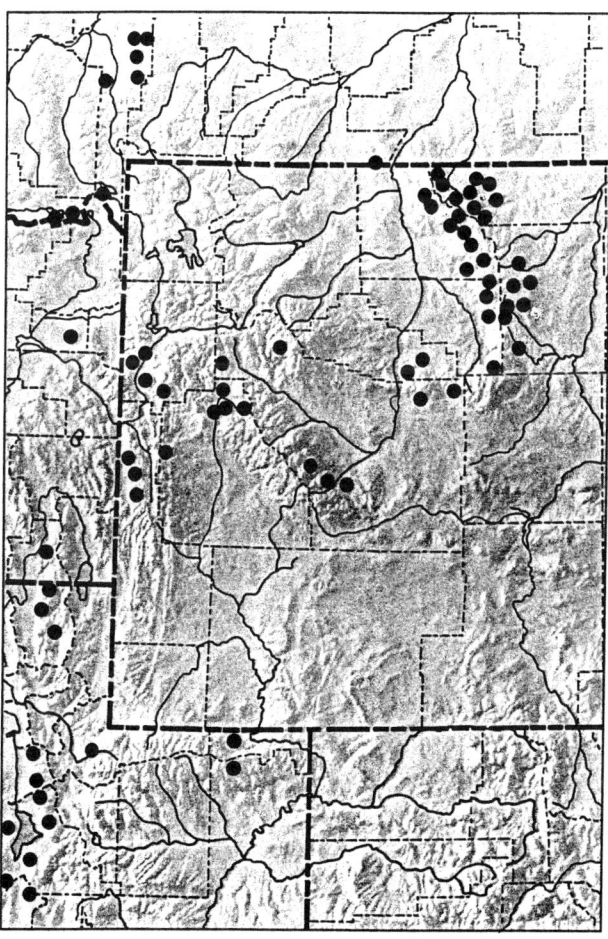

Potentilla L.
Cinquefoil

Annual, biennial, or perennial herbs (shrubby in *P. fruticosa*) with alternate, palmately or pinnately compound leaves, the leaflets incised or toothed. Flowers single and terminal, or numerous and in bracteate or leafy cymes. Sepals 5, with 5 alternating bractlets. Petals 5 and yellow, white, or rarely reddish purple. Stamens commonly 20 or more. Pistils rarely less than 10, the styles terminal and filiform, or lateral and thickened. Fruit a group of light brown akenes.

KEY TO THE SPECIES OF *POTENTILLA*

1. Plants erect, woody shrubs; ovaries and akenes pubescent — *P. fruticosa*
1. Plants herbaceous; ovaries and akenes glabrous
   2. Plants glandular-pubescent, the lower stems rarely glabrous; petals cream to pale yellow; styles basal — *P. glandulosa*
   2. Plants glabrous or pubescent (sometimes finely glandular-puberulent); petals bright yellow; styles terminal or lateral
      3. Basal leaves pinnately compound
         4. Styles lateral; plants finely glandular-puberulent — *P. brevifolia*
         4. Styles terminal or subterminal; plants eglandular
            5. Basal leaves digitately 5-foliolate with a reduced, remote pair of leaflets on the petiole — *P. subjuga*
            5. Basal leaves lacking a reduced, remote pair of leaflets
               6. Styles about the same length as the akenes, thickened and glandular-roughened at the base; akenes often glandular-roughened at the apex
                   7. Stems green or greenish, erect to ascending, more than 20 cm tall; basal leaves with 7 or more leaflets — *P. pensylvanica*
                   7. Stems red, at least near the base, spreading-decumbent, less than 20 cm tall; basal leaves with 5 leaflets — *P. rubricaulis*
               6. Styles much longer than the akenes, neither thickened nor glandular-roughened at the base; akenes smooth
                   8. Plants usually more than 20 cm tall, leaflets more than 2 cm long, oblong, toothed less than halfway to the midrib, usually tomentose on both surfaces — *P. hippiana*
                   8. Plants usually less than 20 cm tall; leaflets less than 2 cm long and oblong, oblanceolate, or cuneate, toothed more than halfway to the midrib, glabrate to tomentose below, rarely tomentose above
                       9. Leaflets 9–13, cleft nearly to the midrib, the ultimate segments less than 1.7 mm wide — *P. ovina*
                       9. Leaflets 5–7, rarely cleft more than halfway to the midrib, the ultimate segments variable in width
                           10. Stems erect to ascending; basal leaves mostly with 7 leaflets, greenish beneath — *P. diversifolia*
                           10. Stems spreading-decumbent; basal leaves with 5 or 7 leaflets, white-tomentose beneath
                               11. Calyx bractlets linear to narrowly lanceolate, acute; stems red, at least near the base; basal leaves mostly with 5 leaflets — *P. rubricaulis*
                               11. Calyx bractlets elliptic to ovate, obtuse; stems green or greenish; basal leaves mostly with 7 leaflets — *P. concinna*
      3. Basal leaves ternate or palmately compound

12. Petioles of basal leaves with a remote, reduced pair of leaflets ... *P. subjuga*
12. Petioles of basal leaves lacking a remote, reduced pair of leaflets
    13. Principal leaves ternate
        14. Leaves equally green on both surfaces, glabrous to sparsely pilose; styles slender, filiform
            15. Plants rhizomatous; stems more than 15 cm tall, 2- to several-flowered; leaflets flabelliform, glabrous or short-pubescent, the margins crenate-dentate ... *P. flabellifolia*
            15. Plants from taproots; stems less than 10 cm tall, mostly 1-flowered; leaflets broadly obovate, pilose, the margins mostly coarsely crenate ... *P. nana*
        14. Leaves green above, gray- to white-pubescent beneath; styles thickened at the base, papillate
            16. Pubescence of petioles and basal stem of 2 kinds, fine puberulence and spreading, long, straight hairs ... *P. hookeriana*
            16. Pubescence of petioles and basal stem of 1 kind, tomentose or pilose
                17. Petioles and basal stem tomentose, the hairs thin, wavy ... *P. nivea*
                17. Petioles and basal stem pilose, the hairs straight or nearly so ... *P. uniflora*
    13. Principal leaves 5–9-foliolate
        18. Stems low, spreading-decumbent, usually less than 10 cm tall; leaves white-tomentose beneath
            19. Calyx bractlets linear to narrowly lanceolate, acute; stems red, at least near the base ... *P. rubricaulis*
            19. Calyx bractlets elliptic to ovate, obtuse; stems green or greenish ... *P. concinna*
        18. Stems erect, often more than 20 cm tall; leaves greenish to gray-tomentose beneath
            20. Leaflets greenish on both sides, the leaflets 1–4 cm long; anthers 0.4–0.6 mm long; akenes 1.3–1.6 mm long ... *P. diversifolia*
            20. Leaflets gray-tomentose beneath, the leaflets 3–6 cm long; anthers 0.6–1.3 cm long; akenes 1.5–2 mm long ... *P. gracilis*

Potentilla brevifolia Nutt. ex T. & G.
Fl. N. Am. 1:442, 1840.

Shortleaf Cinquefoil

Erect, weakly rhizomatous, glandular, perennial herb from a branched caudex covered with old leaf bases. Basal leaves pinnate; leaflets 3–7, the blades broadly obovate to orbicular, mostly 3-parted, the lobes crenate; cauline leaves sessile, reduced upward, pinnately cleft. Flowers few, in open cymes; hypanthium a shallow cup. Sepals 5, lanceolate to ovate, pubescent, spreading. Petals 5, yellow, obovate, often shallowly retuse. Stamens numerous. Pistils numerous, the styles attached laterally above the middle. Akenes greenish, ovate, smooth.

Talus, scree, and fellfields of the Absaroka, Teton, and Wind River Ranges. Oregon to northwestern Wyoming and south to Nevada.

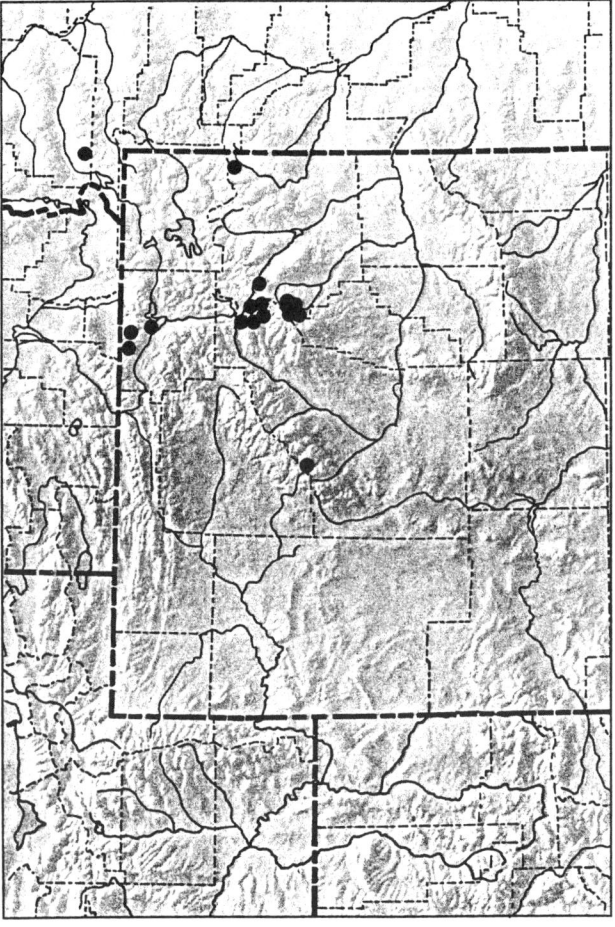

Potentilla concinna Richards.
Bot. App. Frankl. J., 739, 1823.

Elegant Cinquefoil

Potentilla beanii Clokey
Potentilla bicrenata Rydb.
Potentilla divisa (Rydb.) Rydb.
Potentilla humifusa Nutt.
Potentilla intermittens Rydb.
Potentilla macounii Rydb.
Potentilla modesta Rydb.
Potentilla pedersenii Rydb.
Potentilla proxima Rydb.
Potentilla quinquefolia (Rydb.) Rydb.
Potentilla rubricaulis Rydb.
Potentilla rubripes Rydb.
Tomentilla humifusa (Nutt.) G. Don

Spreading to somewhat prostrate, sparsely to densely tomentose, perennial herb from a thick, woody taproot and branching caudex. Basal leaves petiolate, digitately to pinnately 5–9-foliolate; leaflets tomentose beneath, oblanceolate, toothed or cleft into linear-lanceolate segments; cauline leaves similar to the basal ones, reduced upward; stipules mostly cleft, the lobes lanceolate. Flowers few, in bracteate cymes; hypanthium cup-shaped. Sepals 5, lanceolate, villous; bractlets elliptic to ovate. Petals 5, yellow, obovate, retuse. Stamens numerous. Pistils numerous, the styles filiform, attached below the apex. Akenes yellow-brown, glabrous.

Fellfields and rocky slopes in most of the alpine ranges. Not yet known from above timberline in the Big Horn, Hoback, and Salt River Ranges. Alberta and Saskatchewan south to South Dakota, New Mexico, Arizona, and Nevada.

Two varieties occur in the alpine zone of the Middle Rocky Mountains: var. *rupripes* (Rydb.) C.L. Hitchc. has pinnately compound leaves and usually 7 leaflets; var. *concinna* has palmately compound leaves and usually 5 leaflets.

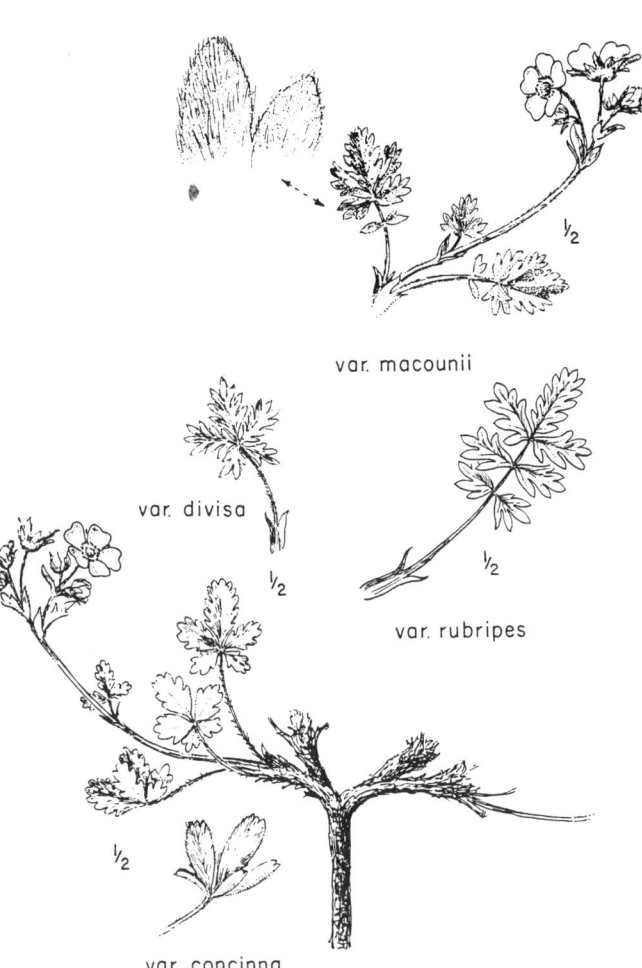

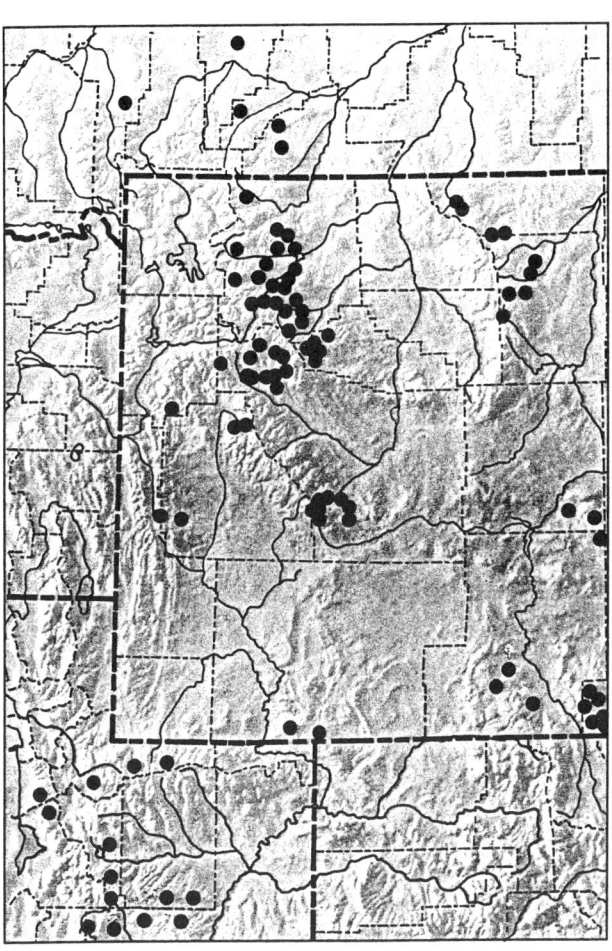

Potentilla diversifolia Lehm.
Stirp. Pug. 2:9, 1830.

Blueleaf Cinquefoil

Potentilla decurrens Rydb.
Potentilla dissecta Nutt.
Potentilla glaucophylla Lehm.
Potentilla multisecta (S. Wats.) Rydb.
Potentilla perdissecta Rydb.
Potentilla ranunculus Lange
Potentilla vreelandii Rydb.

Spreading to erect, perennial herb from a thick, often branched, woody caudex; stems 1 to several, erect or ascending, often decumbent at the base. Basal leaves petiolate, palmately or pinnately compound, glabrous or sparsely pubescent; leaflets oblanceolate to obovate, toothed to deeply dissected into linear or lanceolate segments; cauline leaves 1–2, reduced and becoming sessile or nearly so upward; stipules lanceolate, entire. Flowers few to many, in open cymes; hypanthium saucer-shaped, pubescent. Sepals 5, ovate-lanceolate, spreading to ascending. Petals 5, yellow, obovate, retuse. Stamens numerous. Pistils numerous, the styles filiform, attached below the apex. Akenes yellow-brown, glabrous to weakly reticulate.

Meadows, alp slopes, and fellfields in all alpine ranges of the Middle Rocky Mountains. Greenland; Alaska south to California, New Mexico, and South Dakota.

Two varieties occur in the alpine zone. The var. *diversifolia* with palmately compound leaves and toothed leaflets is the more common. The var. *perdissecta* (Rydb.) C.L. Hitchc. has subpinnate to pinnate leaves and leaflets dissected nearly to the midvein.

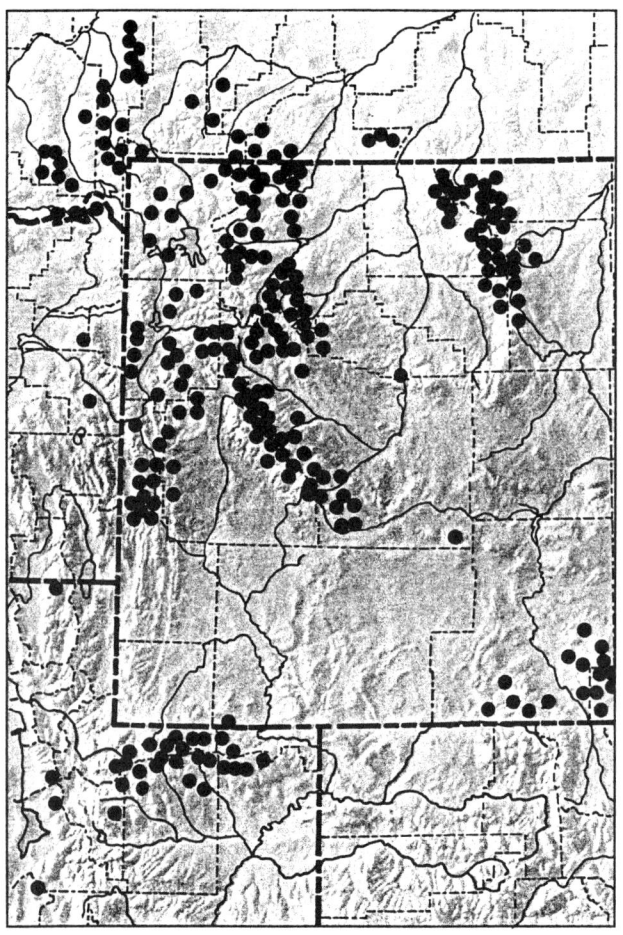

Potentilla flabellifolia Hook. ex T. & G.
Fl. N. Am. 1:442, 1840.

Fanleaf Cinquefoil

Potentilla gelida S. Wats.

Erect to decumbent, pubescent to puberulent, perennial herb from a branched caudex or weakly creeping rhizome. Basal leaves long-petiolate, trifoliolate; leaflets obovate to somewhat cordate, serrate on the margins, dentate at the apex, glabrate; cauline leaves 1 or 2, reduced, sessile or nearly so; stipules ovate, membranous at least below. Flowers mostly 1–3, in short cymes; hypanthium saucer-shaped. Sepals 5, ovate; bractlets entire or 2–3-toothed. Petals 5, yellow, obovate, retuse. Stamens numerous. Pistils numerous, the styles filiform, inserted below the apex. Akenes yellow-brown, glabrous.

Meadows, stream banks, and fellfields of the Teton Range. British Columbia south to California, Idaho, and northwestern Wyoming.

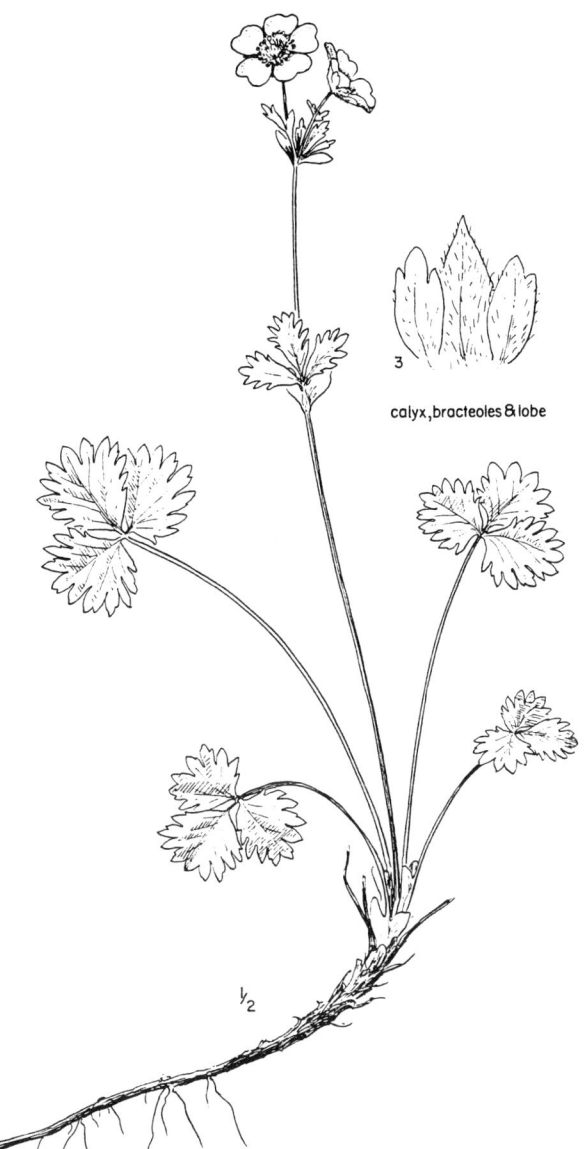

calyx, bracteoles & lobe

Potentilla fruticosa L.
Sp. Pl., 495, 1753.

Shrubby Cinquefoil

Dasiphora fruticosa (L.) Rydb.
Dasiphora riparia Raf.
Fragaria fruticosa (L.) Crantz
Pentaphylloides floribunda (Pursh) A. Löve
Potentilla floribunda Pursh

Spreading to erect, many-branched shrub, the branches pubescent when young, with shredding, brown bark when older. Leaves pinnate, the usually 5 leaflets entire, narrowly elliptic to lanceolate, glabrous to pubescent, often revolute; stipules scarious, brown. Flowers solitary or in few-flowered axillary cymes; hypanthium saucer-shaped. Sepals 5, villous, spreading, acuminate; bractlets lanceolate. Petals 5, yellow, broadly obovate to orbicular, often shallowly retuse. Stamens numerous. Pistils numerous, the styles lateral, attached near the middle. Akenes brownish, densely villous.

Ledges, fellfields, coarse talus, and alp slopes in most alpine ranges of the Middle Rocky Mountains; not yet known from above timberline in the Hoback and Salt River Ranges. Circumboreal; south in western North America to California, Arizona, New Mexico, and South Dakota.

A. Löve et al. (1971) note that North American, Asian, and European plants are 2n=14 and have been called *Pentaphylloides floribunda* (Pursh) A. Löve. They restrict the name *Potentilla fruticosa* to Eurasian plants with 2n=28 chromosomes. I follow Elkington (1969) in recognizing the North American diploids as ssp. *floribunda* (Pursh) Elkington.

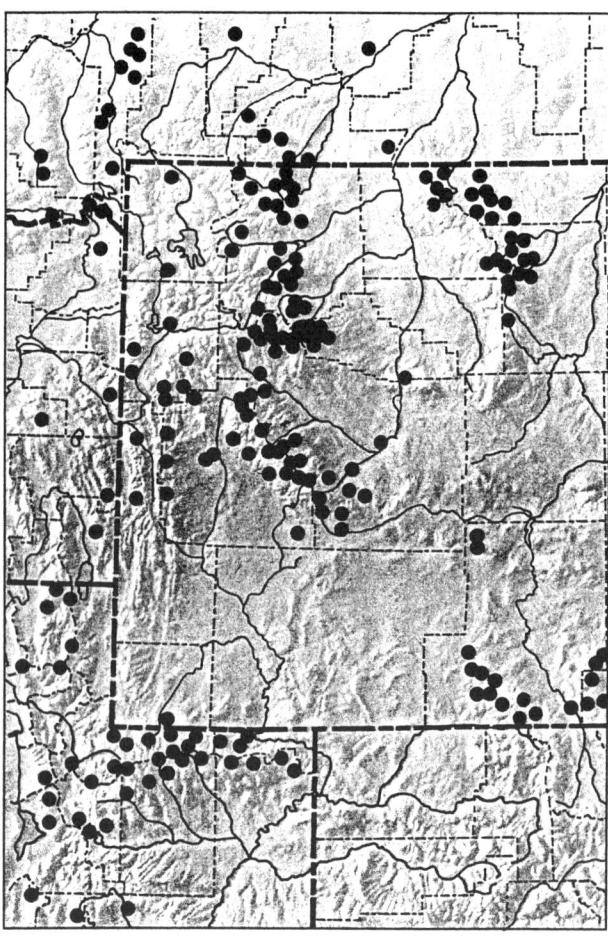

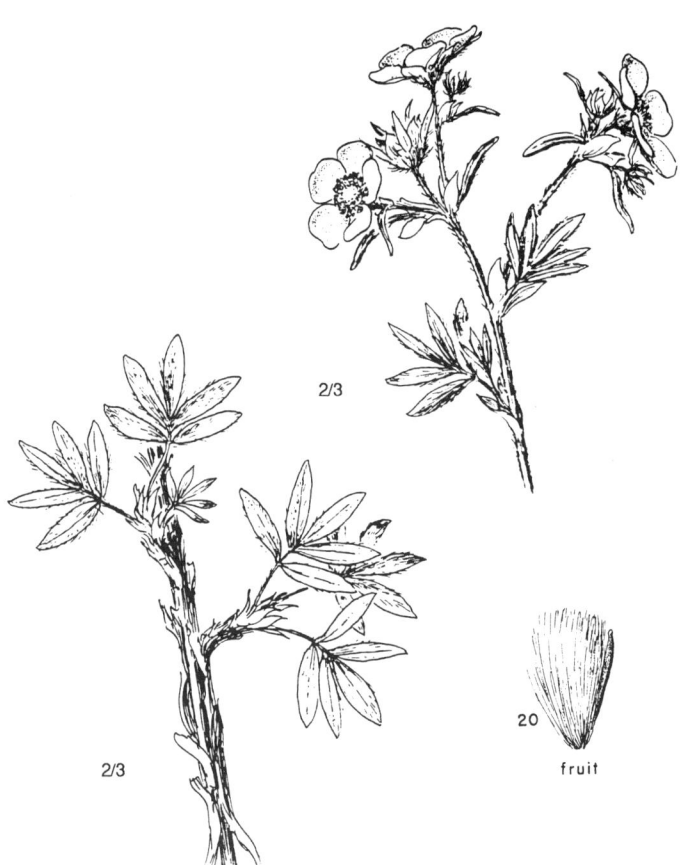

Potentilla glandulosa Lindl.
Bot. Reg. 19:Pl. 1583, 1833.

Sticky Cinquefoil

Drymocallis albida Rydb.
Drymocallis amplifolia Rydb.
Drymocallis ashlandica (Greene) Rydb.
Drymocallis foliosa Rydb.
Drymocallis glabrata Rydb.
Drymocallis glandulosa (Lindl.) Rydb.
Drymocallis glutinosa (Nutt.) Rydb.
Drymocallis monticola (Rydb.) Rydb.
Drymocallis oregana Rydb.
Drymocallis pseudorupestris (Rydb.) Rydb.
Drymocallis pumila Rydb.
Drymocallis reflexa (Greene) Rydb.
Drymocallis rhomboidea (Rydb.) Rydb.
Drymocallis valida (Greene) Piper
Drymocallis viscosa Rydb.
Drymocallis wrangelliana (Fisch. & Ave-Lall.) Rydb.
Potentilla albida (Rydb.) Greene
Potentilla amplifolia (Rydb.) Fedde
Potentilla ashlandica Greene
Potentilla ciliata T.J. Howell
Potentilla hanseni Greene
Potentilla pseudorupestris Rydb.
Potentilla reflexa (Greene) Greene
Potentilla rhomboidea Rydb.
Potentilla valida Greene
Potentilla viscosa (Rydb.) Fedde
Potentilla wrangelliana Fisch. & Ave-Lall.

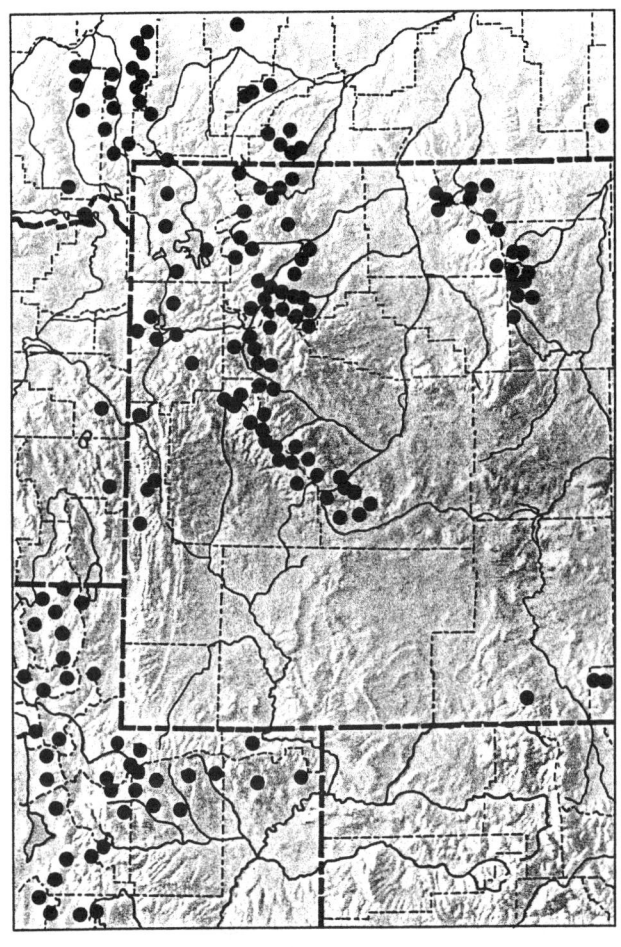

Erect, mostly multistemmed, glandular perennial from a branched, scaly caudex. Basal leaves pinnate; leaflets obovate, cuneate at the base, glandular, glabrate to villous, the margins coarsely sharp-serrate; cauline leaves few, subsessile, reduced upward; stipules lanceolate to ovate. Flowers few to many in an open to compact, leafy-bracteate cyme; hypanthium glabrate to pubescent, cup-shaped. Sepals 5, lanceolate to ovate, spreading; bractlets narrowly lanceolate, shorter than the sepals. Petals 5, white to cream or light yellow, obovate, longer than the sepals in ours. Stamens numerous. Pistils numerous, the styles somewhat clavate, attached laterally near the base. akenes light yellow-brown, glabrous.

Fellfields, ledges, talus, and scree in most Middle Rocky Mountain alpine ranges. Not yet known from above timberline in the Hoback, Medicine Bow, and Wyoming Ranges. British Columbia and Alberta south to California, Utah, Arizona, Colorado, and South Dakota.

Alpine plants of the Middle Rocky Mountains with white or cream petals distinctly longer than the sepals are var. *pseudorupestris* (Rydb.) Breit.

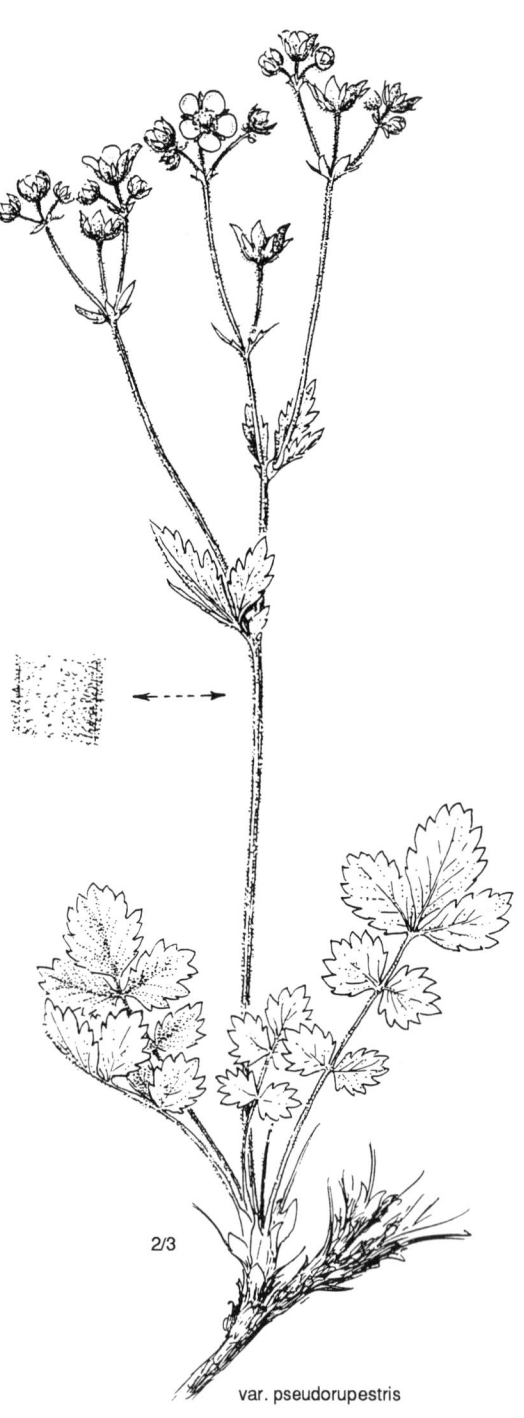

var. pseudorupestris

Potentilla gracilis Dougl. ex Hook.
Curtis' Bot. Mag. 57:Pl. 2984, 1830.

Slender Cinquefoil

Potentilla angustata Rydb.
Potentilla blasckeana Turcz. ex Lehm.
Potentilla brunnescens Rydb.
Potentilla camporum Rydb.
Potentilla candida Rydb.
Potentilla chrysantha Lehm.
Potentilla ctenophora (Rydb.) Rydb.
Potentilla dascia Rydb.
Potentilla dichroa Rydb.
Potentilla elmeri Rydb.
Potentilla etomentosa Rydb.
Potentilla fastigata Nutt.
Potentilla filipes Rydb.
Potentilla flabelliformis Lehm.
Potentilla glabrata (Lehm.) Rydb.
Potentilla glomerata A. Nels.
Potentilla grosseserrata Rydb.
Potentilla hallii Rydb.
Potentilla indiges M.E. Peck.
Potentilla jucunda A. Nels.
Potentilla longiloba Rydb.
Potentilla longipedunculata Rydb.
Potentilla macropetala Rydb.
Potentilla nuttallii Lehm.
Potentilla pecten Rydb.
Potentilla pectinisecta Rydb.
Potentilla permollis Rydb.
Potentilla pulcherrima Lehm.
Potentilla rectiformis Rydb.
Potentilla rigida Nutt.
Potentilla viridescens Rydb.

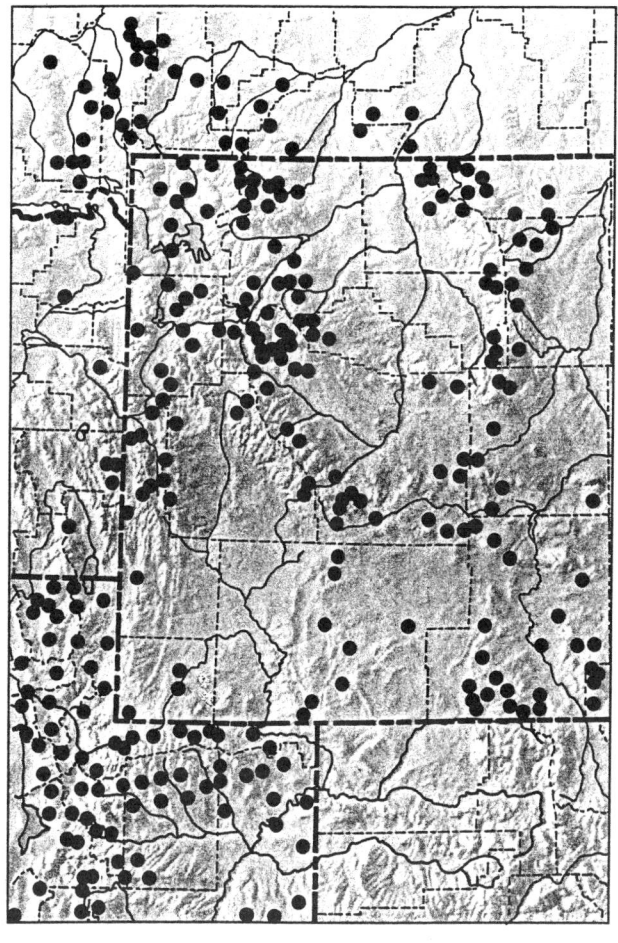

Erect, perennial herb from a stout, often short-branched caudex; stems 1 to several, stout, often decumbent at the base. Basal leaves long-petiolate, palmately compound; leaflets glabrous to white-tomentose, at least on the lower surface, oblanceolate to oblong, serrate to deeply incised or dissected nearly to the midvein; cauline leaves 1–3, subsessile, reduced upward; stipules lanceolate, entire to toothed. Flowers numerous in open, leafy-bracteate cymes; hypanthium cup-shaped, glabrate to pubescent and glandular. Sepals 5, lanceolate to ovate, acuminate. Petals 5, yellow, obovate, usually retuse. Stamens numerous. Pistils numerous, the styles filiform, attached below the apex. Akenes yellowish or greenish, glabrous.

Widely scattered in meadows and fellfields of the lower alpine zone of the Absaroka, Beartooth, Medicine Bow, Salt River, Teton, Uinta, Wasatch, and Wind River Ranges. Alaska south to California, Arizona, New Mexico, Colorado, and South Dakota.

Three varieties of this complex species are known to occur above timberline: var. *pulcherrima* (Lehm.) Fern. has leaflets lobed less than half the distance to the midrib and white-tomentose beneath; var. *nuttallii* (Lehm.) Sheld. and var. *elmeri* (Rydb.) Jeps. both have leaflets parted more than half the distance to the midrib. The latter two are distinguished by the deeper, linear lobes of var. *elmeri*, which are silky to sparsely tomentose beneath, compared with the somewhat triangular, shallower lobes of var. *nuttallii,* which are greenish to sparsely gray-pubescent below.

Potentilla hippiana Lehm.
Stirp. Pug. 2:7, 1830.

Woolly Cinquefoil

Pentaphyllum effusum (Dougl. ex Lehm.) Lunell
Pentahpyllum leucophyllum (Torr.) Lunell
Potentilla argyrea Rydb.
Potentilla diffusa Gray
Potentilla effusa Dougl. ex Lehm.
Potentilla filicaulis (Nutt.) Rydb.
Potentilla leneophylla Torr. & James ex Eaton
Potentilla leucophylla Torr.
Potentilla propinqua (Rydb.) Rydb.
Potentilla rupincola Osterhout

Decumbent-spreading, pubescent, perennial herb from a thick, stout, brown-scaly caudex. Basal leaves long-petiolate, odd-pinnate; leaflets oblanceolate to oblong, white- to grayish-tomentose, at least below, the margins coarse-serrate; cauline leaves short-petiolate, otherwise similar to basal leaves; stipules entire, lanceolate. Flowers several to many in open, branched cymes; hypanthium saucer-shaped, pubescent. Sepals 5, tomentose, lanceolate; bractlets linear. Petals 5, yellow, obovate, rounded to retuse. Stamens numerous. Pistils numerous, the styles filiform, attached below the apex. Akenes yellow-brown, glabrous.

Typical of grasslands and sagebrush foothills, this species occurs in the lower alpine zone above timberline in the Absaroka, Beartooth, and Uinta Ranges. Alaska south to New Mexico, Colorado, Nebraska, and North Dakota.

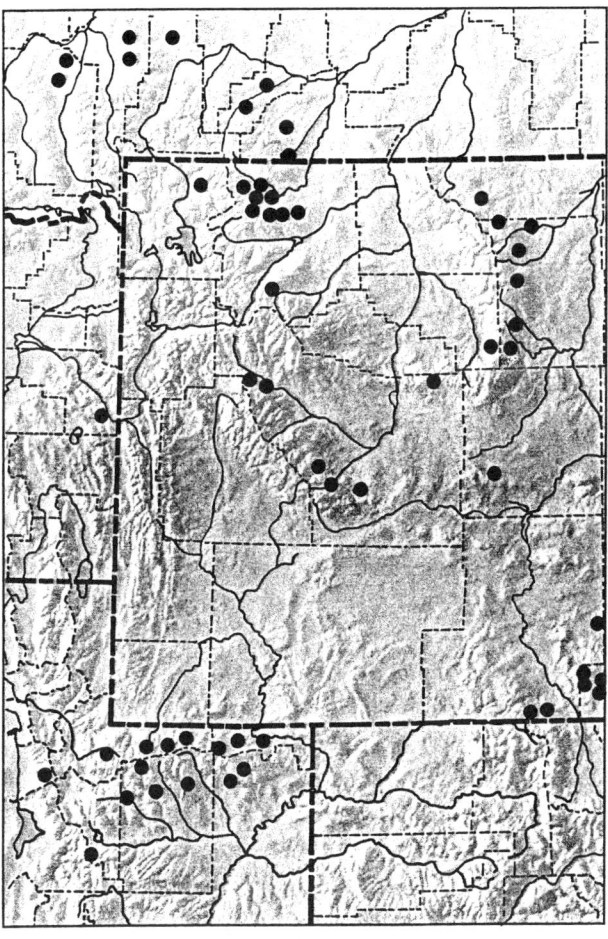

Potentilla hookeriana Lehm.
Del. Sem. Hort. Hamb. 1849:10, 1849.

Hooker Cinquefoil

Potentilla chamissonis Hult.
Potentilla furcata Pors.

Erect to somewhat decumbent, perennial herb from a thick, branched caudex; stems 1 to several, pubescent, the pubescence of fine, short hairs and long, spreading hairs. Basal leaves long-petiolate, trifoliolate in ours; leaflets obovate, coarsely serrate to nearly pinnatifid, green above, white-tomentose below; cauline leaf reduced, subsessile; stipules ovate, entire. Flowers few to several, in an open cyme; hypanthium cup-shaped. Sepals 5, ovate, acute; bractlets lanceolate. Petals 5, yellow, obcordate, longer than the sepals. Stamens numerous. Pistils numerous, the styles thickened and papillate at the base, attached below the apex. Akenes yellowish brown, glabrous.

Fellfields of the Absaroka, Beartooth, Medicine Bow, and Wind River Ranges. Siberia; Alaska south to northwestern Wyoming.

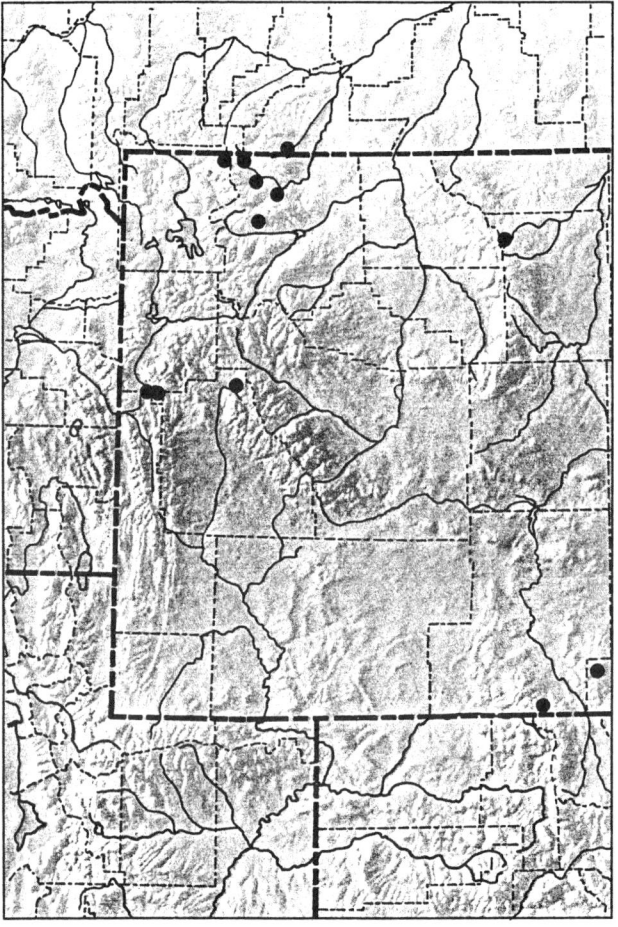

Potentilla nana Willd. ex Schlecht.
Ges. Nat. Fr. Berl. Mag. 7:296, 1813.

Arctic Cinquefoil

Potentilla emarginata Pursh
Potentilla groenlandica R. Br.
Potentilla hyparctica Malte

Low, somewhat tufted, perennial herb from a simple to branched, brownish-scaly caudex; stems few to several, pubescent with short and long, spreading hairs. Basal leaves trifoliolate, petioles equaling or longer than the blade; leaflets obovate, the base cuneate, glabrate to pilose with long hairs, ciliate, the veins sometimes glandular; margins sharp- or crenate-toothed, often with an asymmetric, basal lobe on the lateral leaflets; cauline leaf reduced, sessile. Flowers single, sometimes 2–3 in a short cyme. Sepals 5, pilose, oblong-ovate, acute or obtuse. Petals 5, yellow, obcordate, emarginate, equaling or exceeding the sepals. Stamens numerous. Pistils numerous, the styles terminal or nearly so, filiform to slightly clavate at the base. Akenes glabrous.

Fellfields and patterned ground on Goat Flat, Down's Mountain, and Ram Flat in the northern Wind River Range. Circumpolar; south in western North America to northwestern Wyoming.

Small-flowered plants, closely resembling *P. nana*, from ranges on the Utah-Nevada border have been described as *P. cottamii* N. Holm. (Holmgren, 1987) and might be expected at high elevations along the western margin of our mountains.

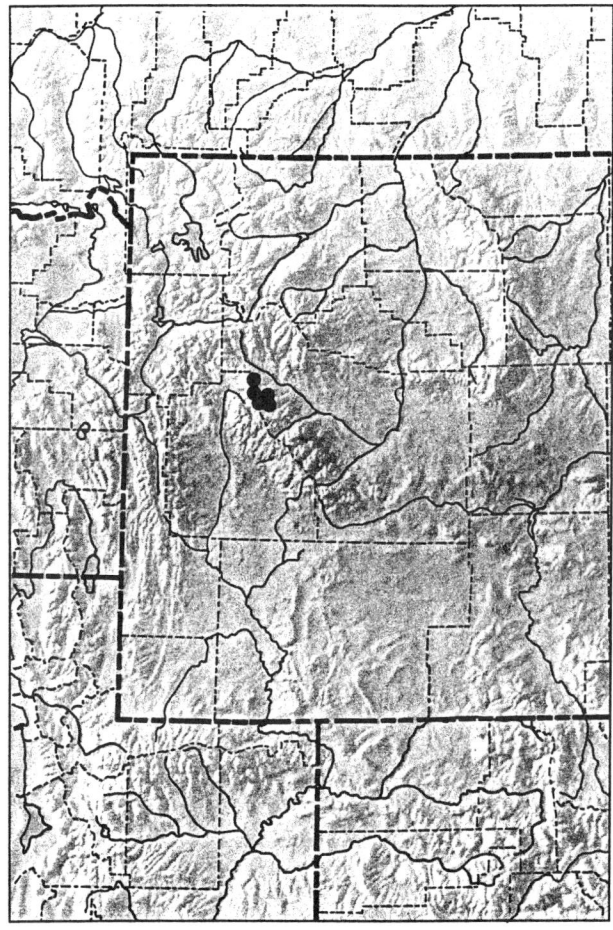

Potentilla nivea L.
Sp. Pl., 499, 1753.

Snowbank Cinquefoil

Fragaria nivea (L.) Crantz
Potentilla altaica Bunge
Potentilla nipharga Rydb.

Low, tufted, pubescent, perennial herb, less than 20 cm tall, from a stout, brownish, much-branched caudex; stems 1 to several, tomentose. Basal leaves trifoliolate; leaflets obovate to elliptic, coarsely serrate to somewhat pinnatifid, green above, tomentose below, the petioles tomentose, usually longer than the blade; cauline leaves 1–2, reduced, subsessile. Flowers 1–5, in leafy-bracteate cymes; hypanthium shallow, saucer-shaped. Sepals 5, ovate; bractlets linear to lanceolate. Petals 5, yellow, obcordate, longer than the sepals. Stamens numerous. Pistils numerous, the styles papillate at the base, attached below the apex. Akenes yellowish brown, glabrous.

Fellfields, rocky slopes, and scree of the Absaroka, Beartooth, Medicine Bow, Uinta, and Wind River Ranges. Circumpolar; south in western North America to Nevada, Utah, and Colorado.

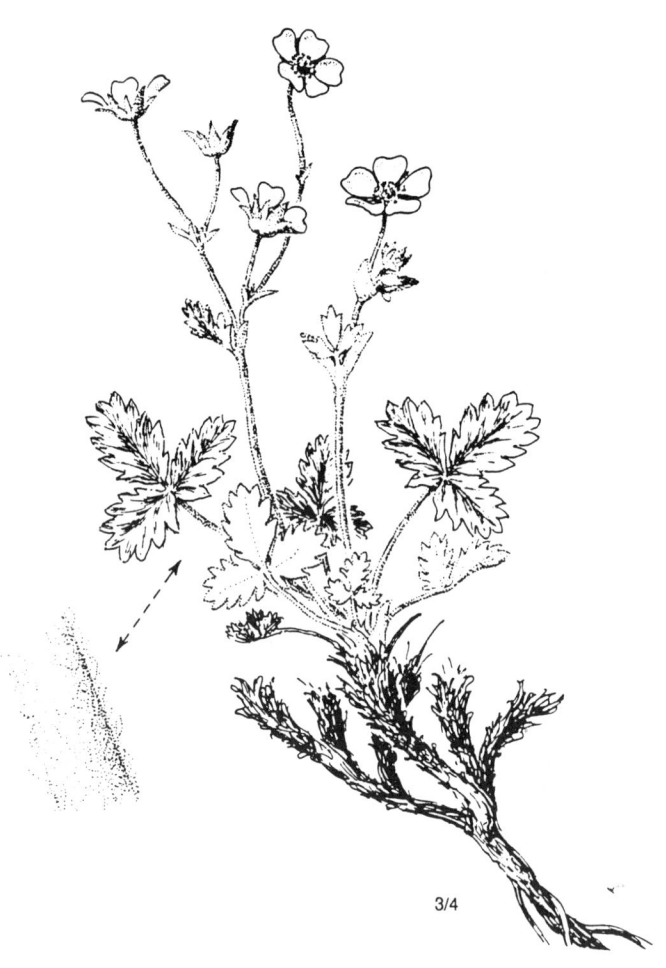

Potentilla ovina Macoun
Can. Rec. Sci. 6:464, 1896.

Sheep Cinquefoil

Potentilla klamathensis Rydb.
Potentilla monidensis A. Nels.
Potentilla nelsoniana Rydb.
Potentilla pinnatisecta (S. Wats.) A. Nels.
Potentilla versicolor Rydb.
Potentilla wyomingensis A. Nels.

Low, tufted, perennial herb, less than 20 cm tall, from a stout, much-branched, brown-scaly caudex and taproot; stems 1 to several, erect to spreading-decumbent. Basal leaves short-petiolate, pinnate; leaflets lobed or cleft nearly to the base into 3–5 linear or narrowly lanceolate segments, sericeous to hirsute. Flowers few to several in an open cyme; hypanthium cup-shaped. Sepals 5, lanceolate, acute. Petals 5, yellow, obcordate, longer than the sepals. Stamens numerous. Pistils numerous, the styles filiform, attached below the apex. Akenes yellow-brown, glabrous.

Fellfields, ledges, and talus of nearly all alpine ranges. Not yet known from above timberline in the Teton Range. Alberta and British Columbia south to California, Utah, and New Mexico.

Alpine plants of the Middle Rocky Mountains with leaflets lobed to the base, or nearly so, and silky-pubescent surfaces are var. *ovina*. Plants with leaflets lobed above, entire or nearly so below, and the surfaces glabrous to sparsely pubescent are var. *decurrens* (S. Wats.) S.L. Welsh and Johnston.

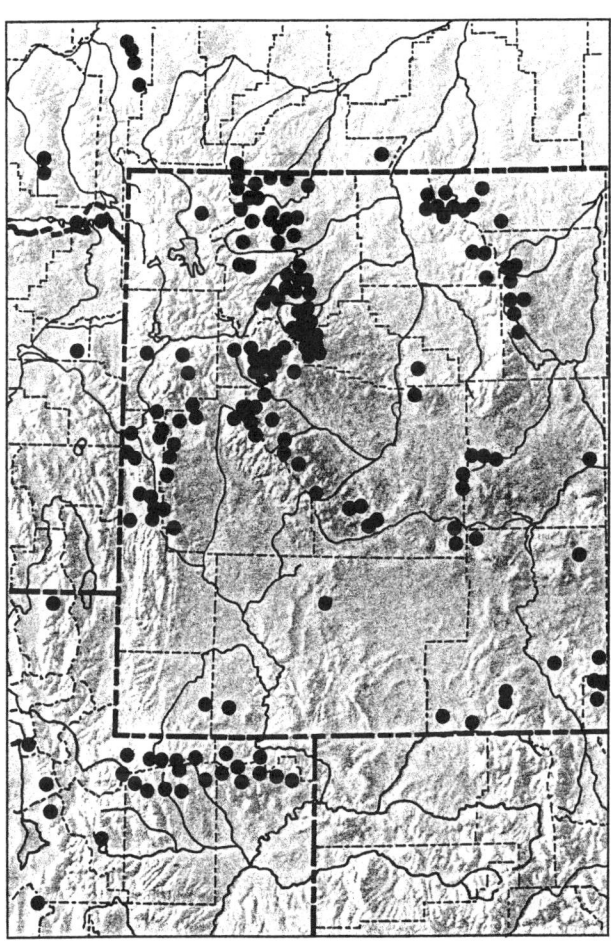

Potentilla pensylvanica L.
Mant., 76, 1767.

Pennsylvania Cinquefoil

Potentilla atrovirens Rydb.
Potentilla bipinnatifida Dougl. ex Hook.
Potentilla finitima Kohli & K.L. Parker
Potentilla glabella Rydb.
Potentilla lasiodonta Rydb.
Potentilla missourica Hornem. ex Lindl.
Potentilla paucijuga Rydb.
Potentilla pectinata Raf.
Potentilla platyloba Rydb.
Potentilla strigosa (Pall. ex Pursh) Pall. ex Tratt.
Potentilla virgulata A. Nels.

Perennial herb from a slender to stout, branched caudex; stems 1 to several, erect or spreading-decumbent, tomentose. Basal leaves pinnate; leaflets elliptic to obovate, pinnately divided into linear segments, green above, sparsely to densely tomentose below; cauline leaves reduced upward, short-petiolate to sessile; stipules ovate, entire to pinnatifid. Flowers several in compact, narrow cymes; hypanthium cup-shaped, pubescent, often glandular. Sepals 5, ovate; bractlets lanceolate. Petals 5, yellow, obovate. Stamens numerous. Pistils numerous, the styles thickened and glandular at the base, attached below the apex. Akenes yellow brown, ovate, glabrous or apically roughened.

Dry meadows and slopes at timberline in the Absaroka, Beartooth, and Uinta Ranges. Alaska south to New Mexico, Colorado, Kansas, Minnesota, and New Hampshire.

The var. *paucijuga* S.L. Welsh & Johnson occurs in the alpine zone of the Uinta Range. It is distinguished from typical var. *pensylvanica* (Welsh, et al., 1993) as follows:

1. Leaflets 5–7, subdigitate; pubescence silvery to yellowish-white, of densely matted, long hairs
   . . . var. *paucijuga*
1. Leaflets 7–17, pinnate; pubescence dull greenish, of sparsely matted, short hairs . . . var. *pensylvanica*

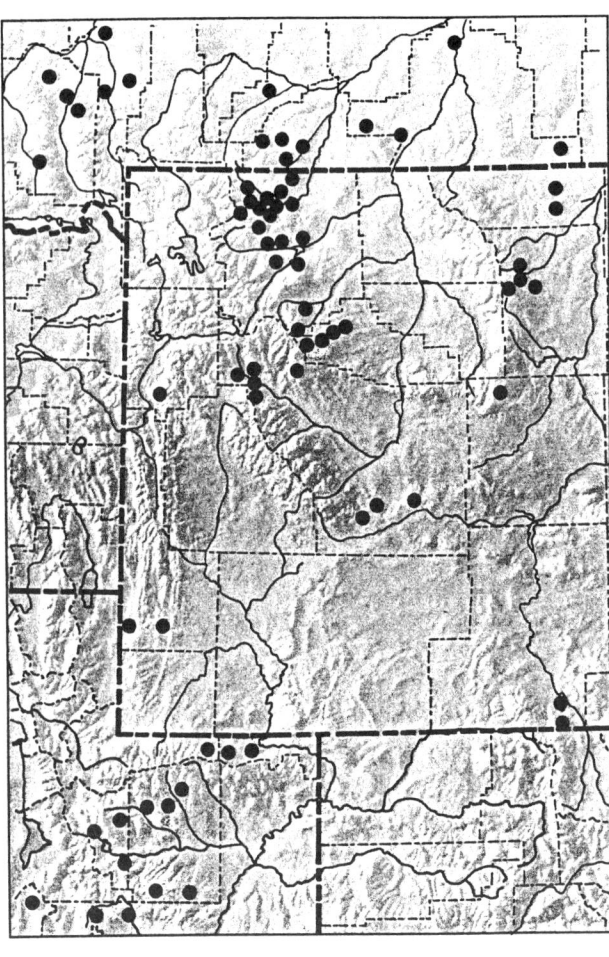

Potentilla rubricaulis Lehm.
Stirp. Pug. 2:11, 1830.

Redstem Cinquefoil

Potentilla saximontana Rydb.

Low, spreading, perennial herb from a simple or branched caudex covered with old leaf bases; stems several, prostrate to ascending, reddish, at least below. Basal leaves petiolate, digitately to pinnately 5-foliolate, occasionally 7-foliolate or ternate; leaflets white-tomentose beneath, glabrous or nearly so above, obovate to oblong, pinnately parted, the segments oblong to linear; cauline leaves much reduced, the leaflets similar in shape to the basal ones. Flowers solitary to few in bracteate cymes; hypanthium cup-shaped. Sepals 5, lanceolate to ovate, villous; bractlets linear to lanceolate. Petals 5, yellow, obovate, emarginate. Stamens numerous. Pistils numerous, the styles thickened and papillose below to filiform and epapillose, attached below the apex. Akenes brownish, glabrous.

Fellfields and exposed rocky places in the Absaroka, Uinta, Wasatch, and Wind River Ranges. Alaska to Greenland; south in western North America to Utah and Colorado.

Potentilla rubricaulis is often difficult to distinguish from alpine forms of *P. pensylvanica* and *P. concinna*. I use the name *rubricaulis* for low, spreading plants with red stems and palmate or subpalmate, mostly 5-foliolate, basal leaves. Most have styles papillose near the base and shorter than the akenes, although some are longer and smooth, or nearly so.

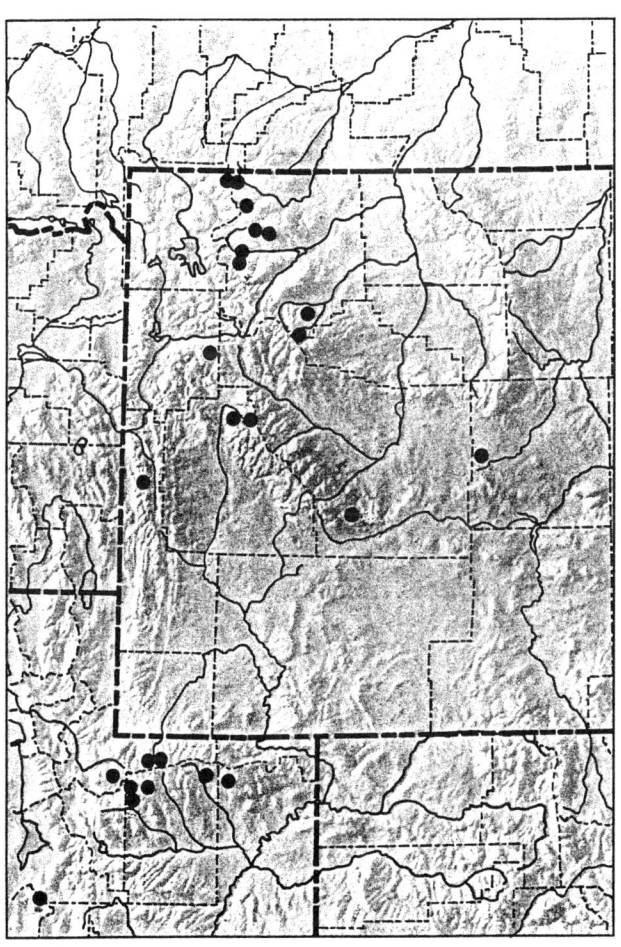

× 1/3

Potentilla subjuga Rydb.
Bull. Torr. Club 23:397, 1896.

Twinleaf Cinquefoil

Potentilla minutifolia Rydb.
Potentilla viridior Rydb.

Tufted, perennial herb from a branched caudex; stems 1 to several, villous. Basal leaves 3–5-foliolate, palmate, oblanceolate to obovate, pinnately cleft, villous; petioles with an additional pair of remote, reduced leaflets; cauline leaves reduced upward, trifoliolate. Flowers in cymes. Sepals 5, strigose; bractlets shorter than the sepals. Petals 5, yellow, longer than the sepals. Stamens numerous. Pistils numerous, the styles filiform, attached terminally. Akenes glabrous.

Meadows near and above timberline in the Absaroka Range. Endemic to Colorado and Wyoming.

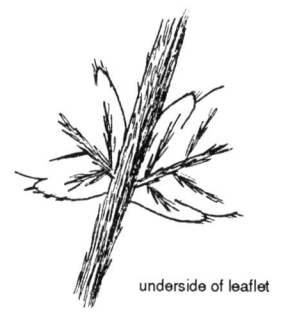
underside of leaflet

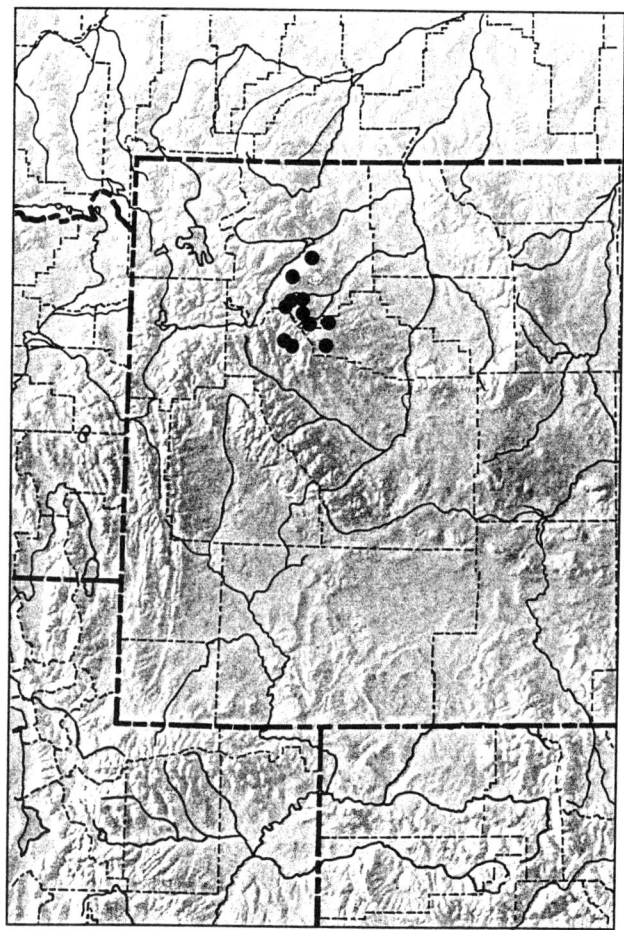

Potentilla uniflora Ledeb.
Mem. Acad. St. Petersb. 5:543, 1812.

Singleflower Cinquefoil

Potentilla ledebouriana Pors.
Potentilla vulcanicola Juz.

Low, tufted, perennial herb, less than 20 cm tall, from a dark brown, scaly, branched caudex; stems 1 to several, pilose, the hairs straight or nearly so. Basal leaves petiolate, trifoliolate; leaflets obovate, incised-pinnatifid, the segments acute, greenish above, tomentose below. Flowers solitary or 2–3 in a bracteate cyme; hypanthium saucer-shaped. Sepals 5, lanceolate to ovate; bractlets narrowly lanceolate. Petals 5, yellow, obcordate. Stamens numerous. Pistils numerous, the style papillate at the base, attached below the apex. Akenes yellow-brown, glabrous or nearly so.

Fellfields of the Absaroka, Gros Ventre, and Beartooth Ranges. Siberia; Alaska south to Montana, Wyoming, and Colorado.

Rubus L.
Raspberry

Low, perennial shrubs, rarely herbs, with prickly, spreading stems. Stems leaf-bearing the first year (primocanes) and floriferous the second year (floricanes). Leaves alternate, stipulate, simple or compound, usually both lobed and serrate. Flowers single, paniculate, cymose, or umbellate. Sepals usually 5, spreading to reflexed, ebracteate, persistent. Petals 5, white or deep rose. Stamens numerous. Pistils numerous, inserted on a hemispheric receptacle which may become elongate and fleshy in fruit. Styles terminal, filiform to clavate. Fruit an aggregate of more or less united drupelets.

Rubus idaeus L.
Sp. Pl., 492, 1753.

Red Raspberry

Batidaea acalyphacea Greene
Batidaea cataphracta Greene
Batidaea filipendula Greene
Batidaea peramoena Greene
Batidaea sandbergii Greene
Batidaea strigosa (Michx.) Greene
Batidaea subcordata Greene
Batidaea unicolor Greene
Batidaea viburnifolia Greene
Rubus acalyphaceus (Greene) Rydb.
Rubus carolinianus Rydb.
Rubus greeneanus Bailey
Rubus melanolasius Dieck
Rubus melanotrachys Focke
Rubus neglectus M.E. Peck
Rubus peramoenus (Greene) Rydb.
Rubus sachalinensis Levl.
Rubus strigosus Michx.
Rubus subarcticus Rydb.
Rubus viburnifolius (Greene) Rydb.

Erect shrub from a branched rhizome; stems biennial, often glaucous, bristly and with slender prickles, glabrous to glandular-pubescent. Leaves 3–5-foliolate; stipules narrowly lanceolate; leaflets lanceolate to ovate, acute to acuminate; irregularly serrate to doubly serrate, often somewhat lobed, green above, sparsely pubescent to tomentose below. Flowers 1–4, in terminal or axillary, glandular-hispid racemes. Calyx 5-lobed, glandular, the lobes ovate-lanceolate, acuminate, reflexed. Petals 5, white, oblong to spatulate, erect or ascending. Stamens numerous. Pistils numerous. Fruit an aggregate of drupelets forming a red, ovoid, tomentulose unit.

Talus slopes, fellfields, and rocky ledges of the Beartooth, Big Horn, Teton, Uinta, Wasatch, and Wind River Ranges. Circumboreal; south in western North America to California, Arizona, and northern Mexico.

Glandular alpine plants with leaves lanate to tomentose below are ssp. *sachalinensis* (Levl.) Focke, var. *sachalinensis*.

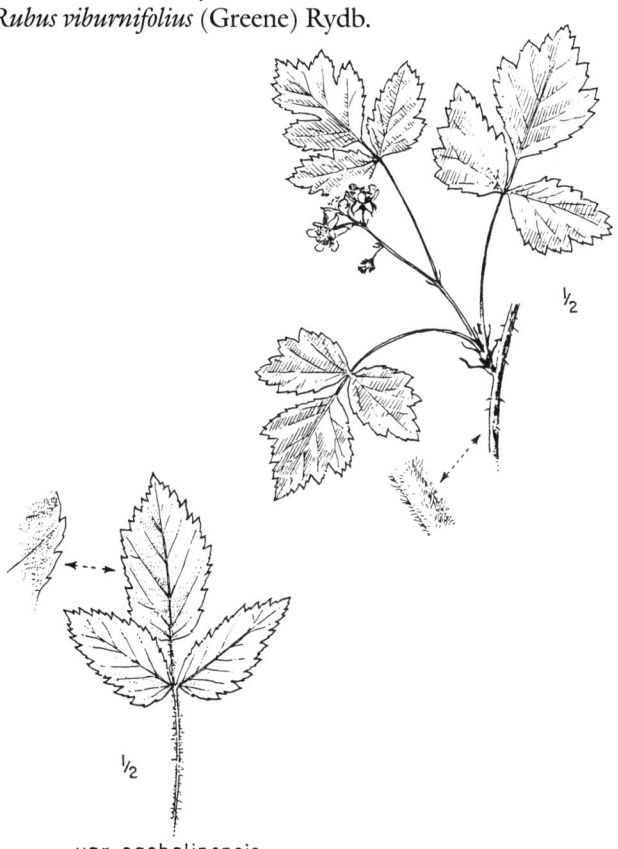

var. sachalinensis

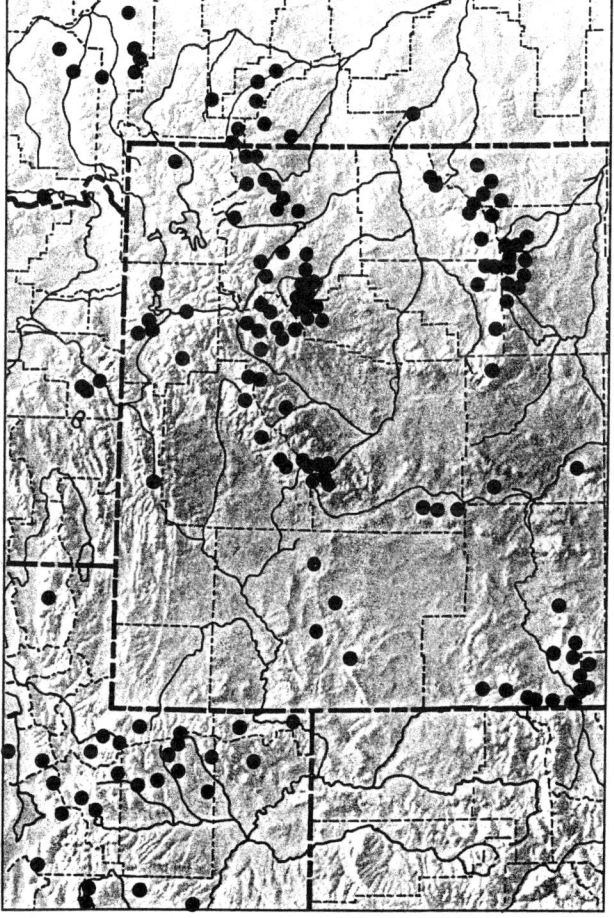

Sibbaldia L.
Sibbaldia

Low, tufted, perennial herbs from a branched, woody caudex. Leaves 3-foliolate, the leaflets truncate, 3-toothed at the apex. Inflorescence cymose on an erect peduncle arising from prostrate, leafy stems. Sepals 5, bracteolate, much exceeding the 5 yellow petals. Stamens 5 in ours. Pistils 5–20. Fruit a group of akenes on a dry receptacle.

Sibbaldia procumbens L.
Sp. Pl., 284, 1753.

Prostrate Sibbaldia

Potentilla procumbens (L.) Clairv.
Potentilla sibbaldii Hall. f.

Densely tufted, perennial herb from a short-creeping rhizome. Leaves trifoliolate, the petioles slender; leaflets oblanceolate to obovate, cuneate, truncate and 3–5-toothed at the apex, sparsely pubescent. Flowers in dense, few-flowered cymes; hypanthium saucer-shaped. Sepals 5, ovate, alternating with the 5 lanceolate bractlets. Petals 5, yellow, linear-oblong to oval, much shorter than the sepals. Stamens 5, alternate with the petals. Pistils 5–15. Akenes ovoid.

Meadows, fellfields, and rocky slopes in all alpine ranges of the Middle Rocky Mountains. Circumpolar; south in western North America to California, Utah, and Colorado, and in the East to New Hampshire.

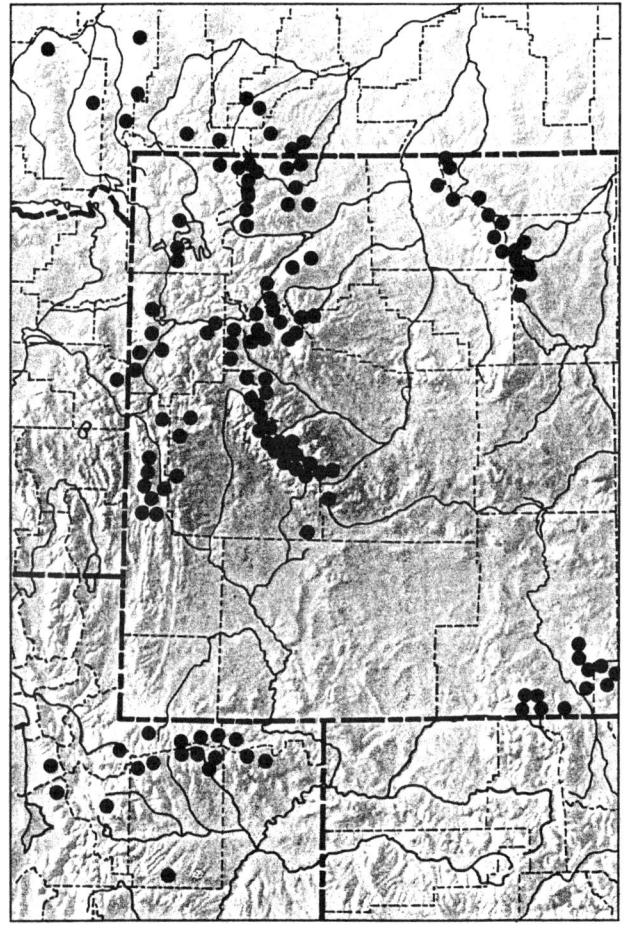

Rubiaceae
Madder Family

Annual or perennial herbs with opposite or whorled, simple and mostly entire, stipulate leaves. Flowers epigynous, regular, mostly perfect, solitary or in terminal cymes or panicles. Calyx 3–5-lobed, or wanting. Corolla 3–5-lobed, sympetalous, rotate to salverform. Stamens 3–5, epipetalous, alternate with the corolla lobes. Pistil 1, the ovary 2-carpellate; locules usually 2; styles 1 or 2. Fruit a capsule or schizocarp.

KEY TO THE GENERA OF RUBIACEAE

1. Leaves whorled — *Galium*
1. Leaves opposite — *Kelloggia*

Galium L.
Bedstraw

Annual to perennial herbs from creeping rhizomes; stems 4-angled, slender, erect or decumbent, sometimes woody at the base. Leaves opposite, appearing whorled because of leaflike stipules. Flowers perfect or unisexual, white, solitary or in terminal or axillary cymes. Calyx lobes obsolete. Corolla rotate, 4-lobed. Stamens 4, occasionally 3. Styles 2; stigmas capitate. Fruit of 2 indehiscent carpels, glabrous to pubescent.

Galium boreale L.
Sp. Pl., 108, 1753.

Northern Bedstraw

Galium hyssopifolium Hoffm.
Galium septentrionale R. & S.
Galium strictum Torr.
Galium utahense Eastw.

Perennial herb from creeping rhizomes; stems erect, simple to branched, 4-angled, glabrous or pubescent at the nodes. Leaves 4 at each node and linear, oblong, or lanceolate, sessile, obtuse to acute, glabrous or scabrous, 3-nerved, the margins entire. Flowers perfect, numerous in dense terminal panicles. Corolla white or cream, 4-lobed. Fruit indehiscent, glabrous or pubescent with short, straight or curled hairs.

Forest and krummholz margins, stream banks, and meadows of the lower alpine zone near timberline in the Beartooth and Big Horn Ranges. Circumboreal; south in western North America to California, Arizona, and Texas.

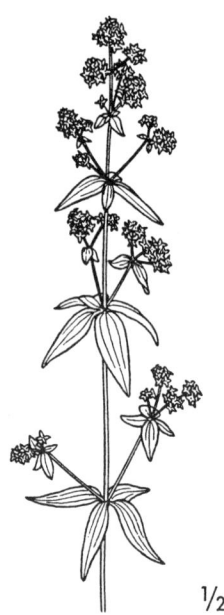

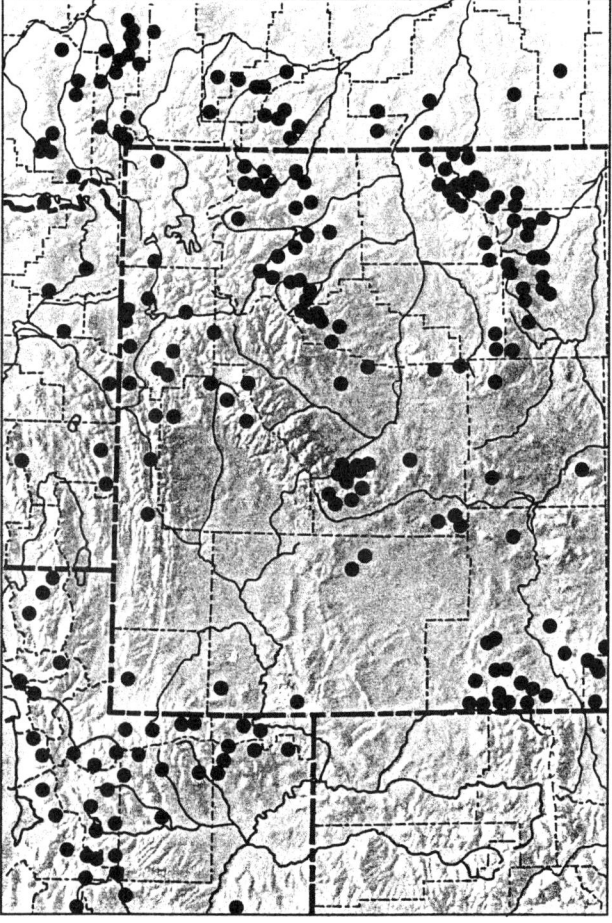

Kelloggia Torr.
Kelloggia

Small, slender, perennial herbs from creeping rhizomes. Leaves opposite, entire, sessile, stipulate, lanceolate. Flowers perfect, white or pinkish, in open, terminal or axillary, cymose panicles. Calyx with 4 minute teeth. Corolla salverform, 4-lobed, hispid. Stamens 4, exerted. Style filiform; stigmas 2. Fruit obovate, indehiscent, with hooked bristles.

Kelloggia galioides Torr.
Bot. Wilkes Exped., 332, 1874.

Kelloggia

Slender, perennial herb from creeping rhizomes. Leaves glabrous, opposite, entire, lanceolate, sessile, stipulate. Flowers small, long-pedicellate, in open panicles. Calyx lobes minute. Corolla funnelform, white or pink, the lobes hispid. Stamens 4, exerted. Ovary 2-loculed; style filiform; stigmas 2. Fruit covered with hooked bristles, dehiscent into 2 schizocarps.

Mountain slopes, moist rocky ledges, woods, or thickets. Known from above timberline on rocky ledges of the lower alpine zone of the Salt River Range. Washington to western Wyoming, south to California, Arizona, and Utah.

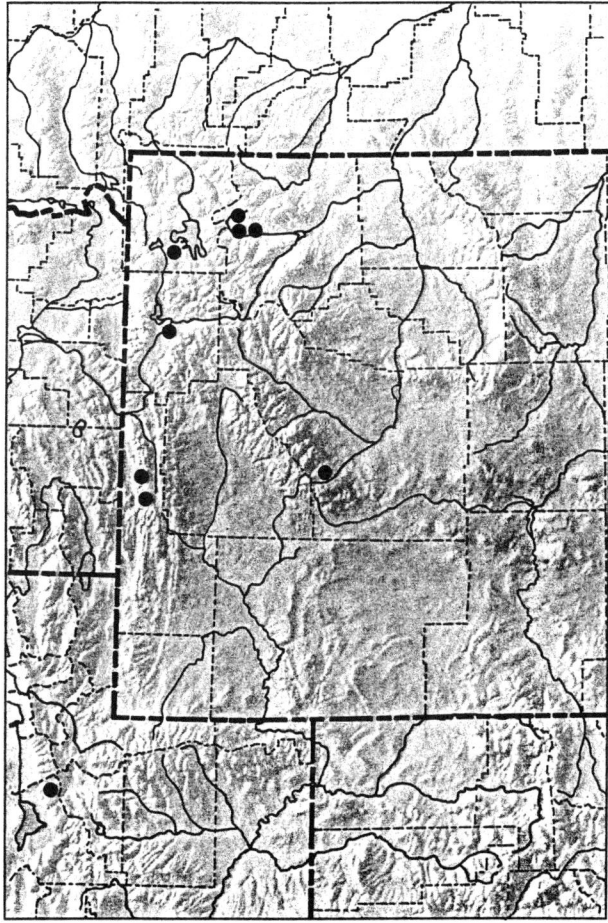

Salicaceae
Willow Family

Trees or shrubs, rarely almost herbaceous, with simple, alternate, stipulate leaves. Plants dioecious, the catkins (aments) erect or pendulous. Each flower subtended by a single scale, the calyx reduced or lacking, the petals lacking. Stamens usually 2 to many. Ovary 1-celled, the stigmas 2–4, often 2-lobed. Fruit a 2–4-valved capsule with many comose seeds.

KEY TO THE GENERA OF SALICACEAE

1. Flowers with a basal, cup-shaped disk; scales fimbriate; buds with several scales — *Populus*
1. Flowers without a basal, cup-shaped disk; scales entire to denticulate buds with a single scale — *Salix*

Populus L.
Cottonwood, Aspen, Poplar

Trees with gray, furrowed bark, the leaves lanceolate to reniform. Buds more or less resinous, with several scales. Catkins precocious, long and pendulous. Each flower with a cup-shaped disk at the base, and subtended by a single, laciniate scale. Stamens several to numerous. Ovary conic or ovoid, the stigmas 2–4. Capsule opening by 2–4 valves.

Populus balsamifera L.
Sp. Pl., 1034, 1753.

Balsam Poplar

Populus candicans Ait.
Populus hastata Dode in part
Populus michauxii Dode
Populus tacamahaca P. Mill.
Populus trichocarpa T. & G. ex Hook.

Trees with glabrous to pubescent branches, the twigs early reddish brown, becoming gray-brown; buds glabrous, resinous. Leaves ovate, acute, rounded at the base, dark green above, whitish-glaucous below; petioles terete for their entire length. Catkins elongate, the rachis pubescent. Capsules ovoid.

Common on stream banks and forming ribbon forest along major drainage systems at lower elevations. A single alpine collection is known from 3385 m (11,000 ft) in the northern Wind River Range, which represents the highest elevational record for this species in the region. *Populus tremuloides* (Quaking Aspen) has also been observed, but not collected, in the same general area at timberline. It might be expected as occasional in dwarfed form on sunny, south-facing talus slopes in the alpine-subalpine ecotone. Transcontinental in northern Canada, south to New York, Nebraska, and Nevada.

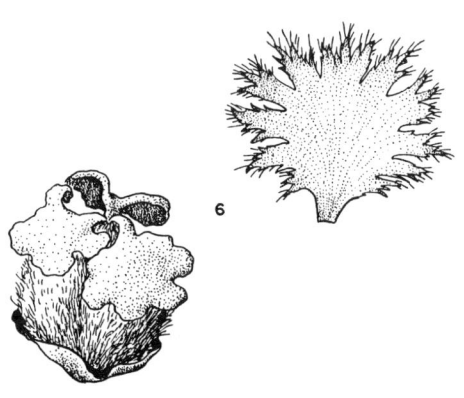

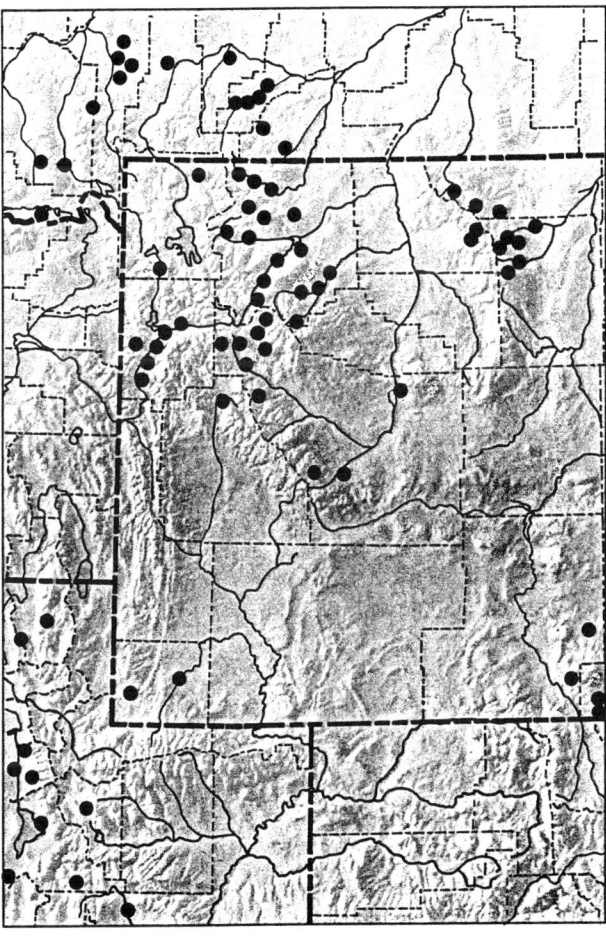

Salix L.
Willow

Shrubs, ours sometimes almost herbaceous, with linear to ovate, short-petioled leaves, the stipules persistent or deciduous. Buds covered by a single scale. Catkins (aments) precocious, cotaneous, or serotinous, erect or spreading, sessile or pedunculate. Flowers each subtended by an entire or, rarely, denticulate scale, which is associated with 1 or 2 small glands. Stamens 2–10. Ovary sessile or pedicellate, glabrous or hairy, the style wanting to elongate, the stigmas short to long, entire or divided. Capsule opening by 2 valves.

KEY TO THE SPECIES OF *SALIX*

1. Plants dwarf, prostrate, often nearly herbaceous, less than 8 cm tall
    2. Leaves 6 mm long or less; capsules glabrous — *S. rotundifolia*
    2. Leaves 7–40 mm long; capsules tomentose to villous
        3. Leaves obtuse or retuse, broadly ovate to suborbicular, strongly reticulate beneath — *S. reticulata*
        3. Leaves acute, lanceolate, pinnately veined beneath
            4. Aments 2–4 cm long, many-flowered; leaves 2–4 cm long, glaucous beneath — *S. arctica*
            4. Aments 0.7–2 cm long, few-flowered; leaves 1–1.2 cm long, green on both surfaces — *S. cascadensis*
1. Plants erect, woody shrubs mostly 3–20 dm tall, often taller at timberline, sometimes prostrate with woody branches on the surface of the ground at high elevations
    5. Pistils and capsules glabrous (rarely puberulent or thinly sericeous)
        6. Aments sessile; leaves glandular-denticulate; branchlets villous — *S. tweedyi*
        6. Aments pedunculate; leaves entire or serrulate; branchlets glabrous
            7. Leaves glabrous — *S. monticola*
            7. Leaves pubescent, at least below
                8. Leaves 20–45 mm long, 8–13 mm wide, gray-sericeous throughout, mostly oblanceolate to elliptical, the margins entire — *S. wolfii*
                8. Leaves 30–70 mm long, 15–30 mm wide, thinly to densely pubescent throughout, mostly ovate-obovate, the margins serrulate to entire — *S. eastwoodiae*
    5. Pistils and capsules pubescent
        9. Buds glutinous; leaves grayish-pubescent, the margins entire and more or less glandular; stipules glandular — *S. barrattiana*
        9. Plants lacking the above combination of characters
            10. Twigs prominently pruinose — *S. drummondiana*
            10. Twigs not pruinose
                11. Pistillate and staminate aments sessile or subsessile; twigs glossy, drying black — *S. planifolia*
                11. Pistillate and staminate aments on short peduncles; twigs yellowish to chestnut brown
                    12. Capsules sessile; bracts yellowish or yellowish white; leaves coarsely pubescent, short-petiolate — *S. brachycarpa*
                    12. Capsules definitely pedicellate; bracts yellowish brown, bicolored, or dark brown; leaves finely pubescent, often long-petiolate — *S. glauca*

Salix arctica Pall.
Fl. Ross. 1(2):86, 1789.

Arctic Willow

Salix anglorum Cham.
Salix arctica R. Br. ex Richards.
Salix brownei (Anderss.) Bebb
Salix caespitosa Kennedy
Salix crassijulis Trautv.
Salix hudsonensis Schneid.
Salix pallasii Anderss.
Salix petrophila Rydb.
Salix tortulosa Trautv.

Prostrate, creeping shrub, the stems on or below the surface of the ground; branches yellowish, becoming brown to purplish or black, glabrous. Leaf blades obovate, oblanceolate, or elliptic, acute to obtuse, entire, dark green above, paler and glaucous below, glabrous at maturity, sparsely villous to pilose when young; stipules small or lacking. Pistillate catkins 2–4 cm long, on leafy peduncles, appearing with the leaves; bracts dark brown, oval, obtuse, villous. Stamens 2, filaments glabrous, the anthers reddish purple. Capsules lanceolate, pubescent, sessile or nearly so; styles reddish, about 1 mm long.

Slopes, meadows, fellfields, and rocky ridges in nearly all alpine areas in the Middle Rocky Mountains; not yet known from above timberline in the Hoback Range. Circumboreal; south in western North America to California, Nevada, and New Mexico.

Rocky mountain plants are var. *petraea* Anderss.

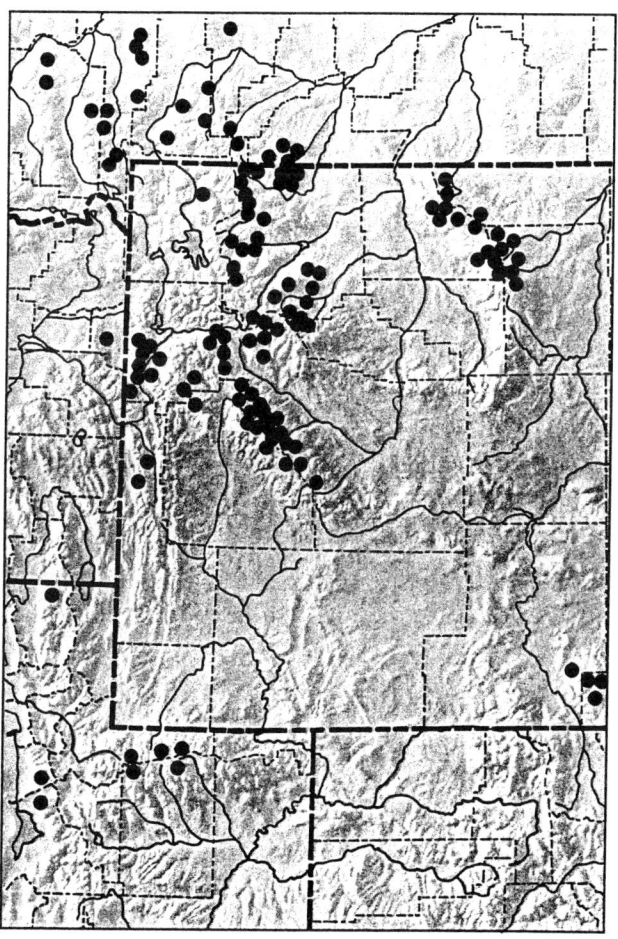

Salix barrattiana Hook.
Fl. Bor.-Am. 2:146, 1838.

Barratt Willow

Salix albertana Rowlee

Erect shrub to 1 m tall, with pubescent, glutinous twigs. Leaf blades elliptic to obovate or oblanceolate, acute, grayish-pubescent, less so above, the margins entire to serrulate; stipules mostly deciduous, glandular-margined. Pistillate catkins sessile, erect, mostly appearing with the leaves; bracts acute, black, pubescent. Stamens 2, the filaments glabrous. Capsules white-tomentose, sessile or nearly so; styles 1–2 mm long.

Stream banks and wet meadows at the head of Wyoming Creek in the Beartooth Range. Alaska and Yukon Territory south to British Columbia, Montana, and northwest Wyoming.

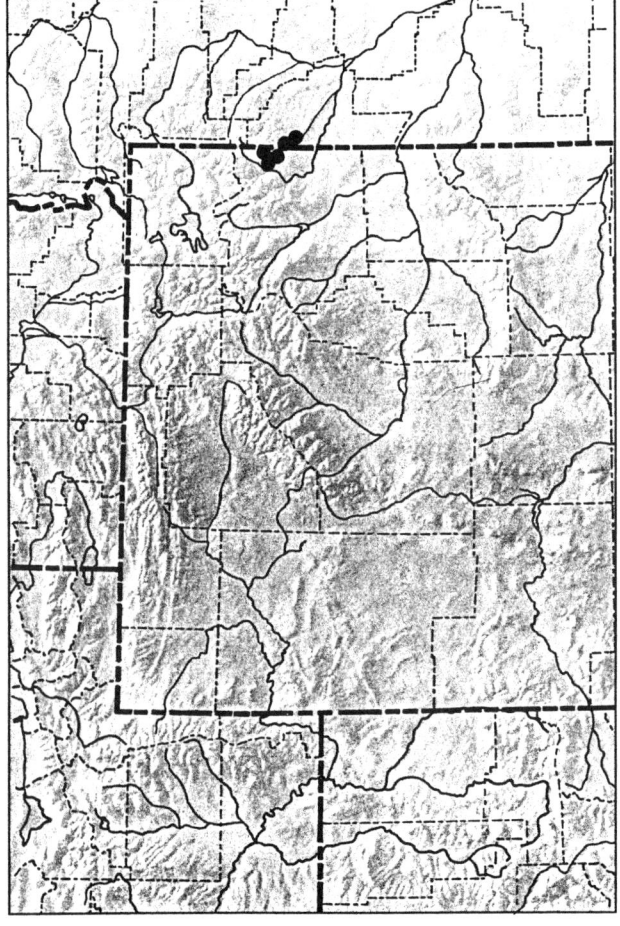

Salix brachycarpa Nutt.
N. Am. Sylva 1:69, 1842.

Barrenground Willow

Salix fullertonensis Schneid.
Salix lingulata Anderss.
Salix muriei Hult.
Salix niphoclada Rydb.
Salix stricta (Anderss.) Rydb.

Erect shrub to 1 m tall, with yellowish to reddish, tomentose twigs. Leaves short-petiolate, the blades elliptic, oval, or obovate, acute, pubescent and usually glaucous beneath, the margins entire; stipules lacking. Pistillate catkins less than 2 cm long, subglobose to short-cylindrical, on short, leafy peduncles appearing with the leaves; bracts oval, yellowish to light brown, short-pubescent. Stamens 2, the filaments glabrous above, pubescent at the base. Capsules sessile or nearly so, white-tomentose; styles 0.5–1 mm long.

Moist meadows and stream banks of the Beartooth, Medicine Bow, Uinta, Wasatch, Wind River, and Wyoming Ranges. Transcontinental in Canada; Alaska south to Oregon, Utah, and Colorado.

Salix brachycarpa and *S. glauca* often hybridize, and plants with characteristics intermediate between the two are to be expected in areas where both species occur.

2/3

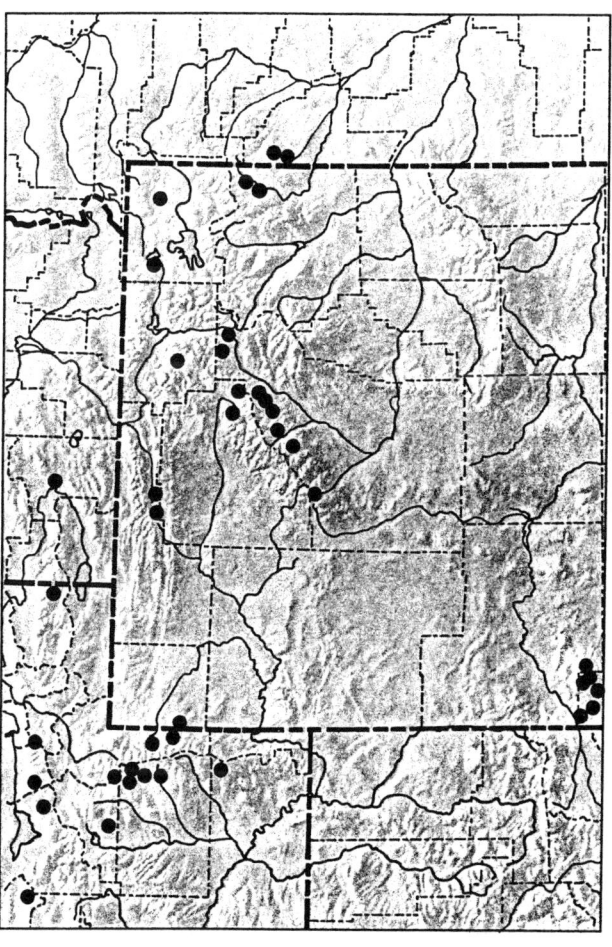

Salix cascadensis Cockll.
Muhlenbergia 3:9, 1907.

Cascade Willow

Salix tenera Anderss.

Prostrate, creeping shrub forming dense mats less than 5 cm tall; twigs yellowish to reddish brown. Leaves short-petiolate, the blades elliptic, acute, pubescent when young, becoming glabrate and shiny with age, green on both surfaces, the margins entire; stipules wanting. Pistillate catkins less than 2 cm long, few-flowered, on short, leafy peduncles appearing with the leaves; bracts oblanceolate, blackish, pubescent at least on the inner surface. Stamens 2, the filaments glabrous. Capsules sessile or nearly so, gray-villous, the styles 0.5–1.2 mm long.

Meadows, rocky slopes, and fellfields in most of the major alpine ranges of the Middle Rocky Mountains. Not yet known from above timberline in the Big Horn, Hoback, Salt River, Wasatch, and Wyoming Ranges. Washington to Montana, south to California, Wyoming, and Colorado.

Middle Rocky Mountain alpine plants are typical var. *cascadensis*.

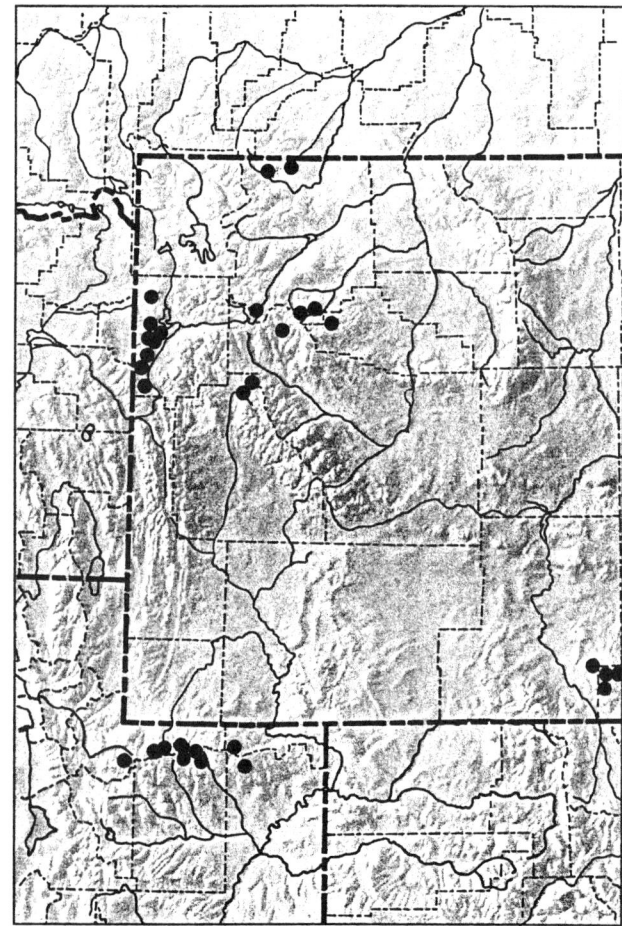

Salix drummondiana Barr. ex Hook.
Fl. Bor.-Am. 2:144, 1838.

Drummond Willow

Salix bella Piper
Salix covillei Eastw.
Salix pachnophora Rydb.
Salix pellita Anderss.
Salix subcoerulea Piper

Erect shrubs, to 1 m tall in the alpine zone. Twigs and older stems glabrous, dark purplish brown, pruinose. Leaves long-petiolate, the blades elliptic to oblanceolate, acute, glabrate above, sparsely silver-pubescent below, the margins entire and mostly revolute; stipules small and narrowly elliptic, or lacking. Pistillate catkins sessile, appearing before or with the leaves; bracts elliptic, dark brown to black, pilose. Stamens 2, the filaments glabrous. Capsules sessile or nearly so, pubescent; styles 0.5–1.5 mm long.

Stream banks in the lower alpine zone of the Wind River Range. Alaska south to California, Nevada, and New Mexico.

Middle Rocky Mountain alpine plants are var. *subcoerulea* (Piper) Ball. They are distinguished from the typical var. *drummondiana* by having pruinose twigs and leaves that are sparsely silver-pubescent beneath.

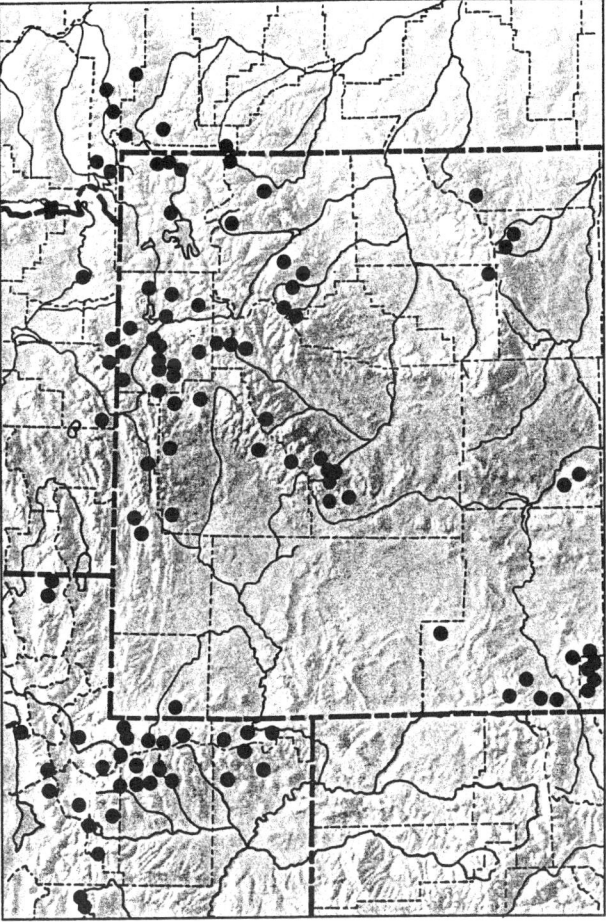

Salix eastwoodiae Cockll. ex A. Heller
Cat. N. Am. Pl. 3:89, 1910.

Eastwood Willow

Salix californica Bebb
Salix commutata Bebb

Erect shrub, to 1.5 m tall in the alpine zone; twigs early yellowish, becoming dark brown with grayish-tomentose pubescence. Leaves short-petiolate, the blades elliptic to obovate, acute to short-acuminate, the margins entire to glandular-serrulate; young leaves densely gray-tomentose on both sides, with glandular margins; older leaves glabrate and nonglaucous; stipules glandular-serrate, villous. Pistillate catkins on leafy peduncles, appearing with the leaves; bracts oblanceolate to oblong, mostly obtuse, light to dark brown, villous. Stamens 2, the filaments glabrous. Capsules short-pedicellate, glabrous or, rarely, sparsely pubescent; styles 0.5–1 mm long.

Stream banks in the Beartooth, Big Horn, Teton, and Wind River Ranges. Alaska and Yukon Territory south to California, Idaho, and Colorado.

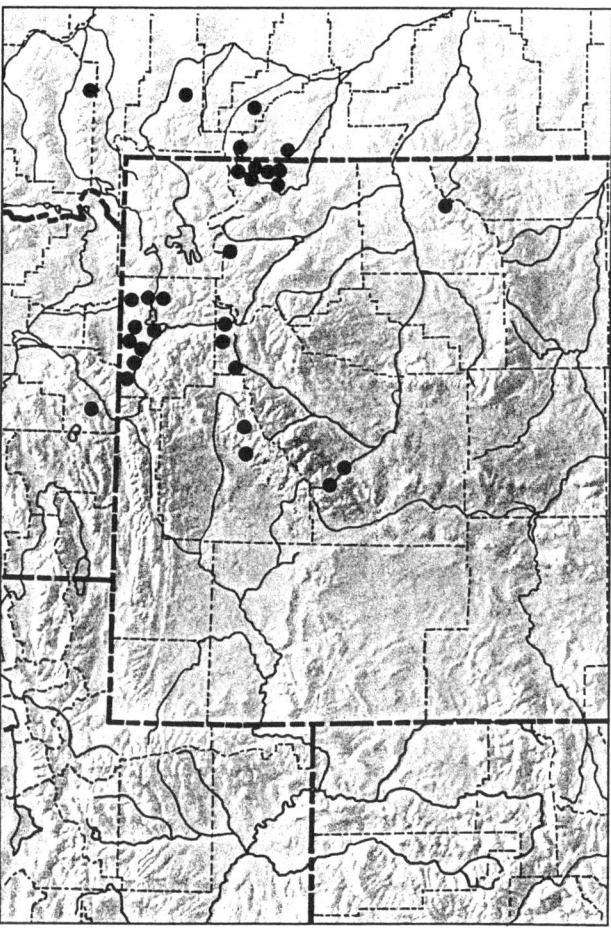

Salix glauca L.
Sp. Pl., 1019, 1753.

Grayleaf Willow

Salix anamesa Schneid.
Salix atra Rydb.
Salix callicarpaea Trautv.
Salix cordifolia Pursh
Salix desertorum Richards.
Salix glaucops Anderss.
Salix labradorica Rydb.
Salix macounii Rydb.
Salix nudescens Rydb.
Salix pseudolapponum von Seem.
Salix rydbergii A. Heller
Salix seemannii Rydb.
Salix vacciniformis Rydb.
Salix villosa D. Don ex Hook.
Salix wyomingensis Rydb.

Erect shrub, to 1 m tall; twigs reddish brown, the younger ones pubescent, the older twigs glabrous, shiny. Leaves mostly long-petiolate, the petioles 5–10 mm long; blades elliptic, obovate, or oblanceolate, acute or obtuse, pubescent on both sides, sometimes glabrate or glabrous, glaucous beneath, the margins entire or sometimes serrulate at the base; stipules inconspicuous and lanceolate, or lacking. Pistillate catkins dense-cylindrical, on leafy peduncles, appearing with the leaves; bracts oblong, brown, often with darker tips, pubescent on both sides. Stamens 2, filaments pubescent at the base. Capsules short-pedicellate, pubescent; styles 0.5–1 mm long.

Meadows, slopes, and stream banks in all alpine areas of the Middle Rocky Mountains. Circumboreal; south in western North America to Nevada, Utah, and New Mexico.

Middle Rocky Mountain plants are var. *villosa* (D. Don ex Hook.) Anderss. See comments under *S. brachycarpa*.

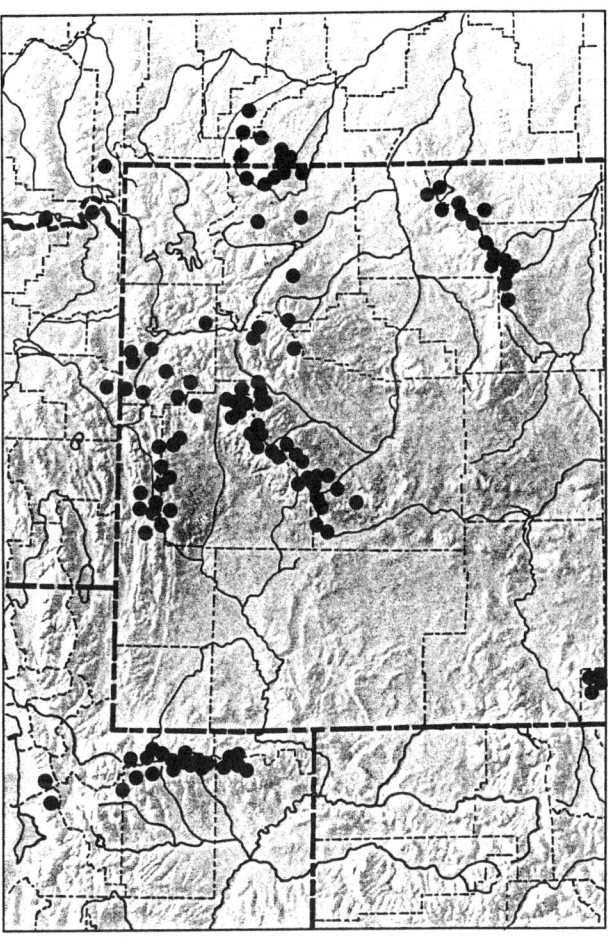

Salix monticola Bebb in Coult.
Man. Bot. Rocky Mt. Reg., 336, 1885.

Mountain Willow

Salix amelanchieroides L. Kelso
Salix dissymetrica L. Kelso
Salix padifolia Rydb.
Salix padophylla Rydb.
Salix pseudomonticola Ball
Salix sawatchicola L. Kelso

Erect shrub, to 1.5 m tall in the alpine zone; twigs yellow or yellowish green, drying dark brown to black, pubescent when young, becoming glabrate and shiny with age. Leaves long-petiolate, the blades ovate to obovate, acute to obtuse, glabrous at maturity, glaucous beneath, the margins crenate-serrulate; stipules small or lacking. Pistillate catkins subsessile on short peduncles, appearing before or with the leaves; bracts acute, dark brown, villous. Stamens 2, filaments glabrous. Capsules short-pedicellate, glabrous; styles 0.8–1 mm long.

Stream banks and wet rocky places in the lower alpine zone of the Medicine Bow Range. Transcontinental in North America; Alaska and Yukon Territory south to Idaho, Utah, and New Mexico.

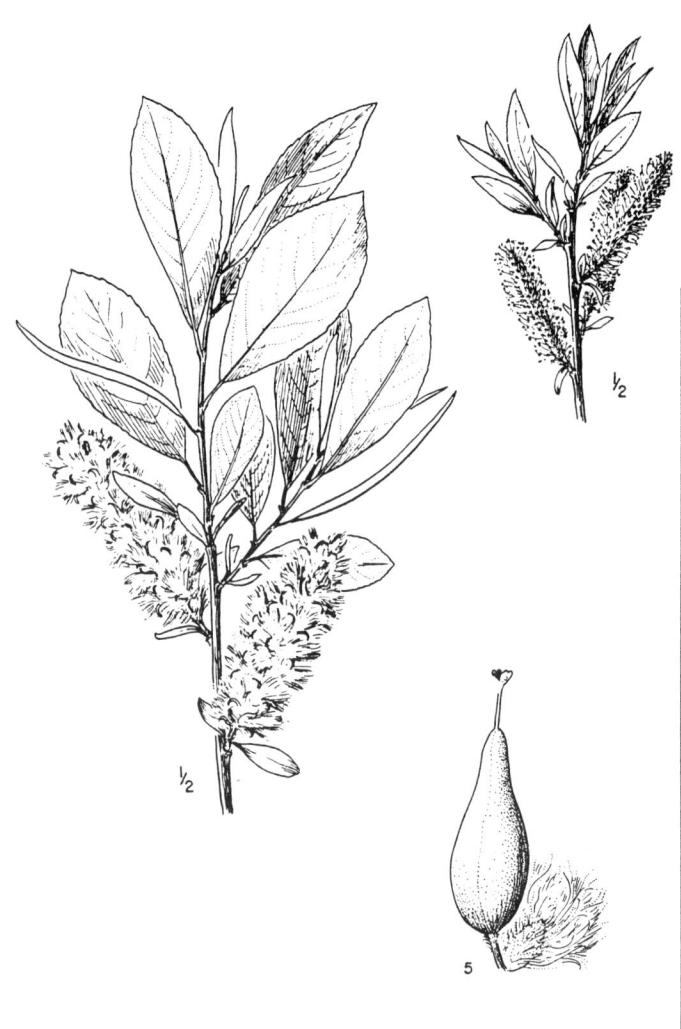

Salix planifolia Pursh
Fl. Am. Sept., 611, 1814.

Planeleaf Willow

Salix chlorophylla Anderss.
Salix monica Bebb
Salix nelsonii Ball
Salix pennata Ball
Salix phylicifolia L.
Salix phylicoides Anderss.
Salix pulchra Cham.
Salix pychnocarpa Anderss.
Salix tyrrellii Raup

Erect shrub, to 1 m in the alpine zone; twigs reddish brown, purplish, or nearly black, glabrous, glossy. Leaves petiolate, the petioles 2–6 mm long; blades elliptic to oblanceolate, acute, glabrous, green above, glaucous below, the margins entire to irregularly and shallowly toothed; stipules lacking, or minute and deciduous. Pistillate catkins sessile to subsessile, appearing before or with the leaves; bracts dark brown to black, villous. Stamens 2, the filaments glabrous. Capsules sessile or nearly so, pubescent; styles 0.6–1.6 mm long.

Slopes and stream banks, often forming dense thickets just above timberline in Middle Rocky Mountain alpine ranges. Not yet known from above timberline in the Gros Ventre, Hoback, Salt River, and Wyoming Ranges. Circumboreal; transcontinental in Canada, south in the western United States to California and New Mexico.

Alpine shrubs rarely exceeding 1 m tall, with leaves less than 5 cm long and averaging 1 cm wide, are var. *monica* (Bebb) Schneider; taller shrubs (1–2 m) of protected timberline sites, with leaves more than 5 cm long and averaging 2 cm wide, are var. *planifolia*.

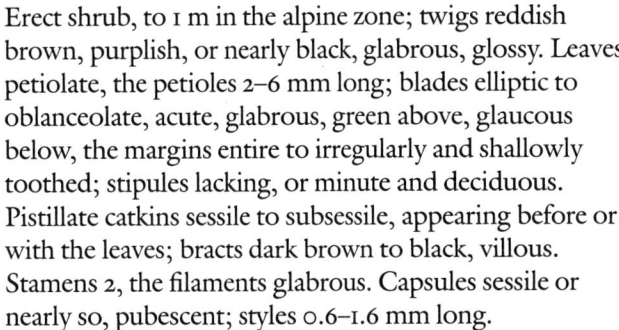

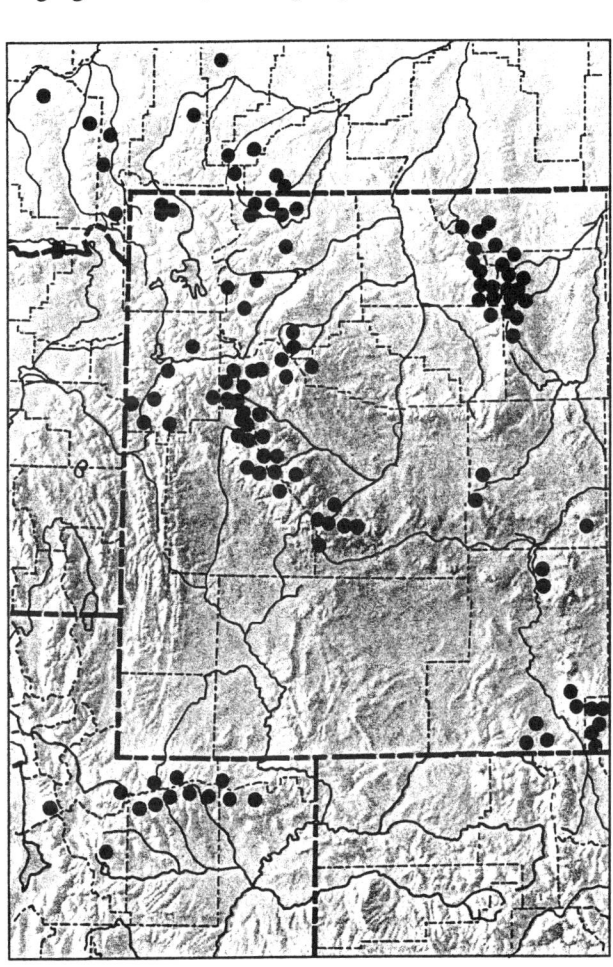

Salix reticulata L.
Sp. Pl., 1018, 1753.

Netleaf Willow

Salix aemulans von Seem.
Salix nivalis Hook.
Salix orbicularis Anderss.
Salix saximontana Rydb.
Salix solheimii E.H. Kelso
Salix venusta Anderss.

Prostrate shrub with creeping, often buried branches, less than 6 cm tall, rarely to 10 cm; twigs greenish to tan or light brown, glabrous. Leaves long-petiolate; blades obovate or broadly elliptic, mostly rounded at the apex, glabrous, shiny green above, glaucous and reticulate below, the margins entire, often somewhat revolute; stipules minute or lacking. Pistillate catkins at the ends of leafy twigs, appearing after the leaves; bracts oblong, greenish, yellowish, or sometimes reddish, pubescent at least on the inner surface. Stamens 2, the filaments pubescent, at least at the base. Capsules sessile, pubescent, styles short, 0.1–0.5 mm long.

Meadows, fellfields, and rocky slopes in nearly all the alpine ranges of the Middle Rockies. Not yet known from above timberline in the Hoback Range. Circumboreal; south in the western United States to California, Nevada, Utah, and New Mexico.

Alpine plants of our region are ssp. *nivalis* (Hook.) Löve et al., distinguished from the more northern and arctic ssp. *reticulata* by having leaves less than 15 mm long and catkins less than 10 mm long.

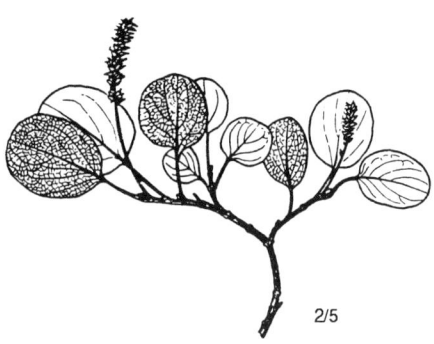

2/5

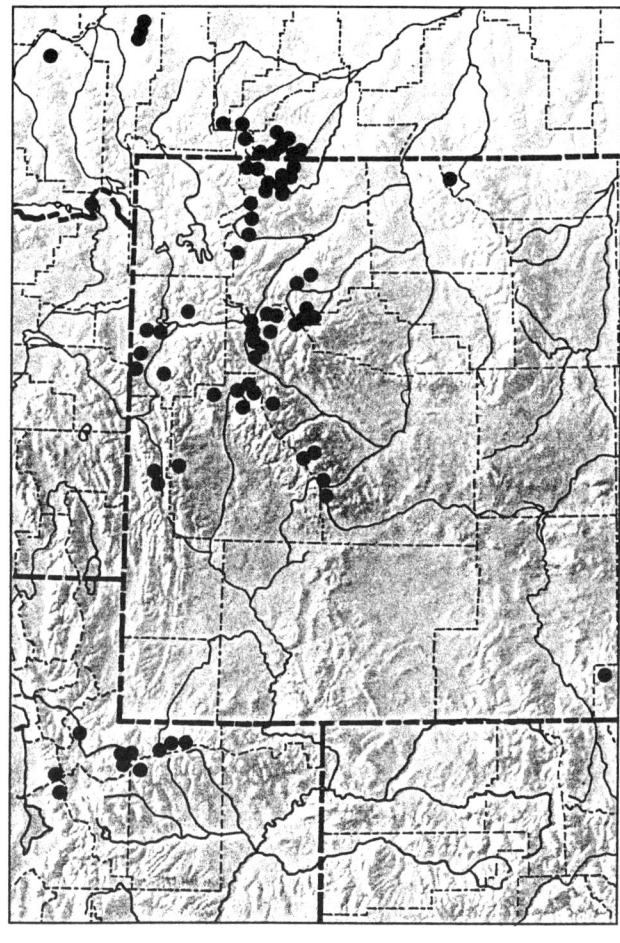

Salix rotundifolia Trautv.
Mem. Soc. Nat. Mosc. 2:304, 1832.

Least Willow

Salix behringica von Seem.
Salix dodgeana Rydb.
Salix leiocarpa (Cham.) Cov.

Densely pulvinate shrub, to 3 cm tall, from a taproot; branches sparsely pubescent, at least when young. Leaves sessile to short-petiolate; blades ovate to slightly obovate, mostly obtuse to rounded at the apex, glabrous, pinnately veined below, the margins entire and often ciliate; stipules lacking. Pistillate catkins sessile at the ends of lateral branches, appearing with the leaves, commonly with fewer than 6 flowers; bracts obovate, light brown to red, somewhat villous within. Stamens 2, the filaments glabrous. Capsules sessile, glabrous; styles mostly less than 1 mm long.

Fellfields of the Absaroka, Beartooth, Big Horn, Gros Ventre, Teton, and Wind River Ranges. Siberia; Alaska and Yukon Territory south to northwestern Wyoming.

Middle Rocky Mountain alpine plants are ssp. *dodgeana* (Rydb.) Argus.

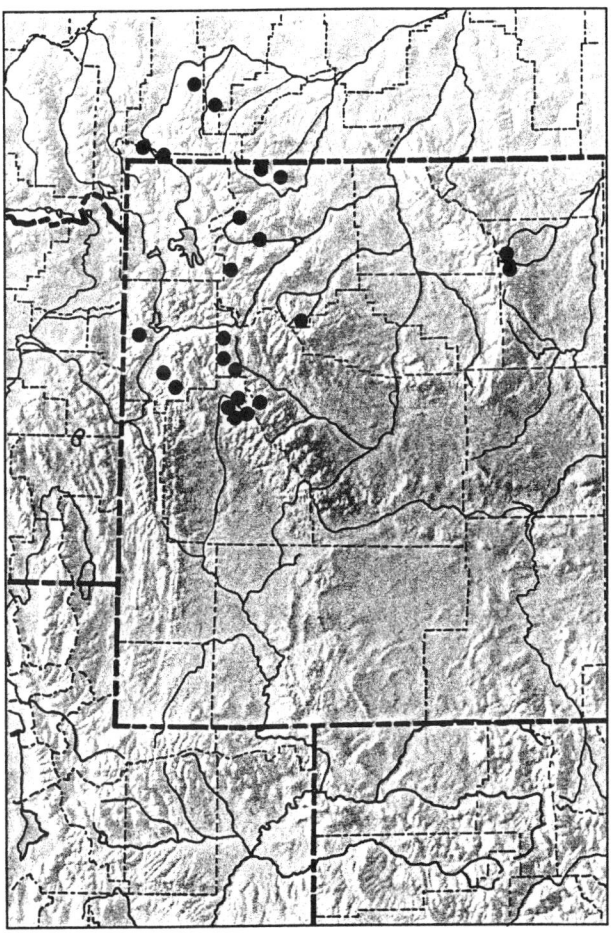

Salix tweedyi (Bebb ex Rose) Ball
Bot. Gaz. 40:377, 1905.

Tweedy Willow

Salix rotundifolia Nutt.

Erect shrub, to 2 m in the lower alpine zone; branches light brown to blackish, villous. Leaves petiolate, the petioles 5–15 mm long; blades ovate to obovate, acute, pubescent when young, becoming glabrate, somewhat glaucescent below, the margins serrulate to denticulate, glandular; stipules foliaceous, ovate to reniform, denticulate. Pistillate catkins sessile on older twigs, appearing before or with the leaves; bracts lanceolate, black, glabrate on the outside, pilose on the inside. Stamens 2, the filaments glabrous. Capsules subsessile, glabrous; styles 1.5–3 mm long.

Sheltered stream banks at timberline and in the lower alpine zone of the Absaroka, Beartooth, Big Horn, and Wind River Ranges. British Columbia south to Idaho and northwestern Wyoming.

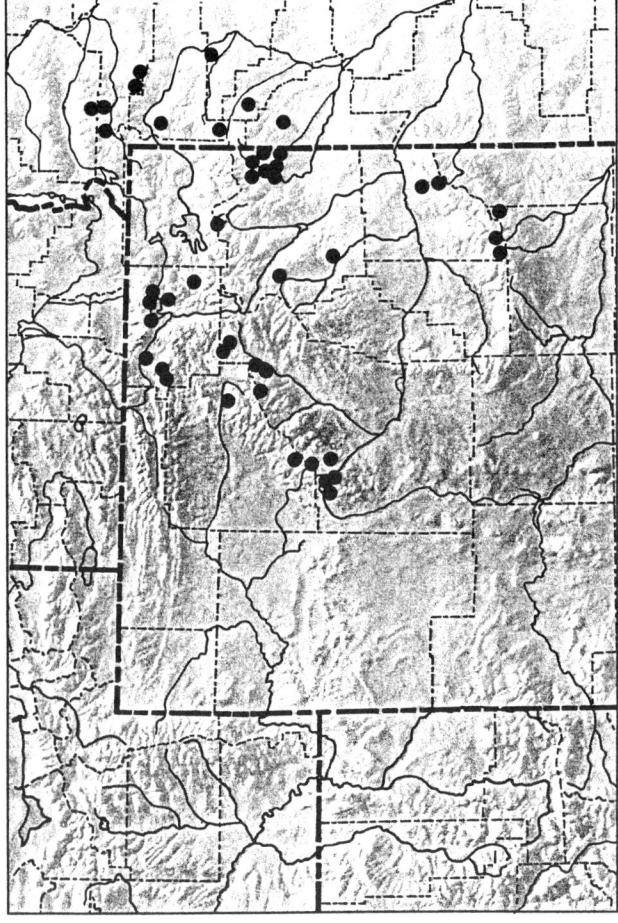

Salix wolfii Bebb in Rothr.
Bot. Wheeler Exped., 241, 1878.

Wolf Willow

Salix idahoensis (Ball) Rydb.

Erect shrub, to 1 m tall; branches yellowish to dark brown, pubescent, especially when young. Leaves short-petiolate, the petioles less than 8 mm long; blades lanceolate to elliptic, acute, gray-pubescent, the margins entire; stipules small, glandular-serrate when young, deciduous. Pistillate catkins on short, leafy peduncles, appearing with the leaves; bracts obovate, black, villous. Stamens 2, the filaments glabrous. Capsules subsessile to pedicellate, glabrous to sparsely pubescent or puberulent; styles 0.5–1 mm long.

Stream banks, moist meadows, and late snow bed margins of the Beartooth and Wind River Ranges. Oregon to Montana and south to Nevada, Utah, and Colorado.

The 2 varieties in the Middle Rocky Mountains are var. *idahoensis* Ball, the more northern form, with puberulent capsules; and var. *wolfii*, of Utah and southwestern Wyoming, with glabrous capsules.

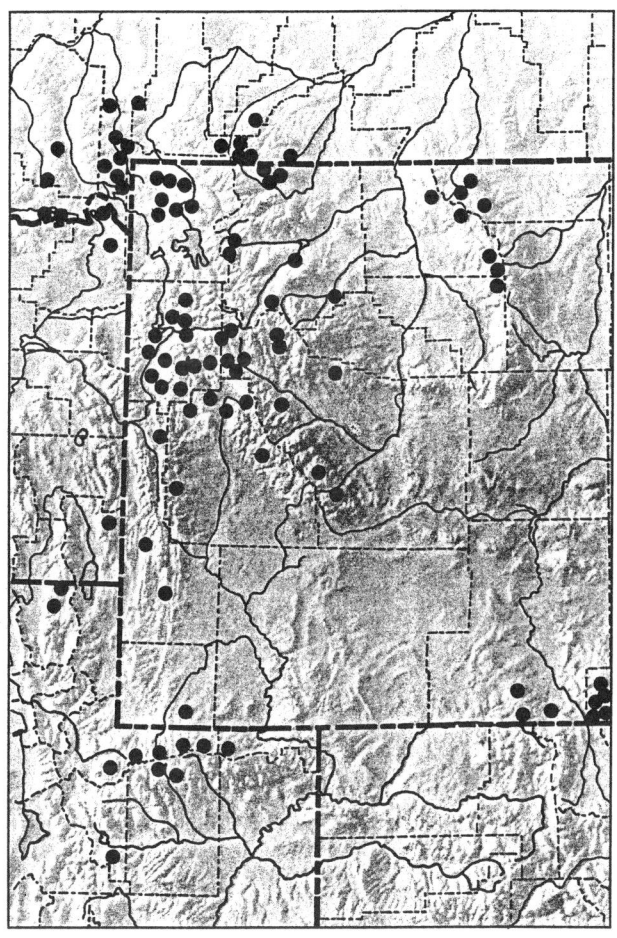

Saxifragaceae
Saxifrage Family

Annual or mostly perennial herbs, the leaves opposite or alternate, often basal, exstipulate. Flowers perfect, perigynous, solitary, racemose, paniculate, or cymose, the hypanthium free or adnate to the ovary. Sepals 5, rarely 4. Petals 5, sometimes reduced or lacking. Stamens 5 or 10, distinct. Pistil 1, of 2 or sometimes 3 or 4 partly united carpels, the ovary 1-celled with parietal placentation, or 2–3-celled with axillary placentation, the styles lacking or well developed. Fruit a many-seeded capsule.

KEY TO THE GENERA OF SAXIFRAGACEAE

1. Stamens 10
   2. Petals palmately cleft; styles 3 — *Lithophragma*
   2. Petals entire; styles 1 or 2
      3. Flowers reddish purple; leaves toothed, alternate and basal, 20–50 mm broad — *Boykinia*
      3. Flowers mostly white or yellow, if reddish purple, then the leaves opposite, entire, and 2–4 mm broad — *Saxifraga*
1. Stamens 5
   4. Petals pinnately cleft or trifid — *Mitella*
   4. Petals entire
      5. Staminodia entire to fimbriate, alternating with the stamens — *Parnassia*
      5. Staminodia lacking — *Heuchera*

Boykinia Nutt.
Boykinia

Low, glandular-pubescent perennials from thick, scaly rhizomes. Leaves mostly basal, with reniform blades and doubly crenate or doubly serrate margins, the few cauline leaves reduced in size. Flowers red or purple, in a narrow, bracteate panicle, the hypanthium partly adnate to the ovary. Sepals 5, ovate-lanceolate. Petals 5, with long, slender claws. Stamens 10. Pistil 1, of 2 carpels, the ovary 2-celled, with axillary placentation. Fruit a capsule, dehiscent between the 2 divergent beaks. Seeds numerous.

Boykinia heucheriformis (Rydb.) Rosend.
Engl. Bot. Jahrb. 37, Beibl. 83:64, 1905.

Boykinia

Telesonix heucheriformis Rydb.
Boykinia jamesii (Torr.) Engl.
Saxifraga heucheriformis (Rydb.) M.E. Jones
Saxifraga jamesiana (Torr.) Engl.
Saxifraga jamesii Torr.
Telesonix jamesii (Torr.) Raf.
Therofon heucheriforme (Rydb.) Rydb.
Therofon jamesii (Torr.) Wheelock

Low, mat-forming, glandular-pubescent, perennial herb from a thick, scaly caudex; stems 1 to several, glandular-pubescent, dark reddish or purplish above. Basal leaves petiolate, the blades reniform, with doubly crenate to crenate-dentate margins, glandular; cauline leaves petiolate to subsessile, reduced upward, the blades resembling those of the basal leaves. Flowers in a crowded, bracteate panicle, the axis and branches usually reddish or purplish. Calyx glandular-pilose, campanulate, the lobes ovate, erect, slightly shorter than the tube. Petals reddish purple, spatulate, long-clawed, about the same length as the calyx lobes. Stamens 10, equaling or exceeding the petals, the filaments subulate. Ovary half-inferior, adnate to the hypanthium below; styles partially connate above, slightly divergent in fruit. Fruit a capsule; seeds oblong, brown, shiny.

Rock crevices, ledges, boulders, and rocky ridgetops of the northern ranges, almost exclusively on limestone. Not known from above timberline in the Hoback, Medicine Bow, Uinta, and Wasatch Ranges. Alberta south to Nevada, Utah, and Colorado.

Alpine plants of the Middle Rocky Mountains with spatulate to obovate petals that are 3–4 mm long and nearly distinct styles are the typical var. *heucheriformis*.

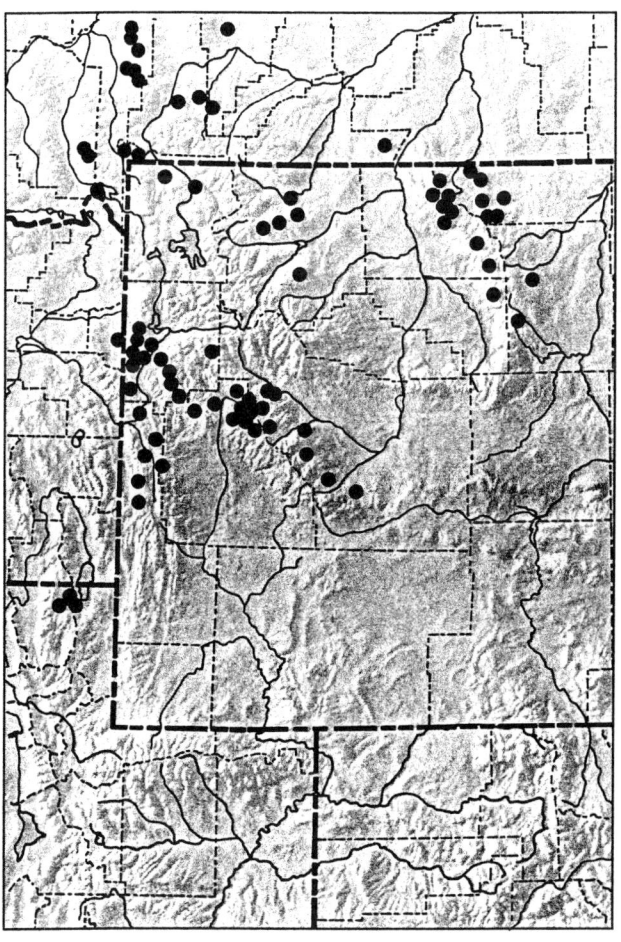

Heuchera L.
Alumroot

Scapose, glandular, perennial herbs from thick, scaly rootstocks with a branched caudex. Leaves basal, densely clustered, long-petioled, the blades cordate to reniform or orbicular, crenate to palmately lobed; stipules membranous, adnate to the petioles. Flowers regular or irregular in a spikelike or open panicle. Sepals 5, green to yellow, often unequal, adnate to the lower half of the ovary. Petals mostly 5, entire, clawed, and greenish, yellowish, or white. Stamens 5, opposite the sepals, inserted near the top of the hypanthium. Ovary 1-celled, with 2 carpels. Styles and stigmas 2. Fruit a many-seeded capsule, dehiscent between the beaks of the carpels.

KEY TO THE SPECIES OF HEUCHERA

1. Calyx and hypanthium turbinate; petals elliptic to ovate, about twice as long as the calyx lobes ..... *H. parvifolia*
1. Calyx and hypanthium campanulate; petals linear, shorter than to about twice as long as the calyx lobes, or lacking
    2. Petals lacking or, if present, less than half as long as the calyx lobes; calyx mostly less than 6 mm long ..... *H. cylindrica*
    2. Petals present, from half as long to nearly twice as long as the calyx lobes; calyx mostly more than 6 mm long ..... *H. grossulariifolia*

Heuchera cylindrica Dougl. ex Hook.
Fl. Bor.-Am. 1:236, 1834.

Roundleaf Alumroot

Heuchera alpina (S. Wats.) Blank.
Heuchera columbiana Rydb.
Heuchera glabella T. & G.
Heuchera ovalifolia Nutt. in T. & G.
Heuchera saxicola E. Nels.
Heuchera suksdorfii Rydb.
Yamala cylindrica (Dougl. ex Hook.) Raf.

Scapose, perennial herb from a stout, branched, scaly caudex. Scapes 1 to several, glandular-puberulent at least above, usually with spreading hairs. Leaves with hirsute, often curved petioles, the blades ovate to suborbicular or reniform, 5–9-lobed, crenate, the bases cordate, glandular-puberulent at least below, often becoming glabrate above. Flowers in a dense to open panicle. Calyx and hypanthium campanulate, greenish below and creamy white above, the lobes oblong, erect. Petals usually lacking, or short-linear and fewer than 5. Stamens included; filaments about the same length as the anthers. Ovary inferior, the top conical; styles very short, less than 0.5 mm long. Capsule oblong, the seeds dark brown, finely spinose.

Rocky places, mostly below timberline, but collected from the alpine zone on the Beartooth Plateau. British Columbia and Alberta south to California, Nevada, and northwestern Wyoming.

Alpine plants of the northern Middle Rocky Mountains with scapes and petioles glandular-puberulent to hirsute throughout are var. *cylindrica*.

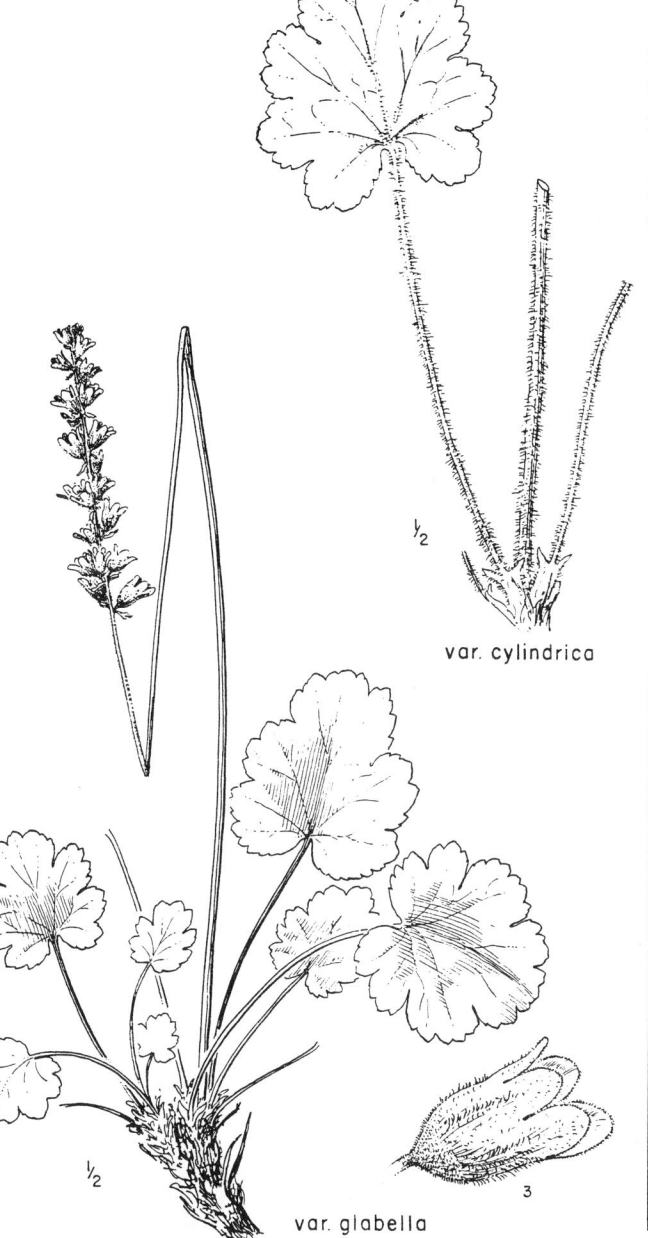

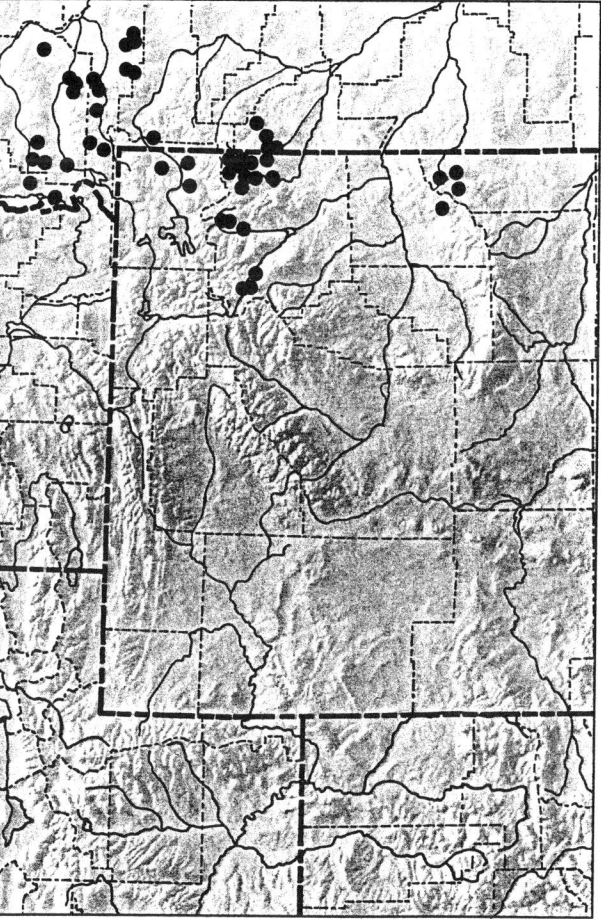

Heuchera grossulariifolia Rydb.
Mem. N.Y. Bot. Gard. 1:196, 1900.

Gooseberryleaf Alumroot

Heuchera cusickii Rosendahl et al.
Heuchera gracilis Rydb.
Heuchera tenuifolia (Wheelock) Rydb.

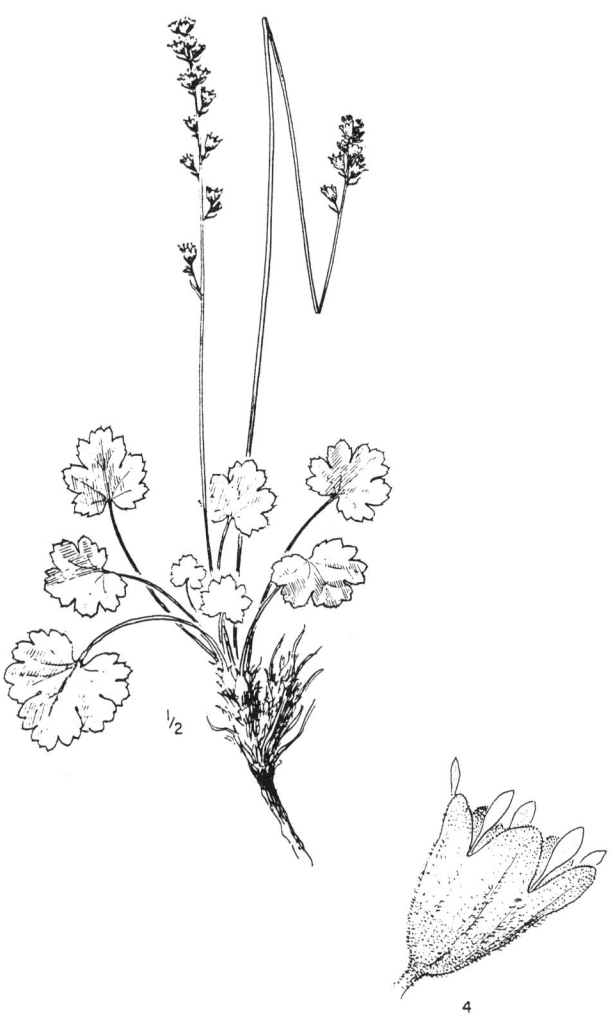

Scapose perennial herb from a stout, branched, scaly caudex. Scapes 1 to several, glandular-pubescent throughout, sometimes glabrate or even glabrous near the base. Leaves glabrous to glandular-puberulent, the blades orbicular or reniform; margins 5–7-lobed, the lobes crenate to dentate; petioles slender, glabrous to glandular-puberulent. Flowers in a dense to open panicle. Calyx and hypanthium campanulate, greenish white, glandular, the lobes erect, oblong, obtuse. Petals white, narrowly spatulate, shorter than to about twice as long as the calyx lobes. Stamens included; filaments shorter than to about the same length as the anthers. Ovary inferior, the top conical. Styles thick, short, less than 1 mm long. Capsule ovoid, the seeds dark brown, finely spinose.

Rocky ledges, scree and talus slopes, and other rocky places. Reported from the Beartooth Range by Lessica (1993). Washington, Oregon, Idaho, and Montana.

Heuchera parvifolia Nutt. ex T. & G.
Fl. N. Am. 1:581, 1840.

Common Alumroot

Heuchera flabellifolia Rydb.
Heuchera flavescens Rydb.
Heuchera missouriensis Rosendahl
Heuchera nivalis Rosendahl et al.
Heuchera puberula (Mack. & Bush) E. Wells
Heuchera utahensis Rydb.

Scapose, perennial herb from a short-branched crown on a thick, scaly rootstock. Scapes 1 to several, glandular-puberulent throughout, or occasionally glabrate below. Leaves long-petiolate, the blades orbicular or reniform; margins 5–9-lobed, the segments crenate or crenate-dentate, glandular-puberulent at least on the lower surface. Flowers in a dense terminal panicle, often interrupted below. Calyx and hypanthium turbinate, greenish, the lobes deltoid, spreading. Petals white or cream, elliptic to ovate, short-clawed, about twice as long as the calyx lobes. Stamens included; filaments about the same length as the anthers. Ovary inferior, the top conical; styles very short, less than 0.5 mm. Capsule ovoid, the seeds dark brown, finely spinose.

Rock crevices and rocky slopes in the Absaroka, Salt River, Uinta, Wasatch, and Wind River Ranges, mostly on sedimentary substrates, especially limestone. Alberta south to Nevada, Arizona, and New Mexico.

Alpine plants of the Middle Rocky Mountains more than 20 cm tall and with leaves 3–6 cm wide are var. *utahensis* (Rydb.) Garrett. In the northern part of our range these intergrade with a smaller plant less than 20 cm tall and leaves 2–3 cm wide, which is var. *dissecta* M.E. Jones.

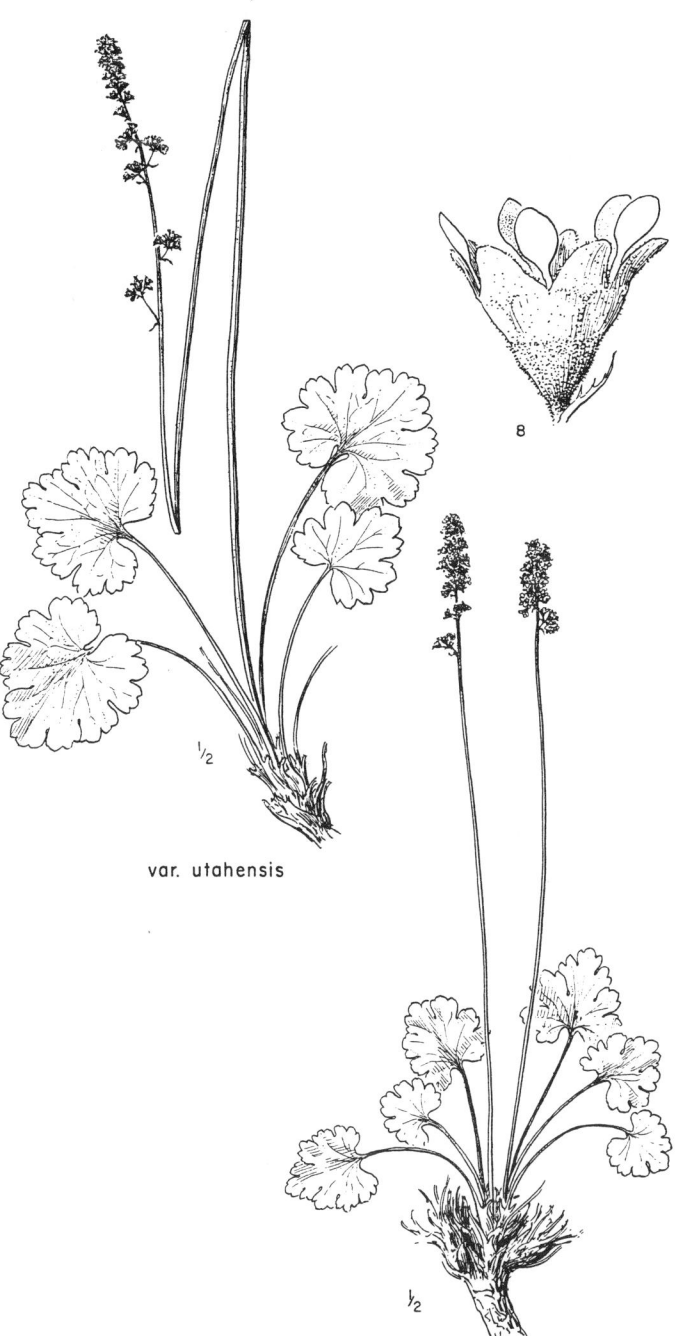

var. utahensis

var. dissecta

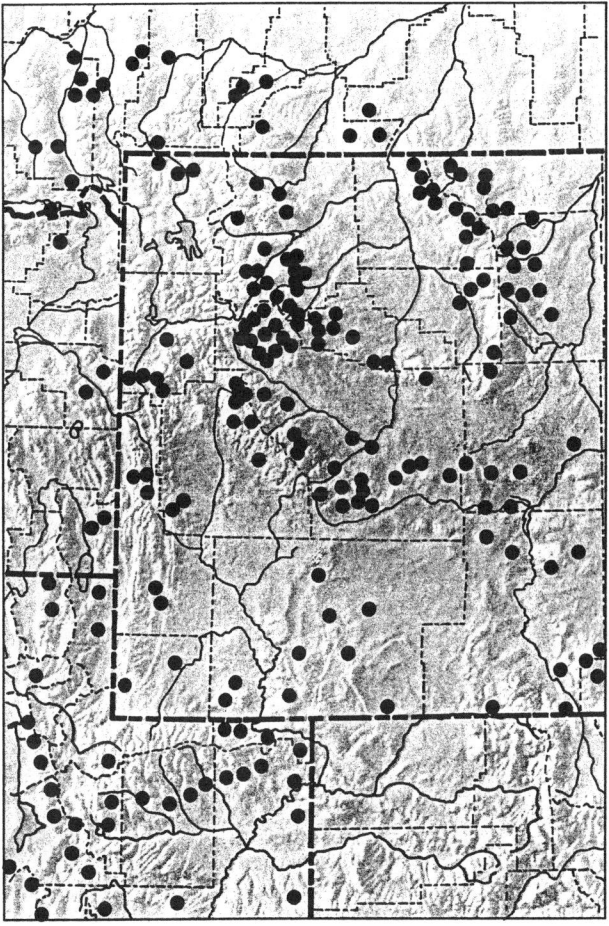

Lithophragma Nutt.
Woodland Star

Perennial, glandular-puberulent herbs from tuberous rootstocks with grainlike bulblets. Leaves mostly basal, orbicular to reniform, palmately lobed or divided, the petioles slender, dilated at the bases; cauline leaves sessile or short-petiolate, usually reduced upward. Flowers white, pink, or lavender, single or in terminal racemes, often replaced by purplish bulblets in the upper leaf axils. Sepals 5, partially fused, the calyx campanulate to funnelform. Petals 5, clawed, palmately 3–7-cleft, rarely shallowly lobed to entire. Stamens 10, included. Pistil 1, of 3 carpels, the ovary 1-celled, with parietal placentation; styles 3. Fruit a 3-beaked, 1-celled capsule.

Lithophragma glabrum Nutt. in T. & G.
Fl. N. Am. 1:584, 1840.

Fringecup Woodland Star

Lithophragma bulbiferum Rydb.
Tellima bulbifera (Rydb.) A. Nels.
Tellima glabra (Nutt.) Steud.

Small, slender, perennial herb with fibrous roots; stems mostly single, often reddish purple, glandular-pubescent. Basal leaves petiolate, the blades 5-cleft, each segment tridentate, glabrous to puberulent; cauline leaves 3-cleft, subsessile, reduced upward, with 1 or more bulblets in the axils. Inflorescence a 2–3-flowered raceme, the lower branches with bulblets replacing flowers. Calyx campanulate, the lobes shallow, broadly triangular, obtuse. Petals white to pink, 3–5-cleft, clawed. Ovary less than half inferior. Fruit capsular; seeds brown, muriculate.

Wet, rocky places on sedimentary rocks of the Beartooth, Big Horn, Salt River, Uinta, Wasatch, and Wyoming Ranges. Washington and Oregon southeast to Montana, Wyoming, and Colorado.

Middle Rocky Mountain plants are var. *ramulosum* (Suksel.) B. Boivin.

Mitella L.
Miterwort

Low, slender, perennial herbs from creeping or ascending rhizomes, the leaves mostly basal, cordate or reniform, with toothed or shallowly lobed margins. Flowers in narrow spikes or racemes, the hypanthium partly adnate to the ovary. Sepals 5, short, entire. Petals 5, green in ours, trifid or pinnatifid. Stamens 5 or 10, sessile or nearly so. Pistil 1, of 2 united carpels, the ovary 1-celled, with 2 parietal placentae, the styles distinct, short, the stigmas often 2-lobed. Fruit a 2-valved capsule, appearing circumscissile. Seeds numerous, shining, brown to black.

Mitella pentandra Hook.
Curtis' Bot. Mag. 56:Pl. 2933, 1829.

Fivestamen Miterwort

Drummondia mitelloides DC.
Mitellopsis drummondiana Meisn.
Mitellopsis pentandra (Hook.) Walp.
Pectiantia latiflora Rydb.
Pectiantia mitelloides (DC.) Raf.
Pectiantia pentandra (Hook.) Rydb.

Slender, scapose, perennial herb from a rhizome, occasionally stoloniferous; stems 1 to several, glabrate to glandular-pubescent. Leaves all basal, long-petiolate, the blades cordate, with 7–11 shallow lobes, the margins coarsely and irregularly serrate, glabrous to pubescent. Inflorescence a lax, few-flowered raceme. Calyx rotate, the lobes deltoid, spreading. Petals green, pectinately lobed, the lobes filiform. Stamens 5, opposite the petals. Ovary mostly inferior; styles lacking or nearly so; stigmas 2, bilobed, the lobes cordate.

Cold, wet, somewhat shaded stream banks near timberline in the lower alpine zone of the Absaroka, Beartooth, Big Horn, Teton, Uinta, Wasatch, and Wind River Ranges. Alaska south to California, Idaho, and Colorado.

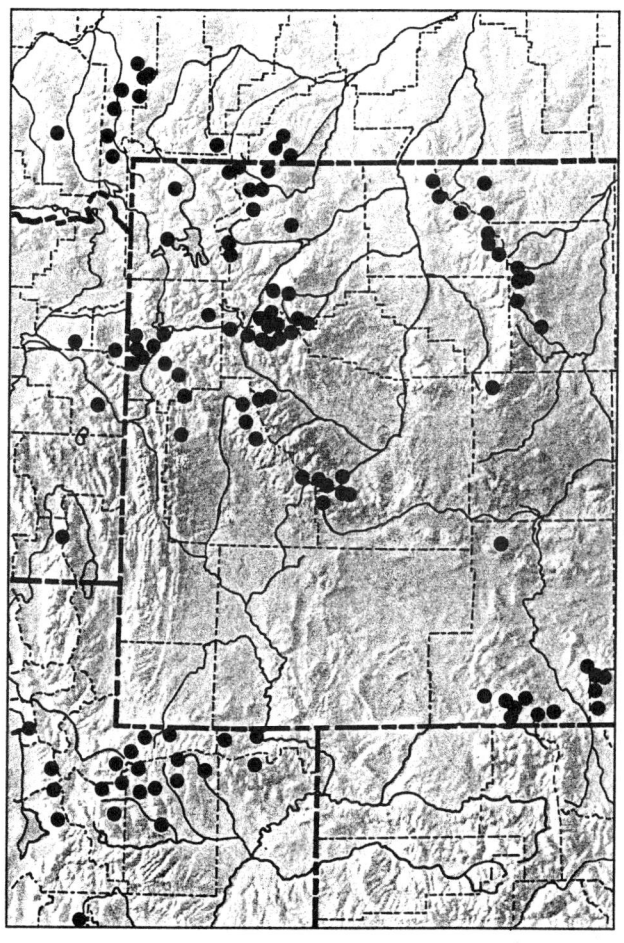
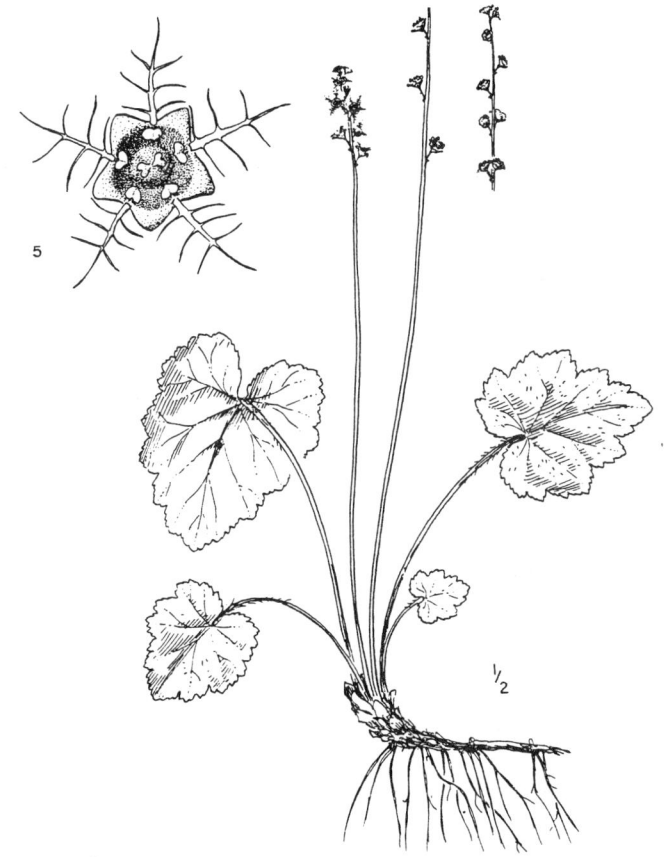

Parnassia L.
Grass of Parnassus

Glabrous, perennial herbs with entire, petiolate, cordate or reniform, usually basal leaves, the stem often bearing a single, sessile bract. Flowers terminal, solitary. Hypanthium short, partly adnate to the ovary. Sepals 5, slightly united at the base. Petals 5, greenish to white, entire or fringed at the base. Stamens 5, alternating with 5 staminodia, the staminodia opposite the petals. Ovary 1-loculed, of 4 united carpels, with 4 parietal placentae, the style short or lacking, the stigmas 4. Fruit a loculicidal, 4-valved, many-seeded capsule.

KEY TO THE SPECIES OF *PARNASSIA*

1. Basal margins of petals fimbriate — *P. fimbriata*
1. Basal margins of petals entire
    2. Flowering stalk with a single leaflike bract near the middle; petals with 5–13 veins — *P. palustris*
    2. Flowering stalk lacking a bract, or the bract near the base; petals with 1–3 veins — *P. kotzebuei*

Parnassia fimbriata Koenig
Ann. Bot. 1:391, 1805.

Fringed Parnassus

Parnassia intermedia Rydb.
Parnassia rivularis Osterh.

Scapose, perennial herb from a short, stout rootstock; stems 1 to several, each with a bractlike leaf at or above the middle. Basal leaves long-petiolate, the blades reniform or cordate, entire. Flowers solitary; sepals elliptic to oval, obtuse; petals white, obovate, clawed, fimbriate at the base, about twice as long as the sepals. Staminodia carnose, with 5–9 terminal lobes. Capsule ovoid.

Stream banks and wet meadows of the Absaroka, Beartooth, Big Horn, and Wind River Ranges. Alaska south to California and New Mexico.

Middle Rocky Mountain alpine plants with short, thick, nonfilamentous lobes of the staminodia are var. *fimbriata*.

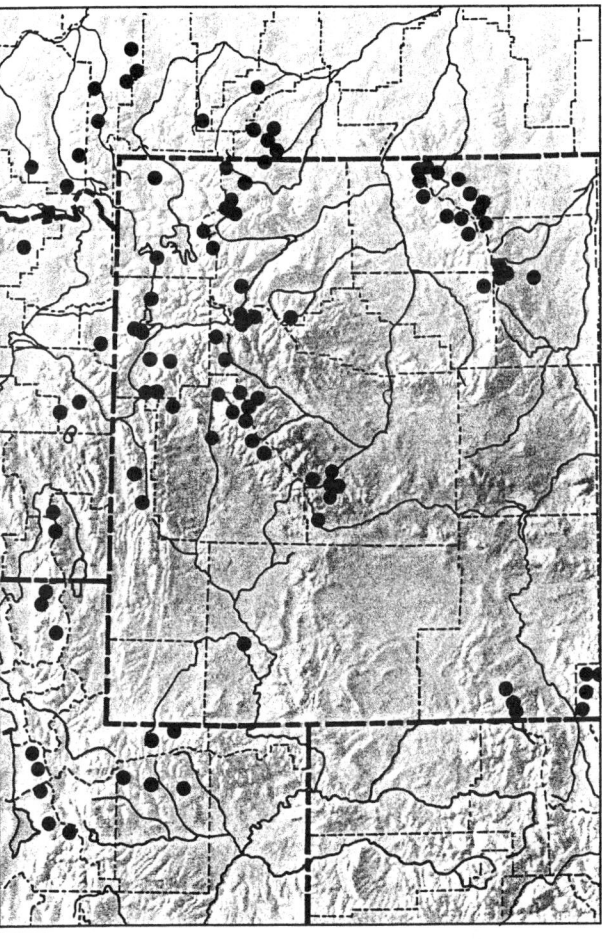

Parnassia kotzebuei Cham. ex Spreng. *Syst.* 1:951, 1825.

Kotzebue Parnassus

Scapose, perennial herb from short rootstocks; scapes single, bractless or with a bract near the base. Leaves petiolate, the blades ovate to elliptic, entire, truncate or broadly cuneate at the base. Flowers solitary; sepals lanceolate or oblong, somewhat acute; petals white, ovate, mostly 3-nerved, about the same length as the sepals. Staminodia (in ours) thin, oblong, with 5 capitate or glandular-tipped, linear segments. Capsule ovoid.

Tundra and moist to wet rocky places in the Beartooth, Big Horn, and Wind River Ranges. Transcontinental in Canada; south to Montana, Wyoming, and Utah.

All Middle Rocky Mountain alpine plants are var. *kotzebuei*.

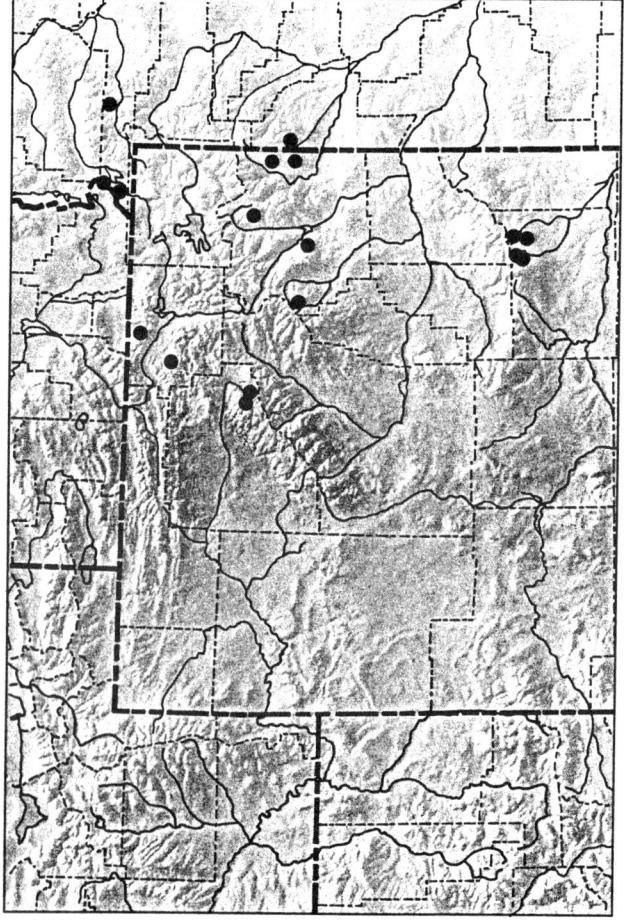

Parnassia palustris L.
Sp. Pl., 273, 1753.

Northern Parnassus

Parnassia californica (Gray) Greene
Parnassia montanensis Fern. & Rydb. ex Rydb.
Parnassia multiseta Fern.

Scapose, perennial herb from short rootstocks; scapes single to clustered, each with a bractlike leaf mostly below the middle. Basal leaves long-petiolate, the blades deltoid to elliptic-ovate, entire, and broadly cuneate, truncate, or somewhat cordate at the base. Flowers solitary; sepals lanceolate to oblong, acute; petals white, ovate to obovate, sessile to somewhat clawed, entire to shallowly undulate, obtuse, 5–13-veined. Staminodia obovate, somewhat clawed, terminally 5–7-fimbriate, the segments capitate or glandular-tipped. Capsule ovoid.

Stream banks of the Teton and Wind River Ranges. Circumboreal; south in western North America to California, Nevada, Utah, and Colorado.

Middle Rocky Mountain plants having 5–7 staminodial segments and petals less than 1.5 times as long as the calyx lobes are var. *montanensis* (Fern. & Rydb. ex Rydb.) C.L. Hitchc.

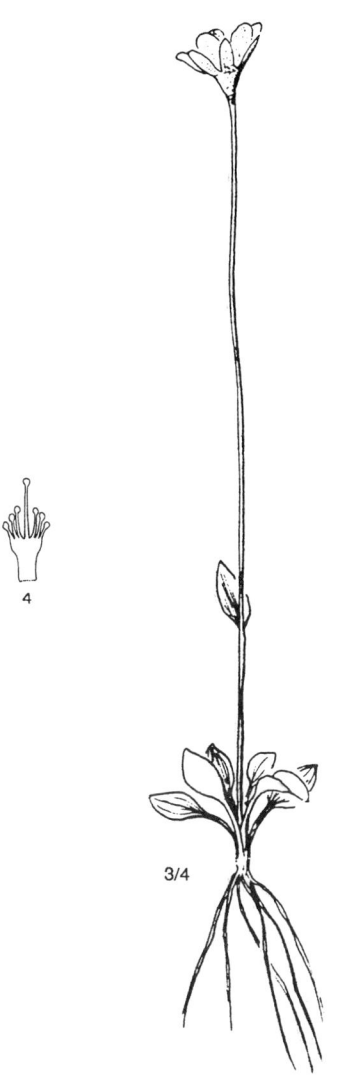

var. montanensis

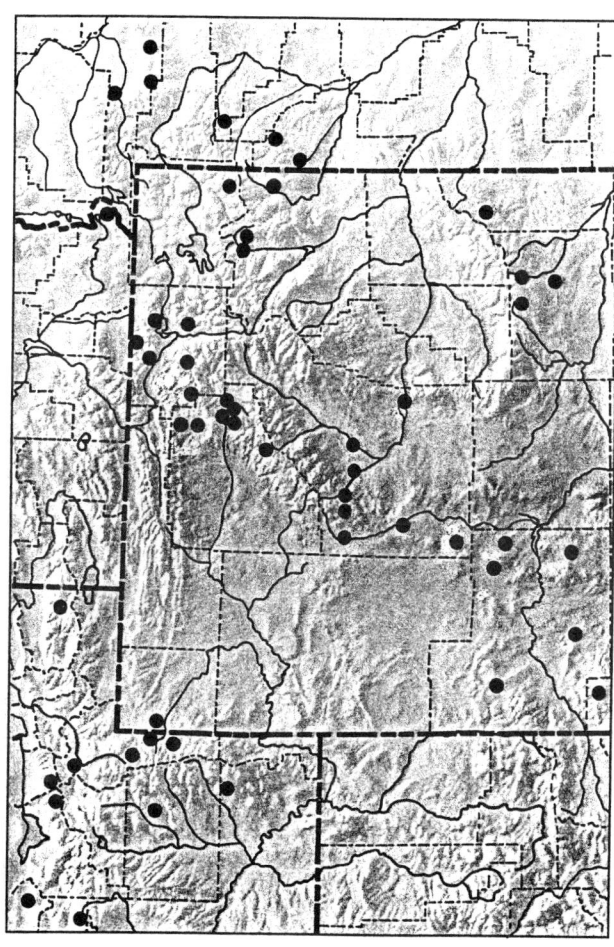

Saxifraga L.
Saxifrage

Scapose or caulescent, perennial herbs with alternate or rarely basal, entire, toothed, lobed, or pinnatifid leaves. Flowers regular to somewhat irregular, perfect, solitary, in paniculate or racemose cymes, the hypanthium free or partly adnate to the ovary. Sepals 5, erect or reflexed. Petals 5, sometimes clawed, and white, yellow, or reddish or purplish, sometimes spotted. Stamens 10. Ovary nearly superior to partly inferior, 2-celled, the carpels united below, with axillary placentation at the base and parietal placentation in the upper, separate parts. Fruit a 2-horned capsule with numerous small seeds.

KEY TO THE SPECIES OF *SAXIFRAGA*

1. Flowers purple or reddish; leaves opposite, imbricate, all cauline — *S. oppositifolia*
1. Flowers white, pinkish, or yellow; leaves alternate, cauline or basal
    2. Leaf blades cordate to reniform, toothed or lobed
        3. Plants scapose, the scape up to 50 cm tall; inflorescence an open, paniculate cyme; blades of leaves mostly 3–7 cm broad, coarsely toothed — *S. odontoloma*
        3. Plants caulescent, the flowering stems 3–15 cm tall; flowers solitary, racemose, or spicate; blades of leaves mostly less than 2 cm broad, lobed or toothed
            4. Inflorescence of a single terminal flower with petals 7–10 mm long, and with sessile bulblets below it — *S. cernua*
            4. Inflorescence not bulbiferous, of 1–3 long-pedicellate flowers with petals 5 mm long or less — *S. rivularis*
    2. Leaf blades linear to obovate, definitely longer than broad
        5. Plants scapose; leaves mostly 2–12 cm long
            6. Inflorescence usually a single, capitate cluster of white flowers, without conspicuous branches; leaves 2–5 cm long, subentire to crenate or dentate — *S. rhomboidea*
            6. Inflorescence definitely branched, the flowers white, pinkish, or greenish; leaves mostly larger, up to 10 cm long, nearly always toothed
                7. Leaves remotely denticulate; flowers glomerate on the branches of an elongate inflorescence; petals in ours minute and greenish or purplish, or lacking — *S. oregana*
                7. Leaves coarsely crenate-dentate; flowers mostly loose on the branches of a somewhat corymbose inflorescence; petals conspicuous, white or sometimes tinged with pink — *S. occidentalis*
        5. Plants caulescent or subscapose, the flowering stem nearly always with 1 or more leaves; leaves less than 15 mm long
            8. Petals yellow; flowers solitary
                9. Plants with long, naked stolons; sepals erect or ascending — *S. flagellaris*
                9. Plants without stolons; sepals reflexed — *S. chrysantha*
            8. Petals white; flowers solitary to many
                10. Leaves entire, linear or lanceolate, the margins ciliate; petals purple-spotted — *S. bronchialis*
                10. Leaves toothed or lobed, cuneate to obovate, the margins not ciliate; petals not spotted
                    11. Leaves with 3 linear lobes; filaments longer than the sepals; petals obovate; plants caespitose — *S. cespitosa*
                    11. Leaves merely toothed or entire; filaments shorter than or equal to the sepals; petals clawed; plants caespitose — *S. adscendens*

Saxifraga adscendens L.
Sp. Pl., 405, 1753.

Wedgeleaf Saxifrage

Muscaria adscendens (L.) Small
Ponista oregonensis Raf.
Saxifraga incompta Peck
Saxifraga oregonensis (Raf.) A. Nels.
Saxifraga petraea Hook.

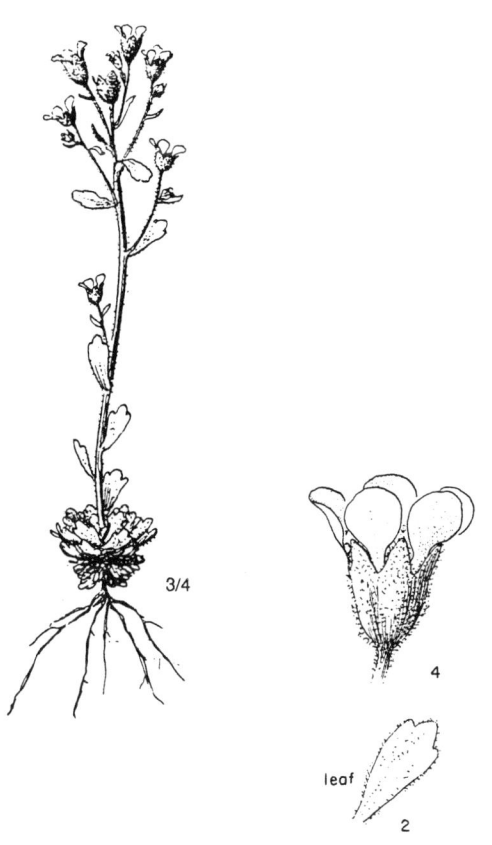

Small, glandular-pubescent, perennial herb, less than 10 cm tall, from a slender caudex; stems simple or branched. Leaves obovate, entire or 3–5-toothed to shallowly lobed, the basal leaves densely imbricate, the cauline leaves mostly 3–8. Inflorescence a leafy-bracteate cyme or raceme. Calyx reddish purple, glandular, campanulate, the lobes deltoid. Petals white, obovate, clawed, about twice as long as the calyx lobes. Stamens inserted on the hypanthium, the filaments shorter than the calyx lobes, not clavate; anthers oval. Ovary more than half inferior. Capsule 3–5 mm long.

Moist, rocky fellfields in the Absaroka, Beartooth, Teton, Uinta, and Wind River Ranges. Eurasia; Alaska south to Oregon, Utah, and Colorado.

The plants of the Middle Rocky Mountains are var. *oregonensis* (Raf.) Breit., which has smaller flowers and broader leaves than the European var. *adscendens*.

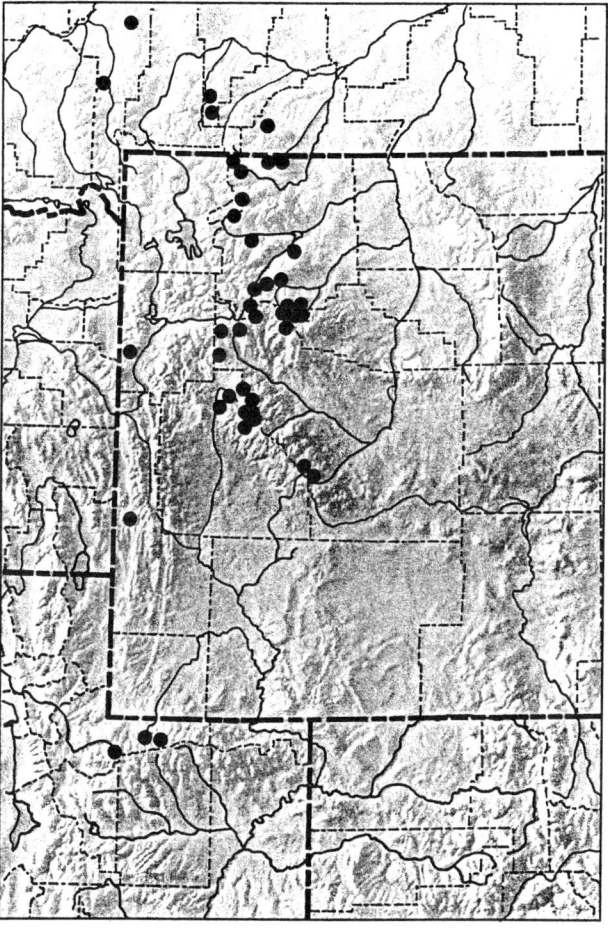

Saxifraga bronchialis L.
Sp. Pl., 400, 1753.

Spotted Saxifrage

Ciliaria austromontana (Wieg.) W.A. Weber
Ciliaria funstonii (Small) W.A. Weber
Ciliaria vespertina (Small) W.A. Weber
Leptasea austromontana (Wieg.) Small
Leptasea vespertina Small
Saxifraga austromontana Wieg.
Saxifraga cherlerioides D. Don
Saxifraga cognata E. Nels.
Saxifraga firma Litv. ex Losinsk.
Saxifraga funstonii (Small) Fedde
Saxifraga vespertina (Small) Fedde

Caespitose, perennial herb from a branched, creeping caudex; stems tufted, procumbent, the terminal flowering portion erect. Leaves imbricate, linear to lanceolate, mostly entire, white-ciliate on margins, acuminate at the tip, marcescent at base of the stem. Flowering stems glabrous to glandular-pubescent, the leaves linear, acuminate, remote; inflorescence paniculate. Calyx rotate, the sepals ovate or lanceolate, glabrous or ciliate, obtuse. Petals white or cream, spotted with purple, oblong to oblong-lanceolate. Stamens inserted at base of the ovary; filaments shorter than petals, about twice as long as sepals, not clavate. Ovary only slightly inferior. Capsule 5–9 mm long, with divergent beaks.

Fellfields, scree, rock ledges, and cirque walls of the Absaroka, Beartooth, and Teton Ranges. Siberia; Alaska south to Oregon, Idaho, Colorado, and New Mexico.

Middle Rocky Mountain plants with linear to lanceolate, entire leaves are var. *austromontana* (Wieg.) G.N. Jones.

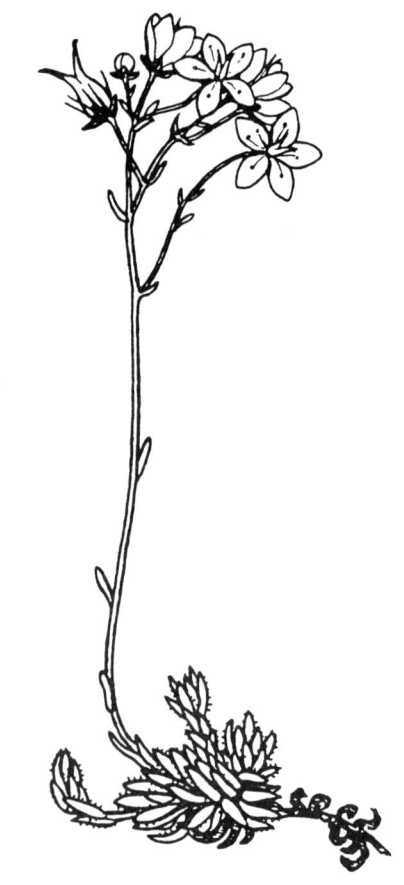

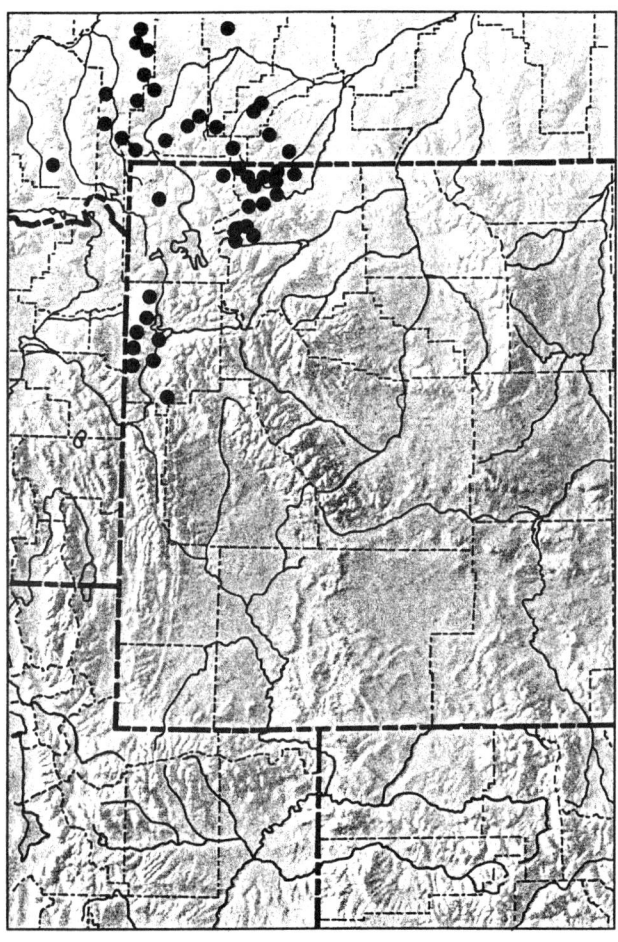

Saxifraga cespitosa L.
Sp. Pl., 404, 1753.

Tufted Saxifrage

Dactyloides cespitosa (L.) Nieuwl.
Muscaria cespitosa (L.) Haw.
Muscaria delicatula Small
Muscaria emarginata Small
Muscaria micropetala Small
Muscaria monticola Small
Saxifraga delicatula (Small) Fedde
Saxifraga emarginata (Small) Fedde
Saxifraga exarata Hook.
Saxifraga groenlandica L.
Saxifraga micropetala (Small) Fedde
Saxifraga monticola (Small) Löve & Löve
Saxifraga sileneflora Sternb. ex Cham.

Densely tufted, caespitose, sparsely pubescent, perennial herb from a branched caudex. Stems with short, leafy, prostrate branches; leaves imbricate, short-petiolate, flabellate, and mostly 3-lobed, marcescent at base of the stem; flowering stems sparsely glandular-pubescent, with mostly 1 or 2 leaves, the upper one often entire; flowers 1–6 in a cymose or paniculate inflorescence. Calyx campanulate, the lobes lanceolate to deltoid. Petals white, oblanceolate, somewhat clawed at the base. Stamens about twice as long as the calyx lobes, shorter than the petals; filaments not clavate. Ovary completely inferior. Capsule 3–7 mm long, nearly beakless.

Fellfields, rocky slopes, and ledges of the Absaroka, Beartooth, Big Horn, Uinta, Wasatch, Wind River, and Wyoming Ranges. Circumboreal; south in western North America to Oregon, Nevada, Arizona, and New Mexico.

Middle Rocky Mountain plants that are sparsely pubescent with oblanceolate petals are var. *minima* Blank.

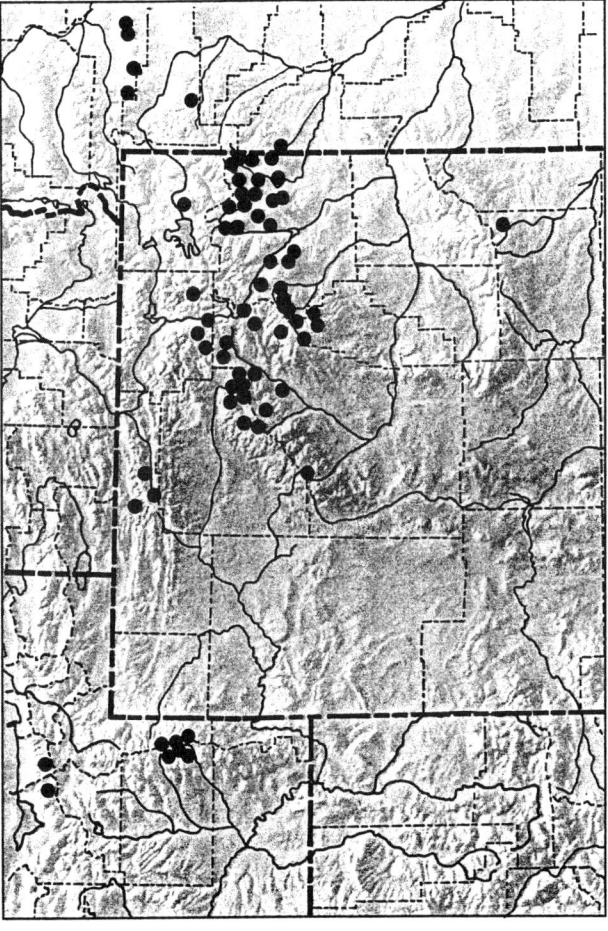

Saxifraga cernua L.
Sp. Pl., 403, 1753.

Bulblet Saxifrage

Lobaria cernua (L.) Haw.
Saxifraga simulata Small

Slender, perennial herb from a compact, bulblet-producing rootstock; stems 1 to several, unbranched, glandular-pubescent above, often rusty-pilose below. Basal leaves petiolate, cordate to reniform, with 5–7 palmate lobes or teeth; cauline leaves reduced upward, the lower ones like the basal leaves, the upper ones becoming sessile and 3-lobed or entire with reddish purple bulblets in the axils. Flowers mostly solitary and terminal. Calyx campanulate, the lobes erect, ovate. Petals white to pinkish, obovate, about 4 times as long as the calyx lobes, clawless. Stamens longer than the calyx lobes; filaments not clavate. Ovary about one-fourth inferior.

Uncommon and sporadic in moist meadows, on scree, and on similar rocky slopes of the Absaroka, Beartooth, Big Horn, Gros Ventre, Medicine Bow, Uinta, and Wind River Ranges. Circumboreal; south in western North America to Washington, Nevada, New Mexico, and South Dakota.

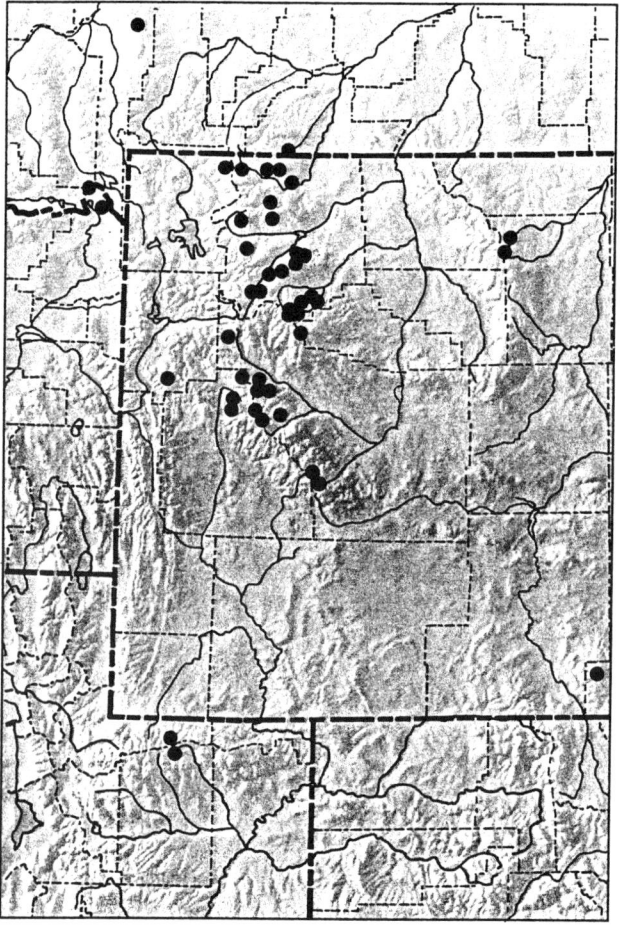

Saxifraga chrysantha Gray
Proc. Am. Acad. 12:83, 1877.

Goldbloom Saxifrage

Leptasea chrysantha (Gray) Small

Small, tufted, glandular-pubescent to glabrate, perennial herb, 2–8 cm tall, from a slender, branched caudex with numerous sterile shoots. Basal leaves oblanceolate to spatulate, entire, obtuse, glabrous or nearly so, in dense rosettes; cauline leaves oblong, entire, often glandular-pubescent. Flowers solitary, occasionally 2 or 3. Sepals ovate, obtuse, reflexed. Petals yellow, oval, narrowed at the base to a short claw. Stamens inserted on the ovary, about half as long as the petals; filaments not clavate. Ovary only slightly inferior. Capsule 3–8 mm long, ovoid.

Uncommon on felsenmeer, rocky slopes, and ridges in the Beartooth, Medicine Bow, Uinta, and Wind River Ranges. Montana, Wyoming, Colorado, Utah, and New Mexico.

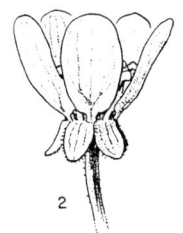

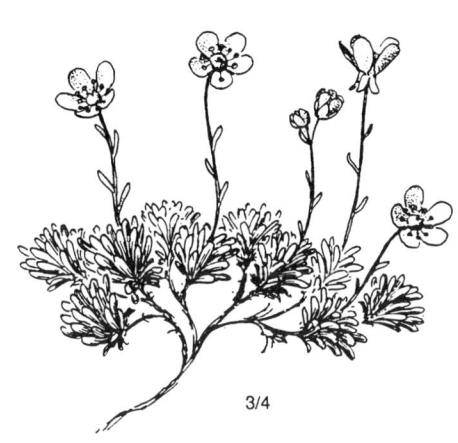

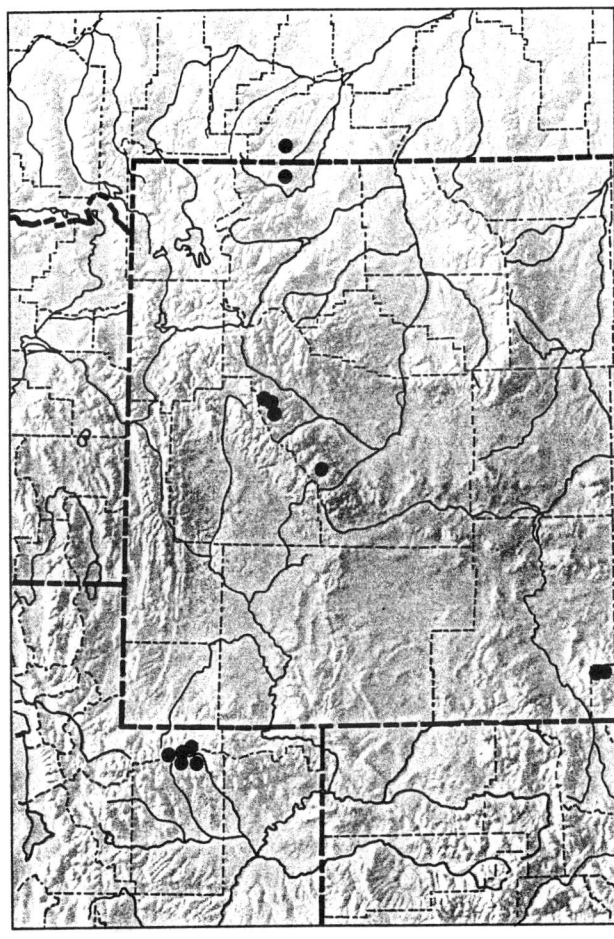

Saxifraga flagellaris Willd. in Sternb.
Rev. Saxifr., 25, 1810.

Alpine Spider Saxifrage

Hirculus flagellaris (Willd.) Haworth
Leptasea flagellaris (Willd.) Small
Saxifraga crandallii Gand.
Saxifraga setigera Pursh

Glandular-pubescent, perennial herb from slender rhizomes, with long, filiform, glabrous, reddish stolons, these terminated by buds or young plantlets. Basal leaves in a dense rosette, oblanceolate, glandular-ciliate on the margins, mucronate; cauline leaves lanceolate, reduced upward. Inflorescence of 1–3 flowers. Calyx campanulate or turbinate, the lobes erect or ascending, lanceolate, about as long as the tube. Petals yellow, obovate, somewhat clawed. Stamens inserted on edge of the ovary; filaments not clavate. Ovary about half inferior. Capsule 4–6 mm, beaked.

Wet meadows, stream gravel, and scree slopes. Its stoloniferous habit allows *S. flagellaris* to survive on active, high altitude cryopedogenic features such as frost boils and stone polygons. Known from all alpine ranges except the Hoback, Salt River, Wasatch, and Wyoming Ranges. Circumboreal; south in the Rocky Mountains to New Mexico and Arizona.

Middle Rocky Mountain alpine plants are ssp. *setigera* (Pursh) Tolm. and are part of a worldwide complex represented in North America by the diploid *S. flagellaris* ($2n=16$) of the Rocky Mountains, and the circumpolar arctic, tetraploid *S. platysepala* (Trautv.) Tolm. ($2n=32$), which has a disjunct population in the southern Rockies (A. Löve et al., 1971).

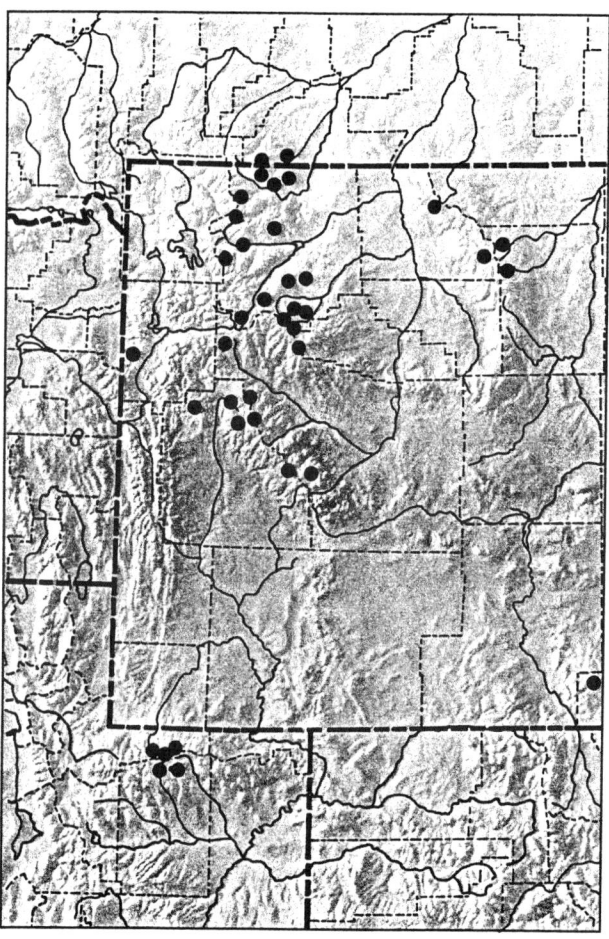

1/2

Saxifraga occidentalis S. Wats.
Proc. Am. Acad. 23:264, 1888.

Alberta Saxifrage

Micranthes lata Small
Micranthes occidentalis (S. Wats.) Small
Micranthes saximontana (E. Nels.) Small
Saxifraga saximontana E. Nels.

Perennial herb from a short, horizontal rhizome; stems 1–3, glandular-pubescent with reddish hairs. Leaves petiolate, the blades ovate to elliptic, coarsely crenate-serrate, reddish-tomentose at least below, somewhat truncate at the base. Inflorescence a compact pyramid of paniculate cymes in ours. Sepals ovate to lanceolate, often purplish, spreading or sometimes reflexed. Petals white to pink, elliptic to ovate, with a very short claw. Stamens inserted around the ovary, about as long as the petals; filaments clavate, reddish. Ovary about one-third inferior. Follicles 3–6 mm long, the beaks long-divergent.

Moist meadows, rocky ledges, and glacial moraines of the Beartooth, Teton, and Wind River Ranges. British Columbia and Alberta south to Oregon, Idaho, Montana, and Wyoming.

Middle Rocky Mountain alpine plants with clavate filaments and small, compact, rounded to pyramidal inflorescences are var. *occidentalis*.

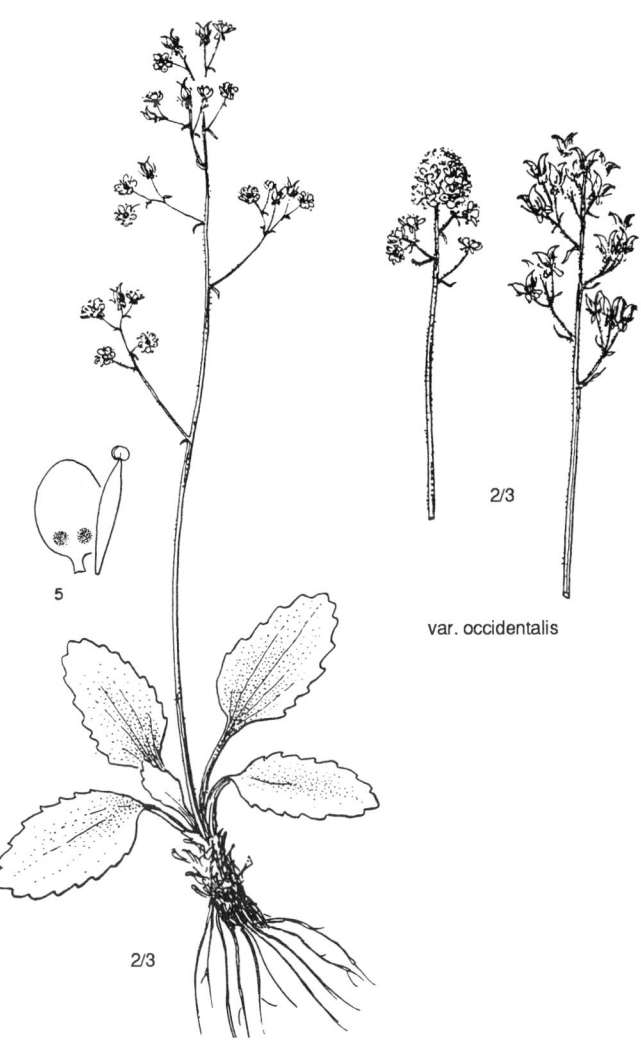
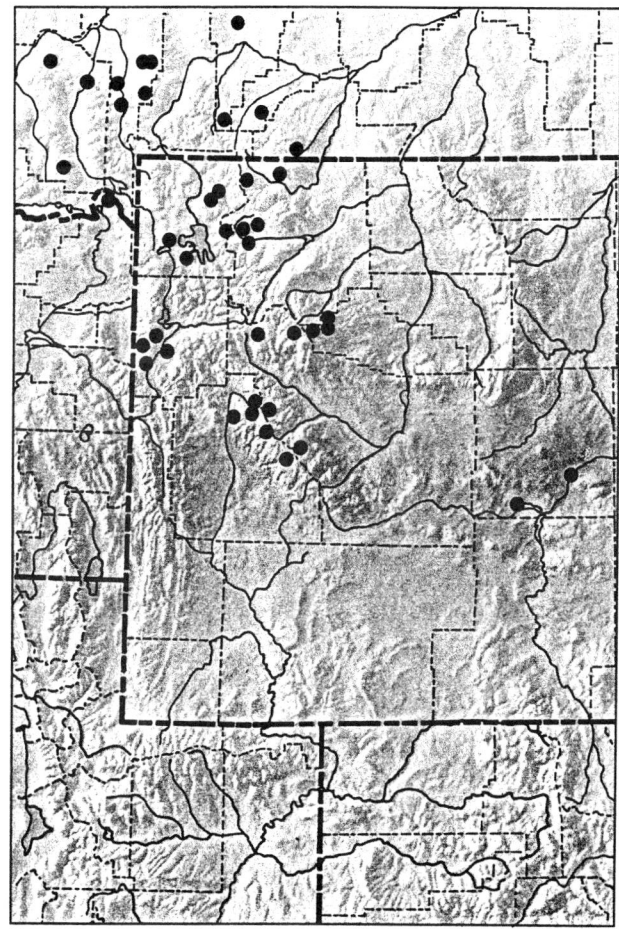

Saxifraga odontoloma Piper
Smithson. Misc. Coll. 50:200, 1907.

Brook Saxifrage

Micranthes arguta (D. Don) Small
Micranthes odontoloma (Piper) A. Heller
Saxifraga arguta D. Don
Saxifraga odontophylla Piper
Saxifraga punctata L.

Scapose, rhizomatous herb; scapes glandular-pubescent above. Leaves petiolate, the blades orbicular to reniform, coarsely dentate or crenate, cordate at the base, glabrous or nearly so. Inflorescence a cymose panicle, the pedicels reddish purple, glandular-pubescent. Sepals purple, oblong to lanceolate, glabrous to glandular-ciliate, spreading to reflexed. Petals white, elliptic to suborbicular, narrowed to a short claw. Stamens shorter than the petals, spreading-ascending; filaments white, clavate. Ovary nearly superior. Capsule 4–10 mm long, with divergent beaks.

Common on shaded and steep-sided, wet stream banks in the lower alpine zone of most Middle Rocky Mountain ranges. Not yet known from above timberline in the Hoback, Salt River, and Wyoming Ranges. Alaska to British Columbia and south to California, Arizona, and New Mexico.

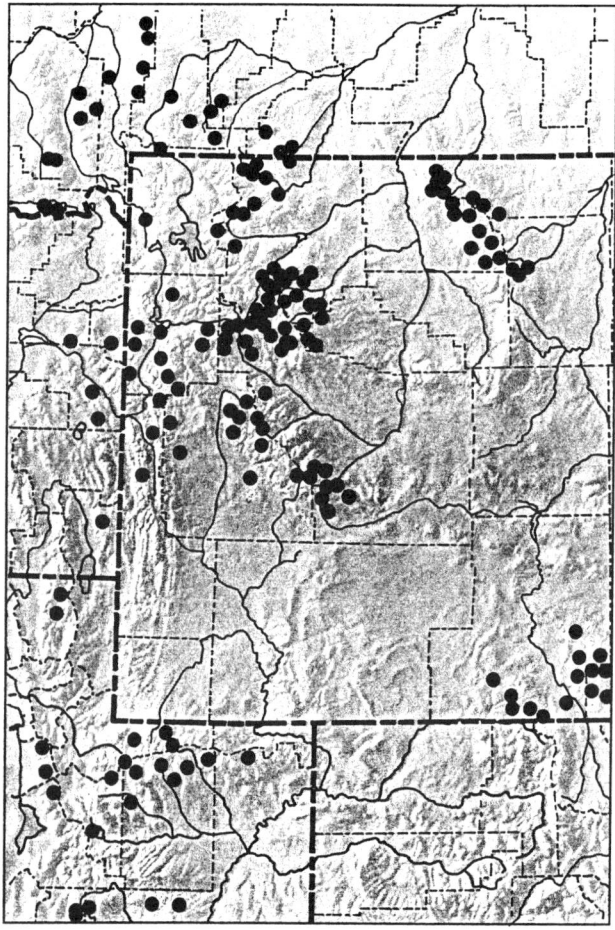

Saxifraga oppositifolia L.
Sp. Pl., 402, 1753.

Twinleaf Saxifrage

Antiphylla oppositifolia (L.) Fourr.
Antiphylla pulvinata (Small) Small
Saxifraga pulvinata Small

Pulvinate, glabrous to slightly pilose, perennial herb, forming dense cushions; stems much branched with opposite, decussate leaves and appearing 4-angled. Leaves sessile, obovate, entire, ciliate, the lower ones marcescent. Flowers solitary at the ends of branches, sessile or nearly so. Calyx campanulate, the lobes oblong to ovate, obtuse, erect, ciliate. Petals purple, obovate, narrowed to a short claw, about twice as long as the calyx lobes, erect. Stamens inserted on the ovary, about twice as long as the calyx lobes; filaments subulate. Ovary nearly superior. Capsule ovoid, 6–8 mm long, long-beaked, the beaks slender, slightly divergent.

Scree and talus slopes, rocky ledges, moraines, and rock crevices of the northern and western ranges of the Middle Rocky Mountains. Not yet known from above timberline in the Hoback, Medicine Bow, Uinta, Wasatch, and Wyoming Ranges. Circumboreal; south in western North America to Washington, Idaho, and Wyoming.

Two subspecies are recognized in western North America: ssp. *smalliana* (Engl. & Irmsch.) Hult., which has densely imbricate leaves, the blades with 2–4 cilia; and ssp. *oppositifolia*, with leaves not densely imbricate, and the blades with more than 4 cilia.

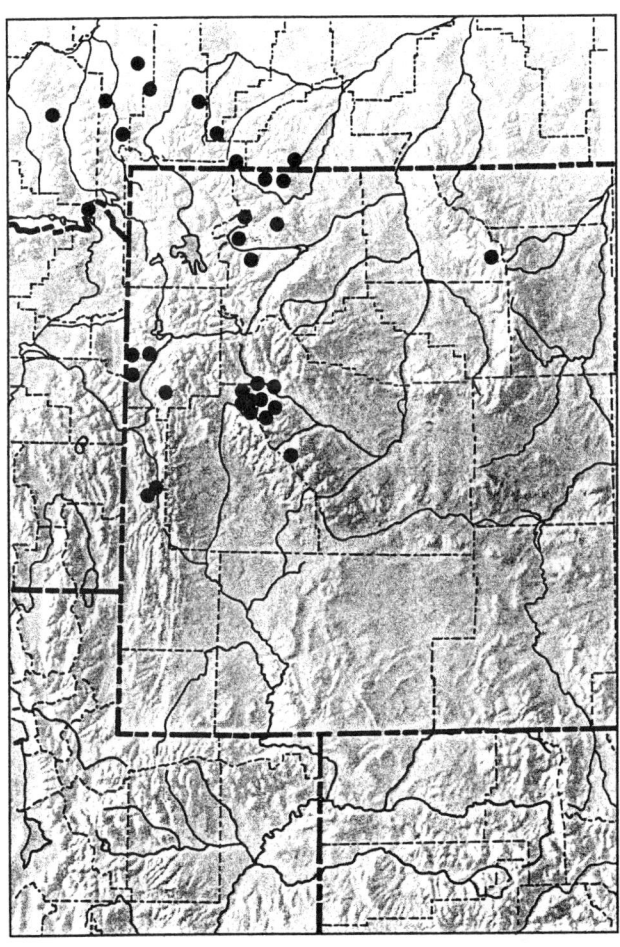

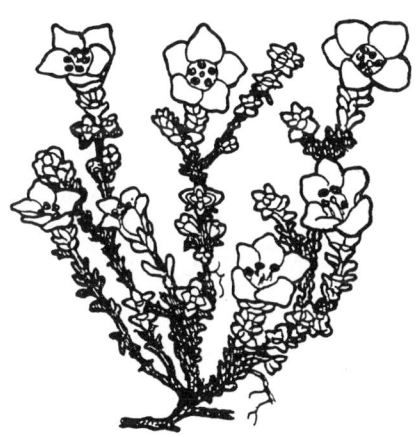

Saxifraga oregana T.J. Howell
Erythea 3:34, 1895.

Oregon Saxifrage

Micranthes arnoglossa Small
Micranthes brachypus Small
Micranthes montanensis (Small) Small
Micranthes oregana (T.J. Howell) Small
Micranthes sierrae (Small) A. Heller
Micranthes subapetala (E. Nels.) Small
Saxifraga arnoglossa (Small) Fedde
Saxifraga montanensis Small
Saxifraga sierrae Small
Saxifraga subapetala E. Nels.

Stout, scapose, perennial herb from a simple or branched caudex; scape glandular-pubescent, at least above. Leaves oblanceolate, lanceolate, or obovate, entire to denticulate in ours, glabrous to sparsely pubescent, the blade narrowed into a short, broad petiole. Inflorescence bracteate, of paniculate cymes. Sepals ovate, becoming reflexed. Petals white or greenish, obovate, obtuse. Stamens inserted on a lobed disk covering the ovary; filaments subulate, purplish in ours. Ovary half inferior or less. Follicles 3–5 mm long, beaked, the beaks divergent.

Wet meadows of the Absaroka, Beartooth, Big Horn, and Wind River Ranges. Washington and Oregon east to Montana, Wyoming, and Colorado.

Our plants, var. *subapetala* (E. Nels.) C.L. Hitchc., have petals less than 1 mm long or lacking, and the filaments, fruits, and calyx lobes are usually purplish-tinged.

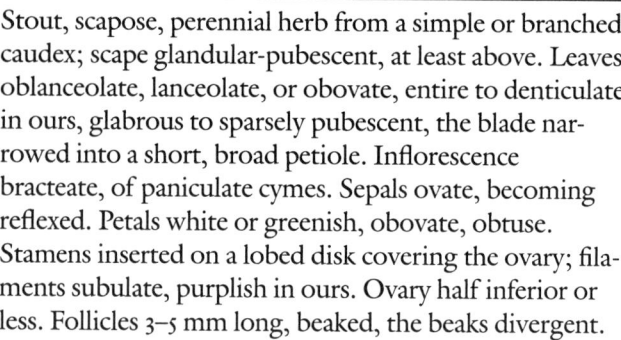

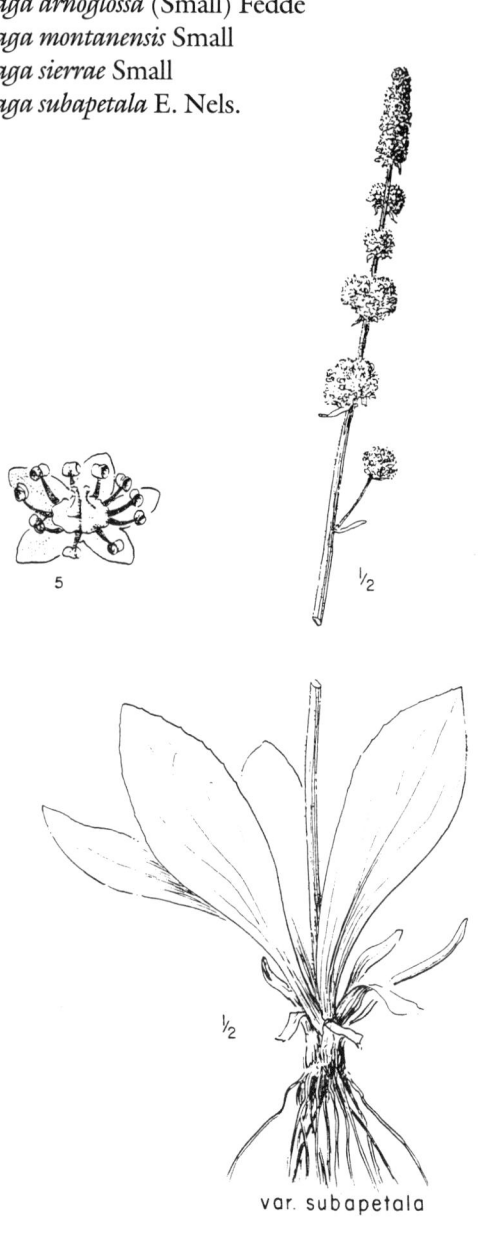
var. subapetala

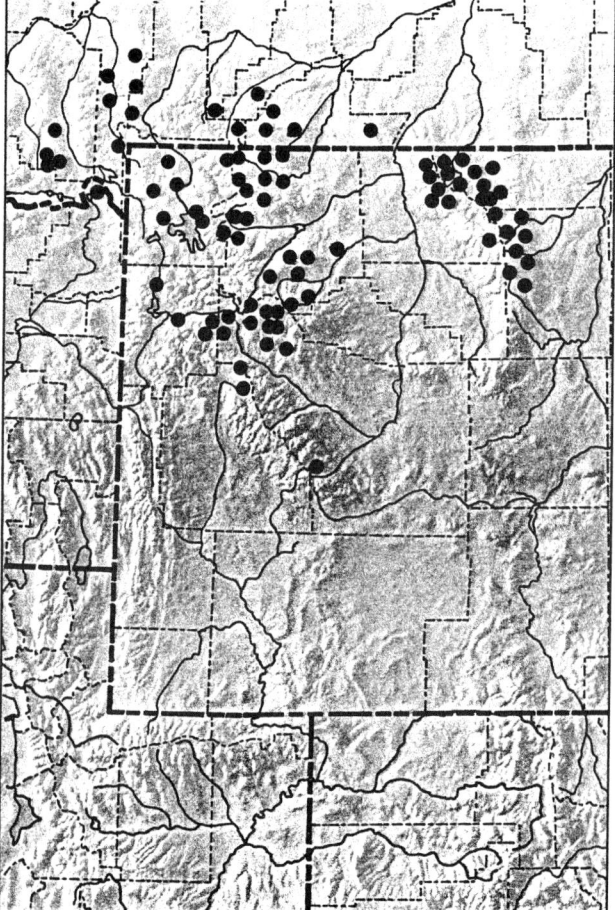

Saxifraga rhomboidea Greene
Pittonia 3:343, 1898.

Snowball Saxifrage

Micranthes austrina (A. Nels.) Rydb.
Micranthes crenatifolia Small
Micranthes franciscana Small
Micranthes greenei Small
Micranthes rhomboidea (Greene) Small
Micranthes rydbergii (Small) Small
Saxifraga austrina (A. Nels.) A. Nels.
Saxifraga greenei Blank.
Saxifraga rydbergii Small

Scapose, perennial herb from a short, erect caudex; scape single, glandular-pubescent. Leaves petiolate, the blades deltoid, rhombic, or ovate, crenate, glabrous above and glabrous to pilose beneath, abruptly narrowed to a ciliate-margined petiole. Inflorescence a dense, capitate, cymose panicle, the lower peduncles sometimes distinct and separated from the upper ones. Sepals oval to ovate, spreading to reflexed. Petals white or cream, obovate to elliptic, clawed, obtuse or emarginate, about twice as long as the sepals. Stamens inserted on a lobed gland; filaments subulate. Ovary about half inferior. Follicles 3–6 mm long, beaked, the beaks divergent, recurved.

Meadows, fellfields, and alpine slopes in all alpine ranges of the Middle Rocky Mountains. British Columbia and Alberta south to Utah and Colorado.

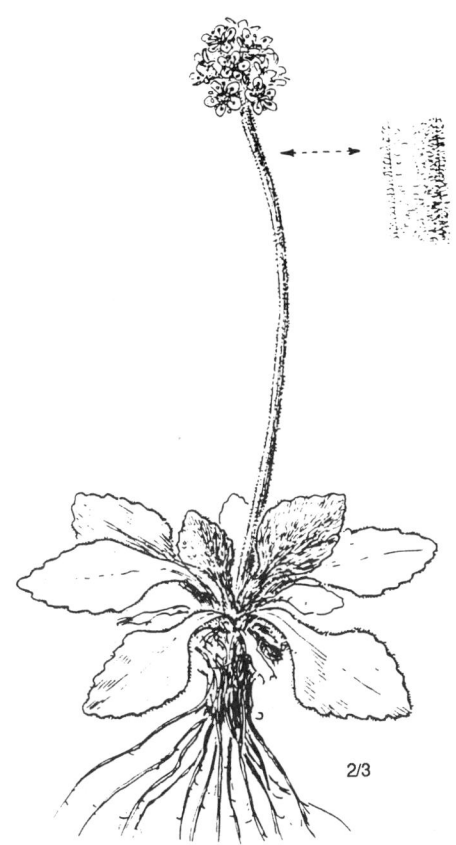

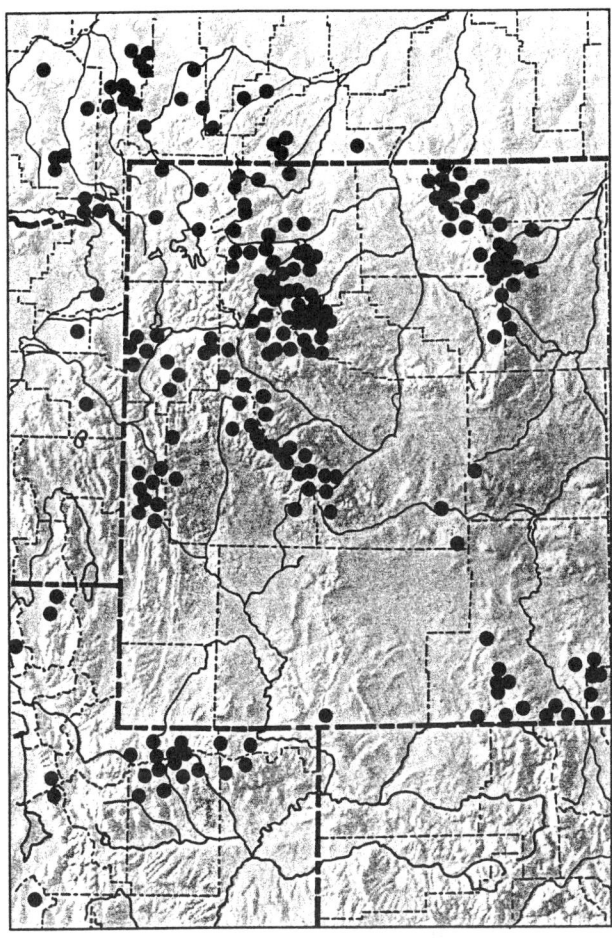

Saxifraga rivularis L.
Sp. Pl., 404, 1753.

Alpine Brook Saxifrage

Saxifraga debilis Engelm. ex Gray
Saxifraga flexuosa Sternb.
Saxifraga hyperborea R. Br.

Small, often tufted, glabrous to glandular-puberulent, perennial herb, to 10 cm tall; stems 1 to several. Basal leaves petiolate, cordate to reniform, 3–7-lobed or toothed; cauline leaves reduced upward, lobed or entire. Inflorescence of 1–3 flowers. Calyx campanulate or turbinate, the lobes erect, ovate, shorter than the tube. Petals white or pinkish, oblong to spatulate, short-clawed, up to twice as long as the calyx lobes. Stamens about the same length as the calyx lobes; filaments not clavate. Ovary about half inferior. Capsule 4–7 mm long, the beaks divergent.

Shaded, protected sites along streams, margins of boulders, ledges, and rock crevices in most alpine ranges of the Middle Rocky Mountains. Not yet known from above timberline in the Hoback, Salt River, and Wyoming Ranges. Circumpolar; south in the United States to California, Arizona, Colorado, and New Hampshire.

Saxifraga rivularis represents a circumpolar, arctic-alpine complex that has traditionally been treated as a mostly arctic group consisting of *S. flexuosa*, *S. hyperborea*, and *S. rivularis* that also occurs in the Rocky Mountains; and an exclusively Rocky Mountain group in southern Canada and the United States consisting of *S. debilis*. There has been wide divergence in authors' applications of these names. Regardless of which names or combinations of names have been used, most authors appear to agree that 2 more or less distinct taxonomic groups exist in the Rocky Mountains. Based on cytotaxonomic evidence, A. Löve et al. (1971) recognized 2 taxa in the Colorado alpine. Tetraploid specimens (2n = 52) were called *S. rivularis* L., and the diploids (2n = 26) were *S. hyperborea* R. Br. ssp. *debilis* (Engelm.) A. Löve, et al. Most recently Dorn (1992) considered the two groups as varieties of *S. rivularis*, which seems to be a practical and workable solution and is used in this treatment as follows:

1. Basal leaves mostly 3–5-lobed; upper stem and pedicels usually pilose, eglandular
   . . . var. *flexuosa* (Sternb.) Engl. & Irmsch.
1. Basal leaves mostly 5–7-lobed; upper stem and pedicels usually glabrate, glandular
   . . . var. *debilis* (Engelm. ex Gray) Dorn

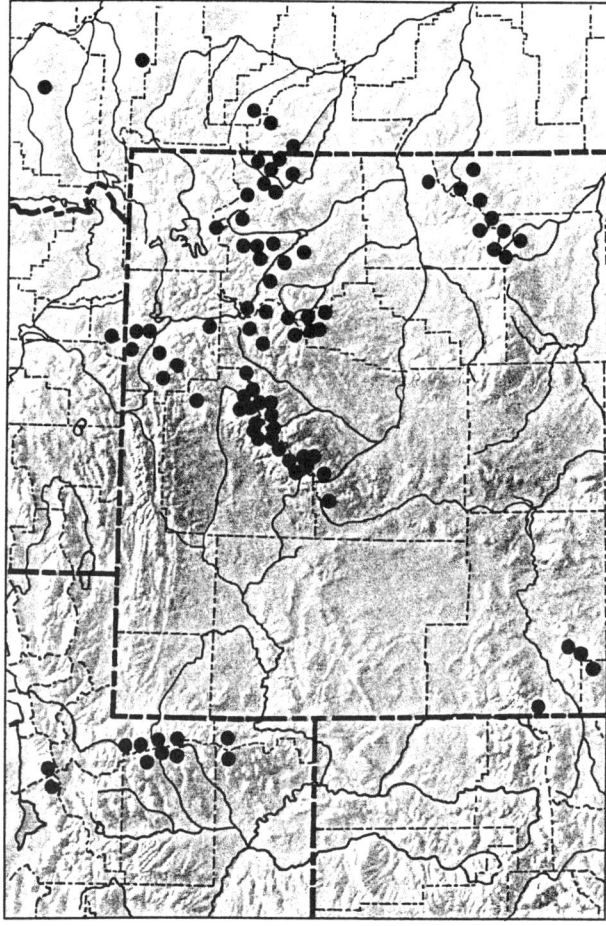

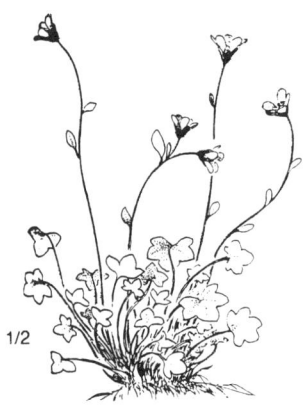

Scrophulariaceae
Figwort Family

Annual or perennial herbs in ours, occasionally shrubby or woody at the base, with alternate or opposite, simple or compound, exstipulate leaves. Flowers hypogynous, 5-merous, perfect, nearly regular to irregular, in terminal spikes, racemes, or cymes. Sepals 4 or 5, distinct or united. Corolla usually zygomorphic and bilabiate, rarely lacking. Stamens 2, 4, or 5, commonly 4 and didynamous, with or without a sterile stamen (filament), inserted on the corolla tube alternate with the lobes. Pistil 1, of 2 united carpels, the ovary 2-celled, with axillary placentation, the style single, the stigmas distinct or united. Fruit usually a many-seeded capsule.

KEY TO THE GENERA OF SCROPHULARIACEAE

1. Stamens 2; corolla rotate or lacking
   2. Leaves all cauline, opposite — *Veronica*
   2. Leaves mostly basal, the cauline ones alternate
      3. Leaves dissected; petals present — *Synthyris*
      3. Leaves finely to coarsely toothed; petals lacking or present — *Besseya*
1. Stamens 4 or 5; corolla bilabiate
   4. Corolla galeate, the upper lip forming a hood or beak
      5. Leaves pinnate or deeply pinnatifid, mostly basal, petiolate — *Pedicularis*
      5. Leaves entire or the upper ones sometimes 3-cleft, mostly cauline, sessile — *Castilleja*
   4. Corolla not galeate, the upper lip 2-lobed
      6. Calyx conspicuously 5-angled; sterile stamens lacking — *Mimulus*
      6. Calyx not 5-angled; sterile stamen present
         7. Plants annual; flowers solitary in the axils of upper leaves — *Collinsia*
         7. Plants perennial; flowers numerous in terminal racemes or panicles
            8. Calyx deeply 5-cleft nearly to the base; corolla blue, purple, or lavender — *Penstemon*
            8. Calyx 5-lobed, not cleft to the base; corolla greenish white to cream — *Chionophila*

Besseya Rydb.
Kittentails

Perennial herbs with mostly basal, petiolate leaves and leafy-bracteate stems, the bracts alternate, at least the upper ones sessile. Flowers small, in terminal, bracteate spikes or racemes. Calyx 2–4-lobed, cleft nearly to the base. Corolla lacking or, if present, purplish or white, bilabiate, the lower lip shorter than the upper. Stamens 2, exserted, the filaments conspicuously purple in ours. Ovary 2-celled, the style filiform, the stigma capitate. Capsule broadly oval, obtuse or emarginate.

KEY TO THE SPECIES OF *BESSEYA*

1. Corolla present, purple — *B. alpina*
1. Corolla lacking — *B. wyomingensis*

Besseya alpina (Gray) Rydb.
Bull. Torr. Club 300:280, 1903.

Alpine Kittentails

Synthyris alpina Gray

Perennial herb, woolly-villous when young, becoming glabrate with age. Basal leaves petiolate, the blades cordate to elliptic, the margins crenate or serrate. Cauline leaves sessile, alternate, reduced. Flowers in terminal spikes. Calyx 2-lobed, the lobes lanceolate, villous. Corolla violet or purple. Stamens not conspicuously colored. Capsule ovoid, somewhat compressed.

Alpine tundra of the Medicine Bow Range. Wyoming, Colorado, Utah, and New Mexico.

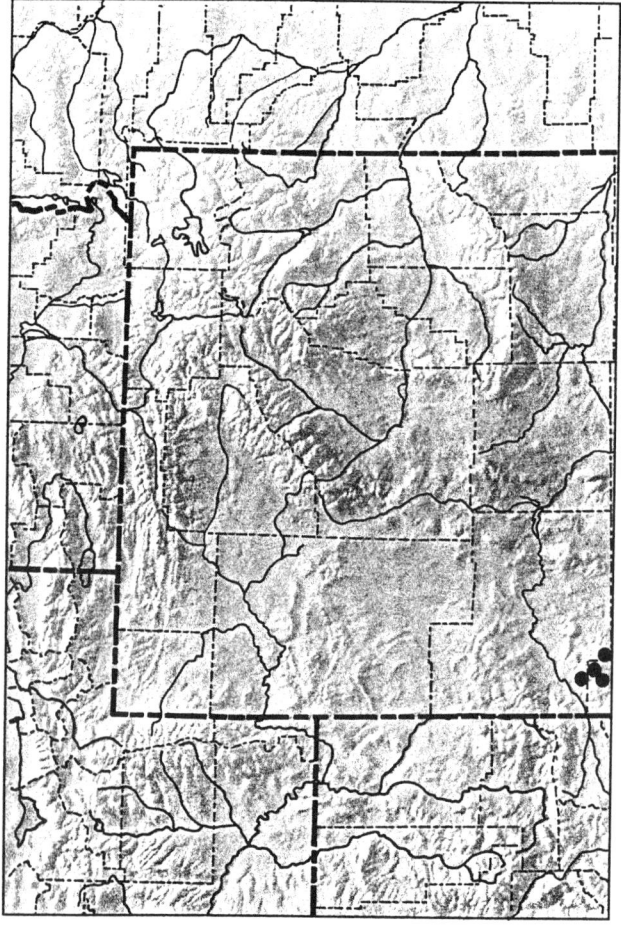

Besseya wyomingensis (A. Nels.) Rydb.
Bull. Torr. Club 30:280, 1903.

Wyoming Kittentails

Wulfenia wyomingensis A. Nels.
Besseya cinerea (Raf.) Pennell
Besseya gymnocarpa (A. Nels.) Rydb.
Synthyris gymnocarpa (A. Nels.) A. Heller
Synthyris wyomingensis (A. Nels.) A. Heller
Wulfenia gymnocarpa A. Nels.

Pubescent, perennial herb from fibrous roots. Basal leaves petiolate, elliptic to ovate, the margins crenate or serrate, cuneate to truncate at the base. Cauline leaves broadly lanceolate to ovate, sessile, toothed to entire. Flowers in dense, terminal spikes. Calyx 2–3-lobed, open on one side of the pistil. Corolla lacking. Stamens conspicuously purple to violet. Capsule ovoid, villous.

Widespread on rocky tundra in most alpine areas of the Middle Rocky Mountains. Not yet known from above timberline in the Hoback, Medicine Bow, Salt River, and Wasatch Ranges. British Columbia to South Dakota, south to Nebraska, Colorado, and Utah.

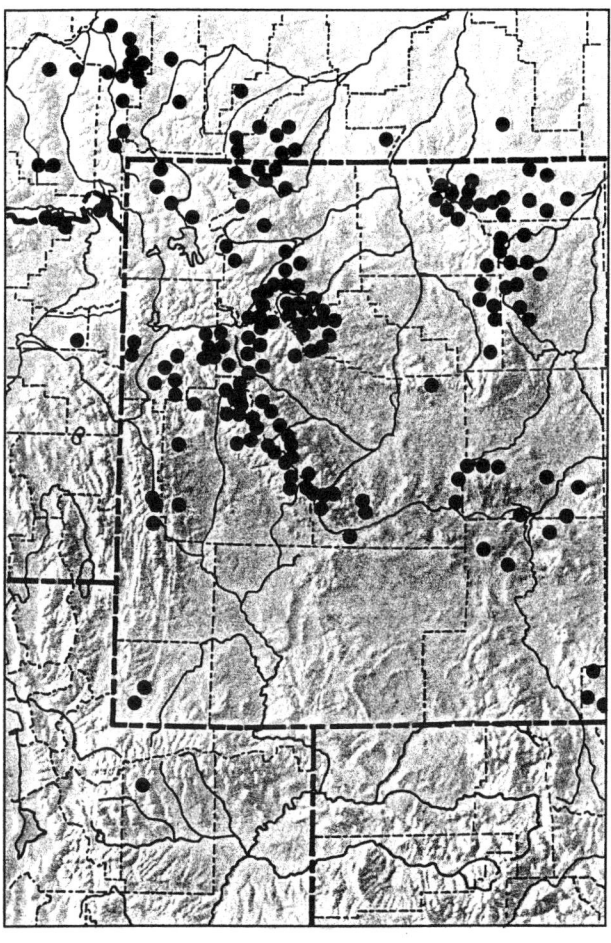

Castilleja Mutis ex L. f.
Paintbrush, Indian Paintbrush

Perennial or rarely annual herbs with alternate, sessile, entire or cleft leaves. Flowers in dense, bracteate spikes, the bracts petaloid, red, purple, yellow, or cream. Calyx petaloid, tubular, 4-lobed, the lobes subequal and usually more deeply cleft above and below than on the sides. Corolla greenish, bilabiate, the upper lip (galea) elongate, entire, the lower lip usually much shorter than the galea, 3-toothed. Stamens 4, didynamous, enclosed by the galea. Style slender, usually exserted, the stigma entire or 2-lobed. Capsule loculicidal, ovoid to cylindric.

KEY TO THE SPECIES OF *CASTILLEJA*

1. Galea less than half the length of the corolla tube; lower lip of corolla more than one-third the length of the galea
   2. Calyx lobes subequal in length; galea villous — *C. nivea*
   2. Calyx lobes unequal, cleft more deeply sagittally than laterally; galea glabrous or puberulent
      3. Calyx lobes acute; plants puberulent with retorse hairs to short-hispid — *C. pallescens*
      3. Calyx lobes obtuse to broadly acute; plants viscid-villous
         4. Bracts of inflorescence equaling or longer than the flowers — *C. cusickii*
         4. Bracts of inflorescence shorter than the flowers
            5. Plants mostly less than 10 cm tall; bracts purplish, at least the lower ones acute; calyx mostly purplish — *C. pulchella*
            5. Plants more than 14 cm tall; bracts yellow, obtuse; calyx mostly yellow — *C. sulphurea*
1. Galea more than half the length of the corolla tube; lower lip of corolla one-fifth to one-third the length of the galea
   6. Bracts yellow or yellowish green — *C. sulphurea*
   6. Bracts red, red-orange, rose, purple, or pink
      7. Plants glandular-viscid; upper leaves mostly cleft into 3 segments — *C. applegatei*
      7. Plants glabrous to villous; upper leaves mostly entire
         8. Flowers mostly 20–30 mm long; bracts rose, purple, or pink, entire, rounded, with 1–2 pairs of short lateral lobes; lower lip one-fourth to one-fifth the length of the galea; galea mostly less than 12 mm long — *C. rhexifolia*
         8. Flowers mostly 30–40 mm long; bracts red to red-orange, toothed, acute; lower lip less than one-fifth the length of the galea; galea mostly more than 12 mm long — *C. miniata*

Castilleja applegatei Fern.
Erythea 6:49, 1898.

Sticky Paintbrush

Castilleja breweri Fern.
Castilleja pinetorum Fern.
Castilleja viscida Rydb.
Castilleja wherryanna Penn.

Erect, perennial herb from a woody caudex; stems single or clustered, glandular-viscid, at least above. Leaves glandular; lower blades linear-lanceolate and entire, the upper ones commonly 3-lobed. Bracts 3–5-lobed, red or occasionally yellowish, equaling or exceeding the flowers. Calyx subequally cleft above and below, the primary lobes divided into 2 acute segments. Corolla exceeding the calyx, the galea puberulent, about equaling the tube and 5 or more times as long as the lower lip.

Rocky slopes of the Uinta, Wasatch, and Wyoming Ranges. Oregon and California east to western Wyoming and Utah.

Middle Rocky Mountain plants with glandular stems and leaves are var. *viscida* (Rydb.) Ownbey. Plants keying here but lacking glandular pubescence may be *C. linariifolia* Benth., Narrowleaf Paintbrush, which has been collected at timberline near the summit of Mount Ogden in the Wasatch Range.

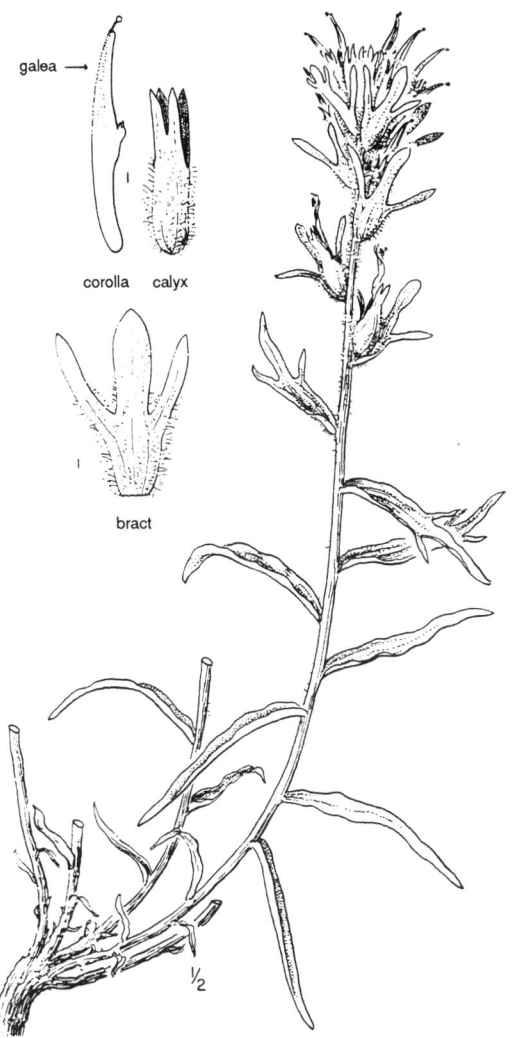

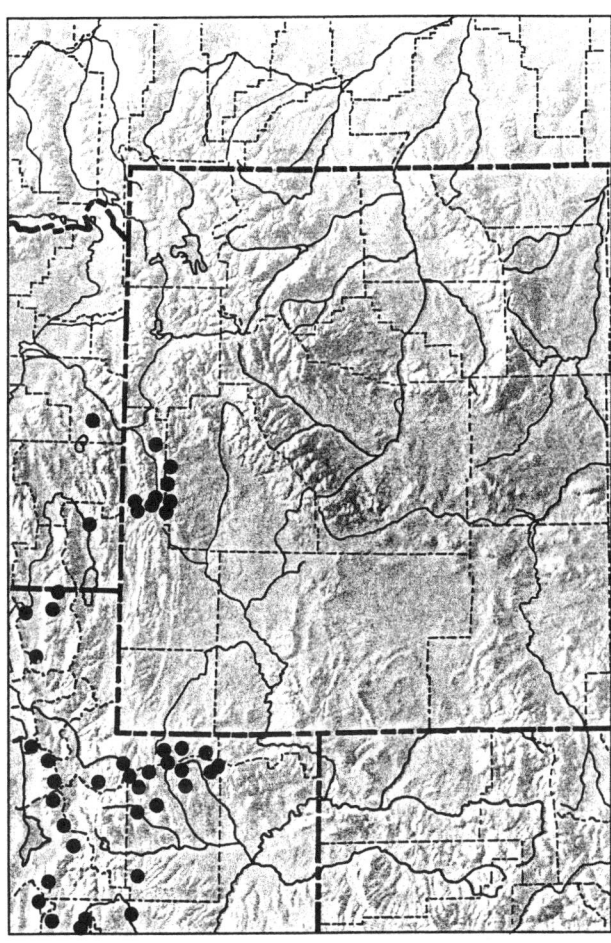

Castilleja cusickii Greenm.
Bot. Gaz. 25:267, 1898.

Cusick Paintbrush

Castilleja camporum T.J. Howell
Castilleja lutea A. Heller
Castilleja pannosa Eastw.
Castilleja pilifera A. Nels.
Castilleja villosa Rydb.

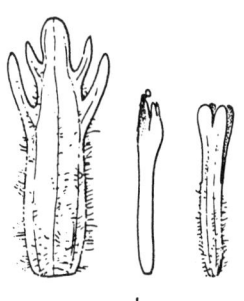

Erect, perennial herb, sometimes decumbent at the base, from a woody crown. Stems clustered, villous or sometimes glabrate. Leaves viscid-villous; lower blades linear-lanceolate, entire, the upper with 1–3 pairs of linear, acute lobes. Bracts longer than the flowers, yellow or purplish, broader than the leaves, oblong, with 1 or 2 pairs of lateral lobes, at least the terminal one obtuse. Calyx subequally cleft above and below, the primary lobes divided into 2 mostly obtuse segments. Corolla shorter or longer than the calyx, the galea puberulent, less than half as long as the corolla tube and about as long as to 3 times longer than the prominent lower lip.

Meadows of the lower alpine zone in the Absaroka, Beartooth, Big Horn, Wind River, and Wyoming Ranges. Alberta south to Oregon, Idaho, and Wyoming.

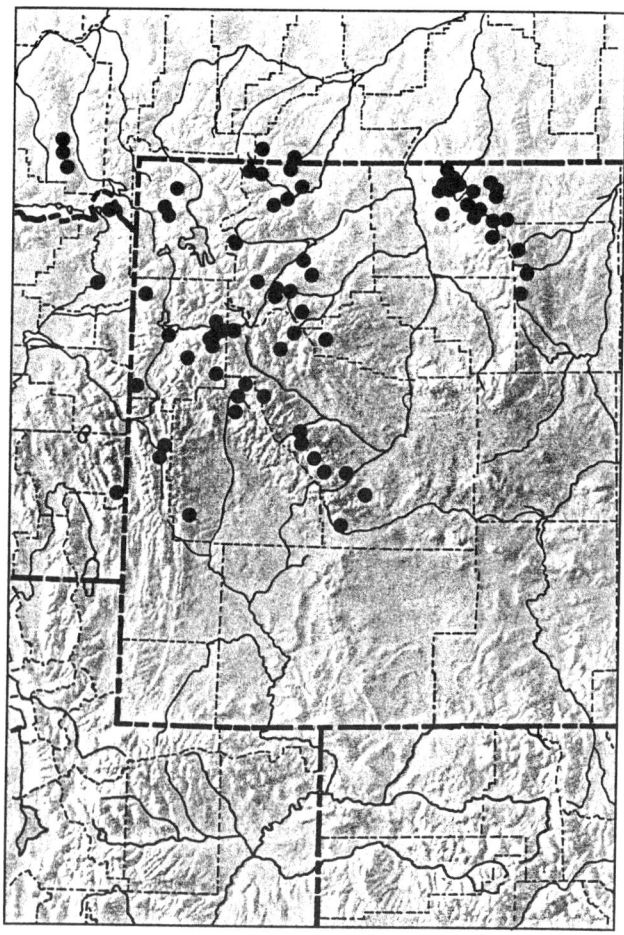

Castilleja miniata Dougl. ex Hook.
Fl. Bor.-Am. 2:106, 1838.

Common Red Paintbrush

Castilleja chrymactis Penn.
Castilleja confusa Greene
Castilleja crispula Piper
Castilleja dixonii Fern.
Castilleja gracillima Rydb.
Castilleja inconstans Standl.
Castilleja lanceifolia Rydb.
Castilleja magna Rydb.
Castilleja oblongifolia Gray
Castilleja peckiana Penn.
Castilleja trinervis Rydb.
Castilleja tweedyi Rydb.
Castilleja variabilis Rydb.
Castilleja vreelandii Rydb.

Erect, perennial herb from a woody caudex. Stems clustered, occasionally single, glabrous to somewhat villous or glandular-villous. Leaves glabrous to short-villous, narrowly lanceolate to lanceolate, entire, the uppermost sometimes 3-lobed. Bracts bright red or orange-red, lanceolate to ovate, entire to variously cleft, the lobes acute. Calyx subequally cleft above and below, the primary lobes divided into 2 linear, acute segments. Corolla much longer than the calyx, the galea puberulent, about equaling the corolla tube and 5 or more times as long as the short lower lip.

Timberline meadows of the lower alpine zone in the Absaroka, Beartooth, Teton, Wasatch, Wind River, and Wyoming Ranges. Alaska south to California, Arizona, New Mexico, and North Dakota. See comments under *C. applegatei,* above.

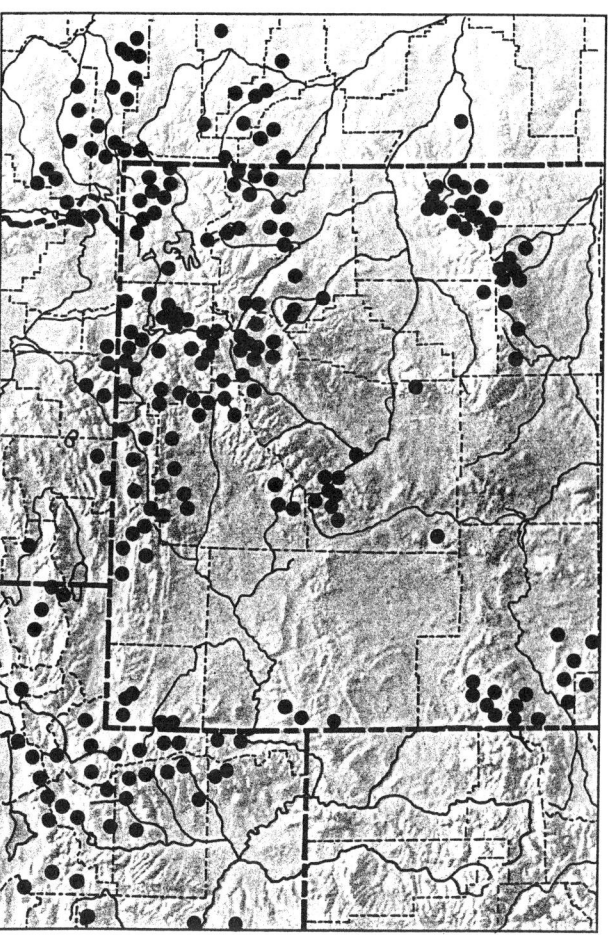

Castilleja nivea Penn. & M. Ownbey
Not. Nat. 227:2, 1950.

Alpine Paintbrush

Low, erect, perennial herb from a woody caudex. Stems several to many, densely to sparsely tomentose. Leaves tomentose; lower blades linear, entire, the upper 3-parted with linear, acute lobes. Bracts yellowish, lanceolate, 3-parted, tomentose, the lobes acute. Calyx subequally cleft into 4 linear, acute segments. Corolla villous, longer than the calyx, the galea about half as long as the corolla tube and about twice as long as the prominent lower lip.

Rocky tundra and fellfields of the Absaroka and Beartooth Ranges. Endemic to southwestern Montana and northern Wyoming.

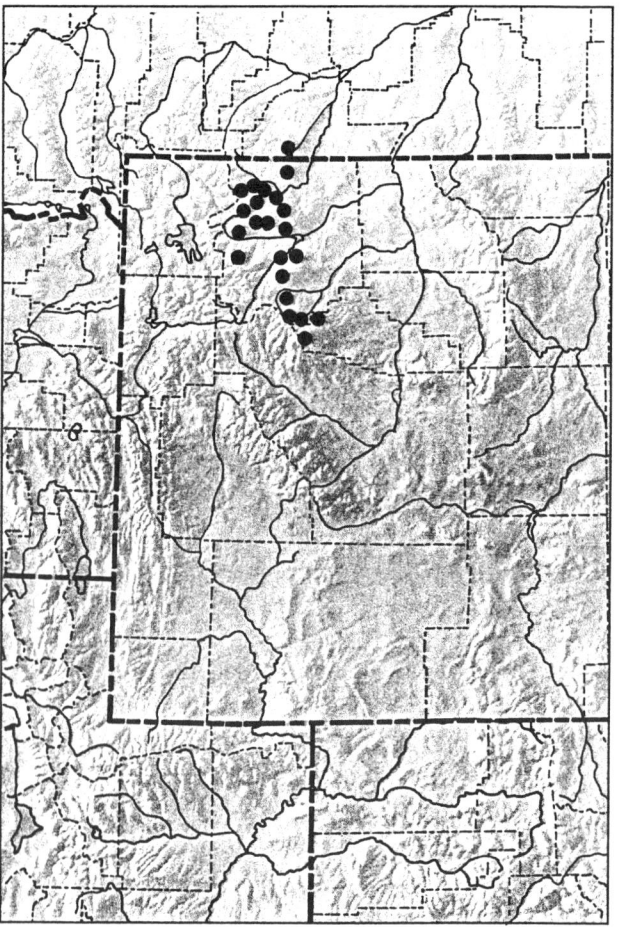

Castilleja pallescens (Gray) Greenm.
Bot. Gaz. 25:266, 1898.

Pale Paintbrush

Orthocarpus pallescens Gray
Castilleja fasciculata A. Nels.
Castilleja inverta (A. Nels. & J.F. Macbr.) Penn. & M. Ownbey in R. Davis
Euchroma pallescens Nutt. ex Gray
Orthocarpus parryi Gray

Erect, perennial herb from a simple to branched, woody caudex. Stems purplish, clustered, occasionally single, often decumbent at the base, puberulent to short-hispid, the hairs often retrorse. Leaves puberulent, the lower ones linear and entire, the upper ones 3–5-parted. Bracts yellow or sometimes purplish, lanceolate, puberulent, ciliate, 3–7-parted, the lobes broadly acute to obtuse. Calyx subequally cleft above and below, the primary lobes divided into 2 triangular, acute segments. Corolla shorter than to slightly exceeding the calyx, the galea less than half as long as the corolla tube and twice as long to about as long as the lower lip.

Typical of slopes and dry plains at lower elevations, this species was reported from the alpine zone of the Beartooth Range by Lessica (1993). Endemic to Montana, Idaho, and Wyoming.

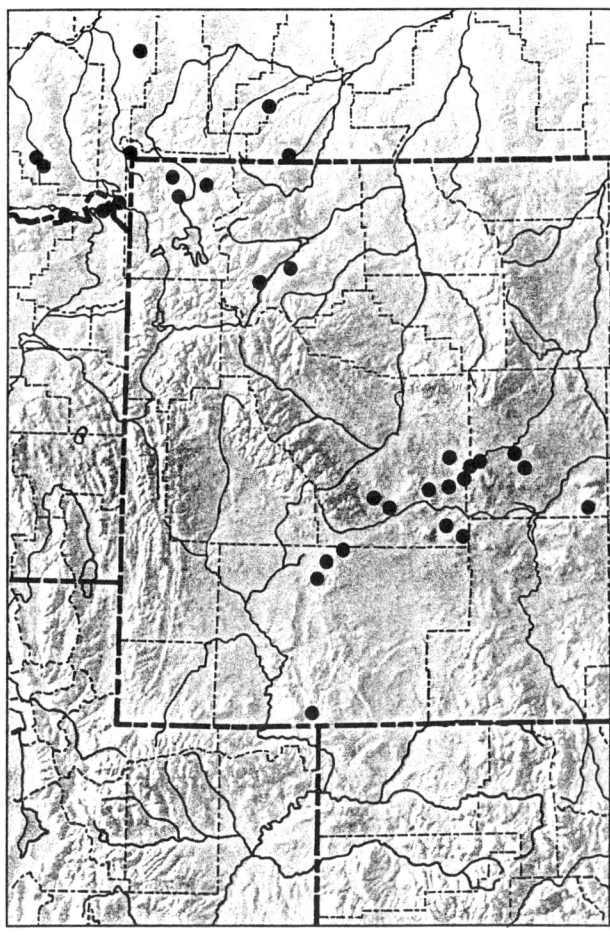

Castilleja pulchella Rydb.
Bull. Torr. Club 34:40, 1907.

Showy Paintbrush

Low, erect, perennial herb from a branched, woody caudex. Stems clustered, often decumbent at the base, viscid-villous. Leaves viscid-villous; lower blades linear, entire, the upper ones usually 3-lobed. Bracts viscid-villous, yellow to purple, ovate, entire to 3-lobed, the terminal lobe usually obtuse. Calyx subequally cleft above and below, the primary lobes entire or divided into 2 broad, obtuse segments. Corolla slightly longer than the calyx, the galea puberulent, about one-third the length of the corolla tube and mostly less than twice as long as the prominent lip.

Widespread and common on alpine tundra in most ranges of the Middle Rocky Mountains. Not yet known from above timberline in the Hoback, Medicine Bow, Salt River, and Wasatch Ranges. Southwestern Montana, western Wyoming, and northeastern Utah.

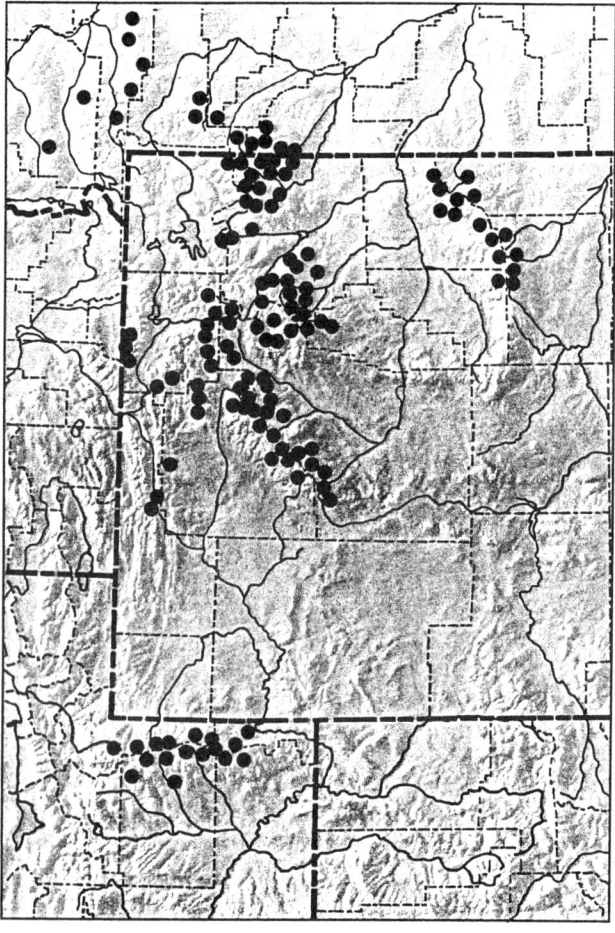

Castilleja rhexifolia Rydb.
Mem. N.Y. Bot. Gard. 1:356, 1900.

Rosy Paintbrush

Castilleja humilis Rydb.
Castilleja lauta A. Nels.
Castilleja leonardii Rydb.
Castilleja obtusiloba Rydb.
Castilleja oregonensis Gand.
Castilleja purpurascens Greenm.
Castilleja purpurascens Rydb.
Castilleja subpurpurascens Rydb.

Erect, perennial herb from a woody caudex. Stems clustered, occasionally single, decumbent at the base, glabrous or sparsely villous. Leaves narrowly lanceolate to ovate, entire, the uppermost sometimes shallowly 3-lobed. Bracts villous and crimson, purple, or rose, ovate, entire or 3-lobed, the lateral lobes much smaller than the broad, obtuse terminal lobe. Calyx subequally cleft above and below, the primary lobes divided into 2 obtuse segments. Corolla longer than the calyx, the galea puberulent, about half the length of the tube and 4 to 5 times longer than the lower lip.

Common and widespread in meadows and moist, rocky slopes in all ranges of the Middle Rocky Mountains. British Columbia and Alberta, south to Oregon, Utah, and Colorado.

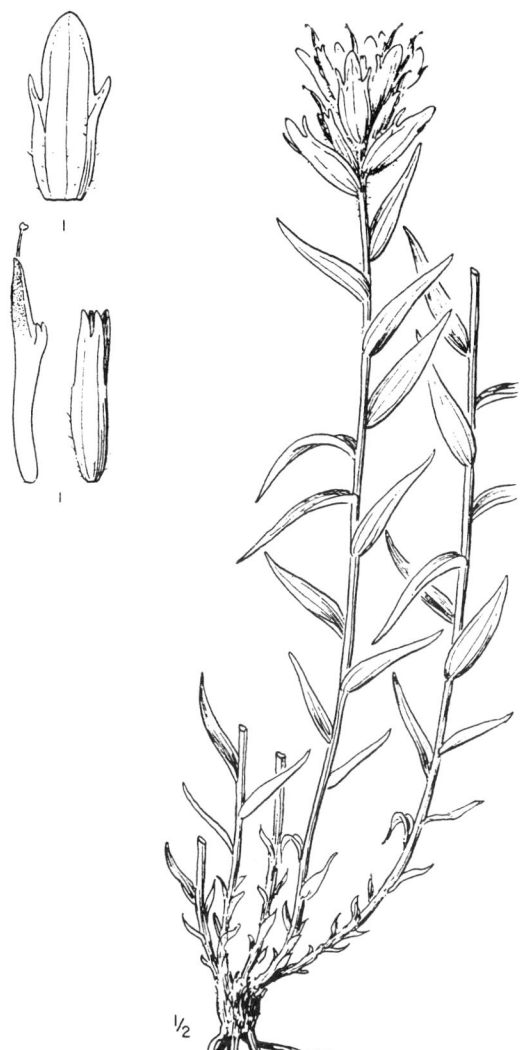
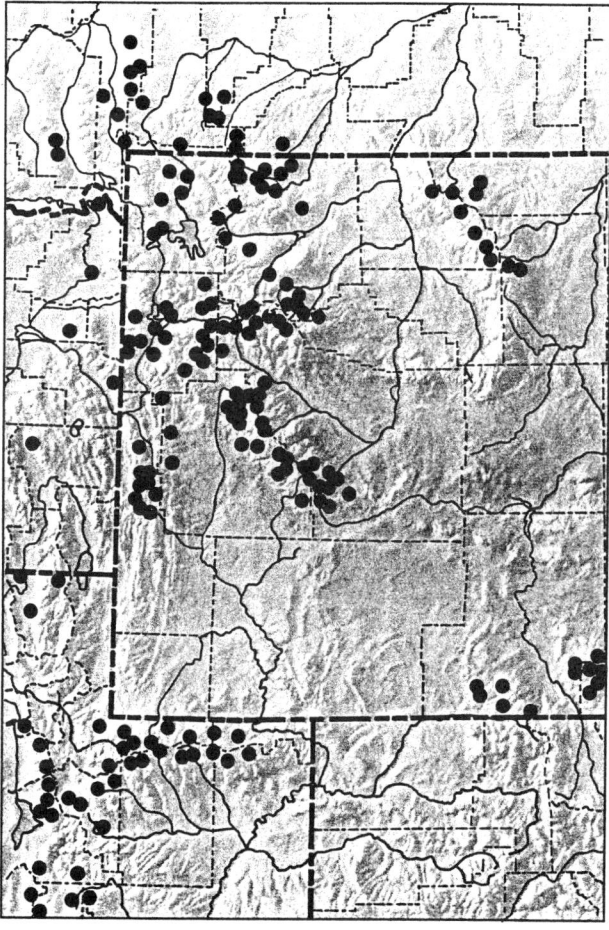

Castilleja sulphurea Rydb.
Mem. N.Y. Bot. Gard. 1:359, 1900.

Sulphur Paintbrush

Castilleja luteovirens Rydb.
Castilleja mogollonica Penn.
Castilleja wyomingensis Rydb.

Erect, perennial herb from a branched, woody caudex. Stems densely clustered, decumbent at the base, glabrous at the base to villous or glandular-puberulent above. Leaves linear to broadly lanceolate, entire, glabrous to puberulent, the upper ones sometimes 3-lobed. Bracts villous, yellowish, broadly lanceolate to ovate, entire, obtuse, sometimes with a short pair of lateral lobes. Calyx subequally cleft above and below, the primary lobes divided into 2 short, obtuse to acute segments. Corolla longer than the calyx, the galea puberulent, less than half as long as the tube, and 3 to 4 times longer than the lip.

Meadows and rocky slopes in nearly all ranges of the Middle Rocky Mountains. Not yet known from above timberline in the Beartooth Range. British Columbia and Alberta south to Utah, New Mexico, and northwestern South Dakota.

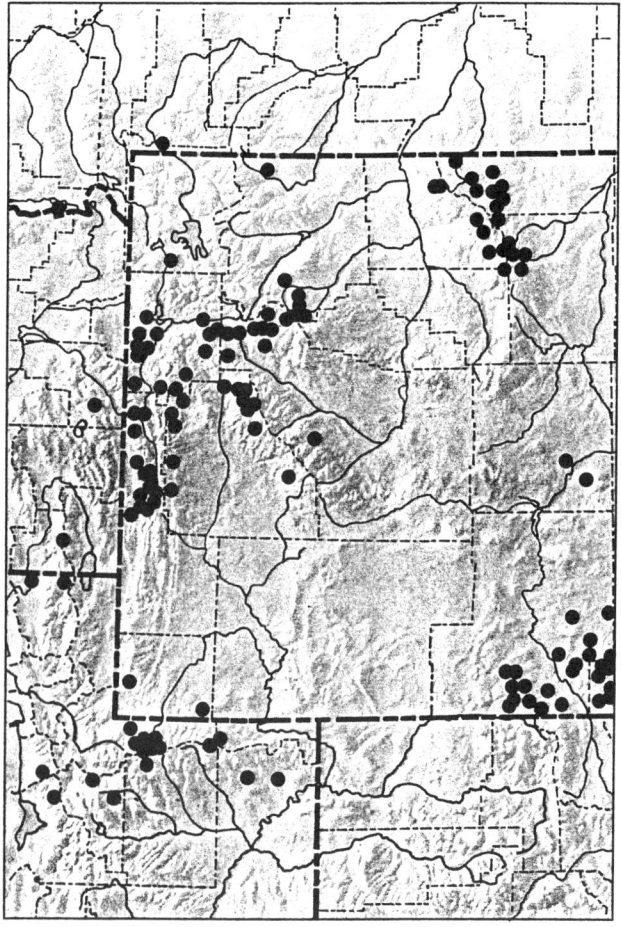

Chionophila Benth.
Snowlover

Slender, perennial herbs with fibrous roots. Leaves mostly basal, petiolate; cauline leaves opposite, reduced, lanceolate. Flowers white, greenish white, or lavender, in secund spikes or racemes. Calyx with 5 lobes. Corolla bilabiate, the upper lip erect, the lower 3-lobed and bearded. Fertile stamens 4, the anther sacs confluent; sterile stamen glabrous, shorter than the fertile ones. Pistil 1, the stigma capitate. Fruit a capsule.

Chionophila jamesii Benth. in DC.
Prodromus 10:331, 1846.

James Snowlover

Small, glabrous to puberulent, perennial herb. Basal leaves oblanceolate, petiolate. Cauline leaves opposite, lanceolate to linear, sessile or nearly so. Flowers in a loose, terminal, secund, bracteate raceme. Calyx 5-cleft, about half the length of the corolla; corolla tubular and white, cream, or greenish white, bilabiate. Fertile stamens 4; sterile stamen short, glabrous. Capsules ovoid to elliptic.

Alpine tundra of the Medicine Bow Range. Endemic to southern Wyoming and Colorado.

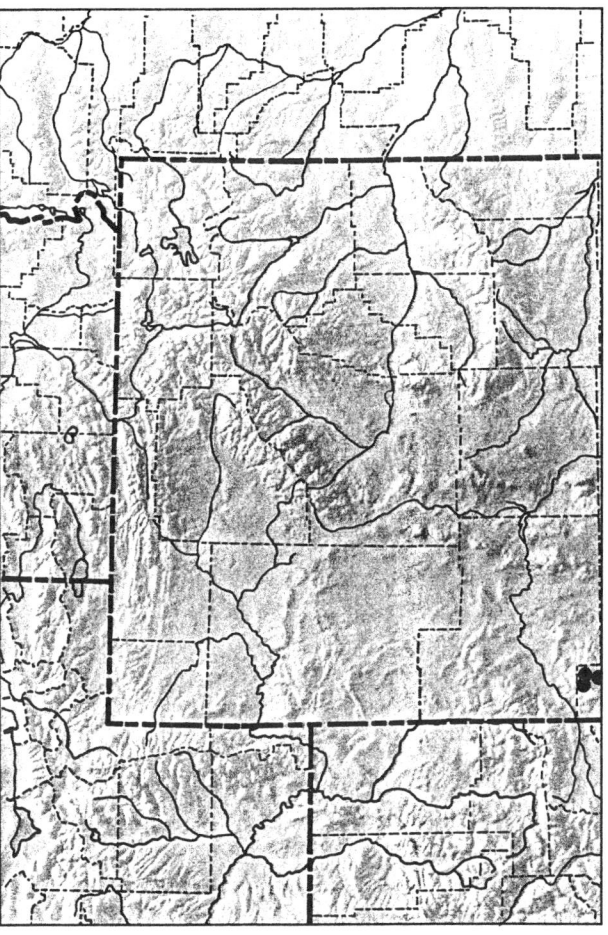

Collinsia Nutt.
Blue-Eyed Mary

Low, annual herbs with slender stems and simple, sessile, opposite or whorled leaves. Flowers white, blue, or bicolored, solitary or several in the axils of the upper leaves. Calyx with 5 subequal lobes. Corolla bilabiate, the upper lip 2-lobed, the lower lip 3-lobed, saccate, and enclosing the 4 didynamous stamens. Style somewhat declined, the stigma capitate or slightly 2-lobed. Capsule ovoid or globose, dehiscing along 4 sutures.

Collinsia parviflora Lindl.
Bot. Reg. 13:Pl. 1082, 1827.

Smallflower Blue-Eyed Mary

Antirrhinum tenellum Pursh
Collinsia grandiflora Lindl.
Collinsia minima Nutt.
Collinsia pusilla (Gray) T.J. Howell
Collinsia tenella (Pursh) Piper

Low, slender, annual herb; stems simple or branched, glabrate to puberulent, sometimes glandular. Leaves simple, entire to serrulate, opposite; lowest blades ovate, petiolate, the upper ones lanceolate, sessile, becoming whorled. Flowers axillary and terminal on slender pedicels. Calyx 5-cleft, the lobes lanceolate, acuminate. Corolla bilabiate, deflexed, blue, upper lip whitish, lower lip blue-violet, the throat yellowish, gibbous on the upper side. Fertile stamens 4. Capsule ovoid, slightly shorter than the calyx lobes; seeds 2–4.

Known in the alpine zone only from meadows and alp slopes in the northern Wind River Range. Alaska and Yukon Territory south to Vermont, Michigan, North Dakota, New Mexico, and California.

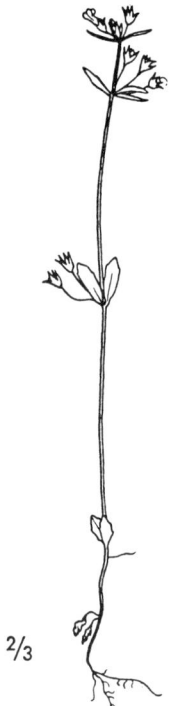

2/3

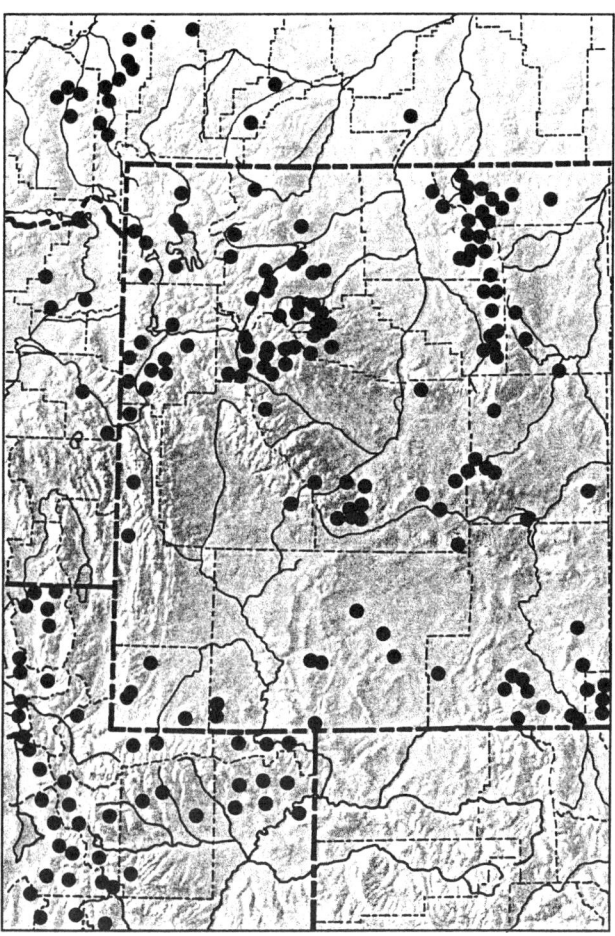

Mimulus L.
Monkeyflower

Annual or perennial herbs with opposite, entire to dentate leaves. Flowers reddish or yellow in ours, usually solitary and axillary, sometimes in terminal, leafy racemes. Calyx tubular, 5-angled, and 5-lobed. Corolla nearly regular to strongly bilabiate, 5-lobed, the throat with a pair of longitudinal ridges on the lower side. Stamens 4, didynamous. Style slender, the stigma usually 2-lobed. Capsule loculicidal.

KEY TO THE SPECIES OF *MIMULUS*		
1. Flowers yellow		*M. guttatus*
1. Flowers pink to magenta		*M. lewisii*

Mimulus guttatus DC.
Cat. Hort. Monspel., 127, 1813.

Common Monkeyflower

Mimulus alpinus (Gray) Piper
Mimulus arvensis Greene
Mimulus bakeri Gand.
Mimulus brachystylis Edwin
Mimulus caespitosus Greene
Mimulus clementinus Greene
Mimulus cordatus Greene
Mimulus corallinus Greene
Mimulus cuspidata Greene
Mimulus decorus (Grant) Suksd.
Mimulus equinnus Greene
Mimulus glareosus Greene
Mimulus grandiflorus T.J. Howell
Mimulus grandis (Greene) A. Heller
Mimulus hallii Greene
Mimulus hirsutus T.J. Howell
Mimulus implexus Greene
Mimulus implicatus Greene
Mimulus langsdorfii Donn ex Greene
Mimulus laxus Penn. ex M.E. Peck
Mimulus longulus Greene
Mimulus lucens Greene
Mimulus luteus Donn ex Sims
Mimulus lyratus Benth.
Mimulus maguirei Pennell
Mimulus marmotatus Greene
Mimulus micranthus A. Heller
Mimulus microphyllus Benth. in DC.
Mimulus minor A. Nels.
Mimulus minusculus Greene
Mimulus nasutus Greene
Mimulus panicolatus Greene
Mimulus pardalis Greene
Mimulus parishii Gand.
Mimulus petiolaris Greene
Mimulus prionophyllus Greene

Mimulus procerus Greene
Mimulus puberulus Gand.
Mimulus puncticalyx Gand.
Mimulus rivularis Nutt.
Mimulus scouleri Hook.
Mimulus subreniformis Greene
Mimulus tenellus Nutt. ex Gray
Mimulus thermalis A. Nels.
Mimulus tilingii Regel
Mimulus unimaculatus Pennell
Mimulus veronicifolius Greene

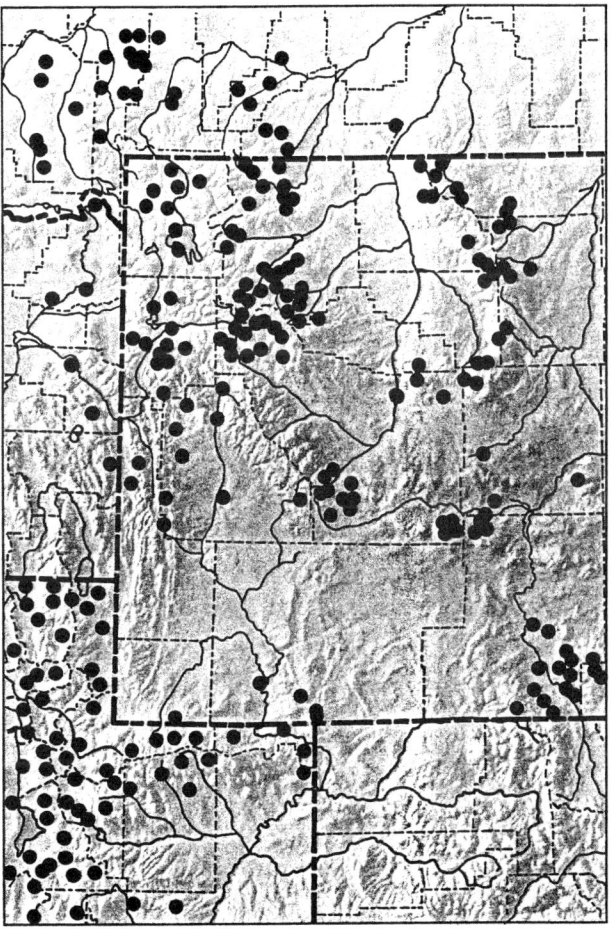

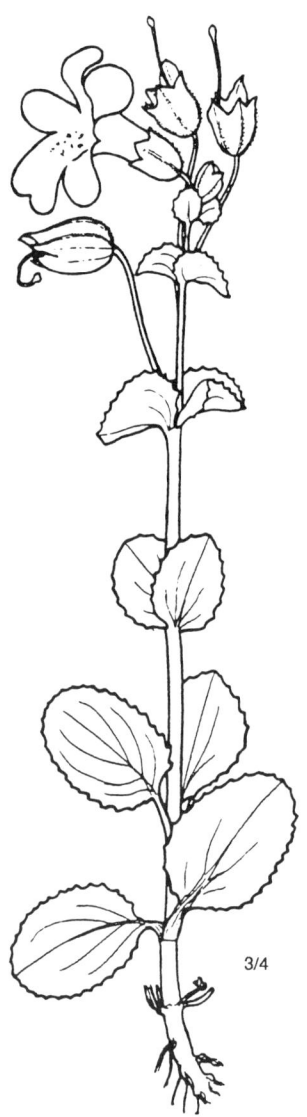

Slender, mat-forming, perennial herb from rhizomes or stolons, or occasionally annual and fibrous-rooted; stems glabrous or sometimes pubescent, solitary, simple or branched, slightly decumbent at the base. Leaves acute, elliptic to ovate, petiolate below, becoming sessile above, the margins dentate, palmately veined. Flowers solitary to many, long-pedicellate in upper leaf axils. Calyx campanulate, irregular, the upper tooth larger than the others, inflated in fruit. Corolla yellow, bilabiate, dotted or splotched with dark red, early deciduous; palate densely yellow-pubescent. Capsule oblong to obovoid, the base stipitate.

Wet, cold stream banks at timberline in the Wasatch Range. Alaska south to California, Arizona, and New Mexico.

Mimulus guttatus is common throughout the Middle Rocky Mountains and should be expected in the vicinity of timberline in all mountain ranges. Plants that are strongly rhizomatous and few-flowered have been called *Mimulus tilingii* Regel, Subalpine Monkeyflower.

Mimulus lewisii Pursh
Fl. Am. Sept., 427, 1814.

Lewis Monkeyflower

Stout, perennial herb from rhizomes; stems erect, simple, often clustered. Leaves acute, viscid-pubescent, sessile, lanceolate to ovate, the margins entire to irregularly dentate, palmately veined. Flowers few to several, axillary, usually on pedicels exceeding the leaves. Calyx tubular, angled, the teeth nearly equal. Corolla rose-purple, strongly bilabiate, splotched or spotted with yellow and dark red. Anthers bearded. Capsule oblong, included within the calyx.

Stream banks and wet rocky places of the Absaroka, Beartooth, Big Horn, Teton, Uinta, Wasatch, and Wind River Ranges. Alaska south to California, Utah, and Colorado.

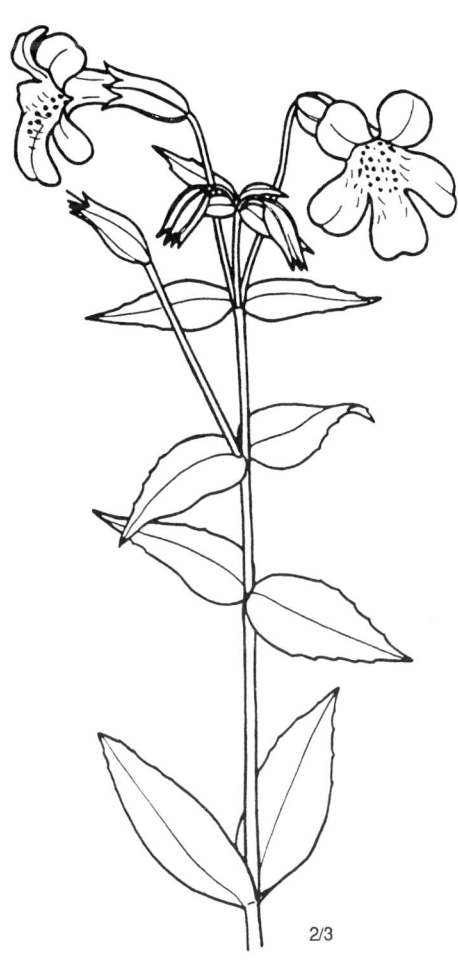

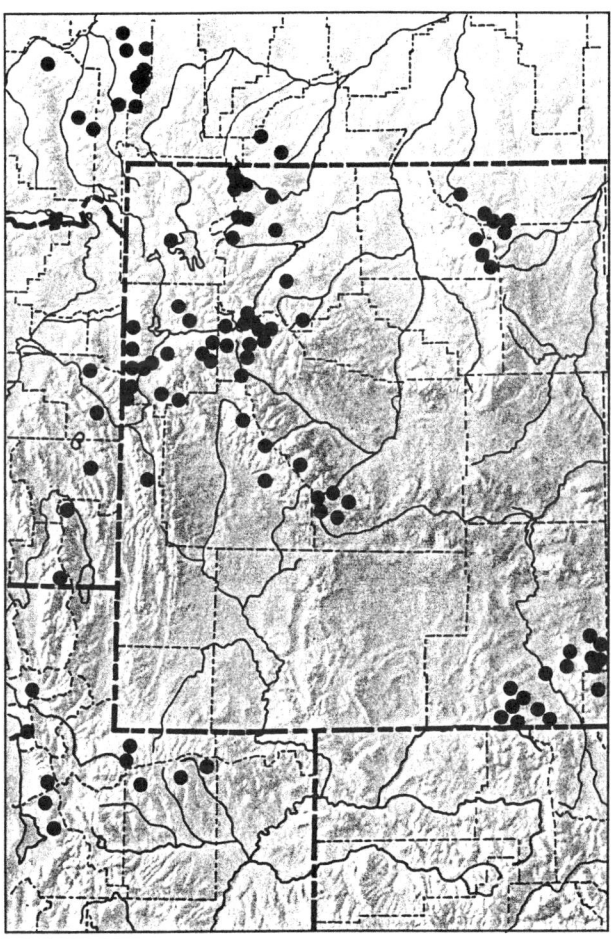

Pedicularis L.
Lousewort

Mostly perennial herbs with erect, usually unbranched stems and alternate, opposite, or basal, toothed to pinnatifid leaves. Flowers white to purple, strongly irregular, in terminal spikes or racemes. Calyx tubular, 2–5-lobed. Corolla bilabiate, the upper lip (galea) hood-shaped or extended into a beak, sometimes with 2 lateral teeth, the lower lip 3-lobed, shorter than the upper lip. Stamens 4, didynamous, included within the galea. Stigma capitate. Capsule ovate to oblong, loculicidal.

KEY TO THE SPECIES OF *PEDICULARIS*

1. Leaves simple, serrate to doubly serrate; calyx lobes 2 — *P. racemosa*
1. Leaves pinnatifid or pinnately compound; calyx lobes 2–5
   2. Galea prolonged into a beak more than 1.5 mm long
      3. Beak straight, 1.5–2 mm long — *P. parryi*
      3. Beak curved, 3–10 mm long
         4. Flowers resembling elephant heads; beak upcurved, well exserted beyond the lower lip of the corolla — *P. groenlandica*
         4. Flowers not resembling elephant heads; beak incurved, nearly equaling the lower lip of the corolla — *P. contorta*
   2. Galea beakless
      5. Corolla purple
         6. Plants less than 15 cm tall; bracts of the inflorescence about the same size and shape as the cauline leaves — *P. pulchella*
         6. Plants more than 15 cm tall; bracts of the inflorescence entire to lobed, not resembling the cauline leaves — *P. cystopteridifolia*
      5. Corolla yellow
         7. Plants with mostly cauline leaves, the largest leaves 5–10 cm broad; stems 3–10 dm tall — *P. bracteosa*
         7. Plants with mostly basal leaves, the largest leaves 1–1.5 cm broad, stems less than 2 dm tall — *P. oederi*

Pedicularis bracteosa Benth. in Hook.
Fl. Bor.-Am. 2:110, 1838.

Bracteate Lousewort

Pedicularis atrosanguinea Penn. & J.W. Thomps.
Pedicularis canbyi Gray
Pedicularis flavida Penn.
Pedicularis latifolia Penn.
Pedicularis montanensis Rydb.
Pedicularis pachyrhiza Penn.
Pedicularis paddoensis Penn.
Pedicularis paysoniana Penn.
Pedicularis siifolia Rydb.
Pedicularis thompsonii Penn.

Stout, perennial herb from fibrous or tuberously thickened roots. Stems single or few, glabrous or nearly so. Basal leaves petiolate, sometimes lacking; cauline leaves sessile to short-petiolate, pinnatifid or bipinnatifid, the divisions often doubly serrate. Flowers in a dense, terminal spike. Calyx 5-lobed, the lobes lanceolate, glandular-puberulent to villous, upper lobe shorter than the others. Corolla yellowish in ours, the galea hooded at the apex, beakless. Capsule ovoid.

Stream banks and meadows in the lower alpine zone of the Absaroka, Beartooth, Teton, Uinta, and Wind River Ranges. British Columbia and Alberta south to California and Colorado.

Middle Rocky Mountain alpine plants with a pubescent calyx, a yellowish corolla, and a beakless galea that is strongly raised above the lower lip are var. *paysoniana* (Penn.) Cronq.

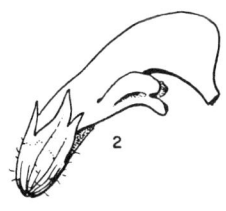
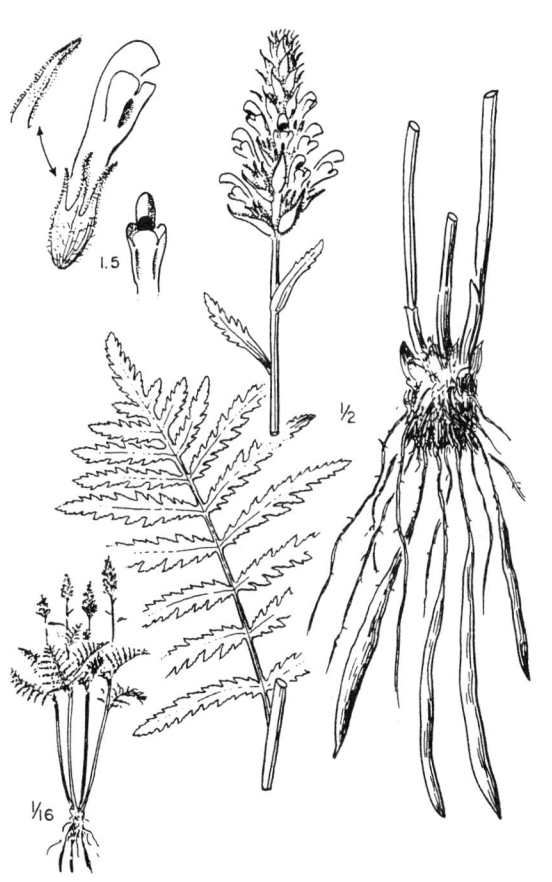
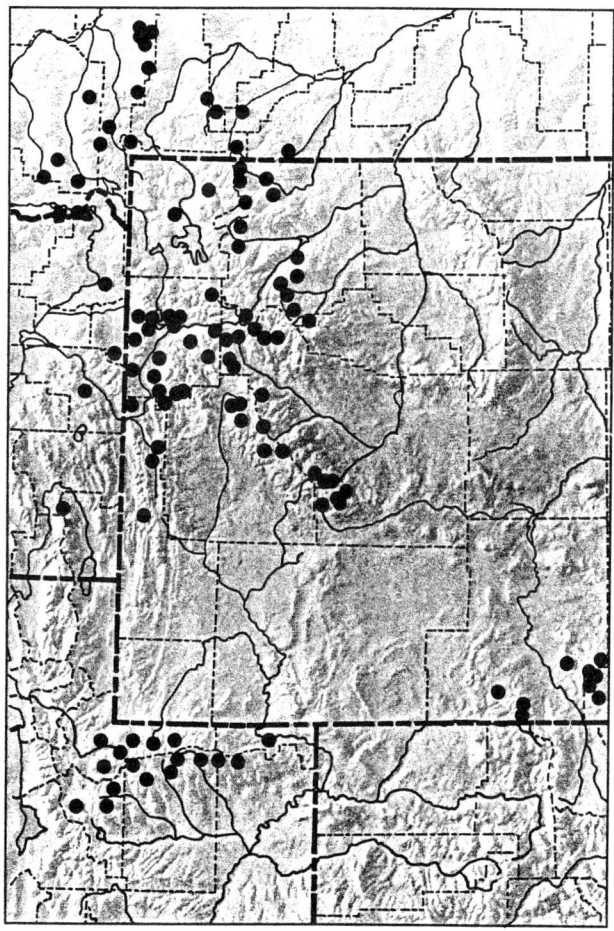

Pedicularis contorta Benth. in Hook.
Fl. Bor.-Am. 2:108, 1938.

Coiled-Beak Lousewort

Pedicularis ctenophora Rydb.
Pedicularis lunata Rydb.

Erect, perennial herb from a stout, woody caudex; stems clustered, glabrous or nearly so. Basal leaves glabrous, petiolate, pinnately divided, the divisions linear to lanceolate, serrate; cauline leaves short-petiolate to sessile, reduced upward. Inflorescence a loose, spikelike raceme. Calyx 5-lobed, the upper lobe shorter than the others, each with a dark midvein. Corolla white, yellowish, or purple, the galea strongly downcurved and enclosed by the lobes of the lower lip. Capsule oval.

Dry meadows of the Beartooth, Big Horn, Gros Ventre, Teton, and Wind River Ranges. British Columbia and Alberta south to northern California, Idaho, and northwestern Wyoming.

Middle Rocky Mountain plants are var. *ctenophora* (Rydb.) Nels. & Macbr., with a pink to purple corolla, and var. *contorta,* with a white or yellowish corolla.

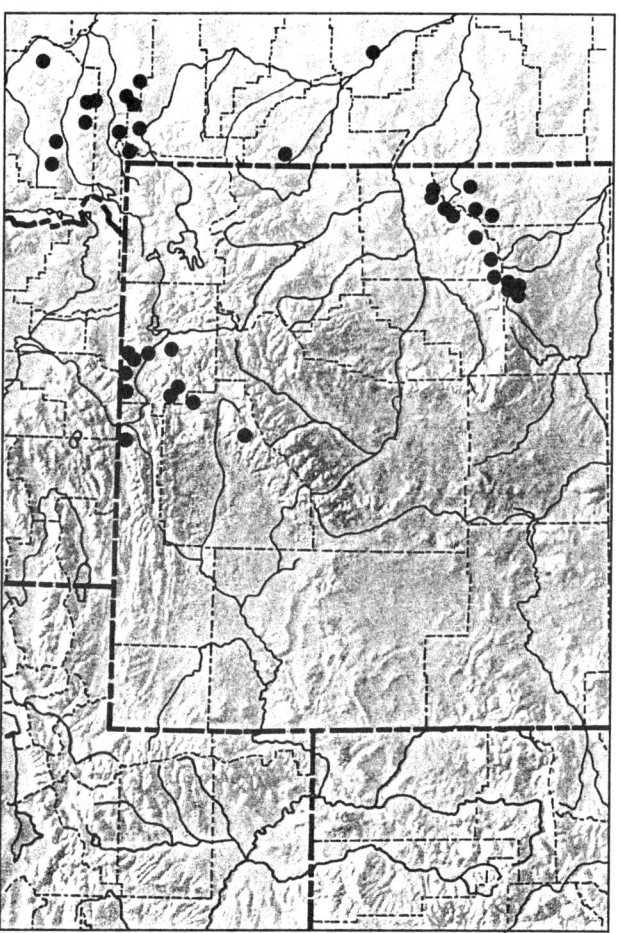

Pedicularis cystopteridifolia Rydb.
Mem. N.Y. Bot. Gard. 1:365, 1900.

Fernleaf Lousewort

Pedicularis elata Pursh

Erect, perennial herb from fibrous roots, often with a short caudex. Stems single, glabrous or puberulent. Basal leaves petiolate, the blades 1–2 times pinnately dissected. Cauline leaves petiolate below, becoming sessile and reduced upward. Inflorescence a villous spike, the bracts strongly differentiated from the leaves. Calyx 5-lobed, the upper lobe smaller than the others. Corolla purple, the galea beakless, minutely bidentate at the tip. Capsule oblong, beaked.

Meadows and rocky slopes of the Absaroka, Beartooth, and Big Horn Ranges. Endemic to southwestern Montana and northwestern Wyoming.

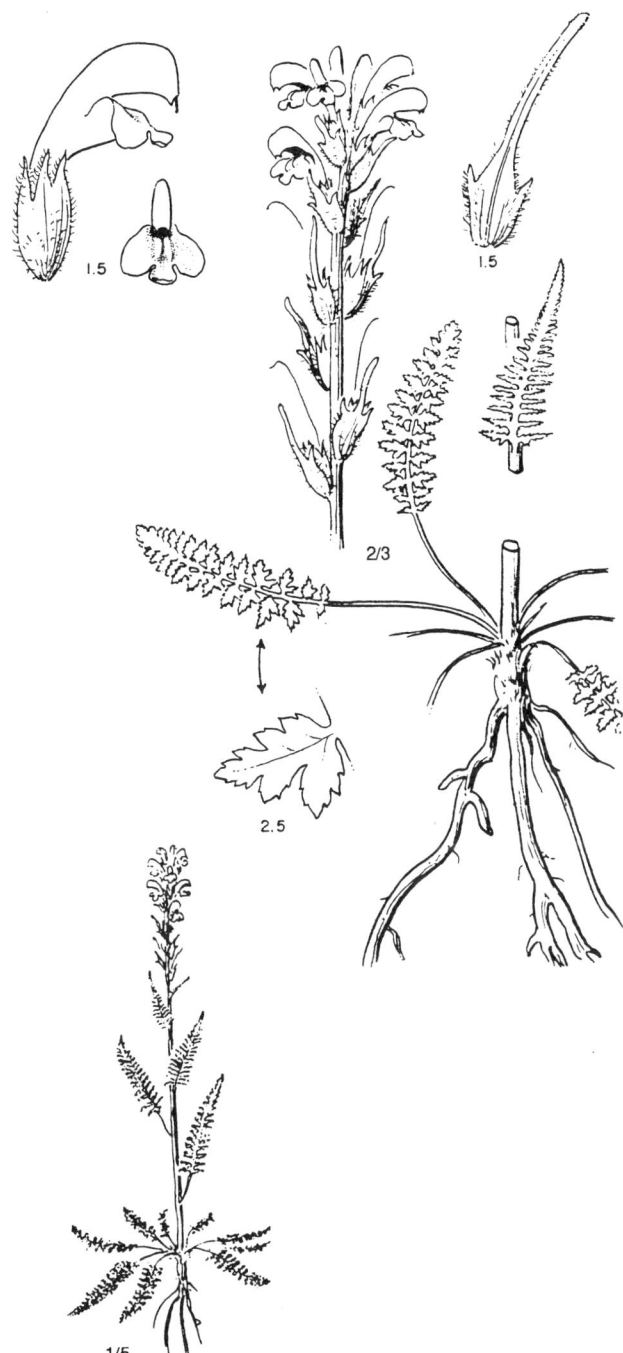

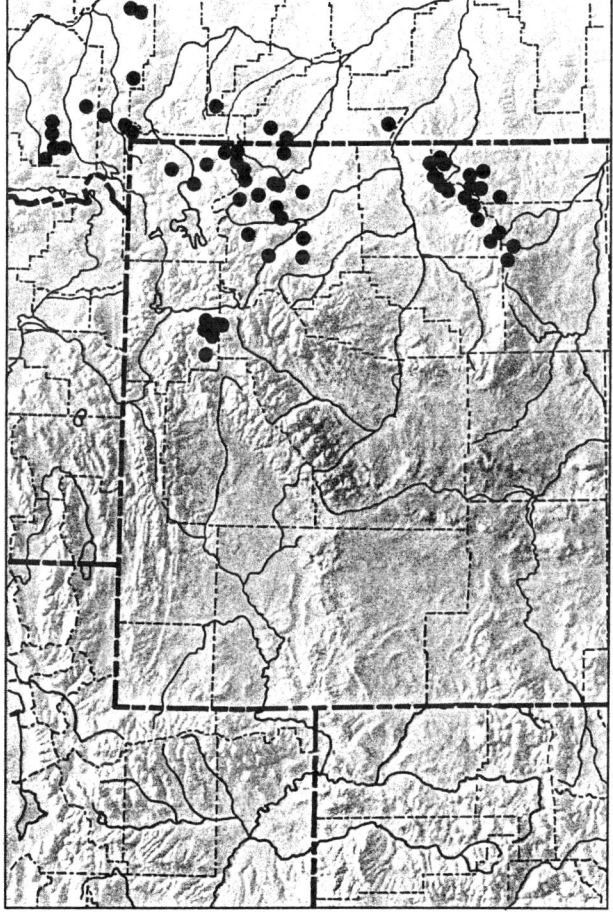

Pedicularis groenlandica Retz.
Fl. Scand. Prodr. 2:145, 1795.

Little Red Elephant

Elephantella groenlandica (Retz.) Rydb.
Pedicularis surrecta Benth. in Hook.

Erect, perennial herb from a stout caudex and fibrous roots. Stems single or clustered, glabrous, often dark reddish purple. Basal leaves long-petiolate, the blades 1–2 times pinnatifid, the pinnae serrate. Cauline leaves pinnatifid, petiolate below, becoming sessile and reduced upward. Inflorescence a dense, elongate spike. Calyx 5-lobed, the lobes short, triangular, entire, subequal. Corolla pink to purple, the galea long-beaked, resembling an elephant head; lower lip small, 3-lobed. Capsule ovoid, mucronate.

Wet meadows, stream banks, and lakeshores throughout most of the Middle Rocky Mountains. Not yet known from above timberline in the Gros Ventre, Salt River, and Wyoming Ranges. Transcontinental in Canada; south in western North America to California and New Mexico.

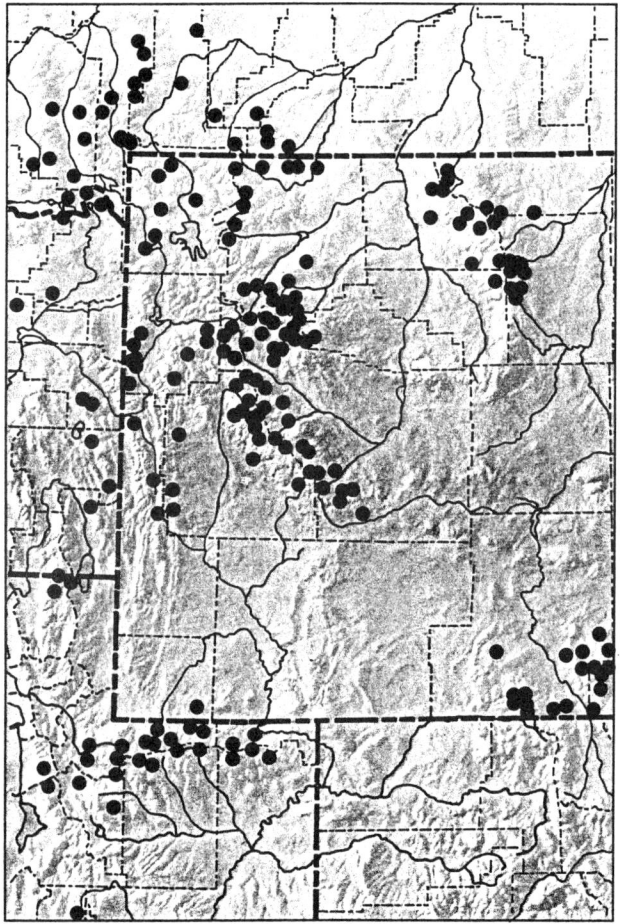

Pedicularis oederi Vahl ex Hornem.
Dansk. Oekon. Pl. 2:580, 1806.

Oeder Lousewort

Pedicularis versicolor Wahlenb.

Short, compact, perennial herb from fibrous roots. Stems single, stout, glabrous at the base, becoming villous-puberulent below the inflorescence. Basal leaves petiolate, 1-pinnatifid, the pinnae somewhat dentate. Cauline leaves petiolate below, becoming sessile and reduced upward. Inflorescence a short, dense spike or spicate raceme, lanate to villous. Calyx 5-lobed, the lobes subequal, lanceolate, sometimes spatulate and toothed. Corolla yellow, the upper lip often purplish or reddish; galea beakless; lower lip 3-lobed. Capsule lance-ovate to lanceolate.

Fellfields of the Beartooth Range. Circumboreal; south in the Rocky Mountains to northwestern Wyoming.

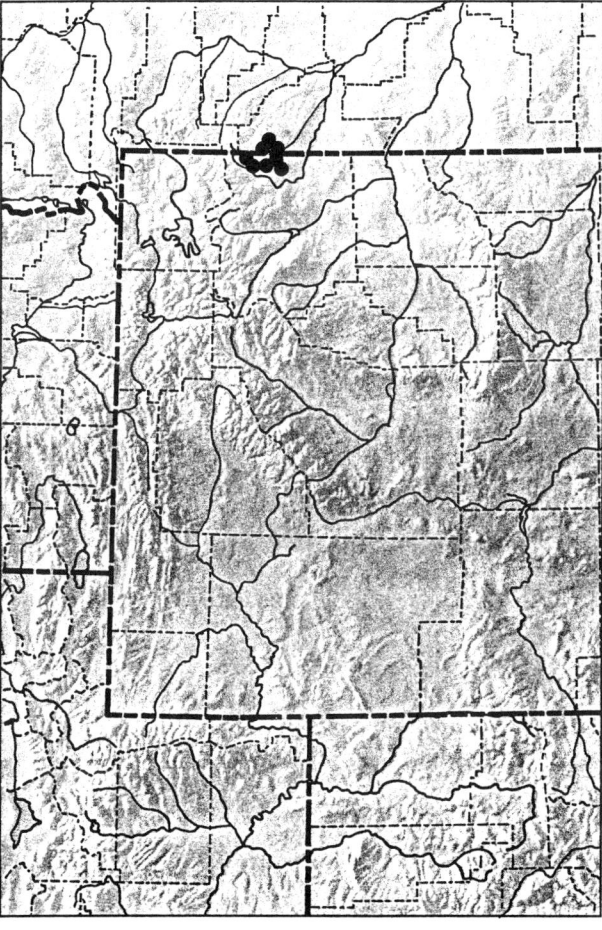

Pedicularis parryi Gray
Am. J. Sci., II, 34:250, 1862.

Parry Lousewort

Pedicularis anaticeps Penn.
Pedicularis hallii Rydb.
Pedicularis mogollonica Greene

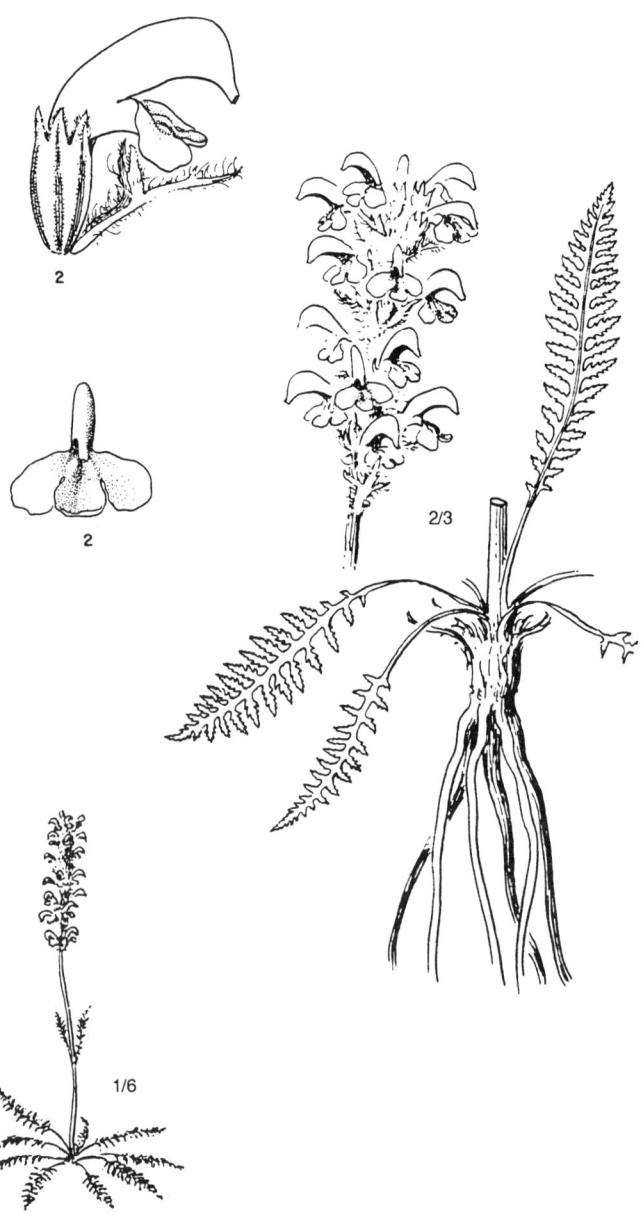

Erect, perennial herb from fibrous roots; stems single, glabrous. Basal leaves petiolate, pinnatifid, the pinnae serrate. Cauline leaves few, reduced, sessile. Inflorescence a dense spike or spicate raceme, villous. Calyx 5-lobed, the lobes subequal, lanceolate, usually with dark medial nerves. Corolla purple or yellow, the galea with a short, straight beak; lower lip 3-lobed.

Meadows, open slopes, and fellfields in most ranges of the Middle Rocky Mountains. Not yet known from above timberline in the Hoback, Salt River, Wasatch, and Wyoming Ranges. Montana and Idaho south to New Mexico and Arizona.

Two varieties occur in our region. Plants of the northern ranges are var. *purpurea* Parry, with purple flowers and villous-ciliate bracts. The typical var. *parryi*, occurring from the Wind River Range southward, has yellow flowers and glabrous bracts. The Wind River Range apparently represents the northern limit of var. *parryi* in our region.

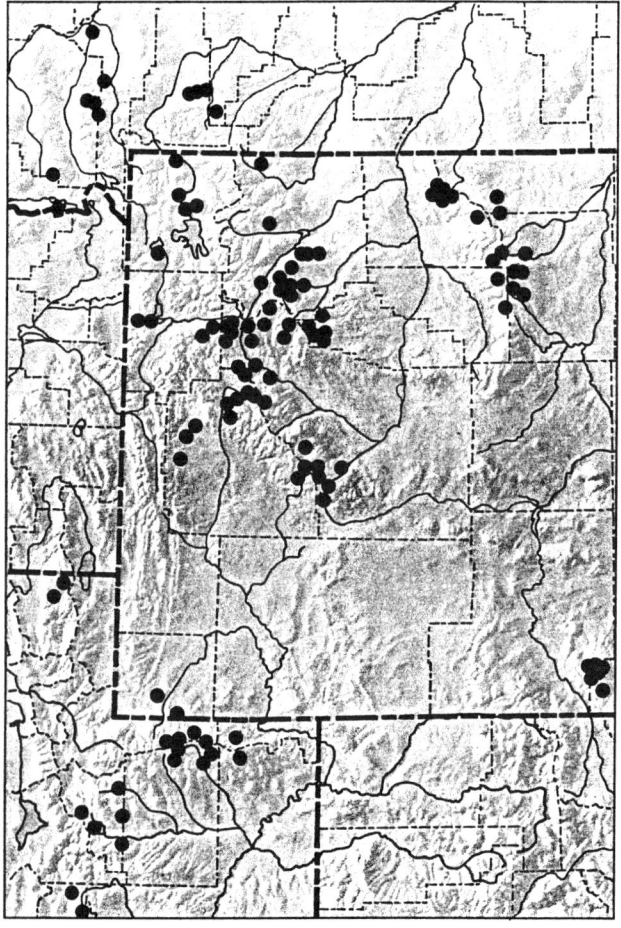

Pedicularis pulchella Penn.
Not. Nat. 95:7, 1942.

Pretty Lousewort

Short, compact, perennial herb from a taproot. Stems single, villous, mostly less than 10 cm tall. Basal leaves petiolate, 1–2 times pinnatifid, the pinnae dentate. Cauline leaves petiolate, reduced upward. Inflorescence a densely compact, villous, spikelike raceme. Calyx 5-lobed, the lobes subequal, lanceolate, entire or toothed. Corolla purple, the galea beakless and minutely bidentate.

Fellfields, scree, and talus of the Absaroka, Beartooth, Big Horn, and Gros Ventre Ranges. Endemic to southwestern Montana and northwestern Wyoming.

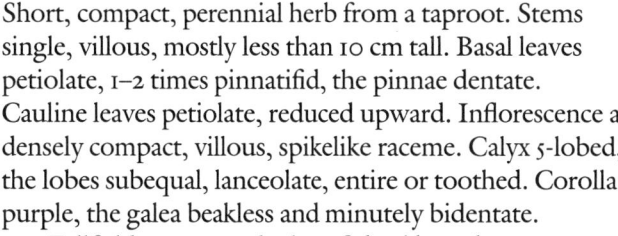

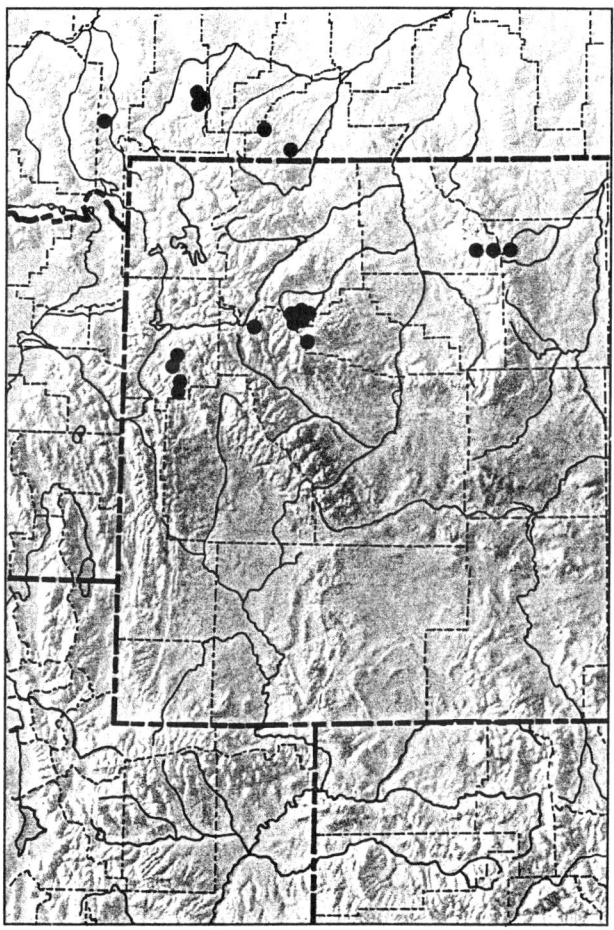

Pedicularis racemosa Dougl. ex Berth. in Hook. *Fl. Bor.-Am.* 2:108, 1838.

Curled Lousewort

Erect, perennial herb from a woody caudex and fibrous roots. Stems clustered, glabrous or nearly so. Leaves mostly cauline, lanceolate to elliptic, the margins crenate or serrate, sometimes doubly serrate. Inflorescence an elongate, sparsely flowered raceme; lower bracts leaflike, becoming reduced upward. Calyx 2-lobed, more deeply cleft below than above, the lobes acuminate. Corolla white or ochroleucous, the galea arched downward into a slender beak; lower lip large, prominent, 3-lobed.

Timberline meadows and willow thickets in the Absaroka, Big Horn, Medicine Bow, Uinta, and Wind River Ranges. British Columbia and Alberta south to California and New Mexico.

Middle Rocky Mountain plants with white or ochroleucous flowers are var. *alba* (Penn.) Cronq.

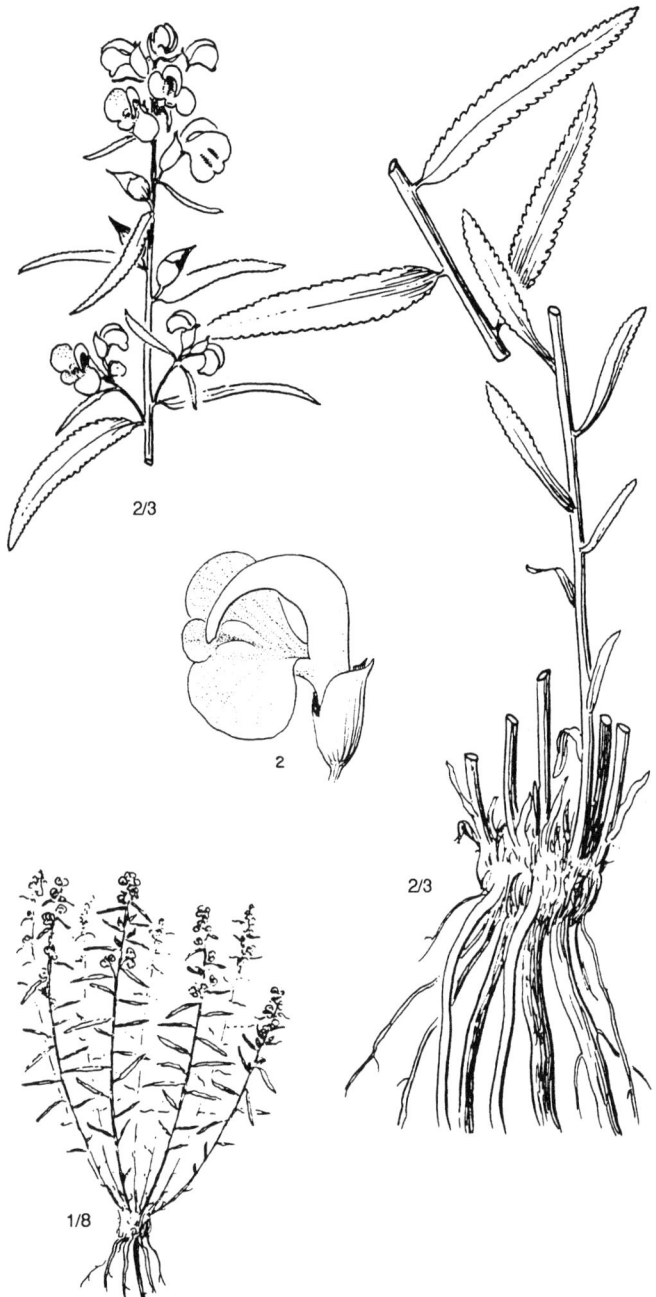
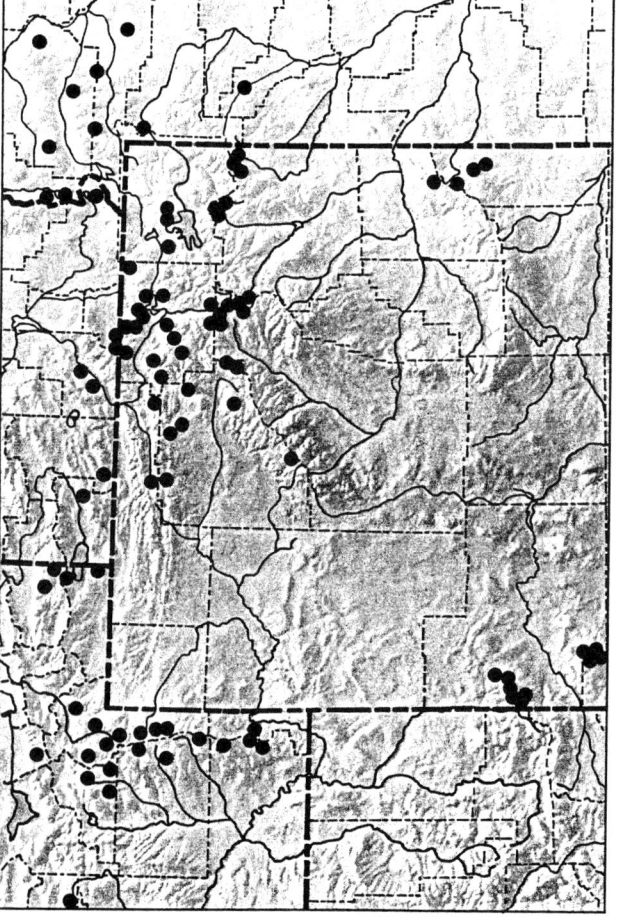

Penstemon Mitch.
Beardtongue, Penstemon

Perennial herbs, sometimes woody at the base, with erect or prostrate stems, the leaves opposite, entire or toothed. Flowers blue to purple in ours, in terminal racemes or panicles. Calyx 5-cleft nearly to the base. Corolla bilabiate, the upper lip 2-lobed, the lower lip 3-lobed. Stamens 4, didynamous, with arched filaments and sagittate anthers, the sterile stamen usually consisting of a spatulate filament, often bearded at the apex. Style elongate, the stigma capitate. Capsule septicidal.

KEY TO THE SPECIES OF *PENSTEMON*

1. Plants woody, at least at the base; anthers densely woolly with tangled hairs
    2. Leaves all cauline, basal tufts of leaves lacking; sterile stamen usually glabrous .... *P. montanus*
    2. Leaves basally clustered, the cauline ones reduced; sterile stamen bearded .... *P. fruticosus*
1. Plants herbaceous throughout; anthers glabrous to puberulent
    3. Inflorescence glandular-pubescent
        4. Anther sacs partially dehiscent, the connective indehiscent .... *P. uintahensis*
        4. Anther sacs completely dehiscent, opening across the connective
            5. Ovary and capsule glandular-puberulent, at least near the summit; corolla 20–35 mm long; calyx more than 7 mm long; staminodium exerted .... *P. whippleanus*
            5. Ovary and capsule glabrous; corolla 8–20 mm long; calyx less than 7 mm long; staminodium included
                6. Leaves glabrous; sepals with erose, scarious margins; anther sacs mostly more than 0.8 mm long .... *P. attenuatus*
                6. Leaves puberulent; sepals with entire margins, scarious or not; anther sacs mostly less than 0.8 mm long .... *P. humilis*
    3. Inflorescence glabrous, sometimes puberulent but not glandular
        7. Corollas more than 15 mm long, medium blue; anthers pubescent .... *P. cyananthus*
        7. Corollas less than 15 mm long, dark purplish blue; anthers glabrous
            8. Corollas 7–10 (–12) mm long, throat 2–3 mm wide; anther sacs 0.3–0.7 mm long .... *P. procerus*
            8. Corollas 11–15 mm long, throat 3–5 mm wide; anther sacs 0.6–1 mm long .... *P. rydbergii*

Penstemon attenuatus Dougl. ex Lindl.
Bot. Reg. 15:Pl. 1295, 1830.

Thinstem Penstemon

Penstemon assurgens Keck
Penstemon cephalanthus Penn.
Penstemon militaris Greene
Penstemon nelsonae Keck & J.W. Thomps.
Penstemon palustris Penn.
Penstemon propinquus Greene
Penstemon pseudohumilis Rydb.
Penstemon pseudoprocerus Rydb.
Penstemon veronicaefolius Greene

Erect, perennial herb from a branched, woody caudex. Stems slender, solitary or a few clumped together, glabrous to puberulent, becoming glandular-puberulent in the inflorescence. Basal leaves entire, petiolate, oblanceolate, glabrous, usually forming tufts; cauline leaves sessile, lanceolate, reduced upward. Inflorescence bracteate, of a few separate, verticillate clusters. Calyx glandular-pubescent, the lobes with erose, scarious margins, acute to somewhat acuminate. Corolla dark blue, or occasionally yellowish or whitish, glandular-puberulent, the lower lip bearded. Anthers glabrous in ours, the pollen sacs explanate and dehiscing completely in ours; sterile stamen bearded at the tip.

Rocky slopes and fellfields in the Beartooth and northern Wind River Ranges. Washington south and east to Oregon, Idaho, and Wyoming.

Middle Rocky Mountain plants with anther sacs completely dehiscent, the sepals erose and prominently scarious are var. *pseudoprocerus* (Rydb.) Cronq.

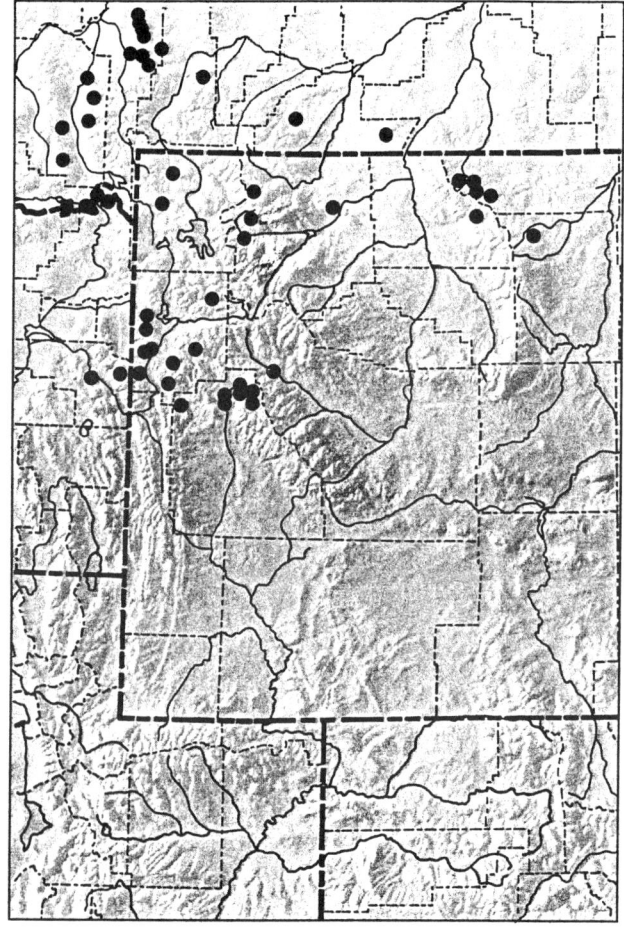

Penstemon cyananthus Hook.
Curtis' Bot. Mag. 75:Pl. 4464, 1849.

Wasatch Penstemon

Penstemon compactus Crosswhite
Penstemon holmgrenii S. Clark
Penstemon longiflorus (Penn.) S. Clark

Erect to spreading, perennial herb from a branched, woody caudex. Stems stout, erect to decumbent, glabrous to puberulent. Leaves glabrous to puberulent, glaucescent; basal leaves petiolate, broadly lanceolate to oblanceolate, acute to obtuse; cauline leaves ovate to subcordate, entire, reduced upward. Inflorescence of 2–9 dense, verticillate clusters. Calyx glabrous to glandular, the lobes ovate to lanceolate, acute to acuminate, with erose, scarious margins. Corolla medium blue, glabrous. Anthers short-pubescent, dehiscing about four-fifths of their length; sterile stamen bearded with yellow hairs about half its length.

Scree slopes and other rocky places above timberline in the Wasatch Range. Endemic to Idaho, Utah, and Wyoming.

Two varieties occur in the Wasatch Range: var. *cyananthus* is more than 30 cm tall and has ascending to erect stems and a glabrous calyx; var. *compactus* (Crosswhite) Neese is less than 25 cm tall, with decumbent or spreading stems and a glandular calyx.

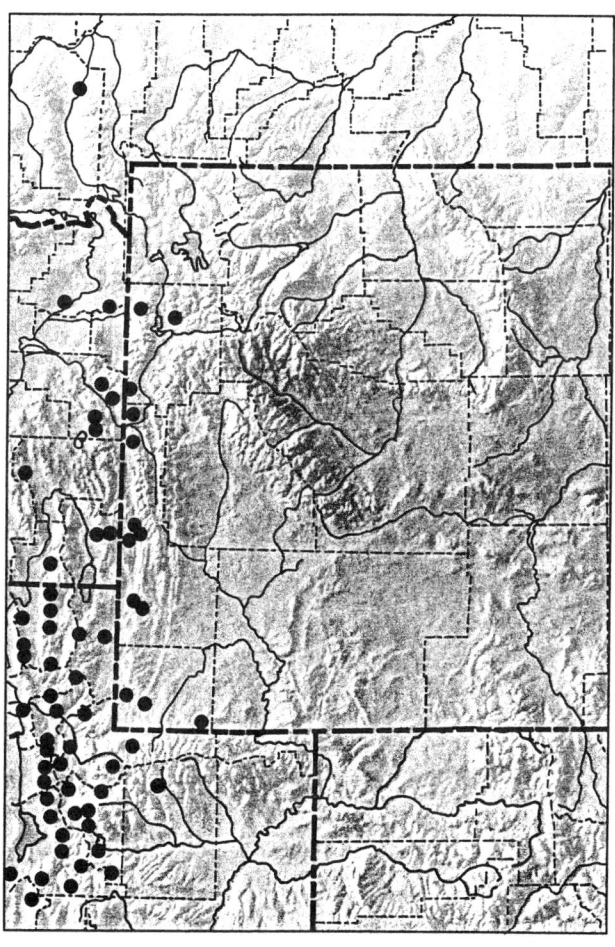

Penstemon fruticosus (Pursh) Greene
Pittonia 2:239, 1892.

Shrubby Penstemon

Gerardia fruticosus Pursh
Dasanthera fruticosa (Pursh) Raf.
Penstemon adamsianus T.J. Howell
Penstemon crassifolius Lindl.
Penstemon douglasii Hook.
Penstemon lewisii Benth. in DC.
Penstemon scouleri Lindl.

Shrub or subshrub forming dense clumps from a branched, woody caudex. Stems decumbent, glabrous to finely puberulent below, becoming glandular above. Leaves sessile or short-petiolate, the blades linear-lanceolate to oblanceolate, entire or serrate, reduced upward on flowering shoots. Inflorescence a short, bracteate raceme. Calyx glandular, the lobes lanceolate, acuminate. Corolla purple or blue, villous on the ventral ridges. Anthers woolly; sterile stamen yellow-bearded near the tip.

Scree, talus, and other rocky habitats in the lower alpine zone of the Absaroka and Beartooth Ranges. British Columbia and Alberta south to Oregon, Idaho, Montana, and northwestern Wyoming.

Alpine plants of the northern portion of the Middle Rocky Mountains are the typical var. *fruticosus*.

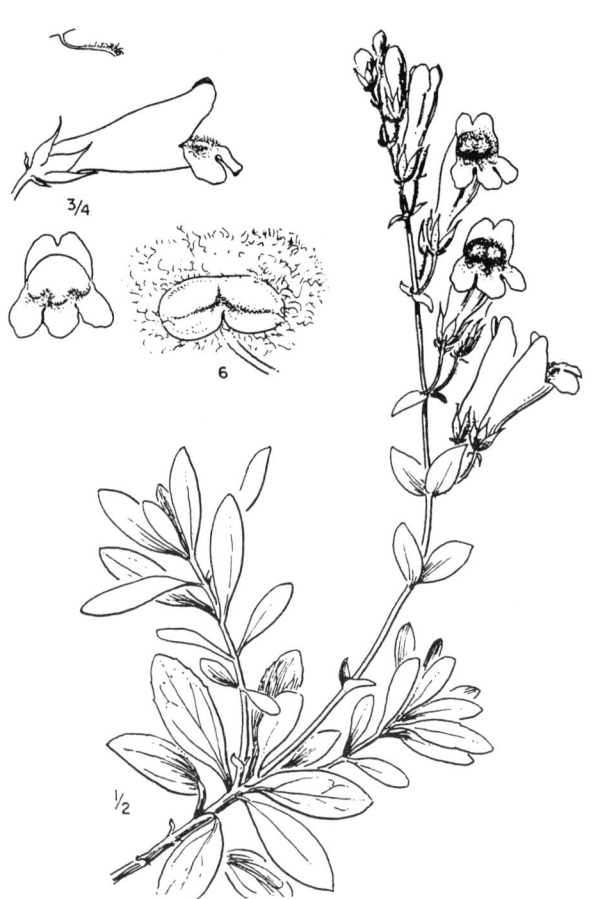

Penstemon humilis Nutt. ex Gray
Proc. Am. Acad. 6:69, 1862.

Low Penstemon

Penstemon brevis A. Nels.
Penstemon cinereus Piper
Penstemon collinus A. Nels.
Penstemon decurvus Penn. ex Crosswhite

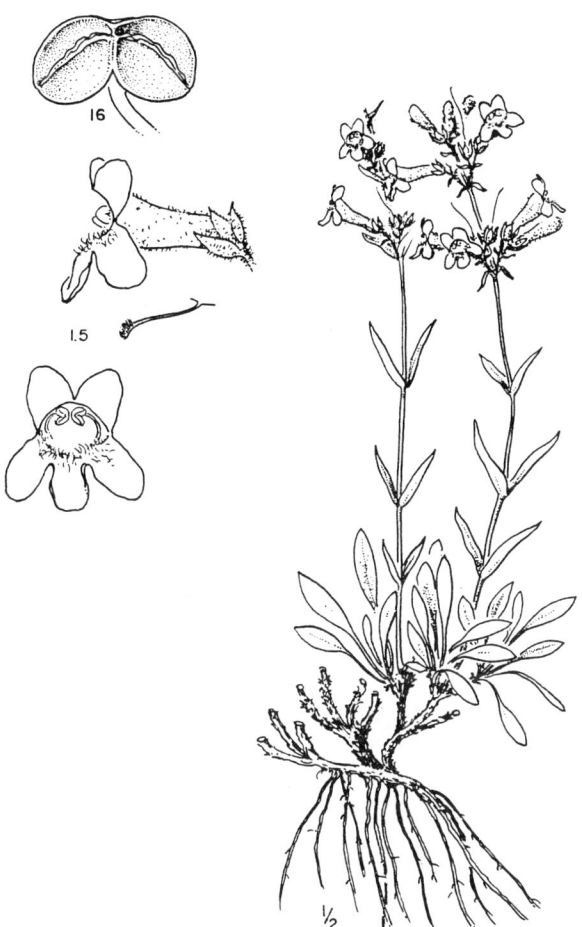

Slender, erect, perennial herb from a branched, somewhat woody caudex. Stems numerous and forming clumps, puberulent, becoming glandular-pubescent above. Basal leaves entire, petiolate, oblanceolate, puberulent to glabrous, usually somewhat tufted; cauline leaves sessile, lanceolate to linear-lanceolate, reduced upward. Inflorescence bracteate, of a few separate, verticillate clusters. Calyx glandular-pubescent, the lobes lanceolate to ovate, acute or short-acuminate, somewhat scarious-margined. Corolla blue to purplish blue, glandular-pubescent, the lower lip bearded. Anthers glabrous, the pollen sacs explanate and dehiscing completely; sterile stamen bearded at the tip with yellowish hairs.

Exposed rocky slopes of the Absaroka, Salt River, Uinta, Wasatch, Wind River, and Wyoming Ranges. Washington south and east to California, Nevada, and Colorado.

Alpine plants of the Middle Rocky Mountains are distinguished as follows: var. *humilis* has puberulent leaves and a corolla usually 12 mm or more long; var. *brevifolius* Gray has glabrous or glabrate leaves and a corolla usually less than 12 mm long.

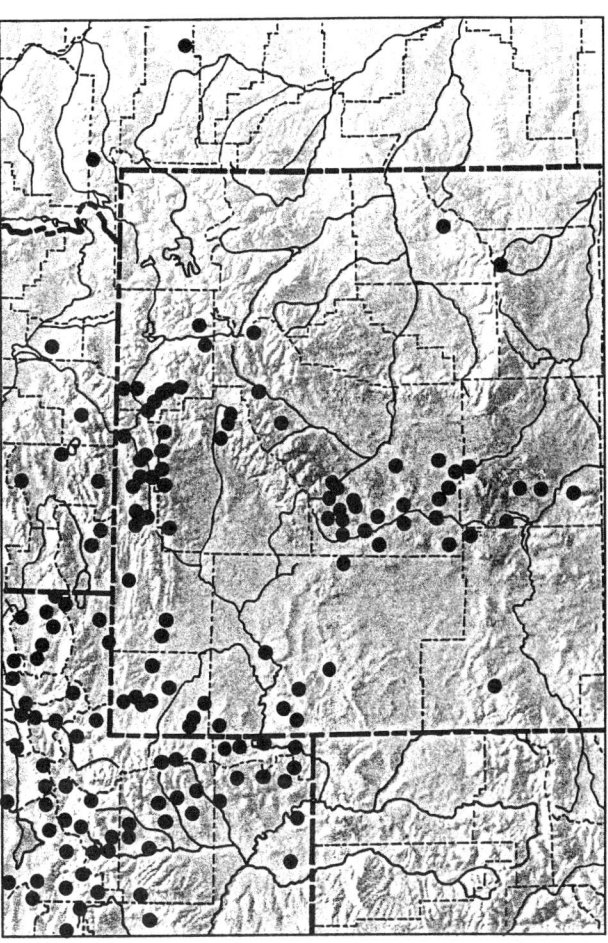

Penstemon montanus Greene
Pittonia 2:240, 1892.

Mountain Penstemon

Penstemon woodsii A. Nels.

Erect to somewhat sprawling, perennial herb or subshrub from a branched, woody caudex. Stems several, usually decumbent, glandular-pubescent, at least above. Leaves sessile, the blades lanceolate, oblanceolate, or ovate, glandular, entire to serrate. Inflorescence a short, compact, few-flowered, bracteate raceme. Calyx with lanceolate, entire lobes. Corolla purple to lavender, the lower lip bearded, at least in the throat. Anthers woolly-villous, dehiscing completely; sterile stamen glabrous to pubescent for most of its length.

Talus, scree, and rocky places in nearly all of the alpine ranges of the Middle Rocky Mountains. Not yet known from above timberline in the Medicine Bow Range. Idaho and Montana south to Utah and Wyoming.

Middle Rocky Mountain alpine plants are the typical var. *montanus*.

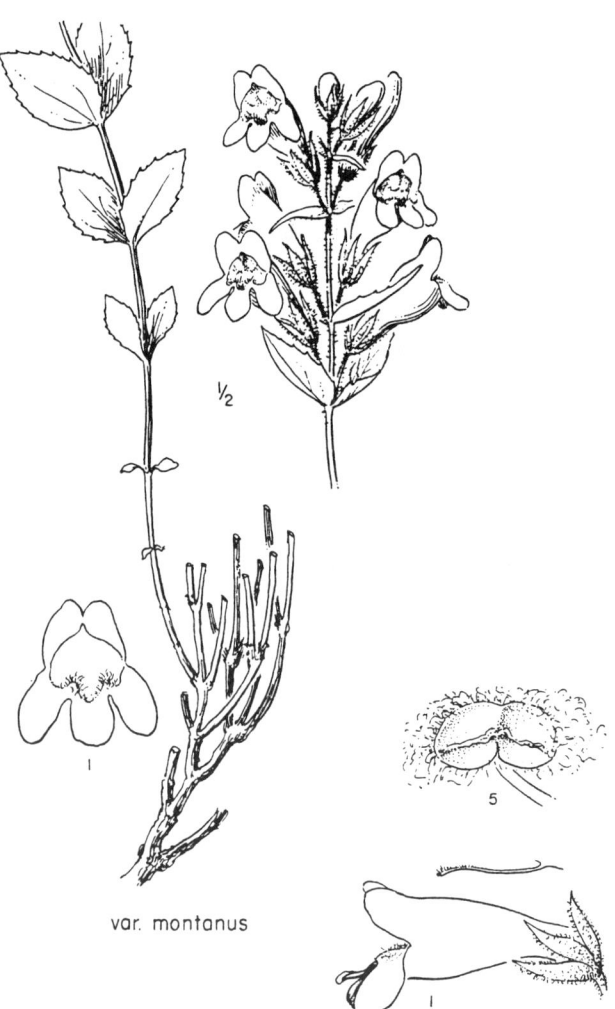

var. montanus

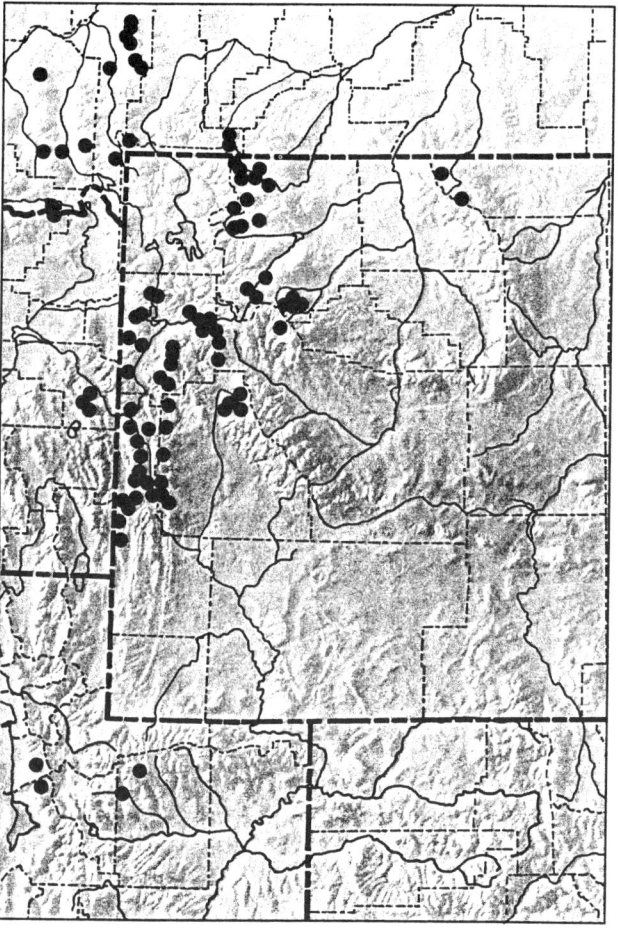

Penstemon procerus Dougl. ex Grah.
Edinb. New Phil. J. 7:348, 1829.

Smallflower Penstemon

Lepteiris parviflora Raf.
Penstemon brachyanthus Penn.
Penstemon cacuminis Penn.
Penstemon formosus A. Nels.
Penstemon micranthus Nutt.
Penstemon modestus Greene
Penstemon pulchellus Greene
Penstemon tolmiei Hook.

Tufted, perennial herb from a woody rhizome. Stems erect or decumbent, glabrous to puberulent. Basal leaves petiolate, oblanceolate, glabrous; cauline leaves sessile, the blades oblanceolate or lanceolate, entire, becoming reduced upward. Inflorescence of interrupted verticillate clusters. Calyx lobes elliptic to obovate, scarious-margined, entire to erose, cuspidate or acuminate. Corolla deep purplish blue, the lower lip bearded within. Anthers glabrous, dehiscing completely; sterile stamen with a few yellowish hairs at the tip.

Common in meadows, fellfields, and on rocky slopes in the Absaroka, Beartooth, Big Horn, Gros Ventre, Medicine Bow, Teton, Uinta, and Wind River Ranges. Alaska and Yukon Territory south to California, Utah, and Colorado.

Middle Rocky Mountain plants are var. *procerus*.

var. procerus

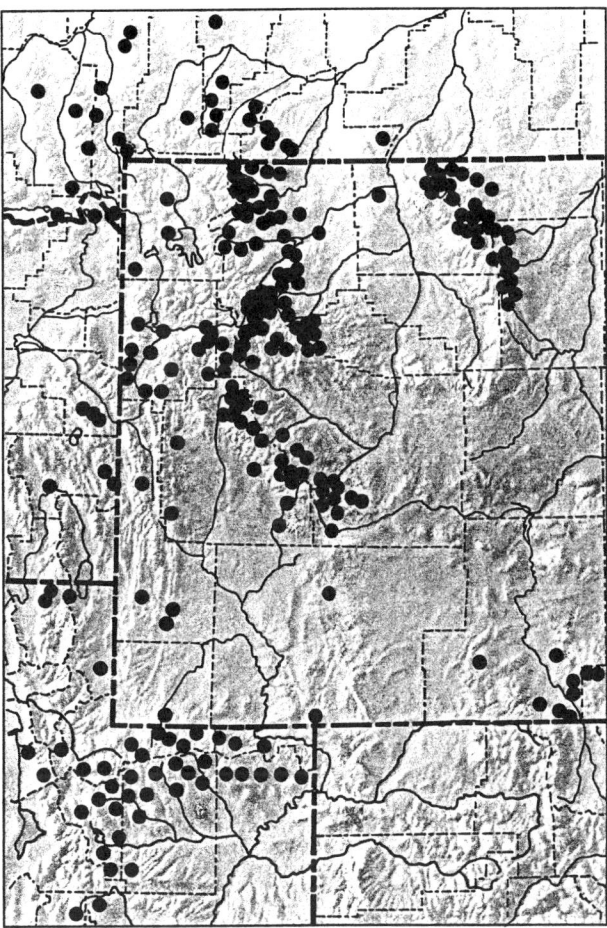

Penstemon rydbergii A. Nels.
Bull. Torr. Club 25:281, 1898.

Rydberg Penstemon

Penstemon aggregatus Penn.
Penstemon hesperius M.E. Peck
Penstemon oreocharis Greene
Penstemon vaseyanus Greene

Erect, perennial herb, sometimes tufted, from a woody caudex. Stems often decumbent at the base, glabrous to puberulent, at least above. Basal leaves petiolate, oblanceolate; cauline leaves sessile, oblanceolate to lanceolate, entire, reduced upward. Inflorescence of interrupted verticillate clusters. Calyx lobes ovate to lanceolate, somewhat acuminate, the margins scarious, erose. Corolla purplish blue, the lower lip bearded within. Anthers glabrous, dehiscing completely; sterile stamen yellow-bearded at the tip.

Meadows near timberline in lower alpine zone of the Big Horn, Teton, and Uinta Ranges. Washington south to California and east to Montana, Wyoming, and Colorado.

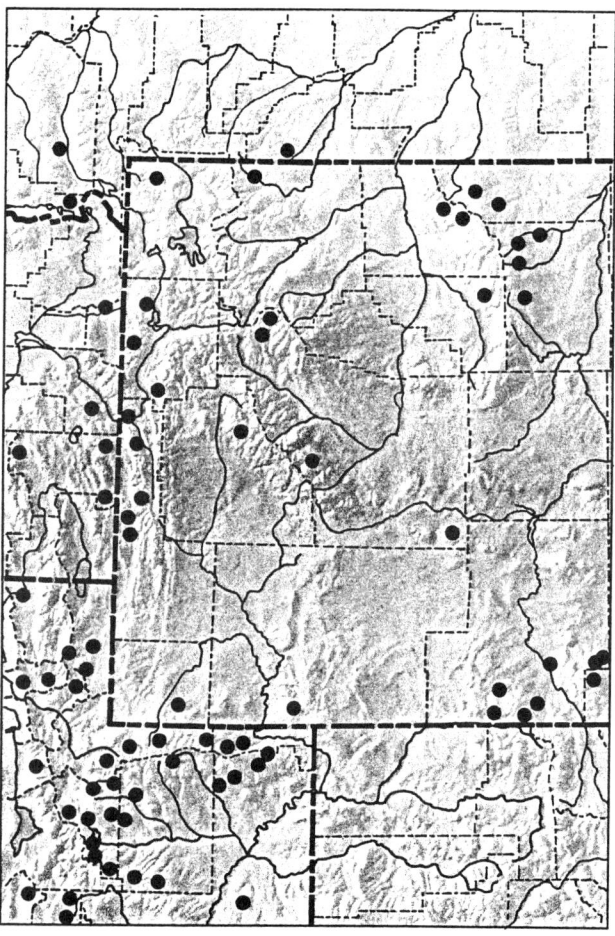

Penstemon uintahensis Penn.
Contr. U.S. Natl. Herb. 20:350, 1920.

Uinta Penstemon

Erect, perennial herb, less than 20 cm tall, from a short, branched caudex. Stems slender, single to several in a clump, glabrous, at least below. Basal leaves petiolate, entire, oblanceolate to somewhat linear; cauline leaves sessile, entire, obtuse. Inflorescence glandular-puberulent, of interrupted verticillate clusters. Calyx glandular-puberulent, the lobes ovate, acute to acuminate; the margins scarious, erose. Corolla glandular-puberulent, blue or bluish violet, the lower lip spreading. Anthers sparsely pubescent, dehiscing incompletely; sterile stamen yellow-bearded at the tip.

Krummholz margins, moraines, and other rocky places above timberline in the Uinta Range. Endemic to this area of northern Utah.

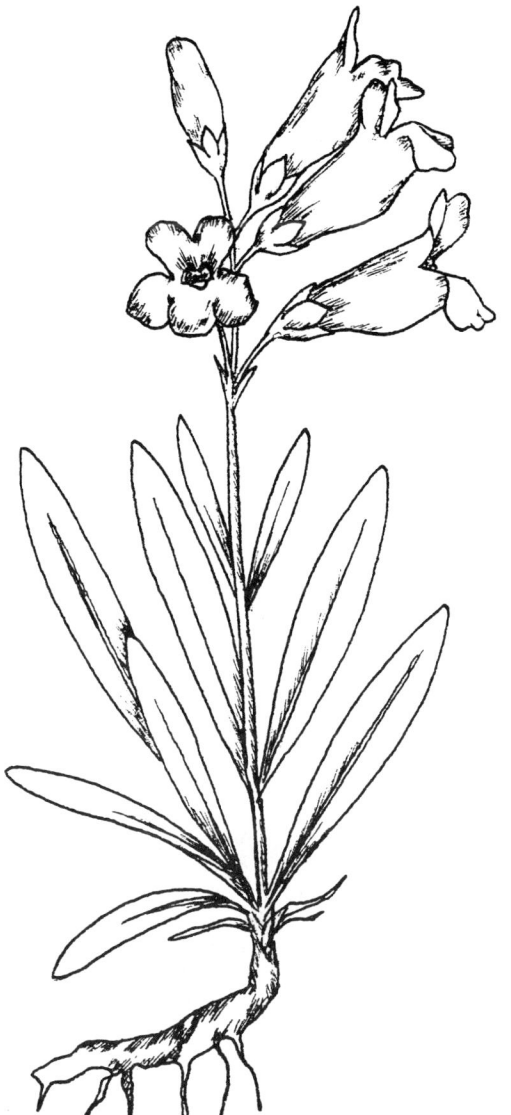

Penstemon whippleanus Gray
Proc. Am. Acad. 6:73, 1862.

Whipple Penstemon

Erect, usually clumped, perennial herb from a branched, somewhat woody caudex. Stems glabrous below, becoming glandular-pubescent above. Basal leaves petiolate, elliptic to ovate; cauline leaves sessile, oblong to lanceolate, entire. Inflorescence of interrupted verticillate clusters, often shortened and nearly capitate. Calyx lobes lanceolate, glandular-pubescent, attenuate, lacking scarious margins. Corolla glandular-puberulent, deep purple to creamy white, the lower lip bearded. Anthers glabrous, dehiscing completely; sterile stamen glabrous or bearded at the tip.

Dry meadows, ledges, and rocky slopes of the lower alpine zone in nearly all ranges of the Middle Rocky Mountains. Not known from above timberline in the Beartooth or Big Horn Ranges. Montana south to Arizona and New Mexico.

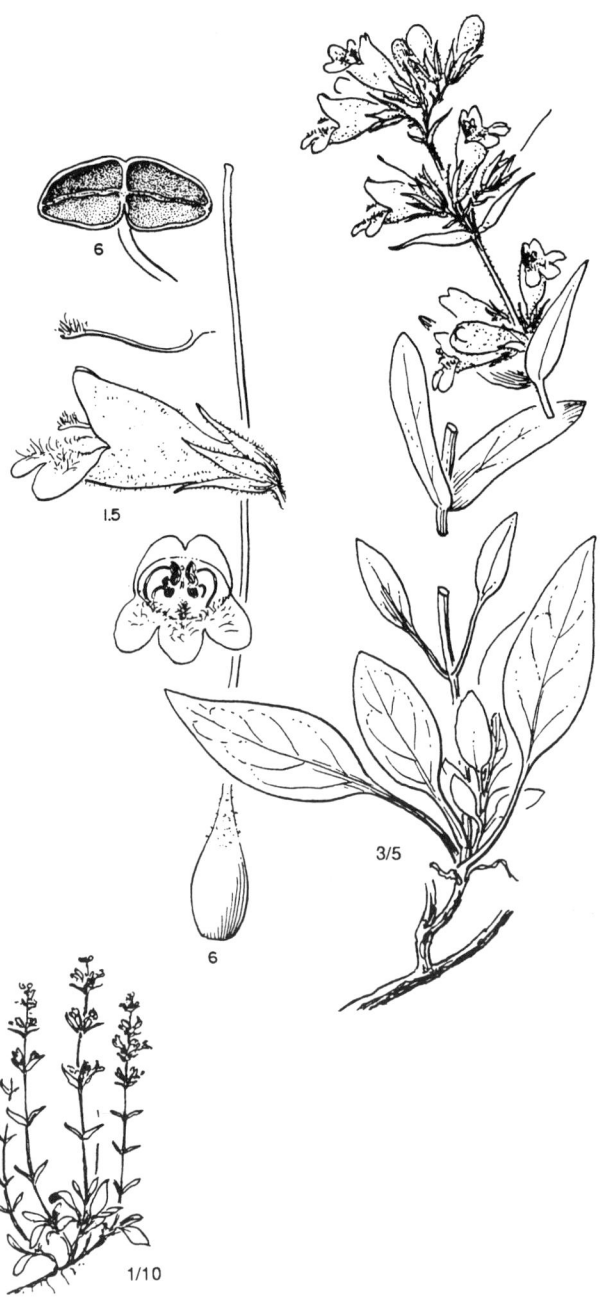

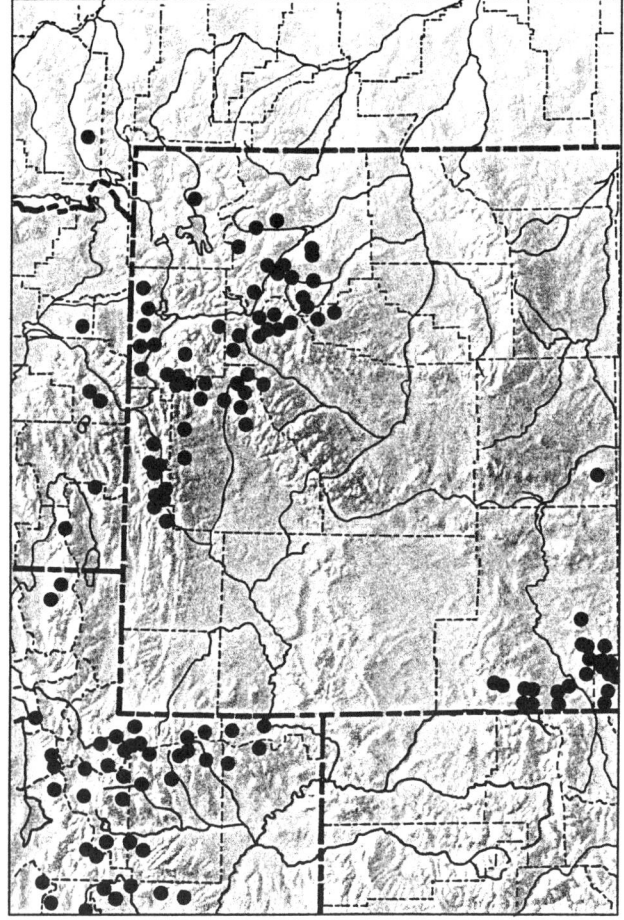

Synthyris Benth.
Kittentails

Perennial herbs from fibrous roots. Leaves mostly basal and pinnately compound in ours; cauline leaves alternate, reduced. Flowers irregular, in narrow, terminal racemes. Calyx 4-parted, the segments nearly distinct; corolla blue, purple, pink, or white, 4-lobed, campanulate to nearly rotate. Stamens 2, exerted. Pistil 1, the style filiform; stigma capitate. Fruit a compressed, loculicidal capsule.

Synthyris pinnatifida S. Wats.
Bot. King Exped., 227, 1871.

Featherleaf Kittentails

Synthyris cymopteroides Penn.
Synthyris hendersonii Penn.
Synthyris lanuginosa (Piper) Penn. & J.W. Thomps.
Synthyris paysonii Penn. & Williams
Wulfenia pinnatifida (S. Wats.) Greene

Low, perennial herb from a short rootstock covered by old leaf bases. Roots long, fibrous. Basal leaves pinnately dissected; leaflet margins deeply incised; cauline leaves alternate, lanceolate to ovate, toothed. Flowers in a dense, terminal, bracteate raceme. Calyx 4-parted; corolla 4-parted, irregular, violet to purple, the lobes entire. Stamens 2, exerted. Pistil 1, the style exerted. Capsule oval.

Tundra, rocky slopes, and fellfields of the Gros Ventre, Hoback, Salt River, Uinta, Wasatch, and Wyoming Ranges. Washington to Montana, south to Wyoming, Idaho, and Utah.

Our plants with glabrous capsules, obovate, obtuse floral bracts, and pinnatifid to pinnately compound leaves are var. *pinnatifida*. Specimens from the extreme southern Wasatch Range in the vicinity of Mount Nebo with simple, laciniate leaves are var. *laciniata* Gray.

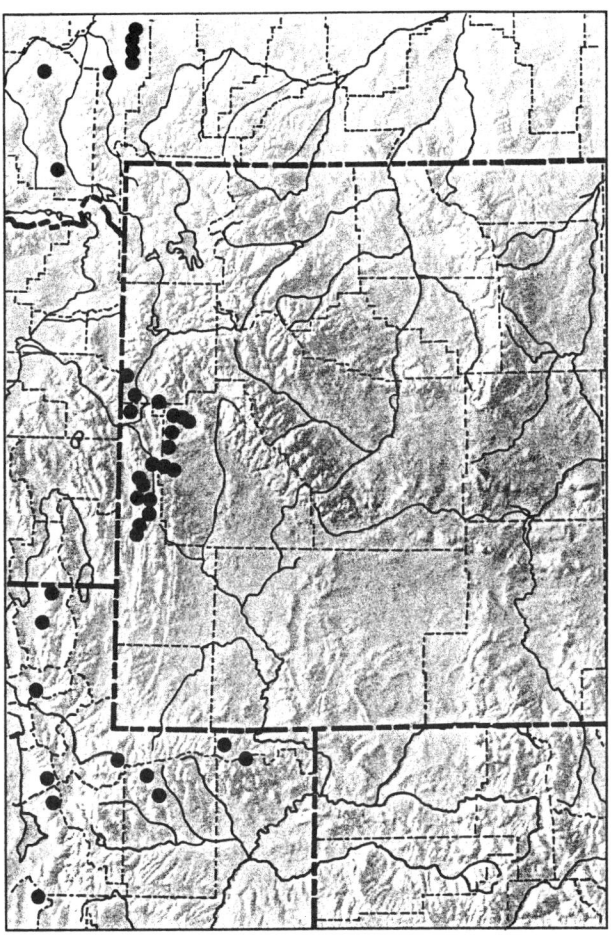

Veronica L.
Speedwell

Annual or perennial herbs with erect stems and sessile or short-petioled, entire or toothed leaves that are opposite, or sometimes alternate above. Flowers small, blue, in terminal or axillary spikes or racemes, or solitary in the upper axils. Calyx of 4 nearly distinct segments. Corolla rotate, 4-lobed, only slightly irregular. Stamens 2, somewhat exserted. Style slender, the stigma capitate. Capsule loculicidal, usually notched or 2-lobed at the apex.

KEY TO THE SPECIES OF *VERONICA*

1. Plants annual, with fibrous roots; flowers white to whitish blue — *V. peregrina*
1. Plants perennial, with slender rhizomes; flowers dark blue or blue-violet — *V. nutans*

Veronica nutans Bong.
Mem. Acad. Petersb. 2:157, 1833.

Alpine Speedwell

Slender, perennial herb from a rhizome. Stems erect or decumbent, glandular-villous, sometimes glabrate at the base. Leaves sessile, cauline, the blades elliptic, lanceolate, or ovate, glabrous to sparsely villous, entire to somewhat crenate, reduced upward, the tips acute to obtuse. Inflorescence a short, crowded, terminal raceme. Calyx lobes glandular-villous. Corolla dark blue or blue-violet, glabrous within, the dorsal lobe wider and more obtuse than the other three. Capsule obovate, emarginate, glandular-pubescent, longer than wide.

Wet meadows, stream banks, and lakeshores in most alpine ranges of the Middle Rocky Mountains. Not yet known from above timberline in the Gros Ventre, Hoback, Salt River, and Wyoming Ranges. Alaska south to California and New Mexico.

Veronica serpyllifolia L. var. *humifusa* (Dickson) Vahl (Thymeleaf Speedwell) has been collected near timberline in the Absaroka and Wind River Ranges and might be expected in the lower alpine zone of any of the Middle Rocky Mountain ranges. It differs from *V. nutans* in having fruits wider than long and stamens with filaments more than 2 mm long. *Veronica nutans* is widely known in the Rocky Mountains as *V. wormskjoldii* R. & S. Löve (1969) argues that since *V. wormskjoldii* of eastern North America is tetraploid, and western plants are both disjunct and diploid, $2n=18$, they are better regarded as *V. nutans* Bong.

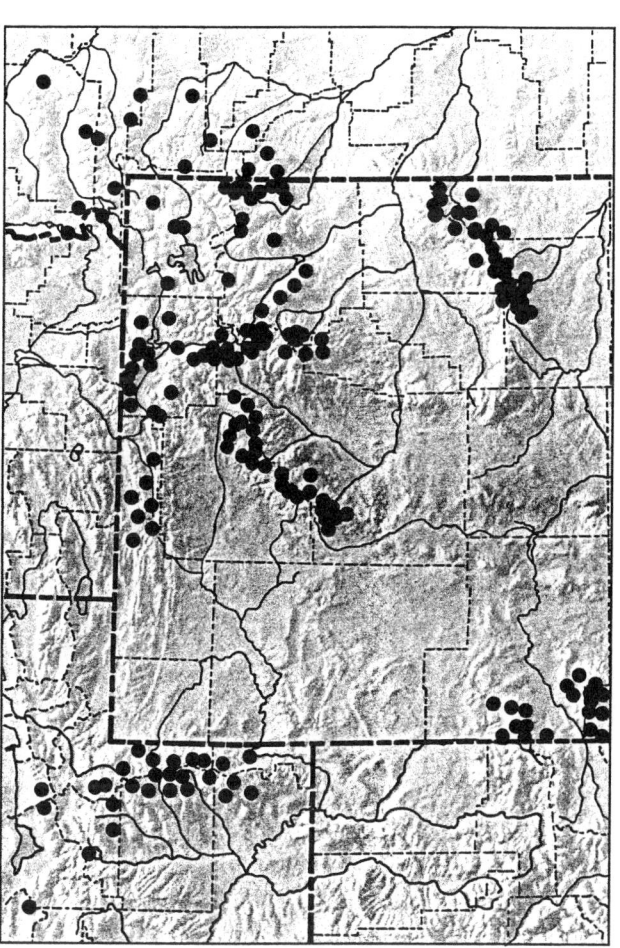

Veronica peregrina L.
Sp. Pl., 14, 1753.

Purslane Speedwell

Veronica sherwoodii M.E. Peck
Veronica xalapensis H.B.K.

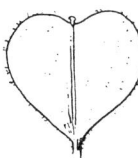

var. xalapensis

Slender annual from fibrous roots. Stems 1 to several, erect or decumbent-ascending, simple or branched, glandular-pubescent. Leaves opposite, at least below, sessile or short-petiolate, cauline, the blades oblanceolate or oblong, glabrous to glandular, entire to shallowly and irregularly toothed, the tips mostly obtuse. Inflorescence an elongate, terminal, bracteate raceme, the bracts alternate, linear-oblong and becoming reduced upward. Calyx lobes lanceolate, glandular. Corolla white or whitish blue, small and inconspicuous, shorter than to about equaling the calyx in length. Capsule obcordate, compressed, with a shallow notch, finely glandular-pubescent.

Wet meadows and stream banks in the lower alpine zone near timberline in the Wasatch Range. Circumboreal; widespread in North America.

Glandular-pubescent plants of western North America are var. *xalapensis* (H.B.K.) St. John and Warren.

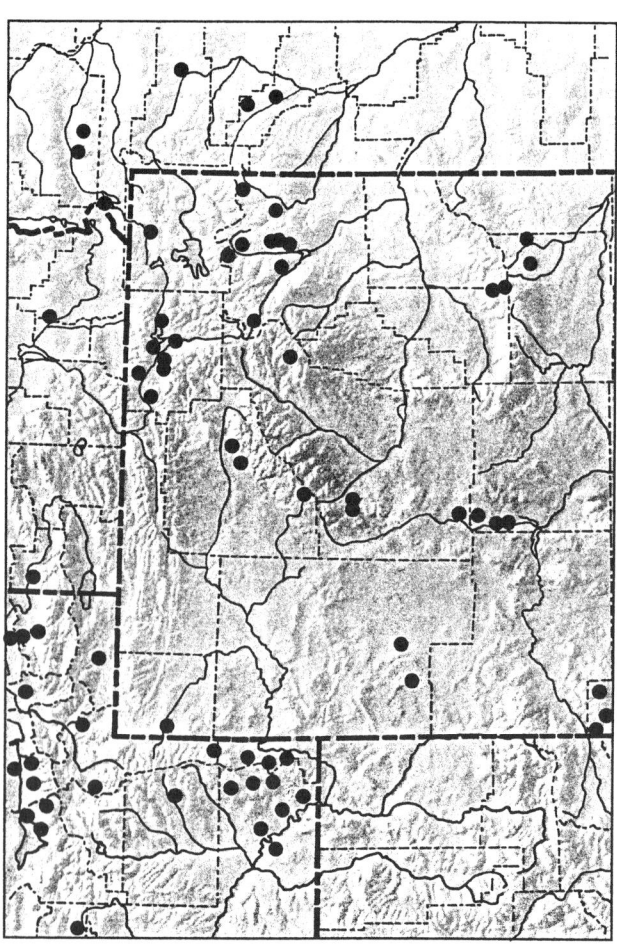

Selaginellaceae
Spikemoss Family

Sporophytes terrestrial, evergreen, mosslike, low and creeping, forming dense tufts or mats. Leaves sessile, imbricate, linear to ovate, 4 mm long or less. Sporophylls larger than the sterile leaves, 4-ranked, aggregated into terminal strobili. Sporangia of 2 types: the megasporangia, producing 1–4 large, yellowish to orange megaspores; and the microsporangia, producing numerous minute, pale-colored microspores. Gametophytes very small, largely retained within the spore walls.

Selaginella Beauv.
Spikemoss

Small, caespitose, perennial herbs; stems erect to prostrate, branched; roots fibrous, adventitious, often on short, down-growing rhizophores. Leaves numerous, crowded, 1-nerved, scalelike, spirally arranged and similar, or 4-ranked and dimorphic. Sporophylls aggregated in terminal strobili; sporangia solitary, axillary, the microsporangia above the megasporangia; microspores numerous; megaspores 1–4.

KEY TO THE SPECIES OF *SELAGINELLA*

1. Leaves tapered, the bristle white, 1–2 mm long — *S. densa*
1. Leaves acute, the bristle greenish yellow, less than 0.5 mm long — *S. watsonii*

Selaginella densa Rydb.
Mem. N.Y. Bot. Gard. 1:7, 1900.

Rydberg Spikemoss

Selaginella engelmannii Hieron.
Selaginella scopulorum Maxon
Selaginella standleyi Maxon

Densely caespitose perennial with prostrate, short-branched stems forming cushionlike mats. Leaves imbricate, appressed, linear-lanceolate to subulate, with a medial dorsal groove, the apex acute, abruptly narrowed to a whitish seta. Strobili quadrangular, erect, the sporophylls triangular-ovate, keeled; setae shorter than those of the sterile leaves. Microsporangia above the larger megasporangia in the strobili; spores pale orange.

Seasonally dry tundra, fellfields, felsenmeer, and exposed rocky ridges in most alpine areas of the Middle Rocky Mountains. Not yet known from above timberline in the Gros Ventre, Hoback, Salt River, and Wyoming Ranges. British Columbia and Alberta south to California, Arizona, Texas, and Oklahoma.

Alpine plants that have plane to abruptly beveled leaf apices, setae of leaves whitish, and sporophylls ciliate toward the apex are var. *scopulorum* (Maxon) Tryon.

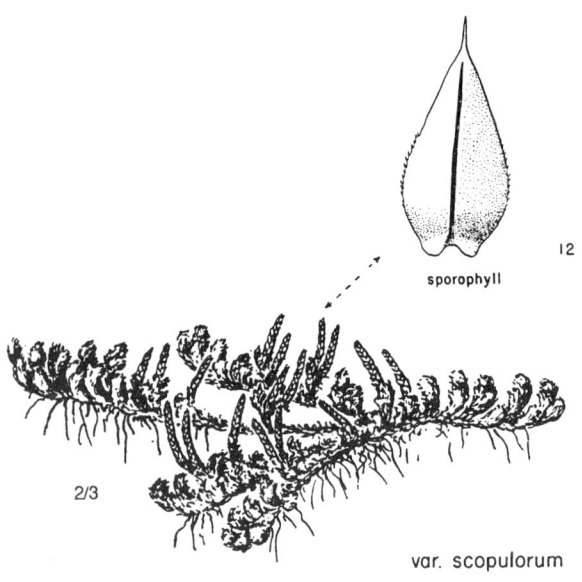

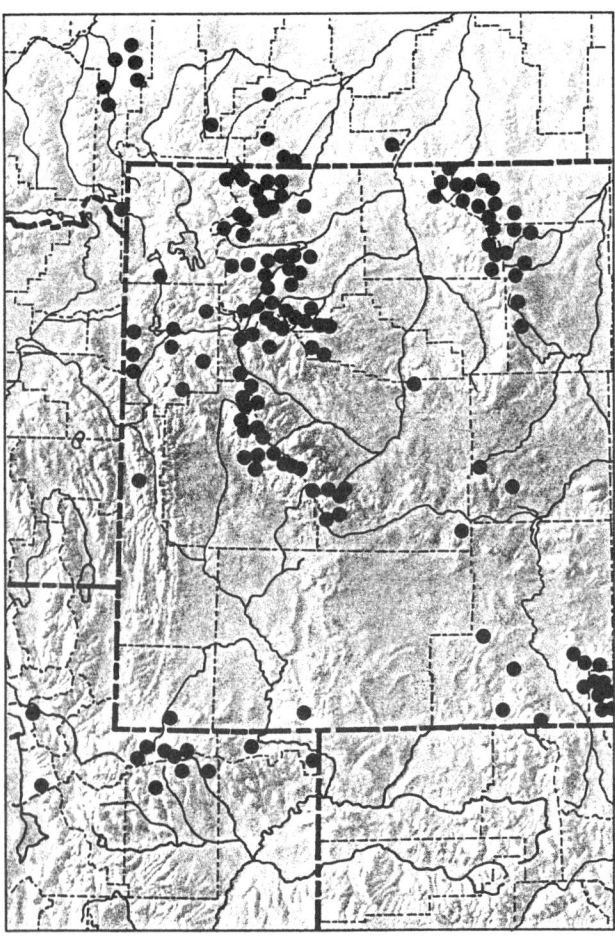

Selaginella watsonii Underw.
Bull. Torr. Club 25:127, 1898.

Watson Spikemoss

Densely matted perennial with creeping stems and short, erect or ascending branches. Leaves imbricate, appressed, lanceolate, somewhat acute, with a narrow dorsal groove, smooth or ciliate-margined, the apex narrowed to a yellowish green seta. Strobili quadrangular, erect or somewhat divergent, the sporophylls triangular-lanceolate to ovate, keeled. Spores pale orange.

Talus and other rocky places in the Beartooth, Uinta, and Wasatch Ranges. Oregon and California east to Montana, Nevada, and Utah.

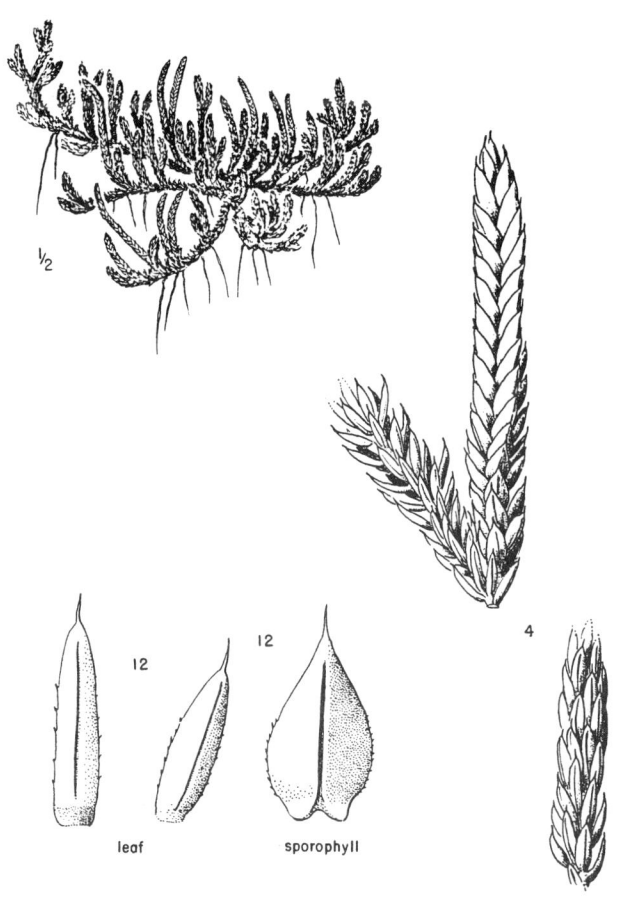

Valerianaceae
Valerian Family

Herbs with opposite, cauline and basal, entire or lobed, exstipulate leaves. Flowers epigynous, perfect or unisexual, regular or irregular, in cymes. Calyx lobes obsolete or inrolled at anthesis and later expanded into several pappuslike structures. Corolla 5-lobed, tubular to funnelform. Stamens 1–4, commonly 3, epipetalous. Pistil 1, the ovary 1–3-celled, 1 cell with a single ovule, the others sterile; style 2–3-lobed. Fruit dry, indehiscent, akenelike.

Valeriana L.
Valerian

Our plants are perennial herbs from strong-scented taproots or rhizomes. Leaves cauline and opposite, or basal, entire to pinnatifid. Flowers irregular, white to pinkish, perfect or unisexual, in paniculate or headlike cymes. Calyx inconspicuous, becoming enlarged and pappuslike in fruit. Corolla 5-lobed, salverform to somewhat funnelform, often saccate or gibbous at the base. Stamens 3, epipetalous, exerted. Ovary 3-carpellate, 2 of the carpels much reduced. Fruit a compressed akene, 3-nerved on the outer surface, 1-nerved on the inner surface, surrounded by the persistent, pappuslike calyx.

KEY TO THE SPECIES OF *VALERIANA*

1. Plants from a taproot; inflorescence paniclelike, elongate — *V. edulis*
1. Plants from a rhizome with fibrous roots; inflorescence corymblike, often capitate
    2. Corolla elongate, tubular, 4–7 mm long, the lobes less than half as long as the tube — *V. acutiloba*
    2. Corolla short, rotate, 2–4 mm long, the lobes about the same length as the tube — *V. occidentalis*

Valeriana acutiloba Rydb.
Bull. Torr. Club 28:24, 1901.

Sharpleaf Valerian

Valeriana puberulenta Rydb.
Valeriana pubicarpa Rydb.

Fibrous-rooted, perennial herb from a stout, branched rhizome; stems erect, glabrous to minutely hirsute. Cauline leaves in 1–3 pairs, the lower pair entire, the upper pinnatifid with lateral lobes smaller than the terminal one; basal leaves petiolate, entire, the blades ovate or obovate. Flowers mostly perfect, in a dense, capitate, corymblike inflorescence. Calyx segments plumose. Corolla white, funnelform, pubescent outside, the tube slightly gibbous. Stamens 3, exerted. Akenes ovate to oblanceolate, pubescent or glabrous.

Alpine meadows, slopes, and krummholz margins of the Beartooth, Gros Ventre, Hoback, Salt River, Teton, Wasatch, and Wyoming Ranges. Oregon to western South Dakota and south to California, Arizona, New Mexico, and Colorado.

Our plants with dense, capitate inflorescences are var. *pubicarpa* (Rydb.) Cronq.

var. pubicarpa

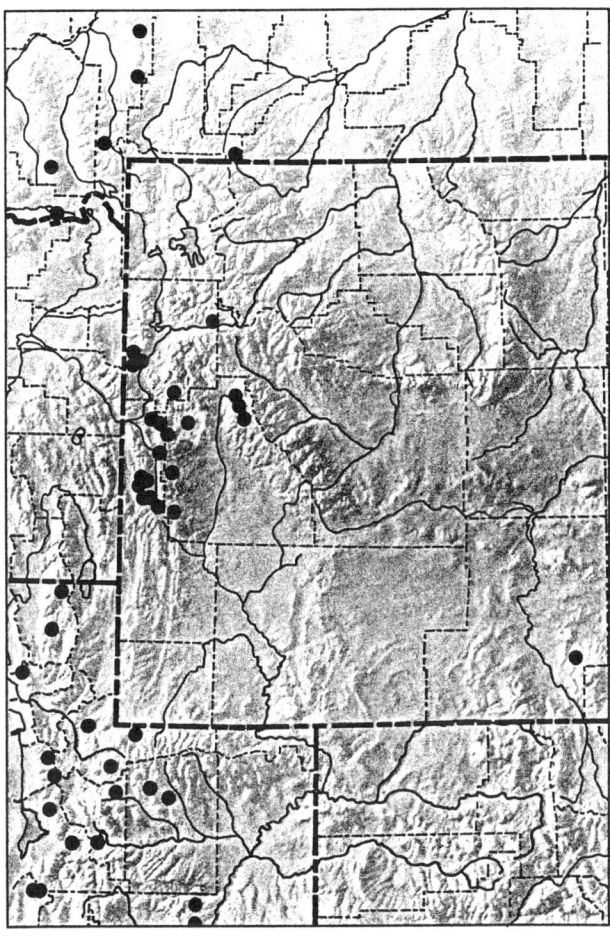

Valeriana edulis Nutt. ex T. & G.
Fl. N. Am. 2:48, 1841.

Edible Valerian

Patrinia ceratophylla Hook.
Valeriana ceratophylla (Hook.) Piper
Valeriana ciliata T. & G.
Valeriana furfurescens A. Nels.
Valeriana obovata (Nutt.) R. & S.
Valeriana trachycarpa Rydb.

Perennial herb from a stout taproot, sometimes with a branched caudex. Cauline leaves in 2–6 pairs, pinnatifid, petiolate below, becoming sessile above; basal leaves pinnatifid or entire, linear to obovate, petiolate. Flowers perfect or unisexual, the plants polygamodioecious; inflorescence paniclelike. Calyx segments plumose. Corollas white to cream, rotate, those of the pistillate flowers less than half the length of the perfect and staminate corollas. Stamens only short-exerted. Akenes ovate to oblong, glabrous or pubescent.

Alpine meadows and moist slopes of the Absaroka, Beartooth, Big Horn, Salt River, Uinta, Wasatch, and Wind River Ranges. Ontario to British Columbia and south to Ohio, Iowa, South Dakota, Colorado, and Mexico.

Our plants with essentially glabrous leaves are ssp. *edulis*.

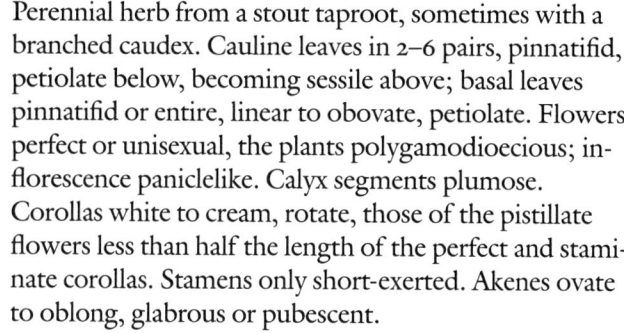
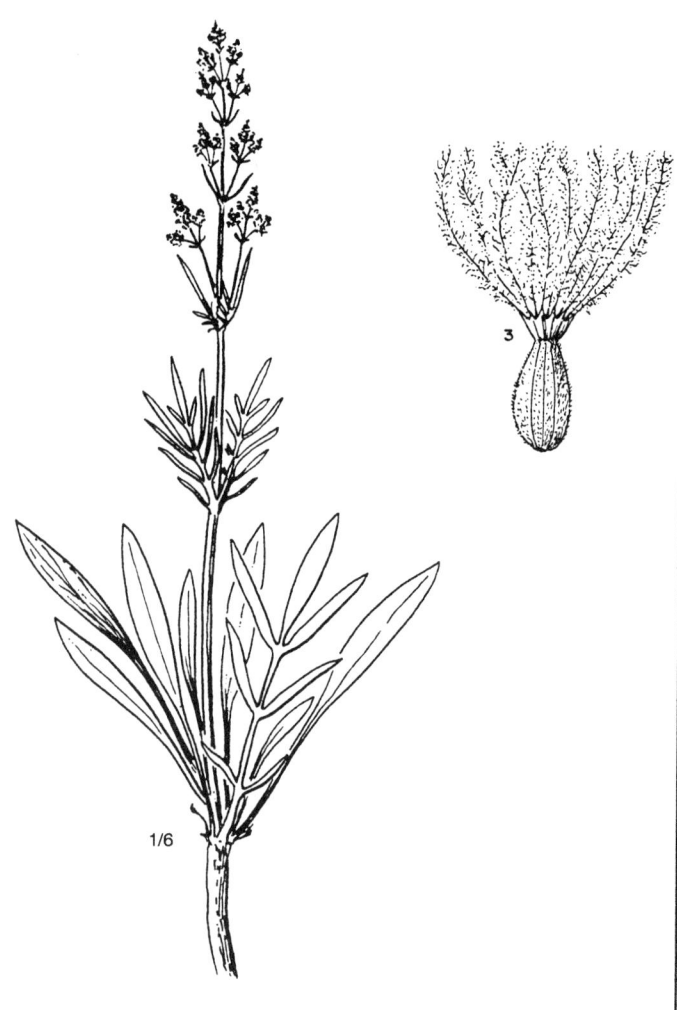
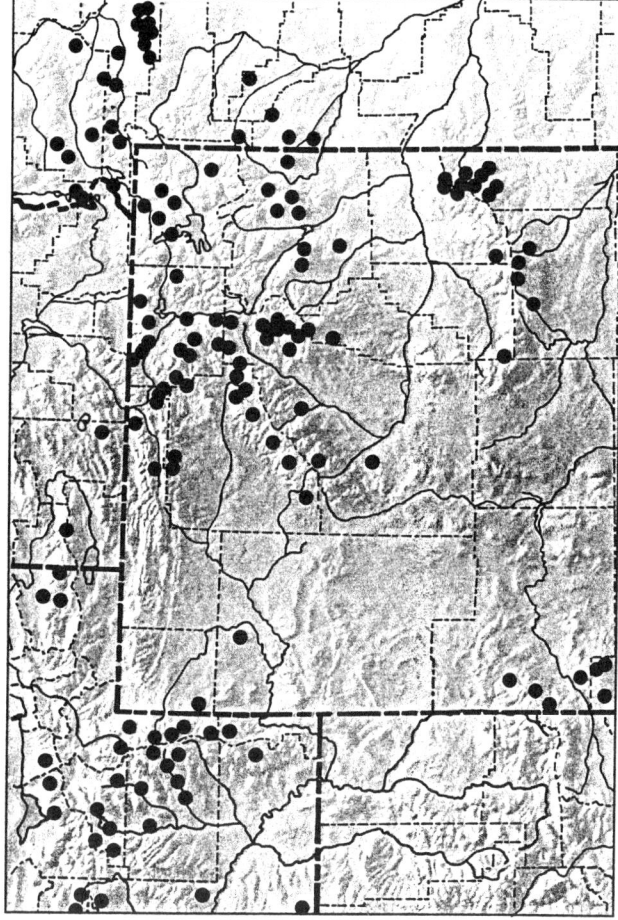

Valeriana occidentalis A. Heller
Bull. Torr. Club 25:269, 1898.

Western Valerian

Valeriana micrantha E. Nels.

Fibrous-rooted, perennial herb from an elongate rhizome. Stems erect, sometimes decumbent at the base, glabrous, pubescent at the nodes. Cauline leaves in 2–4 pairs, pinnatifid, petiolate below, becoming sessile above; basal leaves petiolate, entire and lanceolate to ovate, or sometimes with 1 or 2 pairs of lateral lobes. Flowers perfect or pistillate, the plants gynodioecious; inflorescence corymblike, becoming elongate as fruits develop. Calyx segments plumose. Corollas white, rotate, the lobes about the same length as the tube. Stamens exerted. Akenes lanceolate to ovate, glabrous or pubescent.

Timberline meadows, krummholz margins, and talus slopes of the Uinta, Wasatch, and Wind River Ranges. Oregon south and east to California, Arizona, Colorado, Wyoming, and South Dakota.

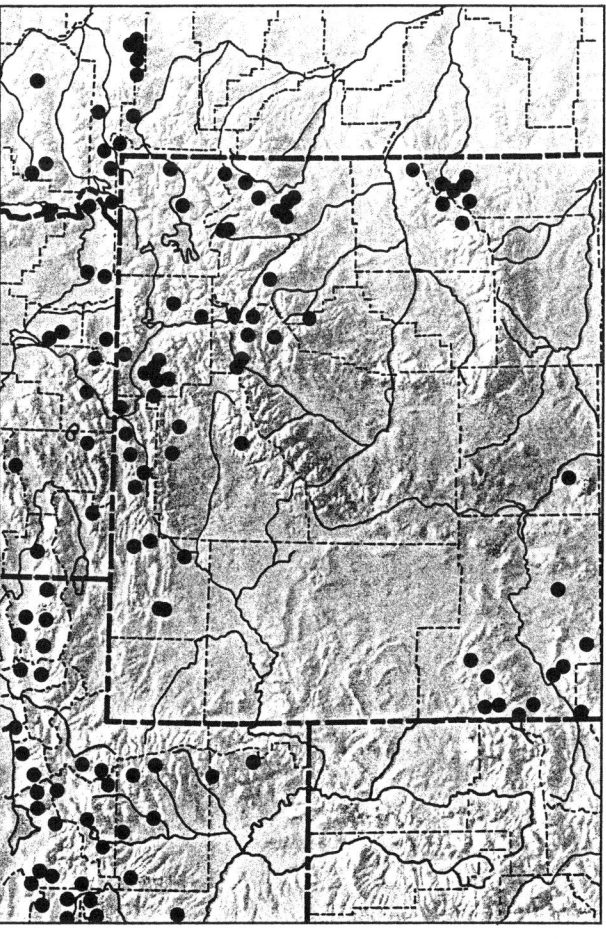

Violaceae
Violet Family

Ours are annual or perennial herbs with simple, prominently stipulate, alternate or basal leaves. Flowers perigynous, perfect, 5-merous, irregular, some of them often cleistogamous, solitary or variously clustered. Sepals 5, persistent, usually distinct. Petals 5, the lower one often spurred or saccate. Stamens 5, connivent around the pistil, the lower two spurred, the filaments sometimes connate. Pistil 1, the ovary 1-celled with 3 parietal placentae, the style single. Fruit a 3-valved, loculicidal capsule with several to many large, shiny seeds.

Viola L.
Violet

Low, perennial, or rarely annual, herbs, with slender rhizomes or stolons. Leaves alternate or basal, blades simple, stipules entire to lobed or toothed. Flowers solitary on axillary or basal peduncles, often of 2 kinds; the earlier ones showy, followed by cleistogamous flowers with the petals reduced or lacking. Sepals 5, distinct or nearly so, sometimes saccate at the base. Petals unequal, the lower one spurred or saccate, the lateral 2 petals often bearded. Stamens 5, slightly connate, the lower 2 spurred. Ovary 1-celled, the style thickened above. Capsule oval to oblong, 3-valved, 1-celled, the seeds ejected at maturity.

KEY TO THE SPECIES OF *VIOLA*

1. Flowers yellow, the petals purplish brown on the back
    2. Leaf margins crenate or serrate — *V. purpurea*
    2. Leaf margins entire — *V. nuttallii*
1. Flowers white, lavender, purple, or blue
    3. Flowers white with purple veins; plants acaulescent — *V. palustris*
    3. Flowers deep blue to purple; plants subcaulescent to caulescent — *V. adunca*

Viola adunca Smith in Rees
Cyclopaedia 37:63, 1817.

Hookspur Violet

Viola aduncoides Löve & Löve
Viola bellidifolia Greene
Viola cascadensis M.S. Baker
Viola mamillata Greene
Viola montanensis Rydb.
Viola monticola Rydb.
Viola odontophora Rydb.
Viola oxyceras S. Wats.
Viola retroscabra Greene
Viola subvestita Greene

Glabrous to puberulent, perennial herb with slender, short to elongate rhizomes. Plants appearing acaulescent early in the season, later developing prostrate to ascending, leafy aerial stems. Leaves alternate, with linear to lanceolate, toothed stipules, the blades ovate to broadly cordate, crenulate; peduncles shorter than to longer than the blades. Petals blue to purple, often whitish at the base, with purple veins; lateral petals bearded; spur prominent, about half as long as the petals. Style slender, bearded.

Occasional to common in moist to wet meadows, willow thickets, and along stream banks in most major ranges of the Middle Rocky Mountains. Not yet known from above timberline in the Absaroka, Gros Ventre, Hoback, and Wyoming Ranges. British Columbia south to California and Colorado, and east to the Atlantic Coast.

Dwarf, caulescent alpine plants with glabrous leaves and petals about 5 mm long are var. *bellidifolia* (Greene) Harrington.

var. bellidifolia

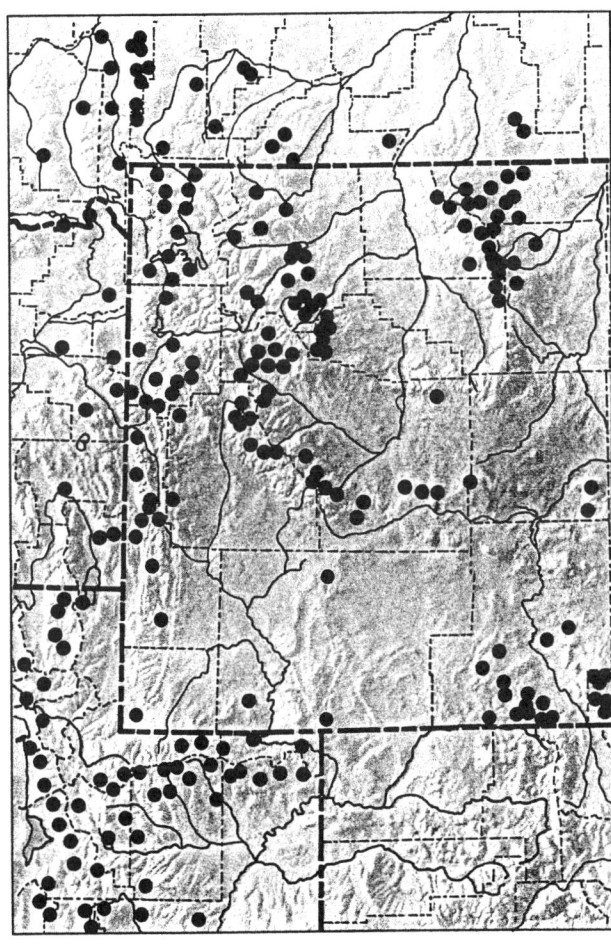

Viola nuttallii Pursh
Fl. Am. Sept., 174, 1814.

Nuttall Violet

Crocion nuttallii (Pursh) Nieuwl. & Lunell
Crocion vallicola (A. Nels.) Nieuwl. & Lunell
Viola erectifolia A. Nels.
Viola flavovirens Pollard
Viola gomphopetala Greene
Viola linguifolia Nutt.
Viola physalodes Greene
Viola praemorsa Dougl. ex Lindl.
Viola russellii B. Boivin
Viola subsagittifolia Suksd.
Viola vallicola A. Nels.
Viola xylorrhiza Suksd.

Subcaulescent, pubescent, perennial herb from short, ascending rootstocks. Leaves with lanceolate, partially adnate stipules; blades narrowly lanceolate to ovate or ovate-deltoid, the bases broadly cuneate to rounded; margins entire, sinuate or weakly serrulate, the bases cuneate to truncate; peduncles shorter than to equaling the leaves. Petals yellow, veined with purple, the upper brownish-backed, the lateral 2 bearded; spur very short. Style bearded.

Timberline stream banks, snow beds, wet meadows, and nivation basins of the Absaroka, Beartooth, Salt River, Uinta, Wasatch, Wind River, and Wyoming Ranges. British Columbia to California and east to Montana, Wyoming, and Colorado.

Alpine plants with cuneate leaf bases, blades glabrous to moderately pubescent, and capsules short-pubescent seem to best fit the weakly differentiated var. *major* Hook.

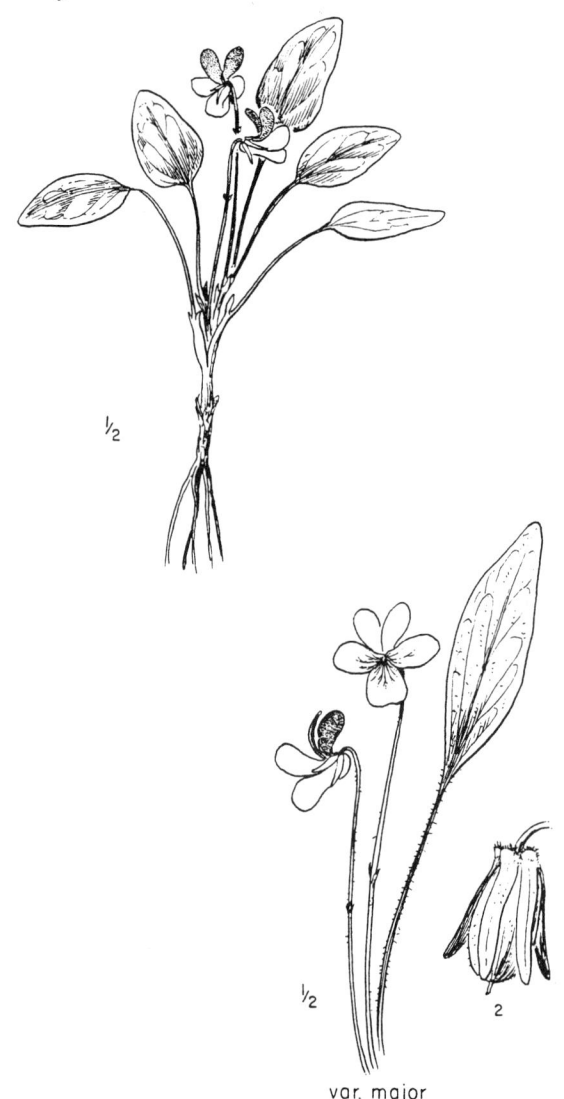

Viola palustris L.
Sp. Pl., 934, 1753.

Marsh Violet

Glabrous, acaulescent, perennial herb from slender, creeping rhizomes and stolons. Stems lacking; peduncles erect, 1-flowered. Leaves petiolate, with lanceolate, toothed stipules; blades cordate to reniform, shallowly crenate; peduncles about equaling, or often surpassing, the leaves. Petals white with purple veins, beardless, or the lateral 2 petals sometimes sparsely bearded; spur short. Style beardless.

Moist, shaded areas along streams of the Wind River Range in the lower alpine zone near timberline. Transcontinental in Canada and south in the western United States to California and Colorado.

White-flowered subalpine and lower alpine plants are var. *brevipes* (M.S. Baker) Davis. *Viola canadensis* L. has been reported from the alpine zone of Carter Mountain in the east-central Absaroka Range (Evert, 1983b). It can be distinguished from *V. palustris* by the presence of leaf-bearing stems, which are lacking in *V. palustris*.

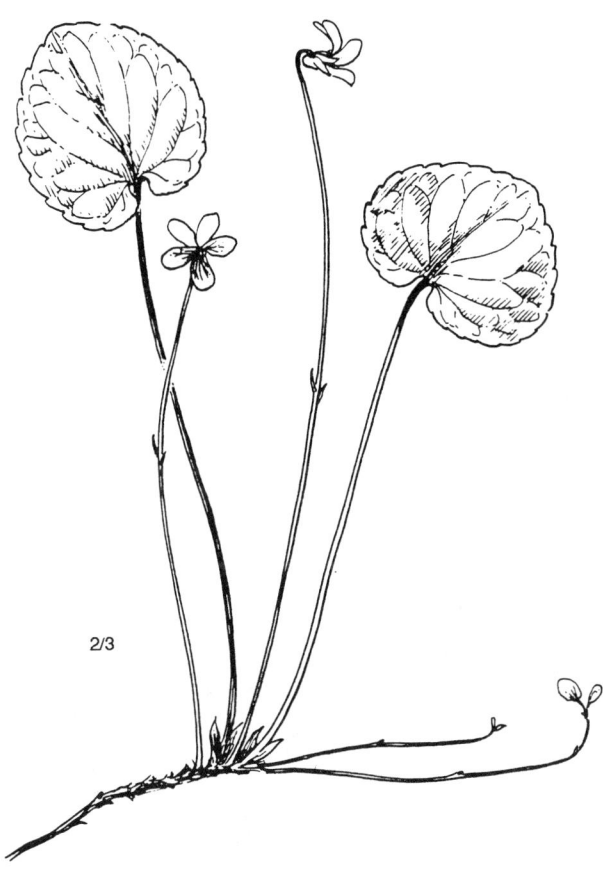

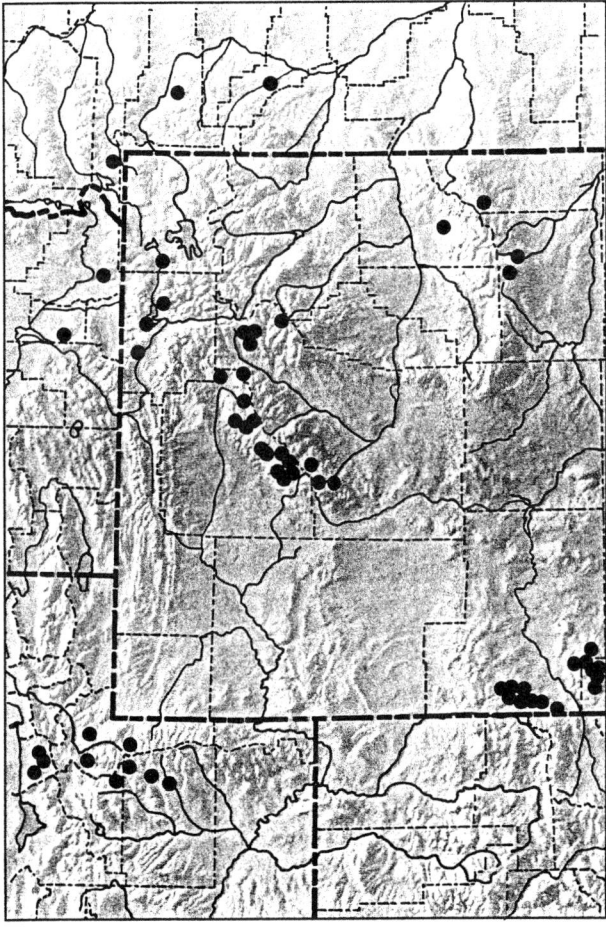

Viola purpurea Kellogg
Proc. Calif. Acad. Sci. 1:55, 1855.

Pine Violet

Viola atriplicifolia Greene
Viola aurea Kellogg
Viola pinetorum Greene
Viola quercetorum M.S. Baker & J.C. Clausen
Viola thorii A. Nels.
Viola utahensis M.S. Baker & J.C. Clausen
Viola venosa (S. Wats.) Rydb.

Puberulent, caulescent, perennial herb from slender, erect, short to elongate rhizomes; stems slender, several to many, ascending to erect. Leaves long-petiolate, with entire to toothed, lanceolate stipules; blades lanceolate to orbicular, crenately toothed to serrulate, the bases cuneate to subcordate, often purple-tinged or purple-veined below, obtuse to acute at the apex; peduncles subequal with the leaves. Petals yellow, veined with purple, often purplish on the dorsal surface, the lateral 2 short-bearded or not; spur very short. Style bearded. Capsules globose, puberulent.

Krummholz and timberline margins in the Salt River and northern Wind River Ranges. Washington south to California, Arizona, and Colorado.

Middle Rocky Mountain plants are var. *venosa* (S. Wats.) Brainerd.

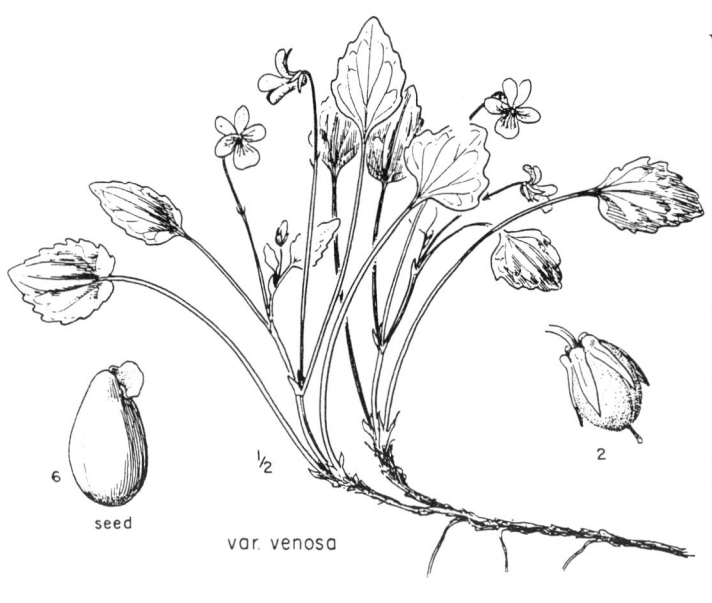

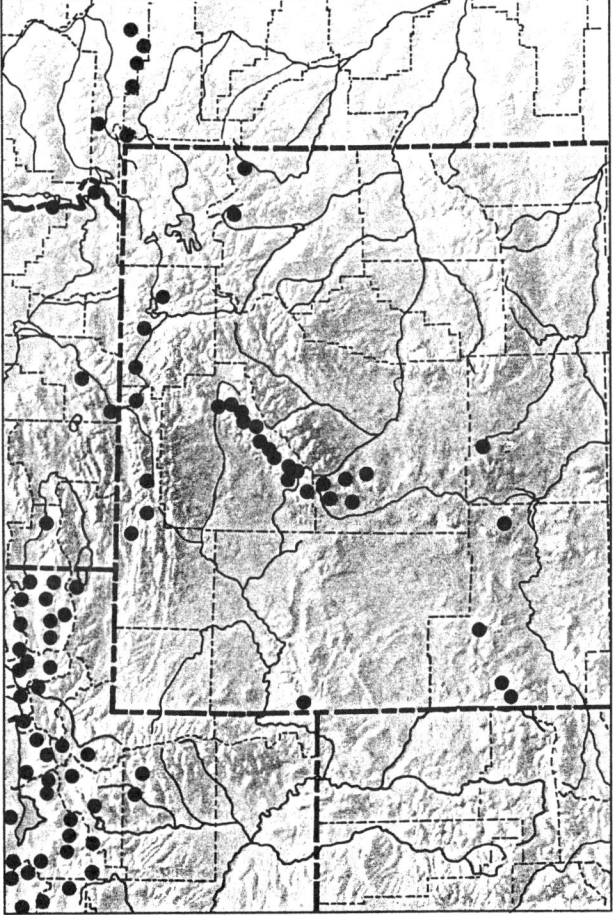

APPENDIX I

Glossary of Descriptive Alpine Terminology

Alp slope	A sloping patch of vegetation among rocks.
Alpine	The zone above timberline.
Alpine bog soil	Dark brown to black, peaty, intrazonal soil in wet areas along streams and below snowbanks.
Alpine desert	A dry, well-drained area with sparse vegetation exposed to winds; occurs on ridgetops and windward slopes.
Alpine meadow	A plant community dominated by grasses, most commonly by species of the genera *Calamagrostis, Deschampsia, Festuca,* or *Poa*.
Alpine meadow soil	Brown to dark brown, intrazonal soil, high in organic matter but not sufficient to be classified as peaty.
Alpine turf soil	Brown to dark brown zonal soil, usually high in organic matter and supporting a dense vegetation mat; cf. *Geum turf*.
Bog	A wet, peaty environment in areas of poor drainage along streams and below snowbanks; dominated by *Salix, Betula,* or *Kalmia*.
Calciphile	A plant restricted to or commonly found on limestone or limestone-derived soils.
Chionophilous	Literally, "snow-loving"; referring to plants associated with snowbanks; often ephemerals that complete flowering and fruiting shortly after snowmelt.
Chionophobous	Literally, "snow-hating"; referring to plants not associated with snowbanks.
Chomophyte	A plant that regularly occurs on rock surfaces such as ledges, scree, talus, and cliffs.
Circle	A patterned ground feature of circular mesh that may be sorted or nonsorted.
Circumboreal	The distribution pattern that includes North America and Eurasia south of the arctic timberline.
Circumpolar	The distribution pattern that includes North America and Eurasia north of the arctic timberline.
Cirque	A steep-walled, armchairlike basin remaining after glacial ice recedes.
Congelifraction	Disruption or wedging of a rock surface by frost action.
Congeliturbation	Disruption of the soil surface by movement due to frost action.
Cryopedogenesis	Soil formation influenced by frost action; it includes congelifraction and congeliturbation.
Cryoplanation	Mass wasting through the processes of frost action.
Dryas island	A clone of *Dryas* (in our area, *D. octopetala*) forming a round or oval mat.

Earth hummock	A patterned ground feature that is raised and knoblike due to frost heaving.
Economic forest line	Closed-canopy forest which can be commercially cut.
Ecotone	The transition area between two communities or ecosystems.
Elfinwood	Dwarfed, stunted conifers growing above timberline.
Entisol	A young, stony soil with little or no evidence of horizons or horizontal development.
Fellfield	Frost-wedged rocks on a level or gently sloping surface stabilized by vegetation mats; or coarse, sandy or gravelly deposits among the rocks.
Felsenmeer	Literally, "sea of rocks"; a level or gently sloping surface covered by frost-wedged rocks and lacking vegetation patches.
Flag	The asymmetrical portion of a timberline tree that has branches missing on the windward side and growth on the leeward side.
Forest limit	The upper limit of closed canopy forest; cf. *Timberline, Economic forest line*.
Frost boil	A patterned ground feature that is an irregular patch of surficial material caused by frost action, often 2–3 m^2 in area.
Frost hummock	Patterned ground feature; see *Earth hummock*.
Frost scar	A patterned ground feature; same as frost boil, but the term is often applied to very small features, less than 0.25 m^2.
Garland	A patterned ground feature resulting from downhill creep that forms the soil into crescents or garlands, which are transitional to terraces.
Geum turf	A dense vegetation mat dominated by *Geum rossii*; common in fellfields.
Historic tree line	An earlier advance of forest or treeline species above the existing treeline.
Hummock	See *Earth hummock*.
Infranival cushion	The "apron" of a krummholz formation; the horizontal branches near to the ground around the perimeter of the krummholz.
Intrazonal soil	Soil with an A-C profile, usually the result of poor drainage.
Island	See *Dryas island*.
Kettle	A small lake or pond in a glacial moraine; fed by groundwater.
Krummholz	Dwarfed, often deformed conifers above timberline.
Kruppelkiefer	Literally, "crippled pines"; dwarfed, often deformed conifers growing above timberline.
Lithosol	Young azonal soil formed in place; cf. *Regosol, Entisol*.
Mass wasting	Downslope movement of material without a suspending or transporting medium such as glacial ice, water, or wind.
Mat	Thick, dense stand of vegetation; often indicates patterned ground features.
Meadow	See *Alpine meadow*.
Mollisol	The zone of seasonal thaw above permafrost.
Moraine	Nonsorted material deposited directly by a glacier.
Multigelation	Repeated freeze-thaw cycles.
Needle ice	Elongate ice crystals that form on or near the surface of wet alpine soils at right angles to the surface; also called brush ice, feather ice, and pipkrake.

Neoglacial	A period of minor glacial advances in the mountains between 8000 B.P. and 100 B.P., reaching a maximum around 5000 B.P.
Net	A patterned ground feature forming an elongate mesh, which may be sorted or nonsorted.
Nival	Pertaining to snow.
Nivation	The denudation of material from around snowdrifts and snowfields.
Nivation basin	A small, usually shallow basin formed by removal of surficial material through erosional processes associated with snowfields.
Nonsorted patterned ground	A patterned ground feature that lacks a border of stones of larger size than the center; often discernible by patterns of vegetation; cf. *Circle, Mat, Polygon*.
Nunatak	An unglaciated bedrock remnant, hill, or peak; in the Middle Rocky Mountains often discernible by felsenmeer surrounded by glacial polish.
Patterned ground	Surface features related to a prevailingly cold climate with permafrost or seasonally frozen ground; see *Frost boil, Frost hummock, Frost scar, Garland, Net, Polygon*, and *Stripe*.
Permafrost	Permanently frozen substrate that has remained below 0°C for two or more years.
Physiognomic forest line	The limit of open-canopy forest.
Pleistocene	Epoch that began 1.65 Ma and ended 10,000 B.P.
Polygon	A patterned ground feature consisting of a 5-sided, polygonal mesh, which may be sorted or nonsorted.
Regosol	Young azonal soil formed on transported rocky material; cf. *Lithosol, Entisol*.
Scree	Fine rock fragments on steep slopes or in deposits at the base of a cliff through the combined action of gravity and physical weathering; cf. *Talus*.
Snow bed	A site occupied by snow for most of the year; recognizable by weathering or vegetational differences compared with the surrounding area.
Snow mat	See *Infranival cushion*.
Snow patch	See *Snow bed*.
Solifluction	Slow, downslope creep of water-saturated soils and rocks covered by a vegetation mat.
Sorted patterned ground	A patterned ground feature with a border of stones of larger size than the center; cf. *Circle, Net, Polygon*.
Stone net	See *Net*.
Stone stripe	See *Stripe*.
Stripe	A patterned ground feature consisting of a striped pattern of parallel lines extending downhill along the fall-line that may be sorted or nonsorted.
Supranival flag	See *Flag*.
Talus	Coarse rock fragments on steep slopes or in deposits at the base of a cliff through the combined action of gravity and physical weathering; cf. *Scree*.
Tarn	A small lake occupying a glacial cirque.
Till	Nonstratified, nonsorted material deposited by a glacier.
Timberline	The upper limit of continuous arborescent growth forms 3 m or more in height.
Tree limit	See *Treeline, Timberline*.

Tree species line	The upper limit represented by any-sized individual of a particular species; also tree species limit.
Treeline	The upper limit of arborescent growth forms 3 m or more in height; also tree limit; cf. *Timberline*.
Tundra	Treeless plains, ridges, slopes, and peaks above alpine timberline or north of arctic timberline.
Turf	Densely matted and rooted vegetation formed by plant species belonging to genera such as *Carex, Geum,* and *Poa;* cf. *Geum turf*.
Tussock	A tufted growth form consisting of a dense cluster of leaves and multiple stems.

APPENDIX 2

Glossary of Botanical Terms

Acaulescent	Lacking or apparently lacking a stem; the leaves, if present, all basal; cf. *Scapose*.
Accrescent	Enlarging with age after flowering, as in the calyx of species of *Lychnis* and *Silene*.
Acicular	Slender with parallel sides; needle-shaped.
Actinomorphic	Having a radially symmetrical structure; regular; usually applied to flowers; cf. *Zygomorphic*.
Acuminate	Gradually tapered to a short, sharp point, the point with concave sides; cf. *Cuspidate*.
Acute	Sharp-pointed, the point with straight sides.
Adnate	Having unlike parts fused.
Adventitious	An organ developing or originating in an unusual location.
Agamospermous	Asexual reproduction in which an embryo develops from maternal tissue.
Aggregate	An apparent single fruit developed from multiple pistils attached to a receptacle, as in *Rubus*.
Aggregated	Clustered.
Akene	A dry, indehiscent fruit from a simple pistil, the pericarp separable from the seed.
Alternate	Situated between, or, as in leaves, one at a node.
Alveolate	Pitted to produce a honeycomb-like surface.
Ament	A spike or raceme of unisexual flowers, such as those in the genera *Betula* and *Salix*; also, a catkin.
Amphibious	Referring to a plant that can grow in water or on land, although usually on wet or damp surfaces.
Androgynous	Having staminate and pistillate flowers in the same inflorescence, with the staminate above the pistillate, as in some sedges.
Angled	Ridged lengthwise.
Annual	A plant that lives for one year.
Annulus	A ring of thick-walled cells in the sporangium of a fern.
Anther	The pollen-producing portion of a stamen.
Anthesis	Opening of a flower.
Apetalous	Lacking petals.
Aphyllopodic	Lacking leaves at the base of the stem, as in species of *Carex*.
Apical	Located at the tip or apex.

Apiculate	Abruptly tapered to a short, sharp point; cf. *Mucronate*.
Apomict	An asexually reproducing plant, or a plant produced asexually.
Appressed	Pressed toward or flattened on a surface.
Approximate	Close together or only slightly spread out.
Aquatic	Growing in water or a water habitat; cf. *Terrestrial*.
Arachnoid	Having fine, cobwebby or tangled hairs.
Arcuate	Curved or arched.
Aril	An often fleshy appendage growing in the area of the hilum of a seed and partially enclosing it.
Aristate	Awned with a fine, slender, bristlelike appendage.
Aromatic	Producing a strong, usually fragrant or pleasant odor, as in *Monardella*.
Articulation	Separation at a point between structures.
Ascending	Growing upward at an oblique angle.
Attenuate	Having a long-tapering tip; cf. *Acuminate*.
Auricle	An earlike structure or lobe.
Auriculate	Having one or more auricles.
Awn	A slender, bristlelike appendage.
Awned	Tipped with an awn.
Axil	The upper angle between an organ, usually a leaf, and a stem.
Axillary	Located in an axil.
Axillary placentation	Attachment of the ovules or seeds in the axils of the locules.
Banner	The large, dorsal petal in the flower of a member of the Fabaceae; also, standard.
Barbellate	With short, stiff hairs (barbs) along an axis.
Basal	Located at the base.
Basal placentation	Attachment of a single ovule or seed at the base of the ovary.
Basifixed	Attached by the base.
Beak	A long, narrow projection, often sharp at the tip.
Bearded	Having a prominent tuft of hairs.
Berry	A simple fruit with a fleshy pericarp, as a tomato, grape, or currant.
Bidentate	With two teeth.
Bidentulate	With two very small teeth.
Biennial	A plant that lives for two years.
Bifid	Forked or once-cleft.
Bilabiate	Two-lipped.
Bilobed	Two-lobed.
Bipinnate	Twice pinnate.

Bipinnatifid	Twice pinnatifid.
Biseriate	In two series or two rows; cf. *Multiseriate, Uniseriate*.
Bivalvate	Having two valves.
Blade	The expanded photosynthetic portion of a leaf.
Boat-shaped	Resembling the prow of a canoe, as in the leaf tips of *Poa*.
Bract	A modified leaf subtending a flower or inflorescence.
Bracteate	With one or more bracts.
Bracteolate	With one or more bracteoles.
Bracteole	A small bract; cf. *Bractlet*.
Bractlet	A small bract on a secondary axis of an inflorescence; the term is often interchangably with *bracteole*.
Bristle	A stiff hair or hairlike appendage.
Bristly	With bristles.
Bulb	An underground stem with a bud covered by fleshy, or otherwise modified, leaves.
Bulbil	A small, bulblike structure produced aboveground in a leaf axil or inflorescence; cf. *Bulblet*.
Bulblet	A small, bulblike structure produced at any location; cf. *Bulbil*.
Caducous	Falling or dropping off early; cf. *Fugacious*.
Caespitose	(also cespitose). Tufted.
Callosity	A hardened thickening or projection.
Callus	The thickened portion of the rachilla at the base of a lemma in some grasses.
Calyx	The outermost whorl of modified leaves in a flower; cf. *Sepal*.
Campanulate	Bell-shaped.
Canaliculate	Channeled along the longitudinal axis.
Canescent	Having a grayish white pubescence.
Capillary	Thin, slender, or hairlike.
Capitate	Spherical, resembling a head; referring to a dense, head-shaped cluster; inflorescence type in members of the Asteraceae; cf. *Head*.
Capsule	Dry, dehiscent fruit from a compound ovary with two or more carpels.
Carnose	Fleshy.
Carpel	One or more modified leaves forming a pistil and producing ovules; megasporophyll.
Carpophore	The slender stalk supporting the fruits in members of the Apiaceae; see *Mericarp*.
Carunculate	Having a caruncle, or outgrowth, near the hilum of a seed.
Caryopsis	A dry, indehiscent fruit with the pericarp fused to the seed; also, grain.
Catkin	A spike or raceme of unisexual flowers such as those in the genera *Betula* and *Salix*; also, ament.
Caudate	Tapered to a tail-like structure.

Caudex	The woody and persistent base of a herbaceous, perennial stem.
Caulescent	Having a stem.
Cauline	On the stem.
Cespitose	See *Caespitose*.
Chaffy	Dry, membranous.
Channeled	Grooved along the longitudinal axis; cf. *Canaliculate*.
Chartaceous	Having a paperlike texture, usually dry and nongreen.
Ciliate	Having a margin with a fringe of hairs.
Ciliolate	Having a margin with a fringe of very small, fine hairs.
Circinate	Coiled downward from the top.
Circumscissile	Dehiscing by a horizontal line around a cylindrical or ovoid fruit or anther, the top falling as a lid.
Clavate	Club-shaped; thickened near the end.
Claw	The abruptly narrowed base of some petals, as in members of the Brassicaceae.
Clawed	Having a claw.
Cleft	Split nearly to the middle.
Cleistogamous	A self-fertilized flower that fails to open, as in members of the Violaceae.
Coherent	Like parts or organs closely joined but not fused together; cf. *Connate*.
Column	A combination of stamens and styles fused into a single unit, as in the column of orchid flowers.
Commissure	A face or surface on which one carpel joins the other in members of the Apiaceae.
Comose	Having a tuft of hairs (coma).
Complete	A flower with all whorls of parts present; i.e., sepals, petals, stamens, and one or more pistils; cf. *Incomplete*.
Complete septum	The transverse septum that extends across a leaf in members of the Juncaceae.
Compound	Having two or more similar parts making up an organ or structure.
Compound leaf	A leaf with two or more leaflets.
Compound umbel	An inflorescence consisting of several to many rays with a common point of attachment, each of these further branched at a common point.
Concave	Having a hollowed-out surface.
Conduplicate	Folded lengthwise.
Cone	An axis bearing microsporophylls or megasporophylls, often woody, as in some members of the Pinaceae.
Confluent	With two parts blending or running together.
Conic	Cone-shaped, broadest near the base.
Connate	Like parts or organs that are fused together; cf. *Coherent*.
Connective	The tissue connecting two anther cells.

Connivent	Converging or close together, but not connected.
Continuous	With a rachis that does not break apart at maturity.
Contracted	An inflorescence with the branches appressed rather than spreading.
Convex	Having a rounded surface.
Convolute	Rolled up lengthwise.
Cordate	Heart-shaped, with the notch at the base and tapered toward the apex.
Coriaceous	Having a leatherlike texture.
Corky	A soft, fluffy texture, like that of cork.
Corm	A solid, underground, bulblike stem covered by papery leaves.
Corniculate	Terminating in a small, hornlike process.
Corolla	The whorl of modified leaves located above or inside the calyx; the floral envelope; cf. *Petal*.
Corona	An outgrowth of the stamens or petals, often crownlike and occurring between them.
Corymb	A flat-topped or convex inflorescence with branches arising at different levels, the outer flowers blooming first.
Corymbose	Resembling a corymb or arranged in corymbs.
Costa	A rib; the midvein of a leaf.
Cotaneous	Flowering as the leaf blades are expanding, as in members of the Salicaceae; cf. *Precocious, Serotinous*.
Cotyledon	An embryonic or seed leaf.
Crenate	Having a scalloped margin with rounded teeth.
Crenulate	Having a margin with very small, rounded teeth.
Crest	An elevated rib or ridge, sometimes toothed, on a surface.
Crisp	Curved, wavy.
Cross-rugulose	Wrinkled at right angles to the axis of a structure.
Crown	A short, scalelike pappus in some flowers of the Asteraceae; top of a tree; top of a root; cf. *Corona*.
Cruciform	Cross-shaped, as the petals of members of the Brassicaceae.
Cucullate	Hood-shaped.
Culm	The flowering stem of grasses and sedges; cf. *Innovation*.
Cuneate	Wedge-shaped.
Cuspidate	Coming to a sharp, rigid point with concave sides; cf. *Acuminate*.
Cylindric	Elongate and circular in cross section; terete.
Cyme	A flat-topped or convex, branched inflorescence in which the innermost or central flowers bloom first.
Cymose	Resembling a cyme, or arranged in cymes.
Deciduous	Falling off during or at the end of each growing season, as many leaves.

Decompound	Several times compound.
Decumbent	Lying on the ground, with the tip ascending.
Decurrent	Bladelike extension of a leaf or stipule downward along the stem from the insertion.
Decussate	With opposite leaves borne at right angles to the preceding pair.
Deflexed	Abruptly bent downward or backward.
Dehiscent	Opening spontaneously along regular lines or sutures.
Deltoid	Triangular.
Dendritic	Treelike.
Dentate	Having a margin with sharp, outward-pointing teeth.
Denticulate	Having a margin with very small, sharp, outward-pointing teeth.
Depressed	Flattened vertically.
Diadelphous	A form in which stamens are united into two groups by their filaments, often in a 9 + 1 arrangement, found in species of Fabaceae; cf. *Monadelphous*.
Dichotomous	Equally twice-branched.
Didymous	Occurring in a pair.
Didynamous	Occurring in two pairs, one long and one short, as the stamens of some members of the Lamiaceae.
Digitate	With fingerlike structures arising from a common point.
Dilated	Expanded or widened, and often flattened.
Dimidiate	Divided into two unequal parts.
Dimorphic	Having two forms.
Dioecious	With staminate and pistillate flowers on different plants; cf. *Monoecious*.
Disarticulate	To separate at a joint or break apart.
Discoid	Having disk flowers.
Disk flower	A tubular flower in members of the Asteraceae.
Dissected	Divided into numerous segments or lobes.
Distinct	Separate.
Divergent	Spreading.
Dolabriform	Pick-shaped; referring to a hair attached at or near the middle; cf. *Malpighian*.
Dorsal	The outer surface, or the surface most distant from the axis; also, abaxial.
Dorsomarginal	Referring to the margins of an outer surface.
Doubly serrate	Having large teeth on a margin with smaller teeth cut into them.
Drupe	A fleshy fruit with a single seed inclosed in a hardened endocarp, as in a cherry.
Drupelet	A small drupe in an aggregate fruit, as in *Rubus*.
Ebracteate	Lacking bracts.
Eglandular	Lacking glands.

Eligulate	Lacking ligules.
Ellipsoid	An elliptic solid.
Elliptic	Having an elongate shape, widest in the middle and gradually tapering toward both ends.
Emarginate	With a shallow notch at the apex.
Embryo	The young sporophyte contained within a seed.
Emergent	A plant with its upper portion raised above the water.
Entire	A smooth margin lacking incisions or lobes.
Epapillose	Lacking papillae.
Ephemeral	Lasting for only a short period.
Epigynous	Having flower parts apparently attached to the top of the ovary, the ovary appearing inferior.
Epipetalous	Having stamens attached to the petals.
Epiphyte	A plant growing on another plant, but not parasitic on it.
Epunctate	Lacking glandular dots.
Equitant	Folded lengthwise in two ranks.
Erect	Upright.
Erose	Having a ragged margin.
Even-pinnate	Pinnately compound and lacking a terminal leaflet.
Evergreen	Remaining green for the entire year.
Excurrent	Extending beyond the apex or margin.
Exerted	Protruding out of an organ, as stamens exerted from the corolla; cf. *Included*.
Exfoliating	Peeling or scaling in thin strips or flakes.
Explanate	Spread out flat.
Exstipulate	Lacking stipules.
Extrorse	Turned outward.
Falcate	Sickle-shaped.
Farinose	With a mealy covering.
Fascicle	A cluster or bundle.
Fasciculate	Growing in a cluster or bundle.
Fertile	Stamens that produce pollen and pistils that produce seed-bearing fruits.
Fibrillose	Having fibers.
Fibrous root	A root with many thin, fiberlike branches.
Filament	The stalk of a stamen that bears an anther.
Filiform	Threadlike.
Fimbriae	Fringes.

Fimbriate	Fringed.
Flabellate	Fan-shaped.
Flabelliform	Fan-shaped.
Fleshy	Succulent, juicy.
Flexuous	Bent back and forth in alternate directions; wavy.
Floccose	Having tufts of woolly hairs.
Floret	The flower of grasses; it includes some combination of lemma, palea, stamens, and pistil; a flower in members of the Asteraceae.
Floricane	A flower-bearing stem (cane); cf. *Primocane*.
Flower	A reproductive structure usually consisting of four whorls of modified leaves: the sepals, petals, stamens, and pistils.
Foliaceous	Leaflike.
Foliose	Leafy.
Follicle	The dry, dehiscent fruit from a single carpel, opening along the ventral suture.
Fornix	A small, arched appendix or scale in the throat of the corolla in members of the Boraginaceae.
Free-central placentation	Attachment of ovules or seeds on a free-standing column in the ovary.
Frond	A fern leaf.
Fruit	One or more ripened ovaries and accessory parts.
Frutescent	Shrubby; also, fruticose.
Fruticose	Shrubby; also, frutescent.
Fugacious	Falling early; cf. *Caducous*.
Funiculus	The stalk of a seed or ovule.
Funnelform	Funnel-shaped; expanding gradually upward, as a funnelform corolla.
Fusiform	Spindle-shaped; widest near the middle and tapering to both ends.
Galea	The upper lip of a bilabiate corolla; see *Hood*.
Galeate	Hood-shaped or helmet-shaped.
Gametophyte	The gamete-producing generation or plant in alternation of generations.
Geniculate	Abruptly bent; resembling a knee joint.
Gibbous	Inflated or swollen on one side.
Glabrate	Becoming glabrous with age; nearly glabrous.
Glabrescent	Glabrate.
Glabrous	Smooth, lacking hairs.
Gland	A structure that secretes a sticky fluid.
Glandular	Having glands.
Glandular-papillate	Having small, dome-shaped glands or tubercles.
Glaucescent	Somewhat glaucous.

Glaucous	Covered with a fine, waxy, whitish to bluish powder that can easily be rubbed off.
Globose	Spherical.
Glochidiate	Pubescent with barbed hairs, often in tufts.
Glomerate	Having dense clusters of flowers.
Glomerulate	Having small, dense clusters of flowers.
Glomerule	A dense, capitate cyme.
Glume	A chaffy bract found at the base of a grass spikelet.
Glutinous	Sticky.
Grain	See *Caryopsis*.
Gynaecandrous	Having staminate and pistillate flowers in the same inflorescence, with pistillate above staminate, as in the sedges.
Gynobase	An elongation of the receptacle, as in members of the Boraginaceae.
Gynodioecious	Plants exhibiting two forms, some with perfect flowers and some with pistillate flowers.
Habit	The general aspect or appearance of a plant.
Hair	An epidermal outgrowth consisting of one to several cells.
Hastate	Halberd-shaped, like an arrowhead but with basal lobes pointed outward.
Head	A dense, spherical or flat-topped cluster of sessile or nearly sessile flowers borne on a common receptacle; loosely applied to any dense inflorescence.
Hemispheric	Consisting of half a sphere, shaped like an open fan.
Herb	A plant lacking woody tissue in the aboveground parts.
Herbaceous	Nonwoody.
Herbage	The leaves and young shoots collectively.
Heterogamous	Having flowers of both sexes, as some members of the Asteraceae.
Heterosporous	Having two kinds of spores: megaspores and microspores.
Hirsute	Having moderately stiff hairs; cf. *Hispid*.
Hirtellous	Minutely hirsute.
Hispid	Having coarse, stiff hairs; cf. *Hirsute, Strigose*.
Homogamous	Having flowers all of one kind, as some members of the Asteraceae, the stamens and carpels maturing at the same time.
Homosporous	Having one kind of spore.
Hood	The upper lip of a bilabiate corolla; a hollow, saccate covering; see *Galea*.
Hyaline	Thin and translucent.
Hypanthium	The fused, cup-shaped, basal portion of floral parts; an outgrowth of the receptacle.
Hypogynous	Having flower parts attached below the ovary, with the ovary superior.
Imbricate	Overlapping.
Imperfect	Having only one sex; a flower lacking either stamens or pistils.
Incised	A margin cut sharply and deeply, resulting in a jagged appearance.

Included	Not protruding beyond an organ, as stamens included in the corolla; cf. *Exerted*.
Incomplete	A flower with one or more whorls of parts lacking; cf. *Complete*.
Incomplete septum	A partially formed septum.
Indehiscent	Not opening along regular lines or sutures.
Indurate	Hardened.
Indusium	An epidermal outgrowth that covers the sorus in ferns.
Inferior	Below.
Inflated	Swollen or bladderlike.
Inflexed	Turned or bent inward.
Inflorescence	A flower cluster, or the arrangement of flowers on a stem.
Inframarginal	Below the margin.
Innovation	A basal, sterile shoot of grasses and sedges; cf. *Culm*.
Inserted	Connected or attached.
Internode	The section of stem between two nodes.
Introrse	Turned inward.
Involucel	A secondary involucre in a compound inflorescence.
Involucrate	Having an involucre.
Involucre	The whorl of bracts subtending an inflorescence.
Involute	Rolled inward from the edges toward the upper side; cf. *Revolute*.
Irregular	Having a bilaterally symmetrical structure (usually applied to flowers); also, zygomorphic; cf. *Regular*.
Keel	A dorsal, longitudinal ridge; the saccate structure formed by fusion of two petals in some members of the Fabaceae.
Lacerate	Having an irregular margin, appearing torn.
Laciniate	Irregularly cut into narrow lobes or segments.
Lanate	Covered with woolly, tangled hairs; cf. *Tomentose*.
Lanceolate	Lance-shaped, widest near the base and tapering to the apex.
Lateral	On or at the side.
Lax	Loose or sparse; separated.
Leaf	The photosynthetic organ; typically it consists of a blade and a petiole.
Leaflet	A division or segment of a compound leaf blade.
Legume	A dry, dehiscent fruit from a single carpel, splitting along two sutures or sometimes breaking crosswise between constrictions.
Lemma	The larger of the two bracts enclosing the stamens and pistil of a grass floret; cf. *Palea*.
Lenticel	A corky spot on young bark that functions in gas exchange.
Lenticular	Lens-shaped.
Ligulate	Strap-shaped; having a ligule.

Ligule	A membranous or fringed appendage at the summit of the leaf sheath in grasses; a strap-shaped corolla in members of the Asteraceae.
Limb	The expanded, spreading portion of a sympetalous corolla above the tube.
Linear	Long and narrow, with parallel sides.
Lithophilous	Literally, "rock-loving"; referring to plants that grow on rock surfaces or in crevices.
Lobed	Divided or indented less than halfway to the midrib or center.
Locule	A compartment or cavity on a pistil or anther.
Loculicidal	Dehiscing on the medial dorsal surfaces, between the septae and in the locules of a capsule.
Lodicule	A small, hyaline scale at the base of the ovary in a grass floret.
Loment	A legume with constrictions between the seeds.
Long-creeping.	Spreading with relatively long internodes; as some rhizomes of grasses and sedges.
Lunate	Crescent-shaped, as a half-moon.
Lyrate	Lyre-shaped; a pinnatifid leaf with a large terminal lobe and lateral lobes becoming smaller toward the base.
Malpighian	Referring to a straight hair attached near the middle and tapered toward both ends; cf. *Dolabriform*.
Marcescent	Withering and persistent, thus remaining attached.
Medial	Pertaining to the middle.
Megasporangium	The sporangium that produces megaspores.
Megaspore	The larger, female spore of heterosporous plants.
Membranaceous	Membranelike, thin and translucent.
Membranous	See *Membranaceous*.
Mericarp	One of two carpels in the Apiaceae that mature independently as fruits; cf. *Schizocarp*.
-merous	Suffix pertaining to parts or multiples of parts; always preceded by a number denoting how many, e.g., 3-merous, 4-merous, 5-merous.
Microsporangium	The sporangium that produces microspores.
Microspore	The smaller, male spore of heterosporous plants.
Midnerve	Middle nerve; see *Nerve*.
Midrib	The middle vein in a leaf or other structure; cf. *Net venation* and *Parallel venation*.
Monadelphous	A form in which the stamens are united into one group or tube by their filaments; cf. *Diadelphous*.
Monoecious	With staminate and pistillate flowers on the same plant; cf. *Dioecious*.
Mucilaginous	Slimy or mucilage-like.
Mucro-umbonate	With a short, sharp point in the center; cf. *Umbonate*.
Mucronate	With a short, sharp point or tip; cf. *Apiculate*.
Multicellular hair	Hair with three or more cells distinguished by clear or colored cross-walls.
Multiseriate	In three or more series or rows; cf. *Biseriate, Uniseriate*.

Multistemmed	With three or more stems originating from the base.
Muricate	Having a surface made rough by numerous, small, hard, sharp projections.
Muriculate	Diminutive of muricate; having very small projections.
Mycorrhizal	Pertaining to a mutualistic relationship between a fungus and a vascular plant root.
Naked	Lacking a perianth or a covering of hairs or scales.
Nectar	Sugary solution produced by a nectary.
Nectary	A plant gland, organ, or part that produces nectar.
Nectiferous	Capable of producing nectar.
Nerve	A vein or thin, narrow rib.
Nerved	Having one or more nerves.
Nerveless	Lacking nerves.
Net venation	A vein pattern characterized by branching and interconnected veins not parallel to the midrib; cf. *Parallel venation*.
Nodal	Pertaining to the node; located or originating at a node.
Node	The point of attachment of a leaf.
Numerous	More than ten.
Nut	A dry, indehiscent fruit from a compound pistil, with a hard, bony pericarp (wall) and one seed.
Nutlet	Diminutive of nut.
Obcordate	Heart-shaped, with the notch at the apex and tapered toward the base.
Oblanceolate	Lance-shaped, widest near the apex and tapered to the base.
Oblique	Asymmetrical, with unequal sides.
Oblong	An elongate shape with the central margins parallel and the ends tapered equally.
Obovate	Egg-shaped, with the widest part near the apex and tapered to the base; also obovoid.
Obovoid	See *Obovate*.
Obscure	Not readily visible; cf. *Obsolete*.
Obsolete	Not evident; vestigial or rudimentary; cf. *Obscure*.
Obtuse	Blunt or rounded at the apex.
Ochroleucous	Yellowish white or cream-colored.
Odd-pinnate	Pinnately compound with a terminal leaflet.
Oil tube	A tube containing aromatic oil found in the mericarps of members of the Apiaceae, visible as surface ribs or nerves.
Operculate	With a lid or cover (operculum) that functions in dehiscence of a fruit.
Opposite	Situated across or against, or as in leaves two at a node.
Orbicular	Flat and circular in shape.
Oval	Broadly elliptic.
Ovary	The basal portion of a pistil that contains one or more ovules.

Ovate	Egg-shaped, with the widest part near the base and tapered to the apex; also ovoid.
Ovoid	See *Ovate*.
Ovulate	Bearing or possessing ovules, as an ovulate cone in members of the Pinaceae.
Ovule	A structure, usually in the ovary, that develops into a seed.
Palate	The raised, convex portion of the lower lip in a bilabiate corolla that closes or nearly closes the throat.
Palea	The smaller of the two bracts that enclose the stamens and pistil of a grass floret; cf. *Lemma*.
Palmate	Having lobes, leaflets, or similar structures arising from a common point, as fingers originating from the palm of the hand.
Palmatifid	Palmately cleft into lobes that extend more than halfway to the base.
Panicle	A compound inflorescence with secondary branching from a vertical axis; a compound raceme.
Paniculate	Borne in a panicle.
Paniculiform	Resembling a panicle.
Papillae	Small, dome-shaped protuberances.
Papillate	See *Papillose*.
Papillose	Having papillae.
Pappus	The calyx of flowers in members of the Asteraceae, consisting of awns, bristles, or scales at the tip of the akene.
Parallel venation	A vein pattern characterized by nonbranched veins arranged parallel to the midrib; cf. *Net venation*.
Parietal placentation	Attachment of ovules or seeds to the wall of the ovary.
Parted	Cleft more than halfway to the midrib or center.
Pectinate	Pinnatifid with narrow, parallel lobes resembling a comb.
Pedicel	The stalk of a flower, or a grass spikelet, in an inflorescence.
Pedicellate	Having a pedicel.
Peduncle	The stalk of a single flower or inflorescence.
Pedunculate	Having a peduncle.
Peltate	Attached in the center of a surface rather than at the edge, as the handle of an umbrella.
Pendent	Hanging down; also, pendulous.
Pendulous	Pendent.
Pentate	In fives, or with five parts.
Perennial	A plant that lives for more than two years.
Perfect	A flower having both stamens and pistil(s).
Perianth	The calyx and corolla collectively; used when both floral whorls are alike in size and color.
Pericarp	The ovary wall in fruit.

Perigynium	A hypogynous sac that covers the ovary and akene in *Carex* and *Kobresia*.
Perigynous	With flower parts attached around the ovary on a cup-shaped hypanthium that is an outgrowth of the receptacle.
Persistent	Remaining attached for a prolonged time rather than falling off.
Petal	A modified and usually colored leaf that is the unit of the corolla.
Petaloid	Resembling a petal or petals.
Petiolate	Having a petiole.
Petiole	The stalk of a leaf blade.
Petiolulate	Having a petiolule.
Petiolule	The stalk of a leaflet in a compound leaf.
Phyllopodic	Having leaves at the base of the stem, as in *Carex*.
Pilose	Covered with soft, thin hairs.
Pinna	The primary division of a pinnate leaf.
Pinnate	Having lobes, leaflets, or similar structures arising on either side of a rachis or axis.
Pinnatifid	Pinnately cleft into narrow lobes that are formed more than halfway to the base or midrib.
Pinnule	The secondary division of a pinnate leaf.
Pistil	The female, seed-bearing structure of a flower, consisting of a stigma, style, and ovary.
Pistillate	Having a pistil or pistils only; female.
Pith	The soft, spongy center of a stem, produced by large, thin-walled parenchyma cells.
Placenta	Area or point of attachment of ovules within an ovary, or seeds within a fruit.
Placentation	The type of attachment of ovules within an ovary, or seeds within a fruit.
Plait	A fold.
Plaited	Folded like a fan, as some corollas in members of the Gentianaceae; plicate.
Planoconvex	Flattened on one side (plano-) and curved outward (convex) on the other; one-half of a lens in cross section.
Plicate	Plaited.
Plumose	Featherlike, with fine hairs spreading at right angles to an axis.
Pollinium	A waxy mass of pollen found in members of the Orchidaceae.
Polygamodioecious	Dioecious but having a few perfect flowers.
Polygamous	Having perfect and unisexual flowers on the same plant.
Polymorphic	Having several to many forms.
Pome	A fleshy fruit from the compound pistil of a perigynous flower, with the hypanthium adnate to the ovary wall, as an apple and other fruits in members of the Rosaceae.
Pore	A small hole.
Poricidal capsule	A capsule that opens, or dehisces, by pores.

Precocious	Flowering before the leaf blades expand, as in members of the Salicaceae; cf. *Cotaneous, Serotinous*.
Prickle	A small, spinelike, epidermal outgrowth.
Primocane	A first-year, nonflowering stem (cane); cf. *Floricane*.
Procumbent	Lying flat on the ground; prostrate.
Prostrate	See *Procumbent*.
Pruinose	With a whitish or bluish, waxy powder on the surface.
Pseudoscape	A false scape; a short stem between the root crown and an apparent basal leaf or whorl of leaves.
Puberulent	Minutely or finely pubescent.
Pubescent	Having hairs; cf. *Glabrate, Glabrous*.
Pulvinate	Cushion-shaped.
Punctate	With glandular, often colored, dots.
Pungent	With a short, sharp point at the tip.
Punticulate	Minutely punctate.
Pustulate	Having hairs with swollen, blisterlike, irregular bases, as in members of the Boraginaceae.
Pyramidal	Pyramid-shaped.
Quadrangular	Square in cross section; four-angled.
Raceme	An elongate, indeterminate inflorescence with a central axis and pedicellate flowers; cf. *Spike*.
Racemose	Racemelike; occurring in racemes.
Rachilla	Axis of a spikelet in the Poaceae that bears florets.
Rachis	The axis of a compound leaf or inflorescence.
Radiate	Having rays; spreading from a central point.
-ranked	Arranged in vertical rows; e.g., two-ranked, three-ranked.
Ray	The primary branch of a compound umbel; also see *Ray flower*.
Ray flower	A strap-shaped (ligulate) flower in members of the Asteraceae.
Receptacle	The expanded portion of an axis that bears flower parts in a single flower, or multiple flowers in a head.
Recurved	Curved backward or downward.
Reflexed	Bent sharply backward or downward.
Regular	Having a radially symmetrical structure (usually applied to flowers); also, actinomorphic; cf. *Irregular*.
Remote	Scattered or spaced some distance away.
Reniform	Kidney-shaped.
Replum	The septum or partition in fruits of members of the Brassicaceae.
Resinous	Having resin, or appearing sticky.

Reticulate	Having a network or a net pattern.
Retrorse	Angled downward or backward.
Retuse	Notched at the tip.
Revolute	Rolled downward from the edges toward the lower side; cf. *Involute*.
Rhizomatous	Having one or more rhizomes.
Rhizome	A horizontal underground stem with buds, scales, nodes, and roots; also rootstock.
Rhizophore	A short, leafless branch that gives rise to roots in members of the Selaginellaceae.
Rhombic	Quadrangular in outline with two obtuse angles; diamond-shaped.
Rib	A prominent nerve or vein.
Ribbed	Having prominent nerves or veins.
Root	A descending, mostly underground organ lacking buds, scales, and nodes.
Rootstock	See *Rhizome*.
Rosette	A cluster of radiating leaves, usually at the base of a stem.
Rostellum	The small, beaklike structure extending from the upper edge of the stigma in members of the Orchidaceae.
Rosulate	Having a rosette.
Rotate	Wheel-shaped; a corolla with a very short tube and spreading limb.
Rugose	Wrinkled.
Rugulose	Finely wrinkled.
Runcinate	Having a toothed or cleft margin with the teeth angled backward.
Saccate	Sack-shaped.
Sagittal	A medial plane dividing a structure into right and left halves.
Sagittate	Arrowhead-shaped.
Salverform	Having a corolla with a long, slender tube and a limb spreading at right angles to the tube.
Saprophyte	A plant that derives its nutrients from dead organic matter.
Scaberulous	Weakly scabrous.
Scabrous	Having short, stiff hairs.
Scale	A thin, dry, reduced leaf.
Scaly	Having scales.
Scape	A leafless flowering stem.
Scapose	Having a scape; cf. *Acaulescent*.
Scarious	Thin, dry, membranous.
Schizocarp	A dry fruit that splits (dehisces) at maturity into two or more seedlike portions (mericarps) in members of the Apiaceae and Malvaceae.
Scorpioid	Curved like the tail of a scorpion, an inflorescence with flowers in two rows along the outer side.

Secund	One-sided.
Seed	The mature or ripened ovule that contains the young embryo.
Semishrub	A plant with a woody base and herbaceous branches.
Sepal	A modified leaf that is the unit of the calyx.
Sepaloid	Resembling a sepal or sepals.
Septate-nodulose	A partition of small nodules.
Septicidal capsule	A capsule that dehisces on the septae and between the locules.
Septum	A partition.
Sericeous	Silky with soft, straight, appressed hairs.
Serotinous	Flowering after the leaf blades have expanded, as in members of the Salicaceae; also applied to cones (of Pinaceae) that remain closed for long periods after ripening; cf. *Cotaneous, Precocious*.
Serrate	Having a margin with sharp, forward-pointing teeth.
Serrulate	Margin with very small teeth pointed forward.
Sessile	Lacking a stalk and attached directly at the base.
Seta	A bristle.
Sheath	A tubular structure surrounding a plant part, as a leaf sheath surrounds the stem.
Short-creeping	Short and nonspreading with very short internodes, as some rhizomes of grasses and sedges.
Shrub	A low, woody plant, branched from the base.
Sigmoid	Shaped like the letter S.
Silicle	A short silique, generally less than two times longer than wide; cf. *Silique*.
Silique	A dry, dehiscent, elongate fruit with two valves splitting away from a persistent septum (replum), as in members of the Brassicaceae; cf. *Silicle*.
Silky	See *sericeous*.
Simple	Composed of an unbranched or undivided unit; not compound.
Simple pubescence	Unbranched hairs.
Sinuate	Wavy-margined, the waves in a horizontal plane; cf. *Undulate*.
Sinus	The space between two lobes of a corolla or leaf.
Sorus	A group of sporangia in ferns.
Spathe	A large, modified leaf that subtends or surrounds an inflorescence (spadix).
Spatulate	Spatula-shaped; cf. *Oblanceolate, Obovate*.
Spicate	Spikelike; occurring in spikes.
Spiciform	Spikelike.
Spike	An elongate, indeterminate inflorescence with a central axis and sessile flowers; cf. *Raceme*.
Spikelet	A small or secondary spike; in grasses, a unit of the inflorescence consisting of two glumes and one or more florets; in sedges, a portion of an inflorescence subtended by a bract.

Spine	A short, modified branch that is sharp at the tip.
Spinose	Having spines.
Spinulose	Having short spines.
Sporangium	A sac or case containing spores.
Spore	A haploid reproductive cell capable of developing into a new individual.
Sporophyll	A spore-bearing leaf.
Sporophyte	The spore-producing generation or plant in alternation of generations.
Spur	A slender, tubular outgrowth on a petal or sepal, often containing nectar.
Spurred	Having a spur.
Squamose	Having scales (squamae).
Stamen	The male, pollen-producing structure of a flower, consisting of an anther and a filament.
Staminate	Having a stamen or stamens only; male.
Staminode	A sterile stamen, often lacking an anther, or modified as a nectary or a petal.
Standard	The large, dorsal petal in the flower of a member of the Fabaceae; cf. *Banner*.
Stellate	Star-shaped; a stellate hair usually has five rays originating from a central point.
Stem	The aboveground or underground axis of a plant that bears nodes and buds; several modifications of the typical aboveground, erect, leafy axis exist; cf. *Bulb, Corm, Rhizome, Stolon, Tuber*.
Sterigma	A short, peglike projection at the base of a leaf in *Picea*.
Sterile	Infertile; not producing a seed, or lacking reproductive structures.
Stigma	The portion of a pistil specialized for receiving pollen.
Stigmatic	Pertaining to the stigma.
Stipe	The stalk of a pistil or fruit above the receptacle; a stalk.
Stipitate-glandular	Having stalked glands.
Stipular sheath	The sheath formed around a stem from the fused stipular margins at the base of a leaf.
Stipulate	Having stipules.
Stipule	One or a pair of usually leaflike appendages at the base of a petiole.
Stolon	A horizontal, aboveground stem that roots at the nodes.
Stoloniferous	Having stolons.
Striate	Having fine, parallel lines or grooves.
Strigillose	Minutely strigose.
Strigose	Having sharp, stiff, straight hairs appressed to the surface; cf. *Hispid*.
Strobilus	A cone or conelike structure with an axis and imbricate scales.
Stylar beak	A tapered style shaped like a sharp-pointed beak.
Style	The stalklike or elongate portion of a pistil between the stigma and ovary.
Stylopodium	An enlargement at the base of the styles in the Apiaceae.

Sub-	A prefix meaning nearly, somewhat, or almost.
Subtend	To occur below or near the base of a structure.
Subterranean	Under the ground.
Subulate	Awl-shaped; tapering to a fine point from a thick base.
Succulent	Juicy, fleshy.
Superior	Above.
Suture	A seam of fusion between two parts, or a line of dehiscence.
Sympetalous	With a corolla of fused petals; also, gamopetalous.
Taproot	The enlarged primary axis of a root system.
Tawny	Brownish yellow.
Terete	Elongate and circular in cross section.
Terminal	At the apex or end.
Ternate	Occurring in threes.
Terrestrial	Growing on land or a land habitat; cf. *Aquatic*.
Tetrad	A group of four objects or structures.
Tetradynamous	Having four long stamens and two short stamens, as in members of the Brassicaceae.
Three-ranked	Arranged on three sides of an axis, as in three-ranked leaves.
Tomentose	Densely woolly with soft, matted hairs; cf. *Lanate*.
Tomentulose	Slightly or finely tomentose.
Tomentum	A woolly, matted pubescence.
Toothed	Having small marginal teeth.
Torulose	Cylindrical with constrictions at regular intervals between nonconstricted areas.
Tridentate	Three-toothed.
Trifid	Three-lobed or twice-cleft.
Trifoliate	With three leaves.
Trifoliolate	With three leaflets.
Trigonous	Three-angled; also, triquetrous.
Tripinnate	Three times pinnately compound.
Triquetrous	Three-angled; also trigonous.
Truncate	A base or apex that ends abruptly as if cut off at a right angle to the longitudinal axis.
Tuber	A short, thickened underground stem modified for storage, as a potato.
Tubercle	A small tuber or tuberlike structure; the persistent base of the style in some members of the Cyperaceae.
Tuberculate	Having tubercles.
Tuberous	Having a tuber or tubers.
Tubular	Cylindrical; shaped like a tube.

Tufted	Clumped, or growing in a clump; caespitose.
Tunicated	Having concentric layers.
Turbinate	Top-shaped.
Turgid	Swollen or plump.
Turion	A scaly, subterranean bud or shoot arising from a rhizome.
Two-ranked	Arranged on two sides of an axis, as two-ranked leaves.
Umbel	An inflorescence, usually flat-topped, consisting of several to many flower-bearing pedicels with a common point of attachment.
Umbellet	A secondary umbel.
Umbellate	Having one or more umbels, or occurring in umbels.
Umbonate	Bearing an umbo, or projection, in the center.
Unarmed	Lacking prickles, spines, or thorns.
Undulate	Wavy-margined, the waves in a vertical plane; cf. *Sinuate*.
Uniseriate	In a single series or row; cf. *Biseriate, Multiseriate*.
Unisexual	A male (staminate) or female (pistillate) flower; cf. *Perfect*.
Urceolate	Urn-shaped.
Utricle	A dry, thin-walled, indehiscent, one-seeded fruit, often inflated.
Valvate	Opening by valves.
Valve	A separable unit of a dehiscent fruit, as a capsule or legume.
Vascular bundle	An internal group of tissues and cells specialized for the conduction of water and nutrients, and for support in the stem.
Velum	The membranous indusium in *Isoetes*.
Ventral	The inner surface, or the surface nearest the axis; also adaxial.
Versatile	Attached near the middle.
Verticillate	Arranged in verticils, or whorls.
Villous	Having long, soft hairs, giving a shaggy but not matted appearance.
Viscid	Sticky.
Viviparous	Sprouting from a seed or bulblet while still attached to the parent plant.
Wanting	Lacking.
Web	A tuft of hairs at the base of the lemma in grasses.
Webbed	Having a tuft of hairs.
Whorl	Three or more leaves or branches at a node; a verticil.
Whorled	Arranged in whorls.
Wing	A membranous appendage of a fruit or seed; a lateral petal in a flower in members of the Fabaceae.
Winged	Having a wing, as a winged fruit.

Wiry	Wirelike.
Woody	Having secondary growth; nonherbaceous.
Woolly	Having long, matted hairs; cf. *Lanate, Tomentose*.
Zygomorphic	Having a bilaterally symmetrical structure, irregular; usually applied to flowers; cf. *Actinomorphic*.

APPENDIX 3

Authors of Accepted Species Names

Abrams	LeRoy Abrams, 1874–1956, professor of botany at Stanford University; author of California and Pacific state floras.
Adans.	Michel Adanson, 1727–1806, French botanist; author of *Familles des Plantes.*
Agardh	Carl Adolf Agardh, 1785–1859, professor of botany in Lund; student of the algae; author of *Synopsis Algarum Scandinaviae* and *Classes plantarum.*
Ait.	William Aiton, 1731–1793, English botanist and royal gardener at Kew; author of *Hortus Kewensis.*
All.	See Allioni.
Allioni	Carlo Allioni, 1725–1804, professor of botany in Turin; author of *Flora Pedemontana.*
Arn.	George Arnold Walker Arnott, 1799–1868, Scottish botanist; coauthor of *The Botany of Captain Beechey's Voyage,* with W.J. Hooker.
Asch.	Paul Friedrich August Ascherson, 1834–1913, professor of botany in Berlin; author of *Flora der Provinz Brandenburg.*
B. & H.	See Benth. & Hook.
L.H. Bailey	Liberty Hyde Bailey, 1858–1954, horticulturist at Cornell University; treatments of cultivated plants, *Carex;* author of *Standard Cyclopedia of Horticulture.*
Ball	Carleton Roy Ball, 1873–1958, agronomist at the U.S. Department of Agriculture; student of *Salix.*
Barneby	Rupert Charles Barneby, 1911–, New York Botanical Garden; student of the Great Basin flora; author of *Atlas of North American Astragalus.*
Barr.	Joseph Barratt, 1796–1882, Connecticut geologist and physician; student of *Salix;* author of *The Indian of New England.*
Beal	William James Beal, 1833–1924, botanist at Michigan State University; author of *Michigan Flora.*
Beauv.	Ambroise Marie François Joseph Palisot de Beauvois, 1752–1820, French naturalist; student of Poaceae; author of *Essai d'une nouvelle agrostographie, ou nouveaux genres des graminées.*
Bebb	Michael Schuck Bebb, 1833–1895; student of *Salix;* author of *Willows of California.*
Benson	Lyman David Benson, 1909–, professor of botany at Pomona College; author of *Cacti of the United States and Canada,* and coauthor of *Trees and Shrubs of Southwestern Deserts,* with Robert Darrow.
Benth.	George Bentham, 1800–1884, English botanist; author of *Handbook of the British Flora* and *Flora Australiensis.*

Benth. & Hook. (B. & H.)	G. Bentham and J.D. Hooker, coauthors of *Genera Plantarum;* see Benth., and also Hook.
Bernh.	Johann Jacob Bernhardi, 1774–1850, professor of botany in Erfurt, Germany; author of *Systematisches Verzeichniss der Pflanzen um Erfurt.*
S. Berthelot	Sabin Berthelot, 1794–1880, French consul on Teneriffe; author of *Histoire Naturelle des Îles Canaries.*
Bess.	Wilibald Swibert Joseph Gottlieb von Besser, 1784–1842, Austrian botanist; student of the flora of Galicia and southwest Russia; author of *Primitiae Florae Galiciae Austriacae utriusque.*
Blake	Sidney Fay Blake, 1892–1959, U.S. Department of Agriculture; coauthor of *Geographical Guide to the Floras of the World,* with A.C. Atwood.
Blank.	Joseph William Blankinship, 1862–1938, botanist at Montana State University, Bozeman (formerly Montana State Agricultural College); student of the flora of Montana.
Boiss.	Pierre Edmond Boissier, 1810–1885, Swiss botanist; author of *Flora Orientalis.*
B. Boivin	Joseph Robert Bernard Boivin, 1916–1985, Department of Agriculture, Ottawa, Canada; author of *Flora of the Prairie Provinces.*
Bong.	August Heinrich Gustav Bongard, 1786–1839, professor of botany in St. Petersburg, Russia; student of Brazilian and Alaskan plants; author of *Observations sur la végétation de l'Îsle de Sitcha.*
Bonpland	Aimé Jacques Alexandre Bonpland, 1773–1858, French botanist; author of *Nova genera et species plantarum.*
F. Boott	Francis Boott, 1792–1863, physician and botanist; author of *Illustrations of the Genus Carex.*
W. Boott	William Boott, 1805–1887; student of *Carex* and Polypodiaceae.
Börner (also Boerner)	Carl Julius Börner, 1880–1953; German botanist; student of *Gentianella,* author of *Eine Flora für das Deutsche Volk.*
R. Br.	Robert Brown, 1773–1858, Scottish botanist, British Museum; student of arctic plants; author of *Prodromus Florae Novae Hollandiae et insulae VanDiemen.*
Briq.	John Isaac Briquet, 1870–1931, Swiss botanist, director of the Conservatoire Botanique in Geneva; student of Apiaceae, Asteraceae, and Lamiaceae. Author of *Notes floristiques sur les Alpes lemaniennes.*
Britt. & Rusby	N.L. Britton and H.H. Rusby; see Britton, and also Rusby.
Britton	Nathaniel Lord Britton, 1859–1934, director of the New York Botanical Garden; author of *An Illustrated Flora of the Northern United States.*
Buckl.	Samuel Botsford Buckley, 1809–1884, American naturalist; state geologist of Texas.
Buek	Heinrich Wilhelm Buek, 1796–1879; author of *Genera, species et synonyma Candolleana alphabetico ordine disposita.*
Carr.	Elie Abel Carrière, 1818–1896, French horticulturist, Muséum d'Histoire Naturelle, Paris; student of Coniferae; *Traite general des Coniferes.*
Cass.	Alexandre-Henri Gabriel, Comte de Cassini, 1781–1832, French botanist; student of Asteraceae; *Opuscules phytologiques.*

Cham.	Adalbert Ludwig von Chamisso (de Boncourt), 1781–1838, curator of the Royal Botanical Gardens at Berlin; poet-naturalist and early botanist in North America; *Uebersicht der nutzbarsten und schadlichsten Gewächse.*
Clairville	Joseph Phillippe de Clairville, 1742–1830, Swiss botanist; author of *Manuel d'herborisation en Suisse et en Valais.*
Cockll.	Theodore Dru Alison Cockerell, 1866–1948, professor of zoology, University of Colorado; student of southwestern natural history and the flora of Colorado.
Coult.	John Merle Coulter, 1851–1928, professor of botany at the University of Chicago; author of *Manual of the Botany of the Rocky Mountain Region* and coauthor of *Gray's Manual of Botany of the Northern United States,* 6th ed., with Sereno Watson.
Coult. & Rose	J.M. Coulter and J.N. Rose; see Coult., and also Rose.
Cov.	Frederick Vernon Coville, 1867–1937, curator of the U.S. National Herbarium; studied Death Valley and Alaska botany; author of *Botany of the Death Valley Expedition.*
Cronq.	Arthur John Cronquist, 1919–1992, curator of the New York Botanical Garden; student of Asteraceae, North American flora; author of *An Integrated System of Classification of Flowering Plants* and coauthor of *Vascular Plants of the Pacific Northwest.*
DC.	Augustin Pyramus de Candolle (also Decandolle), 1778–1841, professor of botany in Geneva; first author of *Prodromus systematis naturalis regni vegetabilis.*
Degl.	Jean Vincent Yves Degland, 1773–1841, professor of botany, Rennes; student of *Kobresia;* author of *De Caricibus Galliae indigenis tentamen.*
Desv.	Niçaise Auguste Desvaux, 1784–1856, professor of botany in Angers; editor of *Journal de Botanique;* author of *Observations sur les plantes des environs d'Angers.*
Dewey	Chester Dewey, 1784–1867, professor at the University of Rochester; student of *Carex,* plants of Massachusetts; author of *Caricography.*
A. Dietr.	Albert Gottfried Dietrich, 1795–1856, curator of the Berlin Botanical Garden; author of *Flora regni Borussici.*
D. Don	David Don, 1800–1841, British botanist, brother of George, professor at Kings' College, London; author of *Prodromus florae Nepalensis.*
G. Don	George Don, 1798–1856, brother of David, plant collector for the Royal Horticultural Society in South America and Africa; author of *A General History of the Dichlamydeous Plants.*
Dorn	Robert Dorn, 1942–; student of western U.S. plants, *Salix;* author of *Vascular Plants of Wyoming, Vascular Plants of Montana,* and *Flora of the Black Hills.*
Dougl.	David Douglas, 1799–1834, plant collector for the Royal Horticultural Society in western North America, discoverer of many species new to science; author of *An Account of the Species of Calochortus.*
Duchn.	Antoine Nicolas Duchesne, 1747–1827, French botanist and horticulturalist; student of *Fragaria;* author of *Histoire naturelle des Fraisiers.*
Duhamel	Henri Louis Duhamel du Monceau, 1700–1782, French botanist and forester; author of *Traité des arbres fruitiers,* and *La physique des arbres.*
D'Urv.	Jules Sébastian César Dumont d'Urville, 1790–1842, French admiral; author of *Flora des îles Malouines.*
Eastw.	Alice Eastwood, 1859–1953, curator of botany at the California Academy of Sciences; flora of western North America.

Eat.	Amos Eaton, 1776–1842, American botanist; author of *Manual of Botany for North America* and *North American Botany*.
D.C. Eat.	Daniel Cady Eaton, 1834–1895, grandson of Amos, professor of botany at Yale University; author of *The Ferns of North America*.
Ehrh.	Friedrich Ehrhart, 1742–1795, German botanist; author of *Beiträge zur Naturkunde*.
Engelm.	George Engelmann, 1809–1884, St. Louis physician; botany of early western American expeditions; author of *Revision of the North American species of the genus Juncus*.
Evert	E.F. Evert, 1940–, Wyoming botanist; *Antennaria, Shoshonea;* student of the northwestern Wyoming flora.
Fern.	Merritt Lyndon Fernald, 1873–1950, director of the Gray Herbarium; author of *Gray's Manual of Botany*, 8th. ed.
Fisch.	Friedrich Ernst Ludwig von Fischer, 1782–1854, director of the Botanic Garden at St. Petersburg, Russia, author of *Sertum Petropolitanum*.
Fisch. & Trautv.	F. von Fischer and E. von Trautvetter; see Fisch., and also Trautv.
Forbes	John Forbes, 1773–1861, English gardener at Woburn Abbey; worked with garden plants; author of *Hortus Woburnensis*.
Fourn.	Eugène Pierre Nicolas Fournier, 1834–1884, Paris physician; student of Asclepiadaceae; author of *Mexicanas plantas nuper a collectoribus expeditionis scientificae allatas*.
Franco	João Manuel Antonio do Amaral Franco, 1921–, Portuguese botanist; author of *Nova flora de Portugal* and *On the Nomenclature of the Douglas Fir*.
Franklin	John Franklin, 1786–1847, Arctic explorer; author of *Narrative of a Journey to the Shores of the Polar Sea in the Years 1819, 20, 21, and 22,* and *Narrative of a Second Expedition to the Shores of the Polar Sea in the Years 1825, 1826, and 1827*.
S.V. Fraser	Samuel Victorian Fraser, 1890–1972; flora of Kansas; author of *American Fruits, Their Propagation, Cultivation, Harvesting, and Distribution*.
Freedman	N.J. Freedman, fl. 1956, botanical collector in Yukon Territory, Canada.
Fries	Elias Magnus Fries, 1794–1878, professor of botany in Uppsala; student of *Hieracium;* author of *Summa vegetabilium Scandinaviae*.
Gaertn.	Joseph Gaertner, 1732–1791, German physician and botanist; author of *De fructibus et seminibus plantarum*.
Gaudin	Jean Francois Gottlieb Philippe Gaudin, 1766–1833, Swiss clergyman (in Nyon) and agrostologist; author of *Agrosographica alpina oder Beschreibung schweizerischer Gräser*.
Gentry	Howard Scott Gentry, 1903–, botanist for the U.S. Department of Agriculture; worked with economically important plants; author of *Agaves of Continental North America*.
Geyer	Carl (Charles) Andreas Geyer, 1809–1853, Austrian botanist; member of the Nicollet Expedition, early collector in the United States.
Goldie	John Goldie, 1793–1886; author of *Description of Some New and Rare Plants Discovered in Canada in the Year 1819*.
Gould	Frank Walton Gould, 1913–1981, professor of range and forestry at Texas A&M University; student of grasses; author of *The Grasses of Texas*.
Grah.	Robert Graham, 1786–1845, professor of botany in Edinburgh, Scotland; author of *Characters of Genera, Extracted from the British Flora of W.J. Hooker*.

Grant	Adele Lewis Grant, 1881–1969, professor of botany at the University of Southern California; student of *Mimulus*.
V. Grant	Verne Edwin Grant, 1917–, Rancho Santa Ana Botanic Garden, Claremont, California, and professor of botany at the University of Texas; student of Polemoniaceae and *Ipomopsis*.
Gray	Asa Gray, 1810–1888, professor of botany at Harvard University; author of *Manual of the Botany of the Northern States*, co-author of *A Flora of North America*, with J. Torrey.
Greene	Edward Lee Greene, 1843–1915, professor of botany at the University of California, Catholic University of America, and the Smithsonian Institution; author of *Flora franciscana* and editor of *Pittonia*.
Greenm.	Jesse More Greenman, 1867–1951, curator of the herbarium at the Missouri Botanical Garden; student of *Senecio*.
Griseb.	August Heinrich Rudolph Grisebach, 1814–1879, professor of botany in Göttingen, Germany; student of Gentianaceae; author of *Die Vegetation der Erde nach ihrer klimatischen Anordnung*.
Gunn.	Johann Ernst Gunnerus, 1718–1773, Norwegian bishop and botanist; author of *Flora norvegica*.
H. & A.	W.J. Hooker and G.A. Arnott; see Hook., and also Arn.
Hackel	Eduard Hackel, 1850–1926; student of European grasses; author of *Monographia Festucarum Europaearum*.
Haenke	Thaddaeus Peregrinus Xaverius Haenke, 1761–1817, Czech botanist; phytographer for the King of Spain, member of the *Malaspina* expedition; author of *Ejus Botanische Beobachtungen im Riesengebirge*, in Jirasek, *Beobachtungen*.
Hall	Harvey Monroe Hall, 1874–1932, professor of botany at the University of California, Carnegie Institution; student of Asteraceae; author of *The Genus Happlopappus: A Phylogenetic Study in the Compositae*.
Hartman	Ronald Hartman, 1945–, curator of the Rocky Mountain Herbarium, University of Wyoming; student of Apiaceae, Caryophyllaceae, the Rocky Mountain flora.
Hartman & Kirkpatrick	R. Hartman and R. Kirkpatrick; see Hartman, and also Kirkpatrick.
F.X. Hartmann	Franz Xaver von Hartmann, 1737–1791; author of *Primae linae institutionum botanicarum Crantzii*.
Hausskn.	Heinrich Karl Haussknecht, 1838–1903, German botanist; explorer of the Orient, student of *Epilobium*; author of *Monographie der Gattung Epilobium*.
Hayek	August von Hayek, 1871–1928, Austrian botanist; student of *Anemone*, *Senecio*; author of *Flora von Steiermark*.
H.B.K.	F.W.H.A. von Humboldt, A.J.A. Bonpland, and C.S. Kunth, coauthors of *Nova genera et species plantarum*.
A. Heller	Amos Arthur Heller, 1867–1944, Pennsylvania botanist; student of western North American plants; author of *Catalog of North American Plants North of Mexico*, editor of *Muhlenbergia*.
Henders.	Louis Forniquet Henderson, 1853–1942, Oregon botanist; author of *The Early Flowering Plants in Lane County, Oregon*.
Henrard	Jan Theodoor Henrard, 1881–1974, Dutch agrostologist; student of Poaceae; author of *A Monograph of the Genus Aristida*.

Hiern	William Philip Hiern, 1839–1925, British Museum; student of the flora of Africa and India; author of *Botany of Lynton and neighbourhood*.
Hill	John Hill, 1716–1775. London physician and pharmacist; herbals and books on nature; author of *The British Herbal*.
A.S. Hitchc.	Albert Spear Hitchcock, 1865–1935, botanist at Kansas State University and the U.S. Department of Agriculture; author of *Manual of Grasses of the United States*.
C.L. Hitchc.	Charles Leo Hitchcock, 1902–1986, professor of botany at the University of Washington; student of the Pacific Northwest flora; senior author of *Vascular Plants of the Pacific Northwest*.
Hitchc. et al.	Charles Leo Hitchcock, Arthur Cronquist, Marion Ownbey, and J.W. Thompson, coauthors of *Vascular Plants of the Pacific Northwest*.
Hitchc. & Maguire	Charles Leo Hitchcock and Bassett Maguire; see C.L. Hitch. and also Maguire.
Holm	Herman Theodore Holm, 1854–1932, Danish botanist; student of Cyperaceae; author of *The Vegetation of the Alpine Region of the Rocky Mountains in Colorado*.
Hook.	William Jackson Hooker, 1785–1865, director of Kew; author of *Flora boreali-americana, Genera Filicum, Species Filicum, The British Flora*.
Hook. f.	Joseph Dalton Hooker, 1817–1911, son of W.J. Hooker, director of Kew; student of Himalayan, Indian, and New Zealand floras; author of *Flora antarctica, Flora Novae Zelandiae, Illustrations of Himalayan plants*.
Hoppe	David Heinrich Hoppe, 1760–1846, student of sedges; coauthor of *Caricologica Germanica*, with J. Sturm.
Hornem.	Jens Wilken Hornemann, 1770–1841, professor of botany in Copenhagen; author of *Flora Danica* and *Hortus regius botanicus Hafniensis*.
House	Homer Doliver House, 1878–1949, New York botanist; author of *Wild Flowers of New York*.
T.J. Howell	Thomas Jefferson Howell, 1842–1912, Oregon botanist; author of *A Flora of Northwest America*.
Huds.	William Hudson, 1730–1793, London apothecary; author of *Flora anglica*.
Hult.	Oskar Eric Gunnar Hultén, 1894–1981, professor of botany in Stockholm; student of northern floras; author of *Flora of Alaska and Neighboring Territories*.
Humboldt	Friedrich Wilhelm Heinrich Alexander von Humboldt, 1769–1859, German explorer and zoologist; author of *Cosmos: A Sketch of a Physical Description of the Universe*.
Iltis	Hugh Hellmut Iltis, 1925–, professor of botany at the University of Wisconsin; student of the Wisconsin flora, Gentianaceae, and Capparidaceae.
Jacq.	Nicolaus Joseph von Jacquin, 1727–1817, professor of botany in Vienna; author of *Flora austriacae* and *Icones plantarum rariorum*.
James	Edwin James, 1797–1861, surgeon and naturalist, early botanical collector in western North America; author of *Account of an Expedition from Pittsburgh to the Rocky Mountains, Performed in the Years 1819 and 1820*.
I.M. Johnston	Ivan Murray Johnston, 1898–1960, professor of botany at Harvard University; student of Boraginaceae, desert plants of the Southwest.
M.E. Jones	Marcus Eugene Jones, 1852–1934, Utah mining consultant; student of Great Basin plants, western botany.

Juss.	Antoine Laurent de Jussieu, 1748–1836, professor at the Jardin du Roi, Paris; characterization of natural families; author of *Genera plantarum*.
Kar.	Grigorij Silyč Karelin, 1801–1872, Russian botanist; explorer of Siberia; coauthor of *Enumeratio plantarum in desertis Songoriae orientalis* with I.P. Kirilov.
Kar. & Kir.	G.S. Karelin and I.P. Kirilov; see Kar., and also Kir.
Karst.	Gustav Karl Wilhelm Hermann Karsten, 1817–1908, professor of botany in Vienna; author of *Flora von Deutschland, Oesterreich und der Schweiz*.
Kearney	Thomas Henry Kearney, 1874–1956, U.S. Department of Agriculture, California Academy of Sciences; student of Malvaceae.
Kearney & Peebles	T.H. Kearney & R.H. Peebles, coauthors of *Flowering Plants and Ferns of Arizona* and *Arizona Flora*.
Keck	David Daniels Keck, 1903–, New York Botanical Garden, National Science Foundation; student of Asteraceae, *Penstemon, Orthocarpus,* and the California flora.
Kell.	Albert Kellogg, 1813–1887, San Francisco physician and botanist, founder of the California Academy of Sciences; author of *Forest Trees of California*.
L. Kelso	Leon Kelso, 1907–; student of western North American plants, *Salix,* and Poaceae.
Kir.	Ivan Petrovič Kirilov, 1821–1842, Russian botanist; coauthor of *Enumeratio plantarum in desertis Songoriae orientalis,* with G.S. Karelin.
Kirkpatrick	Robert Kirkpatrick, 1957–; student of mountain floras; author of *A Flora of the Southeastern Absarokas, Wyoming*.
Koch	Wilhelm Daniel Joseph Koch, 1771–1849, German botanist; student of Apiaceae; author of *Synopsis Florae germanicae et helveticae*.
Koel.	Georg Ludwig Koeler, 1765–1807, German botanist; studies on the grasses of Germany and France; author of *Descriptio graminum in Gallia et Germania*.
König	Karl Dietrich Eberhard Charles König (Koenig), 1774–1851, German geologist.
Kükenth.	Georg Kükenthal (Kuekenthal), 1864–1956, bishop, Coburg, Germany; author of *Das Pflanzenreich, Cyperaceae, Heft 38*.
Kunth	Carl Sigismund Kunth, 1788–1850, professor of botany in Berlin; coauthor of *Flora Berolinensis,* and *Handbuch der Botanik*.
L.	Carolus Linnaeus (Carl von Linné), 1707–1778, professor of botany, Uppsala; author of *Species Plantarum*, binomial nomenclature.
L. f.	Carl von Linné, the son, 1741–1783, also professor of botany in Uppsala; author of *Decas prima (et secunda) plantarum rariorum horti Upsaliensis*.
Lag.	Mariano Lagasca y Segura, 1776–1839, director of the Botanic Garden of Madrid, student of Antonio José Cavanilles; author of *Genera et species plantarum*.
Lam.	Jean Baptiste Antoine Pierre Monnet de Lamarck, 1744–1829, French naturalist; early evolutionary theory; coauthor of *Flore Française,* with A.P. de Candolle.
Lam. & DC.	J.B. de Lamarck and A.P. de Candolle; see Lam., and also DC.
Ledeb.	Carl Friedrich von Ledebour, 1785–1851, Estonian botanist; author of *Flora altaica* and *Flora rossica*.

Lehm.	Johann Georg Christian Lehmann, 1792–1860, director of the Botanic Garden of Hamburg; student of *Potentilla;* author of *Monographia generis Primularum* and *Monographia generis Potentillarum.*
Leiberg	John Bernhard Leiberg, 1853–1913, collector and explorer for the U.S. Department of Agriculture and the U.S. Geological Survey; author of *General Report on a Botanical Survey of the Coeur d'Alene Mountains.*
Leyss.	Friedrich Wilhelm von Leysser, 1731–1815, German administrator and botanist; author of *Flora Halensis.*
Lightf.	John Lightfoot, 1735–1788; author of *Flora Scotica.*
Lilj.	Samuel Liljeblad, 1761–1815, Swedish botanist and economist; author of *Utkast til en Svensk flora.*
Lindl.	John Lindley, 1799–1865, professor of botany in London, horticulturist; student of Orchidaceae; author of *Rosarum monographia.*
Link	Johann Heinrich Friedrich Link, 1767–1851, director of the Botanic Garden in Berlin; author of *Grundlehren der Anatomie und Physiologie der Pflanzen.*
Loisel.	Jean Louis Auguste Loiseleur-Deslongchamps, 1774–1849, French physician and botanist; author of *Flora gallica.*
Loud.	John Claudius Loudon, 1783–1843, English horticulturist; author of *An Encylopaedia of Trees and Shrubs.*
D. Löve	Doris Löve, 1918–, University of Colorado; student of arctic and alpine plants, Caryophyllaceae, nomenclature, and cytotaxonomy.
Ma	Yu-Chuan Ma, 1916–, Chinese botanist; student of Apiaceae, Brassicaceae, and Gentianaceae.
Macbr.	James Francis Macbride, 1892–1976, American botanist, Chicago Natural History Museum; floras of western America and Peru; author of *Common Wild Flowers, Flora of Peru.*
Mack.	Kenneth Kent Mackenzie, 1877–1934, New York attorney; student of *Carex;* author of Cyperaceae in *North American Flora*, 1931.
Macoun	John Macoun, 1831–1920, Canadian botanist; author of *Catalogue of Canadian Plants.*
Maguire	Bassett Maguire, 1904–1991, New York Botanical Garden; student of the flora of the Great Basin and South America, Asteraceae.
Malte	Malte Oscar Malte, 1880–1933, agrostologist, chief botanist at the National Herbarium of Canada; student of Poaceae.
Maxim	Carl Johann Maximovich, 1827–1891, director of the Botanic Garden, St. Petersburg; author of *Primitiae Florae Amurensis.*
McClatchie	Alfred James McClatchie, 1861–1906, American horticulturist, professor of botany at Throop Polytechnic Institute, Pasadena; author of *Flora of Pasadena.*
Merr.	Elmer Drew Merrill, 1876–1956, director of the Arnold Arboretum; student of the floras of China, Indo-Malaysia, and the Philippines; author of *Plant Life of the Pacific World* and *Botany of Cook's Voyages.*
Mert.	Franz Karl Mertens, 1764–1831, German botanist; coauthor of *Deutschlands Flora*, 3rd ed., with W.D.J. Koch.

Mey.	Georg Friedrich Wilhelm Meyer, 1782–1856, German botanist; author of *Flora hanoverana excursoria*.
C.A. Mey.	Carl Anton von Meyer, 1795–1855, director of the St. Petersburg Botanic Garden; author of *Florula provinciae Tambov, Florula provinciae Wiatka,* and other treatises in *Beiträge zur Pflanzenkunde der Russischen Reichs*.
E. Mey.	Ernst Heinrich Friedrich Meyer, 1791–1858, German botanist; student of Juncaceae; author of *Synopsis Juncorum, Synopsis Luzularum*.
Michx.	André Michaux, 1746–1802, French botanist; student of *Quercus*; author of *Flora boreali-americana*.
Mill.	Philip Miller, 1691–1771, British gardener; author of *The Gardeners Dictionary*.
Mirbel	Charles François Brisseau de Mirbel, 1776–1854, French botanist at the Musée d'Histoire Naturelle, Paris; author of *Traité d'anatomie et de physiologie végétales*.
Mitch.	John Mitchell, 1711–1768, physician and botanist; author of *Dissertatio brevis de principiis botanicorum et zoologorum*.
Moench	Conrad Moench, 1744–1805, German botanist; student of *Juniperus*; author of *Einleitung zur Pflanzenkunde*.
Mulligan	Gerald Alfred Mulligan, 1928–, Canadian cytotaxonomist; student of Brassicaceae.
Munro	William Munro, 1818–1880, English agrostologist; student of Poaceae; author of *Hortus agrensis*.
Murray	Johann Andreas Murray, 1740–1791, Swedish botanist, professor of botany at Göttingen, Germany; student of Linnaeus; author of *Prodromus designationis stirpium Gottingensium*.
Mutis	José Célestino Bruno Mutis y Bosio, 1732–1808, Spanish botanist; explorer and collector in Colombia; author of *La flora de la real expedición botánica del Nuevo Reino de Granada*.
Nash	George Valentine Nash, 1864–1921, head gardener at the New York Botanical Garden; horticulturalist and agrostologist; editor of *Addisonia,* contributor to *North American Flora*.
A. Nels.	Aven Nelson, 1859–1952, professor of botany and president of the University of Wyoming; student of the western North American flora; revised Coulter's *New Manual of Botany of the Central Rocky Mountains*.
E. Nels.	Elias Emanuel Nelson, 1876–1949, horticulturist, U.S. Department of Agriculture; author of *The Shrubs of Wyoming*.
Newberry	John Strong Newberry, 1822–1892; physician and naturalist for U.S. Army expeditions in the trans-Mississippi West.
Nutt.	Thomas Nuttall, 1786–1859, Philadelphia naturalist; early collector of western North American plants; author of *The Genera of North American Plants* and *The North American Sylva*.
Oakes	William Oakes, 1799–1848; student of the Massachusetts and Vermont floras.
Olney	Stephan Thayer Olney, 1812–1878, Rhode Island; student of *Carex*; author of *Algae rhodiaceae*.
Opiz	Philipp Maximilian Opiz, 1787–1858, Czech botanist; author of *Böheims phanerogamische und kryptogamische Gewächse*.

Osterh.	George Everett Osterhout, 1858–1937, Colorado lumberman and amateur botanist; student of the flora of Colorado.
M. Ownbey	Francis Marion Ownbey, 1910–1975, professor of botany at Washington State University; student of *Allium* and *Calochortus;* coauthor of *Vascular Plants of the Pacific Northwest.*
Packer	John G. Packer, 1929–, Canadian taxonomist, University of Alberta; author of *Flora of Alberta,* 2nd revised edition.
Pall.	Peter Simon Pallas, 1741–1811, German botanist; student of the Siberian flora, *Astragalus;* author of *Flora Rossica.*
K.L. Parker	Kittie Lucille (Fenley) Parker, 1910–, George Washington University; student of Asteraceae; author of *An Illustrated Guide to Arizona Weeds.*
Parry	Charles Christopher Parry, 1823–1890, botanist with the U.S.-Mexico Boundary Survey and other early surveys.
Pax	Ferdinand Albin Pax, 1858–1942, German botanist, professor of botany and director of the Breslau Botanical Garden; student of Aceraceae and Primulaceae; author of *Beiträge zur Morphologie und Systemtik der Cyperaceen.*
Pays.	Edwin Blake Payson, 1893–1927, professor of botany at the University of Wyoming; student of Boraginaceae, Brassicaceae, Asteraceae, and the flora of the Rocky Mountains.
Peeb.	Robert Hibbs Peebles, 1900–1956, agronomist for the U.S. Department of Agriculture; coauthor of *Flowering Plants and Ferns of Arizona* and *Arizona Flora,* with T.H. Kearney.
Penn.	Francis Whittier Pennell, 1886–1952, curator of Botany at the Academy of Natural Sciences of Philadelphia; author of *The Scrophulariaceae of Eastern Temperate North America.*
Pers.	Christian Hendrik Persoon, 1761–1836, Paris physician; author of *Synopsis Plantarum.*
Phil.	Rudolph Amandus Philippi, 1808–1904, director of the Museo Nacional and professor of botany in Santiago, Chile; author of *La flora de Nueva Zelanda comparada con la Chilena, Florula Atacamensis,* and *Descripción de las nuevas plantas.*
Phipps	Constantine John Phipps, 1744–1792, arctic explorer; author of *A Voyage to the North Pole.*
Piper	Charles Vancouver Piper, 1867–1926, agrostologist for the U.S. Department of Agriculture; author of *Flora of Washington.*
Poir.	Jean Louis Marie Poiret, 1755–1834, French botanist; author of *Flora of the Palatinate.*
Polunin	Nicholas Vladimir Polunin, 1909–, British botanist; author of *Circumpolar Arctic Flora* and *Botany of the Canadian Eastern Arctic.*
Porter	Thomas Conrad Porter, 1822–1901, professor of botany at Lafayette College; author of *Flora of Pennsylvania.*
Presl	Karel Boriwag Presl, 1794–1852, professor of natural history in Prague; author of *Reliquiae Haenkeanae.*
J. Presl	Jan Swatopluk Presl, 1791–1849, professor in Prague, the brother of Karel; coauthor of *Flora čechica,* with Karel Presl.
Pritz.	Georg August Pritzel, 1815–1874, Academy of Sciences, Berlin; student of Lycopodiaceae and *Anemone;* author of *Thesaurus literaturae botanicae.*

Pursh	Frederick Traugott Pursh, 1774–1820, Philadelphia; author of *Flora Americae septentrionalis*.
R. & S.	J.J. Roemer and J.A. Schultes; see Roem., and also Schult.
Raf.	Constantine Samuel Rafinesque (or Rafinesque-Schmaltz), 1783–1840, naturalist, botanical explorer; author of binomials, *Medical Flora, Florula Ludoviciana*.
Rech.	Karl Rechinger, 1867–1952. Austrian botanist; author of *Beiträge zur Flora von Oesterreich*.
Rech. f.	Karl Heinz Rechinger, 1906–, son of Karl Rechinger; author of *Flora aegaea* and *Flora iranica*.
Rees	Abraham Rees, 1743–1825; author of *Cyclopaedia*.
Regel	Eduard August von Regel, 1815–1892, scientific director of the St. Petersburg Botanical Garden; editor of *Gartenflora;* author of *Bemerkungen über die Gattungen Betula und Alnus*.
Reichenb.	Heinrich Gottlieb Ludwig Reichenbach, 1793–1879, German naturalist, professor in Dresden; author of *Flora Lipsiensis pharmaceutica*.
Retz.	Anders Johan Retzius, 1742–1821, Swedish botanist, professor of natural history at the University of Lund; author of *Florae Scandinaviae Prodromus*.
Richards	John Richardson, 1787–1865, Scottish botanist with the Franklin Expeditions to arctic America; author of the *Botanical appendix to John Franklin's Narrative*.
Richter	Karl Richter, 1855–1891, Vienna, collector; author of *Taschenbuch der Botanik*.
Robins.	Benjamin Lincoln Robinson, 1864–1935, curator of the Gray Herbarium; student of Asteraceae, Gentianaceae, and Rubiaceae.
Roem.	Johann Jacob Roemer, 1763–1819, Swiss botanist; produced an edition of Linnaeus's *Systema vegetabilium* with J.A. Schultes.
Rollins	Reed Clark Rollins, 1911–, director of the Gray Herbarium; student of Brassicaceae; coauthor of *The Genus Lesquerella* (*Cruciferae*) *in North America*, with Elizabeth Shaw; author of the *Cruciferae of continental North America*.
Rose	Joseph Nelson Rose, 1862–1928, botanist at the U.S. National Herbarium; student of Apiaceae, Cactaceae, and Crassulaceae.
Rosend.	Carl Otto Rosendahl 1875–1956, professor of botany at the University of Minnesota; student of the flora of Minnesota, *Heuchera*.
Rostk.	Friedrich Wilhelm Gottlieb Theophil Rostkovius, 1770–1848, Polish physician; student of Juncaceae; author of *Dissertatio botanica de Junco*.
Roth	Albrecht Wilhelm Roth, 1757–1834, German physician and botanist; author of *Tentamen florae germanicae* and *Manuale botanicum*.
Rothr.	Joseph Trimble Rothrock, 1839–1922, professor of botany at the University of Pennsylvania; botanist on the Wheeler Expedition to California, Colorado, and New Mexico.
Rottb.	Christen Friis Rottboell, 1727–1797, professor of botany and director of the Copenhagen Botanical Garden; author of *Plantas horti universitatis rariores*.
Rupr.	Franz Joseph Ruprecht, 1814–1870, Russian botanist, curator of the Herbarium at St. Petersburg, Russia; student of Apiaceae, Poaceae, and cryptogams; author of *Tentamen Agrostographiae universalis*.

Rusby	Henry Hurd Rusby, 1855–1940, dean of Columbia College of Pharmacy; student of the flora of South America, *Senecio*.
Rydb.	Per Axel Rydberg, 1860–1931, curator of the New York Botanical Garden; author of *Flora of Colorado, Flora of the Rocky Mountains and Adjacent Plains* and *Flora of the Prairies and Plains of Central North America*.
Salisb.	Richard Anthony Salisbury, 1761–1829, British botanist; author of *Prodromus stirpium*.
Scheele	Georg Heinrich Adolf Scheele, 1808–1864, German botanist; student of the flora of Texas.
Schlecht.	Diederich Franz Leonhard von Schlechtendal, 1794–1866, German botanist, University of Halle; student of Elaeagnaceae; cofounder of *Botanische Zeitung,* with H. von Mohl.
F. Schmidt	Friedrich Schmidt, 1832–1908; student of the Siberian flora; author of *Flora der Insel Moon*.
F.W. Schmidt	Franz Wilibald Schmidt, 1764–1796, professor of botany in Prague; author of *Flora boëmica inchoata*.
Schrad.	Heinrich Adolph Schrader, 1767–1836, German botanist, student of *Verbascum,* author of *Flora germanica* and *Analecta ad floram Capensem*.
Schult.	Joseph August Schultes, 1773–1831, Austrian botanist; produced an edition of Linnaeus's *Systema Vegetabilium,* with J.J. Roemer; author of *Oestreichs Flora*.
F.W. Schultz	Friedrich Wilhelm Schultz, 1804–1876, German physician; student of the flora of the upper Rhine River; author of *Archives de la flore de France et d'Allemagne*.
Schulz	Otto Eugen Schulz, 1874–1936, German botanist; student of Brassicaceae and Erythroxylaceae; author of *Monographie der Gattung Cardamine*.
K. Schum. (or K. Sch.)	Karl Moritz Schumann, 1851–1904, curator of the Berlin Herbarium; student of Cactaceae; author of *Blühende Kakteen*.
O. Schwarz	Otto Karl Anton Schwarz, 1900–1983, Vienna; author of *Flora des tropischen Arabiens*.
Scop.	Johann Anton (Giovanni Antonio) Scopoli, 1723–1788, Austrian physician and botanist; author of *Flora carniolica*.
Scribn.	Frank Lamson Scribner, 1851–1938, agrostologist for the U.S. Department of Agriculture; student of Poaceae, grasses of Yellowstone National Park.
Scribn. & Sm.	F.L. Scribner and J.G. Smith; see Scribn., and also J.G. Smith.
Ser.	Nicolas Charles Seringe, 1776–1858, French physician and botanist, director of the Lyon Botanical Garden; collaborator in de Candolle's *Prodromus;* author of *Flore des jardins et des grandes cultures*.
Shinners	Lloyd Herbert Shinners, 1918–1971, professor of botany at Southern Methodist University; student of the flora of Texas and the southeastern United States; author of *Spring Flora of the Dallas–Fort Worth Area, Texas*.
Small	John Kunkel Small, 1869–1938, curator of the New York Botanical Garden; author of *Flora of the Southeastern United States* and *Manual of the Southeastern Flora*.
Smith	James Edward Smith, 1759–1828, British botanist; founder of the Linnaean Society; author of *Exotic Botany* and *A Grammar of Botany*.
J.G. Smith	Jared Gage Smith, 1866–1925, agrostologist for the U.S. Department of Agriculture; student of Poaceae and *Sagittaria;* author of *North American Species of Sagittaria and Lophotocarpus*.

Sobol.	Gregor Fedorovitch Sobolewski, 1741–1807, Russian physician and botanist; author of *Flora Petropolitana*.
Soland.	Daniel Carl Solander, 1733–1782, Swedish botanist; student of Linnaeus, member of the Cook Expedition; editor of *Elementa botanica*.
Sowerby	James Sowerby, 1757–1822, English botanist; author of *English Botany, or coloured figures of British plants*.
Spreng.	Kurt Polykarp Joachim Sprengel, 1766–1833, professor of medicine and botany at the University of Halle, Germany: history of botany, Apiaceae; author of *Anleitung zur kenntniss der Gewächse*.
Stacey	John William Stacey, 1871–1943, American businessman and botanist, research associate of the California Academy of Sciences; author of *Notes on Carex*, in the series Leaflets of Western Botany.
Steph.	Christian Friedrich Stephan, 1757–1814, Russian botanist and physician, director of the Forestry Institute at St. Petersburg; author of *Icones plantarum Mosquensium*.
Sternb.	Caspar Maria (Graf) von Sternberg, 1761–1838, Czech botanist; author of *Revisio Saxifragarum iconibus illustrata* and *Versuch einer geognostisch-botanischen Darstellung der Flora der Vorwelt*.
Steud.	Ernst Gottlieb Steudel, 1783–1856, German physician and botanist; student of grasses and sedges; author of *Synopsis plantarum glumacearum*.
St. John	Harold St. John, 1892–1991, professor of botany at the University of Hawaii; author of *Flora of Mt. Baker* (Oregon).
Stokes	Jonathan Stokes, 1755–1831, British botanist; author of *A Botanical materia medica*.
St. Yves	Alfred Marie Augustine St. Yves, 1855–1933, French botanist; student of agrostology, *Festuca*; author of *De l'utilité des algues marines*.
Sw.	Olof Peter Swartz, 1760–1818, Swedish botanist; botany of the West Indies; author of *Flora Indiae occidentalis*.
Sweet	Robert Sweet, 1783–1835, British horticulturist and ornithologist; student of Geraniaceae; coauthor of *British botany*, with H. Weddell.
Swezey	Goodwin Deloss Swezey, 1851–1934, American astronomer and botanist, professor of astronomy at the University of Nebraska; author of *Nebraska Flowering Plants*.
T. & G.	John Torrey and Asa Gray; see Torr., and also Gray.
Tausch	Ignaz Friedrich Tausch, 1793–1848, Czech botanist; student of *Hieracium* and *Salsola*; author of *Hortus Canalius*.
Thompson	Zadock Thompson, 1796–1856; author of *Geography and History of Lower Canada*, and *History of Vermont*.
Thurb.	George Thurber, 1821–1890, New York botanist with the U.S.-Mexico Boundary Survey, professor of botany and horticulture at Michigan State University; author of *American Weeds and Useful Plants* (revision of *Agricultural Botany*).
Timm	Joachim Christian Timm, 1734–1805, German pharmacist and botanist; student of the flora of Mecklenburg; author of *Florae Megapolitanae Prodromus*.
Torr.	John Torrey, 1796–1873, physician and professor of chemistry and botany at the College of Physicians and Surgeons in New York, professor of chemistry at Princeton; coauthor of *A Flora of North America*, with A. Gray.

Trautv.	Ernst Rudolf von Trautvetter, 1809–1889, Russian agronomist, director of the St. Petersburg Botanical Garden; author of *Flora riparia Kolymensis* and *Flora Terrae Tschuktschorum*.
Trel.	William Trelease, 1857–1945, director of the Missouri Botanical Garden, professor of botany at the University of Illinois; student of *Agave, Epilobium, Piper, Quercus,* and *Yucca*.
Trin.	Carl Bernhard von Trinius, 1778–1844, German physician and botanist, agrostologist; author of *Species graminum iconibus et descriptionibus illustravit* and *Phalaridea*.
Turcz.	Nicolaus Turczaninow (Nikolai Stepanovich Turchaninov), 1796–1864, Russian botanist; author of *Flora Baikalensi-Dahurica*.
Underw.	Lucien Marcus Underwood, 1853–1907, professor of botany at Columbia University; author of *Our Native Ferns and Their Allies*.
Vahl	Martin Hendriksen Vahl, 1749–1804, professor of botany in Copenhagen; coeditor of *Flora Danica*; author of *Icones illustrationi plantarum americanarum*.
Vasey	George Vasey, 1822–1893, agrostologist for the U.S. Department of Agriculture, curator of the U.S. National Herbarium; author of *Monograph of the Grasses of the United States and British America*.
Vill.	Dominique Villars, 1745–1814, physician and professor in Grenoble and Strasbourg; flora of the western Alps; author of *Histoire des plantes du Dauphiné*.
Wahlenb.	Göran (Georg) Wahlenberg, 1780–1851, professor of botany in Uppsala; author of *Flora lapponica* and *Flora suecica*.
Walp.	Wilhelm Gerhard Walpers, 1816–1853, German botanist; author of *Repertorium botanices systematicae* and *Annales botanices systematicae*.
Walt.	Thomas Walter, 1740–1789, American botanist; author of *Flora Caroliniana*.
Wats. (also S. Wats.)	Sereno Watson, 1826–1892, curator of the Gray Herbarium; flora of western North America; coauthor of *Gray's Manual of Botany of the Northern United States,* 6th ed., with John Coulter.
Webb & Berth.	P.B. Webb and S. Berthelot; coauthors of *Histoire Naturelle des Îles Canaries;* see P.B. Webb, and also S. Berthelot.
P.B. Webb	Philip Barker Webb, 1793–1854, British naturalist; see Webb & Berth.
F. Weber	Friedrich Weber, 1781–1823, German botanist; author of *Beiträge zur Naturkunde*.
G.H. Weber	George Heinrich Weber, 1752–1828, German physician and botanist, professor at the University of Kiel; author of *Commentatio botanico-medica sistens*.
W.A. Weber	William Alfred Weber, 1918–, professor of botany, University of Colorado; student of the flora of Colorado and southern Rocky Mountains; author of *Rocky Mountain Flora, Colorado Flora, Western Slope* and *Eastern Slope*.
Weinm.	Johann Anton Weinmann, 1782–1858, Russian botanist; director, Botanic Garden, St. Petersburg; author of *Hymeno- et Gasteromycetes, Elenchus plantarum horti imperialis Pawlowskiensis*.
Welsh	Stanley Larson Welsh, 1928–, professor of botany, Brigham Young University; author of *A Utah Flora*.
Wherry	Edgar Theodore Wherry, 1885–1982, professor of botany, University of Pennsylvania; Polemoniaceae, author of *Guide to Eastern Ferns, The Southern Fern Guide*.

Wiggers	Friedrich Heinrich Wiggers, 1746–1811, author of *Flora of Holstein*.
Wikst.	Johann Emanuel Wikstrom, 1789–1856, director, Botanical Museum, Stockholm; author of *Juncus, Conspectus litteraturae botanicae in Sueciae, Stockholms Flora*.
Willd.	Carl Ludwig Willdenow, 1765–1812, German botanist; director, Botanical Garden, Berlin; author of *Grundriss der Krauterkunde, Flora Berolinensis*.
Williams	Frederic Newton Williams, 1862–1923, British physician and botanist; student of Caryophyllaceae; author of *The High Alpine Flora of Britain*.
T.A. Williams	Thomas Albert Williams, 1865–1900, agrostologist for the U.S. Department of Agriculture; author of *Grasses and Forage Plants of the Dakotas*.
With.	William Withering, 1741–1799, British physician and botanist; student of the flora of Britain, sedges; author of *Account of the Foxglove, and Some of Its Medical Uses*.
S.J. Wolf	Steven J. Wolf, fl. 1979, University of Missouri; student of *Minuartia*.
Wulf.	Franz Xaver von Wulfen, 1728–1805, Austrian botanist; author of *Flora norica phanerogama* and *Plantarum rariorum descriptiones*.

APPENDIX 4

Chromosome Numbers of Alpine Plants in the Middle Rocky Mountain Flora

Taxa are listed alphabetically by family, genus, and species, as they appear in the text. The numbered sources are documented at the end of this appendix. Different chromosome numbers from separate sources are indicated by parentheses. For example, in the case of *Arabis microphylla* the diploid numbers of 14,15 were reported in source no. 34; the diploid number of 28 was reported in source no. 20. Two sources separated by a comma indicate agreement in chromosome numbers.

Taxon	n	2n	Source
ADIANTACEAE			
Cryptogramma acrostichoides	—	60	1
Pellaea breweri	29	—	18
APIACEAE			
Angelica			
grayi	11	—	32
roseana	11	—	32
Bupleurum americanum	—	14, c. 28	3
Cymopterus			
evertii	—	—	—
hendersonii	—	—	—
longipes	11	—	32
nivalis	11	—	25
terebinthinus	—	22	20
Heracleum sphondylium	—	22	3
Ligusticum			
filicinum	—	66	20
porteri	11	—	14
tenuifolium	—	—	—
Lomatium			
cous	—	22	3
graveolens	—	—	—
triternatum	—	22, 44	3
Oreoxis alpina	—	60	1
Osmorhiza depauperata	—	22	20
ASPLENIACEAE			
Asplenium trichomanes-ramosum	—	72	3
Athyrium			
alpestre	—	80	3
filix-femina	—	80	3
Cystopteris fragilis	—	168, 252	3
Polystichum lonchitis	—	82	3

Continued on next page

Appendix 4 *continued*

Taxon	n	2n	Source
Woodsia scopulina	—	76	3
ASTERACEAE			
Achillea millefolium	—	18, 27, 36, 45, 54, 72	3
Agoseris			
aurantiaca	—	36, 72	3
glauca	—	18, 36	20
Anaphalis margaritacea	—	14, 28	3
Antennaria			
alpina	—	56, 63, 70, 84	3
anaphaloides	—	28	3
aromatica	—	28, 56, 84	17
corymbosa	—	28	14
lanata	—	28	3
microphylla	—	56, 126	3
monocephala	—	56, 63, (56, 70)	3 (17)
umbrinella	—	56	3
Arnica			
cordifolia	—	38, 57, 76, 95	3
fulgens	—	38, 57	3
latifolia	—	38, 57, 76, 112	3
longifolia	—	57, 76	3
mollis	—	38, 57, 76, 95, 114, 133, 152	3
rydbergii	—	38, 76	3
Artemisia			
campestris	—	18, 36	3
frigida	—	18	3
ludoviciana	—	18, 36, 54	3
michauxiana	—	18	3
norvegica	—	18, 36	3
scopulorum	—	18	1
tridentata	9, 18	(18, 36)	10 (3)
Aster			
alpigenus	—	18, 36	17
foliaceus	8, 25, 32	(16, 32, 48, 64, 80, c.96)	14 (30)
glaucodes	—	18	17
integrifolius	—	18	17
occidentalis	—	16, 27 (32, 64)	14 (30)
sibiricus	—	18	3
Chaenactis			
alpina	6	—	16
douglasii	—	12, 37	14
Chrysothamnus viscidiflorus	—	18, 36, 54	20

Continued on next page

Appendix 4 *continued*

Taxon	n	2n	Source
Cirsium			
eatonii	—	34	20
hookerianum	—	34	3
pulcherrimum	—	34	20
subniveum	—	34–36	14
Crepis			
nana	—	14	3
runcinata	—	22	3
Erigeron			
acris	—	18	3
caespitosus	—	18, 36	3
compositus	—	54	1
eatonii	9, 18	—	14
flabellifolius	—	—	—
garrettii	—	—	—
goodrichii	—	—	—
gracilis	—	—	—
humilis	—	36	3
lanatus	—	36	3
leiomerus	—	18	14
melanocephalus	—	18	1
ochroleucus	—	18 (54)	3 (30)
peregrinus	—	18	3
pinnatisectus	—	18	1
radicatus	—	36	3
rydbergii	—	18	22
simplex	—	18	1
speciosus	—	18	1
tener	—	—	—
ursinus	9	—	14
Eriophyllum lanatum	8, 9, 10, 16, 24, 32	—	14
Gutierrezia sarothrae	—	8, 16	3
Happlopappus			
acaulis	—	36	14
lyallii	—	18	3
macronema	—	18	20
pygmaeus	—	—	—
suffruticosus	9	—	31
uniflorus	—	—	—
Heterotheca villosa	—	18, 36	3
Hieracium gracile	9	—	14
Hulsea algida	19 II	—	26
Hymenoxys			
acaulis	—	28, 30, 56, 60, 70	3
grandiflora	15	—	10
Machaeranthera			
canescens	4	—	14

Continued on next page

Appendix 4 *continued*

Taxon	n	2n	Source
kingii	—	18	14
Nothocalais nigrescens	—	—	—
Petradoria pumila	—	18	14
Saussurea weber	13	—	24
Senecio			
amplectens	—	c.40, c. 175, 180 ± 10	33
atratus	—	40, 46, 90+	14
canus	—	46, 92, 138	3
crassulus	20	—	26
cymbalarioides	—	46	3
dimorphophyllus	23	—	23
fremontii	—	40	1
fuscatus	—	46, 48	7
integerrimus	—	40, 80	3
lugens	—	40, 80	3
serra	20	—	14
sphaerocephalus	—	40	14
streptanthifolius	—	23	14
triangularis	—	40, 80	3
werneriifolius	23	—	14
Solidago			
multiradiata	—	18, 36	3
parryi	9	(18)	14 (30)
simplex (as *spathulata*)	—	18, 36	3
Sphaeromeria diversifolia	—	—	—
Taraxacum			
ceratophorum	—	24, 32, 40	3
eriophorum	—	—	—
lyratum	—	16, 24, 32	3
officinale	—	16, 24, 32, 40	3
Townsendia			
alpigena	—	18, 36	20
condensata	—	18, c.36	3
leptotes	—	18, 27–36?	20
parryi	—	18, 36	3
spathulata	—	36?	20
Tragopogon dubius	—	12	3
BETULACEAE			
Betula nana	—	28	3
BORAGINACEAE			
Cryptantha celosioides	—	18	3
Eritrichum nanum	—	44, 46	20
Hackelia			
floribunda	—	24	3
micrantha	12	—	14
patens	12	—	14
Mertensia			
alpina	—	—	—

Continued on next page

Appendix 4 *continued*

Taxon	n	2n	Source
arizonica	—	—	—
ciliata	12, 24	—	8
oblongifolia	—	24 (48)	26 (29)
viridis	—	—	—
Myosotis alpestris	—	24, 48, 72	3
BRASSICACEAE			
Arabis			
drummondii	7	14, 20, 28	34
holboellii	—	13+2B, 14 20 +2B, 21, 21+2B	34
lemmonii	—	14	3
lyallii	—	c.20, 21	34
microphylla	7	14, 15 (28)	34 (20)
nuttallii	—	—	—
williamsii	14	—	34
Descurainia			
incana	7	14, 28, 42	3
torulosa	—	—	—
Draba			
albertina	12	(24)	34 (3)
aurea	32, 37, 36 or 37, 41,	40+1, 64, (74)	34 (3)
breweri	16	32	34
crassa	—	24	34
crassifolia	20	40	34
densifolia	—	36	3
fladnizensis	(8)	16, 32	(27) 34
glabella	32, 40	64, 80	34
incerta	56	112	34
lonchocarpa	8	16	34
oligosperma	32	64	34
paysonii	—	42	34
porsildii	(16)	32	(27) 34
praealta	28	(56)	34 (3)
ventosa	—	36	34
Erysimum asperum	18	36	34
Lesquerella			
alpina	5	10	34
garrettii	—	—	—
occidentalis	5	10	28
paysonii	—	—	—
utahensis	—	—	—
Parrya nudicaulis	—	14, 28	34
Physaria			
acutifolia	5, 8, 12	10, 16, 24	34
didymocarpa	—	8, 16, 24	34
Rorippa curvipes	—	16	3
Smelowskia calycina	—	12, 22, 24	3

Continued on next page

Appendix 4 *continued*

Taxon	n	2n	Source
Thlaspi			
montanum	7, 14	—	14
parviflorum	7	—	33
CALLITRICHACEAE			
Callitriche palustris	—	20	4
CAMPANULACEAE			
Campanula			
rotundifolia	—	34, 68, 102	3
uniflora	—	34	3
CAPRIFOLIACEAE			
Sambucus racemosa	—	36, 38	3
Symphoricarpos oreophilus	—	—	—
CARYOPHYLLACEAE			
Arenaria			
capillaris	—	22	3
congesta	—	22	3
Cerastium			
arvense	—	36, 72	3
beeringianum	—	72	1
Lychnis			
apetala	—	24	12, 14
drummondii	—	48	3
Minuartia			
austromontana	—	30	7
dawsonensis	—	22, 26	20
macrantha	—	—	—
nuttallii	—	36	3
obtusiloba	—	26, c.52, 78	3
rubella	—	24	3
Paronychia pulvinata	—	32	14
Sagina saginoides	—	22	1
Silene			
acaulis	—	24	3
douglasii	—	48	12
menziesii	—	24, 48	3
parryi	—	48, 96	3
Spergularia rubra	—	36	12
Stellaria			
calycantha	—	26, 52	3
crassifolia	—	36	3
longipes	—	52, 78–104	3
obtusa	—	c.78	3
umbellata	—	26	3
CELASTRACEAE			
Paxistima myrsinites	—	—	—
CHENOPODIACEAE			
Chenopodium			
capitatum	—	18	3
fremontii	—	18	14

Continued on next page

Appendix 4 *continued*

Taxon	n	2n	Source
Monolepis nuttalliana	—	18	14
CRASSULACEAE			
Sedum			
debile	—	14–18	12
lanceolatum	—	16, 32, 48	3
rhodanthum	—	14	12
rosea	—	36	1
stenopetalum	—	16, 48	12
CUPRESSACEAE			
Juniperus			
communis	—	22	3
horizontalis	—	22	3
CYPERACEAE			
Carex			
albonigra	—	52	3
aquatilis	—	76	4
atrata	24–28		9
aurea	—	52	3
bipartita	—	58, c.60, 62, 64	3
breweri	—	—	—
brunnescens	—	56	4
canescens	—	54, 56, 62	3
capillaris	—	52, 54	3
capitata	—	50	1
douglasii	—	60	14
ebenea	42	—	9
echinata	—	58	22
elynoides	—	—	—
filifolia	—	50	14
geyeri	—	—	—
hoodii	—	58, 60	3
illota	—	64	3
jonesii	—	—	—
leporinella	—	—	—
luzulina	—	—	—
macloviana	—	74, 76, 78, 80, 82, 86, 90	3
maritima	—	60	4
misandra	—	40	3
nardina	—	68, 70	3
neurophora	—	—	—
nigricans	—	c.72	3
norvegica	—	54, 56	3
nova	—	—	—
obtusata	—	52	3
petasata	—	—	—
phaeocephala	—	c.84	3

Continued on next page

Appendix 4 *continued*

Taxon	n	2n	Source
podocarpa	—	56, 60, c.64, 70, c.84	3
praeceptorum	—	—	—
praticola	—	64, 70? (70, 76, 78)	14, (3)
pyrenaica	—	62	20
raynoldsii	—	70	3
rossii	—	36	14
rostrata	—	76, 82	3
rupestris	—	50, 52	3
saxatilis	—	80	3
scirpoidea	—	62, 64	3
scopulorum	36–40	—	17
stenophylla	—	60, 62	3
straminiformis	—	—	—
subnigricans	—	—	—
vesicaria	—	74 (82, 86)	4, (3)
Eleocharis			
acicularis	—	20	3
quinqueflora	—	20, 132, 134, 136	3
Eriophorum			
callitrix	—	60	3
polystachion	—	58, 60	3
scheuchzeri	—	58	3
Kobresia			
myosuroides	—	58	1
schoenoides	—	58	1
ERICACEAE			
Arctostaphylos uva-ursi	—	26, 52	1
Gaultheria humifusa	—	—	—
Kalmia microphylla	—	24	1
Phyllodoce			
empetriformis	—	24	3
glanduliflora	—	24	3
Pyrola			
minor	—	46	3
picta	—	46	3
secunda	—	38	3
Vaccinium			
cespitosum	—	24	3
myrtillus	—	24	1
scoparium	—	24	3
uliginosum	—	24, 48, 72	20
FABACEAE			
Astragalus			
adsurgens	—	32	3
alpinus	—	16, 32	3
australis	—	16, 32	3

Continued on next page

Appendix 4 *continued*

Taxon	n	2n	Source
kentrophyta	—	24	3
miser	—	22	3
molybdenus	—	—	—
Hedysarum			
boreale	—	16	3
occidentale	—	—	—
sulphurescens	—	—	—
Lupinus			
argenteus	—	48	3
lepidus	—	48	3
leucophyllus	—	24, 48	11
wyethii	24	—	11
Oxytropis			
besseyi	—	16	20
campestris	—	48	3
deflexa	—	16	3
lagopus	—	—	—
parryi	—	16	20
podocarpa	—	16	3
sericea	—	48	3
viscida	—	32	3
Trifolium			
dasyphyllum	—	16	1
haydenii	—	16	15
longipes	—	16, 32, 48	14
nanum	—	16	14
parryi	—	16,	1
FUMARIACEAE			
Dicentra uniflora	—	—	—
GENTIANACEAE			
Frasera speciosa	—	78	1
Gentiana			
affinis	—	—	—
algida	13	—	8
calycosa	—	26	3
prostrata	—	36	3
Gentianella			
amarella	—	36	3
tenella	—	10	7
Gentianopsis			
barbellata	—	—	—
detonsa	—	44	3
Swertia perennis	9, 12, 14	—	8
GERANIACEAE			
Geranium			
richardsonii	—	26, 52	3
viscosissimum	—	52	3
GROSSULARIACEAE			
Ribes			

Continued on next page

Appendix 4 *continued*

Taxon	n	2n	Source
cereum	8	—	11
lacustre	—	16	20
montigenum	—	16	20
oxyacanthoides	8	(16)	11, (19)
HYDROCHARITACEAE			
Elodea canadensis	—	24, 48	9
HYDROPHYLLACEAE			
Phacelia			
hastata	—	22, 44	3
sericea	—	22	3
HYPERICACEAE			
Hypericum scouleri	—	16	3
ISOETACEAE			
Isoetes bolanderi	—	—	—
JUNCACEAE			
Juncus			
biglumis	—	120	1
castaneus	—	60	4
drummondii	—	120	3
mertensianus	—	40, 80	3
nevadensis	—	—	—
parryi	—	—	—
triglumis	—	c.130, c.134	9
Luzula			
glabrata	—	24	3
parviflora	—	22, 24, 36	3
spicata	—	12, 14, 18, 24, 36	3
wahlenbergii	—	24	7
LAMIACEAE			
Monardella odoratissima	—	42	20
LILIACEAE			
Allium			
brandegei	—	14	26
brevistylum	7		9
cernuum	—	14	3
schoenoprasum	—	16	3
Calochortus gunnisonii	9		9
Erythronium grandiflorum	—	24	3
Lloydia serotina	—	24	7
Tofieldia glutinosa	—	30	3
Zigadenus elegans	—	32	3
LINACEAE			
Linum perenne	—	18	3
LYCOPODIACEAE			
Lycopodium selago	—	264	3
ONAGRACEAE			
Epilobium			
alpinum	—	36	3

Continued on next page

Appendix 4 *continued*

Taxon	n	2n	Source
angustifolium	18, 36	(36, 72, 108)	11, (3)
ciliatum	—	36	3
halleanum	—	36	3
latifolium	—	36, 72	3
Gayophytum			
diffusum	7, 14	(28)	11, (14)
racemosum	14	—	11
Oenothera cespitosa	—	14	3
OPHIOGLOSSACEAE			
Botrychium lunaria	45	(90)	13, (1)
ORCHIDACEAE			
Habenaria dilatata	—	42	3
Spiranthes romanzoffiana	—	30	3
PAPAVERACEAE			
Papaver kluanense	—	42, 56	3
PINACEAE			
Abies lasiocarpa	12	—	13
Picea engelmannii	—	24	3
Pinus			
albicaulis	12		24
contorta	—	24	3
flexilis	—	24	3
Pseudotsuga menziesii	—	26, 27	14
PLANTAGINACEAE			
Plantago tweedyi	—	12	14
POACEAE			
Agropyron cristatum	—	14, 28, 42	14
Agrostis			
humilis	—	14	3
idahoensis	—	28	9
mertensii	—	42, 56	3
scabra	—	42	3
thurberiana	—	14, 28	3
variabilis	—	28	3
Alopecurus			
aequalis	—	14	3
alpinus	—	112	1
Bromus			
anomalus	—	14, 28	3
carinatus	—	28, 42, 56, 70	3
ciliatus	—	14, 28	3
inermis	—	28, 42, 49, 54, 56, 58, 70, 84	3
Calamagrostis			
canadensis	—	28, 42–66	2
montanensis	—	28	3
purpurascens	—	28, 40, 42, 58, 84	3

Continued on next page

Appendix 4 *continued*

Taxon	n	2n	Source
scopulorum	—	28	9
stricta	—	28, 56, 70, 84	3
Danthonia intermedia	—	18, 36, c.96	9
Deschampsia			
atropurpurea	—	14	13
cespitosa	—	26, 27, 28	3
elongata	—	26	3
Elymus			
elymoides	—	28	3
glaucus	—	28	3
lanceolatus	—	28, 42	3
scribneri	—	28	3
smithii	—	56	3
spicatus	—	14, 21, 28	3
trachycaulus	—	28	3
Festuca			
baffinensis	—	28	1
ovina	—	14, 21, 28, 35, 42, 49, 56, 70	3
rubra	—	14, 21, 28, 42, 49, 56, 63, 70	3
Helictotrichon			
hookeri	—	14	1
mortonianum	—	—	—
Hierochloe odorata	—	28, 42, 56	2
Koeleria macrantha	—	14, 28	3
Leucopoa kingii	—	56	9
Phippsia algida	—	28	5
Phleum alpinum	—	14, 28, 56	3
Poa			
alpina	—	14, 21, 39, 74, (42)	3, (4)
annua	—	28	3
arctica	—	36	1
fendleriana	—	28, 42, 56, 59, 84	3
glauca	—	42–78	3
leptocoma	—	42	3
lettermanii	—	14	17
nervosa	—	28, 56, 61, c.74, 91	3
pattersonii	—	42	1
pratensis	—	18, 22–147	13
reflexa	—	28	14
secunda	—	42, 56, 70, c.78, 80, 82, c.90, 106	3

Continued on next page

Appendix 4 *continued*

Taxon	n	2n	Source
Stipa			
lettermanii	—	32, 66, 68	9
occidentalis	—	36, 44	13, 14
Trisetum			
spicatum	—	14, 28, 42	9
wolfii	—	—	—
POLEMONIACEAE			
Collomia			
debilis	8	—	22
linearis	—	16	3
Gymnosteris parvula	6	—	22
Ipomopsis			
aggregata	7	—	14
spicata	7	—	22
Linanthus nuttallii	—	18	14
Phlox			
hoodii	—	28	3
multiflora	—	14	14
pulvinata	—	14	14
Polemonium			
pulcherrimum	—	18	1
viscosum	—	18	1
POLYGONACEAE			
Eriogonum			
acaule	—	—	—
brevicaule	—	—	—
flavum	—	38, 76–80	3
ovalifolium	—	40	3
umbellatum	—	76, 80	3
Koenigia islandica	—	28	1
Oxyria digyna	—	14, (42)	1, (3)
Polygonum			
bistortoides	—	44	20
douglasii	—	40	3
kelloggii	—	—	—
minimum	—	—	—
sawatchense	—	—	—
viviparum	—	120	1
Rumex			
acetosa	—	14 ♀, 15 ♂	3
densiflorus	—	120	22
paucifolius	—	14, 28	3
salicifolius	10, 20	—	12
PORTULACACEAE			
Claytonia			
lanceolata	—	16, 24, 32, 48, 64, 74, c.90	3
megarhiza	—	32, 34, 36	3

Continued on next page

Appendix 4 *continued*

Taxon	n	2n	Source
Lewisia			
pygmaea	—	c.66	3
triphylla	—	—	—
Spraguea umbellata	—	—	—
PRIMULACEAE			
Androsace			
chamaejasme	—	20	3
septentrionalis	—	20	3
Dodecatheon			
alpinum	22	—	8
conjugens	—	44	3
pulchellum	—	44, 88, 132	3
Douglasia montana	—	—	—
Primula parryi	—	44	14
RANUNCULACEAE			
Anemone			
lithophila	—	48	3
multifida	—	32	3
narcissiflora	—	14	1
parviflora	—	16	3
patens	—	16	1
Aquilegia			
coerulea	—	14	1
flavescens	—	—	—
jonesii	—	14	3
Caltha leptosepala	—	48	1
Delphinium			
bicolor	—	16	3
occidentale	—	16	14
Ranunculus			
adoneus	—	16	18
alismifolius	—	16	14
aquatilis	—	16, 32, 48	3
eschscholtzii	—	32	3
gelidus	—	16	1
glaberrimus	—	64, 80, 128	3
hyperboreus	—	16, 24, 32, 64	14
inamoenus	—	32, 48	3
pedatifidus	—	32, 48	3
pygmaeus	—	16	3
verecundus	—	—	—
Thalictrum			
alpinum	—	14	1
fendleri	—	28, 56, c.70	12
occidentale	—	56	3
venulosum	—	—	—
Trollius laxus	—	32	1
ROSACEAE			
Cercocarpus ledifolius	—	18	23

Continued on next page

Appendix 4 *continued*

Taxon	n	2n	Source
Dryas octopetala	—	18	3
Fragaria vesca	—	14, 56	3
Geum			
macrophyllum	—	42	3
rossii	—	70	1
triflorum	—	42	3
Holodiscus dumosus	18	—	23
Ivesia			
gordonii	—	—	—
utahensis	—	—	—
Kelseya uniflora	—	18	23
Petrophyton caespitosum	9	—	22
Potentilla			
brevifolia	—	—	—
concinna	—	70	24
diversifolia	—	42–101, c.140	3
flabellifolia	14	—	31
fruticosa	—	14, 28, 42	6
glandulosa	—	14	3
gracilis	—	51–109	3
hippiana	—	42, 70, 77, 84, 98	20
hookeriana	—	42, 49	3
nana	—	42, 49	2
nivea	—	14, 28, 49, 56	3
ovina	—	—	—
pensylvanica	—	28	3
rubricaulis	—	56	21
subjuga	—	—	—
uniflora	—	28	3
Rubus idaeus	—	14, 21, 28, 35, 42	3
Sibbaldia procumbens	—	14	3
RUBIACEAE			
Galium boreale	—	22, 44, 55, 66	3
Kelloggia galioides	—	—	—
SALICACEAE			
Populus balsamifera	—	38	3
Salix			
arctica	—	76, 114	3
barrattiana	—	—	—
brachycarpa	—	38	3
cascadensis	—	—	—
drummondiana	—	38, 57, 76	3
eastwoodiae	—	38	3
glauca	—	76, 95, 114	3
monticola	—	38	3
planifolia	—	57, 76	3
reticulata	—	38	1

Continued on next page

Appendix 4 *continued*

Taxon	n	2n	Source
rotundifolia	—	38	7
tweedyi	—	—	—
wolfii	—	38	14
SAXIFRAGACEAE			
Boykinia heucheriformis	—	14	3
Heuchera			
cylindrica	—	14, 28	3
grossulariifolia	—	—	—
parvifolia	—	14	1
Lithophragma glabrum	—	14, 28	3
Mitella pentandra	—	14	1
Parnassia			
fimbriata	—	36	3
kotzebuei	—	18, (36)	3, (4)
palustris	—	18, 27, 36, 54	3
Saxifraga			
adscendens	—	22	3
bronchialis	—	26	3
caespitosa	—	80	3
cernua	—	36–72	3
chrysantha	—	—	—
flagellaris	—	16	1, 3
occidentalis	—	38	3
odontoloma	—	48	3
oppositifolia	—	26, 52	3
oregana	—	38	3
rhomboidea	—	40	20
rivularis	—	26, (52)	(1), 2
SCROPHULARIACEAE			
Besseya			
alpina	—	24	20
wyomingensis	—	—	—
Castilleja			
applegatei	—	24	14
cusickii	—	24	3
miniata	—	24, 48, 72, 96, 120, c.144	3
nivea	—	—	—
pallescens	—	—	—
pulchella	12	—	14
rhexifolia	—	(12), 24, 48, 96	(14), 3
sulphurea	12, 24	—	26
Chionophila jamesii	—	16	20
Collinsia parviflora	—	14, 28	3
Mimulus			
guttatus	14	(28)	8, (14)
lewisii	—	16	3
Pedicularis			

Continued on next page

Appendix 4 *continued*

Taxon	n	2n	Source
bracteosa	—	16	3
contorta	8	—	23
cystopteridifolia	16	—	33
groenlandica	—	16	3
oederi	—	16	20
parryi	—	16	20
pulchella	—	—	—
racemosa	—	16	3
Penstemon			
attenuatus	—	c.24, 48	20
cyananthus	—	16	20
fruticosus	—	16	3
humilis	8	—	8
montanus	—	—	—
procerus	—	16, 32	3
rydbergii	8, 16	—	8
uintahensis	—	—	—
whippleanus	—	16	14
Synthyris *pinnatifida*	—	24	20
Veronica			
nutans	—	18	14
peregrina	—	52, (54)	14, (3)
SELAGINELLACEAE			
Selaginella			
densa	—	18	3
watsonii	—	—	—
VALERIANACEAE			
Valeriana			
acutiloba	—	—	—
edulis	—	64	14
occidentalis	—	64	25
VIOLACEAE			
Viola			
adunca	—	20, 30, 40	3
nuttallii	—	12, 24, 36, 48	3
palustris	—	48	3
purpurea	—	12, 24, 30	20

Sources

1. Löve, A., D. Löve, and B.M. Kapoor. 1971. Cytotaxonomy of a century of Rocky Mountain orophytes. Arct. & Alp. Res. 3:139–165.

2. Löve, A., and D. Löve. 1965. Taxonomic remarks on some American alpine plants. Univ. Colo. Studies, Series in Bio., no. 17. Univ. Colorado Press, Boulder. 43 pp.

3. Moss, E.H. 1983. Flora of Alberta. 2nd ed., revised by John G. Packer. Univ. Toronto Press, Toronto. 687 pp.

4. Löve, A., and J.C. Ritchie. 1966. Chromosome numbers from central northern Canada. Can. J. Bot. 44:429–439.

5. Bowden, W.M. 1960. Chromosome numbers and taxonomic notes on northern grasses. II. Tribe Festuceae. Can. J. Bot. 38:117–131.

6. Elkington, T.T. 1969. Cytotaxonomic variation in *Potentilla fruticosa* L. New Phytol. 68:151–160.

7. Johnson, A.W., and J.G. Packer. 1968. Chromosome numbers in the flora of Ogotoruk Creek, N.W. Alaska. Bot. Notiser. 121:403–456.

8. Hitchcock, C.L., et al. 1959. Vascular plants of the Pacific Northwest. Part 4: Ericaceae through Campanulaceae. Univ. Washington Press, Seattle. 510 pp.

9. Cronquist, A., et al. 1977. Intermountain flora, vol. 6. N.Y. Bot. Gard., Columbia Univ. Press, New York. 584 pp.

10. Hitchcock, C.L., et al. 1955. Vascular plants of the Pacific Northwest. Part 5: Compositae. Univ. Washington Press, Seattle. 343 pp.

11. Hitchcock, C.L., et al. 1961. Vascular plants of the Pacific Northwest. Part 3: Saxifragaceae to Ericaceae. Univ. Wash. Press, Seattle. 614 pp.

12. Hitchcock, C.L., et al. 1964. Vascular plants of the Pacific Northwest. Part 2: Salicaceae to Saxifragaceae. Univ. Washington Press, Seattle. 597 pp.

13. Hitchcock, C.L., et al. 1969. Vascular plants of the Pacific Northwest. Part 1: Vascular cryptogams, gymnosperms and monocotyledons. Univ. Washington Press, Seattle. 914 pp.

14. Welsh, S.L., et al. 1987. A Utah flora. Great Basin Nat. Mem. 9, Brigham Young Univ., Provo. 894 pp.

15. Mosquin, T., and J.M. Gillett. 1965. Chromosome numbers in American *Trifolium* (Leguminosae). Brittonia 17:136–143.

16. Mooring, J.S. 1965. Chromosome studies in *Chaenactis* and *Chamaechaenactis* (Compositae, Helenieae). Brittonia 17:17–25.

17. Goldblatt, P., and D.E. Johnson, eds. 1991. Index to plant chromosome numbers 1988–1989. Monogr. Syst. Bot. 40, Missouri Bot. Gard. St. Louis. 238 pp.

18. Kapoor, B.M., and A. Löve. 1970. Chromosomes of Rocky Mountain *Ranunculus*. Caryologia 23:575–594.

19. Darlington, C.D., and A.P. Wylie. 1956. Chromosome atlas of flowering plants. Macmillan, New York. 519 pp.

20. Bolkovskikh, Z., et al. 1969. Chromosome numbers of flowering plants. Izdatel'stvo Nauka, Leningrad. 926 pp.

21. Cave, M.S., ed. 1958. Index to plant chromosome numbers, vol. 1. Univ. North Carolina Press, Chapel Hill. (No page nos.)

22. Goldblatt, P., and D.E. Johnson, eds. 1990. Index to plant chromosome numbers 1986–1987. Monogr. Syst. Bot. 30, Missouri Bot. Gard. St. Louis. 243 pp.

23. Goldblatt, P., ed. 1988. Index to plant chromosome numbers 1984–1985. Monogr. Syst. Bot. 23, Missouri Bot. Gard. St. Louis. 264 pp.

24. Goldblatt, P., ed. 1985. Index to plant chromosome numbers 1982–1983. Monogr. Syst. Bot. 13, Missouri Bot. Gard. St. Louis. 224 pp.

25. Goldblatt, P., ed. 1984. Index to plant chromosome numbers 1979–1981. Monogr. Syst. Bot. 8, Missouri Bot. Gard. St. Louis. 423 pp.

26. Goldblatt, P., ed. 1981. Index to plant chromosome numbers 1975–1978. Monogr. Syst. Bot. 5, Missouri Bot. Gard. St. Louis. 553 pp.

27. Mulligan, G.A. 1974. Cytotaxonomic studies of *Draba nivalis* and its close allies in Canada and Alaska. Can. J. Bot. 52:1793–1801.

28. Rollins, R.C. and E.A. Shaw. 1973. The genus *Lesquerella* (Cruciferae) in North America. Harvard Univ. Press, Cambridge. 288 pp.

29. Milek, J.A. 1988. A genetic and taxonomic study of *Mertensia oreophila* and three varieties of *Mertensia oblongifolia* in Wyoming and southern Montana. M.S. thesis, Univ. Northern Colorado, Greeley. 94 pp.

30. Semple, J.C. 1985. Chromosome number determinations in fam. Compositae, Tribe Astereae. Rhodora 87:517–527.

31. Moore, R.J. 1977. Index to plant chromosome numbers 1973–74. Regnum Vegetabile 96. Utrecht, Netherlands. 257 pp.

32. Moore, R.J. 1974. Index to plant chromosome numbers 1972. Regnum Vegetabile 91. Utrecht, Netherlands. 108 pp.

33. Moore, R.J. 1973. Index to plant chromosome numbers 1967- 71. Regnum Vegetabile 90. Utrecht, Netherlands. 539 pp.

34. Rollins, R. 1993. The Cruciferae of continental North America. Stanford Univ. Press, Stanford, Calif. 976 pp.

Bibliography

Albee, B.J., et al. 1988. Atlas of the vascular plants of Utah. Utah Museum of Natural History, Logan. 670 pp.

Allred, K.W. 1976. The plant family Gentianaceae in Utah. Great Basin Nat. 36:483–495.

Andrews, J.P. 1961. Flora of Alaska and adjacent parts of Canada. Iowa State Univ. Press, Ames. 543 pp.

Anderson, J.T. 1975. Glacial systems, an approach to glaciers and their environments. Duxbury Press, North Scituate, Mass. 191 pp.

Arno, S., and R. Hammerly. 1984. Timberline: mountain and arctic forest frontiers. The Mountaineers, Seattle. 304 pp.

Arnow, L.A. 1981. *Poa secunda* Presl versus *P. sandbergii* Vasey (Poaceae). Syst. Bot. 6:412–421.

Arnow, L.A., et al. 1980. Flora of the central Wasatch Front, Utah. 2nd ed. Univ. Utah Printing Service, Salt Lake City. 663 pp.

Atwood, D., et al. 1990. Idaho and Wyoming endangered and sensitive plant field guide. USDA Forest Service, Intermountain Region, Ogden, Utah, 192 pp.

Atwood, W.W., and W.W. Atwood, Jr. 1938. Working hypothesis for the physiographic history of the Rocky Mountain region. Bull. Geol. Soc. Am. 49:957–980.

Babcock, E.B., and G.L. Stebbins. 1938. The American species of *Crepis*. Carnegie Institute, Wash. 199 pp.

Barkley, T.M., et al. 1986. Flora of the Great Plains. Univ. Press Kansas, Lawrence. 1392 pp.

Barneby, R. 1964. Atlas of North American *Astragalus*, parts 1 and 2. New York Botanical Garden, New York. 1188 pp.

Barneby, R.C. 1981. *Dragma hippomanicum* VII. A new alpine *Astragalus* (Leguminosae from western Wyoming. Brittonia 33:156–158.

Barry, R.G. 1973. A climatological transect along the east slope of the Front Range, Colorado. Arct. & Alp. Res. 5:89–110.

Bayer, R.J. 1984. Chromosome numbers and taxonomic notes for North American species of *Antennaria* (Asteraceae: Inuleae). Syst. Bot. 9:74–83.

Bayer, R.J. 1987. Morphometric analysis of western North American *Antennaria* (Asteraceae: Inuleae). I. Sexual species of sections *Alpinae*, *Dioicae*, and *Plantaginifoliae*. Can. J. Bot. 65:2389–2395.

Bayer, R.J. 1989a. A systematic and phytogeographic study of *Antennaria aromatica* and *A. densifolia* (Asteraceae: Inuleae) in the western North American Cordillera. Madroño 36:248–259.

Bayer, R.J. 1989b. A taxonomic revision of the *Antennaria rosea* (Asteraceae: Inuleae: Gnaphaliinae) polyploid complex. Brittonia 41:53–60.

Bayer, R.J. 1990. A systematic study of *Antennaria media*, *A. pulchella*, and *A. Scabra* (Asteraceae: Inuleae) of the Sierra Nevada and White Mountains. Madroño 37:171–183.

Bayer, R.J., and G.L. Stebbins. 1981. Chromosome numbers of North American species of *Antennaria* Gaertner (Asteraceae: Inuleae). Am. J. Bot. 68:1342–1349.

Bayer, R.J., and G.L. Stebbins. 1987. Chromosome numbers, patterns of distribution, and apomixis in *Antennaria* (Asteraceae: Inuleae). Syst. Bot. 12:305–319.

Beaman, J.H. 1957. The systematics and evolution of *Townsendia* (Compositae). Contr. Gray Herb. 183. Harvard Univ. Press, Cambridge. 151 pp.

Beetle, A.A. 1970. Recommended plant names. Univ. Wyoming Agric. Exp. Sta. Res. J. 31. 124 pp.

Beetle, A.A., and M. May. 1971. Grasses of Wyoming. Univ. Wyoming Agric. Exp. Sta. Res. J. 39. 151 pp.

Bell, K.L., and R.E. Johnson. 1980. Alpine flora of the Wassuk Range, Mineral County, Nevada. Madroño 27:25–35.

Billings, W.D., and L.C. Bliss. 1959. An alpine snowbank environment and its effects on vegetation, plant development, and productivity. Ecology 40:388–397.

Blackstone, D.L. 1971. Traveler's guide to the geology of Wyoming. Geol. Surv. Wyo. Bull. 55. 90 pp.

Bliss, L.C. 1956. A comparison of plant development in microenvironments of arctic and alpine tundras. Ecol. Monogr. 26:303–337.

Bliss, L.C. 1962. Adaptations of arctic and alpine plants to environmental conditions. Arctic 15:117–144.

Bliss, L.C. 1963. Alpine plant communities of the Presidential Range, New Hampshire. Ecology 44:678–697.

Bliss, L.C. 1966. Plant productivity in alpine microenvironments on Mount Washington, New Hampshire. Ecol. Monogr. 36:125–135.

Bliss, L.C. 1985. Alpine. *In* B.F. Chabot and H.A. Mooney, eds., Physiological ecology of North American plant communities. Chapman & Hall, New York. 351 pp.

Bolkovskikh, Z., et al. 1969. Chromosome numbers of flowering plants. Izdatel'stvo Nauka, Leningrad. 926 pp.

Boivin, B. 1967. Enumeration des plantes du Canada. VI. Monopsides, pt. 2. Nat. Can. 94:471–528.

Bonney, O.H., and L.G. Bonney. 1977. Guide to the Wyoming mountains and wilderness areas. 3rd rev. ed. Swallow Press, Chicago. 701 pp.

Borror, D.J. 1971. Dictionary of word roots and combining forms. Mayfield Pub. Co., Palo Alto, Calif. 134 pp.

Bowden, W.M. 1960. Chromosome numbers and taxonomic notes on northern grasses. II. Tribe Festuceae. Can. J. Bot. 38:117–131.

Brummitt, R.K. 1971. Relationship of *Heracleum lanatum* Michx. of North America to *H. sphondylium* of Europe. Rhodora 73:578–584.

Brummitt, R.K., and C.E. Powell, eds. 1992. Authors of plant names. Royal Botanic Gardens, Kew. Whitsable Litho, Whitsable, Kent. 732 pp.

Caldwell, M.M. 1970. The wind regime at the surface of the vegetation layer above timberline in the central Alps. Zentralbl. Gesamte Forstwesen 87:65–74.

Caldwell, M.M., R. Robberecht, and W.D. Billings. 1980. A steep latitudinal gradient of solar ultraviolet-B radiation in the arctic-alpine zone. Ecology 61:600–611.

Cave, M.S., ed. 1958. Index to plant chromosome numbers, vol. 1. Univ. North Carolina Press, Chapel Hill.

Chabot, B.F., and W.D. Billings. 1972. Origins and ecology of the Sierran alpine flora and vegetation. Ecol. Monogr. 42:163–199.

Chabot, B.F., and H.A. Mooney. 1985. Physiological ecology of North American plant communities. Chapman & Hall, New York. 351 pp.

Chinnappa, C.C., and J.K. Morton. 1991. Studies on the *Stellaria longipes* complex (Caryophyllaceae)—taxonomy. Rhodora 93:129–135.

Chmielewski, J.G., and C.C. Chinnappa. 1988. Taxonomic notes and chromosome numbers in *Antennaria* Gaertner (Asteraceae: Inuleae) from arctic North America. Arct. & Alp. Res. 20:117–124.

Correll, D., and M.C. Johnston. 1970. Manual of the vascular plants of Texas. Texas Research Foundation, Renner. 1818 pp.

Coulter, J.M. and A. Nelson. 1909. New manual of botany of the Central Rocky Mountains. Amer. Book Co., New York. 646 pp.

Cronquist, A. 1943. Revision of the western North American species of *Aster* centering about *Aster foliaceus* Lindl. Am. Midl. Nat. 29:429–468.

Cronquist, A., et al. 1972. Intermountain flora, vol. 1. N.Y. Bot. Gard. Hafner Pub. Co., New York. 270 pp.

Cronquist, A., et al. 1977. Intermountain flora, vol. 6. New York Botanical Garden. Columbia Univ. Press, New York. 584 pp.

Cronquist, A., et al. 1978. Compositae tribe Senecioneae. North American Flora, ser. 2, pt. 10:14–179.

Cronquist, A., et al. 1984. Intermountain flora, vol. 4. New York Botanical Garden, New York. 573 pp.

Cronquist, A., et al. 1989. Intermountain flora, vol. 3, pt. B. New York Botanical Garden. New York. 279 pp.

Cronquist, A., et al. 1994. Intermountain flora, vol. 5. New York Botanical Garden. New York. 496 pp.

Darlington, C.D., and A.P. Wylie. 1956. Chromosome atlas of flowering plants. Macmillan, New York. 519 pp.

Daubenmire, R.F. 1941. Some ecological features of subterranean organs of alpine plants. Ecology 22:370–378.

Davis, R.J. 1952. Flora of Idaho. Wm. C. Brown, Dubuque, Iowa. 836 pp.

Dorn, R.D. 1977. Manual of the vascular plants of Wyoming, vols. 1 and 2. Garland, New York. 1498 pp.

Dorn, R.D. 1984. Vascular plants of Montana. Mountain West Publishing, Cheyenne, Wyo. 276 pp.

Dorn, R.D. 1988. Vascular plants of Wyoming. Mountain West Publishing, Cheyenne, Wyo. 340 pp.

Dorn, R.D. 1992. Vascular plants of Wyoming, 2nd ed. Mountain West Publishing, Cheyenne, Wyo. 340 pp.

Dorn, R.D., and R. Hartman. 1988. Nomenclature of *Lomatium nuttallii*, *L. kingii*, and *L. megarrhizum* (Apiaceae). Madroño 35:70–71.

Douglas, G.W., K.E. Denford, and I. Karas. 1977. A contribution to the taxonomy of *Antennaria alpina* var. *media*. *A. microphylla*, and *A. umbrinella* in western North America. Can. J. Bot. 55:925–933.

Drury, W.H., and R.C. Rollins. 1952. The North American representatives of *Smelowskia* (Cruciferae). Rhodora 54:85–119.

Duft, J.F., and R.K. Moseley. 1989. Alpine wildflowers of the Rocky Mountains. Mountain Press, Missoula, Mont. 200 pp.

Duman, M.G. 1941. The genus *Carex* in eastern arctic Canada. Catholic Univ. America Press, Washington, D.C. 84 pp.

Eardley, A.J. 1962. Structural geology of North America. 2nd ed. Harper & Row, New York. 743 pp.

Elkington, T.T. 1969. Cytotaxonomic variation in *Potentilla fruticosa* L. New Phytol. 68:151–160.

Engler, H.G.A., ed. 1909. Das Pflanzenreich vol. 38 pt. IV:20, Cyperaceae-Caricoideae by Georg Kükenthal. Leipzig. 824 pp.

Evert, E.F. 1983a. A new species of *Lomatium* (Umbelliferae) from Wyoming. Madroño 30:143–146.

Evert, E.F. 1983b. Some notable collections for 1983. Wyo. Nat. Pl. Soc. Newsletter 3:3.

Evert, E.F. 1984a. A new species of *Antennaria* (Asteraceae) from Montana and Wyoming. Madroño 31:109–112.

Evert, E.F. 1984b. Notes on some plants occurring in the Beartooths I. Wyo. Nat. Pl. Soc. Newsletter 3:1–3.

Evert, E.F. 1995. Personal communication.

Evert, E.F., and R.L. Hartman. 1984. Additions to the vascular flora of Wyoming. Great Basin Nat. 44:482–483.

Fenneman, N.M. 1931. Physiography of western United States. McGraw-Hill, New York. 534 pp.

Fernald, M.L. 1942. Critical notes on *Carex*. J. New Eng. Bot. Club 44:281–331.

Fertig. W. 1992. A floristic survey of the west slope of the Wind River Range, Wyoming. Unpub. M.S. thesis, Univ. Wyoming, Laramie. 188 pp.

Flint, R.F. 1957. Glacial and Pleistocene geology. Wiley & Sons, New York. 553 pp.

Flora of North America Editorial Committee (eds.) 1993. Flora of North America North of Mexico. vol. 1. Introduction. Oxford Univ. Press, New York. 372 pp.

Flora of North America Editorial Committee (eds.) 1993. Flora of North America North of Mexico. vol. 2. Pteridophytes and Gymnosperms. Oxford Univ. Press, New York. 475 pp.

Foose, R.M. 1960. Secondary structure associated with vertical uplift in the Beartooth Mountains, Montana. Proc. Internat. Geol. Congress, Part XVIII, pp. 53–61.

Foose, R.M., D.U. Wise, and G.S. Garbarini. 1961. Structural geology of the Beartooth Mountains, Montana and Wyoming. Geol. Soc. Am. Bull. 72:1143–1172.

Geiger, R. 1965. The climate near the ground. Harvard Univ. Press, Cambridge. 611 pp.

Gillett, J.M. 1965. Taxonomy of *Trifolium*: five American species of the section *Lupinaster* (Leguminosae). Brittonia 17:121–136.

Goldblatt, P., ed. 1981. Index to plant chromosome numbers 1975–1978. Monogr. Syst. Bot. 5, Missouri Botanical Garden. St. Louis. 553 pp.

Goldblatt, P., ed. 1984. Index to plant chromosome numbers 1979–1981. Monogr. Syst. Bot. 8, Missouri Botanical Garden. St. Louis. 423 pp.

Goldblatt, P., ed. 1985. Index to plant chromosome numbers 1982–1983. Monogr. Syst. Bot. 13, Missouri Botanical Garden. St. Louis. 224 pp.

Goldblatt, P., ed. 1988. Index to plant chromosome numbers of 1984–1985. Monogr. Syst. Bot. 23, Missouri Botanical Garden. St. Louis. 264 pp.

Goldblatt, P., and D.E. Johnson, eds. 1990. Index to plant chromosome numbers 1986–1987. Monogr. Syst. Bot. 30, Missouri Botanical Garden. St. Louis. 243 pp.

Goldblatt, P., and D.E. Johnson, eds. 1991. Index to plant chromosome numbers 1988–1989. Monogr. Syst. Bot. 40, Missouri Botanical Garden. St. Louis. 238 pp.

Goldblatt, P. and D.E. Johnson, eds. 1994. Index to plant chromosome numbers 1990–1991. Monogr. Syst. Bot. 51, Missouri Botanical Garden, St. Louis. 267 pp.

Goodrich. S. 1983. Utah flora: Salicaceae. Great Basin Nat. 43:531–550.

Goodrich. S. 1986. Utah flora: Juncaceae. Great Basin Nat. 46:366–377.

Grant, V., and D.H. Wilken. 1988. Racial variation in *Ipomopsis tenuituba* (Polemoniaceae). Bot. Gaz. 149:443–449.

Greuter, W., et al. 1988. The International Code of Botanical Nomenclature: adopted by the Fourteenth Institutional Botanical Congress, Berlin, July–August 1987. Koeltz Scientific Books, Konigstein. 328 pp.

Hadley, J.L., and W.K. Smith. 1986. Wind effects on needles of timberline conifers; seasonal influence on mortality. Ecology 67:12–19.

Hadley, J.L., and W.K. Smith. 1987. Influences of krummholz mat microclimate on needle physiology and survival. Oecologia 73:82–90.

Hallsten, G.P., et al. 1987. Grasses of Wyoming. 3rd ed. Univ. Wyoming Agric. Exp. Sta. Res. J. 202. 432 pp.

Hamerlynck, E.P., and W.K. Smith. 1994. Subnivean and emergent microclimate, photosynthesis, and growth in *Erythronium grandiflorum* Pursh, a snowbank geophyte. Arc. & Alp. Res. 26:21–28.

Harrington, H.D. 1964. Manual of the plants of Colorado. 2nd ed. Swallow Press, Chicago. 666 pp.

Harshberger, J.W. 1929. Preliminary notes on American snow patches and their plants. Ecology 10:275–281.

Hartman, E.L., and M.L. Rottman. 1988. The vegetation and alpine vascular flora of the Sawatch Range, Colorado. Madroño 35:202–225.

Hartman, R.L., et al. 1985. Noteworthy collections, Wyoming. Madroño 32:125–128.

Hartman, R.L., and R.S. Kirkpatrick. 1986. A new species of *Cymopterus* (Umbelliferae) from northwestern Wyoming. Brittonia 38:420–426.

Hartman, R.L., and R.W. Lichvar. 1980. Additions to the vascular flora of Teton County, Wyoming. Great Basin Nat. 40: 408–413.

Helm, D. 1982. Multivariate analysis of alpine snow-patch vegetation cover near Milner Pass, Rocky Mountain National Park, Colorado, U.S.A. Arctic and Alpine Research 14:87–95.

Hermann, F.J. 1970. Manual of the Carices of the Rocky Mountains and Colorado Basin. Agric. Handb. 374, USDA Forest Service. U.S. GPO, Washington, D.C. 397 pp.

Heuberger, H. 1974. Alpine quaternary glaciation. Pp. 319–338 *in* J.D. Ives and R.G. Barry, eds., Arctic and alpine environments. Methuen, London. 999 pp.

Hickman, J.C., ed. 1993. The Jepson manual: higher plants of California. Univ. Cal. Press, Berkeley. 1400 pp.

Hitchcock, A.S. 1950. Manual of the grasses of the United States. 2nd ed., rev. by Agnes Chase. U.S. GPO, Washington, D.C. 1051 pp.

Hitchcock, C.L. 1941. A revision of the Drabas of western North America. Univ. Washington Pub. Biol. 11. 132 pp.

Hitchcock, C.L., et al. 1955. Vascular plants of the Pacific Northwest. Part 5: Compositae. Univ. Washington Press, Seattle. 343 pp.

Hitchcock, C.L., et al. 1959. Vascular plants of the Pacific Northwest. Part 4: Ericaceae through Campanulaceae. Univ. Washington Press, Seattle. 510 pp.

Hitchcock, C.L., et al. 1961. Vascular plants of the Pacific Northwest. Part 3: Saxifragaceae to Ericaceae. Univ. Washington Press, Seattle. 614 pp.

Hitchcock, C.L., et al. 1964. Vascular plants of the Pacific Northwest. Part 2: Salicaceae to Saxifragaceae. Univ. Washington Press, Seattle. 597 pp.

Hitchcock, C.L., et al. 1969. Vascular plants of the Pacific Northwest. Part 1: Vascular cryptogams, gymnosperms, and monocotyledons. Univ. Washington Press, Seattle. 914 pp.

Hitchcock, C.L., and A. Cronquist. 1973. Flora of the Pacific Northwest. Univ. Washington Press, Seattle. 730 pp.

Hitchcock, C.L., and B. Maguire. 1947. A revision of the North American species of *Silene*. Univ. Washington Pub. Biol. 13:1–73.

Holm, D. 1982. Multivariate analysis of alpine snow-patch vegetation cover near Milner Pass, Rocky Mountain National Park, Colorado, U.S.A. Arct. & Alp. Res. 14:87–95.

Holm, T. 1900. Studies in Cyperaceae. Am. J. Sci. 10:266–284.

Holmgren, N. 1987. Two new species of *Potentilla* (Rosaceae) from the intermountain region of western U.S.A. Brittonia 39:340–344.

Hooker, W.J. 1829–34. Flora Boreali-Americana, Vol. 1. Treuttel and Wurtz. 351 pp.

Hooker, W.J. 1838–40. Flora Boreali-Americana, Vol. 2. Treuttel and Wurtz. 328 pp.

Hopkins, D.M. 1975. Time—stratigraphic nomenclature for the Holocene Epoch. Geology 3:10.

Houston, R.S., et al. 1968. A regional study of rocks of Precambrian age in that part of the Medicine Bow Mountains lying in southeastern Wyoming—with a chapter on the relationship between Precambrian and Laramide structure. Geol. Surv. Wyo. Mem. 1. 167 pp.

Howard, A.D., J.W. Williams, and I. Raisz. 1972. Physiography. *In* Geologic atlas of the Rocky Mountain region. Rocky Mountain Association of Geologists. A.B. Hirschfeld Press, Denver. 29–31.

Hultén, E. 1937. Outline of the history of arctic and boreal biota during the Quaternary Period. Bok. Aktieb. Thule, Stockholm. 168 pp.

Hultén, E. 1956. The *Cerastium alpinum* complex. A case of

Hultén, E. 1959a. A new species of *Saussurea* from Colorado. Svensk Bot. Tidskr. 53:200–202.

Hultén, E. 1959b. Studies in the genus *Dryas*. Svensk Bot. Tidskr. 53:507–542.

Hultén, E. 1959c. The *Trisetum spicatum* complex. Svensk Bot. Tidskr. 53:203–228.

Hultén, E. 1962. The circumpolar plants. I. Vascular cryptogams, conifers, monocotyledons. Almqvist & Wiksell, Stockholm. 275 pp.

Hultén, E. 1967. Comments on the flora of Alaska and Yukon. Almqvist & Wiksell, Stockholm. 147 pp.

Hultén, E. 1968. Flora of Alaska and neighboring territories. Stanford Univ. Press, Stanford, Calif. 1008 pp.

Hunt, C.B. 1967. Physiography of the United States. W.H. Freeman, San Francisco. 480 pp.

Hustich, I. 1966. On the forest-tundra and the northern treelines. Ann. Univ. Tur. 36:7–47.

Iltis, H. 1965. The genus *Gentianopsis* (Gentianaceae): transfers and phytogeographic comments. SIDA Contr. Bot. 2:129–154.

Jackson, B.D. 1964. Guide to the literature of botany; being a classified selection of botanical works. Hafner Pub. Co., New York. 626 pp.

Jensen, E.R. 1987. Flowers of Wyoming's Big Horn Mountains. Basin Republican Rustler Printing, Basin, Wyo. 251 pp.

Johnson, A.W., and J.G. Packer. 1968. Chromosome numbers in the flora of Ogotoruk Creek, N.W. Alaska. Bot. Notiser. 121:403–456.

Johnson, D.A., ed. 1979. Special management needs of alpine ecosystems. Range Sci. Ser. no. 5. Society for Range Management, Denver. 100 pp.

Johnson, P.L. 1962. The occurrence of new arctic-alpine species in the Beartooth Mountains, Wyoming-Montana. Madroño 16:229–223.

Johnson, P.L., and W.D. Billings. 1962. The alpine vegetation of the Beartooth Plateau in relation to cryopedogenic processes and patterns. Ecol. Monogr. 32:105–135.

Johnson, W.M. 1964. Field key to the sedges of Wyoming. Univ. Wyoming Agric. Exp. Sta. Bull. 419. 239 pp.

Kapoor, B.M., and A. Löve. 1970. Chromosomes of Rocky Mountain *Ranunculus*. Caryologia 23:575–594.

Kartesz, J. 1994. A synonymized checklist of the vascular flora of the United States, Canada, and Greenland, 2nd ed. Vol. 1: Checklist. Timber Press, Portland, Ore. 622 pp.

Kartesz, J. 1994. A synonymized checklist of the vascular flora of the United States, Canada, and Greenland, 2nd ed. Vol. 2: Thesaurus. Timber Press, Portland, Ore. 816 pp.

Kartesz, J., and R. Kartesz. 1980. A synonymized checklist of the vascular flora of the United States, Canada, and Greenland. Vol. 2: The Biota of North America. Univ. North Carolina Press, Chapel Hill. 498 pp.

Kellogg, E.A. 1985. A biosystematic study of the *Poa secunda* complex. J. Arnold Arb. 66:201–242.

Kirkpatrick, R.S. 1987. A flora of the southeastern Absarokas, Wyoming. Unpub. M.S. thesis, Univ. Wyoming, Laramie. 166 pp.

Knight, S.H. 1990. Illustrated geologic history of the Medicine Bow Mountains and adjacent areas, Wyoming. Geol. Surv. Wyo. Mem. 4. 48 pp.

Komarkov, V.L. 1934. Flora of the USSR, vol. 2. English translation, Israel Program for Scientific Translations. IPST Press, Jerusalem, 1963. 622 pp.

Komarkov, V.L. 1935. Flora of the USSR, vol. 3. English translation, Israel Program for Scientific Translations. IPST Press, Jerusalem, 1964. 512 pp.

Lackschewitz, K.H. 1976. Montana mountain flora: new records. Madroño 23:361–362.

Lawrence, H.M., et al. 1968. Botanico-Periodicum-Huntianum. Hunt Botanical Library, Pittsburgh. 1063 pp.

Lessica, P. 1993. Vegetation and flora of the Line Creek Plateau area, Carbon County, Montana. Unpub. report to USDA Forest Service. Montana Natural Heritage Program, Helena. 30 pp.

Lichvar, R.W. 1979. The flora of the Gros Ventre Mountains. Unpub. M.S. thesis, Univ. Wyoming, Laramie. 384 pp.

Lichvar, R.W. 1983. Evaluation of *Draba oligosperma, D. pectinipila,* and *D. juniperina* complex (Cruciferae). Great Basin Nat. 43:441–444.

Lipschitz, S. 1979. Genus *Saussurea* DC. (Asteraceae). Nauka Sectio Leninopotitana, Leningrad. 282 pp.

Löve, A. 1975. Cytotaxonomical atlas of the arctic flora. J. Cramer. 598 pp.

Löve, A. 1970. Emendations in the Icelandic flora. Taxon 19:298–302.

Löve, A. 1969. The plants of the Queen Charlotte Islands (review). Taxon 18:225–229.

Löve, A., and V. Evenson. 1967. The taxonomic status of *Rumex paucifolius*. Taxon 16:423–425.

Löve, A., and D. Löve. 1965. Taxonomic remarks on some American alpine plants. Univ. Colo. Stud. Ser. Biol. 17. Univ. Colorado Press, Denver. 43 pp.

Löve, A., D. Löve, and B.M. Kapoor. 1971. Cytotaxonomy of a century of Rocky Mountain orophytes. Arct. & Alp. Res. 3:139–165.

Löve, A., and J.C. Ritchie. 1966. Chromosome numbers from central northern Canada. Can. J. Bot. 44:429–439.

Löve, D., and N.J. Freedman. 1956. A plant collection from SW Yukon. Bot. Notiser. 109:153–211.

Love, J.D. 1939. Geology along the southern margin of the Absaroka Range, Wyoming. Geol. Soc. Am. Spec. Pap. 20. 134 pp.

Love, J.D., and J.M. Love. 1983. Road log, Jackson to Dinwoody and return. Geol. Surv. Wyo. Public Info. Circ. 20. 34 pp.

MacArthur, R.H., and E.O. Wilson, 1967. The theory of island biogeography. Princeton Univ. Press, Princeton. 203 pp.

Mackenzie, K.K. 1931. North American Flora: Poales, Cyperaceae. Vol. 18, pt. 1, pp. 1–478. New York Botanical Garden.

Maguire, B. 1950. Studies in the Caryophyllaceae. IV. A synopsis of the North American species of the subfamily Silenoideae. Rhodora 52:233–245.

Maguire, B. 1958. *Arenaria rossii* and some of its relatives in America. Rhodora 60:44–58.

Mahaney, W.C., and J. Spence. 1984. Glacial & periglacial sequence and floristics in Jaw Cirque, central Teton Range, western Wyoming. Am. J. Sci. 284:1056–1081.

Mallory. W.W., ed. 1972. Geologic atlas of the Rocky Mountain Region. A.B. Hirschfeld Press, Denver. 331 pp.

Markow, S. 1994. A floristic survey of the Targhee National Forest and vicinity, east-central Idaho and west-central Wyoming. Unpub. M.S. Thesis, Univ. of Wyoming, Laramie. 164 pp.

Marr. J.W. 1977. The development and movement of tree islands near the upper limit. Ecology 58:1159–1164.

Marriott, H. 1991. Status report for *Descurainia torulosa* (Wyoming tansymustard). Nature Conservancy, Wyoming Natural Diversity Database. 12 pp.

Marriott, H. 1992. Unpub. field notes, summer 1991. 7 pp.

Marsh, V.L. 1952. A taxonomic revision of the genus *Poa* of the United States and southern Canada. Am. Midl. Nat. 47:202–250.

Mathias, M.E., and L. Constance. 1944. North American flora: Umbelliferae. Vol. 28B, pt. 1, pp. 43–160. New York Botanical Garden, New York.

Mathias, M.E., and L. Constance. 1945. North American flora: Umbelliferae. Vol. 28B, pt. 2, pp. 161–397. New York Botanical Garden, New York.

Matthes, F.E. 1942. Glaciers. *In* O.E. Meinzer ed., Physics of the earth. Vol. 9: Hydrology. Dover, New York. 712 pp.

McArthur, E.D., and B.L. Welch. 1984. Proceedings of symposium on the biology of *Artemisia* and *Chrysothamnus*. Gen. Tech. Rep. Int-200. Intermountain Research Station, Ogden, Utah.

McDougall W.B., and H.A. Bagley, 1956. Plants of Yellowstone. Yellowstone Library and Museum Association, Yellowstone National Park, Wyo. 186 pp.

McNeill, J. 1978. *Silene alba* and *S. dioica* in North America and the generic delineation of *Lychnis, Melandrium,* and *Silene* (Caryophyllaceae). Can. J. Bot. 56:297–308.

Mears Jr., B. 1981. Periglacial wedges and the late Pleistocene environment of Wyoming's intermontane basins. Quat. Res. 15:171–198.

Michaux, A. 1803. Flora boreali-americana, vols. 1 (330 pp.) and 2 (340 pp.). Facsimile of the 1803 edition, ed. J. Ewan. Hafner Pub. Co., New York, 1974.

Milek. J.A. 1988. A genetic and taxonomic study of *Mertensia oreophila* and three varieties of *Mertensia oblongifolia* in Wyoming and southern Montana. Unpub. M.S. thesis, Univ. Northern Colorado, Greeley. 94 pp.

Montagne, J.M. 1972. Quaternary system, Wisconsin glaciation. *In* W.W. Mallory, ed., Geologic atlas of the Rocky Mountain region. A.B. Hirschfeld Press, Denver. 331 pp.

Mooney, H.A., and W.D. Billings. 1960. The annual carbohydrate cycle of alpine plants are related to growth. Am. J. Bot. 47:594–598.

Moore, R.J. 1973. Index to plant chromosome numbers 1967–1971. Regnum vegetabile 90:1–539.

Mooring, J.S. 1965. Chromosome studies in *Chaenactis* and *Chamaechaenactis* (Compositae, Helenieae). Brittonia 17:17–25.

Mosquin, T., and J.M. Gillett. 1965. Chromosome numbers in American *Trifolium* (Leguminosae). Brittonia 17:136–143.

Mosquin, T., and D.E. Hayley. 1966. Chromosome numbers and taxonomy of some Canadian arctic plants. Can. J. Bot. 44:1209–1218.

Moss, E.H. 1983. Flora of Alberta. 2nd ed., rev. by John G. Packer. Univ. Toronto Press, Toronto. 687 pp.

Mueller, P.A., et al. 1988. Age and composition of a late Archean magmatic complex, Beartooth Mountains, Montana-Wyoming. Mont. Bur. Mines Spec. Pub. 96:7–22.

Mulligan, G.A. 1971. Cytotaxonomic studies of the closely allied *Draba cana, D. Cinerea,* and *D. groenlandica* in Canada and Alaska. Can. J. Bot. 49:89–93.

Mulligan, G.A. 1972. Cytotaxonomic studies of *Draba* species in Canada and Alaska: *D. oligosperma* and *D. incerta*. Can. J. Bot. 50:1763–1766.

Mulligan, G.A. 1974. Cytotaxonomic studies of *Draba nivalis* and its close allies in Canada and Alaska. Can. J. Bot. 52:1793–1801.

Mulligan, G.A. 1975. *Draba crassifolia, D. albertina, D. nemorosa,* and *D. stenoloba* in Canada and Alaska. Can. J. Bot. 53:745–751.

Mulligan, G.A. 1976. The genus *Draba* in Canada and Alaska: key and summary. Can. J. Bot. 54:1386–1393.

Munz, P.A., and D.D. Keck. 1959. A flora of California. Univ. California Press, Berkeley. 1681 pp.

Murray, D.F. 1969. Taxonomy of *Carex* sect. *Atratae* (Cyperaceae) in the Southern Rocky Mountains. Brittonia 21:55–76.

Murray, D.F. 1970. *Carex podocarpa* and its allies in North America. Can. J. Bot. 48:313–324.

Nelson, B.E. 1984. Vascular plants of the Medicine Bow Range. Jelm Mountain Press, Laramie, Wyo. 357 pp.

Nesom, G.L. 1989. *Solidago simplex* (Compositae: Astereae), the correct name for *S. glutinosa*. Phytologia 67:155–157.

O'Neill, H., and M. Duman. 1941. A new species of *Carex* and some notes on this genus in arctic Canada. J. New Eng. Bot. Club 43:413–425.

Ornduff, R. 1967. Index to plant chromosome numbers for 1965. Regnum Vegetabile 50:1–128.

Ornduff, R. 1968. Index to plant chromosome numbers for 1966. Regnum vegetabile 55:1–126.

Parsons, W.H. 1978. Field guide: Middle Rockies and Yellowstone. Kendall-Hunt, Dubuque, Iowa, 233 pp.

Polunin, N. 1959. Circumpolar arctic flora. Oxford Univ. Press, London. 514 pp.

Porsild, A.E. 1966. Contributions to the flora of southwestern Yukon Territory. Nat. Mus. Can. Bull. 216, Contr. Bot. 4. Ottawa, Canada. 86 pp.

Porter, S.C., and G.H. Denton. 1967. Chronology of neoglaciation in the North American Cordillera. Am. J. Sci. 265:177–210.

Price, L. 1981. Mountains and man. A study of process and environment. Univ. California Press, Berkeley. 506 pp.

Pritzel, G.A. 1871. Thesarus literaturae botanicae. Reprint, ed. G.G. Gorlich. Off. Grafiche Ricordi, Milan, Italy, 1950.

Rambert, E. 1869. Les Alpes Suisses. Bâle et Genève, Genève. 298 pp.

Raven, J., and M. Walters. 1956. Mountain flowers. Collins, London. 240 pp.

Reid, R.R., W.J. McMannis, and J.C. Palmquist. 1975. Precambrian geology of North Snowy Block, Beartooth Mountains, Montana. Geol. Soc. Am. Spec. Pap. 157. 135 pp.

Reveal, J. 1973. *Eriogonum* (Polygonaceae) of Utah. Phytologia 25:169–217.

Reveal, J. 1989. On the modern death of David Douglas. Madroño 36:137–140.

Reznicek, A.A., and P.W. Ball. 1980. The taxonomy of *Carex* section *Stellulatae* in North America north of Mexico. Contr. Univ. Mich. Herb. 14:153–203.

Richmond, G.M. 1984. Surficial deposits of the Bridger

Wilderness. Pp. 76–85 *in* Workshop proceedings, air quality and acid deposition potential in the Bridger and Fitzpatrick Wildernesses. USDA Forest Service, Intermountain Region (Air Quality Group). Odgen, Utah.

Richmond, G.M. 1986. Stratigraphy and correlation of glacial deposits of the Rocky Mountains, the Colorado Plateau and the ranges of the Great Basin. Quat. Sci. Rev. 5:99–127.

Richmond, G.M., and D.S. Fullerton. 1986. Summation of Quaternary glaciations in the United States of America. Quat. Sci. Rev. 5:183–196.

Robbins, G.T. 1944. North American species of *Androsace*. Am. Midl. Nat. 32:137–163.

Rollins, R.C. 1981a. Studies in the genus *Physaria* (Cruciferae). Brittonia 33:332–341.

Rollins, R.C. 1981b. Studies on *Arabis* (Cruciferae) of western North America. Syst. Bot. 6:55–64.

Rollins, R.C. 1983a. Interspecific hybridization and taxon uniformity in *Arabis* (Cruciferae). Am. J. Bot. 70:625–634.

Rollins, R.C. 1983b. Studies in the Cruciferae of western North America. J. Arnold Arb. 64:491–501.

Rollins, R.C. 1993. The Cruciferae of continental North America. Stanford Univ. Press, Stanford, Calif. 976 pp.

Rollins, R.C., and E.A. Shaw. 1973. The genus *Lesquerella* (Cruciferae) in North America. Harvard Univ. Press, Cambridge. 288 pp.

Russell, R.J. 1933. Alpine land forms of western United States. Bull. Geol. Soc. Am. 44:927–950.

Rydberg, P.A. 1912. Studies on the Rocky Mountain flora, XXVII. Bull. Torr. Bot. Club 39:301–328.

Rydberg, P.A. 1917. Flora of the Rocky Mountains and adjacent plains. Pub. by the author, N.Y. 1110 pp.

Schröter, L., and and C. Schröter. 1904. Taschenflora des Alpenwanderers. Albert Raustein. Zürich. 26 pls.

Scott, R.W. 1966a. The alpine flora of northwestern Wyoming. Unpub. M.S. thesis, Univ. Wyoming, Laramie. 219 pp.

Scott, R.W. 1966b. The alpine flora of northwestern Wyoming. Spec. Pub. Rocky Mtn. Herb., Univ Wyoming, Laramie. 219 pp.

Scott, R.W. 1974. The effect of snow duration on alpine plant community composition and distribution. *In* V. Bushnell and M.G. Marcus, eds., Icefield Ranges Research Project. Scientific Results, vol. 4:307–318. Amer. Geogr. Soc., N.Y. and Arctic Instit. of North Amer., Montreal.

Scoggan, H.J. 1978. The flora of Canada, pts. 1–4. National Museums of Canada, Ottawa. 1711 pp.

Semple, J.C. 1985. Chromosome number determinations in fam. Compositae, Tribe Astereae. Rhodora 87:517–527.

Shaw, R.J. 1952. A cytotaxonomic study of the genus *Geranium* in the Wasatch region of Idaho and Utah. Madroño 11:297–304.

Shaw, R.J. 1968. Vascular plants of Grand Teton National Park, Wyoming. SIDA Contr. Bot. 4:1–56.

Shaw, R.J. 1976. Field guide to the vascular plants of Grand Teton National Park and Teton County, Wyoming. Utah State Univ. Press, Logan. 301 pp.

Shaw, R.J. 1989. Vascular plants of northern Utah. Utah State Univ. Press, Logan. 412 pp.

Shaw, R.J. 1992. Vascular plants of Grand Teton National Park and Teton County: an annotated checklist. Grand Teton Natural History Association. Moose, Wyoming. 92 pp.

Siems, B.A., and E.A. Neely. 1990. *Braya glabella* var. *glabella* in Colorado. Madroño 37:145–146.

Snow, N., et al. 1990. Additions to the vascular flora of Yellowstone National Park, Wyoming. Madroño 37:214–216.

Soreng, R.J. 1985. *Poa* L. in New Mexico, with a key to Middle and Southern Rocky Mountain species (Poaceae). Great Basin Nat. 45:395–417.

Spence, J.R., and R.J. Shaw. 1981. A checklist of the alpine vascular flora of the Teton Range, Wyoming, with notes on biology and habitat preferences. Great Basin Nat. 41:232–242.

Spence, J.R., and R.J. Shaw. 1983. Observations on alpine vegetation near Schoolroom Glacier, Teton Range, Wyoming. Great Basin Nat. 43:483–491.

Stafleu, F.A., and R.S. Cowan. 1976a. Taxonomic literature, vol. 1: A–G. 2nd ed. Bohn, Scheltema & Holkema, Utrecht, Netherlands. 1136 pp.

Stafleu, F.A., and R.S. Cowan. 1976b. Taxonomic literature, vol. 2: H–Le. 2nd ed. Bohn, Scheltema & Holkema, Utrecht, Netherlands. 991 pp.

Stafleu, F.A., and R.S. Cowan. 1981. Taxonomic literature vol. 3: Lh–O. 2nd ed. Bohn, Scheltema & Holkema, Utrecht, Netherlands. 980 pp.

Stafleu, F.A., and R.S. Cowan. 1983. Taxonomic literature vol. 4: P–Sak. 2nd ed. Bohn, Scheltema & Holkema, Utrecht, Netherlands. 1214 pp.

Stafleu, F.A., and R.S. Cowan. 1985. Taxonomic literature vol. 5: Sal-Ste. 2nd ed. Bohn, Scheltema & Holkema, Utrecht, Netherlands. 1066 pp.

Stafleu, F.A., and R.S. Cowan. 1986. Taxonomic literature vol. 6: Sti-Vuy. 2nd ed. Bohn, Scheltema & Holkema, Utrecht, Netherlands. 926 pp.

Stafleu, F.A., and R.S. Cowan. 1988. Taxonomic literature vol. 7: W-Z 2nd ed. Bohn, Scheltema & Holkema, Utrecht, Netherlands. 653 pp.

Standley, P.C. 1921. Flora of Glacier National Park, Montana. Contr. U.S. Natl. Herb. 22, pt. 5. U.S. GPO, Washington, D.C. 438 pp.

Stearn, W.T. 1993. Botanical Latin, 4th ed. Redwood Books, Trowbridge, England. 546 pp.

St. John, H. 1941. Revision of the genus *Swertia* (Gentianaceae) of the Americas and the reduction of *Frasera*. Am. Midl. Nat. 26:1–29.

Strickler, D. 1990. Alpine wildflowers. Flower Press, Columbia Falls, Mont. 112 pp.

Swan, L.W. 1968. Alpine and aeolian regions of the world. *In* H.E. Wright and W.H. Osburn, eds., Arctic and alpine environments. Indiana Univ. Press, Indianapolis. 308 pp.

Swift, L.H. 1970. Botanical bibliographies: a guide to bibliographic materials applicable to botany. Burgess Pub. Co., Minneapolis, Minn. 804 pp.

Thilenius, J.F., and G.R. Brown. 1987. Herded vs. unherded sheep grazing systems on an alpine range in Wyoming. USDA Forest Service, Gen. Tech. Rep. RM-147. 8 pp.

Thilenius, J.F., and D. Smith. 1985. Vegetation and soils of an alpine range in the Absaroka Mountains, Wyoming. USDA Forest Service, Gen. Tech. Rep. RM-121. 18 pp.

Tidestrom, I. 1925. Flora of Utah and Nevada. Contr. U.S. Natl. Herb. 25. U.S. GPO, Washington, D.C. 665 pp.

Tolmachev, A.I. 1960. Arkticheskaya Flora SSSR [Flora arctica URSS]. Vol. 1: Polypodiaceae–Butomaceae. Editio Nauk, Moscow.

Tolmachev, A.I. 1963. Arkticheskaya Flora SSSR [Flora arctica URSS]. Vol. 4: Lemnaceae–Orchidaceae. Editio Nauk, Moscow.

Tolmachev, A.I. 1964. Arkticheskaya Flora SSSR [Flora arctica URSS]. Vol. 2: Graminae. Editio Nauk, Moscow.

Tolmachev, A.I. 1966a. Arkticheskaya Flora SSSR [Flora arctica URSS]. Vol. 3: Cyperaceae, Editio Nauk, Moscow.

Tolmachev, A.I. 1966b. Arkticheskaya Flora SSSR [Flora arctica URSS]. Vol. 5: Salicaceae–Portulacaceae. Editio Nauk, Moscow.

Tolmachev, A.I., ed. 1965. Distribution of the flora of the USSR. English translation, Israel Program for Scientific Translations. IPST Press, Jerusalem. 178 pp.

Torrey, J. and A. Gray. 1838–40. A flora of North America, Vol. 1 (711 pp), Vol. 2 (504 pp). Wiley and Putnam, New York.

Tutin, T.G., et al., eds. 1964. Flora Europaea. Vol. 1: Lycopodiaceae to Platanaceae. Cambridge Univ. Press, Cambridge. 464 pp.

Tutin, T.G., et al., eds. 1968. Flora Europaea. Vol. 2: Rosaceae to Umbelliferae. Cambridge Univ. Press, Cambridge. 455 pp.

Tutin, T.G., et al., eds. 1972. Flora Europaea. Vol. 3: Diapensiaceae to Myoporaceae. Cambridge Univ. Press, Cambridge. 370 pp.

Tutin, T.G., et al., eds. 1976. Flora Europaea. Vol. 4: Plantaginaceae to Compositae (and Rubiaceae). Cambridge Univ. Press, Cambridge. 505 pp.

Tutin, T.G., et al., eds. 1980. Flora Europaea. Vol. 5: Alismataceae to Orchidaceae. Cambridge Univ. Press, Cambridge. 452 pp.

Ugborogho, R.E. 1973. North American *Cerastium arvense* L. 1. Cytology. Cytologia 38:559–566.

United States Department of Agriculture. 1982a. National list of scientific plant names. Vol. 1: List of plant names. Soil Conservation Service SCS-TP-159. 416 pp.

United States Department of Agriculture. 1982b. National list of scientific plant names. Vol. 2: Synonymy. Soil Conservation Services SCS-TP-159. 438 pp.

Voss, E.G. 1966. Nomenclatural notes on monocots. Rhodora 68:435–463.

Voss, E.G. 1972. Additional nomenclatural and other notes on Michigan monocots and gymnosperms. Mich. Bot. 11:26–37.

Wardle, P. 1968. Engelmann spruce (*Picea engelmannii* Engel.) at its upper limits on the front Range, Colorado. Ecology 49:483–495.

Wardle, P. 1971. An explanation for alpine timberline. New Zeal. J. Bot. 9:371–402.

Watson, T.J. 1978. Chromosome numbers in *Xylorhiza* Nuttall (Asteraceae-*Astereae*). Madroño 25:205–210.

Watson, T.J., and K.H. Lackschewitz. 1980. The genus *Saussurea* (Asteraceae-*Cynareae*) in Montana. Northwest Sci. 54:106–108.

Webber, P.J., and D.E. May. 1977. The magnitude and distribution of belowground plant structures in the alpine tundra of Niwot Ridge, Colorado, Arct. & Alp. Res. 9:157–174.

Weber, W.A. 1955. Additions to the flora of Colorado, II. Univ. Colo. Stud., Ser. Biol. 3:65–108.

Weber, W.A. 1961. Handbook of plants of the Colorado Front Range. Univ. Colorado Press, Boulder. 232 pp.

Weber, W.A. 1967. Rocky Mountain flora. Univ. Colorado Press, Boulder. 437 pp.

Weber, W.A. 1976. Rocky Mountain flora. Colorado Assoc. Univ. Press, Boulder. 479 pp.

Weber, W.A. 1989. New names and combinations, principally in the Rocky Mountain Flora, VII. Phytologia 67:425–428.

Weber, W.A. 1990. Colorado flora: eastern slope. Univ. Press Colorado, Niwot. 396 pp.

Weber, W.A., and R.C. Whittmann. 1992. Catalog of the Colorado flora; a biodiversity baseline. Univ. Press Colorado, Niwot. 215 pp.

Welsh, S.A. 1974. Anderson's flora of Alaska and adjacent parts of Canada. Brigham Young Univ. Press, Provo. 724 pp.

Welsh, S.L. 1983. A bouquet of daisies (*Erigeron*, Compositae). Great Basin Nat. 43:365–368.

Welsh, S.L., et al., eds. 1987. A Utah flora. Great Basin Nat. Mem. 9, Brigham Young Univ., Provo. 894 pp.

Welsh, S.L., et al., eds. 1993. A Utah flora, *2nd* edition, revised. Brigham Young Univ. Provo. 986 pp.

Welsh, S.L., D. Atwood, and J.L. Reveal. 1975. Endangered, threatened, extinct, endemic, and rare or restricted Utah vascular plants. Great Basin Nat. 35:327–376.

Welsh, S., and J.L. Reveal. 1977. Utah flora: Brassicaceae (Cruciferae). Great Basin Nat. 37:279–365.

Went, F.W. 1953. Annual plants at high altitudes in the Sierra Nevada, California. Madroño 12:109–114.

Wiggins, I.L., and J.H. Thomas. 1962. A flora of the Alaskan arctic slope. Arc. Inst. N. Am. Spec. Pub. 4. Univ. Toronto Press, Toronto. 425 pp.

Wight, W.F. 1902. The genus *Eritrichum* in North America. Bull. Torr. Bot. Club 29:407–414.

Wilken, D.H. 1975. A systematic study of the genus *Hulsea* (Asteraceae). Brittonia 27:228–244.

Williams, J., ed. 1986. Rocky Mountain alpines. Timber Press, Portland, Ore. 309 pp.

Woolridge, G. et al. 1992. Airflow patterns in a small subalpine basin. Theor. Appl. Climatol. 45:37–41.

Zwinger, A.H., and B.E. Willard. 1972. Land above the trees: a guide to American alpine tundra. Harper & Row, New York. 489 pp.

Index to Common Names

A
ADDER'S TONGUE
  FAMILY, 500
Agoseris
  Orange, 72
  Pale, 73
Alberta
  Draba, 232
  Saxifrage, 747
Alp Lily, 482
Alpine
  Arnica, 93
  Aster, 102
  Avens, 679
  Bentgrass, 519
  Bladderpod, 250
  Bluebells, 211
  Bluegrass, 599
  Bistort, 618
  Bog Swertia, 444
  Brook Saxifrage, 752
  Buttercup, 655
  Campion, 279
  Clover, 424
  Collomia, 590
  Dandelion, 192
  Dryad, 674
  Dusty Maiden, 109
  Fescue, 557
  Forget-Me-Not, 205
  Foxtail, 527
  Goldenrod, 187
  Groundsel, 169
  Harebell, 268
  Hulsea, 157
  Kittentails, 754
  Lady Fern, 62
  Locoweed, 420
  Meadowrue, 667
  Milkvetch, 397
  Nerved Sedge, 350
  Oatgrass, 561
  Oreoxis, 58
  Paintbrush, 760
  Pussytoes, 77
  Sagewort, 100
  Shootingstar, 635
  Smelowskia, 261
  Sorrel, 611
  Speedwell, 791
  Spider Saxifrage, 746
  Springbeauty, 626
  Timothy, 566
  Townsendia, 195
  Willowherb, 490
  Wintergreen, 383
Alpinesedge, 378
  Mousetail, 379
  Northern, 380
Alumroot, 730
  Common, 733
  Gooseberryleaf, 732
  Roundleaf, 731
American
  Bistort, 613
  Rockbrake, 36
  Thorowax, 41
Anemone, 640
  Cutleaf, 642
  Fellfield, 641
  Narcissus, 643
  Northern, 644
Angelica, 38
  Gray's, 39
  Rose, 40
Annual
  Bluegrass, 570
  Gentian, 439
Arctic
  Bentgrass, 521
  Bluegrass, 571
  Buttercup, 661
  Cinquefoil, 699
  Draba, 238
  Gentian, 435
  Pearlwort, 289
  Poppy, 505
  Sandwort, 286
  Willow, 715
Arizona Bluebells, 212
Arnica, 86
  Alpine, 93
  Broadleaf, 89
  Hairy, 92
  Heartleaf, 87
  Longleaf, 91
  Orange, 88
  Rydberg, 93
  Sticky, 92
Aromatic Pussytoes, 80
Arrowleaf Groundsel, 182
Aspen, 713
ASTER FAMILY, 67
Aster, 102
  Alpine, 103
  Blueleaf, 105
  Leafybract, 104
  Siberian, 107
  Thickstem, 106
  Western, 108
Avens, 677
  Alpine, 679
  Largeleaf, 678

B
Baffin Fescue, 555
Ballhead Sandwort, 274
Balsam Poplar, 713
Barratt Willow, 716
Barrenground Willow, 717
Beach Dock, 623
Beaked Sedge, 363
Bear River Daisy, 143
Bearberry, 382
Beardtongue, 779
Beauty
  River, 494
Bedstraw, 710
  Northern, 710
BELLFLOWER FAMILY,
  266
Bentgrass, 518
  Alpine, 519
  Arctic, 521
  Idaho, 520
  Mountain, 524
  Thurber, 523
Bering Chickweed, 277
Big Sagebrush, 101
Bigflower Groundsmoke, 496
Bigroot Daisy, 138
BIRCH FAMILY, 201
Birch, 201
  Bog, 202
Birdfoot Buttercup, 663
Biscuitroot, 54
  Cous, 55
  King, 56
Bistort
  Alpine, 618
  American, 613
Bitter Fleabane, 123
Bitterroot, 627
  Pygmy, 628
  Threeleaf, 629
Black
  Alpine Sedge, 351
  Groundsel, 170
  Nothocalais, 164
Black-and-White Sedge, 323
Blackhead Daisy, 134
Blackroot Sedge, 337
Blackscale Sedge, 325
Blacktip Groundsel, 178
Bladder Fern, 64
  Brittle, 64
  Sedge, 371
Bladderpod, 249
  Alpine, 250
  Fremont, 253
  Garrett, 251
  Payson, 253
  Utah, 254
  Western, 252
Bleeding Heart, 429
Blite
  Strawberry, 305
Blue-Eyed Mary, 766
  Smallflower, 766
Blue
  Flax, 486
  Wildrye, 547
Bluebells, 210
  Alpine, 211
  Arizona, 212
  Greenleaf, 215
  Mountain, 213
  Oblongleaf, 214
Blueberry, 390
  Bog, 394
  Dwarf, 391
  Low, 392

Bluebunch Wheatgrass, 551
Bluegrass, 567
   Alpine, 569
   Annual, 570
   Arctic, 571
   Bog, 575
   Greenland, 574
   Kentucky, 579
   Letterman, 576
   Nodding, 580
   Patterson, 578
   Sandberg, 581
   Wheeler, 577
Bluejoint Reedgrass, 534
Blueleaf
   Aster, 105
   Cinquefoil, 690
Blunt
   Sedge, 354
   Starwort, 300
Bog
   Birch, 202
   Blueberry, 394
   Bluegrass, 575
   Laurel, 384
Bog Orchid
   White, 501
Bolander Quillwort, 459
BORAGE FAMILY, 203
Boykinia, 728, 729
Bracteate Lousewort, 771
Brandegee Onion, 476
Braya
   Dwarf, 228
Brewer
   Cliffbrake, 37
   Sedge, 329
Brittle Bladder Fern, 64
Broadleaf
   Arnica, 89
   Knotweed, 616
Brome
   Fringed, 531
   Mountain, 530
   Nodding, 529
   Smooth, 532
Bromegrass, 528
Brook Saxifrage, 748
Broom Snakeweed, 146
Brown
   Pussytoes, 85
   Sedge, 330
BUCKWHEAT FAMILY, 604
Buckwheat
   Cushion, 608
   Shortstem, 606
   Stemless, 605
   Sulfur, 609
   Yellow, 607

Bulblet Saxifrage, 744
BUTTERCUP FAMILY, 640
Buttercup, 654
   Alpine, 655
   Arctic, 661
   Birdfoot, 663
   Dwarf, 664
   Plantainleaf, 656
   Sagebrush, 660
   Snowbed, 658
   Timberline, 665
   Tundra, 659
   Unpleasant, 662
Butterweed Groundsel, 179

C

Campion, 278, 290
   Alpine, 279
   Douglas, 292
   Drummond, 280
   Moss, 291
Canada
   Gooseberry, 452
   Violet, 803
   Waterweed, 454
Capitate Sedge, 333
Cascade Willow, 718
Catchfly, 290
   Menzies, 293
   Parry, 294
Chestnut Rush, 462
Chickweed, 275
   Bering, 277
   Field, 276
Chive, 479
Cinquefoil, 686
   Arctic, 699
   Blueleaf, 690
   Elegant, 689
   Fanleaf, 691
   Hooker, 698
   Pennsylvania, 702
   Redstem, 703
   Sheep, 701
   Shortleaf, 688
   Shrubby, 692
   Singleflower, 705
   Slender, 695
   Snowbank, 700
   Sticky, 693
   Twinleaf, 704
   Woolly, 697
Cleftleaf Groundsel, 173
Cliff Fern, 66
Cliffbrake, 37
   Brewer, 37
Clover, 423
   Alpine, 424
   Dwarf, 427
   Hayden, 425

   Longstalk, 426
   Parry, 428
Cloverhead Horsemint, 474
CLUBMOSS FAMILY, 487
Clubmoss, 487, 488
Cobwebby Goldenweed, 150
Coiled-Beak Lousewort, 772
Collomia, 589
   Alpine, 590
   Narrowleaf, 591
Colorado Columbine, 647
Columbine, 646
   Colorado, 647
   Jones, 649
   Yellow, 648
Common
   Alumroot, 733
   Dandelion, 193
   Juniper, 315
   Lady Fern, 63
   Monkeyflower, 767
   Red Paintbrush, 759
   Sagewort, 95
   Timothy, 566
   Twinpod, 258
   Yarrow, 70
Cotton Grass, 375
Cottonsedge, 375
   Tall, 376
   Tufted, 375
   White, 377
Cottonwood, 713
Cous Biscuitroot, 55
Cowparsnip, 48
   Hogweed, 49
Cream Mountain Heath, 386
Creeping
   Juniper, 316
   Wintergreen, 382
Crested Wheatgrass, 517
Curled Lousewort, 778
Curly Sedge, 364
Currant, 448
   Wax, 449
Cushion
   Buckwheat, 608
   Townsendia, 196
Cusick Paintbrush, 758
Cutleaf Anemone, 642
CYPRESS FAMILY, 314

D

Daisy, 121
   Bear River, 143
   Bigroot, 138
   Blackhead, 134
   Eaton, 126
   Fanleaf, 127
   Fernleaf, 125
   Garrett, 128

   Goodrich, 129
   Low, 131
   Oneflower, 140
   Oregon, 141
   Pale, 135
   Peregrine, 136
   Pinnateleaf, 137
   Rydberg, 139
   Slender, 130
   Smooth, 133
   Thin, 142
   Tufted, 124
   Woolly, 132
Dandelion, 189
   Alpine, 192
   Common, 193
   Rocky Mountain, 191
   Tundra, 190
Dark Alpine Sedge, 370
Darkthroat Shootingstar, 637
Death Camas, 484
   Mountain, 484
Delicate Gentian, 440
Denseflower Dock, 621
Dock, 619
   Beach, 623
   Denseflower, 621
Douglas
   Campion, 292
   Dusty Maiden, 110
   Fir, 513
   Knotweed, 614
   Rabbitbrush, 111
   Sedge, 334
Douglasia, 638
   Mountain, 638
Draba, 230
   Alberta, 232
   Arctic, 238
   Fleshy, 235
   Golden, 233
   Longfruit, 241
   Nuttall, 237
   Payson, 243
   Porsild, 244
   Smooth, 239
   Snowbank, 242
   Tall, 245
   Thickleaf, 236
   White, 234
   Wind River, 246
   Yellowstone, 240
Drummond
   Campion, 280
   Rockcress, 219
   Rush, 463
   Willow, 719
Dryad, 674
   Alpine, 674
   Mountain, 674

Duncecap Larkspur, 653
Dunhead Sedge, 356
Dusky Groundsel, 176
Dusty Maiden
    Alpine, 109
    Douglas, 110
Dwarf
    Blueberry, 391
    Braya, 228
    Buttercup, 664
    Clover, 427
    Hawksbeard, 119
    Locoweed, 418
    Lupine, 410
    Rockjasmine, 632

E

Early Sedge, 358
Eastwood Willow, 720
Eaton
    Daisy, 126
    Thistle, 114
Ebony Sedge, 335
Edible Valerian, 798
Elderberry, 270
    Red, 270
Elegant Cinquefoil, 689
Elk Sedge, 339
Elkweed, 431, 432
Engelmann Spruce, 508
Eriophyllum, 144
    Woolly, 144
EVENING PRIMROSE FAMILY, 489
Evening Primrose, 498
Everlasting
    Pearly, 75
Evert's Springparsley, 43
Explorer Gentian, 436

F

Fairy Trumpet, 593
False
    Dandelion, 71
    Sagebrush, 188
Falsephlox, 592
    Leafless, 592
Fanleaf
    Cinquefoil, 691
    Daisy, 127
Featherleaf Kittentails, 789
Fellfield Anemone, 641
Fendler Meadowrue, 668
Fern
    Alpine lady, 62
    Bladder, 64
    Brittle Bladder, 64
    Holly, 65
    Rock Holly, 65
    Shield, 65

Fernleaf
    Daisy, 125
    Lousewort, 773
    Lovage, 51
Fescue, 554
    Alpine, 557
    Baffin, 555
    Idaho, 557
    Red, 558
    Rydberg, 557
    Sheep, 556
    Spike, 564
Field Chickweed, 276
FIGWORT FAMILY, 753
Fir, 506
    Douglas, 513
    Subalpine, 507
Fireweed, 489, 491
Fiveflower Spikerush, 374
Fivestamen Miterwort, 735
Flattop Pussytoes, 81
FLAX FAMILY, 485
Flax, 485
    Blue, 486
Flaxflower, 596
    Nuttall, 596
Fleabane, 121
    Bitter, 123
Fleshy Draba, 235
Flower
    Pasque, 645
Forget-Me-Not, 205, 216
    Alpine, 205
    Mountain, 216
Foxtail, 525
    Alpine, 527
    Shortawn, 526
Fremont
    Bladderpod, 253
    Goosefoot, 306
    Groundsel, 175
Fringecup Woodland Star, 734
Fringed
    Brome, 531
    Gentian, 441
    Parnassus, 737
    Sagewort, 96
Fringed Gentian, 441
    Perennial, 442
    Rocky Mountain, 443
FROGBIT FAMILY, 453
FUMITORY FAMILY, 429

G

Garden Sorrel, 620
Garrett
    Bladderpod, 251
    Daisy, 128
GENTIAN FAMILY, 431

Gentian, 433
    Annual, 439
    Arctic, 435
    Delicate, 440
    Explorer, 436
    Fringed, 441
    Moss, 437
    Perennial Fringed, 442
    Rocky Mountain Fringed, 443
    Rocky Mountain Pleated, 434
    Star, 444
GERANIUM FAMILY, 445
Geranium, 445
    Richardson, 446
    Sticky, 447
Gilia, 593
Glacier Lily, 481
Globeflower, 671
Goatsbeard, 200
Goldbloom Saxifrage, 745
Golden
    Draba, 233
    Sedge, 327
Goldenaster, 154
    Hairy, 154
Goldenrod, 184
    Alpine, 187
    Northern, 185
    Parry, 186
    Rock, 165
Goldenweed, 147
    Cobwebby, 150
    Lyall, 149
    Pygmy, 151
    Shrubby, 152
    Singlehead, 153
    Stemless, 148
Goldflower, 158
    Stemless, 158
Goodrich Daisy, 129
GOOSEBERRY FAMILY, 448
Gooseberry, 448
    Canada, 452
    Mountain, 451
    Swamp, 450
Gooseberryleaf Alumroot, 732
GOOSEFOOT FAMILY, 304
Goosefoot, 304
    Fremont, 306
Gordon Ivesia, 682
Grape Fern, 500
GRASS FAMILY, 515
Grass of Parnassus, 736
Grayleaf Willow, 721
Gray's Angelica, 39
Green
    Gentian, 431

Spleenwort, 61
Greenland Bluegrass, 574
Greenleaf Bluebells, 215
Groundsel, 167
    Alpine, 169
    Arrowleaf, 182
    Black, 170
    Blacktip, 178
    Butterweed, 179
    Cleftleaf, 173
    Dusky, 176
    Fremont, 175
    Lambstongue, 177
    Mountain, 183
    Roundhead, 180
    Thickleaf, 172
    Twistedleaf, 181
    Twoleaf, 174
    Woolly, 171
Groundsmoke, 495
    Bigflower, 496
    Rocky Mountain, 497
Gunnison Sego Lily, 480

H

Hair Sedge, 332
Hairgrass, 540
    Mountain, 541
    Slender, 543
    Tufted, 542
Hairy
    Arnica, 92
    Goldenaster, 154
Hall's Willowherb, 493
Harebell, 266
    Alpine, 268
    Mountain, 267
Haresfoot Locoweed, 417
Hastate Phacelia, 456
Hawksbeard, 118
    Dwarf, 119
    Meadow, 120
Hawkweed, 156
    Slender, 156
Hayden Clover, 425
Heartleaf Arnica, 87
HEATH FAMILY, 381
Henderson Springparsley, 44
Hepburn Sedge, 349
Hoary Tansyaster, 161
Hogweed Cowparsnip, 49
Holboell Rockcress, 220
Holly Fern, 65
Holodiscus, 681
HONEYSUCKLE FAMILY, 269
Hood
    Phlox, 598
    Sedge, 340
Hooded Ladies' Tresses, 503

Hookspur Violet, 801
Hooker
    Cinquefoil, 698
    Oatgrass, 560
Horsemint, 473
    Cloverhead, 474
Huckleberry, 390
Hulsea, 157
    Alpine, 157
Hyemnoxys, 158

I
Icegrass, 565
Iceland Koenigia, 610
Idaho
    Bentgrass, 520
    Fescue, 557
Incurved Sedge, 347
Indian
    Milkvetch, 398
    Paintbrush, 756
Ipomopsis, 593
    Spike, 595
Ivesia, 682
    Gordon, 682
    Utah, 683

J
Jacob's Ladder, 601, 602
James Snowlover, 765
Jessica Tickweed, 208
Jones
    Columbine, 649
    Reedgrass, 537
    Sedge, 342
Junegrass, 563
Juniper, 314
    Common, 315
    Creeping, 316

K
Kellogg Knotweed, 615
Kelloggia, 711
Kelseya, 684
    Oneflower, 684
Kentucky Bluegrass, 579
King Biscuitroot, 56
King Tansyaster, 163
King's Crown, 312
Kinnikinnick, 382
Kittentails, 753, 789
    Alpine, 754
    Featherleaf, 789
    Wyoming, 755
Knotweed, 612
    Broadleaf, 616
    Douglas, 614
    Kellogg, 615
    Sawatch, 617
Kobresia, 378

Koenigia, 610
    Iceland, 610
Kotzebue Parnassus, 738

L
Ladies' Tresses, 503
    Hooded, 503
Lady Fern, 62
    Alpine, 62
    Common, 63
Lamb's Quarter, 304
Lambstongue Groundsel, 177
Lanceleaf
    Springbeauty, 625
    Stonecrop, 310
Largeflower Sandwort, 284
Largeleaf Avens, 678
Larkspur, 651
    Duncecap, 653
    Little, 652
Leadville Milkvetch, 402
Leafless Falsephlox, 592
Leafybract Aster, 104
Least Willow, 725
Lemmon Rockcress, 222
Letterman
    Bluegrass, 576
    Needlegrass, 584
Lewis Monkeyflower, 769
Liddon Sedge, 355
LILY FAMILY, 475
Lily
    Alp, 482
    Glacier, 481
    Mariposa, 480
    Sego, 480
    Wand, 484
Limber Pine, 512
Little
    Gentian, 438
    Larkspur, 652
    Red Elephant, 774
Locoweed, 413
    Alpine, 420
    Dwarf, 418
    Haresfoot, 417
    Nodding, 416
    Parry, 419
    Plains, 414
    Silky, 421
    Sticky, 422
Lodgepole Pine, 511
Lomatium, 57
    Nineleaf, 57
Longfruit Draba, 241
Longleaf Arnica, 91
Longstalk
    Clover, 426
    Springparsley, 45
    Starwort, 299

Lousewort, 770
    Bracteate, 771
    Coiled-Beak, 772
    Curled, 778
    Fernleaf, 773
    Oeder, 775
    Parry, 776
    Pretty, 777
Lovage, 50
    Fernleaf, 51
    Porter, 52
    Slenderleaf, 53
Lover
    Mountain, 302, 303
Low
    Blueberry, 392
    Daisy, 131
    Penstemon, 783
Lupine, 407
    Dwarf, 410
    Silvery, 408
    Velvet, 411
    Wyeth, 412
Lyall
    Goldenweed, 149
    Rockcress, 223

M
Maclovian Sedge, 345
MADDER FAMILY, 709
MAIDENHAIR FAMILY, 35
Mariposa Lily, 480
Marsh, 650
    Marigold, 650
    Violet, 803
Matchbrush, 145
Meadow
    Hawksbeard, 120
    Sedge, 359
Meadowrue, 666
    Alpine, 667
    Fendler, 668
    Veiny, 670
    Western, 669
Menzies Catchfly, 293
Michaux Sagewort, 98
Milkvetch, 395
    Alpine, 397
    Indian, 398
    Leadville, 402
    Prickly, 399
    Standing, 396
    Timber, 400
Millet Woodrush, 470
Miner's Candle, 203
    Northern, 204
MINT FAMILY, 473
Miterwort, 735
    Fivestamen, 735
Monkeyflower, 767

    Common, 767
    Lewis, 769
    Subalpine, 768
Moonwort, 500
Moss
    Campion, 291
    Gentian, 437
Mountain
    Bentgrass, 524
    Bluebells, 213
    Brome, 530
    Death Camas, 484
    Douglasia, 638
    Dryad, 674
    Forget-Me-Not, 216
    Gooseberry, 451
    Groundsel, 183
    Hairgrass, 541
    Lover, 302, 303
    Mahogany, 673
    Harebell, 267
    Pennycress, 263
    Penstemon, 784
    Snowberry, 271
    Sorrel, 622
    Spray, 681
    Tansymustard, 227
    Willow, 722
Mountain Heath, 385
    Cream, 386
    Pink, 385
Mountain Sorrel, 611
Mousetail Alpinesedge, 379
MUSTARD FAMILY, 217
Muttongrass, 572

N
Nailwort, 288
    Rocky Mountain, 288
Narcissus Anemone, 643
Narrowleaf
    Collomia, 591
    Paintbrush, 757
Narrowpetal Stonecrop, 313
Needlegrass, 583
    Letterman, 584
    Western, 585
Needleleaf Sedge, 368
Netleaf Willow, 724
Nevada Rush, 465
New Sedge, 353
Nineleaf Lomatium, 57
Nodding
    Bluegrass, 580
    Brome, 529
    Locoweed, 416
    Onion, 478
Northern
    Alpinesedge, 380
    Anemone, 644

Bedstraw, 710
 Goldenrod, 185
 Miner's Candle, 204
 Parnassus, 739
 Reedgrass, 538
 Starwort, 297
 Sweetvetch, 404
Norway Sedge, 352
Nothocalais, 164
 Black, 164
Nuttall
 Draba, 237
 Flaxflower, 596
 Povertyweed, 307
 Rockcress, 225
 Sandwort, 285
 Violet, 802

O

Oatgrass, 539, 559
 Alpine, 561
 Hooker, 560
 Timber, 539
Oblongleaf Bluebells, 214
Obtuse Yellowcress, 260
Oeder Lousewort, 775
Old-Man-of-the-Mountain, 160
Old Man's Whiskers, 680
One-Sided Wintergreen, 389
Oneflower
 Daisy, 140
 Kelseya, 684
Onion, 475
 Brandegee, 476
 Nodding, 478
 Shortstyle, 477
Orange
 Argoseris, 72
 Arnica, 88
ORCHID FAMILY, 501
Orchid
 White Bog, 502
Oregon
 Daisy, 141
 Saxifrage, 750
Oreoxis, 58
 Alpine, 58
ORPINE FAMILY, 308

P

Paintbrush, 756
 Alpine, 760
 Common Red, 759
 Cusick, 758
 Narrowleaf, 757
 Pale, 761
 Rosy, 763
 Showy, 762
 Sticky, 757
 Sulphur, 764
Painted Pyrola, 387
Pale
 Agoseris, 73
 Daisy, 135
 Paintbrush, 761
 Sedge, 331
Parnassus
 Fringed, 737
 Kotzebue, 738
 Northern, 739
Parry
 Catchfly, 294
 Clover, 428
 Goldenrod, 186
 Locoweed, 419
 Lousewort, 776
 Primrose, 639
 Rush, 466
 Townsendia, 198
Parrya, 255
 Smoothstem, 255
PARSNIP FAMILY, 38
Patterson Bluegrass, 578
Pasque Flower, 645
Payson
 Bladderpod, 253
 Draba, 243
PEA FAMILY, 395
Pearlwort, 289
 Arctic, 289
Pearly
 Everlasting, 75
 Pussytoes, 79
Pennsylvania Cinquefoil, 702
Pennycress, 262
 Mountain, 263
 Smallflower, 264
Penstemon, 779
 Low, 783
 Mountain, 784
 Rydberg, 786
 Shrubby, 782
 Smallflower, 785
 Thinstem, 780
 Uinta, 787
 Wasatch, 781
 Whipple, 788
Peregrine Daisy, 136
Perennial Fringed Gentian, 442
Phacelia, 455
 Hastate, 456
 Silky, 457
PHLOX FAMILY, 589
Phlox, 597
 Hood, 598
 Rocky Mountain, 599
 Tufted, 600
PINE FAMILY, 506

Pine, 509
 Limber, 512
 Lodgepole, 511
 Violet, 804
 Whitebark, 510
PINK FAMILY, 272
Pink Mountain Heath, 385
Pinnateleaf Daisy, 137
Plains
 Locoweed, 414
 Reedgrass, 535
Planeleaf Willow, 723
PLANTAIN FAMILY, 514
Plantain, 514
 Tweedy, 514
Plantainleaf Buttercup, 656
Poplar, 713
 Balsam, 713
POPPY FAMILY, 504
Poppy, 504
 Arctic, 505
Porsild Draba, 244
Porter Lovage, 52
Povertyweed, 307
 Nuttall, 307
Prairie Sagewort, 97
Pretty Lousewort, 777
Prickly Milkvetch, 399
PRIMROSE FAMILY, 631
Primrose, 639
 Parry, 639
Prostrate Sibbaldia, 708
Purple Reedgrass, 536
Purpleleaf Willowherb, 492
PURSLANE FAMILY, 624
Purslane Speedwell, 792
Pussypaws, 630
 Umbellate, 630
Pussytoes, 76
 Alpine, 77
 Aromatic, 80
 Brown, 85
 Flattop, 81
 Pearly, 79
 Rosy, 83
 Singlehead, 84
 Woolly, 82
Pygmy
 Bitterroot, 628
 Goldenweed, 151
Pygmyflower Rockjasmine, 633
Pyrenees Sedge, 360
Pyrola, 387
 Painted, 387
 Small, 387

Q

Quaking Aspen, 713
QUILLWORT FAMILY, 459

Quillwort, 459
 Bolander, 459

R

Rabbitbrush, 111
 Douglas, 111
Raspberry, 706
 Red, 707
Raynold Sedge, 361
Red
 Elderberry, 270
 Fescue, 558
 Raspberry, 707
 Sand Spurry, 295
 Sandwort, 287
Redstem Cinquefoil, 703
Reedgrass, 533
 Bluejoint, 534
 Jones, 537
 Northern, 538
 Plains, 535
 Purple, 536
Richardson
 Geranium, 446
River Beauty, 494
Rock
 Goldenrod, 165
 Holly Fern, 65
 Rose, 499
 Sandwort, 283
 Sedge, 367
Rockbrake, 35
 American, 36
Rockcress, 218
 Drummond, 219
 Holboell, 220
 Lemmon, 222
 Lyall, 223
 Nuttall, 225
 Small-Leaf, 224
 Williams, 226
Rockjasmine, 631
 Dwarf, 632
 Pygmyflower, 633
 Rockmat, 685
 Tufted, 685
Rockspiraea, 685
Rocky Mountain
 Dandelion, 191
 Fringed Gentian, 443
 Groundsmoke, 497
 Nailwort, 288
 Phlox, 599
 Pleated Gentian, 434
 Woodsia, 66
ROSE FAMILY, 672
Rose
 Angelica, 40
Rosecrown, 311
Roseroot, 312

Ross
    Sandwort, 282
    Sedge, 362
Rosy
    Paintbrush, 763
    Pussytoes, 83
Roundhead Groundsel, 180
Roundleaf Alumroot, 731
Rough Sedge, 336
RUSH FAMILY, 460
Rush, 460
    Chestnut, 462
    Drummond, 463
    Nevada, 465
    Parry, 466
    Subalpine, 464
    Threeflower, 467
    Twoflower, 461
Russet Sedge, 365
Rydberg
    Arnica, 93
    Daisy, 139
    Fescue, 557
    Penstemon, 786
    Spikemoss, 794
    Twinpod, 257

S

Sagebrush, 94
    Big, 101
    False, 188
Sagebrush Buttercup, 660
Sagewort
    Alpine, 100
    Common, 95
    Fringed, 96
    Michaux, 98
    Prairie, 97
    Spruce, 99
Sailorcaps Shootingstar, 636
Sandberg Bluegrass, 581
Sand Spurry, 295
    Red, 295
Sandwort, 272, 281
    Arctic, 286
    Ballhead, 274
    Largeflower, 284
    Nuttall, 285
    Red, 287
    Rock, 283
    Ross, 282
    Threadleaf, 273
Saussurea, 166
    Weber, 166
Sawatch Knotweed, 617
SAXIFRAGE FAMILY, 728
Saxifrage, 740
    Alberta, 747
    Alpine Brook, 752
    Alpine Spider, 746

Brook, 748
Bulblet, 744
Goldbloom, 745
Oregon, 750
Snowball, 751
Spotted, 742
Tufted, 743
Twinleaf, 749
Wedgeleaf, 741
Scorpionweed, 455
Scouler St. John's Wort, 458
Scribner Wheatgrass, 549
SEDGE FAMILY, 317
Sedge, 317
    Alpine Nerved, 350
    Beaked, 363
    Black Alpine, 351
    Black-and-White, 323
    Blackroot, 337
    Blackscale, 325
    Bladder, 371
    Blunt, 354
    Brewer, 329
    Brown, 330
    Capitate, 333
    Curly, 364
    Dark Alpine, 370
    Douglas, 334
    Dunhead, 356
    Early, 358
    Ebony, 335
    Elk, 339
    Golden, 327
    Hair, 332
    Hepburn, 349
    Hood, 340
    Incurved, 347
    Jones, 342
    Liddon, 355
    Maclovian, 345
    Meadow, 359
    Needleleaf, 368
    New, 353
    Norway, 352
    Pale, 331
    Pyrenees, 360
    Raynold, 361
    Rock, 367
    Ross, 362
    Rough, 336
    Russet, 365
    Shasta, 369
    Sheep, 341
    Shortleaf, 348
    Shortstalk, 357
    Sierra Hare, 343
    Singlespike, 366
    Threadleaf, 338
    Twotip, 328
    Water, 324

Woodrush, 344
Sego Lily, 480
    Gunnison, 480
Sharpleaf Valerian, 797
Shasta Sedge, 369
Sheep
    Cinquefoil, 701
    Fescue, 556
    Sedge, 341
Shield Fern, 65
Shootingstar, 634
    Alpine, 635
    Darkthroat, 637
    Sailorcaps, 636
Shortawn Foxtail, 526
Shortleaf Sedge, 348
Shortleaf Cinquefoil, 688
Shortstalk Sedge, 357
Shortstem Buckwheat, 606
Shortstyle Onion, 477
Showy
    Paintbrush, 762
    Thistle, 116
Shrubby
    Cinquefoil, 692
    Goldenweed, 152
    Penstemon, 782
Sierra Hare Sedge, 343
Sibbaldia, 708
    Prostrate, 708
Siberian Aster, 107
Silky
    Locoweed, 421
    Phacelia, 457
Silvery Lupine, 408
Singleflower Cinquefoil, 705
Singlehead
    Goldenweed, 153
    Pussytoes, 84
Singlespike Sedge, 366
Sky Pilot, 603
Slender
    Cinquefoil, 695
    Daisy, 130
    Hairgrass, 543
    Hawkweed, 156
    Spikerush, 373
    Townsendia, 197
    Wheatgrass, 552
Slenderleaf Lovage, 53
Small
    Pyrola, 387
    Sweetcicely, 59
Smallflower
    Blue-Eyed Mary, 766
    Pennycress, 264
    Penstemon, 785
Small-Leaf Rockcress, 224
Smelowskia, 261
    Alpine, 261

Smooth
    Brome, 532
    Daisy, 133
    Draba, 239
    Woodrush, 469
Smoothstem Parrya, 255
Snakeweed, 145
    Broom, 146
Snow Thistle, 117
Snowball Saxifrage, 751
Snowbank
    Cinquefoil, 700
    Draba, 242
    Springparsley, 46
Snowbed Buttercup, 658
Snowberry, 271
    Mountain, 271
Snowlover, 765
    James, 765
Sorrel, 619
    Alpine, 611
    Garden, 620
    Mountain, 611, 622
Speedwell, 790
    Alpine, 791
    Purslane, 792
    Thymeleaf, 791
Spike
    Fescue, 564
    Ipomopsis, 595
    Trisetum, 587
    Woodrush, 471
Spikegrass, 564
SPIKEMOSS FAMILY, 793
Spikemoss, 793
    Rydberg, 794
    Watson, 795
Spikerush, 372
    Fiveflower, 374
    Slender, 373
SPLEENWORT FAMILY, 60
Spleenwort, 61
    Green, 61
Spoonleaf Townsendia, 199
Spotted Saxifrage, 742
Spreading Tickweed, 209
Springbeauty, 624
    Alpine, 626
    Lanceleaf, 625
Springparsley, 42
    Evert's, 43
    Henderson, 44
    Longstalk, 45
    Snowbank, 46
    Turpentine, 47
Spruce, 508
    Engelmann, 508
    Sagewort, 99
Squirreltail, 545
STAFF TREE FAMILY, 302

Standing Milkvetch, 396
Star Gentian, 444
Starwort, 296
   Blunt, 300
   Longstalk, 299
   Northern, 297
   Thickleaf, 298
   Umbrella, 301
   Water, 265
Steershead, 430
Stemless
   Buckwheat, 605
   Goldenweed, 148
   Goldflower, 158
Sticky
   Arnica, 92
   Cinquefoil, 693
   Geranium, 447
   Locoweed, 422
   Paintbrush, 757
   Tofieldia, 483
ST. JOHN'S WORT FAMILY, 458
St. John's Wort, 458
   Scouler, 458
Stonecrop, 308
   Lanceleaf, 310
   Narrowpetal, 313
   Weakstem, 309
Strawberry, 675
   Wild, 675
Strawberry Blite, 305
Subalpine
   Fir, 507
   Monkeyflower, 768
   Rush, 464
Sulfur
   Buckwheat, 609
   Sweetvetch, 406
Sulphur Paintbrush, 764
Swamp Gooseberry, 450
Sweetcicely, 59
   Small, 59
Sweetgrass, 562
Sweetvetch, 403
   Northern, 404
   Sulfur, 406
   Western, 405
Swertia
   Alpine Bog, 444

T
Tall
   Cottonsedge, 376
   Draba, 245
Tansymustard, 227
   Mountain, 228
   Wind River, 229
Tansyaster, 161
   Hoary, 161

King, 163
Thickleaf
   Draba, 236
   Groundsel, 172
   Starwort, 298
Thickspike Wheatgrass, 548
Thickstem Aster, 106
Thin Daisy, 142
Thinstem Penstemon, 780
Thistle, 113
   Eaton, 114
   Showy, 116
   Snow, 117
   White, 115
Thorowax, 41
   American, 41
Threadleaf
   Sandwort, 273
   Sedge, 338
Threeflower Rush, 467
Threeleaf Bitterroot, 629
Thurber Bentgrass, 523
Thymeleaf Speedwell, 791
Ticklegrass, 522
Tickweed, 206
   Jessica, 208
   Spreading, 209
   Western, 207
Timber
   Milkvetch, 400
   Oatgrass, 539
Timberline Buttercup, 665
Timothy, 566
   Alpine, 566
   Common, 566
Tofieldia, 483
   Sticky, 483
Townsendia, 194
   Alpine, 195
   Cushion, 196
   Parry, 198
   Slender, 197
   Spoonleaf, 199
Trisetum, 586
   Spike, 587
   Wolf, 588
Trumpet
   Fairy, 593
Tufted
   Cottonsedge, 375
   Daisy, 124
   Hairgrass, 542
   Phlox, 600
   Rockmat, 685
   Saxifrage, 743
Tundra
   Buttercup, 659
   Dandelion, 190
Turpentine Springparsley, 47

Tweedy
   Plantain, 514
   Willow, 726
Twinleaf
   Cinquefoil, 704
   Saxifrage, 749
Twinpod, 256
   Common, 258
   Rydberg, 257
Twistedleaf Groundsel, 181
Twoflower Rush, 461
Twoleaf Groundsel, 174
Twotip Sedge, 328

U
Uinta Penstemon, 787
Umbellate Pussypaws, 630
Umbrella Starwort, 301
Unpleasant Buttercup, 662
Utah
   Bladderpod, 254
   Ivesia, 683

V
VALERIAN FAMILY, 796
Valerian, 796
   Edible, 798
   Sharpleaf, 797
   Western, 799
Veiny Meadowrue, 670
Velvet Lupine, 411
VIOLET FAMILY, 800
Violet, 800
   Canada, 803
   Hookspur, 801
   Marsh, 803
   Nuttall, 802
   Pine, 804

W
Wahlenberg Woodrush, 472
Wallflower, 247
   Western, 247
Wand Lily, 484
Wasatch Penstemon, 781
WATER STARWORT FAMILY, 265
Water
   Sedge, 324
   Starwort, 265
Watercrowfoot
   White, 657
WATERLEAF FAMILY, 455
Waterweed, 453
   Canada, 454
Watson Spikemoss, 795
Wax Currant, 449
Weakstem Stonecrop, 309
Weber Saussurea, 166
Wedgeleaf Saxifrage, 741

Western
   Aster, 108
   Bladderpod, 252
   Meadowrue, 669
   Needlegrass, 585
   Sweetvetch, 405
   Tickweed, 207
   Valerian, 799
   Wallflower, 247
   Wheatgrass, 550
Wheatgrass, 516, 544
   Bluebunch, 551
   Crested, 517
   Scribner, 549
   Slender, 552
   Thickspike, 548
   Western, 550
Wheeler Bluegrass, 577
Whipple Penstemon, 788
Whiskers
   Old Man's, 680
White
   Bog Orchid, 502
   Cottonsedge, 377
   Draba, 234
   Thistle, 115
   Watercrowfoot, 657
Whitebark Pine, 510
Whitlow Grass, 230
Whitlow-wort, 288
Wild
   Buckwheat, 604
   Strawberry, 675
Wildrye, 544
   Blue, 547
Williams Rockcress, 226
WILLOW FAMILY, 712
Willow, 714
   Arctic, 715
   Barratt, 716
   Barrenground, 717
   Cascade, 718
   Drummond, 719
   Eastwood, 720
   Grayleaf, 721
   Least, 725
   Mountain, 722
   Netleaf, 724
   Planeleaf, 723
   Tweedy, 726
   Wolf, 727
Willowherb, 489
   Alpine, 490
   Hall's, 493
   Purpleleaf, 492
Wind River
   Draba, 246
   Tansymustard, 229
Wintergreen, 387
   Alpine, 383

One-Sided, 389
Wolf
    Trisetum, 588
    Willow, 727
Woodland Star, 734
    Fringecup, 734
Woodrush, 468
    Millet, 470

Sedge, 344
    Smooth, 469
    Spike, 471
    Wahlenberg, 472
Woodsia
    Rocky Mountain, 66
Woolly
    Cinquefoil, 697

Daisy, 132
Eriophyllum, 144
Groundsel, 171
Pussytoes, 82
Woollyleaf, 144
Wormwood, 94
Wyeth Lupine, 412
Wyoming Kittentails, 755

Y
Yarrow, 69
    Common, 70
Yellow
    Buckwheat, 607
    Columbine, 648
Yellowcress, 259
    Obtuse, 260
Yellowstone Draba, 240

Index to Latin Names

Accepted names are in bold-face type.

A

Abies, 506
  arizonica, 507
  bifolia, 507
  douglasii, 513
  engelmannii, 508
  lasiocarpa, 507
  lindleyana, 513
  menziesii, 513
  mucronata, 513
  subalpina, 507
  taxifolia, 513
  trigona, 513
Absinthium frigidum, 96
Acetosa
  alpestris, 620
  gracilescens, 622
Acetosella gracilescens, 622
  paucifolius, 622
Achillea, 70
  alpicola, 70
  angustissima, 70
  arenicola, 70
  asplenifolia, 70
  borealis, 70
  californica, 70
  eradiata, 70
  fusca, 70
  gigantea, 70
  gracilis, 70
  lanulosa, 70
  laxiflora, 70
  megacephala, 70
  millefolium, 70
  nigrescens, 70
  occidentalis, 70
  pacifica, 70
  puberula, 70
  rosea, 70
  subalpina, 70
  tomentosa, 70
Acomastylis
  depressa, 679
  gracilipes, 679
  humilis, 679
  rossii, 679

  sericea, 679
  turbinata, 679
Acrostichum crispum, 36
Actinea
  acaulis, 158
  arizonica, 158
  eradiata, 158
  epunctata, 158
  grandiflora, 160
  herbacea
  incana, 158
  lanata, 144
  simplex, 158
Actinella
  acaulis, 158
  argentea, 158
  epunctata, 158
  eradiata, 158
  grandiflora, 160
  incana, 158
  lanata, 144, 158
  simplex, 158
Actinocyclus secundus, 389
Adenolinum lewisii, 486
ADIANTACEAE, 35
Aegilops hystrix, 545
Aetopteron lonchitis, 65
Agoseris, 71
  agrestis, 73
  altissima, 73
  arachnoidea, 73
  arizonica, 73
  aspera, 73
  attenuata, 73
  aurantiaca, 72
  carnea, 72
  caudata, 73
  confinis, 72
  dasycarpa, 73
  deas leonis, 73
  eisenhoweri, 73
  glauca, 73
  gracilens, 72
  gracilenta, 72
  graminifolia, 72
  greenei, 72
  howellii, 72
  isomeris, 73
  lacera, 73

  laciniata, 73
  lanulosa, 73
  lapathifolia, 73
  leontodon, 73
  longirostris, 72
  longula, 73
  maculata, 73
  microdonta, 73
  monticola, 73
  nana, 72
  parviflora, 73
  procera, 73
  pubescens, 73
  pumila, 73
  purpurea, 72
  rosea, 73
  roseata, 73
  rostrata, 72
  scorzoneraefolia, 73
  subalpina, 72
  taraxacifolia, 73
  taraxacoides, 73
  turbinata, 73
  vestita, 73
  villosa, 73
Agropyron, 516
  albicans, 548
  alaskanum, 552
  andinum, 552
  bakeri, 549, 552
  biflorum, 552
  boreale, 552
  brevifolium, 552
  caninoides, 552
  caninum, 552
  cristatiforme, 517
  cristatum, 517
  dasystachyum, 548
  desertorum, 517
  divergens, 551
  elmeri, 548
  fragile, 517
  gmelini, 552
  griffithsii, 548
  inerme, 551
  lanceolatum, 548
  latiglume, 552
  molle, 550
  novae-angliae, 552

  occidentale, 550
  palmeri, 550
  pauciflorum, 552
  pectiniforme, 517
  psammophilum, 548
  pseudorepens, 548, 552
  richardsonii, 552
  riparium, 548
  scribneri, 549
  sibiricum, 517
  smithii, 550
  spicatum, 551
  subsecundum, 552
  subvillosum, 548
  tenerum, 552
  teslinense, 552
  trachycaulon, 552
  trachycaulum, 552
  unilaterale, 552
  vaseyi, 551
  violaceum, 552
  violescens, 552
  yukonense, 548
Agrostemma apetalum, 279
Agrostis, 518
  algida, 565
  atrata, 523
  bakeri, 521
  borealis, 521
  clavata, 520
  caespitosa, 542
  filicumis, 520
  geminata, 522
  heimalis, 522
  humilis, 519
  idahoensis, 520
  mertensii, 521
  michauxii, 522
  nootkaensis, 522
  nutkaensis, 522
  rossae, 524
  rupestris, 521
  scabra, 522
  scabriuscula, 522
  tenuiculmis, 520
  tenuis, 520
  thurberiana, 523
  variabilis, 524
  varians, 524

Aira
  alpicola, 542
  ambigua, 542
  atropurpurea, 541
  brevifolia, 581
  cespitosa, 542
  canbyi, 581
  elongata, 543
  gracilis, 563
  holciformis, 542
  latifolia, 541
  macrantha, 563
  pungens, 542
  spicata, 587
  subspicata, 587
  sukatschewii, 542
  vaseyana, 543
Airochloa gracilis, 563
Airopis brevifolia, 581
Allium, 475
  alleghemense, 478
  brandegei, 476
  brevistylum, 477
  cernuum, 478
  diehlii, 476
  minimum, 476
  natans, 478
  neomexicanum, 478
  recurvatum, 478
  schoenoprasum, 479
  sibiricum, 479
Allosorus
  acrostichoides, 36
  brewerii, 37
  crispus, 36
Alopecurus, 525
  aequalis, 526
  alpinus, 527
  aristulatus, 526
  beeringianus, 527
  borealis, 527
  caespitosus, 526
  fulvus, 526
  glaucus, 527
  macounii, 526
  occidentalis, 527
  stejnegeri, 527
Alpinus
  albicaulis, 510
  flexilis, 512
Alsinanthe
  biflora, 286
  macrantha, 284
  stricta, 283
Alsine
  baicalensis, 301
  biflora, 286
  borealis, 297
  brachypetala, 297
  crassifolia, 298

  hirta, 287
  laeta, 299
  linnaei, 289
  longipes, 299
  michauxii, 283
  obtusa, 300
  oxyphylla, 297
  palmeri, 299
  polygonoides, 300
  rossii, 282
  rubella, 287
  rubra, 295
  simcoei, 297
  stricta, 299
  strictiflora, 299
  subvestita, 299
  uliginosa, 283
  validus, 299
  viridula, 300
  washingtoniana, 300
Alsinella saginoides, 289
Alsinopsis
  dawsonensis, 283
  macrantha, 284
  obtusiloba, 286
  occidentalis, 285
  propinqua, 287
  quadrivalvis, 287
  rossii, 282
  stricta, 283
  tenella, 283
Alyssum alpinum, 250
Amadea
  diffusa, 633
  puberulenta, 633
Amarella
  acuta, 439
  amarella, 439
  anisosepala, 439
  californica, 439
  cobrensis, 439
  conferta, 439
  copelandi, 439
  distegia, 439
  heterosepala, 439
  lemberti, 439
  macounii, 439
  monantha, 440
  plebeja, 439
  revoluta, 439
  scopulorum, 439
  strictiflora, 439
  tenella, 440
  tortuosa, 439
Amelia minor, 387
Amellus villosus, 154
Amerosedum
  debile, 309
  lanceolatum, 310
  nesioticum, 310

  subalpinum, 310
  stenopetalum, 313
Ammogeton scorzonerae-
    folius, 73
Anacharis
  canadensis, 454
  planchonii, 454
Anaphalis, 75
  angustifolia, 75
  lanata, 75
  margaritacea, 75
  occidentalis, 75
  sierrae, 75
  subalpina, 75
Androsace, 631
  albertina, 632
  arguta, 633
  bungeana, 632
  carinata, 632
  chamaejasme, 632
  diffusa, 633
  glandulosa, 633
  gormani, 633
  lehmanniana, 632
  puberulenta, 633
  septentrionalis, 8, 633
  subulifera, 633
  subumbellata, 633
  uniflora, 638
Anemone, 640
  drummondii, 641
  globosa, 642
  hirsutissima, 645
  hudsoniana, 642
  lithophila, 641
  ludoviciana, 645
  multifida, 642, 645
  narcissiflora, 643
  nuttalliana, 645
  parviflora, 644
  patens, 645
  tetonensis, 642
  wolfgangiana, 645
  zephyra, 643
Angelica, 38
  grayi, 26, 39
  roseana, 40
Anotites
  dorrii, 293
  halophila, 293
  latifolia, 293
  macilenta, 293
  menziesii, 293
  nodosa, 293
  picta, 293
  tereticaulis, 293
  viscosa, 293
Antennaria, 76
  acuminata, 83
  aizoides, 85

  alaskana, 77
  albescens, 85
  albicans, 85
  alborosea, 83
  alpina, 76
  anaphaloides, 79
  angustata, 84
  angustifolia, 77, 83
  arenicola, 77
  arida, 83
  aromatica, 80
  atriceps, 77
  austromontana, 77
  bayardii, 77
  bocheria, 77
  borealis, 77
  bracteosa, 83
  breitungii, 83
  brevistyla, 77
  brunnescens, 77
  burwellensis, 84
  cana, 77
  candida, 77
  canescens, 77
  chlorantha, 77
  columnaris, 85
  compacta, 77
  concinna, 83
  confinis, 85
  confusa, 77
  congesta, 84
  corymbosa, 81
  crymophila, 77
  densa, 77
  densifolia, 77
  ekmaniana, 77
  elegans, 83
  ellyae, 85
  erigeroides, 83
  exilis, 84
  fernaldiana, 84
  flavescens, 85
  foggii, 77
  foliacea, 83
  formosa, 83
  friesiana, 77
  fusca, 85
  glabrata, 84
  gormanii, 85
  groenlandica, 83
  hansii, 83
  hendersonii, 83
  hudsonica, 84
  hygrophila, 81
  imbricata, 83
  incarnata, 83
  intermedia, 85
  isolepis, 85
  laingii, 83
  lanata, 82

lanulosa, 85
leontopodioides, 83
leuchippii, 83
macounii, 77
maculata, 85
margaritacea, 75
media, 77
megacephala, 77
microphylla, 83
modesta, 77
monocephala, 25, 84
mucronata, 85
nardina, 81
neoalaskana, 77
nitens, 84
nitida, 83
oxyphylla, 83
pallida, 85
parvifolia, 83
peasei, 85
pedunculata, 77
philonipha, 84
pulchella, 77
pulvinata, 77
pygmaea, 84
reflexa, 77
rosea, 83
rousseaui, 77
sansonii, 85
scabra, 77
scariosa, 83
sedoides, 85
shumaginensis, 84
solitaria, 84
solstitialis, 83
sorida, 83
sornborgeri, 77
speciosa, 83
stolonifera, 77
straminea, 85
subcanescens, 77
subviscosa, 83
tansleyi, 84
tomentella, 77
tweedsmurii, 84
umbrinella, 85
vexillifera, 77
viscidula, 85
wiegandii, 85
Anthericum serotinum, 482
Anthopogon
  barbata, 443
  barbellata, 442
  detonsa, 443
  elegans, 443
  macounii, 443
  thermalis, 443
Anthoxanthum nitens, 562
Anticlea
  alpina, 484

chlorantha, 484
coloradensis, 484
elegans, 484
longa, 484
Antiphylla
  oppositifolia, 749
  pulvinata, 749
Antirrhinum tenellum, 766
APIACEAE, 38
Aplopappus alpigenus, 103
  falcatus, 148
  howellii, 153
  inuloides, 153
  nelsonii, 148
  parryi, 186
Aquilegia, 646
  coerulea, 647
  flavescens, 648
  jonesii, 649
  leptocera, 647
  oreophila, 647
  piersoniana, 647
  pinetorum, 647
  rubicunda, 648
  scopulorum, 647
Arabis, 218
  acutina, 220
  albertina, 219
  armerifolia, 223
  bourgovii, 220
  brachycarpa, 219
  bracteolata, 222
  brevisiliqua, 220
  bridgeri, 225
  caduca, 220
  collinsii, 220
  confinis, 219
  connexa, 219
  consanguinea, 220
  dacotica, 220
  densa, 223
  densicaulis, 224
  depauperata, 222
  divaricarpa, 220
  drepanoloba, 222
  drummondii, 219
  egglestonii, 222
  exilis, 220
  holboellii, 220
  kennedyi, 222
  kochii, 220
  latifolia, 222
  lemmonii, 222
  lignifera, 220
  lignipes, 220
  lyallii, 223
  mcdougalii, 220
  macella, 225
  macounii, 224
  microphylla, 224

multiceps, 223
nubigena, 223
nudicaulis, 255
nuttallii, 225
oblanceolata, 220
oreocallis, 222
oreophila, 223
oxphylla, 219
paupercula, 223
pendulocarpa, 220
philonipha, 219
pinetorum, 220
polyantha, 220
polyclada, 222
pratincola, 220
retrofracta, 220
rhodanthus, 220
secunda, 220
semisepulta, 222
spathulata, 225
spatifolia, 220
stokesiae, 220
tenuicula, 224
tenuis, 220
williamsii, 26, 226
Aragallus
  albertinus, 414
  albiflorus, 421
  alpicola, 414
  argophyllus, 418
  atropurpureus, 417
  besseyi, 418
  blankenshipii, 417
  campestris, 414
  cervinus, 414
  collinus, 418
  deflexa, 416
  dispar, 414
  foliolosus, 416
  gracilis, 414
  hallii, 420
  hudsonicus, 422
  inflatus, 420
  lagopus, 417
  luteolus, 414
  macounii, 414, 421
  majusculus, 421
  melanopontus, 421
  monticola, 414
  nanus, 418
  parryi, 419
  pinetorum, 421
  podocarpa, 420
  podocarpus, 420
  saximontanus, 421
  sericea, 421
  spicatus, 421
  ventosus, 418
  villosus, 414
  viscidulus, 422

viscidus, 422
Arbutus uva-ursi, 382
Arctostaphylos, 382
  adenotricha, 382
  media, 382
  officinalis, 382
  procumbens, 382
  uva-ursi, 382
Arenaria, 272
  aequicaulis, 287
  alpestris, 297
  biflora, 286
  burkei, 274
  calycantha, 297
  capillaris, 273
  cephaloidea, 274
  congesta, 274
  dawsonensis, 283
  filiorum, 282
  formosa, 273
  gregaria, 285
  glabrescens, 274
  lithophila, 274
  litorea, 283
  macra, 283
  macrantha, 284
  nardifolia, 273
  nuttallii, 285
  obtusa, 286
  obtusiloba, 286
  propinqua, 287
  pungens, 285
  quadrivalvis, 287
  rossii, 282
  rubella, 287
  rubra, 295
  sajanensis, 286
  stricta, 283
  subcongesta, 274
  tenella, 283
  uliginosa, 283
Armeria
  canescens, 598
  hoodii, 598
Arnica, 86
  abortiva, 87
  alpina, 26
  amplifolia, 92
  andersonii, 87
  angustifolia, 93
  aphanactis, 89
  aprica, 89
  arachnoidea, 92
  arcana, 91
  aurantiaca, 93
  austinae, 87
  betonicaefolia, 89
  caespitosa, 93
  cascadensis, 93
  caudata, 91

chionophila, 87
coloradensis, 92
columbiana, 89
cordifolia, 87
crocea, 92
crocinea, 92
diversifolia, 92
eriopoda, 89
evermannii, 87
flodmanii, 89
fulgens, 88
gracilis, 89
grandifolia, 89
granulifera, 89
hardinae, 87
humilis, 87
jonesii, 89
laevigata, 89
lasiosperma, 93
latifolia, 89
latucina, 89
leptocaulis, 89
longifolia, 91
macrophylla, 87
menziesii, 89
merriami, 92
mollis, 92
monocephala, 88
multiflora, 89
myriadenia, 91
ovalifolia, 89
ovalis, 87
ovata, 92
parvifolia, 87
pedunculata, 88
platyphylla, 89
polycephala, 91
puberula, 89
pumila, 87
rivularis, 92
rydbergii, 93
scaberrima, 92
sylvatica, 92
subcordata, 87
subplumosa, 92
sulcata, 93
sylvatica, 92
tenuis, 93
teucrifolia, 89
ventorum, 89
whitneyi, 87
Artemisia, 94
  albula, 97
  angusta, 101
  angustifolia, 101
  arctica, 99
  argophylla, 97
  atomifera, 97
  borealis, 95
  bourgeauana, 95

brittonii, 97
campestris, 95
camporum, 95
canadensis, 95
candicans, 97
caudata, 95
chamissoniana, 99
cuneata, 97
discolor, 98
diversifolia, 97
falcata, 97
floccosa, 97
flodmanii, 97
forwoodii, 95
frigida, 96
gnaphalodes, 97
gracilenta, 97
graveolens, 98
herriottii, 97
incompta, 97
latiloba, 97
lindheimeriana, 97
longipedunculata, 99
ludoviciana, 97
maccallae, 95
mexicana, 97
michauxiana, 98
microcephala, 97
norvegica, 99
pabularis, 97
pacifica, 95
parishii, 101
paucicephala, 97
platyphylla, 97
potens, 97
pudica, 97
purshiana, 97
rhizomata, 97
richardsoniana, 95
ripicola, 95
rothrockii, 101
saxicola, 99
scopulorum, 100
scouleriana, 95
spiciformis, 101
spithamea, 95
subglabra, 98
tenuis, 98
tridentata, 101
underwoodii
vaseyana, 101
Arundo
  canadensis, 534
  langsdorfii, 534
  neglecta, 538
  purpurascens, 536
  stricta, 538
Askellia nana, 119
Asphodeliris glutinosa, 483

Aspidium
  alpestre, 62
  angustum, 63
  filix-femina, 63
  fragile, 64
  lonchitis, 65
ASPLENIACEAE, 60
Asplenium, 61
  cyclosorum, 63
  filix-femina, 63
  trichomanes-ramosum, 61
  viride, 61
Aster, 102
  alpigenus, 103
  amplexifolius, 106
  amplissimus, 104
  andersonii, 103
  andinus, 108
  apricus, 104
  attenuatus, 161
  bakerensis, 107
  burkei, 104
  caespitosus, 148
  campestris, 26
  canbyi, 104
  candollei, 187
  canescens, 161
  ciliomarginatus, 104
  cusickii, 104
  delectabilis, 108
  delectus, 108
  diabolicus, 104
  elatus, 103
  engelmannii, 26
  eriocaulis, 104
  espenbergensis, 107
  foliaceus, 104
  fremontii, 108
  frondeus, 104
  glacialis, 136
  glastifolius, 104
  glaucodes, 105
  glaucus, 105
  glossophyllus, 161
  haydenii, 103
  hendersonii, 104
  incertus, 104
  inornatus, 161
  integrifolius, 106
  jamesii, 149
  kingii, 163
  kootenayi, 104
  leiodes, 161
  leucanthemifolius, 161
  macronema, 150
  meritus, 107
  minor, 186
  misellus, 108
  montanus, 107
  multiradiatus, 185

occidentalis, 108
paludicola, 108
peregrinus, 136
phyllodes, 104
pulchellus, 103
pumilus, 165
pygmaeus, 107
richardsonii, 107
rubricaulis, 161
salsuginosus, 136
sibiricus, 107
spathulatus, 108
stenotus, 151
subintegerrimus, 107
subspathulatus, 108
subspicatus, 104
suffruticosus, 152
tweedyi, 104
uniflorus, 153
vaccinus, 104
vallicola, 108
viscidiflorus, 111
wasatchensis, 105
williamsii, 108
yosemitanus, 108
ASTERACEAE, 67
Astragalus, 395
  aboriginorum, 398
  aboriginum, 398
  acauleatus, 399
  adsurgens, 396
  albertinus, 414
  albiflorus, 421
  alpestris, 397
  alpicola, 414
  alpinus, 397
  andinus, 397
  arcticus, 397
  astragalinus, 397
  australis, 398
  blankenshipii, 417
  brunetianus, 397
  campestris, 414
  carltonii, 400
  centrophytus, 399
  chandonnetti, 396
  crandellii, 396
  decumbens, 400
  deflexus, 416
  divergens, 400
  forwoodii, 398
  garrettii, 400
  giganteus, 397
  glabriusculus, 398
  gracilis, 400
  grayanus, 414
  griseopubescens, 400
  hylophilus, 400
  impensus, 399
  jessiae, 399

kentrophyta, 399
labradoricus, 397
lagopus, 417
lapponicus, 397
laxmanni, 396
lepagei, 398
linearis, 398
mazama, 414
miser, 400
molybdenus, 26, 402
montanus, 399
nitidus, 396
palliseri, 400
parryanus, 419
pauciflorus, 397
phacinus, 397
plumbeus, 402
retroflexus, 416
richardsonii, 398
rydbergianus, 414
rydbergii, 400
saximontanus, 421
scrupulicola, 398
serotinus, 400
shultziorum, 402
sordidus, 414
spatiosus, 398
striatus, 396
strigosus, 400
subpolaris, 397
sulphurescens, 396
tananaicus, 396
tegetarius, 399
vaginatus, 398
varians, 414
viciifolius, 396
viridis, 399
viscidus, 422
Atelophragma
aboriginorum, 398
alpinum, 397
forwoodii, 398
glabriuscula, 398
herriotii, 398
labradoricum, 397
lineare, 398
wallowense, 398
Athyrium, 62
alpestre, 62
americanum, 62
angustum, 63
asplenioides, 63
distentifolium, 62
filix-femina, 63
fragile, 64
Atropis
canbyi, 581
fendleriana, 572
laevis, 581
lettermani, 576

nevadensis, 581
pauciflora, 581
scabrella, 581
tenuifolia, 581
Aulospermum
angustum, 45
longipes, 45
Avena
airoides, 587
americana, 560
atropurpurea, 541
caespitosa, 542
cristata, 517
hookeri, 560
mollis, 587
mortoniana, 561
odorata, 562
ovina, 556
spicata, 587
subspicata, 587
versicolor, 560
Avenochoa hookeri, 560
Avenula hookeri, 560

B
Bahia
achillaeoides, 144
gracilis, 144
integrifolia, 144
lanata, 144
Barkhausia nana, 119
Batanthes
aggregata, 593
attenuata, 593
formosissima, 593
pulchella, 593
Batidaea
acalyphacea, 707
cataphracta, 707
filipendula, 707
peramoena, 707
sandbergii, 707
strigosa, 707
subcordata, 707
unicolor, 707
viburnifolia, 707
Batrachium
aquatile, 657
bakeri, 657
confervoides, 657
drouetii, 657
flaccidum, 657
porteri, 657
trichophyllum, 657
Besseya, 753
alpina, 754
cinerea, 755
gymnocarpa, 755
wyomingensis, 755

Betula
crenata, 202
exilis, 202
glandulifera, 202
glandulosa, 202
hallii, 202
michauxii, 202
nana, 202
terra-novae, 202
BETULACEAE, 201
Bicuculla uniflora, 430
Bigelovia macronema, 150
Bigelowia
douglasii, 111
glauca, 111
lanceolata, 111
viscidiflora, 111
Bistorta
bistortoides, 613
littoralis, 618
macounii, 618
ophioglossa, 618
vivipara, 618
Blechnum crispum, 36
Blitum
capitatum, 305
chenopodioides, 307
hastatum, 305
nuttallianum, 307
Boechera
collinsii, 220
drummondii, 219
fendleri, 220
holboellii, 220
lemmonii, 222
retrofracta, 220
tenuis, 220
BORAGINACEAE, 203
Botrychium, 500
lunaria, 500
minganense, 500
onondagense, 500
Botrypus lunaria, 500
Botrys fremontii, 306
Boykinia, 728
heucheriformis, 729
jamesii, 729
Brachyachyris euthamiae, 146
Brachyris euthamiae, 146
BRASSICACEAE, 217
Braxilia minor, 387
Braya humilis, 228
Bromopsis
anomala, 529
canadensis, 531
ciliata, 531
dicksonii, 532
inermis, 532
porteri, 529
pumpelliana, 532

Bromus, 528
anomalus, 529
arcticus, 532
breviaristatus, 530
canadensis, 531
carinatus, 530
ciliatus, 531
cristatus, 517
dudleyi, 531
flodmannii, 530
frondosus, 529
hookerianus, 530
inermis, 532
latior, 530
marginatus, 530
maritimus, 530
multiflorus, 530
oregonus, 530
ovinus, 556
paniculatus, 530
parviflorus, 530
pauciflorus, 530
polyanthus, 530
porteri, 529
pumpellianus, 532
richardsonii, 531
scabratus, 529
secundus, 558
subvelutinus, 530
richardsonii, 531
virens, 530
Bryanthus
empetriformis, 385
glanduliflorus, 386
Buda rubra, 295
Bulbocodium serotinum, 482
Bupleurum, 41
americanum, 41
angulosum, 41
purpureum, 41
ranunculoides, 41

C
Calamagrostis, 533
alaskana, 534
americana, 538
angustifolia, 534
anomala, 534
arctica, 536
atropurpurea, 534
blanda, 534
borealis, 538
californica, 538
canadensis, 534
chordorrhiza, 538
dubia, 534
elongata, 538
expansa, 538
fernaldii, 538
hyperborea, 538

inexpansa, 538
labradorica, 538
lactea, 534
lacustris, 538
langsdorfii, 534
laricina, 536
laxiflora, 538
lepageana, 536
lucida, 538
macouniana, 534
maltei, 536
michauxii, 534
micrantha, 538
montanensis, 535
neglecta, 538
nubila, 534
oregonensis, 534
pallida, 534
poluninii, 536
purpurascens, 536
purpurea, 534
robusta, 538
scabrua, 534
scribneri, 534
scopulorum, 537
stricta, 538
vaseyi, 536
wyomingensis, 538
yukonensis, 536
Calamaria bolanderi, 459
Calandrinia
   grayi, 628
   nevadensis, 628
   pygmaea, 628
Callistachys pyrenaica, 360
Callisteris
   arizonica, 593
   attenuata, 593
   formosissima, 593
   pulchella, 593
   violacea, 593
CALLITRICHACEAE, 265
Callitriche, 265
   palustris, 265
   verna, 265
Calochortus gunnisonii, 480
Caltha, 650
   auriculata, 650
   biflora, 650
   chelidonii, 650
   howellii, 650
   leptosepala, 7, 650
   rotundifolia, 650
   uniflora, 650
Calyptridium umbellatum, 630
Campanula, 266
   alaskana, 267
   dubia, 267

heterodoxa, 267
intercedens, 267
macdougalii, 267
parryi, 26
petiolata, 267
rotundifolia, 267
sacajaweana, 267
scabrella, 26
uniflora, 268
CAMPANULACEAE, 266
Campella caespitosa, 542
Cantua aggregata, 593
Capnorchis uniflora, 430
CAPRIFOLIACEAE, 269
Cardamine
   articulata, 255
   nudicaulis, 255
Carduus
   butleri, 115
   hookeriana, 115
   leiocephalus, 114
   nevadensis, 117
   olivescens, 114
   polyphyllus, 114
   pulcherrimus, 116
   tweedyi, 114
Carex, 8, 25, 317
   ablata, 344
   accedens, 367
   acutina, 324
   acutinella, 324
   affinis, 379
   albonigra, 323
   alpina, 352
   altior, 324
   ambusta, 365
   amphilogos, 347
   angarae, 352
   angustior, 336
   apoda, 325
   aquatilis, 324
   arctiformis, 331
   arctogena, 333
   arctica, 347
   athabascensis, 366
   atrata, 325
   atrosquama, 325
   aurea, 327
   banata, 347
   behringensis, 357
   bella, 325
   bellardi, 379
   bipartita, 328
   brevipes, 362
   breweri, 329
   brunnescens, 330
   bucculenta, 347
   camptotropa, 347
   campylocarpa, 367
   canescens, 331

capillaris, 332
capitata, 333
cephalantha, 336
chalciolepis, 325
chimaphila, 367
chlorostachyas, 332
constanceana, 355
crandallii, 360
curatorum, 366
curta, 331
danaensis, 347
diversistylis, 362
douglasii, 334
drummondiana, 364
duriuscula, 368
eastwoodiana, 356
ebenea, 335
echinata, 336
elbertiana, 353
eleocharis, 368
elyniformis, 338
elynoides, 337
engelmannii, 329
epapillosa, 325
estesiana, 353
exsiccata, 371
farwellii, 362
festiva, 345
festivella, 345
filifolia, 338
fissuricola, 344
fuscidula, 332
garberi, 327
geyeri, 339
gymnoclada, 367
halleri, 352
hassei, 327
hawaiiensis, 336
haydeniana, 29, 345
hepburnii, 349
hermaphroditica, 379
heteroneura, 325
hoodii, 340
howellii, 324
hyalinolepis, 347
illota, 341
incondita, 345
incurva, 347
incurviformis, 347
inflata, 363
interimus, 324
irrasa, 334
jacob-peteri, 360
jonesii, 342
josselynii, 336
jucunda, 347
juncifolia, 347
krausei, 332
lachenalii, 328
lagopina, 328

lapponica, 331
laricina, 336
leersii, 336
leporinella, 343
liddoni, 355
luzulina, 344
lyallii, 361
macloviana, 29, 345
maritima, 347
media, 352
melanocephala, 353
michauxii, 366
micropoda, 360
microptera, 29, 345
miliaris, 365
misandra, 348
miserabilis, 367
monile, 371
montanensis, 357
multimoda, 345
muricata, 336
myosuroides, 379
nardina, 349
nelsonii, 353
neurophora, 350
nigella, 357
nigricans, 351
norvegica, 352
nova, 353
nubicola, 345
nuttallii, 334
obtusata, 354
olympica, 345
ormantha, 336
orthocaula, 347
pachystachya, 29, 345
paddoensis, 329
panda, 324
paysonis, 357
pelocarpa, 353
petasata, 355
phaeocephala, 356
phyllomanica, 336
physocarpa, 365
piperi, 359
platylepis, 345
podocarpa, 357
praeceptorum, 358
pratensis, 359
praticola, 359
preslii, 345
prionophylla, 367
procerula, 365
psammogaea, 347
pseudofoetida, 347
pseudoscirpoidea, 366
psychroluta, 347
pulla, 365
pyrenaica, 360
pyrophila, 345

884   Index to Latin Names

rachillis, 370
raeana, 371
raynoldsii, 361
rhomalea, 365
rigida, 367
rossii, 362
rostrata, 363
rufovariegata, 355
rupestris, 26, 364
saxatilis, 365
scirpiformis, 366
scirpina, 366, 379
scirpoidea, 366
scopulorum, 367
setina, 347
sitchensis, 324
soperi, 345
spectabilis, 357
spreta, 367
stans, 324
stantonensis, 349
stellulata, 336
stenochlaena, 366
stenophylla, 368
stenoptera, 345
sterilis, 336
stevenii, 352
straminiformis, 369
subfusca, 345
subnigricans, 370
substricta, 324
suksdorfii, 324
svensonis, 336
tolmiei, 357
transmarina, 347
tripartita, 328
uncompahgre, 325
utriculata, 363
vahlii, 352
variabilis, 324
venustula, 357
vesicaria, 371
vitilis, 330
Caricina
  incurva, 347
  stellulata, 336
Caricinella rupestris, 364
CARYOPHYLLACEAE, 272
Cassiope mertensiana, 26
Castilleja, 756
  applegatei, 757
  breweri, 757
  camporum, 758
  chrymactis, 759
  confusa, 759
  crispula, 759
  cusickii, 758
  dixonii, 759
  fasciculata, 761

gracillima, 759
humilis, 763
inconstans, 759
inverta, 761
lanceifolia, 759
lauta, 763
leonardii, 763
linariifolia, 26
lutea, 758
luteovirens, 764
magna, 759
miniata, 759
mogollonica, 764
nivea, 760
oblongifolia, 759
obtusiloba, 763
oregonensis, 763
pallescens, 761
pannosa, 758
peckiana, 759
pilifera, 758
pinetorum, 757
pulchella, 762
purpurascens, 763
rhexifolia, 763
subpurpurascens, 763
sulphurea, 764
trinervis, 759
tweedyi, 759
variabilis, 759
villosa, 758
viscida, 757
vreelandii, 759
wherryanna, 757
wyomingensis, 764
Catabrosa algida, 565
CELASTRACEAE, 302
Cephalophora acaulis, 158
Cerastium, 275
  alsophilum, 276
  angustatum, 276
  arvense, 26, 276
  beeringianum, 277
  bialynickii, 277
  bracteatum, 276
  buffumae, 277
  campestre, 276
  confertum, 276
  earlei, 277
  effusum, 276
  elongatum, 276
  fuegianum, 276
  graminifolium, 276
  latifolium, 276
  leibergii, 276
  nitidum, 276
  oblongifolium, 276
  occidentale, 276
  oreophilum, 276
  patulum, 276

pilosum, 277
pulchellum, 277
scammaniae, 277
scopulorum, 276
sonnei, 276
strictum, 276
subulatum, 276
tenuifolium, 276
terrae-novae, 277
thermale, 276
variabile, 276
velutinum, 276
vestitum, 276
viride, 276
villosum, 276
Ceratochloa
  breviaristata, 530
  carinata, 530
  grandiflora, 530
  marginata, 530
Cercocarpus, 673
  hypoleucus, 673
  intricatus, 673
  ledifolius, 26, 673
Cerinthodes alpinum, 211
Cerophyllum inebrians, 449
Chaenactis, 109
  alpina, 109
  achillaeafolia, 110
  angustifolia, 110
  brachiata, 110
  cheilanthoides, 110
  cineria, 110
  douglasii, 110
  humilis, 110
  imbricata, 110
  leucopsis, 109
  miniscula, 109
  panamintensis, 110
  rubella, 109
  rubricaulis, 110
  suksdorfii, 110
Chamaenerion
  angustifolium, 491
  exaltatum, 491
  latifolium, 494
  spicatum, 491
  subdentatum, 494
Chamerion
  angustifolium, 491
  danielsii, 491
  latifolium, 494
  platyphyllum, 491
  spicatum, 491
  subdentatum, 494
Cheiranthus
  alpestris, 247
  angustatus, 247
  argillosus, 247
  aridus, 247

arkansanus, 247
asper, 247
asperrimus, 247
bakeri, 247
californicus, 247
capitatus, 247
elatus, 247
nivalis, 247
oblanceolatus, 247
pacificus, 247
perennis, 247
radicatus, 247
wheeleri, 247
Cheirinia
  amoena, 247
  argillosa, 247
  arida, 247
  arkansana, 247
  aspera, 247
  asperrima, 247
  bakeri, 247
  brachycarpa, 247
  cockerelliana, 247
  desertorum, 247
  elata, 247
  nevadensis, 247
  nivalis, 247
  oblanceolata, 247
  radicata, 247
  wheeleri, 247
CHENOPODIACEAE, 304
Chenopodium, 304
  aridum, 306
  atrovirens, 306
  capitatum, 8, 305
  fremontii, 306
  hians, 306
  incanum, 306
  incognitum, 306
  overi, 305
  watsonii, 306
  wolfii, 306
Chionophila jamesii, 765
Cholocrepis tristis, 156
Chondrophylla
  americana, 437
  aquatica, 437
  fremontii, 437
  prostrata, 437
Chrysoma pumila, 165
Chrysopsis
  acaulis, 148
  alpicola, 154
  amplifolia, 154
  angustifolia, 154
  arida, 154
  asprella, 154
  bakeri, 154
  ballardii, 154
  barbata, 154

butleri, 154
caespitosa, 148
caudata, 154
columbiana, 154
compacta, 154
cooperi, 154
depressa, 154
foliosa, 154
fulcrata, 154
grandis, 154
hirsuta, 154
hirsutissima, 154
hispida, 154
horrida, 154
imbricata, 154
mollis, 154
pedunculata, 154
pumila, 154
resinolens, 154
villosa, 154
viscida, 154
wisconsinensis, 154
Chrysothamnus, 111
　axillaris, 111
　douglasii, 111
　elegans, 111
　glaucus, 111
　humilus, 111
　lanceolatus, 111
　latifolius, 111
　levocladus, 111
　linifolius, 111
　marianus, 111
　puberulus, 111
　pumilus, 111
　serrulatus, 111
　stenolepis, 111
　stenophyllus, 111
　tortifolius, 111
　viscidiflorus, 26, 111
Ciliaria
　austromontana, 742
　funstonii, 742
　vespertina, 742
Ciminalis
　fremontii, 437
　prostrata, 437
Cineraria lewisii, 125
Cirsium, 113
　butleri
　davisii, 117
　eatonii, 114
　hookerianum, 115
　humboldtense, 117
　nevadensis, 117
　olivescens, 114
　polyphyllum, 114
　pulcherrimum, 116
　subniveum, 117
　tweedyi, 114

wallowense, 117
Cistanthe umbellata, 630
Clavula acicularis, 373
Claytonia, 624
　aurea, 625
　bellidifolia, 626
　chrysantha, 625
　flava, 625
　lanceolata, 625
　megarhiza, 8, 26, 626
　multicaulis, 625
　multiscapa, 625
　nivalis, 626
　rosea, 625
　triphylla, 629
Clementsia rhodantha, 311
Clinelymus glaucus, 547
CLUSIACEAE, 458
Cnicus eatonii, 114
Cnicus hookeriana, 115
Cogswellia
　alata, 57
　anomala, 57
　brevifolia, 57
　circumdata, 55
　cous, 55
　lapidosa, 45
　leptophylla, 57
　montana, 55
　platycarpa, 57
　robustior, 57
　simplex, 57
　triternata, 57
Collinsia, 766
　grandiflora, 766
　minima, 766
　parviflora, 8, 766
　pusilla, 766
　tenella, 766
Collomia, 589
　aggregata, 593
　debilis, 590
　hurdlei, 590
　larsenii, 590
　linearis, 8, 591
　parviflora, 591
Colpodium monandrum, 565
Colutea
　astragalina, 397
　australis, 398
Comastoma tenellum, 440
Comandra umbellata, 26
COMPOSITAE, 67
Corion rubrum, 295
Corniveum uniflorum, 430
Costia cristata, 517
Cotyledon debilis, 309
Coulterina didymocarpa, 258
CRASSULACEAE, 308

Crepidium
　glaucum, 120
　runcinatum, 120
Crepis, 118
　alpicola, 120
　andersonii, 120
　barberi, 120
　denticulata, 120
　glauca, 120
　glaucella, 120
　nana, 119
　neomexicana, 120
　obtusissima, 120
　pallens, 120
　perplexans, 120
　petiolata, 120
　platyphylla, 120
　riparia, 120
　runcinata, 120
　tomentulosa, 120
Crinitaria viscidiflora, 111
Crocion
　nuttallii, 802
　vallicola, 802
Cronyxium serotinum, 482
Cryptantha, 203
　bradburiana, 204
　celosioides, 204
　confusa, 204
　hypsophila, 204
　interrupta, 204
　macounii, 204
　nubigena, 204
　sheldonii, 204
　sobolifera, 204
　spiculifera, 204
　subretusa, 204
Cryptogramma, 35
　acrostichoides, 36
　crispa, 36
Cucubalis acaulis, 291
CUPRESSACEAE, 314
Cyathea
　filix-femina, 63
　fragilis, 64
Cyclopteris fragilis, 64
Cymopterus, 42
　alpinus, 58
　bipinnatus, 46
　calcareus, 47
　evertii, 26, 43
　hendersonii, 44
　lapidosus, 45
　longilobus, 44
　longipes, 45
　nivalis, 46
　terebinthinus, 47
Cynoglossum glomeratum, 204
Cynomarathrum

alpinum, 56
macbridei, 46
CYPERACEAE, 317
Cystea fragilis, 64
Cystopteris
　dickieana, 64
　filix-femina, 63
　fragilis, 64

D
Dactyloides cespitosa, 743
Danthonia, 539
　canadensis, 539
　cusickii, 539
　intermedia, 539
Dasanthera fruticosa, 782
Dasiphora
　fruticosa, 692
　riparia, 692
Dasystephana
　affinis, 434
　bracteosa, 434
　calycosa, 436
　forwoodii, 434
　interrupta, 434
　monticola, 436
　obtusiloba, 436
　oregana, 434
　parryi, 434
　romanzovii, 435
Delphinium, 651
　abietorum, 653
　bicolor, 652
　cucullatum, 653
　glaucum, 26
　multiflorum, 653
　occidentale, 653
　reticulatum, 653
Delphinastrum occidentale, 653
Deschampsia, 8, 540
　alpicola, 542
　alpina, 542
　ambigua, 542
　atropurpurea, 541
　beringensis, 542
　bottonia, 542
　brevifolia, 542
　cespitosa, 542
　ciliata, 543
　congestiformis, 536
　confinis, 542
　curtifolia, 542
　elongata, 543
　glauca, 542
　holciformis, 542
　hookeriana, 541
　hudsonica, 542
　komarovii, 542
　latifolia, 541

pacifica, 541
paramushirensis, 542
pumila, 542
pungens, 542
sukatschewii, 542
Descurainia, 227
  incana, 9, 227–28
  incisa, 228
  richardsonii, 228
  rydbergii, 228
  serrata, 228
  sophia, 26
  torulosa, 9, 26, 229
Deyeuxia
  americana, 538
  canadensis, 534
  dubia, 534
  lactea, 534
  macouniana, 534
  montanensis, 535
  neglecta, 538
  purpurascens, 536
  scabra, 534
Dicentra uniflora, 430
Diclytra uniflora, 430
Diemisia capitata, 333
Dieteria
  canescens, 161
  divaricata, 161
  incana, 161
  pulverulenta, 161
  sessiliflora, 161
  viscosa, 161
Diplogon villosum, 154
Diplopappus
  canescens, 124
  grandiflorus, 124
  hispidus, 154
  incanus, 161
  villosus, 154
Dodecatheon, 634
  acuminatum, 636
  albidum, 636
  alpinum, 635
  amethystinum, 637
  campestrum, 636
  conjugens, 636
  cusickii, 637
  cylindrocarpum, 636
  glastifolium, 636
  meadia, 637
  multiflorum, 637
  pauciflorum, 637
  philoscia, 637
  puberulentum, 637
  puberulum, 637
  pubescens, 636
  pulchellum, 637
  pulchrum, 636
  radicatum, 637

salinum, 637
sinuatum, 637
superbum, 637
thornense, 637
uniflorum, 637
viscidum, 636
watsonii, 637
zionense, 637
Donia uniflora, 153
Douglasia, 638
  biflora, 638
  montana, 638
Draba, 25, 230
  albertina, 232
  andina, 242
  apiculata, 237
  aurea, 233
  aureiformis, 233
  bakeri, 233
  breweri, 234
  caeruleomontana, 237
  cana, 234
  canadensis, 239
  cascadensis, 245
  chrysantha, 235
  crassa, 235
  crassifolia, 8, 236
  columbiana, 245
  daurica, 239
  decumbens, 233
  deflexa, 232
  densifolia, 237
  dolichocarpa, 245
  fladnizensis, 25, 238
  glabella, 25, 239
  globosa, 237
  henneana, 239
  hirta, 239
  incerta, 240
  juniperina, 242
  juvenilis, 239
  laevicapsula, 240
  lanceolata, 234
  lapilutea, 245
  laurentiana, 239
  lonchocarpa, 241
  luteola, 233
  mccallae, 233
  megasperma, 239
  minganensis, 233
  mulfordae, 237
  nelsonii, 237
  neomexicana, 233
  nitida, 232
  novolympica, 243
  oligosperma, 242
  parryi, 236
  pattersonii, 238
  paysonii, 243
  peasei, 240

pectinata, 237
pectinipila, 242
porsildii, 25, 244
praealta, 9, 245
saximontana, 242
sphaerula, 237
stenoloba, 232
streptocarpa, 26
stylaris, 234
subsessilis, 242
surculifera, 233
tschuktschorum, 238
uber, 233
valida, 234
ventosa, 26, 246
vestita, 243
yellowstonensis, 245
Drosace
  albertina, 632
  carinata, 632
Drummondia mitelloides, 735
Dryas, 674
  alaskensis, 674
  drummondii, 26
  hookeriana, 674
  kamtschatica, 674
  octopetala, 8, 674
  punctata, 674
Drymocallis
  albida, 693
  amplifolia, 693
  ashlandica, 693
  foliosa, 693
  glabrata, 693
  glandulosa, 693
  glutinosa, 693
  monticola, 693
  oregana, 693
  pseudorupestris, 693
  pumila, 693
  reflexa, 693
  rhomboidea, 693
  valida, 693
  viscosa, 693
  wrangelliana, 693
Dryopteris lonchitis, 65

E

Echeveria debilis, 309
Echinospermum
  floribundum, 207
  subdecumbens, 209
Edritria rupestris, 364
Eleocharis, 372
  acicularis, 373
  bernardina, 374
  pauciflora, 374
  quinqueflora, 374
  reverchonii, 373

suksdorfiana, 374
Elephantella groenlandica, 774
Elodea, 453
  brandegae, 454
  canadensis, 454
  ioensis, 454
  linearis, 454
  planchonii, 454
Elymus, 8, 544
  albicans, 548
  americanus, 547
  brevifolius, 545
  edentatus, 547
  elymoides, 545
  glaber, 545
  glaucus, 547
  griffithsii, 548
  hispidulus, 547
  howellii, 547
  hystrix, 545
  lanceolatus, 548
  longifolius, 545
  marginalis, 547
  nitidus, 547
  pauciflorus, 552
  petersonii, 547
  pubescens, 547
  riparius, 548
  rydbergii, 548
  scribneri, 549
  sierrus, 552
  sitanion, 545
  smithii, 550
  spicatus, 551
  strigatus, 547
  subsecundus, 552
  subvillosus, 548
  trachycaulus, 552
  virescens, 547
Elyna
  bellardi, 379
  schoenoides, 380
  scirpina, 379
  sibirica, 380
  spicata, 379
Elytrigia
  dasystachya, 548
  riparia, 548
  smithii, 550
  spicata, 551
Epilobium, 489
  adenocaulon, 492
  affine, 492
  alpinum, 490
  americanum, 492
  anagallidifolium, 490
  angustifolium, 491
  behringianum, 490
  boreale, 492

brevistylum, 492
californicum, 492
ciliatum, 492
cinerascens, 492
clavatum, 490
ecomosum, 492
franciscanum, 492
glandulosum, 492
glareosum, 490
griseum, 492
halleanum, 493
hornemannii, 490
lactiflorum, 490
latifolium, 494
macdougalii, 492
nutans, 490
occidentale, 492
oregonense, 490
parishii, 492
perplexans, 492
praecox, 492
pringleanum, 493
pulchrum, 490
sertulatum, 490
spicatum, 491
treasianum, 490
ursinum, 492
watsonii, 492
Eragrostis fendleriana, 572
Eremogone
   americana, 273
   capillaris, 273
   congesta, 274
ERICACEAE, 381
Ericala acuta, 439
Ericameria
   discoidea, 150
   suffruticosa, 152
Ericoilea prostrata, 437
Erigeron, 25, 121
   acre, 123
   acris, 123
   acutatus, 136
   andersonii, 103
   angustifolius, 136
   asteroides, 123
   caespitosus, 124
   callianthemus, 136
   canescens, 124
   ciliolatus, 136
   compositus, 125
   controversus, 128
   coulteri, 26
   debilis, 123
   divergens, 26
   droebachensis, 123
   eatonii, 126
   elatus, 123
   elongatus, 123

eucephaloides, 141
flabellifolius, 127
garrettii, 26, 128
glacialis, 136
goodrichii, 26, 129
gormani, 125
gracilis, 130
grandiflorus, 141
hesperocallis, 136
humilis, 131
juncundus, 123
kamtschaticus, 123
laetevirens, 135
lanatus, 25, 132
lapiluteus, 123
leiomerus, 133
leiophyllus, 141
leucotrichus, 140
loratus, 136
macounii, 135
macranthus, 141
melanocephalus, 134
membranaceus, 136
microlonchus, 126
minusculus, 133
montanensis, 135
multifidus, 125
nivalis, 123
obtusatus, 136
ochroleucus, 135
pacificus, 126
parryi, 135
pedatus, 125
peregrinus, 136
pinnatisectus, 137
plantagineus, 126
podolicum, 123
politus, 123
radicatus, 138
regalis, 136
robertianus, 126
rydbergii, 139
salsuginosus, 136
scribneri, 135
simplex, 140
sonnei, 126
spathulifolius, 133
speciosus, 141
subcanescens, 124
subtrinervis, 141
suksdorfii, 136
tener, 142
thompsoni, 136
trifidus, 125
tweedyanus, 135
uintahensis, 141
unalaschkensis, 131, 136
ursinus, 143
yellowstonensis, 123

Eriogonum, 604
   acaule, 605
   aridum, 609
   biumbellatum, 609
   brevicaule, 606
   caespitosum, 26
   campanulatum, 606
   chloranthum, 607
   chrysocephalum, 606
   chrysops, 26
   cognatum, 609
   covillei, 609
   crassifolium, 607
   croceum, 609
   davisianum, 608
   desertorum, 606
   dichroanthum, 608
   ellipticum, 609
   ephedroides, 606
   flavum, 607
   glaberrimum, 609
   grayi, 606
   hausknechtii, 609
   lagopus, 606
   latum, 609
   loganum, 606
   medium, 606
   micranthum, 606
   montanum, 609
   nanum, 606
   neglectum, 609
   nivale, 608
   nudicaule, 606
   ochroleucum, 608
   orendense, 606
   orthocaulon, 608
   ovalifolium, 608
   piperi, 607
   polyanthum, 609
   polyphyllum, 607
   purpureum, 608
   rhodanthum, 608
   roseiflorum, 608
   rubidum, 608
   rydbergii, 609
   sericeum, 607
   stellatum, 609
   subalpinum, 609
   tolmieanum, 609
   torreyanum, 609
   umbellatum, 609
   umbelliferum, 609
   vineum, 608
   viridulum, 606
   wasatchense, 606
Eriogynia
   caespitosa, 685
   uniflora, 684
Eriophorum, 375
   angustifolium, 376

   callitrix, 375
   capitatum, 377
   komarovii, 376
   leucocephalum, 377
   ochreatum, 376
   polystachion, 376
   scheuchzeri, 377
   subarcticum, 376
   triste, 376
Eriophyllum, 144
   achillaeoides, 144
   caespitosum, 144
   gracile, 144
   harfordii, 144
   integrifolium, 144
   lanatum, 26, 144
   leucophyllum, 144
   monoense, 144
   multiflorum, 144
   nevadense, 144
   pedunculatum, 144
   ternatum, 144
   trichocarpum, 144
   watsonii, 144
Eritrichum, 205
   aretioides, 205
   argenteum, 205
   chamissonis, 205
   elongatum, 205
   howardii, 205
   nanum, 205
Erocallis triphylla, 629
Erxlebenia minor, 387
Erysimum, 247
   alpestre, 247
   amoenum, 247
   angustatum, 247
   argillosum, 247
   aridum, 248
   arkansanum, 248
   asperrimum, 248
   asperum, 247
   bakeri, 248
   californicum, 248
   capitatum, 248
   cockerellianum, 248
   desertorum, 248
   drummondii, 219
   elatum, 248
   holboellii, 220
   moniliforme, 248
   nevadense, 248
   nivale, 248
   nuttallii, 225
   oblanceolatum, 248
   perenne, 248
   pumilum, 248
   radicatum, 248
   tilimi, 248
   wheeleri, 248

Erythrocoma
  affinis, 680
  campanulata, 680
  canescens, 680
  ciliata, 680
  flavula, 680
  grisea, 680
  triflora, 680
Erythronium
  giganteum, 481
  grandiflorum, 7, 481
  idahoense, 481
  leptopetalum, 481
  nuttallianum, 481
  obtusatum, 481
  pallidum, 481
  parviflorum, 481
  utahense, 481
Eucephalus
  formosus, 105
  glaucus, 105
Euchroma pallescens, 761
Eucycla
  ovalifolia, 608
  purpurea, 608
Eulophus triternatus, 57
Eutoca sericea, 457
Exinia pulchella, 637

F

FABACEAE, 395
Facolos brunnescens, 330
Fasciculus ruber, 295
Festuca, 8, 554
  arenaria, 558
  arizonica, 556
  aucta, 558
  baffinensis, 555
  brachyphylla, 556
  brevifolia, 556
  calligera, 556
  capillata, 556
  confinis, 564
  densiuscula, 558
  duriuscula, 556
  earlei, 558
  fallax, 558
  heterophylla, 558
  idahoensis, 556
  inermis, 532
  ingrata, 556
  kingii, 564
  kitaibeliana, 558
  lanuginosa, 558
  minutiflora, 556
  multiflora, 558
  nervosa, 577
  oregana, 558
  ovina, 556
  patagonica, 581
  prolifera, 558
  pubescens, 558
  richardsonii, 558
  roemeri, 556
  rubra, 558
  saximontana, 556
  spaniantha, 581
  spicata, 551
  supina, 556
  vallicola, 558
Filix fragilis, 64
Fonna hoodii, 598
Forasaccus
  breviaristatus, 530
  ciliatus, 531
  inermis, 532
  marginatus, 530
  pumpellianus, 532
Fragaria, 675
  americana, 675
  australis, 675
  bracteata, 675
  californica, 675
  canadensis, 675
  crinita, 675
  fruticosa, 692
  glauca, 675
  grayana, 675
  helleri, 675
  latiuscula, 675
  multicipita, 675
  nivea, 700
  ovalis, 675
  pauciflora, 675
  platypetala, 675
  prolifica, 675
  pumila, 675
  retrorsa, 675
  sibbaldifolia, 675
  suksdorfii, 675
  truncata, 675
  vesca, 675
  virginiana, 675
Frasera, 431
  angustifolia, 432
  macrophylla, 432
  scabra, 432
  speciosa, 9, 432
  stenosepaia, 432
FUMARIACEAE, 429

G

Gagea serotina, 482
Gaillardia acaulis, 158
Galium, 710
  boreale, 710
  hyssopifolium, 710
  septentrionale, 710
  strictum, 710
  utahense, 710
Gastrolynchis
  apetala, 279
  drummondii, 280
  kingii, 279
  uralensis, 279
Gaultheria, 383
  humifusa, 383
  myrsinites, 383
Gayophytum, 495
  caesium, 497
  decipiens, 497
  diffusum, 496
  eriospermum, 496
  helleri, 497
  intermedium, 496
  lasiospermum, 496
  nuttallii, 496
  racemosum, 497
Genersichia obtusata, 354
Gentiana, 433
  acuta, 439
  affinis, 434
  algida, 435
  amarella, 439
  anisosepala, 439
  aquatica, 437
  barbata, 443
  barbellata, 442
  bigelovii, 434
  bracteosa, 434
  calycosa, 436
  cusickii, 436
  detonsa, 443
  distegia, 439
  elegans, 443
  forwoodii, 434
  fremontii, 437
  frigida, 435
  glacialis, 440
  gormani, 436
  heterosepala, 439
  humilis, 437
  idahoensis, 436
  interrupta, 434
  macounii, 443
  menziesii, 434
  monantha, 440
  myrsinites, 436
  oregana, 434
  parryi, 434
  plebeja, 439
  polyantha, 439
  prostrata, 9, 437
  raupii, 443
  remota, 434
  richardsonii, 443
  romanzovii, 435
  rusbyi, 434
  saxicola, 436
  scopulorum, 439
  strictiflora, 439
  tenella, 440
  thermalis, 443
  tortuosa, 439
  wrightii, 439
GENTIANACEAE, 431
Gentianella, 438
  acuta, 439
  amarella, 9, 439
  barbellata, 442
  clementis, 439
  detonsa, 443
  heterosepala, 439
  strictiflora, 439
  tenella, 8, 440
  tortuosa, 439
Gentianodes
  algida, 435
  romanzovii, 435
Gentianopsis, 441
  barbata, 443
  barbellata, 442
  detonsa, 8, 443
  elegans, 443
  raupii, 443
  thermalis, 443
GERANIACEAE, 445
Geranium, 445
  albiflorum, 446
  canum, 447
  gracilentum, 446
  hookerianum, 446
  incisum, 447
  loloense, 446
  nervosum, 447
  richardsonii, 446
  strigosius, 447
  strigosum, 447
  viscosissimum, 447
Gerardia fruticosus, 782
Geum, 677
  campanulatum, 680
  canescens, 680
  ciliatum, 680
  gracilipes, 679
  macrophyllum, 678
  oregonense, 678
  perincisum, 678
  rossii, 7, 679
  sericeum, 679
  triflorum, 680
  turbinatum, 7, 679
Gilia
  aggregata, 593
  arizonica, 593
  attenuata, 593
  bridgesii, 593
  candida, 593

cephaloidea, 595
debilis, 590
formosissima, 593
globularis, 595
howardii, 590
larseni, 590
linearis, 591
nuttallii, 596
parvula, 592
pulchella, 593
spicata, 595
tenuituba, 593
texana, 593
tridactyla, 595
Glyerica canbyi, 581
Gnaphalium
  alpinum, 77
  margaritaceum, 75
Gnomonia ovina, 556
Gormania debilis, 309
Graphephorum
  brandegei, 588
  muticum, 588
  wolfii, 588
Gregoria montana, 638
Grossularia
  cognata, 452
  irrigua, 452
  neglecta, 452
  nonscripta, 452
  oxyacanthoides, 452
  setosa, 452
GROSSULARIACEAE, 448
Gutierrezia, 145
  divaricata, 146
  diversifolia, 146
  euthamiae, 146
  fasciculata, 146
  filifolia, 146
  ionensis, 146
  juncea, 146
  lepidota, 146
  linearis, 146
  longifolia, 146
  myriocephala, 146
  sarothrae, 146
  scoparia, 146
  tenuis, 146
Gymnodes spicata, 471
Gymnogramma acrostichoides, 36
Gymnosteris, 592
  leibergii, 592
  parvula, 8, 592
  rydbergii, 592
Gyrostachys
  porrifolia, 503
  romanzowiana, 503
  stricta, 503

H
Habenaria, 501
  borealis, 502
  dilatata, 502
  dilatatiformis, 502
  gracilis, 502
  graminifolia, 502
  leptoceratitis, 502
  leucostachys, 502
Hackelia, 206
  floribunda, 206
  jessicae, 208
  leptophylla, 206
  micrantha, 208
  patens, 209
Haplopappus, 147
  acaulis, 148
  discoideus, 150
  gossypinus, 153
  lanceolatus, 153
  lanuginosus, 26
  lyallii, 149
  macronema, 150
  parryi
  pygmaeus, 151
  suffruticosus, 152
  uniflorus, 153
Hedysarum, 403
  albiflorum, 406
  boreale, 404
  canescens, 404
  carnosulum, 404
  cinerascens, 404
  flavescens, 406
  gremiale, 404
  lancifolium, 405
  mackenzii, 404
  marginatum, 405
  occidentale, 405
  pabulare, 404
  sulphurescens, 406
  uintahense, 405
  utahense, 404
Helenium lanatum, 144
Helictotrichon, 559
  hookeri, 560
  mortonianum, 561
Heracleum, 48
  douglasii, 49
  lanatum, 49
  maximum, 49
  montanum, 49
  sphondylium, 49
Hesperis incisa, 228
Hesperochloa kingii, 564
Heterotheca, 154
  foliosa, 154
  fulcrata, 154
  horrida, 154
  pumila, 154
  villosa, 154
  viscida, 154
  wisconsinensis, 154
  zionensis, 154
Heuchera, 730
  alpina, 731
  columbiana, 731
  cusickii, 732
  cylindrica, 731
  flabellifolia, 733
  flavescens, 733
  glabella, 731
  gracilis, 732
  grossulariifolia, 732
  missouriensis, 733
  nivalis, 733
  ovalifolia, 731
  parvifolia, 733
  puberula, 733
  saxicola, 731
  suksdorfii, 731
  tenuifolia, 732
  utahensis, 733
Hieraciodes
  nanum, 119
  rucinatum, 120
Hieracium, 156
  gracile, 156
  hookeri, 156
  runcinatum, 120
  triste, 156
  utahense, 156
Hierochloe, 562
  arctica, 562
  nashii, 562
  odorata, 562
Hirculus flagellaris, 746
Holcus
  atropurpureus, 541
  odoratus, 562
Holodiscus, 681
  dumosus, 26, 681
  glabrescens, 681
  microphyllus, 681
Homalobus
  aboriginorum, 398
  aboriginum, 398
  aculeatus, 399
  camporum, 400
  decumbens, 400
  decurrens, 400
  divergens, 400
  hitchcockii, 400
  humilis, 400
  hylophilus, 400
  microcarpus, 400
  miser, 400
  montanus, 399
  oblongifolius, 400
  palliseri, 400
  paucijugus, 400
  serotinus, 400
  strigosus, 400
  tegetarius, 399
  tenuifolius, 400
  wolfii, 399
Homopappus
  inuloides, 153
  spathulatus, 187
  uniflorus, 153
Hoorbekia
  acaulis, 148
  lyallii, 149
  uniflora, 153
Hordeum elymoides, 545
Horkelia
  gordonii, 682
  utahensis, 683
Hulsea, 157
  algida, 157
  caespitosa, 157
  carnosa, 157
  nevadensis, 157
Huperzia selago, 488
Hutchinsia calycina, 261
HYDROCHARITACEAE, 453
HYDROPHYLLACEAE, 455
Hymenopappus douglasii, 110
Hymenoxys, 158
  acaulis, 158
  argentea, 158
  grandiflora, 160
  ivesiana, 158
Hyopeltis lonchitis, 65
HYPERICACEAE, 458
Hypericum, 458
  formosum, 458
  nortonae, 458
  scouleri, 458

I
Ibidium
  porrifolium, 503
  romanzoffianum, 503
  strictum, 503
Ilex myrsinites, 303
Inula villosa, 154
Ipomeria aggregata, 593
Ipomopsis, 593
  aggregata, 9, 593
  bridgesii, 593
  candida, 593
  spicata, 595
  tenuituba, 593
ISOETACEAE, 459
Isoetes, 459
  bolanderi, 459
  pygmaea, 459

Isolepis acicularis, 373
Ivesia, 682
  alpicola, 682
  gordonii, 682
  utahensis, 683

J
JUNCACEAE, 460
Juncodes
  glabratum, 469
  parviflorum, 470
  spicatum, 471
Juncoides
  glabratum, 469
  majus, 469
  parviflorum, 470
  piperi, 472
  spicatum, 471
Juncus, 8, 460
  albescens, 467
  badius, 465
  biglumis, 461
  castaneus, 462
  columbianus, 465
  drummondii, 463
  duranii, 465
  glabratus, 469
  inventus, 465
  leucochlamys, 462
  melanocarpus, 470
  mertensianus, 464
  nevadensis, 465
  parryi, 466
  parviflorus, 470
  pauperculus, 463
  schischkini, 467
  slwookoorum, 464
  spicatus, 471
  subtriflorus, 463
  suksdorfii, 465
  triglumis, 467
  truncatus, 465
Juniperus, 314
  alpina, 315
  canadensis, 315
  communis, 315
  depressa, 315
  horizontalis, 316
  hudsonica, 316
  nana, 315
  prostrata, 316
  repens, 316
  sabina, 316
  sibirica, 315

K
Kalmia, 384
  glauca, 384
  microphylla, 8, 384
  occidentalis, 384
  polifolia, 384
Kelloggia galioides, 711
Kelseya uniflora, 684
Kentrophyta
  aculeata, 399
  coloradoensis, 399
  impensa, 399
  minima, 399
  montana, 399
  rotunda, 399
  tegetaria, 399
  ungulata, 399
  viridis, 399
  wolfii, 399
Kobresia, 378
  arctica, 380
  bellardii, 379
  hyperborea, 380
  macrocarpa, 380
  myosuroides, 379
  schoenoides, 380
  scirpina, 379
  sibirica, 380
  simpliciuscula, 26
Koeleria, 563
  canescens, 587
  cristata, 563
  elegantula, 563
  gracilis, 563
  latifrons, 563
  macrantha, 563
  nitida, 563
  pyramidata, 563
  robinsoniana, 563
  spicata, 587
  subspicata, 587
  yukonensis, 563
Koenigia islandica, 8, 610

L
LABIATAE, 473
LAMIACEAE, 473
Lapathum
  alpestre, 620
  mexicanum, 623
Lappula
  caerulescens, 209
  floribunda, 207
  jessicae, 208
  subdecumbens, 209
Larix lyallii, 26
LEGUMINOSAE, 395
Leontodon
  ammophilum, 191
  angustifolium, 191
  ceratophorus, 190
  dumetorum, 190
  eriophorum, 191
  latiloba, 193
  leiospermum, 190
  lyratus, 192
  mexicanum, 193
  monticola, 190
  rupestre, 192
  scopulorum, 192
  taraxacum, 193
  vulgare, 193
Lepidium calycinum, 261
Lepigonum rubrum, 295
Leptasea
  austromontana, 742
  chrysantha, 745
  flagellaris, 746
  vespertina, 742
Lepteiris parviflora, 785
Leptodactylon nuttallii, 596
Lesquerella, 249
  alpina, 250
  condensata, 250
  curvipes, 250
  cusickii, 252
  fremontii, 253
  garrettii, 26, 251
  goodrichii, 252
  nodosa, 250
  occidentalis, 252
  parvula, 250
  paysonii, 26, 253
  spathulata, 250
  utahensis, 26, 254
Leucopoa kingii, 564
Lewisia, 627
  aridorum, 628
  exarticulata, 628
  glandulosa, 628
  minima, 628
  nevadensis, 628
  pygmaea, 628
  triphylla, 629
Lidia obtusiloba, 286
Ligularia
  amplectens, 169
  holmii, 169
Ligusticum, 50
  affine, 52
  apiifolium, 51
  filicinum, 51
  goldmanii, 52
  madrense, 52
  nelsonii, 52
  oreganum, 53
  porteri, 52
  scopulorum, 51
  simulans, 52
  tenuifolium, 53
LILIACEAE, 475
Limnobotrya
  lacustris, 450
  montigena, 451
  parvula, 450
Limnorchis
  borealis, 502
  dilatata, 502
  dilatatiformis, 502
  foliosa, 502
  gracilis, 502
  graminifolia, 502
  leptoceratitis, 502
  leucostachys, 502
LINACEAE, 485
Linagrostis polystachia, 376
Linanthastrum nuttallii, 596
Linanthus nuttallii, 596
Linosyris
  lanceolata, 111
  pumila, 111
  serrulata, 111
  viscidiflora, 111
Linum, 485
  kingii, 26, 486
  lepagei, 486
  lewisii, 486
  perenne, 486
Lithophragma, 734
  bulbiferum, 734
  glabrum, 734
Lloydia serotina, 482
Lobaria cernua, 744
Lomatium, 54
  alatum, 57
  alpinum, 56
  anomalum, 57
  attenuatum, 55
  brevifolium, 57
  circumdatum, 55
  cous, 55
  graveolens, 56
  kingii, 56
  lapidosum, 45
  montanum, 55
  platycarpum, 57
  purpureum, 55
  robustius, 57
  simplex, 57
  triternatum, 57
Lomatogonium tenellum, 440
Luciola spicata, 471
Luetkea
  caespitosa, 685
  pectinata, 26
  uniflora, 684
Lupinus, 407
  abortivus, 410
  achilleaphilus, 408
  adscendens, 408
  agropyrophilus, 411
  alcis-temporis, 410
  alicanescens, 408
  aliumbellatus, 408

alpestris, 408
alsophilus, 408
alturasensis, 408
amnicoli-cervi, 410
amniculi-salicis, 412
andersonianus, 411
argenteus, 408
aridus, 410
brachypodus, 410
caespitosus, 410
candicans, 412
canescens, 411
capitis-amnicoli, 408
cariciformis, 408
charlestonensis, 408
christianus, 408
clarkensis, 408
comatus, 412
corymbosus, 408
cusickii, 410
cyaneus, 411
danaus, 410
davisianus, 408
decumbens, 408
depressus, 408
diversalpicola, 412
edward-palmeri, 408
enodatus, 411
equi-coeli, 408
erectus, 411
evermannii, 408
falsoerectus, 411
flavescens, 412
floribundus, 408
foliosus, 408
forslingii, 411
fremontensis, 408
fruticulosus, 410
garrettianus, 408
hellerae, 410
hullianus, 408
humicola, 412
ingratus, 408
jonesii, 408
lacuum-trinitatum, 408
lanatocarinatus, 408
laxiflorus, 408
laxus, 408
lenorensis, 410
lepidus, 410
leptostachyus, 408
leucophyllus, 411
longivallis, 410
lucidulus, 408
lyallii, 410
lysichitophilus, 411
macounii, 408
macrostachys, 411
maculatus, 408
markleanus, 410

minearanus, 408
minimus, 410
minutifolius, 410
monticola, 408
montis-cookii, 408
montis-liberatatis, 408
myrianthus, 408
parviflorus, 408
paulinus, 410
perditorum, 410
perplexus, 408
piperi, 410
plumosus, 411
pulcherrimus, 408
pureriae, 411
retrorus, 411
roseolus, 408
rubricaulis, 408
rydbergii, 412
salicisocius, 411
seclusus, 408
sericeus, 26
serradentum, 408
sinus-meyersi, 410
sitgreavsii, 408
sparhawkianus, 408
spathulatus, 408
stenophyllus, 408
summae, 408
tenellus, 408
tenuispicus, 411
torreyi, 410
varneranus, 408
volutans, 410
watsonii, 410
wyethii, 412
Luzula, 468
cusickii, 471
fastigiata, 470
glabrata, 469
hitchcockii, 469
labradorica, 470
melanocarpa, 470
orestera, 471
parviflora, 470
piperi, 469, 472
spicata, 471
wahlenbergii, 472
Lychnis, 278
apetala, 279
attenuata, 279
drummondii, 280
elata, 294
kingii, 279
montana, 279
parryi, 294
pudica, 280
striata, 280
LYCOPODIACEAE, 487
Lycopodium, 487

appressum, 488
arcticum, 488
porophilum, 488
selago, 4, 488

M
Machaeranthera, 161
angustifolia, 161
asteroides, 161
attenuata, 161
canescens, 9, 161
divaricata, 161
glabella, 161
incana, 161
inornata, 161
kingii, 163
laetevirens, 161
latifolia, 161
leptophylla, 161
leucanthemifolia, 161
linearis, 161
magna, 161
montana, 161
paniculata, 161
pinosa, 161
pulverulenta, 161
ramosa, 161
rubricaulis
scoparia, 161
sessiliflora, 161
spinulosa, 161
subalpina, 161
superba, 161
verna, 161
viscosa, 161
Macounastrum islandica, 610
Macrocarpus
achilleaefolius, 110
douglasii, 110
Macronema
discoidea, 150
grindelifolium, 152
lineare, 150
obtusum, 150
pygmaeum, 151
suffruticosum, 152
Macrorhynchus
glaucus, 73
laciniatus, 73
purpureus, 72
troximoides, 72
Madronella
discolor, 474
glauca, 474
nervosa, 474
oblongifolia, 474
odoratissima, 474
parvifolia, 474
sessilifolia, 474
Mairania uva-ursi, 382

Mariscus acicularis, 373
Meadia
cusickii, 637
salina, 637
Melandrium
apetalum, 279
drummondii, 280
kingii, 279
Melargyra rubra, 295
Menziesia
empetriformis, 385
glanduliflora, 386
grahamii, 385
Merathrepta intermedia, 539
Mertensia, 210
alpina, 211
amoena, 215
arizonica, 212
bakeri, 215
cana, 215
canescens, 215
ciliata, 213
coriacea, 215
coronata, 214
cusickii, 215
cynoglossoides, 215
explicata, 214
foliosa, 214
incongruens, 213
intermedia, 214
lateriflora, 215
leonardii, 212
lineariloba, 215
muricata, 215
myosotifolia, 215
nevadensis, 214
nivalis, 215
nutans, 214
oblongifolia, 214
obtusiloba, 211
oreophila, 214
ovata, 215
pallida, 213
parryi, 215
perplexa, 215
picta, 213
polyphylla, 213
praecox, 214
pubescens, 214
punctata, 213
sampsonii, 212
stenoloba, 214
subpubescens, 213
tubiflora, 214
tweedyi, 211
viridis, 215
Micranthes
arguta, 748
arnoglossa, 750
austrina, 751

brachypus, 750
crenatifolia, 751
franciscana, 751
greenei, 751
lata, 747
montanensis, 750
occidentalis, 747
odontoloma, 748
oregana, 750
rhomboidea, 751
rydbergii, 751
saximontana, 747
sierrae, 750
subapetala, 750
Micropetalon lanceolatum, 297
Microseris nigrescens, 164
Mimulus, 767
  alpinus, 767
  arvensis, 767
  bakeri, 767
  brachystylis, 767
  caespitosus, 767
  clementinus, 767
  corallinus, 767
  cordatus, 767
  cuspidata, 767
  decorus, 767
  equinnus, 767
  glareosus, 767
  grandiflorus, 767
  grandis, 767
  guttatus, 767
  hallii, 767
  hirsutus, 767
  implexus, 767
  implicatus, 767
  langsdorfii, 767
  laxus, 767
  lewisii, 769
  longulus, 767
  lucens, 767
  luteus, 767
  lyratus, 767
  maguirei, 767
  marmotatus, 767
  micranthus, 767
  microphyllus, 767
  minor, 767
  minusculus, 767
  nasutus, 767
  panicolatus, 767
  pardalis, 767
  parishii, 767
  petiolaris, 767
  prionophyllus, 767
  procerus, 767
  puberulus, 767
  puncticalyx, 767
  rivularis, 767
  scouleri, 767
  subreniformis, 767
  tenellus, 767
  thermalis, 767
  tilingii, 767
  unimaculatus, 767
  veronicifolius, 767
Minuartia, 8, 281
  austromontana, 282
  biflora, 286
  dawsonensis, 8, 283
  macrantha, 284
  michauxii, 283
  nuttallii, 285
  obtusiloba, 26, 286
  propinqua, 287
  pungens, 285
  quadrivalvis, 287
  rolfii, 282
  rossii, 282
  rubella, 287
  sajanensis, 286
  stricta, 283
  tenella, 283
Minuopsis
  michauxii, 283
  nuttallii, 285
Mitella pentandra, 735
Mitellopsis
  drummondiana, 735
  pentandra, 735
Monardella, 473
  discolor, 474
  elegantula, 474
  glauca, 474
  nervosa, 474
  odoratissima, 474
  parvifolia, 474
  purpurea, 474
Monolepis, 307
  chenopodioides, 307
  nuttalliana, 26, 307
Morocarpus capitatus, 305
Muscaria
  adscendens, 741
  cespitosa, 743
  delicatula, 743
  emarginata, 743
  micropetala, 743
  monticola, 743
Myginda myrtifolia, 303
Myosotis, 216
  alpestris, 216
  nana, 205

N

Nacrea lanata, 75
Narthecium glutinosum, 483
Nasturtium
  obtusum, 260
  sphaerocarpum, 260
Navarretia
  debilis, 590
  linearis, 591
  nuttallii, 596
Nephrodium asplenioides, 63
Neskiza
  aquatilis, 324
  aurea, 327
Neuroloma
  nudicaule, 255
  rydbergii, 255
Noccaea
  cochleariformis, 263
  montana, 263
Nothocalais nigrescens, 26, 164

O

Oenothera, 498
  cespitosa, 26, 499
  idahoensis, 499
  marginata, 499
  montana, 499
Oligosporus
  borealis, 95
  campestris, 95
  pacifica, 95
Olotrema
  filifolia, 338
  juncifola, 347
ONAGRACEAE, 489
OPHIOGLOSSACEAE, 500
Ophioglossum pinnatum, 500
Orchiastrum
  porrifolium, 503
  romanzoffianum, 503
ORCHIDACEAE, 501
Orchis dilatata, 502
Oreastrum
  alpigenum, 103
  andersonii, 103
  elatus, 103
  haydenii, 103
Oreobroma
  aridorum, 628
  exarticulatum, 628
  grayi, 628
  minima, 628
  nevadensis, 628
  pygmaea, 628
  triphylla, 629
Oreocarya
  affinis, 204
  celosioides, 204
  cilio-hirsuta, 204
  glomerata, 204
  interrupta, 204
  macounii, 204
  nubigena, 204
  perennis, 204
  sheldonii, 204
  spiculifera, 204
  subretusa, 204
Oreochrysum parryi, 186
Oreophila myrtifolia, 303
Oreastrum
  alpigenum, 103
  andersonii, 103
  elatus, 103
  haydenii, 103
Oreostemma
  alpigenum, 103
  andersonii, 103
  haydenii, 103
Oreoxis alpina, 58
Orthilla secunda, 389
Orthocarpus
  pallescens, 761
  parryi, 761
Osmorhiza, 59
  depauperata, 59
  obtusa, 59
Osmunda
  crispa, 36
  lunaria, 500
  lunata, 500
Oxyria digyna, 611
Oxytropis, 413
  alaskana, 414
  albertina, 414
  albiflora, 421
  alpicola, 414
  argentata, 417
  bessyi, 418
  blankenshipii, 417
  campestris, 414
  cascadensis, 414
  chartacea, 414
  collina, 418
  columbiana, 414
  condensatus, 421
  cusickii, 414
  deflexa, 416
  dispar, 414
  foliolosa, 416
  foliosa, 416
  gaspensis, 422
  glutinosa, 422
  gracilis, 414
  hallii, 420
  hudsonica, 422
  hyperborea, 414
  ixodes, 422
  johannensis, 414
  jordalii, 414
  lagopus, 417
  lunelliana, 418
  luteola, 414

macounii, 414, 421
mazama, 414
monticola, 414
nana, 418
obnapiformis, 418
okanoganea, 414
olympica, 414
parryi, 419
paysoniana, 414
pinetorum, 421
podocarpa, 420
retrorsa, 416
rydbergii, 414
saximontanus, 421
sericea, 421
sheldonensis, 422
spicata, 421
terra-novae, 414
varians, 414
verruculosa, 422
viliosa, 414
viscida, 422
viscidula, 422

P
Pachylophus
  cespitosus, 499
  canescens, 499
  crinitus, 499
  cylindrocarpus, 499
  glabra, 499
  hirsutus, 499
  macroglottis, 499
  marginatus, 499
  montanus, 499
  psammophilus, 499
Packera
  cana, 171
  cymbalarioides, 173
  dimorphophyllus, 174
  oodes, 181
  streptanthifolia, 181
  werneriifolia, 183
Paneion
  interius, 574
  longiligulum, 572
  pratense, 579
  sandbergii, 581
Panicularia
  fendleriana, 572
  nuttaliana, 581
  scabrella, 581
  thurberiana, 581
PAPAVERACEAE, 504
Papaver, 504
  alaskanum, 504
  alpinum, 504
  kluanense, 504
  lapponicum, 504
  radicatum, 504

Paneion
  interius, 574
  longiligulum, 572
  pratense, 579
Panicularia fendleriana, 572
Panicum spicatum, 587
Parnassia, 736
  californica, 739
  fimbriata, 737
  intermedia, 737
  kotzebuei, 738
  montanensis, 739
  multiseta, 739
  palustris, 739
  rivularis, 737
Paronychia pulvinata, 288
Parrya, 255
  macrocarpa, 255
  nudicaulis, 25, 255
  platycarpa, 255
  rydbergii, 255
  turkestanica, 255
Pascopyrum smithii, 550
Pastinaca lanata, 49
Patrinia ceratophylla, 798
Paxistima
  macrophylla, 303
  myrsinites, 303
  schaefferi, 303
Pectiantia
  latiflora, 735
  mitelloides, 735
  pentandra, 735
Pedicularis, 770
  anaticeps, 776
  atrosanguinea, 771
  bracteosa, 771
  canbyi, 771
  contorta, 772
  ctenophora, 772
  cystopteridifolia, 773
  elata, 773
  flavida, 771
  groenlandica, 774
  hallii, 776
  latifolia, 771
  lunata, 772
  mogollonica, 776
  montanensis, 771
  oederi, 775
  pachyrhiza, 771
  paddoensis, 771
  parryi, 776
  paysoniana, 771
  pulchella, 777
  racemosa, 778
  siifolia, 771
  surrecta, 774
  thompsonii, 771
  versicolor, 775

Pellaea brewerii, 37
Penstemon, 779
  adamsianus, 782
  aggregatus, 786
  alpinus, 26
  assurgens, 780
  attenuatus, 780
  brachyanthus, 785
  brevis, 783
  cacuminis, 785
  cephalanthus, 780
  cinerus, 783
  collinus, 783
  compactus, 781
  crassifolius, 782
  cyananthus, 781
  decurvus, 783
  douglasii, 782
  ellipticus, 26
  formosus, 785
  fruticosus, 782
  hesperius, 786
  holmgrenii, 781
  humilis, 783
  leonardii, 26
  lewisii, 782
  longiflorus, 781
  micranthus, 785
  militaris, 780
  modestus, 785
  montanus, 784
  nelsonae, 780
  oreocharis, 786
  palustris, 780
  procerus, 785
  propinquus, 780
  pseudohumilis, 780
  pseudoprocerus, 780
  pulchellus, 785
  rydbergii, 786
  scouleri, 782
  tolmiei, 785
  uintahensis, 787
  vaseyanus, 786
  veronicaefolius, 780
  whippleanus, 788
  woodsii, 784
Pentameris intermedia, 539
Pentaphylloides floribunda, 692
Pentaphyllum
  effusum, 697
  leucophyllum, 697
Petradoria, 165
  graminea, 165
  pumila, 165
Petrophyton caespitosum, 685
Peucedanum
  circumdatum, 55

  cous, 55
  graveolens, 56
  kingii, 56
  lapidosum, 45
  montanum, 55
  nuttallii, 57
  simplex, 57
  triternatum, 57
Phaca
  aboriginorum, 398
  adsurgens, 396
  alpina, 397
  arctica, 397
  astragalina, 397
  australis, 398
  decumbens, 400
  glabriuscula, 398
  lapponica, 397
  minima, 397
  misera, 400
  parviflora, 400
  serotina, 400
  viridis, 399
Phacelia, 455
  alpina, 456
  ciliosa, 457
  frigida, 456
  hastata, 456
  leucophylla, 456
  lyallii, 26
  magellanica, 456
  nervosa, 456
  sericea, 457
Philotria
  canadensis, 454
  ioensis, 454
  linearis, 454
Phippsia, 565
  algida, 565
  monandra, 565
Phleum, 566
  alpinum, 566
  commutatum, 566
  haenkeanum, 566
  pratense, 566
Phlox, 597
  caespitosa, 600
  canescens, 598
  condensata, 600
  costata, 599
  depressa, 599
  glabrata, 598
  hoodii, 598
  lanata, 598
  multiflora, 599
  patula, 599
  pulvinata, 8, 600
  scleranthifolia, 598
Phorolobus
  acrostichoides, 36

crispus, 36
Phyllodoce, 385
   empetriformis, 8, 385
   glanduliflora, 386
   grahamii, 385
   intermedia, 385
Physaria, 256
   acutifolia, 257
   australis, 257
   didymocarpa, 258
   lanata, 258
   macrantha, 258
   repanda, 257
   stylosa, 257
Physematium scopulinum, 66
Picea, 508
   columbiana, 508
   engelmannii, 508
Picradenia acaulis, 158
Pilosella gracillis, 156
   tristis, 156
PINACEAE, 506
Pinus, 509
   albicaulis, 510
   cembroides, 510
   contorta, 511
   douglasii, 513
   flexilis, 512
   lasiocarpa, 507
   longaeva, 510
   murrayana, 511
   taxifolia, 513
PLANTAGINACEAE, 514
Plantago tweedyi, 514
Plantanthus
   patens, 488
   selago, 488
Plantinia alpina, 566
Platanthera
   dilatata, 502
   gracilis, 502
   graminea, 502
   leucostachys, 502
Pneumonanthe
   affinis, 434
   bracteosa, 434
   calycosa, 436
   forwoodii, 434
   parryi, 434
Poa, 8, 25, 567
   acuminata, 580
   acutiglumis, 581
   agassizensis, 579
   albescens, 572
   alcea, 581
   algida, 565
   alpicola, 571
   alpigena, 579
   alpina, 569
   ammophila, 574

ampla, 29, 581
andina, 581
anadyrica, 574
angustifolia, 579
annua, 9, 570
aperta, 571
arctica, 571
brachyglossa, 581
brandegei, 576
brevipaniculata, 572
brintnellii, 571
bryophila, 574
buckleyana, 581
callichroa, 571
canbyi, 29, 581
capillaris, 581
cenisia, 571
columbiensis, 577
confusa, 581
cottoni, 572
curta, 577
cusickii, 572
cuspidata, 577
eatonii, 572
englishii, 581
epilis, 572
fendleriana, 572
filifolia, 572
glauca, 574
glaucantha, 574
gracillima, 29, 581
grayana, 571
helleri, 581
idahoensis, 572
incurva, 29, 581
interior, 574
invaginata, 581
juncifolia, 29, 581
kingii, 564
laeviculmis, 581
laevigata, 581
laevis, 581
lanata, 571
laxa, 571
leckenbyi, 581
leptocoma, 575
lettermanii, 26, 576
limosa, 581
longiculmis, 571
longiligula, 572
longipedunculata, 572
longipila, 571
lucida, 581
montevansii, 576
multnomae, 581
nematophylla, 572
nervosa, 577
nevadensis, 581
nudata, 581
olneyae, 577

orcuttiana, 581
paddensis, 572
pattersonii, 578
pauciflora, 581
paucispicula, 575
peckii, 579
phoenicea, 571
pratensis, 579
pudica, 580
purpurascens, 572
reflexa, 580
rupestris, 574
rupicola, 574
sandbergii, 29, 581
saxatilis, 581
scaberrima, 572
scabrella, 581
scabrifolia, 572
scabriuscula, 572
scopulorum, 574
secunda, 29, 581
spillmani, 572
subaristida, 572
subpurpurea, 572
subreflexa, 577
subtrivialis, 574
tenerrima, 581
tenuifolia, 581
thurberiana, 581
tolmatchewii, 571
tricholepis, 571
truncata, 581
vaseyana, 577
vivipara, 569
wheeleri, 577
williamsii, 571
wyomingensis, 581
POACEAE, 515
Podagrostis
   humilis, 519
   thurberiana, 523
Podionapus caespitosus, 542
POLEMONIACEAE, 589
Polemonium, 601
   berryi, 602
   confertum, 603
   delicatum, 602
   fasciculatum, 602
   grayanum, 603
   haydenii, 602
   humile, 602
   lindleyi, 602
   mexicanum, 602
   orbiculare, 602
   oreades, 602
   parvifolium, 602
   pilosum, 602
   pulcherrimum, 602
   rotatum, 602
   shastense, 602

   speciosum, 603
   viscosum, 603
Polyantherix hystrix, 545
POLYGONACEAE, 604
Polygonum, 612
   austinae, 614
   bistortoides, 7, 613
   cephalophorum, 613
   confertiflorum, 615
   douglasii, 8, 614
   emaciatum, 614
   engelmannii, 614
   fugax, 618
   glastifolium, 613
   kelloggii, 8, 615
   linearifolium, 613
   macounii, 618
   microspermum, 614
   minimum, 8, 616
   minutissimum, 615
   montanum, 614
   pannosum, 614
   sawatchense, 8, 617
   unifolium, 615
   viviparum, 618
   vulcanicum, 613
Polypodium
   alpestre, 62
   filix-femina, 63
   fragile, 64
   lonchitis, 65
Polystichum, 65
   lonchitis, 65
   scopulinum, 65
Ponista oregonensis, 741
Populus, 713
   balsamifera, 26, 713
   candicans, 713
   hastata, 713
   michauxii, 713
   tacamahaca, 713
   tremuloides, 26
   trichocarpa, 713
PORTULACACEAE, 624
Potentilla, 25, 686
   albida, 693
   altaica, 700
   amplifolia, 693
   angustata, 695
   argyrea, 697
   ashlandica, 693
   atrovirens, 702
   beanii, 689
   bicrenata, 689
   bipinnatifida, 702
   blaschkeana, 695
   brevifolia, 688
   brunnescens, 695
   camporum, 695
   candida, 695

chamissonis, 698
chrysantha, 695
ciliata, 693
concinna, 689
cottamii, 699
ctenophora, 695
dascia, 695
decurrens, 690
dichroa, 695
diffusa, 697
dissecta, 690
diversifolia, 690
divisa, 689
effusa, 697
elmeri, 695
emarginata, 699
etomentosa, 695
fastigata, 695
filicaulis, 697
filipes, 695
finitima, 702
flabellifolia, 691
flabelliformis, 695
floribunda, 692
fruticosa, 692
furcata, 698
gelida, 691
glabella, 702
glabrata, 695
glandulosa, 693
glaucophylla, 690
glomerata, 695
gordonii, 682
gracilipes, 679
gracilis, 695
groenlandica, 699
grosseserrata, 695
hallii, 695
hanseni, 693
hippiana, 697
hookeriana, 698
humifusa, 689
hyparctica, 699
indiges, 695
intermittens, 689
jucunda, 695
klamathensis, 701
lasiodonta, 702
ledebouriana, 705
leneophylla, 697
leucophylla, 697
longiloba, 695
longipedunculata, 695
macounii, 689
macropetala, 695
minutifolia, 704
missourica, 702
modesta, 689
monidensis, 701
multisecta, 690

nana, 25, 699
nelsoniana, 701
nipharga, 700
nivalis, 679
nivea, 700
nuttallii, 695
ovalis, 675
ovina, 701
paucijuga, 702
pecten, 695
pectinata, 702
pectinisecta, 695
pedersenii, 689
pensylvanica, 702
perdissecta, 690
permollis, 695
pinnatisecta, 701
platyloba, 702
procumbens, 708
propinqua, 697
proxima, 689
pseudorupestris, 693
pulcherrima, 695
quinquefolia, 689
ranunculus, 690
rectiformis, 695
reflexa, 693
rhomboidea, 693
rigida, 695
rubricaulis, 703
rubricaulis, 689
rubripes, 689
rupincola, 697
saximontana, 703
sibbaldii, 708
strigosa, 702
subjuga, 704
uniflora, 705
utahensis, 683
valida, 693
versicolor, 701
vesca, 675
virgulata, 702
viridescens, 695
viridior, 704
viscosa, 693
vreelandii, 690
vulcanicola, 705
wrangelliana, 693
wyomingensis, 701
Prenanthes pygmaea, 119
Primula, 639
angustifolia, 26
montana, 638
mucronata, 639
parryi, 639
PRIMULACEAE, 631
Pseudathyrium alpestre, 62
Pseudocymopterus
bipinnatus, 46

hendersonii, 44
nivalis, 46
Pseudopteryxia
hendersonii, 44
longiloba, 44
Pseudoreoxis
bipinnatus, 46
nivalis, 46
Pseudoroegneria spicata, 551
Pseudotsuga, 513
caesia, 513
flahaulti, 513
glauca, 513
globulosa, 513
lindleyana, 513
menziesii, 513
merrillii, 513
mucronata, 513
rehderi, 513
taxifolia, 513
vancouverensis, 513
Psilochaenia runcinata, 120
Psychrophila auriculata, 650
leptosepala, 650
Psyllophora pyrenaica, 360
Pteris crispa, 36
Pteryxia
hendersonii, 44
terebinthacea, 47
terebinthina, 47
Ptilepida
acaulis, 158
grandiflora, 160
Puccinellia
canbyi, 581
laevis, 581
lettermanii, 576
nevadensis, 581
scabrella, 581
Pulmonaria
alpina, 211
ciliata, 213
oblongifolia, 214
Pulsatilla
hirsutissima, 645
ludoviciana, 645
nuttalliana, 645
patens, 645
Pyrola, 387
aphylla, 388
blanda, 388
conardiana, 388
conferta, 387
dentata, 388
minor, 387
pallida, 388
paradoxa, 388
picta, 388
secunda, 389
septentrionalis, 388

sparsifolia, 388
Pyrrocoma
cheiranthifolia, 153
gossypina, 153
howellii, 153
inuloides, 153
linearis, 153
lyallii, 149
plantaginea, 153
sericea, 153
uniflora, 153

R
Radicula
alpina, 260
curvipes, 260
integra, 260
obtusa, 260
sphaerocarpa, 260
underwoodii, 260
Ramischia
elatior, 389
secunda, 389
secundiflora, 389
RANUNCULACEAE, 640
Ranunculus, 25, 654
adoneus, 655
affinis, 663
alismifolius, 656
alismellus, 656
alpeophilus, 662
apetalus, 663
aquatilis, 657
arizonicus, 662
calthaeflorus, 656
capillaceus, 657
drummondii, 659
ellipticus, 660
eschscholtzii, 658
eximus, 658
gelidus, 659
glaberrimus, 660
grayi, 659
hartwegii, 656
helleri, 658
hookeri, 659
hyperboreus, 661
inamoenus, 662
intertextus, 661
lemmonii, 656
micropetalus, 662
natans, 661
ocreatus, 658
oxynotus, 658
pedatifidus, 659
pedatifidus, 663
porteri, 657
pygmaeus, 664
ramulosus, 659
reconditus, 660

sabinii, 664
saxicola, 658
suksdorfii, 658
trichophyllus, 657
trisectus, 658
unguiculatus, 656
utahensis, 662
verecundus, 665
waldronii, 660
Rhabdocrinum serotinum, 482
Rhodiola
　alaskana, 312
　atropurpurea, 312
　integrifolia, 312
　polygama, 312
　roanensis, 312
　rosea, 312
Ribes, 448
　camporum, 452
　cereum, 449
　cognatum, 452
　echinatum, 450
　grossularioides, 450
　hendersonii, 452
　inebrians, 449
　irriguum, 452
　lacustre, 450
　lentum, 451
　leucoderme, 452
　molle, 451
　montigenum, 451
　nonscripta, 452
　nubigenum, 451
　oxyacanthoides, 452
　parvulum, 450
　pumilum, 449
　reniforme, 449
　saximontanum, 452
　saxosum, 452
　setosum, 452
　spathianum, 449
　viscidulum, 449
Rochelia patens, 209
Roegneria
　albicans, 548
　borealis, 552
　canina, 552
　latiglumis, 552
　pauciflora, 552
　scandia, 552
　spicata, 551
　trachycaula, 552
　violaceum, 552
　virescens, 552
Rompelia roseana, 40
Rorippa, 259
　alpina, 260
　curvipes, 9, 260
　integra, 260
　obtusa, 260
　sphaerocarpa, 260
　underwoodii, 260
ROSACEAE, 672
RUBIACEAE, 709
Rubus, 706
　acalyphaceus, 707
　carolinianus, 707
　greeneanus, 707
　idaeus, 707
　melanolasius, 707
　melanotrachys, 707
　neglectus, 707
　peramoenus, 707
　sachalinensis, 707
　strigosus, 707
　subarcticus, 707
　viburnifolius, 707
Rumex, 619
　acetosa, 620
　alpestris, 620
　californicus, 623
　crassus, 623
　densiflorus, 621
　digynus, 611
　geyeri, 622
　hesperius, 623
　lacustris, 623
　mexicanus, 623
　paucifolius, 622
　praecox, 621
　pycanthus, 621
　quadrangulivalvis, 623
　salicifolius, 623
　thyrsiflorus, 620
　transitorius, 623
　triangulivalvis, 623
　utahensis, 623
Rupestrina pubescens, 587
Rydbergia grandiflora, 160

S

Sabina
　horizontalis, 316
　prostrata, 316
Sabulina
　dawsonensis, 283
　propinqua, 287
　stricta, 283
Sagina, 289
　linnaei, 289
　micrantha, 289
　saginoides, 26, 289
SALICACEAE, 712
Salix, 8, 25, 714
　aemulans, 724
　albertana, 716
　amelanchieroides, 722
　anamesa, 721
　anglorum, 715
　arctica, 8, 715
　arctica, 715
　atra, 721
　barrattiana, 716
　behringica, 725
　bella, 719
　brachycarpa, 717
　brownei, 715
　caespitosa, 715
　californica, 720
　callicarpaea, 721
　cascadensis, 718
　chlorophylla, 723
　commutata, 720
　cordifolia, 721
　covillei, 719
　crassijulis, 715
　desertorum, 721
　dissymetrica, 722
　dodgeana, 725
　drummondiana, 719
　eastwoodiae, 720
　fullertonensis, 717
　glauca, 8, 721
　glaucops, 721
　hudsonensis, 715
　idahoensis, 727
　labradorica, 721
　leiocarpa, 725
　lingulata, 717
　macounii, 721
　monica, 723
　monticola, 722
　muriei, 717
　nelsonii, 723
　niphoclada, 717
　nivalis, 724
　nudescens, 721
　orbicularis, 724
　pachnophora, 719
　padifolia, 722
　padophylla, 722
　pallasii, 715
　pellita, 719
　pennata, 723
　petrophila, 715
　phylicifolia, 723
　phylicoides, 723
　planifolia, 8, 723
　pseudolapponum, 721
　pseudomonticola, 722
　pulchra, 723
　pychnocarpa, 723
　reticulata, 8, 724
　rotundifolia, 725
　rotundifolia, 726
　rydbergii, 721
　saximontana, 724
　sawatchicola, 722
　seemannii, 721
　solheimii, 724
　stricta, 717
　subcoerulea, 719
　tenera, 718
　tortulosa, 715
　tweedyi, 726
　tyrrellii, 723
　vacciniformis, 721
　venusta, 724
　villosa, 721
　wolfii, 727
　wyomingensis, 721
Sambucus, 270
　callicarpa, 270
　melanocarpa, 270
　microbotrys, 270
　pubens, 270
　racemosa, 270
Saussurea weberi, 25, 166
Savastana
　nashii, 562
　odorata, 562
Saxifraga, 25, 740
　adscendens, 741
　arguta, 748
　arnoglossa, 750
　austrina, 751
　austromontana, 742
　bronchialis, 742
　cespitosa, 26, 743
　cernua, 744
　cherlerioides, 742
　chrysantha, 26, 745
　cognata, 742
　crandallii, 746
　debilis, 752
　delicatula, 743
　emarginata, 743
　exarata, 743
　firma, 742
　flagellaris, 26, 746
　flexuosa, 752
　funstonii, 742
　greenei, 751
　groenlandica, 743
　heucheriformis, 729
　hyperborea, 752
　incompta, 741
　jamesiana, 729
　jamesii, 729
　lyallii, 26, 27
　micropetala, 743
　montanensis, 750
　monticola, 743
　occidentalis, 747
　odontoloma, 748
　odontophylla, 748
　oppositifolia, 8, 749
　oregana, 750
　oregonensis, 741

petraea, 741
platysepala, 746
pulvinata, 749
punctata, 748
rhomboidea, 7, 751
rivularis, 752
rydbergii, 751
saximontana, 747
setigera, 746
sierrae, 750
sileneflora, 743
simulata, 744
subapetala, 750
vespertina, 742
SAXIFRAGACEAE, 728
Schedonorus
   inermis, 532
   spicatus, 551
Schizonotus dumosus, 681
Scirpidium aciculare, 373
Scirpus
   acicularis, 373
   angustifolius, 376
   bellardi, 379
   nanus, 374
   pauciflorus, 374
   quinqueflorus, 374
SCROPHULARIACEAE, 753
Sedum, 308
   alaskanum, 312
   atropurpureum, 312
   debile, 309
   douglasii, 313
   frigidum, 312
   integrifolium, 312
   lanceolatum, 310
   nesioticum, 310
   polygamum, 312
   rhodanthum, 311
   rhodiola, 312
   rhodioloides, 312
   rosea, 312
   rupicolum, 310
   stenopetalum, 313
   subalpinum, 310
   uniflorum, 313
Selaginella, 793
   densa, 794
   engelmannii, 794
   scopulorum, 794
   standleyi, 794
   watsonii, 795
SELAGINELLACEAE, 793
Selago vulgaris, 488
Selinum
   grayi, 39
   sphondylium, 49
   terebinthinum, 47
Senecio, 167

acutidens, 181
adamsi, 181
admirabilis, 179
alpicola, 183
altus, 180
amplectens, 169
andinus, 179
aquariensis, 181
arachnoideus, 177
atratus, 170
atriapiculatus, 177
bivestitus, 176
blitoides, 175
canus, 171
carthamoides, 175
caulanthifolius, 177
chapacensis, 181
cognatus, 181
columbianus, 177
condensatus, 177
convallium, 171
cordatus, 177
crassulus, 172
crocatus, 26
cymbalaria, 26
cymbalarioides, 173
cymbalarioides, 181
denalii, 176
dileptiifolius, 181
dimorphophyllus, 174
dispar, 177
ductoris, 175
exaltatus, 177
farriae, 181
flintii, 177
foliosus, 177
fondinarum, 177
fraternus, 181
fremontii, 175
fulgens, 181
fuscatus, 176
gibbonsii, 182
glaucescens, 178
hallii, 171
harbourii, 171
heterodoxus, 174
holmii, 169
hookeri, 177
howellii, 171
imbricatus, 178
integerrimus, 177
invenustus, 175
jonesii, 181
kernensis, 171
lactucinus, 169
laetiflorus, 181
lanceolatus, 179
lapathifolius, 172
laramiensis, 171
latus, 180

leibergii, 177
leonardi, 181
lindstroemii, 176
longidentatus, 182
longipetiolatus, 181
lugens, 178
majus, 177
mendocinensis, 177
mesadenia, 177
milleflorus, 170
molinarius, 183
moresbiensis, 173
muirii, 183
occidentalis, 175
ochraceus, 177
oodes, 181
oreganus, 180
oreopolus, 171
ovinus, 173
pammelii, 181
pentodontus, 183
perennans, 183
perplexus, 177
petraeus, 183
petrocallis, 183
petrophilus, 183
platylobus, 181
prionophyllus, 182
purshianus, 171
rubricaulis, 181
rydbergii, 181
saliens, 182
saxosus, 183
scaposus, 183
scribneri, 177
semiamplexicaulis, 172
seridophyllus, 169
serra, 179
solidago, 179
solitarius, 177
sonnei, 177
speculicola, 183
sphaerocephalus, 180
streptanthifolius, 181
subcuneatus, 181
subnudus, 173
subvestitus, 182
suksdorfii, 181
triangularis, 182
trigonophyllus, 182
tundricola, 176
turbinatus, 183
variifolius, 182
vaseyi, 177
wardii, 181
werneriifolius, 183
whippleanus, 177
willingii, 181
Sericotheca
   concolor, 681

dumosa, 681
glabrescens, 681
microphylla, 681
Seriphidium tridentatum, 101
Seseli triternatum, 57
Sibbaldia procumbens, 708
Sieversia
   campanulata, 680
   canescens, 680
   ciliata, 680
   flavula, 680
   gracilipes, 679
   grisea, 680
   humilis, 679
   rosea, 680
   rossii, 679
   sericea, 679
   triflora, 680
   turbinata, 679
Silene, 290
   acaulis, 291
   attenuata, 279
   dilatata, 292
   dorrii, 293
   douglasii, 292
   drummondii, 280
   exscapa, 291
   hitchguirei, 279
   kingii, 279
   lyallii, 292
   macounii, 294
   macrocalyx, 292
   menziesii, 293
   monantha, 292
   multicaulis, 292
   obovata, 293
   oraria, 292
   parryi, 294
   stellarioides, 293
   tetonensis, 294
   tetragyna, 294
   uralensis, 279
   wahlbergella, 279
   williamsii, 293
Siphonella
   montana, 596
   nuttallii, 596
   parviflora, 596
Sisymbrium
   hartwegianum, 228
   incanum, 228
   incisum, 228
   pauciflorum, 220
   procerum, 228
   richardsonianum, 228
   richardsonii, 228
   viscosum, 228
Sitanion
   albescens, 545
   basalticola, 545

breviaristatum, 545
brevifolium, 545
californicum, 545
ciliatum, 545
cinereum, 545
elymoides, 545
glabrum, 545
horteoides, 545
hystrix, 545
insulare, 545
latifolium, 545
longifolium, 545
marginatum, 549
molle, 545
montanum, 545
pubiflorum, 545
rigidum, 545
strigosum, 545
velutinum, 545
Smelowskia, 261
americana, 261
calycina, 261, 26
lineariloba, 261
lobata, 261
porsildii, 261
Solidago, 184
bellidifolia, 187
ciliosa, 185
confertifolia, 187
corymbosa, 185
cusickii, 185
decumbens, 187
dilatata, 185
gillmanii, 187
glutinosa, 187
graminea, 165
hesperius, 187
laevicaulis, 185
multiradiata, 185
neomexicana, 187
oreophila, 187
parryi, 186
petradoria, 165
pumila, 165
racemosa, 187
randii, 187
sarothrae, 146
scopulorum, 185
simplex, 187
spathulata, 187
spiciformis, 187
vespertina, 187
yukonensis, 187
Sophia
brevipes, 228
californica, 228
hartwegiana, 228
incisa, 228
leptophylla, 228
procera, 228

purpurascens, 228
ramosa, 228
richardsonii, 228
serrata, 228
sonnei, 228
viscosa, 228
Spergella saginoides, 289
Spergula
rubra, 295
saginoides, 289
Spergularia rubra, 8, 295
Spergulastrum lanceolatum, 297
Sphaeromeria diversifolia, 188
Sphondylium
lanatum, 49
vulgare, 49
Spiesia
campestris, 414
inflata, 420
lagopus, 417
monticola, 414
parryi, 419
podocarpa, 420
viscida, 422
Spiraea
caespitosa, 685
dumosa, 681
splendens, 26
uniflora, 684
Spiranthes, 503
porrifolia, 503
romanzoffiana, 503
stricta, 503
Spraguea, 630
multiceps, 630
umbellata, 9, 630
Sporobolus bolanderi, 581
Stellaria, 296
alpestris, 297
americana, 26
arenicola, 299
biflora, 286
borealis, 297
brachypetala, 297
calycantha, 297
crassifolia, 298
dulcis, 299
edwardsii, 299
gonomischa, 301
hultenii, 299
laeta, 299
laxmannii, 299
longifolia, 297
longipes, 299
monantha, 299
obtusa, 300
oxyphylla, 297
simcoei, 297

sitchana, 297
stricta, 299
strictiflora, 299
subvestita, 299
uliginosa, 283
umbellata, 301
viridula, 300
washingtoniana, 300
weberi, 301
Stellularia
borealis, 297
longipes, 299
umbellata, 301
Stenactis speciosa, 141
Stenanthium occidentale, 26
Stenotus
acaulis, 148
caespitosus, 148
falcatus, 148
latifolius, 148
lyallii, 149
pygmaeus, 151
rudis, 148
scaber, 148
Stipa, 583
californica, 585
columbiana, 585
elmeri, 585
lettermanii, 584
minor, 585
nelsonii, 585
occidentalis, 585
oregonensis, 585
stricta, 585
williamsii, 585
Stipularia rubra, 295
Streptanthus
angustifolius, 219
cordatus, 26
virgatus, 220
Struthiopteris crispa, 36
Stylopappus laciniatus, 73
Swertia, 444
congesta, 444
fritillaria, 444
obtusa, 444
occidentalis, 444
palustris, 444
parallela, 444
perennis, 444
radiata, 432
scopulina, 444
Symphoricarpos, 271
oreophilus, 271
tetonensis, 271
utahensis, 271
vaccinioides, 271
Synthyris, 789
alpina, 754
cymopteroides, 789

gymnocarpa, 755
hendersonii, 789
lanuginosa, 789
paysonii, 789
pinnatifida, 789
wyomingensis, 755

T
Talinum pygmaeum, 628
Tanacetum diversifolium, 188
Taraxacum, 189
ambigens, 190
ammophilum, 191
amphiphron, 190
angustifolium, 191
arctogenum, 190
atroglaucum, 193
brachyceras, 190
campylodes, 193
carneocoloratum, 190
carthamopsis, 190
ceratophorum, 190
croceum, 193
curvidens, 193
cylocentrum, 193
dahlstedtii, 193
davidssonii, 193
devians, 193
dilutisquameum, 193
dumetorum, 190
eriophorum, 191
eurylepium, 190
firmum, 193
hyperboreum, 190
ingratum, 190
islandiciforme, 193
kamtschaticum, 192
kok-saghyz, 193
lacerum, 190
lapponicum, 190
latilobum, 193
latispinulosum, 190
laurentianum, 190
leiospermum, 190
longii, 190
lyratum, 192
malteanum, 190
maurolepium, 190
mexicanum, 193
mitratum, 190
montanum, 190
multissimum, 190
naevosum, 190
officinale, 26, 193
olympicum, 191
ovinum, 190
palustre, 193
paucisquamosum, 190
pellianum, 190
plentiflorum, 193

pseudonorvegicum, 190
purpuridens, 190
retroflexum, 193
rhodolepis, 193
rupestre, 192
scopulorum, 192
sibiricum, 192
taraxacum, 193
torngatense, 190
trigonolobum, 190
turforsum, 193
umbrinum, 190
undulatum, 193
vulgare, 193
xanthostigma, 193
Telesonix
heucheriformis, 729
jamesii, 729
Tellima
bulbifera, 734
glabra, 734
Tephroseris
fuscata, 176
lindstroemii, 176
Terrellia glauca, 547
Tesseranthium
angustifolium, 432
macrophyllum, 432
radiatum, 432
scabrum, 432
speciosum, 432
stenosepalum, 432
Tetraneuris
acaulis, 158
arizonica, 158
brevifolia, 158
crandallii, 158
epunctata, 158
eradiata, 158
grandiflora, 160
herbacea, 158
incana, 158
ivesiana, 158
lanata, 158
lanigera, 158
pygmaea, 158
septentrionalis, 158
simplex, 158
trinervata, 158
Thalictrum, 666
alpinum, 667
breitungii, 669
columbianum, 670
confine, 670
fendleri, 668
fissum, 670
leiophyllum, 667
megacarpum, 669
occidentale, 669
polycarpum, 668

propinquum, 669
rainierense, 669
stipitatum, 668
turneri, 670
venulosum, 670
Therofon
heucheriforme, 729
jamesii, 729
Thlaspi, 262
australe, 263
californicum, 263
cochleariforme, 263
coloradense, 263
fendleri, 263
glaucum, 263
hesperium, 263
montanum, 263
nuttallii, 263
parviflorum, 26, 264
prolixum, 263
purpurascens, 263
stipitatum, 263
Tissa rubra, 295
Tium
alpinum, 397
misera, 400
miserum, 400
Tofieldia, 483
glutinosa, 483
intermedia, 483
occidentalis, 483
Tolmachevia integrifolia, 312
Tomentilla humifusa, 689
Tonestus
kingii, 163
lyallii, 149
pygmaeus, 151
Torresia odorata, 562
Townsendia, 194
alpigena, 195
alpina, 198
anomala, 196
condensata, 196
dejecta, 195
leptotes, 197
minima, 195
montana, 195
parryi, 198
spathulata, 199
Tozzettia fulva, 526
Tragacantha
aboriginorum, 398
adsurgens, 396
alpina, 397
decumbens, 400
glabriuscula, 398
misera, 400
montana, 399
serotina, 400
tegetaria, 399

Tragopogon, 200
dubius, 26, 200
major, 200
Trasus
atratus, 325
capillaris, 332
vesicarius, 371
Triantha glutinosa, 483
Trianthella glutinosa, 483
Trichodium
album, 522
algidum, 565
scabrum, 522
Trichophyllum
integrifolium, 144
lanatum, 144
multiflorum, 144
Trifolium, 423
anemophilum, 424
atrorubens, 426
brachteolatum, 424
brachypus, 426
caurinum, 426
confusum, 426
covillei, 426
dasyphyllum, 424
elmeri, 426
hansenii, 426
haydenii, 425
inaequale, 428
lividum, 424
longipes, 426
montanense, 428
multipedunculatum, 426
nanum, 8, 26, 427
oreganum, 426
parryi, 428
pedunculatum, 426
rusbyi, 426
rydbergii, 426
salictorum, 428
scariosum, 424
shastense, 426
stenolobum, 424
uintense, 424
Trimorpha acris, 123
Triorchis
romanzoffiana, 503
stricta, 503
Tripolium occidentale, 108
Trisetaria airoides, 587
Trisetum, 586
airoides, 587
alaskanum, 587
americanum, 587
brandegei, 588
congdoni, 587
majus, 587
molle, 587
muticum, 588

spicatum, 587
subspicatum, 587
triflorum, 587
villosissimum, 587
williamsii, 539
wolfii, 588
Triticum
boreale, 552
caninum, 552
cristatum, 517
dasystachyum, 548
desertorum, 517
divergens, 551
pauciflorum, 552
pectinatum, 517
richardsonii, 552
sibiricum, 517
subsecundum, 552
trachycaulum, 552
Trollius, 671
albiflorus, 671
americanus, 671
laxus, 671
Troximon
arachnoideum, 73
arizonicum, 73
aurantiacum, 72
glaucum, 73
gracilens, 72
montanum, 72
parviflorum, 73
pubescens, 73
pumilum, 73
purpureum, 72
roseum, 73
taraxacifolium, 73
villosum, 73
Tryphane rubella, 287
Tsuga lindleyana, 513
Turritis
brachycarpa, 220
drummondii, 219
retrofracta, 220
stricta, 219

U
UMBELLIFERAE, 38
Uncinia filifolia, 338
Urostachys selago, 488
Uva-ursi
buxifolia, 382
procumbens, 382
uva-ursi, 382

V
Vaccinium, 390
arbuscula, 391
cespitosum, 391
erythrococcum, 393
humifusum, 383

myrtillus, 392
nivictum, 391
occidentale, 394
oreophilum, 392
paludicola, 391
scoparium, 393
uliginosum, 394
Vahlodea
atropurpurea, 541
flexulosa, 541
latifolia, 541
Valeriana, 796
acutiloba, 797
ceratophylla, 798
ciliata, 798
edulis, 798
furfurescens, 798
micrantha, 799
obovata, 798
occidentalis, 799
puberulenta, 797
pubicarpa, 797
sitchensis, 26
trachycarpa, 798
VALERIANACEAE, 796
Veronica, 790
nutans, 791
peregrina, 792
serpyllifolia, 791
sherwoodii, 792
wormskjoldii, 791
xalapensis, 792
Vesicaria
alpina, 250
didymocarpa, 258
occidentalis, 252

Vignea
canescens, 331
capitata, 333
incurva, 347
stellulata, 336
stenophylla, 368
Vilfa
algida, 565
scabra, 522
monandra, 565
Viola, 800
adunca, 801
aduncoides, 801
atriplicifolia, 804
aurea, 804
bellidifolia, 801
canadensis, 803
cascadensis, 801
erectifolia, 802
flavovirens, 802
gomphopetala, 802
linguifolia, 802
mamillata, 801
montanensis, 801
monticola, 801
nuttallii, 802
odontophora, 801
oxyceras, 801
palustris, 803
physalodes, 802
pinetorum, 804
praemorsa, 802
purpurea, 804
quercetorum, 804
retroscabra, 801

russellii, 802
subsagittifolia, 802
subvestita, 801
thorii, 804
utahensis, 804
vallicola, 802
venosa, 804
xylorrhiza, 802
VIOLACEAE, 800

W
Wahlbergella
apetala, 279
attenuata, 279
drummondii, 280
kingii, 279
montana, 279
parryi, 294
striata, 280
Wasatchia kingii, 564
Washingtonia obtusa, 59
Woodsia, 66
oregana, 26
scopulina, 66
Wulfenia
gymnocarpa, 755
pinnatifida, 789
wyomingensis, 755
Wyomingia tweedyana, 135

X
Xanthocephlum sarothrae, 146
Xerophyllum tenax, 26

Y
Yamala cylindrica, 731
Youngia nana, 119

Z
Zeia
albicans, 548
canina, 552
cristata, 517
dasystachya, 548
griffithsii, 548
mollis, 550
occidentalis, 550
pseudorepens, 52, 548
richardsonii, 552
riparia, 548
smithii, 550
spicata, 551
tenera, 552
Zerna
anomala, 529
arctica, 532
ciliata, 531
inermis, 532
pumpelliana, 532
richardsonii, 531
Zigadenus, 484
alpinus, 484
chloranthus, 484
coloradensis, 484
elegans, 484
glaucus, 484
longus, 484
washakianus, 484